THE OFFICIAL DATACAD™ USER'S GUIDE

The Official DataCAD™ User's Guide

Michael Smith

Richard Morse

McGraw-Hill
New York San Francisco Washington, D.C.
Auckland Bogotá Caracas Lisbon London
Madrid Mexico City Milan Montreal New Delhi
San Juan Singapore Sydney Tokyo Toronto

Library of Congress Cataloging-in-Publication Data

Smith, Michael R.
The official DataCAD user's guide. / Michael R. Smith and Richard C. Morse
p. cm.
Includes index.
ISBN 0-07-136356-4 (set)
1. DataCAD 2. Architectural drawing—Data processing. 3. Computer-aided design. I. Morse, Richard C. II. Title.

NA2728 .S57 2000
720'.28'402855369—dc21

00-051124
CIP

McGraw-Hill

A Division of The **McGraw-Hill** Companies

1 2 3 4 5 6 7 8 9 0 DOC/DOC 0 6 5 4 3 2 1 0

ISBN 0-07-136357-2
PART OF ISBN 0-07-136356-4

The sponsoring editor for this book was Wendy Lochner and the production supervisor was Pamela Pelton. It was set in Century Schoolbook by D&G Limited, LLC.

Printed and bound by R. R. Donnelley & Sons Company.

This book is printed on recycled, acid-free paper containing a minimum of 50 percent recycled de-inked fiber.

CONTENTS

ACKNOWLEDGMENTS

They say that there is a first time for everything. But there is also only one first time, and as a first time author I will certainly not forget the experiences of creating this book. The difficult but rewarding process was punctuated by input and encouragement from several sources that I would like to acknowledge. The first of those would certainly have to be my wife, who has put up with, and never complained about, all the late evenings and lost weekends. She has offered me nothing but the greatest of encouragement, not only for the writing of this book, but also for our first year of marriage, as well.

I would also like to thank the people who helped me with the myriad of technical details that are such a large part of a book such as this. Evan Shu is the author of the monthly DataCAD "Cheap Tricks" newsletter, whose articles have been such a tremendous resource for me. If this book is written even half as well as Evan's articles then I will be happy indeed. I must also mention Mark Madura, the president and CEO of DataCAD LLC, Bill D'Amico, Patrick McConnell, James Horecka, and everyone on the DataCAD internet forum (DBUG). More thanks go to Bill D'Amico, who always took the time to give me so much technical information, and offer his insights into the inner workings of DataCAD. My partner, William Mitropoulos, deserves a great deal of thanks as well, for allowing me the time for phone calls, e-mail and frequent trips to the Post Office.

I also extend special thanks David Giesselman, senior programmer extraordinaire at DataCAD LLC, and one of the primary reasons that this book turned out as well as it has. He is not only an incredible programmer, but also one of the nicest and most helpful people one could hope to work with.

Additional thanks must go to Wendy Lochner and Margaret Webster-Shapiro, at McGraw-Hill, and the rest of the support crew, all of whose expertise and patience were much appreciated by Rick Morse and myself.

Michael Smith

I would like to express my gratitude to the many persons who made this book possible. Thank you to DATACAD LLC, particularly Mark Madura, who introduced me to DataCAD in 1989 and David Giesselman, one of the original authors of the program, and a key innovator who never seems to tire of a challenge. Special thanks to Mark Toce, who assembled, collated, and transcribed much of the book's raw data.

A number of persons provided critical resource materials for the 3D portion of the book: indispensable macros from Patrick McConnell and David Henderson deserve recognition, and sample renderings provided by Ernest Burden III of O'Really, Inc. and Donald Sanders of Learning Sites, Inc.

Finally, I owe a considerable debt to the many persons who bear the solitude imposed by the endless hours of polygon and pixel twiddling. Thanks indeed!

It is customary to dedicate a book in a person's honor. The 3D section of this book is dedicated to the memory of the late William Riseman, the original Voxelator, mentor and 3D magician.

Richard Morse

How To Use This Book

Text Conventions

We will be using the following typefaces and fonts to clarify input and output:

MENU BUTTONS AND TYPED INPUT

- DataCAD Menu Buttons. **Move** is the "Move" button in the menu. Multiple selections are shown with forward slashes between commands:
- Move/ToLayer means to select the "Move" button, then the "ToLayer" button.
- Enter an angle of **90 <Enter>**, means to type in the number 90, then press the Enter key on your keyboard.

MOUSE COMMANDS

Mouse commands. *Right-click* means to click on the right mouse button, as explained below.

Using the Keyboard

DataCAD makes extensive use of the keyboard and keyboard "shortcuts." We'll cover this in more detail in Chapter 5, but you will find that the keyboard shortcuts give you a huge advantage in drawing speed over a mouse-driven CAD program.

UPPER/LOWER CASE:

Some shortcuts are implemented with either the upper case or the lower case letter on the keyboard, and others are implemented with only one or the other. Remember that having the CAPS LOCK on or off will impact how you choose upper or lower case letters.

- **Nn**—means press either the upper case "N" or the lower case "n" key.
- **F**—means press the upper case "F" key.
- **m**—means press the lower case "m" key.

KEY COMBINATIONS (KEY+KEY):
When you encounter these phrases you will hold down the first key while simultaneously pressing the second key. Let go of both keys and the command is complete.

- **Shift+5**—means press the Shift key, then the 5 key simultaneously.
- **Alt+G**—means press the Alt key, then the G key simultaneously.

Note that when using the Shift and Alt key combinations it does NOT matter whether you use an upper case or a lower case letter. They are interchangeable. The convention used in this book will always show upper case letters for readability.

FUNCTION KEYS:

- Some of DataCAD's shortcuts are accessed by the function keys. These are the keys with **F1**, **F2**, etc. on them, usually located across the top or along the left side of your keyboard.
- Sometimes you access a shortcut by pressing Shift, then a function key. This would be represented by something like **Shift+F4**.

OTHER SHIFT KEY COMBINATIONS:

- Just as with any typewriter or computer keyboard, some characters on the keyboard are accessed by pressing and holding the Shift button, then pressing another key. For instance, the "**&**" character is accessed by pressing **Shift+7**.
- Since these are standard keyboard characters and not a DataCAD shortcut, we will not tell you to to press the Shift key. In the example above we would simply tell you to press the **&** key, not Shift+&.

THE ENTER KEY:

- When you are to press this key you will see this: **<Enter>**.

Mousing Around

- To *click* means to click the left mouse button once and then release it. This is the primary button you will use in DataCAD.
- To *middle click* assumes you have a 3-button mouse (highly recommended), and means to click the middle mouse button once,

then release it. In DataCAD the middle mouse button accomplishes the same thing as pressing the "**Nn**" (object snap) key; it "snaps" to points on objects. So if you do not have a 3-button mouse, use the "**Nn**" key instead.

- To *right click* means to click the right mouse button once and then release it. In DataCAD this will do one of three things, depending on what you are currently doing. It will either:
  - Toggle back and forth between the Edit and Utility menus (in both 2D & 3D)
  - Exit or Cancel a current selection, or
  - Function as the **<Enter>** key.

  Note that although DataCAD is now a Windows program, its roots (and concurrently its strength) lie in the DOS world. DataCAD does not support the standard Windows95/NT right click function, which usually brings up special menus, commands or options. You'll actually come to love this function in DataCAD; it's a real time saver.
- To *drag* means to hold down the left (primary) mouse button and simultaneously move the mouse without releasing the mouse button. Once the *drag* is completed the mouse button is released. Releasing the mouse button will automatically end the function.

CHAPTER 1

CAD Concepts

The first thing we should do is clarify the acronym CAD. You may actually see it spelled one of two ways: CAD or CADD. CADD stands for *Computer Aided Design and Drafting*. Depending on your own personal take, CAD can mean either *Computer Aided Design* or *Computer Aided Drafting*. Take your pick. Better yet, if you want an animated discussion among your CAD peers, just ask them what it means. We use the conventional acronym of CAD, but we'll let you decide what it means.

CAD is first and foremost a tool. As the A in CAD implies, it is an aid to the designer, whether it be to assist in drafting construction drawings, to visualize the project in 3D, to help maintain office drawing standards, or to track the time and personnel working on a project. Just as a word processor can't write the document for you, a CAD program cannot design the building for you. That is still, thankfully, the role of the designer. CAD is simply a means to an end, in this case the visualization and documentation of a building.

Buildings are collections of three-dimensional spaces and entities. Anything that can help the designer and the client visualize the project will ultimately lead to a better understanding of its form and function. The CAD design process in the past was usually to create the 2D drawings first and then create a 3D model of the finished product. Today, however, designers and clients alike are demanding easy 3D visualization at the front end of a project before a lot of time is wasted on extraneous drawing. And more and more CAD programs are beginning to make the 2D aspects of the software indistinguishable from those of the 3D. You'll hear this referred to as *parametrics*.

Parametric entities can be thought of as "smart" objects. Instead of a wall being made up of a series of individual lines, a wall is a single object that has properties such as thickness, height, and materials. If you want to change any of those properties, you don't have to redraw the lines that make up the wall. All you must do is tell the program what the new thickness, height and materials are, and then the program makes all the appropriate adjustments automatically. If there were other parametric objects in that wall, such as doors and windows, the CAD program would immediately know what changes must be made to those objects due to the changes in the wall. This is almost certainly the wave of the future and is quite promising, but the implementation of this concept still has a number of shortcomings to be worked out. Although the use of 3D CAD is growing, for the time being the bread and butter of any design firm is still the 2D depiction of construction drawings.

In this book we will take three approaches to CAD. The first is the strictly 2D approach. That is, whatever you want to depict has to be drawn line by line, curve by curve. Fortunately, DataCAD excels at this task, as its 2D drafting capabilities are arguably the best on the market. This is very important because even if you design your building in 3D, you will eventually need to create 2D drawings for bidding and construction. But even the strictly 2D approach is not truly limited to 2D in DataCAD, since all 2D entities in DataCAD can have a Z-height. This $2^1/_2$ D feature of DataCAD enables you to get quick 3D images of your 2D work with little or no effort at all.

The second approach is the 3D method. You design the building in 3D, in as much detail as possible or appropriate, and then use the resulting 3D model to create your 2D construction drawings. The resultant drawings depend a great deal on the level of detail placed in the model. Since it is usually impractical to draw every last object that makes up a building, you must decide which 3D elements will give you the most bang for the buck. For instance, drawing all the window flashings, termite shields, and anchor bolts would take an inordinate amount of time to figure out and draw in 3D. It is therefore usually better to concentrate on the important elements, leaving the hidden minutia to the 2D drawing phase.

The final approach is a combined method of 3D and 2D, using the best of both to create the desired output. For instance, we often use the $2^1/_2$ D properties of our floor plan elements to create the elevations, rather than drawing the elevations from scratch. We might use the 3D roof-building tool to create a roof that can be saved as a 2D roof plan. We also often use the 3D stair tool to model a basic stair. From that, we can save 2D plan, elevation, and section views of the stair, saving a great deal of calculation and drawing time. We will be using this approach in the tutorial project in Chapter 4.

With all the great 2D and 3D tools available in DataCAD and other CAD programs, they can still only do so much for an office. CAD is not a panacea. Any computer program can only do what it is told or to extrapolate as best as it can with what limited information it has. However, given good input, you can get excellent output, and the end result can truly be greater than the sum of its parts.

The Basics of CAD

So how does CAD work? What are the keys to its efficiency and how does it automate the drafting process? There is a long list of answers to those questions, to which you will develop your own answers as you work through this book, but we're going to focus on seven of them in this chapter: layers, linetypes, symbols, architectural functions, **Stretch**, **Copy**, and **Change**.

First, a quick definition. In CAD, the word "entity" refers to any single, whole drawing element. A line is one entity. Three lines are three entities. A circle is one round entity. A window is made up of a lot of different entities. Everything displayed in the drawing window is an entity of one type or another. Simple entities include lines, circles, and arcs. Complex entities are made up of any number of simple entities and include polygons, symbols, linetypes, and bezier curves among others (see Figure 1-1).

Layers

Layering is the foundation of CAD and is invariably the first thing that is discussed when learning about it. Think of each layer in a CAD drawing as

Figure 1-1 Simple and compound entities

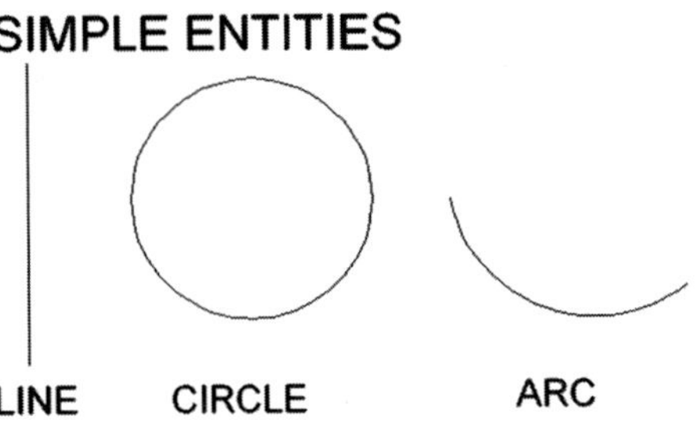

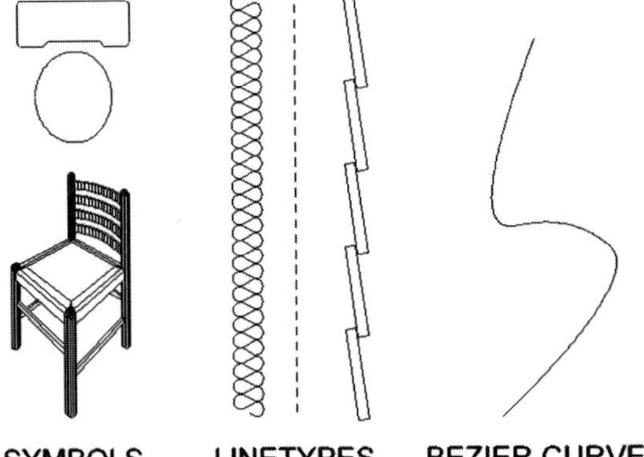

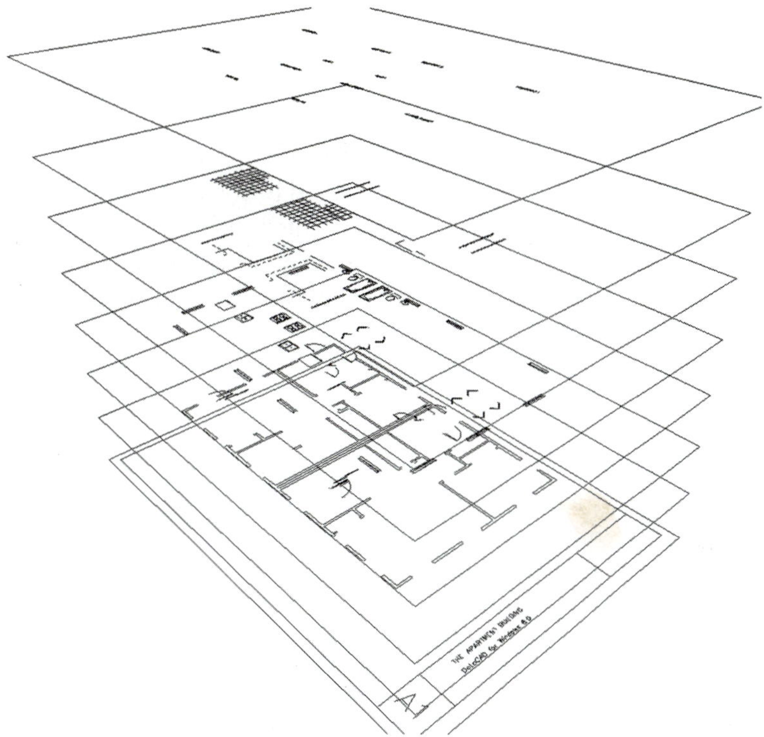

Figure 1-2
An example of layering

a sheet of clear drafting film. Different building entities or groups of entities are drawn on each sheet, as in Figure 1-2. Individual sheets are aligned one on top of the other to form a complete drawing or series of drawings. For instance, you could place a title block on one layer, the exterior walls on one layer, windows and doors on another, and plumbing fixtures on a fourth. By displaying all the sheets, or layers, at the same time, you see a complete drawing with walls, doors, windows, and plumbing fixtures.

So why put all these things on different layers instead of only one? For starters, you usually don't need to display all the layers in a drawing at the same time, so layering enables you to keep your drawing area uncluttered by displaying only the layers you need to see at any one time. For instance, you usually don't need to see the site plan while you are placing bathroom fixtures (see Figure 1-3).

Secondly, what happens if you try to erase a chair from a floor with a tile pattern on it, and both are on the same layer? You may accidentally erase the tile pattern instead of the chair because the two are displayed so closely together. By keeping them on separate layers, it is easier to edit them individually.

Figure 1-3
Select only certain layers, rather than all of them, for clarity

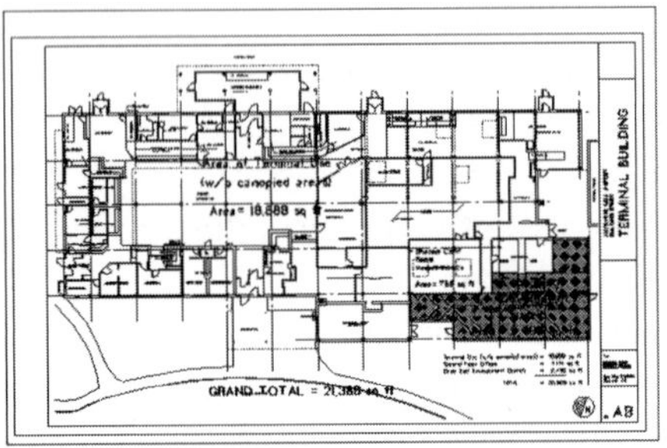

ALL LAYERS TURNED ON
(Notice how cluttered the drawing is)

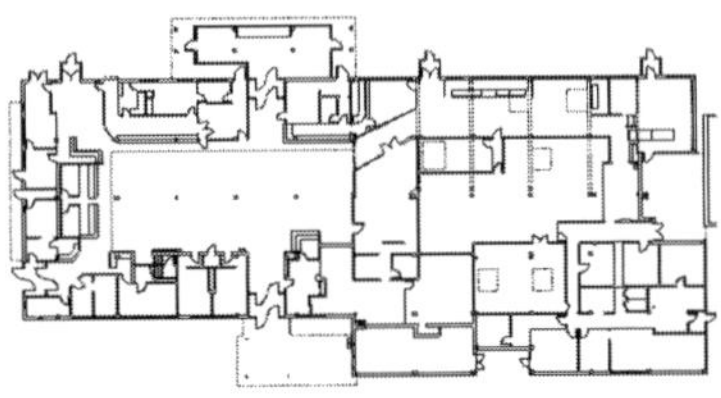

SELECT LAYERS TURNED ON

Thirdly is alignment. In keeping related drawing elements aligned over the top of one another, it is easier to make and keep track of changes that affect those drawing elements. For instance, suppose the first-floor walls are on one layer, the second-floor walls are on another layer, and both are aligned directly over the top of one another. If the first-floor plan changes, you can turn on the second-floor plan layer and make the same changes to that floor as well. If the two drawings were not on separate layers and aligned, it would be difficult to make precisely the same changes to the second floor.

Linetypes

Linetypes in DataCAD are single entities that appear to be multiple entities. By using linetypes (like the batt insulation, dashed line, and wood siding shown in Figure 1–1), these complex graphics can be drawn and edited quickly and easily, saving a great deal of time and effort. Because they are considered to be only one entity by DataCAD, the size of your drawing file will be kept smaller, and that allows DataCAD to run faster.

Symbols

Symbols are one of the most powerful features in any CAD program, and the way in which DataCAD implements them makes symbols doubly efficient.

Think of symbols as elements on a traditional plastic drawing template. In fact, DataCAD symbols are likewise grouped together into templates.

Placing symbols such as furniture and plumbing fixtures with a single mouse click relieves you from having to redraw or tediously copy each of them, saving a great deal of time and effort. Also, if you place 100 2x2 fluorescent light fixture symbols in your drawing and then later decide to change all of those fixtures to 2x4s, you can have DataCAD replace all of them in one step without having to tediously replace each of them one at a time.

2D and 3D symbols exist for cars, furniture, trees, bathroom fixtures, and many more. DataCAD comes with over 3,000 2D and 3D symbols (see Figure 1-4). But best of all, you can create your own symbols. This will all be covered in greater depth in Chapter 14 "Templates and Symbols."

Architectural Functions

DataCAD is, by its very nature, CAD software made *specifically* for architecture. With its functionality and ease of use, it could just as well be used to draw machine parts or electrical circuits, but its commands and functions are centered on the construction industry. This is what makes it such a powerful and efficient tool for designers. Functions like Doors, Windows, Roofs, Columns, Walls, Stairs, and Elevator are purely architectural, and DataCAD has many more similar automatic drawing features.

Doors are one example of how a complex object can be simplified with automatic drawing tools. Something as simple as a door may be made of 13 or more lines (eight for the two jambs, four for the door, and one for the door swing). You could create a symbol for every door in your project, but this method has at least two drawbacks. One, think of how many doors you

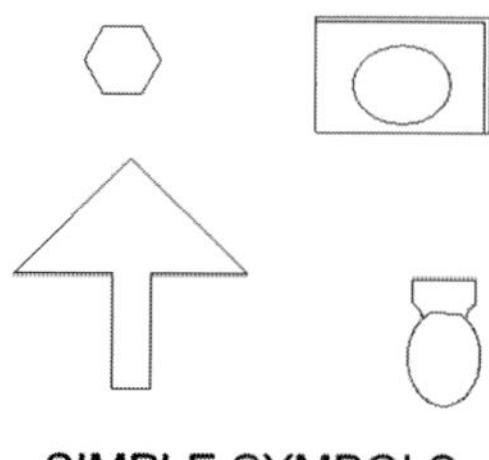

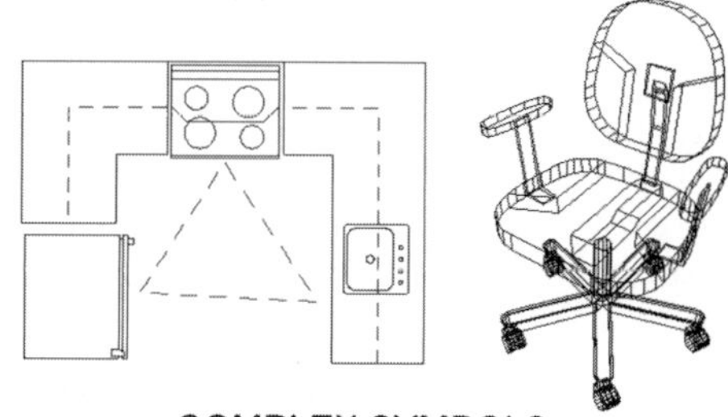

Figure 1-4
Simple and complex symbols

might have on a project. There might be 2′ [610], 2′-6″ [762], 2′-8″ [813], 3′-0″ [914], and 3′-6″ [1066] doors; there may be pocket doors, roll-up garage doors, and sliders; and of all those, there may be many different jamb sizes, adding to the number of door symbols to keep track of. Secondly, if you place a door using a symbol, you still have to manually cut the wall opening. If the door size changes, you will have to trim the wall opening again after placing a new door symbol.

With DataCAD, you can pick the type of door you want, pick the location of the two jambs, and the direction of the door swing. Then DataCAD will cut the wall as required and place the door with the proper jamb dimensions. If you later decide to change the door size, DataCAD will remove the door and repair the wall. You can then place the new door and DataCAD will cut the wall again. Figure 1-5 shows the three points you would have to pick in order for DataCAD to draw a swinging door.

DataCAD will also create more complex items in the same manner. Stairs, especially complex ones like spirals or L- and U-shapes, can take a great deal of time to calculate and draw. DataCAD can automate the process and do the calculations for you. By changing options like floor height, stair width, and rails, you can quickly try out different designs. These architectural features can save an extraordinary amount of time and can provide the kind of accuracy that could not be achieved by manual drafting methods.

Copy

Perhaps the most useful CAD command is **Copy**. It is a feature that helps ensure both efficiency and accuracy. As we pointed out earlier, CAD excels at repetitive tasks, and **Copy** enables the user to repeat an entity or groups of entities quickly and at a specific or uniform distance. This is especially useful for complex forms like window elevations, schedules, or stairs. And

Figure 1-5 Three points to create a door

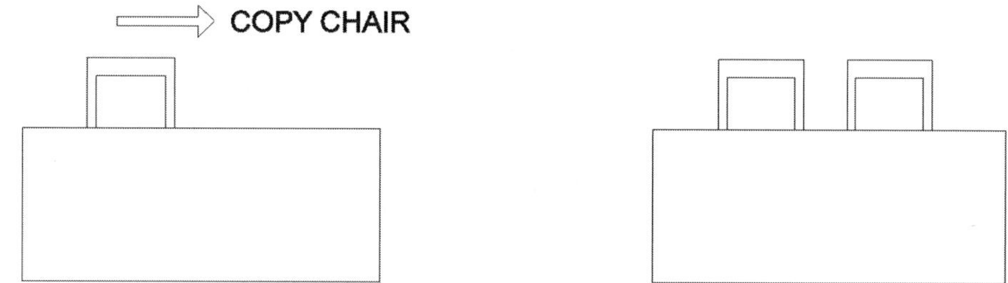

Figure 1-6
An example of **Copy** at work

by copying entities at a specific distance, accuracy is ensured. Anything that can be drawn can be copied, including symbols and 3D elements. Entities, even 2D entities, can even be copied in the third (Z-axis) dimension as well (see Figure 1-6).

Move

Perhaps the second most useful CAD command is **Move**. It is much like the **Copy** command, except that, as the name implies, this command is used to relocate entities from point A to point B. Like the **Copy** command, you can move entities along any axis, X, Y and Z.

Change

No, we're not talking about nickels and dimes here. We're talking about changing entities from one state to another. The beauty of this command is that you can change the appearance or properties of entities without having to redraw them.

Consider a straight, solid line drawn with a pencil on paper. To make that line appear as a dashed line, you would either have to erase many small parts of the line or erase the whole line and redraw it. If you wanted to make that line appear as batt insulation, then you would certainly have to erase the line and redraw it from scratch. In DataCAD, all we have to do is change the line to a dashed line type or a batt insulation line type, saving a great deal of time and effort (see Figure 1-7).

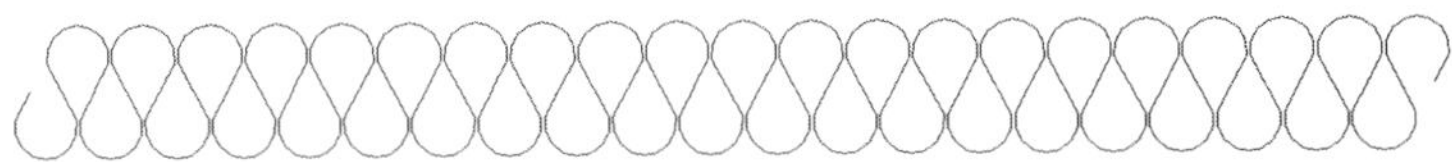

Figure 1-7
Linetype changes

Figure 1-8
Font changes in DataCAD

Text can also be easily changed in DataCAD. We can change the font, size, color, and most importantly the contents of a line of text (see Figure 1-8).

Change can also be used to change the location or size of entities in the Z-axis in both 2D and 3D. DataCAD draws its 2D entities with an extruded Z-height (DataCAD calls this $2^{1}/_{2}$ D) to enable the designer to quickly view their drawing in 3D. Change can be used to alter the location of the top and/or bottom of the extruded entities. This affects how the drawing is viewed in 3D but does not change anything in 2D.

Stretch

Last but not least is the **Stretch** command. You can think of this command like Silly Putty. Once we form the putty into a shape, we can stretch it in any direction to make the shape longer or shorter, or to change it into a new shape. A square can be made into a rectangle by stretching one side, or it can be made into a trapezoid by stretching one or more corners (see Figure 1-9).

Stretch is a natural with DataCAD's associative dimensions and associative hatching. *Associative* means that these entities are smart enough to

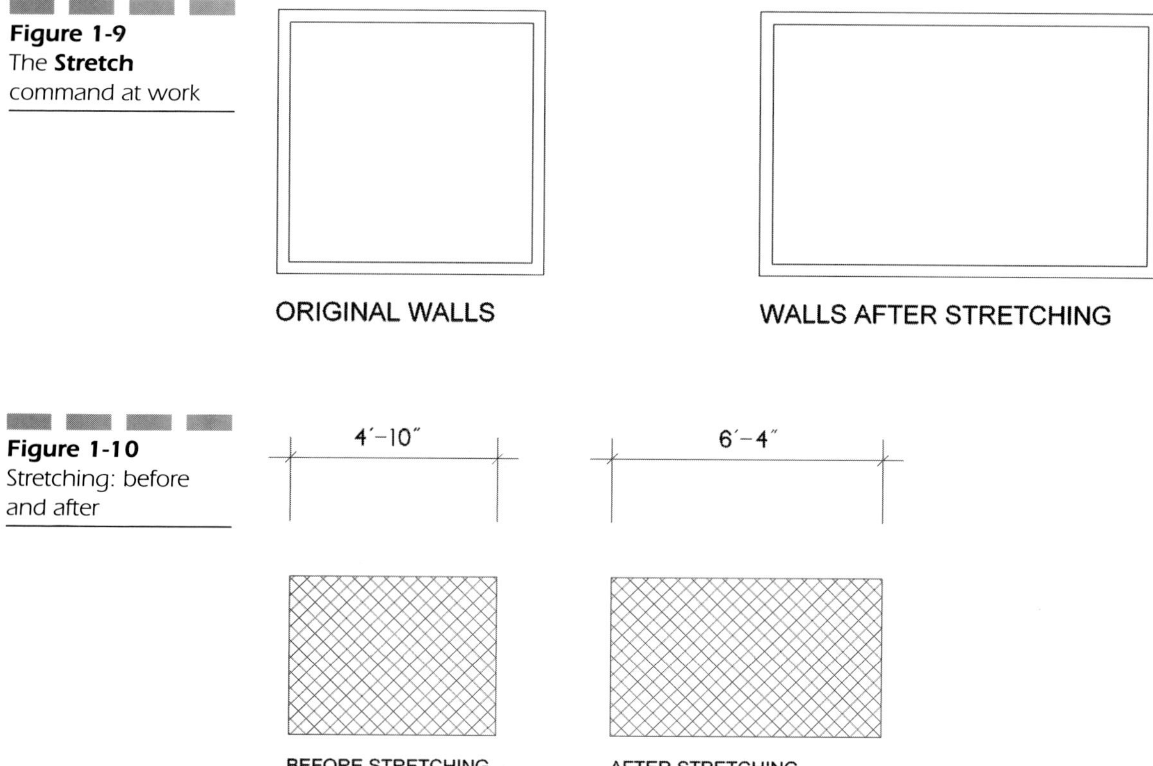

Figure 1-9 The **Stretch** command at work

Figure 1-10 Stretching: before and after

properly change the way they display themselves when, among other things, they are stretched. If a horizontal associative dimension displays a dimension of 3′0″ [914] and you stretch it 1′0″ [305] to the right, the dimension will stretch the leader and the arrowhead 1′0″ [305] to the right, and the numeric dimension display will update itself to show that the dimension is now 4′0″ [1219]. Associative hatching will likewise update the way it displays when stretched (see Figure 1-10).

Reusability

Part of the efficiency of CAD is its capability to reuse any or all elements from one project to the next. Symbols are one example of its reusability. You can draw a symbol once and then reuse it over and over again. The **Copy** command is another example. If we were building a chain of 10 stores on different sites, we could draw the project once, copy the entire project 10 times, and then alter each one to fit the particular site it is to be located on.

Now many of us would cringe at the proposition of "cookie cutter" buildings like this. Perhaps this is an extreme example, but even if the building's form was to change, most of the details, schedules, and basic building elements (in plan, elevation, and section) would remain identical or at least so similar that they would warrant being copied and then modified, rather than redrawn from scratch. Even in traditional pencil drafting, we trace or copy details from one project to the next. It only makes sense not to have to reinvent the wheel on every project.

You might say, "But my projects are all different. How, then, could I benefit from the reusability of CAD?" Again, a building project consists of many different parts, even vastly different ones that are repeated from one project to another. How about schedules (door, window, mechanical, electrical, and finish), door and window details, partition types, stair details, elevator details, roofing details, bathroom plans, landscaping details, ceiling details, furniture, fixtures, equipment, and on and on. I'm sure you can think of even more.

Backing It All Up

Paper drawings can be very fragile. Paper can be ripped or damaged by water. Even mylar drafting film can be badly creased or burn in a fire. We usually protect our paper drawings by keeping them neatly rolled up or put away in flat files. We file our documents in folders in filing cabinets. If you have a special degree of foresight, you may even keep the originals stored away from your office in a fireproof safe. Paper drawings tend to take up a great deal of space, however, especially if you've been in business for any length of time, and making multiple sets of paper drawings can get expensive.

CAD files likewise have their own Achilles' heels. CAD drawings are stored on a magnetic medium such as a computer hard drive or individual floppy disks. Your CAD drawings are subject to accidental erasure, a failed hard drive, crushed or lost floppies, or a fire in the office. In their favor, however, they are easily and cheaply copied, take up very little space, are cheap, and are easily stored in a small fireproof safe. But none of these factors do us any good if we don't use them.

Copying or backing up your CAD files should be considered an essential part of maintaining a successful practice. What would happen to your business right now if all your drawings from the past 10 years went up in flames? And with those projects your two current projects, 25 percent and 75 percent complete, went up in flames as well? What if you office was broken into and all your computers stolen? How would you recover? Could you recover? Would the clients give you an extra few months to do the project over again? It's a scary proposition. With CAD, something as simple as a lightning strike that spikes your computers could put you out of business in a flash (pun intended).

A variety of backup mediums are available today, including tape drives, removable hard drives, Zip and Jaz drives, and rewritable CD-ROM drives. You can get any of these for $100 to $250 U.S. That's the cheapest insurance you could ever hope to buy. But remember that all the backup hardware in the world won't do you any good if you don't use it. The best backup strategy involves daily, weekly, and monthly backups. At the very least, you should back up your data files once a week.

Where you store your backup files is just as important as backing up the information in the first place. Because your home or office is subject to fire, theft, and other unforeseen accidents, it is highly recommended that you store a primary set of backups in your home or office and to also store a duplicate copy of all your backups off-site and in a fireproof safe.

Backup, backup, backup!

This chapter was more of an introduction to CAD in general with some specific references to DataCAD. In Chapter 2, we will discuss what DataCAD can do, get it up and running, and see how to get around in the program. Chapter 4 is a tutorial to teach you the basics of DataCAD. Working through the tutorial is highly recommended as it will make the rest of the book much easier to use and understand.

CHAPTER 2

Getting Started

About DataCAD

To quote the DataCAD Reference Manual, "DataCAD is a powerful computer-aided design (CAD) system specifically developed for architectural, engineering, and construction applications." What is significant about that statement is the part about being developed for construction industry professionals. In contrast, programs like AutoCAD are generic CAD programs that are built to do a wide variety of things, none of them extraordinarily well. DataCAD was developed from the ground up to do one thing well, to provide a program that would intuitively cater to the design professional. The party line at DataCAD LLC has always been "DataCAD: designed *for* architects *by* architects."

It is useful to know that until Version 8, DataCAD had been a DOS program only. With a few exceptions, what you see on your monitor is just what has been seen in the DOS version for years. For those of you accustomed to standard Windows program dialogs, toolbars, and scrollbars, the look of the basic DataCAD screen may seem somewhat archaic. You will come to find, however, that this is actually one of DataCAD's strengths. The interface is refreshingly uncluttered and simple, making navigation and execution that much easier.

Here's my personal top 10 list of reasons why DataCAD is better than other CAD programs for the design professional:

1. *A short learning curve* Days or weeks instead of months
2. *Keyboard shortcuts* Customizable, fast access to dozens of functions. No more digging through endless drop-down menus that obscure your view of the drawing.
3. *Thousands of symbols* A huge array (over 3,000) of architectural and engineering symbols are a standard part of the program and are displayed visually for selection, not just by name.
4. *Defaults to drawing lines* DataCAD defaults at all times to drawing lines and walls. You never have to access a menu to draw a line or wall.
5. *Customizability* DataCAD is easily customized by any end user. No programming or complicated script writing is involved in creating custom shortcuts and toolbars.
6. *3D* Some of the easiest-to-use 3D modeling around. DataCAD will even build complicated roofs, stairs, doors, and windows with a few clicks of the mouse. This goes for 2D entities as well, since DataCAD enables all of them to have a height and to be viewed three-dimensionally.
7. *3D GLShader* Fill the solids of your model with nearly instant color. Instead of waiting minutes for a hidden line drawing or hours for a full

rendering, the GLShader will give you quick color feedback in only seconds.

8. *Third-party support* The power and flexibility of DataCAD's fully documented DCAL programming language has ensured a large base of third-party software support.
9. *Grass roots support* Perhaps more than any other CAD program, DataCAD has a large grass roots following, with numerous user groups throughout the world, and a very active and helpful Internet discussion group.
10. *Middle mouse button snapping* This has got to be one of the greatest features ever! This feature alone will endear you to DataCAD.

Why Use DataCAD?

Ease of use, price, and performance. These are hallmarks of the DataCAD program. Other CAD programs may have an edge in some of these categories, but DataCAD is a clear leader when you consider all of them together.

Surely, I'm somewhat biased. I use DataCAD every day in my practice as an architect and have for the past six years. But that's exactly why I have come to regard DataCAD as an exceptional CAD program for architects. It allows me to practice architecture rather than continually struggling to learn how to operate the software. DataCAD operates like architects think. It is a very intuitive program, requiring much less time to master than competing CAD programs. When new employees with no exposure to DataCAD come into our office, they are very quick to pick it up, becoming proficient in a matter of only one to two weeks. Conversely, even avid users of programs such as AutoCAD will tell you that the learning curve for their software is very steep, often requiring months to become proficient.

DataCAD does architecture right out of the box, at a cost of under $1,000 per station. No architectural add-ons are required. At a cost of over $3,000 per station, AutoCAD can't even draw a simple door out of the box. It requires a third-party architectural front-end program to perform even basic architectural tasks at an additional cost of up to $1,000 per station. Simple math shows us that based on cost alone the DataCAD office can equip four persons for the cost of equipping one person in the AutoCAD office! On the average, AutoCAD version upgrades have cost around four times as much as a DataCAD upgrade.

The typical third-party DataCAD software macro costs between $5 and $30. A macro is like a mini-program that works within DataCAD to accomplish tasks that DataCAD does not do on its own.

Since DataCAD requires no third-party architectural add-ons, a DataCAD user from one office can begin work in a new DataCAD office with no retraining. If an AutoCAD user changes offices, and the new office uses a different architectural front-end program, he must be retrained to use the new interface. The differences can be considerable, requiring a lot of time to learn.

Don't let DataCAD's ease of use and low price fool you. Pound for pound, it is a very powerful program. In 1994, a "Top Gun" CAD contest was sponsored by Oran Woody and Houston CADD of Spring, Texas. The task was to monitor a small project for the number of keystrokes it took for a number of different CAD programs (with experienced operators) to produce the exact same drawing of a floor plan. These were the results:

- DataCAD: 620 strokes
- CADvance: 845 strokes
- AutoCAD (with third party add-on): 940 strokes
- VersaCAD: 975 strokes
- Integraph: 1090 strokes
- AutoCAD (with no third party add-on): 1,180 stokes
- Other CAD programs: 1,310 strokes

Another point worth noting, DataCAD LLC, the company that produces DataCAD, has traditionally maintained a close relationship with its users. DataCAD LLC has shown a strong reluctance to add features for the sake of adding features. Too much software these days contains features that the user really doesn't need or want, adding to bloated code, poor performance, and cluttered menus. Recent additions to DataCAD have come directly from end users. In 1996, the company put out a long laundry list of potential software features that DataCAD was considering adding. Users responded by checking the items that they wanted and returning the survey to DataCAD LLC. This list has led to new and improved features that architects and engineers really want and need.

What DataCAD Does Not Do

Every software program has its strong and weak points. You can't please everyone, and you can't anticipate everything that a program needs to do.

That's true of DataCAD just as any other CAD program. Knowing what a program cannot do is nearly as important as knowing what it can.

DataCAD can do great tricks such as automatically inserting windows and doors of varying types, building complicated roofs and stairs, and creating multiple-line walls with multiple line types in one easy step. But it is not a true "parametric" CAD program. Parametrics is the CAD buzzword of the late 1990s, and although this is the direction that many leading programs are leaning toward, some difficult problems must be worked out before the programs become truly workable.

A truly parametric program is one that enables the user to define the pieces of architectural elements, such as the frame, muntins, sill, and size of a window, as well as have the program automatically insert them into the drawing and "database" that information. If the user redefines any of the pieces in the database (sizes, shapes, or quantities), the inserted element (window, door, or roof) will automatically be updated throughout the drawing. DataCAD can do this with symbols, but not for automatically drawn elements like windows, doors, stairs, and so on. In DataCAD, if we want to change our hypothetical window, we must first remove the window and then redefine its elements and reinsert the window in its previous location. If we have multiple windows to change in the project, we must erase each of them and reinsert the new windows in all those locations.

Parametric CAD programs have been know to be more "gee-whiz" than truly useful. They often make impressive 3D demonstrations, but the bread and butter of any architectural office is in the production of 2D construction drawings. Parametric programs can generally only produce parametric building elements that the programmers were wise enough to see and be able to program in. If you want to create a custom window, storefront, or detail that the programmers hadn't counted on, then you'll be out of luck and will have to draw them the "old-fashioned way." If parametrics can be perfected, then they are truly the wave of the future, but for now I will place my money on the productivity advantages of programs like DataCAD.

Many of those same parametric programs do not make a separation between 2D and 3D drawing elements, and the smarter ones even know that they must display elements differently in 2D than in 3D. For better or worse, DataCAD 9 does separate the two, yet maintains a very close relationship between its 2D and its 3D modules. It's so close, in fact, that sometimes you will find yourself in 3D when you thought you were in 2D, and some functions you will use in producing 2D drawings will come from the 3D module. DataCAD's 2 $^1/_2$ D capabilities also blur the line, even within DataCAD itself, between what is 2D and what is 3D.

With all that said, I should mention that DataCAD LLC has recently introduced their own parametric CAD program, called *DataCAD Plus.*

What I like about this new program is that it is the standard DataCAD program at its core. So, most of the functionality that is in DataCAD 9 is also in DataCAD Plus, but with more bells and whistles, albeit with a higher price tag. It looks like a great integration of the ease and speed of DataCAD, with the added functionality of parametric entities and additional rendering options. For more information, stop by the DataCAD Web site (`www.datacad.com`).

Like it or not (and if I liked it I wouldn't be writing this book), AutoCAD is by far the most commonly used CAD program for architecture in the world (which is somewhat humorous given that AutoCAD by itself does not do architecture). But since AutoCAD is currently the apparent industry standard by measure of the sheer volume of users, anyone using DataCAD will invariably encounter the problem of translating drawings to and from the AutoCAD drawing format (represented by the .dwg file extension). DataCAD, however, does provide a decent conversion utility inside of DataCAD to enable users to read in and save out any version of AutoCAD up to ACAD 2000. Generally the utility works with little or no intervention, but other times you will be confronted with incompatibilities that must be sorted out. DataCAD does not support some of AutoCAD's key features, namely Paper Space and XREFs, though DataCAD does have its own version of these features.

Perhaps DataCAD's biggest shortcoming is its perception as a fringe program, even though a recent *American Institute of Architects* (AIA) survey showed DataCAD was the second most used CAD program within the construction industry in the United States.

Metrics in DataCAD

DataCAD implements metrics quite well and, in fact, can convert between metric and imperial units within the drawing with the click of a button (as long as your dimensions are associative). In an effort to appease all the world's various metric users, DataCAD has six different metric implementations (see Chapter 3, "Settings and Display Options," for a description of each):

- Metric
- Meters
- Centimeters
- Millimeters

- DIN (German DIN meters)
- AS 1100 (Australian standard meters)

Since the authors of this book reside in the United States, the units used throughout the book are Imperial. Their metric equivalents are shown in parentheses for the rest of the world. However, since various implementations of metric dimensioning exist throughout the world, we had to pick a standard and go with it. For better or worse, we chose to go with millimeters, as represented by numbers displayed in brackets, like this: 2′-6″ [762].

New Features

In your main DataCAD directory is a file called *ReadMe.win*. It contains a section outlining the various features and fixes for each version since the original release of DataCAD for Windows 8.0. This file contains a great deal of useful information, and we consider it required reading. The following are just the highlights of the new features found in DataCAD 9:

- A *multiple document interface* (MDI), giving users the ability to have an unlimited number of DataCAD files open simultaneously
- Cut and paste capabilities of private vector and public metafile data between DataCAD and other computer applications
- Global unlimited Undo and Redo
- Context-sensitive toolbars for navigation that can be repositioned anywhere on your screen
- External reference files (XREFs) providing users with the ability to reference an unlimited number of DataCAD drawings within a single DataCAD file
- Entity-to-view mapping, which enables users to link drawing data to specific drawings or details. Click on a linked entity, and the linked view will appear in the Drawing Window.
- Batch plotting, which supports batch plotting of single or multiple drawings with multiple plot layouts. Users can queue up entire print sets using a simple, Explorer-style interface to access any number of plot layouts and then automatically plot these to multiple devices or files.
- A new OpenGL-based shader for fast rendering and hidden line removal. The new shader has also been enhanced to support multiple light sources and control over lighting effects.

- Integrated 32-bit Windows estimating providing users with the ability to assign materials and costs to drawing entities and symbols. This information can then be exported directly to Estimator, where additional information such as overhead, labor costs, and profit can be added. DataCAD users can also export material take-offs directly from the 3D FrameIt macro included with DataCAD. The Estimator for Windows will output to popular business accounting programs, such as Quicken® and Quickbooks®.
- The font previews are back in this version. They were lost in the conversion from DOS to Windows.
- The linetype and hatch previews do not used saved bitmaps to display a preview. They now display a preview based on the linetype and hatch data in the DCADWIN.LIN and DCADWIN.PAT files, respectively.
- A new and more functional Directory dialog
- A plot preview window with full pan and zoom functionality
- Plot previews can be saved in Adobe Acrobat (PDF) format.
- 3D *GotoViews* (GTVs) can now be accessed from the 2D menus.

The DOS Version versus the Windows Version

For those users who may be making the leap from the DOS version of DataCAD to the Windows version, first read the new features list for each of the Windows versions in the Readme.win file. Then review the following outline for additional major features of the Windows version that are the same as, different from, or not available in the DOS version.

The DataCAD for Windows features consist of the following:

- It is a true 32-bit application and runs native under Windows and NT.
- It contains a standard Windows menu bar with pull-down selections. Some are standard DataCAD commands, while others are new to DataCAD, such as **New**, **Open**, **Close**, **Save As**, and **Import/Export** DWG/DXF. These pulldown windows menus can be accessed with **Alt**+ and **Ctrl**+ key combinations.
- The last four files accessed in DataCAD are available under the File menu.
- A very complete, Windows-standard Help system has been added.

- DataCAD can be minimized, maximized, or run in a window.
- Now you can toggle or click between DataCAD and other Windows applications.
- Bitmaps and metafiles of the Drawing Window and plot previews can be exported.
- All toolbar icons are standard .BMP images rather than proprietary .POF images.
- The user interface is extremely customizable and uses True Type fonts. Most settings are alterable under the **Tools/Program Preferences** menu from the Windows menu bar.
- All printing and plotting is now done with standard Windows printer/plotter drivers.
- Fractional dimensions can now be optionally stacked.
- More metric options support various worldwide standards.
- The scale type for area/perimeter calculations is now remembered independently from the current drawing scale type.
- The new setting **AreaPrec** controls the number of digits of precision to be displayed for area calculations when the **ScaleTyp** option in the Area/Per menu is set to **Decimal**, **InchDec**, **Meters**, **Centimtr**, **Millimtr**, **DIN**, or **AS1100**.
- The 0 in dimensions less than 1″ (such as 0 $^{3}/_{16}$″) does not appear anymore, only the fraction.
- A SetPens menu is available to map pen colors, thicknesses, and screening. All 255 colors can now be assigned to any of 15 pens. All settings in this dialog are stored within each drawing file.
- A new Display List function has been implemented in the Windows version. It is far better and much faster than the old DOS implementation.
- The DWG/DXF Import/Export function works for versions R12 through R2000:
  - Colors translate properly in both directions.
  - DCAD's custom linetypes are translated to ACAD format.
  - You can export all layers or just the activated ones.
  - The translator does not run in a DOS window.
  - It is much more crash-resistant.
  - Imports have a user interface to assign unidentified linetypes, hatches, and fonts.

- You can read DataCAD 9 files in older versions of DataCAD 5, 6, 7, or 8, but no backward compatability is available in *multi-scale plotting* (MSP) from DataCAD 9 to pre-DataCAD 8 versions (in most cases, you would lose your multi-scale plot layouts). To get around this, you can start a drawing in any one of the DOS versions of DataCAD (assuming you have it) and then continue working on it in the Windows version. The resulting version 9 file should then be backward-compatible.
- You can transfer your DOS linetypes, hatches, keyboard macro files, and other customized files to DataCAD 9, but it has to be done manually. In most cases, the existing files need only have the prefix name changed from DCAD to DCADWIN. For instance, your DCAD.LIN file from the DOS version would be renamed to DCADWIN.LIN to be referenced under DataCAD 9. The Readme.win file on the DataCAD CD gives all the details.
- Long file names are supported by DataCAD 9, but it only supports continuous-character file names (no spaces are allowed).
- The 12-MB file size limitation of the DOS version has been increased to 20 MB for the Windows version.
- Support for Windows cut, copy and paste is available.
- All version 5 through 9 DCAL macros work in DataCAD 9, and you can still use DCAL for writing macros.
- Many options are now accessed through standard Windows dialog boxes rather than through the side menu window, such as fonts and toolbox macros.
- When you select **Change/Text** and select a string of text, the cursor appears at the end of the message line. In the DOS version, the cursor appears at the beginning of the line:
  - Pressing **Home** will send the cursor to the beginning of the line.
  - Pressing **End** will send the cursor to the end of the line.
- The process of printing and plotting has been completely changed.

System Requirements

You should ensure that you have the following minimum hardware in order for DataCAD 9 for Windows to run properly:

- Windows 95 or higher, or Windows NT 4.0 or higher
- An IBM-compatible 80486 or Pentium-based computer
  - A Pentium 166 or higher is recommended.
  - A Pentium II or III type processor will greatly enhance performance.
- 32 MB of system RAM (64 MB or more will improve performance)
  - 32 MB is recommended for Windows 95/98.
  - 64 MB is recommended for NT.
- A minimum of 20 MB of free disk space. 200 MB is required for a full installation.
- A CD-ROM drive
- A VGA or SVGA graphics card supporting a minimum of 16-bit (High) color. It is not recommended to run the software at only 256 colors, since many colors will not display properly at such low color depth. A minimum of two MB of video card RAM is recommended.
- A Windows-compatible mouse or compatible digitizer/tablet. A three-button mouse is highly recommended. The Logitech Mouse (with three or four buttons) and the MS Intellimouse work well.

Installation

When the installation prompts you for the name of a directory to install DataCAD in, the default choice will be shown as DATACAD. We suggest you name it DCADWIN. That is the name we will be using throughout this book to denote DataCAD's "root" directory. To begin installation, follow these steps:

1. Insert the CD-ROM in your computer.
2. Click on the Windows **Start** button.
3. Click on **Run**.
4. Click on **Browse**.
5. Double-click on the My Computer icon.
6. Your CD-ROM drive icon should say something like "**DataCAD9**." Double-click on your CD-ROM drive's icon to open it.
7. Double-click on the **Setup.exe** icon by clicking on it.

8. Click on the **OK** button in the Run dialog box.
9. The installation program will begin.
10. When prompted, input the **Name** and **Company** that you want to appear in the About tab in DataCAD.
11. In the **Serial:** box, type in the unique software serial number that comes with the software (not required for the demo version). This number can be found on the DataCAD registration card.
12. Click on **Next**.
13. You will be asked where you want to install the program and all its associated files. By default, it selects C:\DATACAD. If you want to install it to a different directory, use the **Browse** button and type in a new path name. We suggest you call the folder DCADWIN.
14. Once you have the correct path chosen, click the **Next** button.
15. Choose which type of installation you want: **Typical**, **Custom**, or **Complete**.

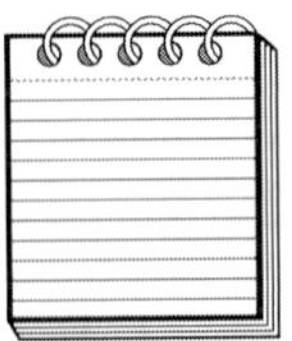

NOTE *You should choose the* ***Custom*** *installation option for one particular reason. When installing the templates and symbols, you will have the option of installing them in a more user-friendly CSI format.*

16. If selecting the **Custom** option, check all the boxes that you want to install.
17. Select **Next**.
18. Choose whether you want DataCAD to default to using **Imperial** or **Metric** units. You can change this setting later in DataCAD if you want to. Select **Next**.
19. Read the remainder of the installation prompts, pressing **Next** between each one. DataCAD and all the options you chose will be installed
20. At the end of the process, you will be prompted to install the hardware key (often called a *dongle*). Plug it into the parallel port of your computer (this does not apply to the demo version).
21. You will then be asked if you want to read the ReadMe file. We suggest that you do read it!
22. Click **Finish** and the installation will be completed.

23. Three new shortcut icons will be added to your Windows Desktop: DataCAD 9, DCViewer, and Estimator.
24. If you are running Windows NT and you get an error message when trying to run DataCAD, there may be a conflict with the hardware key. On the DataCAD CD-ROM, open the folder called **HardLock** and read the **Readme.txt** file for help.

A typical installation consists of the following:

DataCAD 9 for Windows	3.3 MB
Standard templates & symbols	16.6 MB
NEW! Templates & symbols	36.8 MB
DataCAD 9 online help	4.8 MB
DCAL Macros	2.1 MB
	Total 63.6 MB

A Complete installation includes the following:

DataCAD 9 for Windows	3.3 MB
Standard templates & symbols	16.6 MB
NEW! Templates & symbols	36.8 MB
DataCAD 9 online help	4.8 MB
DCAL Macros	2.1 MB
DataCAD Estimator for Windows	14.1 MB
Productivity Pack v.3.0 program files	1.3 MB
Productivity Pack v.3.0 linetypes	.6 MB
Productivity Pack v.3.0 keyboard macros	.1 MB
Sample drawings	19.1 MB
DC Viewer	7.8 MB
Renderize Live	17.4 MB
DataCAD Applications Language (DCAL)	1.5 MB
	Total 125.5 MB

In a Custom installation, all of the applications listed in the Complete installation are available for installation individually or in any combination. Use this if you want to install applications that you opted not to install originally or if you need to reinstall one or more applications.

Software Protection

Software piracy is unfortunately a rampant problem for all software developers, DataCAD LLC included. It is illegal to buy a single license or seat of a piece of commercial software and then install it on multiple computers for use by multiple individuals. Even so, this has been common practice for many companies and individuals, which seriously cuts into the profits of the software companies who cannot afford to stay in business or to continue research and development without that income.

To combat the problem, DataCAD employs a two-level software protection scheme (not in the demo version). First of all, with every copy of DataCAD, a small hardware lock (or dongle) is shipped along with the software. It plugs into the computer's parallel port (the port typically hooked up to your printer or plotter). You can install DataCAD without it, but you cannot run the program. The dongle has a pass-through feature that enables you to place the dongle between your computer and any cable that is connected to your parallel port. Secondly, the DCADWIN.EXE file is encrypted with a software code that reads information off of the dongle in order to start DataCAD. This is why DataCAD takes a few seconds to start after clicking on the program icon; the encryption code is being decrypted.

A few parallel port hardware devices apparently do not get along too well with the dongle, even though it is supposed to be a passive, pass-through device. If the device doesn't work after installing the dongle, you will probably have to take the dongle off temporarily to allow the device to work.

Performance Enhancements

You can enhance the performance of DataCAD, and probably your other applications as well, by doing some or all of the following:

- Use a Pentium II or III computer.
- In Windows, go to **My Computer\Control Panel\System\Performance:**
  - Under the Performance Status heading status, File System and Virtual Memory should be **32-bit**.
  - Select **File System**. Under "Typical role of this computer:," select **Network server** (even if you are not on one, as it increases the cache settings). Make sure "Read-ahead optimization:" is set to **Full**.
  - Select **Graphics** and set the slider to **Full**.

- Select **Virtual Memory**. Check the option "Let me specify my own virtual memory settings." Set the **Minimum** and **Maximum** boxes (the permanent swap file size) to 2.5 to three times the memory you have on your computer. For instance, if you have 32 MB of RAM in your computer, set the minimum and maximum to between **80** and **96**. You will get a warning message. Ignore it and continue on. Restart your computer for the settings to take effect.

DataCAD Directory Tree

This section covers the subdirectories located in the DCADWIN directory and a description of what is located in each.

\DCADWIN is the main (root) directory of DataCAD. Besides containing the following subdirectories, it contains the DataCAD executable program:

BMP\	Saved GL Shader BMP images
CHR\	Text and font files
DCAL\	DCAL files
DCVIEWER\	DC Viewer program files
DCX\	Macro files
DEFAULT\	Default drawings
DWG\	Drawing files
ESTWIN\	DataCAD Estimator program files
FRM\	Template report form files
HELP\	Online help and online documentation
LYR\	Layer files from the Layers menu and LyrUtil macro
PLT\	Plot and text files
RAY\	DC Render drawing files
SUP\	Program support files and default files
SUP\MENUPOF\	Support files for icon toolbars
SYM\	Symbol files
TEMP\	DataCAD temporary files
TPL\	Template files
XFER\	Exported DXF and DWG drawings

Configuration

Now that you have successfully installed DataCAD, you will need to configure the program to work properly and efficiently. Here are a few simple settings to get you started. Chapters 3 and 21 give more detailed information on file and system settings to enhance speed and performance. Start DataCAD, enter any name for a new drawing file, and select **Open**. Go to the Windows menu bar at the top of your screen and perform the following steps:

1. Select **Tools/Program Preferences/Misc**. Check the box next to **Display List/On/Off**.
2. Select **Tools/Program Preferences/Interface Settings**. You will see a graphic image that looks like a small version of the DataCAD program (see Figure 2-1).

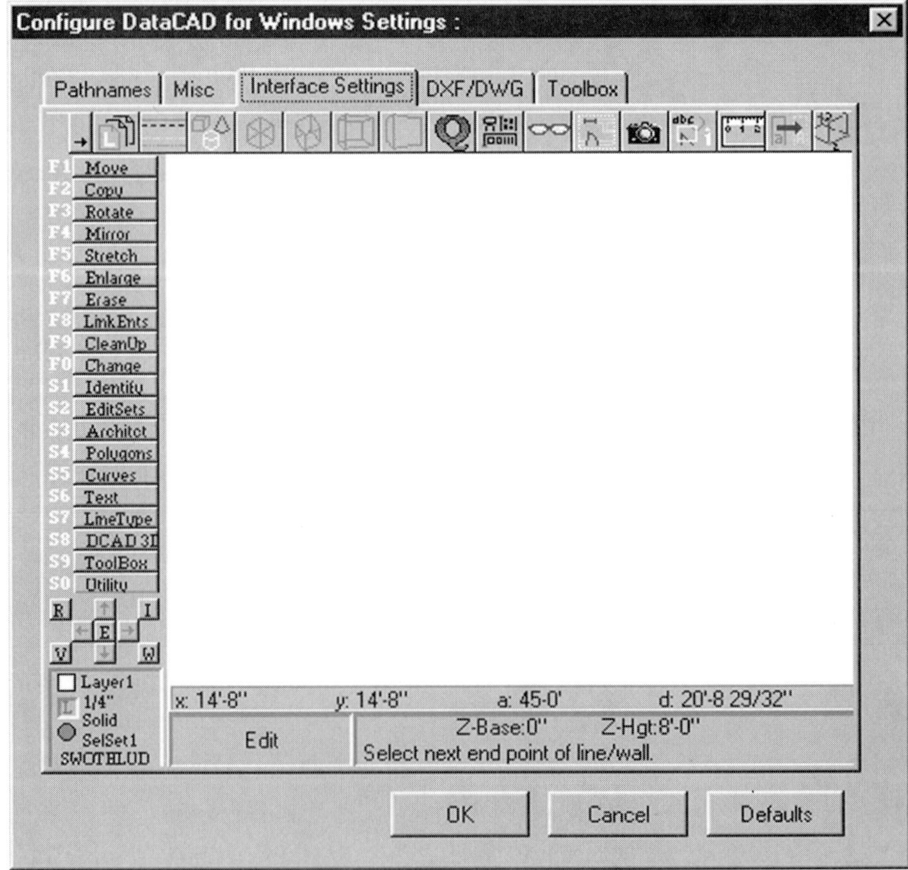

Figure 2-1 The **Interface Settings** dialog

a. Move the cursor around the graphic, pausing occasionally to see the help tags that tell you which DataCAD settings the various areas in the graphic control.
b. Click on the image of the vertical menu buttons. A Windows dialog box appears. Here you can pick the True Type font, the font style, and the font size to appear on the DataCAD menu buttons. This is especially useful if some of the menu button text doesn't quite fit in the width of the menu buttons. Click **Cancel** or **OK** to exit the dialog.
c. Now click in the area below the image of the menu buttons. The menu buttons will jump to the other side of the graphic, allowing you to choose which side you want the menus on.
d. Click on the image of the message area; the horizontal gray area along the bottom. Like the menu buttons, you can now pick the True Type font, the font style, and the font size to appear in the message area. Click **Cancel** or **OK** to exit the dialog.
e. Click in the large open area (the Drawing Window) of the graphic. A Windows color selection dialog box appears. With this, you can pick the color of the Drawing Window. We suggest only black or white. For now leave the Drawing Window black.
f. Now click on the horizontal bar across the top of the image. This turns the icon bar on or off. Make sure it is on; small graphic icons should be displayed.

3. Select **Tools/Program Preferences/Pathnames** There is nothing to set here for now, but take note of all the various pathnames shown. They all assume that each folder shown is located within the root \DCADWIN directory. You will want to change some of these later if you plan to place things like drawing files, templates, symbols, and DXF and DWG files on a network file server (see Chapter 26, "Networking").
4. If you have made any setting changes, be sure to click on **OK** to enable and save the changes. Select **Cancel** if you don't want to make any changes.
5. Select **Utility/Input Mode** and check the **Relative Polar** setting.

Now let's set a few things in the DataCAD menus, the vertical buttons on either side of your DataCAD screen:

1. *Right-click* your mouse a few times. Notice that the vertical menus change each time, but also notice that there are only two menus, an **Edit** menu and a **Utility** menu. The menu you are in is shown in the

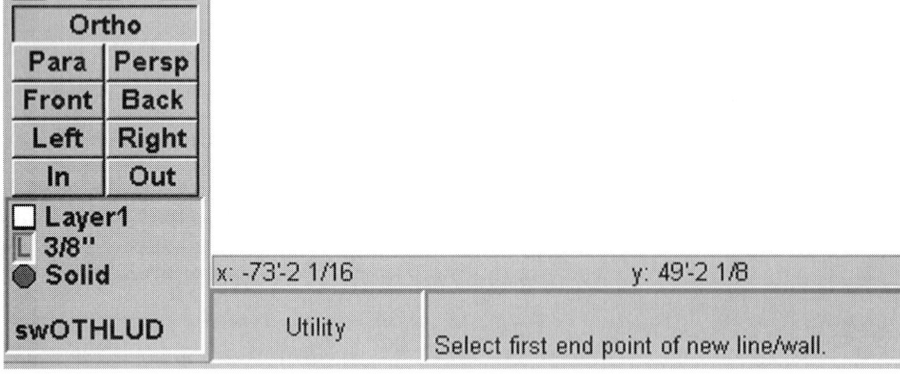

Figure 2-2 `Utility` is the current menu here.

lower left corner of the screen, to the right of the word "SWOTHLUD" (if your menus are on the left—see Figure 2-2).

2. *Right-click* until the **Edit** menu is displayed. Select **Text** and turn on the **TxtScale** button (see Chapter 5, "Basic Drawing") by clicking on it. It is on if the button text turns green.
3. *Right-click* until the **Utility** menu is displayed. Select **ScaleTyp.** Choose the scale type that is appropriate for your geographical location. For architectural drawings in Imperial units, select **Arch**. This book is based on that setting with millimeters shown in parenthesis. *Right-click* to back out of the menu until you reach the main **Utility** menu again.
4. Select **Utility/Grids/SnapGrid**. Make sure the **SnapGrid** button turns green to indicate it is on. Now click on the **GridSize** button and select **SetSnap**. Pick the **1**″ [25.4] setting. *Right-click* to back out of the menu until you reach the main **Utility** menu again.
5. Select **Utility/Display**. Turn the following buttons on if they are not already. The rest should be turned off:
   - **a.** *ShowTxt*
   - **b.** *ShowDim*
   - **c.** *ShowHtch*
   - **d.** *ShowWgt*
   - **e.** *UserLine*
   - **f.** *ShowIns*
   - **g.** *CurveCtr*
   - **h.** *DimPoint*
6. From the **Utility** menu, select **Utility/ObjSnap**. Turn the following buttons on if they are not already. The rest should be turned off:

a. *EndPnt*
b. *MidPnt*
c. *Center*
d. *Quadrant*
e. *Intsect*
f. *LyrSnap*

7. *Right-click* back to either the **Edit** or **Utility** menus.

You are done for now. To read more about the settings you just selected, see Chapter 3.

Mice

DataCAD works well with a two-button mouse but is best suited to a three-button mouse. The middle button of a three-button mouse is used for object snapping in DataCAD (see Chapter 3) and greatly simplifies and speeds up the drawing process. You can accomplish the same thing by pressing the **Nn** key, but it is not nearly as quick and efficient. It is highly recommended that you get a three-button mouse for use with DataCAD.

See Chapter 21, "Techno-Files," for information about particular mice that users have had good results with. Note that in some cases you cannot assign the middle button to double-click in Windows and still have the middle button operate for snapping in DataCAD. You may be forced to pick one or the other.

Starting DataCAD

In this chapter and later in Chapter 3, we will walk you through some of the basics of DataCAD operation. You'll learn how to open and close files and how to get around in DataCAD.

Well, let's get drawing, shall we? The first thing we need to do is fire up DataCAD. Start Windows and then *double-click* on the DataCAD icon.

The Startup Screen

A blank DataCAD screen appears when you start DataCAD or when you close all the drawings you are working on (See Figure 2-5). Select **File/ Open**, and the **File/ Open** dialog box appears (see Figure 2-3).

Figure 2-3 The **File/Open** dialog

Just as with any other Windows application, you can do one of two things:

- Select an existing drawing from the dialog box by *clicking* on the file and selecting **Open** or by *double-clicking* on the file.
- Type in the name of a new drawing that you want to create and press **Open**. DataCAD will create a new drawing using the default drawing indicated in the **Tools/Program Preferences** dialog.

One other option is available to you at the Startup screen. If you want to start a new drawing using a default drawing other than the one indicated in the **Tools/Program Preferences** dialog, you can select that default file. From the **File** menu in the Menu bar, select the **New** option and then click on the **Default** button. The drawing files in the \Default directory will appear in a Windows dialog box. Pick the file you want and press **Open**. Now you can type a new name for your new drawing and select **Create** (see Figure 2-4).

The DataCAD Screen

Eight major elements make up the DataCAD screen (see Figure 2-5):

- *Windows Bars* At the top of the screen are the Windows **Title Bar** and **Menu Bar**. The title bar is the topmost bar and shows the directory path and the name of the current drawing. The menu bar is directly below the title bar and contains standard Windows program options like **File** and **Help**.
- *Drawing Window* This is where all of the elements of your drawings are created and displayed. When you move the mouse, the cursor moves

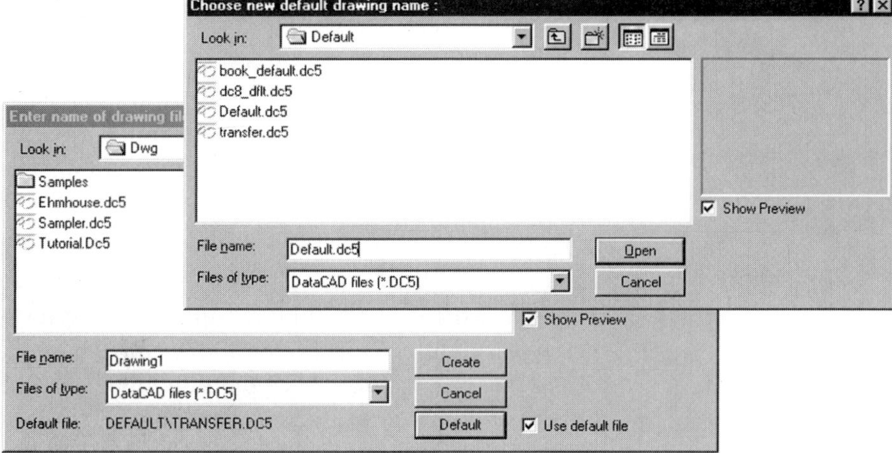

Figure 2-4
Picking a default drawing

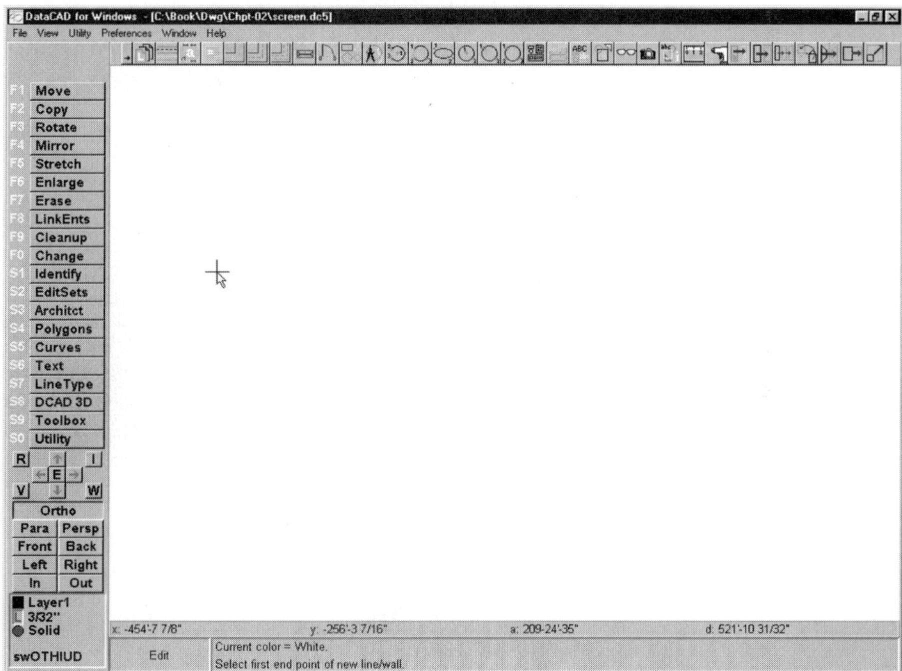

Figure 2-5
The DataCAD screen

around the Drawing Window. When the cursor moves out of the Drawing Window, it changes its appearance to an arrow. In DataCAD, you have the option of displaying a small cursor or a big cursor that is displayed from edge to edge of the Drawing Window. This is useful for aligning things as you draw them.

- *Icon Toolbar* This contains a user-customizable group of graphic icons that enable quick access to DataCAD commands without having to dig through multiple menus. When you hold the cursor over one of the icons, a flyout message will appear below the icon, telling you what the icon does. If you prefer to maximize your Drawing Window, the toolbars can be turned off. Every icon command can be accessed in the menus. No hidden commands exist in DataCAD.
- *Menu Window* All of DataCAD's commands are accessed from the buttons displayed in this menu window. When you hold the cursor over one of the buttons, it will be highlighted with a green border and a message will appear at the bottom of the screen in the Message window telling you what it does. The menu window buttons will change as different options are selected. Some buttons are just on/off switches (you toggle these switches on and off like an electric toggle switch). A button that displays as a depressed, green button is toggled on, while one that displays red is toggled off.

  Two primary menus are available here: the Utility menu and the Edit menu. To switch between the two, you can press the **Utility** or **Edit** buttons at the bottom of each menu or just *right-click* your mouse. Now don't get confused by this. If you see the **Utility** button at the bottom of the menu, then you are in the **Edit** menu (a look at the lower-left box of the prompt line will confirm this). The **Utility** and **Edit** buttons are buttons, not labels, and if you press one, then you will go to that menu. Try it out.

  Next to each menu button is the letter **F** or **S** followed by a number from **1** to **0**. The **F**'s indicate that you can execute that menu button by pressing the corresponding function key (F1, F2, and so on) on your keyboard. To execute the **S**'s, press the Shift key and the function key together. For instance, to execute a button with **S4** next to it, you would press **Shift+F4**. You will find that using these shortcuts can speed up your drawing and decrease the amount of repetitive mouse movements. It is one of the keys to speed in DataCAD.
- *Navigation pad* These buttons allow you to quickly view and navigate around your drawing. The red arrows are used to pan around the Drawing Window left, right, up, and down. Clicking on the **R** (Recalculate Extents) button recalculates the extents of the current drawing and displays it from edge to edge. The **I** (Identify) button is used to identify the properties of any entity. The **V** (Views) button places you in the 3D Views menu. Clicking on the **W** (Window In) button allows you to define

an area of the drawing to zoom into. The **E** button in the center of the arrows displays the drawing at its most recently calculated extents. Since no recalculation is done, this option works more quickly than the Recalc option. Pressing the **Esc** key will do the same thing.

- *Projection pad* This set of nine buttons provides a quick way to access different 2D and 3D views of your drawing.
- *Status area* As you change views, layers, scales, and other settings, DCAD constantly updates their status in the Status area so they can be reviewed at a glance. Most of these status displays can also be changed by selecting them directly. The current layer, scale, selection set, and line type can be viewed and changed in the four windows. Layer Search is displayed by the square with the L in it and can be toggled on and off by clicking on it or by pressing the apostrophe (') key. The colored square above the Layer Search button shows you the active layer color. At the bottom of this area there appears to be a strange word called SWOTHLUD. It's actually not a word. Each letter corresponds to a different DataCAD setting. An upper case letter means the setting is on, and a lower case letter means it is off. Various methods can be used to turn these features on and off. More on this later.
- *Message window* The message window is comprised of four subwindows. These elements of this window are constantly changing to inform the user of what is currently happening or to prompt the user for input information.
    - *Coordinate readout line* Current cursor coordinates are displayed here. All readouts are relative to the last point picked with the cursor (via a mouse button click or keyboard input), even if no entities were drawn as a result of the point picked. The **x:** and **y:** readouts give the horizontal and vertical locations of the cursor. The **a:** readout gives the angle of the cursor, and the **d:** readout gives the distance of the cursor from the last point picked. Some DataCAD messages temporarily display themselves in place of the coordinate readouts, such as when DataCAD informs you that an autosave has just taken place.
    - *Information line* Two major items are displayed here. The first is the current Z-location of your cursor (Z-Base) along with the $2^1/_2$ D extrusion height at which all entities will be drawn (Z-Hgt). Secondly, when an entity or entities are moved, stretched, or copied, DataCAD will display the number of entities that have been edited. Like the coordinate readout line, some DataCAD messages temporarily display themselves in place of the information line readouts.

- *Prompt line* This is where DataCAD will prompt you to enter information or to verify that a current default setting is correct. The feedback that DataCAD gives you on this line is very important. If you are ever confused about what to do next or ask yourself, "what did I just do?," you can often find the answers simply by looking at this line.
- *Menu name box* This small box will show the menu, function, or toolbox macro you are currently in, such as Edit, Utility, Toolbox, Text, and so on.

Windows Menu Options

Nearly every option in the menus from **Edit** through **Help** is a duplicate of the commands that are accessed from the standard menus on the side of the screen. The **File** menu offers the following:

- *New* Select this to start a new drawing file
- *Open* Select this to open an existing drawing file
- *Close* Closes the current drawing file. You will be asked if you want to save the current drawing.
- *Close All* Closes all open drawing files. You will be asked if you want to save each drawing as they are closed.
- *Save* Select this to save the current drawing. Alternatively, you can press **Ctrl+S** (for save) or the capital letter **F** (for forced save) to do the same thing
- *Save As* Use this to save the current drawing under a new file name. The file dialog box will appear, so you can enter a new name. Press **Save** to save the file. The Windows title bar will now show the new file name. Warning: if you do not want to lose any of your current changes in the current drawing, you must first **Save** the current drawing. Otherwise, the current file will not be saved and any editing that was done since the last save will be lost.
- *Save All* Use this to save all open drawing files at once
- *Import* Use this to import a .DWG or .DXF format drawing. The .DWG or .DXF drawing will be converted to entities that can be displayed and edited in DataCAD. DataCAD will automatically detect the version number of the file (R12, R14, and so on).
- *Export* This option offers the following selections:

- *Dwg* This enables you to export to all the layers or only the layers that are turned on:
  - *All Layers* Use this to export all the layers in the current DataCAD drawing file, whether they are on or off, to a .DWG format drawing.
  - *On Layers* Use this to export only the layers that are currently turned on in the current DataCAD drawing file to a .DWG format drawing.
- *Dxf* This also enables you to export to all the layers or only the layers that are turned on.
  - *All Layers* Use this to export all the layers in the current DataCAD drawing file, whether they are on or off, to a .DXF format drawing.
  - *On Layers* Use this to export only the layers that are currently turned on in the current DataCAD drawing file to a .DXF format drawing.
- *Ray* This saves all the layers of the current drawing file as a .RAY format drawing. This is the format that the third-party DC Render program uses.

- *Print* This opens up the DataCAD **Plotter** menu where you can then select Plot to print a drawing. If you select this option in a new drawing file prior to selecting a printer/plotter and paper size, a message will appear telling you this must be done first. The Printer/Plotter Settings menu will then appear so you can make the selection.
- *Print Setup* This accesses the Printer/Plotter Settings menu. You should make sure the information in this menu is correct prior to laying out your drawings to make sure the right paper size and margins are selected.
- *File History* The last four drawing files accessed by DataCAD are displayed here. When you select one, the selected file will be opened.
- *Exit* Select this to exit the current drawings and close down DataCAD. Alternatively, you can press the **X** in the upper-right corner of your screen (the standard Windows shortcut to Exit).

The **Edit** menu contains the following options:

- *Undo* This undoes the last action or edit command.
- *Redo* This redoes the last action or edit command.
- *Cut* Once you select entities with the **Clipboard Select** option, you can cut them out of their current location to place them elsewhere.

- *Copy* This acts like Cut but leaves the original entities in place while allowing you to copy the selected entities elsewhere.
- *Paste* This pastes in text or entities from the Windows clipboard or after using the cut or copy options from a DataCAD drawing.
- *Paste Special* This works like the paste option but allows more options.
- *Clipboard Select* Before using the cut or copy options, you must first use this option to pick the entities to be acted upon.
- *Move* through *Change* These are duplicates of the commands found in the Edit menu on the side of the screen.
- *Partial Erase* This enables you to erase part of a 2D line or arc. You can access this same command with **Edit/Erase/Partial**.
- *Line CleanUp* This accesses 2D line cleanup functions (the same ones found in the **Edit/Cleanup** menu).
- *Wall CleanUp* This accesses 2D wall cleanup functions (the same ones found in the **Edit/Cleanup** menu).

The **View** menu offers the following:

- *Orthographic* Select this to view the current Drawing Window as a flat, 2D view. This is the only projection view that 2D entities can be drawn in, so this is the view option where all your 2D drafting takes place. You can also select this option in the Projection Pad at the bottom of the DataCAD menus.
- *Elevation* This option contains the following choices:
  - *Front* This displays a front view in 3D space. While in an Orthographic view, front is defined as viewing from the bottom of the screen to the top.
  - *Back* This displays a back view in 3D space. While in an Orthographic view, back is defined as viewing from the top of the screen to the bottom.
  - *Left* This displays a left view in 3D space. While in an Orthographic view, left is defined as viewing from the left of the screen to the right.
  - *Right* This displays a right view in 3D space. While in an Orthographic view, right is defined as viewing from the right of the screen to the left.
  - *New* This enables you to define a new view in 3D space.

- *Isometric* This displays the current drawing in the last user-defined isometric view. If you haven't yet defined one, DataCAD displays a default isometric view.
- *Parallel* This displays the current drawing in the last user-defined parallel view. A parallel view (all lines parallel) is any view other than a perspective view (all lines converging or diverging). This includes the four elevation views, an isometric view, and an oblique view. If you haven't yet defined one, DataCAD displays a default parallel view.
- *Set Oblique* This takes you to the **DCAD_3D/3DEdit/3DViews/SetObliq** menu to set the oblique viewing options.
- *Oblique* This displays the current drawing in the last user-defined oblique view. If you haven't yet defined one, DataCAD displays a default oblique view.
- *Set Perspective* This enables you to access and set the options for a perspective view. This is a shortcut to the **DCAD_3D/3DViews/SetPersp** command.
- *Perspective* This displays the current drawing in the last user-defined perspective view. If you haven't yet defined one, DataCAD displays a default perspective view.
- *Icon Toolbar* This turns the Icon Toolbar on and off.
- *Viewer Toolbar* Use this to turn on the context-sensitive toolbars for navigating around the Drawing Window. Only one toolbar is displayed at one time. Which one is displayed depends on how you are viewing the Drawing Window. Three toolbars are available: Orthographic, Parallel, and 3D Views.
- *Undo/Redo Toolbar* This turns the Undo/Redo toolbar on and off.
- *Go-to-View* This displays the first page of the 3D GotoView menu. This is the same as selecting the **Utility/GotoView** command.
- *Multi-View* This toggles the four vertical multi-view windows on or off. This is a shortcut to the **Utility/Display/MltiView** option.
- *WindowIn* This enables you to define a zoom window. Draw a bounding box with your cursor. After picking the second point, DataCAD will zoom or window in to the drawing to best fit the screen. You can also window in by pressing the **W** button on the Navigation Pad or pressing the **/** key on the keyboard.
- *WindowIn Recalc* This command will first cause DataCAD to find the farthest extents (left/right, and up/down) of all the visible entities in

the activated layers or of the current *Clip Cube* (CC). Then DataCAD will zoom or window in/out to those extents, as they best fit the screen. You can accomplish the same thing by pressing the **R** button on the Navigation Pad.

- *To Scale* This displays the current drawing at one of the predefined scales.
- *Refresh* This causes all the visible entities in the Drawing Window to be freshly redrawn. This is useful when overlapping entities have been edited to a point where the pixels of many entities are no longer visible. You can accomplish the same thing by pressing the **Esc** key on your keyboard.
- *Hidden Line Removal* This visually removes the lines that would be hidden from view if the model were a solid. This may take only a few seconds or several minutes, depending on the size and complexity of the model. This is a shortcut to the **DCAD_3D/Hide/Begin** command.
- *Shade* This is similar to Hidden Line Removal, but instead of hiding lines, the model's surfaces are shaded with a solid color. This is a much faster process than the hidden line removal process. This is a shortcut to the **DCAD_3D/3DEdit/GlShader/Shade** command.

The **Insert** menu contains the following:

- *Symbol* This brings up a dialog box to allow direct viewing and selection of symbols without having to first open a template file.
- *Reference File* This inserts an external .DC5 drawing file (also called an XREF) into the current drawing file.
- *Text File* This inserts a .TXT format text file into the Drawing Window. The **ToFile** option will also be available to export a text file.
- *Windows MetaFile* This inserts a Windows metafile (.WMF) format image into the Drawing Window.
- *Reference File Management* This option offers the following choices:
  - *Manager* This displays the Reference File Manager dialog, enabling you to manage previously inserted .DC5 reference files.
  - *Refresh All* This will re-read all the current XREFed files to update the entities while maintaining the layer on/off settings that are currently set for that XREF.
  - *Reload All* This will re-read the XREFed file to update the entities but will use the layer on/off settings that were last saved in that

XREFed file, rather than using the settings that were set in the Reference File Manager dialog box.

- *Orphans* This displays the Orphans dialog box, enabling you to reestablish lost links to referenced .DC5 drawing files.

The **Create** menu offers all the entity creation options that are also available in the menus on the side of the screen. In general, these commands are all self-explanatory, so we won't describe each one here.

The **Tools** menu contains the following:

- *Ortho Mode* This turns Ortho mode on and off. You can do the same thing by pressing the **Oo** key.
- *Snap Grid* This turns the Snap Grid on and off. You can do the same thing by pressing the **x** key.
- *Identify* Select this and then click on any entity in the Drawing Window. All the information about that entity will be displayed in the vertical menu window, including the entity type, layer name, entity color, linetype, spacing, line weight, Z-base, and Z-height. You can accomplish the same thing by pressing the **I** button on the Navigation Pad or the **I** key on your keyboard.
- *SetAll* After identifying an entity, use this option to change all the current settings to match those of the selected entity, including the layer name, entity color, linetype, spacing, line weight, Z-base, Z-height, associative hatch settings, text settings, and dimension settings.
- *Coordinate Identification* Select this option and then click or object snap anywhere in the Drawing Window. The message line will display the cursor location in X, Y, and Z coordinates relative to absolute zero.
- *Reference Point* This enables you to select a point in the Drawing Window, either by left-clicking or object snapping, in order to draw or place an entity relative to that point (a very useful feature). You can accomplish the same thing by pressing the apostrophe key (on the same key as the tilde (~).
- *Snap Point* Use this to place a non-printing snap point in the Drawing Window. You can then snap to this point by locating the cursor near the point, and then snapping with the middle mouse button, or pressing the **Nn** key. This is a shortcut to the **Utility/Measures/SnapPnt** command.
- *Directory* This enables you to enter and print project billing information, view layer information, check system paths, and review

and purge drawing symbols. This is a shortcut to the **Utility/Directry** command.

- *Layers* This accesses the Layers menu.
- *Grids* This accesses the Grids menu.
- *Geometry* This accesses the Geometry menu options.
- *Measure* This accesses the Measures menu options.
- *Link Entities* This accesses the LinkEnts menu.
- *Selection Sets* This accesses the EditSets menu.
- *Input Mode* This option contains the following input modes that can be chosen by pressing the **Insert** key on your keyboard:
  - *Relative Polar* This defines lines as a distance and angle from the last point picked. (This is what I use nearly all the time.)
  - *Absolute Polar* This defines lines as a distance and angle relative to the absolute zero point of the drawing.
  - *Relative Cartesian* This defines lines as a distance and angle by X and Y coordinates from the last point picked (other users swear by this method).
  - *Absolute Cartesian* This defines lines as a distance and angle by X and Y coordinates relative to the absolute zero point of the drawing.
- *Options* This accesses the Display, Object Snap, 2D Settings, and 3D Settings menus.
- *Program Preferences* This option contains the following five choices:
  - *Pathnames* This is where you can view and change the directory paths used by DataCAD. To change paths, click on the yellow folder to the right. You can also type a path, but if you type in an invalid path, DataCAD will not let you leave the input box until a valid path is entered.
  - *Misc* Here you can change your default drawing, toolbar .KEY file, default character (font) name, GLShader background color, DC Render materials, Copy to Clipboard options, and symbol preview options. You can also select whether the display list is on, off, or at maximum acceleration.
  - *Interface Settings* Selecting this will display a graphic representation of the DataCAD program. You can turn the toolbars on/off, change the menu and message window font characteristics, change the location of the vertical menu window, and change the Drawing Window color.

- *DXF/DWG* Here you can set the various import and export options for .DXF and .DWG filetypes (see Chapter 25, "Converting File Formats," for more info).
- *Toolbox* Here you can add toolbox macros to the **Macros** drop-down menu.

The **Macros** menu offers the following:

- *Configure* Use this to open the Macros dialog in the Tools/Program Preferences menu. Here you can add and delete toolbox macros from the drop-down list.
- *Macro List* This displays the list of available macros. Click on one to run it.

The **Window** menu offers the following:

- *Save as Bitmap* This saves the image in the current Drawing Window to the Windows clipboard in a bitmap (.BMP) format. Once there, you can paste the image into other documents or graphic paint programs. The resolution of the image will match the resolution at which your computer display is set. The higher the resolution, the better the .BMP image.

- Tile Horizontally This displays all the open drawing files at once. To make them fit, this option displays their Drawing Windows at their full horizontal widths with equal spacings vertically.
- *Tile Vertically* This displays all the open drawing files at once. To make them fit, this option displays their Drawing Windows at their full vertical heights with equal spacings horizontally.
- *Cascade* This arranges multiple drawing files in an overlapping format like a fan of playing cards.
- *Next* With multiple drawing files open, this option makes the next drawing in the list the active drawing.
- *Previous* With multiple drawing files open, this option makes the previous drawing in the list the active drawing.
- *Open Drawing List* This lists all of the currently open drawing files. Click on one to make it the active drawing.

Finally, the **Help** menu offers the following:

- *DataCAD Help* This displays the DataCAD Help file dialog box. Here you can get help regarding DataCAD via a contents or index menu, or by using the Find function.
- *Add-On Products* This displays Help files for third-party macros.
- *Using Help* This displays the standard Windows Help file dialog box. Here you can get information regarding the Windows Help system via a contents or index menu, or by using the Find function.
- *About DataCAD . . .* Click on this to view information about the version of DataCAD that you are running. You will see the DataCAD version number, your company name, your name, and the program's product serial number. If any of these items are incorrect or blank, you can edit the information in the DCADWin.ini file in your Windows or WinNT directory folder.

Drawing File Options

In this section, we'll examine the drawing file options including saving, opening, and creating default drawing files.

Multiple Document Interface

The multiple document interface is a long-awaited new feature to DataCAD. Now you can have an unlimited number of drawing files open at one time. The only limit is your computer's available RAM. Like other Windows programs, you can switch quickly between all the open drawing files, and you can cut and paste drawing entities between all of them. You can also save text, vector, and graphic information from DataCAD to the Windows Clipboard where you can then paste that data into another Windows program (you cannot yet paste bitmap images into DataCAD, only vector graphics). The standard Windows shortcut of **Ctrl+Tab** will scroll through all the open drawing files.

Opening an Existing Drawing

Open a drawing just as you would any other document in Windows:

1. Go to **File/Open**.
2. Browse to the file you want.
3. *Double-click* on it or highlight it and then select **Open**.

WARNING: *If you are on a network and you try to open a drawing file that is already open on someone else's computer, you will get the following error message:*

```
ATTENTION...
This file is already marked In Use.
Continuing with this file will result in data loss.
Are you sure you wish to continue?
```

*If you choose **Yes** to continue, the file will open and you can add to and edit the drawing. However, that means two of you are working in the same drawing file at the same time. If you **Save** the drawing, you will overwrite all the changes the other user has made, and likewise when he **Saves** the drawing, he will overwrite all the changes that you made.*

How does DataCAD know the file is already open? Whenever an existing file is opened, or a new drawing is created and saved, DataCAD creates a semaphore file, *a dummy file with the same name as the drawing file, but with three dollar signs at the end, instead of the .DC5 extension:*

File name:	*Myfile.dc5*
Semaphore file name:	*Myfile.$$$*

*If you want to see which files are open prior to selecting one, select **File/Open** and then click on the down arrow at the right of the "**Files of type:**" dialog box and select the **All files (*.*)** option. Now all the files, including the .$$$ semaphore files, will be displayed.*

Starting a New Drawing

Unless otherwise directed, DataCAD will use the current default drawing (as selected in **Tools/Program Preferences/Misc/Default Drawing File**) whenever starting a new drawing file. A default drawing is simply a seed drawing. In other words, it is a DataCAD drawing file that has been set up ahead of time with predefined settings and information. Creating default drawings can save a great deal of time whenever a new drawing file is started. It can be an empty drawing or can contain things like a title block.

If you just opened Data CAD:

1. When you first start DataCAD, a dialog box will be displayed, from which you can select an existing drawing file. But you can also start a new drawing from this dialog.
2. Navigate to the folder where you want the new drawing to be saved to.

3. In the File name: box, type in a new name for your drawing them select **Open**, or press the **Enter** key. You do not need to add the .dc5 file extension. Data CAD will do that for you.
4. A **Confirm** dialog box will be displayed, asking you, "Do you want to create the drawing file?"
5. The Title Bar will show the name of the new, current drawing file.

If you just opened DataCAD, but want to use a different default drawing file:

1. Close the current dialog box by selecting **Cancel**.
2. Follow the next three steps below, selecting a new default drawing in Step 2.

If you already have DataCAD running:

1. Select **File/New**.
2. If you want to use a different default drawing:
   a. Select **Default**. The default drawings in the DataCAD\DEFAULT directory will be displayed.
   b. Select the default drawing you want to use then press **Open**. You will be returned to the main File to Create dialog.
3. Navigate to the folder where you want the new drawing to be saved to.
4. In the File name: box, type in a new name for your drawing then select **Open**, or press the **Enter** key. You do not need to add the .dc5 file extension. DataCAD will do that for you.
5. The Title Bar will show the name of the new, current drawing file.

NOTE: *You can name your DataCAD drawings with anything acceptable to the Windows operating system, including long file names. Certain characters cannot be used, including asterisks, plus signs, forward/backward slashes, brackets, colons, periods, and spaces* ***(* + / \ [ ] : . [space])****.*

Saving a Drawing

You can save your current drawing in one of three ways:

- **File/Save**
- **Ctrl+S**
- **Shift+F** (notice that only an upper case F works). The letter F stands for forced save. It's an old DataCAD term that has stuck.

Using Save As . . .

Use this to save the current drawing as a new file with a new name:

1. Go to **File/Save As** . . . .
2. The Windows **Save file as:** dialog box will appear.
3. Type in a new name for your drawing and then select **Open** or press the **Enter** key. You do not need to add the .dc5 file extension. DataCAD will do that for you.
4. The title bar will show the name of the new drawing file.

WARNING: *If you do not* ***Save*** *your current drawing before saving the drawing with* ***Save As*** *. . . , you will lose all the changes in the original drawing that you made since the last* ***Save****. This is a convention of the Windows operating system.*

Autosave

As with most modern programs, DataCAD will save a backup of your current drawing at regular intervals. This is called *Autosave*. By default, DataCAD will execute an autosave every 15 minutes, but you can change that setting in the **Utility/Settings/SaveDlay** menu. DataCAD saves the backup with an .ASV file extension, and by default it is saved to the \DWG directory. You can change where DataCAD saves these backup files by selectingp a new path in **Tools/Program Preferences/Pathnames/Autosave** Files.

If you are working with large files, you may be tempted to set the **SaveDlay** setting to something higher, such as 60 minutes, since saving a large file may take several seconds, and the program will stop in its tracks while the save takes place. But look at it this way, how much of your drawing do you want to have to redraw if something happens to your file? Let's say it's been 45 minutes since you last did a full save (**F** or **File/Save**), and your **SaveDlay** is set to 60 minutes. All of a sudden you get a power surge that knocks out your computer, or your dog accidentally unplugs it. You have just lost 45 minutes of work, all of which will have to be redrawn. But if you have your **SaveDlay** set to 15 minutes, then you will only lose 15 minutes of work. It's up to you.

If something does happen and you need to recover the .ASV file, see Chapter 21 for more information about backup files and file recovery. For now, just remember to always save, save, save.

Exiting DataCAD

When you are ready to leave or exit a drawing, you have a number of choices in how you do that:

- *Close* This closes the current drawing only. You will be asked if you want to save the drawing prior to closing it.
- *Close All* This closes all current drawing files. You will be asked if you want to save each drawing prior to closing it.
- *Exit* Select this to close all the current drawings and exit DataCAD. Alternatively, you can press the **X** in the upper-right corner of your screen (standard Windows shortcut to Exit). Each drawing that is open will ask you if you want to save it before exiting.

Default Drawings

DataCAD comes with its own default drawings that are located in the ***Dcadwin\Default*** directory. If you do not choose an alternate default file, DataCAD will start with the Transfer.dc5 file. This will then be the base from which the rest of your drawing will be created.

You should understand two key concepts about using default files:

- To save a default file, save the drawing to the ***Dcadwin\Default*** directory. You can have an unlimited number of default drawings.
- You must configure DataCAD to tell the program which default drawing file you want to open every time DataCAD starts. DataCAD will automatically open this file every time until you choose a new one.

Personally, I don't like most of the settings any of the standard default files that come with DataCAD, so I have come up with my own. As you continue to work with DataCAD, you will invariably want to create you own default drawing files. Let's create one right now with the settings we will be using for the tutorial in the next chapter.

Creating a New Default Drawing

In order to create a new default drawing, follow these steps:

1. Start DataCAD (always a good place to begin . . .).
2. The Drawing File dialog box will pop up in the Drawing Window.
3. In the File name box, type ***Default*** and then select **Open**, or press **Enter**. The screen will change to display the standard Drawing Window and menus.
4. *Right-click* or select the **Utility** button at the bottom of the menu. The DataCAD **Utility** menu will be displayed (the first button at the top of the menu should be **ToScale**).
5. Select the following buttons and change the settings as shown (remember that you can *right-click* to move back or exit out of menus). Buttons not specifically indicated should remain in their current on or off position. Chapter 3 and other chapters will explain what all of this means, but for now we just want to get something useable for our projects:
   a. **Hatch:**
      - **Associat = on**
      - ***HtchType/NoOutLin***

   b. ***Dmension/Linear:***
      - ***Horizntl* = on**
      - ***Assoc* = on**
      - ***TextStyl/TextSize/..$^1/_8$ (That's two periods before the fraction. This is DataCAD shorthand for entering $^1/_8$ of an inch.):***

- ***TxtScale* = on**
- ***Auto* = off**
- ***FontName/ARCHWY2HC***

- ***DimStyl:***
  - ***Offset/..1/8 [3.2]*** (remember the brackets are metrics)
  - ***Overlap/..1/8 [3.2]***
  - ***Overrun/..1/8 [3.2]***
  - ***FixdDist* = off**

- ***ArroStyl:***
  - ***TickMrks* = on**
  - ***Size/.5***
  - ***Color/Green***

c. ***Settings:***
- ***ScaleTyp/Arch*** (sets Imperial units)
- ***ScaleTyp/StakFrac***
- ***AngleTyp/Normal***
- ***SaveDlay/15***
- ***BigCursr* = on**

d. ***Grids:***
- ***SnapGrid* = on**
- ***DspGrid1* = on**
- ***DspGrid2* = off**
- ***GridSize/SetSnap/1" [25.4]***
- ***SnapAng/8***

e. ***Display:***
- ***ShowTxt* = on**
- ***ShowDim* = on**
- ***ShowHtch* = on**
- ***ShowWgt* = off**
- ***UserLine* = on**
- ***OverSht* = off**
- ***ShowIns* = on**
- ***ShowAttr* = off**
- ***CurveCtr* = on**

f. **ObjSnap:**
- ***Nearest* = off**
- ***EndPnt* = on**
- ***MidPnt* = on**

- ***No. Pnts* = off**
- ***Center* = on**
- ***Quadrant* = on**
- ***Intsect* = on**
- ***Perpend* = off**
- ***Tangent* = off**

6. *Right-click* or select the **Edit** button at the bottom of the menu. The DataCAD Edit menu will be displayed (the first button at the top of the menu should be **Move**).
7. Select the following button and change the settings as shown.
   a. ***Text:***
   - ***Size/..1/8 [3.2]***
   - ***FontName/ARCHWY2HC***
   - ***Dynamic* = on**
   - ***TxtScale* = on**
   - ***Arrows:***
     - ***Size/.5***
     - ***Style/Open* = on**
     - ***Aspect/3***

   b. ***Left* = on**
8. We need to make one setting from the Windows menu bar. Select **Tools/Input Mode/Relative Polar**. This sets the method by which we will select points and directions. It is worth noting that if at any time your input method seems to have gone haywire, you may have accidentally pressed the **Insert** key (the shortcut key for Input Mode). Press it a few times until the **Relative Polar** message shows again, or go back to the Windows menu bar and reselect Relative Polar.
9. Now we need to save the drawing to the DCADWIN\DEFAULT directory. Go to the menu bar and select File/Save As. The File dialog box will pop up.
10. Navigate to the DCADWIN\DEFAULT directory, type in the name ***Default***, and press Save.
11. You have now saved this file to the ***Dcadwin\Default*** directory and named it Default. This file contains all the standard settings we will be starting with for all the tutorial projects. Note that you are still working in the original drawing file. Notice that as with other Windows programs, when you use **Save As**, you will immediately be working in that new file. The original file will be lost unless you first save it. We don't need it, so we didn't save it first.

NOTE: ***Configure for Default*** *If you want to make this new* ***default*** drawing the one that DataCAD will always use by default, go back to the ***Tools/Program Preferences/Misc*** *menu and select the* ***default*** *drawing that you just saved in the* ***Default Drawing File:*** *box. Now whenever you start a new drawing in DataCAD, this file, along with all the settings we made, will automatically be used to start the new drawing.*

Starting with a Different Default Drawing

Let's see how we would use this default drawing (or any other default drawing) to start a new project. We want to start from scratch to try out the whole process, so let's exit this drawing without saving it and then start a new drawing using our new default drawing.

1. Select **File/New**.
2. When the File dialog box pops up (the one that says, *"Enter name of drawing file to create"*), press the **Default** button.
3. Open the file we just created, called **default.dc5** (located in the \Default folder of your root DataCAD directory), and then press **Open**. You have now told DataCAD that you want to create a new drawing using all the settings saved in the **default.dc5** file. Next, you will have to name the new drawing file that you want to start.
4. In the original File dialog box in the **File name:** box, type in a new file name and press Create (you don't have to add the .dc5 extension at the end of the filename; DataCAD will do that for you).

Now that we have a default drawing set up, you can use it to start a new project in Chapter 4, "Tutorial." Completing this tutorial project will build on the basic concepts covered in this chapter and will show you how to draw in DataCAD by creating a small, two-unit apartment.

DataCAD Online Help

DataCAD uses the standard Windows Help module. To access it, select the **Help** option from the menu bar. A drop-down list will appear with three more options (see Figure 2-6).

Figure 2-6 Selecting **Help**

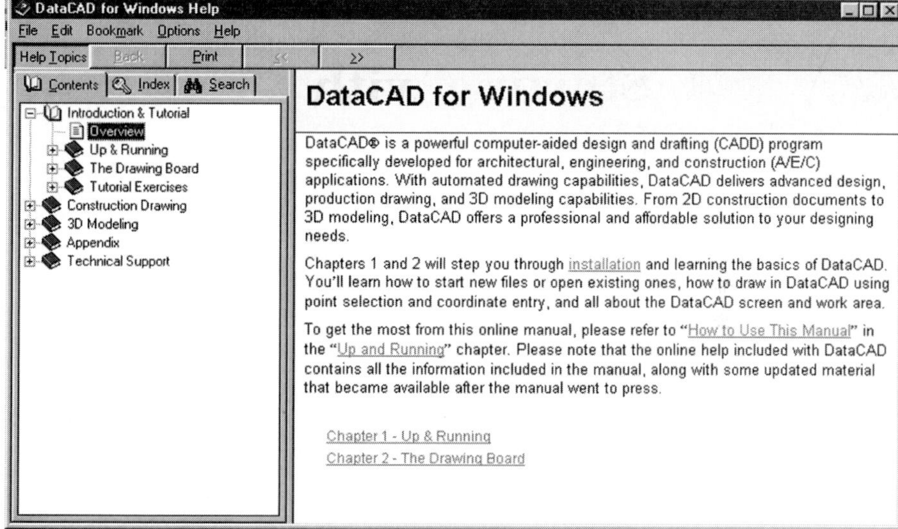

Figure 2-7 The **Contents** tab

DataCAD Help Notice that you can use **Ctrl+F1** at any time to access DataCAD's help. When you select this option, the window shown in Figure 2-7 appears.

For detailed help on how to use the various options displayed, see the following **"Using Help"** section.

Using Help This option allows you to get information about how to use the Windows Help system. It contains the following choices:

Here's what each of the three main tabs will do for you (see Figure 2-8).

- *Contents* This is the default tab that is displayed when you select the **Help** option, as previously shown. To use the Contents tab, just double click on a book or a help topic to read more about each one.
- *Index* In the first window, you can type in the first few letters of a subject you would like to learn about. As you do this, the second window will jump to the alphabetical listing that you are typing.

Figure 2-8 The **Using Help/Contents** tab

Double-click on the topic you want to learn about and a yellow Windows Help dialog box will be displayed, as in Figure 2-9.

- *Find* This option enables you to search for specific words and phrases in Help topics by scanning the entire Help database. To do this, Windows will scan the entire Help file and create its own database for faster searches. Unfortunately, this adds a lot of data to your computer's hard drive and can cause some computers to become sluggish. When you first select the **Find** option, you will get the **Windows Find Setup Wizard**. If you do not want to have Windows set up the Find database, then do not proceed with the **Next** command.

About DataCAD . . . Selecting this option will display information about the version of DataCAD currently running, including the version number, the registered user, and the software serial number.

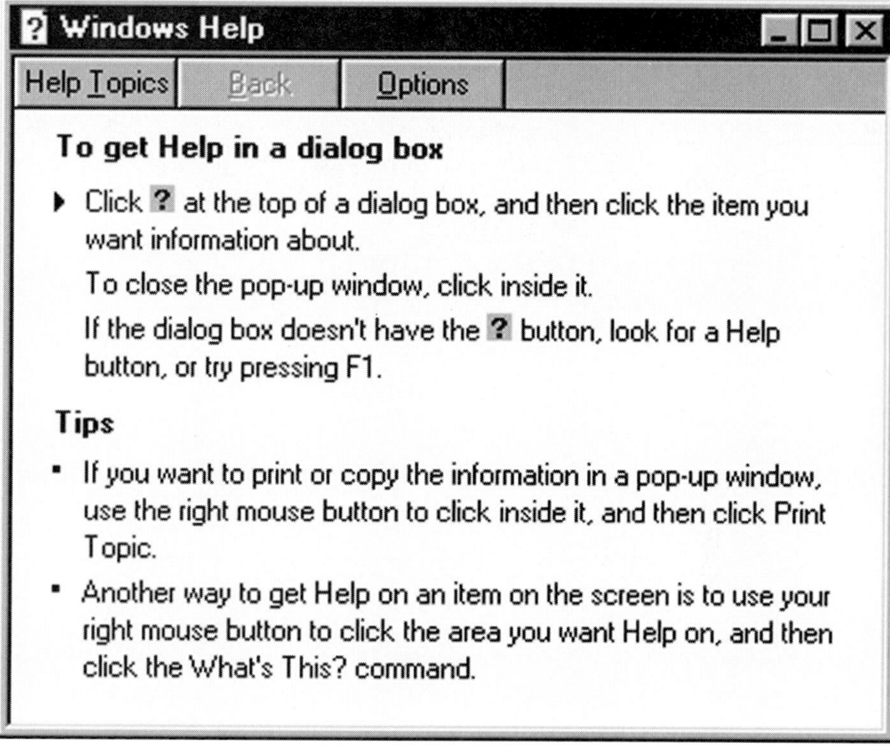

Figure 2-9
A typical **`Using Help`** dialog box

Whenever DataCAD issues a new downloadable software version or patch, they usually also offer updated help files, so be sure to check for those on the DataCAD Web site as well.

CHAPTER 3

Settings and Display Options

This chapter will describe the various settings and display options that can be made within any DataCAD drawing. The settings you choose will have a tremendous impact on how DataCAD operates. Saving the settings in your default drawings will ensure that settings are standardized, since every new drawing that is opened will start with these defaults (see Chapter 2). Standardizing these settings will also ensure that each drawing file acts predictably like all other drawings.

To start from scratch regarding settings and options, you can use the **dc9_dflt.dc5** file, located in the **dcadwin\default** folder. If you have just installed DataCAD, this should already be your default drawing file. Perform the following steps:

1. Start DataCAD or pick **File/New** if you are already in DataCAD. A Windows dialog box will appear (see Figure 3-1).
2. Navigate to the folder that you want to save the drawing in, such as the **dcadwin\dwg** folder.
3. Type a new drawing name, such as Settings, in the **File name:** box and then press **Open**.

Settings

The **Settings** option button is located in the **Utility** menu and contains numerous options that tell DataCAD how to run and what to display to you.

Figure 3-1
Creating a new file

These settings are valid only for the current file and do not affect other drawings. To access this button, follow these steps:

1. *Right-click* until you are in the **Utility** menu (the Menu Name box will say Utility).
2. Press the **Settings** button.

NOTE: *DataCAD saves all current settings with the drawing file when it is saved.*

Password

This option enables you to protect your drawing with a password of up to 10 characters. Once the drawing is saved with a password, to open the drawing you must successfully enter the correct password. When opening the file, you will be given three opportunities to input the correct password.

WARNING: *Be careful with this one! If you forget the password, you will not be able to open that file again.*

To change the password, after opening the drawing, select Password, and then type in a new password and verify it. To delete an existing password so that the file no longer requires a password, type nothing, leaving the password dialog empty.

ScaleTyp (Scale Type)

The **ScaleTyp** option will establish the type of units used throughout the drawing file (see Figure 3-2). If you change the setting, then all associative dimensions in the drawing will change to the new scale type. This is a very

Figure 3-2
Scale type examples

3.096 m	METRIC
3.096 m	METERS
309.562 cm	CENTIMETER
3095.625 mm	MILLIMETER
3.09^{6}	DIN
3 096	AS 1100

significant feature, since many CADD programs cannot do this. Non-associative dimensions are not changed. The types of units used by ScaleTyp are as follows:

- **Arch** (Architectural) Feet, inches, and fractions
- **Engr** (Engineering) Feet and decimal inches
- **Decimal** Decimal feet
- **Inch/Frc** Inch fractions
- **Inch/Dec** Decimal inches
- **Metric** Meters, centimeters, and millimeters (such as 5.105 m)
- **Meters** Decimal meters (such as 5.105)
- **CentiMtr** Decimal centimeters
- **MilliMtr** Decimal millimeters (such as 5105)
- **DIN** German DIN meters (such as 5.10^{5})
- **AS 1100** Australian standard meters (such as 5 105)
- **Units** This displays the dimensions with (when toggled on) or without (when toggled off) the unit abbreviations (m, mm, cm, ′, ″).
- **DoFloat** This option is only available when Decimal, Inch/Dec, Meters, CentiMtr, and MilliMtr are turned on. When DoFloat is toggled on, the distance readouts will use a floating decimal point. When it is toggled off, the Message Windows will show the number of significant

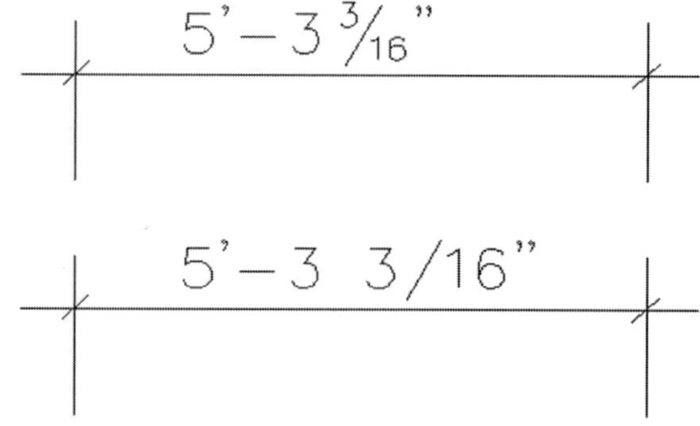

Figure 3-3
A example of **StakFrac** at work

digits to be displayed. You can configure it to set the decimal places from 0 to 3.

- **SigDigit** This will only be displayed if the DoFloat option is toggled off. It changes the number of significant digits that appear. DoFloat and SigDigit are available only with Decimal, Inch/Dec, Meters, CentiMtr, or MilliMtr scale types.
- **StakFrac** This option will only appear if the Arch option is toggled on. It creates stacked fractions (like $^3/_4$″) instead of linear ones (like 3/4″), as shown in Figure 3-3. With StakFrac on, all further associative and non-associative dimensions will be created with stacked fractions. At the same time, all associative dimensions in the current drawing file will be displayed with stacked fractions, since the StakFrac option is a global setting. However, existing non-associative dimensions are not affected. Note that a few fonts may not work with StakFrac. This will become evident if you see the fractions laying over the top of the whole number inches.

AngleTyp (Angle Type)

The **AngleTyp** option will establish the method of entering angles throughout the drawing file (see Figure 3-4). This option includes the various settings:

- **Normal** This sets zero degrees equal to a horizontal line to the right. This is the default setting. Along with the **Compass** and **StartAng** settings, this setting is not a type of angle but rather determines in which direction the angles are drawn.

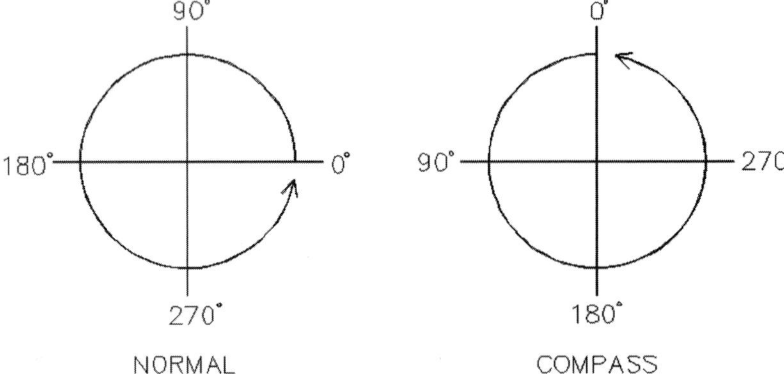

Figure 3-4 The **AngleTyp** option in effect

- **Bearing** When toggled on, you enter angles with bearings. When it is toggled off, all angles are entered in degrees.
- **DecDeg** This enters angles in decimal degrees.
- **Radians** This enters angles in radians.
- **Grads** This enters angles in gradients.
- **Compass** This sets zero degrees to a vertical line pointing straight up or "North". Along with the **Normal** and **StartAng** settings, this setting is not a type of angle but rather determines in which direction angles are drawn (see Figure 3-11).
- **ClkWise** When toggled on, angles increase in a clockwise direction. When toggled off, angles increase in a counterclockwise direction the default setting. When you choose the **Compass** setting, **ClkWise** is automatically toggled on.
- **StartAng** This sets zero degrees equal to two specified points. Along with the **Normal** and **Compass** settings, this setting is not a type of angle but rather determines in which direction angles are drawn. This option is useful if you are working off entities that are at an angle and you want to use those entities as a baseline from which to draw other entities. To do so, follow these steps:

  a. Click on **StartAng**.

  b. Select a starting point with the cursor.

  c. Enter a second point by

  - Selecting with your cursor or
  - Pressing the space bar to enter a distance and angle. Note that the method of inputting distances and angles here is dependent on

which **ScaleTyp**, **AngleTyp**, and **Input Mode** (with **Utility/ Input Mode**, or pressing the **Del** key) are currently selected.

EditDefs (Edit Definitions)

This is a very nice feature of DataCAD. This option allows you to list, add, change, or delete the values for Scales, Angles, and Distances, rather than having to live with the default values that come with DataCAD. Although the default values are probably all that most users will need, occasions may occur when you will want to change them. If you make changes to these definitions, DataCAD will automatically insert them in the proper numerical order, making it easier to find them in their menus. Unfortunately, you can only have a total of 18 values defined (one menu's worth).

Once you change some settings, you can save them to a file (with the appropriate file extensions of .SCL, .ANG, or .DIS) for use later on. These settings are saved with each drawing. If you decide that all your new projects will use your new settings, be sure to resave your default drawing file(s) with the new settings so that you won't have to reload them every time you start a new project. See Chapter 23 for instructions on how to customize these settings.

Click on **EditDefs**, then pick **Scales**, **Angles**, or **Distances**, and the following options will appear:

- **List** Lists the current scale, angle, or distance settings
- **Add** Used to add new values to the list
- **Change** Used to change current valves in the list
- **Delete** Deletes current values from the list
- **Save File** Saves the current list to a specified file with the appropriate .SCL, .ANG or .DIS extension
- **Load File** Loads an existing saved list of values

Scales

Here you can edit the values of the scales found via **Utility/Plotter/Scale**. These are the scales that DataCAD uses for plotting (Multi-Scale Plotting details and **QwkLyout**) and for screen display. Every time you use **PgUp/PgDn** to zoom in and out of your drawing, DataCAD is using the scales in this menu to determine how far to zoom in or out.

To change settings, following these steps:

1. Select **Change**.
2. Select the value you want to change in the list.
3. The Prompt Line will read, "**Enter a new string for this scale:**." The text you type here will be the text displayed on the menu button.
4. Type in a new value, up to eight characters, and then press **Enter**. The Prompt Line will read, "**Enter a new scaling value for this scale:**." The value you type here will be the mathematical value that determines the correct scale.
5. Type a new scaling value, expressed in decimal form, and then press **Enter**.

Architectural Scaling Values To calculate the scaling value, express the fraction in decimal form and then divide by 12 to convert the feet to inches. For example, 1/4″ = 1′-0″ becomes 0.25 (1 divided by 4). Divide that number by 12 and the final scaling value becomes 0.02083333. Here are some other examples:

1/32″=1′-0″	.00260416
1/16″=1′-0″	.00520833
3/32″=1′-0″	.00781250
1/8″=1′-0″	.01041666
3/16″=1′-0″	.01562500
1/4″=1′-0″	.02083333
3/8″=1′-0″	.03125000
1/2″=1′-0″	.04166666
3/4″=1′-0″	.06250000
1″=1′-0″	.08333333
1 1/2″=1′-0″	.12500000
2″=1′-0″	.16666666
3″=1′-0″	.25000000
6″=1′-0″	.50000000

12″=1′-0″	1.00000000
24″=1′-0″	2.00000000

Engineering Scaling Values To calculate the scaling value, express the ratio in decimal form and then divide by 12 to convert the feet to inches. For example, 1:20 becomes 0.05 (1 divided by 20). Divide that number by 12 and the final scaling value becomes 0.00416666. Here are some further examples:

1:10	.00833333
1:20	.00416666
1:30	.00277777
1:40	.00208333
1:50	.00166666
1:80	.00104166
1:100	.00083333
1:200	.00041666
1:400	.00020833
1:600	.00013888
1:1000	.00008333

Metric Scaling Values To calculate the scaling value, simply express the ratio in decimal form, such as the following:

1:1	1.00000000
1:2	0.50000000
1:5	0.20000000
1:10	0.10000000
1:20	0.05000000
1:50	0.02000000
1:100	0.01000000
1:200	0.00500000
1:500	0.00200000
1:1000	0.00100000
1:2000	0.00050000

1:5000	0.00020000
1:10000	0.00010000

*CAUTION: If you use the **Toolbox** macro **LyrUtil/LyrSave** to save your drawing and then use **LyrUtil/LyrLoad** on a default drawing that has different scale definitions than the original drawing, you will get some anomalous results with your Multi-Scale Plotting and **QwkLyout** details. The display of scales, like so many things in DataCD, is dependent upon the order in which they appear in the **Scale** menus. The detail you laid out in the original drawing will assume the scales of the default drawing. For instance, if you changed the standard **1:20** scale to **1:30** in the original drawing but left the standard **1:20** scale in the default drawing, any details that were laid out at 1:30 in the original drawing will now be displayed and printed at a scale of 1:20 in the new drawing.*

Angles

Here you can edit the values of the angles displayed whenever you are prompted to input an angle. Note that, similar to the Scale values, the **Angle** values are saved with each drawing, but since they do not have anything to do with the layout of details, using the **LyrUtil** macro will not adversely affect your drawing.

Distances

Here you can edit the values of the distances displayed whenever you are prompted to input a distance. Note that, like the Scale and Angle values, the Distance values are saved with each drawing. Also similar to the Angle values, since the Distance values do not have anything to do with the layout of details, using the **LyrUtil** macro will not adversely affect your drawing.

MissDist (Miss Distance)

This setting controls how close the cursor must be to a selected entity for DataCAD to find it. It is called Miss Distance because it defines how close the cursor must be to avoid "missing" the entity. The Miss Distance extends in a square box area around the cursor and is measured in pixels. The smaller the Miss Distance, the closer you must place the cursor to an entity to select it. If you want to actually see this box or aperature, you can select **Utility/ObjSnap/Aperture** to turn it on. It is important to note that the **MissDist** function affects a number of DataCAD functions such as L-Intsct and T-Intsct. The default value is **10**. This is probably the best setting for most if not all of your work.

SmalGrid (Display Small Grid)

This setting controls the smallest grid to display on the screen, assuming you use a displayed grid. Use this option to turn off the display grid at a predefined pixel spacing. Without this setting, the grid dots would always be displayed, so that when you **PgUp** away from the grid dots, they would get so tight that you could barely see the screen. A good setting might be between 15 and 20 pixels, but you will have to experiment. Remember that this grid does not print, and its points cannot be snapped to but are instead used as visual references when drawing.

ScrlDist (Scroll Distance)

This setting controls how far the screen will scroll during panning with the arrow keys. The selected value corresponds to a percentage of the visible screen. So if you select a value of 30, then the screen will scroll by 30 percent of its display width or height each time you press an arrow key. The default value is **20** percent.

SaveDlay (Automatic Save Delay)

This is an extremely important setting. It controls how often DataCAD automatically backs up your current drawing file. The default is 15 minutes. I tend to set mine to 30 minutes. This auto-save drawing is saved with the name of the current drawing but with the .ASV file extension instead of the .DC5 extension. When the .ASV file is saved, all layers in the drawing file are saved, including layers that are off or locked. The location where these files are saved is set in the **Tools/Program Preferences/Pathnames** dialog box.

It is also very important to note that the primary .DC5 file is not saved when DataCAD saves the .ASV file. To do that, you must press the key or select **File/Save** from the menu bar.

Although it is very important to save often, if the **SaveDlay** option is set too low and you have a large drawing file, then you will be constantly interrupted by auto-saves, since DataCAD must pause all operations while the drawing is being saved. See Chapter 21, "Techno-Files," for more file recovery information. The suggested setting is 15 minutes.

DrwMarks (Draw Marks)

With **DrwMarks** turned on, each time you select a point in the Drawing Window with the left or middle mouse buttons DataCAD will draw a temporary, non-printing point (in the form of a small cross) centered on that point. These are visual references so that you can see what or where you picked, which can be especially important with middle snapping functions and tightly grouped entities. These marks are erased when the screen is Refreshed (**Esc**), ReCalculated, or when you pan around the drawing.

While you are first learning DataCAD, you will probably want to leave the **DrwMarks** setting on. It is very handy to have DataCAD draw temporary marks when using the Reference Point (`) option. So, here's a great trick for those of you, like myself, who prefer to leave Draw Marks turned off for all operations except Reference Point:

1. First, go to **Utility/Measures/RefPt** and turn **DrwMarks** on.
2. Now go to **Utility/Settings** and turn **DrwMarks** off.
3. Now **DrwMarks** will only display when using Reference Point.

BigCursr (Big Cursor)

This setting controls the size of the cursor crosshairs. With **BigCursr** turned on, the cursor crosshairs will extend from one edge of the Drawing Window to the other. This is especially useful for aligning entities or text. With **BigCursr** tuned off, the crosshair will be displayed as a small cross.

When **BigCursr** is turned from on to off, another menu will appear where you can adjust the size of the small cursor. The menu displays standard integer selections where you can define the cursor size in pixels, or you can type in a value and press **Enter**.

Although I personally use the big cursor at all times, it is worth noting that according to the DataCAD Reference Manual, "Because the big cursor speed has not been optimized for lower resolution monitors, the small cursor may be more efficient." So if you have a low resolution monitor you might want to turn the big cursor off. The suggested setting is on.

SHORTCUT KEY: *You can press the* **+** *key to toggle between a large cursor and a small one.*

NegDist (Negative Distance)

With **NegDist** turned on, the Coordinate Readout area will display the cursor location with both positive and negative values in relation to the last selected point. If **NegDist** is turned off, then all values are displayed as positive numbers. The suggested setting is on.

ShowZ (Show Z-Base and Z-Height)

When toggled on, this option will continuously display the current Z-Base and Z-Height information on the Information Line. If toggled off, Z-Base and Z-Height will not be shown. The suggested setting is on.

FixedRef (Fixed Reference Setting)

This is another one of the most important features in DataCAD. Toggle **FixedRef** off to choose a floating reference setting, or on for a fixed reference setting. The **FixedRef** toggle is usually off, indicating a floating reference setting that lets you input a distance from the last point you entered to the current cursor location, and then you can set all the distance information from zero each time you enter a new point. For the majority of your work, you will probably always keep this option toggled off.

When **FixedRef** is toggled on, you can enter a specific fixed reference point or select an absolute zero point in the drawing field using the **AbsZero** option. The absolute zero point, which appears at the lower-left point of the screen when you created a new drawing, remains constant throughout the drawing process. Your can also enter a fixed reference point to compute all X and Y distances to the current location. This distance is displayed in the Message Window. Toggling **FixedRef** on is sometimes useful for inputting survey data from a fixed reference point. The suggested setting is off.

DistSync (Synchronize the Distance)

Turn **DistSync** on to synchronize the distance displayed in the Message Window with a rotated cursor angle. If you are working with a rotated cursor but want to view the distance relative to normal X and Y coordinates, toggle DistSync off. The suggested setting is on.

TxtScale (Text Scale)

This is one of the truly masterful features of DataCAD. It controls how the size of text is calculated and displayed, greatly simplifying the process. This setting is found not only in this menu, but in the **Text** and **Plotter** menus as well. See the lengthy explanation of Text Scale in Chapter 5, "Basic Drawing."

In a nutshell, without **TxtScale**, offices came up with elaborate charts to remind everyone what the text size was supposed to be for different scale drawings. A $^1/_2$″ scale [1:50] drawing may have used text that, at a $^1/_2$-inch-scale [1:50], was defined as 6 inches [152] high, to create text to be plotted out at $^1/_8$-inches [3.2] high. What **TxtScale** does is figure out all those conversions for you on the fly. The suggested setting is on.

Grids

The **Grids** option button is located in the **Utility** menu. Three different grids are available in DataCAD: a snap grid and two display grids. The two display grids are visual reference points that are displayed in the Drawing Window. They do not print and you cannot snap to them.

Each layer in DataCAD has its own set of snap and display grids. This lets you have independently sized grids on each layer, letting you move quickly between drawing layers without having to change grid settings or sizes each time you change to a new active layer. Like the **Settings** options, DataCAD saves all current grid settings with the drawing file when it is saved.

SnapGrid

The snap grid does not display on the screen. With **SnapGrid** turned off, your cursor moves pixel by pixel around the screen. With **SnapGrid** turned on, your cursor will jump from one point to another by the increment selected in the **GridSize** option. How large or small you set the grid depends on the smallest size entities you draw the majority of the time. For instance, perhaps you draw a lot of wood framing details and need to draw 1x and 2x framing members. You would perform the following steps:

1. Turn **SnapGrid** on.
2. Select **GridSize/SetSnap** and select a setting of a $^1/_2$″ [12.7].
3. Now you can draw exact 1x and 2x framing without having to input distances with the keyboard. While drawing entities, just watch the Coordinate Readout line. You will see that the cursor now only draws in $^1/_2$″ [12.7] increments.

Even if you don't draw entities with fixed increments all the time, in most cases you will probably want to leave **SnapGrid** on to avoid allowing the cursor to move in increments that are smaller than average construction tolerances. For instance, I set my snap grid to $^1/_8$″ [3.2] since there are very few if any entities which I draw that are required to be drawn down to an accuracy of $^1/_{16}$″ [1.6] or $^1/_{32}$″ [0.8]. The suggested setting here is on.

SHORTCUT KEY: *You can toggle the **SnapGrid** option on and off by pressing the **x** key. Note that the "S" in SWOTHLUD in the Status Area appears lowercase when **SnapGrid** is toggled off and is uppercase when toggled on.*

DspGrid1 (Display Grid 1)

Turn this option on to display a uniform grid of reference dots in the Drawing Window. You set the grid distance with the **GridSize/SetGrid1**.

DspGrid2 (Display Grid 2)

Turn this option on to display a uniform grid of reference dots in the Drawing Window. You set the grid distance with the **GridSize/SetGrid2**.

GridSize

Use this option to set the size of the snap grid and the two display grids. It is usually a good idea to keep your grid sizes as multiples of each other to avoid ending up with oddly spaced grids due to fractional divisions between the display grid dots and the snap grid. When you press GridSize, the following options appear:

- **SetSnap (Set Snap Grid Size)** Use this to select the size of the X and Y values for the snap grid. The suggested setting is 1″ or $^1/_2$″ [12.7] (this depends on the kind of drawings you do).

SHORTCUT KEY: *Press the* **s** *key to go instantly to the* **Setsnap** menu.

- **SetGrid1 (Set Display Grid 1 Size)** Use this to select the size of the X and Y values for Display Grid 1.
- **SetGrid2 (Set Display Grid 2 Size)** Use this to select the size of the X and Y values for Display Grid 2.

GridColr (Grid Color)

Use this option to set the color of the dots that make up the display grids. The default color for Grid1 is **LtGrey**. The default color for Grid2 is **Yellow**.

Pick the grid (**SetGrid1** or **SetGrid2**) that you want to change, and then pick the new color. Black works well for a white Drawing Window, but keep in mind that if you want any of your grids to use black, then you must select the color **White** from the color menu since DataCAD automatically substitutes black for white when a white Drawing Window is being used.

SnapAng (Snap Grid Angle)

This option sets the number of cursor snap segments (like slices of a pie) to be used when **Ortho** is turned on after placing the first point of an entity. This allows you to move your cursor and draw entities at exact angles by simply moving your mouse, and without having to type any coordinates or angles.

By default, DataCAD uses 45-degree segments for a total of eight segments (45 degrees × 8 segments = 360 degrees). But you can change that as you see fit. Figure 3-5 shows the default eight-segment SnapAng setting and a five-segment setting.

Try changing the **SnapAng** and seeing what your cursor does. Watch the Coordinate Readout line to see the angle display jump from angle to angle as you move the cursor. The suggested setting is **8**.

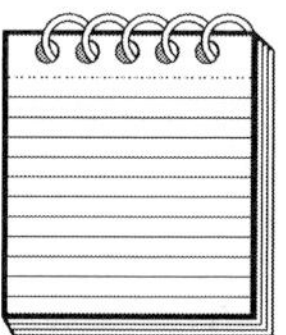

NOTE: *With **SnapGrid** turned on and **Ortho** mode off, the cursor's rubberband does not follow the **SnapAng** setting. It moves per the **SnapGrid** setting only.*

Figure 3-5
SnapAng examples

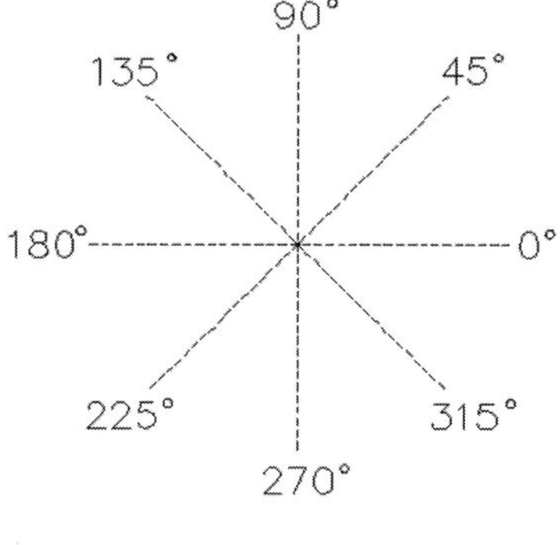

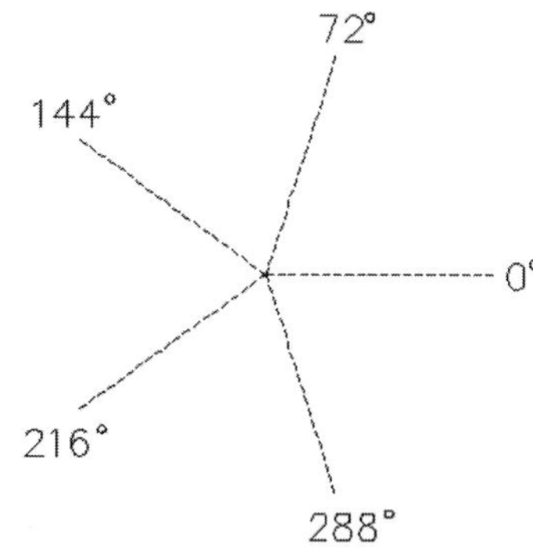

Angle (Grid Angle)

Use this option to rotate all three grids (**SnapGrid**, **DspGrid1**, and **DspGrid2**) simultaneously on the current layer. Choose or type an angle value from 0 to 90 degrees and press **Enter**. To match the grid angle with an existing entity in the drawing, choose the **Match** option in the **Value** menu and select an existing line or a line defined by any two points. The suggested setting is **0.0.0**.

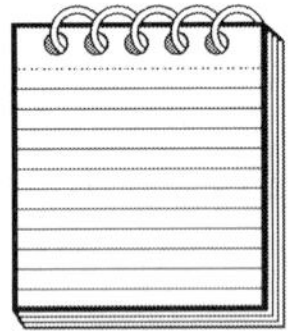

NOTE: *Notice that the cursor crosshairs rotate along with the grids. Also note that your X and Y axis do not rotate along with the grids. X remains along the horizontal and Y remains along the vertical axis no matter which Grid Angle you choose.*

GridOrg (Grid Origin)

Use this to change the origin of all three grids (**SnapGrid**, **DspGrid1**, and **DspGrid2**) simultaneously. By default, it will reset the origin on every layer in the drawing file whether currently displayed or not (remember that DataCAD enables each layer to have its own set of snap and display grids). To use GridOrg

- To select a new grid origin after selecting **GridOrg**, simply click wherever you want the origin to be located in the Drawing Window. Use the middle snap button to snap to existing objects.
- To display the new grid, refresh the Drawing Window by pressing **Esc**.
- To change the origin on the active layer only, toggle **SetAll** off before entering the new grid origin.

Display

The **Display** option button is located in the **Utility** menu. It controls what DataCAD displays and how it is displayed. It is a somewhat eclectic mix of settings, many of which might be better off in the menus to which they are directly related, but here they are nonetheless. More than half of the display settings are toggle buttons; they are either on (green) or off (red). The remainder of display settings have various options associated with them; they are selected or toggled on. Like **Settings** and **Grids**, DataCAD saves all current display settings with the drawing file when it is saved.

ShowTxt (Show Text)

Displaying text onscreen, especially fonts with complicated strokes or large amounts of text, takes a lot of computer resources (like RAM, video RAM, processor cycles, and so on). This can sometimes slow the refresh of the Drawing Window considerably, even in a powerful computer.

To help speed up the display of text, you can use the **ShowTxt** toggle to show text normally in its entirety (when toggled on) or to display text as bounding boxes (when toggled off) that show the extents of your text rather than showing the actual text (see Figure 3-6). This toggle affects all text, including text within dimensions and text within symbols. With **ShowTxt** toggled off, the bounding boxes display much faster than standard text, although you won't be able to see the words themselves (see the section on **BoxColor** for a related setting.)

Note that with **ShowTxt** toggled off to display bounding boxes, new text entered dynamically on the screen will display as text rather than bounding boxes until the Drawing Window is refreshed (using Esc or ReCalc), at which point the text will revert to a bounding box. This is the kind of setting (along with **ShowHtch**) that many people create a keyboard shortcut or icon for in order to be able to quickly toggle the setting on or off. The suggested setting is on.

WARNING: *If you fail to turn **ShowTxt** back on prior to printing, then the text will not be printed; only the bounding boxes will be printed.*

Figure 3-6
ShowTxt toggled on and off

TEXT

ShowDim (Show Dimension)

This toggle is much like the **ShowTxt** button. Toggling the button on will show all associative dimensions. Toggling it off will cause all associative dimensions not to be displayed. With **ShowDim** off, no bounding boxes are displayed as with the **ShowTxt** option. The dimensions are simply not displayed at all: no text, no leaders no arrows. Again, the reason for this is to speed up screen displays.

Note that this only affects the display of associative dimensions. Since non-associative dimensions are simply a disassociated collection of lines and text, they are not affected by this toggle. The suggested setting here is on.

ShowHtch (Show Hatch)

Again, this setting is much like the **ShowTxt** button. Toggling the button on will show all associative hatching, while toggling it off will cause all associative hatching not to be displayed. The display of hatching in DataCAD can be even more taxing on a computer's resources than the display of text. When toggled off, all associative hatches are displayed as closed polygons.

This is the kind of setting (along with **ShowTxt**) that many people create a keyboard shortcut or icon for in order to be able to quickly toggle the setting on or off. Note that this only affects the display of associative hatching. Since non-associative hatches are simply a disassociated collection of lines, they are not affected by this toggle. The suggested setting is on.

TIP: *Toggling **ShowHtch** off is a useful way to edit the boundaries of associate hatches. With **ShowHtch** on, it is often difficult or impossible to find all the vertices that make up the boundaries of each hatched area. With **ShowHtch** off, however, the polyline that is displayed in place of the hatching shows exactly where each vertex is. Now it is quite simple to move, add, or delete vertices with the Polyline macro found in the **Toolbox**.*

WARNING: *If you fail to turn **ShowHtch** back on prior to printing, then the hatch patters will not be printed; only the polyline outline will be.*

ShowWgt (Show Line Weight)

If you use Line Weights for your entities (not recommended for most of your work) rather than mapping line colors to pen thicknesses, this toggle will determine whether those lineweights (or widths) will be displayed in the Drawing Window or not. With **ShowWgt** off, all entities are displayed and printed at a width of one pixel. With **ShowWgt** on, all entities are displayed and printed with their assigned lineweights.

I personally don't recommended using Line Weights (press **Ww** or **Shift+Ww** to toggle Line Weights up and down) because of the way DataCAD displays and prints them. To make entities wider, DataCAD simply adds identical lines and arcs adjacent to the first ones. Thicker Line Weights mean more lines added. As more lines are added, display times increase and finding the actual center of lines and arcs soon becomes an effort in futility. However, some people like the Line Weights option, so by all means try it out and see if you like it. The suggested setting here is off.

WARNING: *If you fail to turn **ShowWgt** on prior to printing, then all entities will be printed at only one lineweight.*

UserLine (Show User Lines)

User lines are all the "complex" linetypes (**Edit/LineType**) available to you in DataCAD. Every linetype, except for the first four, are *user linetypes* in DataCAD. The first four are native to DataCAD and cannot be changed or edited. Toggling the **UserLine** button on will display all user linetypes. Toggling it off will display all linetypes as solid.

Toggling **UserLine** off is useful when trying to trim or edit complex linetypes that don't have a well-defined centerline, such as **LapSidR**, **PShngl_R**, **Marble**, and so on. The suggested setting is on.

WARNING: *If you fail to turn **UserLine** back on prior to printing, complex linetypes will not be printed; all lines will be printed solid instead.*

Figure 3-7 Toggling **OverSht** off and on

OverSht (Show Line Overshoot)

Use **OverSht** to display line overshoots, giving a looser, more hand-drawn look to your drawings. To set the size of line overshoots, go to the **Edit/LineType** menu and select **OverSht** (more on this later in the chapter). Toggling the **Display/OverSht** button on will display all the lines with the overshoot defined in the **LineType/OverSht** menu. Toggling it off will display all the lines without any overshoot (see Figure 3-7). The suggested setting is off.

WARNING: *If you fail to turn **OverSht** back on prior to printing, overshoots will not be printed. All the line intersections will be printed without overshoots.*

ShowIns (Show Symbol Insertion Point)

Toggling this button on will display an X at the insertion point of all symbols in the drawing. Toggling it off will not display anything at the symbol insertion points. The suggested setting is on.

ShowAttr (Show Attributes)

Use this to display visible attributes added to entities by a *DataCAD Applications Language* (DCAL) macro. Toggling this button on will not change the way any standard entities created by DataCAD will be displayed. However, when a programmer creates a macro (the mini-programs found in your \DCX directory), they have an option in the DCAL programming language to attach attributes (like text) to entities created by the macro. Toggling this button on will cause these attributes to be displayed. If you purchase or use such a macro, the instructions will tell you if attached attributes exist that can be displayed. The suggested setting here is off.

CurveCtr (Show Center of Curves)

Turn this toggle on to show a small dot at the centerpoint of all 2D curves, including arcs, circles, bezier curves, and so on. Toggling it off will not display anything at the centerpoint of the curves. This is especially useful since you can middle-snap to the centerpoint when this toggle is on and the centerpoint is visible. With it toggled off, you cannot see or snap to a curve centerpoint. The suggested setting is on.

DimPoint (Show Dimension Control Points)

Turning this toggle on will display control points for associative dimensions. Displaying these control points makes it much easier to select the ends of dimension leaders in order to edit the leaders or the dimensions (see Figure 3-8).

Associative dimensions and their leaders can be stretched only by selecting the actual points from which the dimensions were created. If you do not have your dimension leaders set with an offset from the object you are dimensioning, then the dimension points, the ends of the dimension leaders, and the object you are dimensioning all coincide with one another. But if your dimension leaders are offset from the object you are dimensioning,

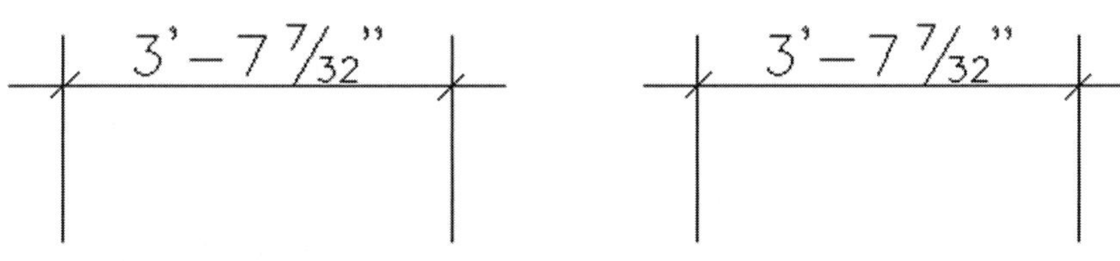

Figure 3-8 Control points off and on

then the ends of the leaders do not coincide with the dimension points, as in Figure 3-8.

Turning **DimPoint** on enables you to more easily see and find these control points on the associative dimensions, making editing easier. The suggested setting here is on.

MltiView (Multi-View Windows)

This toggle turns on and off the display of four Multi-View windows, displayed on the opposite side of the Drawing Window from your menu buttons. They are thumbnail or snapshot views of whatever was displayed in the Drawing Window when the In button is pressed under each window. They act very much like mini-Drawing Windows, except that you cannot draw or edit entities in them. You can, however, refresh and quick-shade the entities within the views (see Figure 3-9).

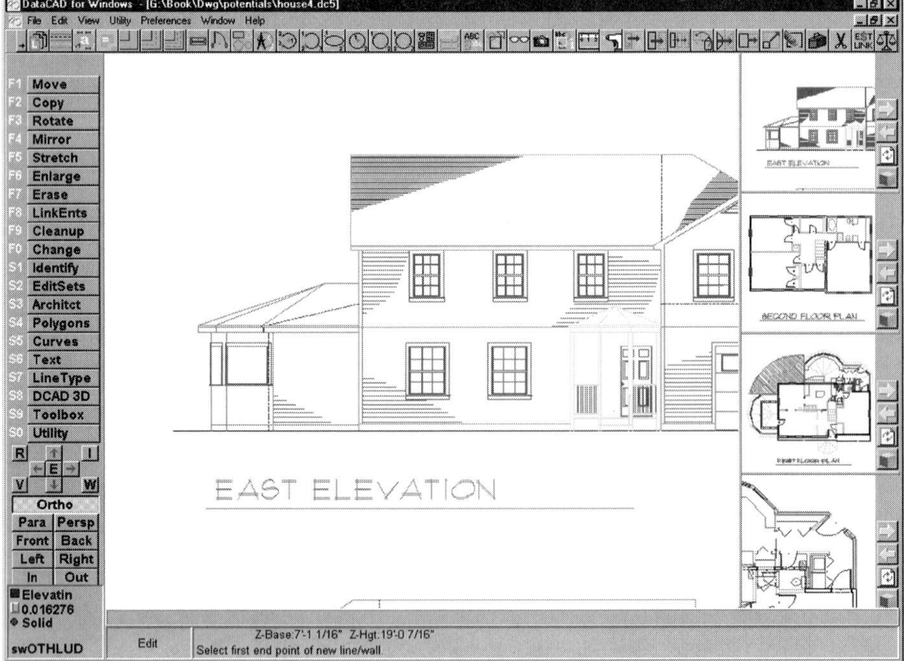

Figure 3-9 The four **MltiView** windows

SHORTCUT KEY: ***MltiView*** *can also be selected by pressing* ***Ctrl+Ww.***

If you want to know how DataCAD determines how to save each view, read about 3D GotoViews in Chapter 10. Multi-View windows work in the same manner, remembering the layers that were on and the extents of the Drawing Window when the view was saved. In fact, as soon as you save a Multi-View, it is added to the list of 3D GotoViews. The first Multi-View window (at the top) is always named Window1. The next three views are likewise named Window2, Window3, and Window4 in the 3D GotoView menu. If you rename them, the 3D GotoViews will be retained, but the views will no longer work in or be displayed in the Multi-View windows.

Under each Multi-View window are four buttons:

Read In This reads the current image in the Drawing Window into the Multi-View window. If an image is already in the window, it is replaced by the new view. Placing the cursor in a Multi-View window and clicking the mouse will do the same thing as pressing the In button. The view is created by DataCAD "remembering" which layers were turned on when the view was saved, and what the extents of the Drawing Window were.

Read Out This reads the image in the Multi-View window back out into the Drawing Window. Again, it works by turning on the layers and showing the extents of the Drawing Window that were saved when the view was read In.

Refresh This refreshes the image in the Multi-View window. The extents and the layers of the view do not change, even if a different view and different layers are currently displayed in the main Drawing Window. Instead, the view is retained, but all changes made to the entities since the image was read In are updated.

Shade When toggled the first time, this button will use the GLShader to shade the current image in the Multi-View window. When toggled the next time, all entities will be displayed without shading. If the button is toggled on and the Out button is pressed, the view will be transferred to the main Drawing Window but without shading. Toggling Sh on will automatically refresh the view in the window prior to shading the entities.

Views in the Multi-View windows, including shading, are retained even when **MltiView** is toggled off.

SmallTxt (Small Text)

Since text at a certain size is too small to read, and because text can cause screen refreshes to slow considerably, this setting is used to display text as text boxes once you zoom out so far that the text appears smaller than the **SmallTxt** setting. A setting of **zero** pixels will display all text and no text boxes. You can only pick a number from 1 to 10, even though more numbers are available in the menu. You will have to experiment a little to see what works for you (see the following section on **BoxColor** for a related setting). The suggested setting is **4**.

BoxColor (Text Box Color)

Use this option to set the color of the text boxes that appear when the text is too small to display (refer to the previous **SmallTxt** section) or when **ShowTxt** is toggled off. By default, the color of all text boxes is the same color as the text itself. You can change the color of the text boxes to a single, global color by selecting one of the displayed colors. The suggested setting is NoChange. BoxColor offers the following options:

- **Adjust** Selecting this enables you to alter the color, hue, saturation, and luminance of whatever color you select by using the standard Windows color dialog box. By selecting the **Palettes** option, it also enables you to save the current RGB file or to load a previously saved RGB file.

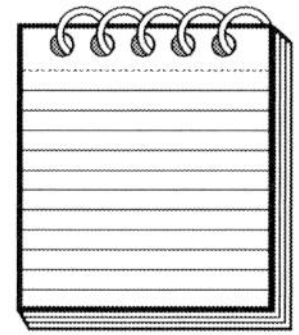

NOTE: *These settings are "trap doors" to the* ***Edit/Change/Color*** *menu and the* ***Utility/Display/Palette*** *option. Adjusting colors or loading a new RGB file will affect the entire drawing file, not just the text box colors.*

- **Custom** Selecting this option enables you to pick any custom color from 1 to 255 in the current Windows color palette
- **Match** After selecting this, pick an entity in the Drawing Window. The color of that selected entity will be used for the text box color.
- **NoChange** Selecting this option causes DataCAD to return to the default state where text boxes are displayed in the same color as the text

SmallSym (Small Symbol)

Since symbols at a certain size are too small to read, and because a large number of symbols can cause screen refreshes to slow down, this setting is used to display symbols as bounding boxes once you zoom out so far that they get smaller than the **SmallSym** setting. A setting of **0** pixels will display all symbols and no bounding boxes. You will have to experiment a little to see what works for you. The suggested setting here is **2**.

ArcFactr (Arc Factor)

Use this to change the precision factor for drawing and plotting circles and arcs. A low setting like **1** results in curves being drawn with more facetting and less smoothness. Screen refresh times will be faster, and plot file sizes will be smaller. A higher setting results in curves being drawn with less facetting and greater smoothness. It will also result in slower screen refresh times and larger plot file sizes. The suggested setting here is one.

*NOTE: Although the menu allows you to pick numbers greater than 10, DataCAD only recognizes an **ArcFactor** setting from 1 to 10. You will also find that a setting of 1 is usually acceptable for nearly all of your work. A setting of 4 is usually as high as you will need to go. Numbers from 5 to 10 don't generally show a noticeable improvement, but feel free to experiment.*

LyrOrder (Layer Refresh Order)

This setting determines the order in which the drawing layers are refreshed whenever the Drawing Window is refreshed. The following options are available here:

- **First** will cause the active layer to be refreshed first.
- **Last** will cause the active layer to be refreshed last.
- **InOrder** will cause all layers to be refreshed in order, starting with the first on layer in the drawing file.

Selecting the **Last** option is probably the most useful setting for most of your work. When entities on the current, active layer overlap one another, it may be difficult or impossible to see them since only one color is displayed at one time. By selecting **Last**, you will be able to see the entities on the active layer when the screen is refreshed since they will be the last ones drawn.

The **First** option is useful if you are working in a large drawing file, are only editing on the active layer, and don't want to keep waiting for all the entities in the Drawing Window to redraw. With **First** selected, you can start the refresh (via Esc, ReCalc, PgUp/Dn, and so on) and then press **Del** to stop the refresh once the entities on the active layer have displayed. The suggested setting is Last.

Palettes (RGB Color Palettes)

Palettes enables you to load and save custom RGB color palettes to alter the way entities and GLShades displayed. RGB color palettes are used to define the 256 colors available for use in DataCAD and affect all entities in the drawing file. See Chapter 23 for a more in-depth explanation.

You can create your own custom RGB palettes, but DataCAD comes with several that are ready for you to use. Located in the **\SUP** directory you will see a number of files with the .rgb file extension. To select one of them, select **Palettes/LoadRGB**. To save the current color settings (if you have made changes that you might want to use in other drawings) to an RGB file, select **Palettes/SaveRGB**, and save the file with a new name in the **/SUP** directory. Here is what you will find in some of these files:

- **Default.rgb** This is the standard, default RGB file.
- **Greyscal.rgb** A monochromatic, grey RGB file
- **Sepia.rgb** A monochromatic, sepia-toned RGB file

DispList (Display List)

This button will only be displayed if the *Display List* (D/L) option (explained later in this chapter) is currently on. Select **DispList** and then **Status** to display the status of the D/L. It will tell you the number of entities saved in the D/L database, how many line segments are in those entities, and how many megabytes of RAM are currently being used to store those elements.

Other Settings

A number of settings and display options are located in various places throughout the DataCAD menus, which are best covered here since they are related more to this chapter than to the others.

Edit/LineType/OverSht (LineType Overshoot)

Use this option to set the Line Type Overshoot dimension. Usually, a dimension of an inch or a fraction of an inch is sufficient, but you will have to experiment to see what you like. Refer to the explanation of **OverSht** in the **Display** menu earlier in this chapter for more information.

Utility/WindowIn/FreeZoom (Free Zoom)

Toggling **FreeZoom** on enables DataCAD to freely zoom in or out when using **WindowIn** and **PgUp/Dn**. Toggled off, DataCAD is restricted to zooming in or out by using the fixed scales under Utility/Plotter/Scale (such as 12′, $^3/_4$″, 1:20, 1:1000, and so on). Most users find the limitations of these fixed scales completely unnecessary. It is therefore recommended that you always leave **FreeZoom** toggled on.

DCADWin.INI File Settings

The DCADWin.INI (DataCAD initialization) file is a simple text file that contains a number of settings that determine how DataCAD operates and how certain user interface elements are displayed. Some of the settings are altered via the **Tools/Program Preferences** dialog, but others can only be changed by manually altering the .INI file yourself.

If you are running Windows 95 or 98, you will find the DCADWin.INI file in your C:\WINDOWS folder. If you are running Windows NT, you will find the DCADWin.INI file in your C:\WINNT folder. See Chapter 23 for a full description of the **Tools/Program Preferences** dialog and the DCADWin.INI file.

Display List: What Is It and How Does It Work?

The *Display List* (D/L) is a speed enhancer for DataCAD. D/L can be turned on by selecting **Tools/Program Preferences/Misc/Display List/On/Off** and checking the box. Essentially, the D/L scans all the layers that are currently turned on and stores an image of them in your computer's RAM. That memory is allocated to D/L dynamically. The memory shrinks and grows as needed, and gets cleared and reallocated every time entities are regenerated or when you close a drawing so that there is no maintenance necessary on the user's part and no need to put the D/L in a RAM drive.

Note that D/L will run faster and faster the tighter you zoom into your drawing. D/L will ignore all the entities outside the Drawing Window since they aren't in view, thereby reducing the number of entities that have to be recalled from D/L memory.

With D/L turned off, DataCAD displays and interacts with the current drawing file directly from the hard disk drive. With D/L turned on, DataCAD saves the current drawing image to RAM, and then displays and interacts with the current drawing file from there. Accessing RAM is much, much faster than accessing the hard drive. Because of this, with D/L turned on, DataCAD can accomplish nearly instantaneous pans and zooms. Using D/L also allows for much faster entity selections (up to 5 to 10 times faster) and object snapping, since DataCAD does not have to scan the entire drawing database from the hard disk to find the proper selection; that information is already saved in the D/L database. You have very few reasons *not* to turn D/L on, but you should be aware of certain aspects of its operation.

D/L and ReGen/Refresh

When DataCAD regenerates a drawing, the program searches through all the layers that are currently turned on, compiles a new list of all the entities found, and then saves them to the D/L in RAM. Depending on the number of entities displayed and the speed of your computer, this may take only a split second or it may take several seconds. Once this is done, however, DataCAD will not need to do another time-consuming search of the drawing database until you turn new layers on or current layers off (see the following list).

In some instances, the D/L updates itself, and occasionally you will want to do it yourself. To do this yourself, simply press the **Uu** key. DataCAD will tally up all the currently displayed entities, save the image to the D/L, and then display them in the Drawing Window.

DataCAD will ReGen when

- The **Uu** key is pressed.
- The **ReCalc** button in the Navigation Pad is pressed.
- Turning layers on and off via **On/Off**, **ActvOnly**, or **AllOn** in the Layers menu.
- Selecting a 3D GotoView (because layers are turned on and off).

With the D/L on the **Esc** (or Screen Refresh) key still behaves the same as it does with D/L turned off, producing a quick refresh of the displayed graphics, without first scanning for new entities. The entities already saved in the D/L memory are cleanly redisplayed in the Drawing Window. This is faster than pressing the **Uu** key since DataCAD is simply redrawing the vectors the D/L has already captured for each entity that has been drawn.

NOTE: *If you stop the screen from displaying all the current entities by pressing **Delete** or **End**, then this partial image is the one saved to the D/L. Thus, pressing **Esc** (Refresh) afterward may not display all the current entities. If you use **Delete** or **End** to stop the screen refresh, then use the **Uu** (ReGen) key afterward to redisplay the entire drawing.*

Max Acceleration

Turn this on by checking the **Max Acceleration** box. It causes the D/L to run even faster, but with a couple of minor drawbacks:

- The **Max Acceleration** option of D/L means that it will not take the extra time to fix the following entities when you zoom and pan around the drawing:
  - Symbols
  - Arcs
  - Circles
  - Ellipsis

  This means that you will get a visual pixel "bloom" (larger-than-life screen garbage) on these entities, but they will refresh faster. Pressing the **Uu** key will regenerate them and get rid of the pixel bloom, but will take a little time to do it, depending on the size of the drawing and the speed of your computer.
- When **Max Acceleration** is turned on, curves will appear more chunky as you zoom in. Regenerating the D/L (press the **Uu** key) will cause DataCAD to resave a smoother version of the curves to the D/L. But on the up side, since the D/L contains the entire drawing saved at the moment of regeneration, if you zoom in close, regenerate the D/L, and then pan out, the curve, text, and marker resolutions will remain high.

Different Behaviors in D/L

A few things will behave differently with D/L turned on. This is because D/L saves an image of the current drawing. When you add to or edit the drawing on your screen, you are actually editing an image that simulates your drawing. The nuances of changing an image versus the actual drawing database is what causes some of these anomalies. Here are some of them:

- With D/L turned on, text can be selected by any visual entity within it. With D/L turned off, text can only be selected by one its six handles (see Chapter 5).
- With D/L turned on, complex linetypes can be selected by any visual entity within it. With D/L turned off, linetypes can only be selected by the center or edge that defines the line.

- The selection of symbols differs depending whether D/L is turned on or off.
  - **Entity and Group** With D/L turned on, the **Entity** and **Group** commands enable a symbol to be selected by any visual entity within it. With it turned off, the **Entity** and **Group** commands can only select symbols by their insertion points.
  - **Area** With the **Area** command, a symbol must be completely enclosed within the area bounding box to be selected. This is true even when D/L is on.
  - **Area/Crossing** With D/L turned on, the **Area/Crossing** bounding box shown in Figure 3-10 will select the symbol, even though the bounding box does not encompass the whole symbol, and the symbol's insertion point (the small X in the bottom-left corner) is not enclosed by the bounding box. With D/L turned off, the **Area/ Crossing** bounding box will not select the symbol at all, since the entire symbol must be within the bounding box, even with the **Crossing** option enabled.

Before D/L came along, some longtime DataCAD users may have used the inability to choose symbols with Area/Crossing except by their insertion point as a feature and incorporated it into their editing practices. Be aware, however, with D/L on, you will no longer be able to work this way, as selecting anywhere on a complex entity such as text, linetypes, and symbols is considered valid as far as D/L is concerned.

Fence With the **Fence** command, a symbol must be completely enclosed within the fence to be selected. This is true even when D/L is on.

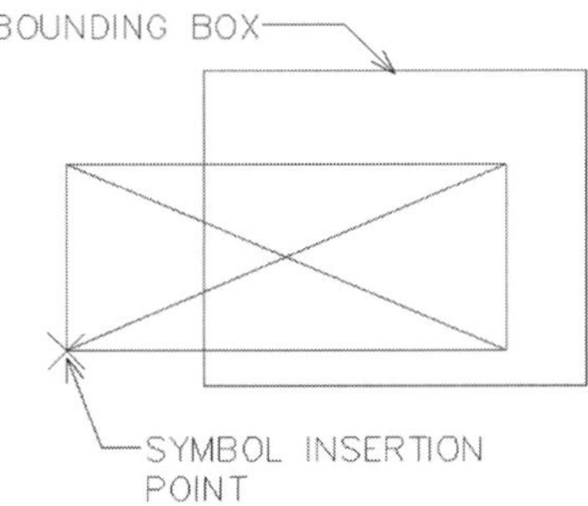

Figure 3-10 The Area/Crossing bounding box

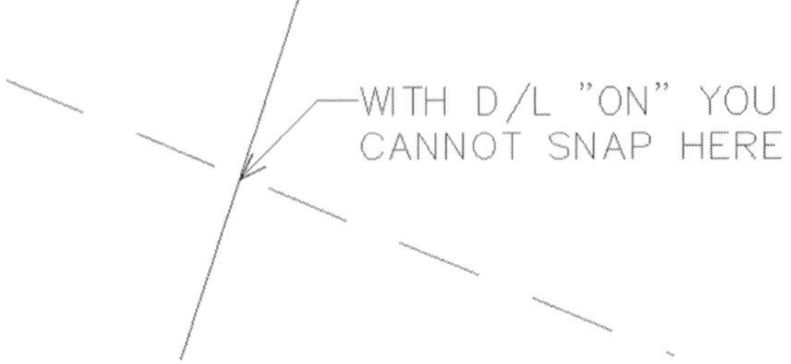

Figure 3-11
An "unsnappable" situation

Linetypes If a blank area of a linetype intersects another entity when D/L is turned on, the cursor will not snap to the intersection (see Figure 3-11). Two solid parts of intersecting linetypes must actually touch one another in order to snap to the intersection.

The workaround is to temporarily either turn D/L OFF, or to toggle **Utility/Display/UserLine** off. The recommended settings are for D/L to be on and for Max Acceleration to be on as well.

ObjSnap (Object Snap Settings)

Object snapping is used to exactly select a particular point on an existing entity or on the intersections of entities. You snap to these points by using the middle button of your three-button mouse or by pressing the **Nn** key.

NOTE: *For Logitech mouse users, the middle button works for snapping with the Microsoft Intellimouse and the Logitech three-button mice. The Intellimouse will concurrently support Windows double-clicking (**click** once on the middle button to perform the action of **double-clicking** on a folder or file). However, the Logitech drivers will not support both Windows double-clicking and DataCAD middle-snapping concurrently. You can only have one or the other, but if you have a Logitech four-button mouse, the fourth button can be set to snap in DataCAD, while the middle button can concurrently provide Windows double-clicking. See Chapter 2 for more information about various mice.*

These settings are dynamic and global. They affect the entire drawing. Whichever of the currently selected Object Snap settings best matches the condition closest to your cursor position when you middle snap or press the **Nn** key is where your cursor will snap to.

More than one toggle may be toggled on at one time. In fact, this is what makes these setting so powerful; you don't have to pick them each time you want to snap to something different. Note that if more than one snappable point is near the cursor, the point closest to the center of the cursor is the one that will be snapped to. For this reason, if two snap points are very close to one another, you may want to **WindowIn** closer to the entities to make sure you select the correct one.

Nearest Toggle this on to snap to the nearest point of any object. Be warned that this can be a dangerous setting at times. If you have other snap settings turned on along with this one, you may think you are snapping to a mid-point or to a tangent when you are actually snapping to a point that is only near your intended snapping point. I have only ever used this setting for placing telephone and electrical outlet symbols on walls. The suggested setting is off.

EndPnt (End Point) Toggle this on to snap to the nearest endpoint of a line or arc (see Figure 3-12). The suggested setting is on.

Figure 3-12
An **EndPnt** example

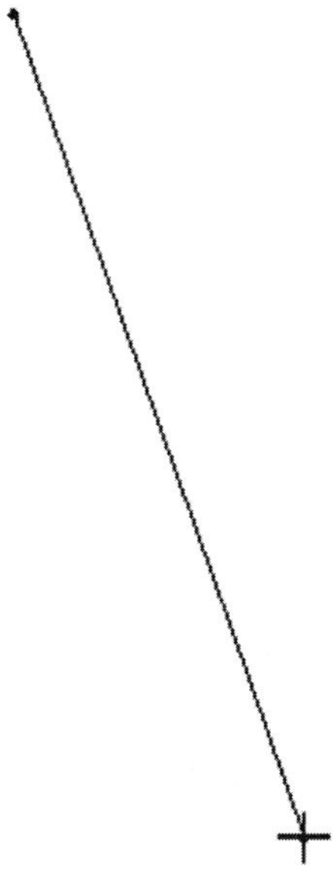

Figure 3-13
A **MidPnt** example

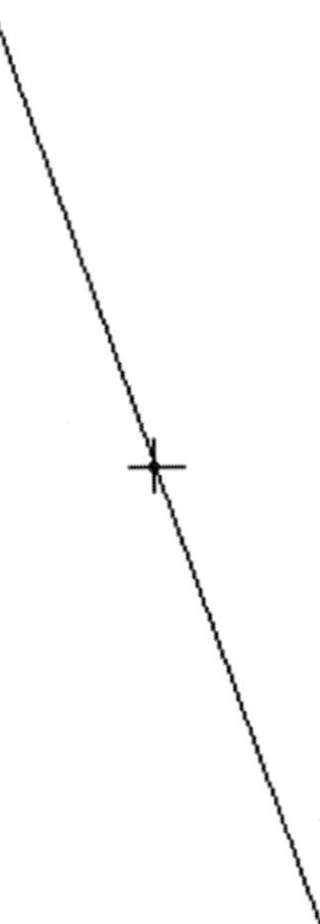

MidPnt (Mid-Point) Toggle this on to snap to the mid-point of a line (see Figure 3-13). The suggested setting is on.

No.Pnts (Number of Points) With this setting, you can select an equidistant number of snappable points along any line (see Figure 3-14). The two endpoints of a line count as the first two points. Toggle this button on to snap to the closest one of these points on a line. The points are not visible; they are implied and they affect all lines in the drawing file. When you toggle this button on, you will be prompted to select the number of points. The suggested setting is off.

Center Toggle this on to snap to the center of an arc or circle (see Figure 3-15). Note that if **Utility/Display/CurveCtr** is toggled off, the center points of circles and arcs will not be visible and will therefore not be snappable. The suggested setting is on.

Quadrant Toggle this on to snap to one of the four quadrants of a circle or arc (see Figure 3-16). The quadrants are at 0, 90, 180 and 270 degrees. The suggested setting is on.

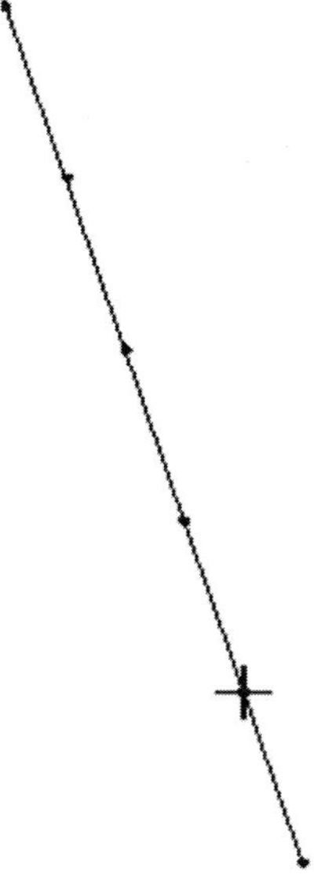

Figure 3-14
A **No.Pnts** example

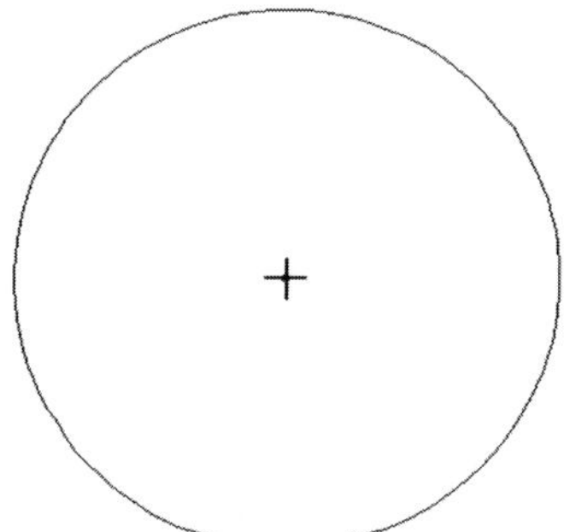

Figure 3-15
Snapping to a center

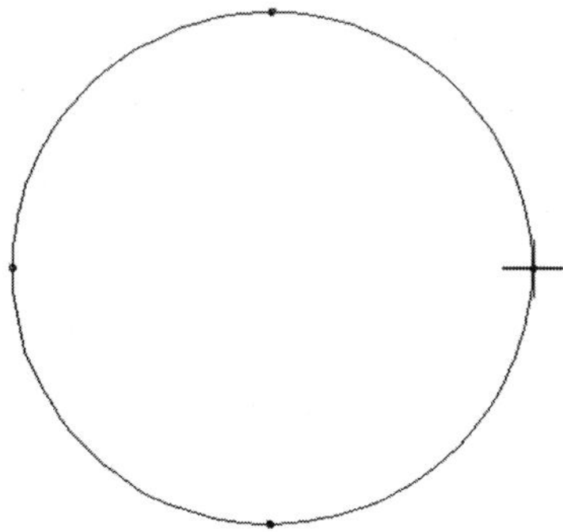

Figure 3-16
Snapping to a quadrant

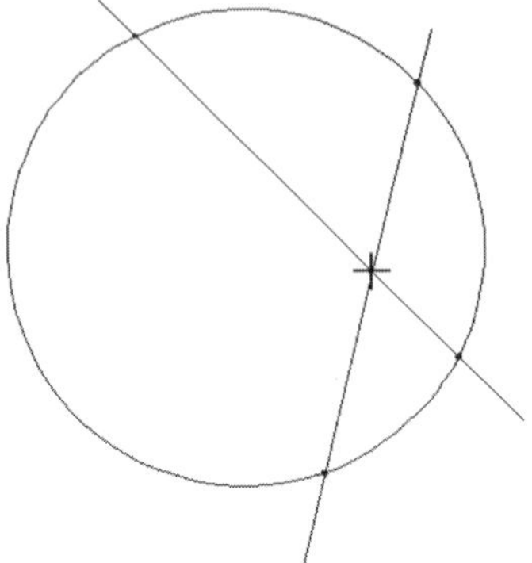

Figure 3-17 Snapping to an intersection

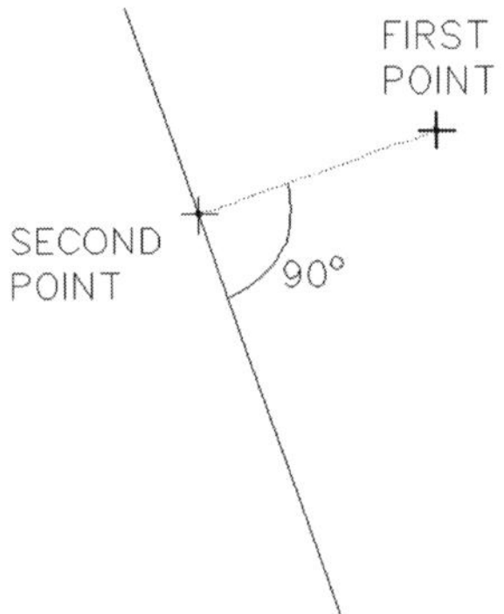

Figure 3-18 The **Perpend** option

Intsect (Intersection) Toggle this on to snap to the intersection of lines, arcs, or circles (see Figure 3-17). The suggested setting is on.

Perpend (Perpendicular) Toggle this on to snap to the point on the nearest line, circle, or arc that forms a perpendicular line through the previous point. As in Figure 3-18, click at point 1 and then click at point 2. The perpendicular line will be drawn from point 1 to the closest point to the cursor on the line (point 2). Note that the **Nearest** toggle does not also need to be toggled on for this to work. The suggested setting is off, but I know a number of DataCAD users who almost never turn it off. Try it both ways to see what is right for you.

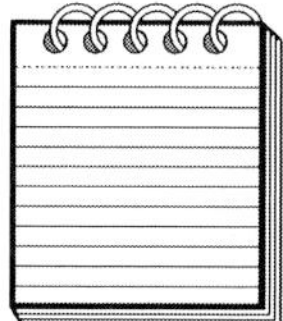

NOTE: A word of caution on this setting: it is very useful, but be aware that leaving it turned on can give you some erroneous results if you are not careful. For instance, you might be trying to snap to a perpendicular point on a line but accidentally snap to the midpoint of a line because it is very close to the perpendicular point.

Tangent Toggle this on to snap to the point on the nearest line, circle, or arc that forms a tangent line through the previous point. As in Figure 3-19, click at point 1 and then click at point 2. The line will be drawn from point 1 to the closest point tangent to the cursor on the circle (point 2). Note that the **Nearest** toggle does not need to be toggled on for this to work. The suggested setting is off.

None **Click** on this button to quickly turn off most of the Object Snap settings. The settings that will not be turned off by this button are as follows:

- **FastSym**
- **Fast3D**
- **LyrSnap**
- **SrchHtch**
- **Aperture**

FastSym (Fast Snap to Symbols) Toggle this on to snap only to the insertion point of a symbol, rather than to any other object snap point within the symbol. This will increase the speed of snapping to symbols, since DataCAD will not have to search all the entities in a symbol. With

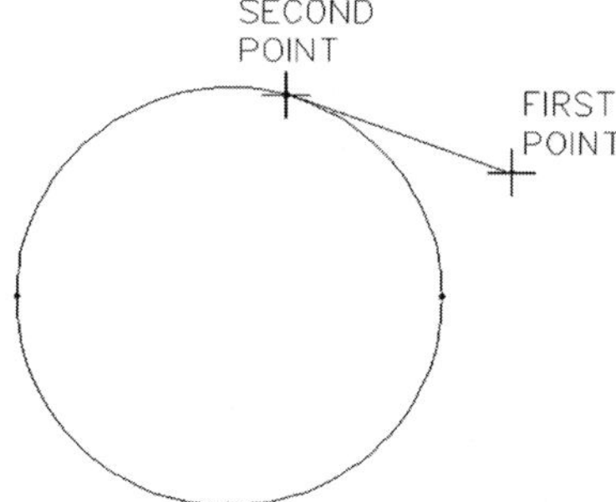

Figure 3-19
A Tangent example

FastSym toggled off, you can snap to any valid object snap point in a symbol. The suggested setting is off.

*NOTE: With **FastSym** toggled off, you can move a symbol within the Drawing Window by any point other than the symbol's insertion point. If you pick a symbol by simply clicking with the left mouse button, it will be picked by its insertion point. But if you snap to an object snap point within a symbol, you can move a symbol by that point. This is a very handy feature.*

Fast3D (Fast Snap to 3D Entities) Toggle this on to select 3D circular entities (arcs, cylinders, cones, truncated cones, domes, tori, or contour curves) only by their center axis markers or control points. This will increase the speed of snapping when many circular 3D entities exist in the drawing. The suggested setting is off.

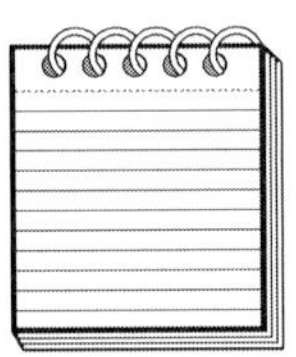

*NOTE: With **Fast3D** toggled on, if you select 3D circular entities using **Area** or **Fence**, DataCAD will select those entities if their center axis markers or control points are located anywhere within the area or fence, even if the entire 3D entity is not fully enclosed by the area or fence.*

MissDist (Miss Distance) Use this to specify the search area in pixels around your cursor center point for all snapping functions. See the following description of the related **Aperture** setting. The suggested setting is **10** (the default).

LyrSnap (Layer Snap) Toggle this on to enable snapping to entities on all the displayed layers. Toggle it off to enable snapping only to the entities on the active layer. The suggested setting is on.

SrchHtch (Snap to Hatch Lines) Toggle this on to enable snapping to any lines in a non-associative hatch pattern. Toggling **SrchHtch** on will enable you to snap to hatch lines just like any other line in DataCAD, but snapping throughout DataCAD will slow down, sometimes considerably, because of the many snappable points DataCAD has to search through.

The reason you will nearly always want to leave **SrchHtch** off is to keep DataCAD from searching that multitude of snappable points in non-associative hatches. This is especially important if the drawing has a lot of non-associative hatching and/or very complex hatch patterns (like patterns made up of lots of small vectors). The suggested setting is off.

NOTE: The Hatch attribute value will be stripped from a non-associative hatch entity if you do any of the following to an entity:

- *Erase a hatch entity and then undo Last Entity Erase.*
- *Erase a group of hatch entities andthen undo Last Group Erase.*
- *Copy*
- *Mirror*

They do not lose the Hatch attribute when trimming functions are run on them (like 2-line trim).

Quick Toggle this on to snap to the first entity (the first one drawn) in the drawing database that is eligible for object snapping. With this setting toggled off, your cursor will instead snap to the nearest point of any entity within the miss distance around the cursor. The suggested setting is off.

*NOTE: Because the cursor snaps to the first entity within the miss distance, you should use **Quick** only when the entity you want is the only entity within the Miss Distance.*

SelSet Toggle this on to snap to entities only within the active selection set. This increases the snapping speed since DataCAD only has to search for entities in the active selection set instead of throughout the entire drawing. The suggested setting is off (until needed).

Aperture (Aperature) Toggle this on to graphically display the Miss Distance around the center of your cursor. When toggled off, the Miss Distance box is still valid but is not displayed visually (see Figures 3-20 and 3-21). The suggested setting is off.

Directry

The **Directry** menu has a number of functions within it. Its overall purpose is to enable you to save and display information about the drawing file and

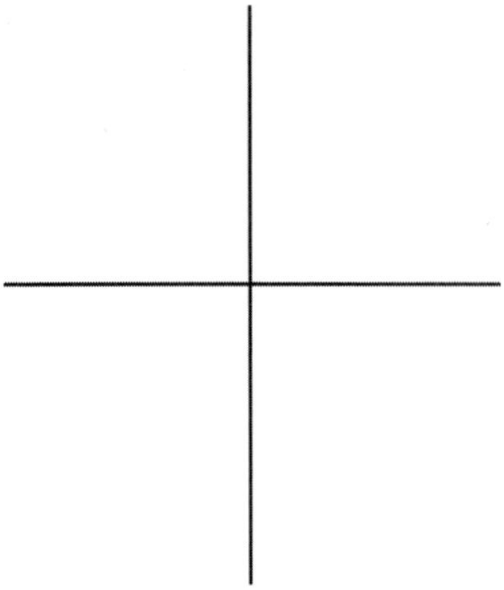

Figure 3-20
Aperature toggled off

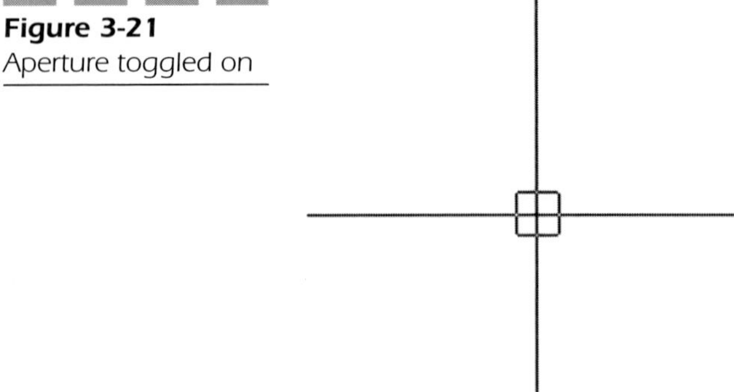

Figure 3-21
Aperture toggled on

what's contained within it. When you select the **Directry** button, a dialog box will be displayed (see Figure 3-22).

The Directory Input These are the rectangular input boxes at the top of the **Directry** dialog box. Use these input boxes to change the items seen in the directory screen:

- **Project Number** Select this to input a user-defined project identifier of up to nine alphanumeric characters
- **Employee** Select this to input a user-defined employee identifier of up to seven alphanumeric characters

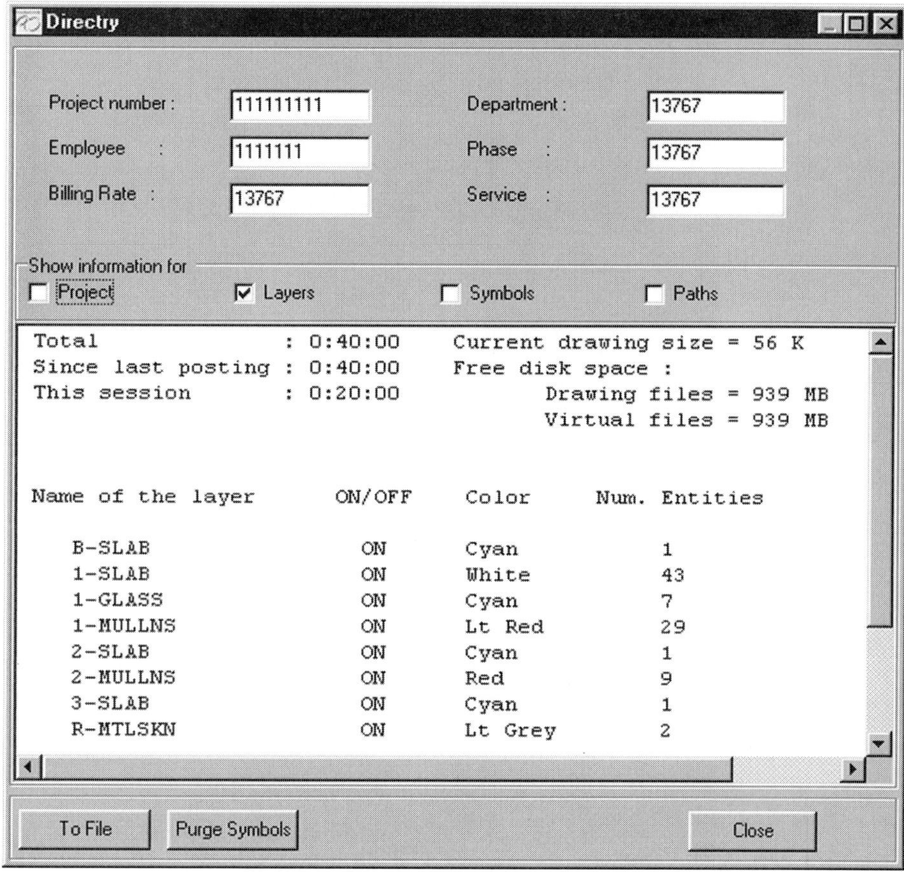

Figure 3-22 The **Directry** menu

- **Billing Rate** Select this to input a user-defined billing rate of up to five numbers
- **Department** Select this to input a user-defined office department identifier of up to five numbers
- **Phase** Select this to input a user-defined project phase identifier of up to five numbers
- **Service** Select this to input a user-defined service identifier of up to five numbers

Show Information For These four options determine what is displayed in the Directry Screen:

- **Project** This displays all the information about the project, including all the Directory Input information, the drawing file times, the current drawing path, and the current file size.
- **Layers** This displays all the information about the layers in the drawing file.
- **Symbols** This displays all the symbols in the drawing database. Since DataCAD retains all the symbols in its database, even when all of them are erased from the Drawing Window, you will still see the symbols listed here, since they are still in the drawing file. Use the **Purge Symbols** command to delete all references to non-existent symbols.
- **Paths** This displays all the current DataCAD path settings (the same ones you see under **Tools/Program Preferences/Pathnames**).

The Directry Screen None of the information shown in this screen, except the "Free disk space" information, is linked to anything outside of the drawing file. You can move the drawing file from directory to directory, and different persons can open the drawing; the information will not change. The Free disk space information is the only exception. It displays current information about the computer that DataCAD is currently running on and/or the computer on which the drawing file is located.

- **Project number** This displays a user-defined project identifier of up to nine alphanumeric characters.
- **Employee** This displays a user-defined employee identifier of up to seven alphanumeric characters.
- **Billing Rate** This displays a user-defined billing rate of up to five numbers.
- **Department** Thisdisplays a user-defined office department identifier of up to five numbers.
- **Phase** This displays a user-defined project phase identifier of up to five numbers.
- **Service** This displays a user-defined service identifier of up to five numbers.
- **Drawing name** This is the name of the current drawing file. Except as noted, no path or .DC5 file extensions are shown with the name. If you have not done a **SaveAs** during the current drawing session, then the name is minus the .DC5 extension and the file path. If you do a **SaveAs** and select **Directory**, then the full path and the .DC5 file

extension will be shown. When the file is closed and then reopened, the file name will revert back to the name only (with no path or .DC5 extension).

- **Total** This value is the total time that the drawing has been open since the file was initially created.
- **Since last posting** This value is the total time that the drawing has been worked on since the last time the **Directory** file information was saved using **ToFile** and the "Since last posting" value was cleared. If the "Since last posting" value has not been cleared since the drawing file was created, then this number will match the "Total" number. See the description of the **ToFile** menu option.
- **This session** This value is the total time that the drawing has been open since the file was reopened. If the file has not been exited since it was created, then this number will match the "Total" number. For "Total," "Since last posting," and "This session," time will continue to be accrued from the time the drawing file is opened until it is exited. To get these counters to temporarily stop accruing time, use the **Pause** button in the **Directry** menu.
- **Current drawing size** This displays the size of the current drawing file in *kilobytes* (k).
- **Free disk space**
  - **Drawing files** This displays the amount of space available on the hard disk of the computer where the current drawing file is stored. If the drawing is stored on your local computer, then that is the hard drive information that is displayed. If the file is located on another computer on a network, then the hard drive information that is displayed comes from that computer, not your own.
  - **Virtual files** This displays the amount of space available on the hard disk of the computer where DataCAD stores its temporary files. The path for this location is set under **Tools/Program Preferences/Pathnames/Temporary Files**. By default, they are located in your DataCAD directory in the /TEMP directory folder. This is where DataCAD stores temporary information if it runs out of RAM for storage. (You can view this location by selecting the **Paths** in the **Directry** dialog and then scrolling down to the bottom of the display until you see the paths.)
- **Layer Information** If the Layers option is checked, then layer information will be displayed:

- **Name of the layer** This displays, in order, all the layer names in the drawing file.
- **On/off** This tells you if each layer in the drawing file is currently turned on or off.
- **Color** This displays the currently selected color for each layer. Even if 255 different colored entities are in that layer, only the currently selected color is displayed here.
- **Num. Entities** This displays the number of entities in each layer. Remember that each instance of a complex linetype is only one entity, as is each symbol.

- **Symbol Information** If the Symbols option is checked, then all the symbol paths in the current drawing database will be displayed, but only if you have placed one or more symbols in the drawing file. Note that depending on how the symbols were originally created and named, you may see the symbol references only by the symbol name or by the symbol's full path and name.
- **Path Information** If the Paths option is checked, then all the current DataCAD path settings (the same ones you see under **Tools/Program Preferences/Pathnames**) will be displayed.
- **To File** Select this to save selected directory information to a text file. By default, the file is saved to the **Text Output Files** path, and it will have a .DIR file extension. Here is a sample of the output:

```
Project number       : 111111111       Department : 13767
 Employee            : 1111111         Phase      : 13767
 Billing Rate        : 13767           Service    : 13767
 Drawing name        :          G:\Book\Dwg\Cover\3DTess2.dc5
 Total               : 0:37:00  Current drawing size = 56 K
 Since last posting  : 0:37:00  Free disk space :
 This session        : 0:17:00  Drawing files = 939 MB
                                Virtual files = 939 MB
Name of the layer    ON/OFF          Color            Num. Entities
B-SLAB               ON              Cyan                        1
  1-SLAB             ON              White                      43
  1-GLASS            ON              Cyan                        7
  1-MULLNS           ON              LtRed                      29
  2-SLAB             ON              Cyan                        1
  2-MULLNS           ON              Red                         9
  3-SLAB             ON              Cyan                        1
  R-MTLSKN           ON              LtGrey                      2
  ConeGls            ON              LtBlue                      1
  ConMulln           ON              LtMgta                     25
  StairTwr           ON              Magenta                     2
```

All the information that is displayed in the Directory window when you pressed ToFile is saved to the .DIR file.

If you had previously saved a .DIR file with information about this drawing file, then when you press **ToFile**, DataCAD will tell you the file already exists and asks if you want to "Overwrite it?" Answer yes to save the new information. You will then be asked, "Do you wish to clear the time since last posting?" If you answer yes, then the Since last posting value in the .DIR file will go back to zero (0). If you answer no, then the Since last posting value in the .DIR file be updated with the current Since last posting value.

Purge Symbols This option will only be available if you have placed one or more symbols in the drawing file. Use this to delete symbol references that are still in the drawing database but are not still in the drawing. To do so, perform these steps:

1. Select the **Purge Symbols** button. You will be asked if you are sure you want to do this.
2. Select **Yes**. All nonexistent symbols will be purged from the database.

You will likely see the message shown in Figure 3-23. The Undo buffer is the database that DataCAD keeps of all your past actions. DataCAD uses this database to remember what you did in a drawing session so that you can use the Undo and Redo functions. If you think that you will need to use Undo/Redo, then do not purge your symbols at this time. Wait until you are sure there is nothing that you need to Undo/Redo and then purge them. Once the buffer is cleared, a new one will be started. It is good practice to get rid of (purge) databased symbols that are no longer in the drawing file.

The information contained in the **Directry** menu is cumulative and persistent. It doesn't matter who opens and works on the drawing file, even over a network. The information accumulates every time the drawing file is opened. However, running the **LyrUtil** macro (**LyrSave/LyrLoad**) on the drawing file will delete all the information in the following categories, starting over again from scratch:

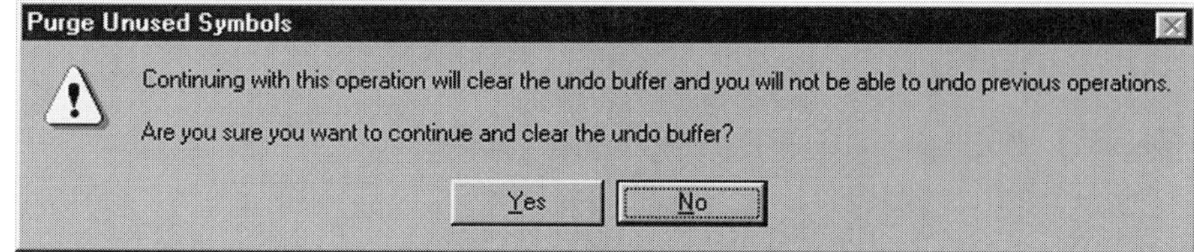

Figure 3-23 The **Purge Unused Symbols** dialog box

- Project number
- Employee
- Billing rate
- Department
- Phase
- Service
- Total
- Since last posting
- This session

SUMMARY

How you set and use the various settings and display options in DataCAD will have a profound impact on how easily and efficiently you can operate the program. For this reason, we stress again how important it is to set most of these options by making them part of your default drawing file(s). In this way, you will only have to set these settings once and they will be carried over into every future drawing you start in DataCAD.

CHAPTER 4

Tutorial

Tutorial: The Apartment

Here we will start by drawing the floor plan of a simple apartment and then copying it to form two apartments. When we're done, it should look like Figures 4-1 and 4-2.

Objectives

The objective is to introduce you to the basic workings of DataCAD. We'll give you some information on what is happening step by step, but don't be too concerned if you don't always understand exactly what's happening. More in-depth explanations will come in the next chapters. We'll introduce a lot of material here so that you can get started drawing in DataCAD, and so that you will understand the basics when reading the other chapters.

Screen Colors

Although you can change the color of the main Drawing Window to any color you want, while you are learning about DataCAD you will probably want to keep the default color of black. That's only because the default entity colors tend to wash out against a bright white background. They

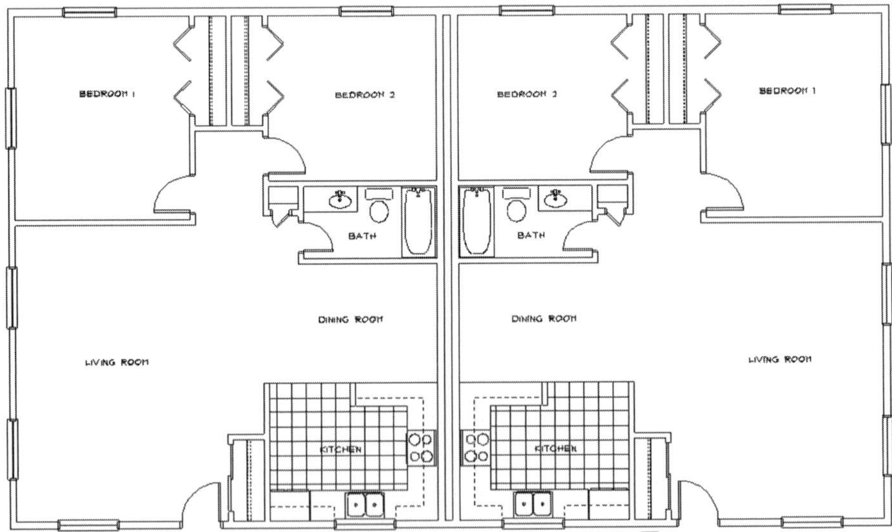

Figure 4-1 The apartment floor plans

Figure 4-2
The apartment with a roof

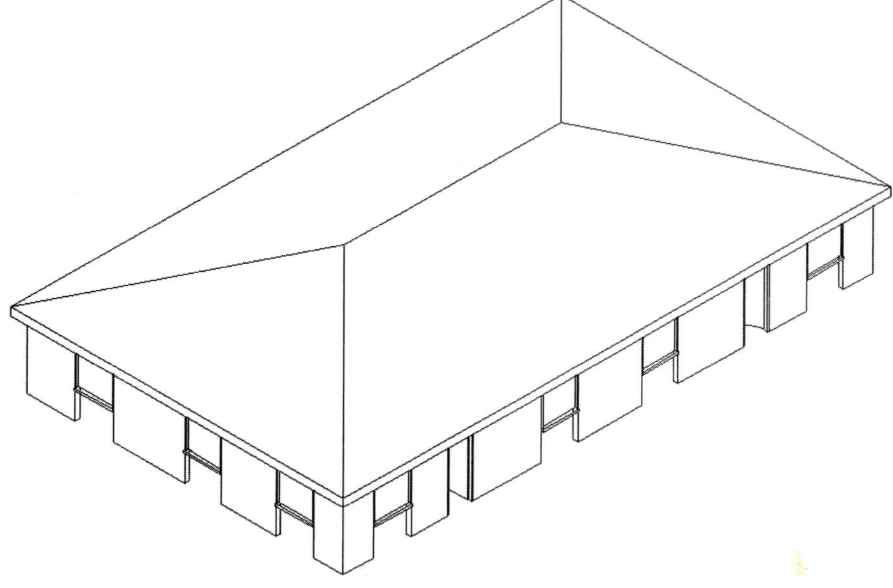

show up pretty well against black, however. In this chapter, and indeed throughout the book, you will see black line drawings on a white background. This is simply due to the fact that black lines on white pages show up better than white lines on a black page. In Chapter 23, "Customizing DataCAD," you can find out how to customize the drawing window and entity colors.

Samples

If you want to jump around a little in the tutorial to look at specific task, the CD-ROM has three drawing file examples of this tutorial in various stages of completion. The name of the file is indicated at the equivalent section of the tutorial.

Let's Start

First, we need to review some high school trigonometry (hey, no moaning!). Look at the following figures. Three dimensions in space are at work here

Figure 4-3
A two-dimensional graph with only horizontal and vertical planes

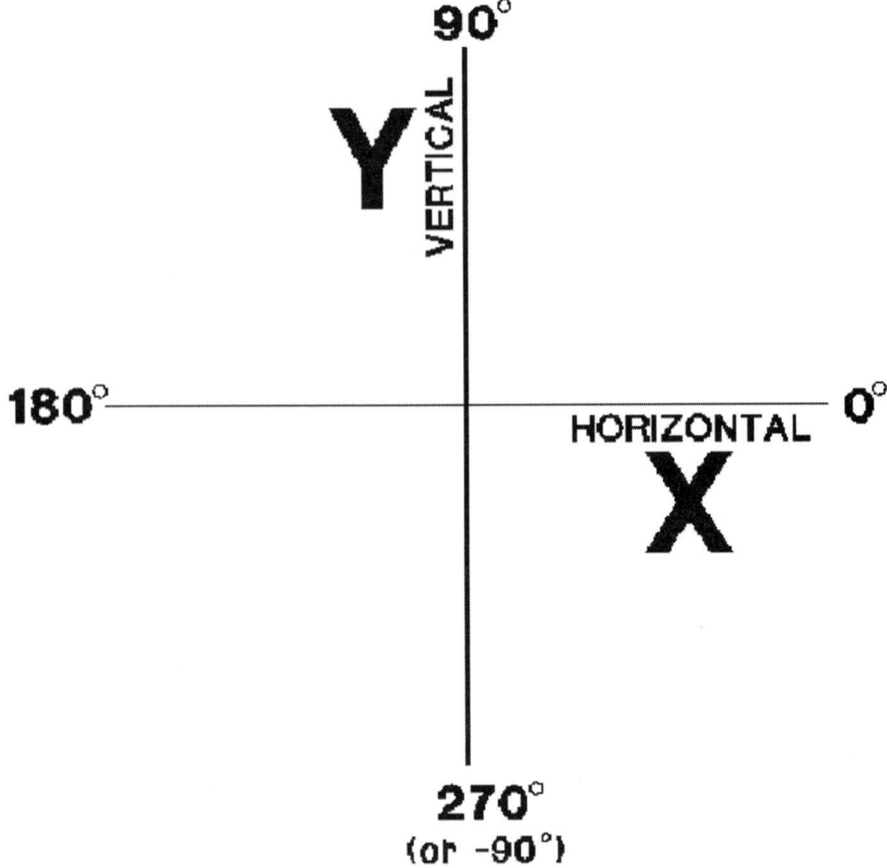

(quantum physics not withstanding) instead of just two, as in Figure 4-4. Figure 4-3 shows the X/Y plane we are familiar with when creating 2D drawings. The X dimension is, on our computer screen, the left/right, or horizontal direction. The Y dimension is the up/down, or vertical dimension. The Z dimension is the height, or depth dimension. Think of this as the dimension that comes out of the screen at you. If you place the eraser of a pencil on your computer screen and point the sharp end toward your face, that's the line that represents your Z dimension (see Figure 4-4). This is pretty straightforward while drawing 2D lines, as we will be here. It gets a little more confusing when we draw in 3D, but let's not worry about that for now.

After starting DataCAD or exiting from a previous drawing, we need to set a default drawing to use for our new project. Here we will use the default drawing that we just created in the previous chapter. If you haven't done so, we recommend that you do so before continuing, since this default

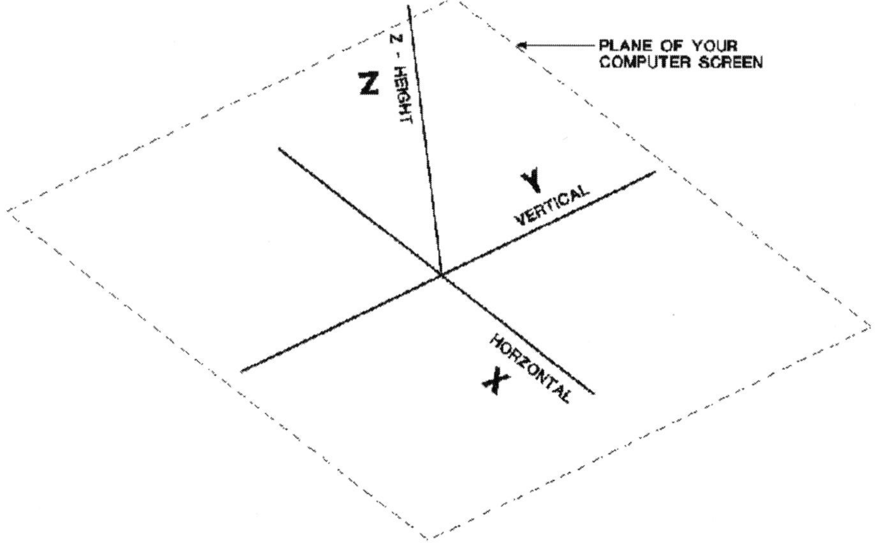

Figure 4-4
A graph with three dimensions

Figure 4-5
Choosing a new default drawing

file will have a lot of settings saved in it that will help you with this tutorial. When you have the default drawing ready, follow these steps:

1. Close the opening dialog box (which is used to open a file that already exists) by clicking on **Cancel**.
2. From the drop-down menu at the top-left of your screen, select **File/New** and then press the **Default** button. A dialog box appears, as shown in Figure 4-5.

3. Highlight the **Default.dc5** file. This is the default drawing we saved in the previous chapter. Press **Open** or **Enter**. You will then be taken back to the drawing file dialog box. The file folder shown in the **Look in:** box is where your new drawing will be saved.
4. Now we need to give our new project a name. In the **Filename:** box, type in **Apartment** and then press **Enter**. A file called Apartment.DC5 is saved. The Windows title bar at the top of your screen will display the computer path and name of the drawing file you are working in. Now we can start drawing!

Draw Marks

An optional setting is available in DataCAD that you might want to use called *Draw Marks*. With this option turned on, DataCAD will place a temporary X on the screen whenever you pick a point. This can be helpful if you are a new user to make sure you have chosen the point you intended. The Xs are only displayed on your screen and will not be saved or printed. If your screen gets too cluttered with these marks, just press the **Esc** key (explained later) and the marks are removed from your screen. Let's turn this option on for this tutorial by following these steps:

1. Go to the **Utility** menu, the one with **ToScale** at the top and **Edit** at the bottom. The name of the current menu is always displayed in the Menu Name box. If you don't see it, *right-click* the mouse until you do.
2. Press the **F0 Settings** button, and turn **S1 DrwMarks** on (the button will turn green).
3. *Right-click* to get out of the **Settings** menu.

Refresh

At times, you will want to "refresh" the display of your drawing window. By pressing the **Esc** key, DataCAD will clean up your drawing window. You'll want to do this when your screen becomes cluttered with extraneous temporary marks, like the Xs from the Draw Marks setting. Also, entities will not show up on your screen if you have done something like erased one entity from on top of another. To see all of your entities again, just press the **Esc** key.

Walls

Now let's draw some walls. We'll use a couple of different methods. You can decide for yourself which way you like best:

1. Go to the **Edit** menu. That's the one with **Move** at the top and **Utility** at the bottom. The name of the current menu is always displayed in the Menu Name box.
2. Select the **S3 Architct** button either by clicking on the button with the mouse or by pressing **Shift+F3** (remember, that's what the **S3** means on the **Architct** button). Alternately, you can press the lower case **a** on the keyboard to get to this menu. That's one of DataCAD's many shortcut keys.
3. **F1 Walls, F2 2LnWalls**, and **S1 Outside** should all be turned on (the buttons should look recessed and be green).
4. Select **F6 Width**. The command line will say something like *"Enter new wall width: 0.4.1/2."* [114] This means that unless you change the value, the walls will be drawn at 4 1/2″ [114 mm] wide. If it says 1.0 [305], then the walls will be drawn at 1′ [305 mm] wide. We're going to change the width of the wall to 6″ [152 mm] wide. To do this, type **.6 [152]** and then hit **Enter**. The command line will now say *"Width* = 6″ [152]."

NOTE: *In DataCAD, you do not have to type in the feet (') or inch (") symbols. Instead, you separate numbers with a period (.), a forward slash (/), or a space. You also don't have to place zeros as placeholders between the . or /.*

The dimension 3' and 6 3/16" takes a long time to type. Instead, all you have to type is one of these:

- ***3.6.3/16***
- ***3.6.3.16***
- ***3 6 3 16***
- ***.42.3/16***

To enter 4", you could type

- ***.4***
- ***(space) 4***

To enter a $^1/_4$", you could type

- ***..1/4***
- ***..1.4***
- ***(space)(space) 1 (space) 4***

DataCAD is smart enough to know that the periods, slashes, or spaces represent dimension placeholders for feet, inches, and fractions. It makes entering distances much faster than in other CAD programs.

6. *Right-click* the mouse once. You will be back in the **Edit** menu.
7. Select **Polygons** either by clicking on the button or by pressing **Shift+F4.**
8. Select **F0 RectAngl** so that it is turned on. Turn off **F4 CntrPnt** (the button will be red and will look raised).
9. Place the cursor anywhere in the upper-left corner of the Drawing Window.
10. Drag the cursor down and to the right. Look at the X and Y values in the Status Line. Move the cursor until the X value reads 30′-0″ [9144] and the Y value reads 36′-0″ [10973]. If at any time you run out of room on the screen, you do not have to stop, zoom out, and redraw the lines. Simply use the **PageUp** key on your keyboard to zoom out, the **PageDn** key to zoom in, or use the arrow keys to pan up, down, left, or right.
11. *Click* the left mouse button. You will be prompted with *"Select a point to define the inside of the wall"* (see Figure 4-6).
12. Place the cursor anywhere inside the rectangle and *click* the mouse. The lines representing the inside of the walls are drawn, and all the corners have been cleaned up! (See Figure 4-7.) Notice that the Command Line says, *"Select first corner of rectangle."* DataCAD is still in the rectangle mode and will remain there until you *right-click* the mouse or press the **S0 Exit** button. DataCAD operates this way in nearly every case. In this way, you can continue drawing without having to constantly go back to the menus.
13. *Right-click.* You will be back in the **Edit** menu.

Figure 4-6
Creating the rectangle

14. Press the upper case **F**. This will save your drawing. The **F** stands for forced save, as opposed to an automatic save. You can also pick **File/Save** from the drop-down menu or press **Ctrl+S**. Do this often! If your computer suddenly crashes or you lose power, you want to make sure that you have a copy of the drawing saved. If not, you'll have to draw it all over again!
15. Now let's make sure we can see the whole floor plan within the Drawing Window. You can quickly recalculate and display the extents

of the drawing by using the **R** button (for Recalculate or Recalc) in the Navigation Pad. If the drawing appears too close to the edge of your screen, just press **PageUp** once to back away from the plan a little.

NOTE: *In the Navigation Pad, two buttons that will enable you to see your entire drawing: Recalc and Extents. The Recalc button, as we mentioned, is the* ***R*** *button. The Extents button is the* ***E*** *in the center of the four arrows in the Navigation Pad. On the surface, they seem to do the same thing, but they are different.*

The Recalc button recalculates the extents of all the layers that are currently turned on. That is, it finds the outermost entities and adjusts the display of the drawing so that it will fit entirely within the drawing window. On a drawing with a lot of entities in it, this may take some time, since DataCAD has to search through every entity to see which ones lie the furthest away.

The Extents button displays the extents of your drawing based on the last Recalc that you did. Until you press Recalc, DataCAD will zoom to the same extents every time, regardless of which layers are on or off. If you don't need to do a full Recalc, then Extents is a better option. Extents does not search through any entities, so it takes much less time to redisplay your drawing.

As you play around with these options, you'll understand it better. Pressing them cannot harm your drawing in any way, so feel free to experiment.

NOTE: *Zooming in and out of your drawing is called* windowing *in DataCAD. The quick and simple way to do this in DataCAD is to press the* ***PageUp*** *key to window (zoom) out away from your drawing to see a larger extent of it, or to press the* ***PageDn*** *key to window in closer to your drawing to see greater detail. The word window comes from the next method of zooming: by area, using a bounding box or window to define the area to be windowed into. The bounding box looks and acts like a rectangular window into your drawing, hence the name. You access it by pressing the* ***W*** *button on the Navigation Pad or by pressing the / (forward slash) key. You can then define a bounding box around the area to be windowed into.*

In the WindowIn menu, one button is labeled ***FreeZoom****. It is a toggle that either remains on or off. If it is turned off, then DataCAD will window in and out by the fixed scales indicated in your Plotter/Scales menu, like*

*$^{1}/_{4}$" or 1:40. If **FreeZoom** is toggled on, then DataCAD will window in and out freely. We suggest you leave it on.*

Erasing

Now I'm sure that with the outstanding step-by-step instructions in this book, you will never make a mistake. But just in case, let's try a few methods of erasing entities. You can erase whole entities or just select parts of them. Use the **right arrow** key to move your view to the right of the four walls you've just drawn. From the **Edit** menu, choose **Polygons/RectAngl** again and draw a smaller version of your four walls. It should look something like Figure 4-7.

In the **Edit** menu, choose **F7 Erase**. The menu will change to display the standard DataCAD Selection menu with

- **F1 Entity**
- **F2 Group**
- **F3 Area**
- **F4 Fence**

Figure 4-7 Creating a smaller version of the rectangle

- **F5 SelSet**
- **F6 LyrSrch**
- **F9 Partial**
- **S1 Clr Undo**

We won't go over **F2 Group** or **F5 SelSet** for now, since those are a little more advanced features, and since we only have one layer in our drawing right now, we'll save the explanation of **F6 LyrSrch** for later in our apartment project.

NOTE: *This Selection menu is not just for the* ***Erase*** *function. You will see this menu every time you select a function that requires you to pick an entity or group of entities. Later in the book you will see references to "EGAFS", meaning to select either Entity, Group, Area, Fence or SelSet.*

When we draw walls in DataCAD, each line that makes up the walls is a separate entity. So if you want to erase a wall, you have to erase each of the lines that make up the wall. To see how this works, follow these steps:

1. Select **Entity**. The Command Line will tell you *"Select entity to <ERASE>"*.
2. Place your cursor close to just one line and *click* the mouse. That line should disappear. Notice that only one line was erased, not both lines that make up the wall (see Figure 4-8).
3. Now select **Area**. The Command Line will tell you *"Select first corner of area to <ERASE>."*
4. Place your cursor just above and to the left of the walls that remain and *click* the mouse once. Do not hold the mouse button down. Now drag the mouse down and to the right until all the walls are completely within the bounding box created by the cursor. Click the mouse to complete the box. Only entities that are completely within the bounding box will be erased.
5. All the walls are erased.
6. Now let's say you erased all those lines accidentally and you want to get them back. Select **Edit/Undo** from the drop-down menu (or press **Ctrl+Z,** the Windows shortcut for Undo). All the lines should reappear.

Figure 4-8
Erasing a wall

7. Now let's see what the **F4 Fence** option does. Essentially, it is just like **Area**, except that you can define from 3 to 36 edges of your bounding box and they can be at any angle. Follow these steps:
   a. Select **Erase/Fence**.
   b. Press the **Oo** (upper or lowercase O) key on your keyboard. Notice that the SWOTHLUD acronym at the bottom of the Status Area now has a lowercase **o** (SWoTHLUD), meaning that DataCAD is not in orthographic mode now. Now you can draw lines freely at any angle.

c. Locate your cursor above and to the left of the remaining walls and *click* the mouse. Drag the mouse to the right of the walls and click the mouse again. The Command Line will read *"Select the 3rd point of the fence to <ERASE>."*

d. Notice also that in addition to the single line you just drew, two other lines are attached to the cursor. This is the fence bounding box.

e. Click on another five or so points around your walls, so that all the walls are within the fence. To finish selecting the vertices of the fence, you can either select **S0 Close** from the menu or, more practically, just right-click the mouse to close the fence.

f. All the walls within the fence disappear (see Figure 4-9).

9. Before going on, you might want to draw a few more lines and experiment with the Entity, Area, and Fence options. See what happens when not all of an entity is located within the bounding box.

Undo and Redo

As long as we're talking about correcting mistakes, DataCAD 9 has globally accessible and unlimited **Undo** and **Redo**. As the name implies, the **Undo** feature will enable you to back up through your previous actions, reversing each one as you go, and it can be performed at any time. If you go too far back with **Undo**, then **Redo** will move you forward through the previously Undone actions, restoring them as you go.

You can access Undo and Redo in one of two ways:

- Select **Edit/Undo** or **Edit/Redo** from the drop-down menu

or

- Press **Ctrl+Z** to Undo, and **Ctrl+Y** to Redo (these are standard Windows shortcuts)

Undo

Undo will back up through any previous commands you have issued to DataCAD, even the deletion of drawing layers. Each action you take in DataCAD makes changes to the drawing database. Selecting **Undo** simply steps backwards through those changes, reversing each action sequentially. When you select the Undo command from the drop-down **Edit** menu, you

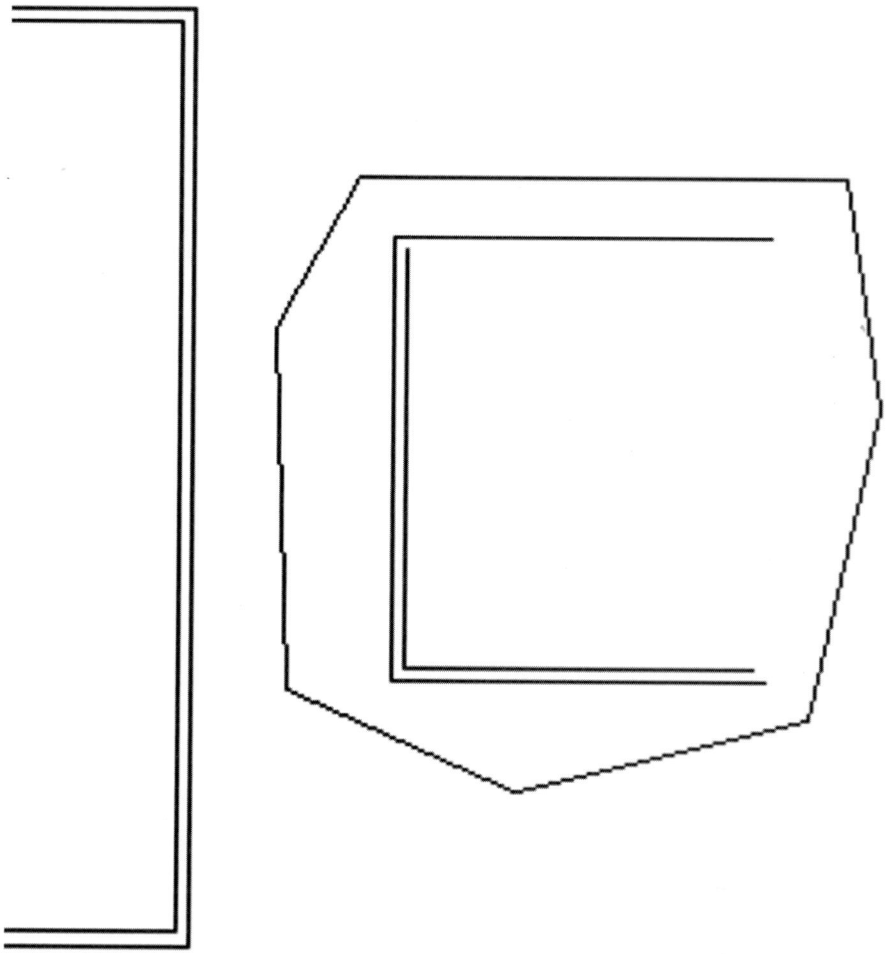

Figure 4-9
"Fencing" in the walls

will see a second word after the word Undo (Draw, Erase, Move, Hatch, Explode, and so on). This tells you quickly what the last command was and what will be undone.

Redo

Redo is the opposite of **Undo**. It is only available once something has been acted on by the **Undo** command. It will move forward through any previous Undo commands that you have issued to DataCAD. Like **Undo**, selecting Redo simply moves through the sequence of drawing database changes, reversing each Undo action sequentially. When you select the Redo command from the drop-down **Edit** menu, similar to Undo, you will see a second word after the word Redo (Draw, Erase, Move, Hatch, Explode, and so on). This tells you quickly what the last command was and what will be redone.

More Walls

We've used just one method for drawing walls. The next method will involve picking start points for our walls and then defining in which direction they should be drawn and at what distance. You may recall from your old math classes that this is the definition of a *vector*. You may also remember entering those directions according to the degrees on a compass. These are referred to as *polar coordinates*. In our default drawing, we made the default input mode Relative Polar. This means that where we start the line is the starting point relative to the cursor, and the direction is defined by polar (compass) coordinates. To begin using this next method, follow these steps:

1. Make sure you are drawing in orthographic mode (uppercase **O** in SWOTHLUD). If the **o** is lowercase, press the **Oo** key on your keyboard to change it.
2. First, we're going to change the width of our interior walls to 4$^1/_2$″ [114]. One way to do this is to go back to the **Architect** menu, select **Width**, then type **.4.1/2** [114](or **.4.1.2**). The other method is as follows:
   a. Press the equals (=) key on the keyboard. This turns DataCAD's wall function on and off. This should be easy to remember since it looks like the two lines of a wall. (Neat, huh? You'll find a lot of these "common sense" shortcuts throughout the program. That's one thing that makes DataCAD so easy to use.)
   b. Pressing the = key once again will turn off the walls function so that DataCAD will only draw single lines. This kind of function is called a toggle, meaning that it has only two settings: on or off. Keep pressing the = key and you will continue to turn the Walls function on and off.
3. At any time, to make sure that the wall function is indeed on, look at the **SWOTHLUD** letters at the bottom of the Status Area. The **W** should be upper case. If it is, then the wall function is on. If it is a lowercase **w**, then it is off.

4. If you press the = key so that the Walls function is on, you will see the prompt *"Enter new wall width : 0.6." [152].*
5. At the prompt type **.4.1/2** and then press **Enter**.

Reference Point

Now a new concept: the *reference point* (RefPt). If we were drafting by hand, we would place our scale on the drawing with the zero at a reference point like a corner or the end of a line and measure to another point where we would start drafting a new line. In DataCAD, we set a RefPt with the ` key (the un-Shifted ~ key) and then enter a distance and angle for the starting point of our new line or wall:

1. Press the RefPt (`) key.

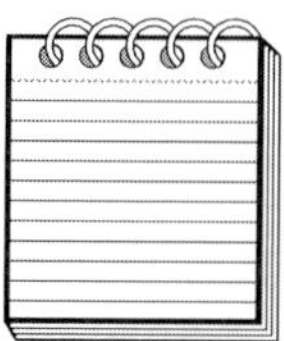

NOTE: *Since this key is easy to confuse for the single quote key ('), and because it is just so small, we'll use the following abbreviation to denote this key: RefPt (').*

2. Snap to the upper-left, outside corner of the wall by placing the cursor near that corner point and *middle-clicking* the mouse (or pressing **Nn**). The Command Line will read *"Select first end point of new line / wall."*
3. To enter a dimension in DataCAD, rather than using the mouse you will always press the Space bar . Pressing the **Space** bar will always tell DataCAD that you want to enter coordinate information. The Command Line will now read *"Enter relative distance : 0.0."* Now you can enter a distance.
4. Type **17** and then hit **Enter** for 17′-0″.
5. The Command Line now reads *"Enter relative angle : 0.0.0."*
6. Press **Enter** to accept the default of 0 degrees.

NOTE: *The default polar method of input in DataCAD places 0 degrees at the three o'clock position and all subsequent angles added counter-clockwise. Ninety degrees is at the 12 o'clock position, 180 degrees at nine o'clock, and 270 degrees at six o'clock.*

Note that angles can be entered as negative numbers as well. Negative numbers go in a clockwise direction. So entering 270 degrees (270 degrees counter-clockwise from the three o'clock position) is the same as entering −90 degrees (90 degrees clockwise from the three o'clock position).

7. Drag the mouse down until the Y coordinate reads 15′-0″. Don't click any mouse buttons yet.
   a. Look at the Status Line. Reading from the left, the first item is the X coordinate. Notice that it reads 0″. That's because we are only drawing our wall in the up/down or Y direction. The Y coordinate reads 15′-0″, the length of the wall we are drawing. The next coordinate reads a:270-0′. This is the angle coordinate. It means that, relative to our starting point, the line we are now drawing is being drawn at 270 degrees, 0 minutes. The final coordinate reads d:15′-0″ This is the Distance coordinate. We will see later that this is more useful to us when we're drawing lines at angles other than 0, 90, 180, or 270. It will read the same as the a: coordinate for this wall.
   b. Before picking the end of your wall, move the cursor back and forth a little. Notice that even though the cursor moves back and forth, the end point of our wall does not. This is because our default drawing is currently set to draw lines in Orthogonal mode. More about this will be discussed later, but for now just realize that you don't have to keep your cursor precisely at the exact end point of your wall. This makes drawing with a mouse much easier, especially if you've got shaky hands (see Figure 4-10).
8. Finally, we can *click* the mouse to pick the end of our wall. Make sure the Y coordinate still reads 15′-0″, and *click* the mouse.
9. The Command Line reads *"Select a point to define the Inside of the wall."* Place the cursor anywhere to the left of the line you just drew and *click* the mouse. This tells DataCAD which side of the line to draw the second line of the wall.
10. Notice that a line is attached now from the cursor to the end of the wall. This is so that you can keep drawing walls without stopping. Do not click the mouse.
11. Drag the mouse to the right until the cursor crosses to the right of the right-hand wall of the apartment and then *click* the mouse. Don't worry about the fact that the walls cross each other. We'll learn how to clean this up shortly. For now, *right-click* the mouse to stop drawing walls. The line that was attached from the cursor to the wall will disappear, and your drawing will look like Figure 4-11.

Figure 4-10
Beginning a wall within a room

12. Press the RefPt (′) key again and snap to the same upper-left corner of the apartment that we used last time.
13. Press the **Space** bar to enter a coordinate, type **12** [3658], and then hit **Enter**. At the next prompt, type in **270** and then press **Enter** again. We will now be drawing a new wall beginning at 12′-0″ [3658] and 270 degrees from our reference point.
14. We're not sure how long we want this wall to be yet, so drag the cursor over to the right and end it just short of the previous vertical wall we

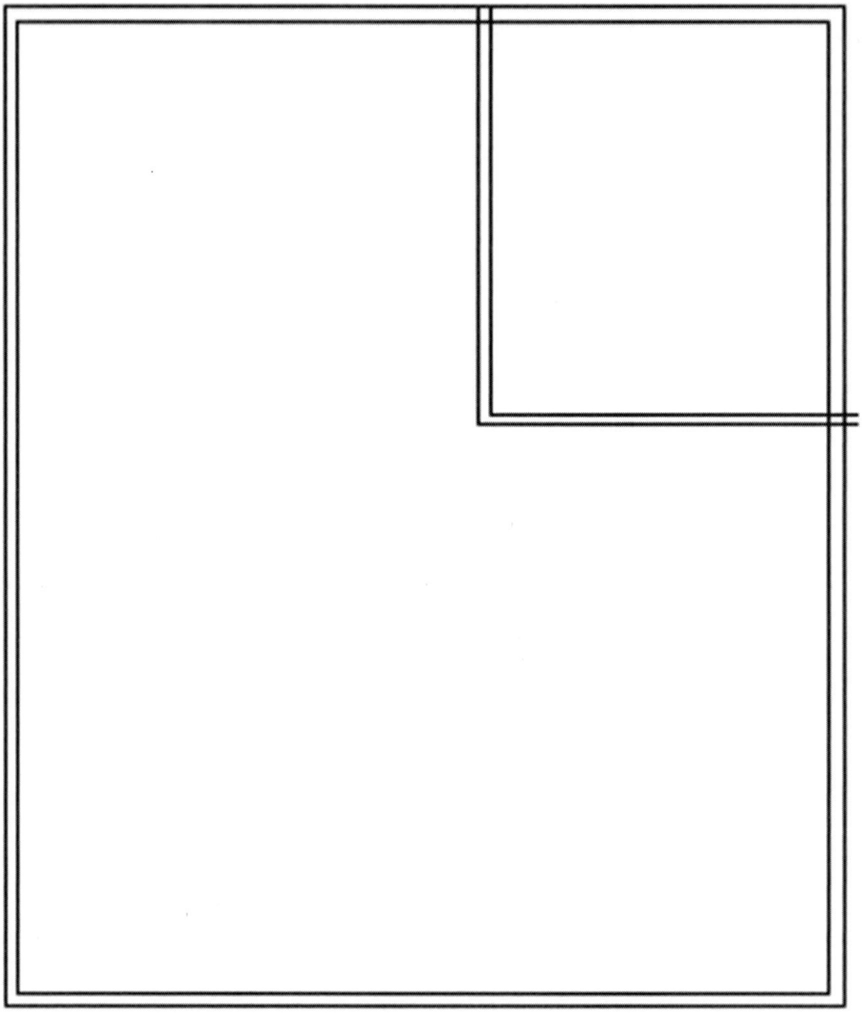

Figure 4-11
The completed wall

drew. When prompted, select a point below the line you just drew. *Right-click* to stop drawing the wall. The drawing will look like Figure 4-12.

Let's keep drawing some more walls. To draw the next walls, we will specify their length with the keyboard rather than the mouse. We'll need to use the RefPt (') key again, so make sure you review how to use it in the earlier example.

Figure 4-12
Creating a second interior wall

15. Again, press the RefPt (′) key. Now we need to snap to a new point. Snap to the intersection with the last wall we drew, as shown in Figure 4-13.
16. Press the **Space** bar and type in a distance of **5.4.1/2** [1638] and an angle of **270** (remember that typing in −90 is the same thing as 270. I like −90 better since all three keys are next to each other and that makes it easy to type in).

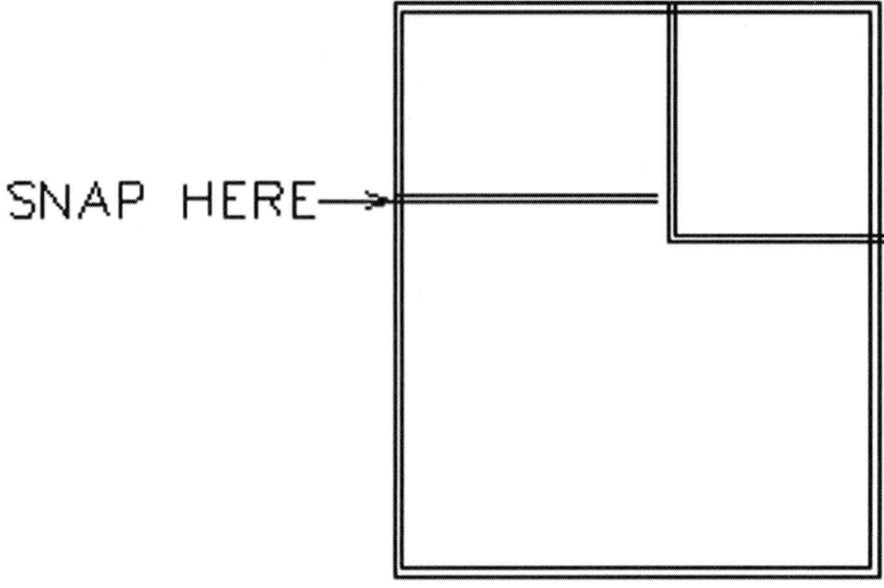

Figure 4-13 Snapping to a new point

17. Now we will draw the length of the wall using the keyboard instead of the mouse. Earlier we learned that to enter a distance from the keyboard you press the **Space** bar. Go ahead and press the **Space** bar now. Enter a distance of **10** [3048] and an angle of **0** (the default is **0**, so all you have to do is press **Enter**).
18. Pick a point above the line you just drew to define the other side of the wall.
19. To complete the fourth wall of the bathroom, drag the mouse up until the cursor crosses the bedroom wall above. *Right-click* to stop drawing the wall.

We'll draw a few more walls and then learn how to clean up all the walls that are crossing. You'll be on your own here for the next few walls. See if you can use what you've learned to make your drawing look like Figure 4-14. The dimension lines are there so you know where to place the walls.

Here's one more little trick for drawing a wall in DataCAD. Look at the finished illustration of the apartment at the beginning of this section. See the little piece of wall at the end of the counter? It is essentially two parallel lines for the wall with closed ends. DataCAD can automatically draw these closed ends, called *caps*, when the walls are drawn. We don't really need to use it to make this one, small wall, but let's do it anyway so you can

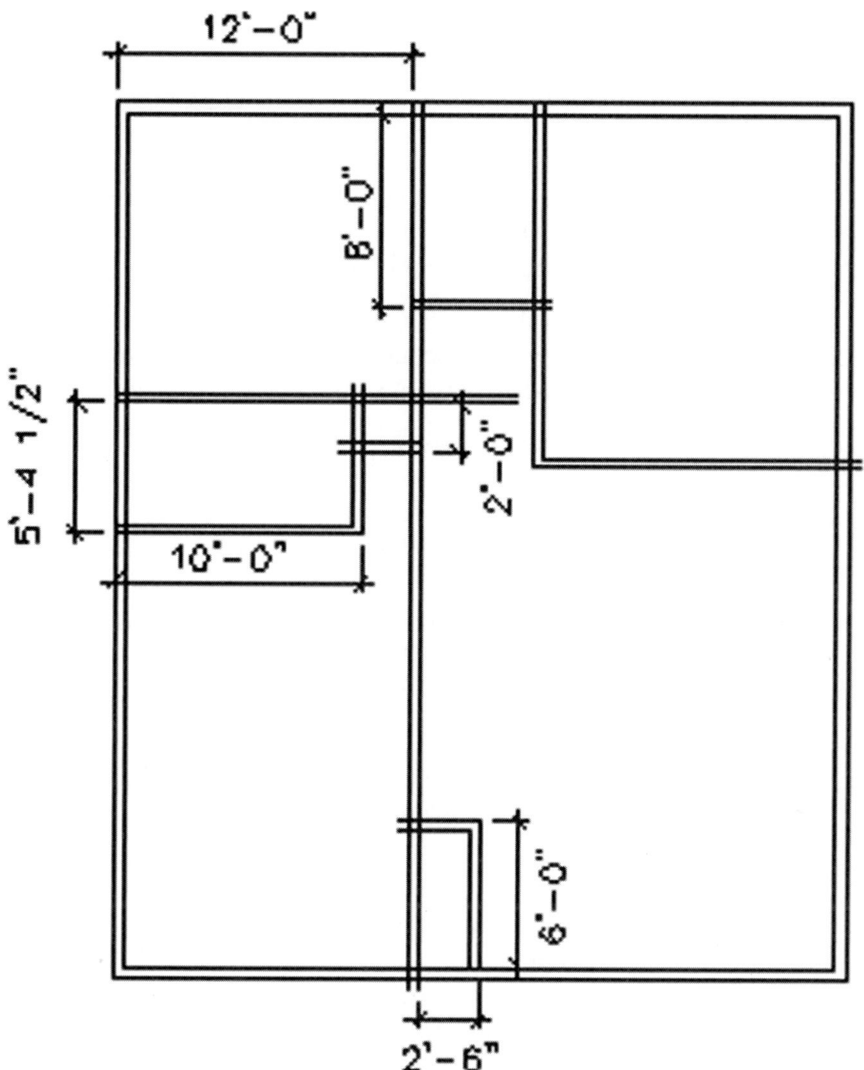

Figure 4-14
Use these dimensions for the next exercises

add it to your bag of tricks. The Cap Walls function can be turned on in a couple of ways:

- This function is an option back in the **Architect** menu (**Edit/Architct** or the **a** key). Press the **a** key to go to the **Architect** menu. Press the **S6 Cap** button to turn on the Cap Walls function. *Right-click* to go back to the **Edit** menu.

- Pressing the pipe (|) key toggles the Cap Walls function on and off. Every time you press the key, the Command Line will tell you *"Walls will be capped."* or *"Walls will NOT be capped."* This is another example of a toggle command. This key looks different in print than it does on your keyboard. Here it looks like a solid vertical line, but on your keyboard it should look like two smaller vertical lines, one over the other. It's on the same key as the backslash (\), and you use it by simultaneously pressing (**Shift + **).

First, let's draw a guideline using a single line, so we have something to snap to when placing this little piece of wall. Drawing guidelines is sometimes just as necessary in CAD as it is when drafting with a pencil:

1. To draw a single line, we need to turn off the Walls function. You could go back to the Architect menu, but the easiest method (and our goal here should always be to do things easier and faster) is to simply toggle off the Walls function by pressing the = key. The Command Line will read *"Walls will NOT be drawn."*
2. Now use the RefPt (′) to snap to the lower-left-inside corner and draw a vertical line, as shown in Figure 4-15.
3. Turn the Walls function back on by pressing the = key. Make sure the width is still set to $4^1/_2$″ [114].
4. Make sure the Cap Walls function is turned on by pressing the pipe (|) key until the Command Line reads *"Walls WILL be capped."*
5. Using RefPt (′) at the lower end of the guideline, locate the top end of the wall at 9′-6″ [2896], 90 degrees upward, as shown in Figure 4-16. Locate the second point of the wall 1′-6″ [483] down (270 degrees) from that point.
6. Pick a point to the left of the line to locate the other side of the wall. *Right-click* to finish drawing the wall. Voila! You have drawn a wall with both ends capped while only having had to pick two points.
7. Erase the guideline by selecting **Ee/Entity** and then selecting the guideline to be erased.
8. Now try a little experiment. Draw an L-shaped wall in the middle of the living room: Pick one point, go right, and pick a second point. Pick to locate the other side of the wall. Now go up and pick a third point. Right-click to finish. Your wall should look something like Figure 4-17.
9. Notice that the corner is cleaned up automatically for you, and that the walls are capped only at the ends, as they should be. Pretty slick, huh?

Figure 4-15 Snapping to the left corner and drawing a guideline

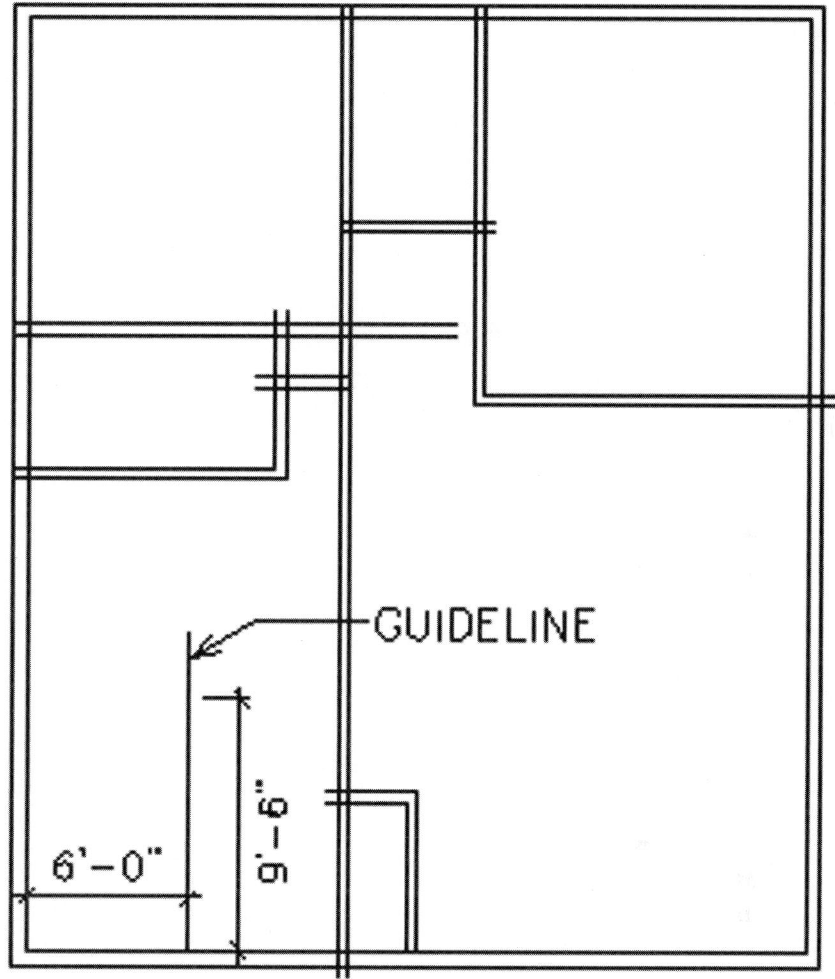

10. Now erase the walls from the middle of the living room (they'd just be in the way of the television).

Cleanup

Well, our apartment is coming along nicely, but a lot of walls are crossing each other and are making a mess. So let's clean them up, shall we? DataCAD has

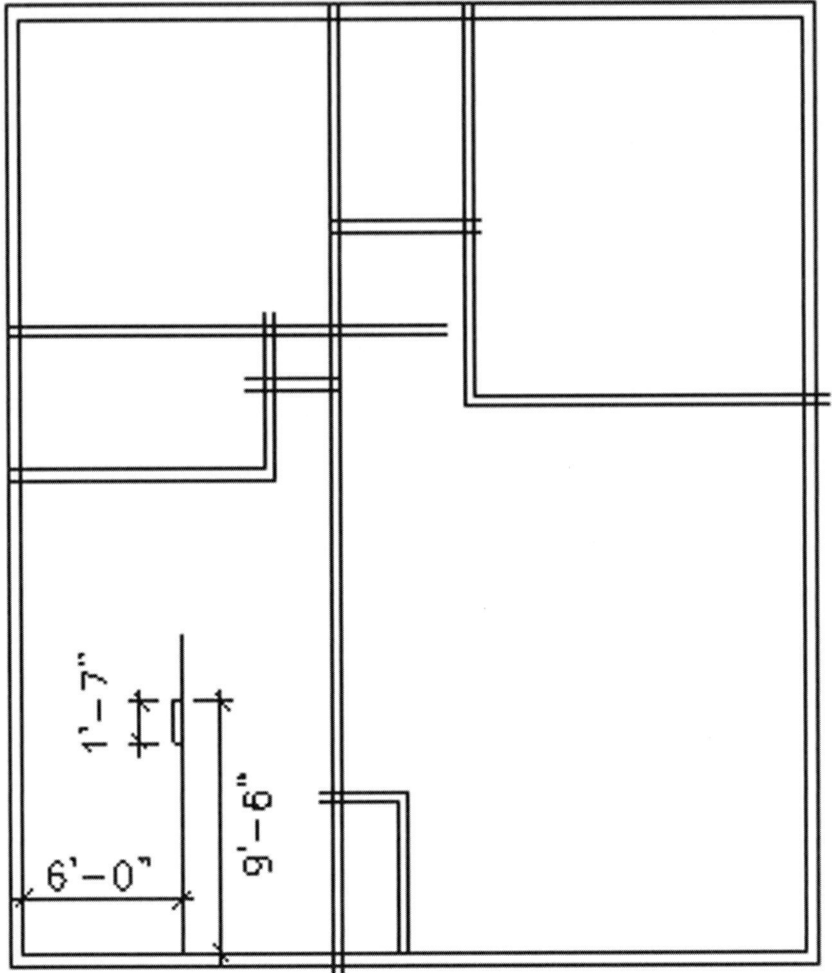

Figure 4-16 Drawing a small piece of wall

some cool functions to make cleaning up easier. To get to them, we need to go to the Cleanup menu in the Edit menu:

1. Go to **Edit/F9 Cleanup**. There you will find 10 different options.
2. We want to clean up all of our T intersections, so pick the **F0 TIntsct** button. The Command Line will read *"Select 1st corner around "T" intersect (wall line ends only)."* What we need to do is place a bounding box around the end of the wall we want to trim and then tell DataCAD

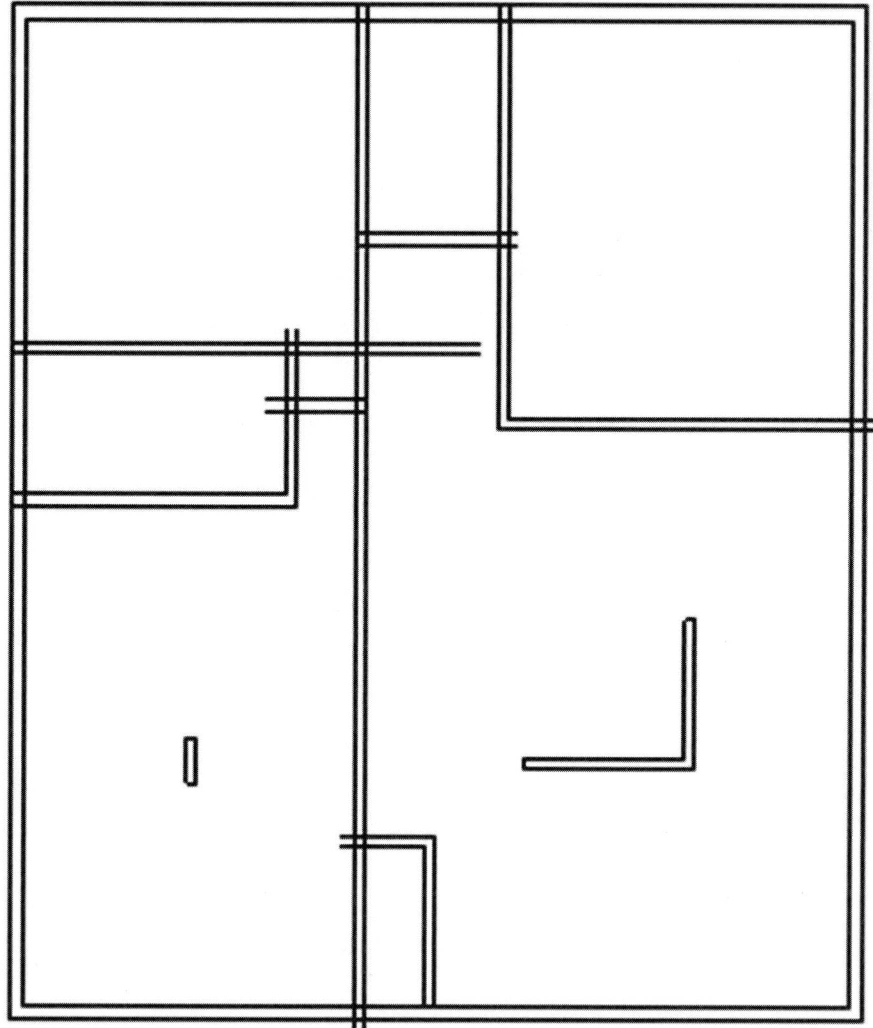

Figure 4-17
An L-shaped wall completed

which wall we want to trim to in order to form a nice clean intersection (see Figure 4-18).

3. As shown in the figure, we must first pick two points to define a bounding box around the ends of the wall. After doing this, the Command Line will read *"Point to wall line to trim to."*
4. Now place your cursor anywhere on the wall you want to trim to and click the mouse. The T intersection will be cleaned up automatically.

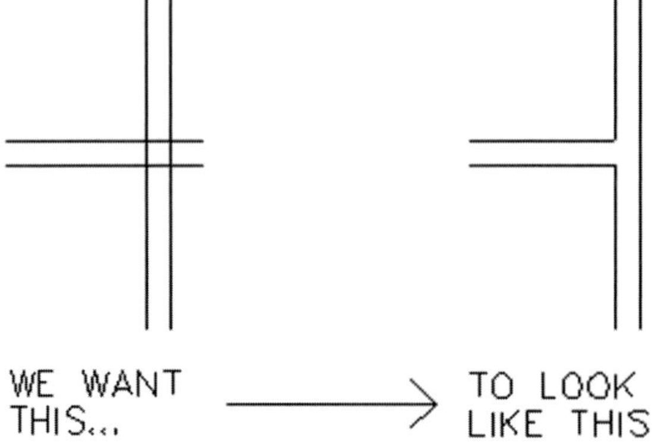

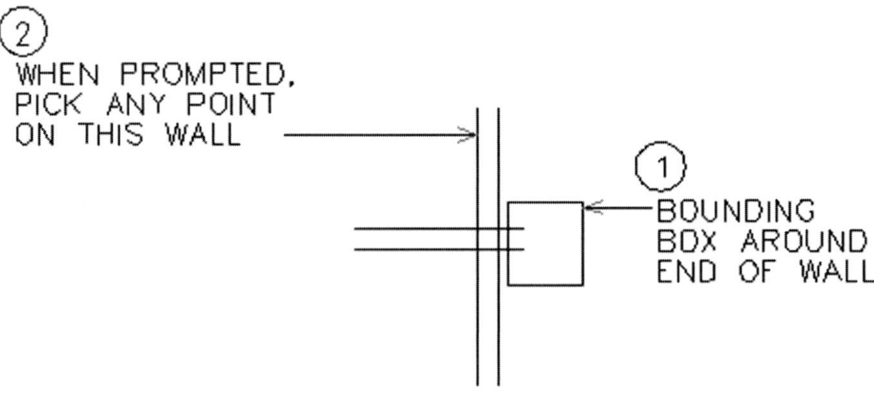

Figure 4-18 Cleaning up an intersection

This method will work for walls that don't actually cross each other too, as in Figure 4-19.

It will also work for walls that are not perpendicular to one another (see Figure 4-20).

One thing you have to be careful of is to make sure that only the end of one wall is inside the bounding box. If you include the ends of two or more walls, as in Figure 4-21, DataCAD won't know which ones you want to trim.

If you pick the wrong wall to trim to, you'll get some weird results (see Figure 4-22).

Now use the T intersection cleanup function to fix as many of the T intersections as you can, so that your drawing looks something like Figure 4-23.

We have one more wall to add and then one to clean up. Let's add the middle wall to the closet between the two bedrooms. We'll do that in a slightly different manner:

Figure 4-19
An example of connecting walls that don't actually cross

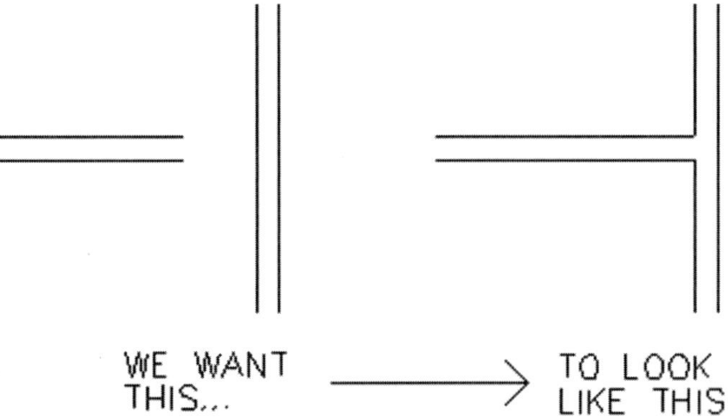

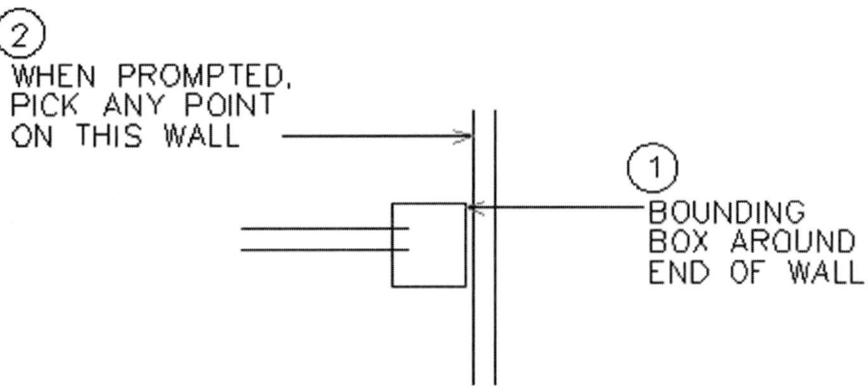

1. Press the **a** key to go to the **Architect** menu. Select **S3 CntrWall** and turn off **Cap**. Now we will draw our wall by defining its centerline, instead of its side.
2. Make sure **Walls** is turned on (press either the **Walls** or the = buttons if it is not on). *Right-click* the mouse.
3. Locate the cursor close to the center of the upper horizontal wall of the closet. *Middle-snap* to that point.
4. Now *middle-snap* to the lower wall and then *right-click* to stop drawing the wall.
5. Use the **Cleanup** function (**Edit/Cleanup**) to cleanup the two ends of the wall. Your closets will look like Figure 4-24.

Figure 4-20
Non-perpendicular walls being joined

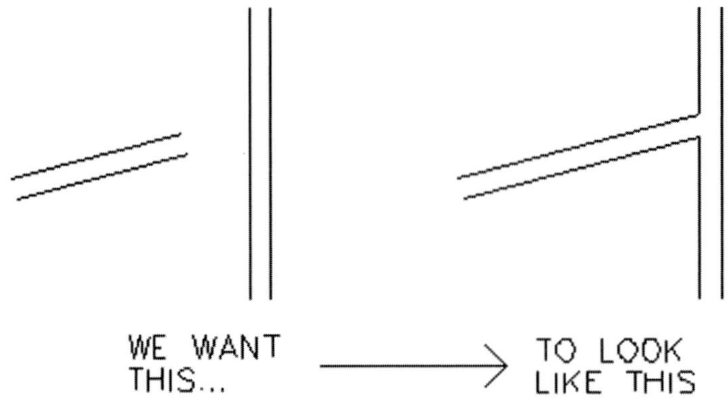

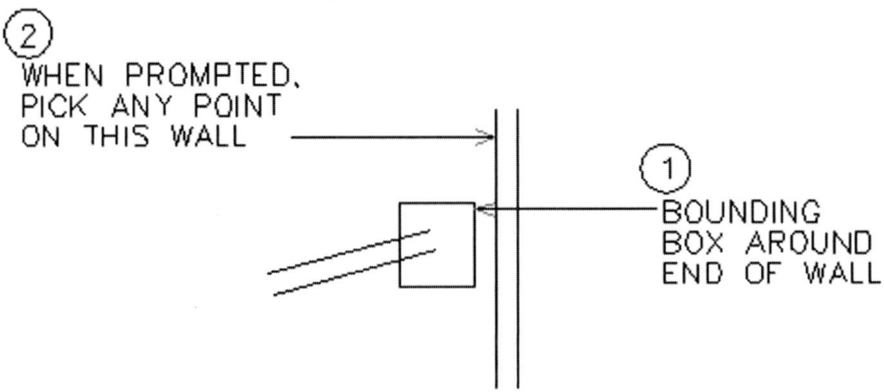

Figure 4-21
The bounding box can only contain one set of endpoints.

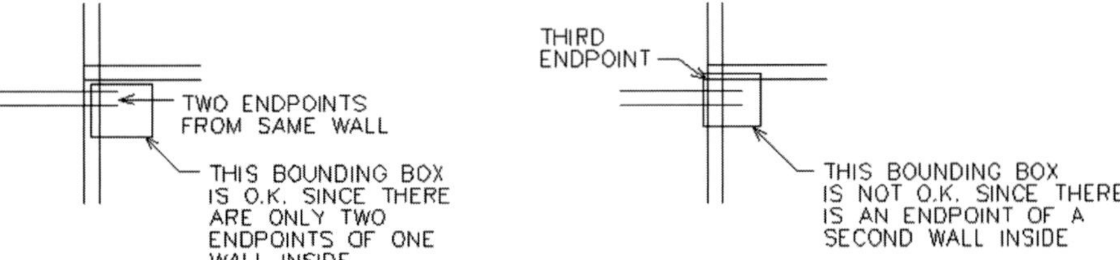

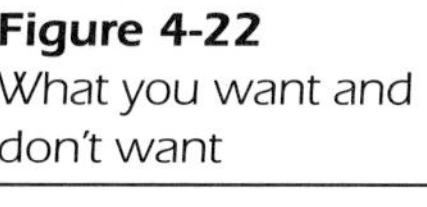

Figure 4-22
What you want and don't want

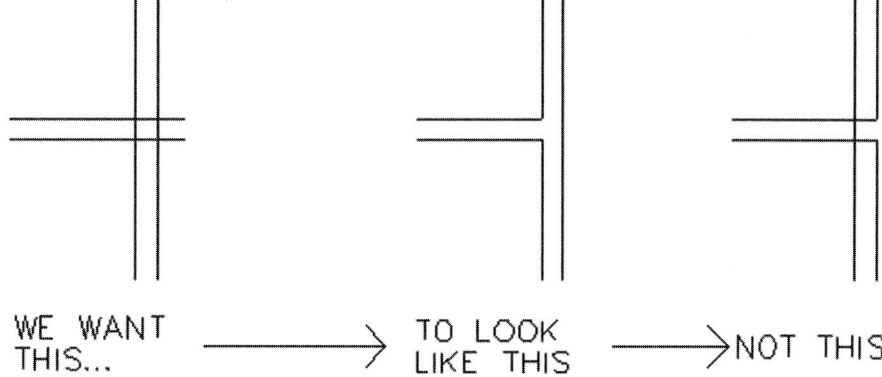

Sample

This drawing is available on the CD-ROM, with the file name "APT-01.DC5"

Partial Erase

To clean up the last wall, we will use the Partial Erase function. This function enables you to erase only selected portions of an entity, but first we will need to draw another guideline:

1. Turn the Walls function off by pressing the = key. Make sure Ortho is on (press **Oo** if necessary).
2. *Middle-snap* to the upper-right corner of the little kitchen wall we drew earlier.
3. Drag the mouse to the right until it crosses the wall we are going to clean up. It should look like Figure 4-25.
4. Go to **Edit/Erase** (or **Ee**) and press the **F9 Partial** button. The Command Line will read *"Select entity to modify."*
5. Select the rightmost vertical line that we are going to trim. The line will turn to a dashed line. This confirms that you have selected the correct line.
6. *Middle-snap* to Point 1.

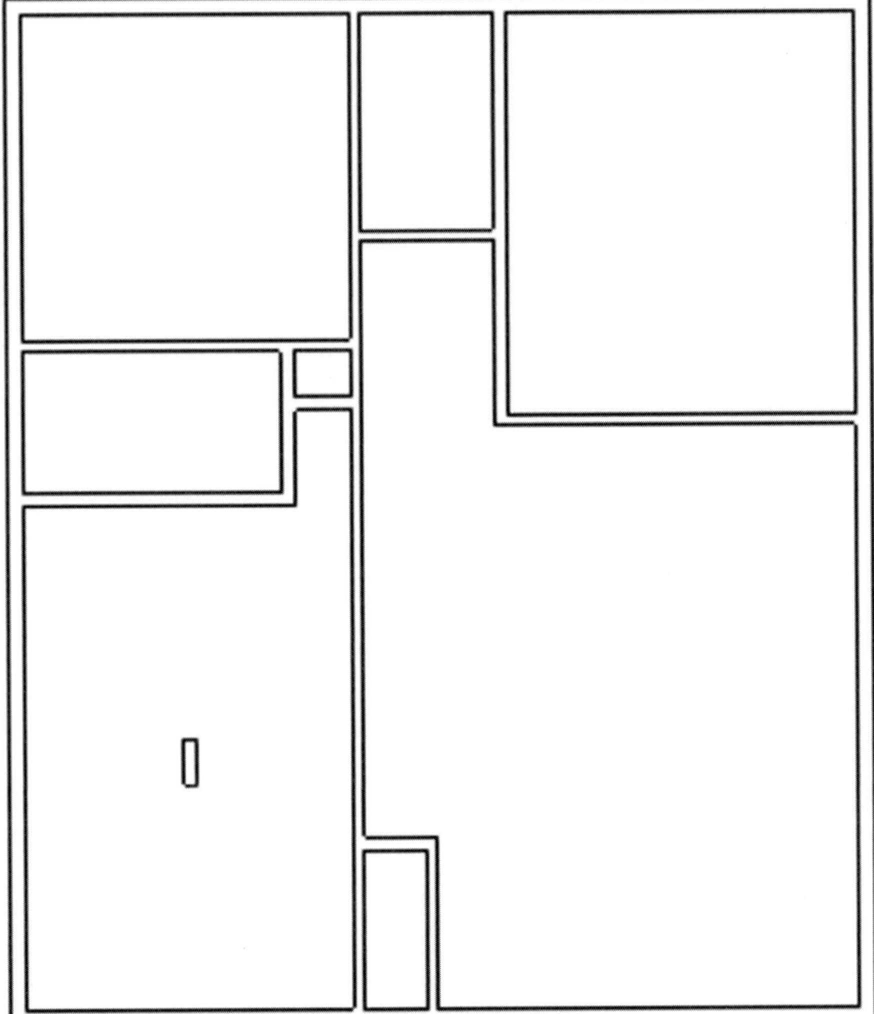

Figure 4-23 The plan after cleanup

7. Now *left-click* to pick Point 2. Pick anywhere in this area. Since there is nothing to snap to (an intersection or endpoint), we are simply selecting any point on the line that is closest to the cursor. You will soon see why the exact location is not important here (see Figure 4-26).
8. The section of the line between Point 1 and Point 2 will be erased.
9. Now we'll use another couple of cleanup functions to finish up the walls.

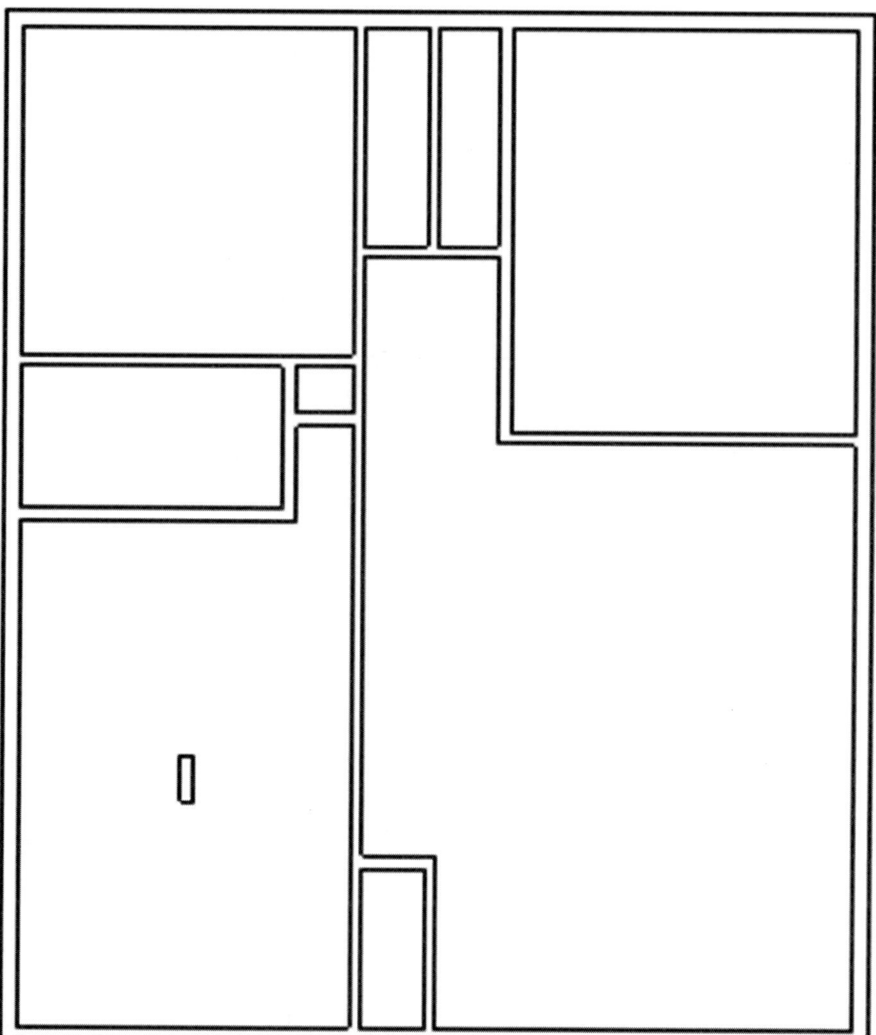

Figure 4-24 The addition of the closet wall

Cleaning Up L Intersections

In order to perform a clean up on L intersections, follow these steps:

1. Go to the **Cleanup** menu (**Edit/Cleanup**). Optionally, you can go straight to the **Cleanup** menu by pressing the Icon Bar button at the far right that looks like a spray bottle (if you are using the standard

Figure 4-25
Drawing a guideline

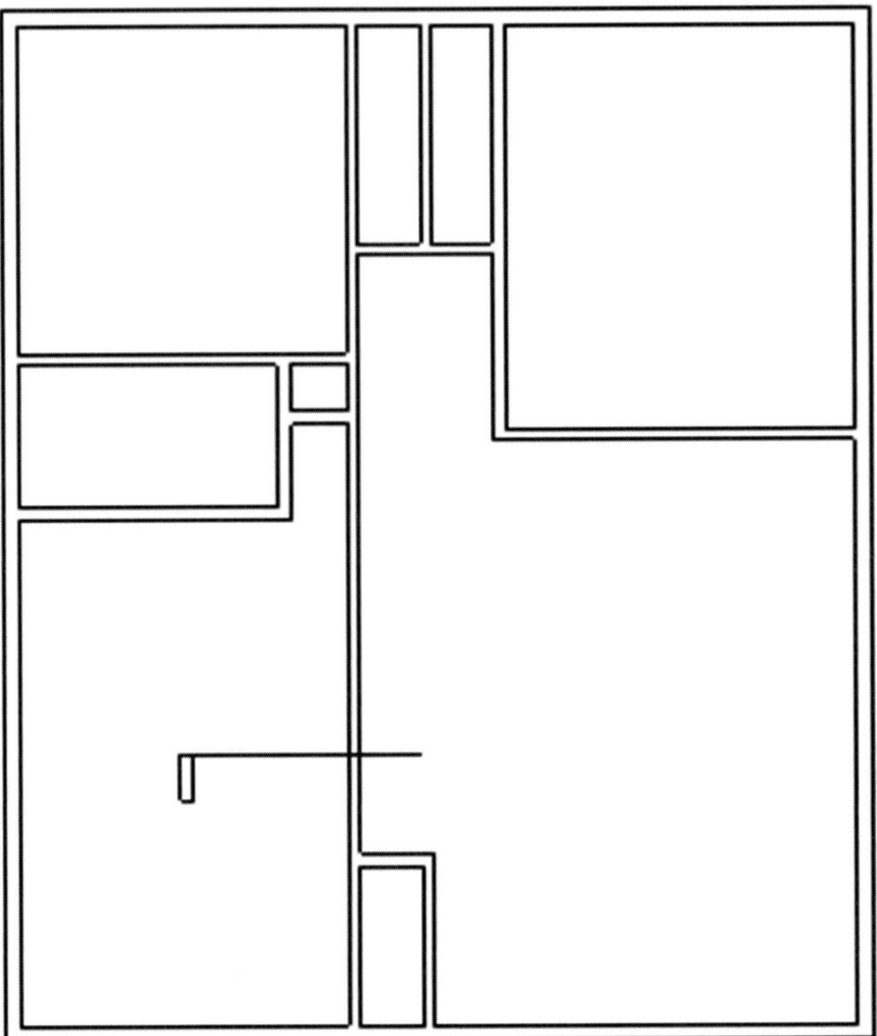

icons). When you place the cursor over the icon, the Information Line will read *"Cleanup menu."*

2. Once in the **Cleanup** menu, select **F5 2LnTrim**. The Command Line will read *"Select first line to trim."*
3. Select a point on the horizontal guideline you drew earlier, somewhere around Point 1.
4. Now select a point on the vertical line shown in Figure 4-27, somewhere around Point 2.

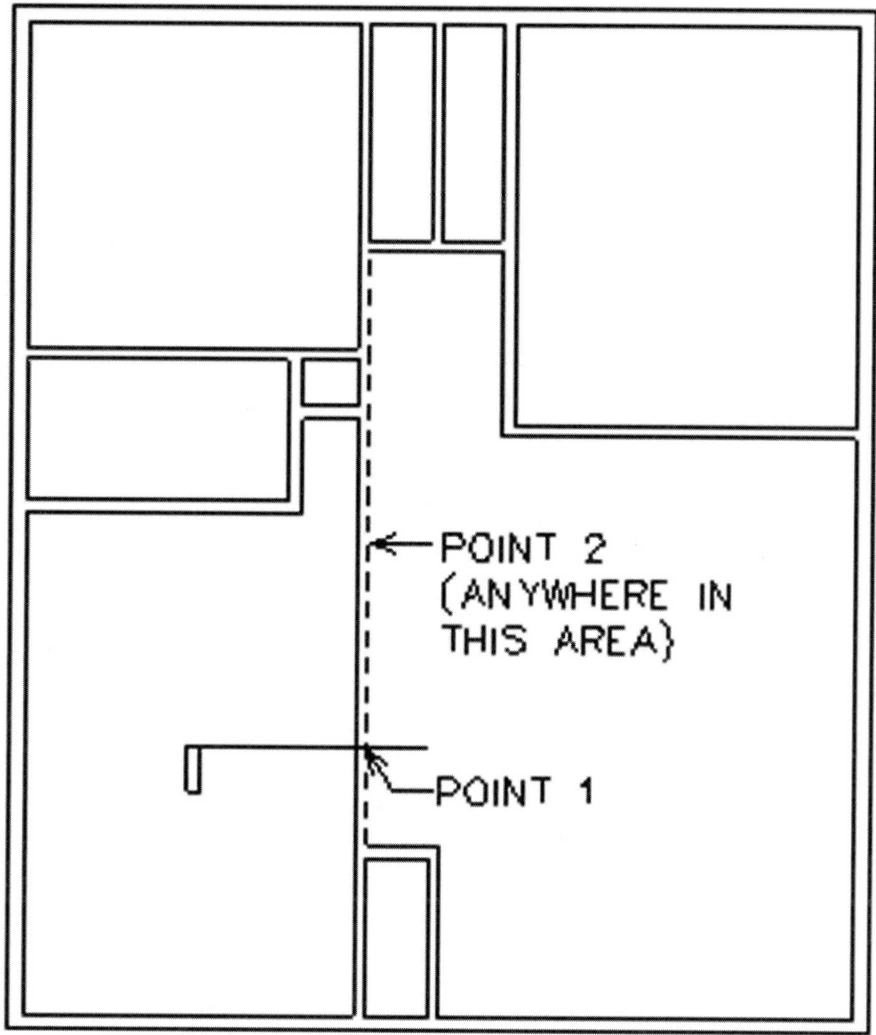

Figure 4-26 Selecting the two points

5. The intersection of those two lines is now cleaned up. Note that the way this feature works is that you place the cursor and pick a point ON THE SECTION OF THE LINE THAT YOU WANT TO RETAIN. This is an important step. If you pick the wrong side of the line, the wrong section of line will be retained. The next cleanup, using the same 2 Line Trim feature, is a little tricky because you will be picking a small section of line to retain. You may want to WindowIn to make the intersection larger and easier to pick.
6. Go to the **Cleanup** menu and pick **2LnTrim** again.

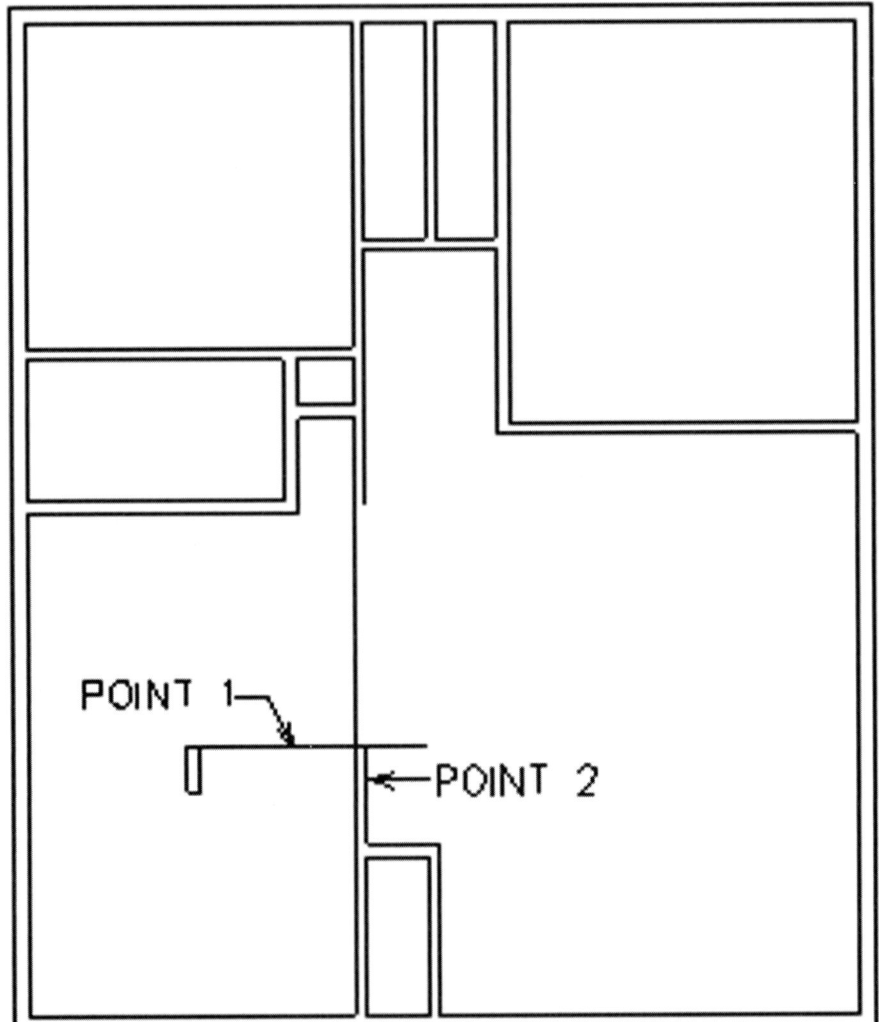

Figure 4-27
Selecting two points

7. Pick the horizontal line at Point 1. Be sure to place the cursor to the right of the vertical line since this is the section of the line you want to keep.
8. Now pick the vertical line as shown in Figure 4-28, somewhere around Point 2.
9. Now we have a nice clean end to our wall. It should look like Figure 4-29.

   Now let's use one more cleanup function to finish off the final wall corner at the hall closet.

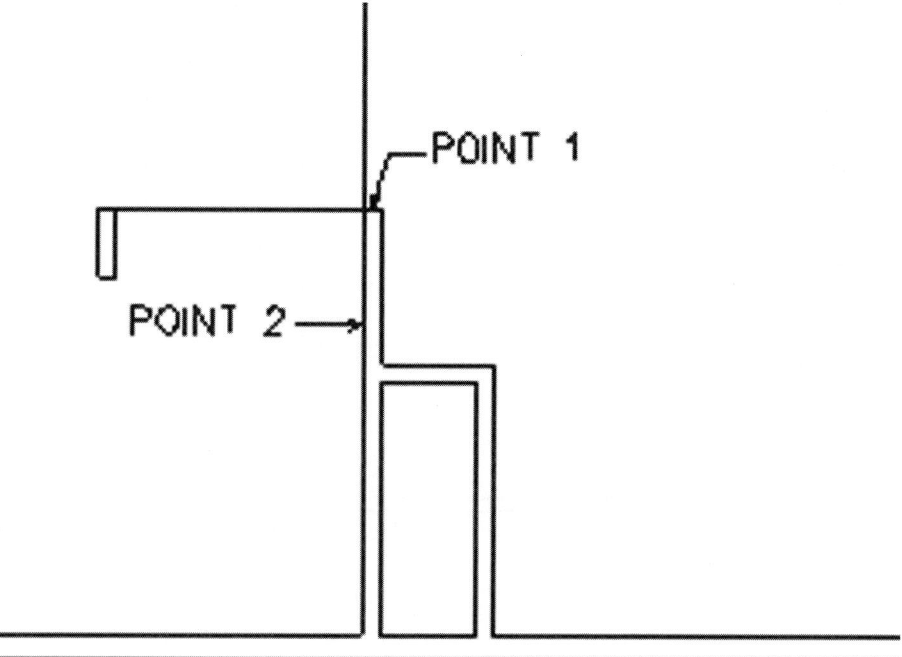

Figure 4-28 Choosing two new points

10. Go to the Cleanup menu once again. This time select **S1 LIntsct**. The Command Line will read, *"Select 1st corner around "L" intersect (wall line ends only)."* Like the T intersection function we used earlier, you must be sure that only the endpoints of the walls you are going to clean up can be within the bounding box we are going to draw.
11. Place a bounding box around all four endpoints of the two walls you are cleaning up (see Figure 4-30). The corner is cleaned up. Cool, huh?

Now that we have all of our walls drawn, it's time to put in the doors. But first, this is a good time to learn about layers. We are going to rename our current layer from Layer1 to Walls and then add a few more layers for some of the other entities we're going to add to our drawing. Finally, we will learn what the **LyrSrch** button does. You've probably seen it quite a few times by now.

Layers

Layers are an important aspect of CAD. Good layering practices can mean the difference between CAD heaven and CAD hell. By placing entities on

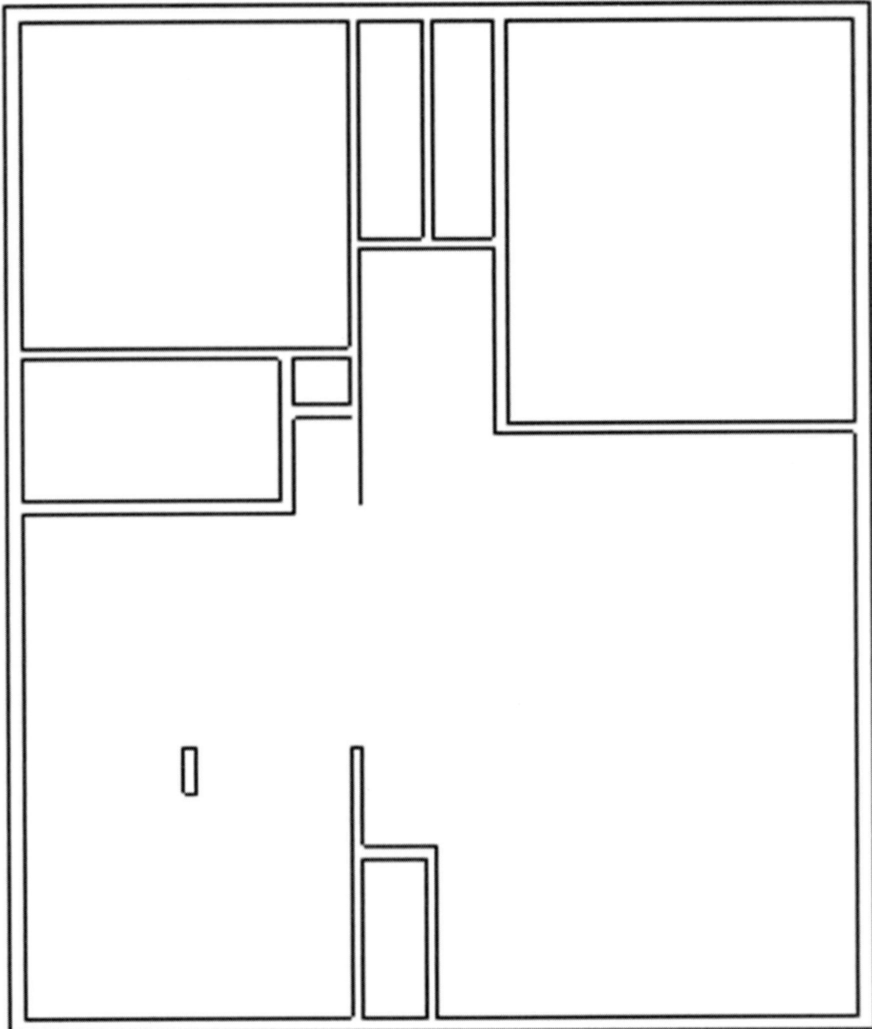

Figure 4-29
The completed wall

different layers, it is much easier to be selective about what you edit and how you edit it. First, let's add six new layers and name them:

1. Go to the **Utility** menu (*right-click* until **F1 ToScale** appears at the top and **S0 Edit** appears at the bottom).
2. Pick **F3 Layers**. Alternately, you can get there by pressing **Ll**.
3. Pick **F9 NewLayer**. The Command Line will read *"Enter the number of new layers to create : 1."* The number **1** is the default value. If you press Enter now, you will only add one layer. We want to add six.

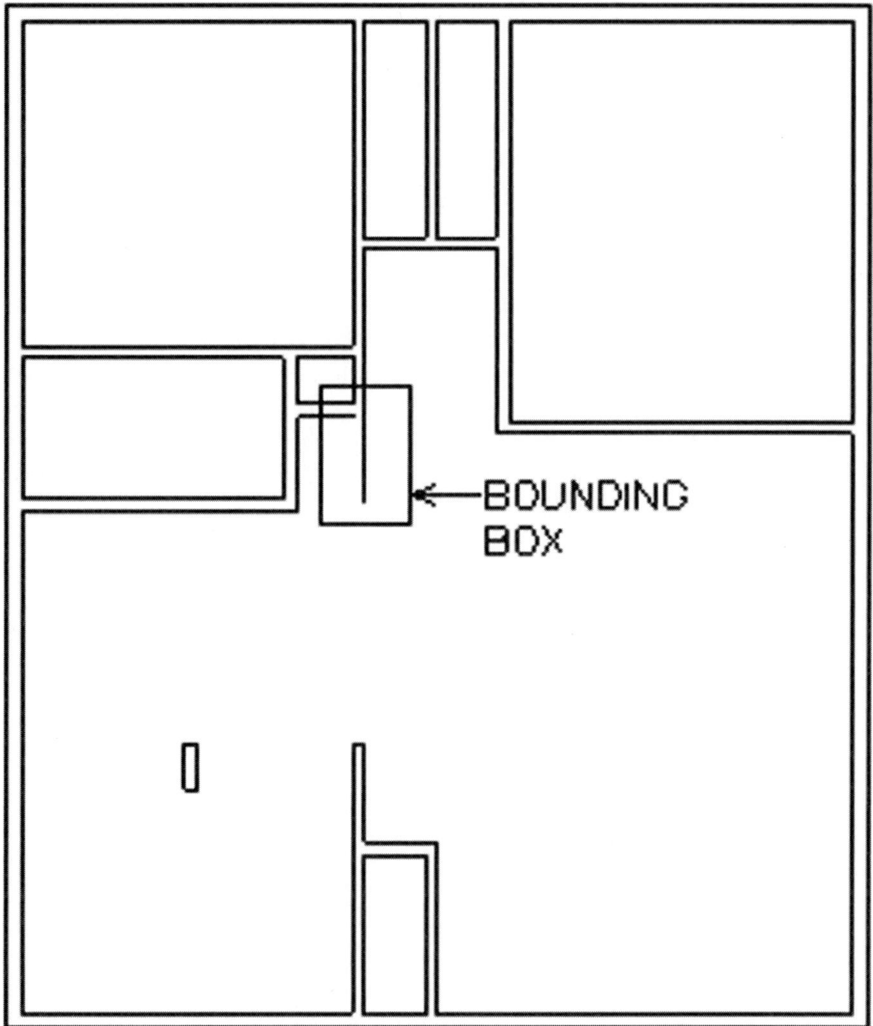

Figure 4-30
Adding a bounding box around the endpoints

4. To add six new layers, we can either pick the **F6 6** button or type in the number **6** and then press **Enter**.
5. The Information Line will read *"6 Layers have been created. Total layers : 7."*
6. While still in the Layers menu, select **F6 Name** so that we can give names to our layers. Names are so much more descriptive than Layer 1, 2, 3, and so on.

7. All the layers in the drawing will be displayed in the menu area. The Command Line reads *"Select layer to be renamed :."*
8. Select **F1 Layer1**. This is the layer that our walls are drawn on. The Command Line will read *"Enter new name : Layer1."* Layer1 is the current layer name. All we have to do to change it is type right over it.
9. Type **Walls** and then press **Enter**. Notice that the first layer in the menu has changed to read **F1 Walls**.

NOTE: *Unlike some other CAD programs, DataCAD enables you to enter names, such as layer names, in upper and/or lowercase letters. You can also use certain characters like the dash (-),the underscore (_), or the ampersand (&). What you type is up to you. I prefer to use both upper and lower case (like* ***1stFloor****,* ***Plan-1****, and so on), while others prefer only upper case.*

10. The Command Line is again asking you for a layer to be renamed. Select **F2 Layer2** and rename it to **Doors**.
11. Use the same process to change the remaining five layers to read:
    a. **Windows**
    b. **Plumbing**
    c. **Cabinets**
    d. **Hatching**
    e. **Text**
12. *Right-click* twice to return to the Utility menu. Notice that down in the Status Area the word **Walls** is displayed to the right of a white box. This means that the Walls layer is currently the active layer, and if you started drawing an entity, it would be on that layer.

    To change the active layer to, say, Doors, so that you are drawing entities on the Doors layer, you can go through the menus (**Utility/ Layers/SetActiv/Doors**, then *right-click*), or you can do it the quick and easy way.

    Instead, you can scroll through all the layers that are currently on by pressing the **Tab** key. Try this now. Look at the Status Area and watch the name change sequentially from **Walls** to **Doors** to **Windows**, and so on.
13. Now press **Shift+Tab** (press the **Shift** key while simultaneously pressing the **Tab** key). You will scroll backward through the layers that are turned on. Try this a few times.

14. Now press the **Tab** key until the word **Walls** appears again, so that it is the active layer.

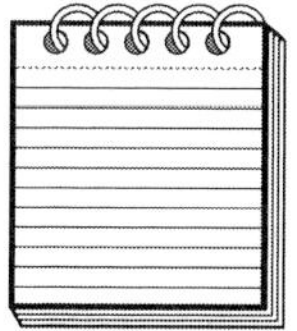

NOTE: *You will see this function in many of DataCAD' s shortcuts. Anytime DataCAD enables you to scroll through a list by pressing a shortcut key on the keyboard, you can press and hold the* ***Shift*** *key to go backward through the list. You'll see this again in the next example on colors.*

The white square to the left of the word **Walls** tells you that unless you change something, all entities on this layer will be drawn as white entities. And, of course, I want you to change something.

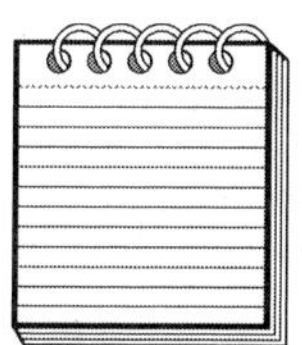

NOTE: *From here on in, we will not be placing the* ***F#*** *or* ***S#*** *characters before the menu choices (like* ***F7 Erase****). I think you probably have the idea by now.*

Layer Colors

O.K., I lied. We're going to learn a few more things before we put in the doors. One advantage of CAD is that we can draw with many different line colors to keep track of what we are drawing. In DataCAD, all the entities on one layer can be one color or many colors. For now, we will keep all the entities on each layer as just one color.

First, let's change the color of all the walls we drew. Let's make them cyan instead of white, and then make sure that all future entities drawn on that layer will be drawn as cyan as well:

1. Go to the **Edit** menu and select **Change**. Make sure **Area** is turned on.
2. Select **Color/Cyan** and then draw a bounding box around all the walls of the apartment. All the walls will change to cyan in color. *Right-click* to go back to the **Edit** menu.
3. To change the default color of the current Walls layer to cyan, press the **Kk** key to change colors.

4. Notice that the Information Line reads *"Current color* = Red." and notice also that the square to the left of the word Walls in the Status Area has changed to the color red to indicate the current color.
5. Notice again that your cursor is also changing color. This is the easiest way to determine which color you are currently drawing with. Just look at the cursor.
6. Press the **Kk** key a few more times and note the changes in the Information Line and the colored square.
7. Let's try the **Shift** key thing again. Press **Shift+Kk** (remember the **Kk** means either the upper case or lower case **K** will work). Notice that the colors are scrolling backwards through the list.
8. Press the **Kk** key until the color in the Information Line reads Cyan. Now everything you draw on the Walls layer will be drawn as cyan.
9. Press the **Tab** key until the active layer is the Doors layer. The default color is currently white. We want to change it to magenta.
10. Press **Kk** until the Information Line reads *"Current color* = Magenta."
11. Press **Tab** to make the active layer "Windows." Change its default color to **Red**.
12. Follow the same procedure to change the default colors of the last four layers:
    a. Plumbing to **LtGreen**
    b. Cabinets to **LtBlue**
    c. Hatching to **White**
    d. Text to **Brown**
13. Finally, press **Tab** until the active layer is the Doors layer. Your cursor should show up as magenta in color.

Adding Doors

It's awfully hard to get around a building without doors, so let's add swinging doors to our apartment project. We'll start in Bedroom 1:

1. Press the **a** key to go to the **Architect** menu and then press **DoorSwng**. You can also select the icon that looks like a door swing. It's 10 icons in from the left if you're using the default icon bar.
2. Select **DoorStyl/Single** to create a single swinging door. *Right-click* back to the **DoorSwng** menu.

3. Select **SwngStyl/Arc** and then *right-click*.
4. In the **DoorSwng** menu, make sure the buttons are set as follows:
   a. **Sides** = On
   b. **CntrPnt** = Off
   c. **Cutout** = On
   d. **InWall** = On
   e. **DrawJamb** = On
   f. **MtchWall** = On
5. Now press **LyrSrch** to turn it on. You will be presented with a list of all the layers in your drawing. You need to tell DataCAD which layer contains the walls that the doors will be cut into. To do this, select Walls.
6. The prompt line reads *"Select hinge side of door."* We want to place the hinge side of the door 6″ [152] from the inside corner of the room. Rather than drawing a reference line and later erasing it, we can simply use the RefPt (′) key to pick the inside corner and then place the wall 6″ up from that point.
7. Press the RefPt (′) key. *Middle-snap* to the lower-left, inside corner of Bedroom 1.
8. Press the **Space** bar so you can input a coordinate.
9. Enter a distance of **.6** [152]and then press **Enter**. Enter an angle of **90** and press **Enter**. A line will be located on the bedroom wall, attached to the cursor. The Command Line will read *"Select strike side of door."*
10. Press the **Space** bar to enter another coordinate. Enter a distance of **2.6** [762] and an angle of **90**. The Prompt Line will read *"Select direction of door swing."*
11. Move the cursor to a point anywhere to the right of the bedroom wall and *left-click* the mouse. The prompt line will read *"Select any point on the outside of the wall."*
12. Keep the cursor to the right of the bedroom wall and *left-click*. The door is inserted, as shown in Figure 4-31. There is a reason for this second step. You can read about it in Chapter 11, "Basic Construction Drawings."

Removing Doors

Now let's see how to quickly remove a door and repair the wall that was cut. We'll use the door we just inserted:

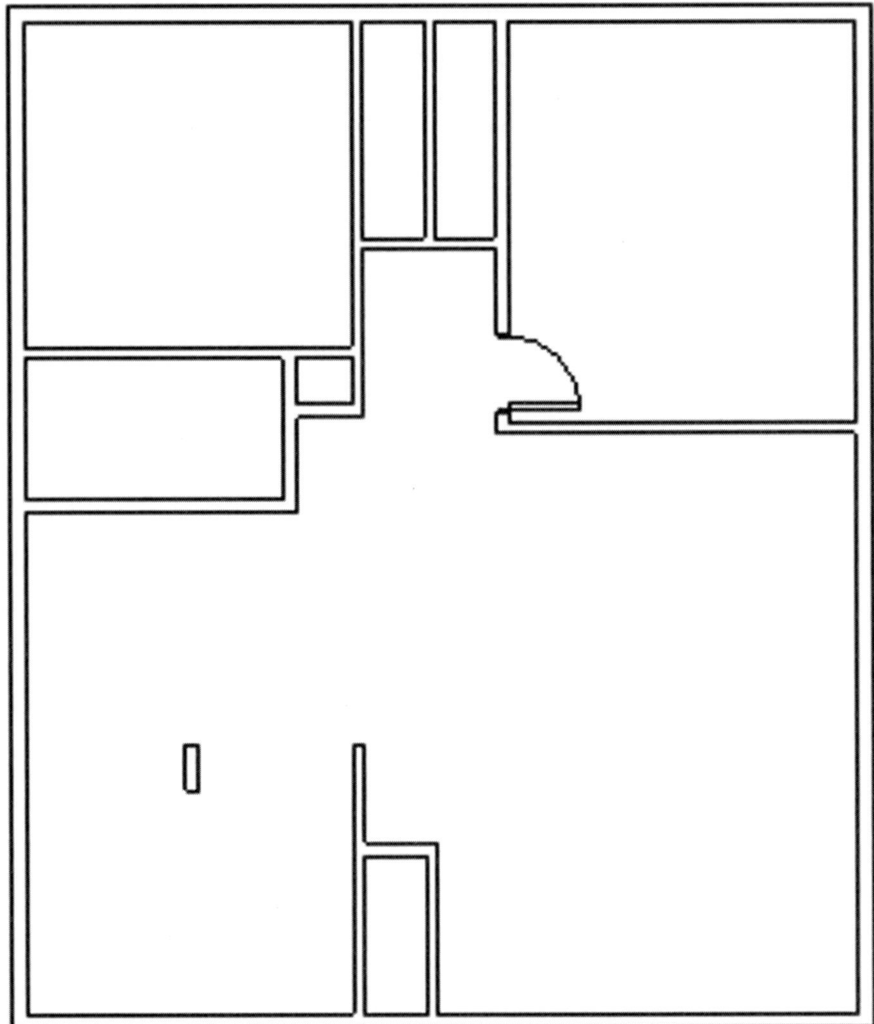

Figure 4-31
Insertion of a door

1. While still in the **DoorSwng** menu, select **Remove**. The prompt line will read *"Select first point of box around door or window to remove."*
2. Pick two points to place a bounding box around the door and both jambs. Be sure to have all the endpoints of the wall cuts and all the entities that make up the door within the bounding box. Also, be sure that no extraneous endpoints or entities are within the bounding box (see Figure 4-32).

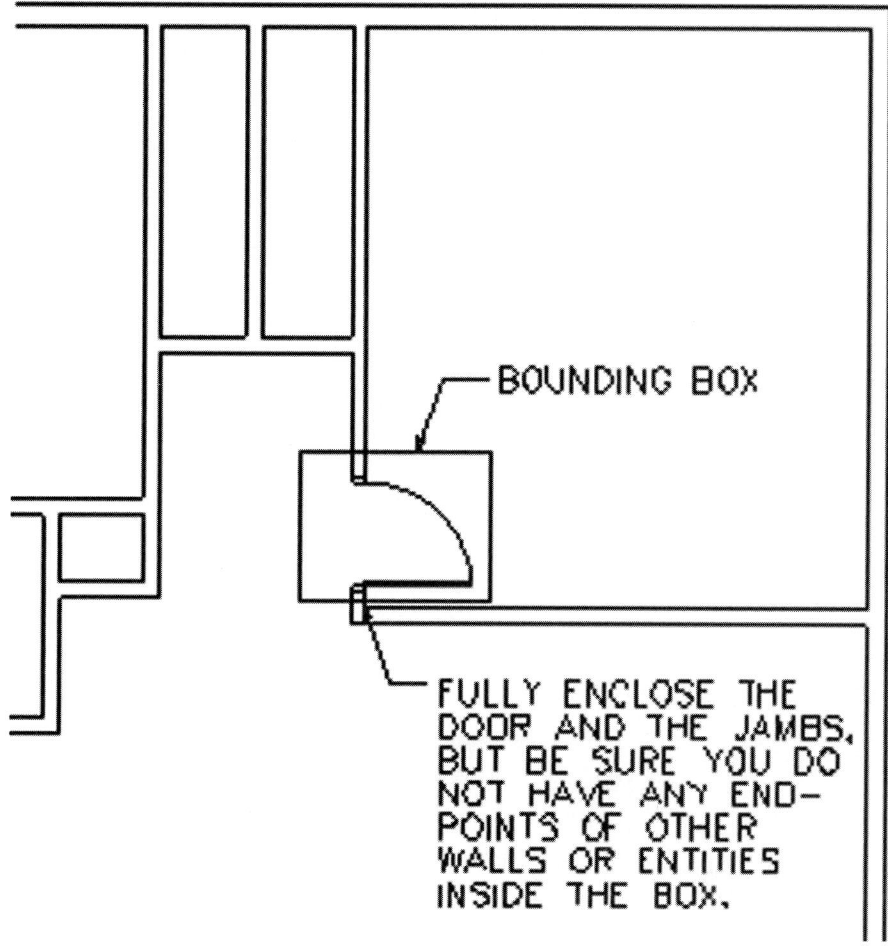

Figure 4-32 Enclosing the door in a bounding box

3. The door will be erased and the wall is reconnected! Select **Edit/Undo** from the drop-down menus to get the door back (more on this later).
4. *Right-click* back to the **DoorSwng** menu.

Identify

Quite often, you will need to see the values and settings of an entity. You may need to remind yourself which layer the entity is on, or what linetype it is, or you'll need to see its dimensions. To quickly do this in DataCAD, an Identify function is available. Four methods exist for activating it:

- Select **Edit/Identify**.
- Press the **I** button in the upper-right corner of the Navigation Pad.
- Press the **I** key (uppercase I).
- Select **Tools/Identify** from the drop-down menu.

Let's see what happened to the wall that was repaired after removing the swinging door:

1. Select Identify using either previous method. I prefer using the **I** on the Navigation Pad. The Command Line will read *"Point to entity to identify."*
2. *Left-click* to select the inside wall of the bedroom, the same one that we cut the door into. Again, you only need to get the cursor close to the entity, not right on top of it. The selected entity will temporarily change to a dashed line to show that it is the one selected.

Look at the menu, which has nine buttons displayed. Each one is telling you something about the entity you just selected. By themselves, they will tell you nearly everything you need to know, but, if not, you can press each one for additional information. You'll learn more about all of them later, but for now look at the first four. Here is what each one is telling you:

- **LINE** The entity type (line, arc, symbol, and so on)
- **Walls** The layer name (the name of the layer on which the entity resides)
- **Cyan** The color of the entity
- **Solid** The line type (solid, dashed, dotted, insul, and so on)

Now look down at the Information Line and the Command Line. You will see additional information there including the length and angle of the line.

The important point to take note of is that DataCAD is smart enough to repair the lines of the walls back to their original state without having to draw separate lines to "infill" the cut in the wall.

More Doors

In this section, we'll add some more doors to the apartment.

While still in the **DoorSwng** menu, see if you can place three more swinging doors for Bedroom 2, the bathroom, and the front door to the

Figure 4-33
The completed set of doors

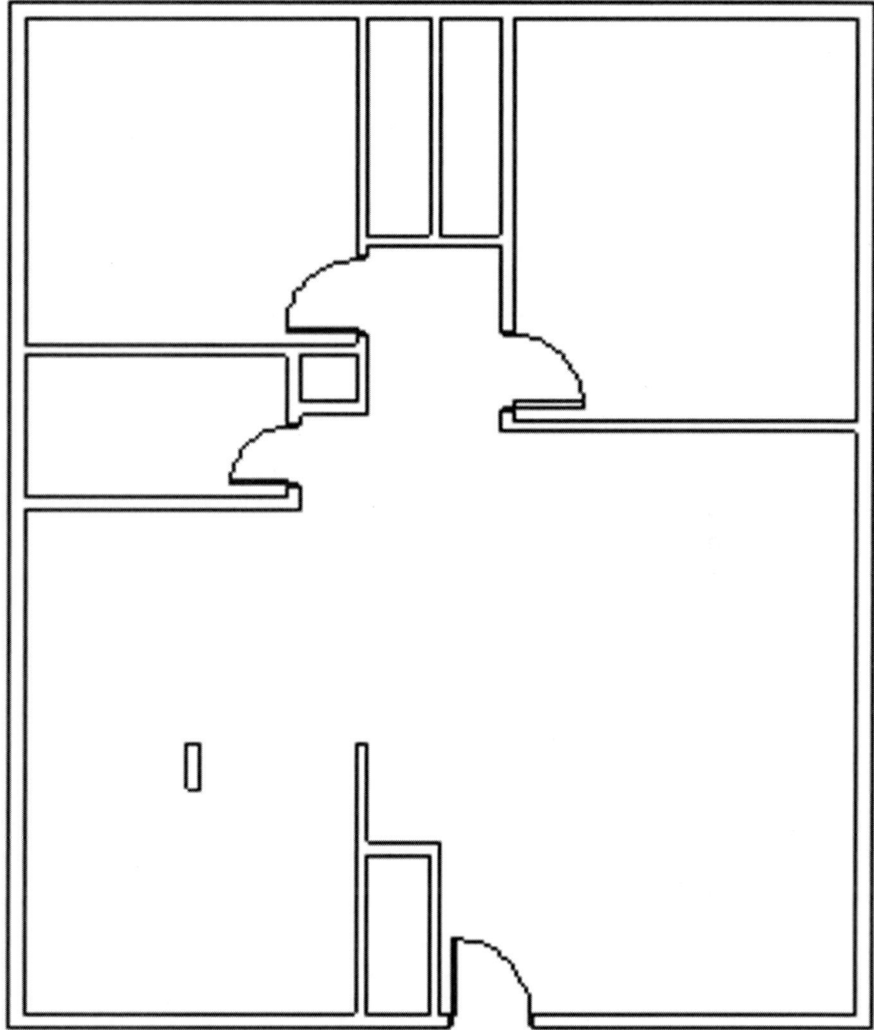

apartment. Do not place the door to the linen closet yet. Since we are retaining all of the settings used to draw the first door, all you need to do is select **a/DoorSwng** and begin to place the door by starting with your RefPt ('). Remember that the first point you pick must be the hinge side of the door. When you are done, your drawing should look like Figure 4-33.

1. To place the linen closet door, we'll use a slightly different technique to get it visually centered in the wall. We will also show the door at a partially open angle of 45 degrees.

2. Zoom into the closet so you can see it better. Select **a/DoorSwng** if you're not already there.
3. Pick **Angle/45-0′** and then *right-click* or press **Enter** to accept it.
4. Turn the **Sides** button to off. The Command Line will read, *"Doors defined by center and strike side. Select center of door."* You can probably decipher its meaning on your own.
5. To center the door on the closet wall, set the cursor near the center of the outside wall of the closet and *middle-click* to snap to it. Read the Command Line.
6. Since we want to place a 1′-2″ [355] door, and half of that is 7″, press the **Space** bar and enter a distance of **.7** [178] and an angle of **0** to pick the strike side of the door.
7. Select a point below the closet to define the direction of swing. Do it again to define the outside of the wall. The door is inserted (see Figure 4-34).

   Now let's add the bifold closet doors in the two bedrooms, starting with Bedroom 1.

9. WindowIn closer to the Bedroom 1 closet if you need to.
10. While still in the **DoorSwng** menu, leave **Sides** turned to off and **Angle** set to 45 degrees.
11. Pick **DoorStyl/Bi-Fold**.
12. Place the cursor close to the center point of the inside closet wall and *middle-snap* to pick the center point of your bifolds.
13. The doors are a total of 6′-0″ [1828] wide, so we need to pick a point 3′-0″ [914] from the center point. Press the **Space** bar and enter a distance of 3′-0″ [914] and an angle of **90**.
14. Pick a point outside the closet for the direction of the door swings. Pick the same point to define the outside wall.
15. Now do the same for the closet doors in Bedroom 2. Your drawing should look like Figure 4-35.

    We're cruising now! Can you see how the automation of this kind of repetitive task can save you a great deal of time? Now let's finish installing our doors by adding the final set of sliders to the closet next to the apartment entrance.

16. While in the **DoorSwng** menu, select **DoorStyl/Sliding**, then **Angle/0-0′**, and press **Enter**, set **DrawJamb** to off. This will give us a pair of sliders that appear closed without jambs drawn in.

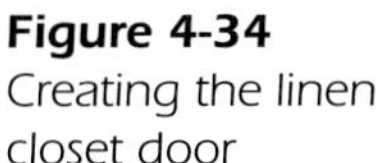

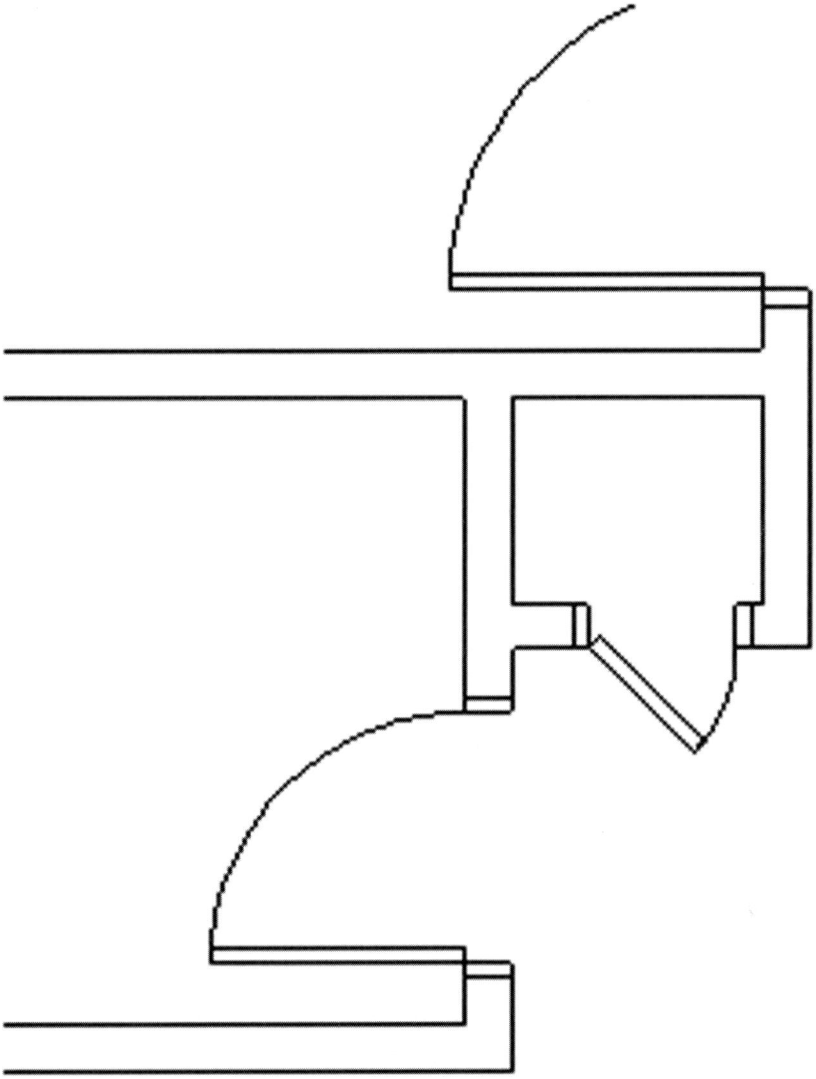

Figure 4-34
Creating the linen closet door

17. Let's select the visual center of the closet wall like we did at the bathroom. *Middle-snap* to the outside center of the closet wall.
18. Press the **Space** bar and then enter a distance of **2.4** [712] and an angle of **90**. Select a point to the right of the closet to finish adding the doors. Notice that with sliders you have to select only a single point outside the closet to define the outside. Your drawing should now look like Figure 4-36.

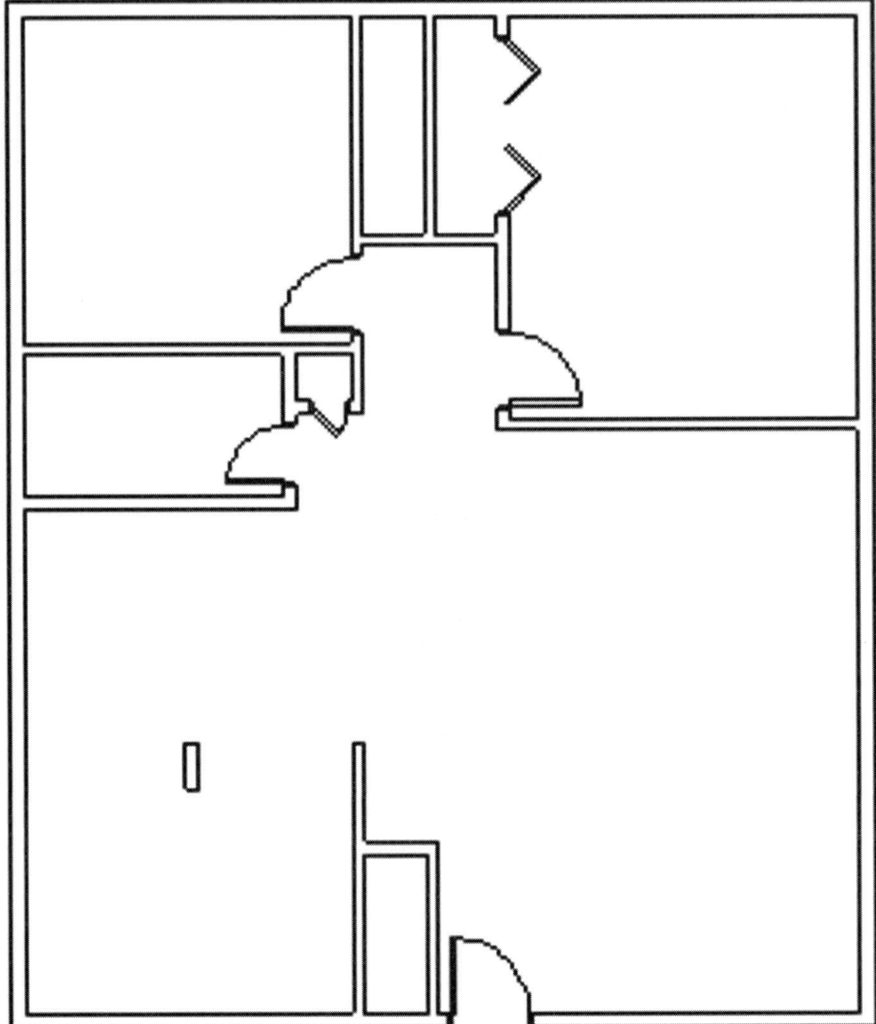

Figure 4-35 The completed bedroom closet doors

OK, we're done with the doors. Now it's onto the windows. You will find the steps and concepts for windows to be very similar to that of the doors.

Adding Windows

Just like with the doors, we can add windows by their jambs or centers. We will use both methods. The Windows menu is also located in the Architect

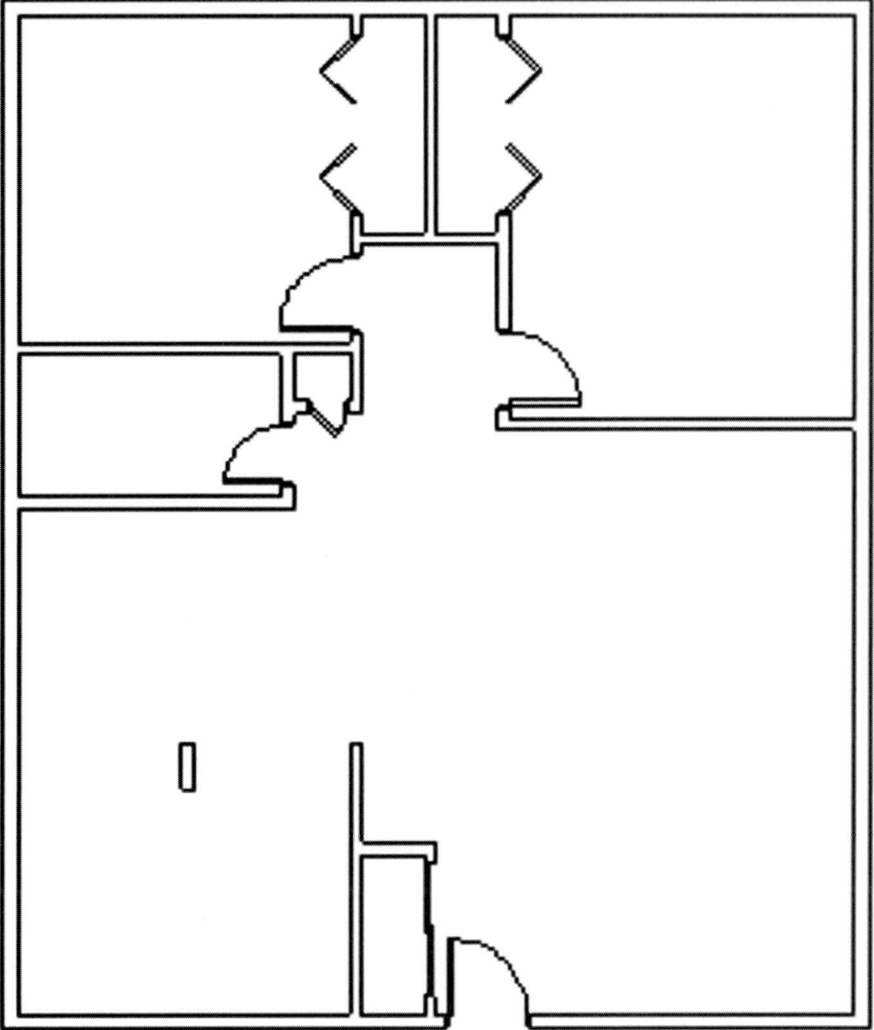

Figure 4-36 Completion of the sliders

menu, accessible with the **a** key. Press the **a** key now and then the **Windows** button, or you can use the icon at the top of your screen that looks like a window. It's the ninth one from the left if you are using the default icon bar. Notice that the **F1** through **F0** buttons are exactly the same as those in the Doors menu. They function the same as the buttons in the Doors menu too, so you will find them very familiar.

First, we need to make the Windows layer the active layer so that all of our windows will be drawn in that layer:

1. Press the **Tab** key. The Status Area will tell you that the Windows layer is the active layer, and your cursor will now be red (the color that we made the Windows layer earlier). Let's put windows in Bedroom 1 first. We'll start with the window at the topmost wall. We'll place it by defining the width between the two jambs. First, we'll have to change some settings.
2. While in the **Windows** menu, set the buttons as follows:
   a. **Sides** = On
   b. **CntrPnt** = Off
   c. **Cutout** = On
   d. **InWall** = On
   e. **DrawJamb** = On
   f. **JambIn** = Off
   g. **MtchWall** = On
3. Now make the following additional settings:
   a. Pick **LyrSrch** and then select the **Walls** layer (if **LyrSrch** is already turned on, turn it off and then back on to get to the Layers menu).
   b. **OutSill / .1.1/2** [38.1](gives us an outside sill of $1^1/_2''$ [38.1])
   c. **InSill / .1** [25.4](you guessed it; it gives us an inside sill of 1″ [25.4mm])
   d. **GlassThk / ..1/2** [12.7]
   e. **SillHgt / 3.4** [1016]
   f. **HeadHgt / 6.8** [2032]
4. Press the RefPt (') key and snap to the inside, upper-right corner of Bedroom 1.
5. Press the **Space** bar and enter a distance of **3.6** [1066] and an angle of **180**.
6. Now just like for a door, define the other side of the window by pressing the **Space** bar and entering a distance of **3.6** [1066] and an angle of **180**. The Command Line will read *"Select any point on the outside of the wall."* Because we have set different dimensions for the outside and inside sills, it is important to pick the correct side of the wall.
7. Pick a point above the wall (the outside). The wall is cut and the window is added. Just like in the **Doors** menu, you can use the **Remove** command to remove the door and repair the wall (see Figure 4-37).

   Now let's place the second window in Bedroom 1 by defining the center and one jamb.
8. In the Windows menu, turn **Sides** to off. The Command Line will read, *"Windows defined by center and jamb. Select center of window."*

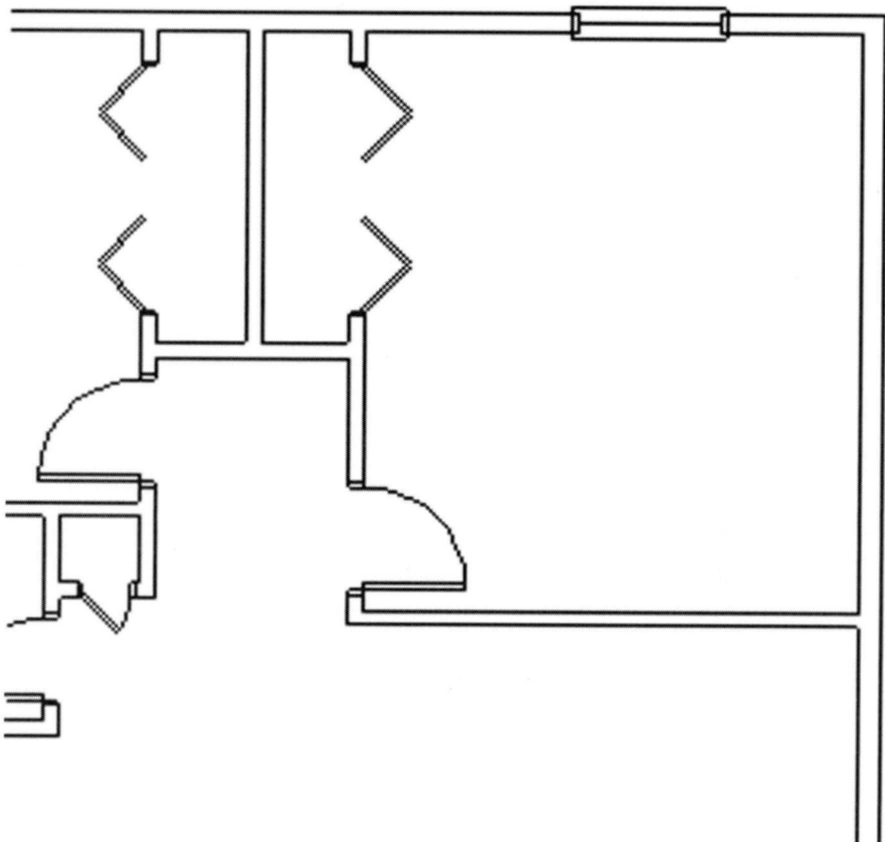

Figure 4-37
Adding the bedroom window

9. Press the RefPt (′) key and snap to the inside, upper-right corner of Bedroom 1.
10. Locate the window in the middle of the wall by *snapping* to the center of the inside wall (by placing the cursor near the center of the line and *middle-clicking*).
11. This window will be a 4′-0″ [1219] window, so define the other side of the window by pressing the **Space** bar and entering a distance of **2** [610] and an angle of **270** (or **−90**). The command line will read, *"Select any point on the outside of the wall."*
12. Pick a point to the right of the wall (the outside). The wall is cut and the window is added (see Figure 4-38).

    Now let's place the two living room windows in the same wall using the same center-of-window method.

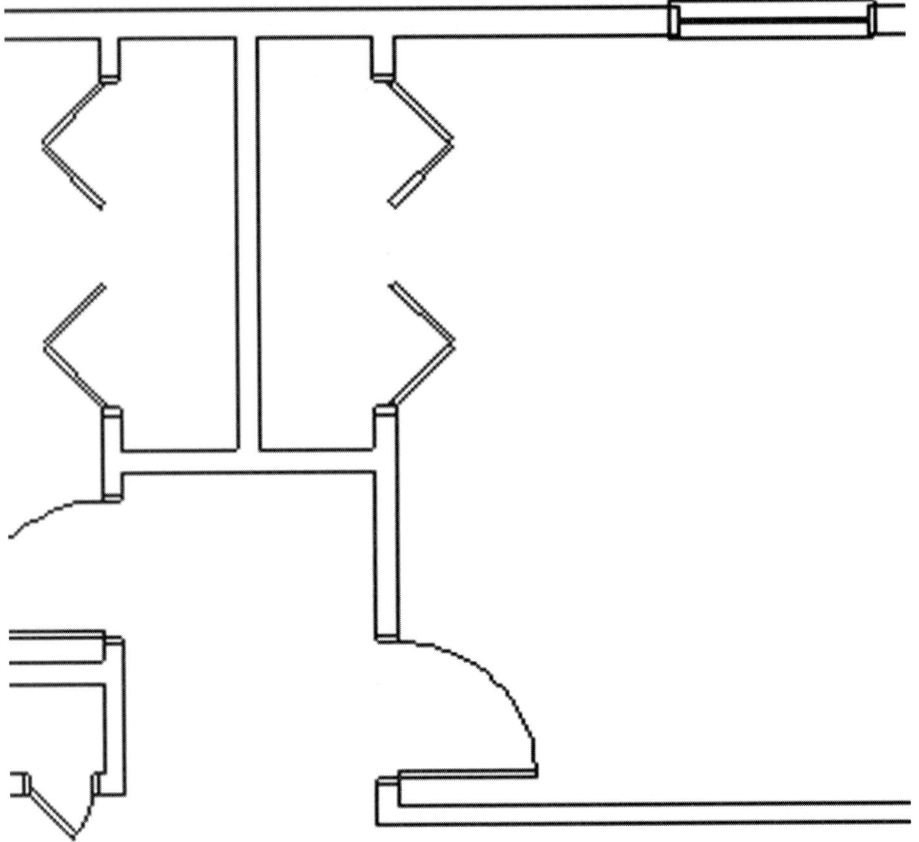

Figure 4-38
Adding another bedroom window

13. Press the RefPt (') by snapping to the inside, upper-right corner of the living room.
14. Press the **Space** bar and enter a distance of **5** [1524] and an angle of **270** (or **−90**).
15. This window will be 4′-0″ [1219] wide. Press the spacebar and enter a distance of **2** [610] and an angle of **90** (we could have used **270**, but I thought we'd be different).
16. Pick a point to the right, outside of the wall. The window is added (see Figure 4-39).

Now that you know how to place windows by using **Sides** and **Center**, add in the remaining four windows yourself. When you are finished, it should look like Figure 4-40.

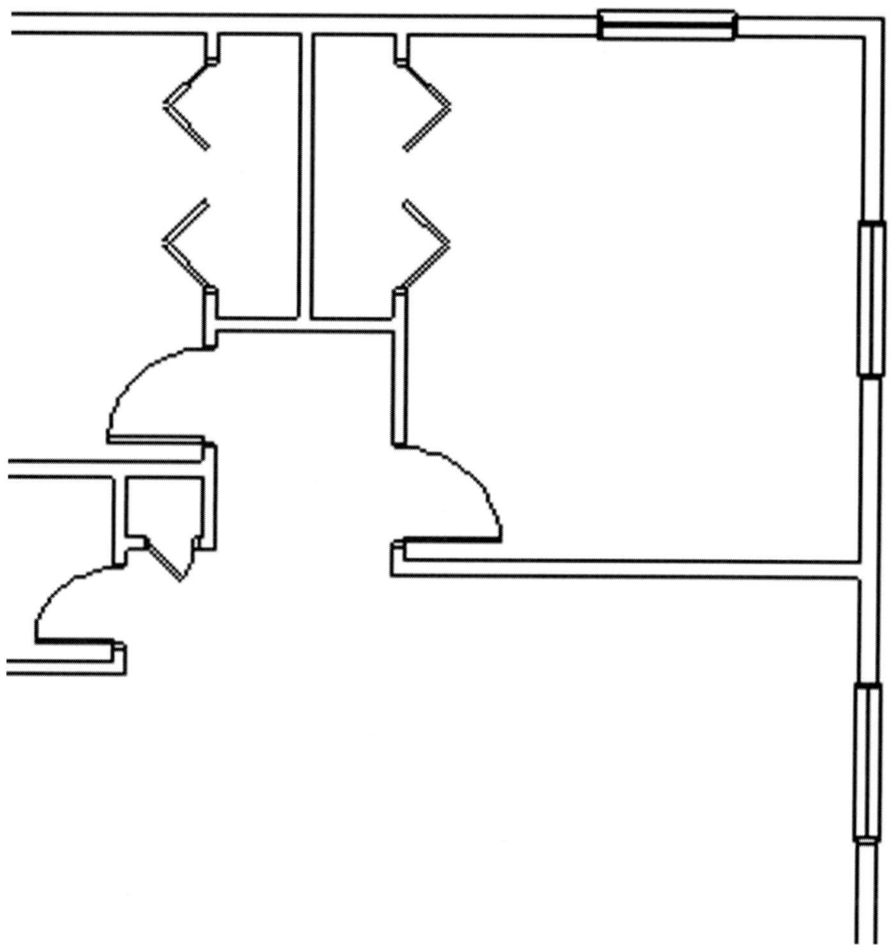

Figure 4-39
Adding a window to the living room

Sample

This drawing is available on the CD-ROM, with the file name "APT-02.DC5."

Isometric View

Well, you've diligently been drawing lots of 2D stuff, but the really eye-catching part of CAD is 3D. One of DataCAD's strong points is the ease with which you can flip back and forth between your 2D drawing and a

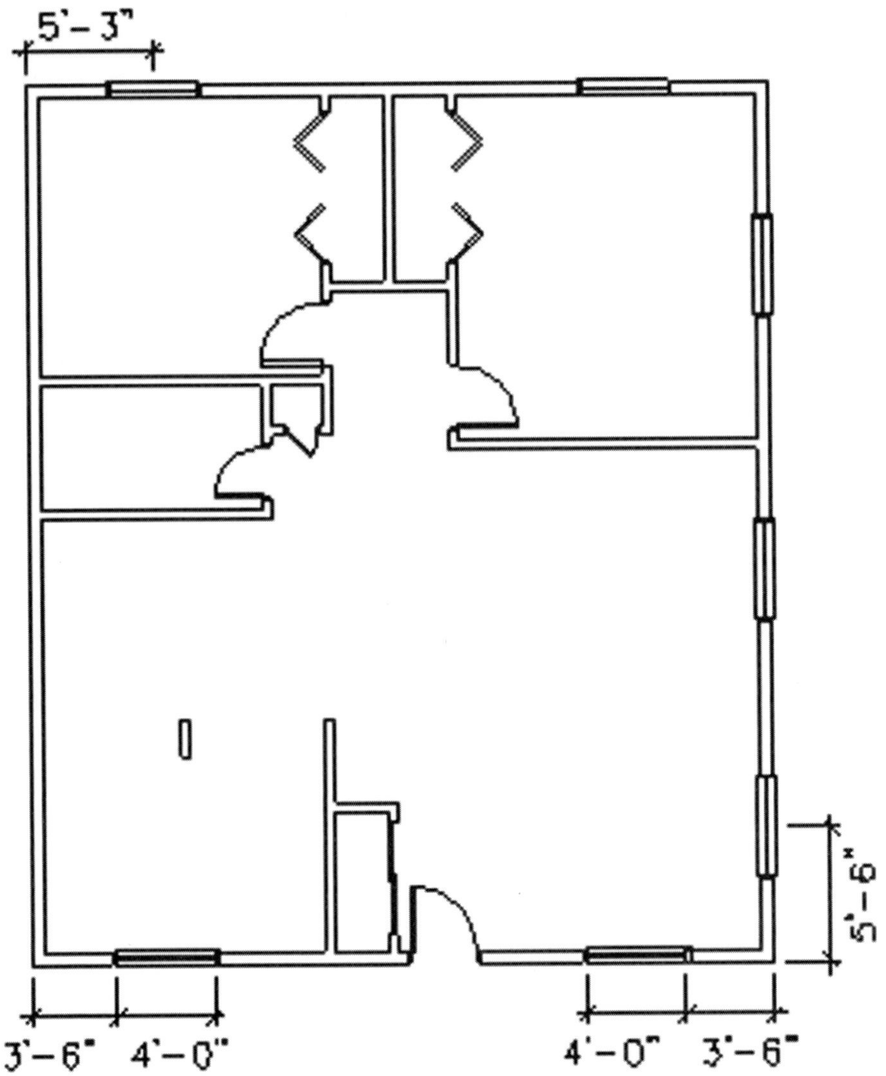

Figure 4-40
The floor plan with all the windows added

simplified 3D view using DataCAD's $2^1/_2$ D entities (all of DataCAD's entities are $2^1/_2$ D). You can read more about this in Chapter 5, "Basic Drawing," but let's at least see what you've got right now without explaining a lot about how you got there.

With your floor plan fully inside the Drawing Window (press Recalc if you have to), follow these steps:

1. Select **Edit/DCAD 3D** to get to the 3D menus. Notice that although most of the menu options here are different than the 2D menus, two primary menus are still available here: **3DEdit** and **3DEntity**. You access each by simply *right-clicking*.
2. Select **3DViews** (available in both the **3DEdit** and **3DEntity** menus).
3. Select **Isometrc** to see an isometric view of your building (see Figure 4-41). If you don't see it, or you only see part of it, just press the **R**ecalc button in the Navigation Pad.
4. While you are still in isometric view, try out these options:
   a. Press the **Ortho** button in the Projection Pad. You will return to the 2D, orthographic view of your building.
   b. Press the **Para** button in the Projection Pad to return to the isometric view. An isometric view is a "parallel" view because none of the entities are in perspective. All entities in the same plane are parallel to one another.
5. Make sure you are back in isometric view, and then *right-click* back to the **3DEdit** menu. Press **Hide/Begin** and a hidden line view of the building will be generated (see Figure 4-42).

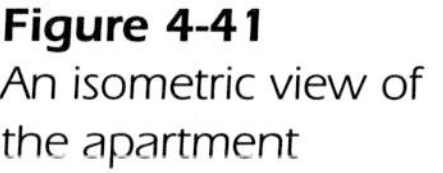

Figure 4-41
An isometric view of the apartment

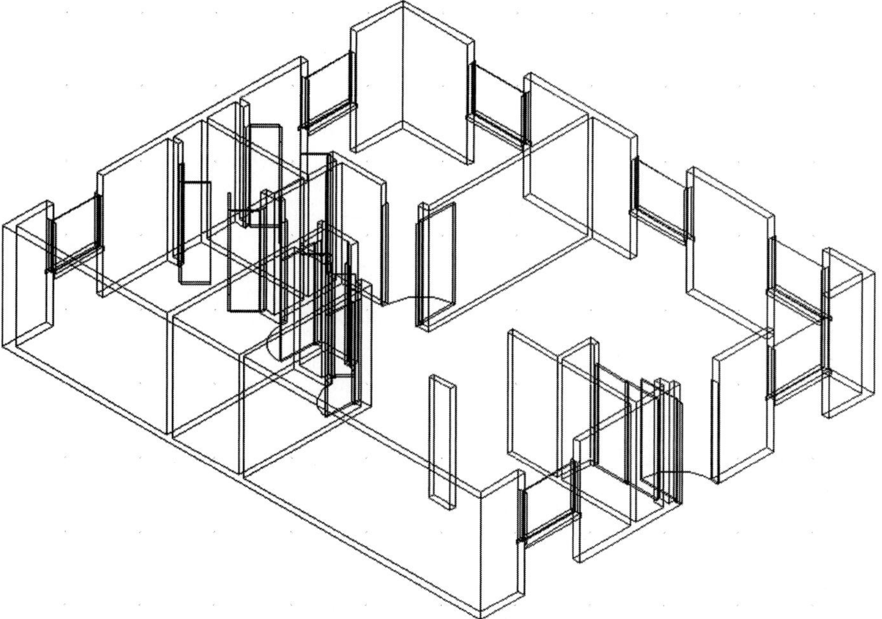

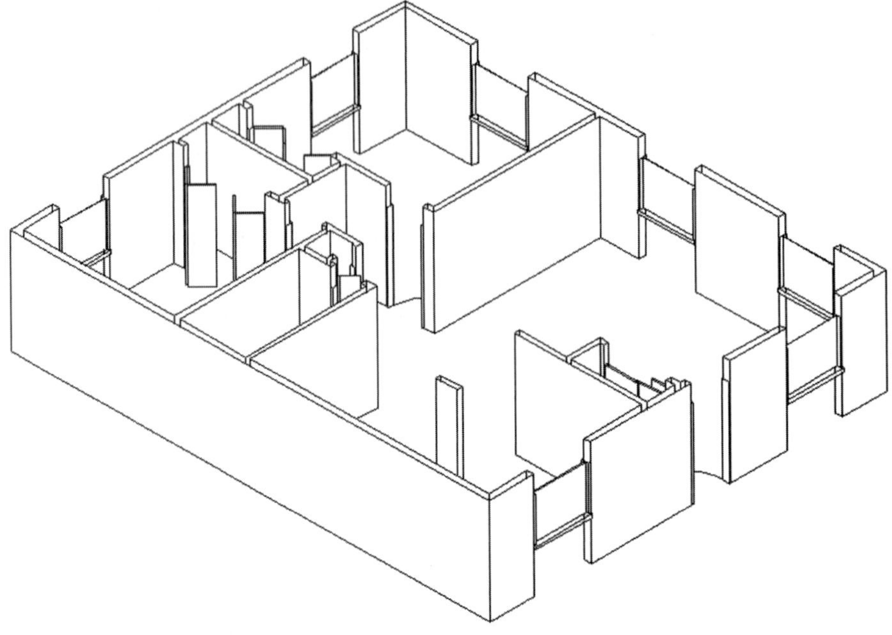

Figure 4-42
A 3D, hidden line view of the apartment

6. *Right-click* back to the **3DEdit** menu and then press the **Esc** key to get back to the wire frame image. Now press the **Ortho** button, and then the **DCAD 2D** button (or the **;** key) to get back to the 2D menus.

We'll do some more work with these isometric and hidden line views at the end of the chapter.

Counters and Closets

Now that the shell of the building is coming along, let's work on the interior. We'll add the countertops in the kitchen and in the bathroom, and then the shelves and closet poles in the closets:

1. *Right-click* until you are back in the **Edit** or **Utility** menu (it doesn't matter which).
2. Press the **Tab** key until the Cabinets layer is the active layer. The cursor should now be LtBlue.

3. Make sure the Walls function is off.
4. In the kitchen, draw a line to start the countertop. Press the RefPt (') key and snap to the inside, lower-right corner of the kitchen.
5. Press the **Space** bar and enter a distance of **.33** [838](that's 33″) and an angle of **180**.
6. Drag the mouse straight up until the Y dimension in the Message Line equals 2′-1″ [635] and click the mouse.
7. Drag the mouse to the left until it crosses the leftmost wall. *Right-click* to stop drawing the line (see Figure 4-43).
8. Press the RefPt (') key and snap to the intersection of the blue counter line and the inside kitchen wall.
9. Press the **Space** bar and enter a distance of **2.1** [635] and an angle of **0**.
10. Drag the mouse straight up until the cursor is aligned with the bottom of the short wall in the kitchen and *click* the mouse.
11. Drag the mouse to the right and snap to the corner of the small wall to finish the line of the counter. *Right-click* to stop drawing the line.
12. Now snap to the upper-left corner of the small kitchen wall.

Figure 4-43
Creating the kitchen counter

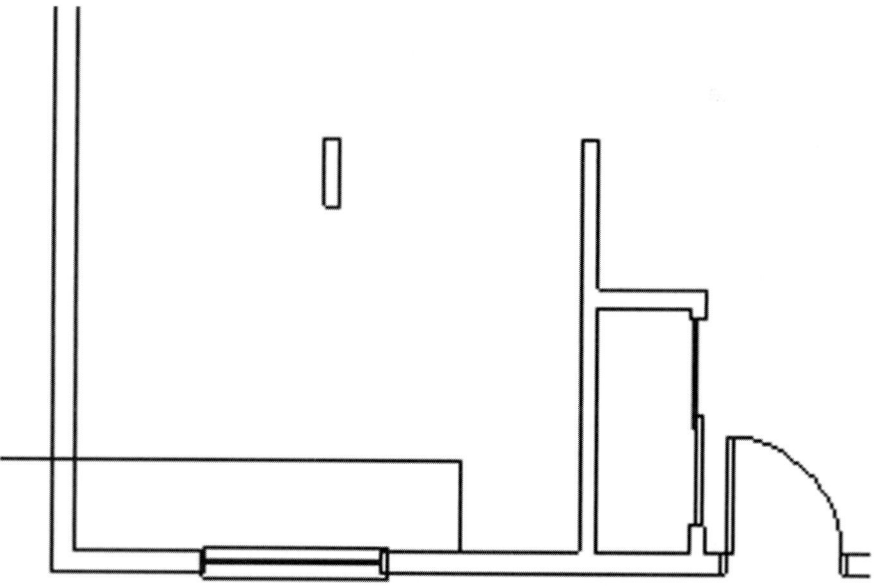

13. Drag the mouse to the left until the cursor lines up on the vertical line of the wall. *Click* the mouse to place the end of the line and then *right-click* to stop drawing the wall.
14. Let's clean up the corner of the counter now. Do you remember how to do it? Hint: **Edit/Cleanup/2LnTrim** (see Figure 4-44).

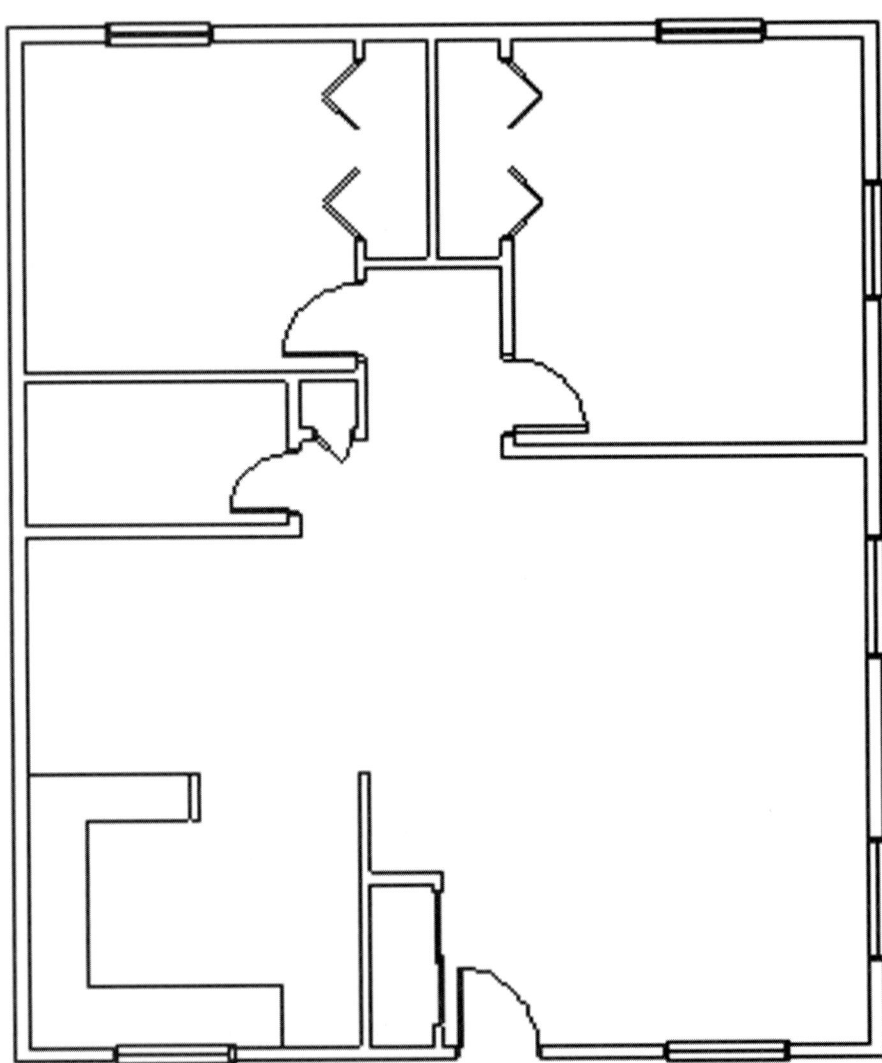

Figure 4-44 Cleaning up the counter

Linetypes

To place dashed lines over the counters to represent overhead cabinets, we need to learn about another feature of DataCAD: linetypes. It would be counterproductive to draw a dashed line as a series of short lines, and drawing something as complicated as batt insulation would likewise be quite futile. To solve the problem, DataCAD comes with 26 pre-made linetypes. You will find some additional linetypes on the accompanying CD-ROM. You can also create your own linetypes, which will be covered in a later chapter. Currently, DataCAD supports the use of up to 175 different linetypes.

A linetype is a representation of a complicated graphic that acts as a single entity. Confused? I don't blame you. So let's draw a few to see what this means:

1. Pick **Edit/LineType** (or **Alt+L**) and you will see 10 menu buttons with names ranging from **Solid** to **Plywood1**. You will also notice that a preview window pops up in the Drawing Window, allowing you to view each linetype as you pass the cursor over them.
2. Now drag your cursor over each of the buttons and notice that a graphic depiction of each linetype is displayed in the Preview Window. Obviously, these graphics make selecting a linetype much easier than by the text alone.
3. To see the rest of the linetypes available, press the **ScrlFwrd** (Scroll Forward) button. You will see 15 more linetypes, from **Plywood2** to **ShingleL**.
4. Press **ScrlFwrd** to see the remaining linetype (**LapSidL**). Now press **ScrlBack** (Scroll Back) to go back one linetype menu and then once more to get to the first linetype menu.

NOTE: *Throughout DataCAD, you will see the **ScrlFwrd** and **ScrlBack** buttons. This is how you move forward and backward through multiple menu choices. If you do not see a **ScrlFwrd** or **ScrlBack** button, then no further menus are available in that direction.*

Solid is, of course, the most often used linetype for any drawing. What you have to remember is that DataCAD will continue to draw with whatever linetype you choose until you choose another, even if you switch to drawing on a different layer. Before we choose the **Dashed** linetype for our cabinets, look just above the **SWOTHLUD** acronym at the bottom of the Status Area. You should see the word "**Solid**." This tells you the current linetype that you are drawing with. When you change linetypes, this readout will change as well.

5. Now press the **Dashed** button in the **LineType** menu. The linetype readout in the Status Area changes to "**Dashed**," and the Prompt line reads, "*Linetype* = Dashed. Current color = LtBlue."
6. Some options exist at the bottom of the **LineType** menu. The only one we are concerned with at the moment is the one that says **Spacing**. Press it now. The Command Line will read *"Enter new line spacing (.01 to 9999): 1.0."* **1.0** stands for 1′-0″ [305] and means that the dashes in the linetype are spaced apart at 1′-0″ [305] on center. We want to make them appear only 6″ [152] apart.
7. Type in **.6** [152] and press **Enter**. Then *right-click* to return to the **Edit** menu.
8. Press RefPt (′) and snap to the upper-left corner or the small kitchen wall.
9. Press the **Space** bar and enter a distance of **1** [305] and an angle of **270** (or **−90**).
10. Drag the mouse to the left and end the line at the kitchen wall by *clicking* the mouse. *Right-click* to stop drawing lines.
11. Now press RefPt (′) and snap to the intersection of the left-hand kitchen wall and the line you just drew. Press the spacebar and enter a distance of **1** [305] and an angle of **0**. Drag the mouse down and end the line at the lower wall of the kitchen and *right-click* (see Figure 4-45).

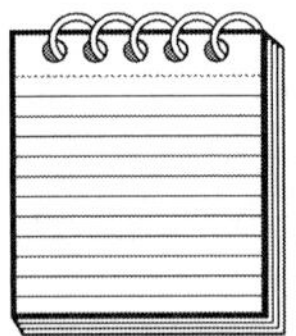

NOTE: *In some cases, it may look like the ends of your dashed lines do no actually touch the walls you drew them to. That's just because the endpoint of the line coincidentally ends at one of the spaces instead of one of the dashes in the line. To prove that the line truly does end at the wall and is still just as long as you originally drew it, use the Identify function to see the full extent of the line:*

- *Press the **I** button in the Navigation Pad.*
- *Click on the vertical dashed line you just drew or any other line that may look like it doesn't end at the wall you drew it to.*

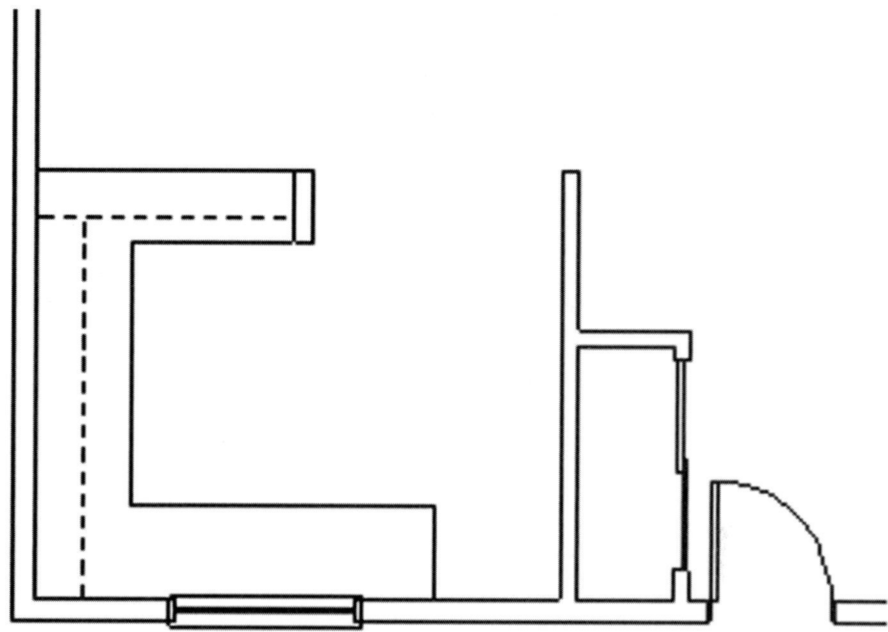

Figure 4-45
Drawing a dashed linetype

- *The line will temporarily change to a tighter, dashed white line, showing you the true dimensions of the line. If you ever have a question about an entity, try using the Identify option to check it out.*
- *Right-click to exit the Identify function.*

Back to drawing our cabinets.

12. Use the RefPt (') key and the *middle-snap* function to create three more dashed lines to complete the cabinet lines to look like this illustration. Notice that I drew the two small lines from the kitchen window so that they extended past the long horizontal line above the window. This will make it easier to clean up the corners to interrupt the cabinets at the kitchen window (see Figure 4-46).
13. Use the Partial Erase function (**Ee/Partial**) and *middle-snapping* at the corners to remove the section of horizontal dashed line in front of the window (see Figure 4-47). Remember to read the Command Line for information on what DataCAD needs you to do for each step. When all else fails, read the directions!

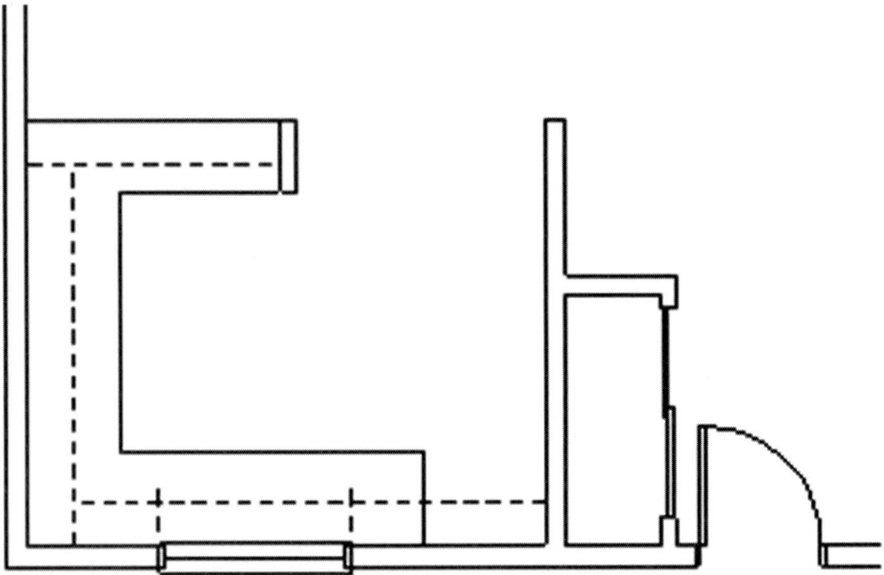

Figure 4-46
Adding three more dashed lines

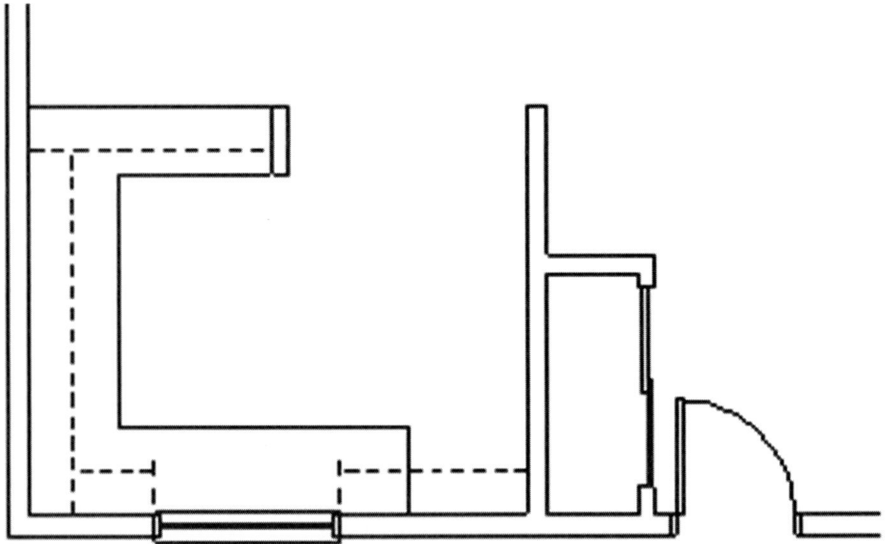

Figure 4-47
Removing the dashed line in front of the window

14. Now use the 2 Line Trim function (**Edit/Cleanup/2LnTrim** or the Cleanup icon in the icon bar) to cleanup the corners of the cabinets (see Figure 4-48).

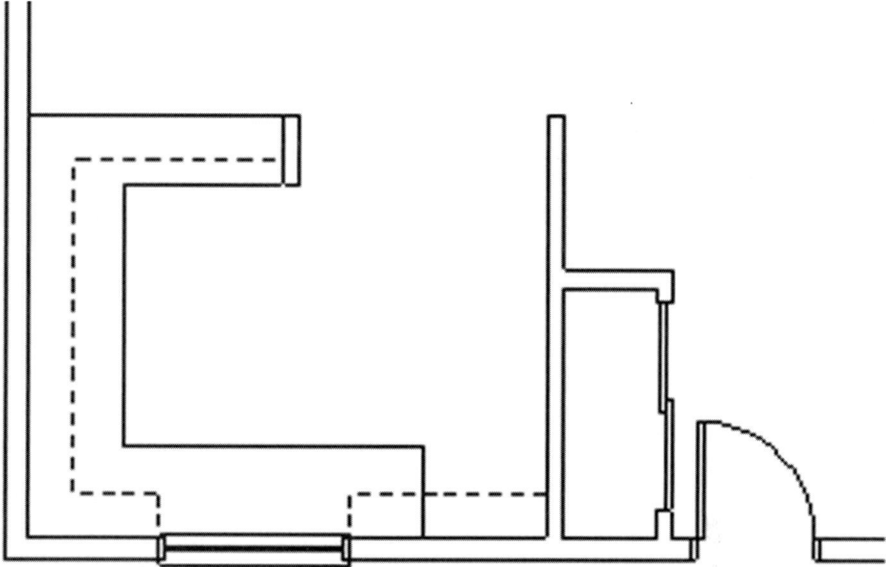

Figure 4-48
Cleaning up the cabinet corners

Remember that when using the **2LnTrim** function you select the part of the lines that you want to keep. Let's add the counter in the bathroom now, but first we need to change back to drawing a solid linetype.

15. We can do it by choosing **Edit/LineType/Solid** or by another little shortcut. Try this instead.
16. By pressing the **Qq** key, you can toggle through the various linetypes available. By using the **Shift+Qq** trick, you can go backward through the linetypes. Press the **Shift** key and hold it down. Now press the **Qq** key twice to go backward through the linetypes. Look at the Linetype readout in the Status Area as you press the **Qq** each time. You will see the linetype change from Dashed to Dotted to Solid. Now you will be drawing solid lines again.
17. Now zoom into the bathroom so it will be easier to work on (use the **W** key in the Navigation Pad or press the / key). *Right-click* to exit the **WindowIn** menu.
19. Use RefPt ('), 2 Line Trim, and/or Partial Erase to draw a counter that is 3′-8″ [111] long and 1′-9″ [533] deep (see Figure 4-49).

Now use what you know about linetypes, RefPt, and cleanup to add the shelves in the four closets. Locate the front edges of the shelves at one foot,

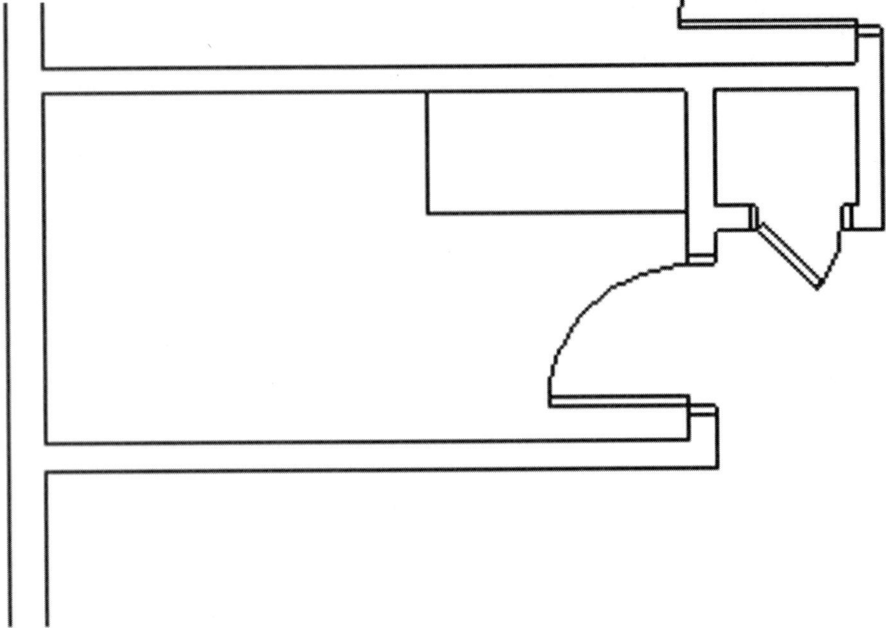

Figure 4-49 Adding the bathroom counter

two inches off the back wall using a solid, light blue line on the Cabinets layer. Don't draw in the closet poles (dashed lines) yet. When you are done, your closets should look like the ones at the apartment entrance in Figure 4-50.

Offsetting Lines

Here's another little trick for creating copies of lines at an offset distance. We'll use it, along with the Change function, to create the closet poles:

1. Zoom into the closet at the apartment entrance.
2. Go to the **Geometry** menu by either pressing **Utility/Geometry** or by pressing **Alt+G**.
3. Now pick **Offset**. If the **Dynamic** button is on, turn it off by pressing it.
4. Press **PerpDist**. The Command Line will read, *"Enter perpendicular distance to offset: 1.0"*. Type in **.1** [25.4] and then press **Enter**. The Command Line will now read, *"Select object to offset."*
5. Click on the LtBlue line of the closet shelf. It will temporarily change to a dashed line, and the Command Line will read, *"Offset to which side?"*

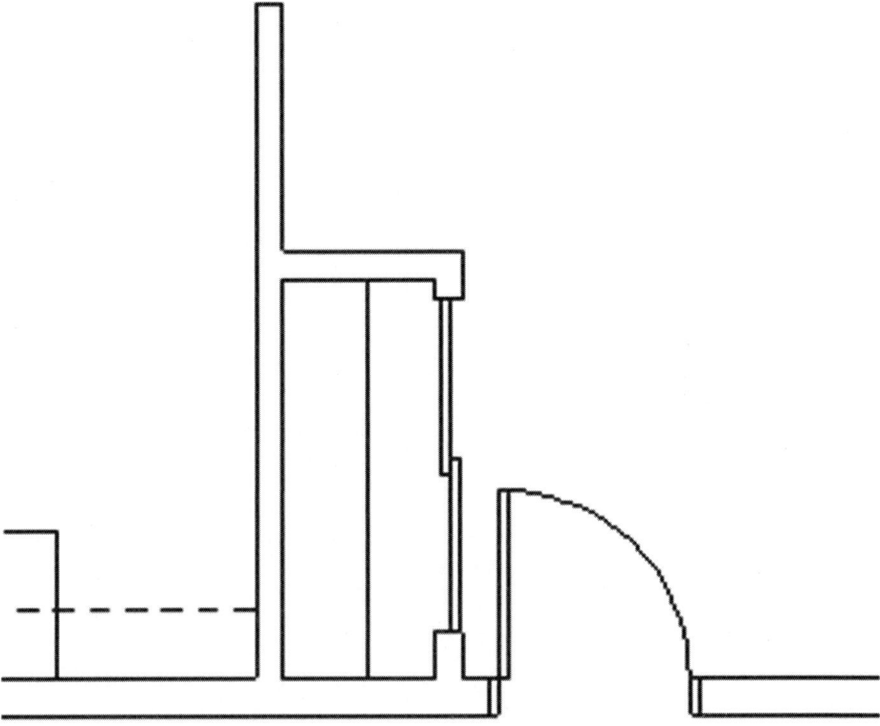

Figure 4-50 *Adding shelves in the closets*

6. Locate the cursor anywhere to the left of the shelf line and click the mouse. A new line matching the shelf line will be drawn 1″ [25.4] to the left of the shelf (see Figure 4-51).
7. The Command Line will read *"Select object to offset."* We want to draw another offset line from the new line but at a distance of $1^1/_2$″ [38] instead.
8. To do this, press the **NewDist** button and then the **PerpDist** button. The Command Line will read *"Enter perpendicular distance to offset: .1."*

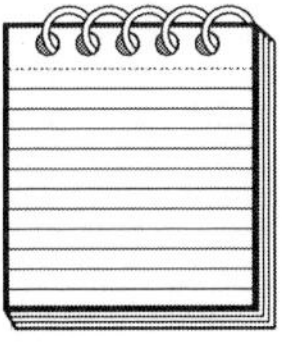

NOTE: *Notice that the default distance is now* ***.1*** *(1″ [25.4]). The last time you selected it the default was* ***1.0*** *[305]. This is because the default automatically becomes the last distance that you entered. If you had entered* ***6.7*** *[2007] instead of* ***.1*** *[25.4], then* ***6.7*** *[2007] would now be the default. Most of DataCAD's default functions work this way.*

9. Type in .1.1/2 [38] and then press **Enter** (for $1^1/_2$″). The Command Line will now read *"Select object to offset."*

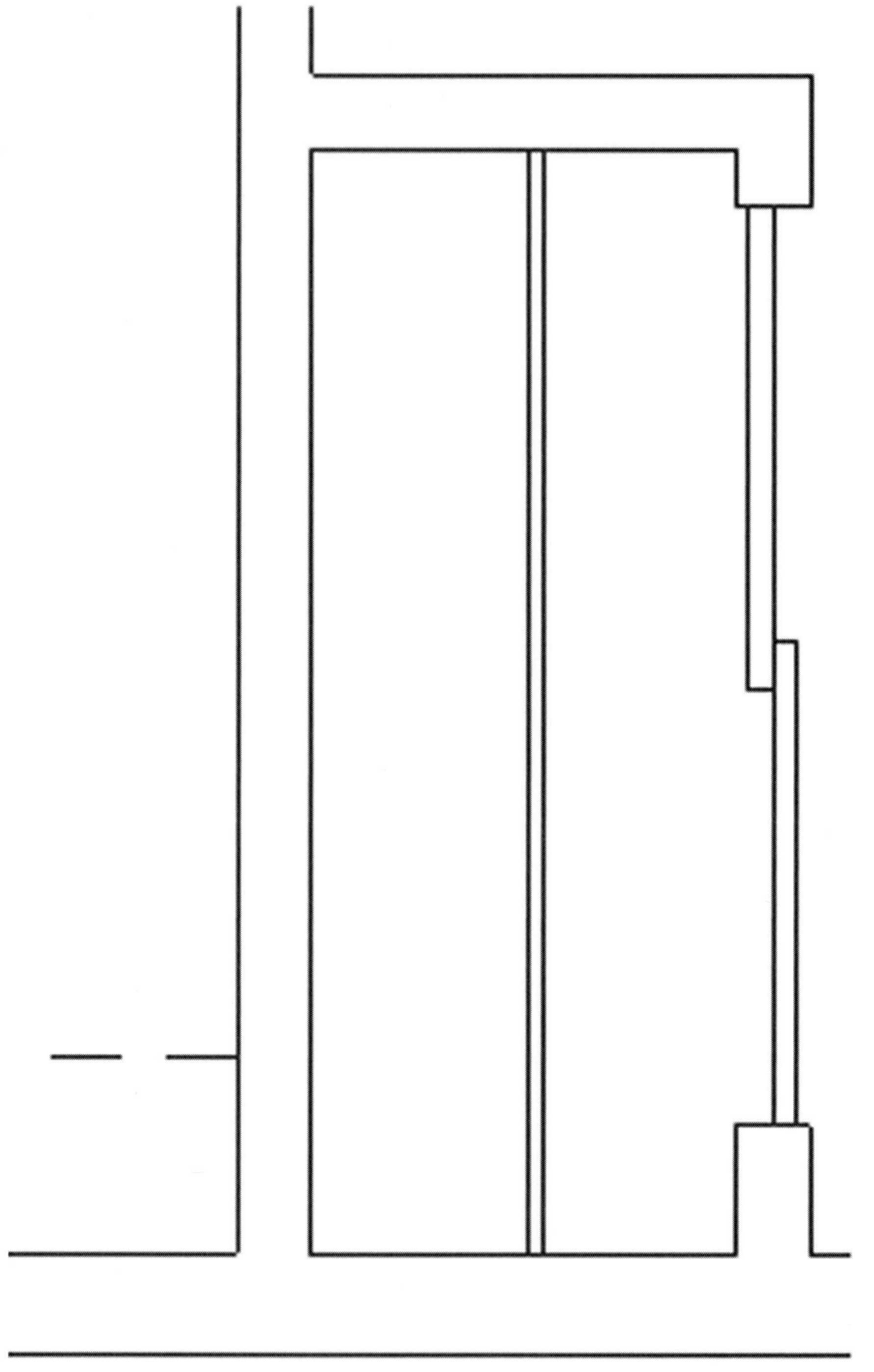

Figure 4-51
Creating the closet pole

10. Click on the LtBlue line of the new line we just drew. It will temporarily change to a dashed line and the Command Line will read *"Offset to which side?"*
11. Locate the cursor anywhere to the left of the line and click the mouse. A new line matching the new line will be drawn $1^1/_2$″ [38] to the left of the shelf (see Figure 4-52).
12. *Right-click* back to the **Edit** menu.
13. Select **Change/LineType/Dashed**. This will enable us to change the closet pole lines to dashed lines. But we want to graphically make the dashed lines tighter together than the 6″ [152] spaced dashed lines we use on our kitchen cabinets. Let's use 3″ [76] instead.
14. To do this, press the **Space** bar. Type in **.3** [76] and then press **Enter**. The Command Line will read *"Select entity to <CHANGE>."*
15. *Click* on the two lines we just drew to represent the closet pole. *Right-click* to stop changing entities (see Figure 4-53).

Now look down at the Linetype readout in the Status Area. Notice that it still says **Solid**. Using the **Change** menu to change lines to a dashed linetype did not affect the current linetype setting. Remember that the **Change** options only change the entities you are working on and do not change your current settings. Use these same techniques to add closet poles to the two bedroom closets as well (see Figure 4-54).

Symbols

We could draw the tub, toilet, sinks, refrigerator, and stove from scratch, but that would waste a lot of time. When you draw plans with a pencil, you use plastic templates to quickly draw those fixtures because it's much easier and faster. Likewise, it is much easier and faster to place ready-made symbols into your CAD drawings as well.

First, you must understand the concept of a template versus a symbol in DataCAD. It's the same concept as the plastic template. The template is the thing that holds all the symbols, like toilets, sinks, and so on. In DataCAD, you will display various templates that contain various symbols.

A symbol exists as a single entity, just like a single line. Unless you place a symbol in the drawing "exploded" (more on this in later chapters), then you cannot edit any of the lines that make up that symbol.

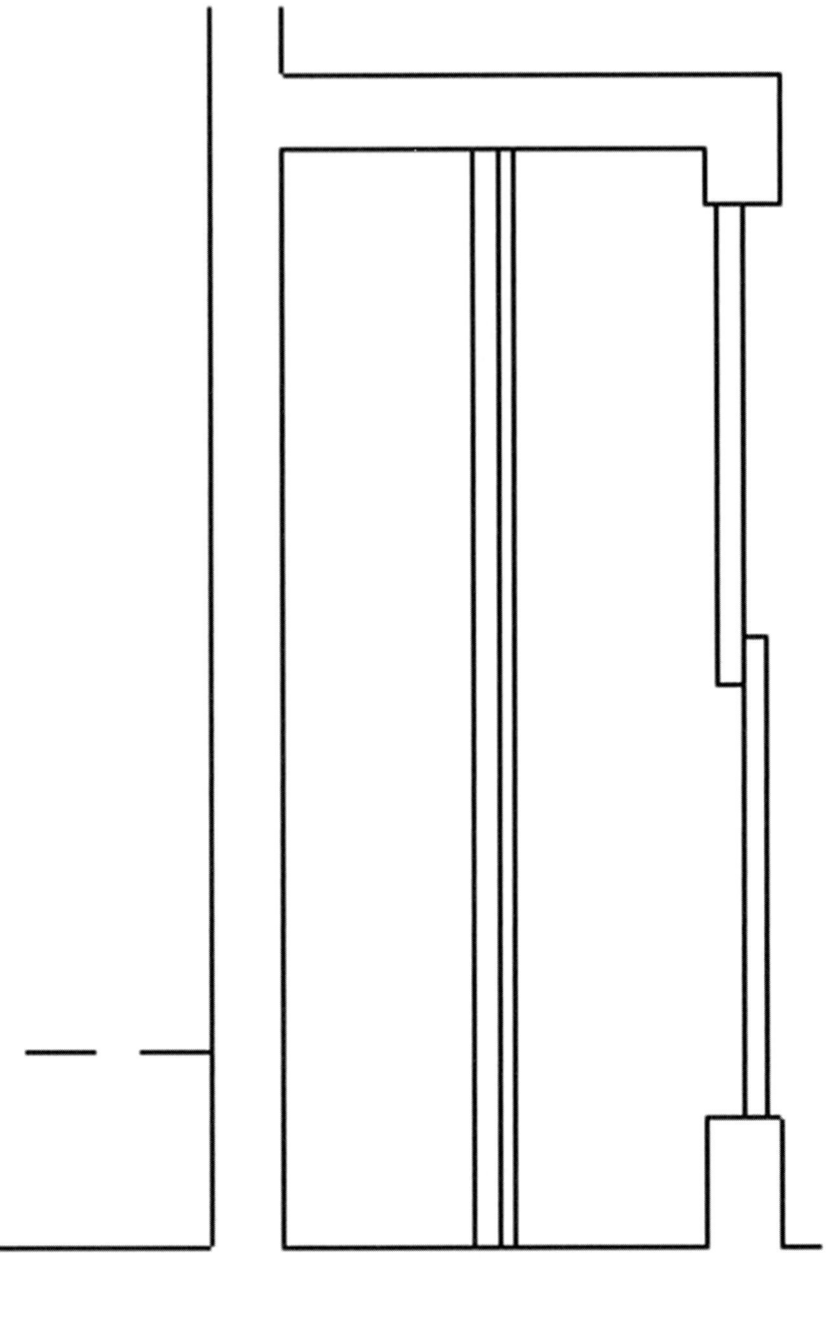

Figure 4-52
Offsetting the second line of the closed pole

Figure 4-53
Finishing up the closet pole

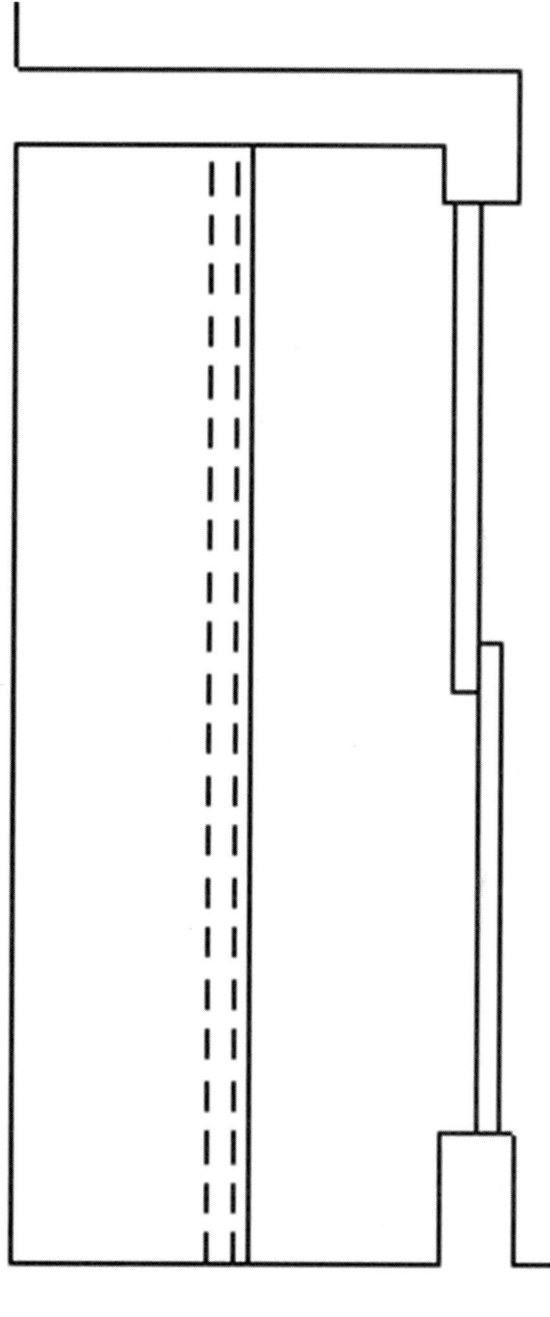

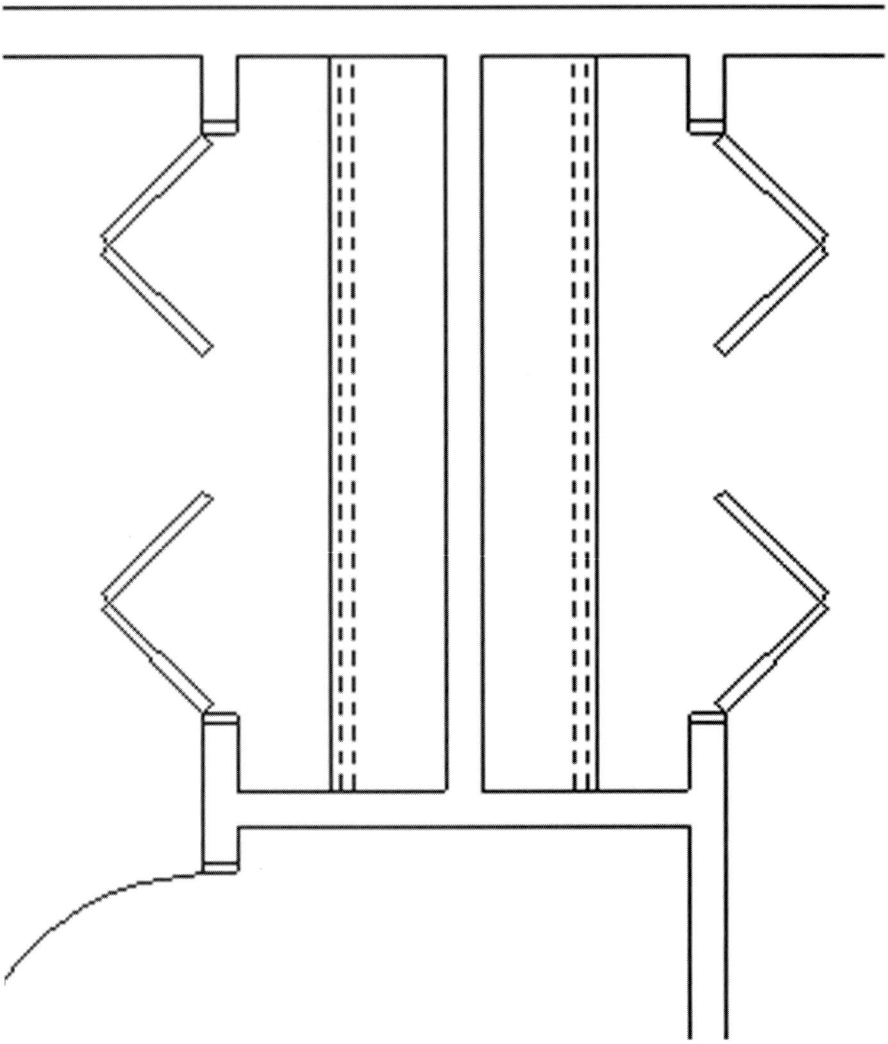

Figure 4-54
Adding more closet poles to the bedroom closets

NOTE: *One really great feature of DataCAD is the way it "instances" symbols. Once a symbol is placed in the drawing, the program holds this symbol in its database. If you add 2, 20, or even 200 instances of the same symbol in your drawing, the size of your drawing file will not get larger. This is because DataCAD simply references the first instance of that one symbol already stored in its database and just displays another copy of it in all the additional locations. This helps keep the size of your drawings down, making it faster and easier to work with.*

We want to place all our new plumbing symbols on the Plumbing layer, so we need to make it the active layer:

1. Hold down the **Shift** key and press **Tab**. The cursor will change to LtGreen and the name of the layer will be shown in the Status Area as **Plumbing**. Let's add the tub first.
2. Press **Utility/Template** (or you can use the shortcut **Tt** to get there). A dialog box like the one shown in Figure 4-55 will be displayed:

   Hundreds of templates and thousands of symbols come with DataCAD, and you will invariably create more of your own later, so it's not very practical to remember the names of all your templates. Instead, we'll search through the list of templates on your hard drive.
3. Press the down arrow to the right of the **Look in:** box. A directory tree of your computer will be displayed.
4. Click on the \TPL folder in your DataCAD directory. Depending on whether you installed the CSI format templates or not, you should see something like Figure 4-56.

   These folders are all of the main template files available to you. You might recognize the numbers and headings of these directories as those used by CSI and Sweet's Catalog. That makes it much easier to find the

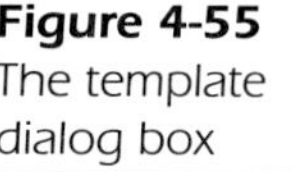

Figure 4-55
The template dialog box

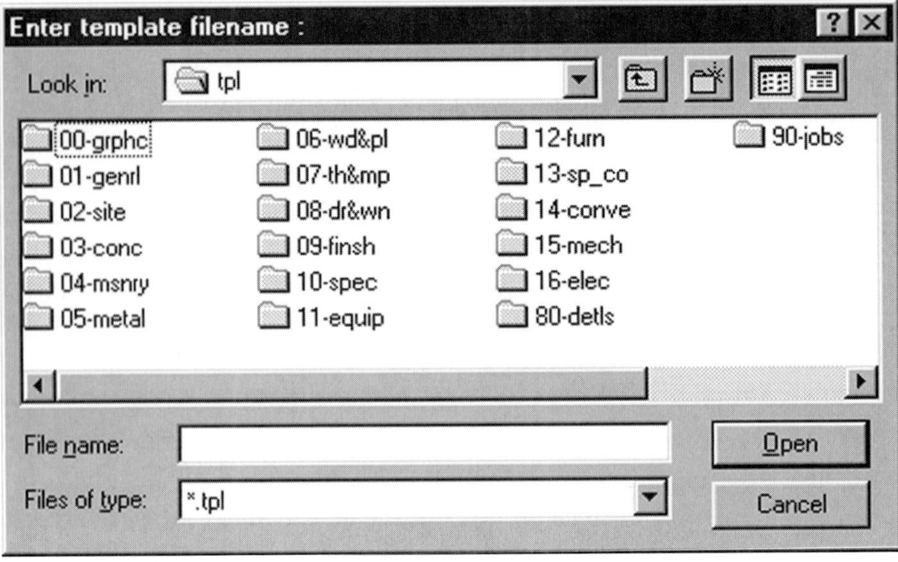

Figure 4-56 The various folders available in your directory

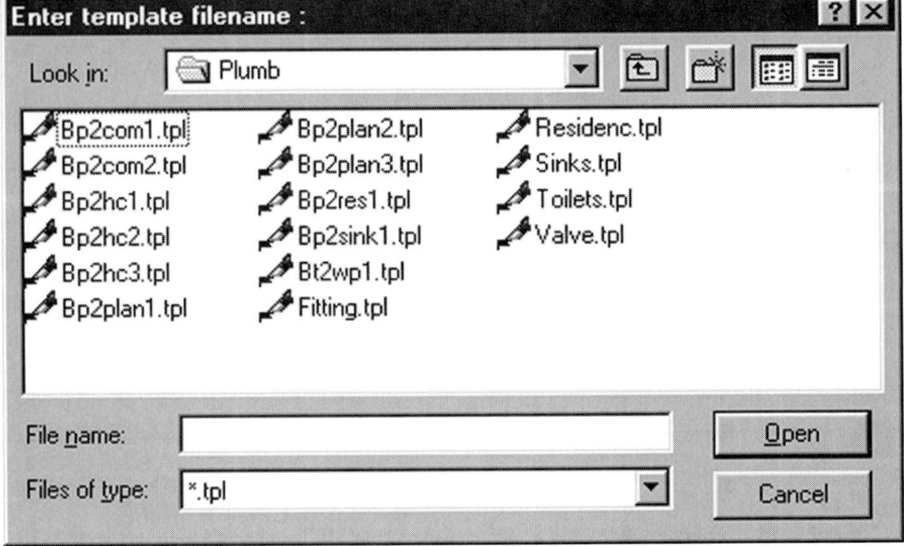

Figure 4-57 The template files in the PLUMB folder

template you are looking for. We want a tub, so we need to go to the **15-MECH** (Section 15 - Mechanical) directory.

5. *Double-click* on the **15-MECH** folder. You will see two more folders. *Double-click* on the **PLUMB** folder. A list of the template files in that folder are displayed, as shown in Figure 4-57.

The .TPL files you see correspond to various plumbing templates. They are a bit cryptic (we have a custom program on the CD to fix that), so you'll have to experiment a little to get the idea of what is in each one. For now, you'll have to take my word for which one to look in.

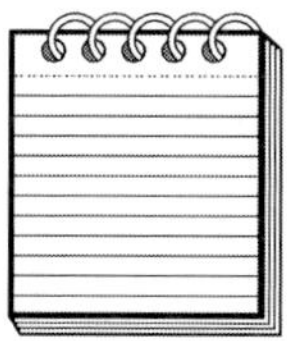

NOTE: *In Figure 4-57, each template filename is preceded by an image of a pen. Your icon will be different. It is a pen on my computer because I have associated the template (.TPL) files with the Windows Write word processing program so that I can easily edit my templates.*

6. *Double-click* on **BP2RES1.TPL.** A visual template full of symbols will appear at the far right of your screen, as shown in Figure 4-58.

   You have now opened up a template that contains 14 symbols. It is very important to distinguish between these two words.
7. A new group of buttons is also displayed in the menus. Don't worry about the majority of these buttons for now. Just make sure these ones are set as follows:
   a. **DynamRot** = Off
   b. **Explode** = Off
   c. Down in the Information Line, you should see a message that says, among other things, *"Enlargement*=1.0 × 1.0 × 1.0." If not, then press **Enlarge/SetAll** and type **1.** Then press **Enter** and *right-click*.
8. Place the cursor over the lower-left bathtub symbol. The Information Line will tell you the name of the symbol: **BATHTUB**.
9. *Click* on it. You will see a box attached to your cursor. This represents the outside limits of the tub symbol. For this symbol, the insertion point is the upper-left corner of the tub, so you can snap your cursor to the upper-left corner of the bathroom to place it.
10. Snap (*middle-click*) to the upper-left corner of the bathroom and the tub is inserted (see Figure 4-59). Notice that the cursor still has the tub symbol attached to it so that you can continue to place more of them if you want to.
11. *Right-click* to stop placing tub symbols.

To place the toilet, we'll draw a guideline that we can snap to. You do not have to leave the Template menu to draw a line. Remember, DataCAD always defaults to drawing lines. It wants to let you draw lines whenever possible. In order to place the toilet, follow these steps:

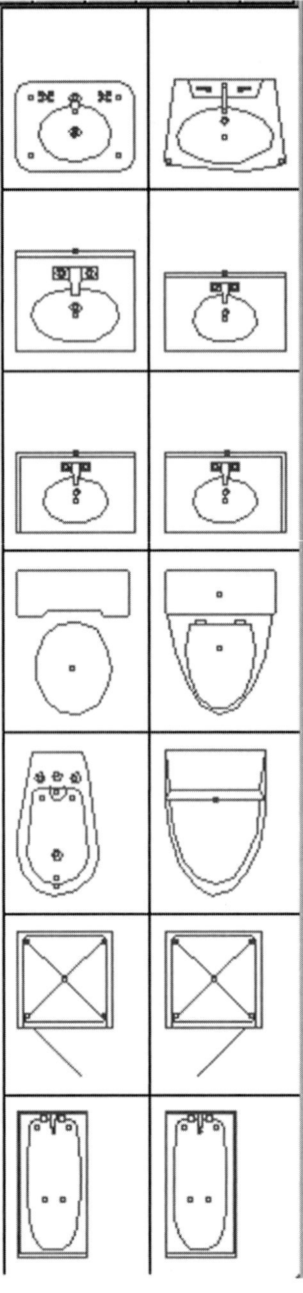

Figure 4-58
The symbols within the BP2RES1.TPL file

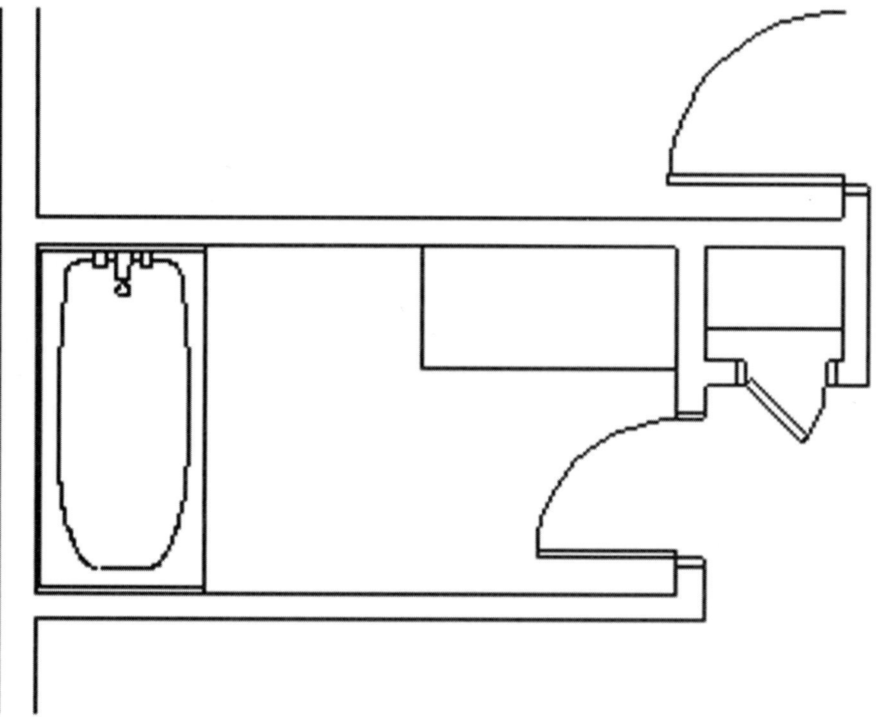

Figure 4-59
Inserting the tub

1. Use RefPt (′) and snap to the upper-left corner of the counter where it intersects with the bathroom wall.
2. Press the **Space** bar and enter a distance of **1.6** [457]and an angle of **180**.
3. Drag the line down about a foot, *click*, then *right-click*. It's just a temporary guideline so that you have a point to snap to when placing the toilet symbol.
4. In the same template, click on the Standard Toilet symbol, the fourth one down on the left side of the template (remember to look at the Information Line for the name of the symbol).
5. *Snap* to the point where your guideline intersects the wall (see Figure 4-60). *Right-click* to stop placing toilets.
6. Press **Ee** and then pick by **Entity** to erase the guideline. To place the sink, we'll draw another guideline for snapping.
7. Use RefPt (′) and snap to the upper-left corner of the counter where it intersects with the bathroom wall.
8. Press the **Space** bar and enter a distance of **1.2** [356] and an angle of **0**.

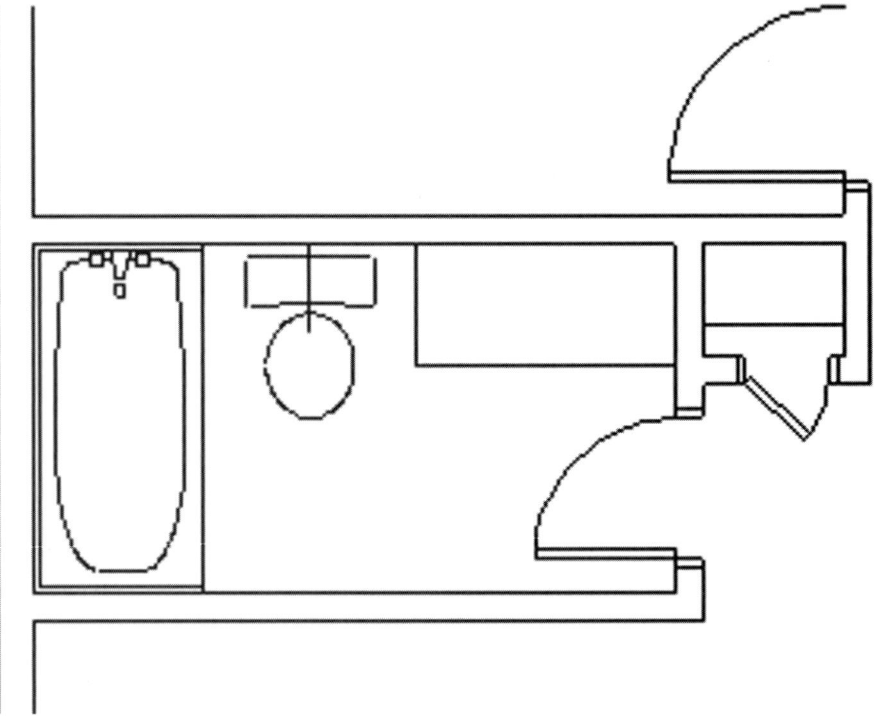

Figure 4-60
Placing the toilet

9. Drag the line down about a foot past the edge of the counter (this will make it easier to erase), *click*, and then *right-click*.

   Now we have to open a new template file to pick the sink.

1. In the **Template** menu, press **NewFile**. The list of templates will appear. It is the same list you chose the previous template from.
2. Now *double-click* on **BP2SINK1.TPL** and a new set of symbols will appear.
3. *Click* on the one in the upper-right corner called **BIPS1-2**.
4. The Information Line will read something like **SYM\BIPS1\BIPS1-2**. Whoever made this particular set of symbols saved them with the full directory path name. \SYM is the DataCAD symbol directory, and BIPS1 is a subdirectory within it. The last item, BIPS1-2, is the actual name of the symbol.
5. *Snap* to the intersection of the guideline and the wall (see Figure 4-61). *Right-click* to stop placing sinks.
6. Use **Ee/Entity** to erase the guideline.

Figure 4-61
Placing a sink

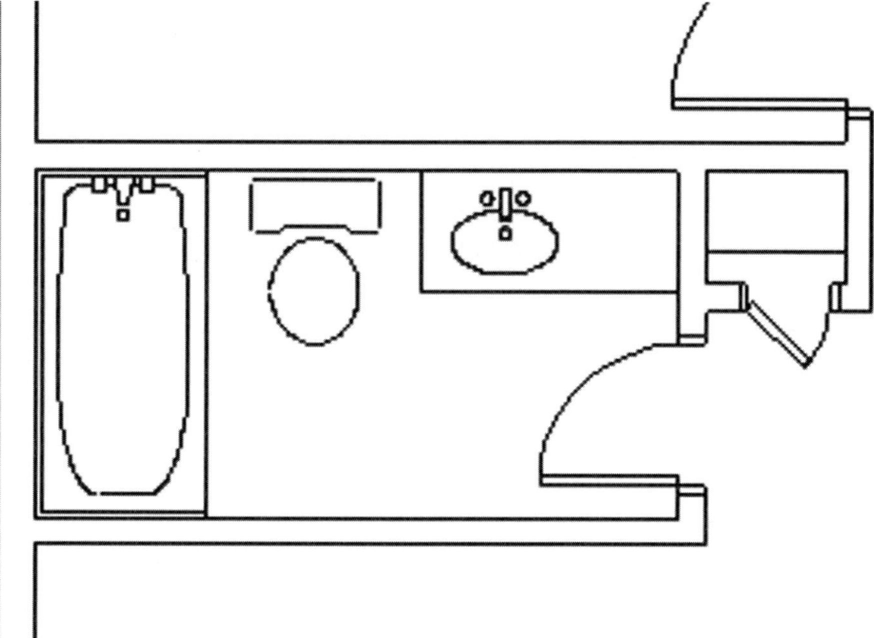

Now we'll add the kitchen symbols.

1. Zoom into the kitchen area.
2. In the **Template** menu, pick **NewFile** and *double-click* on **RESIDENC.TPL.** A new set of symbols is displayed.
3. Now pick **DynamRot** (Dynamic Rotate) in the menu. This will enable you to freely rotate your symbol before finally placing it.
4. We will be using the fourth symbol down from the top. The symbol is called **Wardrobe**. Why? Beats the heck out of me, since it's a refrigerator (though I suppose it could double for a freestanding wardrobe closet).
5. The normal insertion point for the symbol is the center of the back of the refrigerator. To see this, pick the refrigerator symbol (fourth from the top) and drag it into the Drawing Window. The cursor is attached to the center back of the refrigerator symbol. This is the insertion point.
6. *Right-click* to cancel the symbol placement.

To more accurately place this symbol, we need to do two things: pick a different insertion point and turn the symbol around 180 degrees in the kitchen plan.

In DataCAD, you are not forced to use the insertion point that was defined when the symbol was created. You can use the *middle-snap* function to snap to any of the object-snap points currently defined in your **Utility/ObjSnap** menu (centerpoints, midpoints, intersections, and so on) in the symbol. This becomes the new temporary insertion point for the symbol (it does not change the original insertion point).

The Dynamic Rotate option will enable you to change the orientation of the symbol on the fly (dynamically) so that you don't have to worry about a symbol's orientation when you place it. Place the insertion point where it should be, then rotate it dynamically around that point:

1. *Snap* to the upper-left corner of the refrigerator symbol. You will see that the insertion point is now at the corner of the symbol instead of the center back.
2. To place the refrigerator, snap to the lower-right corner of the kitchen to place it.
3. Drag the mouse just to the right of the symbol and you will see a line forming a handle attached to the symbol, extending to the right, and ending at your cursor. This handle enables you to turn the symbol. Think of your cursor as a hand on the handle. Rotate the handle and the symbol will rotate with it (see Figure 4-62).
4. Make sure **Ortho** mode is turned on (press **Oo**).
5. Rotate the handle 180 degrees so that the symbol is correctly oriented.
6. *Click* to finish placing the symbol. *Right-click* to stop placing refrigerators. The insertion point for the next symbol, the double sink, is at the center back, which is where we want it, but we still have to rotate it, so leave **DynamRot** on. We'll use the center of the window sill to locate the symbol.
7. Pick the double sink symbol, the second symbol from the bottom. It's appropriately called **Sink**.
8. Snap to the center of the window sill, rotate the symbol 180 degrees, and *click* to place it (see Figure 4-63). *Right-click* to stop placing sinks.

To place the final symbol, the stove (third from the bottom), we need to draw a guideline to

1. *Snap* to the center of the vertical line of the counter and draw a short line to the right.
2. *Snap* to the bottom center of the stove symbol; that's the line that makes up the front handle of the oven door. This will be the insertion point.
3. *Snap* to the intersection of the guideline and the counter.

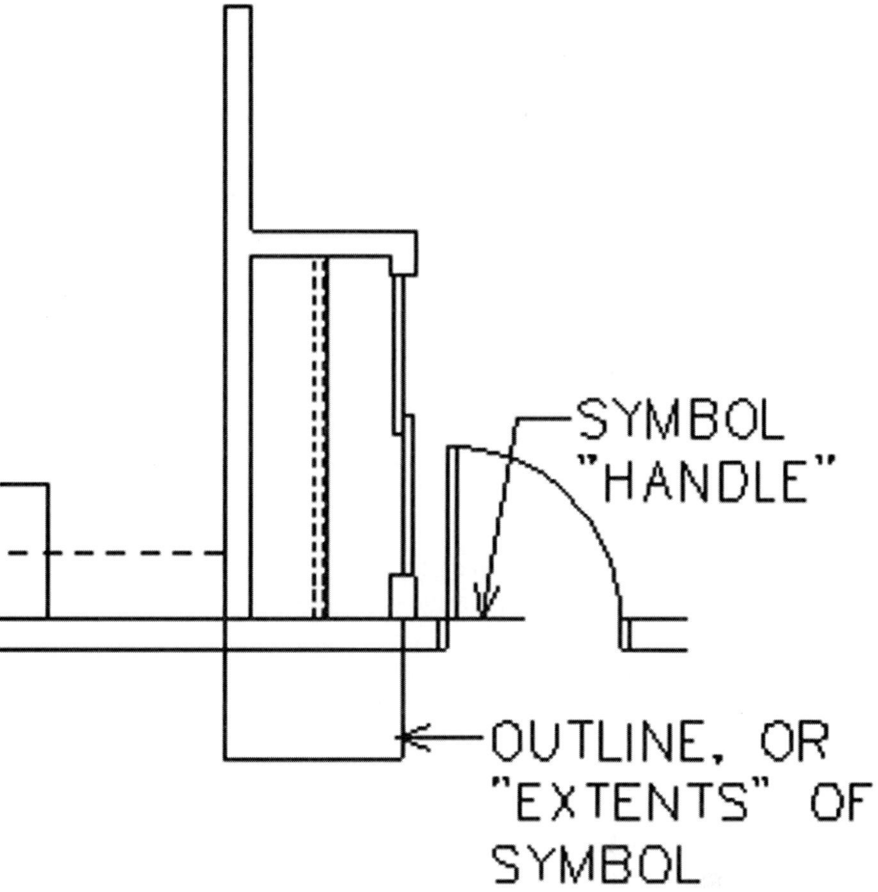

Figure 4-62
Using the handle to turn the symbol

4. Rotate the symbol into place and click to place it. *Right-click*.
5. Use **Ee/Entity** to erase the guideline (see Figure 4-64).

   We're done placing symbols, so we need to turn off the display of the symbol template.
6. Press **TemplOff** (or **Alt+B**) to turn off the display of the template and exit the Template menu.

Hatching

Hatching has more than one meaning in CAD. If you are used to drawing by hand, you probably think of hatching as a graphic convention to show

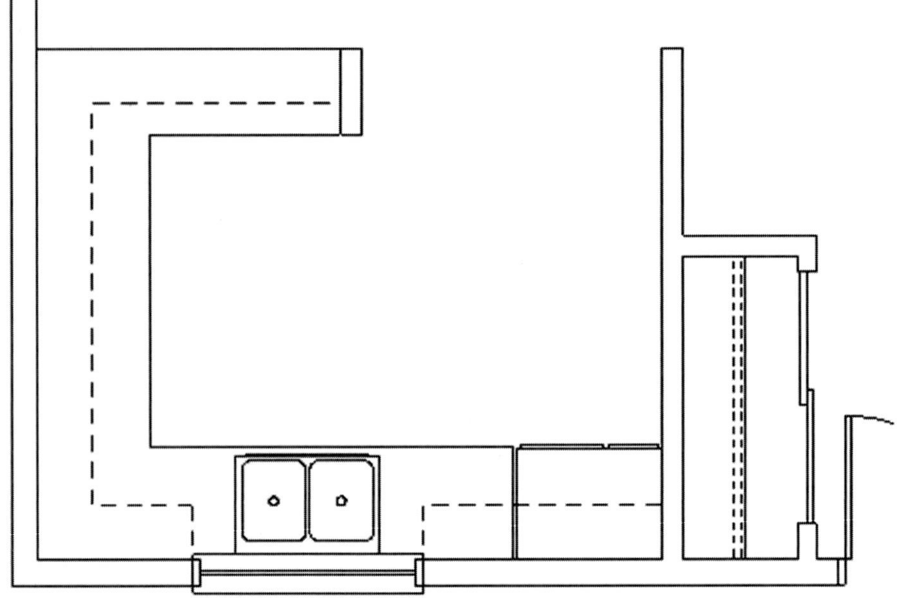

Figure 4-63
Placing a double sink symbol

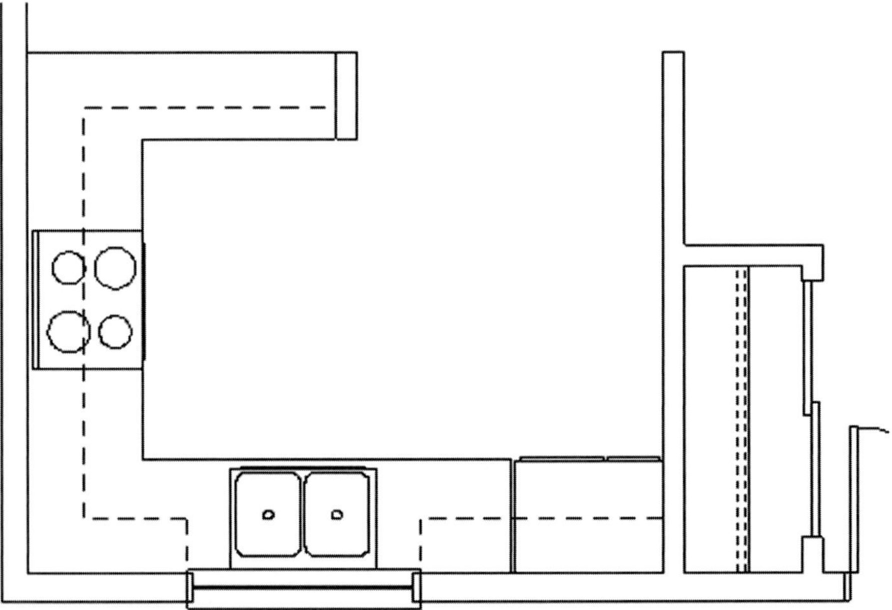

Figure 4-64
Last but not least, the stove is placed.

materials in section, like brick, CMU, steel, or earth. In CAD, hatching is used for that same purpose, but is also used to make literal representations of materials, like ceramic tile in plans and clapboard siding in elevation.

Hatching in CAD can save you a lot of tedious drawing time. For instance, we are going to add some 4″ × 4″ [102 × 102] ceramic tile to the floor of our kitchen plan. Drawing the pattern of ceramic tile on the floor involves drawing quite a few horizontal and vertical lines. By using the Hatching function in DataCAD, you can simply define the outside edges of the ceramic tile area and the program will draw all the lines for you. Now think of how long it would take you to draw a brick pattern by hand on an elevation. By using hatching, you just define the outline of the area to be shown as brick, and then the program draws all the brick. Two kinds of hatching exist in DataCAD: associative hatching and non-associative hatching.

Associative hatching is very much like a symbol. It is treated as one whole entity, rather than a bunch of separate entities. Every line within the hatch is associated with every other line, and you cannot edit the individual lines and arcs within the hatch. This has some distinct advantages. Let's say you decide to replace the 4″ × 4″ [102 × 102] tile that is currently in your kitchen plan with a 2″ × 2″ [51 × 51] pattern. All you have to do is point to the current 4″ × 4″ [102 × 102] associative hatch and tell DataCAD to replace it with a 2″ × 2″ [51 × 51] pattern and the whole hatch will be updated in one step with no erasing or redrawing. Think of how much time that could save if you wanted to change all the brick in the previous example to clapboard siding.

Non-associative hatching is drawn the same way as associative hatching; you define the edges and then the program draws all the lines for you. The difference is that all the lines of the hatching are drawn as separate entities. To erase the hatching, you must erase each line individually. The lines are not associated with one another, so you lose the ability to change or update the hatching like you can with associative hatching.

Let's add an associative hatch pattern of 4″ × 4″ [102 × 102] tile to our kitchen plan:

1. Zoom into the kitchen area.
2. Press the **Tab** key until your active layer is the **Hatching** layer. Your cursor will turn to white.
3. Pick **Utility/Hatch** (or **Hh**). Make sure **Associat** is on.
4. Pick **HtchType/NoOutLin** and *right-click*.
5. Choose **Pattern**. A menu full of hatch patterns will be displayed along with a Hatch preview window, as shown in Figure 4-65.

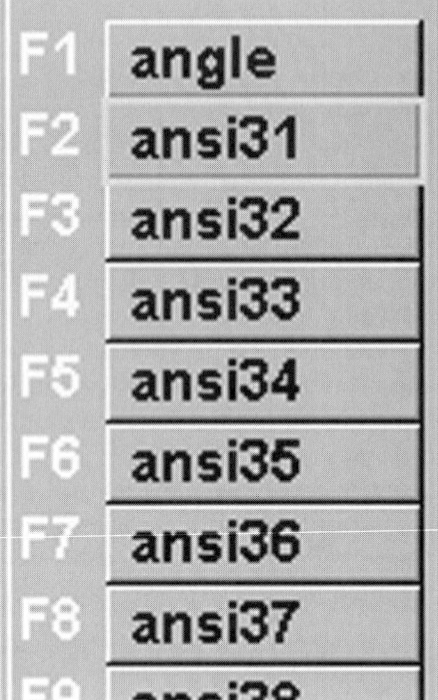

Figure 4-65 The Hatch preview window

As you locate your cursor over a hatch pattern button, the Preview Window will show you what the pattern looks like.

   a. If the pattern appears too large or too small, go back to the main **Hatch** menu, pick **Scale**, and select a larger number, like 100.

6. Press **ScrlFwrd** until you see the hatch pattern called **4x4til**. Click on it.
7. Pick **Scale** and type in **1** [305]. Then press **Enter**.

NOTE: *Selecting an appropriate scale for each hatch pattern is very important. In this case, the 4″ × 4″ [102 × 102] tile pattern is designed to be 4″ × 4″ [102 × 102] at a scale factor of 1. If you choose a scale factor of 2, then the tile will be displayed at twice the size, or 8″ × 8″ [204 × 204]. A factor of 3 would be three times as large, and so on.*

Not all hatch patterns have an obvious scale factor, however. For instance, the ***brick*** *hatch pattern requires a scale factor of 32 in order to display the true dimensions of standard brick.*

To avoid some of the trial and error of figuring out which scale factors to use, the Appendix shows you not only all of DataCAD's standard hatches, but many suggested scale factors for displaying various materials at various scales.

8. Press **Boundary**. The Command Line reads *"Select first point on boundary to hatch to."* This is the command you will use most often when creating a new hatch area. The boundaries of the tile hatch pattern are the boundaries of the kitchen floor, so we can use the *middle-snap* button to latch onto most of the points we need in order to define the pattern. A hatch pattern must be defined by three or more sides, forming a closed polygon.
9. Make sure that **Ortho** (**Oo**) is on. It will make placing the hatch easier.
10. Snap and click to the six points shown in Figure 4-66. Do not snap back on Snap 1.
11. After completing Snap 6, you need to close the polygon (from Snap 6 to Snap1). You can do this in one of two ways. Either *right-click* or press the **Close** button. I prefer to *right-click* since it's quick and easy.
12. Your hatch pattern is not visible yet. The final step is to press **Begin**. The hatch will appear.

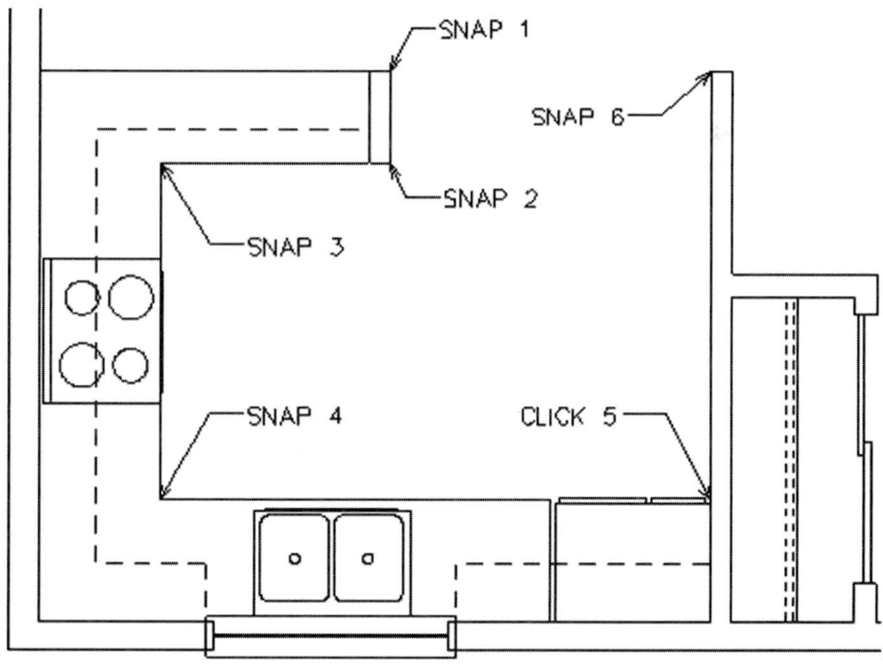

Figure 4-66 Defining the hatch boundary

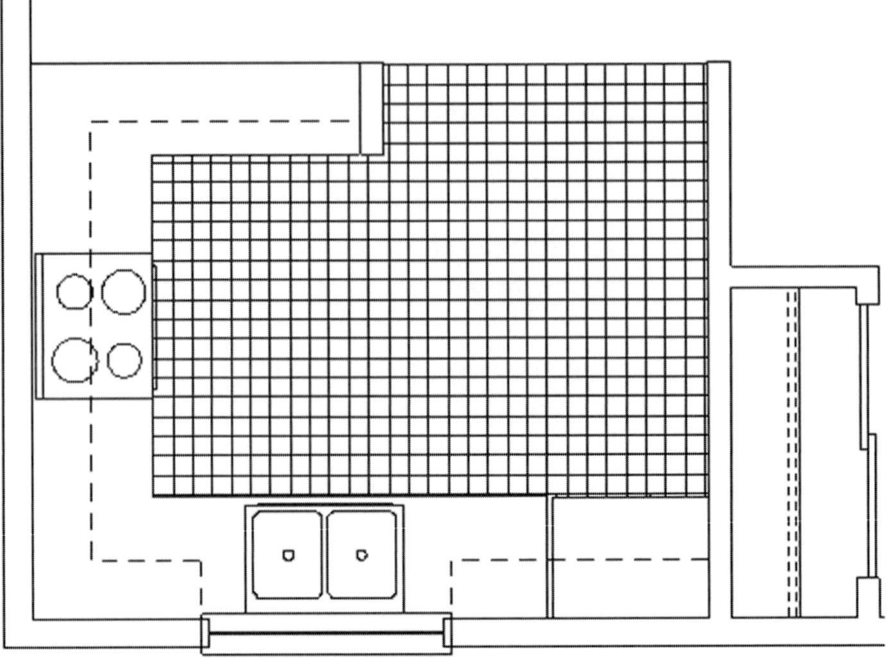

Figure 4-67 The completed tile floor

13. Press **Esc** to refresh the screen (see Figure 4-67).

    To see how useful an associative hatch can be, let's change the tile pattern to a 12″ × 12″ [305 × 305] pattern. If you had drawn all of this tile on paper with pencil, or if you had drawn it in CAD with a non-associative hatch, then you would have to erase each and every line of the tile, and then redraw the new pattern.
14. While in the **Hatch** menu, pick **Pattern**, scroll forward to the **12x12til** button, and click on it.
15. Now pick **Entity** from the **Hatch** menu. The Command Line reads, *"Select an entity to <Hatch>."*
16. Click on any part of the tile floor hatch pattern. The pattern will change to dashed lines to indicate that you have selected it.
17. Press **Begin** and the hatch will be updated.
18. *Right-click* to exit the **Hatch** menu and press **Esc** to refresh the screen.
19. Make sure the Walls function is off, and snap a single line from Snap 6 to Snap 1 to close off the edge of the tile pattern (see Figure 4-68).

    You probably noticed that your tile pattern may not start exactly where you would want it to. You would probably prefer to see a row of full tiles

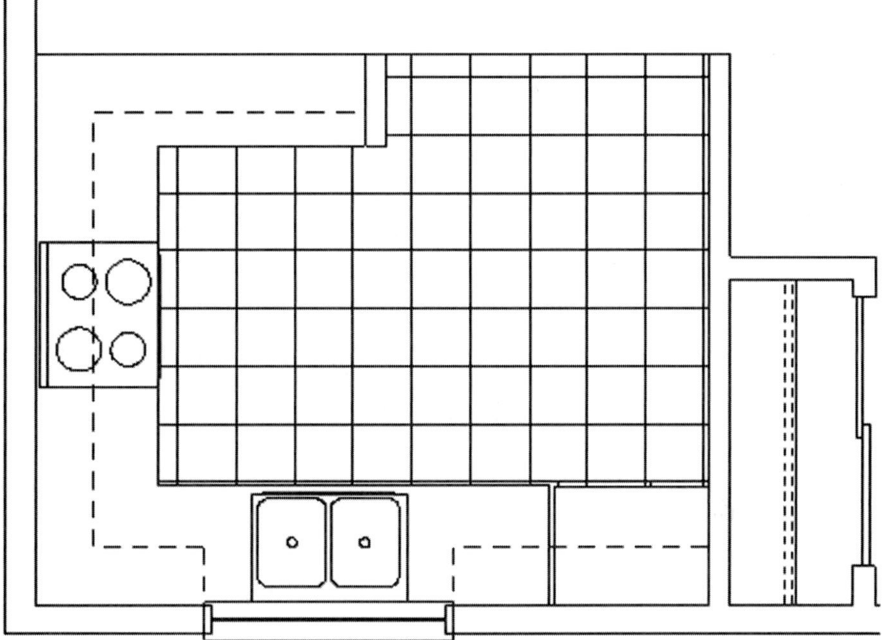

Figure 4-68 Placing a 12-by-12-inch pattern

at the top edge of the kitchen floor where it meets the dining room. We can do that.

20. Go back to the **Hatch** menu (**Hh**) and select the **Origin** button. It will enable us to redefine the starting point (origin) of the hatch pattern. The command line reads, *"Select hatch pattern origin."*
21. *Snap* to the upper-right corner of the kitchen floor (where Snap point 6 was). The command line reads *"Select entity to <Hatch>."*
22. Click on any part of the tile floor hatch pattern. The pattern will change to dashed lines to indicate that you have selected it.
23. Press **Begin** and the hatch will be updated again.
24. *Right-click* to exit the **Hatch** menu and press **Esc** to refresh the screen (see Figure 4-69).

Text

We have one last thing to do to finish off this first apartment. We need to add some text to identify the spaces in the apartment. Let's begin:

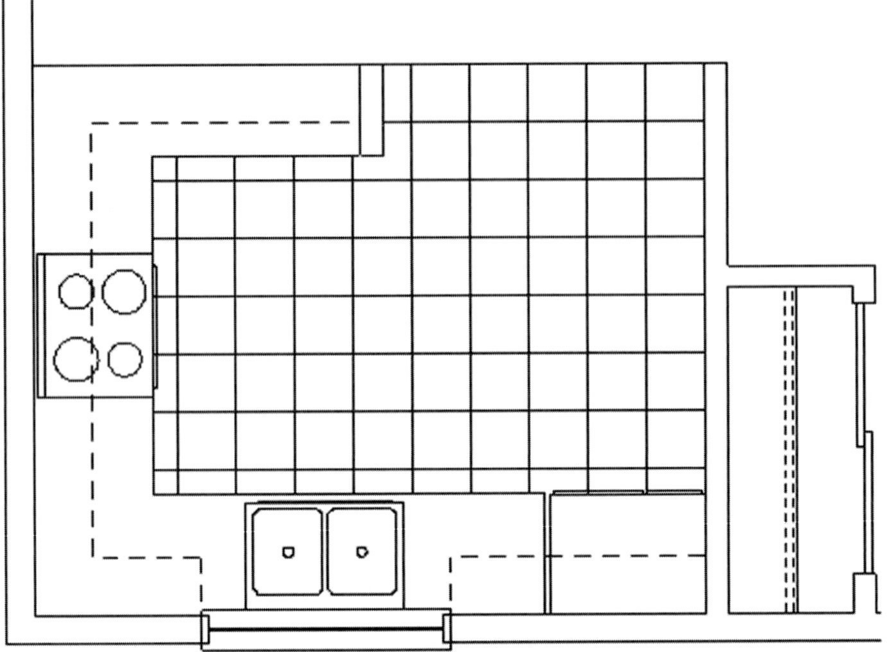

Figure 4-69 Changing the hatch origin

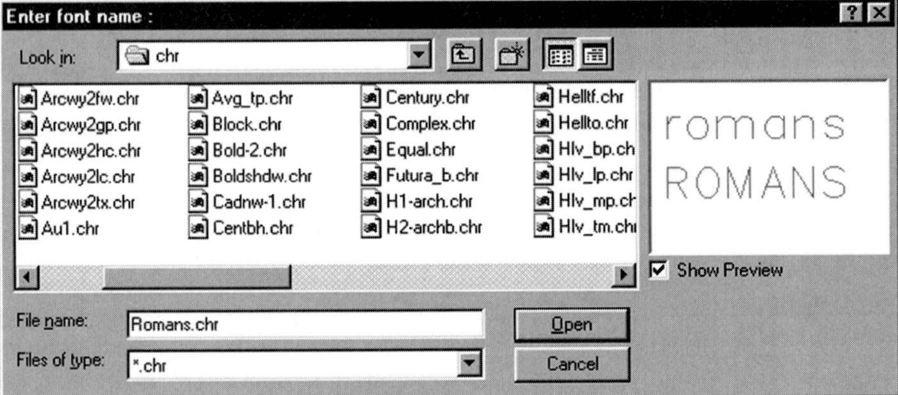

Figure 4-70 Choosing the text font

1. Press the **Tab** key until your active layer is the **Text** layer. Your cursor will turn to red (if not, just press the **Kk** key until it does).
2. Pick **Edit/Text**.
3. Make sure **Dynamic**, **TxtScale**, and **Left** are all turned on.
4. Click on **FontName**. A font dialog box will appear, as shown in Figure 4-70.

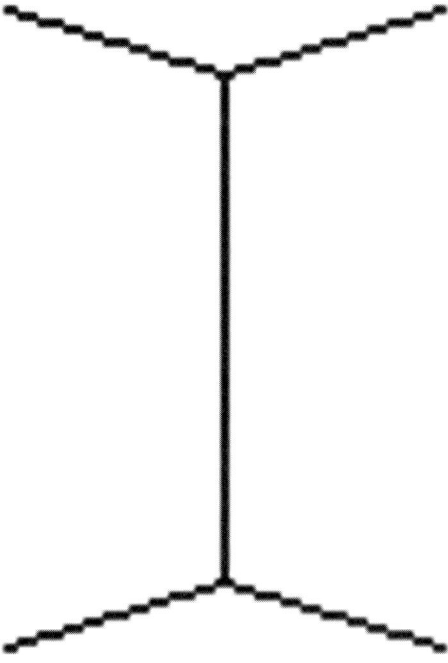

Figure 4-71
The text cursor

Click once on each font name and the preview window will show what that font looks like. You will normally see both upper and lower case letters in the preview, but some fonts only support upper case letters. Those fonts will therefore show only upper case letters in the preview window. The default location for DataCAD fonts is the \CHR folder in your main DataCAD directory.

6. Select the font called **Arcwy2hc** by double-clicking on it or highlighting it and pressing **Open**. Your cursor then changes to a funny little symbol that looks like back-to-back brackets. This is your text cursor (see Figure 4-71).
7. Place the cursor in Bedroom 2 at the left side of where you want to start the text. You should be careful of one particular text feature in DataCAD . With **TxtScale** turned on, as we have it now, if you press **PageUp** or **PageDn** before clicking to place the text cursor, the cursor will get larger or smaller as you do. Down in the Information Line, you will see a message that reads "Current plotting scale is now xxx" where xxx is a scale like 3/4″ [1:20], or 1-$^1/_2$″ [1:10]. This is so that you can dynamically change the size of your text depending on what scale you intend to plot your drawing at. With most other DataCAD features, the **PageUp** and **PageDn** keys zoom in and out of your drawing. This is a

feature that you will most likely want to change. See Chapter 23, "Customizing DataCAD," to learn how to do this. For now, just be aware of it.

8. Since we intend to print this drawing at a 1/4″ [1:50] scale, press **PageUp** or **PageDn** until the Information Line tells you the plotting scale is at 1/4″ [1:50] (or select **Text/Size** and type **..1/4** [6.4] and then press **Enter**).
9. Click the mouse to place the cursor. The text cursor will stay in place in the room while you get your standard cursor back. The Command Line will read *"Enter text:."*
10. Turn on **Caps Lock** on your keyboard and then type **BEDROOM 2**. *Right-click* to finish adding the text.
11. Do the same for the other five spaces: BEDROOM 1, BATH, DINING, LIVING ROOM, and KITCHEN.
12. *Right-click* to exit the **Text** menu and press the Recalc button (**R** in the projection pad) and **PageUp** to view your whole drawing (see Figure 4-72).

Mirror

Now that we have the first apartment drawn, we need to create the second one, which is just a mirror image of the first apartment. Appropriately enough, the Mirror function in DataCAD will make short work of this task for us. Let's assume we want to leave a 6″ [152] chase space between the two walls of the apartments:

1. Make sure you save the drawing before making major changes like these just in case (**File/Save** or **F**).
2. Press **Edit/Mirror** (**Alt+M**). The Command Line will read *"Select first point along the line of reflection."* The line of reflection is the plane about which the apartment will be mirrored.
3. Press the RefPt (′) key and snap to the lower-left, outside corner of the apartment. Press the spacebar and enter a distance of **.3** [76]and an angle of **180**. The Command Line will read, *"Select second point along the line of reflection."*
4. Move the cursor up a few feet to form a vertical line (it doesn't matter at what distance) and click the mouse to finish the line. Make sure the line is perfectly vertical. (If you are not in Ortho mode, press the **Oo** key to get there.)

Figure 4-72
The completed floor plan

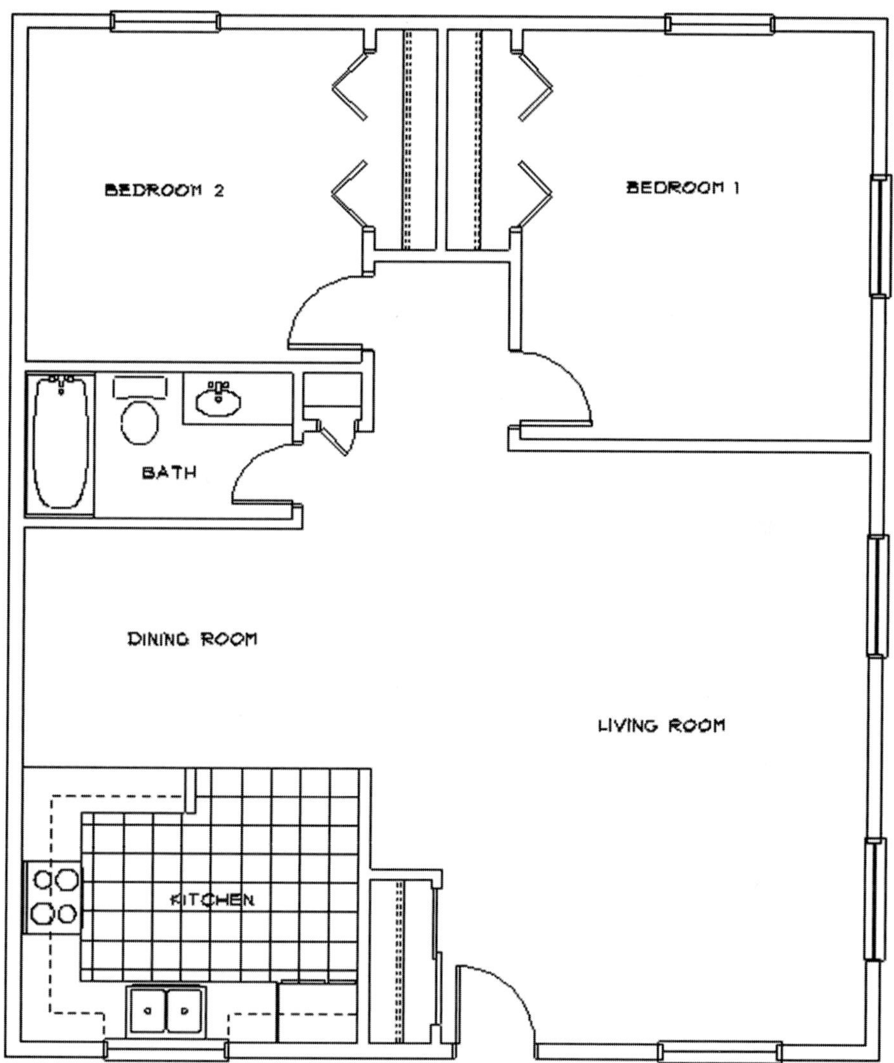

5. The Command Line will read *"Select entity to <MIRROR>."*
6. Make sure that **AndCopy** is on (otherwise, the apartment will be mirrored to the other side of the line you define, but without leaving a copy behind). Then turn **LynSrch** and **Area** on.
7. Draw a bounding box around the entire apartment and click the mouse.
8. If you made a mistake, you can use **Edit/Undo** (**Ctrl+Z**) to erase the mirrored apartment.

9. If everything looks correct, then *right-click* to exit the **Mirror** menu. If something is wrong, select **Edit/Undo** (**Ctrl+Z**) and try again. Press Recalc and **PageUp** to get a view of the entire drawing (see Figure 4-73).

   Notice that DataCAD is intelligent enough to know that the text should be moved to the correct position but not be displayed with the letters mirrored.

10. Now zoom into the lower part of the drawing where the two apartments meet.
11. Press the **Tab** key until Walls is the active layer.
12. Draw a single line between the two inside corners of the kitchen walls by *middle-snapping*.
13. Use the 2 Line Trim function (**Edit/Cleanup/2LnTrim**) to finish off the cavity, as shown in Figure 4-74.

Weld

To finish off the exterior wall we could just draw a short line to connect the two, but that would be very inefficient. Every extra entity in a drawing file increases the size of the drawing, decreasing speed and efficiency. If you add

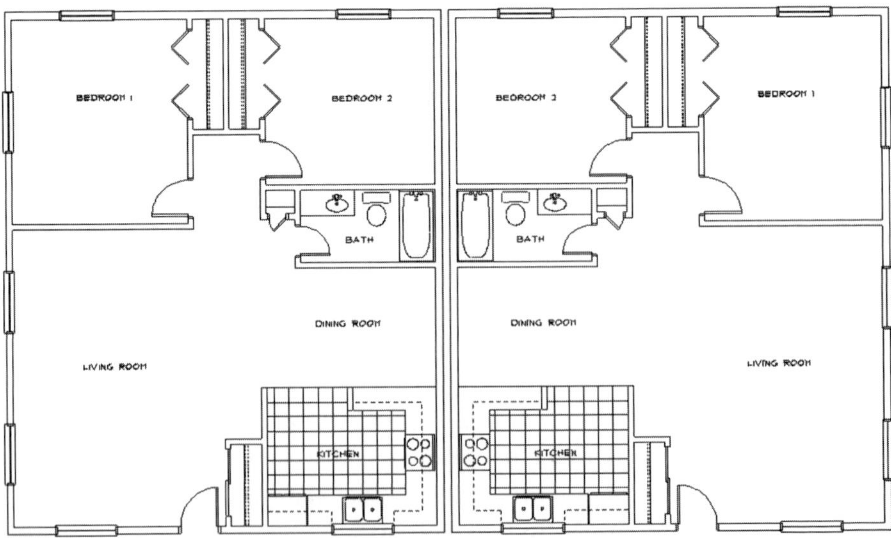

Figure 4-73
Nearly completed plans for both apartments

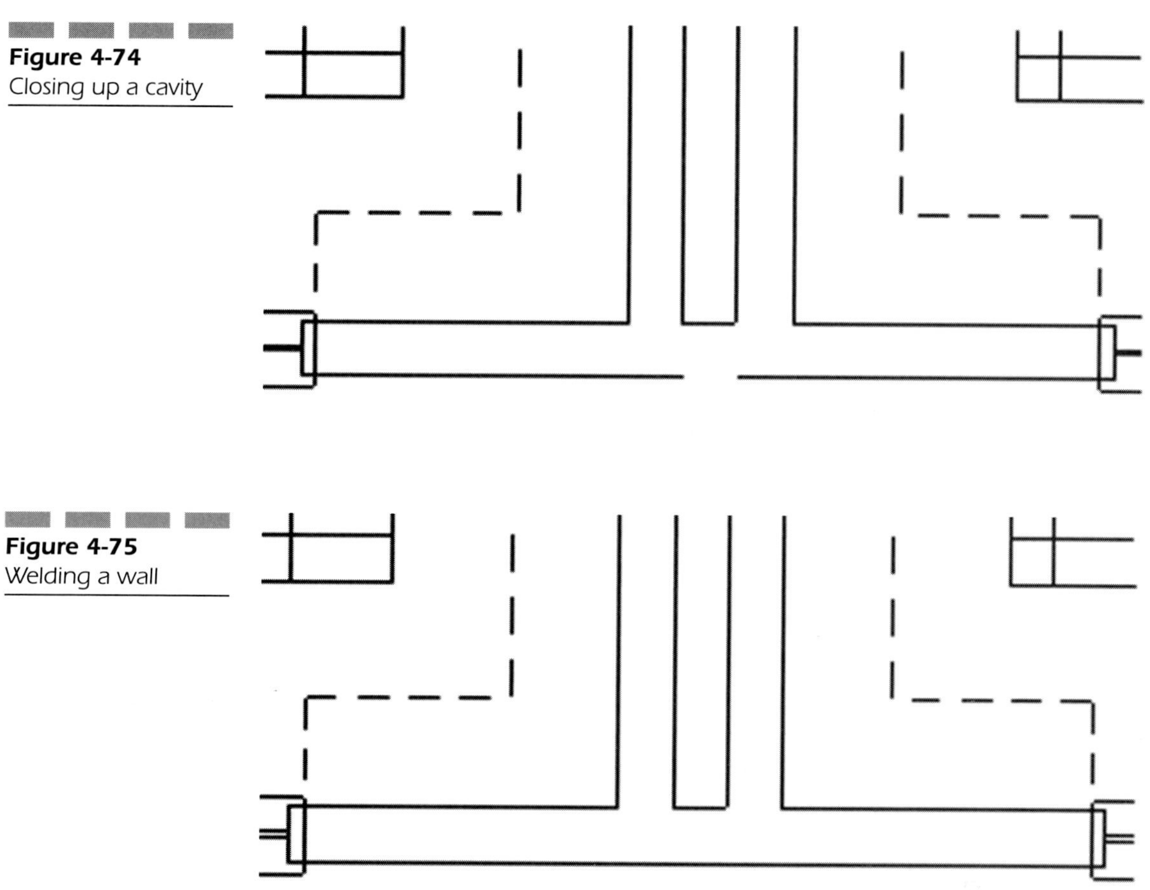

Figure 4-74
Closing up a cavity

Figure 4-75
Welding a wall

small lines to a drawing often, especially in a large file, editing the drawing becomes a nightmare due to all the little bits and pieces that have to be individually edited, cleaned up, copied, and so on. For now, you'll just have to take my word for it that it really does become a big hassle later on.

To get around this problem, DataCAD has two functions called *Weld Line* and *Weld Wall*. As the names imply, they enable you to weld two lines or two walls together to form one entity instead of three. We'll use the Weld Line function here, but you can read more about it in Chapter 5:

1. While still in the **Cleanup** menu, press the **WeldLine** button. The Command Line reads *"Select first line to weld."*
2. Pick the first line, then the second line, and the wall line will be welded together (see Figure 4-75).

3. To prove it, press Identify (the **I** button in the Projection Pad), and click on the newly welded line. The line is highlighted and you can see that it is now one whole line instead of three.
4. *Right-click* to get back to the **Cleanup** menu.
5. Now use this same technique to clean up the wall intersections at the top of the apartment.
6. *Right-click* until you get back to the **Edit** menu.
7. Press **Recalc** and **PageUp** to see the finished drawing (see Figure 4-76).

Add a Roof

We're going to deviate a bit from the 2D aspects of this chapter to use one of the 3D features that will put a 3D roof on the building. In order to do this, we need to access another DataCAD feature called the **Toolbox**. Here you will find an array of various mini-programs called "macros." DataCAD comes with a lot of these macros, and hundreds more are available elsewhere (see Chapters 24, "Third-Party Macros," and Chapter 27, "DataCAD Resources"). We're going to use one called ROOFIT. This macro contains a lot of settings

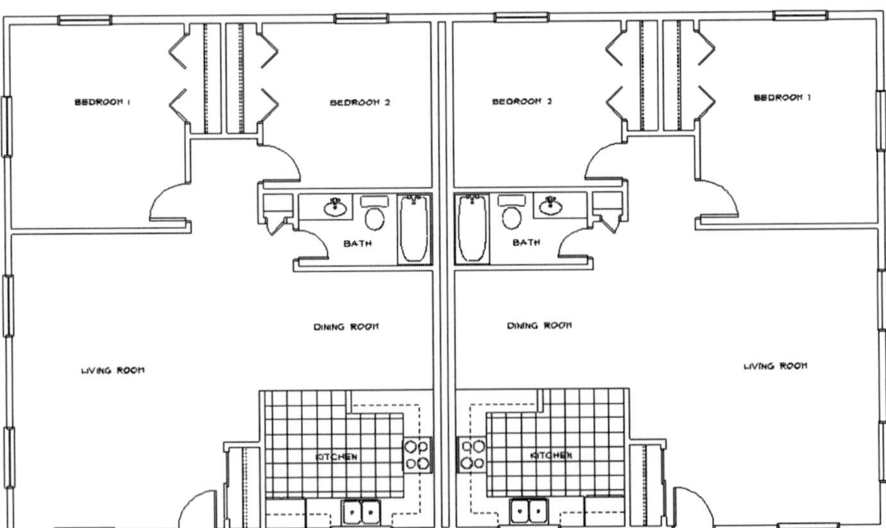

Figure 4-76
The finished plans

and features, which we won't get into in this example, but we will show you some of what it can do:

1. Create a new layer called Roof (**Utility/Layers/NewLayer/1/Enter/Name/Roof** and *right-click* twice).
2. Press **Tab** until **Roof** is the active layer. Press **Kk** until the color is **LtRed**.
3. Go to the **Edit** menu and select the **Toolbox** option (or pick the red Toolbox icon near the end of the standard DataCAD icon toolbar). A dialog box with all the available macros is displayed (see Figure 4-77).

   The default directory for the macros is the \DCX folder, and all DataCAD macros have a file extension of .DCX.
4. Find the **RoofIt.dcx** macro. Select it by *double-clicking* on it or by highlighting it and then pressing **Open**. You will see a menu like the one in Figure 4-78.

   If the first option at the top of the menu is **PolyRoof** instead of **RectRoof**, just click on the button and it will change to **RectRoof**. This enables you to draw simple rectangular roofs with a single plate height, as we're going to do here.
5. We're going to draw a Hip roof, so select the **Hip** button.

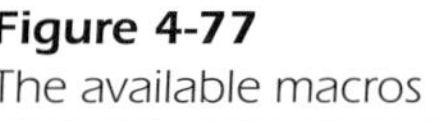

Figure 4-77
The available macros

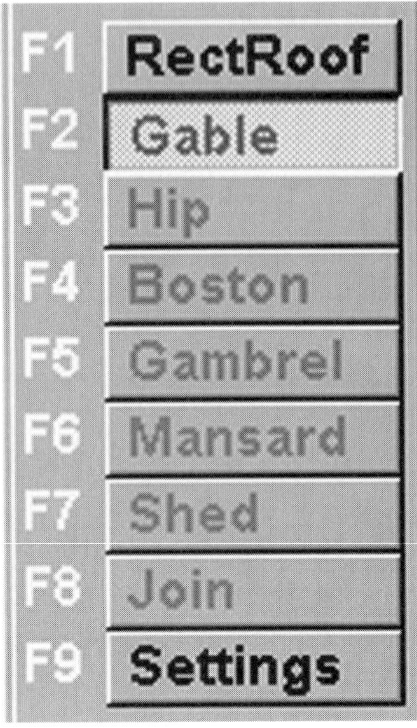

Figure 4-78
The Rooflt.dcx macro menu

6. Now select **Settings**. Click on **PlatHgt** (plate height). Make sure the Command Line says that the current plate height is 8.0 (8′-0″ [2438]).
7. Click on **Pitch** and select **4:12**.
8. Click on **RoofThck**, type in **1** (for a roof thickness of 1′-0″ [305]), and then press **Enter**.
9. Click on **WallThck**, type in **.6** [152] (the thickness of our exterior walls), and then press Enter.
10. *Right-click* to exit the **Settings** menu.
11. The Message Line will tell you the current settings. The Command Line will be prompting you to *"Pick the 1st point of the roof boundary."* You need to define two adjacent outside walls on the perimeter of the building so that DataCAD knows the boundaries of the roof. It is very important to pick these points in a clockwise direction. If you don't, then your roof will be created upside down.
12. *Middle-snap* to the upper-left, outside corner of the building and then *middle-snap* to the upper-right, outside corner of the building. You have just defined the long direction of the roof.

13. The Command Line will now be prompting you to *"Enter second point of roof width." Middle-snap* to the lower-right, outside corner of the building. The roof will be created. Because you are drawing a simple rectangular roof, you only need to define three corners of the roof boundaries (to define length and width).
14. To see the roof in 3D, *right-click* to exit the RoofIt macro and then select **Edit/DCAD 3D** to get to the 3D menus. Notice that although most of the menu options here are different than the 2D menus, two primary menus are still available, **3DEdit** and **3DEntity**, and you access each by simply *right-clicking*.
15. Select **3DViews** (available in both the **3DEdit** and **3DEntity** menus).
16. Select **Isometrc** to see an isometric view of your building. If you don't see it, or you only see part of it, just press the **R**ecalc button in the navigation pad (see Figure 4-79).
17. While you are still in isometric view, try out these options:
    a. Press the **Ortho** button in the Projection Pad. You will return to the 2D, orthographic view of you building.
    b. Press the **Para** button in the Projection Pad to return to the isometric view. An isometric view is a "parallel" view because none

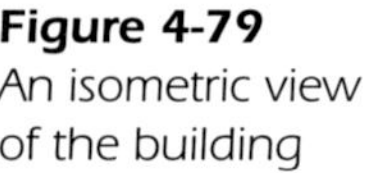

Figure 4-79
An isometric view of the building

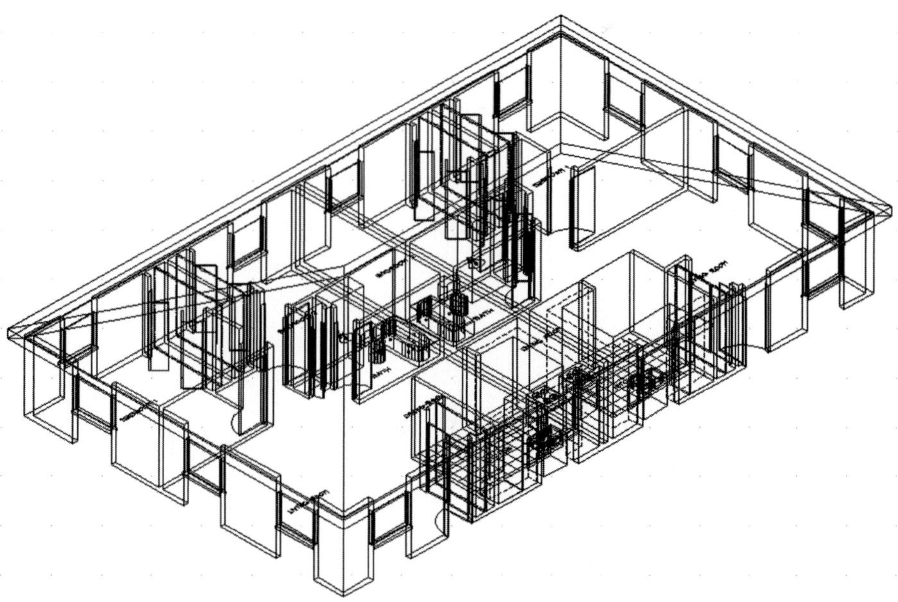

of the entities are in perspective. All entities in the same plane are parallel to one another.

c. Now press **Front**, **Back**, **Left**, and **Right** in the Projection Pad to see views of the building elevations. If you now press the **Para** button again, notice that you do not go back to the isometric view. That's because elevation views are also parallel views, and DataCAD always returns to the last parallel view that was selected. In this case, this is the elevation view. To get back to an isometric view, you will have to go back and select **3DViews/ Isometrc**.

d. In the **3DEdit** menu, select **Layers/On/Off**. Try turning on and off various layers. Notice that the layers will dynamically turn on and off in the isometric view. *Right-click* out of the **Layers** menu.

18. Now let's see what the building looks like without all its insides showing. Go back to an isometric view. In the **3DEdit** menu, select the **Hide** option. Make sure **SaveImg**, **CropImg**, and **All** are turned on.

19. Press **Begin** to start the hidden line removal process. The Command Line will tell you the progress of the process. The image on the screen is a new 2D image drawn by DataCAD in the Drawing Window (see Figure 4-80).

Figure 4-80
A hidden line view of the building

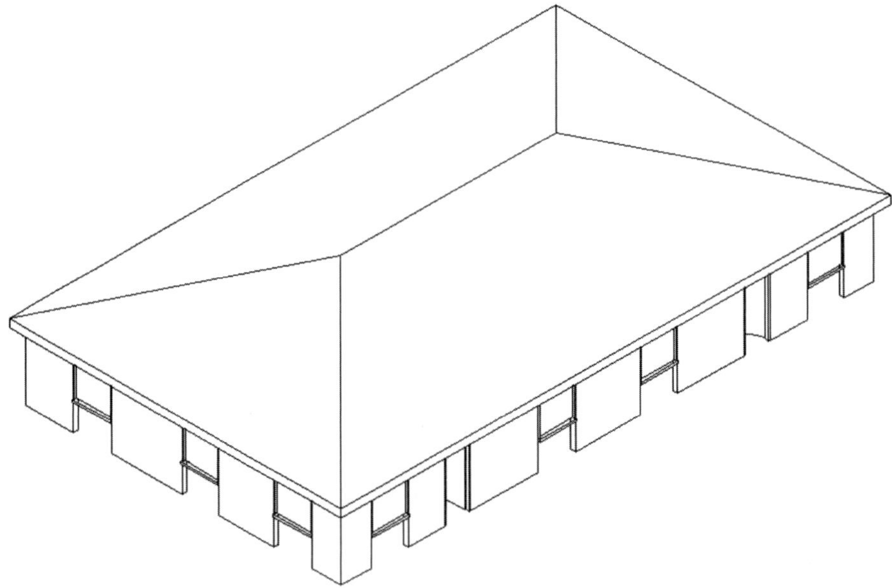

To see that it is only a 2D image, press the **Esc** key. The hidden line image will disappear and the full isometric image will again be displayed. The hidden line view is only temporary, unless you save it.

20. If you want to save a 2D image of the hidden line view, perform the following steps:
    a. *Right-click* back to the **Hide** menu. Press **Begin** to create a hidden line view.
    b. Select the **NewLyr** button in the menu. You will be prompted to *"Enter name of new layer to be added: Layer9."*
    c. Type **3DRoof** and then press **Enter**. You have now saved a 2D image of the building to a new layer called **3DRoof**.
    d. Select **Off**. This will turn off the new layer you just created.
    e. *Right-click* back to the **3DEdit** menu.
    f. Select **3DEdit/Layers/ActvOnly/3DRoof**. Your 2D image will be displayed, but it's been rotated! What is going on is that you are still in isometric view. Press the **Ortho** button to get back to a 2D, orthographic view.
21. Select **3DEdit/DCAD 2D** or press the semicolon (;) key to get back to the 2D menu.

SUMMARY

Whew! If you've never used CAD before, then you may be thinking that it took an awfully long time to draw this simple floor plan. With a little practice, though, you will find yourself drawing more in less time with greater accuracy. Really!

We've touched on a good number of DataCAD's features, but it has a great deal more. This chapter was meant to be an introduction to basic concepts such as entities, bounding boxes, menus, keyboard shortcuts, moving around, hatching, Move, Copy, Erase, and Mirror. We also touched on some of the $2^1/_2$ D and 3D aspects of DataCAD. These concepts should give you a good foundation for understanding and using the information in the following chapters.

We didn't cover printing in this chapter since that is more of a specific task than a concept. You can read about printing in Chapter 7, "Printing and Plotting," which will explain how to do so in greater detail.

CHAPTER 5

Basic Drawing

In this chapter, we will explore the very basics of drawing in DataCAD, that is, how to draw lines, arcs, and text, and how to navigate the keyboard and menus. One of the benefits of DataCAD over all other CAD programs is that its menu system is primarily text-based, making menu navigation a much easier task. So if all else fails, read the menus!

Drawing Lines

Drawing a line in DataCAD literally requires a total of two mouse clicks; nothing more. Unlike most other CADD programs, DataCAD is designed to draw a line by default. You don't have to type or select any commands or icons at all:

1. Click once anywhere in the Drawing Window.
2. Drag the mouse anyplace else in the Drawing Window then click again to select the second point of the line. A line is drawn from the first point to the second.

Pretty easy, huh! But now you may ask, "How do I draw a more precise line, in both distance and direction? Essentially, two methods can be used: using a snap grid or entering a distance and an angle via the keyboard. Keep in mind that the steps for drawing walls are essentially the same as those for drawing lines, so what you read about here will be applicable to the drawing of walls, which is covered later in this chapter.

The keyboard method further breaks down into four methods for input (Relative Polar, Absolute Polar, Relative Cartesian, and Absolute Cartesian). Each of these methods has its merits for different situations, but just the same, a friendly debate exists in the DataCAD community over whether the Relative Polar or Relative Cartesian method is best for day-to-day use. The Relative Polar method is the predominant method described in this book, but you should try out the Relative Cartesian method as well to see which suits you best:

- **Relative Polar** defines lines as a distance from the last point picked at a specified angle.
- **Absolute Polar** defines lines as a distance at a specified angle, relative to the absolute 0 (zero) point of the drawing file.

- **Relative Cartesian** defines lines as increments along the X- and Y-axis, relative to the last point picked. This is the default DataCAD entry method.
- **Absolute Cartesian** defines lines as increments along the X- and Y-axis, relative to the absolute 0 (zero) point of the drawing file.

You can use the **Insert** key to toggle between each of these input methods.

Lines with Snap Grids

Chapter 3, "Settings and Display Options," describes the various settings for snap grids. In a nutshell, a snap grid is a non-visual, regular X/Y grid that forces the cursor to move at fixed distances vertically and horizontally. Depending on how closely or how far away you are zoomed into the Drawing Window, as the cursor moves around the screen, rather than moving smoothly, it will appear to skip, or "snap" by a fixed distance. By watching the coordinate readout at the bottom of the Drawing Window, you can draw lines of precise angles and distances.

The following are the settings we will work with in this example. Keep in mind that you can and should alter these settings as required for your particular situation.

In the top menu bar, select the **Utility/Input Mode/Relative Polar**. In the **Utility/Settings** menu, set the following settings:

- **ScaleTyp/Arch**
- **AngleTyp/Normal**
- **BigCursr** = on

In the **Utility/Grids** menu, set the following grids:

- **SnapGrid** = on
- **DspGrid1** = on
- **DspGrid2** = off
- **GridSize/SetSnap/1′-0″** [305] (It is important to set this properly for this example)
- **SnapAng/8**

Start the first line by *clicking* anywhere in the Drawing Window. Now move the cursor to the right, watching the Coordinate Readout at the bottom of the screen. Keep the angle (**a:**) at 0 degrees while dragging the cursor to a distance (in the **x:** direction) of 10′-0″ [3050]. Notice that the cursor "jumps" in one-foot increments as you move it to the right. This is because your snap grid is currently set to 1′-0″ [305]. To finish the line, make sure the coordinates are still 0 degrees and 10′-0″ [3050]; then *click* the mouse. The first line is completed and a new line is begun from the end of the first line to the cursor (see Figure 5-1).

Move the cursor in a slow circle around the end point of the first line. Watch the coordinate readout change. The angle (**a:**) will change by 45-degree increments because your snap angle (**SnapAng**) is set to eight increments ($^{360}/_{8}$ = 45 degrees). If you set the **SnapAng** to **16**, then the cursor would snap to 22.5-degree increments ($^{360}/_{16}$ = 22.5 degrees).

Now drag the new line, still attached between the end of the first line and the cursor, upward and to the right with the cursor at an angle of 45 degrees. Notice that the distance readout (**d:**) does not give you an even distance divisible by 1′-0″ [305], but that the **x:** and **y:** coordinate readouts do. That's because the **SnapGrid** is only an X/Y grid, and the distance from one corner of the grid to the next (a 45-degree angle) is a little less than 1′-5″ [432] (see Figure 5-2).

So a snap grid setting is usually not very helpful for drawing anything not on the X/Y axis, but for X/Y distances it is very useful. Don't forget that you can change the snap grid quickly by pressing **Gg** and changing the **SnapGrid** to 1″ [25.4], 6″ [152], 21′-6$^{1}/_{2}$″ [6566], and so on. But to draw a line at an angle, such as 45 degrees, at a specified distance, you can use the next method of typing in the distance and angle. Whenever you want to stop drawing lines, simply *right-click* the mouse. You might need to *right-click* a couple of times to return to the **Utility** or **Edit** menus.

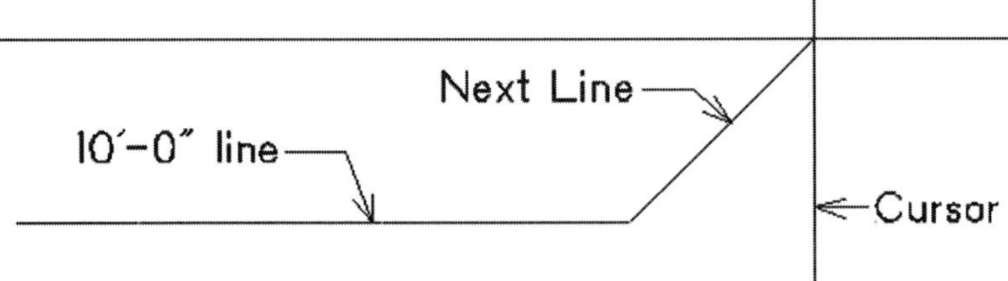

Figure 5-1
Drawing two lines in DataCAD

Figure 5-2
Coordinate and angle readouts

Lines by Distance and Angle

Here you will find two basic options: Relative Polar and Absolute Polar. Both of these options are available from the menu bar via **Utility/Input Mode**.

Entering Distances

In DataCAD, you do not have to type in the feet (′) or inch (″) symbols. Instead you separate numbers with a period (.), a forward slash (/), or a space, and you don't have to place 0s as place holders between them. Let's look at some examples.

The dimension 3′-6$^3/_{16}$″ inches takes a long time to type. Instead, all you have to type is one of the following:

- **3.6.3/16**
- **3.6.3.16**
- **3 6 3 16**
- **.42.3/16**

To enter 4″, you could type

- **.4**
- **(space) 4**

To enter $^1/_4$″, you could type

- **..1/4**
- **..1.4**
- **(space)(space) 1 (space) 4**

As you can see, DataCAD is smart enough to know that the periods, slashes, or spaces represent dimension placeholders for feet, inches, and fractions. It makes entering distances much faster than in other CAD programs.

Entering Angles

The default orientation for defining angles is shown in Figure 5-3. Note that it is not oriented like a compass. Zero degrees starts to the right, and all positive angles are described in a counterclockwise direction. You can

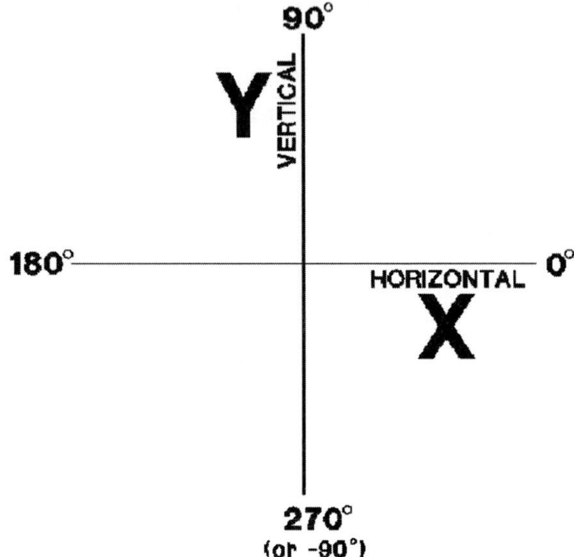

Figure 5-3 The default orientation for defining angles

change this orientation if you'd like. See Chapter 3 for information on changing settings.

Relative Polar

Polar means that the new line is drawn based on the angles of a compass. *Relative* means that a line is drawn relative to the last point picked. So if you *click* the mouse in the Drawing Window to define the first point of a line, that point becomes the last point picked. The second point of the line will then be drawn at an angle and distance from that point. To draw a line with this method follow these steps:

1. Make sure the current input mode is set to Relative Polar (select **Utility/Input Mode** from the drop-down menu, or press the **Insert** key until you see the Relative Polar input mode in the Message Line).
2. Click in the Drawing Window to pick the first point of the line.
3. Press the **Space** bar. This tells DataCAD you want to define a new line by typing something in.

4. You are prompted for a distance. Type in a distance like 5′-0″ [1524] (just type the number **5** and press **Enter**).
5. Now you are prompted for an angle. Since the default angle is **0**, just press **Enter** to accept it and the line is completed (see Figure 5-4).

If you draw a complete line from point 1 to point 2, then the end of that line (point 2), becomes the last point picked. As long as there is an active line attached from Point 2 to the cursor, a new line can be defined relative to Point 2 (see Figure 5-5).

If you *right-click* the mouse, then any lines in the process of being drawn will cease to be drawn and will be disconnected from the cursor. A new line will no longer be drawn starting from Point 2. If you want to continue to draw a new line from Point 2, you must middle-snap (or press **Nn**) to snap to Point 2. From there, you can define a distance and angle for the new line.

But even if you *right-click* after drawing a line from Point 1 to Point 2, DataCAD still knows that Point 2 was the last point picked and will enable you to use this as a reference point from which to start a new line. Try drawing a line and then *right-clicking*, then try this:

1. Press the **Space** bar. This tells DataCAD you want to define a new line. Since you didn't define the start point of a new line, DataCAD assumes the last point picked (Point 2) as a reference point.
2. You are prompted for a distance. Type in a distance like 5′-0″ [1524] (just type the number **5** and press **Enter**).
3. Now you are prompted for an angle. Type in **90** and press **Enter**.

Figure 5-4 Drawing a line with Relative Polar

Figure 5-5 Point 2 becomes the starting point for a new line.

Notice that the first point of the new line is drawn starting at 5 feet and 90 degrees from Point 2. DataCAD uses the last point picked as a reference point from which to define the start of a new line. Point 2 is not used as the first point for the new line. For that, you must always middle-snap to redefine that point as the point from which to start the new line.

Absolute Polar

As stated earlier, polar means that the new line is drawn based on the angles of a compass. *Absolute* means that a line is drawn relative to the absolute zero (0,0) point of the drawing. Every CADD drawing has a single absolute zero point that cannot be moved or changed. With the Absolute Polar method, if you *click* the mouse in the Drawing Window, that point will be the first point of the line (Point 1). If you then press the **Space** bar, you can input the second point (Point 2) of the line in reference to the absolute zero point, rather than in reference to Point 1.

1. Make sure the current input mode is set to Absolute Polar (select **Utility/Input Mode** from the drop-down menu, or press the **Insert** key until you see the Absolute Polar input mode in the Message Line).
2. Click the mouse anywhere in the Drawing Window (Point 1).
3. Press the **Space** bar. This tells DataCAD you want to define the second point of the line from the absolute zero point.
4. You are prompted to *"Enter distance from the origin."* Press **Enter** to accept the default distance of 0.
5. Now you are prompted to *"Enter angle from the origin."* Press **Enter** to accept the default angle of 0.

The line is drawn from your first mouse click (Point 1) to the absolute zero point (Point 2). Depending on how far away Point 1 is from the absolute zero point, the new line could be quite long (see Figure 5-6).

You can also draw the second point of the line to any point referenced from, but not on, the absolute zero point:

1. Click the mouse anywhere in the Drawing Window (Point 1).
2. Press the **Space** bar. This tells DataCAD you want to define the second point of the line from the absolute zero point.

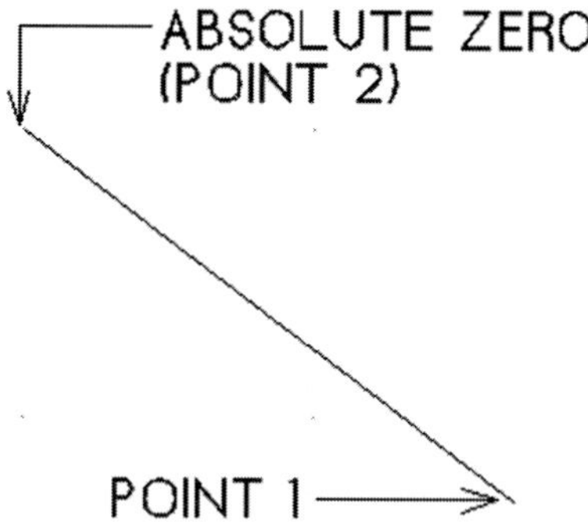

Figure 5-6
A line drawn from Point 1 to the absolute zero point

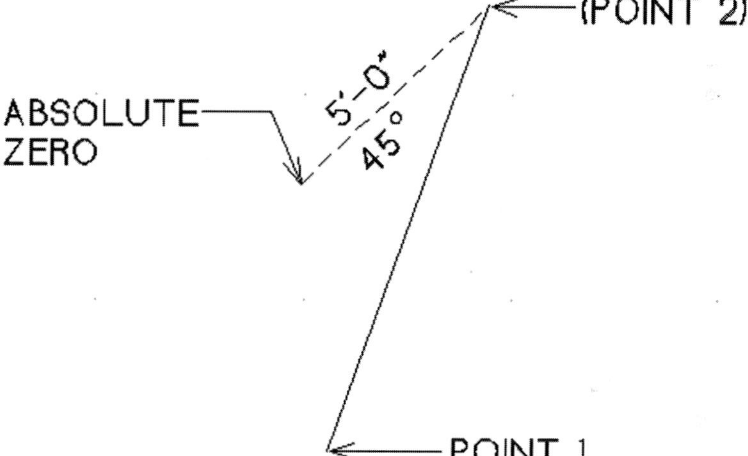

Figure 5-7
An Absolute Polar line drawn in reference to absolute zero

3. You are prompted to *"Enter distance from the origin."* Press **5** and then **Enter** to enter a distance of 5′-0″ [1524] from absolute zero.
4. Now you are prompted to *"Enter angle from the origin."* Press **45** and then **Enter** to enter an angle of 45 degrees from absolute zero.

The line is drawn from your first mouse click (Point 1) to a point 5 feet and 45 degrees from absolute zero (Point 2), as shown in Figure 5-7.

If you do not *right-click* after drawing the first line, then a second line will be displayed from Point 2 to the cursor. If you *click* the mouse, the second line will be drawn from Point 2 to Point 3. But if you do not *right-click*, and instead press the **Space** bar, then you can define Point 3 by typing the distance and angle. So the location of Point 3 was "referenced" off of absolute zero, not off of Point 2.

If you again do not *right-click*, then continue drawing another line by pressing the space bar, the next line will be drawn from Point 3 to Point 4, with Point 4 being again defined as a distance and angle from absolute zero. Let's try drawing three contiguous lines using this method:

1. *Click* the mouse anywhere in the Drawing Window (Point 1).
2. Press the **Space** bar. This tells DataCAD you want to define the second point of the line off of the absolute zero point.
3. You are prompted to *"Enter distance from the origin."* Press **5** and then **Enter** to enter a distance of 5′-0″ [1524] from absolute zero.
4. Now you are prompted to *"Enter angle from the origin."* Press **45** and then **Enter** to enter an angle 45 degrees from absolute zero.
5. A line is drawn from Point 1 to Point 2.
6. Press the **Space** bar.
7. You are prompted to *"Enter distance from the origin."* Press **12** and then **Enter** to enter a distance of 12′-0″ from absolute zero.
8. Now you are prompted to *"Enter angle from the origin."* Press **270** and then **Enter** to enter an angle 270 degrees from absolute zero.
9. A line is drawn from Point 2 to Point 3.
10. Press the **Space** bar.
11. You are prompted to *"Enter distance from the origin"*. Press **14** and then **Enter** to enter a distance of 14′-0″ [4267] from absolute zero.
12. Now you are prompted to *"Enter angle from the origin."* Press **190** and then **Enter** to enter an angle 190 degrees from absolute zero.
13. A line is drawn from Point 3 to Point 4 (see Figure 5-8).

Lines by X and Y Coordinates

The next two Cartesian methods of input are related to the Polar methods, but instead of entering a single distance followed by an angle to determine the position of the second point of the line, you enter an X and then a Y distance. This can be quite useful when drawing lines that are traditionally described by X and Y coordinates, like a roof pitch of 4 in 12.

All X distances to the right of the Y-axis are described by positive numbers, while all X distances to the left of the Y-axis are described by negative

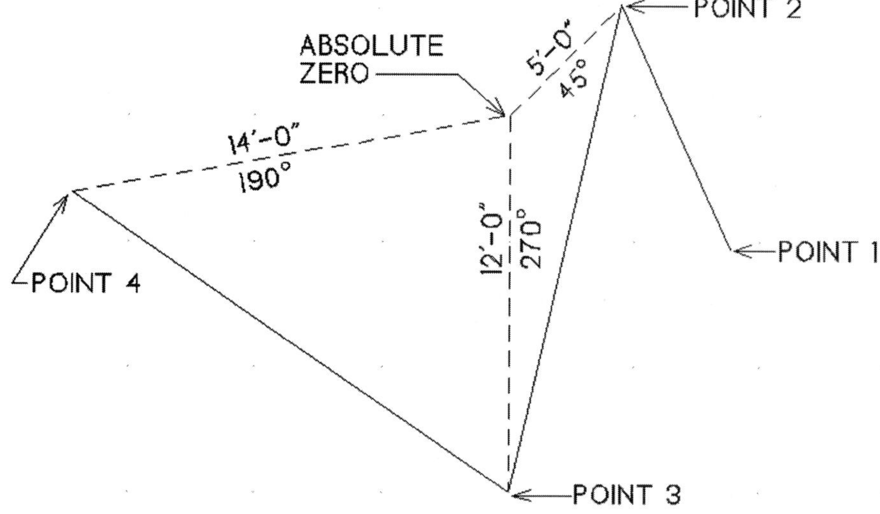

Figure 5-8 Three contiguous lines drawn with Absolute Polar

numbers. Likewise, all Y distances above the X-axis are described by positive numbers, and all Y distances below the X-axis are described by negative numbers (see Figure 5-9). Take a look at the following examples.

Relative Cartesian

Cartesian means that the new line is drawn based on X (horizontal) and Y (vertical) coordinates. Relative means that a line is drawn relative to the last point picked. So if you *click* the mouse in the Drawing Window to define the first point of a line, that point becomes the last point picked. The second point of the line will then be drawn at X and Y coordinates from that point. Here's how you would draw a 5′-0″ [1524] line horizontally to the right:

1. Make sure the current input mode is set to Relative Cartesian (select **Utility/Input Mode** from the drop-down menu, or press the **Insert** key until you see the Relative Cartesian input mode in the Message Line).
2. Click in the Drawing Window to pick the first point of a line.
3. Press the **Space** bar. This tells DataCAD you want to type in the second point of a new line.
4. You are prompted for an X distance. Type in a distance like 5′-0″ [1524] (just type the number **5** and press **Enter**).

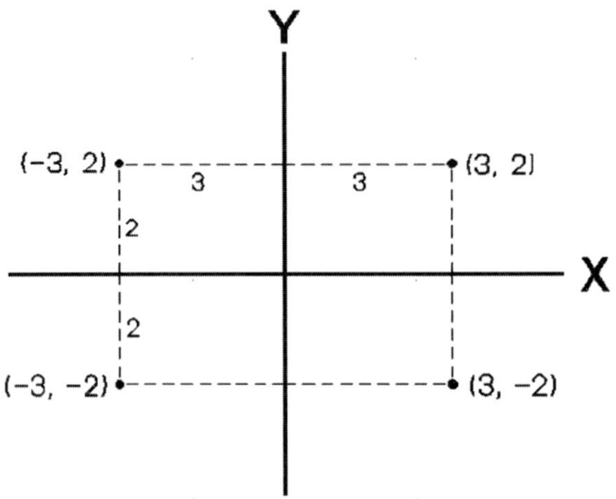

Figure 5-9 Examples of Cartesian coordinates

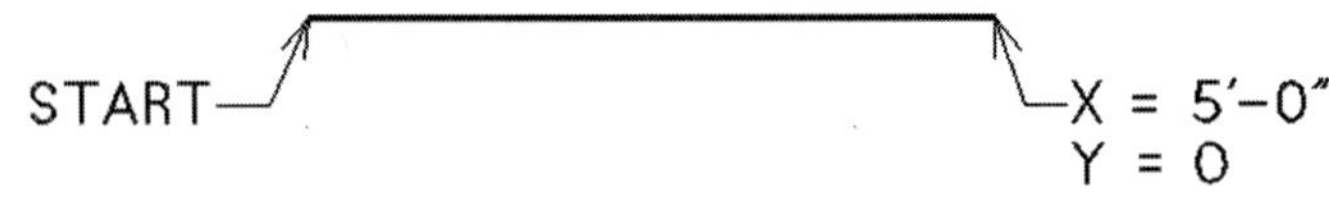

Figure 5-10 Drawing a horizontal line using Cartesian coordinates

5. Now you are prompted for a Y distance. Since the default distance is **0**, just press **Enter** to accept it and the line is completed (see Figure 5-10).

Let's say you want to draw an angled line. How you draw that line using the Relative Cartesian method depends on what kind of line you want to draw. If you want a line that is drawn to a specific point, you need only know the X and Y coordinates of that point without needing to know the actual length of the line. Let's say you want to draw a line from the lower-left corner of an imaginary box to the upper right corner as shown in Figure 5-10b. Since we know the X and Y distances, this is a simple matter:

1. Select a start point (the lower-left corner of the imaginary box).
2. Press the **Space** bar. This tells DataCAD you want to type in the second point of the line.
3. You are prompted for an X distance. Type in the X distance of the line, which is 10′-6″ [3200] (type **10.6** and then press **Enter**).

4. Now you are prompted for a Y distance. Type in the Y distance of the line, which is 6′-2″ [1880] (type **6.2** and then press **Enter**), and the line is completed.

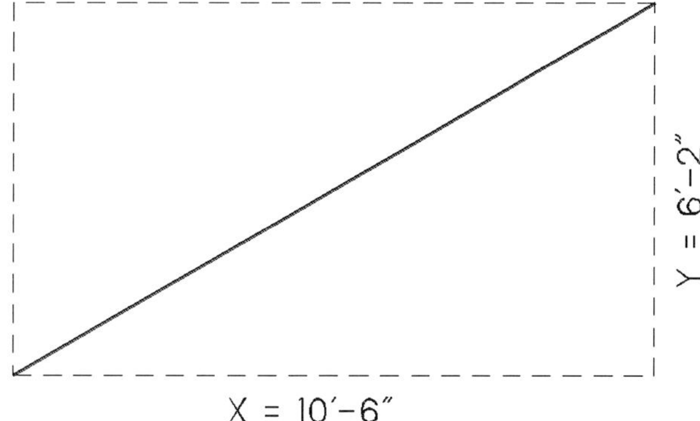

Figure 5-10b
Drawing a diagonal line using Cartesian coordinates

But if you want a line of a specific length and angle, you would need to know trigonometrically what the X and Y distances would be for that line. You would probably be better off switching temporarily to Relative Polar coordinates to do this. Otherwise, if you wanted to draw a 30-degree line 6′-6″ [1981] long using the Relative Cartesian input method, first you would need to calculate the X distance, and then the Y distance using trigonometry. The end result is shown in Figure 5-11.

After making those calculations you would then use the original four steps at the beginning of this section to input the X and Y distances for the line.

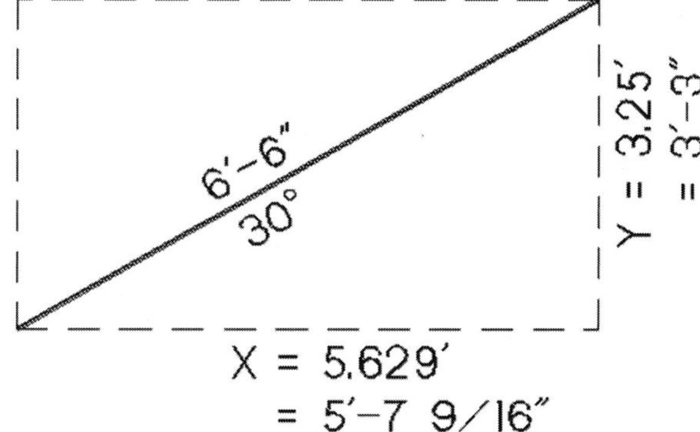

Figure 5-11
The X and Y distances had to be figured with trigonometry.

Absolute Cartesian

Cartesian means that the new line is drawn based on X (horizontal) and Y (vertical) coordinates. Absolute means that a line is drawn relative to the 0.0 (absolute zero) point of the drawing. Every CAD drawing has a 0,0 point which can not be moved or changed. In the absolute Cartesian method, if you *click* the mouse in the Drawing Window, that point will be the first point of the line (Point 1). If you then press the space bar, you can input the second point (Point 2) of the line in reference to the absolute zero point, rather than in reference to Point 1. Let's try an example. See Figure 5-12.

1. Make sure the current input mode is set to Absolute Cartesian (select **Utility/Input Mode** from the drop-down menu, or press the **Insert** key until you see the Absolute Cartesian input mode in the Message Line).
2. *Click* the mouse anywhere in the Drawing Window (Point 1).
3. Press the **Space** bar. This tells DataCAD you want to define the second point of the line off of the absolute zero point.
4. You are prompted for an X distance. Type in the X distance of the line, which is 7′-6″ [2286] (type **7.6** and then press **Enter**).
5. Now you are prompted for a Y distance. Type in the Y distance of the line, which is 3′-2″ [1117] (type **3.2** and then press **Enter**), and the line is completed.

The line is drawn from your first mouse click (Point 1) to a point referenced from absolute zero (Point 2), as shown in Figure 5-12.

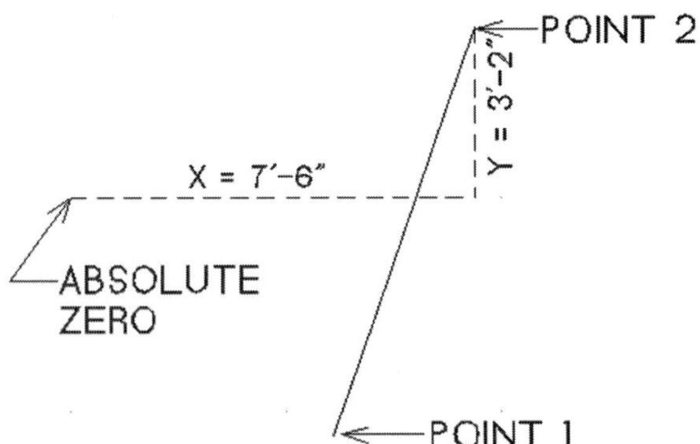

Figure 5-12 Drawing a line with the Absolute Cartesian method

The Absolute Cartesian input method is very similar to the Absolute Polar method, except for using X/Y references. For more examples of its use, see the section regarding the Absolute Polar input method.

Drawing Arcs and Circles

For arcs and circles, just as for lines, you can pick the points of the entities based on any one of the four input methods (Relative Polar, Absolute Polar, Relative Cartesian, and Absolute Cartesian). A line needs only two points to describe it: a start point and end point. For describing arcs and circles, however, a wide variety of methods can be used, depending on what kind of curved entity you want to draw and how you want to draw it.

The various curves and arcs are all accessible through the **Edit/Curves** menu. Figure 5-13 displays the 17 options available in that menu.

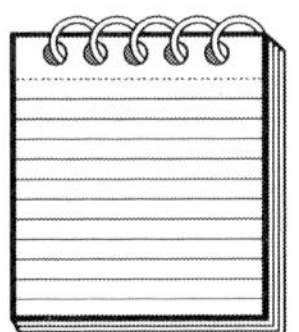

NOTE: *Whenever you pick one of the curve options, you will see a button labeled **Dynamic** (toggled on in this example).*

If this button is toggled off, you will not see the curve dynamically update as you pick points, so you won't see the curve until you have finished picking all the input points.

*With **Dynamic** toggled on, you will see the curve dynamically update as you pick points and move the cursor. In most cases, you will want to use this option. You can pick the **Dynamic** button with your mouse at any time, or you can press the **F1** key.*

2PtArc (Two-Point Arc)

Use this option to draw arcs by entering an arc center point and then selecting its two end points (see Figure 5-14):

1. Select **2PtArc** and turn **Dynamic** on.
2. Select the center (radius) point of the arc.

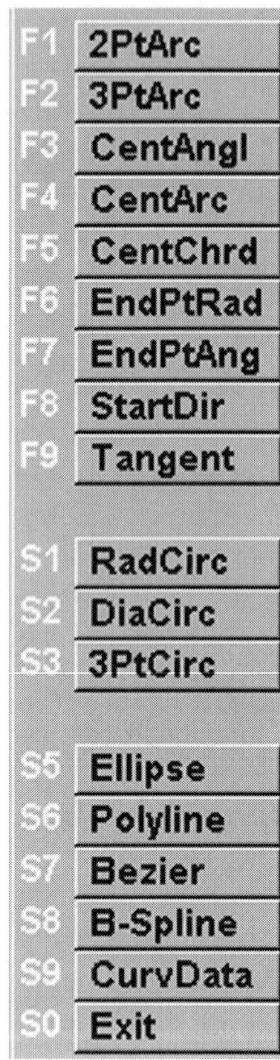

Figure 5-13
The curves menu

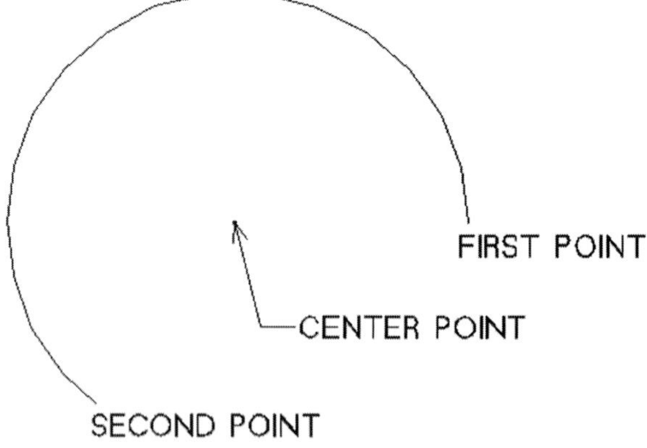

Figure 5-14
`2PtArc` in action

3. Select the start point of the arc. The arc is drawn about the center point in a counterclockwise direction unless **ClkWise** is turned on.
4. Select the second point of the arc. This point only defines the end of the angle of the arc, so it does not have to actually be located on or near the arc.

3PtArc (Three-Point Arc)

Use this option to draw arcs by entering the two end points and a third point anywhere on the arc (see Figure 5-15):

1. Select **3PtArc** and turn **Dynamic** on.
2. Select the start point of the arc.
3. Select the end point of the arc.
4. Select a point anywhere along the arc between the two end points. DataCAD draws the arc in a counterclockwise direction.

This type of curve is very useful for pointing to things, like small dimensions. Figure 5-16 is an example.

CentAngl (Center Angle Arc)

Use this to draw an arc of a specified angle after specifying the start point and then the center point of the arc:

1. Select **CentAngl**.
2. Select the starting point for the arc.

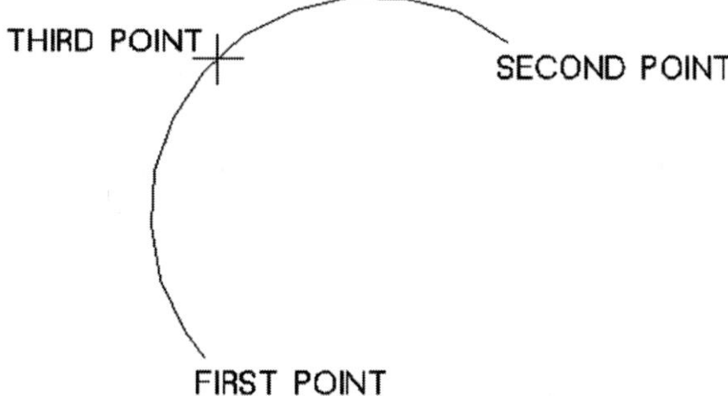

Figure 5-15
Using **3PtArc**

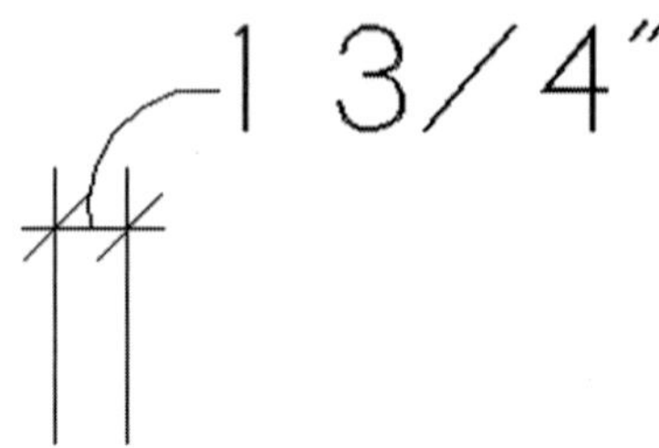

Figure 5-16
`3PtArc` example

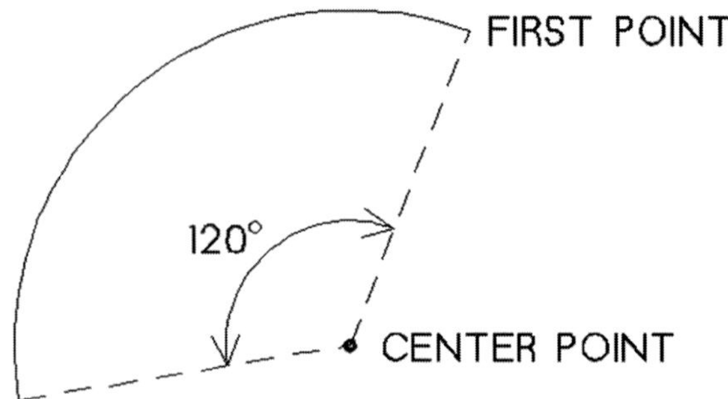

Figure 5-17
Using **`CentAngl`** to draw an angle

3. Select the center (radius) point of the arc.
4. Type or select an angle value. The arc will be drawn in a counterclockwise direction. In the following example, the angle value selected is 120 degrees (see Figure 5-17).

CentArc (Center Arc)

Use this to draw an arc of a specified length after specifying the center point and starting point of the arc:

1. Select **CentArc**.
2. Select the starting point for the arc.
3. Select the center (radius) point of the arc.

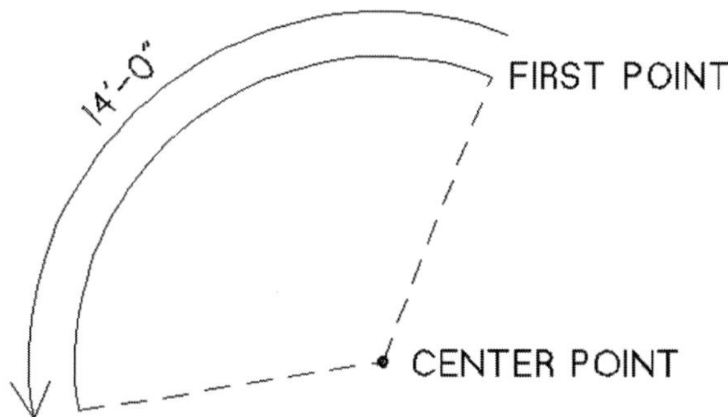

Figure 5-18 Creating the curve by arc length

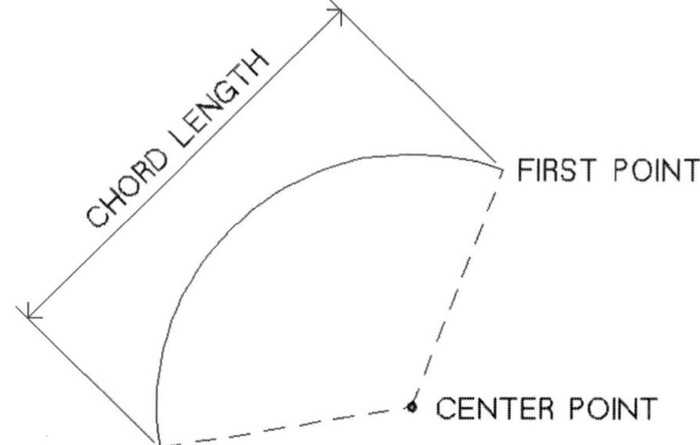

Figure 5-19 Using **CentChrd** to create the arc of a specific chord distance

4. Type or select an arc length value. The arc will be drawn in a counterclockwise direction. The arc length value selected here is 14 feet (see Figure 5-18).

CentChrd (Center Chord Arc)

Use CentChrd to draw an arc of a specified chord distance after selecting the start point and the center point of the arc (see Figure 5-19):

1. Select **CentChrd**.
2. Select the starting point for the arc.

3. Select the center (radius) point of the arc.
4. Type or select an arc length value. The arc will be drawn in a counterclockwise direction. Here the arc chord value selected is 14 feet.

EndPtRad (End Point Radius Arc)

Use this option to draw an arc of a specified radius after selecting a start point and an end point. Note that the radius must be larger than half the distance between the two endpoints of the arc:

1. Select **EndPtRad**.
2. Select the starting point for the arc.
3. Select the end point for the arc.
4. Type or select a radius length value. The arc will be drawn in a counterclockwise direction. In Figure 5-20, the arc radius value selected is 14 feet.

EndPtAng (End Point Angle Arc)

Use this to draw an arc with a specified angle after selecting a start point and an end point:

1. Select **EndPtAng**.
2. Select the starting point for the arc.
3. Select the end point of the arc.

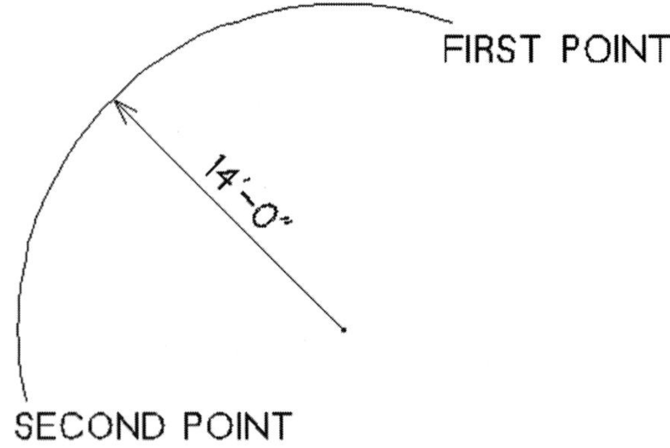

Figure 5-20
EndPtRad in action

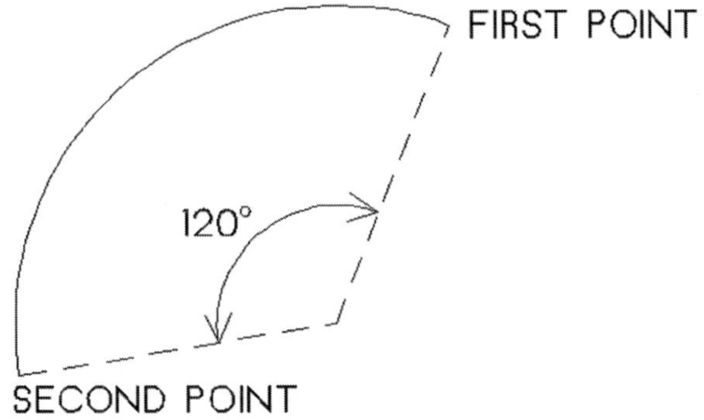

Figure 5-21
An arc with a specified angle

4. Type or select an angle value. The arc will be drawn in a counterclockwise direction. In the following example, the angle value selected is 120 degrees (see Figure 5-21).

StartDir (Start Direction of Arc)

Use StartDir to draw an arc that is tangent to the **first** and **third** points entered:

1. Select **StartDir**.
2. Select the starting point of the arc.
3. Select the end point of the arc. As you move the cursor, a rubberband line extends out from the first point selected. This represents the line that the curve will be drawn tangent to.
4. Locate the rubberband line as you want it in relation to the first point selected and then click the mouse. The curve will be drawn, as shown in Figure 5-22.

Tangent (Tangent Arc)

Use this option to draw an arc segment that is tangent to an existing line or arc segment. You can draw additional tangent arc segments by entering additional endpoints. When you do, the cursor remains connected to the last point you entered:

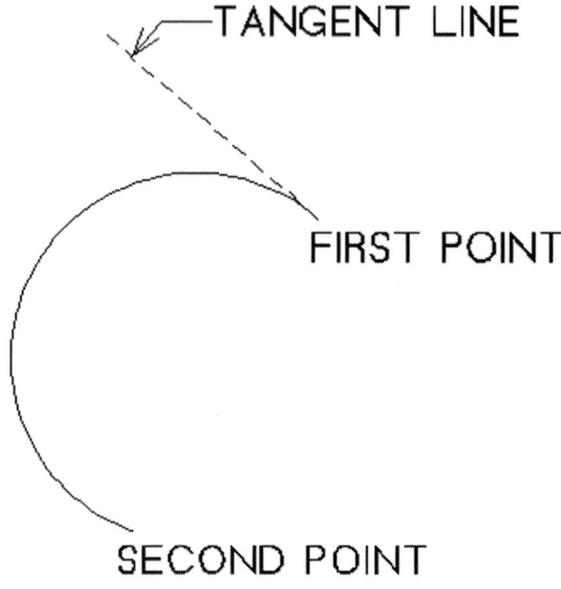

Figure 5-22
`StartDir` at work

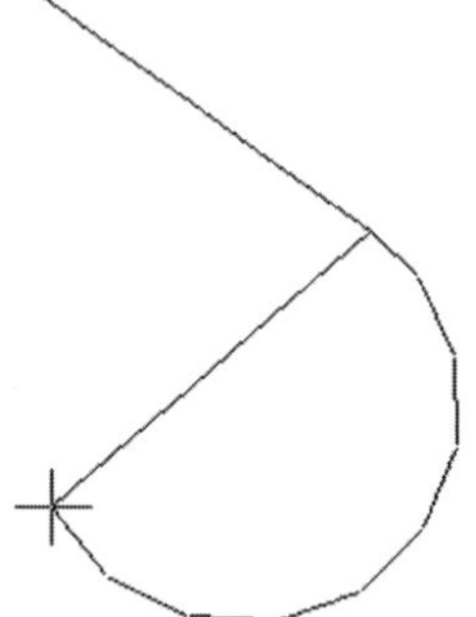

Figure 5-23
An arc that is drawn tangent to a particular line

1. Select **Tangent** and turn **Dynamic** on.
2. Click on the line or arc to draw tangent to. A rubberband line denoting the chord of the new arc is displayed. The arc will expand and contract as the cursor is moved, always tangent to the line or arc selected. Figure 5-23 is an example of an arc being drawn from a selected line.
3. Select the end point of the arc. The arc is drawn (see Figure 5-24).
4. Once an arc is drawn, you can continue to draw arcs from the last point selected. In this example, we will continue to draw a series of tangent arcs, one after the other (see Figure 5-25).

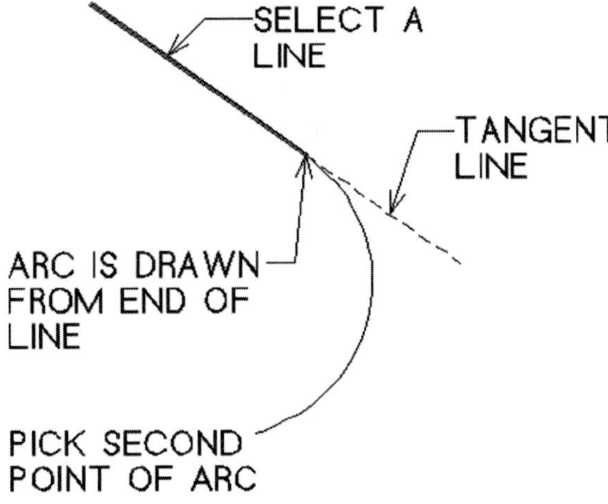

Figure 5-24
Completing the arc

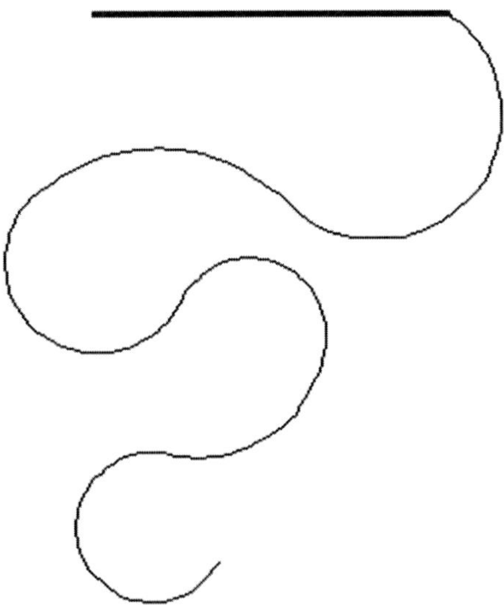

Figure 5-25
Carrying on with the process

5. After selecting the line or arc to draw tangent to, a **NewLine** option will appear in the menu. This option allows you to stop drawing the current arc and select a new line or arc to draw tangent to. You can accomplish the same thing by *right-clicking* the mouse to stop drawing the current arc.

RadCirc (Radius Circle)

Use this to draw a circle by defining the center point and then any other point on the circle (see Figure 5-26):

1. Select **RadCirc** and turn **Dynamic** on.
2. Select the center of the circle.
3. Select any point on the circle to define the radius distance.

DiaCirc (Diameter Circle)

Use this to draw a circle by defining two points on opposite sides of a circle (see Figure 5-27):

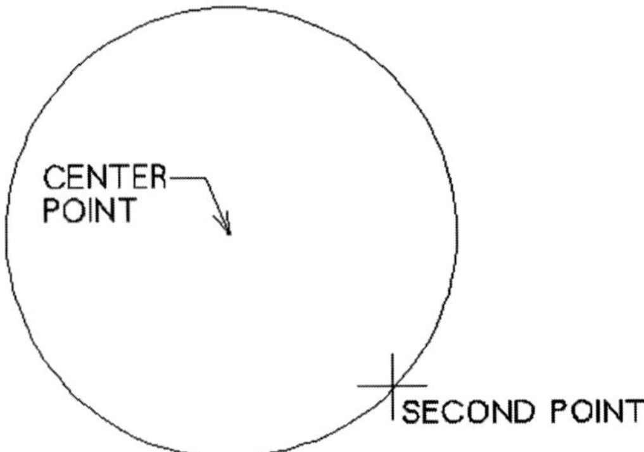

Figure 5-26 Drawing a circle by its radius

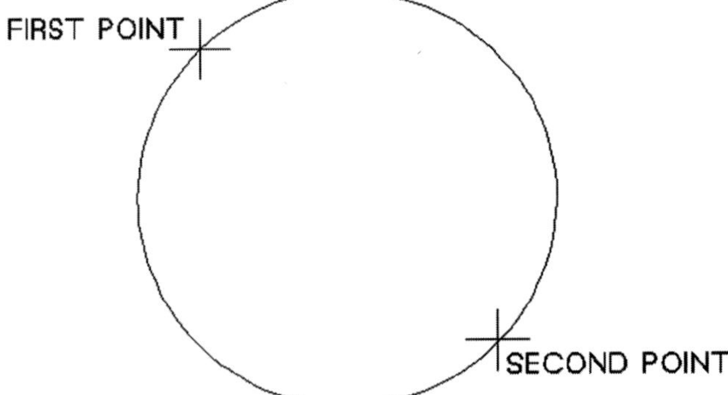

Figure 5-27 Drawing a circle diameter by its two opposite points

1. Select **DiaCirc** and turn **Dynamic** on.
2. Select any point on one side of the circle.
3. Select any point on the opposite side of the circle.

3PtCirc (3-Point Circle)

Use this to draw a circle by defining three points along the perimeter. This option is useful when the endpoints are predefined, as in the example of the two sets of walls in Figure 5-28.

1. Select **3PtCirc** and turn **Dynamic** on.
2. Select the first point of the circle.
3. Select the second point of the circle.
4. Select the third point of the circle.

Ellipse

Use this to draw an ellipse by using two opposite diagonal points that define the width and height of the ellipse. The extents of the ellipse will be

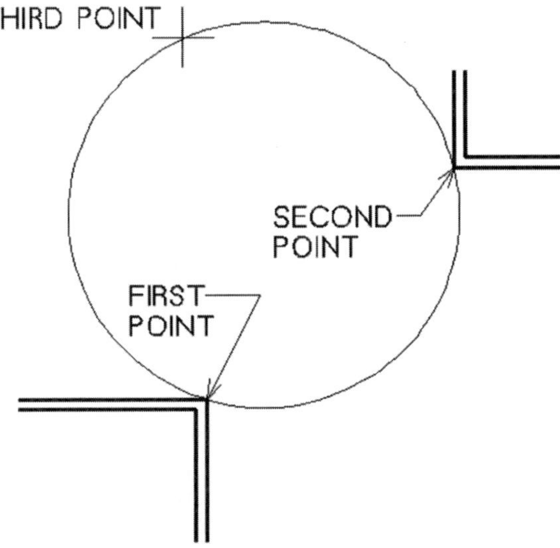

Figure 5-28 Drawing a circle using three points

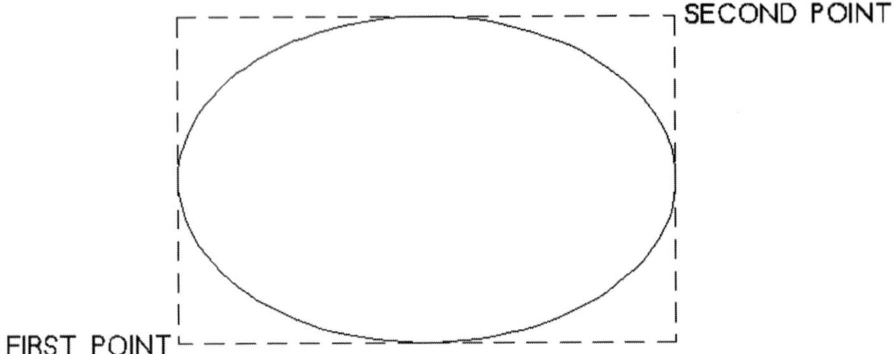

Figure 5-29 Drawing an ellipse

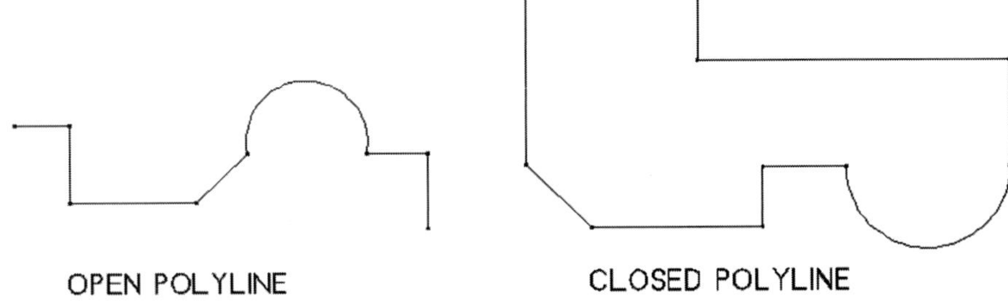

Figure 5-30 Open and closed polylines

displayed differently depending on whether **Dynamic** is turned on or off (see Figure 5-29):

1. Select **Ellipse**. If **Dynamic** is off, then the ellipse will be defined by a rubberband box defining the width and height of the ellipse. If **Dynamic** is on, then the ellipse will be displayed dynamically, though it is still defined by two opposite corners of an imaginary box.
2. Select the first corner of the ellipse box.
3. Select the second corner of the ellipse box.

Polyline

Use this to create polylines. A polyline is a continuous series of lines and/or arcs that are treated as a single entity. A polyline can be a single line or arc, or many lines or arcs. They can be open or closed depending on whether you toggle **Closed** on or off (see Figure 5-30). For information about editing existing polylines, see the section called "More On Polylines" after this section. To create a polyline, follow these steps:

1. Select **Polyline**.
2. Turn **Closed** on or off. When turned on, this creates a closed polyline. When turned off, the polyline is open at its end point.
3. Select the first point of the polyline. Two new menu options appear (see the previous curve options for instructions on how to use them):
   a. **2PtArc** draws a two-point arc segment of the polyline.
   b. **3PtArc** draws a three-point arc segment of the polyline.
4. Select the next point of the polyline. Another option will appear, **Tangent**, which draws a tangent arc segment of the polyline.
5. Continue selecting points until you complete the polyline.
6. Finish drawing your polyline:
   a. *Right-click* to stop drawing.
   b. Select **Exit** to stop drawing.
   c. Select **Close** to automatically close the polyline from the first point selected to the last point selected.

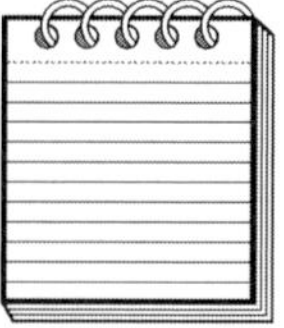

NOTE: *Polylines can be filled in with a hatch pattern, either associative or non-associative. In fact, when you create a hatch in DataCAD, you are actually creating a closed polyline with a hatch pattern inside. For more on this, see the section on "Hatching" later in the chapter. Be aware that if you hatch an open polyline, DataCAD will automatically close the polyline between the first point and the last point.*

You can also calculate a polyline's area, perimeter, and volume using the ***Utility/Measures*** *option or better yet by using the* ***Polyline*** *macro in the* ***Toolbox****.*

Bezier (Bezier Curve)

This option enables you to create Bezier curves. Bezier curves were originally developed by Pierre Bézier in the 1970s for CAD/CAM operations. In its simplest form, a Bezier curve is defined by two endpoints and two control points (you can read more about them at `www.moshplant.com/direct-or/bezier/simple_curve.html`). DataCAD currently only allows a total of two endpoints and six control points. The distance between each

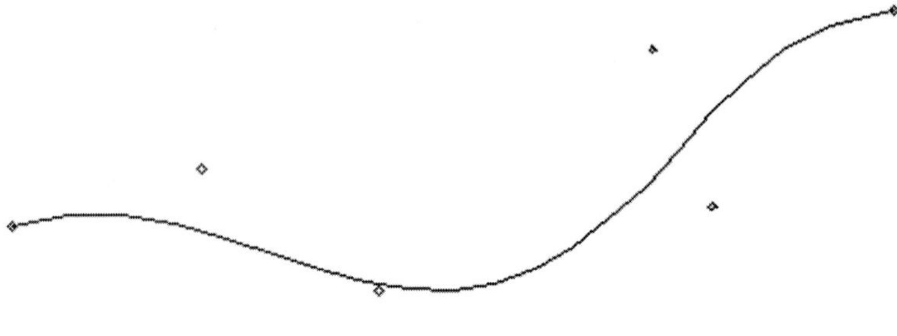

Figure 5-31
A Bezier curve

endpoint and their associated control points determines the degree of curvature between the points. Bezier curves can be used to give you continuous, flowing curves, as shown in Figure 5-31:

To draw a Bezier curve, perform the following steps:

1. Select **Edit/Curves/Bezier**. Make sure **Dynamic** is turned on.
2. Click in the Drawing Window to select the first endpoint of the Bezier curve. A marker is drawn on the screen (these markers will not print, but they can be snapped to).
3. Pick a second point. This will be the first control point.
4. Move the cursor around and you will see the curve dynamically bend and change its shape. The location of the control point determines how the curve will be shaped. Think of the control point as a kind of magnet that tugs at the curve, causing it to bulge.
5. Select a second control point. If you *right-click* now, the second control point also becomes the endpoint, and you are done creating this curve.
6. If you do not *right-click* in Step 5, then you can continue to place control points and move the cursor to change the shape of the curve. Notice that even the previous sections of the Bezier curve may be affected by the current control point and the movement of the cursor. At this step. you have the following options:
   a. **Backup** erases the last control point and the last curve defined by it.
   b. **Cancel** erases the whole curve and its control points. It is available after you select the third point.

c. **Add** detaches the cursor from the curve. The curve is drawn to the final control point selected prior to the **Add** button being selected. *Right-clicking* or pressing **Exit** will do the same thing.

DataCAD's Bezier curves are unfortunately not very useful in terms of construction, since you cannot draw them as walls, nor can they be offset to form walls. There is also no way of dimensioning the curves in DataCAD, so constructing them in the real world would be difficult at best.

B-Spline (B-Spline Curve)

Use this to create B-spline curves. A B-spline curve is very similar to a Bezier curve, but each successive control point has more impact on its own section of the spline curve and less impact on previous sections of the spline curve. Like Bezier curves, DataCAD currently only allows a total of two end-points and six control points for B-splines. The distance between each end-point and their associated control points determines the degree of curvature between the points. The steps for creating a B-spline curve are the same as those for Bezier curves. In Figure 5-32, we have used the same two end-points and the same four control points to create both a Bezier curve and a B-spline curve, so that you can see the relative difference between the two.

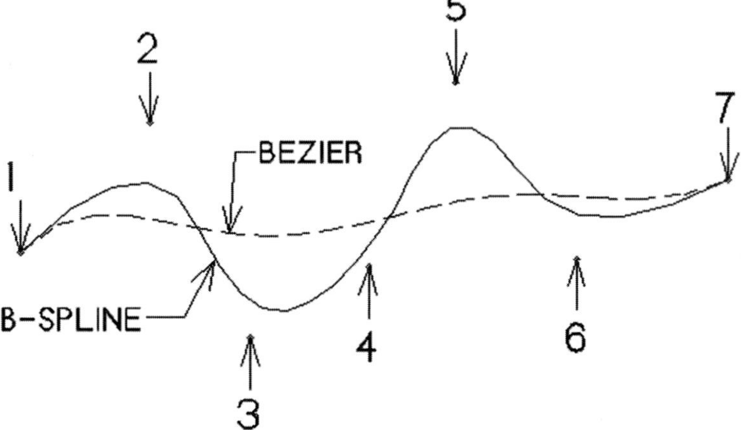

Figure 5-32 Comparing a Bezier curve and a B-spline curve

- **Backup** erases the last control point, and the last curve defined by it.
- **Cancel** erases the whole curve and its control points. It is available after you select the third point.
- **Add** detaches the cursor from the curve. The curve is drawn to the final control point selected prior to the **Add** button being selected. *Right-clicking* or pressing **Exit** will do the same thing.

CurvData (Curve from Data)

Use this to enter curves using surveyors' data. **CurvData** is DataCAD's built-in coordinate geometry routine. Here is the complete list of available toggles. Based on the toggles you choose and the information you enter, additional data is calculated. When you select the displayed attributes, any known or calculated information appears in the Message Window (see Figure 5-33).

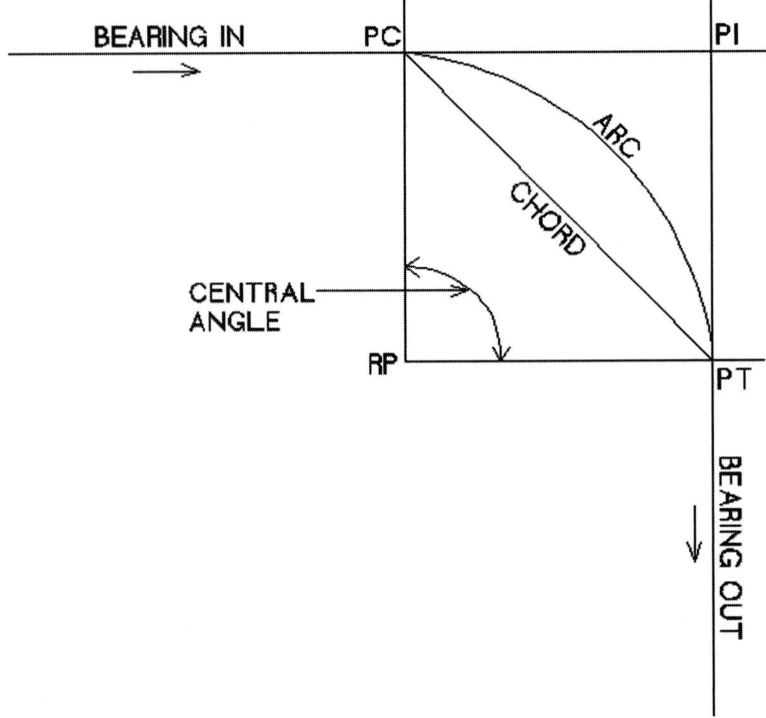

Figure 5-33
The Curve Data options

RP	Use the Radius Point to define the center of the circle or arc.
PC	Use the Point of Curve to define the starting point of the arc.
PT	Use the Point of Tangency to define the end point of the curve.
PI	Use the Point of Intersection to define the point where the tangent line intersects the Point of Curve.
Angle	Use the Angle to define the central angle about the center Radius Point defined by the Point of Curve and the Point of Tangency.
Radius	Use the Radius to define the radius of the central angle, or the distance from the Radius Point to the Point of Curve or the Point of Tangency.
Tangent	Use the Tangent to define the distance to the intersection of the tangent lines from the Point of Curve or the Point of Tangency.
Chord	Use the Chord to define the straight line distance from the Point of Curve to the Point of Tangency.
ArcLnth	Use the Arc Length to define the distance along the curve from the Point of Curve to the Point of Tangency.
BrngIn	Use Bearing In to define the bearing of the line going into the Point of Curve.
BrngOut	Use Bearing Out to define the bearing of the line going away from the Point of Tangency.
Brnge	Radius Point to the Point of Curve
BrngPT	Use Bearing to Point of Tangency to define the bearing of the line from the Radius Point to the Point of Tangency.

More on Polylines

The following section contains some more useful information about editing polylines. For whatever reason, DataCAD does not have polyline editing tools in the **Polyline** menu. Instead you have to use the Polyline macro in the **Toolbox**.

Add, Delete, and Move Polyline Vertices

Use the **Polyline** macro found in the **Toolbox** (**Edit/Toolbox**, or **M**) to add, delete, or move vertices in an existing polyline. Because a hatch pattern is a filled-in polyline, you can use this option to edit your hatches as well (see the section about Hatching later in this chapter):

1. Select **Edit/Toolbox**.
2. Select the **Polyline.dcx** macro and then press **Open**. The top of the menu will look like Figure 5-34.

Polyline Use this to draw a new polyline, just as you would in the standard **Polyline** macro. If you use this to draw a new closed polyline, you could then go to the **Hatch** menu, select **Entity**, pick the polyline, and the polyline will be hatched inside.

AddVertx Use this to add a vertex to an existing polyline by performing the following steps:

1. Select **AddVertx**.
2. You will be prompted to select a polyline to add the vertex to. *Click* on or near one of the boundary lines or vertices of the hatch polyline.
3. Pick a point between two existing vertices. A new vertices will be created, attached by a rubber-banding line to your cursor.
4. *Click* or snap this new vertices wherever you need to place it. Remember to avoid crossing over any of the other lines of your hatch polyline.

DelVertx Use this to delete an existing polyline vertex:

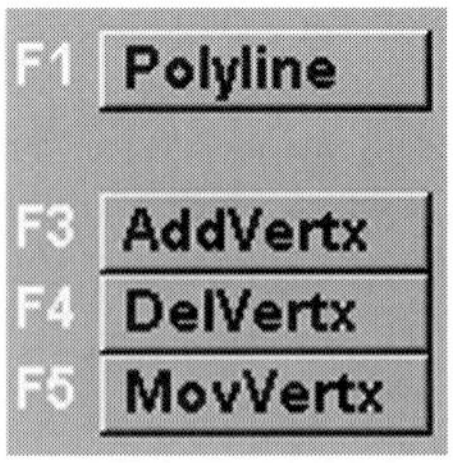

Figure 5-34
The Polyline macro menu

F1 Polyline

F3 AddVertx
F4 DelVertx
F5 MovVertx

1. Select **DelVertx**.
2. You will be prompted to select a polyline to add the vertex to. *Click* on or near one of the boundary lines or vertices of the hatch polyline.
3. *Click* on or near the vertex you want to delete and it will be removed.

MovVertx Use this to move an existing vertex. This is especially useful if two vertices from two different polylines overlap one another, but you only want to move one of them:

1. Select **MovVertx**.
2. You will be prompted to select a polyline to add the vertex to. *Click* on or near one of the boundary lines or vertices of the hatch polyline.
3. *Click* on or near the vertex you want to move. The selected vertices will be attached by a rubber-banding line to your cursor.
4. *Click* or snap the vertices wherever you need to place it. Remember to avoid crossing over any of the other lines of your hatch polyline.

FreeHand

In the Utility menu is another drawing function called **FreeHand**, which allows you to draw freehand lines and polylines without having to hold down a mouse button. It works pretty much like a pencil on paper. To use it, follow these steps:

1. Select **Utility/FreeHand**. Four options will appear.
2. Choose to draw your FreeHand entities as **2D Lines**, **3DLines**, or a **PolyLine** (see Figure 5-35).
3. Turn the **Closed** option on or off as appropriate.

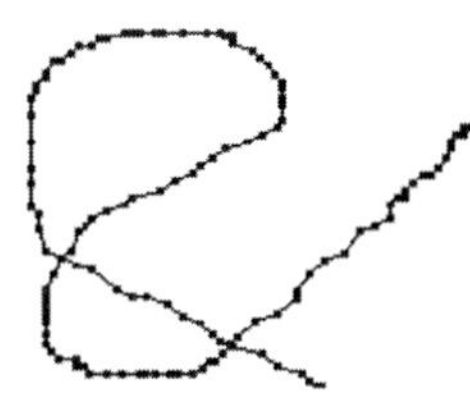

Figure 5-35 The various options provided by **Freehand**

4. Place the cursor in the Drawing Window where you want to start the line, and then *click* the mouse to start drawing.
5. Every move of your cursor will now draw a series of lines forming a continuous FreeHand shape.
6. *Click* either the right or left mouse button to stop drawing FreeHand lines.

The **Lines**, **3DLines**, and **Polyline** commands are mutually exclusive. If one is on, the others are off. Here is a brief description of each one:

Lines	Draw contiguous 2D lines.
3DLines	Draw contiguous 3D lines.
PolyLine	Draw a contiguous polyline. Note that each vertices of the polyline is denoted by a non-printing node.
Closed	This is a toggle that can be on or off, no matter which line type is selected. With **Closed** turned on, a line will automatically be added between the first point of the FreeHand line, and the last point, forming a closed shape. With **Closed** turned off, no line will be drawn between the first and last points.

Simple and Complex Linetypes

Simple linetypes are basically straight lines like the solid, dashed, dotted, or similar linetypes. Complex linetypes in DataCAD can be described as single entities that display like multiple entities. By using linetypes (like the batt insulation, plywood, and wood siding shown in Figure 5-36), these complex graphics can be drawn and edited quickly and easily, saving a great deal of time and effort. And because they are considered to be only one entity by DataCAD, the size of your drawing file will be kept smaller, thereby allowing DataCAD to run faster.

Selecting Linetypes

All linetypes are selected from the **Edit/LineType** menu. Once a linetype is selected, DataCAD will continue to draw everything with that linetype until you choose a new one. Let's try drawing a few:

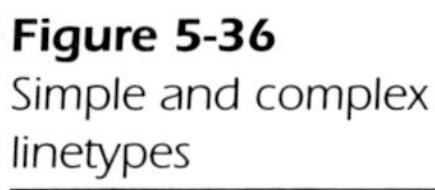

Figure 5-36
Simple and complex linetypes

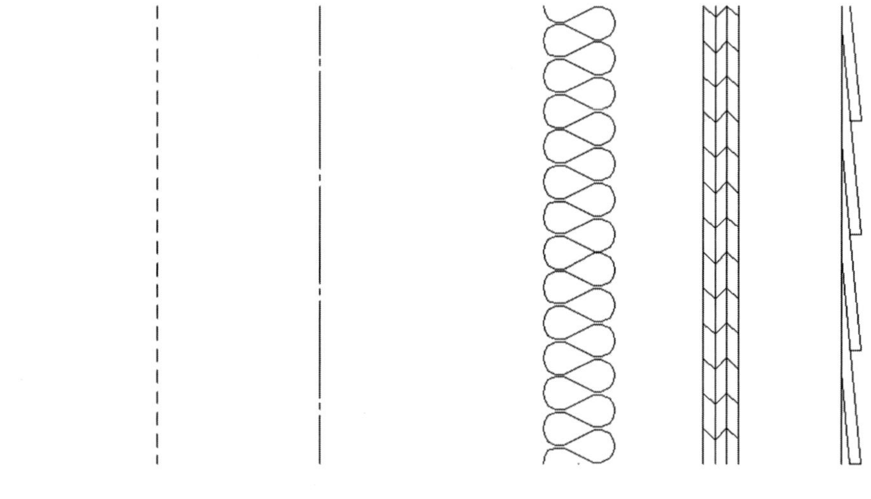

1. Go to the **Edit/LineType** menu. You will see a menu something like the one in Figure 5-37.
2. If you place your cursor over one of the LineType menu buttons, like the **Plywood2** linetype shown in Figure 5-38, you will see a graphic image of the linetype that is displayed in the right preview window.
3. *Click* on the **Plywood2** linetype and the Status Area will change to show that the **Plywood2** linetype is now the current linetype.
4. *Right-click* once to exit the **LineType** menu.
5. Place your cursor anywhere in the Drawing Window and draw a line (refer to the beginning of this chapter if you need help with this). A line that looks like plywood is drawn (more on this later in this section).
6. Draw a few more lines so that you can see how you can continue to draw more **Plywood2** lines without having to select any other options.
7. Go back to the **LineType** menu and press the **ScrlFwrd** button to see some of the other linetypes available. Notice that at the bottom of this new list of linetypes are both a **ScrlFwrd** button and a **ScrlBack** button so that you can move forward and backward through the list of linetypes.
8. Pick another linetype, such as the **Brick** pattern, and draw another line in the Drawing Window. You may now have something that looks like the line on the right in Figure 5-39.

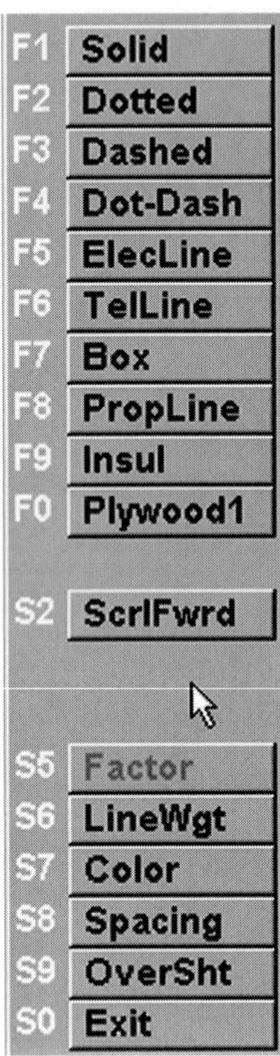

Figure 5-37
The LineType menu

Notice that both linetypes shown in Figure 5-39 display at different widths, even though you didn't change any settings. You might ask how DataCAD determines how wide to draw a line. This is controlled by the **Spacing** option in the **LineType** menu. We'll cover this later in this section about *Spacing*.

Linetype Toggle

Instead of going to the **Edit/Linetype** menu each time you want to draw a different linetype, you can toggle through the list simply by pressing the **Qq** key. You do not have to be in the **LineType** menu. Each press of the key will move you forward through the list of linetypes. Pressing **Shift+Q** will move you backward through the list. This works pretty well unless you have a lot

Figure 5-38
The LineType preview window

F1 Plywood2
F2 Hedge1
F3 CentrLin
F4 NewSectn
F5 DShngl_R
F6 LapSidR
F7 Shiplap
F8 Brick
F9 4Block
F0 MsnryBlk
S1 12Block
S2 RigdIns1
S3 RndmLite
S4 Wiggle
S5 EvShuWgl
S6 ScrlBack
S7 ScrlFwrd

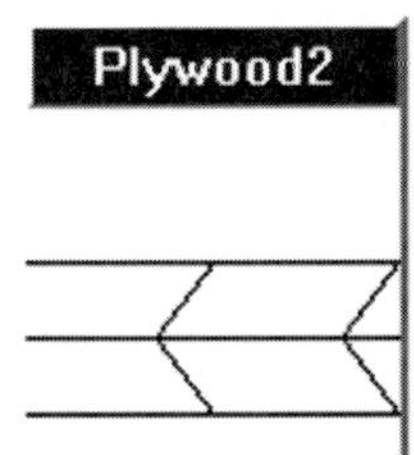

Figure 5-39
Two different complex linetypes

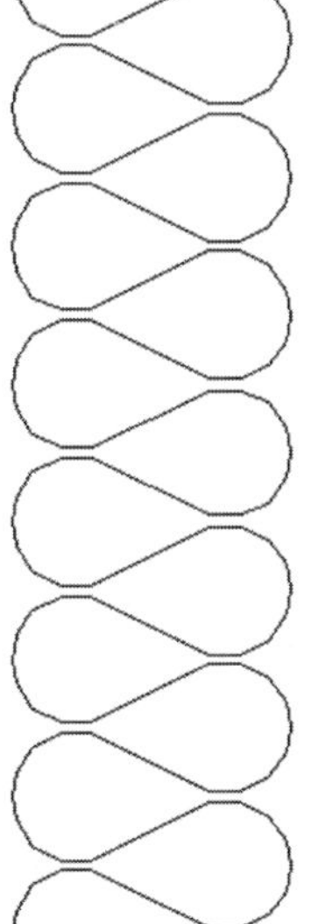

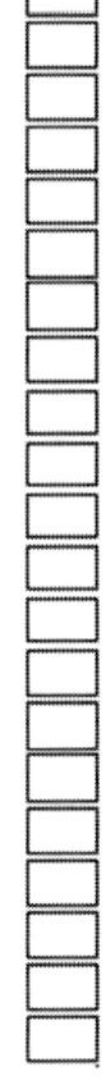

of linetypes and are not sure whether you should go forward or backward, in which case you might want to revert to the **Edit/LineType** menu.

Each time you press **Qq**, you will see the current linetype change in the Status Area. Figure 5-40 shows that the current linetype is called ***Dot-Dash***. Pressing **Qq** will cause the current linetype to change from Dot-Dash to whatever is the next linetype in your LineType menu.

Each of the lines shown in Figure 5-41 were drawn by selecting a start point and an end point. DataCAD took care of drawing all the complex linework. Look at each of the complex linetypes and try to imagine having to draw all that linework from scratch. Think of how much time you are saving by using these linetypes, instead of drawing them from scratch.

A variety of materials would be time-consuming if done by hand. To prove that each of these complex lines is actually only one line, let's turn off the display of complex linetypes via the **Display** menu (see Chapter 3, "Settings and Display Options," for more information about this menu):

1. Select **Utility/Display** and turn **UserLine** off.
2. All complex linetypes are now displayed in their basic form: simple lines with a start point and an end point, as shown in Figure 5-42.
3. Turn **UserLine** back on by picking it.

Figure 5-40 The linetype displayed in the Status Area

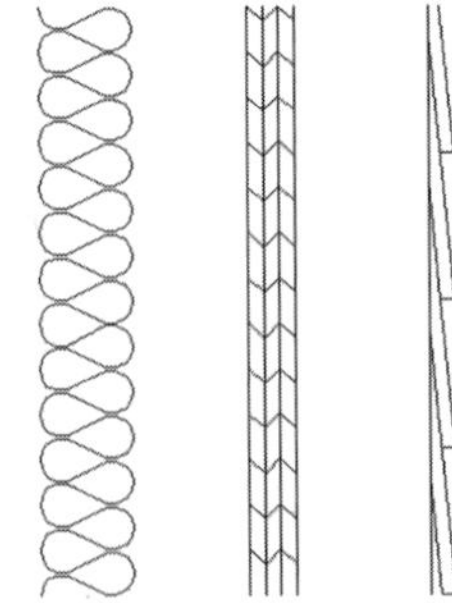

Figure 5-41 Each line was drawn by selecting only a start point and an end point.

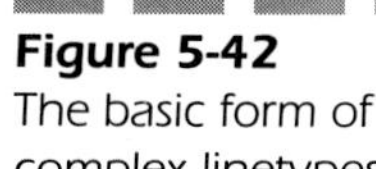

Figure 5-42
The basic form of complex linetypes

Figure 5-43
Examples of complex linetypes of various geometries

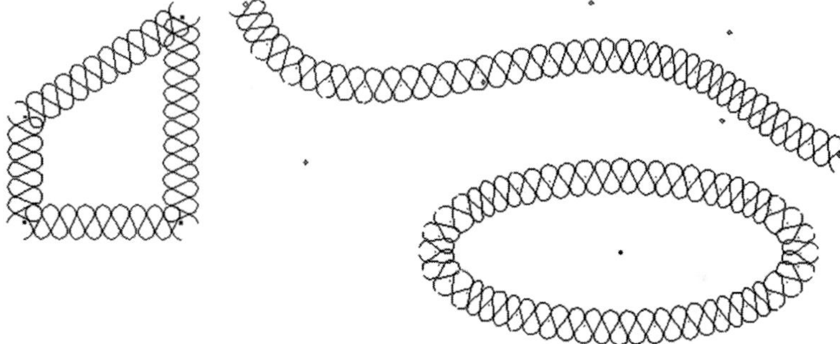

All these complex linetypes will work on any 2D lines, circles, arcs, and curves, including ellipses, polylines, Bezier curves, and B-spline curves, such as those shown in Figure 5-43.

How Linetypes Are Defined

The full explanation of how a linetype is defined is fairly complex. Without going into a lot of technical explanations, linetypes can be defined by their centers, their left edge, their right edge, or along any edge, actually. It is therefore important to know how each linetype is defined before placing it

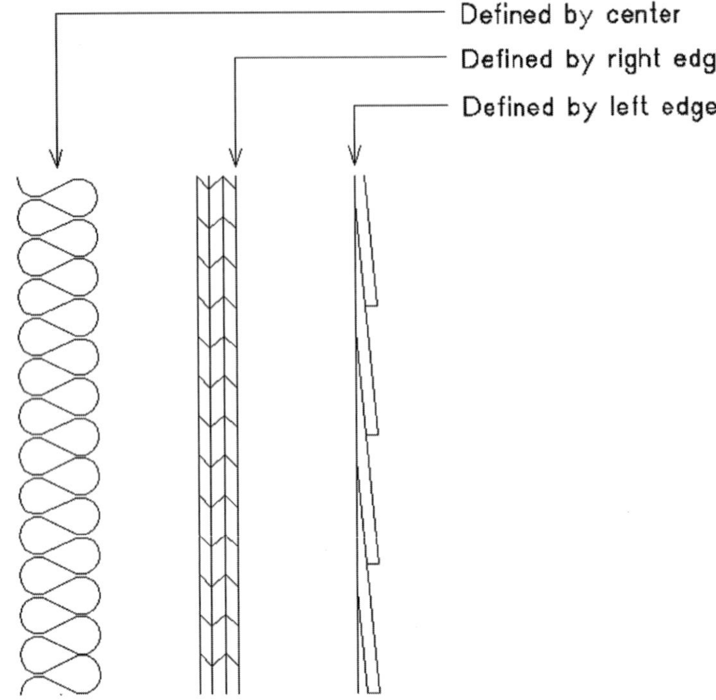

Figure 5-44 Three linetypes defined by different edges

in your drawing. If you're not sure, draw a test line and see. The three examples in Figure 5-44, are all defined differently. Look in the Appendix of this book for a complete list of the default DataCAD linetypes and how they are defined. A printable PDF chart is also included on the CD-ROM.

Some of the linetypes are directional too, such as the **ShingleR** linetype at the far right in Figure 5-44. The "R" in the linetype name means that the wood siding is displayed facing to the right when the line is drawn from top to bottom, as shown in Figure 5-45.

So if you draw it from the bottom up instead, then it would look like Figure 5-46.

When the author of **ShingleR** created the linetype, she had to decide if the user was likely to draw the line from the top down or from the bottom up. Most linetypes that are directional, such as ShingleR, ShingleL, Grass, and GroundLn, are designed to be drawn in either one of two ways:

- From top to bottom for typically vertical linetypes like building sidings
- From left to right for typically horizontal linetypes like **Grass** and **GroundLn**

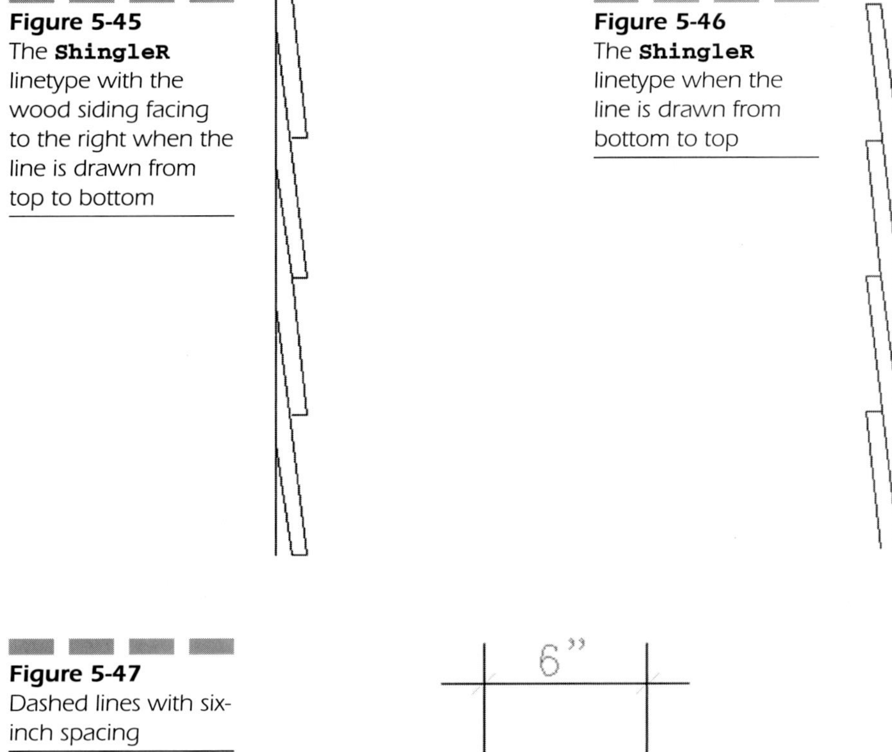

Figure 5-45
The **ShingleR** linetype with the wood siding facing to the right when the line is drawn from top to bottom

Figure 5-46
The **ShingleR** linetype when the line is drawn from bottom to top

Figure 5-47
Dashed lines with six-inch spacing

Spacing

Spacing describes the "scale" of the linetype and often, but not always, describes the space between each repeating element of a linetype. For instance, a spacing of 6″ [152] for the **Dashed** linetype would draw a dashed line with a spacing of 6″ [152] from the center of one dash to the center of the next (see Figure 5-47):

So for some linetypes, like the simple Dotted, Dashed, and Dot-Dash linetypes, the **Spacing** option controls how far apart the dots and dashes are. But for many of the more complex linetypes, the spacing determines the width of the linetype rather than the spacing between elements. For instance, a spacing of $^3/_4$″ [19.1] for the **Plywood1** linetype would give you a line that displays as $^3/_4$″ [19.1] wide, as shown in Figure 5-48.

In some of the previous linteype examples, you probably noticed that different linetypes displayed with different widths, even if the spacing is

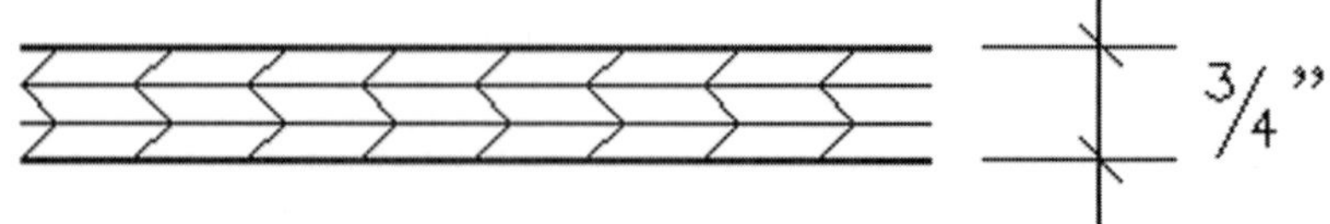

Figure 5-48
The **Plywood1** linetype with a spacing of 3/4″

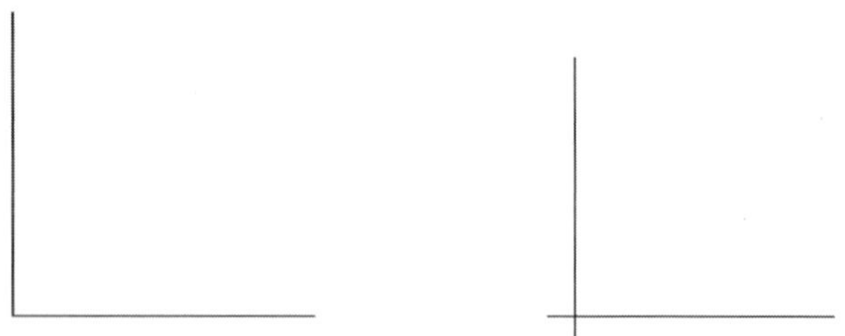

Figure 5-49
Displaying the line overshoots

unchanged. Or you may have asked yourself how to draw the **Brick** pattern so that the bricks are shown at the correct size. All of these factors are controlled by two things, both of which are related to one another:

1. How the linetype was defined by the person who created the linetype
2. The spacing of the line in the **LineType** menu

The first factor, how the linetype was defined, is described in very basic terms in the previous heading (if you want to find out more about it, you can read about creating custom linetypes in Chapter 23). However, it is worth mentioning that when linetypes are created, the authors usually try to create something that makes sense and works with the **Spacing** option. For instance, to get batt insulation that is 6″ [152] wide, you need to set the **Spacing** to 3″ [76]. Since the **Insul** linetype is defined by its centerline, this gives you 3″ [76] on either side of the centerline for a total width of 6″ [152]. The author of the **Insul** linetype could have defined it so that a **Spacing** of 6″ [152] would yield 6″ [152] insulation, but he/she did not.

So how do you know what to set the **Spacing** to for each linetype? A complete list of the default DataCAD linetypes, along with their spacings, is included in the Appendix of this book, and there is a printable chart on the CD.

OverSht (Overshoot)

Use **OverSht** to display line overshoots, giving a looser, more hand-drawn look to your drawings (see Figure 5-49). Usually, a dimension of an inch or a fraction of an inch is sufficient, but you will have to experiment to see what you like. Toggling the **Utility/Display/OverSht** button on will display the overshoots of all the lines for which you defined an overshoot. Toggling **Utility/ Display/OverSht** off will display all lines without any overshoot.

Use **OverSht** to set the linetype overshoot dimension. Usually, a dimension of an inch or a fraction of an inch is sufficient, but you will have to experiment to see what you like. Refer to the explanation of **OverSht** in the **Display** menu earlier in this chapter for more information. Alternatively, instead of going through the menus, you can toggle **OverSht** on and off by pressing the dash (-) key on your keyboard.

Factor

This is a toggle button that lets you switch between using the default spacing factor (the one used when the linetype was created) and the spacing factor set by the **Spacing** option. Selecting the **Spacing** option will automatically toggle the **Factor** button off.

The default spacing of a linetype is set within the definition of the linetype itself. It is there to ensure that the line will form the desired real scale size. For instance, the **Plywood1** linetype was created so that its default width would be $^3/_4''$ [19.1] wide. With the **Factor** button turned on, the plywood will always be drawn at $^3/_4''$ [19.1] wide.

If you have been drawing other linetypes using different **Spacing** settings, turning on the **Factor** button will temporarily override the current **Spacing** setting and revert back to the default spacing for the linetype you are drawing with. Turn the **Factor** button off and you will be back to whatever spacing you were working with previously.

LineWgt (Line Weight)

In DataCAD, you can vary the thickness of your displayed and plotted lines in one of two ways:

1. Using a **LineWgt** factor
2. Mapping specific line widths to each color throughout your drawing

The second option is by far the better option for the majority of users. The only real drawback to this method is that all the lines on your screen display at the same width, so you don't have a WYSIWYG (what you see is what you get) display of all the various line thicknesses. You don't actually see the results until you select **Plot/Preview**, or print the drawing. To map lineweights to colors, use the **PenTable** option in the **Plotter** menu. You can read more about this in Chapters 3 and 7.

In this section, we will describe how line weights work and then list their drawbacks, so you can decide for yourself if you want to use the LineWgt option. In Chapter 3, we suggested that you do not use line weights because of the way DataCAD displays and prints them. To make entities wider, DataCAD simply adds identical lines and arcs adjacent to the first ones, as shown in Figure 5-50. Notice that due to the computer screen's resolution, horizontal, vertical, and diagonal lines often do not look the same on your screen, even though they are of the same line weight.

The reason for this line weight methodology dates back to when all plotters drew lines with a real ink pen. Since the ink pen had a fixed width, to get a fatter line the plotter would have to draw a number of lines close together to output lines that were wider than the pen itself. The LineWgt function is a carry over from those days.

Drawing thicker lines in this manner means there are a great deal more lines on your screen, giving you correspondingly slower redraw and refresh times (because the computer has to draw many more lines). Because so many lines overlap one another, finding the snap points of intersections, lines, and arcs becomes difficult. Also notice that the corners of the rectangle do not intersect cleanly when zoomed into.

If you zoom out away from the lines, they do appear more like single, fat lines, but the image still leaves a lot to be desired, as Figure 5-51 will show you.

Figure 5-50
Line weights display differently depending on angle.

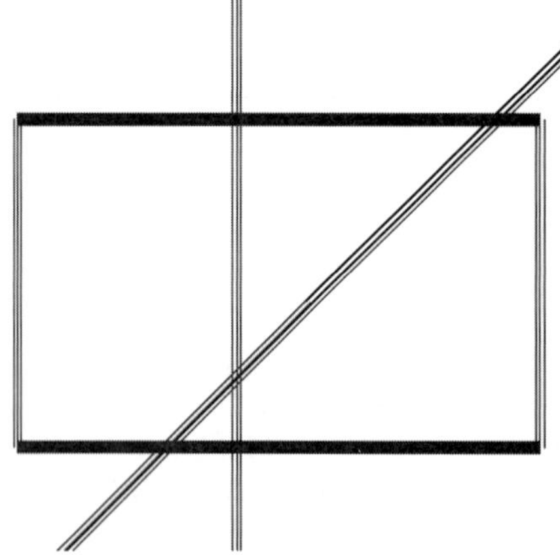

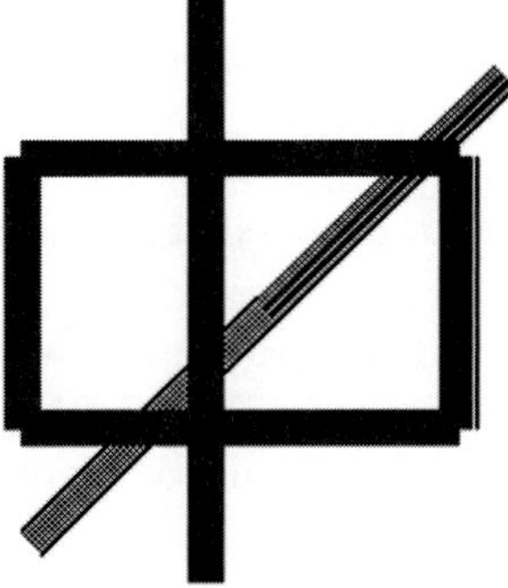

Figure 5-51
Zoomed-out view of the increased line weights

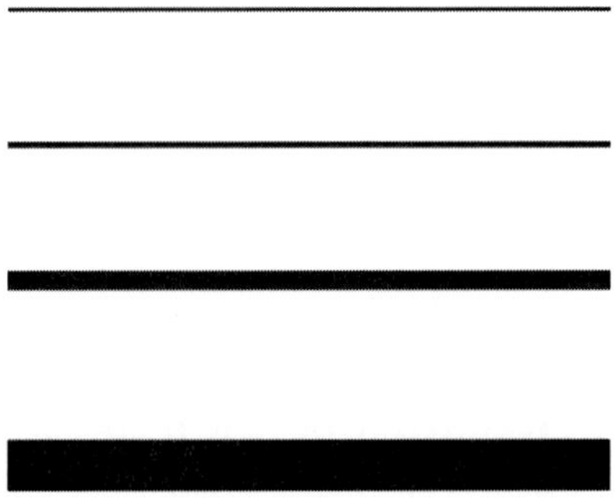

Figure 5-52
Line weights of 1, 2, 4, and 10

To change line weights, perform the following steps:

1. Select **Edit/LineType**.
2. Pick a new linetype or keep the current one.
3. Select the **LineWgt** button.
4. Type a whole number between 1 and 99 and then press **Enter**.
5. The selected line weight will remain in effect until you change it.

Alternatively, instead of going through the menus, you can increase line weights, one number at a time, by pressing the **Ww** key on your keyboard (use **Shift+W** to decrease line weights).

Figure 5-52 illustrates what line weights of 1, 2, 4, and 10 might look like on your screen when you are zoomed out (so that you don't see the multitude of lines that make them look wider).

A line weight of 1 will display and print as a single line. Each additional weight adds a stroke of the "pen" to each side (think pen plotter days). A line weight of 2 will display and print a total of three lines: the primary line down the center, plus one additional stroke on each side of it.

Warning: Technical Jargon Is Imminent

A line weight of 1 is one stroke, a single line. A line weight of 2, as noted earlier, draws the primary line down the center and then adds one stroke on either side. To find out how many lines will be displayed and printed, you can therefore use the following formula:

$$\text{Total strokes} = (\text{numerical weight} \times 2) - 1, \text{ or } (W \times 2) - 1$$

So a **LineWgt** of 5 would have a total of nine strokes to it: $(5 \times 2) - 1 = 9$.

The interstitial space (or gaps) between the individual strokes is set by going to **Utility/Plotter/PenWidth**. The value is set as a whole number that represents 100th of $^1/_{32}$″ [0.8] (because DataCAD's basic unit of measurement is $^1/_{32}$″ [0.8]). Confused? Here's an example:

1. A **LineWgt** of 17 would yield a fat line with 32 interstitial spaces: the total spaces = (numerical weight − 1) × 2, or $(17 - 1) \times 2 = 32$.
2. If you then set the **PenWidth** to 50 (the highest allowable value is 99), you would wind up with 32 spaces at $^{50}/_{100}$ ths of $^1/_{32}$ of an inch, or a total of half-inch.
3. If you print or plot that line at full scale (12″ scale 1:1), you will get a line that measures a $^1/_2$″ wide.
4. If you then try a **LineWgt** of 33 (which yields 64 interstitial spaces) still at a **PenWidth** of 50, then the printed line will be 1″ [25.4] wide.

NOTE: *With **Utility/Display/ShowWgt** off, all entries are displayed and printed at a width of one pixel. With **ShowWgt** on, all entries are displayed and printed with their assigned line weights.*

Drawbacks to Using LineWgt

The drawbacks to the LineWgt option are mainly due to the large number of lines that are displayed and printed in order to show wider lines. These drawbacks include

- Slower refresh and redraw times
- A confusing number of lines when zoomed in
- Difficulties finding snap points and intersections, especially on dense drawings
- An inconsistent display of horizontal, vertical, and diagonal lines (they don't look the same on your screen)
- Increased plotting times
- Available line widths are limited because each weight is limited by the **PenWidth** setting. Although you can theoretically create a nearly infinite number of line weights by using the **PenWidth** setting, doing so would be extremely time-consuming and would have to be done for every drawing file.
- Without also using the pen mapping technique you cannot make screened or half-tone lines
- If you need to print a smaller scale version (like an $8\frac{1}{2}'' \times 11''$ check plot) of a larger sheet or detail, you cannot easily change the line weights, causing the printed drawing to bleed together with a lot of fat lines
- Corners show a gap and do not meet cleanly. This is more apparent with thicker line weights.

The majority of current DataCAD users do not use the **LineWgt** option due to all the drawbacks, but others swear by it. You will have to decide for yourself.

Color

As the name suggests, use this button to change the current color of the active layer. Simply select the **Color** option, then pick a color. As you change colors, you will see both the cursor crosshairs and the colored square in the upper-left corner of the Status Area change to the selected color (see Figure 5-53).

Figure 5-53 The square displays the current color.

Alternatively, instead of going through the menus, you can toggle through the 15 standard DataCAD colors by pressing the **Kk** key on your keyboard (press **Shift+K** to go backward through the colors). This is pretty darn fast, so I never end up using the **Edit/LineType/Color** option.

The four buttons at the bottom of the menu are for choosing custom colors beyond the 15 standard colors and for adjusting the way colors are displayed on your screen. This is helpful for persons who are color blind or for adjusting colors that may be hard to see, like Yellow lines on a white background. See Chapter 3 for more information, especially regarding the **Palettes** option. The four buttons are as follows:

- **Adjust** Selecting **Adjust/Custom** and then *right-clicking* will bring up the standard Windows color picker that allows you to alter the color, hue, saturation and luminance of whatever color you select (see Figure 5-54). By selecting **Adjust/Palettes**, you can save the current RGB file or load a previously saved RGB file.

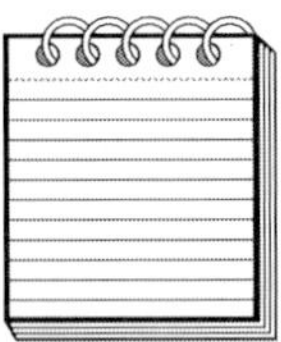

NOTE: *These **Adjust** settings are "trap doors" (shortcuts) to the **Edit/Change/ Color** menu and the **Utility/Display/Palette** option. Adjusting colors or loading a new RGB file will affect the entire drawing file, not just the line colors.*

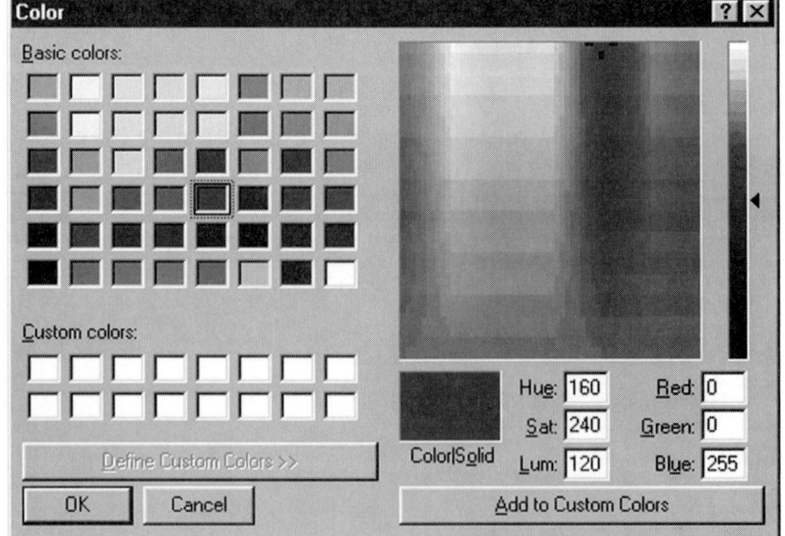

Figure 5-54 The Color picker dialog box

- **Custom** Selecting this option allows you to pick any custom color, from 1 to 255, from the current Windows color palette. Don't confuse this **Custom** option with the **Adjust/Custom** option described earlier. This custom option does not use the Windows color picker but instead picks from the standard 255 colors available to DataCAD.
- **Match** Select **Match** and then pick an entity in the Drawing Window to match its color.
- **NoChange** Selecting this option is the same as picking the **Exit** button in other menus. It simply stops all current selections and backs out of the Color menu. You can do the same thing by *right-clicking* your mouse instead.
- **Exit** Selecting this button exits you from the **LineType** menu back to the **Edit** menu.

Z-Heights and $2^1/_2$ D

Beside having a length and a width, all 2D entities in DataCAD have an extruded Z-height as well. DataCAD refers to this as $2^1/_2$ D (somewhere between 2D and 3D). By default, most DataCAD entities have a $2^1/_2$ D height of 8′-0″ [2438], starting at a Z-base of 0. But you can make them any height you want, and they can start at any Z-height, not just at zero. By default, 2D doors are drawn with a Z-base of 0 and a Z-height of 6′-8″ [2032]. 2D windows are drawn with sills at a Z-base of three feet, 3′-4″ [1016]. and a header height of 6′-8″ [2032]. These and other door and window settings can be changed to suit your needs.

None of this will affect how or what you see in **Ortho** view in DataCAD, but all of this can be very useful for creating quick views of the building you are working on. Figure 5-55 shows a 2D building plan.

Figure 5-56 shows that same plan viewed in perspective (from the DataCAD 3D menu).

Notice the heights of the doors and windows. Also notice that there are no solid walls or headers over or under the doors and windows. That's because the 2D lines that make up the walls are simply extruded upward. If you want headers and walls to fill in the gaps around the windows and doors, you can draw 2D lines to fill in across the doors and windows, and change their Z-bases and Z-heights to match the tops and bottoms of the doors and windows.

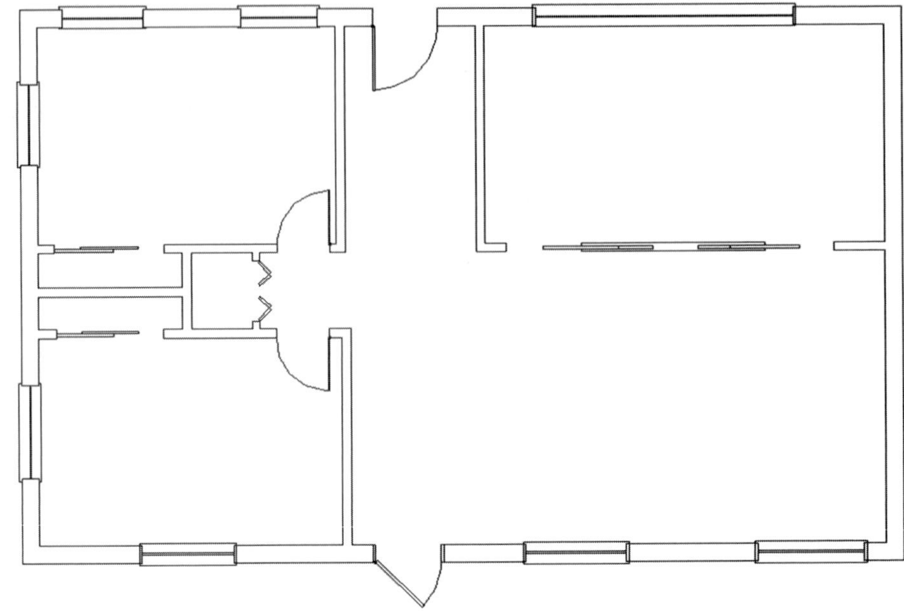

Figure 5-55
A 2D building plan

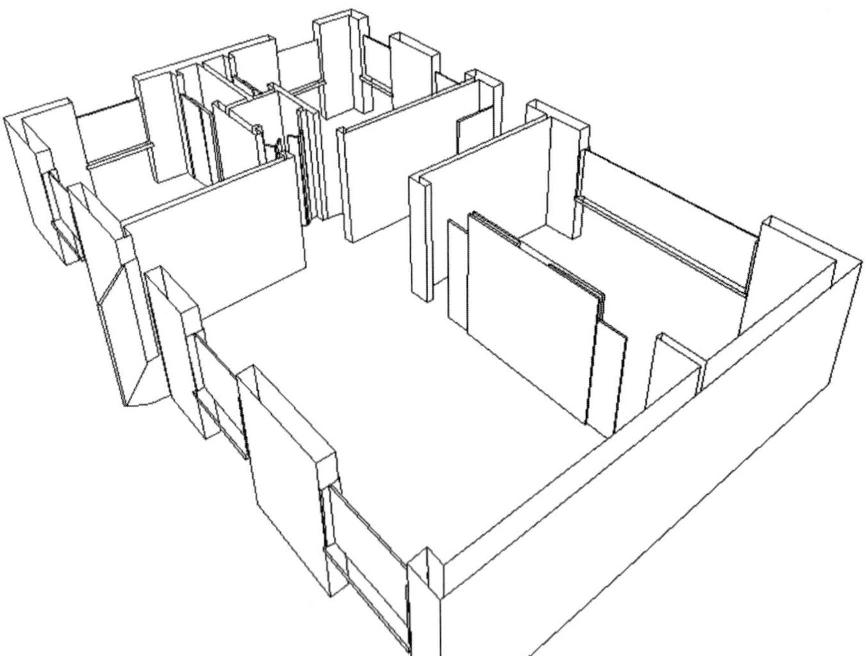

Figure 5-56
A 3D perspective of $2^1/_2$ D entities

Lines and arcs are drawn with whatever the current Z-base and Z-height are set to. By default, these are 0 and 8′-0″ [2438]. If you want to change these defaults, go to **Edit/DCAD 3D/Settings/Z-Base** and **Z-Height** and set them to whatever you want. These settings will remain until you change them and will only affect the file you are working in.

If you want to change the Z-base or Z-height of existing entities, perform the following steps:

1. Go to **Edit/Change** and select **Z-Base** or **Z-Height**. You can select one at a time, or both at the same time.
2. Select the entity or entities to be changed.

If I add the door and window headers and sills, add a horizontal 3D slab for a ceiling, and then a 3D roof to top it all off, we have a pretty fair representation of our building, inside and out, without a lot of additional effort (see Figures 5-57 and 5-58).

The $2^{1}/_{2}$ D aspect of DataCAD is one of its strengths and one that can really help you and your client quickly visualize the building.

Undo and Redo

Perhaps the most long-awaited new feature in DataCAD 9 is the addition of global, unlimited **Undo** and **Redo**. In the past, DataCAD could only do one

Figure 5-57
A building composed of $2\,^{1}/_{2}$ D entities and a 3D roof

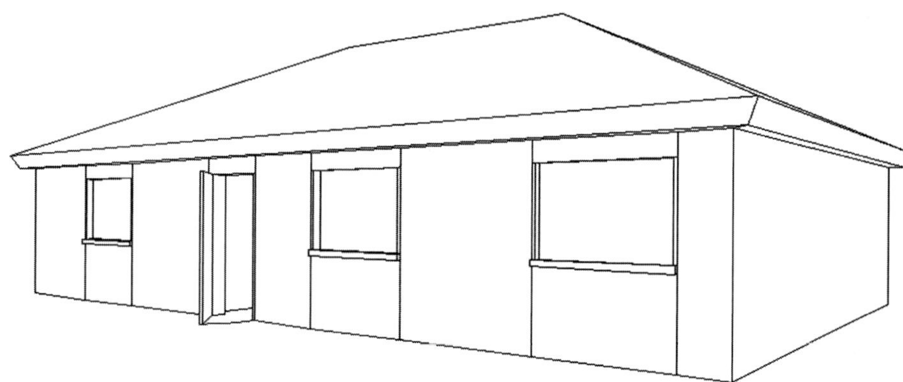

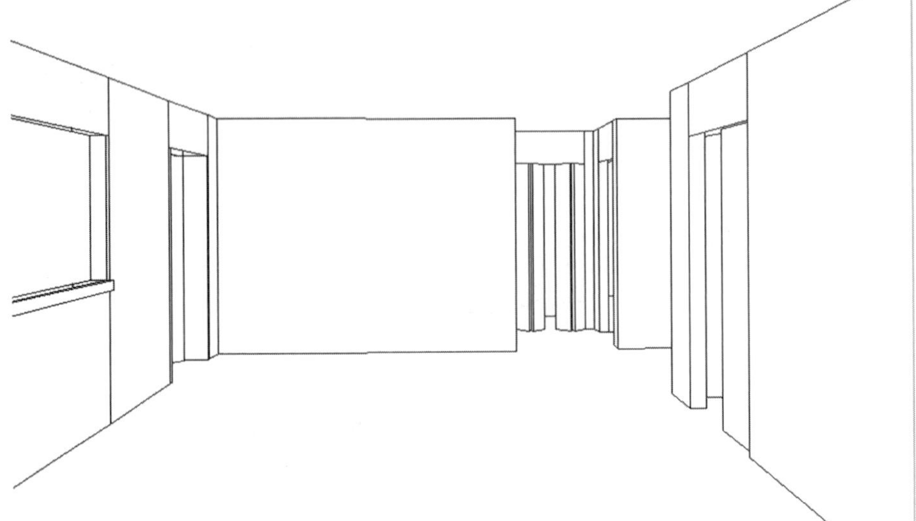

Figure 5-58 Interior perspective of 2 1/2 D building entities

level of Undo on some editing commands like **Move**, **Copy**, **Rotate**, or **Change**. Many other commands had no undo capability at all, including all of the 3D editing tools. In addition, the **Undo** commands that were available were local to the command being executed and had to be performed immediately after executing the command or the chance to undo was lost. DataCAD 9 changes all that with the addition of globally accessible and unlimited **Undo** and **Redo**.

As the name implies, the **Undo** feature allows you to back up through your previous actions, reversing each one as you go, and it can be performed at any time. If you go too far back with **Undo**, then **Redo** will move you forward through the previously Undone actions, restoring them as you go. You can access Undo and Redo in one of three ways:

- Select **Edit/Undo** or **Edit/Redo** from the drop-down menu.
- Press **Ctrl+Z** to Undo, and **Ctrl+Y** to Redo.
- Turn on the Undo/Redo toolbar (**View/Undo/Redo** Toolbar), then click on the Undo or Redo buttons.

Undo

Undo will back up through any previous commands you have issued to DataCAD, even the deletion of drawing layers. Each action you take in DataCAD makes changes to the drawing database. Selecting **Undo** simply steps backwards through those changes, reversing each action sequentially. When you select the Undo command from the drop-down **Edit** menu, you will see a second word after the word Undo (Draw, Erase, Move, Hatch, Explode, and so on). This tells you quickly what the last command was and what will be undone.

Redo

Redo is the opposite of **Undo**. It is only available once something has been acted on by the Undo command. It will move forward through any previous Undo commands that you have issued to DataCAD. Like **Undo**, selecting **Redo** simply moves forward through the sequence of drawing database changes, reversing each Undo action sequentially. When you select the Redo command from the drop-down **Edit** menu, you will see a second word after the word Redo (Draw, Erase, Move, Hatch, Explode, and so on). This tells you quickly what the last command was and what will be Redone.

If you have used DataCAD prior to version 9, then you will notice that there are no longer any **Undo** options in the Menu Windows, so you will now have to get used to using **Ctrl+Z** and **Ctrl+Y**. However, the following old keyboard shortcuts still work as they did before and can perform the following tasks:

,	Erase the last entity
.	Restore the last entity erased
<	Erase the last group
>	Restore the last group erased

These shortcuts are very limited, however, since they generally only go back one step, and only act upon the currently active layer. You are therefore better off not using these shortcuts and getting used to using **Ctrl+Z** and **Ctrl+Y** instead.

Undo and Redo only act upon the drawing you are currently working in, so if you have more than one drawing open at one time, Undo and Redo will only act on the currently active drawing. When you switch to one of the other drawings, Undo and Redo will then only act on that drawing file.

Undo/Redo Buffer

All the actions you take in DataCAD are stored in a buffer in your computer's RAM memory. As long as the information in the buffer remains intact, then DataCAD can Undo and Redo any commands you have given to DataCAD. The information in this buffer will stay intact until one of the following happens:

- You exit the drawing normally.
- You exit the drawing abnormally (DataCAD or your computer crashes).
- You clear the Undo/Redo buffer when purging symbols (**Utility/Directry/Purge** Symbols or **Utility/Template/PurgeSym**).
- You save the drawing under a new name using the **File/Save As** option.

What Will Not Undo and Redo

As of the writing of this book, Undo and Redo do not work on a certain things.

- The Undo and Redo commands do not undo and redo the views you were in (pan, zoom, and so on) as the Undo and Redo commands are acted on.
- The Undo command will not restore deleted 3D GotoViews.
- When a layer is deleted from the drawing file, it is also deleted from any 3D GotoViews that it was in. Undo will not reinsert that layer into the 3D GotoViews that it was removed from.
- Undo does not currently work on any of the Multi-Scale Plotting functions.
- Most Toolbox macros will work fine with Undo/Redo, but some macros may experience problems if Undo is invoked while the macro is running. When using macros that were written prior to the implementation of version 9's Undo/Redo, you should save your work frequently and use Undo/Redo sparingly until you gain confidence that any particular macro interacts with Undo/Redo without unexpected consequences.

The Technical Side of Undo/Redo

The Undo system does its work by recording the modifications, and only the modifications, made to the entities in the drawing database (the .SWP file) and then reversing those modifications upon the user's request (when the Undo or Redo command is executed). In other words, and in a simple case, if the user changes a line's color from white to red, DataCAD executes this request by modifying that line's record in the database (.SWP file). The Undo system sees this action taking place and records the entity affected and the color before the operation (white) and the color afterwards (red). It also records the notion that DataCAD did an undraw and then a redraw of the entity to show the change. If and when the user asks for this operation to be undone, the actions recorded are reversed. The recorded actions that are performed by the user using the **Change** command and picking the line 'X' would be

1. Undraw entity X (draw it in the background color so it disappears).
2. Change the data value of entity X (the color value from white to red).
3. Draw entity X (in its current color).

When undoing this operation, the Undo system simply performs the reverse of each operation and in reverse order. If the user hits Undo, then the following steps are performed and in this order:

1. Undraw entity X (and draw it in the background color so it disappears). This is the reverse of Step 3 earlier.
2. Change the data value of entity X (the color value from red to white), which is the reverse of Step 2 above.
3. Draw entity X (in its current color), which is the reverse of Step 1 above.

A Redo after the Undo would end up executing the original three steps in their original order and therefore produce the same results as the original **Change** command.

Thus, the undo buffer is really like a recorded script of database changes and drawing operations, and it is smart enough to know and understand each operation's inverse. Once again, this script or undo buffer resides in memory and outside of the DataCAD temporary file system.

Reference Point

A *reference point* is a point that you measure from in order to accurately place or draw another entity. If you were drafting with pencil and paper, you would place your scale on the drawing with the zero at a reference point like a corner or the end of a line, and measure to another point where we would start drafting a new line. In DataCAD, you get to the **RefPnt** option in one of three ways:

- By pressing the ` key (the un-Shifted ~ key above the **Tab** key)
- By selecting **Utility/Reference Point** from the drop-down menu
- By selecting **Utility/Measures/RefPnt** from the standard DataCAD menus

To be honest, I don't know anyone who uses the **Utility/Measures/RefPnt** method. It is far easier to just press the ` key. However, one thing you will want to do is go to the **Utility/Measures/RefPnt** menu and turn on the **DrwMarks** (Draw Marks) option. Now a temporary, non-printing × will be displayed on the screen at the reference point. This is great for confirming that you have indeed selected the correct reference point, especially in a crowded drawing (Chapter 3 suggests that you make this setting part of all your default drawings so that you won't have to remember to turn it on in each new drawing). It's interesting to note that some other CAD programs don't have an equivalent feature such as this.

Using the outside corner of the wall in Figure 5-59, let's see how you would draw a line whose first point started 2′-0″ [610] to the right of the intersection of two lines:

1. Press the ` key.
2. Snap to the intersection of the two lines that you want to reference off of. If you have the **DrwMarks** setting turned on, a small × will appear at the point you snapped to.

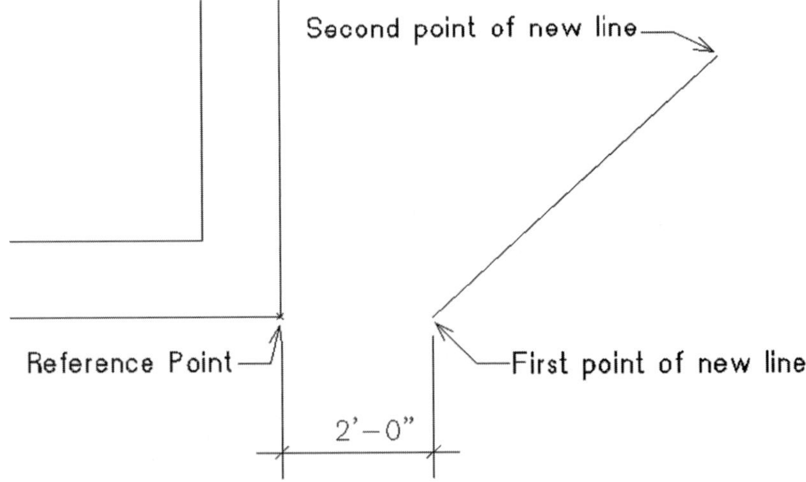

Figure 5-59 Starting a line two feet from the intersection of two lines

3. Press the **Space** bar (so DataCAD will prompt you to type the coordinates in).
4. Type a distance of 2′-0″ [610] (type **2** and then **Enter**).
5. Now type an angle of 0 degrees and then press **Enter**.
6. The cursor crosshairs will snap to a point 2′-0″ [610] to the right (0 degrees) of the intersection of the wall lines. This is the first point of your new line. Note that a line was not drawn from the reference point to this location.
7. Move the cursor anywhere on the screen and click the mouse to select the final point for the line. *Right-click* to quit drawing the line.

The **RefPnt** command can be used any time you need to locate something off of another entity, such as walls, arcs, symbols, and so on. For instance, if you wanted to place a symbol 2′-0″ [610] to the right of the intersection of two lines, you would perform the following steps:

1. Pick the symbol.
2. Press the ` key
3. Snap to the intersection of the two lines that you want to reference off of. If you have the **DrwMarks** setting turned on, a small × will appear at the point you snapped to.
4. Press the **Space** bar (so DataCAD will prompt you to type the coordinates in).
5. Type a distance of 2′-0″ [610] (type **2** and then **Enter**).
6. Now type an angle of 0 degrees and then press **Enter**.
7. The insertion point of the symbol is placed 2′-0″ [610] to the right of the intersection of the wall lines.

RefPnt will also work to select the second point (or third, fourth, and so on) of a line or arc. When DataCAD prompts you to select the second point, just press the ` key and enter the distance and direction.

Viewing Options

Once you have something drawn, you will need methods of navigating around the drawing file. Numerous options are available in DataCAD.

ReCalc and Extents

The Navigation Pad contains two buttons that will enable you to see your entire drawing: Recalc(ulate) and Extents. On the surface, they seem to do the same thing, but they are different. They can be accessed in several ways. For the Extents, you would perform one of the following steps:

1. Press the **E** in the center of the four arrows in the Navigation Pad.
2. Select **Utility/WindowIn/Extents**.

To Recalculate, you would do one of the following:

1. Press the **R** button in the Navigation Pad.
2. Select **View/WindowIn Recalc** from the drop-down menu.
3. Select **Utility/WindowIn/ReCalc**.

The **R**ecalc function recalculates the extents of all the layers that are currently turned on. That is, it finds the outermost entities and adjusts the display of the drawing so that it will fit entirely within the Drawing Window. On a drawing with a lot of entities in it, this may take some time, since DataCAD has to search through every entity to see which ones lie the farthest away.

The **E**xtents function displays the extents of your drawing based on the last **R**ecalc that you did. Until you press **R**ecalc, DataCAD will zoom to the same extents every time, regardless of which layers are on or off, and regardless of whether you have added or deleted entities. If you don't need to do a full **R**ecalc, then **E**xtents is a better option, since it does not search through any entities and therefore takes much less time to redisplay your drawing.

As you play around with these options, you'll understand it better. Pressing them cannot harm your drawing in any way, so feel free to experiment. You can interrupt the **E**xtents or **R**ecalc commands before they have completed their screen refresh by pressing the **Delete** key. This can be useful if you have a complex drawing that takes a while to redraw on the screen.

Refresh (ESC key)

Pressing the **Esc** key (or selecting **View/Refresh** in the drop-down menu) will cause your current view to refresh itself. This is useful if you have been doing a lot of editing, which tends to leave some stray visual debris on your screen and cause some entities to seem to disappear. Pressing the **Esc** key will clean up the visual clutter on your screen.

However, if you have DataCAD's *Display List* (D/L) feature turned on, pressing the **Esc** key will not always refresh some of the entities in the Drawing Window. For instance, some entities that you erased may still appear to be in your drawing. If this happens, then you need to press the **Uu** key, which is how you refresh (or regenerate) the screen when D/L is on. The drawback to this is that pressing the **Uu** key takes longer to refresh the screen than pressing the **Esc** key.

Refresh (Uu Key)

As previously noted, with the D/L feature turned on (see Chapter 3), you need to press the **Uu** key to update the Drawing Window. If you have D/L turned on and then press the **Uu** key, DataCAD regenerates (ReGens) the drawing by searching through all the layers that are currently turned on, compiling a new list of all the entities found and then saving them to the D/L in RAM. Depending on the number of entities displayed and the speed of your computer, this may take only a split second or it may take several seconds.

Window In/Out:

Zooming in and out of your drawing is called *windowing* in DataCAD. The quick and simple way is as follows:

- Press the **PgUp** key to window (zoom) out, away from your drawing to see a larger view of it
- Press the **PgDn** key to window (Zoom) in closer to your drawing to see greater detail.
- Or press the **In** and **Out** buttons in the Projection Pad.

The word "window" comes from the next method of zooming: by "area," using a bounding box or window to define the area to be windowed into. The bounding box looks and acts like a rectangular window into your drawing, hence the name. You access this in one of three ways, after which you can then define a bounding box around the area to be windowed into:

- Press the **W** button on the Navigation Pad, or
- Press the **/** (forward slash) key, or
- Select **View/WindowIn** from the drop-down menu

In the **Utility/WindowIn** menu is a button labeled **FreeZoom**. It is a toggle that either remains on or off. If it is turned off, then DCAD will window in and out by the fixed scales indicated in your Plotter/Scales menu, such as $^1/_4''$ or 1:40. If **FreeZoom** is toggled on, then DCAD will window in and out freely. I prefer to leave it on.

Center View

If you want to center your view on something in the visible Drawing Window, place your cursor where you want the center of your view to be and then press **Ctrl+Home**.

Orthographic and 3D Viewing

In the 3D section of this book, you will learn about the various options for viewing entities in 3D. For now, it is important just to know where these options are and what they do. They are in two basic areas: the View drop-down menu and the Projection Pad.

The View Drop-Down Menu In the **View** drop-down menu you will find the options shown in Figure 5-60:

- **Orthographic** This is the default DataCAD viewing method. All 2D entities must be created and manipulated in this mode.
- **Elevation** This is a quick way to view the front, back, and sides of a 3D model.
- **Isometric, Parallel, Oblique,** *and* **Perspective** Use these to view entities in various 3D projections.
- **Set Perspective** This enables you to set up a 3D view.

The Projection Pad The buttons in this pad do the same thing as some of the previous options, showing Orthographic, Parallel, Perspective, and the four elevation views.

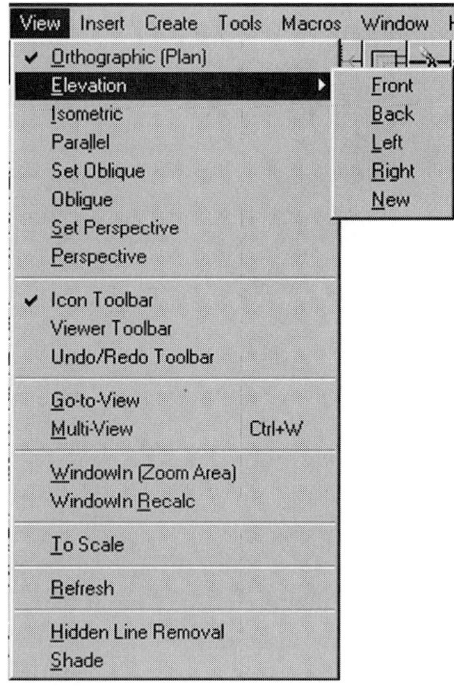

Figure 5-60
The View menu options

Figure 5-61
The Ortho View bar

Context-Sensitive Toolbars

A new feature of DataCAD 9 are the floating, context-sensitive toolbars, accessed by selecting **View/Context-Sensitive Toolbar** from the drop-down menu. A different toolbar will be displayed, depending on what viewing method you are in.

Ortho View Bar Here you have eight buttons to control your movement around an orthographic screen: pan left, right, up, and down; zoom in and out; zoom by area; and recalc the drawing (see Figure 5-61).

Parallel View Bar Here you have 19 buttons to control your movement around a parallel view screen: pan left, right, up and down; zoom in and out; zoom by area; recalc the drawing; rotate about the X, Y, and Z axis; view

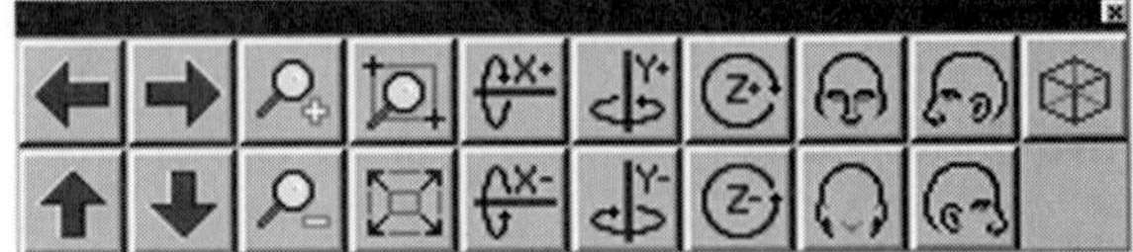

Figure 5-62
The Parallel View bar

Figure 5-63
The 3D View bar

in isometric; and display front, back, left, and right elevations (see Figure 5-62).

3D View Bar Here you have 10 buttons to walk around the 3D screen: walk forward and backward; step left and right; turn left and right; step up and step down; and look up and down (see Figure 5-63).

You can use your cursor to drag the toolbars anywhere on your screen. You can also orient the toolbars horizontally or vertically by *right-clicking* anywhere in the toolbars and selecting **Horizontal** or **Vertical**. A toolbar can be turned off by unselecting it from the View drop-down menu, by clicking on the X in the upper-right corner of the toolbar or by *right-clicking* on the toolbar and selecting **Exit**.

Online Calculator

DataCAD has a neat little calculator feature that is available any time you need to enter a value on the Command Line, such as a distance or an angle. Whenever you are prompted to enter a value, you can type =, followed by your equation. DataCAD will then use the calculated value for your input. To accept the calculated value, press **Enter**.

Unfortunately, for users of imperial units, you must enter distances in true decimal units, since the input function of the online calculator does not understand DataCAD's feet/inch method. For instance, to enter 7′-4″ [2235], you must enter 7.333. However, the calculated value will be converted to DataCAD's current **ScaleTyp** setting, so the output side of the online calculator does understand DataCAD's feet/inch method (go figure!). If you are using the **Arch** scale type, then the result will be the standard DataCAD architectural decimal format. That is, a value of 7′-4″ [2235] will be displayed as 7.4.

Because imperial units must be entered in real decimal units (not DataCAD's feet/inch method), a calculation entered like this will not work:

$$= 2 \times (2.5.11/32)$$

Because of double decimals in the feet/inch values, the online calculator gets confused. In this case, you would need to first convert the feet/inch values into a whole number, such as one of these:

$= 2 \times (29 + (11/32))$	2′-5″ converted to inches
$= 2 \times (2.41666 + (11/32))$	2′-5″ converted to decimal feet

The online calculator assumes that angles are entered as radians. If you want to enter degrees instead, type the letter **d** after the numerical degree value. Like distances, the calculated angle value will be converted to DataCAD's current **AngleTyp** setting. The default output is **Normal**: degrees/minutes/seconds.

Here are some examples of the possible calculation functions:

Operation	Example	Calculated Value
Addition	= 3.5 + 7	10.6 (10′-6″)
Subtraction	= 9 − 3.5	6.5 (6′-6″)
Multiplication	= 4 × 2.333	9.4 (9′-4″)
Division	= $^{15}/_{3}$	5.0 (5′-0″)
Parenthetical	= (4 + 3.5) × (8/2)	30.0 (30′-0″)
Square	= SQR (4)	16.0 (16′-0″)
Square Root	= SQRT (9)	3.0 (3′-0″)
Cosine	= COS (27d)	0.10.11/16 ($10^{11}/_{16}$″)
Sine	= SIN (27d)	0.5.7/16 ($5^{7}/_{16}$″)
Tangent	= TAN (27d)	0.6.1/8 ($6^{1}/_{8}$″)
Arc Tangent	= ATAN (10,18)	0.6.3/32 ($6^{3}/_{32}$″)
Natural Logarithm	= LN (8.2)	2.1.1/4 (2′-$1^{1}/_{4}$″)
Pi	= pi	3.1.11/16 (3′-$1^{11}/_{16}$″)

When entering equations that include multiple math operations, such as addition, multiplication, and so on, standard math rules apply. That is, math operations are calculated from left to right and in this order:

- Parentheses
- Addition and subtraction
- Multiplication and division

Make very sure that you read your equations carefully from left to right. It will make all the difference in the world when obtaining the proper output.

Here are some examples of other possible math equations:

$= 6.666 \times 2 + 3.5$

$= 6.666 \times (2 + 9.5 - 4.333 - 2.5)$

$= 4 \times (3.333 + (1/16))$

$= (3/32)/12$ (read the following section regarding the accuracy of such an expression)

Calculator Accuracy

When using the online calculator, it is important to realize its limitations. For many years, the online calculator was an undocumented feature that was primarily used by the programmers at DataCAD LLC. As such, it is not as easy to use or as accurate as you might expect.

When DataCAD was originally developed in the good ole' days of DOS, computers were extremely slow compared to the machines we typically use today. Maintaining an accuracy of four, five, or more decimal places while running on these machines was prohibitively slow. Thus, the developers of DataCAD decided to limit its accuracy for most functions to the smallest unit that generally occurs in the U.S. construction industry: $^1/_{32}$″ [0.8]. So, the online calculator is incapable of displaying a fractional value lower than a 32nd of an inch, and for that reason, some of the math results you get with the online calculator will be inaccurate if the mathematical result is less than $^1/_{32}$″ [0.8]. Numbers less than that, like $^1/_{64}$″ [0.4], will be rounded up to the nearest $^1/_{32}$″ [0.8]. Further, in most cases, when using a decimal format in DataCAD, it will only display out to three decimal places, so 0.03125″ would be rounded up to 0.031″.

To make things more confusing, it also depends on where you are entering an online calculator value. For instance, if you go to **Utility\Settings\EditDefs\Scales** and choose to change a value by using the online calculator (like entering = (3/32)/12), when you input values, they will be calculated in decimal format out to an accuracy of eight decimal places. But in other locations in DataCAD, when you are entering a distance, it uses fractions (down to $^1/_{32}$″ [0.8]) and will only display decimal values out to three decimal places. So different fields use different expression calculators. Sometimes a good hand held calculator is just the best solution.

Drawing Walls

This topic is covered in more depth in Chapter 11, but it is worth touching on here since the drawing of walls, just like lines, arcs, and circles, is based on the four input methods (Relative Polar, Absolute Polar, Relative Cartesian, and Absolute Cartesian). And like a line, a single wall needs only two points to describe it: a start point and end point. Because walls are often strung together to form a room or a building, however, some additional considerations are important to keep in mind.

What is a "wall" in DataCAD? A wall is simply two to four parallel lines or arcs. The lines that make up the wall can be simple, solid lines, or they can be made up of complex linetypes. Figure 5-64 shows some examples.

Two methods exist for turning on the automatic Walls feature:

- Select **Edit/Architct/Walls**.
- Press the = sign on the keyboard. This should be easy to remember since the = sign looks like a two-line wall.

While in Walls mode, the W in the SWOTHLUD acronym at the lower corner of your screen will be shown in upper case. When you are in single-line mode (with Walls toggled off), the w will be shown in lower case (see Figure 5-65).

You will remain in the Walls mode until you turn it off, either by selecting **Edit/Architct/Walls**, or by pressing the = sign again.

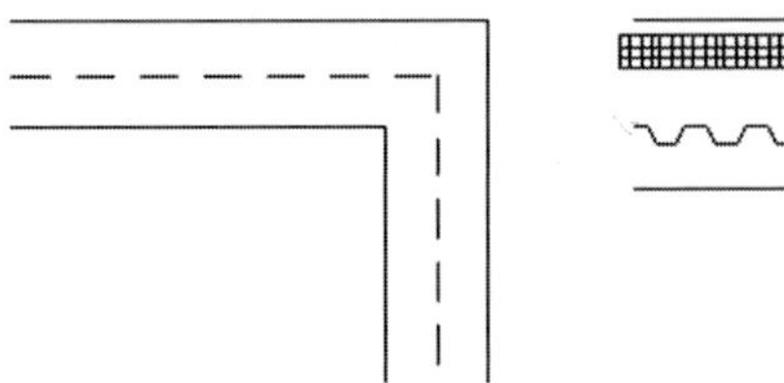

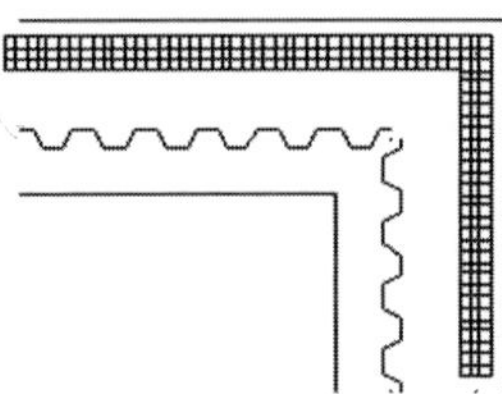

Figure 5-64
Some example of DataCAD walls

SWOTHLUD SwOTHLUD

Figure 5-65
Walls mode on and off

Although a curved wall is still a wall to DataCAD, some of the automatic drawing features in DataCAD, such as the insertion of doors and windows, will only work in straight walls.

NOTE: *It is very much worth nothing that there is really nothing special about a wall in DataCAD. If two or four lines or arcs are parallel to one another, they are automatically considered a wall. DataCAD does not label or treat those lines differently within the program.*

So conversely, you can use the ***Walls*** *command to draw anything that needs to be shown with two to four parallel lines. For instance, you can draw plumbing pipes, a sidewalk or road, a lamp post, or trim on a building elevation. One of the benefits to this methodology is that you can use all the wall cleanup features for these entities.*

Everything in Real Scale

Perhaps the most difficult thing for new CAD users to get used to, after drawing with paper, pencil, and a scale, is drawing everything in real scale. In the old days of paper drawings, if you wanted to draw details of, say, the exterior wall of a house, you perhaps drew a complete wall section, roof to foundation, at a scale of ${}^{3}/_{4}''$ = 1′-0″ [1:20]. You then drew details of the foundation, the windows, and the roof at a larger scale, such as 3″ = 1′-0″ [1:5]. With CAD, however, you draw everything at true scale, as if you were drawing the house on a giant, 20-foot tall piece of paper. You then decide at which scale you want to display your details and then lay them out on “real-size” sheets of paper within DataCAD, telling the program what scale to “display” the details at.

By drawing everything in real scale, you never have to make calculations to scale entities up or down. You never have to pull out your scale to figure out that a line of 3′-9″ [1143] at a${}^{3}/_{4}''$ [1:20] scale would have to be drawn at a length of $2^{13}/_{16}''$, as you would if you were drawing with paper and pencil. Instead, you draw the detail at real scale (3′-9″ [1143] in this case). Then tell the computer to display the detail at a $^{3}/_{4}''$ [1:20] scale when it is printed or plotted. Here is a short example:

1. Draw a line 10′-0″ [3048] long in the Drawing Window.
2. Go to **Utility/Plotter/Scale** and set the current scale to a $^{1}/_{4}''$ [1:50].
3. Print this line on your printer.
4. Measure the line on the printout with an architect's scale and you will see that at a $^{1}/_{4}''$ [1:50] scale the line is 10′-0″ [3048] long.
5. Now go to **Utility/Plotter/Scale** and set the current scale to 3″ [1:5].
6. Print this line on your printer.
7. Measure the line on the printout with an architect's scale and you will see that at a 3″ [1:5] scale the line is again 10′-0″ [3048] long.

Multiple Scales on One Drawing Sheet

Because everything is drawn at real scale, you may be wondering how to show drawings and details of different scales on the same drawing sheet. A couple of methods exist for this, the better of which is called *Multi-Scale Plotting* (MSP) and is covered in depth in Chapter 12. The other method involves enlarging (scaling up or down) your drawings and details so that they are not at real scale anymore (not a good idea). This is touched on in Chapter 11.

Multiple Drawing Technique

DataCAD excels at the Multiple Drawing technique of CAD drawing. This involves keeping the entire project, or at least multiple drawing sheets, in one drawing file, and reusing and copying entities as much as possible to avoid inefficiently wasting time and effort redrawing repetitive information or switching between drawings. The key to this technique is proper layer management and the use of 3D GotoViews (see Chapter 10).

Entities that are exactly the same on multiple drawings or details should only be drawn once whenever possible. For instance, even if you have five drawing sheets, your title block and sheet borders are the same from sheet to sheet, so you should have one layer with one border/title block on it. You can then make five new layers, each with only sheet-specific text information on it, such as the sheet number and sheet title. This keeps the drawing file small while eliminating repetitive information, the very foundation of CAD productivity.

Let's use a medium-size project to demonstrate. You can follow along if you'd like by copying the **Ymcapln2.dc5** file to your hard drive from Chapter 5 of the CD-ROM. You may even have this drawing on your hard disk already if you opted to install the sample drawings when you installed DataCAD.

When you first open the Ymcapln2.dc5 drawing, you will see a Lower-Level plan, sheet number A-2, with dimensions and text. The plan was set up on multiple layers: 14B-BDR, 14B-BLK, A2-BLK, A2-COL, A2-DIM, and so on. These layers represent both the $^1/_4''$ scale titleblock layers and the lower-level plan information displayed on sheet A-2. This one drawing file actually contains a total of 12 drawing sheets! Let's turn some of these layers on and off to see how this drawing file works.

Sheet A-3: Upper-Level Plan

When the file is first opened, all the layers required to display sheet A-2 are turned on. Let's see how some of the layers for sheet A-2 are also used for sheet A-3:

1. Go to the **Utility/Layers** menu. Note which layers are on in the first page of layer names:
   a. **14B-BDR**
   b. **14B-BLK**
   c. **14B-LOG**
   d. **14B-NOR**
   e. **14B-PRE**
   f. **14B-SHT**
2. Go to the next menu of layer names using the **ScrlFwrd** button, and again note which layers are on:
   a. **14B-TXT**
   b. **14G-DIM**
   c. **14G-GRD**
   d. **14G-HCH**
   e. **14G-SYM**
3. **ScrlFwrd** again to the third menu of layers. All the **A2** layers are currently turned on. **ScrlFwrd** to the fourth menu of layers and you will see the remainder of the 15 **A2** layers, all of which are on.

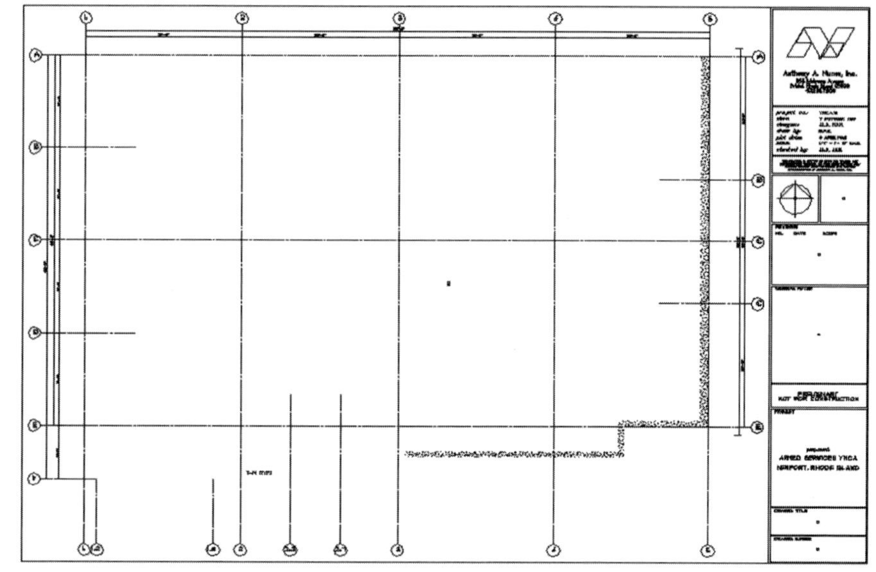

Figure 5-66 The building with only selected layers turned on

4. Select each of the 15 **A2** layer buttons to turn them all off, so that only the **14B** and **14G** layers remain on (see Figure 5-66).

   To see which entities are on each of the remaining layers, turn each one off and then back on. Notice that all the **14B** layers contain all the entities associated with the sheet border and titleblock. As you will see, the remaining **14G** layers contain entities that are shared between both sheets A-2 and A-3.

5. **ScrlFwrd** to the fourth and fifth menus of layers and turn all the **A3** layers on.

Notice how the **14G** layers are applicable to the floor plan on sheet A3. This is a key concept not only for the Multiple Drawing technique, but for CAD in general, *never draw anything twice!* Since it's so important, let me say that again: *never draw anything twice!*

So let's take a quick inventory here. Here are all the entities that are on shared layers between sheets A-2 and A-3:

- Sheet outline
- Titleblock lines, text and north arrow
- Dimensions

- Column grid lines and bubbles
- Concrete wall hatching

The **14B** layers are also used for the A-1, A-4, A-5, A-8, E-1, E-2, P-1, and P-2 sheets. Hopefully, you can see the speed and efficiency gains here. The more you can reuse entities between sheets, plans, elevations, and details, the faster you can finish your drawings, and the better coordinated they will be.

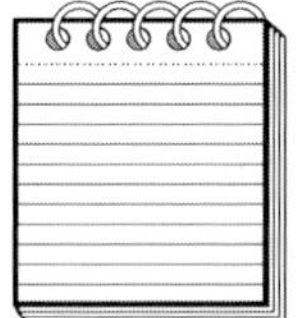

NOTE: *If you are flustered by constantly scrolling forward and backward to turn layers on and off, and getting equally frustrated trying to remember which layers are supposed to be on and off for each drawing sheet, then you are in luck. There is a better way!*

Now might be a good time to skip ahead to Chapter 10 to check out a DataCAD feature called GotoViews. It's a way of saving all the layer information for each drawing sheet so that you can instantly display a complete sheet or a detail with only a few mouse clicks.

Coordination

With the multiple drawing technique, if you stretch the building by 10 feet, you can turn on both the first- and second-floor layers prior to making the stretch. Now you can be assured that the first and second floors remain perfectly aligned. If you want windows to align between the tenth and second floors, simply turn on the tenth and second floor layers so you can be sure they are aligned. If you draw ductwork or plumbing fixtures on one floor, you can immediately see where and how they impact the other floors.

Drawbacks

The idea of the Multiple Drawing technique is to achieve better accuracy, coordination, and speed. A drawback can occur, however, if more than one person in your office is sharing drawings over a network. Take the Ymcapln2.dc5 drawing file, for instance. In DataCAD, two people cannot work in the same drawing file at the same time, so only one person can work in the Ymcapln2.dc5 drawing file at one time. You can see that this would not work at all in an office of more than one person.

You may choose to save your drawings in a hybrid manner. Some drawings, like all the floor plans, for instance, may all reside in one drawing file for better coordination. All the elevations may reside in a second drawing file, a group of details in another, and finally another group of details in the final drawing file. This enables more than one person to work on the project at one time, but it retains many of the benefits of the Multiple Drawing technique. Chapter 13 describes yet another method, called *External File Referencing* (XREF). Based on the size of the project and the number of staff in your office, you will have to decide what is best for you.

External File Referencing (XREF)

XREF is similar to the idea of the multiple drawing technique described previously, but the sharing of layers and other drawing information is done across multiple .DC5 drawing files, rather than keeping all the information in only one drawing file. This is an incredibly useful feature, especially if you have multiple employees in your office and work across a network.

Selection Methods

Five basic methods exist for selecting entities in DataCAD, best remembered by the acronym EGAFS: **Entity**, **Group**, **Area**, **Fence**, and **SelSet** (Selection Set). Since DataCAD defaults to drawing things, rather than picking or moving things, you will only see these options after selecting an option such as **Move**, **Copy**, **Mirror**, **Erase**, **Change**, and so on. You first tell DataCAD which edit function you want to use and then use EGAFS to tell the program which entities you want to edit.

- **Entity** enables you to select individual entities in the Drawing Window. Remember that each of these more complex entities are still just one entity each:
  - Complex linetypes, such as Insul, Plywood1, and ShingleR
  - 2D and 3D polygons
  - Associative hatching

- Strings of text
- 3D solids like slabs, blocks, cones, and spheres

- **Group** enables you to select groups of entities. A group is a set of any number of entities linked together logically but not necessarily graphically. Entities that do not touch may be part of the same group, and entities that are contiguous may be part of different groups.
- Entities are linked as a group in a number of ways. A group of entities drawn in succession without disconnecting the cursor is linked as one group. Similarly, a series of lines of text entered at the same time is linked as one group. You can link entities or groups of entities using the **LinkEnts** command. Entities are also grouped when they are created at the same time with the Copy command. When an existing entity is edited, with the **Change** or **Move** commands, for instance, the integrity of the group is retained. See Chapter 9, "Editing," for a more in-depth description of the **LinkEnts** command.
- **Area** enables you to select one or more entities at one time by drawing a rectangular bounding box fully around the entities to be selected. Only entities that are fully within the bounding box will be selected. If even one pixel of an entity is outside the bounding box, that entity will not be selected.

NOTE: *One shortfall of the Area selection feature is that the sides of the bounding box are always drawn horizontally and vertically. If you rotate your cursor by rotating the Grid Angle, or with the Tangents (**Utility/ Geometry/Tangents**) options, the Area selection box does not rotate with the cursor. This is visually confusing to your brain when trying to select entities in a rotated environment.*

- *Crossing* After selecting the first point of the area bounding box, you will see the **Crossing** option appear in the menu window. This feature allows you to pick not only entities that are fully inside the Area selection bounding box, but all entities that are crossed by one or more of the four sides of the bounding box. To use it, press **F1** or click on the **Crossing** button prior to selecting the second point of the Area selection box.

- **Fence** enables you to select one or more entities at one time by drawing an irregular boundary fully around the entities to be selected. The Fence method also offers a **Crossing** option (see Area) as well as a **Backup** option. Only entities which are fully within the bounding fence will be selected. If even one pixel of an entity is outside the fence that entity will not be selected.

- *Backup* In case you want to modify your fence in the middle of the selection process this option allows you to delete the vertices of your fence by moving successively backward in your selection fence with each selection of the **Backup** option.

- **SelSet** enables you to select entities grouped together in the active selection set. Pressing the **SelSet** option will edit the currently active selection set. Selection sets are much like groups, but you must deliberately group the entities together via the **Edit/EditSets** options. You can then save your selections in eight available selection sets. Entities can exist concurrently in more than one selection set, but only one selection set can be active at one time (**Edit/EditSets/SetActiv**).

  Another feature that sets selection sets apart from groups is that you can discriminate between the types of entities you want to add to the selection set. You can add entities by each of these methods or by any combination of them:

  - *Entity* Line, Circle, Bezier curve, Arc, Point, Ellipse, Text, 3D Line, B-Spline, Associative Dimension, or Symbol
  - *Color* Only the 16 basic colors are available, not custom colors
  - *Linetype* Any of the standard or custom linetypes that you have in your Dcadwin.lin file can be selected.
  - *Weight* Choose by any particular lineweight from 1 to 99.

  See Chapter 9 for a more in-depth description of how to use Selection Sets.

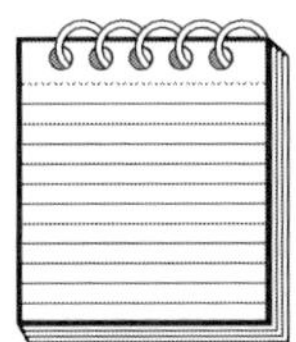

NOTE: *If you enable the D/L setting (**Tools Program/Preferences/Misc/Display List**), the rules for which parts of entities can be selected by each selection method may be slightly altered. Refer to Chapter 3 for a more thorough explanation.*

LyrSrch (Layer Search)

The **LyrSrch** option can be found in nearly every menu in DataCAD. It is one of the most basic, integral, and powerful features in DataCAD. Fortunately, it is also one of the easiest features to understand. With **LyrSrch** on, all visible entities can be selected, edited, and erased. With **LyrSrch** off, only entities on the active layer can be selected, edited, and erased. This is extremely helpful when you are working on a complex or very dense drawing, where many entities may be located in close proximity to one another, making selections difficult.

To see how **LyrSrch** works, open an existing drawing or create two new layers in a new drawing file (see Chapter 6 if you need help creating new layers). In this example, we will call them **Layer1** and **Layer2**:

1. Draw a few entities of one color (like Red) on **Layer1**.
2. Make **Layer2** the active layer (pressing the **Tab** key is the fastest way). Then draw some more entities of another color (like Blue) on that layer.
3. Select **Edit/Erase** and turn both **LyrSrch** and **Area** on.
4. Make sure **Layer2** is the currently active layer (its name will appear in the Status Area).
5. Try erasing all the entities from both layers by drawing a bounding box around all the entities.
6. All the entities, both blue and red, should be erased.
7. Press **Ctrl+Z** to undo the erasure and get the entities back.
8. Now turn **LyrSrch** off.
9. Again, try erasing all the entities from both layers by drawing a bounding box around all of them.
10. This time only the blue entities on **Layer2** (the active layer) were erased. Because **LyrSrch** is off, DataCAD only searched for and acted upon entities found on the active layer.

When You're On, You're On

Pressing the **LyrSrch** button in any menu globally changes the Layer Search setting throughout all the menus in DataCAD, but only in the current drawing file. Whenever **LyrSrch** is turned on or off, the square in the Status Area with the **L** in it (see Figure 5-67) will also toggle on or off (in

Figure 5-67 Layer Search on and off

and green for on; out and red for off) to continually show you the status of Layer Search. The figure on the left shows Layer Search on. The figure on the right shows Layer Search off.

Other Layer Search Toggles

Two other methods can be used for turning Layer Search on and off:

- Press the ' (single quote) key, or
- Click on the square with the **L** in it in the Status Area

Just like pressing the **LyrSrch** button in any menu, pressing either of these toggles will also globally change the Layer Search setting throughout the current drawing file.

Text and Fonts

DataCAD's fonts are proprietary. That means they only exist in DataCAD. You cannot use the True Type fonts installed in your computer like you can with your favorite word processor. To do so is on every user's wish list, but it's not here yet. Also, text is always drawn as solid lines in DataCAD, so you cannot select the dashed linetype and get dashed text.

Since text in any CAD drawing is printed at lots of different scales, even within one drawing sheet, no point sizes exist as in a word processor or any other program that uses True Type fonts. Instead, you have to tell DataCAD literally what size to create and display the font at: 1/8″ [3.2], 1″ [25.4], 6″ [152], and so on. The Text Scale option described in the following section makes that process a lot easier, however. First things first, however. Let's look at how to pick fonts and their settings, and to place text in a drawing.

Picking Fonts

To pick a font, perform the following steps:

1. Go to **Edit/Text**.
2. Make sure **Dynamic**, **Left**, and **TxtScale** are on.
3. Pick Size and type in **..1/8** (1/8″) and then hit **Enter**.

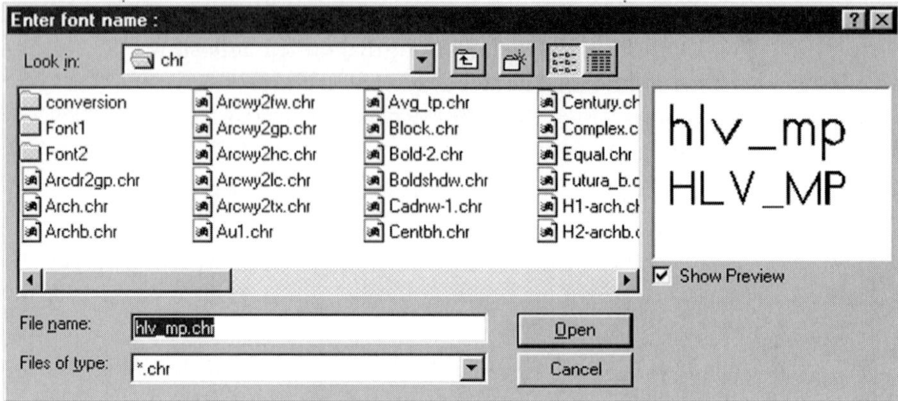

Figure 5-68 The **`Enter font name:`** dialog box

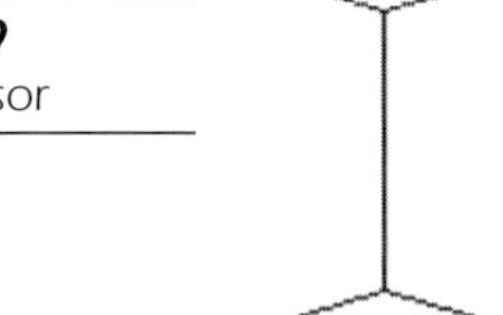

Figure 5-69 The text cursor

4. Select **FontName**. A dialog box like the one in Figure 5-68 will appear.
5. Highlight a font to see what it looks like in the preview window (make sure Show Preview is checked).
6. Press **Open** to select the highlighted font.
7. The text cursor will appear in the Drawing Window.
8. If it is too small to see, *right-click* to exit the text menu and use **PgDn** to move in closer to the Drawing Window. Enter the **Text** menu again and you should see the text cursor (see Figure 5-69).
9. Place the cursor where you want it in the Drawing Window and *click* the mouse to place it.
10. Type in some text. Notice that you can use the **Backspace** key to delete text while typing. Also notice that the crosshair cursor is free to move around the screen once the text cursor is placed in the Drawing Window.
11. When you are finished typing in the text you want, you have two options:

a. *Right-click* or press the **Exit** button. You can now move the text cursor to a new location to start typing text in a new location.
b. Press the **Enter** key. The text cursor will move down one line to enable you to type more text until you *right-click* or press the **Exit** button.

Snapping to Text

If you already have some text in the Drawing Window and want to type more text below it, and you want to make sure the beginning of both lines of text are aligned, you can use the object snap feature in DataCAD to do this quite easily:

1. Select **Text**. The text cursor will appear in the Drawing Window.
2. Locate the text cursor at the beginning of the text already in the drawing window. Now snap to the end of the text by *middle-snapping* or pressing the **Nn** key. The cursor will snap to the beginning of the text (see Figure 5-70).
3. Press the **Enter** key and the text cursor will move down one line, perfectly aligned with the text you snapped to (see Figure 5-71). Each press of the **Enter** key will move the text cursor down one more line. You can accomplish the same thing by *middle-snapping* after you place the text cursor. Each press of the *middle-snap* button will move the text cursor down one more line.

Figure 5-70 Snapping to the beginning of existing text

TEXT STRING

Figure 5-71 Returning the text cursor down one line

TEXT STRING

Text Settings

Like any other good word processor, you will invariably want to adjust your font settings in DataCAD. Here is what each of the settings in the **Text** menu do.

Size Select this option to choose the height of your text. You can select from the menu window or you can type in a height. Remember that whether **TxtScale** is on or off, it has a tremendous impact on the final size of your text.

Angle Like the angle of a line, use this setting to determine at what angle the text will be drawn. If your current **AngleTyp** setting is set to **Normal**, then an angle of zero degrees would give you horizontal text. If your **AngleTyp** is set to **Compass** mode, then an angle of zero degrees would draw text vertically on the screen.

Match Use this option to set the text angle by matching an existing line in the Drawing Window. Select the **Match** option and then click on a line in the drawing window to match. *Right-click* or press Enter to accept the value.

2Points This option is like the **Match** option, but instead of selecting an existing entity, you *click* on two points in the Drawing Window. The angle between these two points is the angle your text will be drawn at.

Weight

Like the **Weight** setting for lines, this will set the weight, or thickness, of the individual strokes of your text. As with lines and other entities in DataCAD, it is suggested that in most cases you do not use the weight setting, that you instead use color mapping to assign weights to your text (but feel free to try using the **Weight** setting to see if it's right for you).

Slant

This setting is like the italic option in your word processor. It adjusts the slant of your text, off of the vertical. Select the **Slant** option and then select an angle from the menu window, or type in an angle.

Subtract Use this option to subtract a number, using the buttons in the menu window, from the number currently entered in the Command Line.

Clear Use this option to clear any numbers currently entered in the Command Line.

Match Use this option to set the text slant by matching an existing line in the Drawing Window. Select the **Match** option and then click on a line in the Drawing Window to match.

2Points This option is like the **Match** option, but instead of selecting an existing entity, you *click* on two points in the Drawing Window. The angle between these two points is the angle your text slant will be drawn at.

Aspect The aspect ratio of text refers to the relative height and width of each character. Select the **Aspect** option and then select a value from the menu window or type one in. An aspect ratio of 1 represents the normal character aspect, while an aspect ratio of 5 produces very tall, slender text. An aspect ratio of 1, however, often produces text that appears just a little too wide to fit comfortably. An aspect of 1.25 is often a better selection. You will have to experiment to see what works for you, based on the font you are using at the time.

Subtract Use this option to subtract a number, using the buttons in the menu window, from the number currently displayed in the Command Line.

Clear Use this option to clear any numbers currently entered in the Command Line.

Factor Use this option to set the line feed, or space, between each line of text. The amount of line feed is calculated by multiplying the text height by the factor. Select the **Factor** option and then select a value from the menu window or type one in.

Subtract Use this option to subtract a number, using the buttons in the menu window, from the number currently entered in the Command Line.

Clear Use this option to clear any numbers currently entered in the Command Line.

FontName Use this option to select any of the available DataCAD fonts installed on your system. When you select this option, a dialog box will appear (Figure 5-68). It should default to the **\CHR** directory, which is where all of DataCAD's fonts are stored. If not, then navigate there through the dialog box. You should see a number of fonts, all with the **.chr** file extension. Choose the font you want and then select **Open**.

NOTE: *You are limited to using a maximum of 32 fonts in any drawing session. A drawing session extends from the time you open DataCAD until the time you close it, whether you open one drawing or 100.*

TxtStyle Because DataCAD offers so many setting options for text, this option will allow you to save the settings you have made for reuse in the future without having to remember or retype any values. As stated in the DataCAD manual, "You can save your text menu settings as a text style, which you can later load back into DataCAD to quickly apply text settings. Also available are options to delete unused text styles or to view text style attributes (**ShowVals**). The text style remains until you load a different style," or until you manually change settings in the text menu. The important concept is that it saves your current text menu settings, calling them back up and applying them when a text style is selected.

NOTE: *In case you're wondering, these text styles are saved in an ASCII text file called DCADWIN.STL, located in the \SUP directory.*

Load Use this option to load previously saved text styles. If you have not saved any, the only one you'll see is called *Default*.

SaveCurr Use this option to save the currently selected text options for future reuse:

1. Select **SaveCurr**.
2. A list of existing text styles appears.
3. Type in a new name of up to eight characters and press **Enter**.

Delete Select this option to remove a text style from a list of available styles.

ShowVals Use this option to display the factor, font name, and coordinate system of previously saved text styles. The information is shown in the Message Line.

NOTE: *Quick Steps*

1. *Go to* ***Utility/Plotter/Scale*** *(or* ***Edit/Text/PltScale*** *if you have the appropriate DCADWIN.INI file settings) and set the scale of the current plan, elevation, or detail.*
2. *Select* ***Edit/Text/TxtStyle****.*
3. *Use* ***Load*** *to load a previously saved text style.*
4. *Use* ***SaveCurr*** *to save your current text menu settings.*
5. *Use* ***Delete*** *to delete previously saved text styles.*
6. *Use* ***ShowVals*** *to display the attributes of any of the previously saved text styles:*
   a. *Factor*
   b. *Font*
   c. *Size*
   d. *Angle*
   e. *Weight*
   f. *Slant*
   g. *Aspect*

Here are some caveats to keep in mind:

- Make sure your current **Plotter/Scale** or **Text/PltScale** is set to the scale of the drawing you are placing text in.
- Text Style names can only be eight characters and are subject to the same character limitations as DOS file names.

Dynamic Toggle this button on to display text in the Drawing Window as you type it. When **Dynamic** is toggled off, the text first appears in the Command Line and will not be added to the Drawing Window until the **Enter** key is pressed. Generally, you will want to leave this option turned on.

TxtScale Toggle this button on to indicate the size of the text relative to the plot scale. If you toggle this button off, the size of the text appears in a real-world scale. See the following section for more information about **TxtScale**. It is a very important option.

FileI/O Use this option to import or export text to and from DataCAD via ASCII text files. Your other option for importing text is via the Windows cut and paste method (described in Chapter 8).

FromFile Use this option to import text from an ASCII text file. When selected, a Windows dialog box will appear. All text options in DataCAD, including plot files and ASCII text files, default to the **\PLT** directory. You do not have to retrieve your ASCII text files from here. If the text file you want to import is somewhere else, navigate to the proper directory in the dialog box, select the file, and then press the **Open** button.

ToFile Use this option to export text to an ASCII text file. When selected, a Windows dialog box will appear. All text options in DataCAD, including plot files and ASCII text files, default to the **\PLT** directory. You do not have to save your ASCII text files to here. If you want to export the text somewhere else, navigate to the proper directory in the dialog box, type in a new name, and then press **Save**.

Arrows This option allows you to draw arrows or pointers from notes or details by selecting the tail of the arrow first and then the head. The size of each arrowhead is relative to the current text size, so it's important to be at the correct text scale prior to picking the **Arrow** option.

The leader lines of the arrows are drawn in the currently active color, with the current lineweight settings. The following arrowhead options affect only the arrowheads (not the leaders) and will remain current in the drawing file until changed.

Size Use this option to adjust the arrowhead size relative to the current text size. Choose a factor from the Menu Window or type in a value. The most common arrowhead factor is .50. Experiment to see what works for you.

- **Subtract** Use this option to subtract a number using the buttons in the Menu Window from the number currently displayed in the Command Line.
- **Clear** Use this option to clear any numbers currently displayed in the Command Line.

Style Five arrowhead styles can be used (see Figure 5-72):

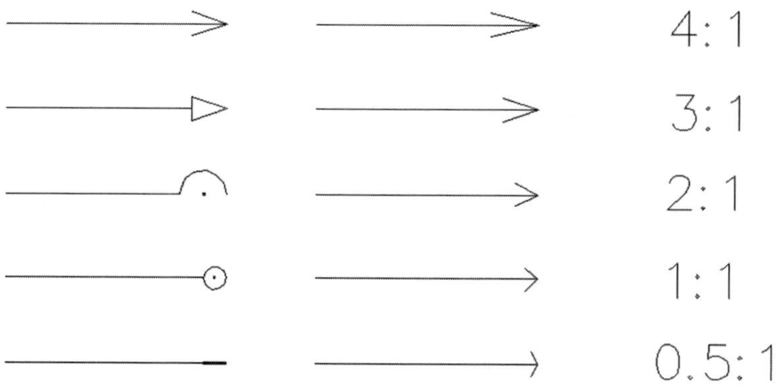

Figure 5-72
The five arrowhead styles and various **Aspect** settings

- **Open** The open style is a standard arrowhead. It can be modified with the **Size** and **Aspect** options.
- **Closed** The closed style is a standard arrowhead. It can be modified with the Size and Aspect options.
- **Bridge** Use this option when two leaders need to cross one another, or use this for symbol diagrams that show the crossing of two pipes or wires.
- **Dot** This is an alternative to the open or closed arrowhead. It can only be modified with the Size option.
- **Tick** This is an alternative to the open or closed arrowhead. It can only be modified with the Size option.

Aspect Use this option to change the aspect ratio of the arrowheads. The aspect is defined as the length of the arrowhead divided by its height (length/height). On the right side of the previous figure, the arrow **Size** is kept constant while the **Aspect** is changed.

Notice how the sizes of the arrowheads decrease as the aspect values decrease. To keep the arrowheads larger, you will need to increase the **Size** value. The ANSI standard arrowhead has an aspect ratio of 6 to 1. The **Aspect** option will only affect the open and closed arrowhead types.

Weight Like the **Weight** setting for lines, this option will set the weight of the individual strokes of the arrowheads. As with lines and other entities in DataCAD, it is suggested that you do not use the weight setting, that you instead use color mapping to assign weights to your arrowheads (but feel free to try using the **Weight** setting to see if it's right for you).

Color This option changes the color of the arrowheads only, allowing the arrowheads to be printed darker or lighter than the leaders. Pick one of the 15 standard DataCAD colors from the menu window, or select the **Custom** option to select any one of 255 colors. Here are the available options:

- **Adjust** This option will affect your entire drawing file and not just the arrowheads! See Chapter 23 for more information. The **Adjust** option will open the Color Palette Editor and enable you to manually adjust the RGB values displayed on your screen. When you press this button, the text will change to read **Palettes**. Pressing this button will then display the **LoadRGB** and **SaveRGB** buttons to allow you to save the current color palette or to load a previously saved palette.
- **Custom** Pick this option if you want to select any color besides the standard 15 DataCAD colors. You can enter a value from 1 to 255. Numbers 1 to 15 are the standard DataCAD colors; all the rest are considered custom colors.
- **Match** Use this option to set the arrowhead color by matching an existing entity color in the Drawing Window. Select the **Match** option, and then click on an entity in the Drawing Window to match.
- **NoChange** If you choose a custom color with the **Custom** option and then decide you don't want to use that color, select the **NoChange** button to return to the original color.

PltScale This option will change the current plot scale of the drawing file. Use this in conjunction with the **TxtScale** option to have your text printed out at the correct size. You'll find the same menu option in the **Utility/Plotter** menu, but with the name **Scale**, instead of **PltScale**. See the following discussion of Text Scale.

Justify With this option, you can justify existing text along a left, center, or right reference line (see Figure 5-73). The line of justification can be any angle, except an angle that matches the angle of the text: a parallel line. To justify text, follow these steps:

1. First, select the justification you want: **Left**, **Center**, or **Right**.
2. Select **Justify**.
3. Select the first and second points along the line of justification, or use object snapping to snap to an existing line or text if you want the line of justification to match that entity.
4. Pick the text you want to justify using EGAFS.

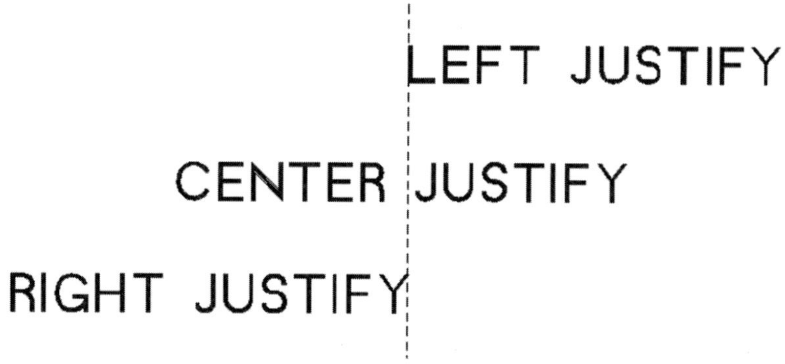

Figure 5-73
Justification options

NewLine Use this option to select a new line of justification without having to back up to the main text menu. The **Left**, **Center**, and **Right** options appear in this menu as well, for the same reason (see Figure 5-73).

Left, Center, and Right These three options are used for two purposes. The first is previously described under the **Justify** option. The second involves picking one of these three options prior to placing new text in the Drawing Window. The new text will be placed in the Drawing Window via whichever option is currently selected. Note, however, that with the **Dynamic** toggle on, the text will not actually display justified by center or right until you finish placing the text by *right-clicking* the mouse or pressing **Enter**.

FitText Use this option to define the width and height of your text. After selecting a horizontal and vertical line to describe the area and then entering the text, the correctly proportioned text will not be displayed until you finish placing the text by *right-clicking* the mouse or pressing **Enter**.

The way this works is that DataCAD uses the horizontal and vertical lines you drew to calculate a new aspect ratio and size for the text. While still in the **FitText** menu, all subsequent text is drawn with these new settings. To return to your original text settings while still in the **FitText** menu, press the **Reset** button. However, if you exit the **Text** menu without first pressing the **Reset** button, the altered aspect ratio and text size is retained.

Reset Use this option to return to your original text settings while still in the **FitText** menu. This option will only appear after selecting **FitText** and defining new text settings.

Text Scale

This feature, (**Edit/Text/TxtScale**), is nothing short of absolutely wonderful! Using **TxtScale** in DataCAD is simple to do, but sometimes difficult to understand at first. First, we will try to explain it, but then it may be easier to understand by doing, so we'll do a little tutorial.

Without **TxtScale**, offices used to come up with elaborate charts to remind everyone what the text size was supposed to be for different scale drawings and details. A 1/4″ [1:50] scale drawing may have used text that was, at a quarter-inch scale, defined as 6″ [152] high in order for the text to plot at 1/8″ [3.2] high. What **TxtScale** does is figure out all those conversions for you on the fly. You tell **TxtScale** what the scale of the detail/drawing will be when you plot it (key concept here) and how big you want the text to be (like 1/8″ [3.2] high) when it is printed out on paper. Then **TxtScale** automatically figures out how big the text should be drawn in the drawing file. Best of all, **TxtScale** carries over into DataCAD's dimensioning as well. If you like your dimension overruns and overlaps to always be 3/32″, then you only have to set it once, make it part of your default drawings, and never have to set those settings again. Let's give it a try:

1. Open a new drawing in DataCAD.
2. Go to **Utility/Plotter** and select the **Scale** option. Select 1″ [1:10] to tell DataCAD that you intend to print this text at a scale of 1″ = 1′-0″ [1:10]. (The end of this section contains an easier method for achieving this same result.)
3. Now go to **Edit/Text**.
4. Turn **TxtScale** on.
5. Go to **Size** and make the text size 1/8″ [3.2] (type in ..**1/8** and then hit **Enter**).
6. Place the text cursor in the drawing screen and type, "**This text will plot at 1/8″ high at 1″ scale**", and then hit **Enter**.
7. Go back to **Utility/Plotter/Scale** and set the scale to 3″ [1:5]. Your plotting scale is now set to 3″ = 1′-0″ [1:5].
8. Place the text cursor just below the first line of text and type, "**This text will plot at 1/8″ high at 3″ scale**", and then hit **Enter**.
9. Now to see how **PgUp** and **PgDn** work. Locate the text cursor below the previous text, but do not click any mouse buttons. Now press **PgUp**. Notice that the text cursor gets larger. Notice also that the

information line at the bottom of the screen is telling you what the new PLOT SCALE is (this is an important concept). This is not the text size. The text size is still an 1/8″ [3.2] high (and always will be until you change it in **Edit/Text/Size**), but by using **PgUp** and **PgDn**, you are defining what plot scale the text will be printed at, just as surely as if you selected **PltScale** and set the scale that way.

10. To prove this, **PgUp** until the info line says, **"Current plotting scale is now 1/4″."**
11. Now place the text cursor below the second line of text and type, "**This text will plot at 1/8″ high at 1/4″ scale**." Then press **Enter**.

NOTE: *At any time while you are typing (after you have placed the text cursor) you can use* ***PgUp*** *or* ***PgDn*** *to Window In/Out without affecting the text cursor or the plot scale.*

12. Go to **Utility/Plotter/Scale** again. You will notice that the **"Current plotting scale is: 1/4."** In effect, the **PgUp** and **PgDn** in the **Text** menu act as a shortcut to this **Utility/Plotter/Scale** function. This is also a key concept. Now we'll print out a few pages of our text to see how it looks at different plot scales.
13. With the Scale still set to a 1/4″ [1:50], print an 8 1/2″ × 11″ [216 × 280] page with the text inside it (all of it should fit).
14. Now go back to **Utility/Plotter/Scale** and set the plot scale to 1″ [1:10]. Print out the page.
15. Go back to **Utility/Plotter/Scale** again and set the plot scale to 3″ [1:5]. Get as much of the text into the sheet as possible. The largest text will not fit completely, but that's OK. Print out the page.

So, if you want to print all the notes of your details at an 1/8″ [3.2] high (no matter what scale), set your text **Size** to an 1/8″ [3.2]. Then, when working on a 3″ [1:5] scale drawing, use **PgUp** and **PgDn** or **PltScale** to set the plot scale to 3″ [1:5] and then type away. When you want to start adding notes to a 1/4″ [1:50] scale drawing, use **PgUp** and **PgDn** or **PltScale** to set the plot scale to a 1/4″ [1:50] and then type away.

Voila! No more trying to remember what size the text should be at 3″ [1:5] or 1/4″ [1:50] scale, and no using antiquated charts to tell everyone in your office what the text sizes and dimension settings should be. Set once, type many.

NOTE: ***TxtScale*** *has no impact on your previously drawn plans and details or on symbols. Remember* ***TxtScale*** *is basically an on-the-fly calculator. It does not affect how your existing text is displayed. Whether* ***TxtScale*** *is turned on or off, your text will always look the same.*

In Step 2, we mentioned that there was a better way to pick the plot scale. By changing one setting in the **Tools/Program Preferences/Misc** menu you can add a new **PltScale** button to the **Text** menu so that you don't have to keep going back to the **Utility/Plotter/Scale** to change the current plotting scale. (This option is only available by installing the 9.01 update available from the DataCAD Web page.)

TxtScale Option

The Text Scale option can function in two ways:

1. Default TxtScale Operation (used in the previous example):
   a. **PgUp/PgDn** causes the plotting scale to change, and therefore the text size increases or decreases with each press of the **PgUp** or **PgDn** keys.
   b. Unfortunately, this means that while the text cursor is on the screen you cannot use **PgUp/PgDn** to zoom in and out of the Drawing Window.
   c. In the text menu, no menu button will be displayed between the **Arrows** and **Justify** buttons.
   d. To maintain this option, make sure the **PgUp/PgDn in Text Menu** option is un-checked in the **Tools/Program Preferences/ Misc** menu.
2. Enhanced TxtScale Operation (the suggested option):
   a. **PgUp/PgDn** will zoom in and out of the drawing.
   b. You will not be able to change the plot scale with the **PgUp/PgDn** buttons.
   c. A new menu button called **PltScale** (plot scale) is displayed between the **Arrows** and **Justify** buttons in the Text menu. Pressing this will allow you to change the current plot scale, just as

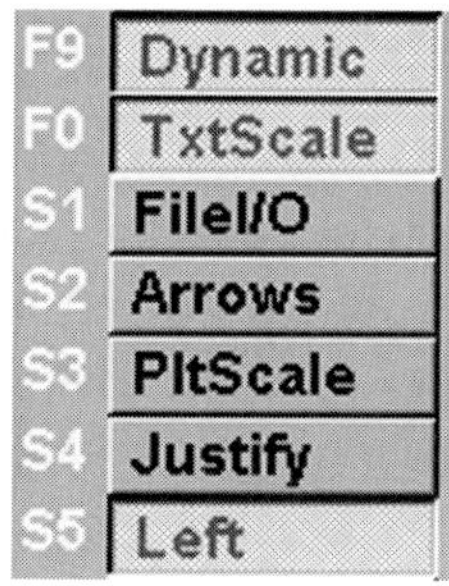

Figure 5-74
A new **PltScale** button added to the Text menu

if you chose it by going to **Utility/Plotter/Scale**, but in fewer steps (see Figure 5-74).

d. To select this option, you need to download the 9.01 upgrade from the DataCAD Web page, then check the **PgUp/PgDn in Text Menu** option in the **Tools/Program Preferences/Misc** menu.

Take your pick. I prefer this second method since it makes changing the plot scale much faster since you can access it within the **Text** menu.

NOTE: *Remember that to get the correct size text when using* ***TxtScale*** *you must have the* ***PltScale*** *set correctly. If the detail you are adding text to will be printed at 1" = 1'-0" [1:10], but you have PltScale set to a* $^1/_2$*" = 1'-0" [1:20], then the text will not be sized properly. Always check your* ***PltScale*** *prior to adding text to a drawing.*

It is highly recommended that you set **TxtScale** on in your default drawing files and never turn it off.

Editing Text

When you type a string of text, whether it be a single letter, one word, or a series of words, DataCAD treats it as one single entity. Each letter in the text string is not treated as an individual entity, as is the case in a word processor and can therefore not be edited individually. After entering one string of text and then pressing the **Enter** key, the cursor moves down one line and you can continue typing more text. The text on this new line is in turn treated as a separate entity from the first line of text. The two lines of text essentially have no relationship to one another (except that the multiple strings of text typed at the same time will be part of the same group. Once you exit the **Text** option, all future text will become part of a separate group). This means that if you want to change those two lines of text later, like changing the content or color, you have to pick and edit each line individually (see Figure 5-75).

Text Handles

Each string of text in DataCAD has six *handles*. When picking or snapping to text, you need to locate your cursor near one of these six points in order for DataCAD to find the text string. Figure 5-76 shows a string of text with an imaginary box around the extents of the text. The small squares at the corners and the top and bottom centers of that box represent the locations of the text handles. No matter how large or small the string of text, the handles are always located in these locations.

If you want to snap to existing text, you can only snap to one of these six handles. If you want to select the text for editing, you must also pick the text by one of these six handles. If your cursor is not close enough to the text to find one of these handles, you will see a message at the bottom of the screen that reads, "The point entered is too far from an entity." Move your cursor closer to one of the six handles and try again.

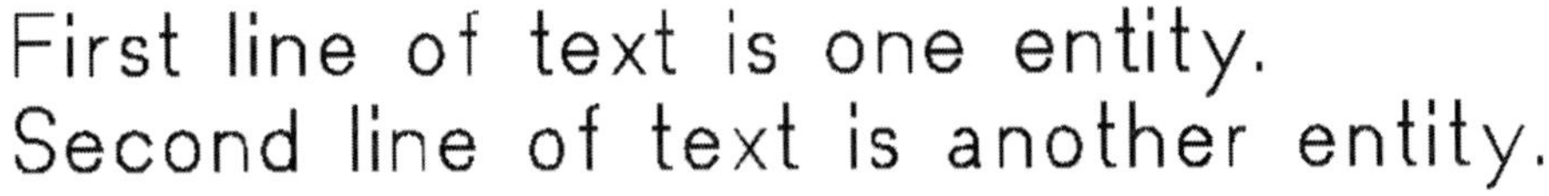

Figure 5-75 Each line of text is a separate entity.

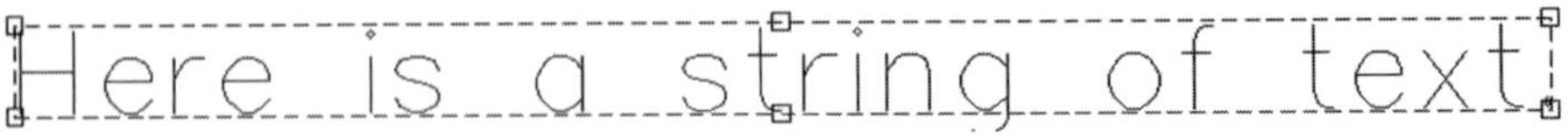

Figure 5-76 A text box with text handles displayed

NOTE: *Chapter 3 describes the use of the D/L (display list) in DataCAD to speed up the refreshing and selection of entities on your screen. With D/L turned on, you will find that these six text handles, although they still exist, are not as relevant as they are with D/L turned off.*

Turning D/L on has the added benefit of allowing you to pick (but not snap to) the text by picking any point on any of the letters in the text. You do not have to get the cursor close to one of the six handles. This makes selecting text for editing much easier.

For snapping, you are still required to snap to one of the six handles. This makes sense since if you could snap to anywhere on the text, you would have a hard time snapping precisely to anything.

Changing Text

To change the content and text properties of a string of text in DataCAD, you use the **Edit/Change/Text** menu. Note that these options are not in the **Text** menu as you might expect. When you select **Edit/Change/Text**, you will see a menu, part of which looks like Figure 5-77.

See the section about Text and Fonts earlier in this chapter for an explanation of **Size**, **Weight**, **Slant**, **Aspect**, **FontName**, and **Match**. You can use the **Change/Text** command to change any of these properties, but in

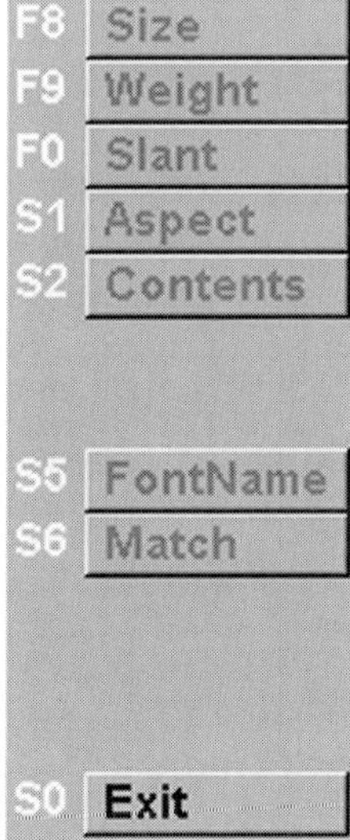

Figure 5-77 A section of the **Change/Text** menu

Lettuce change sum tixt.

Figure 5-78
Some spelling errors that need to be fixed

Figure 5-79
The text to be changed is transferred to the Message Window.

this section, we are primarily concerned with the **Contents** option, which is the one to use if you want to change the contents of the text string itself. Figure 5-78 has some spelling errors, so let's see how to go about changing the text to correct that.

1. Select **Edit/Change/Text/Contents**. The Message Window will read, ***"Select entity to <CHANGE>."***
2. At the top of the menu, make sure the **Entity** option is selected. You could use any of the other options (EGAFS) instead, but since we only need to change one line of text, we only need to use Entity.
3. *Click* near any of the six imaginary handles to select the text (unless D/L is on). Since the position of the handles in the center of the text is more difficult to locate, you will generally want to use one of the four corner handles.
4. The text will disappear from the Drawing Window and will reappear in the Message Window at the bottom of your screen (see Figure 5-79).
5. Notice that all of the text is highlighted. Just as in other Windows programs, if you start typing now, all the highlighted text will be deleted and replaced with the new text you are typing.

NOTE: *If you begin to type something in the Message Line and then realize you goofed and want to get your original text back, you can do one of two things:*

- *Press the **Tab** key on your keyboard. All the new text will be deleted and the old text will return to the Command Line.*
- *Right-click the mouse. All the new text will be deleted and the old text will return to the Drawing Window.*

6. Since we only want to change parts of the text in the Command Line, you can move the cursor to the area to be changed via several methods, most of which are standard Windows conventions. You can use any of these methods as any time during text editing, not just at the start:

Action	Result
Position the mouse and then *click and drag* to highlight.	Just as in a word processor, you can highlight selected text by clicking and dragging the mouse over the text to be changed. Typing will then replace all the highlighted text with new text.
Position the mouse and *click* to place the cursor.	Just as in a word processor, you can locate the cursor anywhere in the line of text. Clicking the mouse will position the cursor. Now you can move the cursor to the correct position by using the left and right keyboard arrow keys.
Press any of the four arrow keys (up, down, left, or right) on your keyboard.	The text highlighting will disappear and the cursor will be displayed at the end of the text. Now you can move the cursor to the correct position by using the left and right keyboard arrow keys.
Press the **End** key.	The text highlighting will disappear and the cursor will be displayed at the end of the text. Now you can move the cursor to the correct position by using the left and right keyboard arrow keys.
Press the **Home** key.	The text highlighting will disappear and the cursor will be displayed at the beginning of the text. This is especially useful if the cursor is at the end of the text line and you want to move it to the beginning of that line. Now you can move the cursor to the correct position by using the left and right keyboard arrow keys.
Press **Tab** or *right-click* the mouse.	All current changes will be nullified and your original text will be displayed back in the Drawing Window.
Press **Delete**.	All current highlighted text in the Message Window will be deleted.
Press **Enter** or *middle-click* the mouse.	The current text displayed in the Command Line will be displayed in the Drawing Window where the original text was located.

The first line of Figure 5-80 shows the corrected line of text. The second line is that same line of text, but notice that we added more text to it. You can add and delete text just as easily as you can correct typos, but remember that as you add text, the line will become longer, so make sure the text is not going to run over some other text or part of your drawing. You can

Figure 5-80
The spelling is fixed and more text has been added.

also add to the text and then use the **Move** command afterward to move the longer line of text out of the way.

NOTE: *For single-text entities, or those that don't lend themselves to using the Area, Fence, or Group options, there is another way. Constantly wading through the menus to select **Edit/Change/Text/Contents** and then selecting the text can get tedious, especially if you have a lot of text to change. To get around this, we altered our **Alt+X** keyboard shortcut to do this automatically (see the Keyboard Shortcut section of this chapter and Chapter 23 for more information). All you have to do is place the cursor on a piece of text to be edited and then press **Alt+X**. The text will appear in the Command Line, ready for editing.*

Multiple Selections

You can continue to change the contents of more than one line of text in one of two ways:

- Once you have changed one line of text and either pressed **Enter** or *middle-clicked* the mouse, you will see that the **Contents** button is still highlighted. You can therefore continue to change a new line of text simply by selecting it, changing it, then pressing **Enter** or *middle-clicking* the mouse, and on and on. This can get tedious if you have a lot of text to change.
- Instead of using the **Entity** option, use the **Area** or **Fence** options to select a lot of text in one step. DataCAD will then display each individual line of selected text, one by one, in the Command Line for editing. DataCAD will display each line of text in the order it was entered into the drawing, not necessarily in the order they are displayed on the screen.

The **Group** option can be used as well, but just be aware that groups of text, like other entities, frequently change groupings depending on how the text was created, and whether or not it was copied or moved at some point in time.

Changing Basic Text Properties

Text is simply another entity, just as a line is. Thus, you can change the color, line weight, Z-base, and Z-height by using the **Edit/Change** command. You cannot, however, change the linetype of the text. Text is always drawn as solid lines, so you cannot select the dashed linetype and get dashed text. You also cannot use the **Change** command to change the **Spacing** or **OverSht** (overshoot) of text, since these properties do not apply to text.

Keyboard Shortcuts

First, we must clarify some terminology. In this book and among DataCAD users, you will hear references to a number of phrases that mean essentially the same thing:

- Keyboard shortcuts
- Keyboard macros
- Immediate mode commands
- Keyboard interrupts
- Hot keys

In this book, we will use the phrase *keyboard shortcuts* to avoid confusion with the Toolbox macros and to avoid using technical jargon like *immediate mode commands*. But they all refer to the same thing: the capability of DataCAD to temporarily interrupt a current action by pressing a particular key or keys in order to cause DataCAD to execute a specific task. For most of these tasks, once they are completed, DataCAD will be returned right back to the point where you left off (with some exceptions).

Some of these keyboard shortcuts are hard-wired into the DataCAD program code and generally cannot be changed by the DataCAD user. This would include all of the single keys, like an **H**, a semi-colon (;), quotation marks ("), **F11**, or **/**. Many of the shortcuts that start with the **Shift** key, and all of the shortcuts that begin with the **Alt** key can be redefined by the DataCAD user. In fact, this is highly recommended once you have some experience with DataCAD and want to begin to work faster and more efficiently.

To see all the default shortcuts, look in the Appendix of the book. You will also find a printable list of these shortcuts on the CD-ROM that comes with this book. If this all sounds confusing, let's try an example. First of all, here is a sampling of some of the default keyboard shortcuts that run on DataCAD when it is first installed. Note that some are case-sensitive, and others are not:

Key or Key Sequence	What It Does
F	Stands for "forced save"; it saves the current drawing just the same as selecting **File/Save** from the drop-down menu
f	Line spacing that enables you to change the current line space setting
F11	Elevation left; it displays the left elevation of a 3D model
Tab	Active layer; each press of the **Tab** key will change the currently active layer by moving forward through the list of layers that are currently on
Shift+Tab	Active layer; while holding down the **Shift** key, each press of the Tab key will change the currently active layer by moving backward through the list of layers that are currently on
K	Color; each press of the **K** (upper case) key will change the currently active color by moving forward through the list of 15 standard colors
k	Color; each press of the k (lower case) key does exactly the same thing as the upper case **K**
Alt+C	**Change** menu; while holding down the **Alt** key and pressing the C key (either upper or lower case), DataCAD will jump to the **Change** menu
Ctrl+N	New file; it does the same thing as selecting **File/New** from the drop-down menu

Single-Key Shortcuts

The first thing you should notice is that all single-letter commands are case-sensitive. Although some commands like **Erase** work whether you press a lower case **e** or and upper case **E**, others like **Edit Sets** (the **S** key) will only work if the proper case is used. For this reason, the convention in this book is to show exactly which key needs to be pressed. If an upper case **S** is required, then that is what will be shown. But if you can use either the upper or lower case letter to accomplish the same thing, as in the **Erase** example, both upper and lower case will be displayed, like this: **Ee**.

If you see a character such as the equals (=) sign, then all you have to do is press that key. But if you see a character such as the plus (+) sign, you will see on your keyboard that in order to select this you have to hold down the **Shift** key and simultaneously press the = key. This book will not tell you that you have to press the **Shift** key.

Keys like the **F11** key refer to the function keys on your keyboard. On some small keyboards found on laptops, you may have to press a "function" key to access those keys. This book assumes that you have a full-size keyboard and only have to press one key.

Alt-Key Shortcuts

The **Alt+C** example in the previous chart means that to execute this shortcut you need to hold down the **Alt** key while simultaneously pressing the **C** key (either upper or lower case). This will immediately take you to the **Change** menu. Note that the **Alt+Key** sequences are not case-sensitive; either upper or lower case will work. For the sake of consistency, this book will always show the letter in upper case.

The **Alt+Key** shortcuts are one of the most powerful features in DataCAD. Not only because they are fast and efficient, but because they are incredibly customizable by even the most non-technical DataCAD user. I know of very few DataCAD users who have not changed some or even all of the default **Alt+Key** shortcuts (those that came with DataCAD when it was first loaded) to reflect their own needs.

In a nutshell, the **Alt+Key** shortcuts can be programmed to duplicate nearly any DataCAD function or series of mouse/keyboard strokes. Tasks that might take you 15 steps to do by selecting menu buttons and *clicking* the mouse can be programmed to be executed by simply pressing the **Alt** key plus one letter on the keyboard. Pretty powerful, huh? Best of all, it will take you about two minutes to create such a shortcut. It's really pretty simple. Chapter 23 details the entire process of customizing your **Alt+Key** shortcuts.

The **Alt+Key** shortcuts only work with letters, and since there are 26 letters in the English alphabet, that's how many shortcuts are available to you. The CD-ROM that came with this book has a large number of sample shortcuts that you can cut and paste for your own use.

Ctrl-Key Shortcuts

These are like the **Alt+Key** shortcuts, but they use the **Ctrl** key instead of the **Alt** key, and in this case, the amount of customization is limited to only certain commands (located in the SUP\DCADWIN.MNU file).

Here are some of the functions that can be accessed by **Ctrl+Key** shortcuts:

- **Ctrl+N** New
- **Ctrl+O** Open
- **Ctrl+S** Save
- **Ctrl+P** Print
- **Ctrl+X** Cut
- **Ctrl+W** Multi-view windows

Chapter 23 details the entire process of customizing your **Ctrl+Key** shortcuts.

Icon Toolbars

The icon toolbar (also called the icon bar) is the strip of graphic buttons (or icons) across the top of your screen. They operate in an almost identical manner to the **Alt+Key** shortcuts described in the previous section, but instead of using a keyboard sequence to execute the shortcuts, you press an icon button instead. A default toolbar is loaded and displayed when you first install DataCAD, but you can alter each button on the toolbar or change them completely if you'd like. Figure 5-81 shows what the default toolbar looks like.

If you have a big enough computer monitor and are running it at a high resolution, then you may be able to see all 42 icons. If not, then you will only see as many buttons as will fit across your screen. To access the rest of the icons, there is a small arrow at the far left of the toolbar (see Figure 5-82).

Figure 5-81
The default toolbar

Figure 5-82
The button to let you access the rest of the toolbar

Pressing this arrow will cause the toolbar icons to shift to the left so that the remainder of the icons are visible. Another small arrow like the one earlier will then point in the other direction. Pressing this arrow will return the icon bar to its original state.

If you don't have an icon bar showing the select **View/Icon Toolbar**. If you want to change to a different icon bar, do the following:

1. Go to the drop-down menu at the top of your screen and select **Tools/Program Preferences/Misc**.
2. Click on the down arrow to the right of the white box in the ***Default .KEY File*** area (Figure 5-83).
3. At least two names should be under the file in the drop-down list: ***DCADWIN3*** and ***DCADWIN***. The ***DCADWIN*** icon bar is the one shown earlier and is the default toolbar for 2D commands and functions.
4. Pick the ***DCADWIN3*** name instead. This is the default toolbar for 3D commands and functions.
5. Press **OK**.
6. Now you should see a toolbar that looks the ones shown in Figure 5-84.
7. If you do not want any toolbars to be displayed in DataCAD, simply uncheck the **On/Off** box (or uncheck the **View/Icon Toolbar** option).
8. There is an easier way to switch between the 2D and 3D icon bars. At the end of the 2D bar is an icon with a picture of a key and the phrase

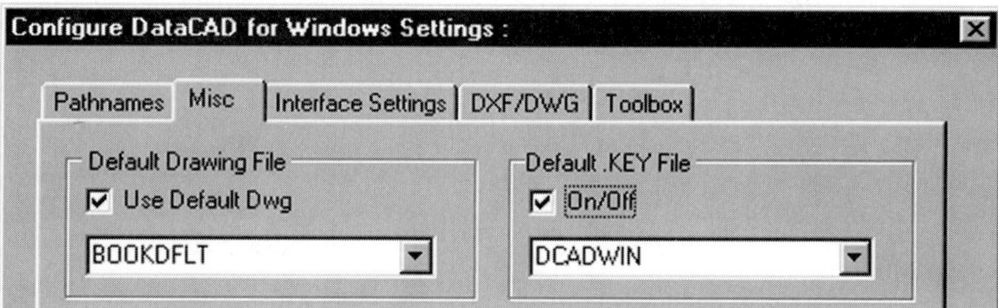

Figure 5-83
The Default.KEY File configuration dialog box

Figure 5-84
The 3D icon toolbar

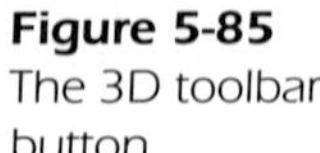

Figure 5-85
The 3D toolbar button

Figure 5-86
The 2D toolbar button

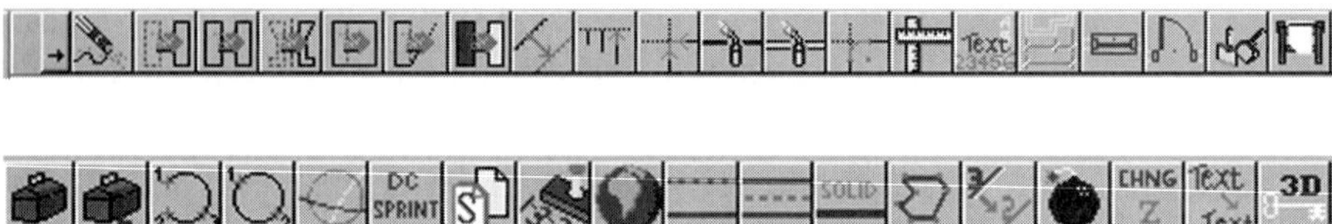

Figure 5-87
My customized toolbar

3D on it (see Figure 5-85). By pressing this button, the 2D toolbar will be replaced with the 3D toolbar.

9. Likewise, at the end of the 3D bar is an icon with a picture of a key and the phrase 2D on it (see Figure 5-86). By pressing this button, the 3D toolbar will be replaced with the 2D toolbar.
10. To show how customized you can make your toolbars, my own toolbar is shown in Figures 5-87.

Notice that a few of the icons remain from the original DataCAD default icon bar, but not many. I custom-created 16 of the icons you see here (they're available on the CD-ROM that comes with this book).

Tag—You're It!

If you want to know what each icon does, simply locate the cursor over one of the icons and a yellow help tag will appear, giving you a short description of what the icon does. The tag only stays visible for about three seconds, so if it disappears and you want to see it again, move the cursor off of and then back onto the icon. Figure 5-88 is an example of a help tag that reads *"Change entity attributes."*

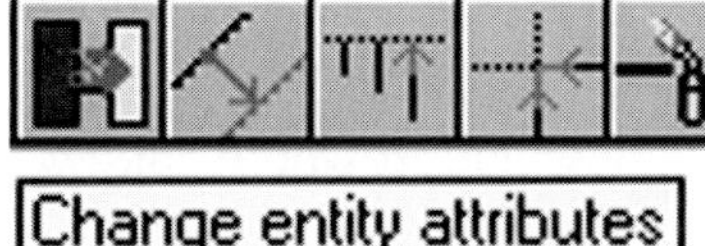

Figure 5-88
An example of a help tag

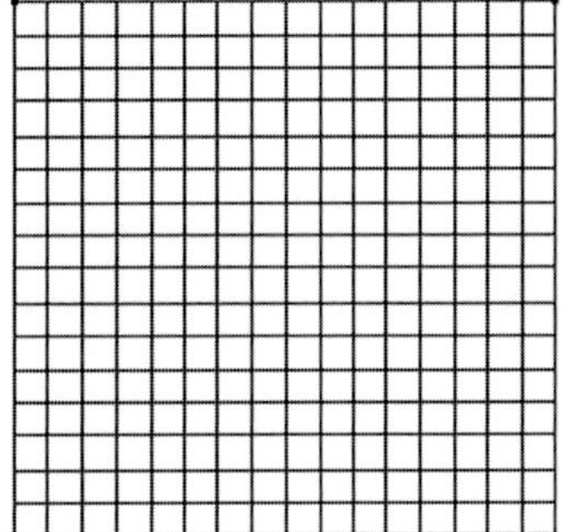

Figure 5-89
Hatching examples

Create Your Own Toolbars

Chapter 23 contains detailed explanations of how to modify and create your own icons and icon toolbars.

Hatching

Use the Hatch command to fill portions of a drawing with a repetitive pattern. If you are used to drafting with paper and pencil, you may think of hatching as infill: a graphic convention to show different materials in a section, such as brick, CMU, steel, or earth. Architects commonly refer to this as *poche*. In CAD, hatching is used for this same purpose, but it is also used to make literal representations of materials as well, such as ceramic tile in a plan and brick in an elevation (see Figure 5-89).

Hatching in CAD can save you a lot of tedious drawing time. For instance, in the tutorial in Chapter 4, we needed to add some 4″ × 4″ ceramic tile to the floor of our kitchen plan. We could have drawn each individual line of the tile pattern on the floor, but this would have involved quite a lot of drawing as well as tediously cleaning up around all the ins and outs of the countertops. By using the **Hatch** function in DataCAD, you can instead define the outside edges of the ceramic tile area (using object snapping will make this even more accurate) and the program will draw all the lines for you. Now think of how long it would take you to draw a brick pattern, like the one shown previously, line by line on an elevation. By using hatching, you just define the outline of the area to be shown as brick and then the program draws all the brick for you. Two kinds of hatching exist in DataCAD: associative hatching and non-associative hatching.

Associative hatching is very much like a symbol. It is treated as one single entity rather than a bunch of separate entities. Every line within the hatch is "associated" with every other line, and you cannot edit the individual lines and arcs within the hatch. You can only edit the boundaries of the hatching. This has some distinct advantages. Let's say you decide to replace the 4″ × 4″ [102 × 102] tile that is currently in your kitchen plan with a 2″ × 2″ [51 × 51] pattern. All you have to do is point to the current 4″ × 4″ [102 × 102] associative hatch and tell DataCAD to replace it with a 2″ × 2″ [51 × 51] pattern and the whole hatch pattern will be updated in one step to show the 2″ × 2″ [51 × 51] pattern; no erasing or redrawing is necessary. Think of how much time that could save if you wanted to change all the brick in the previous example to clapboard siding.

Why Use Associative Hatching?

There are probably few times when you would want to use non-associative hatching over associative hatching. Non-associative hatching is drawn the same way as associative hatching; you define the edges and then the program draws all the lines for you. The difference is that all the lines of the hatching are drawn as separate entities. Although each entity in the hatch is still identified as a hatch and are all part of a common group, each entity must be edited separately, and you can no longer just change the boundaries of the hatching. To erase the hatching, you have to erase every line individually. None of the lines are associated with one another, so you lose the ability to change or update the hatching like you can with associative hatching.

Here are some reasons why you would want to use associative hatching 99.9 percent of the time:

- Because each associatively hatched area is treated as a single entity, it keeps the size of the drawing file down. For instance, an associatively hatched area of a brick pattern might visually contain 200 entities, but as far as the drawing file is concerned, it is only one entity. The same brick pattern hatched with a non-associative hatch would add all 200 of those entities to the drawing file. Do this a dozen times and you have a file with 2,400 entities instead of only 12.

- Also related to file size, when you erase entities from a CAD file, the drawing file does not decrease. Although the entities are gone, the space they took up in the file is not. So if you erase 2,400 non-associative hatch entities from a drawing file, the file size does not get smaller. If you erase an associative hatch pattern, only the space for one entity remains in the drawing database.
- Editing an associative hatch is far easier than a non-associative hatch pattern. You can change the entire hatch pattern with a few button clicks, never having to erase the first pattern or to redraw the original hatch boundary. To change a non-associative hatch pattern, you must first erase the original hatch entities (of which there might be dozens or hundreds) and then retrace the original hatch boundary. If the boundary has a lot of corners, this can get quite tedious, especially if you decide to change the hatch pattern a third or fourth time.

Drawing a Hatch Pattern

Let's add a brick pattern to an elevation of a simple rectangular wall (see Figure 5-90).

Figure 5-90
The wall we will add a brick pattern to

1. Go to the **Utility/Hatch** menu.
2. Make sure **Associat** is on.
3. Select **HtchType** and make sure **NoOutLin** is on. *Right-click* once to get back to the **Hatch** menu.
4. Select **Pattern** and a preview box will display (see Figure 5-91). As you move your cursor over each hatch pattern button, DataCAD will display an image of what the hatch looks like.
5. Pick the **brick** pattern.
6. Now select **Scale** and type in a value of **31.95** (this will give you a standard nominal American brick size of $2\frac{2}{3}'' \times 8''$ [68×203]). Then press **Enter**.
7. Pick **Angle** and make sure it is set to 0.
8. Now select the **Boundary** option so that you can define the edges of the hatch pattern. Use your middle mouse button or the **Nn** key to snap to the four corners of the rectangle. It is very important that you move sequentially, either clockwise or counterclockwise, when defining a hatch pattern.

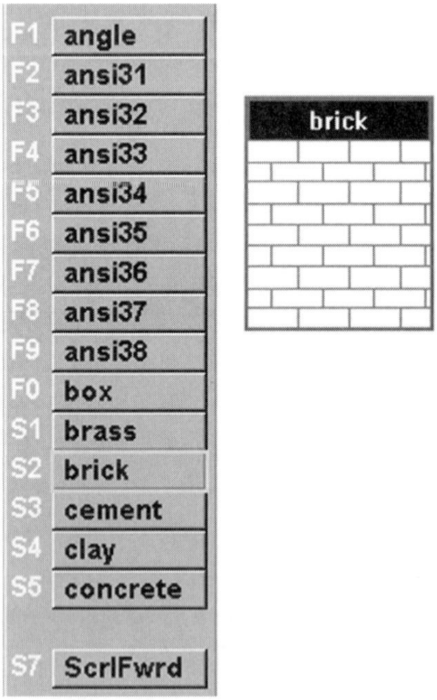

Figure 5-91
The hatch pattern menu and preview window

NOTE: *Selecting an appropriate scale for each hatch pattern is very important. In this case, the brick pattern is designed to display a standard nominal American brick size of* $2^2/_3''$ × *8″ [68 × 203] at a scale factor of 32. If you choose a scale factor of 64 (double the 32), then the brick will be displayed at two times its intended size, or* $5^1/_3''$ × *16″ [136 × 406] inches. A factor of three would be three times as large and so on.*

Many hatch patterns have a more obvious scale factor. For instance, the ***4x4til*** *hatch pattern requires a scale factor of 1 in order to display the true dimensions of the 4″ × 4″ tile.*

To avoid some of the trial and error of figuring out which scale factors to use, the Appendix shows you all of DCAD's standard hatch patterns along with many suggested scale factors for displaying various materials at various scales.

9. After snapping to point 4, you need to close the hatching boundary (from point 4 to point 1). You can do this in one of two ways. Either *right-click* or press the **Close** button. I prefer to *right-click* since it's quick and easy.
10. Your hatch pattern is not visible yet. The final step is to press **Begin**. The hatch will then appear.
11. Press **Esc** to refresh the screen. The drawing should look something like Figure 5-92.
12. Notice that the hatch pattern starts with only half of a brick along the bottom course. If you want the brick pattern to start with a full brick in the lower left-hand corner, you need to adjust the **Origin** (the place where the hatch pattern starts) to start at that point.
13. In the **Hatch** menu, select **Origin**.
14. *Snap* to the lower-left corner (Point 1) of the wall.
15. Pick the existing brick hatch pattern. You do this by placing the cursor on any line within the hatch pattern and *clicking* the mouse button. The hatch pattern will appear shaded, confirming that this is the hatch you are picking.
16. Press **Begin** and the hatch pattern realigns itself to start at the new origin (see Figure 5-93).

All the settings in the **Hatch** menu (from **Boundary** to **Origin**) are described in more detail later in this section, but first you should learn a little about what DataCAD is doing when you create a hatching boundary.

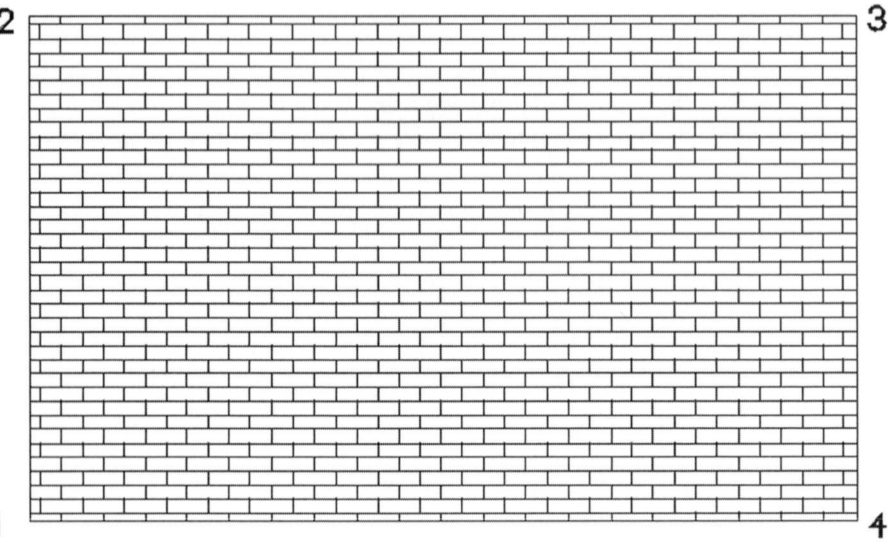

Figure 5-92 The completed hatch pattern

Figure 5-93 The hatch pattern realigning at the new origin

Hatches as Polylines

In DataCAD, a polyline is a single entity made up of a linked series of straight lines or curves. DataCAD uses closed polylines, consisting of at least three vertices, to define the boundaries of a hatch pattern (3D polygons can be hatched as well, but 3D polygons are limited to 32 vertices in DataCAD).

When you draw a hatch pattern, you are drawing a closed polyline. If a polyline is open when you try to hatch it, DataCAD will infer that the point between the first and last vertices should be connected. The hatch pattern will be drawn based on that inference. You can also draw a closed polyline by using the **Polyline** macro found in the **Toolbox** and then hatch it afterward. The shapes in Figure 5-94 are all valid polygons, but notice that the last one has boundaries that cross over one another.

Although this last polygon is valid, you will want to avoid doing this, since if you fill it with a hatch pattern, it would look like in Figure 5-95.

Figure 5-94 Polygons ready for hatching

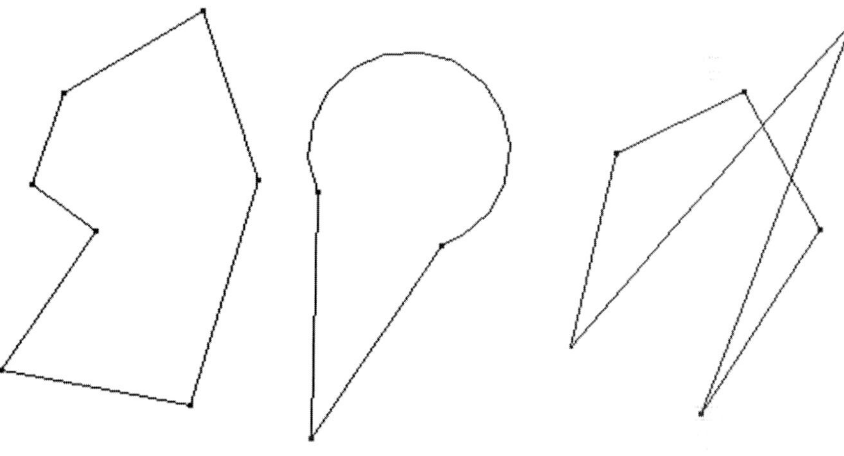

Figure 5-95 The polygons with hatching applied

Boundary

2PtArc and 3PtArc

Hatch boundaries (closed polylines) can be made up of curves as well as straight lines. When you select **Boundary** and then select your first point, you will then see two new buttons appear: **2PtArc** and **3PtArc**. When you select a third point, a **Tangent** button will appear as well. These three buttons enable you to draw various types of curves for your hatch boundaries. The ice cream cone-shaped hatch pattern in the previous figure was drawn in part with the **3PtArc** option. For a complete description of how to use these three options, see the section about **Drawing Arcs and Circles** earlier in this chapter.

Hatch Types

As discussed at the beginning of this section, the two types of hatching in DataCAD are associative and non-associative. The **Associat** button in the **Hatch** menu controls which of the two will be drawn. Each one behaves a little differently and also has some slightly different menu options. The biggest difference between how the two are used is that associative hatching only works with closed polylines, while non-associative hatching can use either closed polylines or regular 2D lines and curves that form a closed area.

Associative Hatching (Associat-On)

With the **Associat** button toggled on, all hatching will be drawn as associative. When you select the **HtchType** button, you will see the following options:

- **OutLine** With this button toggled on, the hatch pattern will display an outline of the polyline defining the perimeter of the hatch.
- **NoOutLin** With this button toggled on, the hatch pattern will not display an outline of the polyline defining the perimeter of the hatch. This will most often be the setting you use since in most cases you will already have some kind of boundary drawn, such as the walls of a room or the edges of the elevation of a building. If **OutLine** is turned on, you

would get another set of lines over the top of the already existing lines, which is needlessly redundant.

Changing One Hatch Pattern to Another

Let's say you want to change the brick pattern in our examples to 12″ × 12″ tile. If you had drawn all that brick with paper and pencil, or if you had drawn it in DataCAD with non-associative hatching, then you would have to erase each and every line of the brick and then redraw the new pattern. With associative hatching (**Associat** on) the task of changing patterns is far easier:

1. While in the **Hatch** menu, pick **Pattern**, scroll forward to the **12x12til** button, and click on it.
2. Now pick **Entity** from the **Hatch** menu. The Prompt Line reads, *"Select an entity to <Hatch>."*
3. Click on any part of the associative brick hatch pattern. The pattern will change to dashed lines to indicate that you have selected it.
4. Press **Begin** and the hatch will be updated to 12x12til.
5. *Right-click* to exit the **Hatch** menu and press **Esc** to refresh the screen.

Non-Associative Hatching (Associat-Off)

Although we said earlier that you will want to use associative hatching 99.9 percent of the time, non-associative hatching does do a few tricks that associative hatching cannot, as illustrated in the following several examples. With the **Associat** button toggled off, all hatching will be drawn as non-associative. When you select the **HtchType** button, you will see several new options. Note that in the figures below the lines defining the hatch areas are regular 2D lines and arcs, not polylines. Although non-associative hatches will also hatch inside of closed polylines, they do not have to be polylines, as associative hatches do. In each of the following hatching examples, the drawing in Figure 5-96 was hatched by drawing an **Area** bounding box around the entire drawing at one time.

The **Fence** option would have worked just as well to select all the previous entities. In fact, all of the EGAFS selection methods will work for non-associative hatching, but no matter which method you use *you must select*

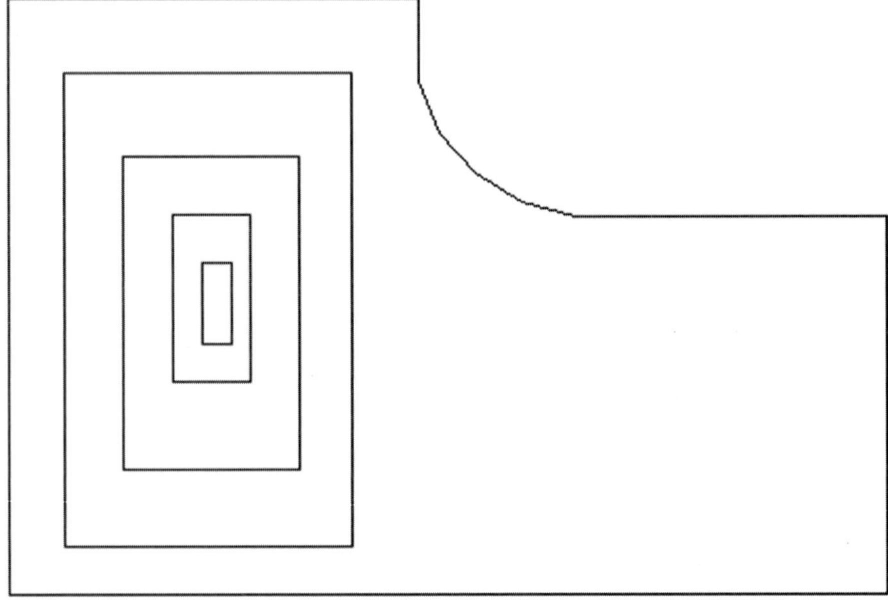

Figure 5-96 This is the drawing example to be hatched with an Area bounding box.

Figure 5-97 A 2D rectangle

entities which form a closed area, even if it takes a number of selections to do so. For instance, Figure 5-97 is a rectangle made up of simple 2D lines.

These four lines form a closed area, and selecting all four lines by using the **Area** or **Fence** commands would be the quickest way to select them. If all four lines are part of the same group, then I can select the **Group** command and click on any one of the four lines. All four lines will be highlighted. Pressing **Begin** will hatch the area inside the four lines. If all four lines are part of the same selection set, I could use the **SelSet** option in the same manner.

To use the **Entity** command, since it only picks entities one by one, I would have to select all four lines individually (in any order) until they were all highlighted. Pressing **Begin** would then hatch the area inside the four lines. If the rectangle were a single, closed polyline, then I would only need to select one point anywhere on the polyline to hatch inside of it.

HtchType (Hatch Type)

Normal Use this to hatch successive boundaries, starting from the outside and working inward. To use this option, you must select all the entities to be hatched prior to pressing **Begin**. As noted earlier, you can use any of the EGAFS selection methods, but **Area** or **Fence** would be the easiest (see Figure 5-98).

OutMost (Outer Most) Use this to hatch only the outermost boundary, starting from the outside, while ignoring all the other boundaries inside. To use this option, you must select all the entities to be hatched prior to pressing **Begin**. As noted earlier, you can use any of the EGAFS selection methods, but **Area** or **Fence** would be the easiest (see Figure 5-99).

Ignore Use this to hatch the entire area within the outermost boundary while ignoring all other boundaries inside. The boundaries inside are

Figure 5-98 The **Normal** option at work

Figure 5-99 Using the **OutMost** option

treated as if they don't exist. To use this option, you must select all the entities to be hatched prior to pressing **Begin** (see Figure 5-100).

Erasing Non-Associative Hatching

Because this type of hatching is made up of dozens or even hundreds of individual lines, you might be wondering how you erase them all if you make a mistake. The first method, of course, is to use the **Undo** command. But if you want to erase those entities later on, when you don't want to Undo a lot of other things just to get back to the hatching entities, you can use this next method. Whenever a hatch pattern is drawn, all the entities in that pattern are grouped together. So select **Edit/Erase** (**Ee**) and then select **Group**. Pick any entity in the hatch pattern and the entire hatch will be erased.

But what if you need to erase a whole bunch of non-associative hatching? Chapter 9 contains a discussion about "masking" entities in DataCAD. This allows you to use **Area** or **Fence** to pick only specific kinds of entities (like hatching) to be added to their own special group called a *Selection Set*. Doing so enables you to pick all the hatching at once. Then use the **SelSet** option in the **Erase** menu to erase all the selected hatching. The reason this

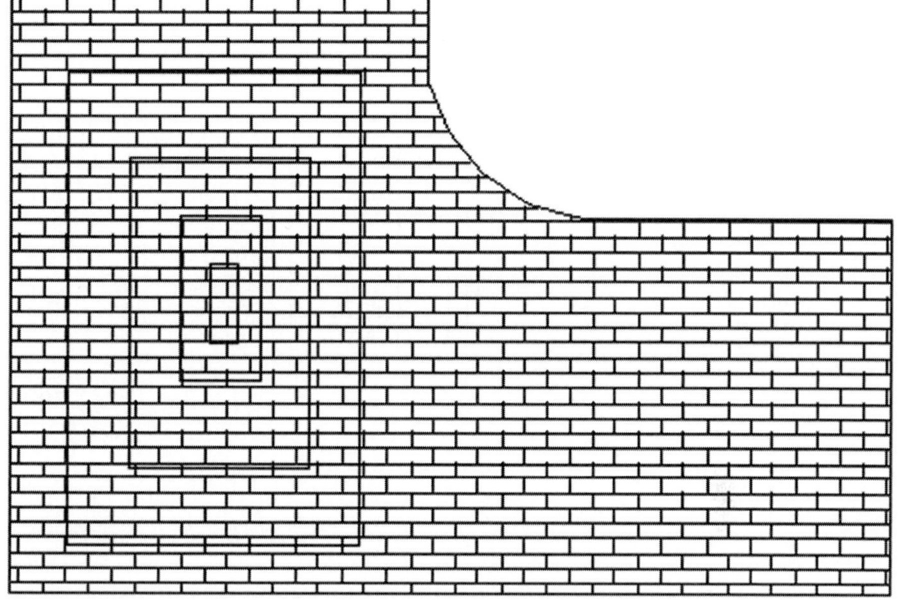

Figure 5-100
The **Ignore** option doesn't pay attention to the interior boundaries.

works is that all the entities in a non-associative hatch will still be identified as an entity type called *Hatch*.

Hatching around Objects

Often you will have something in the middle of the area to be hatched that you don't want to be hatched. For instance, you may have a window in a building elevation that you, of course, do not want to hatch with a brick pattern. Hatching around objects like this is a bit trickier. Unfortunately, DataCAD does not let you fill an area like a paint program, and you cannot cut out an area from inside an existing associative hatch. Instead you have to define all the edges of the area to be hatched, including the edges of any objects that are not to be hatched, by literally drawing around the object.

The easy way to remember how the hatch polyline should be drawn around objects is to think like you were tracing the hatch boundary with a pencil, but the pencil line must be continuous; it cannot be lifted from the paper, and it cannot cross over any of the other lines of the polyline. You can have the lines of the closed polyline lie on top of one another, as you will see, but they should not cross over one another.

Figure 5-101
A hatched brick wall

Figure 5-101 is an example of an associatively hatched elevation with a single window in it, and the following steps show you how we achieved this.

1. Start with the unhatched elevation (see Figure 5-102).
2. Go to the **Utility/Hatch** menu.
3. Make sure **Associat** is on.
4. Select **HtchType** and make sure **NoOutLin** is on. *Right-click* once to get back to the **Hatch** menu.
5. Select **Pattern** and then choose the **Brick** pattern.
6. Now select **Scale** and type in a value of **32** (this will give you a standard nominal American brick size) of $2^2/_3'' \times 8''$ and then press **Enter**.
7. Pick **Angle** and make sure it is set to **0**.
8. Now select the **Boundary** option so that you can define the edges of the hatch pattern.
9. Use your middle mouse button or the **Nn** key to snap to the points shown in Figure 5-103. It is very important that you follow the exact order of the points shown to avoid having your hatch pattern cross over itself, which would give you some strange looking results.
10. When you are finished entering point 10, either *right-click* or select the **Close** button to finish the polyline. Remember, what DataCAD is doing

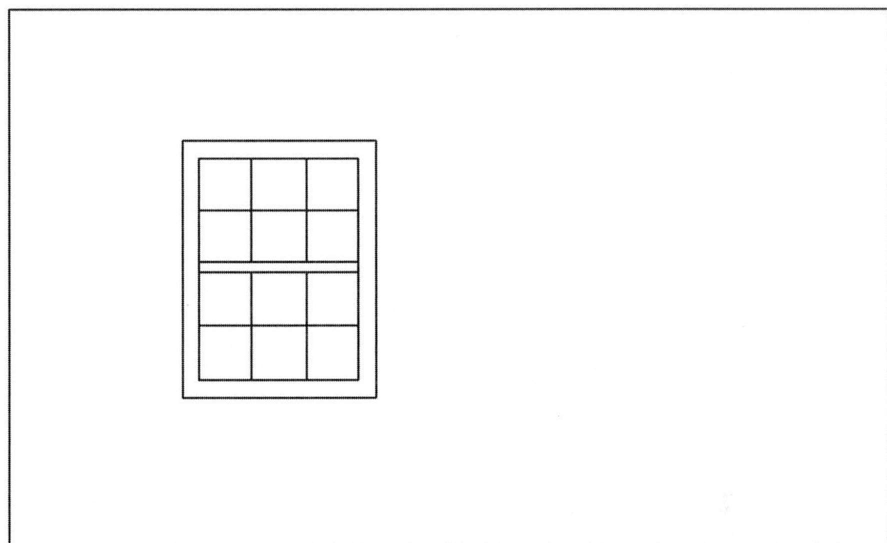

Figure 5-102
The unhatched wall

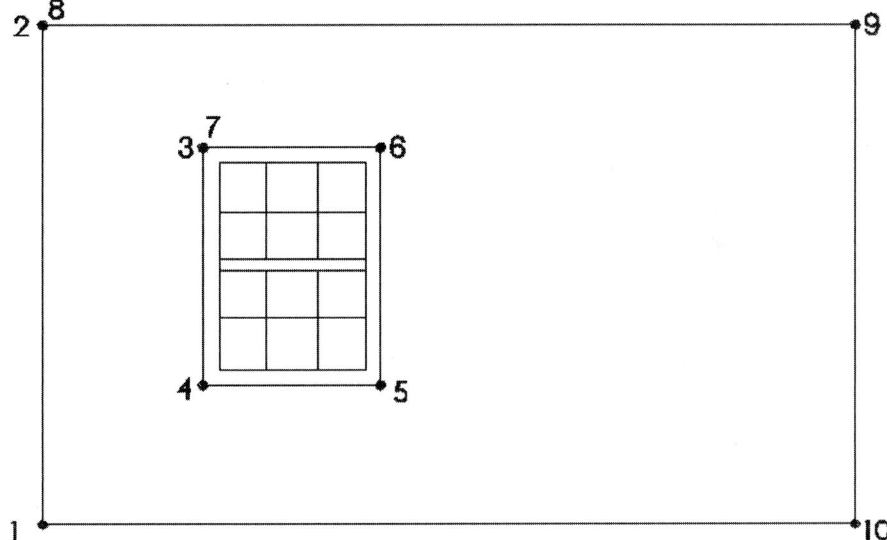

Figure 5-103
A display of all the points to snap to

here is drawing a continuous, closed polyline, which it will then fill with the selected hatch pattern. Notice how we had to maintain a continuous line from point 2 to point 3, and how we had to double back from point 7 to point 8. As long as **HtchType/NoOutlin** is toggled on, no visible line will show where this occurs.

Figure 5-104
The completed wall

11. Press the **Begin** button. Remember that the pattern will not be applied to the polyline until you press this button.
12. The final result should look something like Figure 5-104.
13. Though it may be difficult to see in the figure, notice that the hatch pattern starts with only half a brick along the bottom course. If you want the brick pattern to start with a full brick in the lower left-hand corner, you need to adjust the **Origin** (the place where the hatch pattern starts) to start at that point.
14. In the **Hatch** menu, select **Origin**.
15. *Snap* to the lower-left corner (Point 1) of the wall.
16. Make sure **Entity** is selected, then pick the existing brick hatch pattern. You do this by placing the cursor on any line within the hatch pattern and *clicking* the mouse button. The hatch pattern will appear shaded, confirming that this is the hatch you are picking.
17. Press **Begin** and the hatch pattern realigns itself to start at the new origin (see Figure 5-105).

Angle

You are not confined to drawing hatches that display at the angle shown in the hatch preview window. The **Angle** function in the **Hatch** menu

Figure 5-105
The realigned pattern

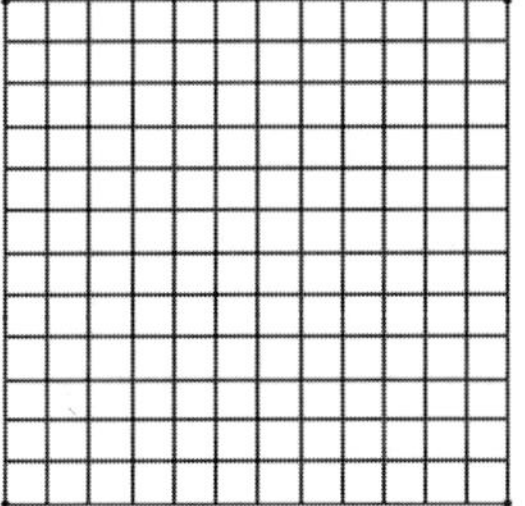

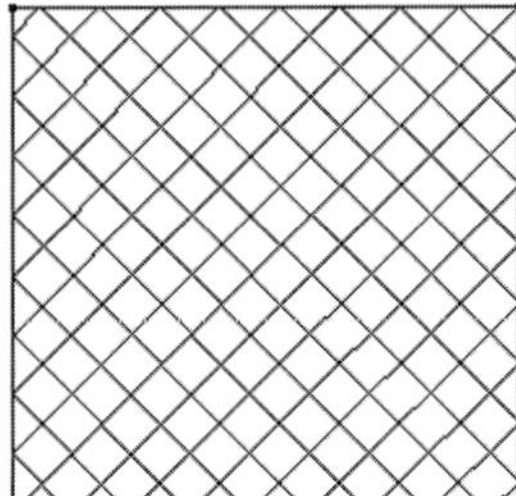

Figure 5-106
Different hatch angles

enables you to display a hatch pattern at any angle. If you draw the **4x4til** hatch pattern with a default angle of **0**, the pattern will draw just as you see it in the preview window, like the first image in Figure 5-106. If you want to rotate the hatch pattern to display diagonally, select the **Angle** button and then type in an angle like **45** and press **Enter**. Pick the existing hatch or draw a new one, press **Begin**, and the hatch will display like the second image.

Origin

When you draw a new hatch, DataCAD does not automatically start the origin of the pattern at one of the vertices of your hatch boundary. Instead,

DataCAD uses the origin from the last hatch pattern drawn. In fact, if you move, mirror, or copy an associative hatch pattern to a new location, the hatch pattern will seem to shift since the origin stays where it was and does not move when the hatch pattern moves. There is actually a very good reason for this.

When hatching complex areas, especially ones with a lot of vertices involved, it is often easier and more efficient to break the complex areas into smaller, easier-to-manage areas. But you still want the hatch pattern to align between adjacent areas so that the hatch pattern displays seamlessly. Here's an example:

1. Draw an area something like Figure 5-107.
2. Select **Pattern** and pick something like **48x24til**.
3. Use the **Associat** and **Boundary** commands to hatch only one of the two boxes (make sure **HtchType/NoOutlin** is on) and the press **Begin**. The hatch will look like Figure 5-108.
4. Now hatch the second box and press **Begin**. The final result should look something like Figure 5-109.

Because the origin remained the same between the two hatches, the two patterns align perfectly. If we then rotate this image 45 degrees using the

Figure 5-107
An unusual area to be hatched

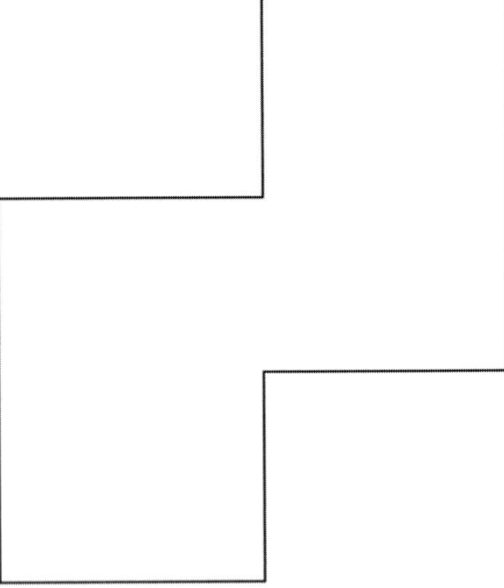

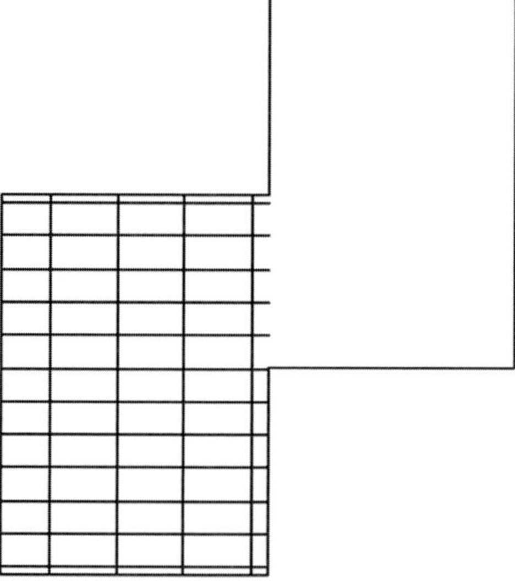

Figure 5-108 Hatching half of the area

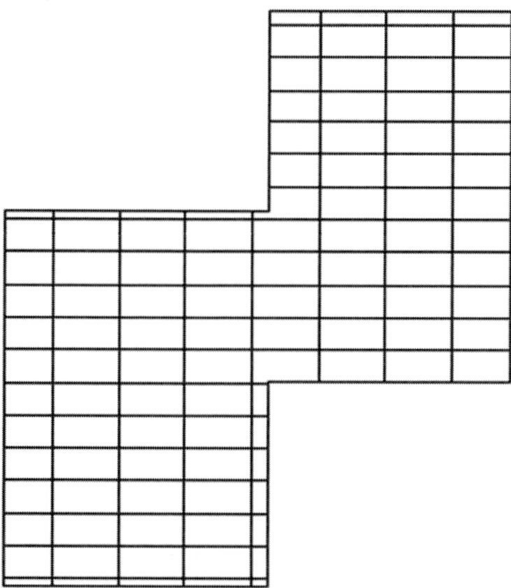

Figure 5-109 Completing the hatch

Edit/Rotate command, the image looks like Figure 5-110. Again you can see that although the boxes rotated, the hatch origin and orientation stay in place.

In the first illustration in Figure 5-111, we took the original image and realigned the origin to the lower-left corner. We then used the **Edit/Move** command to move the whole drawing to the right one foot, as you can see in

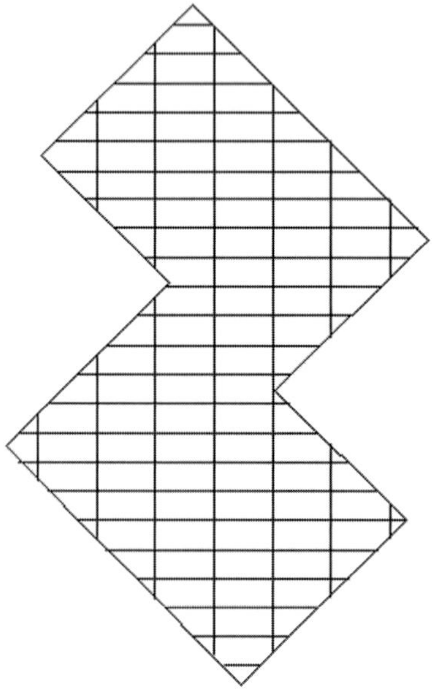

Figure 5-110 The rotated area

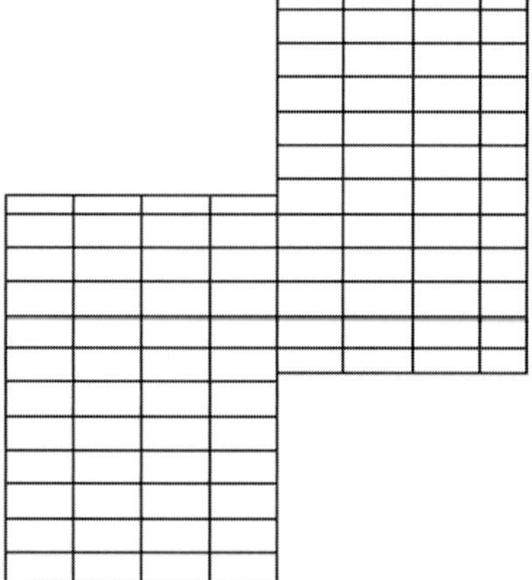

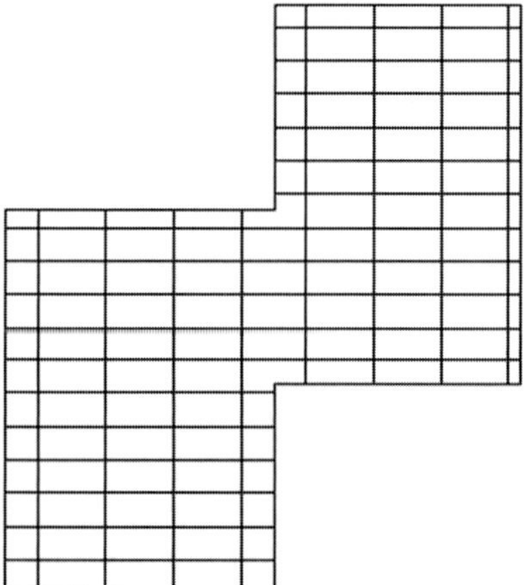

Figure 5-111 Using the **`Origin`** command to make any necessary alterations

the second illustration. Notice how the pattern stays in place, because the pattern's origin did not move, and the boxes did, yet the pattern seemed to have shifted. To fix this, use the **Origin** command to snap to the lower-left corner, realigning the hatch pattern to its original configuration.

Each hatch pattern can have its own origin, or they can all share a common origin. It all depends on how you use the **Origin** command.

Changing Associative Hatch Boundaries

Design is rarely static. Sooner or later you will need to make some changes to that perfectly hatched kitchen floor plan. If you are using associative hatching, then this task is greatly simplified. If your kitchen plan just got longer or wider, all you have to do is use the **Edit/Stretch** command to stretch the kitchen. The associative hatch pattern will stretch with it!

If you need to relocate individual vertices, you can use the **Stretch** command on each vertices as well. If two different vertices are located on top of one another, but you only want to stretch one of them, you will not be able to accomplish this with the Stretch command, since both vertices will be stretched. See the following instructions on how to move a polyline vertices with the Polyline macro.

Viewing Polyline Vertices If you hatched a simple rectangular floor plan with a total of four vertices and then later changed the floor plan to an L-shape, you won't be able to change the shape of the hatch pattern by simply stretching it. You actually need to add vertices to the hatch boundary to define the new shape. Unfortunately, you cannot alter the current polyline boundaries of the hatch from within the **Hatch** menu. You need to use a **Toolbox** macro called *Polyline*, which is installed with DataCAD. But even with this macro, it can be difficult to add or delete vertices because you can't see them directly. Here's a little trick to make the process a lot easier:

1. Go to the **Utility/Display** menu and turn **ShowHtch** off.
2. All your hatch patterns will temporarily disappear so that only the polyline outlines of the hatches are shown. You can now see small dots at each vertices of the hatch outline, which makes editing them much easier.
3. When you are done editing the hatch polylines, go back to the **Utility/Display** menu and turn **ShowHtch** on. The hatch patterns will reappear. If you have **ShowHtch** off when you print the drawing, you will see only the polyline outline, so don't forget to turn it back on.

Add, Delete, and Move Polyline Vertices Use the **Polyline** macro, found in the **Toolbox** (**Edit/Toolbox**, or **M**), to add, delete, or move vertices in an existing polyline, like an associative hatch pattern. This process is greatly facilitated by using the little trick mentioned above prior to running the macro:

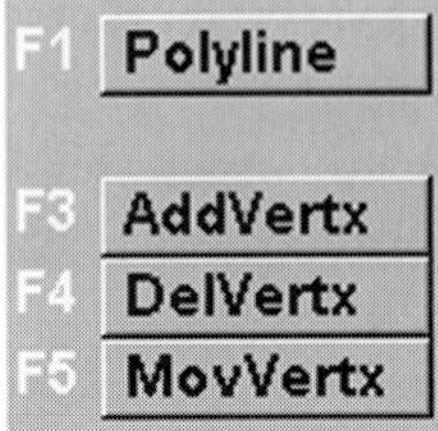

Figure 5-112 The Polyline macro menu

1. Select **Edit/Toolbox**.
2. Select the **Polyline.dcx** macro and then press **Open**. The top of the menu will look like Figure 5-112.

Polyline Use this option to draw a new polyline. If you use this to draw a new closed polyline, you could then go back to the **Hatch** menu, select **Entity**, pick the polyline, and the polyline will be hatched inside.

AddVertx Use this option to add a vertex to an existing hatch polyline:

1. Select **AddVertx**.
2. You will be prompted to select a polyline to add the vertex to. *Click* on or near one of the boundary lines or vertices of the hatch polyline.
3. Pick a point between two existing vertices. A new vertices will be created, attached by a rubber-banding line to your cursor.
4. *Click* or *snap* this new vertices wherever you need to place it. Remember to avoid crossing over any of the other lines of your hatch polyline.

DelVertx Use this option to delete an existing polyline vertex:

1. Select **DelVertx**.
2. You will be prompted to select a polyline to delete the vertex from.
3. *Click* on or near the vertex you want to delete and it will be removed.

MovVertx Use this option to move an existing vertex. This is especially useful if two vertices overlap one another, but you only want to move one of them:

1. Select **MovVertx**.
2. You will be prompted to select a polyline whose vertex is to be moved.

3. *Click* on or near the vertex you want to move. The selected vertices will be attached by a rubber-banding line to your cursor.
4. *Click* or snap the vertices wherever you need to place it. Remember to avoid crossing over any of the other lines of your hatch polyline.

Creating Polylines with DC Sprint

Chapter 24, "Third-Party Macros," describes a third-party macro called *DC-Sprint* by Bill D'Amico. One of the features in this macro is called *Line-to-Polyline* (Lin2Pln). It allows you to convert contiguous 2D entities into a single polyline. This can be a great help when trying to hatch complicated areas with lots of nooks, crannies, and curves, like the walls of a floor plan. Here's how it works.

1. Usually, you will want to create a temporary layer to do this work on, after which you can move the final results to whatever layer you desire.
2. Draw the boundaries for your hatch pattern with 2D lines. The great thing about this is that you can use DataCAD's great 2D tools like **Erase, Erase/Partial**, 2-line trim, and the full **Curves** menu to create your hatch outline. Make sure the final result is a closed boundary.
3. **PgUp** or **R**ecalc so that all the lines that make up the boundary are visible.
4. Go to the DataCAD 3D menus and right-click until you are in the **3DEdit** menu.
5. Select the **Hide** option. Make sure **SaveImg** is on.
6. Select **Begin**. This will create a hidden line image of the boundary lines.
7. Select **NewLyr** and then type in the name of a new, temporary layer. You will be asked if you want the layer to be on or off. Select **Off**.
8. *Right-click* out of the **Hide** menu and then press the ; key to get back to the 2D menus.
9. Go to the **Layers** menu and select **ActvOnly**. Select the new layer that you just created, so that it will be the only layer displayed.
10. Run the **DCSprint/Lin2Pln** macro and select the boundaries (using **Area** or **Fence**) that you just created.
11. The 2D lines will be converted into a single, closed polyline.
12. Now you can go to the **Hatch** menu and hatch the new polyline by **Entity**, **Area**, or **Fence**.

13. Select **Edit/Move** and move the new hatch to the layer that you want the hatch to be located on. You can then delete the temporary hatch layer that you created earlier.

NOTE: *Steps 4 through 9 are not necessarily required, but I almost always do them so that I don't have any problems. The* ***Lin2Pin*** *macro works by scanning the entities sequentially as they were created in the database. But many of the lines that appear visually contiguous were created in no particular order, so they may not be contiguous in the database. This means that sometimes the polyline vertices created by the macro will jump around, creating areas that are not hatched correctly. Running the* ***Hide*** *command will create a new set of lines on a new layer, which will automatically create contiguous lines in the database.*

You don't need to do all this if you have a simple area to hatch, but for complex hatches, like the walls in large floor plans, it will save a great deal of time. If you try this method, I guarantee you'll love it!

Identify

Often you will need to find out some information about an entity that you have already drawn. For instance, you might need to know what layer it's on, or you might need to find out the font name of a piece of text, the name of a symbol, or whether a particular line is a 2D line or a 3D line. All of this and more can be found by using the **Identify** command. It can be selected in a number of ways:

- Pressing the **I** button in the Navigation Pad.
- Pressing the **I** (upper case "I") on the keyboard.
- Clicking on **Edit/Identify**.
- Selecting **Tools/Identify** from the drop-down menu.

Pick the **Identify** command and then click on the entity you want information about. In the vertical Menu Window, you will see most of the available information about that entity. The actual information given depends on the type of entity. Some of the information is given outright. The menu buttons will say something like **LINE** for the entity type, **Walls** for the layer name, **Brown** for the color, and **RigidIns** for the linetype name. Other buttons have to be clicked on for them to display the information you want. For instance, click on the **Spacing** button and the Message Line at the bottom of the screen will say, "***Current line spacing* = 6**".

Note that when you first identify an entity, some of this information is duplicated in the Message Line at the bottom of the screen, such as Spacing, Weight, Overshoot, and Base and Height, so you may not always have to click on a button to display that information. Again, it depends on the type of entity. The rest of the entity information, such as length, angle, and attribute, will be displayed in the Coordinate Readout line, just below the Message Line. You will find the **Identify** command to be one of those commands that you use very, very often.

Identify/SetAll

The **SetAll** function is one of the truly great features of DataCAD. Select the **Identify** option, pick any entity (line, arc, associative hatch, associative dimension, and so on) in the Drawing Window, select the **SetAll** option, and then all the current settings will be changed to match those of the entity you picked, including linetype, layer, color, spacing, Z-heights, hatch settings, associative dimension settings, and so on. Keep in mind that this command only matches entity properties, but not geometries. Thus, picking a circle will not cause an identical radius or diameter to be selected, for instance.

This **SetAll** command is so important that we have a permanent keyboard shortcut (**Alt+I**) set up for it (see Chapter 23). I place the cursor on the entity I want to match and then press **Alt+I**. All my current settings are changed to match. Since one of those settings is the active layer, now I am working on the same layer that the selected entity resides on.

Here is how this can save you a lot of time. Let's say you have a Brown, **RigidIns** linetype entity with a spacing of 6″ [152] on the **Insulatn** layer at a Z-height of 2′-0″ [610]. You want to draw some more of these lines on the same layer with all the same settings, but you are currently drawing on a layer called **Walls**, with a blue, **Plywood1** linetype, a spacing of $1^1/_2$″ [38] and a Z-height of 0. To change all your settings to match the entity on the **Insulatn** layer, perform the following steps. We'll begin with the long method:

1. Go to **Utility/Layers**.
2. Select **SetActiv/Insulatn**.
3. *Right-click* several times to back out of the menu.
4. Select **Edit/LineType**.
5. **ScrlFwrd** and select **RigidIns**.
6. Select **Spacing**.
7. Type in **.6** [152] and press **Enter**.
8. *Right-click* out of the menu.

Here's the short method:

1. Select the **I** (identify) button in the Navigation Pad.
2. *Click* on the brown, **RigidIns** line in the Drawing Window.
3. Select the **SetAll** option. You're done.

Here's the REALLY short method: (see Chapter 23):

1. Place the cursor on the brown, **RigidIns** line in the Drawing Window.
2. Press **Alt+I**. You're Done!

Measures

Besides drawing lines, drafting is all about measurements. Since you can't put a scale to your screen to measure the things you've drawn, however, there have to be other ways. The Identify function will give you a lot of information about whole entities. If you identify a line, DataCAD will tell you its full length, angle, and a number of other things about it. If you identify an arc or circle, DataCAD will tell you, among other things, its radius. This is all fine for a limited amount of information about the whole entity, but the functions in the **Utility/Measures** menu will provide you with a great deal more measurement options.

We'll now examine the **Measures** functions and describe what each one does. Perhaps the most important one is the **PntToPnt** (point-to-point) measurement. We use it so often that we have a toolbar icon that, when selected, skips over all the other measurement options and goes straight to the **PntToPnt** function.

RefPnt (`)

This extremely important measurement option is covered in great detail earlier in this chapter.

SnapPnt

This will allow you to create a non-printing snap point on your screen, one that can be snapped to with the middle mouse button or the **Nn** key. It will appear as a small dot. You can place them anywhere in your drawing, even floating out in space. This is primarily used to mark a point in space off of which other lines and arcs are referenced. To get rid of these points, they must be erased just like any other entity.

Line

Select this option, then click on a line, and DataCAD will give you its length down in the Message Line. A **ToDrwing** button appears at the top of the menu. If you select this option, your crosshair cursor will change to a text cursor. *Clicking* anywhere in the Drawing Window will then place the distance dimension in the Drawing Window. If you do not change any text settings, then the current text settings, including the size and font, will be used.

NOTE: *Whenever the **ToDrwing** option is selected in any DataCAD menu, another set of menu buttons will appear like this:*

These functions will allow you to change the current text settings prior to placing the text in the Drawing Window. If you do not change any of

these settings, then the current text settings including the size and font will be used. But be aware that any changes that are made here will also be applied to the current text settings outside of the ***Measures*** *menu.*

PntToPnt

This option will measure the distance between two points. Snapping to points is not a requirement of this function, but snapping is generally used in order to precisely measure a distance between endpoints, midpoints, intersections, quadrants, and so on. The distance will be displayed in the Message Window. After you pick the second point, a **ToDrwing** button appears at the top of the menu.

Diameter

Select this option, then *click* on an arc or circle, and DataCAD will give you its diameter in the Message Line. A **ToDrwing** button appears at the top of the menu.

Chord

Select this option, then *click* on an arc, and DataCAD will give you its chord dimension (the distance across the two endpoints of the arc) in the Message Line. When you do, a **ToDrwing** button appears at the top of the menu.

ArcLnth

Select this option, then *click* on an arc, and DataCAD will give you its total length (along the curve from endpoint to endpoint) in the Message Line. A **ToDrwing** button appears at the top of the menu.

Circumf

Select this option, then *click* on a circle, and DataCAD will give you its circumference in the Message Line. A **ToDrwing** button appears at the top of the menu.

InclAngl

Select this option, then click on two lines, and DataCAD will give you the included angle between the two in the Message Line. A **ToDrwing** button appears at the top of the menu. The Included Angle is an angle that is less than 180 degrees, formed by the intersection of two lines. In Figure 5-113, the Included Angle is 45 degrees.

ExclAngl

Select this option, then click on two lines, and DataCAD will give you the excluded angle between the two in the Message Line. A **ToDrwing** button appears at the top of the menu. The Excluded Angle is an angle that is greater than 180 degrees, formed by the intersection of two lines. In Figure 5-114, the Excluded Angle is 315 degrees.

Figure 5-113 45 degrees is the Included Angle.

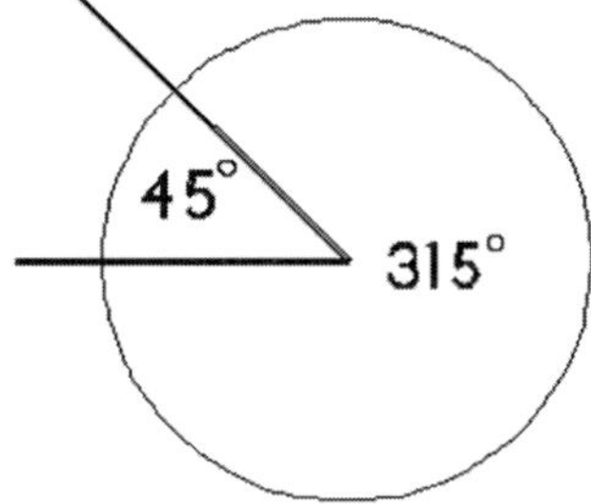

Figure 5-114 315 degrees is the Excluded Angle.

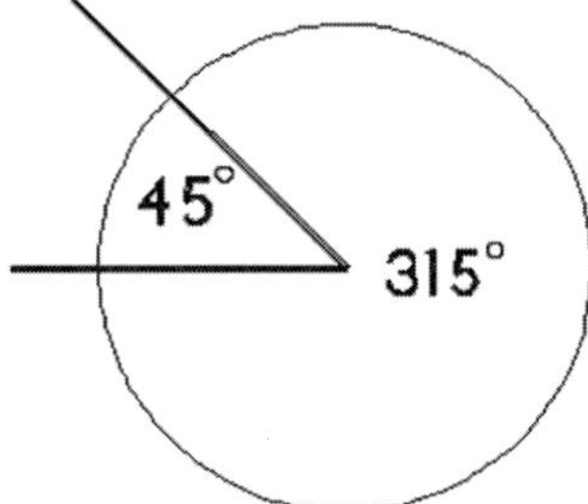

LineAngl

Select this option, then click on a line, and DataCAD will give you the angle it is drawn at in the Message Line. A **ToDrwing** button appears at the top of the menu.

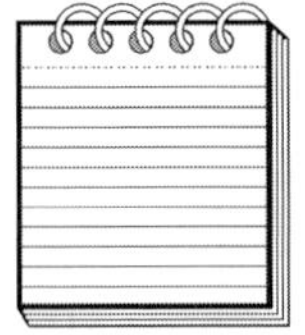

NOTE: *Every line has two possible angles, depending on how it is drawn. DataCAD determines the angle of a line by measuring from the first point to the second point as it was drawn. For instance, a perfectly vertical line would have an angle of 90 degrees if it were drawn from bottom to top, but that same line would have an angle of 270 degrees if it were drawn from top to bottom.*

Area/Per (Area Perimeter)

Use this to compute the area, perimeter, or both of any existing enclosed shape. For a circle or a polyline, all you have to do is select the entity, but for other shapes, you must trace over the shape to get the calculations. Tracing a rectilinear shape is straightforward, but for a complex curved shape like an ellipse, you will find this somewhat tedious and inaccurate. For arcs and simple curves, tools in the **Area/Per** menu can help.

When you press the **Area/Per** button, you will only see four menu options: **ScaleTyp**, **Select**, **Acres**, and **AddSnap**. The Command Line will read, *"Select first point on polyline,"* because DataCAD assumes that you want to start tracing a shape. Five other options will appear once you have started to trace a shape in the Drawing Window.

ScaleTyp (Scale Type) Use this option to change the scale type to be used for the calculation. These are the same scale types as found in the **Utility/Settings/ScaleTyp** menu. See Chapter 3 if you need information about any of these scale types. If you change the scale type here in the **Area/Per** menu, the current scale type in the Settings menu will not be affected.

Select Use this option to select a circle or polyline for an area or perimeter calculation. Once you have clicked on this option, all you have to do is click on the entity and DataCAD will make the calculation. Both the area and the perimeter calculations will be displayed in the Message Line. You will then be presented with the same 11 options as in the **Close** menu, described later in this section.

NOTE: *If you select a polyline that is not closed, DataCAD will assume that the shape should be closed between the first vertex and the last vertex and will make the calculations based on that assumption.*

Acres If **Acres** is toggled on, the area units will be calculated in acres. If **Acres** is toggled off, the area unit appears as the scale type set in the **ScaleTyp** menu. The **Acres** option will not be displayed if a metric **ScaleTyp** is set.

AddSnap Choose this option before you begin to trace a shape and DataCAD will add a snapping point to each vertex of the tracing polyline. For more information, see the **SnapPnt** description at the beginning of this section.

Tracing Shapes with Area/Per

To determine the area of shapes other than circles or polylines, you need to select **Area/Per** and trace over the shape, using object snapping as appropriate. Once you select the first point to begin tracing, the other five menu options will appear: **2PtArc**, **3PtArc**, **Tangent**, **BackUp**, and **Cancel**.

NOTE: *Whenever tracing objects, do not enter the last point (same as the first point) to close the shape. Instead use the* ***Close*** *button or click the right mouse button and DataCAD will close it for you. You will then be presented with another set of menu options, described at the end of this section.*

2PtArc

This button enables you to draw a two-point arc in your tracing boundary. For a complete description of how to use this option, see the section *"Drawing Arcs and Circles"* earlier in this chapter.

3PtArc

This button enables you to draw a three-point arc in your tracing boundary. For a complete description of how to use this option, see the section *"Drawing Arcs and Circles"* earlier in this chapter.

Tangent

After you select a second point, the **Tangent** button will appear, allowing you to draw a tangent arc in your tracing boundary. For a complete description of how to use this option, see the section *"Drawing Arcs and Circles"* earlier in this chapter.

BackUp

During your tracing, you may make a mistake in entering a vertex. Use the **BackUp** button to backup one step. Each press of the **BackUp** button will move you one more point backward, all the way to the first point if you need to.

Cancel

Use this option to cancel all current area/perimeter operations. DataCAD will revert to the main **Area/Per** menu.

Close

Select **Close** or *right-click* to close the shape you are tracing. When you do, you will be presented with 11 more options, as follows:

Volume This option calculates the volume of the polygon and displays the value in the Message Line.

ToDrwing When you select this option, your crosshair cursor will change to a text cursor. More options will appear to allow you to change the text settings. *Clicking* anywhere in the Drawing Window will place the calculation in the Drawing Window. If you do not change any text settings, then the current text settings including the size and font will be used.

Centroid This option locates the centroid of the circle, polyline, or tracing polygon. When selected, you will have two options:

ShowIt This option displays the centroid as a cross that disappears when you exit the **Measures** menu.

AddIt This option adds a snapping point to the drawing at the centroid.

ScaleTyp As described earlier, use this option to change the scale type to be used for the calculations.

Select Use this option to select another circle or polyline for area or perimeter calculations. Once you have clicked on this option, all you have to do is click on the entity and DataCAD will make the calculation.

Acres If **Acres** is toggled on, the area units will be calculated in acres. If **Acres** is toggled off, the area unit appears as the scale type set in the **ScaleTyp** menu. The Acres option will not be displayed if a metric **ScaleTyp** is set.

AddSnap Choose this option and DataCAD will add a snapping point to each vertex of the tracing polygon. For more information, see the **SnapPnt** description at the beginning of this section.

The next four options are toggles used to keep a running total of perimeter and/or area calculations. As long as you remain in the **Measures** menu, the totals will continue to be tracked. Once you exit the **Measures** menu, all current totals will be nullified.

Perim+ When this option is toggled on, the next perimeter measurement will be added to the running total.

Perim– When this option is toggled on, the next perimeter measurement will be subtracted from the running total.

Area+ When this option is toggled on, the next area measurement will be added to the running total.

Area When this option is toggled on, the next area measurement will be subtracted from the running total.

TakeOffs

Use this option to calculate the total length and/or area of all lines, polylines, circles, and arcs in a drawing according to color, linetype, or weight. You can use the **Color**, **LineType**, and **Weight** filters individually or in any combination, and you can turn **LyrSrch** on or off. But other than these four discriminators, the results will be tabulated for the whole drawing file, since no standard EGAFS selections are available.

The readout for the takeoffs is displayed in the Coordinate Readout line. The Message Line will display the status of the currently chosen filters, described later. DataCAD uses the word *Perimeter* to describe the total length of all selected lines. Because all 2D lines in DataCAD are actually $2^1/_2$ D lines (having both length and height), the Area option describes the total area of the selected lines.

Here is one possible use of the **TakeOffs** feature. Perhaps you want to know how much paint it will take to cover the walls of your house design:

1. Make sure you accurately set the $2^1/_2$ D heights of all your walls.
2. Make sure all the walls are on their own layer, or have their own color or weight so that you can filter out all other lines.
3. Run the **TakeOffs** function.
4. Use **Color**, **LineType**, **Weight**, and **LyrSrch** to filter your walls.
5. Press **Begin**.
6. The **Area** calculation will give you the total area of your walls.

If there is only one color, linetype, or weight in the drawing file, then the corresponding menu button will not appear, since there is nothing to select. Here are the takeoffs that are available:

- **ToDrwing** places the takeoff measurement in the Drawing Window (the **ToDrwing** option may not appear until you have selected another menu option or pressed **Begin**). Once you select it, you will get three more options:
  - **Perimeter** places only the total line-lengths measurement in the Drawing Window.
  - **Area** places only the total area of $2^1/_2$ D lines in the Drawing Window.
  - **Both** places both the perimeter and area takeoff measurements in the Drawing Window.

After selecting one of the previous options, your crosshair cursor will change to a text cursor. Clicking anywhere in the Drawing Window will then place the takeoff measurement in the Drawing Window. If you do not change any text settings, then the current text settings including the size and font will be used.

- **Color** calculates the total length of all lines, polylines, circles, and arcs of a specified color.
- **LineType** calculates the total length of all lines, polylines, circles, and arcs of a specified linetype.
- **Weight** calculates the total length of all lines, polylines, circles, and arcs of a specified line weight.
- **LyrSrch** calculates the total length of all specified lines, polylines, circles, and arcs. With LyrSrch off, only the total length of all specified lines on the current active layer will be calculated.
- **Begin** is used to start the calculation of all selected lines, polylines, circles, and arcs in the drawing file, based on the filters selected. The total measurement will be displayed in the Coordinate Readout line.

Using Toolbox Macros

We've saved the best for last in this chapter. Macros are mini-programs that work within DataCAD to accomplish specific tasks. DataCAD comes with numerous macros, but hundreds more are available from third-party sources. All of the installed macros can be found in the DataCAD **\DCX** directory, and if you add more macros, this is where they should go as well. You can find out more about how to find other macros in Chapter 27. Chapter 24 describes the features of some of the best of these macros and Chapter 21 describes the process of creating your own DataCAD macros.

NOTE: DataCAD Applications Language *(DCAL) is the language that all of DataCAD's Toolbox macros are written in. It is a "proprietary" language, meaning it was developed solely for DataCAD. Because of this, you will not find any DCAL programming references on your local computer or bookstore shelves. But DataCAD LLC has provided full*

documentation of the language on your DataCAD installation CD, and other online sources can be used to find information and help if you are interested in programming your own toolbox macros.

DataCAD macros are accessed by selecting the **Edit/Toolbox** button, or pressing **M**. After pressing it, you will see the dialog box shown in Figure 5-115.

All DataCAD macros have a .DCX file extension. To select a macro, double-click on its name, or click once on it and then press **Open**. The macro will run, typically displaying new menu buttons. In this example, we selected the **Dcad_aec.dcx** macro, which gives us a new set of menu options like the ones shown in Figure 5-116.

Here you can see that DataCAD can be made to do a lot of specialized tasks that are not built into the main program. In this case, you can do things like create column grids, draw elevators and stairs, add room labels, and do square foot calculations. Some of the most useful macros that come with DataCAD are described throughout this book as they apply to certain tasks. To find out how the rest of the macros work, you can read about them in the DataCAD Reference Manual.

A great many useful and very powerful macros are available for Data-CAD, and most of them are very inexpensive. Some free ones are even avail-

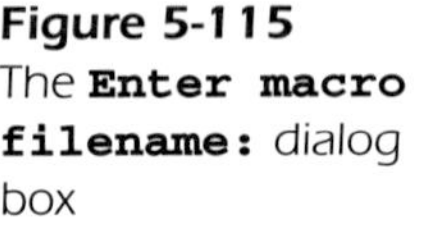

Figure 5-115
The **`Enter macro filename:`** dialog box

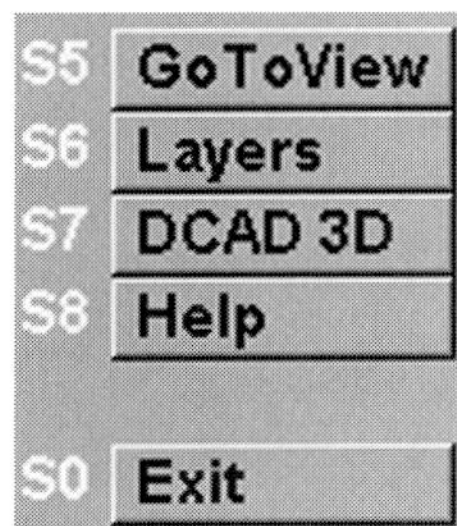

Figure 5-116
The menu from the DCAD-AEC macro

able from the DataCAD Web site and on the CD that comes with this book. You should definitely take the time to learn to use these toolbox macros, and we encourage you to check out all the third-party sources for macros. They will add greatly to your repertoire of drafting tools and are sure to add useful features, speed, and efficiency to your work.

CHAPTER 6

Organizational Concepts

Some topics in this chapter are much more subjective than any others. Although there are usually two or three ways to "skin a cat" in DataCAD, when it comes to file names and filing systems, the options are nearly infinite. Firstly, what we will attempt to do in this chapter is to suggest some possible organizational criteria that you might want to try yourself, after which I'm sure you will develop your own methods. Secondly, we will describe some of the features of DataCAD that will help you find, organize, and more efficiently use your files.

The first consideration is whether or not you are running DataCAD in a networked environment. We will try to describe methods for both network and non-networked situations. Even if you are running DataCAD on a single computer, you can still fool your computer into believing it is networked, offering some distinct organizational advantages. See Chapter 26 for more information regarding networking.

Drawing File Names

Your CAD drawing files should be named carefully and simply so that anyone can understand them, not just you. In general, the sheet number should be used for the file name in some manner. Although you can create drawing file names of up to 255 characters, this can quickly be overdone if you are not careful. A filename that is 40 or 50 characters long loses its usefulness very quickly.

Another consideration is the fact that long filenames are not supported by the DOS versions of DataCAD. Those filenames are limited to only eight characters, a period, and the DC5 file extension (FIRSTFLR.DC5, for example). So if you plan to distribute copies of your drawing files to DataCAD DOS users, you will need to create eight-character filenames, or you could rename the file to a shorter name in Windows prior to distributing the files.

Rules for Naming Files

Some characters cannot be used by Windows in any file name, and some DataCAD cannot use. Nearly any alphanumeric combination will work for drawing file names, with certain exceptions. Windows and DataCAD cannot use the following:

/ a slash

\ a backslash

: a colon
* an asterisk
? a question mark
< a less than sign
> a greater than sign
| a pipe (vertical bar)

DataCAD cannot use the following:

( ) a space
. a period
" a double quote
[a left bracket
] a right bracket
+ a plus sign
= an equal sign
; a semicolon
, a comma

plus any ASCII characters less than 21H

Here are some examples of valid filenames. Note that the .DC5 extension must be present for DataCAD to open the file. When you save a file, DataCAD automatically appends this extension to your file name, so you don't have to type it yourself. But if you are renaming a file via the Windows or DOS interface, be sure to leave the .DC5 extension:

- 1stFloorPlan.DC5
- 1st_Floor_Plan.dc5 (the underscore characters are valid)
- 1stfloorplan&second_floor.dc5 (the ampersand is valid)
- 1stFloorPlan_For_the_Jones_House.dc5

Here are some examples of filenames that are not valid:

- 1st Floor Plan.dc5 (spaces are not valid)
- 1stFloorPlan (no .DC5 extension)
- 1stFloorPlan\for\the\Jones\House.DC5 (the backslashes are not valid)

Well, that was the objective part of this chapter. Here come the more subjective parts. Let's take a look at some of the possible naming conventions you might use. It is important that your file names make sense to you and to anyone else who might open them. If you send drawing files to consultants or others, the filenames should be easily understood without further explanation. Support of long file names should help, but don't overdo it.

Here are examples of some possible file names:

A400.DC5	Sheet A-400
A40X.DC5	Contains multiple sheets, like A-400, A-401, and A-402
A501-A502-A503.DC5	Contains sheets A-501, A-502, and A-503
9801A200.DC5	Project number and sheet A200
PLANS&ELEVATIONS.DC5	Plans and elevations

The combinations are nearly infinite, but the common theme should always remain understandability.

File Locations

It is very important to maintain all your CAD files in sensible locations so that anyone (including yourself) can find files when looking for them. This is true whether or not you operate over a network. Essentially, all drawing, layer, plot, and DWG/DXF files should be in similarly named directories, preferably by project name or number. The method described here is only one possibility, but it has worked well for our office.

If You Do Not Have a Server

By default, DataCAD creates certain directory names for storing various file types. You can create your own directory names in place of these if you'd like, but we'll use the default directories. These directories are as follows:

- **DWG** DataCAD drawing files (.DC5 extension)
- **DEFAULT** DataCAD default drawing files (.DC5 extension)
- **LYR** DataCAD layer files (.LYS and .LYR extensions)
- **PLT** Plot files (.PRN or .PLT extensions)
- **XFER** DXF and DWG import/export files (.DWG and .DXF extensions)

If you use the Sticky Back macro (in the optional DC Sprint macro, available from DataCAD LLC), then you will want to add one more directory: theSTKYBACK, which stores "stickyback" drawings.

Let's assume you have a new project, and that its project number in your office is 9801. Use Windows to add a subdirectory with the project number as its name under each of the above noted directories. Your file structure would then look like this:

```
C:\DCADWIN
  \DWG\9801
  \LYR\9801
  \PLT\9801
  \XFER\9801
  \STKYBACK\9801
```

Within each directory you can add further directories to organize your files as you need, but keep in mind that the best system is the one that is universal in all folders. For instance, let's say you want to have another subdirectory to store files directly related to your engineering consultants. You could add a new folder in each of the previous directories called \ENG. Your directories would then look like this:

```
C:\DCADWIN
  \DWG\9801\ENG
  \LYR\9801\ENG
  \PLT\9801\ENG
  \XFER\9801\ENG
  \STKYBACK\9801\ENG
```

If You Do Have a Server

The same rules apply if you have a file server (a separate computer on the network on which to store all your computer files), but you will create and keep these files on the server instead of your own computer. To avoid having people in our office accidentally saving files to their own hard drives instead of the server, we actually remove the previous directories entirely from their own computers' hard drives.

It is highly recommended that you create separate virtual **drive partitions** on the server for each file type. You do this by **mapping** directories to a drive letter. See Chapter 26 for directions. Here is a quick overview of what it means.

Instead of simply having directories somewhere on the C:\ drive of your server, you will fool all your programs (including DataCAD) into thinking that each of the previous directories resides on its very own hard drive in the server. You must do this on every computer that accesses the server.

First, create the drive partitions and folders you want on the server. Use the directory names listed earlier (DWG, XFER, and so on.). You may also want to do this for the Template and Symbol directories as well (Chapter 14, "Templates and Symbols," explains why). When selecting drive letters, try to use something that makes phonetic sense if possible, and be sure not to use a drive letter that already exists on the workstation. Here is what we use:

F:\	Default directory
G:\	DWG directory
L:\	LYR directory
P:\	PLT directory
S:\	SYM (symbol) directory
T:\	TPL (template) directory
X:\	XFER directory
Y:\	STKYBACK directory

Next, on each networked workstation, go to My Computer on the Windows desktop and then click on the second icon from the left in the toolbar (if the toolbar is not visible, click on **View/Toolbar**). This is the **Map Network Drive** button. Pick a drive letter that makes sense for your files like the ones listed above, and then select or type in the path of the resource you want on the Server, such as **\\server\dwg.** Check the **"Reconnect at logon"** box to make Windows reestablish the connection with the new drive next time you start your computer. Click on **OK**. Some other steps are involved, but this is the basic concept.

Now open DataCAD and select the **Tools/Program Preferences/ Pathnames** tabs. Change the paths to reflect the new virtual drives you just created. For instance, the Drawing Files box should have G:\DWG displayed in it. This is a powerful feature that you will come to really appreciate if you give it a try.

Default Files

Default files are not only big time-savers, but the key to a well-organized drawing file and should be used at all times. Think of a default file as a "seed" drawing: one that contains any repetitive information and settings that you do not want to recreate every time you start a new drawing. You can have one default file or many. If you do not define your own default drawing, then DCAD will choose one for you (called DEFAULT.DC5, unless you specify something different). Chapters 2 and 3, "Settings and Display Options," describe the creation of a default drawing file, and the selection of many of the possible settings in a DataCAD file. By using a predefined default file, you will be assured that your new drawing will start out with the correct settings, such as

- Input mode (Relative Polar, Cartesian, and so on)
- Snap settings
- Display Settings
- Font and dimension settings
- Layers

All the default files are stored in the \Default folder in the DataCAD root directory (or on the server if that's where you have reassigned the \Default folder). When you select a default drawing, DataCAD will use the settings in that file but will name and store the new drawing file according to what you enter in the New File dialog box.

Remember that you can have more than one default file for different occasions. You may have only one default drawing or a dozen. For instance, you may have certain settings you like for 2D drawings and different settings for 3D drawings. You may have one default drawing file for building elevations and another for plans, each with layer names already set up.

Besides standard drawing settings, the use of various layer-naming conventions is very effective for default drawings. Having office standards for layer names is critical to maintaining orderly, useable drawing files. The best way to maintain those layer standards is to create them in default files. For instance, you might have a default file called ELEVATION.DC5 containing all the layers for creating elevations, and one called PLANS.DC5 containing, you guessed it, layers for creating floor plans. Remember that one of CAD's greatest advantages is the reduction of repetitious tasks. By creating default files for every occasion, you eliminate the repetition of setting up new layers every time you start a new drawing.

Layer Concepts

Layers are the lifeblood of any CAD drawing file. DataCAD has some unique features that enable a measure of flexibility in layer settings and layer naming. It also has some unusual quirks that you should understand.

Creating New Layers

If you ran through the tutorial in Chapter 4, then you already know something about creating new layers. If you didn't then here is the process:

1. Press **Ll** to go to the **Layers** menu.
2. Select **NewLayer**. Decide how many new layers you want and select that number in the menu, or type in the number and then press **Enter** or *right click*. That only created the layers. Now you need to name them. You don't have to do that right away, but they won't be very useful until you do.
3. Once back in the **Layers** menu, select **Name**. The current list of layers will appear. Note that all the layers that you just created are set to on by default (the menu buttons are green and appear to be pressed in).
4. The message line reads, *"Select layer to be renamed:."* Pick one of the layers. That layer name will appear in the message line and will be highlighted. As long as the text is highlighted, you can simply type in a new name to replace the old one. If it is not highlighted, then place the cursor where you want it in the text and use **Del** or **Backspace** to delete characters and type in the new name.
5. Press **Enter** and the new layer name will appear in the menu. Continue to select and rename layers. When you are done, just *right click* back to the **Layer** menu.

NOTE: *In DataCAD, unlike some other CAD programs, you can have all uppercase letters, all lowercase letters, or a combination of both. Using both can help with layer name readability, but some users prefer all upper case. To each their own!*

If you are exporting a drawing to DXF or DWG format, DataCAD will automatically convert all layer names to uppercase, since many CAD programs require it.

Layer Names

Like drawing files, it is very important to create layer names that are easily understood, not only by yourself and your office personnel, but for any clients, engineers, or consultants you might send drawings to. The best method is usually to keep them phonetic in some way: WALLS for walls,

DOORS for doors, and so on. You might also use something like 1F as a prefix for all first floor elements or NE as a prefix for all north elevation elements. Unlike DataCAD drawing files, however, the length of a layer name is currently limited to a maximum of eight characters (so that they can fit within a menu button), and just like file names certain characters cannot be used.

Some acceptable layer names would include

- 1-COL-LN
- 6_Doors
- Door&Win

Some unacceptable layer names would include

- 1 Floor (has a blank space in it)
- First-Floor (has more than eight characters)
- 1*Floor (has asterisk)

Because it often makes sense to create large scale ($1\frac{1}{2}''$ [1:10], 3″ [1:5], and so on) details from the smaller scale sections ($\frac{3}{8}''$ [1:30], $\frac{1}{2}''$ [1:20], and so on) by using clip cubes and *Multi-Scale Plotting* (MSP), it is beneficial to try to keep related groups of layers physically adjacent to one another with matching prefixes to identify them. If you set up all the required groups of layers ahead of time then you won't need to add additional layers to the group. Layer management is then relatively easy. However, DataCAD keeps track of layers by their order in the drawing file, not by their name, and does not allow you to insert new layers between existing layers. More on this later in the chapter, but for now just know that it is beneficial to include more layers in each related group than you think you will actually need, since reordering your layers later is problematic at best. This is arguably one of DataCAD's biggest weaknesses at the moment.

Here is one possible layer naming system. You will invariably come up with your own over time, but this is good food for thought.

The general **New Plan** layer-naming guidelines are as follows:

X = Floor number or letter (1, B, and so on)

X-COL-LN	Column lines/grid
X-STRUCT	Special structures (columns, etc.)
X-SITE	Site information
X-EXWALL	Exterior walls (heavier lineweight)

X-INWALL	Interior walls/partitions (lighter lineweight)
X-LOWALL	Half-height objects (not to show up in ceiling plans)
X-RMNAME	Room names
X-RMNUMB	Room numbers (not required if included in X-RMNAME)
X-DOORS	Doors and swings
X-DRNUMB	Door numbers
X-WINDW	Windows
X-WNNUMB	Window numbers
X-FIXTUR	Plumbing, mechanical, and so on
X-MILLWK	Cabinets, counters, millwork, and so on
X-HATCH	Hatching and other pocheing (including linetypes used as poche)
X-NOTES	Notes and text
X-DIMS	Dimensions
X-MISC-1	Miscellaneous #1 (change name as required)
X-MISC-2	Miscellaneous #2 (change name as required)
X-CLHEAD	Headers over doors
X-CLGRID	Ceiling grids
X-CLFIX1	Ceiling fixtures (Electric)
X-CLFIX2	Ceiling fixtures (Mechanical)
X-CLNOTE	Ceiling notes and dimensions
X-CLMIS1	Ceiling miscellaneous #1
X-CLMIS2	Ceiling miscellaneous #2

Here are the general **Renovation Plan** layer-naming guidelines:

X = Floor number or letter (1, B, and so on)

XR-CO-LN	Column lines/grid
XR-STRUC	Special structure (columns, and so on)
XR-SITE	Site information
XR-EXWAL	Exterior walls (heavier lineweight)
XR-INWAL	Interior walls/partitions (lighter lineweight)
X-LOWAL	Half-height objects (not to show up in ceiling plans)
XR-RMNAM	Room names

XR-RMNUM	Room numbers (not required if included in X-RMNAME)
XR-DOORS	Doors and swings
XR-DRNUM	Door numbers
XR-WINDW	Windows
XR-WNNUM	Window numbers
XR-FIXTR	Plumbing, mechanical, and so on
XR-MILWK	Cabinets, counters, millwork, and so on
XR-HATCH	Hatching and other pocheing (including linetypes used as poche)
XR-NOTES	Notes and text
XR-DIMS	Dimensions
XR-MIS-1	Miscellaneous #1 (change name as required)
XR-MIS-2	Miscellaneous #2 (change name as required)
XR-CLHED	Headers over doors
XR-CLGRD	Ceiling grids
XR-CLFX1	Ceiling fixtures (Electric)
XR-CLFX2	Ceiling fixtures (Mechanical)
XR-CLNOT	Ceiling notes and dimensions
XR-CLMS1	Ceiling miscellaneous #1
XR-CLMS2	Ceiling miscellaneous #2

The general **Existing Plan** layer-naming guidelines are the same as renovations, but use **E** in place of **R.**

Here are the general **Interior and Exterior Elevation** layer-naming guidelines:

X-COL-LN	Column lines/grid
X-STRUCT	Special structure (columns, and so on)
X-SITE	Site information
X-WALL1	Exterior walls #1
X-WALL2	Exterior walls #2
X-DOORS	Doors
X-WINDW	Windows
X-WNNUM	Window numbers
X-TRIM	Trim

X-HATCH	Hatching and other pocheing (including linetypes used as poche)
X-NOTES	Notes and text
X-DIMS	Dimensions
X-MIS-1	Miscellaneous #1 (change name as required)
X-MIS-2	Miscellaneous #2 (change name as required)

X = North, South, East, West (N, S, E, W) *or* elevation number (1, 2, and so on)

The **Miscellaneous** drawing information layer-naming guidelines are as follows:

KEY	Key plan (add text after KEY as required)
TTLBLK	Sheet/title block
A-100	Sheet number and sheet-specific information (goes with TTLBLK)
LEGEND	Legends (symbols, and so on)
ROOFDRN	Roof drains

Here are the general **Detail** layer-naming guidelines:

A	**X-WALL**	Entities making up the walls, floors, roofs, and so on
	X-STRUCT	Steel, aluminum, and other structural elements
	X-WINDOR	Windows and doors
	X-HATCH	Hatching and other pocheing (including linetypes used as poche)
	X-NOTES	Notes and dimensions (don't include dimensions if X-DIMS is used)
B	**X-DIMS**	Dimensions
	X-MISC	Miscellaneous
C	**X-2HATCH**	Hatching and pocheing in large-scale details
	X-2NOTES	Notes and text in large-scale details
	X-2MISC	Miscellaneous in large-scale details

X = Number or Letter of Detail (1, 6, A, G, and so on)

A = Simple details

$A + B$ = More complex details

$A + B + C$ = Additional layers for large-scale drawings (like $1^1/_2''$ enlargements of $^3/_4''$ details)

Layer Order

It is important to note that DataCAD's layers remain in the order in which they are created. Currently, layers cannot be reordered. In fact, when DataCAD reads layer information in the drawing file, it does so by order and not by name. This is vastly different than other CAD programs and can cause problems with a few features in DataCAD if you're not careful. It also means that the order in which you create your layers can have a definite impact on the usability of the drawing file. For instance, if you create five layers for the first floor, five for the second floor, and five for the third, and then later decide to add a sixth layer to each floor, the three new layers will be added at the end of the layer list. This makes it a little more difficult to pick all the first floor layers, since you constantly have to go hunting for the last layer.

However, all is not lost here. Two unique features of DataCAD can make the task much easier: 3D GotoViews, and the Filter option. 3D GotoViews are covered in depth in Chapter 10 (think of them as instant snapshots of layer groupings that can be instantly recalled). In this chapter, we will discuss the Filter option.

Layer Filtering

In the **Layer** menu is a button near the bottom labeled **Filter**. Use this option to find and group layers with similar names. You can search, or filter, by a variety of methods: by prefix, by suffix, or by wildcards. If you are familiar with DOS search filters, you will be instantly familiar with DataCAD's. Since DataCAD's layers cannot be reordered once created, the **Filter** feature is a powerful alternate way to group similar layers together, especially when you have a great many layers. For instance, if you have multiple pages of layers having prefixes like 1, 2, 3, and so on, you can view together all the layers with a 1 in front using the **Filter** option. You can group together layers like 1-WALL, 1-WINDOW, 1-DOOR, and 1-HATCH, or ELLIGHT, ELSYMB, and ELFIXT.

You can also use an asterisk (*) as a wildcard in your searches. Like the Joker in a deck of cards, the asterisk can represent any character or series of characters. The way it works is that DataCAD disregards all characters after the asterisk. So if you Filter by typing in **1***, DataCAD will display all the layers that start with the number 1 (1WALL, 1-HTCH, 12FLOOR) but will not display any other layers. And if you **Filter** by typing, **1FL***, DataCAD will only display layers that start with 1FL (1FLWALL, 1FLRDR 1FL-FURN) and will not display any other layers.

The question mark (**?**) can also be used to narrow the display or to help search for particular layers. The question mark is used in place of individual characters rather than a series of characters. Filtering by typing **???WALL** will display only those layers that have any three characters (because of the three question marks) followed by WALL, such as 1FLWALL, 2C-WALL, and NEWWALL. Typing **???WALL** would not find layers such as 1WALL, 1STFWALL, or 2-WALL because these layers have more or less than three characters in front of the letters WALL.

SaveLyr and LoadLyr

SaveLyr and **LoadLyr** are used to, you guessed it, save and load individual layers. These layers will be saved outside of the current drawing file rather than in it. This is useful if you want to keep your file sizes down or if you have layers with corrupted data in them (recovering corrupted data and drawings is covered in depth in Chapter 21, "Techno Files").

DataCAD can save drawing information in several ways. The .DC5 file format is the primary method. Whenever you choose **File/Save** (or **F**), DataCAD saves all the current information on every layer to a single file with the .DC5 file extension. Every time a new layer with drawing information is added to the drawing file, the size of the file increases as well. But deleting layers does not decrease the size of the drawing file (see Chapter 21 for an explanation). For small files, this is generally not a problem, but once your drawing file begins to get larger, you will notice slower screen refreshes, longer layer search times, and a general decrease in performance. By using DataCAD's **SaveLyr** and **LoadLyr** options, you can keep your file sizes smaller while retaining information that you might not use frequently but still want to be available to your drawing file.

Let's say you are working on a 3D drawing and you want to save some perspective views from different angles. Once you have created a perspective view that you want to save, you could use the **SaveImg/NewLyr** option in the **3DViews** menu to save the image to a new layer in the current drawing file, which would increase the file size due to the new layer and entities that would be added. The alternative would be to go to the **3DViews** menu and select the option called **SaveImg/LyrFile**. This will save the current image to a new layer with the file extension of .LYR in the \LYR directory. You can do this for all your various views without adding to your .DC5 file size.

*NOTE: DataCAD saves all layers to the default path you set in the **Tools/Program Preferences/Pathnames/Layer Files** box. If you want to save them somewhere else, you can browse to a new location when the Windows file dialog box appears. If you do save to a new location, that location will replace the old location in the **Layer Files** box.*

Of course, if you have gone to the trouble of saving layers, then you may eventually want to add that layer to another drawing file. Perhaps you have a set of notes or a detail saved in a .LYR file and you want to add one of them to a new drawing file. To add individual saved layers to your drawing, you use the **LoadLyr** option. In the file dialog box, select the layer file you want to load. The layer will be added to your drawing at the end of the list of current layers.

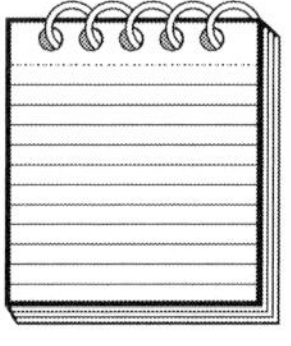

*NOTE: The **LoadLyr** utility has a feature that some would call a bug. You can decide for yourself. If you use **LoadLyr** to read in a layer that has the same name as a layer that already exists in the current drawing file, the new layer will be appended to the end of the list of current layers, thereby leaving you with a drawing file that has duplicate file names. This will usually not cause any immediate problems, but it can later on, especially if you use* Multi Scale Plotting *(MSP) or GotoViews. Duplicate layer names can occasionally cause DataCAD to crash when MSP sheets or GotoViews are selected. To avoid this problem, simply rename the newly added layer, or move all its entities to the layer with the same name and delete the duplicate layer.*

ViewLyr and ViewFile

The **ViewFile** selection is a cousin of the **SaveLyr** and **LoadLyr** options. While in a drawing file, you can use **ViewFile** to temporarily view a previously saved .LYR file. This way you can view a layer before deciding to add it to your drawing via the **LoadLyr** option.

Unlike the **ViewFile** option, **ViewLyr** is used to temporarily view a layer within the current drawing file. When you select **ViewLyr/Select** and then pick a layer name, that layer will be temporarily displayed in the drawing window. When the screen is refreshed, the original drawing view will return.

Besides **Select**, four other options can be found in the **ViewLyr** and **ViewFile** menu. They control how the layer to be viewed will be displayed when it is selected:

- **Extents** Toggling this on will display the layer at its extents, from edge to edge of the drawing window. Toggling it off will display the layer at the current drawing scale.
- **LyrRfsh** Toggling this on will cause the drawing window to refresh as each layer is displayed. Toggling it off will cause each new layer selected to be displayed on top of the previously selected layers.
- **Border** Toggling this on will display a blank border area around the entities of the selected layer by indicating the size of the image as a percentage of screen size. Toggling it off will cause the image to be displayed from edge to edge of the drawing window.
- **ImgSize** This option is only available when the Border option is toggled on. Use it to specify the percentage of the screen to use to display the layer image.

Layer Match Function

The **Match** function appears in many of the layers' submenus. Use it to make a selection by picking an entity in the drawing window. Whichever layer the entity is on is the layer that will be selected. For instance, if you want to turn off your Window layer, you could select **Layers/On/Off** and then scroll through the layer names until you find the Window layer to click on. If you have a drawing file with few layers, this works just fine, but in drawing files with many layers, it would be easier to select **Layers/On/Off** and then **Match**. Now place the cursor on a window (or anything else on that layer) and left-click or middle-click on it. That layer will be turned off.

The **Match** function in conjunction with **SetActiv**, **ActvOnly**, and **Lock** is just one of the features we use on a daily basis in our office. This is a very handy feature and is a must-try option for any DataCAD user. Try them all out and we think you will be glad you did.

Deleting Layers

Layers are deleted one by one in DataCAD, with a confirmation required between each delete command. Using the **Filter** command can help speed up your selection by grouping related layer names together. If you delete a layer, then it and all the entities on it will be erased from the drawing. If you made a mistake in deleting the layer, select **Edit/Undo** from the drop-down menu, or press **Ctrl+Z**. Notice that when you do this the layer is reinserted in its original location in the layer list. You must also be aware that deleting layers can have a detrimental affect on 3D GotoViews and Multi-Scale Plot details. These pitfalls are covered in the respective chapters for those features.

The Layer Utility Macro

Layer Utility (**LyrUtil**) is a macro found in the **Toolbox**. Think of it as the **SaveLyr/LoadLyr** functions on steroids. Rather than saving and loading only one layer at a time, this macro enables you to save and load any number of layers all at once. Additionally, if you are using Multi-Scale Plotting or 3D GotoViews, it will enable you to save your sheet and view information along with the layer information if you so choose. Again, see Chapter 21 for an explanation of how to use **LyrUtil**.

To use the LyrUtil macro, select **Toolbox/LyrUtil.dcx**. You will see three options: **LyrSave**, **LyrLoad**, and **DelLyrs**. Let's start with the easiest one, **DelLyrs**.

Use **DelLyrs** to delete layer files created by the **LyrUtil** macro. When you select it, a file dialog box appears. The dialog box will open up to the directory you set in the **Tools/Program Preferences/Pathnames/Layer Files** box.

NOTE: *Layer files saved with the LyrUtil macro have a file extension of .LYS. Note that this is a different extension than the one saved by the* ***SaveLyr*** *and* ***LoadLyr*** *utilities in the Layer menu that have a .LYR extension. For this reason, only layers saved with the* ***LyrUtil*** *macro will be displayed in the file dialog box here, since by default you will notice that the* ***"Files of type:"*** *box shows that only* ****.lys*** *files are shown.*

An .LYS file is basically a "pointer" that tells DCAD where to look for the actual .LYR files that make up each individual layer. These .LYR files are located in another directory with a name matching the .LYS file. To see this, make sure you have at least one layer file saved with the ***LyrUtil*** *macro. Then go to Windows Explorer and open up the \LYR directory in the DataCAD directory. You will see a file like WINDOW.LYS, and a corresponding directory called WINDOW. Open this directory and you will see all the layer files (with the .LYR extension) saved within.*

To select a layer file to delete, simply highlight it in the file dialog box and then press the **Delete** key or the **Open** button of the dialog box. The file will be sent to the Windows recycle bin, so you can retrieve it later if you find that you shouldn't have deleted it.

LyrSave will save a single layer or group of layers. When you click on it, you will see a submenu with five more options: **SaveAll**, **SaveOn**, **3D Views**, **Details**, and **LyrMenu**. We'll outline them here:

- **SaveAll** Selecting this button will save all the layers in the current drawing file. It is important to turn the **3D Views** and **Details** buttons on or off *before* selecting **SaveAll**.
- **SaveOn** This will save only the layers that are currently turned on in the drawing file. Layers that are currently off will not be saved.
- **3D Views** Toggling this on will save all the drawing file's 3D GotoViews along with the selected layers. Note that even 3D GotoViews containing layers that are not being saved will be included. Toggling this option off will not save any 3D GotoView information. (See Chapter 10 for more information about 3D GotoViews.)
- **Details** Toggling this on will save all the drawing file's *Multi-Scale Plotting* (MSP) details and sheets along with the selected layers. Note that even MSP details containing layers that are not being saved will be included. Toggling this option off will not save any MSP information. (See Chapter 12 for more information about MSP, including potential file corruption and program crashing problems related to **LyrUtil**.)
- **LyrMenu** This is a trap door to the standard DataCAD **Layers** menu. It is a quick way to turn on/off layers or to otherwise manipulate layers without having to exit the **LyrUtil** macro.

To save an entire drawing file, all layers, MSP sheets, and 3D GotoViews, simply select the **SaveAll** button. A file dialog box will appear displaying the default directory indicated in the **Tools/Program Preferences/ Pathnames/Layer Files** box. In the **File name:** box, type in a new name and then press **Open** to save it with that name.

If you do not want to save 3D GotoViews and MSP details, then you must always make sure to first toggle the **3D Views** and **Details** buttons off and then select **SaveAll** or **SaveOn**. This is an important concept to know and understand, since accidentally combining two files with their own MSP and 3D GotoView details can cause problems with the resulting drawing file.

CHAPTER 7

Printing and Plotting

Printing from a CAD program is somewhat different from the more straightforward process of printing from a word processing program, but as long as you understand the process, it becomes just as simple. A quick word of caution: Windows printer and plotter drivers are notoriously finicky when it comes to any CAD program, not just DataCAD, since many printer manufacturers write software drivers for their machines that do not conform to all the Windows standards as they should. If you experience problems, the first thing you should do is make sure you have the most up-to-date driver for your printer. Check the manufacturer's Web site or call them directly.

At times in this chapter we will use two terms, printing and plotting, whose definitions are more a matter of semantics than anything else, but we will use them to separate two different concepts. We will use the word printing to indicate sending the drawing output to a small format printer, one that outputs a smaller paper size like $8^1/_2 \times 11$ [216 × 280], or 11 × 17 [280 × 432]. By this definition, a printer would typically be small enough to sit on a desktop. We will use the word plotting to mean sending the drawing output to a large format plotter, one that outputs a larger paper size like 24 × 36 [610 × 914], or 30 × 42 [762 × 1066]. Most of the time we will use the word printer to refer to either a printer or a plotter.

The process in a nutshell is as follows:

1. Get what you want to print onscreen.
2. Make sure your plotter is set up to get pens and settings "from software."
3. Select **Setup** in the DataCAD **Plotter** menu:
   - **a.** Pick the printer driver.
   - **b.** Pick the paper size.
   - **c.** If you are going to print a "check plot," select **Fit to:** and check the appropriate printer and paper size.
   - **d.** Select **OK**.
4. Load the correct **PenTable** from the **Plotter** menu.
5. If you are using **QwkLyout**, pick the **Scale** you want to print at.
6. Pick **QwkLyout** or pick a **MltLyout** (*Multi-Scale Plot* [MSP]) sheet to choose the area to print.
7. Press **Plot** to print the drawing right away press **ToFile** to save the drawing to a file for later printing, or press **Preview** to see a preview of the drawing layout.

What Is a Printer Driver?

Every printer or plotter speaks its own unique "language." Windows does not inherently know how to speak those languages, so printer manufacturers write a language interpreter called a *software driver*. When you want to print to a new printer on your computer, you must install this driver on your computer so that Windows can take the printing information from DataCAD or any other program and pass it to the driver, which then tells the printer how to print what you see on your screen. You do not have to actually have a particular printer or plotter physically attached to your computer in order to use its driver.

Why Should I Care about Printer Drivers?

Without understanding how a printer driver operates and what it controls, you will not be able to successfully print your drawings. Here are some of the things that a driver will control in DataCAD:

- Paper size
- Paper source
- Output quality
- Color versus black and white output
- Portrait or landscape orientation
- Output language (HPGL/2 or proprietary)

All printing and plotting in DataCAD is controlled from the **Utility/Plotter** menu. If you have not yet picked a printer or plotter driver, DataCAD will display a message on the screen (see Figure 7-1).

This message appears because you cannot create any sheet layouts or print anything without first selecting a printer or plotter driver. Once you select a driver and save the drawing, this message will no longer appear because DataCAD saves the name of the driver along with the drawing. But if you then open this same drawing on a computer that does not have the same printer driver installed, DataCAD will instead select the computer's default driver, since it won't be able to find the driver that was originally

Figure 7-1 DataCAD informing you that a printer or plotter driver has not yet been selected

saved with the drawing. You will then most likely have to pick a new paper size as well, since the default paper size will probably be different than the paper size you originally picked.

Configuring Your Plotter

Many plotters (rather than printers) have at least two methods of determining how to print lines on paper: via hardware or software. The hardware method means that all settings are configured at the plotter via the plotter's own user interface panel. Usually, it enables you to determine settings such as line widths, screening, color, paper orientation, and plot file language. See your plotter's operation manual for further information.

The software method means that the plotter gets all of its settings and takes all its directions from the plot data created by the CAD program in conjunction with the plotter's software driver. In most cases, this is how your plotter will function. Therefore, a crucial step in the plotting process is to make sure that your plotter knows it will be getting its information from the plot file created by DataCAD. In most cases, there should be a setting on your plotter that is called something like "from software." Make sure this is set prior to sending plots to the plotter. See your plotter's operation manual for further information.

Settings Saved with the Drawing

Once you choose your printing/plotting settings, most of them are saved with the DataCAD drawing file when you save the drawing, including

- **Scale**
- **PenTable** name
- **QwkLyout** and **MltLyout** (MSP) views
- All settings in the **Setup** dialog (driver, paper size, orientation, and so on)

If a DataCAD drawing is created with one set of drivers under the **Setup** dialog, and that drawing is opened on another computer that does not have those drivers, DataCAD will simply revert to the default driver and paper size of the current computer. To view and plot the same **QwkLyout** and **MltLyout** (MSP) sheets that were saved with the drawing, you will need to select a new driver and paper size that support the sizes used in the original drawing. When the drawing is saved, these will be the new settings saved with the drawing. The old driver and paper size settings will no longer be available.

Likewise, if the drawing is opened on a computer that does not have a .PEN file with the identical name (like MYLINE.PEN) in the DataCAD \SUP directory, DataCAD will default to the DCADWIN.PEN file. The .PEN file is a simple text file and can be copied to the \SUP directory of any computer running DataCAD.

Printer/Plotter Settings

Accessing and selecting a printer/plotter driver as well as setting up all your printer/plotter options can be done in one of three ways:

- Go to the **Utility** menu and select **Plotter/Setup**.
- In the Windows menu bar, select **File/Print Setup**.
- Press **Ctrl+P**.

The menu in Figure 7-2 will appear.

The **Printer** section of the menu contains these settings:

- **Name** Use this setting to select the current printer/plotter driver.
- **Where** This setting tells you where the printer/plotter driver will send the prints.
- **Properties** Use this setting to access the driver's own property settings. Often this is where custom paper sizes and output quality are selected.

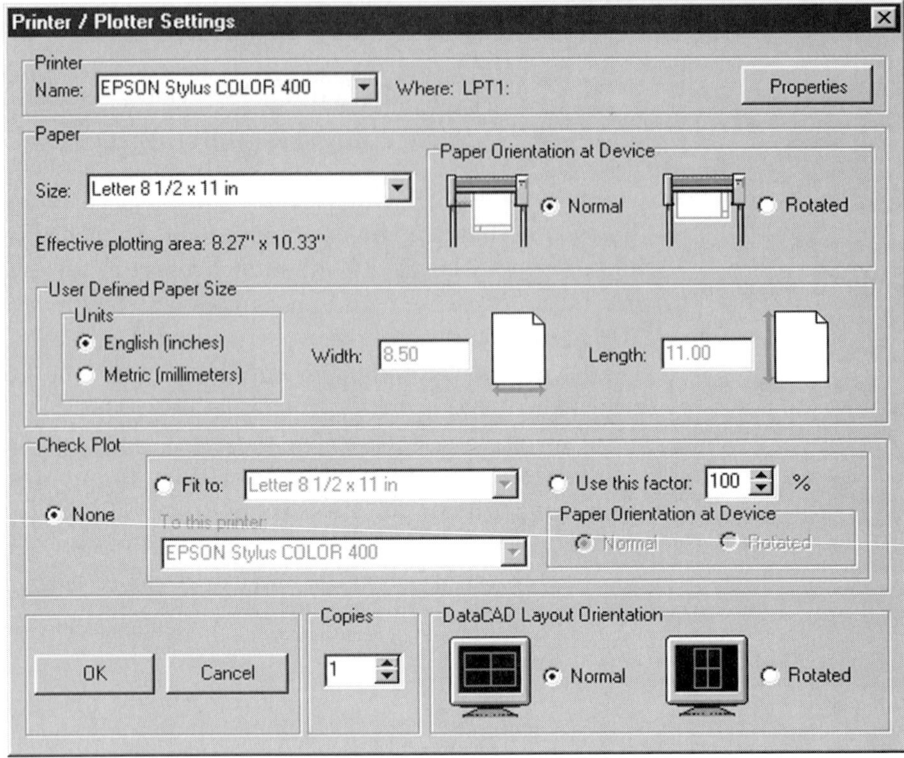

Figure 7-2 The **Printer/ Plotter Settings** menu

The **Paper** section contains the following settings:

- **Size** Here you select the paper size for your prints. The **Effective plotting area:** underneath the Size setting gives you the maximum printable area that your printer's current paper size will support. These are also the dimensions of the drawing limits that DataCAD displays when you select **QwkLyout** or use MSP.
- **Paper Orientation at Device** As the graphic shows, this setting controls the orientation of the long and short edges of the paper when printing. This is especially useful to prevent wasting paper on wide-format, roll-feed plotters. Not all plotters will support this option.
- **User Defined Paper Size** This setting is where you may be able to select custom paper sizes, but not all printer/plotter drivers will support this feature. If you are not getting the **Effective plotting area:**, which you set under **Width** and **Length**, then use the **Properties** setting to pick a custom paper size via the printer driver's own interface. This area

will remain grayed out (unavailable for selection) until you select User defined or Custom in the Paper/Size box.

- **Check Plot** This is where you can select an alternate printer to print a mini-version of your full-size drawing. As the name implies, this is usually used to check drawings for layout and content. This section contains the following settings:
  - **None** This is the default setting. By keeping this circle checked, all your printed output will go to the printer listed at the top of the Setup dialog under **Printer/Name**. This box will automatically uncheck itself when you select the **Fit to** option.
  - **Fit to** By checking this box, all your printed output will go to the printer shown in the **To this printer:** box. The entire drawing will be scaled down to fit the paper size selected in the box to the right of the **Fit to:** option. DataCAD will automatically calculate the required scaling percentage to make it fit and will show that percentage to the right of the **Use this factor**: option.

*NOTE: One great feature of the **Fit to:** option is that it will also automatically scale down the width of all your lines. If it did not, then the lines of the reduced version of your drawing would end up bleeding together into dense lumps of ink, making the output unreadable.*

  - **To this printer** Use this setting to pick which printer/plotter to print the check plot to. (This can be the same printer as the one selected under **Printer/Name**, but usually it is something different.)
  - **Use this factor** This setting determines the scale factor at which to print the check plot. Check the circle and you can type in your own percentage. Otherwise, you can leave it unchecked and DataCAD will automatically calculate the required scaling percentage to make it fit when you select a size under the **Fit to:** option.
  - **Paper Orientation at Device** This setting controls the orientation of the long and short edges of the paper when printing. Not all plotters will support this option.
- **Copies** This setting enables you to print more than one copy of a regular plot or a check plot.

NOTE: The Copies option may not work with some plotter drivers, like that of our CalComp TechJet 720c. If this is the case with your plotter you may still be able to select more than one copy via the driver's own settings in the ***Properties*** *option, or by setting the number of copies at the control panel of the plotter itself.*

- **DataCAD Layout Orientation** This setting controls the orientation of all paper layouts in DataCAD, including QwkLyout and MSP. Most of the time you will leave the **Normal** option checked, but if you have a tall detail you need to print, you might want to temporarily change the orientation to **Rotated** to get it to fit. This option controls both the standard **printer** layout and the **check plot** layout. This option does not work with Clip Cubes (see Chapter 12)

Line Weights and the PenTable

You may notice that when you draw lines in the Drawing Window they all display at the same width, or Line Weight. But the convention for both hand-drafting and CAD drafting is to use lines of varying weights (widths). Although DataCAD is capable of displaying lines onscreen with varying widths, doing so is not the best option for most users, due to certain limitations in how this feature is implemented (see Chapter 3, "Settings and Display Options," regarding the **Utility/Display/ShowWgt** option for more information). Instead, DataCAD, and in fact nearly every CAD program, uses a technique called Pen Mapping to vary line weights. With this method, each color within the drawing file is assigned a line weight by means of a Pen Table. You therefore control the width of your lines in the drawing file by varying the colors of the entities you draw. The downside to this method is that you will not see your actual line weights until you print the drawing, but the upside is that it allows for much finer control of line weights and enables you to change line weights en mass with just a couple of mouse clicks.

If you want to follow along with the following example, you can copy the PRINTING.DC5 file to your hard drive from the Chapter 7 directory of the CD-ROM.

To map lineweights to colors, use the **PenTable** option in the **Plotter** menu. When selected, you will see a dialog box as shown in Figure 7-3.

Figure 7-3
The **Pen Table** dialog box

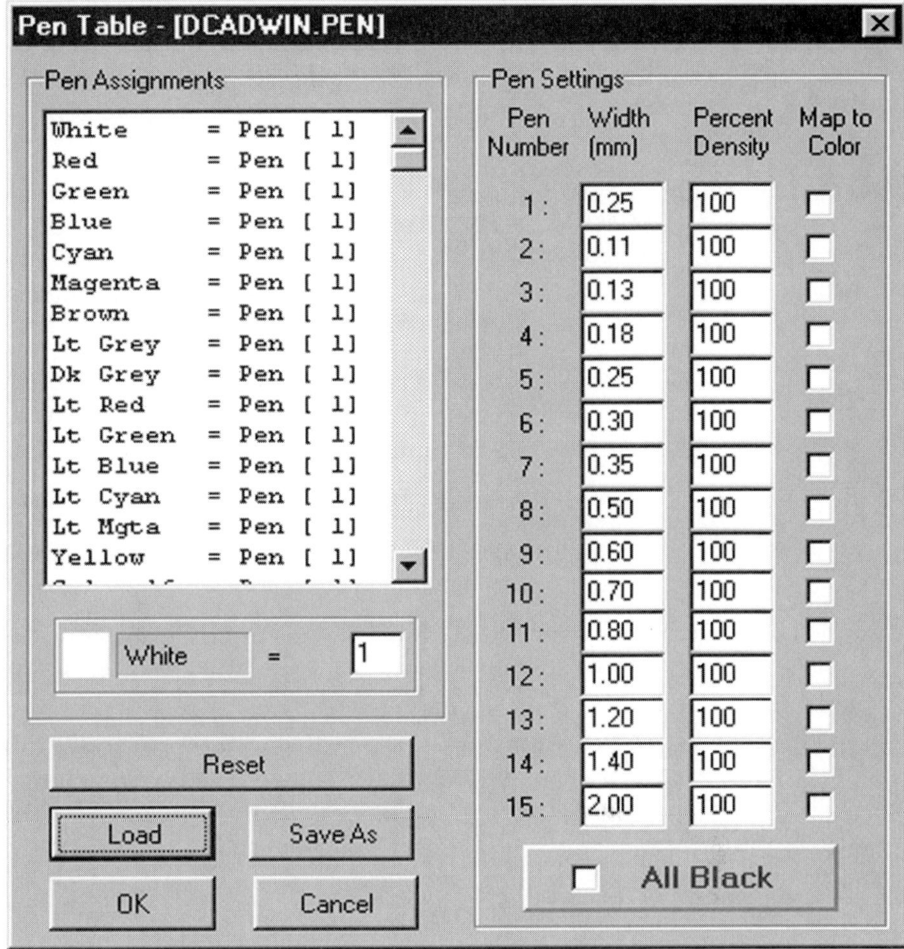

The word Pen is simply a carryover from the days when all plotters drew lines on paper by means of a series of real pens filled with ink. The word now is just a convenient way to describe each line weight that is assigned and then printed.

The general process for mapping pens to colors and then to line weights is as follows:

1. Assign a Pen Number to each drawing color in the **Pen Assignments** area.
2. Assign a **Width** (line weight) to each pen number.

3. As an option, assign a density factor with **Percent Density** (**Map to Color** must be selected for this to work).
4. As an option, assign a different print color to each pen with **Map to Color** (You can have a blue line print as orange if you want).

Pen Assignments

The Pen Assignments window consists of three entities, the Pen window, the Color box, and the Pen Number box:

- **Pen window** The colors displayed on the left correspond to the colors of the entities in your drawing. The 15 standard DataCAD colors are shown first and then the remainder of the 255 possible colors that can be used in DataCAD. After = **Pen**, there is a number between a pair of brackets. This is the pen number assigned to each color. Only numbers one through 15 are valid options, since DataCAD only allows for 15 pen assignments.
- **Color box** Located at the lower-left corner of Pen Assignments box, this small box visually displays the pen color as you select each color in the previous pen window.
- **Pen number box** Located at the lower-right corner of Pen Assignments box, this is where you type in a new Pen Number assignment. The number must be a whole number between 1 and 15.

Pen Settings

The Pen Settings section of the Pen Table dialog box consists of the following settings:

- **Pen Number** These numbers correspond to the pen numbers set in the Pen Assignments dialog.
- **Width** Here you type in the width of each Pen. Note that the CAD industry standard is millimeters. Although you can input any number from 0.01 and up (you don't have to type the zero before the decimal), your printer/plotter may only support certain finite widths. If you type in a width not supported by your printer/plotter, you may see some unexpected line widths as your printer/plotter tries to round up or down to the closest supported width.

- **Percent Density** This setting is used for screened lines. With this option, you can make entities print solid (**100** percent) or at a shaded percentage (like **75**, **50**, or **20** percent). Selecting **0** (zero) percent yields a line that does not print (useful for guidelines). These percentages will only work if you check the **Map to Color** boxes and un-check the **All Black** option.

NOTE: *A 20-percent screened black line will look like a 50-percent value to the human eye. This is known as the* Bezold Effect, *a geometric progression that tends to confuse the eye/mind relationship. So values set greater than a 20-percent density may quickly be taken for black. The screening of colors gets far more tricky, since each color has its value in gray-scale in relationship to full black. You will most likely have to experiment to find what works visually for you.*

Map to Color Use this setting to assign a different print color to each pen. The following guidelines can help the success of your output:

- If you want black and white output and some of your lines need to be less than 100 percent density, check all the **Map to Color** boxes and set each color to black. (If you check the **All Black** button, your **Percent Density** settings will not work.)
- If you want color output and some of your lines need to be less than 100 percent density, do not check any **Map to Color** boxes and do not check the **All Black** option. Your drawing entities will print with the same colors you see onscreen and with whatever percentages you type in under **Percent Density**.
- If you want black and white output with all lines at 100 percent density, do not check any **Map to Color** boxes, but simply check the **All Black** box.

All Black Check this box to print all entities black at 100 percent density. Leave it unchecked if you want your output to print in color or with entities at less than 100 percent density.

Reset This will reset all Pen Table settings to the default DataCAD settings. The default settings are saved in a file called DcadWin.Pen (see the description of the **Load** button).

Load *Click* on this button and the **Load pen file** dialog box will appear. By default, the dialog box opens the DataCAD \SUP directory. To select a .PEN file, highlight it and then press **Open**, or *double-click* on it. At least one file should be called DcadWin.Pen. It contains the default DataCAD Pen Table settings. If other .PEN files are listed, you can select one of those instead.

Save As *Click* on this button and a **Save pen file as** dialog box will appear. By default, the dialog box opens the DataCAD \SUP directory. Use this option to save the current Pen Table settings. To do so, type in a new name and select **Save** to save the current settings. This is very useful if you need one group of settings for your plotter, one for your laser printer, and one for your color printer, or if you deal with different pen width standards from different consultants or clients.

Print Settings

Using the Printing.DC5 sample file in the Chapter 7 folder on the CD, let's see how to print one of the floor plans from this drawing file onto an $8^1/_2 \times 11$ [216×280] sheet of paper. When you first open the file, it looks something like Figure 7-4.

Setup

Before the actual printing, we must go through some preliminary setup options:

1. Select **File/Print Setup** or **Utility/Plotter/Setup** or **Ctrl+P**. The **Printer/ Plotter Settings** dialog box will appear.
2. In the **Printer/Name** box, select a printer driver. In our case, we chose the Epson Stylus Color 400.

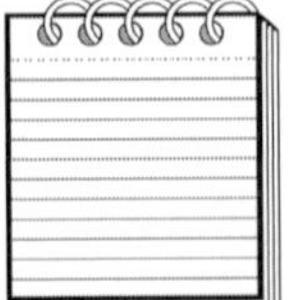

NOTE: *Remember that selecting a printer versus a plotter is more a matter of verbal semantics. This example would work just as well if we chose a large format plotter as the printer to send our drawing to.*

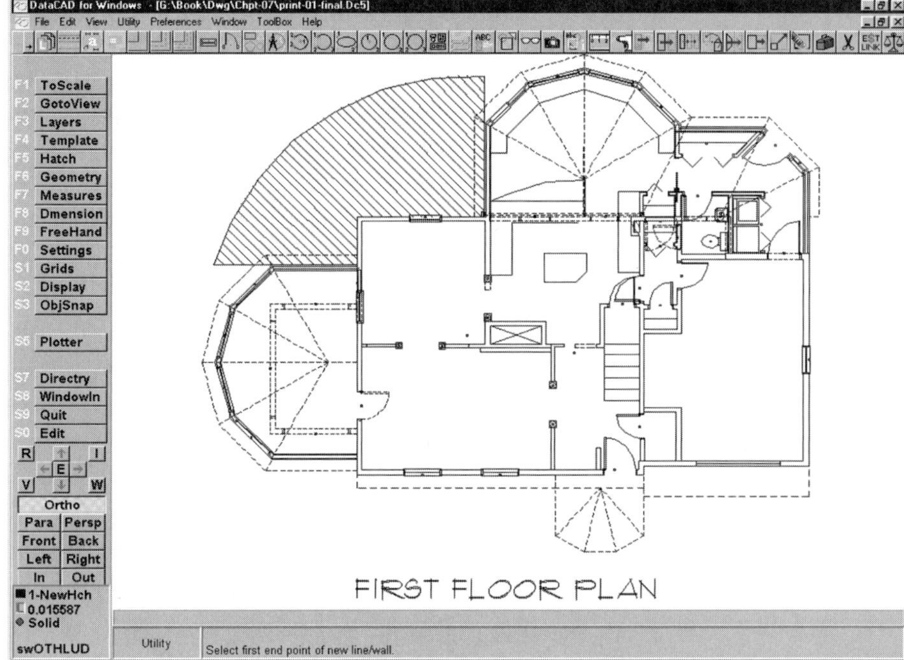

Figure 7-4
The floor plan to be printed

3. Under the **Paper/Size** section, choose a small paper size supported by your printer. In our case, we picked "Letter $8\frac{1}{2}'' \times 11''$" Check the dimensions shown next to the **Effective plotting area:**. In our case, it reads $8.27'' \times 10.33''$. Note that this is slightly smaller than the $8\frac{1}{2}'' \times 11''$ we selected. This is because the Epson printer driver knows that there is a region around the edge of the paper that is unprintable with this printer and automatically displays the resulting dimensions here.
4. Make sure the **Normal** option is checked under both the **Paper Orientation at Device** area and the **DataCAD Layout Orientation** area.
5. Click on **OK**. You will be back at the **Plotter** menu.

Scale

Most of the time you will probably want to print your drawings at a particular scale, like an architectural scale of a $\frac{1}{4}'' = 1'\text{-}0''$ or an engineering scale such as 1″ = 40′. In the **Plotter** menu is a button called **Scale**. If you select this button, you will see DataCAD's default scales, starting with 12″ (1″ = 1′-0″) [1:1] and ending with **1:1000** ″ (1″ = 1000′-0″). To customize the scales DataCAD uses, see "Custom Scales, Angles, and Distances" in Chapter 23, "Customizing DataCAD." To set up the scale, follow these steps:

1. In the **Plotter** menu, select **Scale**.
2. Pick the 1/4″ [1:50] button. This will cause the drawing to print at a scale of 1/4″ = 1′-0″. The Message Window will display a message such as *"Current plotting scale is: 1/4."*

Line Widths (Line Weight)

Now we need to tell the printer how wide to print each entity in the drawing. We will be mapping line weights to colors so that every red line prints at the same width, every blue line prints at the same width, and so on:

1. In the **Plotter** menu, select **PenTable**. The Pen Table dialog box will appear. Note that the currently selected .PEN file is indicated in the colored title bar at the top of the dialog box.
2. Click on the line that says **White = Pen** in the **Pen Assignments** box. It will now be highlighted. The Color Box will change to White, and the Pen Number box will change to the current pen number.
3. Place the cursor in the Pen Number box, highlight the current number, and then change it to the number 4.

NOTE: *This means that all white entities in your drawing are now mapped to Pen Number 4 and will take on the* ***Width, Percent Density,*** *and* ***Map to Color*** *settings for Pen Number 4, shown under the* ***Pen Settings*** *area of the dialog box.*

4. Click on each successive color in the Pen Assignments box and change the pen numbers as follows (note that the same pen number can be mapped to multiple colors):
   a. White = 4
   b. Red = 2

c. Green = 5
d. Blue = 8
e. Cyan = 1
f. Magenta = 1
g. Brown = 6
h. Lt Gray = 3
i. Dk Gray = 7
j. Lt Red = 2
k. Lt Green = 2
l. Lt Blue = 4
m. Lt Cyan = 2
n. Lt Mgta = 3
o. Yellow = 2

Pen Settings

For this exercise, we will leave all the Percent Density settings at 100 percent, and will not remap any colors under Map to Color. The only settings we will change are the line widths under the Width column:

1. At the bottom of the **Pen Settings** area, check the **All Black** box. Note that the **Percent Density** and **Map to Color** columns are now grayed out, indicating none of the settings in those columns will be used. When this box is checked, DataCAD automatically prints all lines black and at 100 percent density.
2. Use the cursor to highlight the number in the **Width** column, adjacent to Pen Number 1. Type in **.06**. All entities mapped to Pen 1 will now be printed at a width of .06 mm. Note that you can type the zero before the decimal point or not.
3. Highlight each successive **Width** and change the widths as follows:
   a. Pen 1 = .06
   b. Pen 2 = .15
   c. Pen 3 = .35
   d. Pen 4 = .50
   e. Pen 5 = .70
   f. Pen 6 = .90
   g. Pen 7 = 1.40
   h. Pen 8 = .35

Note that pen numbers do not have to exist in numeric order, like Pen Number 8. Don't worry about any of the other pen numbers, since we did not define any mapped pens above Pen 8 in the Pen Assignments table.

Save Settings

Let's save these pen settings so that you won't have to retype them in the future when you want to use the same settings:

1. *Click* on **Save As**. The **Save pen file as** dialog box appears.
2. In the **File name:** box, type **Example1**. Press **Enter** or **Save**.
3. These settings are now saved to a file called Example1.PEN in the DataCAD \SUP directory. Note that this new file name is shown in the title bar at the top of the **Pen Table** dialog box (see Figure 7-5).
4. Select **OK** to close the **Pen Table** dialog box and accept the current settings.

Printing with QwkLyout

The **QwkLyout** option is an important printing option in DataCAD. It is the quickest and most instantaneous method of printing entire drawings or, as in this case, selected areas of a drawing. Unlike most CAD programs and word processors, with DataCAD's QwkLyout feature you move the paper around the image on your screen rather than moving the image around the paper. As you'll see, this method is actually easier and much more accurate. To use QwkLyout, follow these steps:

1. Use the **Plotter/Setup** dialog to set your printer and paper size.
2. In the **Plotter** menu, select **Scale** and set the scale that you want to print at (if you are following along with the Printing.DC5 sample file, set the scale to $^1/_8$ ″ [1:100]).
3. Now select **QwkLyout**.
4. The image on your screen will increase or decrease in size, depending on the current Scale setting. Two rectangles will appear, both of which are sized according to the current **Scale** and the maximum effective plotting area of the current printer driver. You will see the current

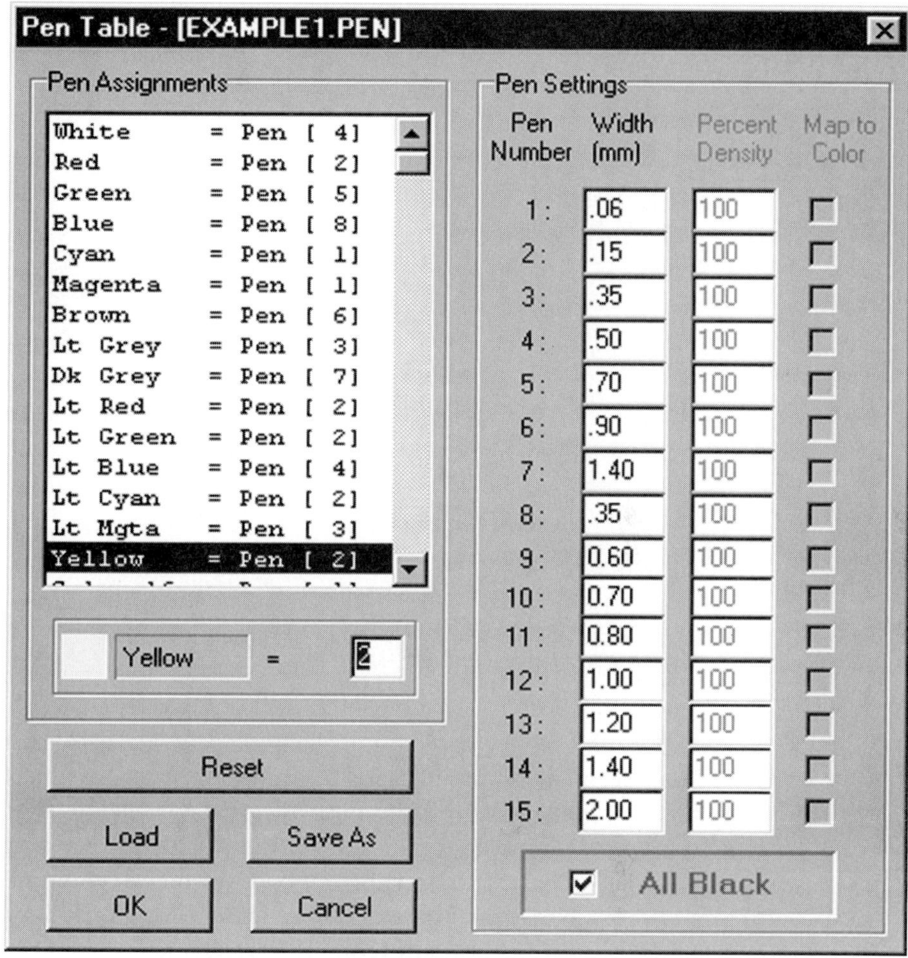

Figure 7-5 The **Example1.PEN** settings

maximum effective plotting area displayed in the Message Window at the bottom of the screen (see Figure 7-6):

a. One layout box has solid lines and moves with your cursor, representing the size of the paper you have selected to print on.
b. One layout box has dashed lines, representing the current paper layout location.

5. The **Layout%** button controls how much of the Drawing Window is filled by the drawing entities.
   a. Click on the **Layout%** button. The Message Window says "Enter layout paper size screen (%):."

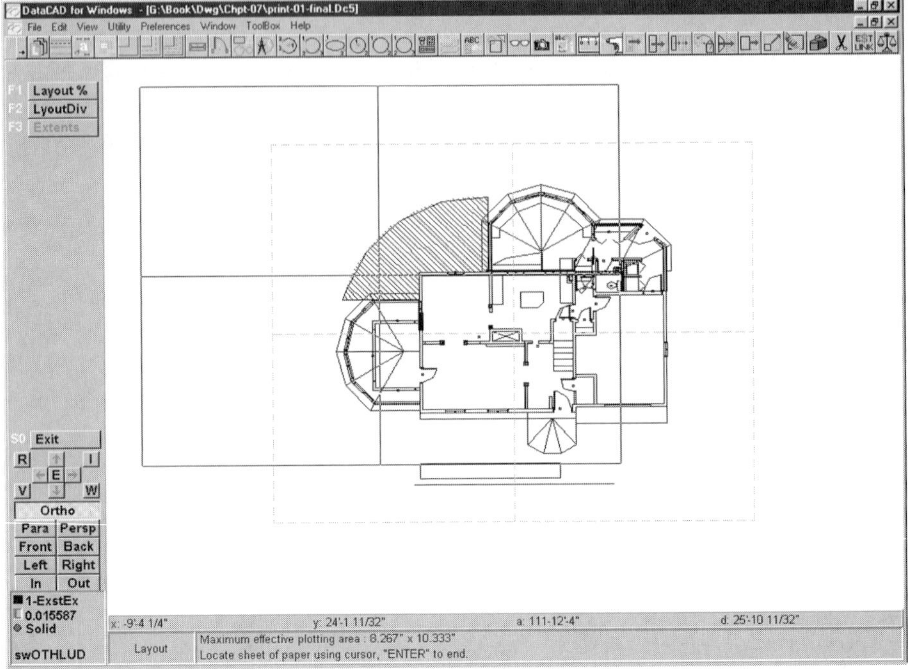

Figure 7-6
The **QwkLyout** paper dimensions are displayed in the Drawing Window.

b. Try typing **90** percent and then press **Enter**. Notice how the floor plan fills the Drawing Window more than it did before. Now pick **Layout%** again and try typing **20**, and see what happens.
c. Change the setting one more time by typing in **75**.
d. Note that you can dynamically change the **Layout%** percent by using the **PageUp** and **PageDn** keys.

NOTE: *While still in the* ***QwkLyout*** *menu, you will also see the* ***LyoutDiv*** *and* ***Extents*** *options.* ***LyoutDiv*** *controls the number of divisions displayed within the layout box. They are simply guidelines to help you lay out your drawings and do not print or otherwise affect the drawing. The default setting is* ***2*** *X-divisions and* ***2*** *Y-divisions. Try different settings to see what works for you.*

Extents *controls whether all the entities of your current drawing are displayed in* ***QwkLyout*** *or whether only a bounding box showing the outermost extents of the drawing is shown. This is useful for drawings with a lot of entities that take a long time to display.*

6. Move the layout box with your cursor until it completely surrounds the floor plan and then *click* the left mouse button to select that layout. The solid-line bounding box remains around the floor plan and your cursor crosshairs return (see Figure 7-7).

   That's all there is to laying out the drawing. All that is left is to send the drawing to the printer.

7. In the **Plotter** menu, select the **Plot** button. This will send the current **QwkLyout** to the currently selected printer or plotter.

Rotating the QwkLyout

Certainly, times will occur when you want to lay out your **QwkLyout** vertically instead of horizontally. All that is required is to select the **Rotated** option in the **Setup** menu. Here are some guidelines as to what this option can do:

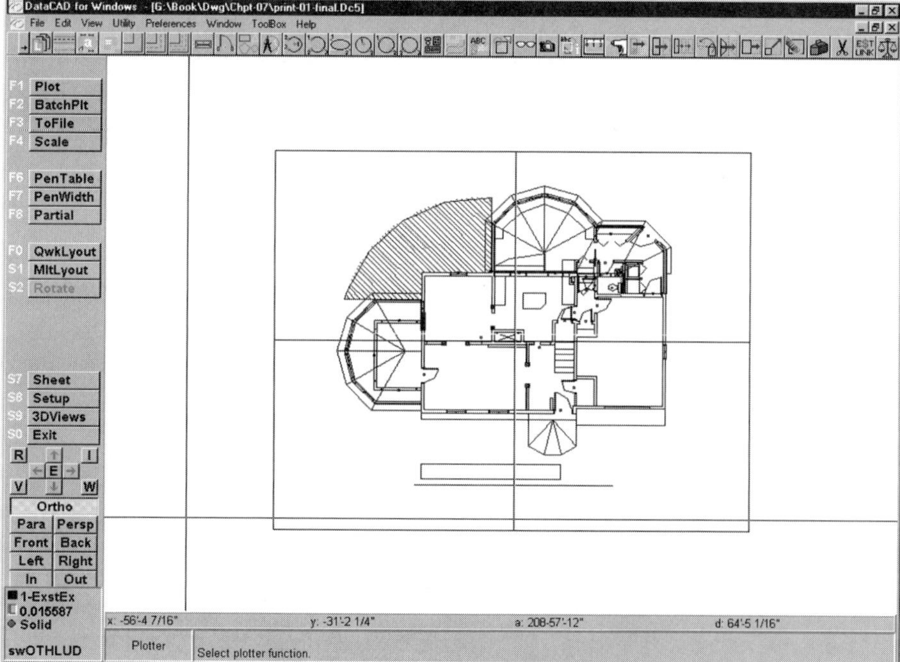

Figure 7-7 The completed Quick Layout

- This method will only let you rotate your printed image at 90 degrees.
- This method enables you to use the standard pan and zoom controls to move the drawing image around the screen.
- When using the MSP feature (see Chapter 12, "Advanced Construction Drawings"), this option will simply rotate the paper in the MSP window; it will not rotate the detail itself.

To rotate the QwkLyout, follow these steps:

1. In the Drawing Window, center the image to be printed. Use **Recalc** or **WindowIn** to zoom in on the image.
2. In the **Plotter** menu, select the **Setup** option.
3. In the **Printer/Plotter Settings** dialog, go to the bottom-right corner of the dialog box and check the **Rotated** option under **DataCAD Layout Orientation**.
4. Select **OK** to return to the main **Plotter** menu.
5. Select **QwkLyout**. Notice that the layout box is now oriented vertically, as in Figure 7-8.

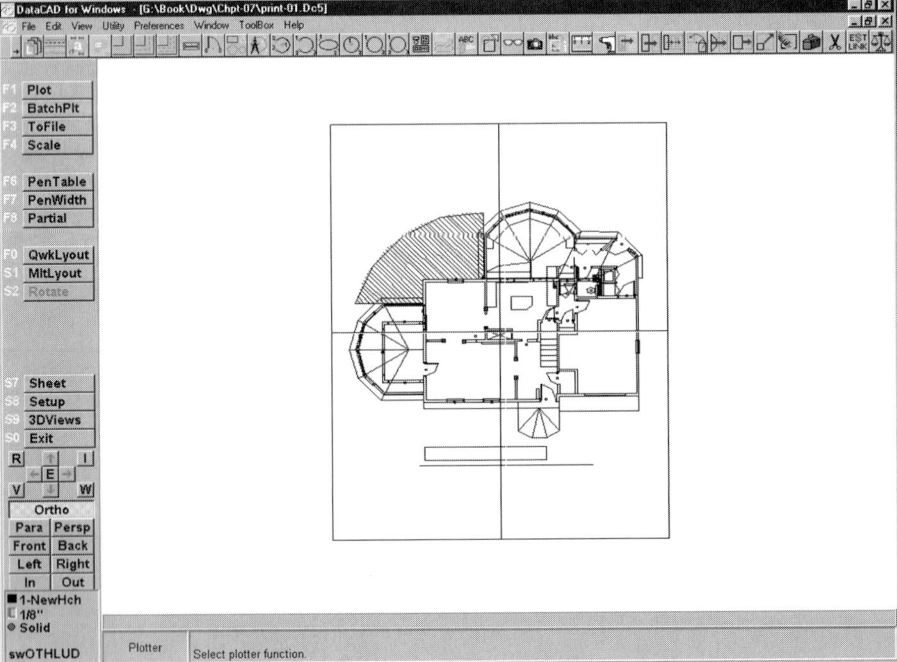

Figure 7-8 Aligning the layout box vertically

6. If you don't see your image right away, you may need to pan and zoom (**PageUp/PageDn** or the up, down, left, and right arrow keys on your keyboard) to get the image where you want it in the Drawing Window.
7. The rest of the steps for printing are the same as previously outlined.

Alternatively, you can leave the **Normal** option checked in the **Printer/Plotter Settings** menu and instead select the **Rotate** option in the **Plotter** menu. Here are some guidelines as to what this option can and cannot do:

- With this option, the drawing image is rotated, rather than the layout box being rotated.
- This option has more steps than the previous method.
- The standard pan and zoom controls don't work with this option. Using the **PageUp/PageDn** or the up, down, left and right arrow keys on your keyboard will cause the image to rotate back to its original orientation.
- With this particular Rotate option, you can select any angle of rotation, not just 90 degrees. That is the one thing you cannot do with the first Rotate method.
- To use the Rotate option in the Plotter menu, follow these steps:
  - In the Drawing Window, center the image to be printed. Use **Recalc** or **WindowIn** to the image.
  - Select **Plotter/Rotate**. You will be prompted to *"Select CENTER of rotation."* Because this Rotate method will rotate the image, rather than the paper, first you must tell DataCAD where in the Drawing Window the center of rotation is to be.
  - Locate the cursor at the center of the image in the Drawing Window and *click* the mouse. You will then be prompted to *"Enter ANGLE of rotation from keyboard: 90.0.0."* Type or select an angle, or press Enter (or just *right-click*) to accept the default of 90 degrees. The image in the Drawing Window will be rotated about the center point selected in Step 2.
  - Locate the paper over the image and click the mouse to place it. The result will be something like Figure 7-9.
  - If you don't see your image right away. you may need to exit **QwkLyout**, recenter the image, and then reselect the **Rotate** option to set the center of rotation.
  - The rest of the steps for printing are the same as outlined earlier.

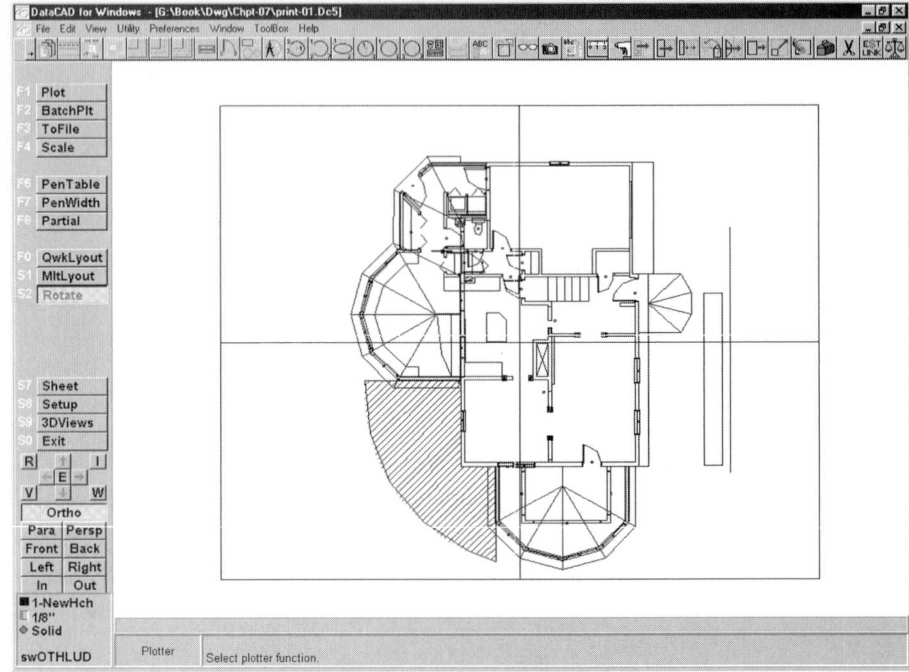

Figure 7-9
The drawing is rotated, rather than the layout box.

Printing with MltLyout

The **MltLyout** feature (Multi-Scale Plotting) in the **Plotter** menu is covered in greater depth in Chapter 12. The biggest difference is that MSP layouts are accessed via the **MltLyout** option rather than the **QwkLyout** option. Otherwise, MSPs are printed the same way QwkLyouts are.

- Settings in the **Setup** menu affect MSPs in the exact same way they do **QwkLyouts**.
  - MSPs can be set up on any size paper, but often require selecting a larger paper size to fit a greater number of details.
  - In the **Setup** menu, in the **Printer/Plotter Settings** dialog you would simply go to the **Printer/Name** option and select a printer or plotter that supports a larger paper size.
- Then choose a larger, supported paper size under the **Paper/Size** dialog.
- The **Rotate** option in the **Setup** menu affects MSP layouts as well.

If you want to follow along with this next example, you can continue to use the PRINTING.DC5 file. Note that selecting a scale is not required for MSP sheets, since each detail is placed with its own scale.

1. Select **Utility/Plotter/PenTable** to open the **Pen Table** dialog box.
2. Pick the **Load** button to access the **Load pen file** dialog. Pick the DcadWin.Pen file (or the Example1.PEN file that we created earlier) and then select **OK**.
3. Select **Setup** in order to pick a plotter and paper size:
   a. In this example, we are picking our **TechJET 720c** plotter under **Printer/Name** and choosing a **Paper/Size** of **ARCH. 4 24x36in**.
   b. If you are going to print an MSP sheet that has already been created, it is important to choose the same size paper (even if it's not the same printer driver) that the MSP sheet was created with in order for it to display and plot properly. To see how MSP sheets are affected by different sizes of paper, simply choose various paper sizes and then go to **Utility/Plotter/MltLyout/Sheet** to see the effects.
4. Press **OK** to exit the dialog.
5. While still in the **Plotter** menu, select **MltLyout/Sheet**. The current MSP sheet will appear in the Drawing Window. If you want to change to a different MSP sheet, simply select a different sheet in the menu. In Figure 7-10, you can see that you can save any number of sheets, and that you can name them as well. Here we have two sheets defined: A-1 and A-2.
6. *Right-click* twice to get back to the **Plotter** menu.
7. To plot the current MSP sheet, simply press the **Plot** button and the MSP sheet will be sent to the plotter.

How Does DataCAD Know to Print a QwkLyout or MSP Sheet?

This is a typical question that will invariably come up when you are learning to print and plot in DataCAD, so I've given the topic its own heading. When the **Plot** button in the **Plotter** menu is pressed, DataCAD will print whichever sheet layout was last chosen: QwkLyout or MSP. It's that simple.

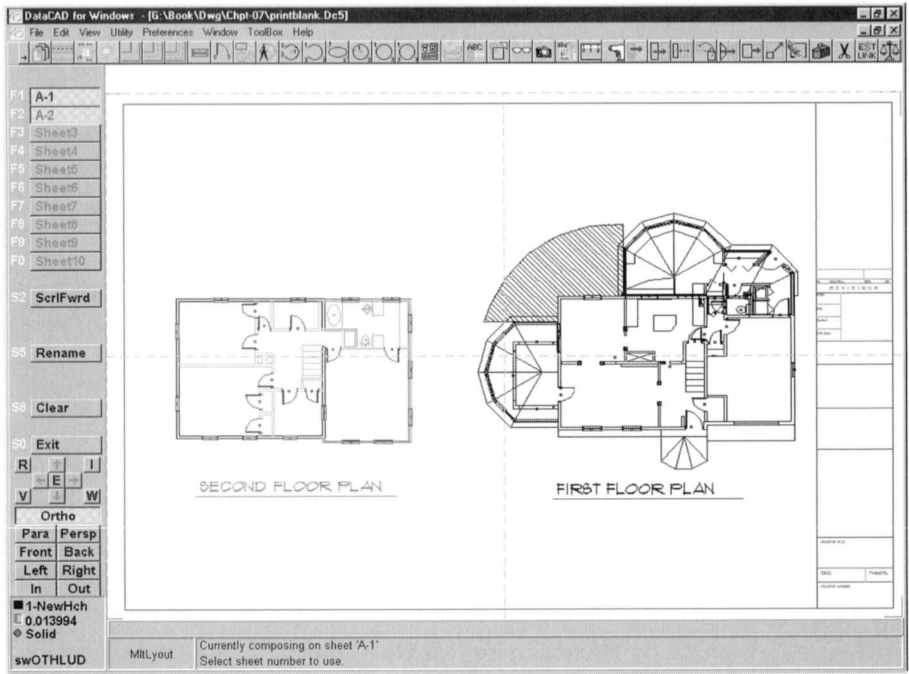

Figure 7-10 MSP Sheet A-1 is displayed.

Plot Preview

To view a plot before sending it to the printer or plotter, you can use the **Preview** option in the **Plotter** menu. Because you can pan and zoom around the preview, this can be especially useful for viewing MSPs, since the MltLyout option does not have that feature. The preview image will display the line widths assigned in the .PEN table or by the **LineWgt** settings. To use the the **Preview** option, perform the following steps:

1. Select all your plotter settings, including the driver, paper size, and .PEN table.
2. Select either a **QwkLyout** or a **MltLyout** sheet.
3. In the **Plotter** menu, select the **Preview** button.
4. Press **Plot**. The drawing will not be sent to the printer or plotter. Instead, you will see a preview screen, as shown in Figure 7-11.
5. Select one of the viewing options or select **Plot** to send the plot to the printer or plotter. You can also select **Cancel** to return to the **Plotter** menu without printing.

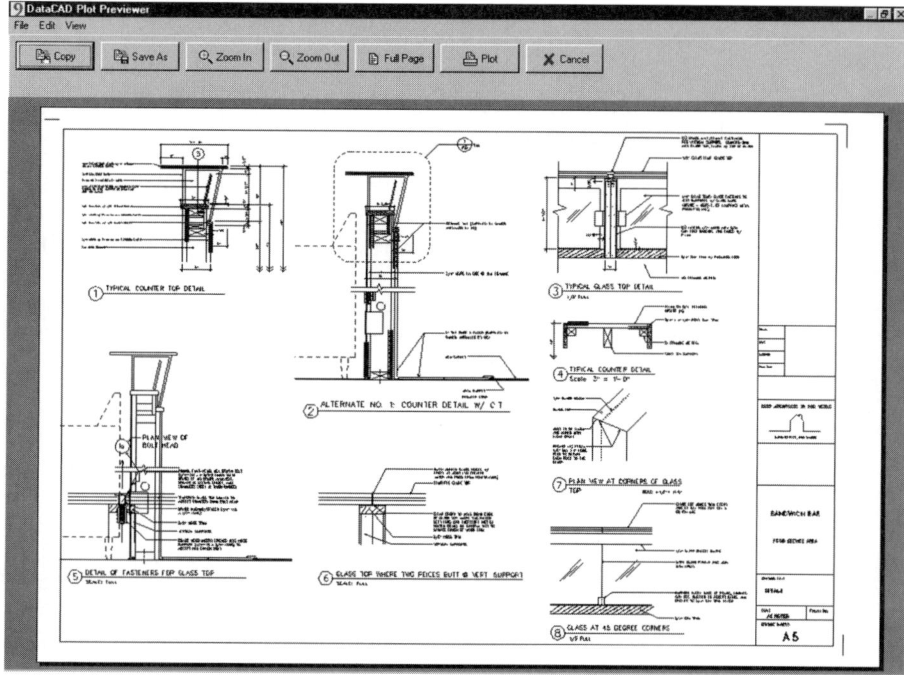

Figure 7-11
The preview screen

Plot Preview Options

The Plot Preview options are as follows:

- **Copy** This option copies the preview image to the Windows clipboard as a *Windows Metafile* (.WMF) object. From there, you can open the image in a paint program or paste the image into any program that accepts metafiles, such as a word processor.
- **Save As** Allows you to save the preview image as an Adobe Acrobat File (.PDF), or a *Windows Metafile* (.WMF) object. A **Save As** dialog box will appear so that you can choose a file type and location to save the file. See the following section for more information about the Adobe Acrobat file format.
- **Zoom In** Use this option to zoom into the center of the current preview image.
- **Zoom Out** Use this option to zoom out from the center of the current preview image.
- **Full Page** Use this option to view the entire drawing sheet.

- **Plot** Selecting this option will send the current plot to the printer or plotter.
- **Cancel** Selecting this option will return you to the **Plotter** menu without printing.
- **Cursor** Use the cursor to pan around the preview image. Place the cursor where you want to zoom into and then *click* the mouse. The image will zoom to that point. Hold down the **Shift** key and *click* the mouse to zoom back out.

You can also use the four keyboard arrow keys, **Page Up**, and **Page Down** to navigate around the preview, just as you can in the DataCAD Drawing Window.

Adobe Acrobat File Format (.PDF)

As previously noted, the **Save As** option in the Preview window has an option to save an Adobe Acrobat File, with a file extension of .PDF. This seemingly innocuous option is actually a VERY powerful new feature. Here is how the Adobe company describes this file format:

> Adobe® Portable Document Format (PDF) is a universal file format that preserves all of the fonts, formatting, colors, and graphics of any source document, regardless of the application and platform used to create it. PDF files are compact and can be shared, viewed, navigated, and printed exactly as intended by anyone with a free Adobe Acrobat® Reader. Adobe PDF is the ideal format for electronic document distribution because it transcends the problems commonly encountered in electronic file sharing.

Common Problem	Adobe PDF Solution
Recipients can't open files because they don't have the applications used to create the documents.	Anyone, anywhere can open a PDF file. All you need is the free Acrobat Reader.
Formatting, fonts, and graphics are lost due to platform, software, and version incompatibilities.	PDF files always display exactly as created, regardless of fonts, software, and operating systems [including PC's vs. Mac's].
Documents don't print correctly because of software or printer limitations.	PDF files always print correctly on any printing device.

Unlike a bitmap graphic image, when you zoom in close to text and lines in a PDF file, the entities do not appear rough and pixelated. Instead, all entities and text are displayed sharp and crisp. This means that if your client wants to see some drawings of the project, you don't have to send paper drawings, or worry that they don't have DataCAD so that they can't view your CAD files. Instead you would simply send them PDF files which they can then open, view, and even annotate with the free Adobe Acrobat Reader.

We have included the free Adobe Acrobat Reader program on the CD at the back of this book. For further information about this format, go to the Adobe website at **`www.adobe.com`**.

Check Plots

As the name implies, this is usually used to "check" drawings for layout and content, and usually consists of printing out a smaller version of your full-size drawing. For instance, you may have a 24″ × 36″ drawing with many details laid out on it. Perhaps you want to print a more manageably sized 11″ × 17″ version of that sheet for when you walk around a project site. Or perhaps you just want to print out an 8½″ × 11″ version just to check that all the details appear to be laid out and aligned properly, rather than wasting paper to print out a full 24″ × 36″ version.

For this example, we will use the PRINT-02.DC5 file from the Chapter 7 directory of the CD. This example does not make use of MSP, but the procedure for MSP sheets is exactly the same. Simply select **MltLyout/Sheet** prior to pressing the **Plot** button, instead of selecting **QwkLyout**, and perform the following steps:

1. Turn on all the layers you want to print. In this PRINT-02.DC5 example, all layers should be turned on.
2. Select **Utility/Plotter/PenTable**. Select an appropriate .PEN file and then select **OK**.
3. Select **Setup** in order to pick a plotter and paper size:
   a. In this example, we are picking our **TechJET 720c** plotter under **Printer/Name** and choosing a Paper/Size of **ARCH. 4 24x36in**.
   b. If you are going to print an MSP sheet that has already been created, it is important to choose the same size paper (even if it's not the same printer driver) that the MSP sheet was created with in order for it to display and plot properly.

*NOTE: The **Printer** and the **Check Plot** areas of the **Printer/Plotter Settings** dialog box are dependent on one another to produce a check plot. It is very important that you understand this concept. For check plots, the paper size you choose under **Paper/Size** at the top of the dialog defines the size of the original QwkLyout or MSP sheet. If your full size drawing is, for example, a 24" × 36" sheet, then you must do the following:*

*a. Select a plotter driver in the **Printer/Name** box that supports a 24" × 36" paper size.*

*b. Select a 24" × 36" paper size in the **Paper/Size** box.*

*The **Check Plot** area of the dialog only controls the printer and paper size that your reduced size check plot will be printed to. If the check plot size is to be 8½" × 11" , for example, then you must perform these steps:*

*a. Select a printer driver in the **To this printer:** box that supports an 8½" × 11" paper size.*

*b. Select an 8½" × 11" paper size in the **Fit to:** box.*

4. Check the **Fit to:** circle. The **None** circle will become unchecked at the same time, and the **Fit to:** and **To this printer:** boxes will become active (they will not be gray any more).
5. In the **To this printer:** box, select a printer to print the check plot to. In this example, we are picking the Epson Stylus COLOR 400.
6. In the **Fit to:** box, select a paper size to print to. In this example, we are picking the "Letter 8½" × 11"" paper size.
7. Notice that when you choose a paper size, DataCAD automatically calculates the reduction factor for you and displays it in the **Use this factor:** box.
8. If you want to print more than one copy of your Check Plot, select the **Copies** option and change it to the value you want (this feature is not supported by some printer drivers).
9. Your dialog box should look something Figure 7-12.
10. Press **OK** to save the changes and exit the dialog.
11. While still in the **Plotter** menu, select **QwkLyout**.
12. Notice that the layout box is displayed at the same dimensions as shown in the **Setup** menu next to the **Effective plotting area:**. To confirm this, DataCAD displays the **Maximum effective plotting**

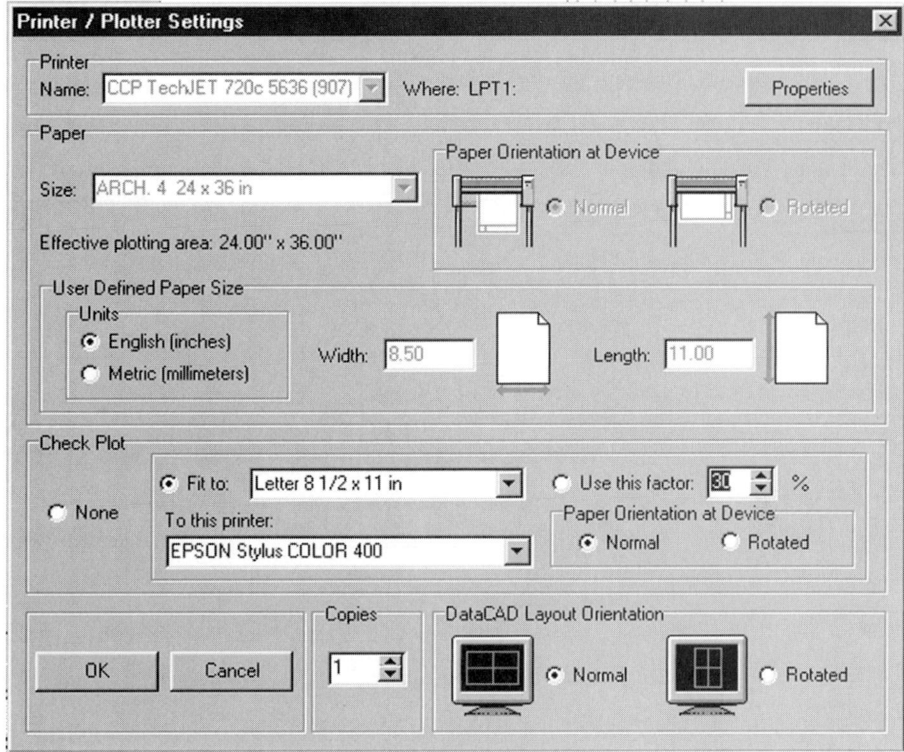

Figure 7-12 The completed **Check Plot** Settings

area: in the Message Window. In this example, it says 24.00″ × 36.00″, which are the dimensions chosen under **Setup/Paper/Size**.

13. Locate the **QwkLyout** box around the area you want to print and click the mouse to select it (see Figure 7-13).
14. In the **Plotter** menu, select **Plot** and a mini-version (check plot) of the full drawing sheet will be sent to the printer selected in the **Setup/ Check Plot** area.

NOTE: *One great feature of the **Fit to:** option is that it will also automatically scale down the width of all your lines. If it did not, then the lines of the reduced version of your drawing would end up bleeding together into dense lumps of ink, making the output unreadable.*

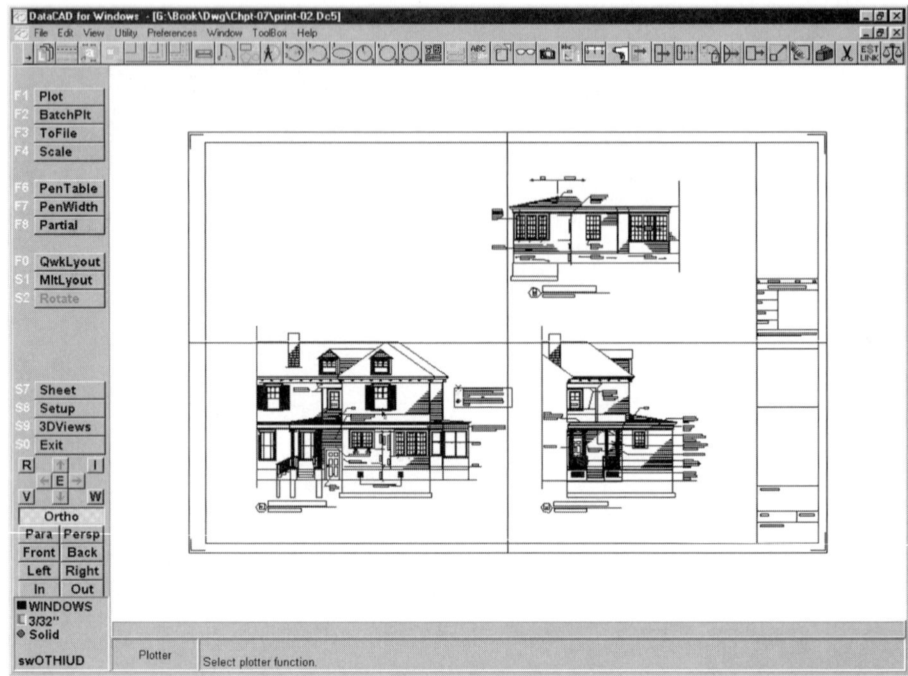

Figure 7-13 Selecting an area to print with **QwkLyout**

Check Plot Properties

Currently, >setting the driver properties of the check plot printer cannot be done from within the Check Plot dialog. If you need to change colors, create custom paper sizes, or set the quality of the prints, then you will have to follow these steps:

1. Minimize DataCAD.
2. Open **My Computer/Control Panel/Printers**.
3. *Right-click* on the printer driver and select **Properties**.
4. Make and save your setting changes.
5. Return back to DataCAD and plot your Check Plot.

Custom Reduction Percentages

There may be times when you will want to print your check plot at a specific percentage. For instance, maybe you want to print the check plot at 50 percent so that all entities are still printed to some definable scale. Or maybe the 43 percent that DataCAD calculated for you doesn't quite fit right, so you want to try 40 percent. The process is exactly the same as a standard check plot except for the following steps:

1. In the **Printer/Plotter Settings** dialog, in the **Check Plot** section, pick **"Use this factor:"** instead of the **Fit to:** option. The box to the right of **Use this factor:** is highlighted. The number in the box is the percentage DataCAD calculates for you to make the drawing fit on the paper size chosen.
2. Use the up/down arrows in the box, or type in your own number.
3. Press **OK** to keep all the options selected.

Keep in mind that if you choose a percentage larger than the one DataCAD calculates, you may experience some anomalies since the drawing won't be able to completely fit on your paper.

Saving to a File

You will probably need to use the **ToFile** option in the **Plotter** menu in at least three instances:

- If you want to send plot files to an outside printing service
- If you want to "batch plot" a group of drawings all at once
- If you want to save, or archive, your plots as a record, or so that you can print more copies of the exact same sheets at a later date

To use the **ToFile** option, follow these steps:

1. To save a drawing **ToFile**, lay out your drawing just as you would if you were going to print it right away. Don't forget to pick the correct **PenTable**.
2. Instead of selecting the **Plot** button, select the **ToFile** option further down.
3. The **Save plotter output to:** dialog box will appear. The default location for saving these files is the \PLT directory. If you want to save them somewhere else, just navigate to the new directory before saving the files. The file name shown in the **File name:** box will default to the name of the current DataCAD drawing. You can either keep it or

rename it. If you are printing an MSP sheet, you will probably want to rename the drawing to match the sheet name or number.

4. The file type shown in the **Save as type:** box will default to "Plot files *.PLT."
   a. .PLT is the standard file extension for plotted drawings. Most plotters will recognize this extension right away. Some printers will also recognize the .PLT extension, but most want to see a .PRN file extension.
   b. Press the down arrow at the right of the **Save as type:** box. You will see three options, as shown in Figure 7-14:
      - **Plot files *.PLT** The default extension
      - **Print files *.PRN** Most Windows-compliant printers, and even a lot of plotters, will recognize this extension as a printer file and will attempt to print the contents of the file (more on this later in this section).
      - **All files *.*** This option enables you to see all the files in a directory, even if they do not have a .PLT or .PRN extension. You cannot save a file with a **.*** extension. This option is simply a convenience so that you can view any file from within this dialog box.

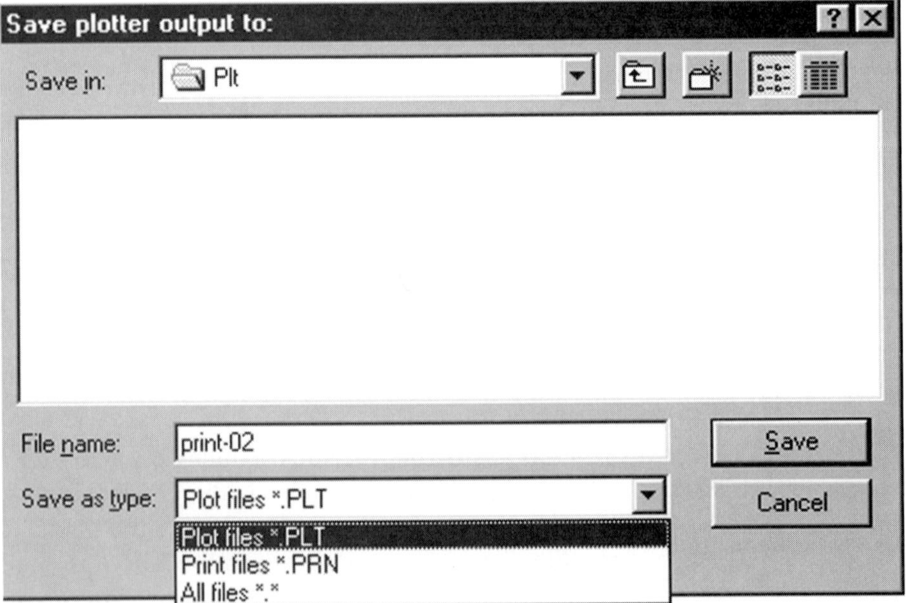

Figure 7-14
Saving plotter output to .PLT or .PRN files

5. Pick the appropriate extension, .PLT or .PRN, depending on the intended output device. When DataCAD saves the drawing shown in the **File name:** box, the chosen extension will automatically be added to the file. Note, however, that the information contained in that file is exactly the same no matter which extension is chosen. The extension is just an indicator to a printer or plotter receiving the file that there may be information in the file that the printer or plotter is capable of printing.
6. The **Open as read-only** option is a Windows convention that adds a read-only attribute to the file so that it can be read and printed but cannot be altered until the read-only attribute is removed. Usually, this is used to prevent someone from accidentally overwriting a file.

Understanding the ToFile Option

When you select a printer or plotter driver, you are selecting a language with which to write the drawing. When any Windows document is saved to a file, the driver saves the exact same information that would have gone to the printer. So the language the file is written in can only be read by a printer that understands that language. If you send an Epson file to an HP printer, you will get a very strange text output that in no way resembles the drawing you saved. Both printers understand different languages. This is why you must select the correct driver via the **Setup** option in DataCAD, whether you are printing to a printer, a plotter, or to a file.

Some printers and plotters are multilingual; they understand more than one language. Years ago, *Hewlett Packard* (HP) developed a standard language for their plotters. The first version was called *Hewlett Packard Graphics Language* (HPGL). A later, more robust version was called *HPGL/2.* This language still remains as a fairly standard language that most HP plotters and printers can understand. But more importantly, many other plotter manufacturers (though not very many printer manufacturers) have made sure that their own plotters can understand this HPGL/2 language in addition to their own languages. This means that if you use a driver that writes true HPGL/2 files, you can make plots on nearly any plotter made today, but these generic drivers appear to be few and far between.

If you plan to send your plot files to an outside source for printing, you must do one of two things:

- Make sure you use a plotter driver that creates real HPGL/2 files

or

- Find out which plotter driver your plotting service is using and install the identical plotter driver on your system. You can get this driver from your plotting service or from the plotter manufacturer's Web site. This is by far the best option and will give the best results with the fewest headaches.

If you don't do this, then your plotting service's plotter will not be able to decipher the language used by your plotter driver. This is a limitation of the plotter, not of DataCAD, and is true of any CAD program's output.

Troubleshooting the ToFile Option

Saving drawings using **ToFile** is a simple process, but printing them out afterward occasionally presents some confusing issues, most of which are related to the idiosyncrasies of specific printer/plotter drivers and not to DataCAD itself. Although printer/plotter manufacturers are supposed to write software drivers that conform strictly to the Windows conventions, they apparently don't always do that. This often creates some problems that show up more often while printing from CAD programs than from other types of software.

As an example, I have never been able to get my Epson Stylus 400 to print using ToFile, even when I follow the Epson instructions for installing the ToFile driver.

If you experience trouble, the first place to turn is the manufacturer of the printer or plotter. If that doesn't help, then you might try calling or e-mailing DataCAD LLC. A third option is to post your problem to the DBUG Forum, an e-mail listserve group (see Chapter 27, "DataCAD Resources"). Your best option is to save the drawings as Adobe PDF files. which can be printed on any printer.

Batch Plotting

This is a new feature in DataCAD 9. It enables you to print or plot one or more drawings all at once without having to select and plot each one individually. Two batch plotting methods exist, each related to the other. One allows you to make all your settings and batch plots from the current drawing, while the other allows you to use the current batch plot settings in each drawing file to batch plot from more than one drawing file at one time.

Whether you want to batch plot from the currently open drawing file or to batch plot many sheets from many drawing files, it all starts by first configuring the plotting parameters within each drawing file.

Batch Plotting from an Open File

While you have a .DC5 drawing file open, you can batch plot only drawings from that file. These are the basic steps:

1. Make all your printer/plotter settings in the **Plotter/Setup** menu. Be sure to select the correct paper size that all your plots are to be printed to.
2. In the **Plotter** menu, select **PenTable** and pick the correct .PEN table.
3. To batch plot a Quick Layout sheet, select **QwkLyout** and layout the detail or sheet before selecting the batch plot utility.
4. Select **Utility/Plotter/BatchPlt**. The **Batch Plotting Setup** dialog box will appear (see Figure 7-15).
5. In the **Batch Plotting Setup** dialog, select all the views and MSP sheets to be plotted from the current drawing file, as will be described later.
6. Select the number of copies, and check the other options as desired.
7. Select **Plot**. Answer **Yes** when asked if you are ready to plot.
8. A **Batch Plotting Session** dialog box will be displayed and the plots will be sent to the printer, plotter, or file. When all the plots are sent, the progress bar will be solid blue. You can then close the session dialog.

Batch Plot Settings

To access the batch plot settings, open the drawing file that you want to plot from. Now go to the **Plotter** menu and pick **BtchPlot**. The dialog box shown in Figure 7-15 appears.

Among other things, with this dialog box you can set up to plot the following elements:

- The current QwkLyout sheet
- One or all of the MSP (Multi-Layout) sheets
- One or all of the 3D GTVs (see Chapter 10)

The 3D GTV option may not be as useful as it first sound, since in most cases it will not actually print your saved views. This option will only work if all the 3D views are layered one on top of the other and are at the same size and scale. We'll explain these issues and more in greater detail at the end of this batch plotting section.

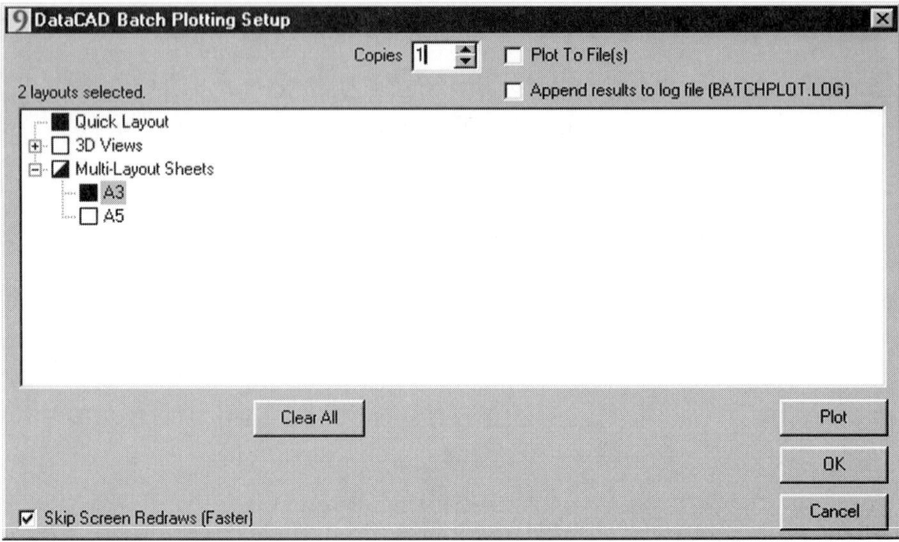

Figure 7-15 The **DataCAD Batch Plotting Setup** dialog box

All plots will be made using the current printer driver and .PEN file settings, so make sure you have the correct **PenTable** selected. The next section will outline what all the settings in the dialog box will do.

Copies The default is 1. If you want to print more than one copy of each drawing, either use the up/down arrows or type in the number of copies that you want.

Plot to File(s) Check this box if you want to save all the prints to individual plot (.PLT) files instead of having them printed on your printer.

Append Results to Log File (BATCHPLOT.LOG) With this box checked, basic information about each batch plotting session will be added to a file called BATCHPLOT.LOG in the DataCAD \PLT directory. Figure 7-16 is a sample.

This .LOG file will keep a running list of all batch plotting sessions. If the box is unchecked, then the .LOG file will remain unchanged. To start with a new, blank .LOG file, you need to either erase everything in the current .LOG file or rename the current one so that a new .LOG file will be created the next time the **Append** box is checked.

Layout Window This is where you select all the views and/or sheets to be printed. *Click* on an empty square box and it will turn solid black, mean-

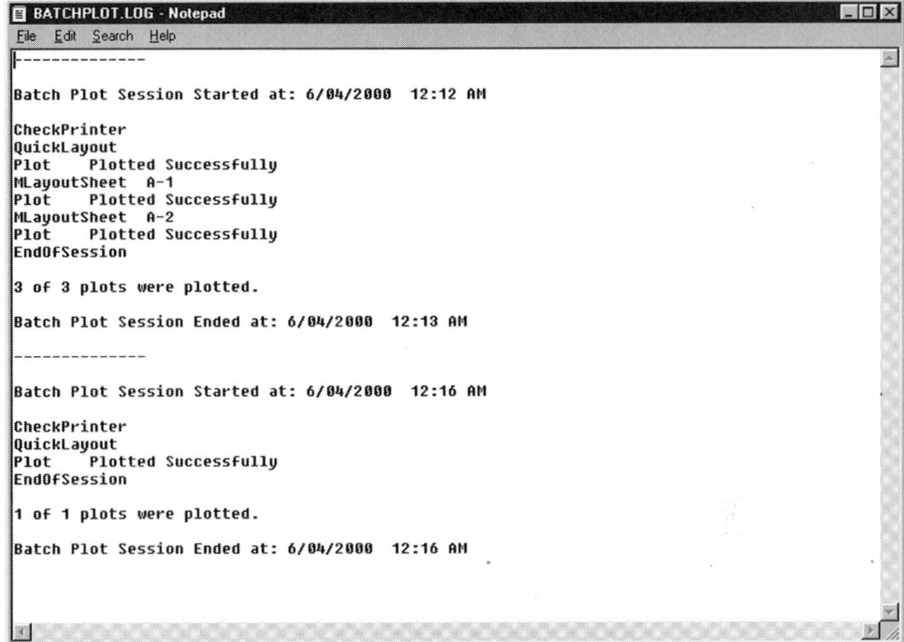
BATCHPLOT.LOG - Notepad

```
--------------

Batch Plot Session Started at: 6/04/2000  12:12 AM

CheckPrinter
QuickLayout
Plot    Plotted Successfully
MLayoutSheet  A-1
Plot    Plotted Successfully
MLayoutSheet  A-2
Plot    Plotted Successfully
EndOfSession

3 of 3 plots were plotted.

Batch Plot Session Ended at: 6/04/2000  12:13 AM

--------------

Batch Plot Session Started at: 6/04/2000  12:16 AM

CheckPrinter
QuickLayout
Plot    Plotted Successfully
EndOfSession

1 of 1 plots were plotted.

Batch Plot Session Ended at: 6/04/2000  12:16 AM
```

Figure 7-16 A sample **BATCHPLOT.LOG** file

ing that every view or detail of that type has been selected. *Click* on a solid black box and it will change to empty, meaning that no view or detail of that type has been selected.

Click on a plus (+) sign to expand the option into a tree view of all the available details or sheets. *Click* on each of the resulting squares to select or deselect each view or sheet. If all the views or sheets of one type are selected, then the parent square will be solid black. If at least one, but not all, of the views or sheets of one type are selected, then the parent square will be half-black and half-white (as in the previous MultiLayout Sheets example).

Selecting **Quick Layout** will cause the last defined QwkLyout view to be printed. Remember that you can only assign one Quick Layout at a time in DataCAD, so you can only print one Quick Layout sheet with the Batch Plot feature. However, you can batch plot any number of 3D GTV's and MultiLayout sheets.

In the previous example, notice that the names that you give to each of the Multi Layout [MSP] sheets are shown, making them easy to identify. This is also true for the 3D views.

Clear All Clicking on this button will deselect all current selections in the Layout Window.

Plot Select this button to send all the selected views and sheets to the selected printer, plotter, or file. A **Confirm** message will appear directly after pressing this button, asking if you are ready to plot all the selected layouts. Selecting **Yes** will cause the **Batch Plotting Session** dialog box to display (described later), and all the selected views and sheets to be sent to the selected printer, plotter, or file.

OK Select this option to save all the current settings without actually plotting. The dialog box will automatically be closed. The second reason for this OK option is because you need to set all of the previous settings in each drawing file before you can use the batch plotting method described later.

Batch Plotting Session Dialog Box

This dialog box appears after selecting **Plot** and **Yes** from the **Batch Plotting Setup** dialog (see Figure 7-17).

Stop Pressing this option will cancel the current plotting session.

More Detail You can select this option during or after the plots have been sent. When selected, you will see an expanded dialog box like Figure 7-18.

The expanded dialog area gives you the current status of all the plots, including whether or not each plot was successfully plotted.

Less Detail This option will only appear once the dialog box has been expanded. When selected, it will cause the expanded dialog box to contract to its original size.

Close Once all sheets have been successfully or unsuccessfully sent, the Close button will become available. When selected, it will close the batch plotting dialog boxes.

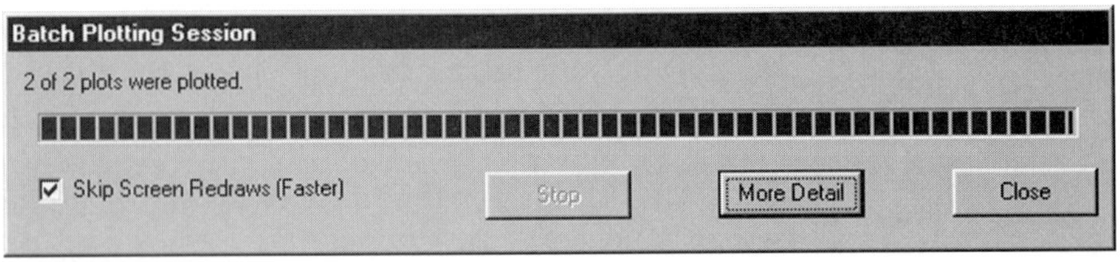

Figure 7-17 The **Batch Plotting Session** dialog box

Figure 7-18
An expanded version of the dialog box

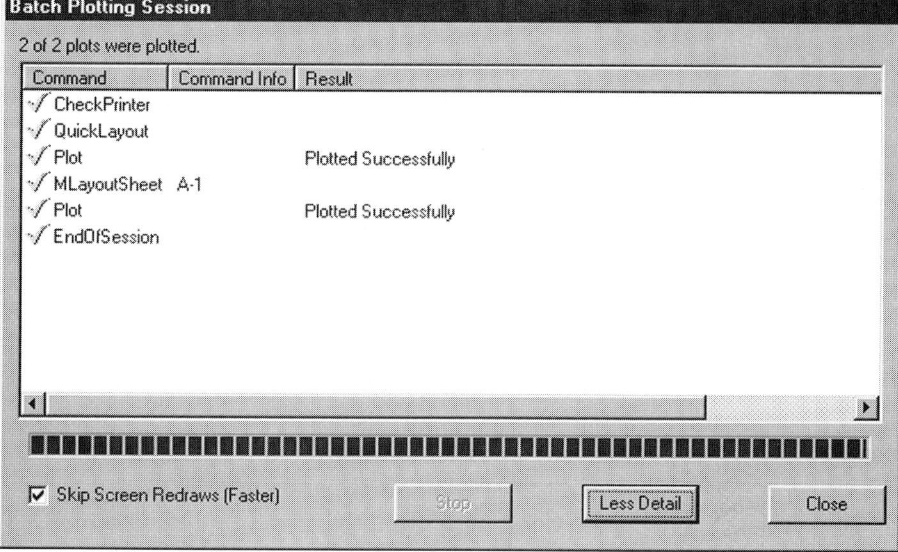

Batch Plotting Quick Layouts

Because you can only assign one Quick Layout at a time in DataCAD, you can only print one Quick Layout sheet with the Batch Plot feature. Therefore, if you routinely set up a drawing file to print multiple drawing sheets using only Quick Layouts, and not Multi-Layout sheets, then you will need to rethink your approach if you want to use Batch Plotting. You have at least two options:

- Use 3D GTVs to save your drawing sheets for batch plotting, as described in the next section

or

- Save all your drawing sheets as Multi-Layout sheets. This doesn't mean that you have to re-lay out each individual detail again just to get the drawing in Multi-Layout format. You can save the whole Quick Layout drawing as one single "detail." in the Multi Layout window. You can read more about the Multi-Layout option in Chapter 12.

The next feature, batch plotting with 3D GTVs, also uses the Quick Layout option to print, but it goes one step further.

Batch Plotting with 3D Goto Views

Earlier we mentioned that the 3D GTV option may not be as useful as it first sounds. If you draw your various details in different locations in the Drawing Window and use 3D GTVs to quickly switch to each detail, then you would probably expect the batch plotting 3D GTV option to print out each view. But these options will only work if all the 3D views are layered one on top of the other and are at the same size and scale, which is generally not the case for most users.

However, these batch plotting options will be useful to some users. They are there because some users layer all the entities of multiple drawing sheets one on top of the other and then use 3D GTVs to bring up each drawing sheet for plotting with the **QwkLyout** option. As long as the title block, sheet border, and all the drawing entities are directly over one another, this option will work in batch plotting.

See Chapter 10 for a thorough explanation of 3D GTVs. For now, here are the steps to set up your Drawing Window to batch plot from these views. After these steps, we'll run through an example to see how it works:

1. Make sure all the entities to be plotted from each GTV are laid out on top of one another.
2. Make sure all your printer/plotter settings are in the **Plotter/Setup** menu. Also be sure to select the correct paper size that all your plots are to be printed to.
3. In the **Plotter** menu, select **PenTable** and pick the correct .PEN table.
4. You need to lay out the QwkLyout box over the area to be printed, so select one of the GTVs to be printed.
5. Select **QwkLyout** and center the Quick Layout box over the detail or sheet. In batch plotting GTVs, DataCAD uses the location of the Quick Layout box to determine what is printed. Only one Quick Layout is available in DataCAD, so that is why the entities to be printed, even if they are in different GTVs, must be located within the same Drawing Window area.
6. Select **Utility/Plotter/BatchPlt**. The Batch Plotting Setup dialog box will appear.
7. In the Layout Window, select all the 3D GTVs to be printed from the current drawing file.
8. Select the number of copies, and check the other options as desired.
9. Select **Plot**. Answer **Yes** when asked if you are ready to plot.

10. A **Batch Plotting Session** dialog box will be displayed and the plots will be sent to the printer, plotter, or file. When all the plots are sent, the progress bar will be solid blue. You can then close the session dialog.

An Example The accompanying CD-ROM contains a sample file called 2d3dExample.DC5. You can copy this file to your computer and follow along if you'd like. Four layers can be found in this file, each with a single string of text. Three of the strings are located very close to one another, and one is farther away, as shown in Figure 7-19.

Each string of text is saved as a separate 3D GTV. Noting where each string of text is located on the screen will be important when you see the final output. Let's begin:

1. Make all your printer/plotter settings in the Plotter/Setup menu. For this example, select an $8^1/_2''$ × 11″ paper size.
2. In the **Plotter** menu, select **PenTable** and pick the correct .PEN table. Then select **Scale** and set the scale to $^1/_8$ ″ [1:100].
3. Turn on the first 3D GTV, called **GTV-1** (**Utility/GotoView**). *Right-click* back to the **Plotter** menu.
4. Select **QwkLyout** and center the Quick Layout box over the detail or sheet, as shown in Figure 7-20.
5. Select **Utility/Plotter/BatchPlt**. The Batch Plotting Setup dialog box will appear.
6. In the Layout Window, select all the 3D GTVs to be printed (see Figure 7-21).
7. Select **Plot**. Answer **Yes** when asked if you are ready to plot.
8. A **Batch Plotting Session** dialog box will be displayed and the plots will be sent to the printer, plotter, or file. When all the plots are sent, the progress bar will be solid blue. You can then close the session dialog.

GOTO VIEW 4

GOTO VIEW 1
GOTO VIEW 2
GOTO VIEW 3

Figure 7-19
The text strings in 2d3dExample.DC5

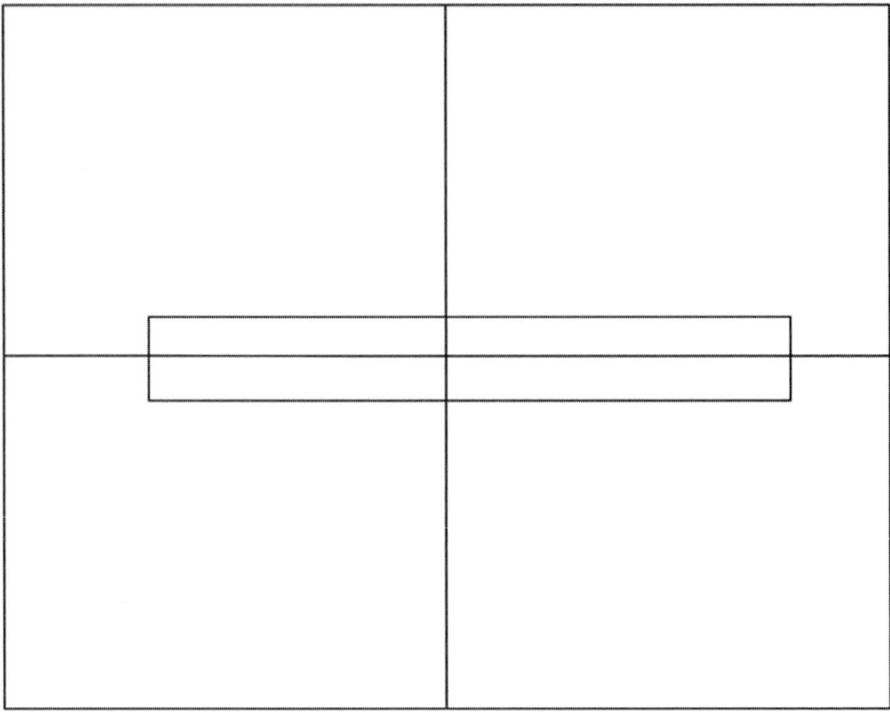

Figure 7-20 Centering the Quick Layout box over the detail or sheet

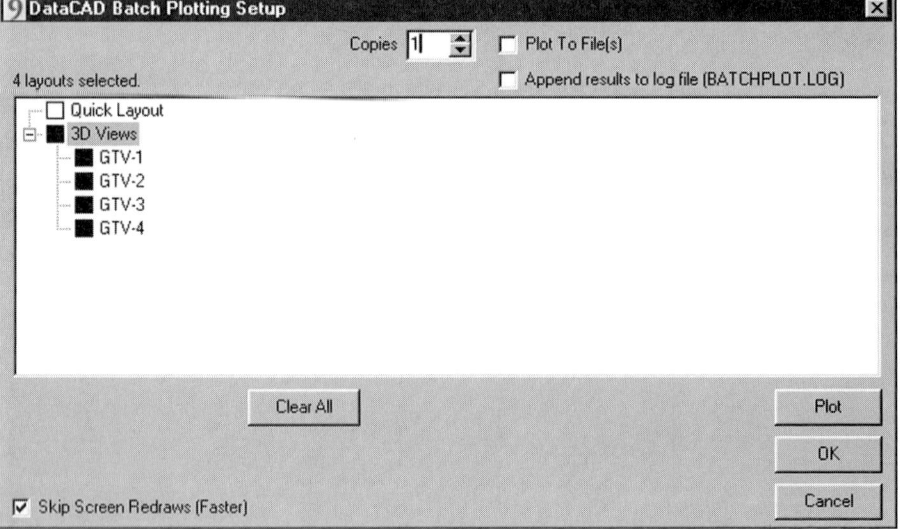

Figure 7-21 Selecting all the 3D GTVs to be printed

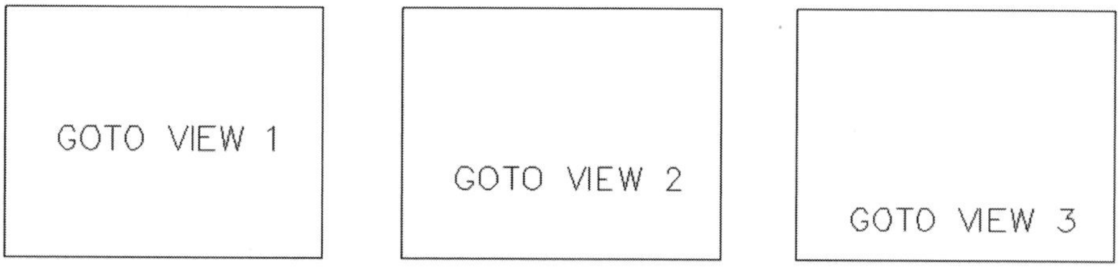

Figure 7-22
The print results

GOTO VIEW 1

GOTO VIEW 2

GOTO VIEW 3

After all four sheets have been printed, you should see that the first three GTVs were printed, though in slightly different locations on the page (see Figure 7-22).

The text in the fourth GTV, however, did not print. That's because the entities were not within the Drawing Window area where the QwkLyout box was placed.

Batch Plotting from Multiple Drawing Files

This batch plotting method can only be accessed when you have DataCAD running, but with no drawing files open. This method allows you to plot sheets and views from multiple drawing files where batch plot settings have already been made. The multiple file batch plotting feature is really just a shortcut for printing the same things that you could batch plot from within the individual drawing files, but it lets you do so without first having to batch plot from each file individually. So you must have first set up all your plot sheet parameters within each drawing file, as described in the previous section, before running this utility.

Because this option is only accessible when no drawings are open in DataCAD, you can select one or dozens of drawing files without ever having to open them. As long as you first set up and save all of these options within each of those .DC5 files, you can print the following elements:

- The last QwkLyout sheet
- One or all of the Multi-Scale Plot (Multi-Layout) sheets
- One or all of the 3D GTVs (see Chapter 10)

Just as the layout and view settings are saved within each drawing file, so are the currently selected printer driver and .PEN file. This means that it is important to ensure you have the correct driver and .PEN file selected in each drawing file before saving and closing the drawing. If you don't pay attention to that, then you can, either by accident or on purpose, send some plots to your plotter and some to your printer, depending on which driver was last selected in each drawing file:

1. Be sure that all your layout, paper, and printer settings are correct within each .DC5 drawing file you are going to batch plot from.
2. Close all open DataCAD drawing files. The only drop-down menu options available then are **File** and **Help**.
3. Select **File/Batch Plot**. A dialog then appears, as shown in Figure 7-23.
4. Click on the **Add** button. A dialog box will appear, allowing you to pick all the .DC5 drawing files that you want to plot from.
5. Select all the drawing files you want to print from. You can use the standard Windows methods of selecting multiple files at once (Ctrl+Click, Shift+Click, and Alt+Click).
6. Press **Open** or **Enter** to accept all your selections.
7. If you want to add more drawing files from other directories, select **Add** again, navigate to the new folder, and then select more files.
8. Select the number of copies and check the other options as desired.
9. If you want to save all these settings for future use, including the list of selected drawings, select **Save As** to save a Project *Set File* (.SET) to the current plot directory (usually \PLT).

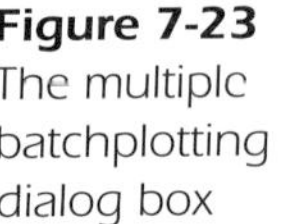

Figure 7-23 The multiple batchplotting dialog box

10. Select **Plot**. Answer **Yes** when asked if you are ready to plot.
11. A **Batch Plotting Session** dialog box will be displayed and the plots will be sent to the printer, plotter, or file. When all the plots are sent, the progress bar will be solid blue. You can then select close the session dialog.

Batch Plotting Setup Options

Most of the options in this dialog box are the same as in the dialog box you saw in the previous section with the addition of a few extra options:

Drawing File Window This is where you view all the drawing files to print from. Click on the **Add** button. A dialog box will appear, allowing you to pick the .DC5 drawing files that you want to plot from. You can use the standard Windows methods of selecting multiple files at once (Ctrl+Click, Shift+Click, and Alt+Click).

Add Use this button to add .DC5 drawing files to the Drawing File Window list. You can select drawing files from multiple directories by selecting **Add** as many times as required.

Remove Highlight a file or files in the Drawing File Window and then use this button to remove those drawing files from the list.

Clear All Clicking on this button will deselect all current selections in the Drawing Window.

Ignore Autosave and File-in-Use Conditions If a drawing file is exited abnormally, when you attempt to open the file again in DataCAD, it will warn you that an *Autosave* (.ASV) file exists and ask you a couple more questions prior to opening the drawing file. When a drawing file is currently in use on another computer on the network, DataCAD will warn you that this is the case and will not let you proceed without first answering a question. Checking the **Ignore** . . . box will enable the batch plot utility to bypass both these messages and plot from those files anyway. In most cases, you will want to leave this box checked.

Skip Screen Redraws (Faster) If you uncheck this box, then you will see DataCAD open each drawing file, display each view or sheet selected for printing, and then send the plot. This is useful if you want to visually monitor the batch plot session, but it will slow down the whole batch plotting process considerably because it takes time for your computer to display each view prior to printing it. If you check the **Skip** . . . box, then DataCAD will open each file and print all the views and sheets in the background without displaying them. This will then speed up the whole batch plotting process.

Save As Use this option to save the current list of .DC5 files displayed in the Drawing File Window for future use. To use it, select **Save As**. A dialog box will appear, allowing you to save a 'Project Set File' (.SET) to the current plot directory (usually \PLT unless you have it set up somewhere else). The number of copies and all of the check box options are not saved.

Load Use this option to retrieve a .SET file saved with the **Save As** command. The drawing file list that was saved in the .SET file is appended to the current list of selected drawings. The **Load** command will also not cause the current selections in the Drawing Window to be overwritten.

Plot Select this button to send all the selected sheets to the selected printer, plotter, or file. A **Confirm** message will appear directly after pressing this button, asking if you are ready to plot all the selected layouts. Selecting Yes will cause the **Batch Plotting Session** dialog box to display, and all the selected views and sheets to be sent to the selected printer, plotter, or file.

OK Select this option to save all the current settings without actually plotting. The dialog box will automatically be closed.

How to Print Saved Drawings

If you have saved .PLT or .PRN files with the **ToFile** option in the **Plotter** menu or the **BatchPlt** dialog, you will eventually need to print them. Oddly enough, Windows does not have any direct way of doing this, so we'll have to do a few tricks in order to print them. If you send your plot files to an outside plotting service, they will have the expensive software to print out your saved .PLT files without having to resort to these tricks. If you want to print those saved files in-house from Windows, you can use Windows' capability to make file-associations. Of course, the best way to do this yourself is to avoid it, and use the Adobe PDF format instead of the .PLT or .PRN formats.

This method uses Windows to associate a file extension, such as .PLT, so that you can print one or a group of plot files by *right-clicking* or *double-clicking* on them and then sending them directly to your plotter. This method involves writing a very simple "batch file" (a mini-DOS program). To do this, you must first know the location (path and name) of your plotter on your own computer or the network. If you don't already know, here is how find out:

1. Open the **Printers** folder in the Windows **Control Panel** (in My Computer on your desktop) and *right-click* on the plotter driver you want to send your plots to.
2. Select the **Properties** option and then the **Details** tab.
3. Look at the **"Print to the following port:"** box.
   a. If your plotter is hooked up directly to your computer, then it probably says **LPT1:** or it might say **LPT2:** or **LPT3:**. This is the location of your plotter.
   b. If your plotter is on a network, then it probably says something like **\\Server\CalComp _720c**. This is the location of your plotter.

To associate all files with the .PLT file extension, do the following:

1. Open a text editor such as Windows Notepad.
2. Type **copy %1 LPT1:**. LPT1: is the location of the plotter in this example. If yours is in a different location, then use that instead (such as copy %1 \\Server\CalComp_720c).
3. Save the file to the C:\ directory with the name PLOT.BAT.
4. Open My Computer from the Windows desktop and select **View/ Options** from the drop-down menu.
5. Select the **File Types** tab.
6. Select the **New Type** button.
7. Next to the **Description of type:**, type **Plot File**.
8. Next to **Associated Extension:**, type **PLT**.
9. Under **Actions**, select the **New** button.
10. In the **Action** field, type **Send to LPT1.** This does not have to be the exact name of your plotter location. What you type here is what will be displayed when you right-click on the .PLT files. You could type something like **Send to Plotter** instead. Just keep it short so it will fit.

11. Select the **Browse** button.
12. In the **Open With** window, browse to the C:\ drive and highlight the PLOT.BAT file you saved earlier. Then press **Open** to select it.
13. Select **OK** to close the New Action dialog. This should return you to the Add New File Type window. Under **Actions**, you should see Send to LPT1 (or whatever you typed in step 10).
14. Highlight **Send to LPT1** and select the **Set Default** button.
15. Select **Close** at the bottom of the Add New File Type window and then select **Close** at the bottom of the Folder Options window. This should bring you back to the My Computer window.
16. From My computer, browse to the PLOT.BAT file in the C:\ drive and *right-click* on PLOT.BAT.
17. Select **Properties** and then select the **Program** tab. In the **Working** field, type **C:\.** This tells Windows where to look for the PLOT.BAT file.
18. In the **Run** drop-down menu, select **Minimized**.
19. Check the **Close On Exit** box and select **OK** at the bottom of the plot properties window.

How It Works

With all that done, now you can send your saved .PLT files to the plotter by making one of the following choices:

- In Windows Explorer, highlight a group of plot (.PLT) files and then *right-click* on them. Select **Send to LPT1** (or whatever you typed in Step 10), and they will all be sent to the plotter

or

- *Double-click* on a single .PLT file and it will be sent directly to the plotter.

Keep in mind that this is just a shortcut for sending files to your output device. The type of file sent must still be saved in a format that the output device understands. If the chosen device does not understand the language of the plot file sent to it, the file will not print properly.

To send drawings to your printer (usually files with a .PRN extension), you can follow the exact same steps as listed earlier, but create a batch file called PRINT.BAT and type in **copy %1 LPT1:** using the location of your printer.

Troubleshooting Tips

Many printing and plotting problems are directly related to problems with the hardware's printer driver. Either the driver is old and out of date, the correct one has not been installed, or it has not been configured properly. The first thing to do if you are having printing problems is to go to the Web site of your printer or plotter manufacturer and download the most up-to-date drivers. Install the driver and see if this corrects the problem.

Screened Lines Are Printing Solid

If you are trying to print lines that are screened (the Percent Density is lower than 100 percent in the Pen Table) but they are instead printing solid, check the following in the **Pen Table** dialog:

- Make sure **All Black** is unchecked. You cannot use **All Black** if you want to print screened lines.
- Ensure the correct **Percent Density** is set for the pen numbers to be screened (see the description of the **Percent Density** option earlier in this chapter).
- If you are printing all in black, but are not using the **All Black** option, then check to make sure the **Map to Color** boxes are checked and the current color for each one is black.
- If you are printing all in color, then check to make sure the **Map to Color** boxes are checked and the current color for each one is the correct color. Remember that colors will generally already print lighter even without a screen setting (see the description of the **Percent Density** option earlier in this chapter).
- Make sure to use **SaveAs** to save the settings for future use.

The Plot Has Been Sent to the Printer or Plotter, but It Does Not Print

This could be any number of problems, most of them probably related to an incorrectly configured printer driver. You should first consult your printer/plotter manual for proper configuration settings. If everything still appears all right, try some of these possible solutions:

- Make sure you have selected the correct driver prior to printing. Perhaps you wanted to print to your laser printer, but you accidentally picked the plotter driver.
- With our HP Laser Jet printer, if we accidentally select a custom paper size that is larger than the printer can handle, the printer will blink with an error message. We have to cancel the print job and reset the printer. Select the proper paper size and try again.
- Give it time. The print may still be spooling in your computer. Open the icon for your printer in the Windows Control Panel to see if the drawing is still spooling. If it is, then the drawing will print when the spooler is ready. Check your printer/plotter's manual for correct driver spool settings.
- If you are on a network, some automatic A/B switch boxes (a box that enables you to connect one computer to more than one printer, or a box that will connect more than one computer to one printer) will not function properly with some printers or plotters. Consult your printer/plotter's manual.
- If you are on a network, your print may be hung up in the server. This happens often to us in Windows NT, but we have not found a solution. Try canceling the print job then try again.

Line Widths Are Not Printing Properly

If some or all of your line widths are wider or thinner than they are supposed to be, check the following:

- Be sure you have the correct driver and Pen Table selected. It is a good idea to check both of these before any print job. I often think I left the correct settings from the last time I printed, only to find that I didn't.
- Some plotters, such as our CalComp TechJet 720c, will only print particular, fixed pen widths. If this is the case with your plotter, look in your plotter manual for the correct line widths and insert them in the Widths column of the Pen Table.
- A conflict is taking place between DataCAD and some printer/plotter drivers that is forcing lineweights to be specified in the Pen Table in ascending order. **Pen Number 1** would have to be set to the thinnest

lineweight and **Pen Number 15** to the widest, with all **Pen Numbers** in between likewise ascending from thinnest to thickest.

- The standard unit of measure of line widths for most (but not all) plotters is millimeters. Therefore, this is how DataCAD's Width settings are defined in the Pen Table. Be sure the widths you are setting are in millimeters. If you are trying to use widths of decimal inches, you will have to convert those inches to millimeters. There are 25.4 mm per inch, so a line width of 0.001 inches is equivalent to 0.0254 mm (0.001 x 25.4).
- On many plotters, you can set line widths either at the plotter itself or via software with DataCAD's Pen Table dialog. In general, you will always want to use the software option. On most plotters, this means you have to manually set a control panel option on the plotter itself in order for it to receive the lineweights, colors, density percentages, and other such information from the software. Check your plotter's operation manual for more information. This is one of the most common solutions to plotting problems, so be sure to read your manual carefully to determine the proper settings.

Not All of My MSP Details Are Printing

There could be several reasons for this. More often than not, I will hear about this problem when users are sending saved plot files to an outside plotting service, but occasionally it happens in-house, too. Thus, make the following checks:

- Check to be sure all your MSP details are actually in the MSP sheet. Remember that if you don't give an MSP detail a name immediately after placing it in the MSP window, the detail will appear to be there, but next time you open the MSP sheet you will see that the detail was not actually saved.
- Make sure you have the most up-to-date driver from your printer/plotter manufacturer.
- One of the most frequent problems involves the timeout settings of your plotter driver. The timeout setting tells the plotter how long to wait for more information from the computer before determining that the plot is finished. Because of the way DataCAD sends MSP details to the plotter, your plotter may be sensing that the plot is done before all the details are finished being sent. The possible solution involves some

fundamental setting changes in the driver. Give these instructions to your outside plotting service as well. They may balk at making changes to their driver settings, but there are no ill effects from the process, so just be insistent.

Non-Networked Printer

These directions are for a plotter that is connected directly to the parallel port in the back of your computer, rather than over a network:

1. Open the Printers folder in the Windows Control Panel and *right-click* on the driver you are trying to use.
2. Select the **Properties** option.
   a. Select the **Details** tab.
   b. In the **"Print using the following driver:"** dialog box, make sure the correct driver is selected. If it is not, click on the down arrow and select the correct driver for your plotter.
3. Now change the **Timeout settings** so that both boxes say **900** seconds.
4. Click on the **Spool Settings** button. Check the **"Spool print jobs so program finishes printing faster"** option, and the **"Start printing after the first page is spooled"** option. Then click **OK**.
5. Click on the **Port Settings** button. Check both boxes and then press **OK**.
6. Press **Apply** and then **OK**.

Networked Computer and Printer

Open the Printers folder in the Windows Control Panel and *right-click* on the driver you are trying to use:

1. Select the **Properties** option.
2. Select the **Details** tab and look at the **Timeout settings** box at the bottom. If the two boxes are grayed out, then you will need to do the following steps, because the timeout settings only work when you capture a printer port.

Figure 7-24 The **Capture Printer Port** dialog box

Figure 7-25 Choosing a network path

a. Click on **Capture Printer Port**. The dialog box in Figure 7-24 will appear.

b. Under **Device**, click on the down arrow to select an unused LPT port, like **LPT2** or **LPT3**. For this example, we will use LPT3.

c. Under **Path**, click on the down arrow to select a network path. You should see the path names of the valid printers and plotters on the network. Pick the one that goes to your plotter. In this example, it is **\\OMEGA\CALCOMP-720C**. Leave the **Reconnect at logon** box checked. Select **OK** (see Figure 7-25).

d. In the **"Print to the following port:"** dialog, click on the down arrow and scroll through the list until you find the LPT port and path you just created. Do not pick the one that only has the plotter path without the LPT port.

e. In the **"Print using the following driver:"** dialog, click on the down arrow and select the correct driver for your plotter.

f. Now change the **Timeout settings** so that both boxes say **900** seconds. The **Details** tab may look something like Figure 7-26 now. Keep in mind that how yours looks will vary slightly with the driver you are using.

g. Click on the **Spool Settings** button. Check the **"Spool print jobs so program finishes printing faster"** option, and

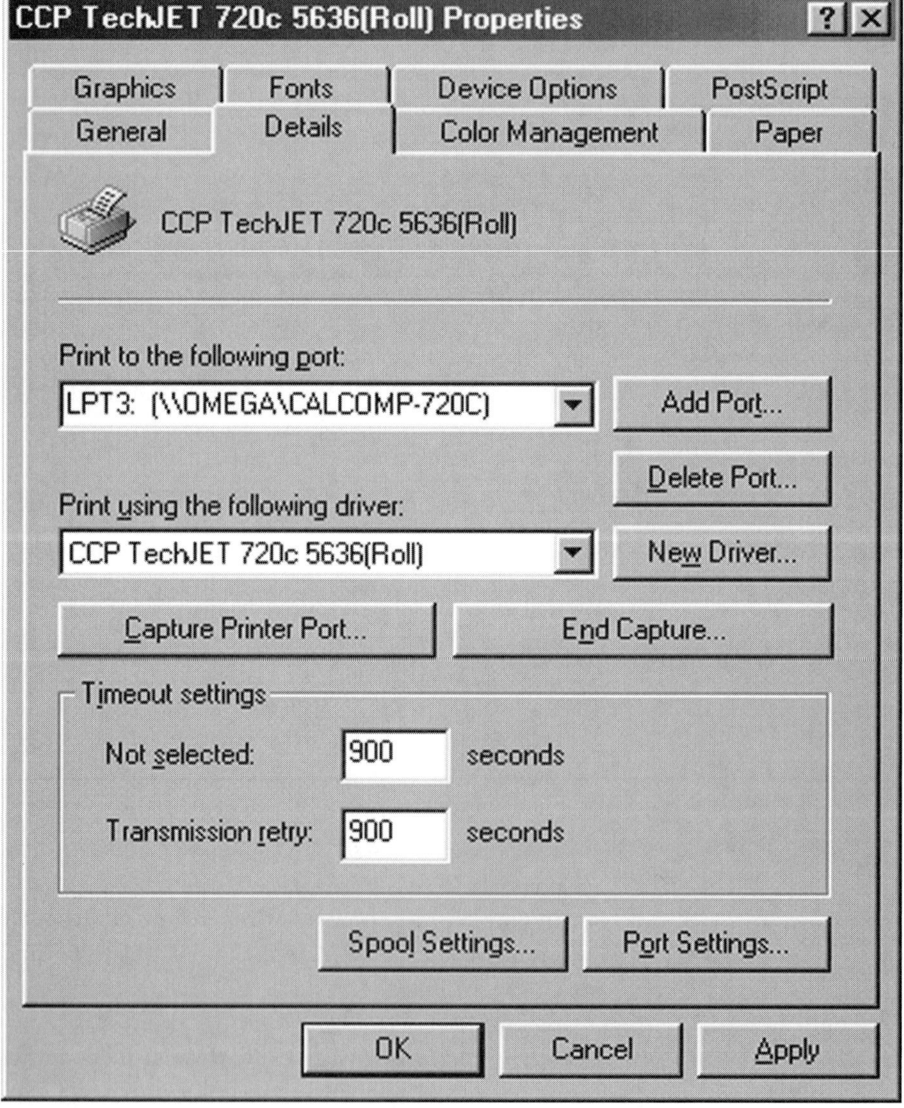

Figure 7-26 The Details tab in the printer properties dialog box

the **"Start printing after the first page is spooled"** option. Then click **OK**.

h. Click on the **Port Settings** button. Check both boxes and then press **OK**.

i. Press **Apply** and then **OK**.

If you are wondering what you just did, read on. If not, then you can skip this explanation. Firstly, the timeout settings of 900 seconds (15 minutes) ensures that the plotter will wait up to 15 minutes after the first batch of plot information before assuming no more MSP details are to be processed. In reality, the plotter will not take any longer to finish plotting, since the "end of file" message in the plot file will tell the plotter when to end the plot. The larger timeout settings just keep the plotter from making an assumption on its own prior to receiving the end of file message.

LPT is the old DOS moniker for a parallel port. This is the port in the back of your computer that attaches to a printer, plotter, or other parallel port device. If your computer does not have a local printer or plotter because they are located elsewhere on the network, then there is no need to use or configure an LPT port for the printer. But to set the timeout settings Windows needs to have an LPT port selected for that printer/plotter driver. What we did here was capture a virtual LPT port. Although no physical third port (LPT3) exists, we have tricked Windows into thinking there is and then assigned the actual plotter path to it.

I Am Using Clip Cubes and Some of My MSP Details Come Out All Wrong

If you are using Clip Cubes with MSP sheets, be aware that currently you cannot rotate Clip Cubes in DataCAD. If you selected the **Plotter/Rotate** option prior to placing an MSP detail, then you may be experiencing some strange output.

Essentially, what is happening is that the entities that make up the detail are being properly rotated in MSP, but the Clip Cube around them is not, so you end up with no entities, or only some entities, being printed because the unrotated Clip Cube is obscuring them. The only solution is to not use Clip Cubes with rotated details in MSP. If you already have such details in your MSP sheets, delete them from each MSP sheet and relay them out without rotating the Clip Cubes.

MSP Details Aren't Printing in the Right Location on the Page, or Some Don't Print at All

This can happen with some printer/plotter drivers if the paper orientation is different between the DataCAD Setup dialog and the printer's own properties settings. Open the DataCAD **Setup** menu. Note the paper orientation shown under **Paper Orientation at Device**. The default is **Normal**. Now select the **Properties** button to directly access the printer driver. In most cases, you will want the paper orientation to be set to **Portrait**, not Landscape. If this doesn't work, then you can try the Landscape option, but in 99.9 percent of the cases the Portrait option will be the correct one.

CHAPTER 8

Moving Things Around

Moving things around can have a number of different meanings in DataCAD. We can move things from point A to point B, move them to a different layer, or even change their location in the third dimension (Z base/Z height). In this chapter we will also discuss the Cut, Copy, and Paste functions. These are new functions to DataCAD but are very familiar to users of other Windows programs.

EGAFS

Just as in other chapters, when we tell you to pick by EGAFS, we are referring to the standard selection methods: Entity, Group, Area, Fence, or Selection Set.

Input Mode

DataCAD has four input modes: Relative Polar, Absolute Polar, Relative Cartesian, and Absolute Cartesian. Most DataCAD users, myself included, use the Absolute Polar method, but plenty of other users swear by the Relative Cartesian method. In this chapter, we will always refer to the Relative Polar method (distance and angle) when moving, copying, or otherwise relocating entities by manually typing in the coordinates. If you like to use the Relative Cartesian input method, just substitute the X/Y input method whenever you are prompted to press the **Space** bar and manually type in coordinates.

Move

This command is used to relocate entities from point A to point B. You can move entities along any axis, be it X, Y, or Z. Use **Move** to move entities to a new location on the screen by any of the EGAFS selection methods. You specify the distance, angle, or direction to move either with the mouse or by typing in coordinates from the keyboard. The distance is a vector that can be entered anywhere on the screen and then be applied to the entities you want to move. Moving entities will not change their attributes (group, color, layer, and so on). They will simply be relocated on the screen. The only exception to this is using **Move** to relocate entities to a new layer.

The steps for moving are as follows:

1. Select **Edit/Move**. You are prompted to *"Select first point of the distance to move."*
2. Select any point in the Drawing Window. You do not have to select a point on or near the entities to be moved.
3. A line representing the move vector is extended from the first point you picked to the cursor. You are then prompted to *"Select second point of the distance to move."*
4. To locate the second point to move to, either
   a. *Click* or snap to the point to move to, or
   b. Press the **Space** bar. This allows you to type in a distance and direction. Type in a distance, then press **Enter**. Type in an angle, then press **Enter**.
5. Select the entities to be moved with one of the EGAFS selection methods.

NewDist

Use this option to select a new distance to move without exiting the **Move** menu.

Invert

Select this option to invert the current move angle. For instance, if the current move angle is 90 degrees, selecting **Invert** will change the move angle to 270 degrees. Selecting **Invert** does not change the move distance.

ToLayer

Use this option to move entities from their current layer to another layer in the drawing file. This option is only available when there is more than one layer in the drawing file:

1. Pick **ToLayer**. The current layer list is displayed.
2. Pick the layer to move the entities to.
3. Select the entities by one of the EGAFS selection methods.

AndCopy

Whenever the EGAFS selection menu is displayed, the **AndCopy** option is also displayed. Select **AndCopy** prior to selecting one of the EGAFS selection options. Use **AndCopy** to leave the original entities in place in the Drawing Window while creating a copy of those entities at the specified distance and angle.

You might ask, "What is the difference between **Move/AndCopy** and the plain old **Copy** command?" The answer is nothing. They do the same thing. But with Move/AndCopy you have one useful option that you don't have with Copy: the **Drag** command.

MoveZ

2D entities, like their 3D counterparts, can exist anywhere in the Z axis. Use **MoveZ** to change the location of entities in the third dimension (the Z axis, in and out of the screen). You can enter positive or negative numbers, and entities will be moved that distance, relative to their current Z location.

Drag

This is one of my favorite DataCAD commands, especially when used in conjunction with **AndCopy**. Use **Drag** to select entities and then dynamically move (or drag) them around the screen with your mouse. Follow these steps:

1. Select **Drag**.
2. Select **AndCopy** if you want the dragged entities to be copied, while leaving the original entities in place.
3. Select one of the EGAFS selection methods.
4. Select the entities to be dragged.

5. Select a reference point to drag from. This point can be anywhere in the Drawing Window; it does not need to be on or near any of the selected entities.
6. Drag the entities to their new location and *click* the mouse to place them.
7. If **AndCopy** is on, continue placing copies, or *right-click* to stop.

MaxLines If a lot of entities are in the drag selection, movement may be slow and jerky because every move of the mouse will cause your screen to try to redraw all of the selected entities. Use this option to limit the number of entities that will be displayed while dragging.

AndCopy Like the standard **AndCopy** option, pick this option prior to selecting one of the EGAFS selection methods. Use **AndCopy** to leave the original entities in place in the Drawing Window while creating a copy of those entities at the point that you drag the entities to. Until you *right-click* the mouse or select **Exit,** every click of the mouse will place new copies of the entities. Think of this feature as a rubber stamp.

This option is a toggle command, meaning that it stays permanently set to on or off until you change it. It's a very common mistake to move entities while forgetting that **AndCopy** is still toggled on.

Multi Use this option if you need to select two or more entities that cannot be selected at once by the one of the EGAFS selection methods. To do so, follow these steps:

1. Select **Multi** prior to selecting EGAFS (you can also select **AndCopy** at the same time if desired).
2. Select one of the EGAFS selection methods.
3. Select the entities to be dragged. While selecting entities, you can switch EGAFS selection methods and continue to select more entities.
4. After picking all the entities, select **Begin**.
5. Select a reference point to drag from. This point can be anywhere in the Drawing Window; it does not need to be on or near any of the selected entities.
6. Drag the entities to their new location and *click* the mouse to place them.

PrevDist (Previous Distance)

DataCAD always remembers the last distance and angle input for the **Move**, **Copy**, or **Stretch** commands (but not **Move/Drag**). Use **PrevDist** to select this previous vector information. Only one distance and angle are remembered, however. For instance, if you moved an entity 20 feet [6096] at 90 degrees and then copied another entity 15 feet [4572] at 180 degrees, the **PrevDist** command will remember only the last distance/angle used, which in this example is 15 feet [4572] at 180 degrees.

Copy

This command is used to copy entities from point A to point B. You can copy entities along any axis, X, Y, or Z. Use **Copy** to copy entities to a new location on the screen by any of the EGAFS selection methods. You specify the distance, angle, or direction to copy either with the mouse or by typing in coordinates from the keyboard. The distance is a vector that can be entered anywhere on the screen and then be applied to the entities you want to copy. Copying entities will not change their attributes (color, layer, and so on), but the copied entities will be part of a new, single group, and not part of their original group. They will simply be relocated on the screen. The only exception to this is using **Copy** to relocate entities to a new layer, since the copied entities will all be on the new layer and not their previous ones.

The steps for copying are as follows:

1. Select **Edit/Copy**. You are prompted to *"Select first point of the distance to copy."*
2. Select any point in the Drawing Window. You do not have to select a point on or near the entities to be copied.
3. A line representing the copy vector is extended from the first point you picked to the cursor. You are prompted to *"Select the second point of the distance to copy."*
4. To locate the second point to copy to, either
   a. Click or snap to the point to copy to

   or

   b. Press the **Space** bar. This allows you to type in a distance and direction. Type in a distance, then press **Enter**. Type in an angle, then press **Enter**.
5. Select the entities to be copied with one of the EGAFS selection methods.

The following sections detail various **Copy** options.

NewDist

Use this option to select a new distance to copy without exiting the **Copy** menu.

Invert

Select this option to invert the current copy angle. For instance, if the current copy angle is 90 degrees, selecting **Invert** will change the copy angle to 270 degrees. Selecting **Invert** does not change the copy distance.

ToLayer

Use this option to copy entities from their current layer to another layer in the drawing file. This option is only available when more than one layer is in the drawing file.

1. Pick **ToLayer**. The current layer list is displayed.
2. Pick the layer to copy the entities to.
3. Select the entities by one of the EGAFS selection methods.

Array

Use an array to repeat an entity or entities a specified number of times at a specific angle and distance:

1. Select **Edit/Copy**. You'll be prompted to *"Select the first point of the distance to copy."*
2. *Click* or snap to a starting point in the Drawing Window. It does not have to be on or near the entities to be copied. You'll be prompted to *"Select the second point of the distance to copy."*
3. Select a copy distance by
   a. *Clicking* or snapping to the point to copy to

   or

   b. Pressing the **Space** bar. This allows you to type in a distance and direction. Type in a distance, then press **Enter**. Type in an angle, then press **Enter**.

4. Select **Array**. You will be prompted to *"Enter the number of repetitions:"*
5. Select or type the number of desired repetitions.
6. Select the entities to be copied with one of the EGAFS selection methods.

The **Array** commands have one quirk about them. The original instance of the entities is counted as one of the members of the array. In this example, we selected an array of 3 (see Figure 8-1).

RectArry (Rectangular Array)

This is similar to the **Array** command, but **RectArry** enables you to repeat entities in both the X and Y directions simultaneously. It works best with the Relative Cartesian input method (press the **Insert** key until the current input mode changes to Relative Cartesian). To use it, follow these steps:

1. Select **Edit/Copy/RectArry**. You'll be prompted to *"Select the first point of the distance to copy."*
2. *Click* or snap to a starting point in the Drawing Window. It does not have to be on or near the entities to be copied. You'll be prompted to *"Select the second point of the distance to copy."*

   You will also notice that two vector arrows are extended from the first point you selected. They move in the X and Y directions as you move

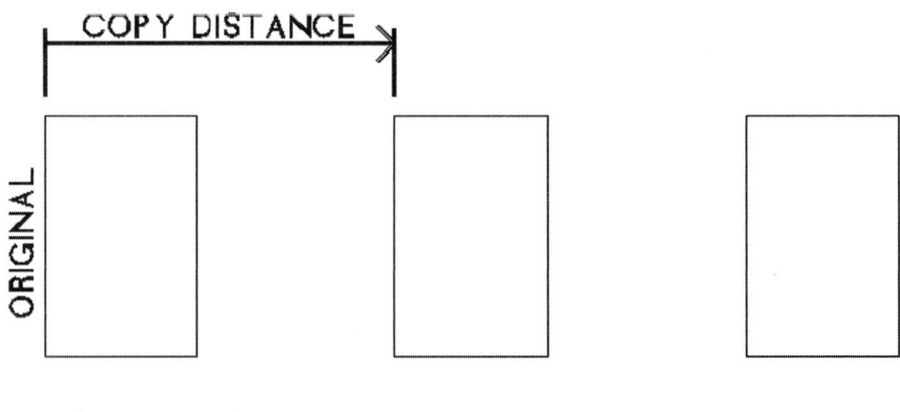

Figure 8-1 An array includes the original entities in the count

your cursor. The rest of these steps assume that you are using the Relative Cartesian input method.

3. Select an X and a Y copy distance by
   a. Watching the X and Y coordinate readouts at the bottom of the screen. When they both equal the X and Y distances desired, click or snap to the point to copy to.

   or

   b. Press the **Space** bar. Here you are prompted to enter the relative X and Y distances. Type in a distance for each, then press **Enter.**
4. You will be prompted to *"Enter number of X repetitions."* After doing so, press **Enter**.
5. You will be prompted to *"Enter the number of Y repetitions."* After doing so, press **Enter**.
6. Select the entities to be copied with one of the EGAFS selection methods.

   The **RectArry** command has two quirks about it:

- The original instance of the entities is counted as one of the members of the array.
- If you are using Polar coordinates, you cannot type in separate X and Y vector distances. The distance and angle you enter in Step 3b earlier is the diagonal distance between the first point and the X/Y intersection. The solution is to simply press the **Insert** key until the current input mode changes to Relative Cartesian. After entering the array, you can press **Insert** to change back to a polar input.

In this example, we used the Relative Cartesian input method and selected an X array of 4 and a Y array of 3 (see Figure 8-2).

Angular This option appears when you select **RectArry**. Angular enables you to do the same thing as a standard Rectangular Array, but the X and Y arrays will be along the angles you enter, rather than being perfectly horizontal and vertical. To use Angular, follow these steps:

1. Select **Edit/Copy/RectArry/Angular**. You'll be prompted to *"Select first point of the distance to copy."*
2. *Click* or snap to a starting point in the drawing window. It does not have to be on or near the entities to be copied. You'll be prompted to *"Select vector indicating angle of array."* You will see a vector arrow

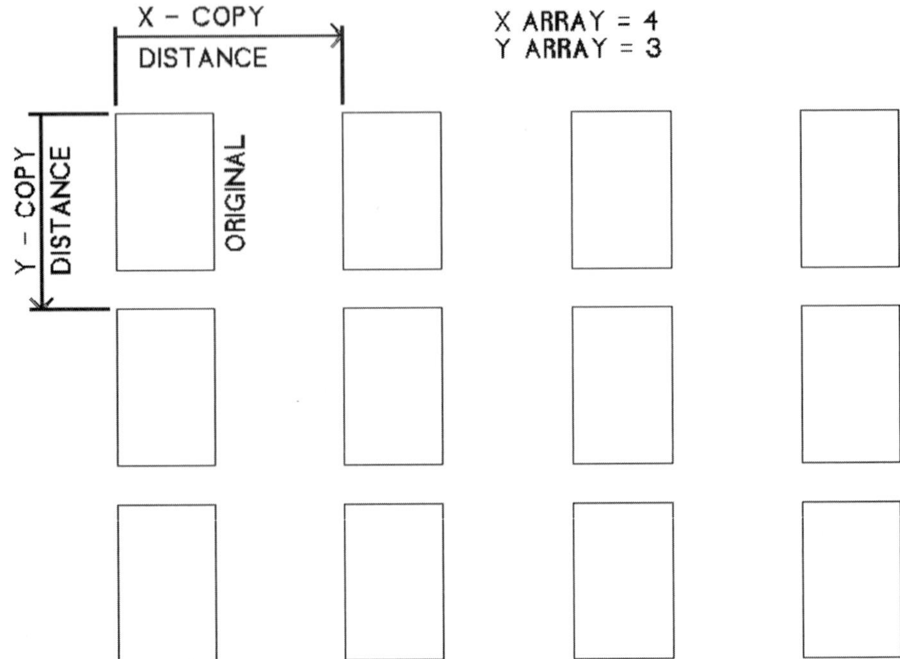

Figure 8-2
An example of a rectangular array

extending from the first point you selected. It moves as you move your cursor.

3. Select a distance and angle for the array angle (using either the Relative Polar or Relative Cartesian input methods). The vector that you just defined establishes the angle along which the X array will be aligned. See Figure 8-3 if you are having trouble following that description. The rest of these steps assume that you are using the Relative Cartesian input method.
4. After entering the vector, you will be prompted to *"Select the second point of the distance to copy."* Select an X and a Y vector distance by
   a. Watching the X and Y coordinate readouts at the bottom of the screen. When they both equal the X and Y distances desired, click or snap to the point to copy to

   or

   b. Press the **Space** bar. Here you are prompted to type in the relative X and Y distances. Type in a distance for each, then press **Enter**.
5. You will be prompted to *"Enter number of X repetitions."* After doing so, press **Enter**.

Figure 8-3
An example of an angular array

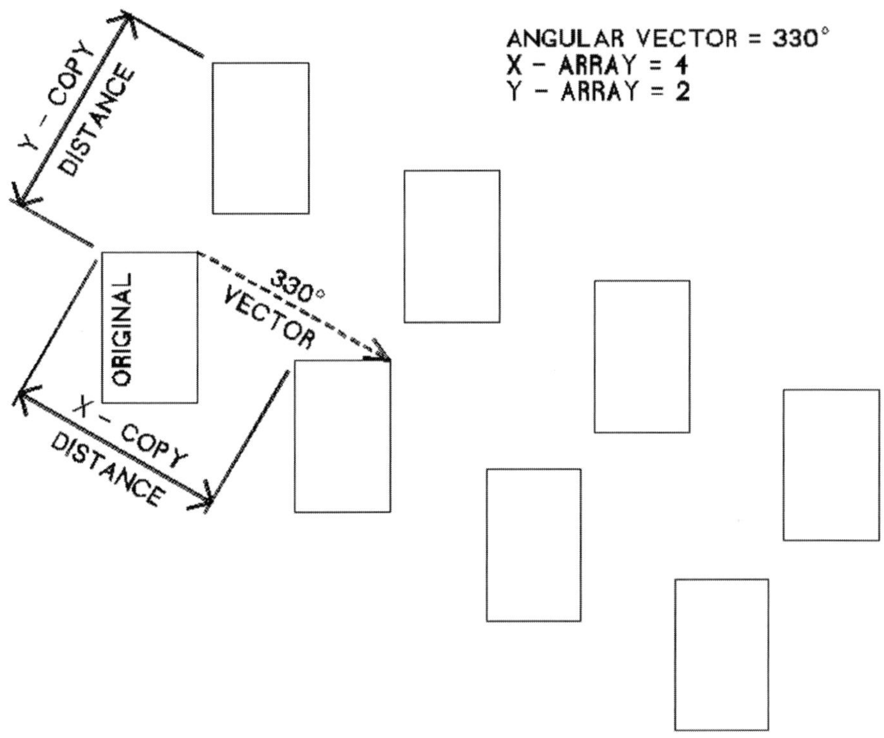

6. You will be prompted to *"Enter number of Y repetitions."* After doing so, press **Enter**.
7. Select the entities to be copied with one of the EGAFS selection methods.

In this example, we began by using the Relative Polar input method to enter an angular vector of 330 degrees. Then we changed to the Relative Cartesian input method to enter the X and Y copy distances, the X array of 4, and the Y array of 2 (see Figure 8-3).

NewArray Select this option to enter the new array settings while remaining in the **Copy** menu.

Z-IncrX Use this option to increment the Z height by a specified amount with each repetition of each entity in the X direction. The increment is reset to 0.0 after each use.

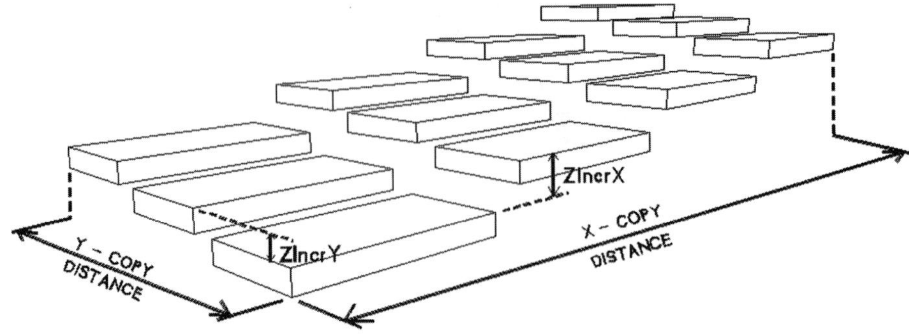

Figure 8-4
An oblique view of a 3D slab copied in a rectangular array with Z-increments

Z-IncrY Use this option to increment the Y height by a specified amount with each repetition of each entity in the Y direction. The increment is reset to 0.0 after each use.

Figure 8-4 is an oblique view example of a 3D slab copied in a rectangular array with both a Z increment in the X direction and a Z increment in the Y direction.

Counter Use this option toggle to turn on and off the display of the status of the copy row and column numbers.

CircArry

Use this option to copy entities a specified number of times around a center of rotation with a specific angle of separation between the copied entities. To use CircArry, follow these steps:

1. Select **Edit/Copy/CircArry**. You will be prompted to select the center point of the array.
2. *Click* or snap on the center point of the array.
3. *Click* on the center of the entities to be copied. You will then be prompted to *"Select the angle between items."*
4. Select or type the angle (in degrees, minutes, and seconds) between the copied entities and then press **Enter**. To copy an array in a

counterclockwise direction, enter a positive angle. To copy an array in a clockwise direction, enter a negative angle.

5. You will be prompted to *"Enter number of objects in array:"* Select or type the number of objects, including the original objects.
6. Select the entities by one of the EGAFS selection methods (see Figure 8-5).

After selecting the number of objects in the array (step 5), the following additional options will appear.

NewCentr This option enables you to select a new center point while still maintaining the previous object center, separation angle, and number of object settings. After selecting **NewCentr**, you will again be prompted to select the entities to copy.

ObjCent This option enables you to select a new object center while still maintaining the previous array center, separation angle, and number of object settings. After selecting **ObjCent**, you will again be prompted to select the entities to copy.

Figure 8-5 An example of a circular array

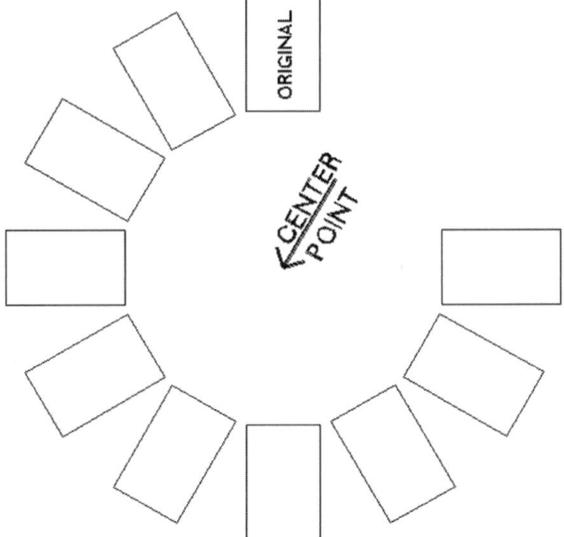

Sep.Angl This option enables you to select a new angle of separation between copied objects while still maintaining the previous array center, object center, and number of object settings. After selecting **Sep.Angl**, you will again be prompted to select the entities to copy.

No.Objct This option enables you to select a new number of copied objects in the array while still maintaining the previous array center, object center, and separation angle settings. After selecting **No.Objct**, you will again be prompted to select the entities to copy.

Z-Incr Use this option to increment the Z height by a specified amount with each repetition of each entity in the array. The increment is reset to 0.0 after each use. This is similar to the **Z-Incr** options of a standard array (refer to Figure 8-5).

Rotate By default, this option is always turned on. If you turn it off prior to selecting the entities to copy, the entities will be copied without changing the X/Y orientation of the entities. The circular array in Figure 8-6 was created with the same settings as the previous array, but with **Rotate** turned off.

Notice that because the selected object center (Step 3 in the CircArray instructions) is at the bottom of the rectangle, that is the point by which the rectangle is rotated, causing the rectangles at the bottom to appear closer to the center point. In actuality, the bottom center of each rectangle is consistently the same distance from the center point.

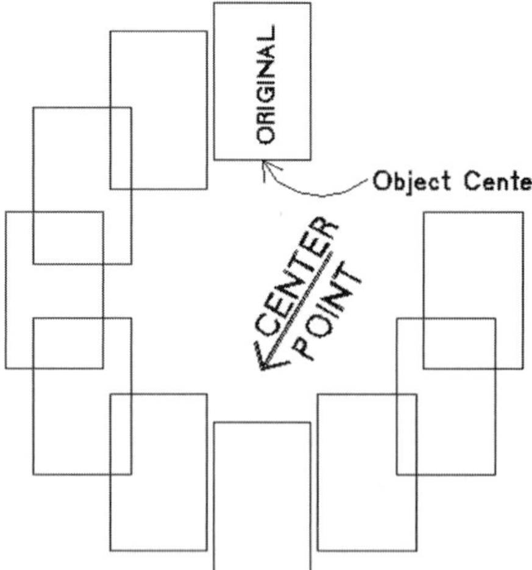

Figure 8-6 A circular array with **Rotate** turned off

PrevDist (Previous Distance)

As stated earlier, **PrevDist** can be used to select the last distance and angle that was entered for the Move, Copy, or Stretch commands, but only one distance and angle are remembered.

Cut/Copy/Paste

Although **Cut**, **Copy**, and **Paste** are standard Windows functions, these features are new to DataCAD. To understand them, let's use the example of a Windows text editor. To cut or copy text, you must first select the text by highlighting it. Once selected, you then pick the Cut or Copy command. The highlighted information is then stored in the Windows Clipboard. When you place your cursor somewhere else in the document and select the Paste command, the information that was saved to the Windows Clipboard is inserted at the location of the cursor.

DataCAD's **Cut**, **Copy**, and **Paste** commands work nearly identically, but they work with all of DataCAD's entities and layers. When you cut or copy and then paste entities from one DataCAD drawing into another, all the selected entities will be pasted into the drawing. If the layers that the entities were on do not already exist in the new drawing, they will automatically be created.

When you use **Cut, Copy** and **Paste** within DataCAD drawings, the pasted entities retain their original groupings (see Chapter 5 for a description of Groups). Contrast this to the standard DataCAD Copy command, which groups all copied entities into one single new group. Each has its advantage.

Select (Ctrl+E)

Before you can cut or copy entities, you must first select them:

1. Press **Ctrl+E** or select **Edit/Clipboard Select** from the drop-down menus.
2. Use the EGAFS selection methods to choose the desired entities. The selected entities will be highlighted. You can make multiple selections and can switch EGAFS selection methods between selections.
3. The selected entities must remain highlighted to use the **Cut** or **Copy** commands.

All selections are placed in a temporary selection set (see Chapter 9, "Editing" for a thorough description). In a nutshell, this allows you to edit or act on your entities as a group. During the selection process, you can add or delete entities from the selection set.

DelFrom (Deleted From) After selecting entities, as described in the three steps previously, all the selected entities are highlighted. You can then use the **DelFrom** option to unselect the highlighted entities, thereby removing them from the temporary selection set:

1. *Click* on **DelFrom**. The button will change to read **AddTo**. The Menu Name box will display the name of the current option: *DelFrom*.
2. Select the entities to be unselected. The entities will be unhighlighted to show that they are no longer selected.
3. Press the **AddTo** button to stop the deletion process and to continue adding entities.

AddTo You only need to select this option after using the **DelFrom** option. Selecting **AddTo** will enable you to continue selecting entities:

1. *Click* on **AddTo**. The button will change to read **DelFrom**. The Menu Name box will display the name of the current option: *AddTo*.
2. Select the entities to be selected. They will be highlighted to show that they are selected.
3. Press the **DelFrom** button to stop the addition process and to allow the deletion of selected entities.

Mask This is another advanced selection feature related to the concept of a selection set (see Chapter 9 for a more thorough explanation). If you have ever done any artwork or even painted your home, you are already familiar with masking. When painting, you use masking tape (now you know where the name comes from) to select areas to be painted, simply by placing the tape in the correct place. Masking enables you to be selective about what you paint. So, masking in DataCAD is a way to be selective about what you edit.

You can mask by Entity, Color, LineType or Weight. To do so, follow these steps:

1. Select **Mask**.
2. Select one of the four masking options:

   a. When selecting the **Entity** option, 11 additional options are available.
   b. When selecting the **Color** option, only the standard 15 DataCAD colors are available for selection.
   c. When selecting the **LineType** option, all the linetypes in the current DCADWIN.LIN file are available.
3. You can select any and all masking options individually or simultaneously. You are not limited to selecting one option at a time.
4. When you are done selecting the masking options, *right-click* to exit the **Mask** menus.
5. Use the EGAFS, **DelFrom**, and **AddTo** selection methods to select entities. Only entities conforming to the current **Mask** settings will be selected. All others will be ignored.
6. If you want to continue selecting entities without the Mask options, simply turn the **Mask** function off by selecting it.

Cut (Ctrl+X)

When you use the **Cut** command, the selected entities are erased from the Drawing Window and added to the Windows Clipboard. Once there, you can paste those entities back into the same drawing, another DataCAD drawing, or even into another Windows program (see the **Paste** command):

1. Use the **Edit/Clipboard Select** option to select the entities to be cut.
2. While the selected entities are still highlighted, select **Edit/Cut** or press **Ctrl+X**.
3. You will be prompted to *"Select reference point for Clipboard Cut." Click* or snap to a reference point (see the following explanation).
4. The selected entities will be erased from the Drawing Window and added to the Windows Clipboard.

The Reference Point In Step 3 above, you are prompted for a reference point. You can pick a random point that is near the selected entities or any distance far away from them. The point you select will be "remembered" along with the entities themselves. If you decide to paste those entities into a DataCAD drawing, this reference point will enable you to define the exact point where the entities are inserted. This point is only relevant within DataCAD and has no bearing on the insertion of those entities into other Windows programs.

If you paste those entities into a DataCAD drawing using the **AbsZero** option, the reference point is ignored, and the absolute zero points of the selected entities and of the target drawing are aligned.

Copy (Ctrl+C)

When you use the **Copy** command, the selected entities remain in their current location in the Drawing Window but are also added to the Windows Clipboard. Once there, you can paste those entities back into the same drawing, another DataCAD drawing, or even into another Windows program (see the **Paste** command):

1. Use the **Edit/Clipboard Select** option to select the entities to be copied.
2. While the selected entities are still highlighted, select **Edit/Copy** or press **Ctrl+C**.
3. You will be prompted to *"Select reference point for the Clipboard Copy." Click* or snap to a reference point (see the previous explanation).
4. The selected entities will remain in place in the Drawing Window and will also be added to the Windows Clipboard.

The Reference Point Refer to the Reference Point discussion in the "Cut" section.

Paste (Ctrl+V)

Use this option to paste information from the Windows Clipboard into the DataCAD Drawing Window. Not all types of information can be pasted into DataCAD; only the following can be inserted:

- *Private vector data* Entities that are native to DataCAD
- *Public metafile data* *Windows Metafile* (.WMF) vector images

- *Public text* Text from a word processor or similar Windows program
- *Spreadsheet data* Text and lines from spreadsheet programs like Excel

All these data types are either vectors or text. You cannot currently import bitmap images, such as photos and paint program images, directly into DataCAD. See the "Importing Bitmaps" section at the end of this chapter for information about some work-around solutions to this shortcoming.

Note that if you copy vector data to the Windows Clipboard from another CAD program, like AutoCAD, some of that data can often be successfully pasted into DataCAD. Some information, most notably text and complex entities, can get muddled in the process. Also, that data will be pasted only into the currently active layer; the original layer structure will not be retained. How well data transfers into DataCAD depends on how the other CAD program saves its data to the Clipboard. For instance, in our testing, we found that IntelliCAD data is saved to the Clipboard differently than AutoCAD. IntelliCAD data brought into DataCAD is improperly scaled and gives some strange results for curves and vertical lines. For information specifically related to pasting text and spreadsheet data, see Chapter 11, Basic Construction Drawings.

Pasting Private Data Here are the steps for pasting private (DataCAD only) data. This would be data that is cut or copied from a DataCAD drawing and then pasted back into another DataCAD drawing.

1. Select **Edit/Paste** or press **Ctrl+V**. A bounding box representing the outside extents of the data will be displayed in the Drawing Window. You may need to use **PgUp** to see the full extents of the bounding box, depending on its size and its proximity to the center of your cursor.
2. If desired, select **AbsZero**, **OrigLyrs**, or **ActivLyr**.
3. *Click* or snap the cursor to paste the clipboard data into the Drawing Window.

AbsZero This option inserts the clipboard data so that the entities are placed with the absolute zero point of the copied entities and the absolute zero point of the target drawing aligned. If you select this option, the data insertion point that you selected with the **Copy** or **Paste** commands will be ignored.

OrigLyrs When you cut or copy DataCAD data, the selected entities remain on their original layers by default. If those layers do not already exist in the current drawing, then they will be created for you. Use the **OrigLyrs** option to paste the clipboard data using their original layers.

ActivLyr This option can be used to paste all of the clipboard data onto the currently active layer only. All the original layers will be ignored and will not be created in the current drawing.

Pasting Public Data The steps for pasting public data (from Windows programs outside of DataCAD) are the same as for private DataCAD data, except that some different options are available:

1. Select **Edit/Paste** or press **Ctrl+V**. A bounding box representing the outside extents of the data will be displayed in the Drawing Window. You may need to use **PgUp** to see the full extents of the bounding box, depending on its size and its proximity to the center of your cursor.
2. If desired, select **Flip**, **Enlarge**, or **FontName**.
3. *Click* or snap the cursor to paste the clipboard data into the Drawing Window.

Flip Selecting this option will cause the clipboard data to be displayed mirrored (in the case of vector data) or in their opposite order (in the case of text; see Chapter 11, "Basic Construction Drawings," for more on this).

Enlarge Select this option to paste the clipboard data at a larger size. Choose an enlargement factor prior to placing the entities.

FontName Select this option to change the current font name for all text to be pasted. If you don't select a new font, the current font will be used.

Paste Special

This option enables you to choose how the clipboard data is pasted into DataCAD. If you are pasting only text, then the only option available will be the Text option. But if you are pasting information like an Excel spreadsheet, you will have the option to paste only the text from the spreadsheet or to paste all the data, including the visible lines that make up the spreadsheet table via the **Picture** (**Windows Metafile**) option. For information specifically related to pasting text and spreadsheet data, see Chapter 11.

Pasting Private Data No difference exists between how private DataCAD data is pasted with the **Paste Special** or the **Paste** commands, other than the dialog box in Figure 8-7.

To use the Paste Private Data command, follow these steps:

1. The **DataCAD Clipboard Data** line will be highlighted. Select **OK** to accept it.
2. A bounding box representing the outside extents of the data will be displayed in the Drawing Window. You may need to use **PgUp** to see the full extents of the bounding box, depending on its size and its proximity to the center of your cursor.
3. If desired, select **AbsZero**, **OrigLyrs**, or **ActivLyr**.
4. *Click* or snap the cursor to paste the clipboard data into the Drawing Window.

See the **Paste** command for information about **AbsZero**, **OrigLyrs**, and **ActivLyr**.

Pasting Public Data The steps for pasting public data are the same as for private DataCAD data, except some different options are available:

Figure 8-7 The `Paste Special` dialog for pasting private data

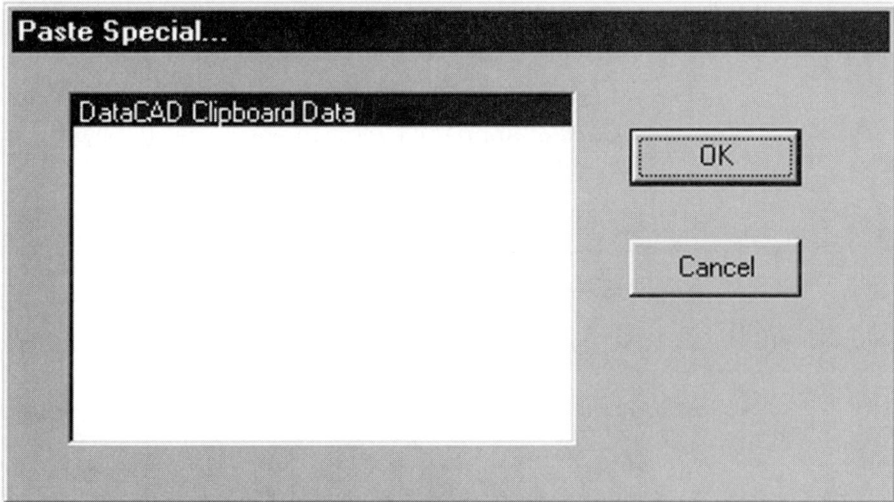

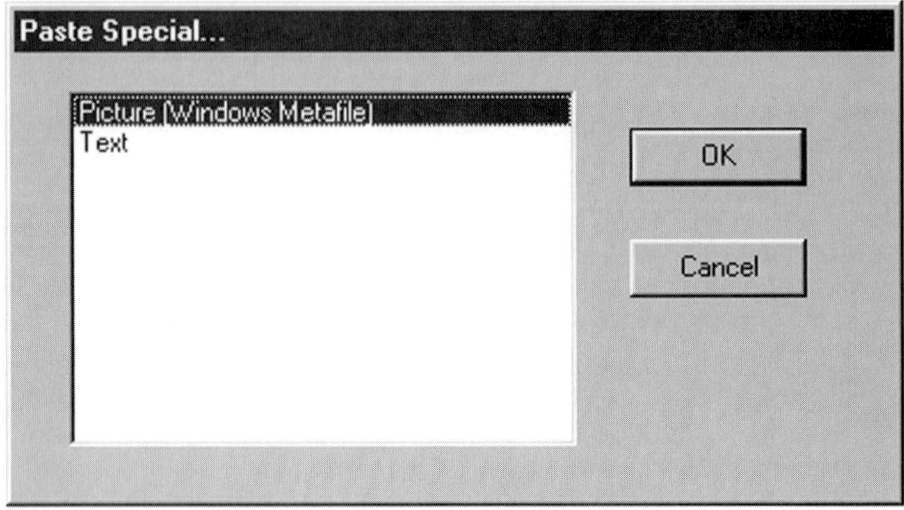

Figure 8-8 The **Paste Special** dialog for pasting public data

1. Select **Edit/Paste Special**. You will see a dialog box, as shown in Figure 8-8.
2. Highlight **Picture** (**Windows Metafile**) if you want to paste the most comprehensive data (vector entities and text). Select **Text** if you only need to paste the text data.
3. Click **OK** to accept the selected option.
4. A bounding box representing the outside extents of the data will be displayed in the Drawing Window. You may need to use **PgUp** to see the full extents of the bounding box, depending on its size and its proximity to the center of your cursor.
5. If desired, select **Flip**, **Enlarge**, or **FontName**.
6. *Click* or snap the cursor to paste the clipboard data into the Drawing Window.

See the section on the **Paste** command for information about **Flip**, **Enlarge**, and **FontName**.

Cut/Copy/Paste Settings

When you select **Tools/Program Preferences/Misc** from the drop-down menus, one option will read "Copy to Clipboard." The settings under this option determine how entities are copied to the Windows Clipboard when you use the **Cut** and **Copy** options. Both options are selected by default. If you unselect **Copy entity color**, then all entities will be copied to the clipboard as black. If you unselect **Copy entity line weight**, then all entities will be copied to the clipboard with the same thin line weight (thickness).

Mirror

Use the **Mirror** options to flip entities about a selected line of reflection. You can leave the original entities in place or mirror them into a new position without leaving the original entities behind. Notice that the SUN ROOM text is not mirrored (see Figure 8-9).

The steps for using the Mirror option are as follows:

1. Select **Edit/Mirror**. You will be prompted to *"Select first point along the line of reflection."*
2. Select the first point by clicking the mouse. You will then be prompted to *"Select the second point along the line of reflection."*

NOTE: *Although the line of reflection you create is a phantom line that disappears after the mirror command is completed, to define it you can use all the standard DataCAD options for drawing lines, such as using the Reference Point option (`), or pressing the* ***Space*** *bar and typing the distance and angle. The length of the line of reflection only needs to be greater than 0.*

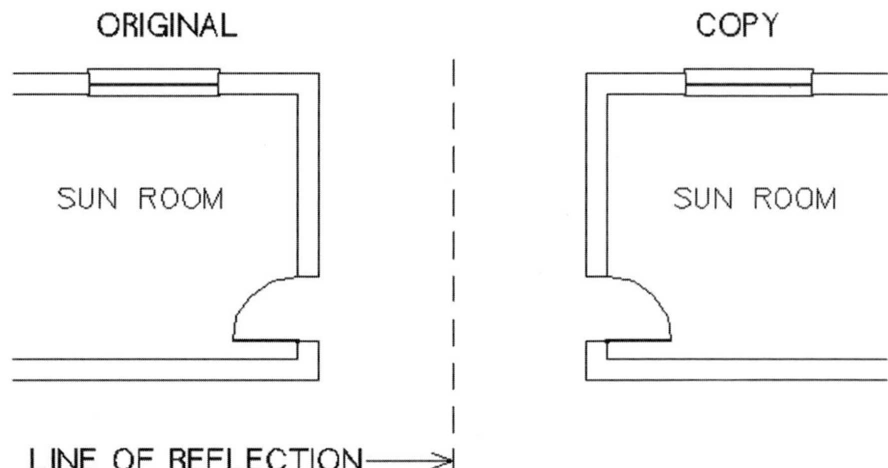

Figure 8-9
The **Mirror** option does not flip the text.

3. *Click* to select the second point of the line or press the **Space** bar to type in the parameters of the line.
4. Select **FitText** and/or **AndCopy**, if desired.
5. Use one of the EGAFS selection methods to select the entities to be mirrored.

NewLine

Use this option to select a new line of reflection without exiting the **Mirror** menu.

FixText

With this option to select a new line of reflection without exiting the **Mirror** me this option turned on, only the position of the text is mirrored, not its orientation. With this option turned off, text will be mirrored and rotated 180 degrees (so it will read upside down and backwards). Note, however, that this only applies if the resulting mirrored text will not be displayed as reverse-reading (see Figures 8-10 and 8-11).

AndCopy

With this option turned off, the selected entities will be mirrored and erased from their original location. With this option turned on, the selected

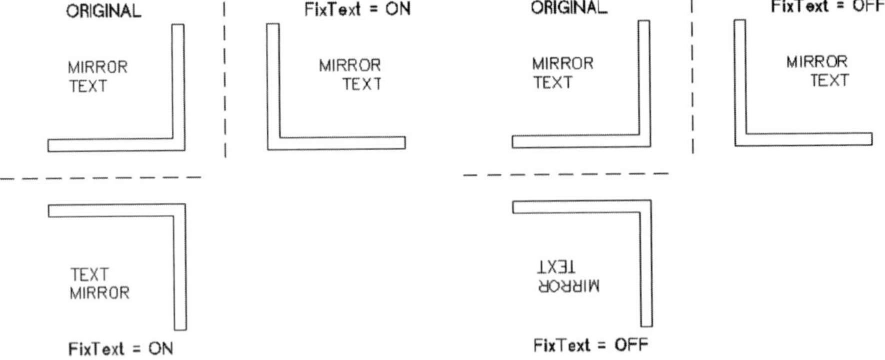

Figure 8-10 Horizontal text mirrored with **FixText** on and off

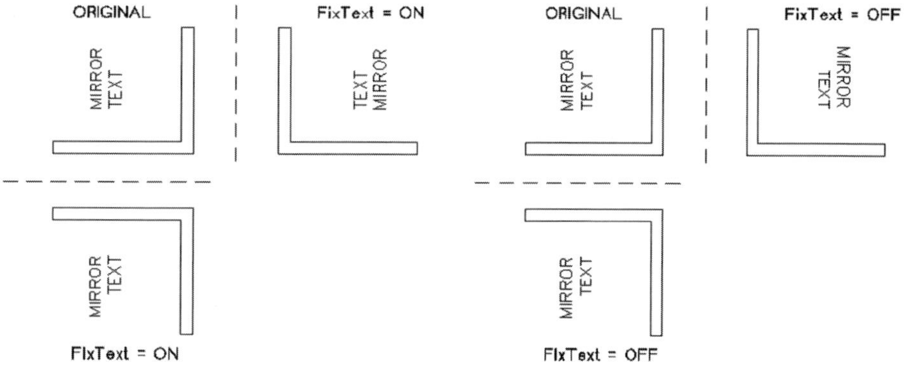

Figure 8-11 Vertical text mirrored with **`Fix Text`** on and off

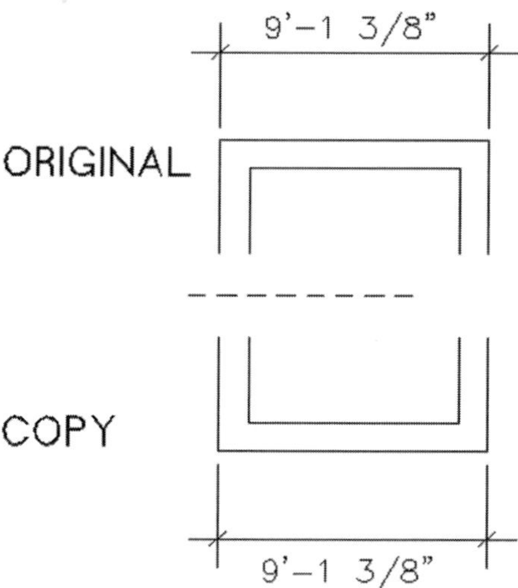

Figure 8-12 A mirrored dimension will relocate the text.

entities will remain in their original location, while a copy of the entities will be mirrored.

Mirroring Dimensions

If the text of your associative dimension is located over the dimension line, when you mirror the dimension, the text will be mirrored to the bottom of the dimension line, as shown in Figure 8-12.

To move the text back to the top of the line, perform the following steps:

1. Select **Utility/Dmension/Linear/Change/TxtPostn**.
2. Select the dimension to change. The dimension text will change to a bounding box attached to the cursor.
3. Locate the text over the top of the dimension line and click the mouse to place it.

Mirroring Symbols

If you **Explode** a symbol after it has been mirrored, DataCAD often mirrors the symbols along the wrong axis, so that the symbol may not actually wind up in the proper orientation. For more information about this, see the "Mirrored Symbols" section of Chapter 14, "Templates and Symbols"

Change/Z Heights

Although these options are located under the **Change** menu, we describe them here because they deal with moving entities, which in this case is in the third, or Z, dimension.

These options only apply to 2D entities, not to 3D entities. Remember that all of DataCAD's 2D entities can have a Z height. If you change both the Z base and the Z height of an entity by the same value, then you will have changed the Z height of the entire entity. For instance, let's say you have an original entity with a Z base of 0′-0″, and a Z height of 8′-0″ [2438]. If you change the Z base to 2′-0″ [610] and the Z height to 10′-0″ [3048], you will have moved the entity by 2′-0″ [610] in the Z dimension.

These options are also used to change the height of the entities themselves. Let's continue with the previous example. If the original entity has a Z base of 0′-0″, and a Z height of 8′-0″ [2438], the entity is 8′-0″ [2438] high. If you leave the Z base at 0′-0″ but change the Z height to 10′-0″ [3048], you will have increased the height of the entity by 2′-0″ [610] in the Z dimension so that it is now 10′-0″ [3048], in height.

The steps to Change Z Heights are as follows:

1. Select **Change**.
2. If you want to change the Z base, then select the **Z-Base** option:

   a. You will be prompted to *"Enter the new Z base."* Select or type a value and then press **Enter**.
   b. If you decide that you really didn't want to change the Z base, you can simply turn the **Z-Base** option off by selecting the button a second time.
3. If you want to change the Z height, then select the **Z-Hgt** option:
   a. You will be prompted to *"Enter new Z height."* Select or type a value and then press **Enter**.
   b. If you decide that you really didn't want to change the Z height, you can simply turn the **Z-Hgt** option off by selecting the button a second time.
4. You will then be prompted to *"Select entity to* <CHANGE>." Use one of the EGAFS selection methods to select the entities to change.

Z-Base

Use this to change the Z base of a 2D entity.

Z-Hgt

Use this to change the Z height of a 2D entity.

Match

Use this to change an entity to match the Z base and Z height of another existing 2D entity:

1. Select **Edit/Change**.
2. Select **Z-Base** to turn it on and then *right-click* (you do not need to select a value).
3. Select **Z-Hgt** to turn it on and then *right-click* (you do not need to select a value).
4. Select **Match**. You will be prompted to *"Select the entity to* <CHANGE>."
5. Use one of the EGAFS methods to select the entities to be changed.

6. You will then be prompted to *"Select the entity to match."*
7. Select the entity whose Z base and Z height you want to match. The entities selected in step 5 will have their Z dimensions changed to match the entity selected in Step 6.

Geometry

Although four basic options can be found in the **Geometry** menu, in this chapter we describe here only the **Divide**, **Intrsect**, and **Offset** commands. For a description of the **Tangents** option, see the section "Working with Rotated Plans" in Chapter 11. If you are wondering why we cover the **Divide** and **Intrsect** commands in this chapter, there are two good reasons: 1) these options are often used as a basis for moving or copying entities, and 2) we couldn't find any other place for them.

Divide

Use Divide to add equally spaced, non-printing snap points along a line, arc, or circle or between two specified points. The snap points remain visible entities in the drawing unless you **Erase** them, but they will not print. These points can then be snapped to, just like any other snappable point on an entity (such as an endpoint, midpoint, intersection, and so on).

The steps for using Divide are as follows:

1. Select **Utility/Geometry/Divide**. The Message Line will tell you the number of divisions currently selected. If you don't need to change this setting, skip to Step 3.
2. To change the number of divisions, select **Divisions**, select or type the new value, and then press **Enter**.
3. If desired, select **GoldSect** to divide the entity or distance into two unequal segments (see the following description).
4. Select the entity to be divided or select two points to divide between
   a. Select **Entity** and then click on the line, arc, or circle to be divided

   or

   b. *Click* or snap to define the first point and then click or snap to define the second point.

Divisions Use this option to set the number of divisions. Select or type a value and then press **Enter**.

Select Two Points By default, you are prompted to select two points to divide between. *Click* or snap to select a first point and then a second point, and the distance between will be divided according the number of divisions currently set (see Figure 8-13).

Entity After setting the number of divisions, select the Entity option if you want to equally divide a 2D line, arc, or circle by that number. The selected entity will be divided equally between the two endpoints (for a line or arc) or along the entire circumference of a circle. In each of these examples in Figure 8-14, the number of divisions was set to 6.

Notice that the first point of the division of the circle is located at the exact point where we selected the circle, so if you want the divisions to start at a particular point, make sure you snap to that exact point.

Div+Dist (Divide by Distance) Use this option to create snap points at a fixed distance and at a specified angle. Note that this can only be done

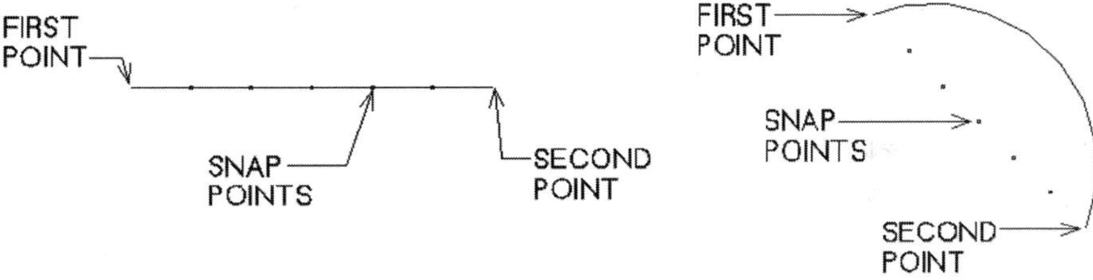

Figure 8-13 Six divisions between two selected points

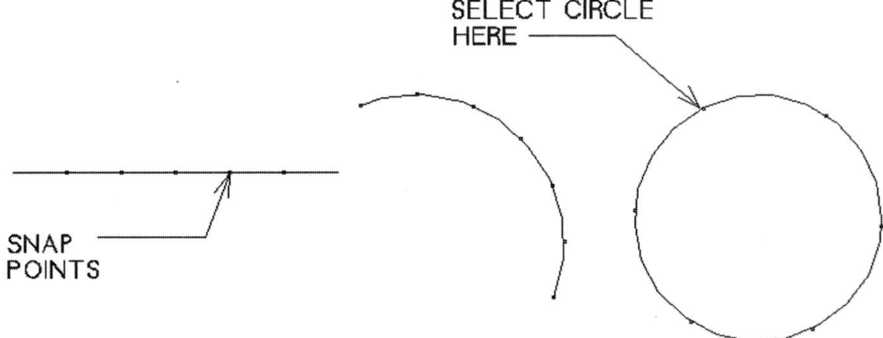

Figure 8-14 Six divisions by selecting the entities themselves

over a straight line distance, and not around a curve or arc. To use **Div+Dist**, follow these steps:

1. Select **Div+Dist**. You will be prompted to *"Enter distance between division points:"*
2. Select or type the distance between division points and then press **Enter**.
3. You will be prompted to *"Select first point." Click* or snap to select the point to start from.
4. You will then be prompted to *"Indicate direction of divisions."* Then you must either
   - **a.** Move the cursor in the direction to create the snap points and then *click* the mouse

   or
   - **b.** Press the **Space** bar, enter a distance (anything greater than zero), and then enter the angle along which the snap points will be drawn.

In this example, we specified a 4′-0″ [1219] distance between points and entered an angle of 0 degrees after selecting the first point to start from (see Figure 8-15).

GoldSect (Golden Section) Use this option to divide a line, or two selected points, into two unequal segments, at a ratio of 1 to 1.618. The ancient Greeks considered this geometric proportion to be the most harmonic of geometries. You can select a line by selecting two points or by selecting an existing line with the **Entity** option (see Figure 8-16).

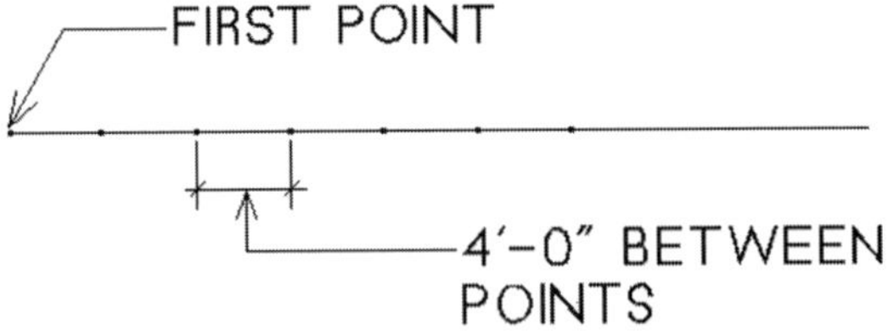

Figure 8-15
Six divisions at 4′-0″, at 0 degrees

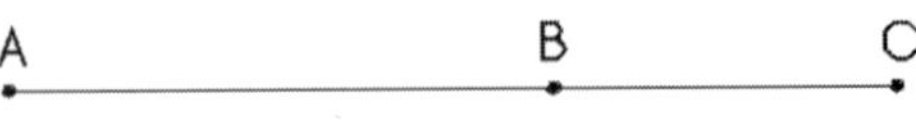

The ratio of AB : AC = 1 : 1.618

Figure 8-16
Dividing a line by the Golden Section

Intrsect (Intersection)

Use this option to place a non-printing snap point at the inferred intersection of two lines. The snap point remains visible in the drawing unless you **Erase** it, but it will not print. Since parallel lines will never intersect, you can only select two lines that are not parallel to one another. To use Intrsect, follow these steps:

1. Select **Utility/Geometry/Intrsect**.
2. Select the first line.
3. Select the second line. A snap point will be added at the intersection of the two lines (see Figure 8-17).

Offset

Use this option to make a parallel copy of a line, arc, circle, or ellipse at a specified perpendicular distance. Besides simple straight lines, DataCAD is very good at offsetting linked curves, as shown in Figure 8-18. Notice how the two original curves are copied so that the offset curves are likewise perfectly linked. This is great for things like sidewalks and streets.

Dynamic Use this to offset an entity by dynamically dragging a copy of the entity with your cursor:

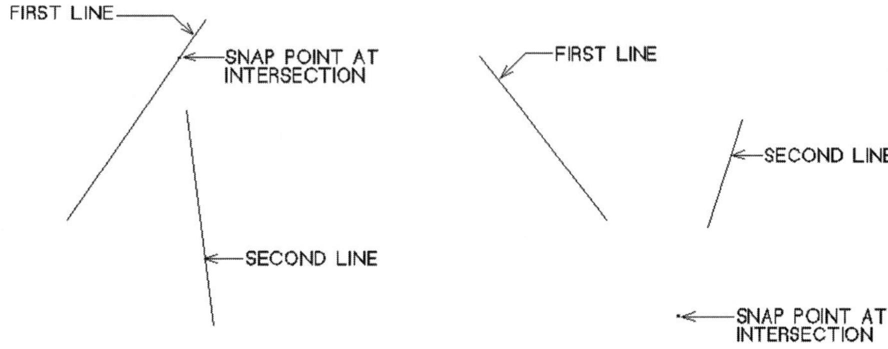

Figure 8-17 Two examples of intersection snap points

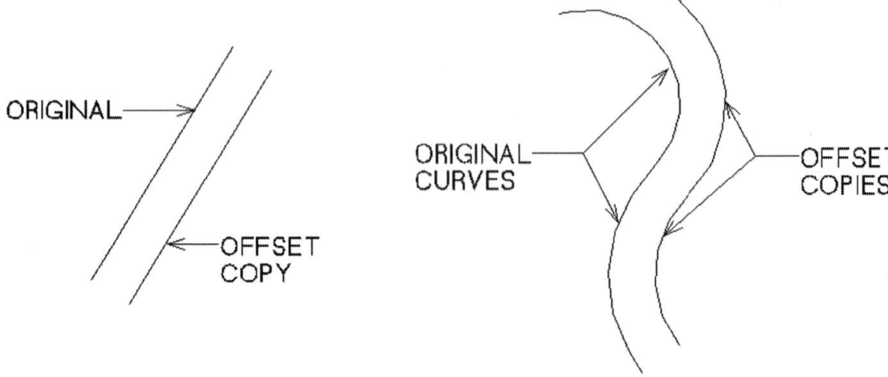

Figure 8-18 Two examples of offset entities

1. Select **Utility/Geometry/Offset/Dynamic**. You will be prompted to *"Select object to offset."*
2. Select the entity to be offset. A copy of the entity will be attached to your cursor.
3. Move the entity to its new location and *click* the mouse to place it.

While attached to your cursor, the offset entity will only move in a direction perpendicular to the original entity. In the case of a curve or circle, this means that the center point of the new offset curve will be the same as the center point of the original curve.

PerpDist This option is only available when **Dynamic** is turned off. Use it to offset an entity by a specific perpendicular distance:

1. Select **Utility/Geometry/Offset/PerpDist**.
2. Select or type the distance to offset by and then press **Enter**.
3. Select the entity to be offset. You will be asked, *"Offset to which side?"*
4. Place the cursor anywhere on the side of the entity to copy to and *click* the mouse. The entity will be offset to that side, by the distance entered in Step 2.

NewDist This option is only available when **PerpDist** is selected. Use it to input a new offset distance without exiting the **PerpDist** menu.

Rotate

Use this option to turn an entity about a center point at a specified angle:

1. Select **Edit/Rotate**. You will be prompted to *"Select center of rotation."*
2. *Click* or snap to select the center point of the rotation.
3. The current rotation angle is displayed in the Message Line:
   a. Select **Dynamic** if you want to dynamically rotate the selected entities with the cursor

   or

   b. To change the rotation angle, select **NewAngle**, then select or type a new angle, and press **Enter**.
4. Turn **AndCopy** on if desired.
5. Use one of the EGAFS selection methods to select all the entities to rotate.
6. If you selected Dynamic in step 3, move the entities with your cursor and then *click* the mouse to place them (see Figure 8-19).

NewAngle

Use this option to select a new angle of rotation without exiting the **Rotate** menu.

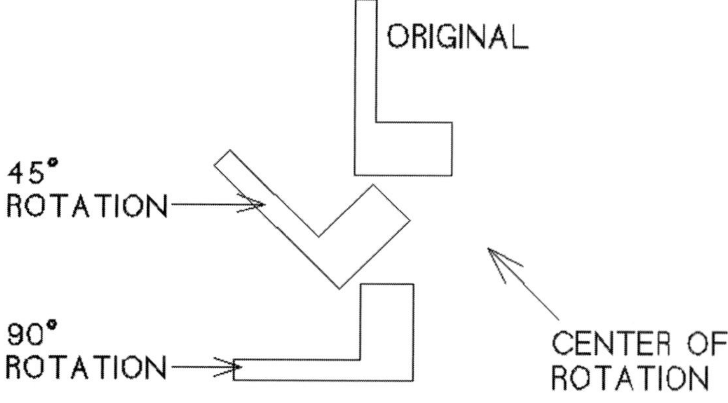

Figure 8-19 Examples of rotations about a center point

Invert

Use this option to select the opposite angle from the current angle. For instance, if the current rotation angle is 90 degrees, selecting **Invert** will change the rotation angle to 270 degrees.

NewCentr

Use this option to select a new center of rotation without exiting the **Rotate** menu.

PrevCent Use this option to select the previous center of rotation.

AndCopy

Use this option to leave the original entities in place while creating a copy of those entities at the specified angle of rotation.

Dynamic

Select this option to dynamically rotate the selected entities with the cursor, rather than entering a specific angle. After selecting the entities to be rotated dynamically, you will be prompted to *"Select the point to rotate by."* Refer to the information on **Copy/Rotate** earlier in this chapter for an explanation of the significance of selecting this point.

MaxLines Use this option to select the maximum number of entities to be displayed while rotating entities with the **Dynamic** option turned on. Depending on the speed of your computer, having too many entities displayed during dynamic rotation could slow down the rotation process considerably.

Multi Select this option prior to picking one of the EGAFS selection methods if you need to select two or more entities that cannot be selected at once by one of the EGAFS options. Once you begin selecting entities, you can switch the EGAFS selection methods to continue picking more entities.

Begin After selecting all the entities to Rotate, select **Begin** to continue the rest of the rotation process.

Stretch

Think of the **Stretch** command like Silly Putty®. Once formed into a shape, it can be stretched in any direction to make the shape longer or shorter, or to change it into a new shape. A square can be made into a rectangle by stretching one side, or it can be made into a trapezoid by stretching one or more corners (see Figure 8-20).

Stretch is a natural with DataCAD's associative dimensions and associative hatching. If a horizontal associative dimension displays a distance of 3′-0″ [914] and you **Stretch** it 1′-0″ [305] to the right, the leader and the arrowhead will be stretched 1′-0″ [305] to the right, and the dimension text will update itself to show that the dimension is now 4′-0″ [1219] long. Associative hatching will likewise update the way it displays when stretched (see Figure 8-21).

You can only stretch linear entities (lines, dimensions, and so on) and polylines. You cannot stretch arcs or circles.

To use Stretch, follow these steps:

1. Select **Edit/Stretch**. You will be prompted to *"Select first point of the distance to stretch."*
2. Select any point in the Drawing Window. You do not have to select a point on or near the entities to be moved.
3. A line representing the stretch vector is extended from the first point you picked to the cursor. You are then prompted to *"Select second point of the distance to stretch."*
4. To locate the second point to stretch to, you can either

   a. *Click* or snap to the point to stretch to

   or

   b. Press the **Space** bar. Enter a distance and angle, then press **Enter**.

Figure 8-20 The result of stretching the right side of a square

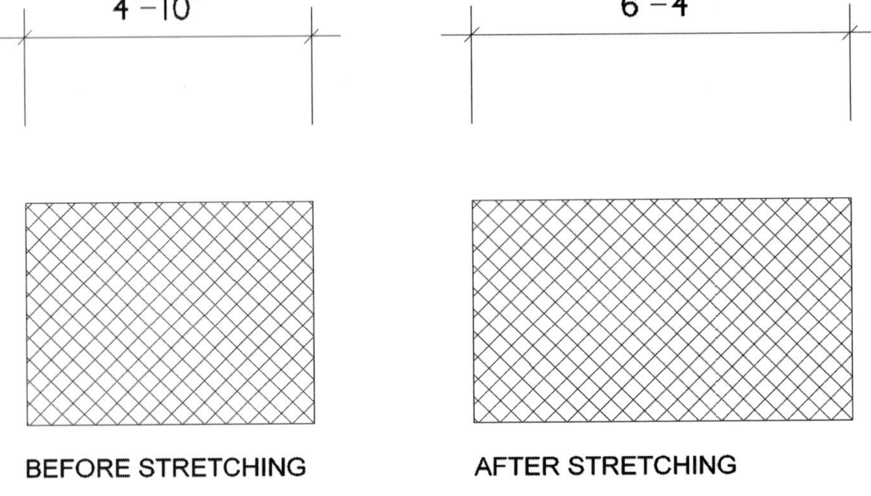

Figure 8-21 Associative hatching and dimensions after stretching

5. Select the entities to be stretched with the **Point**, **Area** or **Fence** options.

Point

Use this option if you only want to stretch a single point such as the end of a line or one polyline vertex. If two valid entities share a common point, both will be stretched (see Figure 8-22).

NewDist (New Distance)

Use this option to select a new distance to stretch without exiting the **Stretch** menu.

Invert

Use this option to select the opposite angle from the current angle. The previous distance will be reused. For instance, if the current stretch angle is 90 degrees, and the distance is 5′-0″ [1524], selecting **Invert** will change the stretch angle to 270 degrees at a distance of 5′-0″ [1524].

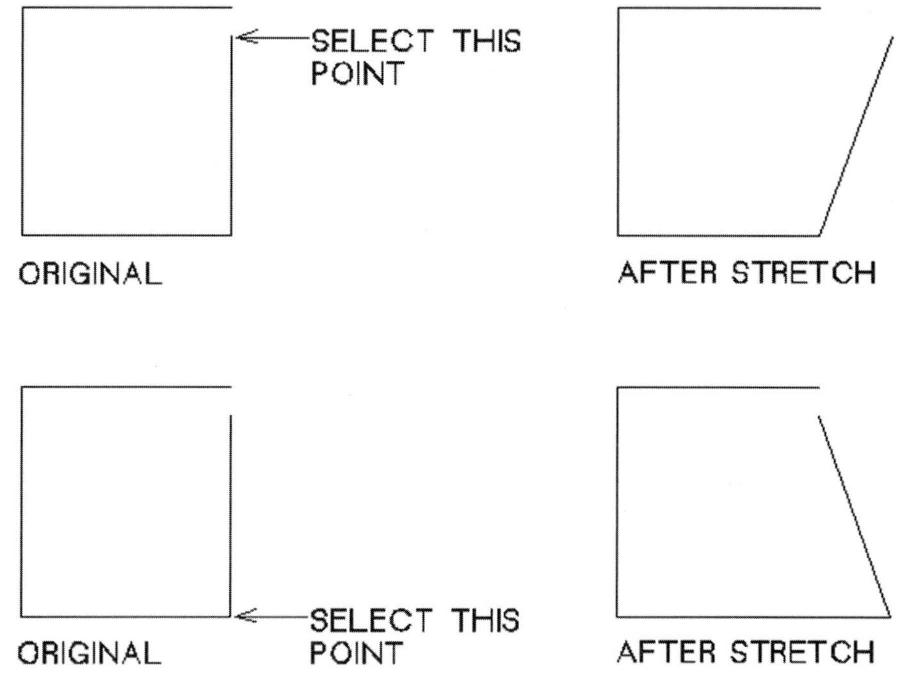

Figure 8-22
Two typical stretching scenarios

PrevDist (Previous Distance)

As stated earlier in the chapter, **PrevDist** enables you to select the last distance and angle that was input for either the **Move, Copy** or **Stretch** commands, but only one distance and angle are remembered.

Stretch Point Macro

This inexpensive third-party macro (available from Cheap Tricks Ware, as discussed in Chapter 27, "DataCAD Resources,") is an excellent supplement to the standard DataCAD **Stretch** options and is highly recommended. It enables you to selectively stretch the endpoint of a line when two or more entities share a common point, such as the corner of a wall in a plan (refer to Figure 8-22). The standard DataCAD **Stretch/Point** option will not enable you to stretch only one of the entity endpoints, but this macro will.

Exporting Bitmaps

Exporting bitmaps and other graphic images from DataCAD to other Windows programs can be done in several ways.

Export Part of the Drawing Window

To save a portion of the image in the current Drawing Window to the Windows Clipboard, where it can then be pasted into another Windows program, perform these steps:

1. Select the entities to be copied with **Edit/Clipboard Select** (**Ctrl+E**) from the drop-down menu.
2. Select **Edit/Copy** (**Ctrl+C**) or **Edit/Cut** (**Ctrl+X**) from the drop-down menu.
3. Select a reference point. Even though this is not required for exporting, it is still a DataCAD requirement to proceed.

The selected entities are now temporarily saved to the Windows Clipboard. To proceed, open the Windows program you want to paste the image into and use the **Paste** or **Paste Special** command in that program to paste the DataCAD image. The image will be in the .WMF format. See the "Cut/Copy/Paste" section of this chapter for specific information about the cut and copy process.

Export the Entire Drawing Window

To export a *Windows bitmap* (.BMP) image of everything currently displayed in the Drawing Window, perform the following steps:

1. Select **Window/Save as Bitmap** from the drop-down menu. A "Save bitmap file as:" dialog box will appear.
2. Navigate to the folder where you want to save the .BMP image. By default, it will be saved to the DataCAD \BMP folder.
3. Type a name for the image and then press **Save**.

Open the Windows program you want to paste the image into and use the **Insert** command (or whatever other command that program uses) to paste the saved .BMP image.

Export a QwkLyout or MSP Sheet

Chapter 7, "Printing and Plotting," describes the new **Plotter/Preview** feature that enables you to save your drawings in a .WMF format. Because the .WMF format is fairly universal, the resulting image can then be pasted into almost any other Windows program.

Export a GL Shader Image

The 3D shaded image that results from using **3DEdit/Shader/Shade**, or **View/Shade** from the drop-down menu, can be saved as a .BMP file. After shading your model, follow these steps:

1. Place the cursor in the Drawing Window and *right-click* the mouse. The dialog box shown in Figure 8-23 will appear.
2. To save the entire Drawing Window:
   a. Select **Save bitmap** to save the image to the DataCAD \BMP folder.
   b. Select **Save bitmap as** to navigate to a different folder to save the image
3. Type a name for the image and then press **Save**.

Save As PDF

This is actually a method for saving a vector file, rather than a bitmap file, but it is an important export option that you should be aware of. Chapter 7

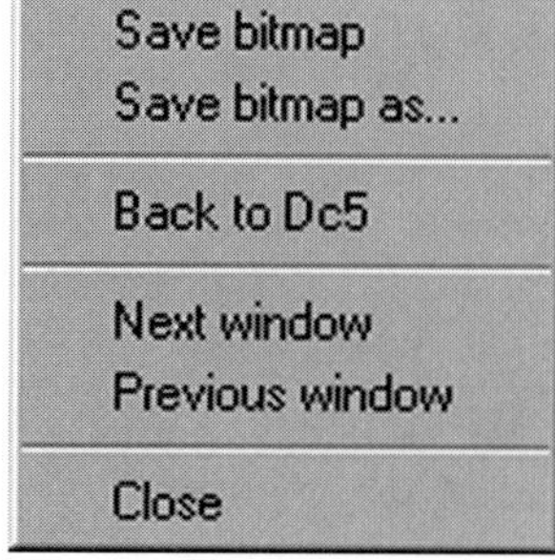

Figure 8-23
The **Save bitmap** options

describes the new **Plotter/Preview** feature that enables you to save your drawings as an Adobe Acrobat .PDF (portable document format) file. .PDF is a universal file format that preserves all the fonts, formatting, colors, and graphics of any source document, regardless of the application and platform used to create it.

Importing Bitmaps

As mentioned earlier, you cannot currently import bitmap images directly into DataCAD, but some work-arounds can help deal with this shortcoming. They all take some time and effort, but they work, and some better than others.

In all the methods described in the following section, the common thread is that they all start with a bitmap image of some type. If all you have is a piece of paper with an image on it, you will first need to use a scanner to scan the image. None of the following methods will enable you to import the colors of an image, so when you scan it, use the Black & White, Gray Scale, or Line Art options in the scanning software. This will provide a better scan and will result in a smaller image file.

Method #1: PCX to DXF

A free conversion utility has been created (which is included on the CD-ROM at the back of this book) to convert a .PCX bitmap image into a .DXF vector image, which can then be imported into DataCAD. If the image you have is in a graphic format other than .PCX (like .BMP, .GIF, .JPG, and so on), any graphics program can convert it to a .PCX format. Once done, you can run the .PCX image through the PCX2DXF.EXE program (follow the directions in the Readme file that comes with the program). Keep in mind that since this is a DOS program, the .PCX file must be limited to a maximum of eight characters and should not contain any spaces or special characters. The resulting image will look something like Figure 8-24 when imported into DataCAD (**File/Import**).

Notice that the image is actually made up of hundreds of horizontal lines. Even the verticals are made up of horizontal lines. If this is all you need, then you could use the image as is, but if you need something else, then you will have to create a new layer in the drawing file and trace the image. Yes, this can be a tedious process, but it does work.

Figure 8-24 A bitmap converted to a .DXF format and imported into DataCAD

More than likely, the imported image will not be at the correct scale. If you are only importing a logo, then this will make little difference, but if you are importing something like a floor plan, you will have to use the **Edit/Enlarge** option to enlarge it to the proper scale. This process can be made easier if you import a graphic scale with the original image prior to running it through the PCX2DXF.EXE program.

Method #2: Corel Draw and Adobe Streamline

Several programs on the market can do the job of tracing a bitmap image for you and can then save the result as an *encapsulated postscript* (.EPS) vector file. That file can be converted to a .DXF vector file, which can then be imported into DataCAD using **File/Import**. Two such programs are *Corel Draw* and *Adobe Streamline*. Here we'll discuss how the *Corel Draw* program works. *Adobe Streamline* works in a similar manner.

Within the *Corel Draw* suite of functions is one called *Corel Trace*. Even very early versions of *Corel Draw* such as version 3.0 have this function and are available at rock bottom prices. Its Centerline option can be used to force the program to trace a single line down the center of each line in the image. Otherwise, a line will be traced on both sides of each line in the image. The resulting image will be in an .EPS vector file format. Now you can export the image in .DXF format and import it into DataCAD.

The resulting vector image, once imported into DataCAD, is often not much more useable than the scanned image imported with the PCX2DXF.EXE program. To get a good, usable vector image with parallel and perpendicular lines, you will probably still have to trace the imported image in DataCAD.

Method #3: Let Someone Else Do It

This is my favorite method because it usually yields the best results and takes practically none of my time. Here all you have to do is to check your local phone book for Computer Scanning or Printing services and look for someone who will scan your image. You need to find a service that can convert a scanned, bitmap image into a vector format image, such as a .DXF or a .DWG file, which can then be imported into DataCAD. For small images such as $8^1/_2'' \times 11''$ [216 × 280], the cost is usually very little. For larger sheets like an old 30″ × 42″ [762 × 1066] floor plan, the cost may be a couple of hundred dollars, but we have always found this to be more cost-effective than the time it takes to redraw the entire sheet by hand. We also consider it a reimbursable cost and charge it back to the client.

Sticky Back Macro

This third-party macro is part of a larger macro called DC Sprint, which you can read about in Chapter 24, "Third-Party Macros." It was created long before the new Cut/Copy/Paste options and enabled you to do many of the same things. However, it also allowed you to permanently save the entities that were cut and copied for use at a later time. In that respect, the macro acts much like DataCAD's templates and symbols, but all selected entities are saved on their current layers (while symbols cannot save layers). Sticky backs are saved in their own directory and can be called up at any time, similar to symbols.

CHAPTER 9

Editing

Editing entities in a drawing is just as much a part of CAD as creating them. You will invariably need to constantly change things that have already been drawn. In this chapter, you will learn about the basic and advanced editing features in DataCAD. This is not a long chapter, but it is a very important one. Although every CAD program will have some identical features like Enlarge, Erase, Move and Copy, you should pay particular attention to some of the more unique and powerful features in this chapter such as Layer Locking, Edit Sets, and Masking.

Undo/Redo

DataCAD can undo and redo nearly any editing that you do within the drawing file. See Chapter 5, "Basic Drawing," for a full explanation.

Enlarge

As the name suggests, this function is used to change the size of entities in the Drawing Window. But just as an enlargement factor of 2 will enlarge entities by two times, using an enlargement factor of 0.5 will size entities down to half-size. So you can use the **Enlarge** function to make entities smaller as well as larger. Another useful feature of this function is that you can enlarge entities in both the X and Y axis, or along only one or the other. To use the **Enlarge** function, follow these basic steps:

1. Select **Edit/Enlarge**. You will be prompted *to "Select CENTER of enlargement."*
2. Pick or snap to a point.
3. Select **Enlrgmnt** and pick the enlargement axis (X, Y, Z, or **SetAll**). In most cases, you will select **SetAll** to enlarge equally in all three axes.
4. Enter an enlargement factor and then *right-click* or press **Enter** to accept the value.
5. *Right-click* or press **Exit** to exit the **Enlrgmnt** menu.
6. Turn **LyrSrch** on or off as appropriate.
7. Turn **AndCopy** on or off as appropriate.

8. Select one of the EGAFS selection methods (**Entity**, **Group**, **Area**, **Fence**, or **SelSet**).
9. Select all the entities to be enlarged. They will be enlarged about the selected center point.

Once you set the enlargement factor in a drawing, that factor will remain until changed to something else, so don't forget to check the current enlargement factor before running the **Enlarge** function again later.

When you first select the **Enlarge** function, DataCAD prompts you to *"Select the center of the enlargement."* This is the most important concept for you to understand. When you tell DataCAD to enlarge something, you also need to locate the point from which it will be enlarged. DataCAD does not automatically assume you want to enlarge from the center of mass of all the selected entities. A different enlargement center will not affect *how* the entities will look after they are enlarged, but it will affect *where* they are displayed in the Drawing Window. Let's see how this works.

After selecting **Enlarge**, pick a center point. In this case, we selected a point somewhere between two chairs (see Figure 9-1).

Select an **Enlrgmnt/SetAll** factor of 2 and then press **Enter**. *Right-click* back to the Enlarge menu. Make sure **AndCopy** is off. Select the **Area** selection method and then draw a bounding box around the chairs to select them both. The chairs will be enlarged, as shown in Figure 9-2.

In the next example, we will enlarge the same two chairs by selecting a different center point, as shown in Figure 9-3.

Notice now that the chairs have still been enlarged by a factor of 2 but are located differently than in the previous example (see Figure 9-4). This is because they were enlarged in reference to a different center point.

Figure 9-1 Choosing the center point for enlargement

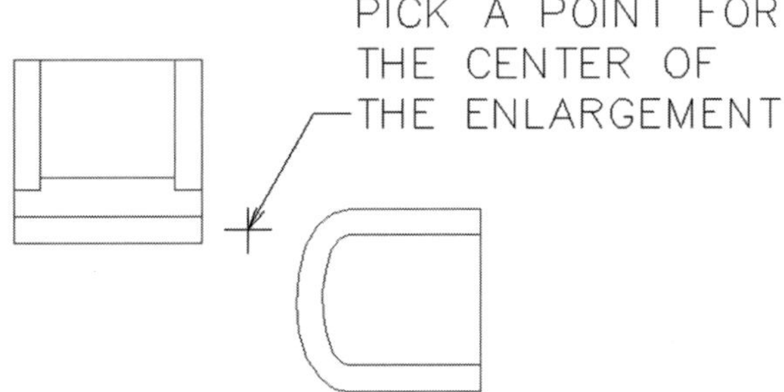

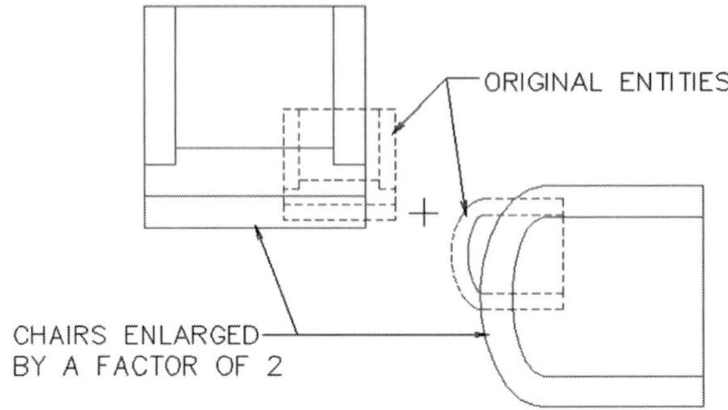

Figure 9-2 The final enlargement compared to the original sizes

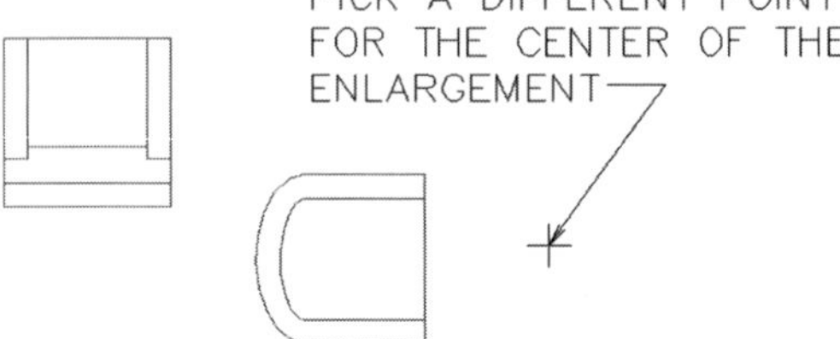

Figure 9-3 Enlarging from a different center point

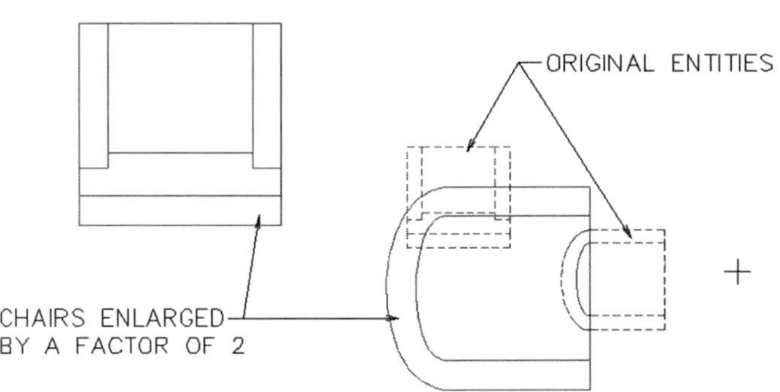

Figure 9-4 The final enlargement along with the original sizes

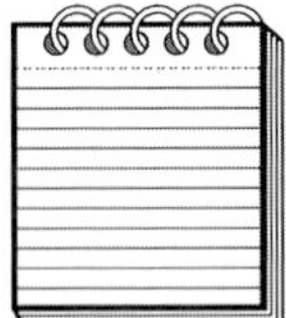

NOTE: *Unless you want to enlarge subsequent entities by the same center point (by selecting* ***PrevCent****), each time you enlarge something you need to pick a new center.*

Once you select an enlargement center, you will see five new enlargement options, Enlrgmnt, Invert, Center, AndCopy, and EnlrgeZ, along with the standard EGAFS selection options.

Enlrgmnt

Use this to set the enlargement factor. You can enlarge by the X, Y, and/or Z axis, and you can change the line spacing of linetypes at the same time. If you simply want to enlarge some 2D entities in a drawing without regard for Z-heights, then all you will normally need to do is press the **SetAll** option, type in the enlargement factor, and then pick the entities you want to enlarge. Here is what each option does:

- **XEnlgmt** Select this option to enter an enlargement factor along the X axis. You can enter a value from 0.01 to 9999.
- **YEnlgmt** Select this option to enter an enlargement factor along the Y axis. You can enter a value from 0.01 to 9999.
- **ZEnlgmt** Select this option to enter an enlargement factor along the Z axis. You can enter a value from 0.01 to 9999.
- **LineFact** Select this option to change the linetype line spacing factor. Do not use this if you want to enlarge your entities and still keep the look (spacing) of the existing linetypes. For instance, if you want to enlarge a detail and you want all your $^3/_4$″ [19] plywood linetypes to continue to display at $^3/_4$″ [18] wide, then leave the **LineFact** at 1.0. But if you want your 3″ [76] batt insulation to display as 6″ [152] wide after enlarging, then enter a **LineFact** of 2.0.
- **SetAll** Use this option to set all four enlargement options to the exact same factor without having to select each one individually.

Invert

Use this option to enlarge by the inverse of the previous enlargement factor. If the previous factor was 2.0, then the inverse will be 0.5. This is useful if you enlarge something and then change your mind. Simply select **Invert**, then select the entities, and they will be returned to their original size. This toggle affects all four of the enlargement options simultaneously.

NOTE: *The number 2 can be expressed as a fraction:* $^2/_1$ *. The inverse of this is described by inverting the fraction, or* $^1/_2$*, which in its decimal for is 0.5. So the inverse of 4 (*$^4/_1$*) would be* $^1/_4$*, or 0.25.*

Center

Use this option to locate a new enlargement center without leaving the **Enlarge** menu.

AndCopy

Use this option to enlarge the selected entities by the selected center while maintaining the original entities in their current location as well.

EnlrgeZ

Select this option to enlarge entities in the Z axis only. You can enter a value from 0.01 to 9999. This is generally used when all you want to change is the Z height of entities and nothing else. You could use the **ZEnlgmt** setting in the **Enlrgmnt** menu, but then you would have to make sure the other three settings are set to 1.0 so that they are not affected.

Erase

Use this option to remove entities from the drawing. When selected, you will see another menu with the standard EGAFS selection methods, along with the three more options described in this section. You can also get to the **Erase** menu by pressing **Ee** on your keyboard. To erase an entity, select **Edit/Erase** (**Ee**), select one of the EGAFS selection methods, and then pick the entity or entities to be erased.

Crossing

After selecting the first point of an **Area** bounding box or the first point of a selection **Fence**, you will see the **Crossing** option appear in the menu window. This feature allows you to pick not only entities that are fully inside the selection area bounding box or fence, but all entities that are crossed by one or more of the sides of the area or fence. To use it, press **F1** or click on the **Crossing** button prior to selecting the second point of the Area selection box or any time prior to selecting the final Fence point.

Partial

Use this option to erase selected portions of a line, arc, or circle, not just the entire entity. However, you cannot partially erase a B-spline or a Bezier curve. The **Partial** erase function is an instance, especially with circles and arcs, where having the **DrwMarks** setting on (in the **Utility/Settings** menu) can be very helpful, since it will visually show you each of the points you have picked.

To partially erase a line, perform the following steps:

1. Select **Partial**. DataCAD will prompt you to *"Select entity to modify."*
2. *Click* on the entity you want to edit. DataCAD will prompt you to *"Select first point of line segment to be removed."*
3. *Click* or snap on the first point. DataCAD will prompt you to *"Select second point of line segment to be removed."*
4. *Click* or snap on the second point. The portion of the line between the first point to the second point will be erased.

To edit a circle or arc, follow these steps:

1. Select **Partial**. DataCAD will prompt you to *"Select entity to modify."*
2. *Click* on the entity you want to edit.
3. DataCAD will prompt you to *"Select first point of line segment to be removed." Click* or snap on the first point.
4. DataCAD will prompt you to *"Select second point of line segment to be removed." Click* or snap on the second point.
5. DataCAD will prompt you to *"Select point on arc to clip out."* Select the arc or portion of the arc to be removed.

If the entity you are editing is a circle, then you will have just divided the circle into two arcs, as shown in Figure 9-5.

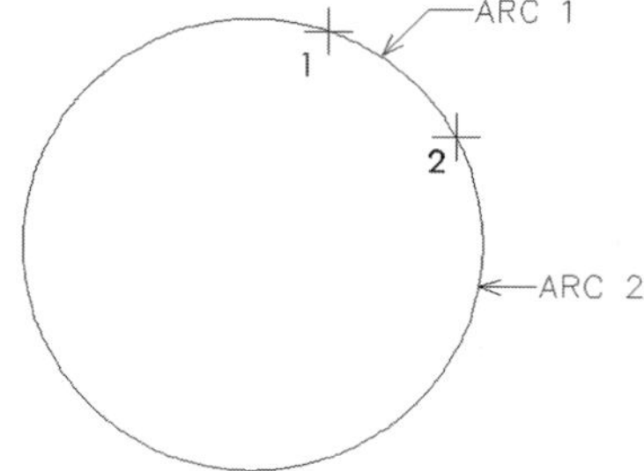

Figure 9-5 A circle being separated into two arcs

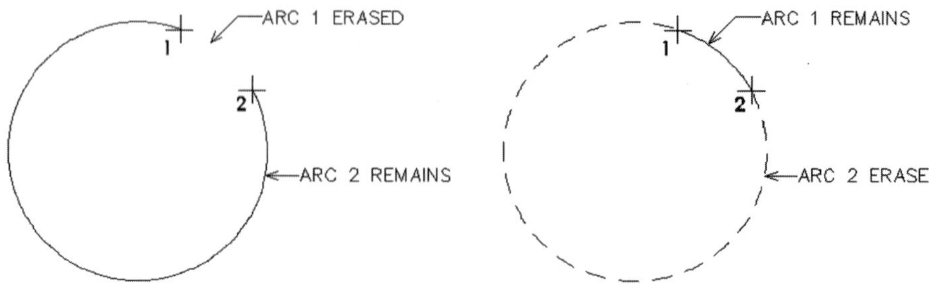

Figure 9-6
The results of erasing either one of the arcs

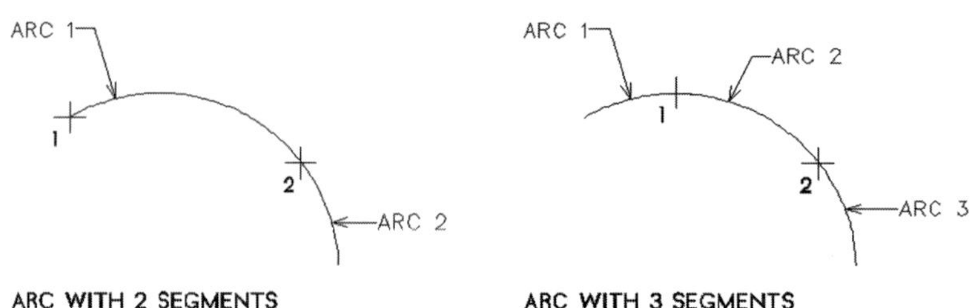

Figure 9-7
Dividing an arc into two and three segments

In Step 5, DataCAD is asking you which of the two arcs you want to erase. Selecting the small arc (Arc 1) will yield the result on the left. Selecting the larger arc (Arc 2) will give you the result on the right (see Figure 9-6).

If the entity you are editing is an arc, then depending on where you select your two points, you can divide the arc into two or three segments, as shown in Figure 9-7. In Step 5, DataCAD is asking you which of the two or three arc segments you want to erase.

The **Partial** function should not be taken lightly. It is a very important editing option in DataCAD, so you should become very familiar with it.

Cleanup

The **Cleanup** menu has several functions used to trim 2D entities in various ways. Many 2D lines and arcs could be "cleaned up" or trimmed using the **Erase/Partial** command, but the Trim, Weld, and Intersection cleanup functions are much quicker and more elegant methods for cleaning up your 2D lines and arcs. The **Fillets** and **Chamfer** options are likewise much faster methods for creating those conditions, rather than manually measuring, erasing, and drawing.

The options in the **Cleanup** menu are an important aspect to quick and efficient drafting in DataCAD, so do take the time to learn them and get into the habit of using them.

Fillets

As the name implies, this option adds an arced fillet between any two lines that are not parallel and, if extended, would intersect. To use this option, do the following:

1. Select **Fillets/Radius**.
2. Pick a fillet radius from the list, or type in your own value. Press **Enter** or *right-click* to accept the value.
3. Turn **Clip** on or off, as required.
4. Turn **LyrSrch** on or off, as required.
5. Pick the two lines to add a fillet between.

Radius Use this option to select the radius of the fillet arc.

Clip With Clip on, the ends of the two selected lines will be clipped off where they extended past the fillet radius. With Clip off, the fillet arc will be added, but the ends of the two selected lines will be retained (see Figures 9-8a through 9-8f.

One important aspect of the **Fillet** and **Clip** commands is that the two selected lines do not have to intersect in order to be acted upon. DataCAD will interpolate where the two lines would cross if they were extended and place the fillet arc at that location.

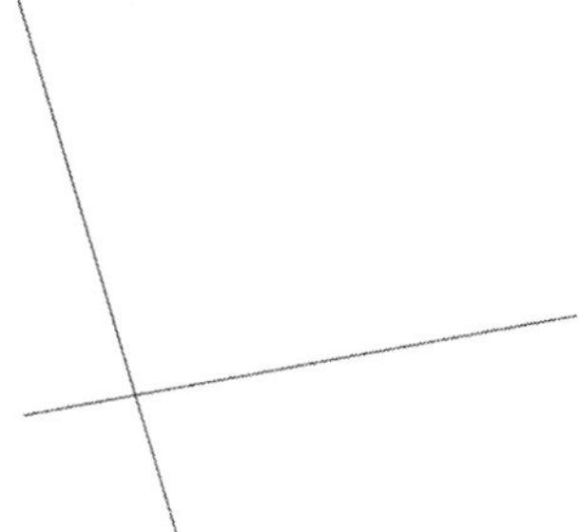

Figure 9-8a
Original two lines

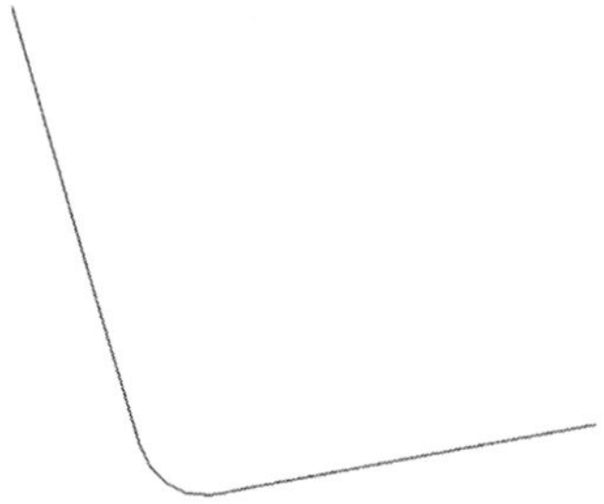

Figure 9-8b
Clip on

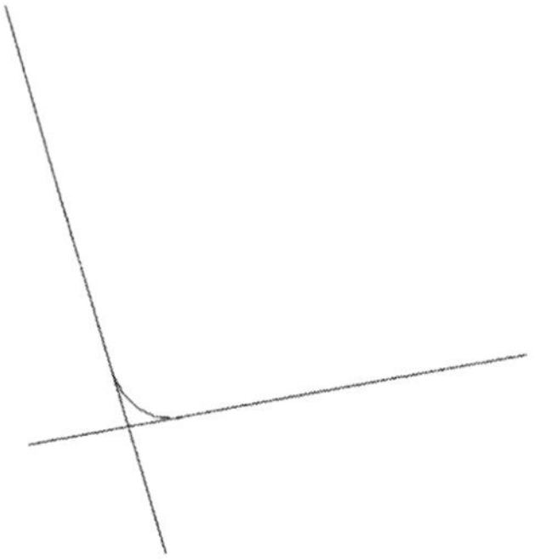

Figure 9-8c
Clip off

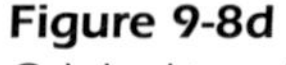

Figure 9-8d
Original two lines

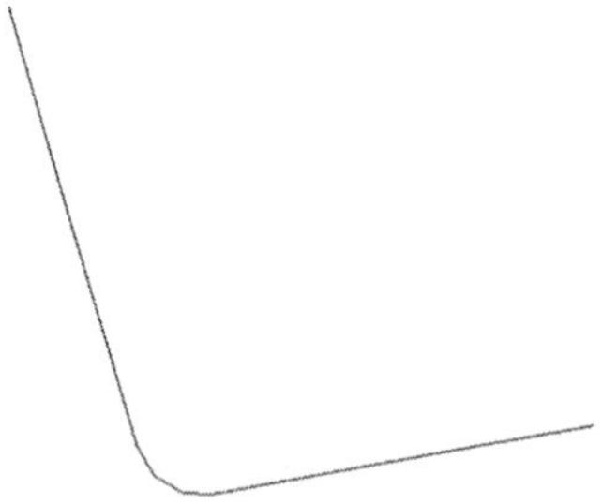

Figure 9-8e
Clip on

Figure 9-8f
Clip off

Chamfer

This option adds a straight-line chamfer between any two lines that are not parallel. To use this option, follow these steps:

1. Select **Chamfer/Distnces**.
2. Select the first chamfer distance. This selection will affect the first line that you pick. Press **Enter** or *right-click* to accept the value.
3. Select the second chamfer distance. This selection will affect the second line that you pick. Press **Enter** or *right-click* to accept the value.
4. Turn **Clip** on or off, as required.
5. Turn **LyrSrch** on or off, as required.
6. Pick the two lines to add a chamfer between them.

Distnces Use this option to select the length of the chamfer line.

Clip With Clip on, the ends of the two selected lines will be clipped off where they extend past the chamfer. With Clip off, the chamfer will be added, but the ends of the two selected lines will be retained (see Figures 9-9a through 9-9f).

One important aspect of the **Chamfer** and **Clip** commands is that the two selected lines do not have to intersect in order to be acted upon. DataCAD will interpolate where the two lines would cross if they were extended and place the chamfer at that location.

1LnTrim (1 Line Trim)

Use this option to trim one or more lines to another line. Trimming will either shorten or lengthen the selected lines. The line to which the entities

Figure 9-9a
Original two lines

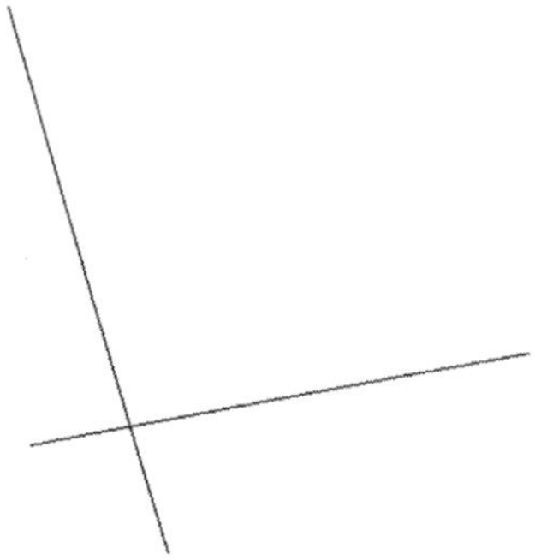

Figure 9-9b
Clip on

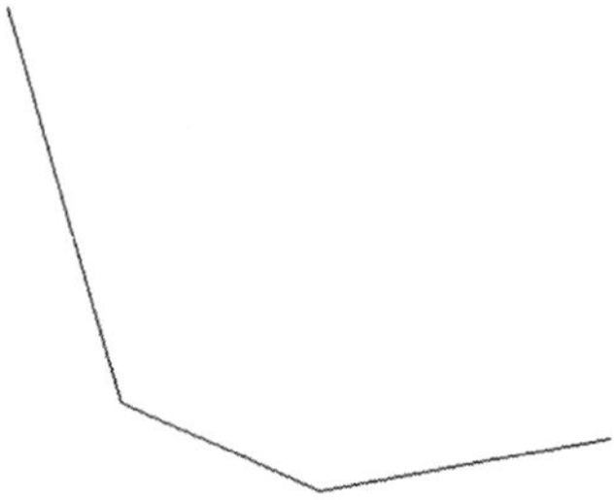

Figure 9-9c
Clip off

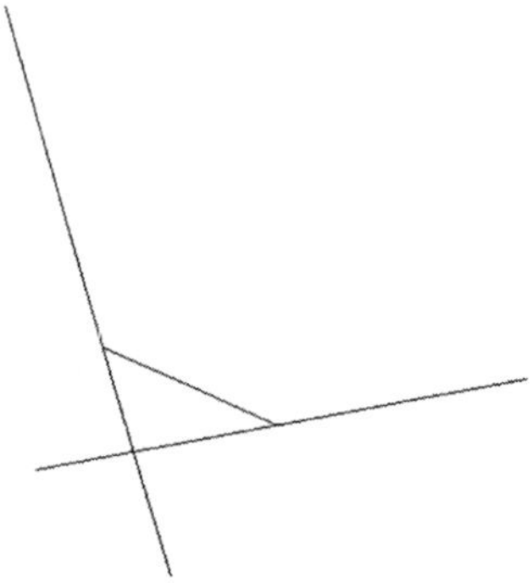

Figure 9-9d
Original two lines

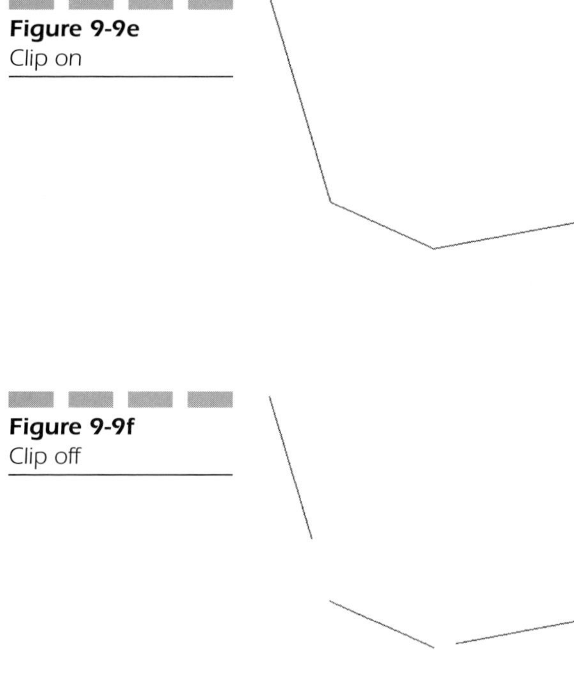

Figure 9-9e
Clip on

Figure 9-9f
Clip off

are trimmed can be a 2D line, or it can be an imaginary line defined by two points.

To utilize this option, perform these steps:

1. Select **1LnTrim**. DataCAD will prompt you to *"Select first point of line to clip to."*
2. Define the line with your mouse by selecting two points, or select **Entity** (a 2D line) if you want to trim to an existing line.
3. Pick a point anywhere on the "outside" of the line to trim to. Selected entities will not appear on the selected side.
4. Select the entities to be trimmed. Since you can select one or more lines to be trimmed, you can select entities by any of the EGAFS selection methods (see Figures 9-10a through 9-10d).

One important aspect of the **1LnTrim** command is that selected lines do not have to intersect in order to be acted upon. DataCAD will interpolate where the lines would cross if they were extended and trim the lines based on that location.

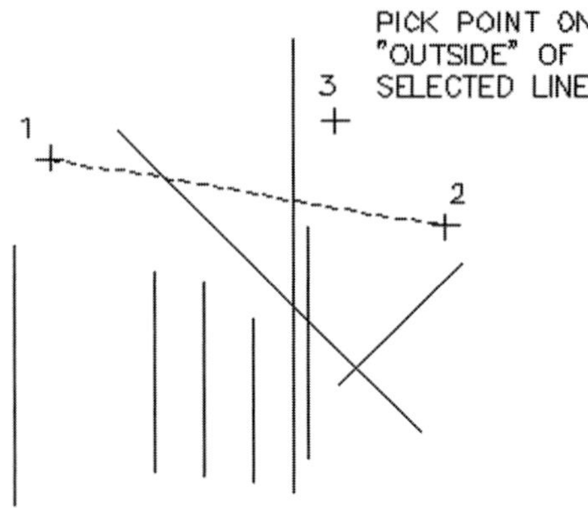

Figure 9-10a
Pick two points to trim to, then a third point to define the "outside"

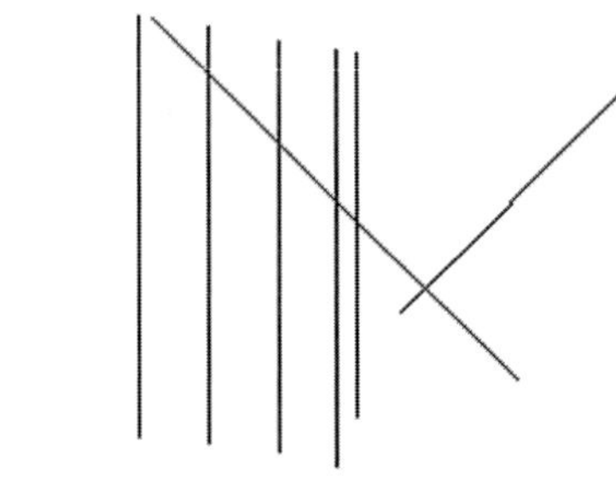

Figure 9-10b
Pick lines to be trimmed

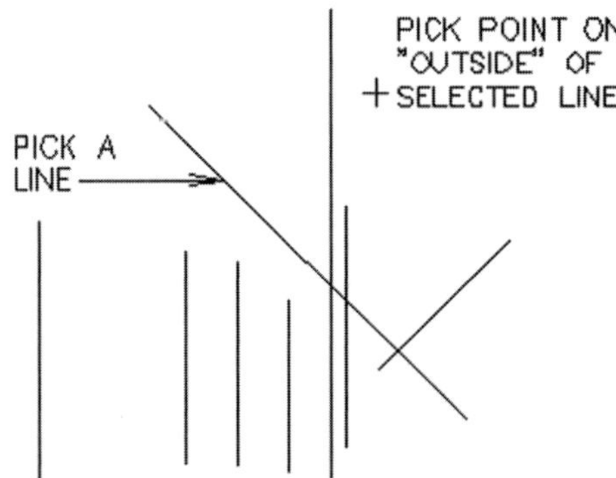

Figure 9-10c
Pick **`Entity`** to trim to, then a third point to define the "outside"

2LnTrim (2 Line Trim)

Use this to trim two lines at their intersection. If the two lines do not intersect, **2LnTrim** will extend and join the lines where they would have intersected (see Figures 9-11a and 9-11b). Follow these steps to use this option:

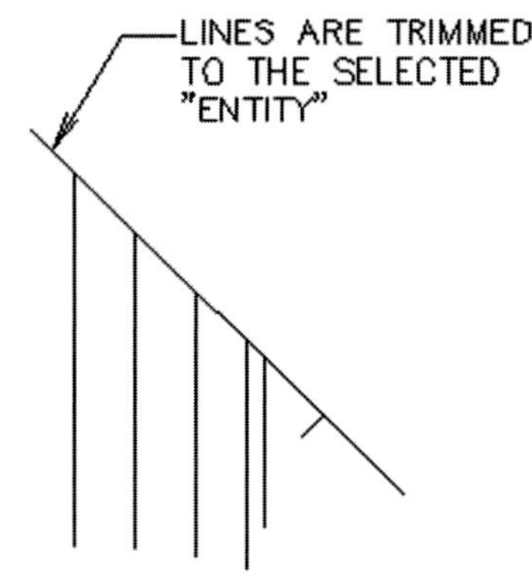

Figure 9-10d Pick lines to be trimmed

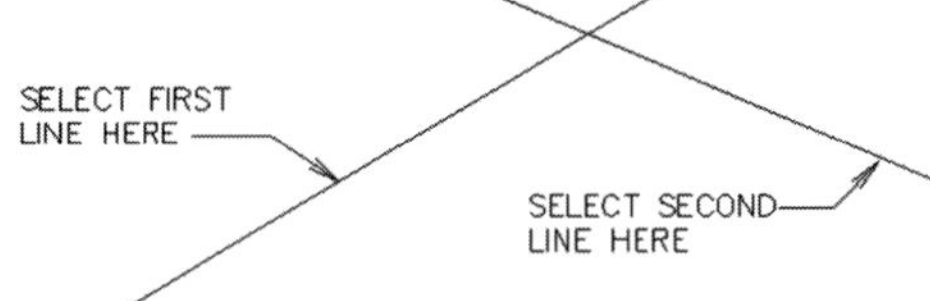

Figure 9-11a Selecting the two lines to trim

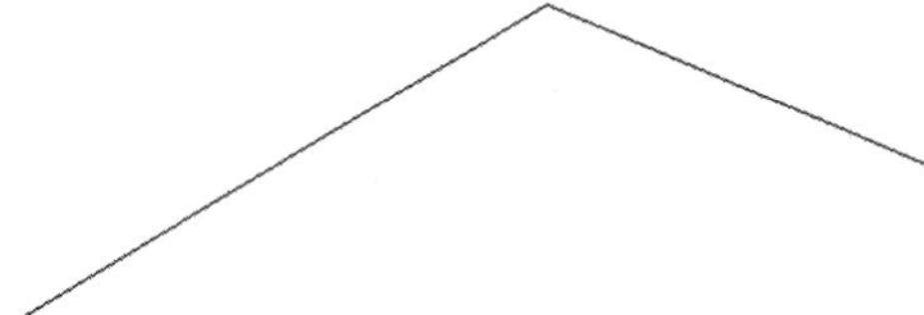

Figure 9-11b The final outcome

1. Select **2LnTrim**.
2. Pick the first line to trim.
3. Pick the second line to trim.

When selecting two intersecting lines, you must be sure to select the side of each line that is to remain.

Look at the previous two-line trim example. Note at which end each line is selected. Now if you trim the same lines, but select the opposite ends of the lines, Figures 9-12a and 9-12b would be the result.

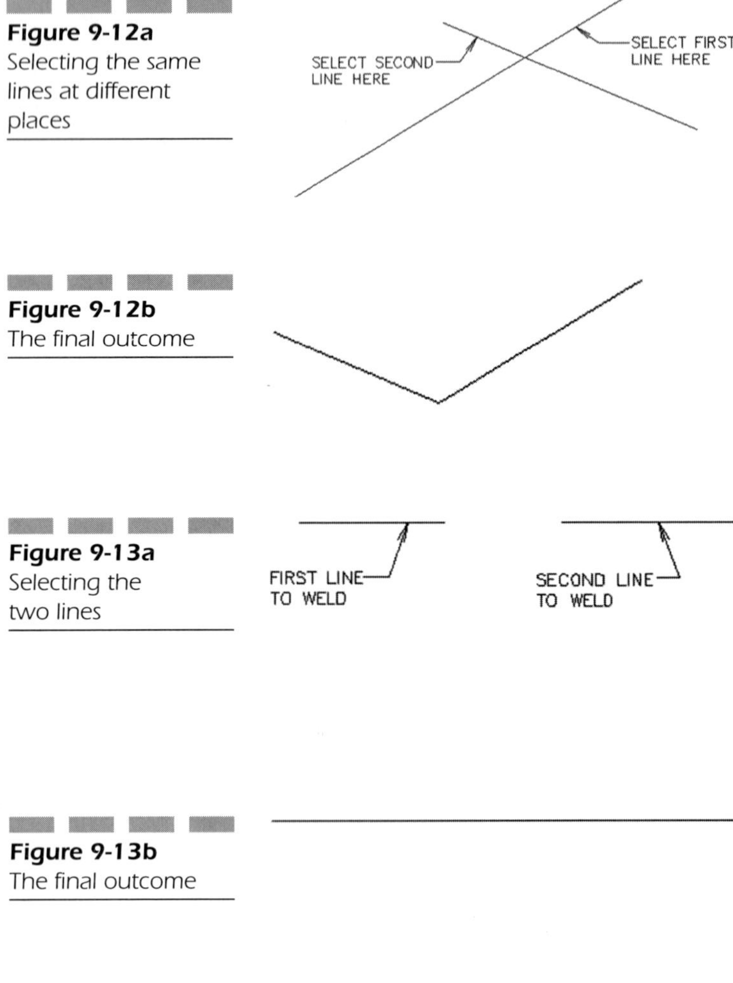

Figure 9-12a
Selecting the same lines at different places

Figure 9-12b
The final outcome

Figure 9-13a
Selecting the two lines

Figure 9-13b
The final outcome

WeldLine

Use this option to repair a broken line to form a single, straight, and continuous line (see Figures 9-13a and 9-13b):

1. Select **WeldLine**.
2. Select the first line to weld.
3. Select the second line to weld.

If you weld together two lines that are not exactly in line with one another, they will still be welded together, but they will be welded between their two outside endpoints. This is one thing you should be careful of when using the **Weld** command, or you may wind up with a new, single line that is at some odd angle that you didn't expect. Figures 9-14a and 9-14b show an extreme example.

If you have ever used AutoCAD, then you may know that it has nothing like the **Weld** function. For drawing efficiency and for your own sanity, it is important to learn and to use this option often. You might wonder why you should weld two lines together rather than just drawing a third line between them, since doing either would have the same visual result. Welding two lines into one should be done for a couple of important reasons.

First, the resulting single entity is far easier to edit, since it is one entity instead of three. If you move the entity, the whole thing moves. If you try to move the three unwelded lines, only one of them will move. If you need to erase the line, you will end up erasing only one of the three lines, having to go back to erase the other two as well.

The second reason to use the weld command is to create a more efficient drawing file. If you simply draw a new line between two lines, rather than welding them, you have three entities in your drawing file, rather than only one. This may not seem like much, but if you multiply this by 10 or a hundred times in a drawing file, you will wind up with a needlessly bloated drawing

Figure 9-14a The two, seemingly incompatible lines

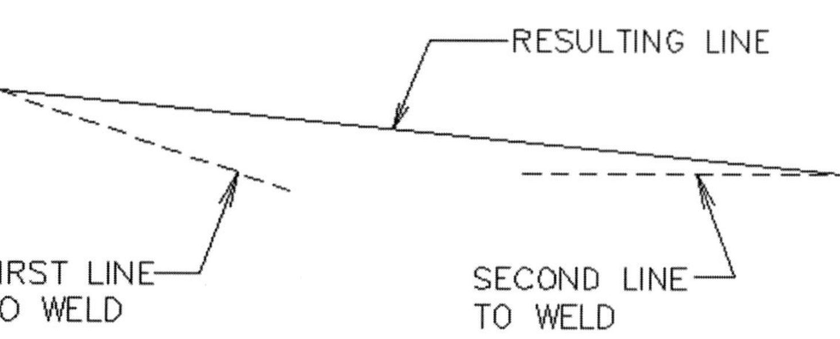

Figure 9-14b The end result of the two lines being welded together

file full of extraneous entities. Too many entities in your drawing will eventually slow down your drawing and editing to a noticeably slower pace.

You can weld together two lines even if they are of different linetypes, but since the resulting single, welded line can only be of one linetype, the final linetype will be changed to whichever line was picked first.

WeldWall

This option is much like the **WeldLine** function. Use it to repair a broken wall to form a single, straight, and continuous wall. You can weld walls made up of two, three, or four lines. It is worth noting that this function is not limited only to walls. It can be used to weld together any matching group of parallel lines, like a handrail, a sidewalk, or a length of countertop.

To use this option, follow these steps:

1. Select **WeldWall**.
2. Draw a bounding box around the ends of the two walls to be joined (see Figures 9-15a and 9-15b).

Because this function joins multiple lines, some important rules must be followed in order for it to work properly:

1. Only two-, three-, and four-line walls can be joined. The function will not work on walls of five or more lines.
2. For a two-line wall, there must be exactly four endpoints (two for each wall) within the bounding box. For a three-line wall, there must be exactly six endpoints (three for each wall) within the bounding box. For

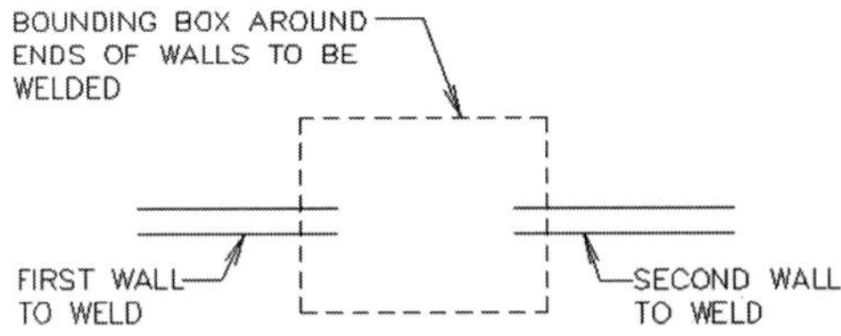

Figure 9-15a The two walls to be welded

Figure 9-15b The completed connection

a four-line wall, there must be exactly eight endpoints (four for each wall) within the bounding box.

3. DataCAD knows which lines to weld by counting the number of endpoints within the bounding box and noting whether or not the two walls are exactly opposite one another. If additional entities exist within the bounding box like a stray line, a door, or a window, **WeldWall** will not work.
4. If the two walls are not of exactly the same width and directly across from one another, then **WeldWall** will not work.

Figure 9-16 provides four examples of elements that can be successfully welded.

Figure 9-17 provides four examples of things that cannot be welded.

Just as it is important to weld two lines into one (refer to the discussion on the **WeldLine** option), it is likewise important to weld two walls into one. Just as if you weld together two lines that are not exactly in line with one another, walls that are not exactly in line with one another will still be welded together, but they will be welded between their two outside endpoints. This is something you should be careful of when using the **WeldWall** command, or you may wind up with a new wall that is at some odd angle that you didn't expect. Refer to the figure in the earlier Weld Wall sections.

Figure 9-16
Four groups that can be welded

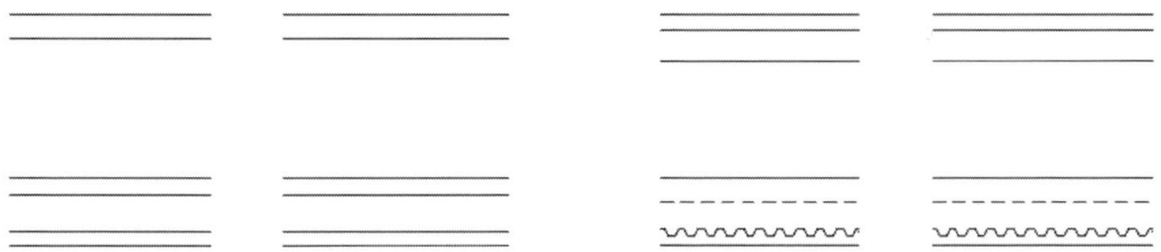

Figure 9-17
Some walls that are unable to be welded

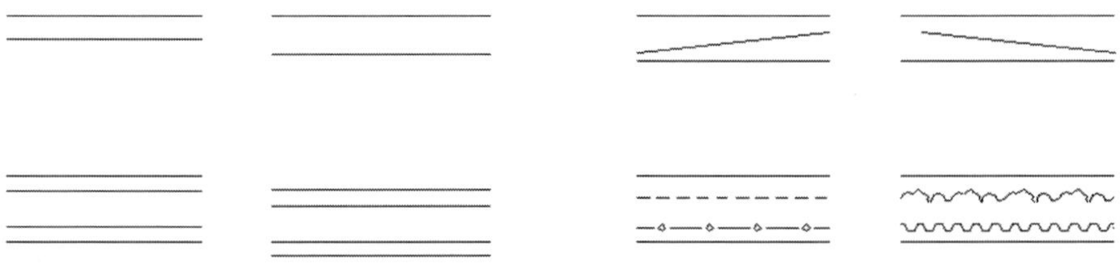

NOTE: *Although any two to four parallel lines can represent a wall in DataCAD, when it comes to the* ***Cleanup*** *functions, DataCAD can be picky. When you use the Walls command in DataCAD, the lines you draw are seen as walls to DataCAD. But if you draw two to four parallel lines without using the Walls function, DCAD may not understand that they are walls that can be cleaned up. If you are having this trouble, you can try running a* ***Toolbox*** *macro called* ***Lin2Wall.DCX****.*

To use the macro, go to the ***Toolbox*** *and run the* ***Lin2Wall*** *macro. Select all the lines that you want to designate as walls. Now you should be able to use all of the wall cleanup functions (****WeldWall, TIntsct, LIntsct,*** *and* ***XIntsct****).*

TIntsct (T Intersection)

Use this to clean wall lines meeting at a T intersection. Intersections with different wall types will also clean up, but with varying results. Follow these steps to use this option:

1. Choose **TIntsct**. You will be prompted to *"Select 1st corner around T intersect (wall line ends only)."*
2. Draw a bounding box around the walls to be cleaned up. Only the walls to be cleaned up can be inside the bounding box.
3. Select the wall line you want to trim to.

The general process is to place a bounding box around the end of the wall we want to trim and then tell DataCAD which wall we want to trim to in order to form a nice clean intersection (see Figure 9-18).

As shown in the figure, Step 1 is to pick two points to define a bounding box around the ends of the wall. After doing this, the Prompt Line will read, *"Point to wall line to trim to."* Step 2 is to place your cursor anywhere on the wall you want to trim to and click the mouse. The T intersection will be cleaned up automatically. This method will work for walls that don't actually cross each other, as in Figure 9-19, and it will work for walls that are not perpendicular to one another, as in Figure 9-20.

One thing you have to be careful of is to make sure that only the end of one wall is inside the bounding box. If you include the ends of two or more walls, as shown in Figure 9-21, DataCAD won't know which ones you want to trim.

And if you pick the wrong wall to trim to, you'll get some weird results, as shown in Figure 9-22.

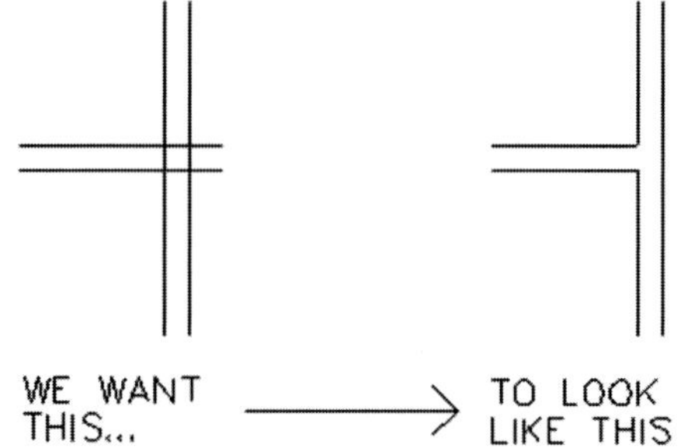

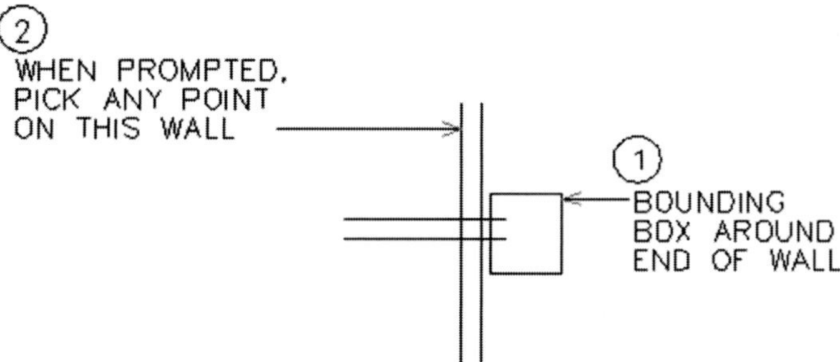

Figure 9-18 Cleaning up a T intersection

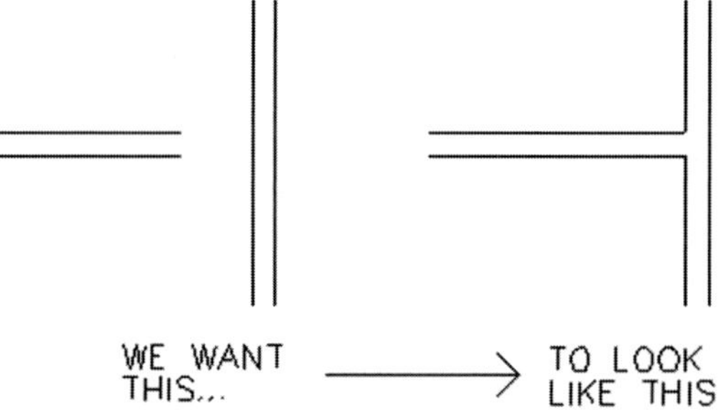

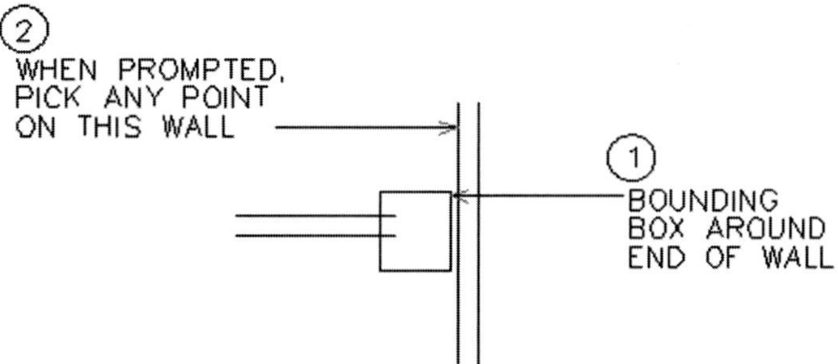

Figure 9-19 Connecting perpendicular walls

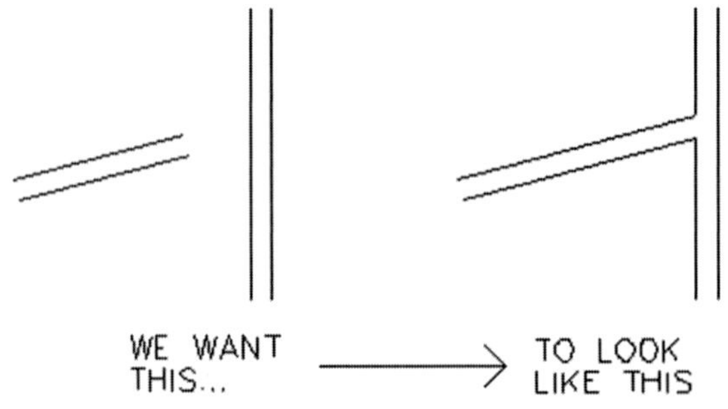

Figure 9-20 Connecting walls that aren't perpendicular

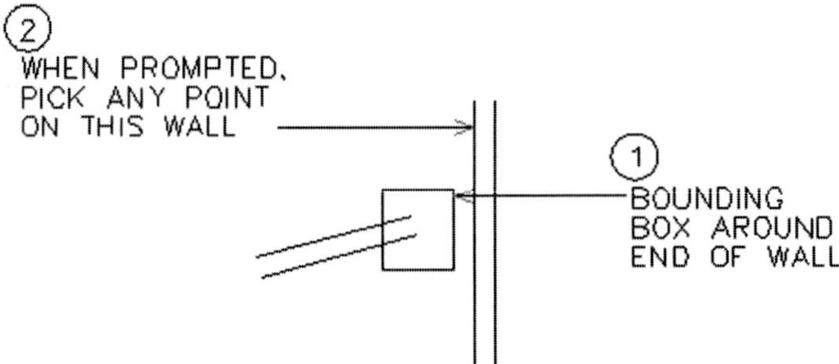

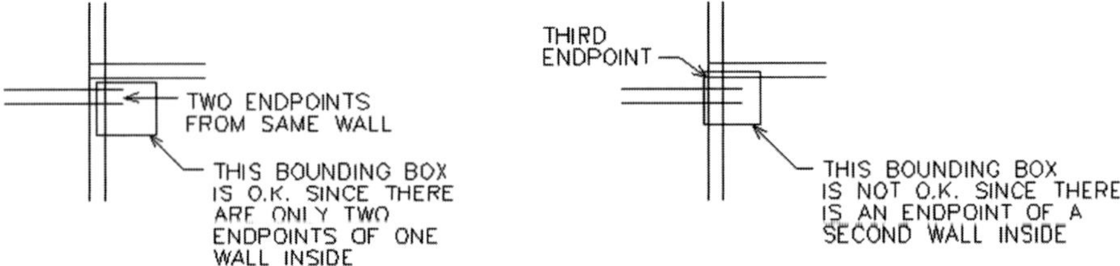

Figure 9-21 Examples of what to do and what not to do when choosing walls to trim

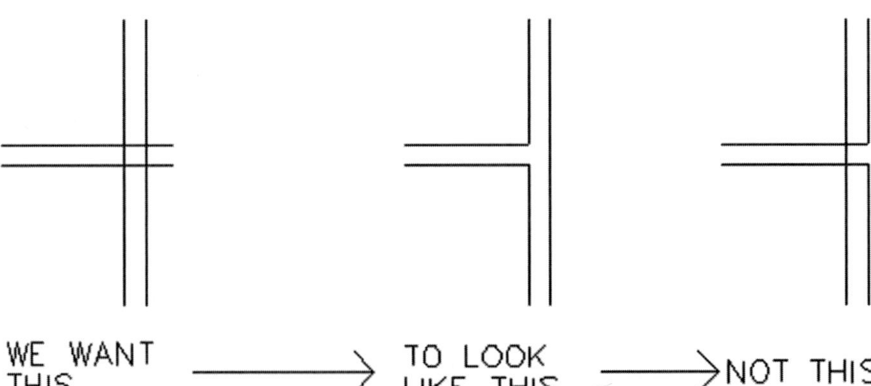

Figure 9-22 The outcome of choosing the wrong wall

Figure 9-23 Cleaned up walls with three to four lines

Figure 9-24 Cleaned up walls with different numbers of lines

You can use the **TIntsct** function to clean up wall types with two, three, or four lines (see Figure 9-23).

You can also use it to clean up walls with different numbers of lines, but the results will look like the examples in Figure 9-24.

If you are having trouble getting the walls to clean up, refer to the note at the end of the WeldWall description earlier in this chapter.

LIntsct (L Intersection)

This option is much like the **2LnTrim** option, but it works on multiple lines such as you would find in a wall. To use it, follow these steps:

1. Select **LIntsct**. You will be prompted to *"Select 1st corner around "L" intersect (wall line ends only)."*
2. Draw a bounding box around the wall ends to be cleaned up. Only the walls to be cleaned up can be inside the bounding box.

As shown in Figure 9-25, to clean up a corner you pick two points to define a bounding box around the ends of the walls. The L intersection will be cleaned up automatically. Only the wall ends to be cleaned can be inside the bounding box.

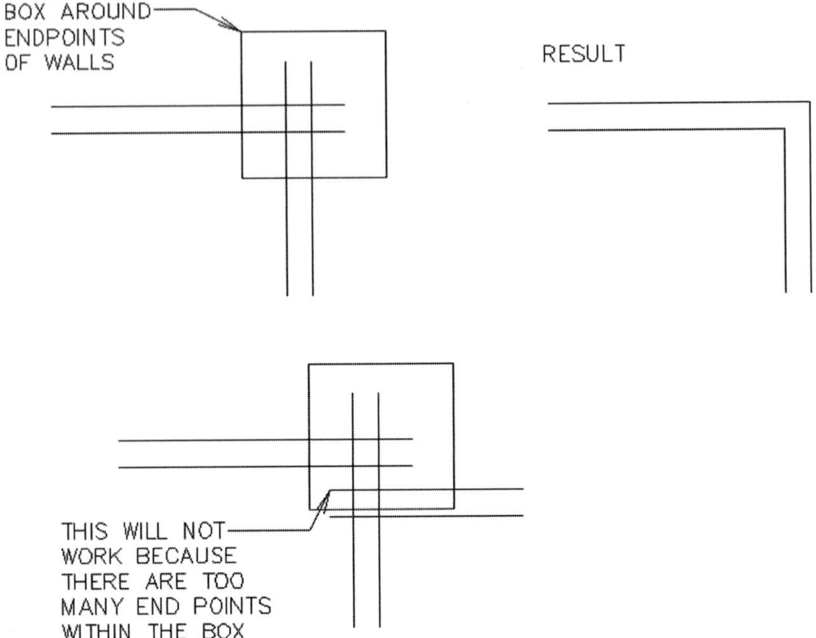

Figure 9-25
Cleaning up a corner

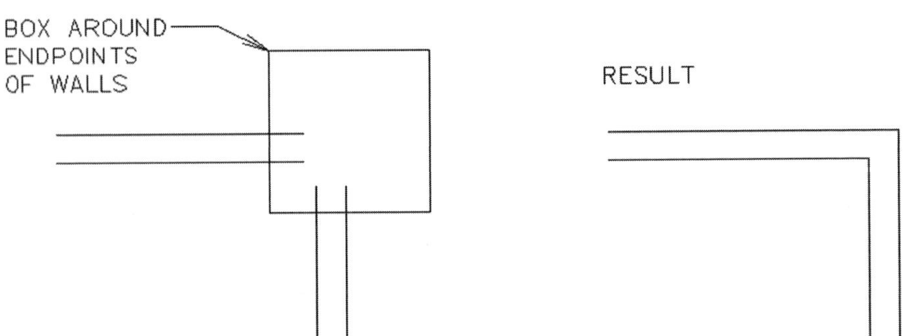

Figure 9-26
Connecting walls that don't cross each other

This method will also work for walls that don't actually cross each other, as shown in Figure 9-26.

LIntsct can also be used on walls that are not perpendicular to one another (see Figure 9-27).

You can use the **LIntsct** function to clean up wall types with two to four lines (no more than four lines), such as the ones shown in Figure 9-28.

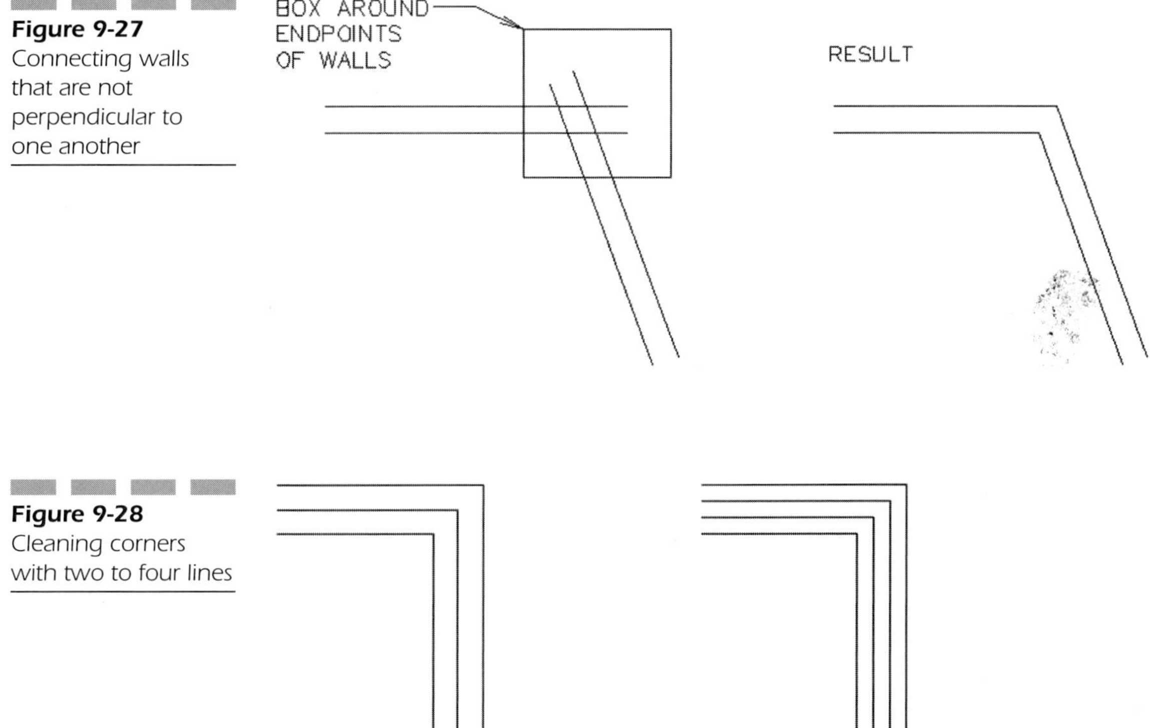

Figure 9-27
Connecting walls that are not perpendicular to one another

Figure 9-28
Cleaning corners with two to four lines

You can also use **LIntsct** to clean up walls with different numbers of lines, but the results will look like the examples in Figure 9-29.

If you are having trouble getting the walls to clean up, refer to the note at the end of the **WeldWall** description earlier in this chapter.

XIntsct (X Intersection)

Like the **LIntsct** option, use **XIntsct** to clean up the intersection of multiple intersecting lines, like when two walls intersect with one another. The steps to use this option are as follows:

1. Select **XIntsct**. You will be prompted to *"Select 1st corner around "X" intersect (0 line ends)."*
2. Draw a bounding box around the wall intersection to be cleaned up. Only the walls to be cleaned up can be inside the bounding box.

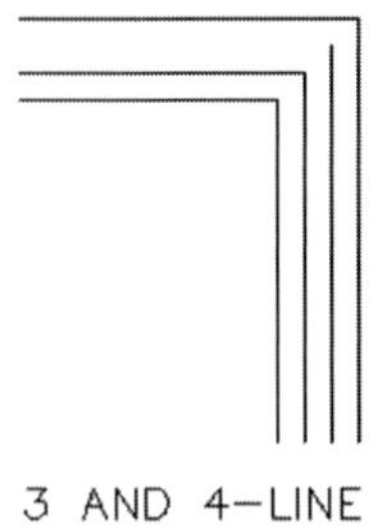

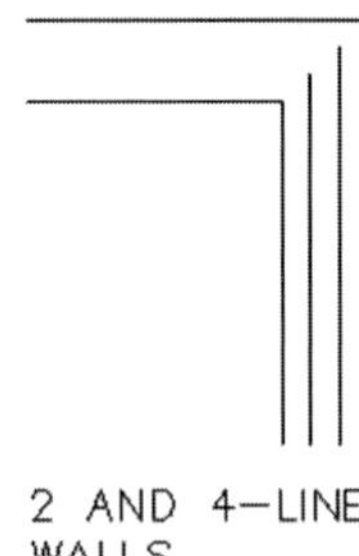

Figure 9-29
The results of cleaning up walls with multiple lines

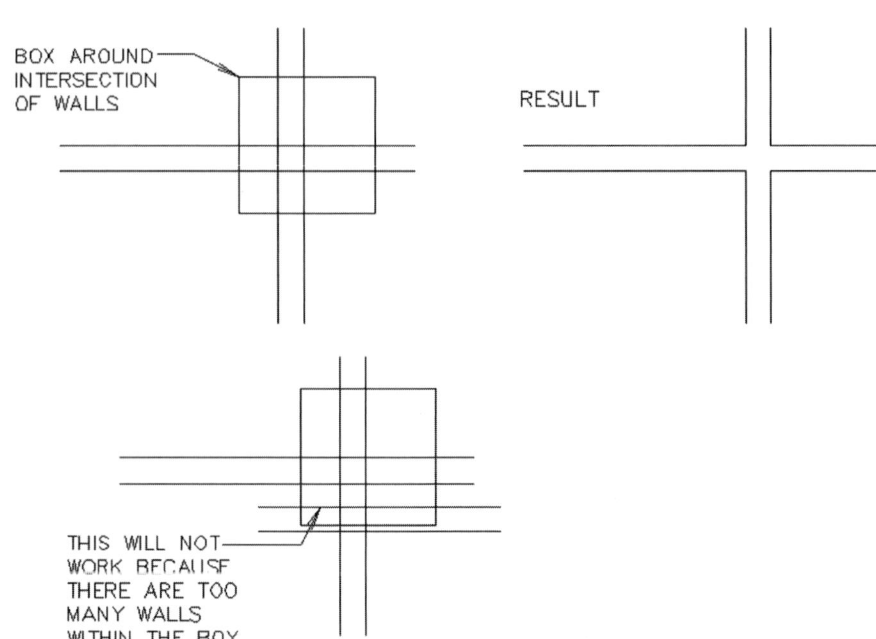

Figure 9-30
As usual, only the walls inside the bounding box will be cleaned up.

As shown in Figure 9-30, to clean up a corner you pick two points to define a bounding box around the intersection of the walls. The X intersection will be cleaned up automatically. Only the walls to be cleaned can be inside the bounding box.

XIntsct can also work for walls that are not perpendicular to one another, as shown in Figure 9-31.

You can use the **XIntsct** function to clean up wall types with two to four lines, like the ones shown in Figure 9-32.

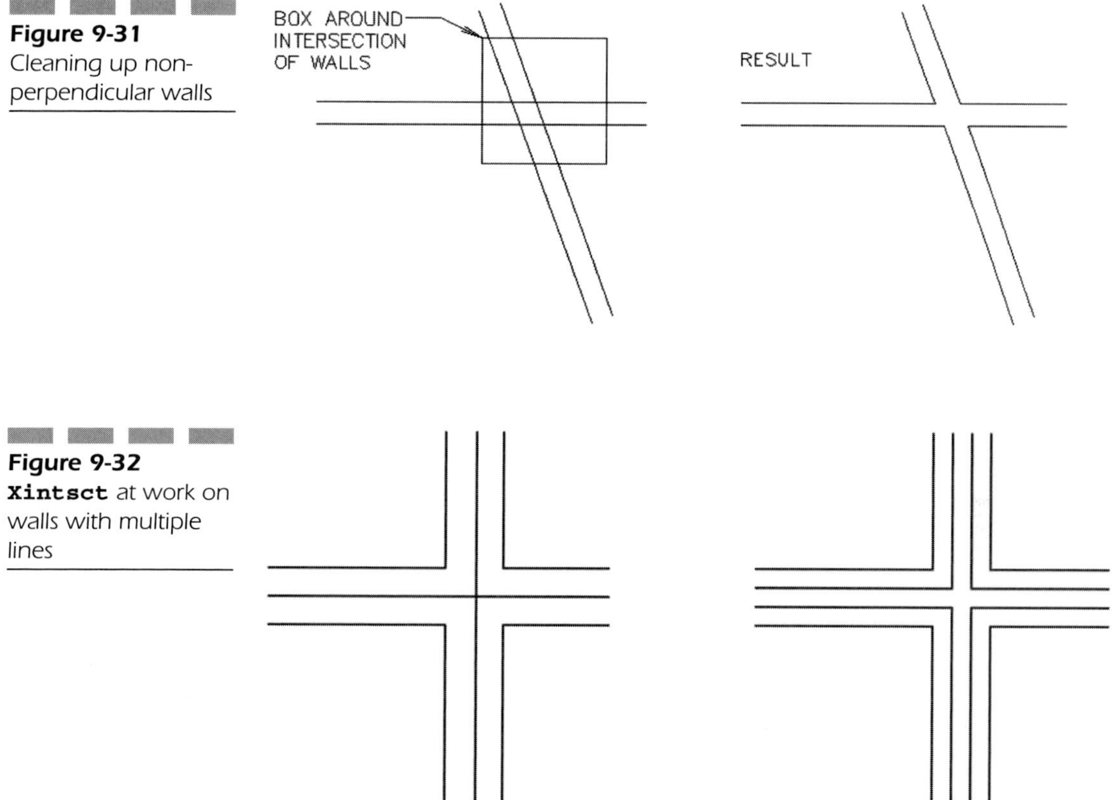

Figure 9-31 Cleaning up non-perpendicular walls

Figure 9-32 **`Xintsct`** at work on walls with multiple lines

You can also use it to clean up walls with different numbers of lines, but the results will look like the examples in Figure 9-33.

If you are having trouble getting the walls to clean up, refer to the note at the end of the **WeldWall** description earlier in this chapter.

FreeTrim

FreeTrim is an interesting feature. It enables you to preselect a group of 2D lines or arcs (not ellipses). Within the selected group you can then pick any entity to be trimmed, and it will be trimmed between its nearest two intersections with any of the other selected entities. This is a great feature for cleaning up something like clapboards behind a downspout. You're probably saying, "What?!" Here are the steps:

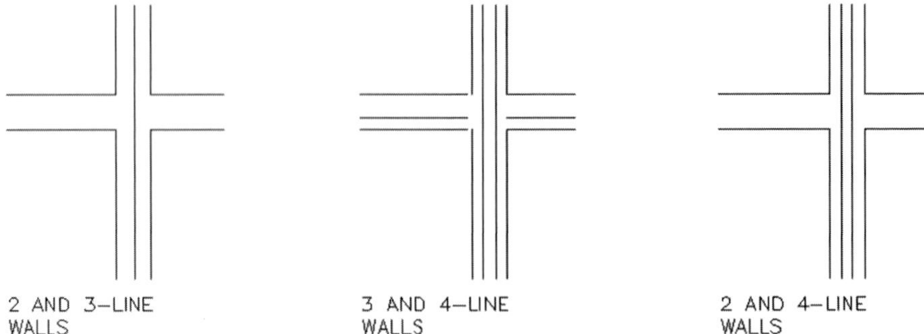

Figure 9-33 The results of cleaning up multi-line walls

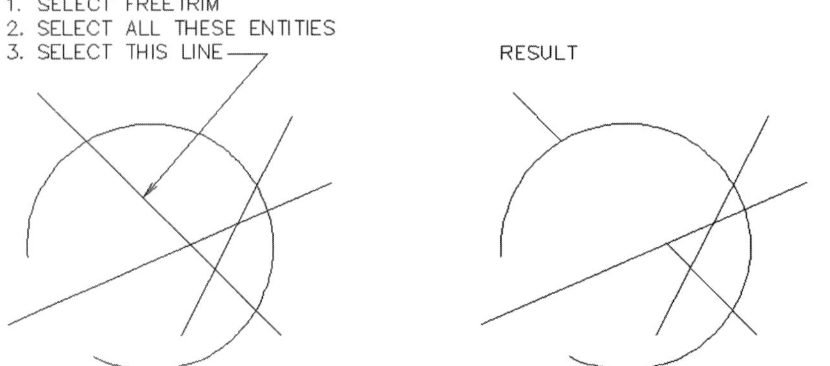

Figure 9-34 `FreeTrim` being used for an unusual trimming

1. Select **FreeTrim**. The EGAFS selection menu appears.
2. Choose a selection method.
3. Pick all the entities to trim to. All selected entities will be highlighted.
4. When you are done selecting entities, press **Begin**.
5. Select the line or arc that you want to trim. DataCAD searches all the highlighted entities for ones that intersect the chosen entity. The section of the entity between the nearest intersections is deleted.
6. Continue selecting lines and arcs to be trimmed.
7. When done, *right-click* or select **Exit** (see Figure 9-34).

Where you select an entity has an effect on how it is trimmed, since DataCAD searches for the nearest intersections from the point you select, as in Figure 9-35.

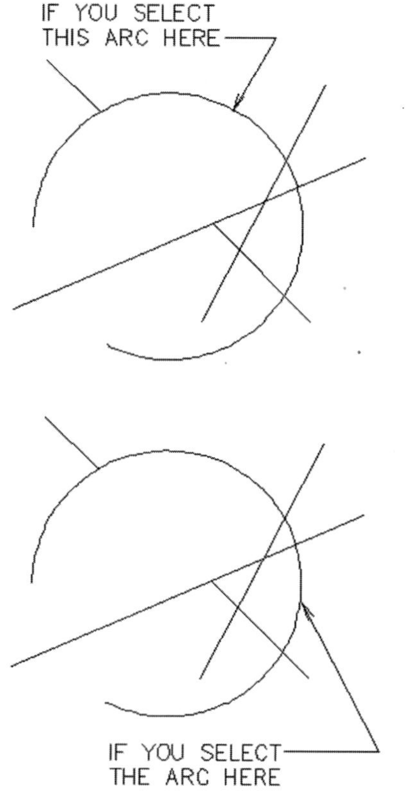

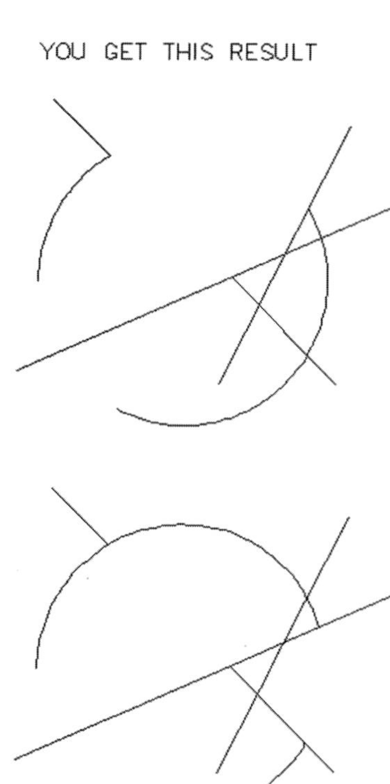

Figure 9-35
Further **FreeTrim** examples

Change

The **Change** menu has several functions used to change the properties or characteristics of various entities in your drawing (see Figure 9-36). When you select **Change**, you will see the standard EGAFS selections and nine more functions.

Note the **Text** option at the bottom. Interestingly, although you change associative dimension properties from within the **Dmension** menu, the **Change** menu is where you make changes to text, not within the **Text** menu, as you might expect.

To change something, perform the following steps:

1. Select **Change**.
2. Select the property you want to change.

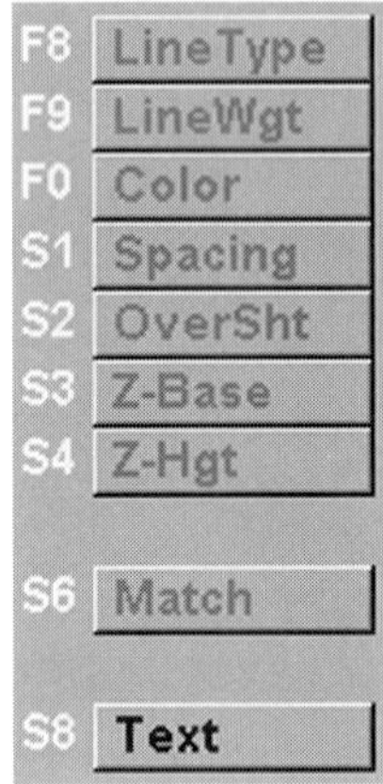

Figure 9-36
The **Change** menu

3. Select the new setting. The **Change** menu will reappear and the property you chose will now be highlighted.
4. Follow Steps 2 and 3 for further property selections.
5. Select by EGAFS. The selected entities are changed to match the new setting.

You can pick one property at a time to change, or you can select multiple properties at one time. So, as an example, you could change an entity's **LineType**, **Color**, and linetype **Spacing** all at once instead of changing each one individually. *Clicking* on any of the nine options (**LineType** through **Text**) when they are already on will subsequently turn them off.

LineType

Use this option to change entities from one linetype to another. When you pick this option, the list of your current linetypes will appear. The first 15 linetypes will appear, along with the **ScrlFwrd** button to move forward in the list.

Pick the new linetype you want to use. The **LineType** menu will disappear and the **Change** menu will reappear. The **LineType** function will be highlighted now, even if you did not select a new linetype.

If the list of linetypes looks a little different from the **Edit/LineType** menu, that's because it is. Although all the linetypes are identical, and in the identical order, the **Edit/LineType** menu only displays the first 10 linetypes in the first menu display. Pressing **ScrlFwrd** causes the next 15 linetypes to display, and so on from there. If you are used to looking for specific linetypes in specific locations within the **Edit/LineType** menu, you may feel a little lost at first in the **Change/LineType** menu.

LineWgt

Use this option to change entities from one line weight to another. Select or type any whole number of 1 or higher. Pick the new line weight you want to use. The **LineWgt** menu will disappear and the **Change** menu will reappear. The **LineWgt** function will be highlighted now, even if you did not select a new line weight.

Color

Use this to change entities from one color to another. Select one of the 15 standard DataCAD colors, or use the **Adjust**, **Custom**, or **Match** functions:

- **Adjust** Selecting this option enables you to alter the color, hue, saturation, and luminance of whatever color you select by using the standard Windows color dialog box. By selecting the **Palettes** option, it also allows you to save the current RGB file or to load a previously saved RGB file.

*NOTE: This setting is a trap door to the **Utility/Display/Palette** option. Adjusting colors or loading a new RGB file will affect the entire drawing file, not just the entity you are changing.*

- **Custom** Selecting this option enables you to pick any custom color, from 1 to 255, from the current Windows color palette.
- **Match** After selecting this, pick an entity in the Drawing Window. The color of that selected entity will be used.
- **NoChange** Selecting this option causes DataCAD to return to the Change menu without selecting a new color. *Right-clicking* will do the same thing.

Pick the new color you want to use. The **Color** menu will disappear and the **Change** menu will reappear. The **Color** function will be highlighted now, even if you did not select a new line color.

Spacing

Use this option to change entities from one line spacing to another (see Chapter 5 if you need an explanation). Select or type any number from .01 to 9999. Pick the new line spacing you want to use. The **Spacing** menu will disappear and the **Change** menu will reappear. The **Spacing** function will be highlighted now, even if you did not select a new line spacing.

OverSht

Use this option to change the overshoot of lines from one setting to another (see Chapter 3, "Settings and Display Options," if you need an explanation). Select or type an overshoot distance. The **OverSht** menu will disappear and the **Change** menu will reappear. The **OverSht** function will be highlighted now, even if you did not select a new overshoot dimension.

Z-Base

Use this option to change the current Z-base of entities (including text) to a new Z-base (refer to Chapter 5 if you need an explanation). Select or type a new positive or negative height. Pick the new Z-base you want to use. The **Z-Base** menu will disappear and the **Change** menu will reappear. The **Z-Base** function will be highlighted now, even if you did not select a new Z-base height.

Z-Hgt

Use this to change the current Z-height of entities to a new Z-height (refer to Chapter 5 if you need an explanation). Select or type a new positive or negative height (changing the Z-height of text entities has no effect on the text). Then pick the new Z-height you want to use. The **Z-Hgt** menu will disappear and the **Change** menu will reappear. The **Z-Hgt** function will be highlighted now, even if you did not select a new Z-height.

Match

Use this option to match any or all of the settings of an existing line in the Drawing Window by following these steps:

1. Select **Match**.
2. Select one or more of the seven options (**LineType** through **Z-Hgt**).
3. DataCAD will prompt you to *"Select entity to* <CHANGE>."
4. Select (via EGAFS) the entity or entities that you want to change.
5. The menu options will disappear and you will be prompted to *"Select entity to match."*
6. Select an entity in the Drawing Window. DataCAD will take note of all its settings (**LineType** through **Z-Hgt**) of the selected entity.
7. The entity or entities you selected in Step 4 will change to match the settings of the entity selected in the Drawing Window in Step 6, but only for the properties you highlighted in Step 2 .

All

This button will appear after you select the Match function. Picking **All** is a quick way to highlight all seven options (**LineType** through **Z-Hgt**).

NOTE: *If you select a property to change (like **LineType**), don't select a new setting, and still pick an entity to change, DataCAD will dutifully change the selected entity. However, since no new linetype is selected in this example, DataCAD will use the current linetype setting (as displayed in the menu Status Area). This is true for all the options, so be careful about which property buttons you leave on when changing entities.*

Layer Locking

Layer Search (**LyrSrch**) is a great tool for weeding out what you want to edit from what you don't, but sometimes it's just not enough. Another method can be used in place of or in conjunction with **LyrSrch:** *layer locking*.

In the **Layers** menu, you will find a button called **Lock**. If a layer is turned on but is locked, then you can see the entities on that layer, but you cannot edit them. This is a great tool when you have to selectively edit entities on specific layers by **Area**, **Group**, or **Fence**. This may sound very much like **LyrSrch**, but with a couple of important differences:

1. With **LyrSrch** turned off, you can freely edit only the active layer.
2. With **LyrSrch** turned on, you can edit all entities on any layer, individually or simultaneously. The drawback is that you often end up accidentally editing entities that you did not want to edit.
3. With layer locking, you can choose exactly which layers you want to be able to edit and which ones you don't while still maintaining the advantage of being able to have any and all layers turned on for viewing.
4. Turn on all the layers you want to view and edit.
5. Click on **Lock** in the Layer menu and select all of the layers that you do not want to edit. A horizontal line (or an X in the DataCAD DOS version) will appear over each layers' button, signifying that it is locked.

6. In the example above, Layers **1-Roof** and **2-NewEx** are locked.

7. Now go to your drawing and move, stretch, erase, and otherwise edit the unlocked layers to your heart's content with no fear of disturbing the locked layers.

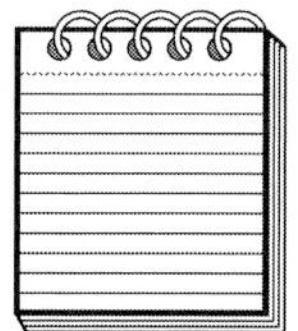

NOTE: *If the active layer is **locked**, you cannot access the **Toolbox** to run macros. Either unlock the active layer or change the active layer (with the **Tab** key or through the **Layers/On/Off** menu) to one that is not locked.*

Editing Multiple Entities

This chapter has so far shown you how to edit individual entities. Often, however, you will want to edit two entities or perhaps 200 entities all at once. DataCAD has two primary means of doing this. One is by the simple linking of entities, and the other is by creating one or more groups of entities, called Edit Sets. Although each method has its differences, the two methods are similar. In fact, you can actually link edit sets together with other entities and groups.

Linking Entities

This function lets you combine entities, groups, areas, or selection sets into a single group that can then be edited together in one step. Let's see how it works and then let's try an example of how you might use it:

1. Select **Edit/LinkEnts**.
2. Choose a selection method (EGAFS).
3. Turn **LyrSrch** on or off, as required.
4. Select each entity or group of entities to be linked together. You can make multiple selections. Note that you can switch back and forth between the various selection methods (EGAFS) at any time during the selection process.

5. Linked entities appear as dashed lines until you exit the **LinkEnts** menu.
6. To finish selecting entities, *right-click* or choose **Exit**.

You can also unlink entities that have been linked. This is useful if you decide later to remove an entity or entities from a linked group, or if you need to unlink an entity that you accidentally linked:

1. Select **Edit/LinkEnts**.
2. Choose a selection method (**EGAFS**).
3. Turn **LyrSrch** on or off, as required.
4. Select each entity or group of entities to be unlinked. You can make multiple selections. Note that you can switch back and forth between the various selection methods (EGAFS) at any time during the selection process.
5. Selected entities appear as dashed lines until you exit **LinkEnts**.
6. To finish selecting entities, *right-click* or choose **Exit**.

Now that you have some linked entities, what can you do with them? Let's try this example. Open the YMCAPLAN.DC5 drawing sample that comes on the DataCAD CD-ROM. If you installed the sample files, then it will be on your hard drive in the \DWG\SAMPLES folder:

1. When the drawing first opens, you will see a door schedule. To view the **A-2** drawing sheet, select **Utility/GotoView** and click on the **A-2** button.
2. Press **Utility/WindowIn/ReCalc** to fill the screen with the drawing.

NOTE: *If you are annoyed with some of the default settings of this drawing, as I am, you can change them before trying these examples. The things I would change are*

- ***Utility/Grids*** *Turning **SnapGrid** off.*
- ***Utility/Display/SmallTxt*** *Set to **0**.*

3. *Right-click* to return to the 2D **Edit** menu. Let's say you wanted to change all the room numbers in this plan. You could select **Change/**

Text/Contents/Entity and select each piece of text, or you could link all the room name labels together to edit them quickly, one after the other. Let's try it.

4. Select **LinkEnts** and make sure the **Entity** option is selected. The message line will read, *"Select entity to* <LINK>."
5. Pick each room number text until they are all selected. By linking the room numbers, they are now part of a common group that can be edited together.
6. Select **Change/Text/Contents** and select the **Group** option.
7. Pick any of the room number texts that you linked. You will now be prompted to enter the new text for each of the linked text entities in turn without having to stop. Now let's say you wanted to change the color of all the text that you linked.
8. Select **Change/Color/Red** and make sure the **Group** option is selected.
9. Pick any of the room number texts that you linked. They will all be changed to the color red all at once.
10. Notice that once the text is linked in Step 5, you do not have to relink the text. They remained linked. That's the power of **LinkEnts**.

More Linking

If you select **LinkEnts** again and pick new entities, the previously linked groups remain linked, but all the newly selected entities become part of a new, completely separate group. Thus, you can still edit previously created **LinkEnts** groups without affecting your new **LinkEnts** group.

If you select **LinkEnts** and one of the entities you select is part of a previously defined group, that entity is now only part of the new group. It ceases to be part of its previous group, yet all the other entities in the previous group remain intact. You can also use the **Group** option to link entities to a previously linked group, and even to link two or more groups together into a new, larger group.

LinkEnts is good for defining temporary groups and is most often used for quick, on-the-fly editing. To create more permanent groups with far more options for defining group members, or to assign individual entities to more than one group at a time, you will want to use the **EditSets** feature instead.

Selection Sets

The **EditSets** option in the **Edit** menu acts much like the **LinkEnts** function, allowing you to place entities into groups, called Selection Sets so that they can be edited or acted on as a group. But unlike **LinkEnts**, **EditSets** enable you to save up to eight Selection Sets and has far more options for the selection of entities. You should be familiar with the EGAFS selection methods in DataCAD, allowing you to select entities by **Entity**, **Group**, **Area**, **Fence** or **SelSet** (Selection Set). Now you know what the **SelSet** option is for. With a selection set activated, you can edit (**Move**, **Copy**, **Change**, and so on) the entities in that group by selecting the **SelSet** option.

To creating a selection set, follow these steps:

1. Pick **Edit/EditSets**.
2. Select **AddTo** so that you can add entities to a selection set.
3. Pick one of the eight selection sets to add entities to. If entities are already in that selection set, the new entities will simply be added to the existing set.
4. Turn **LyrSrch** on or off as you require. Select entities in the drawing file by **Entity**, **Group**, **Area**, **Fence** or **SelSet** (yes, you can add entities from other selection sets into the current selection set).
5. If you want to pick entities with the **Mask** option (explained later), select it prior to selecting by EGAFS.
6. Once entities are placed in a selection set, that set must be made active in order for the **SelSet** selection method to act upon those entities. To do this, go to the main **EditSets** menu and pick **SetActiv**. Pick one of the eight selection sets to become the currently active set. A dollar sign (**$**) will appear in front of the set, indicating that it is now the current selection set.
7. Now when you pick the **SelSet** option, all the entities in the currently active selection set will be acted upon. If you deactivate all the selection sets (by clicking on the active set a second time), none will be acted upon.
8. You can only have one selection set active at one time.

The main **EditSets** menu is shown in Figure 9-37.

The Selection Set menu is shown in Figure 9-38.

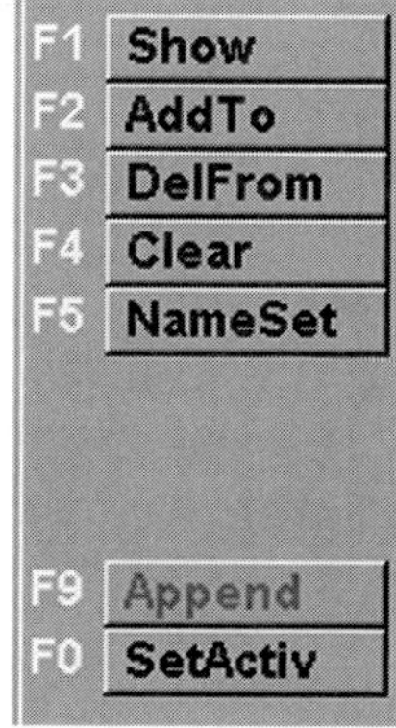

Figure 9-37
The **EditSets** menu

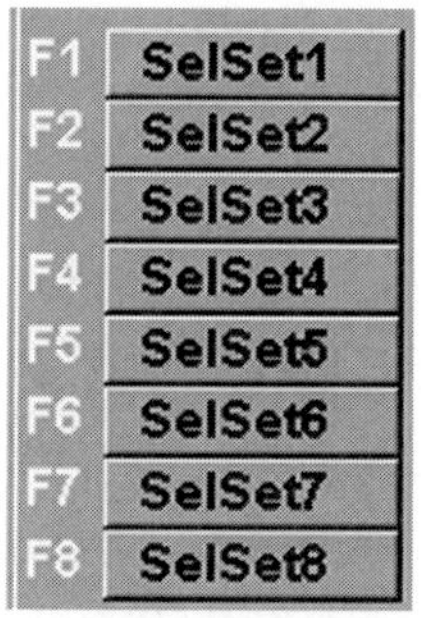

F1 SelSet1
F2 SelSet2
F3 SelSet3
F4 SelSet4
F5 SelSet5
F6 SelSet6
F7 SelSet7
F8 SelSet8

Figure 9-38
The Selection Set menu

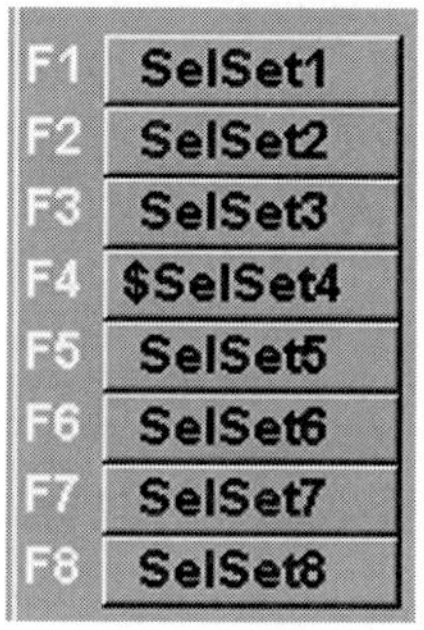

F1 SelSet1
F2 SelSet2
F3 SelSet3
F4 $SelSet4
F5 SelSet5
F6 SelSet6
F7 SelSet7
F8 SelSet8

Figure 9-39
Selection set 4 is the active set here.

In Figure 9-39, selection set number 4 (SelSet4) is the active set, as evidenced by the dollar sign (**$**) in front.

We'll now discuss what the various options are for. The **Mask** function is by far one of the most powerful features in DataCAD and for that reason it is explained in depth after the main options.

Show

This command will show you all the entities in a selection set:

1. Pick **Show** to display the SelSet list.
2. Pick the selection set with the entities you want to see. The entities will be shown as gray and dashed.
3. *Right-click* or press **Exit** to exit the **Show** menu.

AddTo

Use this option to add entities to a new selection set or to add more entities to an existing selection set:

1. Select **AddTo** to display the SelSet list.
2. Pick the selection set to add entities to.
3. Select enties by EGAFS or **Mask**.
4. If you want that selection set to be the active set, *right-click* back to the main **EditSets** menu, pick **SetActiv**, and then pick the selection set.
5. *Right-click* or press **Exit** to exit the **AddTo** menu.

DelFrom

Use this option to remove entities from an existing selection set:

1. Select **DelFrom** to display the SelSet list.
2. Pick the selection set you want to delete entities from. All the entities currently in that selection set will be shown as gray and dashed to make it easy to see which entities you want to delete.
3. Select the enties to be removed by EGAFS or **Mask**. As entities are removed from the set, they will change from gray and dashed to their original colors and linetypes.
4. *Right-click* or press **Exit** to exit the **DelFrom** menu.

Clear

Use this option to remove all entities from an existing selection set:

1. Select **Clear** to display the SelSet list.
2. Pick the selection set you want to delete all the entities from.
3. The message line will ask, *"Are you sure you want to clear it?"*
4. Selecting **Yes** will remove all the entities from that selection set and exit from the **Clear** menu.
5. Selecting **No** will enable you to exit the **Clear** menu without removing any entities. The current selection set will remain intact.

NameSet

This option enables you to name any of the eight selection sets to make it easier to remember what is in each one. You can use up to eight characters. To use this option, follow these steps:

1. Select **NameSet** to display the SelSet list.
2. Pick the selection set you want to rename.
3. Type in the new name and press **Enter**.
4. *Right-click* or press **Exit** to exit the **NameSet** menu.

Append

This option is another way to add entities to a selection set. By activating a selection set and toggling **Append** on, every entity added to your drawing will automatically be added to the currently active selection set. To discontinue adding entities to the set, either toggle the **Append** option off, or deactivate all selection sets:

1. Select **Append**. When on, the button will be depressed, and the text will change to green (when off, the button will not be depressed, and the text will be red).

2. Now select **SetActiv** to display the **SelSet** list.
3. Pick the selection set to be active. Future drawing entities will be added to this set.
4. *Right-click* or press **Exit** to exit the **SelSet** menu.
5. As you draw, mirror, copy, and so on, the new entities will be added to the active selection set until you turn **Append** off or deactivate all selection sets.

SetActiv

Use this option to activate or deactivate any of the eight selection sets. Only one selection set may be active at one time. To use **SetActiv**, follow these steps:

1. Select **SetActiv** to display the **SelSet** list.
2. To make a selection set active, pick it from the list. A dollar sign (**$**) will be placed to the left of the **SelSet** name.
3. To deactivate the currently active set (the one with the dollar sign), pick it again. The dollar sign (**$**) will disappear, indicating that no selection sets are active.
4. *Right-click* or press **Exit** to exit the **SelSet** menu.

Masking

Masking is just one aspect of **EditSets** but is by far the most powerful. **Mask** is an option after selecting **AddTo/SelSet#** or **DelFrom/SelSet#**. If you have ever done any artwork or even painted your home, you are already familiar with masking. When painting, you use masking tape (now you know where the name comes from) to select areas to be painted or not, simply by placing the tape in the correct spots. Masking enables you to be selective about what you paint and thus masking in DataCAD is a way to be selective about what you edit. Refer to the earlier directions for **AddTo** and **DelFrom** for the steps involved in using **Mask** with **EditSets**.

Picking **EditSets/AddTo** or **EditSets/DelFrom** will bring up the main **SelSet** menu. Pick the selection set where you want to add or delete entities. Then the EGAFS menu will appear along with the **Mask** option and either the **DelFrom** or **AddTo** option. Selecting **Mask** will bring up the menu shown in Figure 9-40.

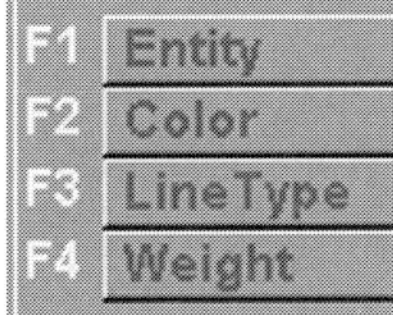

Figure 9-40
The **Mask** menu

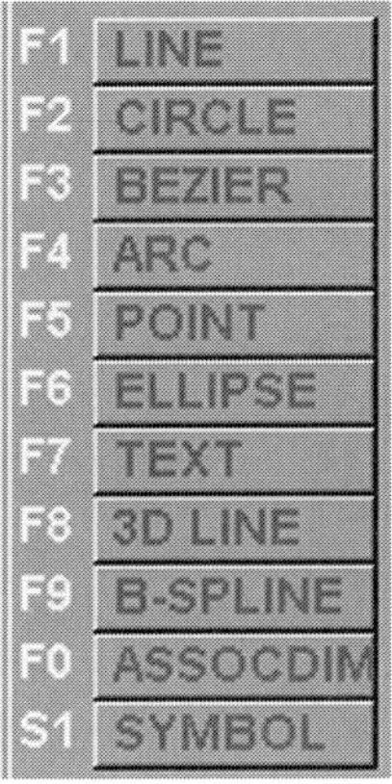

Figure 9-41a
The **Entity** menu

The Entity, Line Type, Weight, and Color submenus are shown in Figures 9-41a through 9-41d.

Instead of meticulously picking individual entities, and sometimes picking ones you didn't mean to pick, you can select exactly the type of entity or entities you want. Looking at the previous four submenus, you can see that these options enable you to be very specific about the entities you pick, especially if you use them in combinations. You can select any or all of the options in any or all of the four menus. For instance, if you select all of these:

- **Entity/Circle/Ellipse**
- **Color/Red/LtRed**
- **LineType/Dashed**
- **Weight/3**

then you will select only circles and ellipses that are **Red** or **LtRed** with a dashed linetype and a lineweight of 3. No other entities will be selected for masking. After selecting those options, if you then select by **Area** or **Fence**, DataCAD will only select the entities matching those parameters, making your job of selecting those entities much easier.

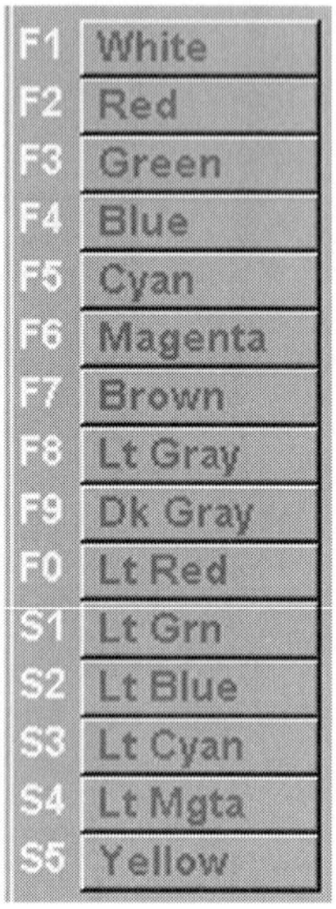

F1 White
F2 Red
F3 Green
F4 Blue
F5 Cyan
F6 Magenta
F7 Brown
F8 Lt Gray
F9 Dk Gray
F0 Lt Red
S1 Lt Grn
S2 Lt Blue
S3 Lt Cyan
S4 Lt Mgta
S5 Yellow

Figure 9-41b
The **`Color`** menu

F1 Solid
F2 Dotted
F3 Dashed
F4 Dot-Dash
F5 ElecLine
F6 TelLine
F7 Box
F8 PropLine
F9 Insul1
F0 NewPlywd
S1 Plywood2
S2 Hedge1
S3 CentrLin
S4 NewSectn
S5 DShngl_R
S7 ScrlFwrd
Solid

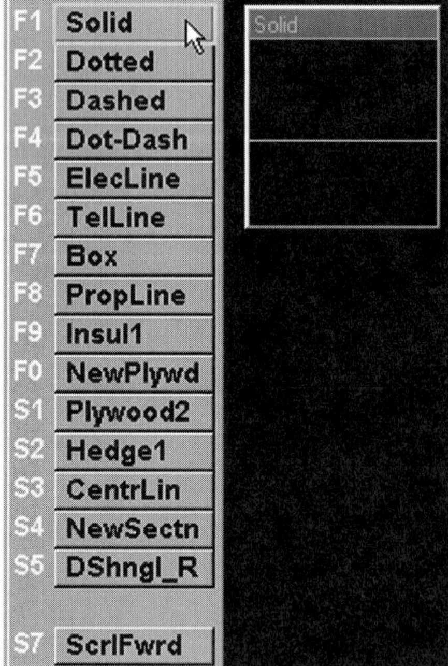

Figure 9-41c
The **`Line Type`** menu

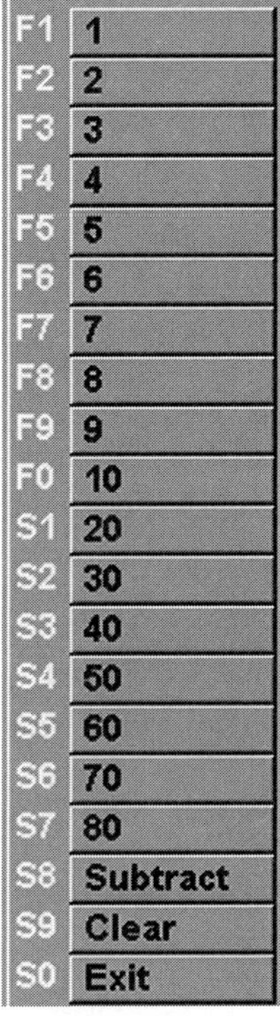

Figure 9-41d
The line **Weight** menu

Would you ever use the selection criteria shown previously? Maybe not, but you could probably think of many other combinations that you would use. For instance, perhaps you want to change all the red text in your drawing to blue. All you would have to do is turn on all the layers of your drawing (or only the ones you want to change), then use the **Mask** option and select:

- **Entity/Text**
- **Color/Red**

Now all the red text could be changed to blue by selecting **Edit/Change/Color/Blue**.

Using a Masking Keyboard Shortcut

Although Chapter 23 will cover the creation of custom *keyboard macros* (also called *keyboard shortcuts*), this following keyboard macro is so highly useful and highly recommended that it is worth noting here. If you're not familiar with keyboard shortcuts in DataCAD, see Chapter 23 first and then return here.

Let's look at one often-used option as an example: masking by **Color**. Let's say you want to change all of your red entities in a drawing to blue, and let's then say that these entities are on different layers. You could pick **Change/Color** and tediously change each entity by EGAFS. Or better yet, you could tell DataCAD to pick (mask) all the red entities and temporarily group them together into a selection set. We use this so often in our office that we have a keyboard macro dedicated to it (see Chapter 23 for information on how to create custom keyboard macros). Here are the commands for that macro:

H^;^S2^F4^F8^F5^F0^F1^F8^S0^F2^F8^F3^F0^F2^S9^

If you run this macro, you will see the **Mask/Color** menu, shown in Figure 9-42.

To run the keyboard macro and group only red entries, follow these steps:

1. Run the keyboard macro (**Alt+H**).
2. Click on the **Red** button to select that color as the color you want to mask.

Figure 9-42 The **Mask/Color** menu

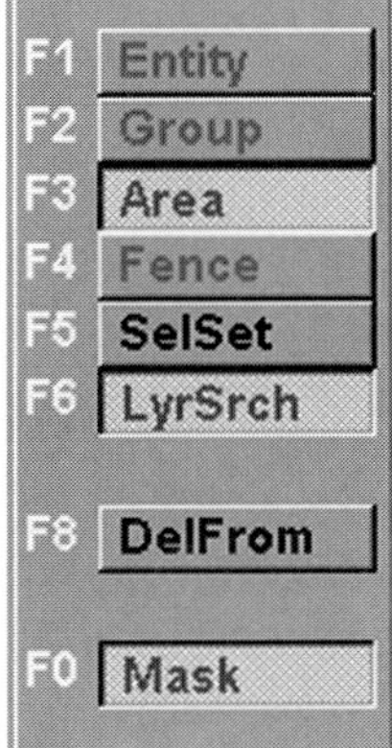

Figure 9-43 EGAFS selection menu

3. *Right-click* twice to get to the EGAFS selection menu (see Figure 9-43).
4. Now select the **Entity**, **Group**, **Area**, or **Fence** option (you cannot use the **SelSet** option, since you are actually in the process of creating a new selection set right now). In Figure 9-43, the **Area** option has been selected. If you will be selecting entities on multiple layers, make sure **LyrSrch** is on before making your selections.
5. More often than not, you will be using the **Area** or **Fence** options. With either of these, you can place a bounding box around your entire drawing of selected areas. In this example, all the Red entities will be selected.
6. You can continue to make multiple selections until you *right-click* the mouse. Once you *right-click*, the selection process ends. All the Red entities you picked have now been grouped into a selection set.
7. Now to change all the selected Red entities to Blue.
8. In the DataCAD **Edit** menu, select **Change/Color/Blue/SelSet** and all the red entities you masked will be changed to blue.

The key here is the **SelSet** command. As noted earlier, eight possible selection sets are available in DataCAD. You can have entities saved simultaneously in all eight of them at the same time, but you can have only one of these selection sets active at any one time. In the case of the keyboard macro we mentioned previously, we used the last selection set, **SelSet8**. Let's see exactly what each item in the keyboard macro does:

- **H** **Alt+H** will initiate this macro
- **;** **Edit** menu
- **S2** **EditSets** (to edit selection sets)

- **F4 Clear**
- **F8 SelSet8**
- **F5 Yes** (to clear it of any previous entities)
- **F0 SetActiv**
- **F1 SelSet1** (If **SelSet8** is already active, pressing it again will turn it off, so **SelSet1** is picked first, and then **SelSet8** is made active in the next step.)
- **F8 SelSet8** (to make it the active **SelSet**)
- **S0 Exit**
- **F2 AddTo** (to add entities to a **SelSet**)
- **F8 SelSet8** (add the entities to **SelSet8**)
- **F3 Area** (choose by **Area**; you can set this to something else if you prefer)
- **F0 Mask**
- **F2 Color**
- **S9 AllOff** (turns off all previous color selections in case there were any)

Whew! Now you know why it's a keyboard macro! From here you can select the color or colors (yes, you can choose more than one at a time) that you want to mask by.

This keyboard macro first makes sure that **SelSet8** is empty (or cleared), makes it the active selection set, and then prompts the user to **Mask** by color. You pick the color and then you are prompted to select entities like you usually do by EGAFS. Only entities of the color you picked will be added to the selection set (but note that entities on locked layers or layers that are turned off cannot be selected). When you run the macro and select entities by color, all the entities you pick are saved in **SelSet8**, which is currently the active **SelSet** because of the macro. **SelSet8** will remain the active selection set until you run the macro again or until you select a new active set with the **EditSets/SetActiv** option.

In the next to last step (**F2 - Color**), you could instead choose by **Entity** (Line, Circle, Bezier, Arc, Point, Ellipse, Text, 3D_Line, B-spline, AssocDim, and/or Symbol), **LineType**, or Line **Weight**. I could have made a keyboard macro to get me only to that step, from which I could choose one of those options, but **Color** is the most often used choice.

If I want to choose by **Entity/Text** instead, then I run the macro and *right-click* the mouse once to back up one menu. The menu will now look like Figure 9-44.

Deselect the **Color** option by clicking on it. The menu will then look like Figure 9-45.

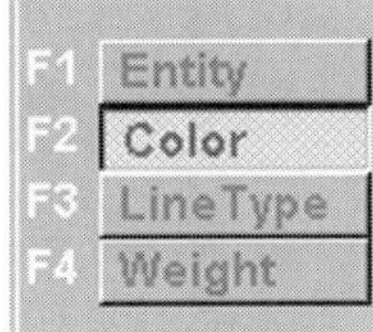

Figure 9-44
`Color` is selected.

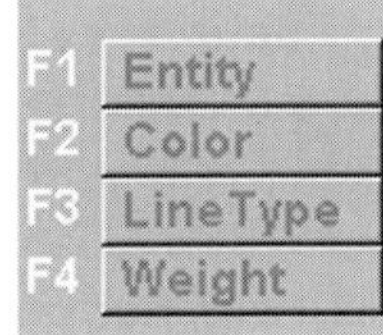

Figure 9-45
`Color` is de-selected.

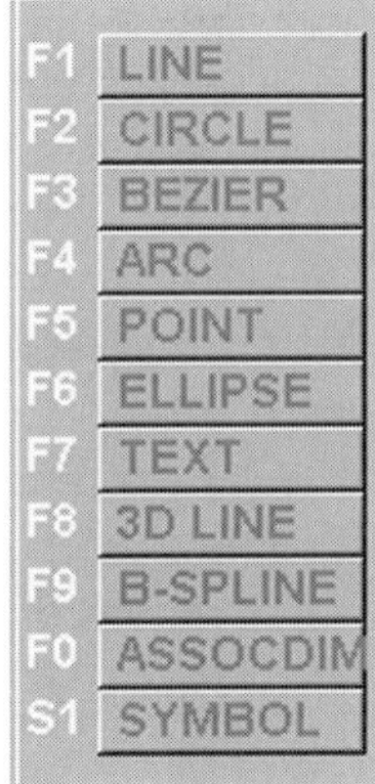

Figure 9-46
The **`Entity`** menu

Now you can choose by something other than **Color**. For instance, if you select the **Entity** option, the menu will now look like Figure 9-46. now you could select any of the **Entity** options instead.

It is important to realize that you can choose multiple, concurrent selections when masking. In the previous example, you do not have to deselect the **Color** option. You caould have choosen all **Text** that is **Red** by selecting both the **Color** and **Text** options, you could deselect the **Color** option and choose only **Text**, you could choose all **AssocDims**, **3D_Lines**, and **Arcs**, or you could choose all **AssocDims**, **3D_Lines**, and **Arcs** that are **Blue** at the same time. Mix and match as you please. That's what makes it so powerful.

So many options exist that we can't possibly go through them all here, but now that you have the idea, go ahead and experiment on your own with all the options in the **Entity**, **Color**, **LineType**, and **Weight** menus. You will probably be amazed at the new tricks you can get DataCAD to do.

ClipIt Macro

In case you are not already aware of it, macros are mini-programs that run within DataCAD to accomplish specialized tasks that the base DataCAD program does not address. All macros are accessed from the **Toolbox** button found in the **Edit** menu. One of those macros is called **ClipIt**.

The **ClipIt** macro is an editing tool that enables you to cut, copy, and paste specific areas of entities, as defined by a rectangular or circular area, or a fence. Think of the **ClipIt** macro like a cookie cutter. You can cut out all the cookie dough within the cookie cutter (clipping), or you can keep the dough inside the cookie cutter and discard all the dough outside of it (cropping). But unlike a cookie cutter, **ClipIt** also allows you to copy the dough within the cookie cutter (for placement elsewhere) without actually removing the dough.

ClipIt only works in an orthographic (plan) view, so if you are in another view, like perspective or oblique, DataCAD will automatically switch to orthographic mode. To use **ClipIt**, perform the following steps:

1. Go to **Edit/Toolbox** and run the **ClipIt** macro by either
   - **a.** Highlighting the **Clipit.dcx** file and then selecting **Open**

     or
   - **b.** *Double-clicking* on the **Clipit.dcx** file.
2. The **ClipIt** menu will appear.
3. Select the type of clipping: **Cut**, **CutCopy**, or **Copy**.
4. Select the method of clipping: **Clip** or **Crop**.
5. Select the types of entities you want to edit. You can select one, two, three, or all four of the options: **2D Line**, **3D Line**, **2D Arc**, and **Circle**.
6. You can use the **Layers** option to temporarily go to DataCAD's **Layers** menu to turn layers on or off.
7. Turn **LyrSrch** on or off as appropriate.
8. Select the type of clipping boundary to use: **Area**, **Fence**, **RadCirc**, or **DiaCirc**.

9. Turn **Boundry** on or off as appropriate.
10. In your drawing, use the cursor to define the boundary to be clipped.
11. If you selected the **CutCopy** or **Copy** options, you will be prompted for where you want to move or copy the clipped selection.

When the **ClipIt** menu appears, it looks like Figure 9-47.
Let's look at what each function does:

- *Selection Methods* You can select only one of these options:
  - **Area** uses a rectangular bounding box to select the area to be clipped or cropped.
  - **Fence** uses a polygon to select the area to be clipped or cropped. Fence polygons can have a maximum of 36 vertices and can include convex and concave arcs, but none of the polygon lines may cross one another.

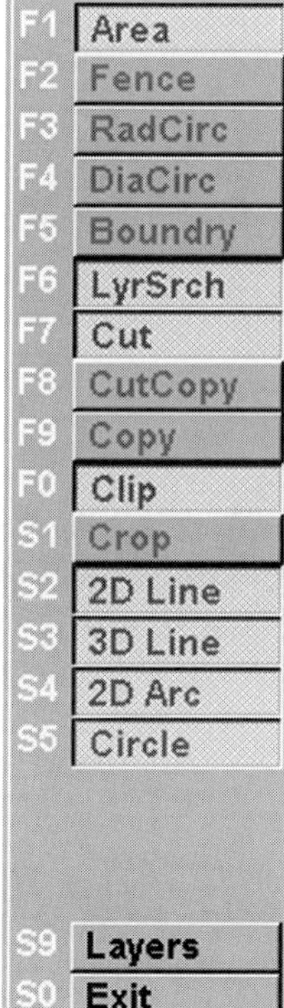

Figure 9-47
The **ClipIt** menu

- **RadCirc** uses a circle to select the area to be clipped or cropped. The circle is defined by selecting the radius point and then the outside boundary of the circle.
- **DiaCirc** uses a circle to select the area to be clipped or cropped. The circle is defined by selecting the opposite sides of the circle.

- *Options* You can turn either of these options on or off, as required:
  - **Boundry** With this option turned on, a visible boundary will be drawn around the clipped or cropped area. The current layer, linetype, and color will be used to create the boundary. With this option turned off, a visible boundary will not be drawn.
  - **LyrSrch** This is a standard DataCAD option. With **LyrSrch** on, DataCAD will clip or crop all the visible layers within the ClipIt selection area. With **LyrSrch** off, DataCAD will only clip or crop entities on the currently active layer.
- *Clipping Type* You can select only one of these options.
  - **Cut** All the selected entities within the **ClipIt** selection area will be erased from the drawing.
  - **CutCopy** This option might be more appropriately called **CutMove**. With this option, all the selected entities within the **ClipIt** selection area will be erased, but then you can move them to another location or layer in your drawing. Think of the cookie cutter analogy, where you cut the dough with the cutter and move it somewhere else.
  - **Copy** All the selected entities within the **ClipIt** selection area will remain uncut but will be copied. You will then be prompted for a location or layer to copy the entities to.

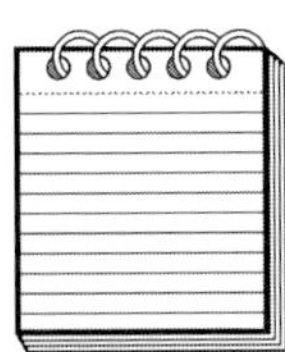

NOTE: *When using **CutCopy** or **Copy** with **LyrSrch** turned on, you can only copy the selected entities to one layer. The selected entities will not be moved or copied to the same layers on which they originally existed.*

- *Clipping Method* You can pick only one of these options:
  - **Clip** This option processes the entities within the **ClipIt** selection area. For instance, with both **Cut** and **Clip** turned on, all the entities within the selection area would be erased.
  - **Crop** This option processes the entities outside of the **ClipIt** selection area. For instance, with both **Cut** and **Crop** turned on, all

the entities outside of the selection area would be erased, while all the entities within the selection area would be retained.

*WARNING: When using the **Crop/Cut** or **Crop/CutCopy** options, especially with **LyrSrch** on, the entire drawing outside the **ClipIt** selection area will be deleted. So make sure this is really what you want to do.*

- *Entities to Clip* You can select one or any number of these options simultaneously:
  - **2D Line** includes 2D line entities in the clipping and cropping processes.
  - **3D Line** includes 3D line entities in the clipping and cropping processes.
  - **2D Arc** includes 2D arc entities in the clipping and cropping processes.
  - **Circle** includes 2D circle entities in the clipping and cropping processes.
- *Other Options:*
  - **Layers** Pressing this button will take you to the standard DataCAD Layers menu where you can turn layers on or off and access all the other options available in the Layers menu.
  - **Exit** Use this option to exit the **ClipIt** macro and return to the standard DataCAD menus.

Things ClipIt Will Not Clip or Crop

Some entities **ClipIt** will not clip, including the following:

- Symbols
- *Text* If the text is fully inside the **ClipIt** boundary, then it will be clipped, but if part of a line of text is inside, and part is outside, **ClipIt** will not clip any of the text.
- Associative hatching
- Associative dimensions
- 3D entities other than 3D lines. For instance, you cannot clip or crop 3D slabs or cylinders.

Clipping Examples

Figures 9-48a through 9-48c are some examples of what the **Clip** option can do with the various clipping types selected.

Cropping Examples

Figures 9-49a through 9-49c show some examples of what the **Crop** option will do with the various clipping types selected.

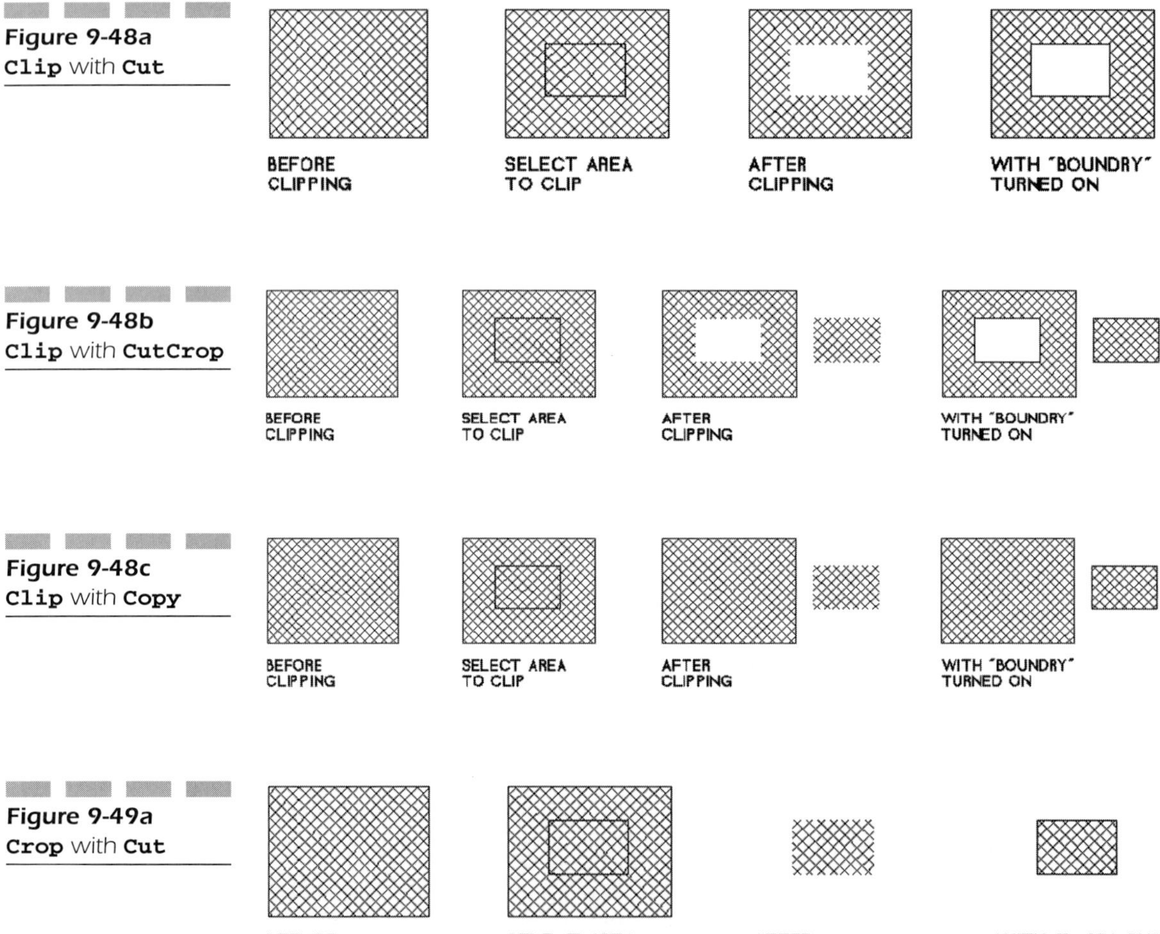

Figure 9-48a Clip with Cut

Figure 9-48b Clip with CutCrop

Figure 9-48c Clip with Copy

Figure 9-49a Crop with Cut

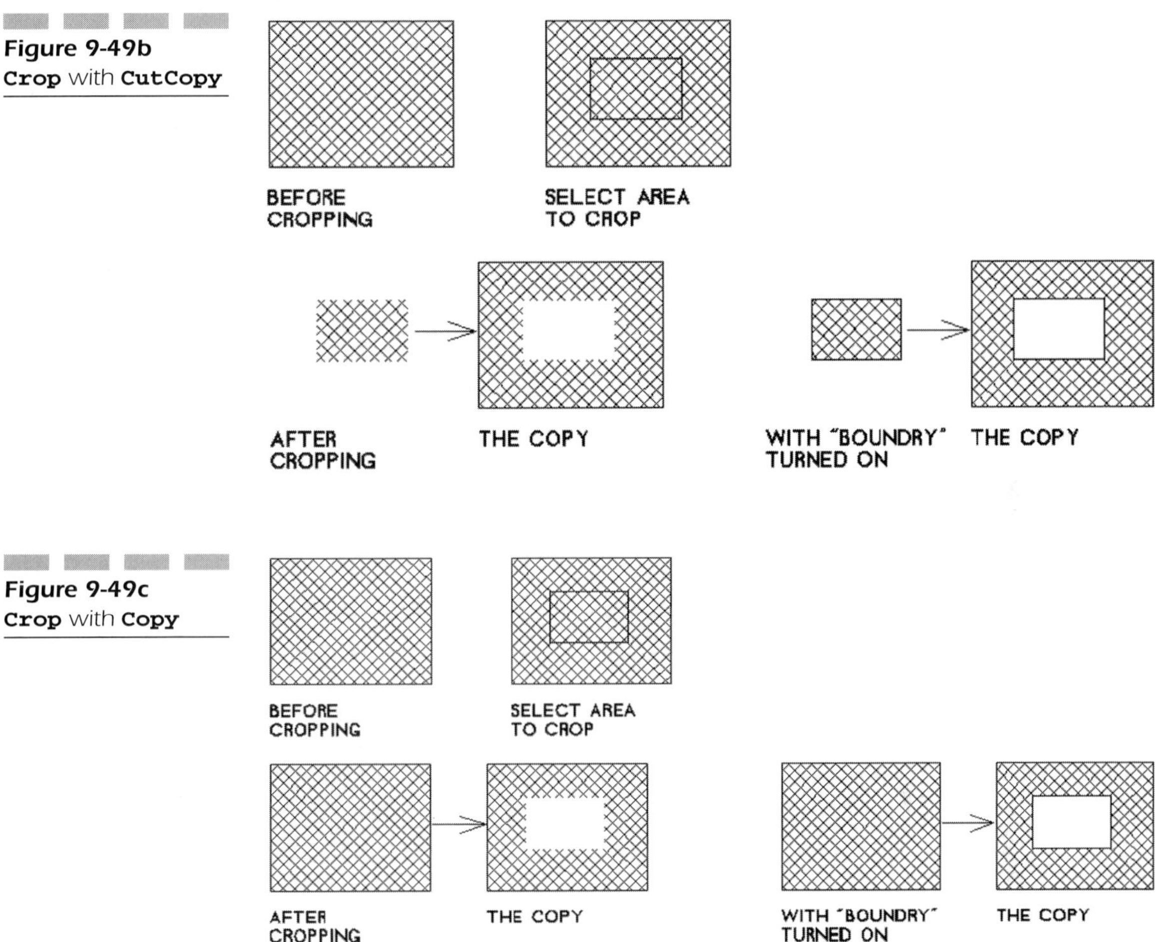

Figure 9-49b
Crop with **CutCopy**

Figure 9-49c
Crop with **Copy**

Notes about ClipIt

Here are some things to keep in mind when using the **ClipIt** macro:

- This macro is sometimes prone to crashing DataCAD. For that reason, in our office we always insist that users save the current drawing prior to running **ClipIt**. That way, if DataCAD crashes, we won't lose any drawing information.
- **ClipIt** generally works fastest when you use a rectangular boundary or a fence with three or four sides. The processing speed of a circular boundary is slower than that of a rectangle. **ClipIt** is slowest when the boundary is a fence of more than four sides.

CHAPTER 10

Creating Efficiency and Order

Although "plain vanilla" DataCAD is by itself very intuitive and efficient, this chapter will describe some of the features and customization options that will help fine-tune your use of the software. If you read nothing else in this chapter, at least make sure you read the section on 3D *GotoViews* (GTVs). This feature is by far *the* most important productivity feature in DataCAD.

3D GotoViews

DataCAD's 3D GTVs are one of its greatest features. It's like taking a Polaroid snapshot of your current Drawing Window that can then be instantly retrieved at any time in the future. You will want to use 3D GTVs as the primary means of retrieving predetermined details and views.

Very simply, 3D GTVs work by memorizing and displaying the following information:

- Which layers should be turned on
- What are the extents of the viewing area when saved
- Whether the Clip Cube is on or off

If you add new layers to a detail or want to add other layers to a view, you must turn them on, WindowIn to your view (if so desired), and then update the GTV.

The GTV menu button is found in the **Utility** menu and is labeled **GotoView**. They are called 3D GTVs because they will save 3D information as well as 2D. Up until Version 9, DataCAD had both 2D and 3D GTVs, each of which had different properties. 2D GTVs were accessed by clicking on the **Utility/GotoView** button, while 3D GTVs were accessed in the 3D menus by selecting **Edit/DCAD_3D/3DViews/GotoView**. Because 2D views were much more limited compared to 3D views, the 2D views were recently eliminated, and the Utility/GotoView button was changed to access the 3D views.

Let's look at an example of why DataCAD's GTVs are so important to efficiency. Let's say you have a drawing file with 100 layers containing a first, second, and third floor plan with four elevations. You want to view and edit only the third-floor plan layers. Without GTVs, you would have to go to the **Layers** menu, turn on only the third-floor plan layers (assuming you even remember which ones are supposed to be on and which are supposed to be off), turn off all the layers not associated with the third floor, and use WindowIn or Extents to get the view you want. With a list of 100 layers, some of which may not be in order, the task of constantly turning on and off the layers to work on a new floor plan or elevation becomes tedious and incredibly time-consuming.

Saving a GTV of the third-floor plan allows you to instantly get the same results as the previous manual method, but in a fraction of the time with a fraction of the effort. It also has the advantage of ensuring that anyone who opens the drawing file and selects the third-floor GTV will instantly have the correct third-floor layers turned on. This is essential to maintaining uniformity and organization within drawing files. And since GTVs can be updated as layers are added, changed, and deleted, they themselves are easy to maintain.

It is important to realize that a GTV is not a static picture of your drawing. It is an internal function that simply tells DataCAD to display the drawing as it currently exists by turning on certain layers and WindowingIn to a particular view. So as entities are added and erased from the third floor plan example, the GTVs will reflect those changes instantly. You do not have to save or update a GTV every time you make a change to the entities in the drawing. You only need to update a GTV if you add or delete layers from it or to change the extents of the view.

The newest addition to the 3D GTV system is the **HyprView** (**Hyper View**) option. It enables you to link any drawing entity, or group of entities, to any 3D GTV in the current drawing file. We'll cover this feature at the end of the description of 3D GTVs.

Save a View

To save a view, perform the following steps:

1. Make sure all the appropriate layers are on or off. Get the view that you want by using **ReCalc**, **PgUp/PgDn**, **WindowIn**, and so on.
2. Select **Utility/GotoView/AddView**.
3. You will be prompted to input a name (up to eight characters). Remember to make it a useful name that anyone can understand. Type in the name for the view and then press **Enter**.
4. *Right-click* to exit the **GotoView** menu and to go back to the 2D menus.

Recall a View

To recall a saved 3D GTV, you have two options:

- Select **Utility/GotoView** and then select the view from the menu, or
- Press **Shift+#** (# is a number from 1 to 10 on the keyboard). This enables you to select any of the first 10 saved 3D GTVs quickly. The downside is that you need to remember which of the 10 views you want. If you have the first three views as floor plans and the next four as elevations, for instance, you might be able to remember them.

Update a View

To update a view, you would do the following:

1. Get the image you want on the screen. Make sure all the appropriate layers are on or off. Get the view that you want by using **ReCalc**, **PgUp/PgDn**, **WindowIn/Out**, and so on.
2. Go to the 3D **GotoView** menu.
3. Select **Update**.
4. From the current list, choose the 3D GTV you want to update. The selected view will now contain all the new parameters as you see them on the screen. The previous parameters have been overwritten.

Warnings about 3D GTVs

Onoccasion, DataCAD has been known to crash if you delete layers and then pick a 3D GTV that contains the deleted layer. For some reason, this seems to vary with computer hardware. In our office, we haven't experienced a crash like this for some years now, but for safety's sake, be careful about deleting layers. If you do delete a layer that is in a 3D GTV, make sure you delete the GTV or update it with the new settings and layers prior to selecting it for viewing.

If you combine two drawings that each have saved 3D GTVs via the **LyrUtil** (Layer Utility) macro, the 3D GTVs will be successfully merged. However, if duplicate layer names exist in the two files, then the 3D GTVs that contain those layers will act erratically. Specifically, when you pick a 3D GTV, DataCAD will scan the layer names starting with the first layer. Whichever duplicate layer names are found first will be the layers that are turned on by the 3D GTV, while the remaining duplicate layers will not be turned on.

NOTE: How 3D GotoViews Are Saved in DataCAD
Like many other things in DataCAD, the layers associated with a 3D GTV are not stored based upon the layers' names. Instead, the layer list is based upon the "Logical Address" (a long pointer (Page / Offset) into memory) of the layer(s). This is why renaming a layer doesn't affect its membership in either 3D GTV's or MSP Details.

As a side note (as long as we're on layers), there is often confusion about renaming Layer files. The confusion arises when a renamed Layer file is loaded into a drawing and still reflects its original layer name. This is due to the layer's name being stored within the header of the file, where it is totally disconnected from the original drawing file's name. This is best illustrated by examining the file names that result from saving out layers with the **LyrUtil** *macro. You will find that the Layer files created by that macro are: 00000001.LYR, 00000002.LYR, 00000003.LYR . . . and so on. So although the layers are saved with numeric file names, when the layers are read back into a new drawing their original layer names reappear because those names were saved in the header of the layers, and DataCAD coughs them back up.*

HyprView (Hyper View)

Link View is one of DataCAD 9's new features and is an extension of the 3D GTV options. It enables you to link any drawing entity, or group of entities, to any 3D GTV in the current drawing file. This can be very useful for tying together groups of related items. For instance, you could associate the following groups with each other so that you could move back and forth between them with just a mouse click:

- Window symbols with a window schedule
- Door symbols with a door schedule
- A section cut with the actual section detail
- An elevation symbol in a plan with the actual elevation view

Here are some examples:

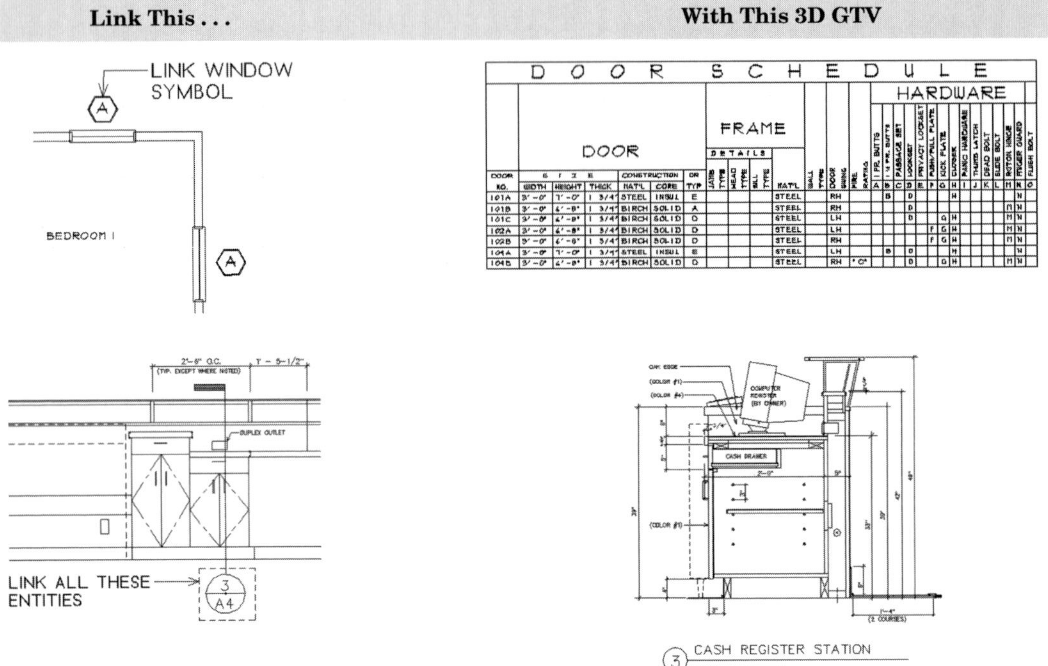

Create a Link: Method #1

If you haven't done so already, create a 3D GTV of the view to be linked to. Then follow these steps:

1. Go to **Utility/GotoView** and select the **HyprView** option. You will be prompted to *"Select the view to link to an entity."*
2. From the current list of views, select the 3D GTV that you want to link to. You will be prompted to *"Select entity to* <HyperView>."
3. Use one of the EGAFS selection methods to select an entity or entities in the Drawing Window to be linked to the previously selected GTV.

Create a Link: Method #2

This method only works if you are going to link a single entity, rather than a group of entities. Follow these steps:

1. Place the mouse on the entity that you want to link to, hold down the **Alt** key, then click the mouse. The current 3D GTV list will be displayed and you are prompted to *"Enter name of new view or select an existing view to link to this entity."*

2. Select an existing 3D GTV from the list or type a new 3D GTV name (up to eight characters). If you type a new viewname, the current Drawing Window will be made into a 3D GTV and the view will be added to the end of the existing 3D GTV list.

Recall a Linked View

To recall a linked view:
Hold down the **Alt** key, place the cursor on a previously linked entity, and click the mouse. The linked 3D GTV will be displayed in the Drawing Window.

Other HyprView Ideas

You might want to try a couple of other ideas. First, you currently have no way to automatically jump back to the view that you were working in when you jumped forward to the linked 3D GTV. So in the above example where we linked the 3/A4 detail title to the section through the cash register station, once your view jumps to the section through the cash register station, you have no way to automatically jump back to the elevation of the station, but a simple work-around can solve that:

1. Create a 3D GTV of the cash register elevation where the section cut title is shown. Let's call it **REGELEV**.
2. Go to the view of the cash register section. Selecting the 3D GTV is the fastest way to get there.
3. Select **HyprView** and pick the previously created **REGELEV** GTV of the cash register elevation.
4. Pick the **Area** selection method, and draw an **Area** bounding box around the whole title of the register section detail. That way you will be able to select any entity in the title.

Now you have dual links for these two views, enabling you to move back and forth between the two details.

The second idea involves the use of a non-printing color. Many offices assign one of DataCAD's colors to be a non-printing color. That is, they pick a color like DkGrey and assign it a pen width of 0.0 (in the **Plotter/PenTable** dialog). That way they can write notes to themselves or draw guidelines in the drawing, but those notes and guidelines will not print. If you use the Internet at all, then you are familiar with hyperlinks. Click on a hyperlink and you are sent to a new Web page. Sound familiar? That's the premise of the **HyprView** feature.

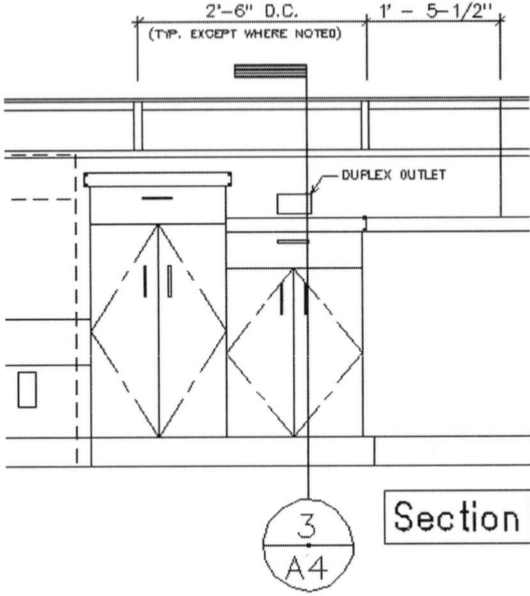

Figure 10-1 Adding a non-printing Hyper View link (the "Section" tag)

So, to put the idea of non-printing entities and hyperlinks together, you could add some non-printing text or entities in each of your details and then create your links using those entities. That way no confusion or ambiguity exists about which entities are linked to which views. In this example, the word Section and the box around it were created with a non-printing color so that they won't print. It, was then linked to the 3D GTV of the cash register section (see Figure 10-1).

The word Section is descriptive enough to tell anyone who views the detail that clicking on that link will take them to a view of the section through the counter.

Currently, the Link View feature is missing a couple of features that it may get in the future:

- You cannot unlink entities and views. You can, however, reassign links. If you try to link an entity that is already linked, you will get a message as shown in Figure 10-2.
- You have no way to tell which entities have been assigned to a linked view. So, you can't tell if I have already associated an entity or group of entities with a view.
- If you copy (or mirror) an entity that has a linked view associated with it, the copied entity retains the same link as the original entity.
- Links cannot be made with views in other drawing files. You are limited to links within the current drawing file.

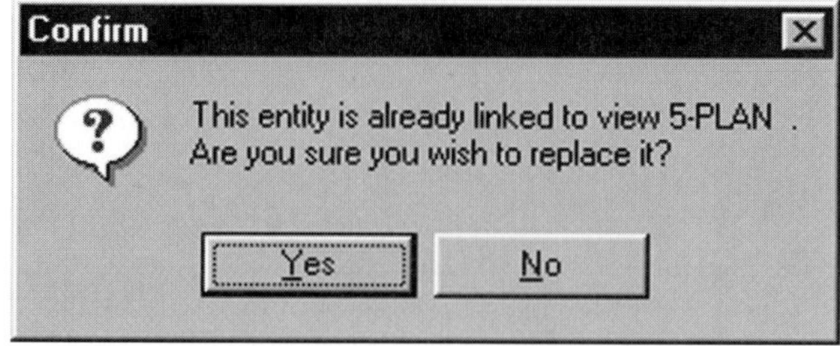

Figure 10-2 Confirm that you want to reassign a link.

Multi-View Windows

You can read more in-depth information about this feature in Chapter 3, "Settings and Display Options." For the purposes of this chapter, just understand that each view in the four windows is directly connected to the 3D GotoView menu. However, these windows cannot be called up using existing 3D GTVs.

You can toggle the display of the four Multi-View windows by selecting **View/Multi-View** from the drop-down menu or by pressing **Ctrl+W**. The windows are displayed on the opposite side of the Drawing Window from your menu buttons. They are thumbnail views of whatever is displayed in the Drawing Window when the top Read In arrow button is pressed. These windows act very much like mini-Drawing Windows, except that you cannot draw or edit entities in them. You can, however, refresh and quick shade the entities within the views (see Figure 10-3).

Multiple .DCX Directories

Chapter 5, "Basic Drawing," describes the use of the **Toolbox** macros. You may find after some time that you are compiling a great deal of these macros. By default, all of these macros reside in the \DCX subdirectory of DataCAD. After a while, you may find that you use some of these macros quite often, while others you may use much less frequently. If you are getting tired of scrolling through all those macros, especially the ones you don't use often, one solution is to divide the macros into two \DCX directories, one for the most often used macros and one for the lesser used ones. Although by default the DataCAD installation creates only one \DCX directory and places all the macros into it, you can create additional subdirectories with any name. In the systems in our office, we have created an additional one called \DCX2. Getting to it is as simple as changing directories after selecting the Toolbox option. Alternately, you can, as I have done, create two toolbar icon buttons, one for each subdirectory, as shown in Figure 10-4. The second icon has a "2" on top of the toolbox.

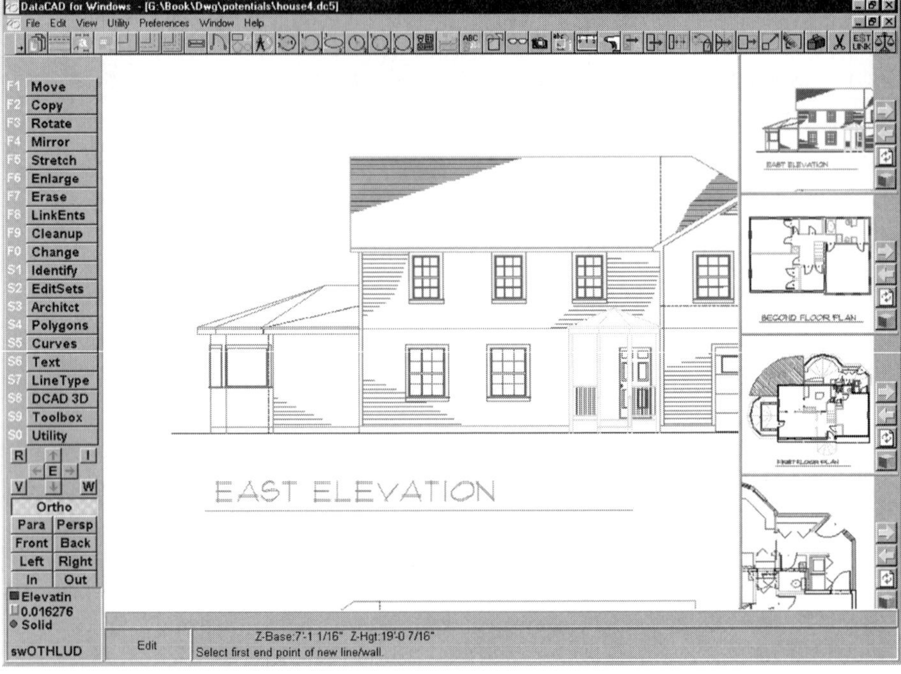

Figure 10-3 The four **Multi-View** Windows

Figure 10-4 Icons for two Toolboxes

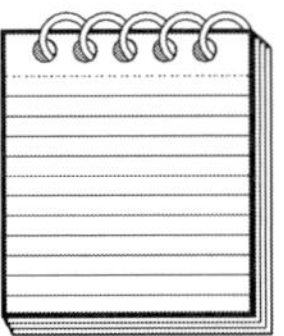

NOTE: *Once you have selected a directory, whether it is the \DCX, \DCX2, or some other directory, any subsequent* ***Toolbox*** *button selections from the main menus will cause DataCAD to return to the last directory selected. So if you press the* ***Toolbox*** *button and don't see the macros you expected to see, check the path to be sure you are in the correct directory. You can get around this issue if you use the icon bar's toolbox icons instead of the* ***Toolbox*** *menu button.*

Interface Customization

Customizing the DataCAD interface is one of the best ways to add to the efficiency and order of the program. Adding, creating, and modifying elements such as the icon toolbar and keyboard shortcuts should be the first items on your list. For more information regarding these and other features, see Chapter 23, "Customizing DataCAD."

CHAPTER 11

Basic Construction Drawings

To understand this chapter effectively, you should have a good, working understanding of the concepts presented in the previous chapters. This chapter will go beyond the basic concepts of simply how to draw in DataCAD and will focus on the ideas you need to understand when putting together basic construction drawing sheets, plans, and elevations.

How Many DC5 Files?

One of the most important decisions you can make when starting a new project is to decide how many DataCAD drawing files (.DC5) to use. If you ask most beginning CAD users, you will usually get one of two answers: make just one .DC5 drawing file that contains all the drawings for the project, or create a separate .DC5 drawing file for each drawing sheet in the project. Neither of these two answers is necessarily incorrect, but other options are available depending on what you want to achieve, and DataCAD provides you some very powerful tools for accomplishing this.

The first thing you must realize is that DataCAD has a method to enable you to place details of different scales on one drawing sheet. It is called *Multi-Scale Plotting* (MSP) and is covered in depth in the following chapter. But in this chapter we will assume that all the details will be located on one drawing sheet and will be of the same scale.

NOTE: *If you are at all familiar with AutoCAD's Paper Space function, then you will immediately grasp the concept of MSP. Although the implementation of these two features is very different, the underlying concepts and the final printed results are essentially the same.*

Here are some of the various issues you will want to consider when deciding how many .DC5 drawing files you will need for a project, and how many drawing sheets (like A1, A2, and so on) you might want to have inside each .DC5 drawing file.

Separating Files for Multiple Users

If you are a one-person design firm, then you don't have to worry about splitting up your drawing files, but if two or more people in your office need to work on drawings at the same time, then you will need to consider how to split up your drawings so that more than one person can work on the project. If you have several people working on a project simultaneously, then you almost can't have too many .DC5 drawing files making up the CAD project. Some of the following concepts will help you decide how your CAD project files should be handled.

Separating Files for Speed and Efficiency

Although DataCAD can save files of up to 20 megabytes in size, files that large will slow down nearly every aspect of DataCAD, including screen refreshes, entity identification, object snapping, and drawing saves. So even if it is possible to keep your entire project in one .DC5 file, it may not be very efficient to do so. By separating your project into multiple, smaller .DC5 drawing files, you will increase the efficiency and speed of your drawings.

Another consideration is the medium you are using: a computer, running software, and saving to magnetic media. Like it or not, these things are not infallible. Occasionally, you may find that a file has been deleted from your computer, or that the drawing file has become corrupted to the point where you can't save all of it, or even any of it. By separating your project into multiple .DC5 drawing files, you may only lose some of your work instead of all of it.

Separating Files by Content

Chapter 5, "Basic Drawing," discussed the Multiple Drawing File technique, where related information and drawings are kept in one drawing file, thereby taking advantage of information that can be shared between different drawings. For instance, before XREFs (Chapter 13), we nearly always kept all of our plans in one .DC5 drawing file: multiple floor plans, reflected ceiling plans, roof plans, and enlarged room plans. We create layers for each floor (first, second, and so on), and more layers for different floor elements (the ceiling grid, lights, roof elements, and so on). All of the floors are aligned directly on top of one another. All of this makes it easy to use common elements (like exterior walls, stairs, and so on) between the floors and to ensure the alignment of elements between the floors.

Some of our projects may include several floors, each of which has a corresponding reflected ceiling plan. Obviously, not all of those plans will fit on one drawing sheet. We might have the first-, second-, and third-floor plans on sheets A1, A2, and A3 respectively. The roof plan might be on A4. The reflected ceiling plans for each of those three floors might then be on sheets A5, A6, and A7. So here we have one .DC5 drawing file with a total of seven drawings sheets laid out.

Likewise, we tend to keep all of our interior elevations in one .DC5 drawing file, all the exterior elevations in another, and all the wall sections in yet another. And just as in the example of the floor plans, each of these .DC5 drawing files may have several drawing sheets laid out inside of them.

Separating Drawing Sheets by Scale

Notice that here we are talking about drawing sheets, not drawing files. One of the first questions I am invariably asked by someone who has learned to draw in DataCAD and is beginning to lay out drawing sheets is, "How do I lay out details of different scales?" Two basic methods exist.

The first method is to use DataCAD's very powerful MSP feature, which is discussed in great detail in the next chapter. It enables you to lay out multiple details of multiple scales on one drawing sheet.

The second and most basic method is the one we will limit our discussion to in this chapter. It involves laying out multiple details of only one identical scale on one drawing sheet. Without using MSP, this is the only option you have, since everything in CAD is drawn at "real" scale, and not drawn at larger or smaller scales, as you do with hand-drafted details.

Here is an example of one project whose drawing files were divided up in various ways. Notice that four of the drawing files have only one drawing sheet in them, while all of the floor plans and reflected ceiling plans are contained in only one drawing file. This project was worked on by two to three people at one time.

File Name	Contents
T1.DC5	Title Sheet and Legends
AllPlans.DC5	Sheet A100, 5th Floor Plan
	Sheet A101, 6th Floor Plan
	Sheet A102, 7th Floor Plan
	Sheet A103, 8th Floor Plan
	Sheet A200, 5th Floor Ceiling Plan

File Name	Contents
	Sheet A201, 6th Floor Ceiling Plan
	Sheet A202, 7th Floor Ceiling Plan
	Sheet A203, 8th Floor Ceiling Plan
A300.DC5	Interior Elevations and Partition Types
A400.DC5	Door Schedule and Details
A401.DC5	General and Finish Notes

Sheet and Title Block Layout

You don't have to lay out a drawing sheet or title block before you start drawing your project. You can just start drawing and worry about a sheet layout later. However, having a sheet layout ahead of time will help you in composing your drawings and in seeing what will fit on each sheet.

Figure 11-1 is an example of a 24″ × 36″ [610 × 914] drawing sheet with a title block. Its true dimensions are 24″ [610] tall by 36″ [914] wide.

Now, setting aside this sheet and title block for a moment, the following concept is one that you can forget when you move to the next chapter regarding MSP, but it is an integral part of this chapter. Everything you draw in CAD is drawn to its real and full scale. An eight-foot stud is drawn eight feet long, a $^3/_4$″ [19.1] thick piece of plywood is drawn $^3/_4$″ [19.1] thick, and a 50-foot-long floor plan is drawn 50 feet long. So with this in mind, how long does a drawing sheet need to be to display that 50-foot-long floor plan completely within it? Well, it's got to be at least 50 feet long. If your drawing sheet were only 24″ × 36″, then it would be far too small to contain the floor plan as in Figure 11-2.

Remember how we said that everything in CAD should be drawn at a real scale? Well, like so many things in life, there are exceptions to every rule, and here comes one now. Once you have drawn your 24″ × 36″ title block at a real 24″ × 36″, you need to use the **Enlarge** function to increase its size so that it will fit the floor plan. But we can't just enlarge it arbitrarily, since we want to be sure that when the drawing sheet is printed, the floor plan will be at a $^1/_4$″ [1:50] scale (for this example), and the title block will fit perfectly on a 24″ × 36″ piece of paper.

Figure 11-1
A 24″ × 36″ drawing sheet example

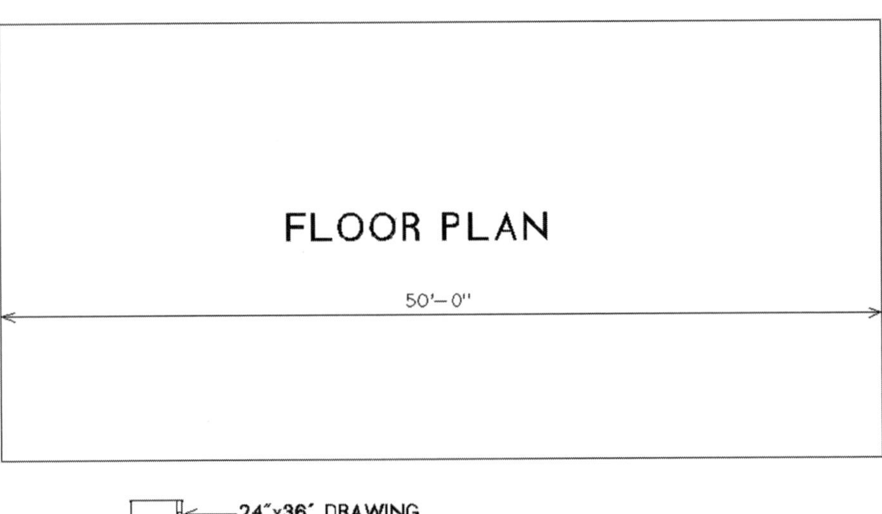

Figure 11-2
A mismatch between the floor plan and the drawing sheet

We'll be kind (because that's the kind of people we are) and actually give you the enlargement factors by which to enlarge your title blocks so they will be at the right scale when printed with **Plotter/QwkLyout**:

Scale	Enlargement Factor
12″	× 1
6″	× 2
3″	× 4
2″	× 6
$1^1/_2$″	× 8
1″	× 12
$^3/_4$″	× 16
$^1/_2$″	× 24
$^3/_8$″	× 32
$^1/_4$″	× 48
$^3/_{16}$″	× 64
$^1/_8$″	× 96
$^3/_{32}$″	× 128
$^1/_{16}$″	× 192
1:20	× 240
1:40	× 480
1:100	× 1,200
1:1000	× 12,000

If we want our 50-foot floor plan to be displayed in our drawing sheet at a $^1/_4$″ [1:50] scale, then we have to **Enlarge** the title block by 48 times. To do so, follow these steps:

1. Select **Edit/Enlarge**.
2. Select the center point of the enlargement.
3. Press **Enlrgmnt/SetAll**.
4. Type in **48** and then press **Enter**.
5. *Right-click* back to the main **Enlarge** menu.
6. Turn on **Area** and **LyrSrch**.
7. Draw a bounding box around the title and sheet border to be enlarged.

Make sure that with the title block you also enlarge any text that belongs with the title block, like the project name, sheet number, sheet title, and so on. Also make sure that you only enlarge the title block information, not the floor plan. After you enlarge it, use the **Move** function to move the title block until it's located around the floor plan (see Figure 11-3).

Now just to prove that everything is still drawn to real scale in CAD, if you measure the dimension of the title block from top to bottom and then left to right, you will find that it is 96′ [29261] tall (24″ × 48 = 1,152″ or 96′) by 144′ [43891] wide (36″ × 48 = 1,728″ or 144′). This is how large a sheet of paper would really have to be to lay underneath the building represented by the floor plan, in Figure 11-3.

Multiple Drawings, One Title Block

You might ask, if I am going to have more than one drawing sheet (A1, A2, and so on) per .DC5 drawing file, do I need to create a separate title block for each drawing? The answer is no and yes. If all your details will be drawn at one scale, then the answer is no. But if you have details at different scales, then you will have to create a title block for each scale of details. We'll cover that last scenario in the next section.

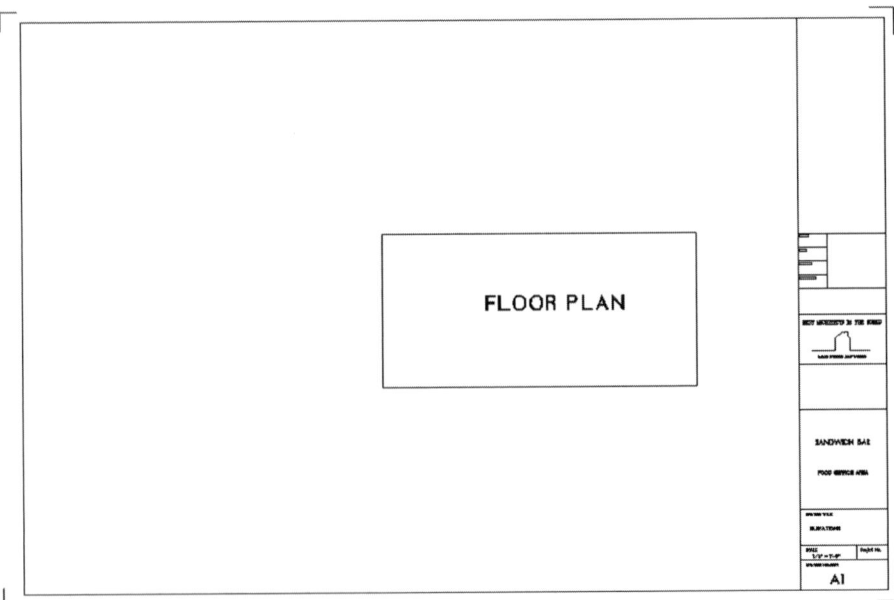

Figure 11-3 The enlarged title block around the plan

If all your details will be of one scale but are located on more than one drawing sheet, then all you need is one title block (appropriately scaled) with different layers for the text that goes on each sheet. Let's say you have a series of $1^{1}/_{2}''$ [1:10] details, and that it will take two drawing sheets, A1 and A2, to fit all of them. You would then follow these steps:

1. Create three layers: TTLBLOCK, A1-TEXT, and A2-TEXT.
2. Make the TTLBLOCK layer the active layer. Using the 24″ × 36″ [610 × 914] title block example again, (you do have your standard title blocks already saved as symbols, don't you?) place it in the drawing.
3. Now make the A1-TEXT layer the active layer. Add the appropriate text within the title block: the A1 sheet number, sheet name, scale, and so on.
4. Now make the A2-TEXT layer the active layer. Add the appropriate text within the title block: the A2 sheet number, sheet name, scale, and so on.
5. Enlarge the title block and the A1 and A2 text by 8x (for $1^{1}/_{2}''$ details).
6. When you want to use the A1 title block, turn on layers TTLBLOCK and A1-TEXT. When you want to use the A2 title block, turn on layers TTLBLOCK and A2-TEXT.
7. Using this method, you will create all your details for both drawing sheets within the boundaries of the single title block, meaning that details will overlap (but only when all layers are turned on). So layer management is important.

If you don't want details to overlap, then the alternative method would be to do the following:

1. Create only two layers, called something like A1-SHEET and A2-SHEET.
2. Place the title block on layer A1-SHEET. You can place all the title block text on this layer too, since there is no reason to separate the two with this method.
3. Enlarge the title block and text.
4. Make a copy of the first title block and place it next to the first one.
5. Create all your details within the appropriate title block.

Multiple Scales, Multiple Title Blocks

As mentioned previously, if you have details at different scales within the same drawing file, then you will have to create a title block for each set of details. Let's say you have a $^1/_4''$ [1:50] floor plan along with that previous series of $1^1/_2''$ [1:10] details. You cannot put both the floor plan and the details on the same drawing sheet since the details are of different scales. So you need to have two title blocks: one scaled to fit the $^1/_4''$ [1:50] floor plan, and one scaled to fit the $1^1/_2''$ [1:10] details.

Start with your standard title block. Let's use the 24″ × 36″ [610 × 914] example again. With your 24″ × 36″ title block placed in the drawing, use the **Copy** command to make a copy of it. Now use the **Enlarge** command to enlarge both title blocks. Enlarge the first one by a factor of 48 (for the $^1/_4''$ [1:50] plan), and the second one by eight (for the $1^1/_2''$ [1:10] details).

Now simply move the title blocks or the details as required. Make sure that only the $^1/_4''$ [1:50] plan is within the first title block, and only the $1^1/_2''$ [1:10] details in the other. If you are going to move the details to fit within the title blocks, be very careful to turn all the drawing layers on before you do, to make sure you move all the entities on all the layers associated with each detail. You don't want to leave anything behind.

Layering for Success

As many opinions about layering systems exist as there are people using CAD. There really are no right or wrong answers. What's most important is that the layer structure and names work for you. Chapter 6, "Organizational Concepts," covers the basics of layering as well as making some suggestions for naming. In this section, we will give you some more suggestions as they relate to construction related drawings. We will also discuss the U.S. government's and the American Institute of Architects' attempts to create a universal layering standard.

What is most important is to avoid random layer naming and disorganized layer ordering. Remember that in DataCAD you cannot change the order of layers once they are created. You can rename them, yes, but they cannot be reordered. This is certainly a shortcoming of DataCAD, but for now it's one that you will have to actively work with.

Default Drawings

The best method for keeping your drawing files organized is to create good default drawings in the first place. If you currently don't make extensive use of default drawings, then you should go right back to Chapter 2, review default drawings, and then get to using them right away. Using default files eliminates the drudgery of creating new layers from scratch each time you start a new drawing. It also ensures that you, your office, and its employees stick to a concise set of layering standards. It is unarguably the single best organizational tool for CAD drawings.

To give you some ideas of the kinds of default .DC5 drawings you might want to consider, Table 11-1 outlines some of the ones that our office uses.

Layer Names

It is very important to create layer names that are easily understood, not only by yourself and your office personnel, but for any clients, engineers, or consultants you might send drawings to. The best method is usually to keep them phonetic in some way: WALLS for walls, DOORS for doors, and so on.

Table 11-1

Some examples of default layer names

Floor Plans	
1NPLAN	Layers for first floor plans (new construction)
2NPLAN	Layers for second floor plans (new construction)
1XPLAN	Layers for first floor plans (existing/renovation construction)
2XPLAN	Layers for second floor plans (existing/renovation construction)
1-2NPLAN	Layers for first and second floor plans (new construction)
1-5NPLAN	Layers for first through fifth floor plans (new construction)
1-2XPLAN	Layers for first and second floor plans (existing/renovation construction)
1-5XPLAN	Layers for first through fifth floor plans (existing/renovation construction)
Elevations	
ELEVATIONS	Kind of says it all, doesn't it!
Sections and Details	
WALLSECT	Layers for multiple details of complex wall sections
DETAILS	Layers for multiple miscellaneous details

You might also use something like 1F as a prefix for all first-floor elements or NE as a prefix for all north elevation elements.

Besides naming considerations, the order in which layers appear in your drawing file is also important. Although DataCAD has a fairly powerful layer-filtering feature (see Chapter 6), it is usually still advisable to try to keep groups of related layers physically next to one another.

Keep in mind that one of DCAD's shortcomings is its incapability to rearrange layers once they have been added to the drawing file. They remain in the same order in which they were created in the drawing file. So whole groups of layers should be added at the same time, even if you don't think all of them will be used. Table 11-2 outlines some layer naming guidelines.

$2^1/_2$ D

Before you look at drawing walls, doors, and other entities, it's worth reviewing the section in Chapter 5 regarding DataCAD's capability to create $2^1/_2$ D entities. With practically no additional effort at all, you can create quick 3D drawings from your 2D construction plans, as in Figure 11-4

Walls

Just like lines, arcs, and circles, drawing walls is based on the four input methods (Relative Polar, Absolute Polar, Relative Cartesian, and Absolute Cartesian). Like a line, a single wall needs only two points to describe it: a start point and an end point. A wall in DataCAD is simply two, three, or four parallel lines or arcs. The lines that make up the wall can be simple, solid lines, or they can be made up of complex linetypes. All of the wall options can be found in the **Edit/Architct** menu (see Figure 11-5).

Two methods can be used to turn on the automatic Walls feature:

1. Select **Edit/Architct/Walls**.
2. Press the equals (=) sign on the keyboard. This is meant to be easy to remember since it looks like a two-line wall.

While in Walls mode, the W in the SWOTHLUD acronym at the lower corner of your screen will be shown in upper case. When you are in single-line mode (Walls toggled off), the w will be shown in lower case (see Figures 11-6 and 11-7).

Table 11-2

Various layer naming guidelines

General New Plan	Layer Naming Guidelines
X-COL-LN:	Column lines/grids
X-STRUCT:	Special structures (columns and so on)
X-SITE:	Site information
X-EXWALL:	Exterior walls (heavier lineweight)
X-INWALL:	Interior walls/partitions (lighter lineweight)
X-RMNAME:	Room names
X-RMNUMB:	Room numbers (not required if included in X-RMNAME)
X-DOORS:	Doors and swings
X-DRNUMB:	Door numbers
X-WINDW:	Windows
X-WNNUMB:	Window numbers
X-FIXTUR:	Plumbing, mechanical, and so on
X-MILLWK:	Cabinets, counters, millwork, and so on
X-HATCH:	Hatching and other pocheing (including linetypes used as poche)
X-NOTES:	Notes and text
X-DIMS:	Dimensions
X-MISC-1:	Miscellaneous #1 (change name as required)
X-MISC-2:	Miscellaneous #2 (change name as required)
X-DRHEAD:	Headers over doors
X-CLGRID:	Ceiling grids
X-CLFIX1:	Ceiling fixtures (Electric)
X-CLFIX2:	Ceiling fixtures (Mechanical)
X-CLNOTE:	Ceiling notes and dimensions
X-CLMIS1:	Ceiling Miscellaneous #1
X-CLMIS2:	Ceiling Miscellaneous #2

X = Floor number or letter (1, B, and so on)

continues

Table 11-2

Continued.

General Renovation Plan	Layer Naming Guidelines
XR-CO-LN:	Column Lines/Grid
XR-STRUC:	Special structures (columns, and so on)
XR-SITE:	Site information
XR-EXWAL:	Exterior walls (heavier lineweight)
XR-INWAL:	Interior walls/partitions (lighter lineweight)
XR-RMNAM:	Room names
XR-RMNUM:	Room numbers (not required if included in X-RMNAME)
XR-DOORS:	Doors and swings
XR-DRNUM:	Door numbers
XR-WINDW:	Windows
XR-WNNUM:	Window numbers
XR-FIXTR:	Plumbing, mechanical, and so on
XR-MILWK:	Cabinets, counters, millwork, and so on
XR-HATCH:	Hatching and other pocheing (including linetypes used as poche)
XR-NOTES:	Notes and text
XR-DIMS:	Dimensions
XR-MIS-1:	Miscellaneous #1 (change name as required)
XR-MIS-2:	Miscellaneous #2 (change name as required)
XR-DRHED:	Headers over doors
XR-CLGRD:	Ceiling grids
XR-CLFX1:	Ceiling fixtures (Electric)
XR-CLFX2:	Ceiling fixtures (Mechanical)
XR-CLNOT:	Ceiling notes and dimensions
XR-CLMS1:	Ceiling Miscellaneous #1
XR-CLMS2:	Ceiling Miscellaneous #2

X = Floor number or letter (1, B, and so on)

General Existing Plan	Layer Naming Guidelines

Same as Renovations, but use **E** in place of **R**

General Interior and Exterior Elevation`Layer Naming Guidelines	
X-COL-LN:	Column lines/grid
X-STRUCT:	Special structure (columns and so on)
X-SITE:	Site information
X-WALL1:	Exterior walls #1
X-WALL2:	Exterior walls #2
X-DOORS:	Doors
X-WINDW:	Windows
X-WNNUM:	Window numbers
X-TRIM:	Trim
X-HATCH:	Hatching and other pocheing (including linetypes used as poche)
X-NOTES:	Notes and text
X-DIMS:	Dimensions
X-MIS-1:	Miscellaneous #1 (change name as required)
X-MIS-2:	Miscellaneous #2 (change name as required)

X = North, South, East, West (N, S, E, W) **or** elevation number (1, 2, and so on)

Miscellaneous Drawing Information	Layer Naming Guidelines
KEY:	Key plan (add text after KEY as required)
SHEET:	Sheet/title block
A-100:	Sheet # and sheet-specific information (goes with TTLBLK)
LEGEND:	Legends (symbols and so on)
ROOFDRN:	Roof drains

continues

Table 11-2
Continued.

	General Detail	Layer Naming Guidelines
	X-WALL:	Entities making up the walls, floors, roofs, and so on
A	**X-STUCT:**	Steel, aluminum, and other metals
	X-WINDOR:	Windows and doors
	X-HATCH:	Hatching and other pocheing (including linetypes used as poche)
	X-NOTES:	Notes and dimensions (don't include Dims if **X-DIMS** is used)
B	**X-DIMS:** Dimensions	
	X-MISC: Miscellaneous	
C	**X-2HATCH:** Hatching and pocheing in large-scale details	
	X-2NOTES: Notes and text in large-scale details	
	X-2MISC: Miscellaneous in large-scale details	
	X = Number or letter of detail (1, 6, A, G, and so on)	
	A = Simple details	
	B = More complex details	
	C = Additional layers for large-scale drawings (only)	

Figure 11-4
A 3D version of a 2D plan

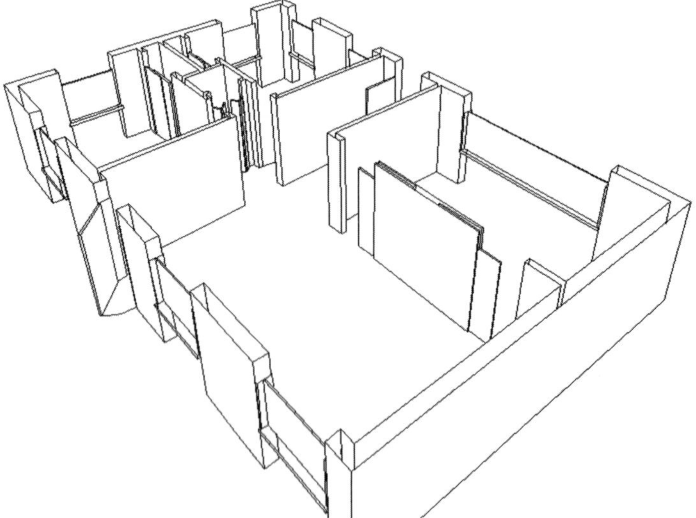

You will remain in Walls mode until you turn it off, either by selecting **Edit/Architct/Walls** or by pressing the equals (=) sign again. When the Walls mode is turned off, DataCAD will draw single lines rather than walls.

NOTE: *Although a curved wall is still a wall to DataCAD, some of the automatic drawing features in DataCAD, such as the insertion of doors and windows will only work in straight walls.*

DataCAD enables you to draw 2-line walls, 3-line walls, or 4-line walls. Each of these wall types has their own set of settings and options, but the basic concept is that 2-line walls draw an outside and an inside line, 3-line

Figure 11-5
One way to draw a wall is to offset a rectangle.

SWOTHLUD

Figure 11-6
Walls mode is on.

Figure 11-7 Walls mode is off (single-line mode).

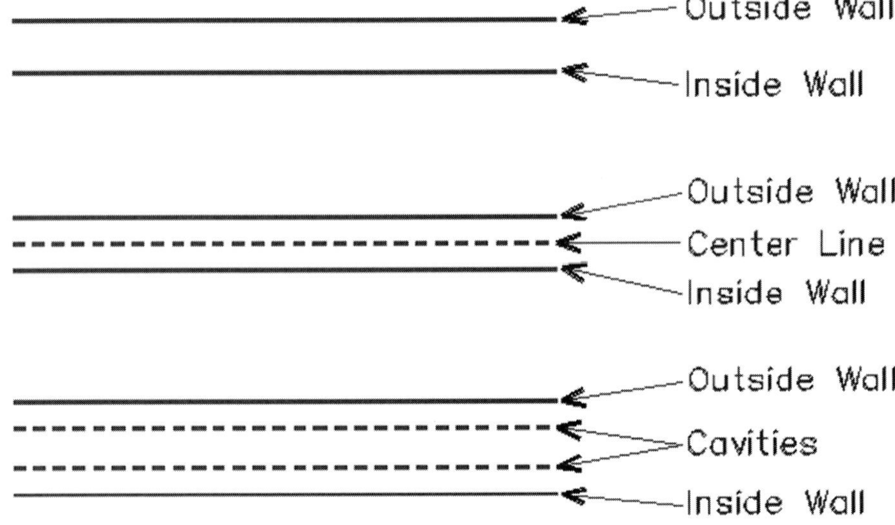

Figure 11-8 The 2, 3, and 4-line walls offered by DataCAD

walls draw a centerline between the inside and outside lines, and 4-line walls draw four lines to create two cavities (see Figure 11-8).

Each of the three wall types are mutually exclusive, meaning that when one is turned on, the others are turned off. Walls can only be drawn in Ortho mode, so press the **Ortho** button in the Projection Pad if you are in any other view mode.

Four different methods are available for drawing walls: Offset Rectangle, Freeform, Distance and Angle, and Offset Lines. They each have their uses in different circumstances. In the first two, Offset Rectangles and Freeform, you will want to make sure that a couple of settings are set to make the task easy and accurate.

Sides versus Centers

As you will see later in this section, you can define walls by three basic "sides": **Outside**, **Inside**, and **CntrWall**. The **Outside** and **Inside** options are interchangeable as long as both sides of the wall are made of the same linetype and the same color. But once you choose the **Hilite**, **3LnWalls**, or **4LnWalls** option, you will need to be careful about choosing the correct side by which to define your walls. All of the various Walls options will be discussed later in this section.

Settings for Freeform and Rectangular Walls

Drawing walls in a rectangular or freeform manner means that you move the cursor around the screen, placing your walls visually, rather than typing in exact distances and dimensions. To get accurate distances and angles, you will want to use Ortho mode and a Snap Grid, as described later.

Press the **Oo** key until the O in SWOTHLUD changes to upper case. The Message Line will read, *"Orthomode is on."* Your cursor will now draw walls by snapping to the angles set in **Utility/Grids/SnapAng** (the default setting of eight divisions means the cursor will snap at 45-degree increments). Now you can be assured of drawing accurate horizontal and vertical walls.

Next, you will want to force the cursor to move in increments that are not too small and not too big. If your Snap Grid is set to $^1/_{32}$″ [0.8] for instance, then the cursor will move in $^1/_{32}$″ [0.8] increments. This is too small, however, for drawing walls. In most cases, your walls will probably only need to be accurate down to 1″ [25.4], or $^1/_2$″ [12.7] at the smallest. To set the Snap Grid to 1″ [25.4], go to **Utility/Grids/GridSize/SetSnap** and set the grid to 1″ [25.4]. *Right-click* twice back to the **Grids** menu and make sure the **SnapGrid** button is turned on. *Right-click* to exit the **Grids** menu.

Walls by Offset Rectangle

With this method, you create a rectangle that represents one side of the walls and then in one step DataCAD offsets those four lines to create the other side of the walls. Follow these steps:

1. Press the equals (=) key until the W in SWOTHLUD is an uppercase W. If the current wall type is set to 2-line walls, then the command line will read something like, *"Enter new wall width: 0.4.1/2."* This tells you that the walls will be drawn 4½″ [114] wide. If this width is what you want to draw, *right-click* or press **Enter** to accept it. If not, go to Step 2.

NOTE: *If the current wall type is set to 3-line or 4-line walls, then you will not see a prompt to change the wall width, but the W in SWOTHLUD will still change to show you that you are in Walls mode.*

2. To change the wall width, type in a new width, like **.6** for 6″ [152] walls. Press **Enter** to accept the new value. *Right-click* if you want to retain the previous value.
3. Go to the **Edit/Architct** menu. Alternately, you can press the lower case **"a"** on the keyboard to get to this menu.
4. **Walls**, **2LnWalls**, and **Outside** should all be turned on. *Right-click* to exit the **Architct** menu.

NOTE: *You can use the* ***Inside*** *setting instead if you want to define your walls by their inside face instead of their outside face. In this case, the prompt in Step 7 will say "define the Outside . . . " instead of "define the Inside."*

5. Select **Edit/Polygons**. Select **RectAngl** so that it is turned on. Turn **CntrPnt** off (leaving this on will draw a non-printing snap point at the center of the rectangle).
6. Place the cursor anywhere in the upper-left corner of the Drawing Window. Drag the cursor down and to the right. Look at the X and Y values in the Status Line. Move the cursor until the X and Y values read what you want them to be. Remember that if at any time you run out of room on the screen, you do not have to stop, zoom out, and redraw the lines. Simply use the **Page Up** key on your keyboard to zoom out, or use the arrow keys to move up, down, left, or right (see Figure 11-9).

Figure 11-9
The initial rectangle

7. *Click* the mouse. You will be prompted with *"Select a point to define the Inside of the wall."*
8. Place the cursor anywhere inside the rectangle and click the mouse. The lines representing the inside of the walls are drawn, and all the corners have been cleaned up (see Figure 11-10).
9. Notice that the Prompt Line says, *"Select first corner of rectangle."* DataCAD is still in Rectangle mode and will remain there until you *right-click* the mouse or press **Exit**. DataCAD operates this way in nearly every case. In this way, you can continue drawing without having to constantly go back to the menus.

Freeform Walls

In most instances, you will not be able to draw your walls by drawing a rectangle. You will draw them in a freeform manner, as described here, or by entering distances and angles, as described next. Drawing walls in a freeform manner means that you move the cursor around the screen, placing your walls visually, rather than typing in exact distances and dimensions. To get accurate distances and angles, you will want to use Ortho

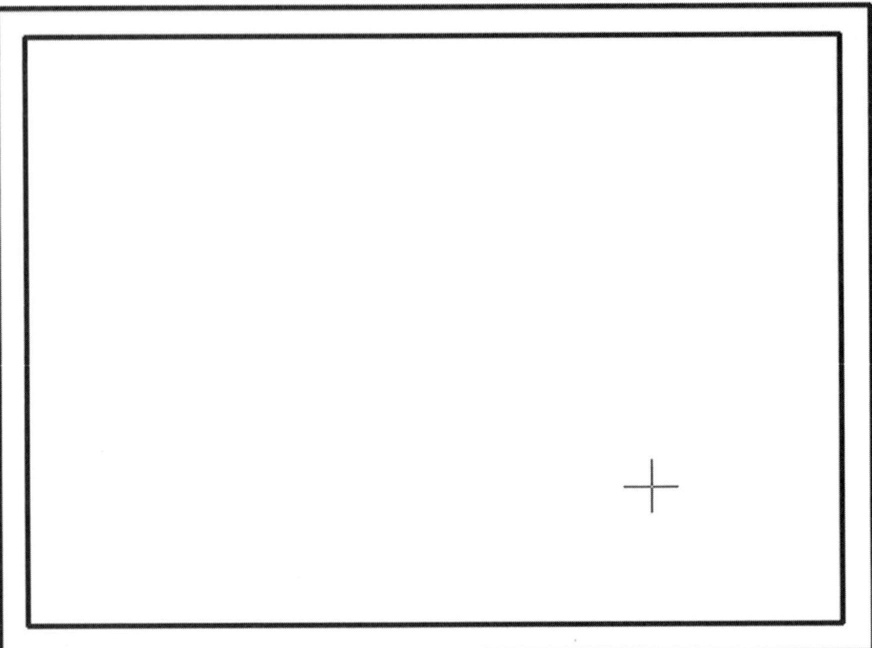

Figure 11-10 Creating the inside of the walls

mode and a Snap Grid, as described previously. With all those settings in place, you are ready to draw freeform walls:

1. Make sure the Walls setting is on (either press the equals (=) key or select **Edit/Architct/Walls**). The W in SWOTHLUD should be an upper case W.
2. Set the wall Width (after turning Walls on with the equals (=) key or with **Edit/Architct/Width**).
3. *Click* anywhere in the Drawing Window to set the start point of the new wall. The Coordinate Readout line will reset all values to 0.
4. Move the cursor in the desired direction for the wall. Watch the Coordinate Readout line and stop the cursor at the desired distance. In this example, we've moved the cursor 12′ [3658] to the right (an angle of 0 degrees), as shown in Figure 11-11.
5. *Click* the cursor to set the endpoint of the new wall. You are then prompted to *"Select a point to define the Inside of the wall."*
6. *Click* to the side of the line where you want the inside line of the wall to be drawn. DataCAD will draw the second line of the wall, and you

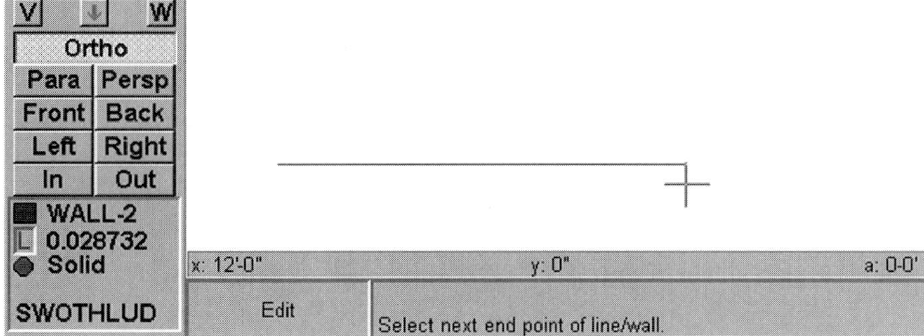

Figure 11-11 Moving the cursor 12 feet to the right

Figure 11-12 Automatically creating the second line of the wall

will be prompted to *"Select next end point of the line/wall"* (see Figure 11-12).

7. If you do not want to draw any more walls, then *right-click* to finish. If you want to continue drawing more walls from the end of the previous wall, go on to Step 8.
8. Notice that a line extends from the end of the first outside wall to the cursor. This represents the next outside line of the next wall. Move the cursor in the desired direction for the wall. Watch the Coordinate Readout line and stop the cursor at the desired distance. In this example, we've moved the cursor 10′-6″ [3200] upward (an angle of 90 degrees), as shown in Figure 11-13.
9. *Click* the mouse to set the endpoint of the second new wall. DataCAD will draw the second line of the wall and you will be prompted to *"Select next end point of the line/wall"* (see Figure 11-14).

Notice that the L-intersection between the two walls is automatically cleaned up. As long as you continue to draw contiguous walls without *right-clicking*, all the corners and intersections will be automatically cleaned up like this. If the **Clean** option is turned on in the **Architct**

Figure 11-13
Moving the cursor up 10 feet, six inches

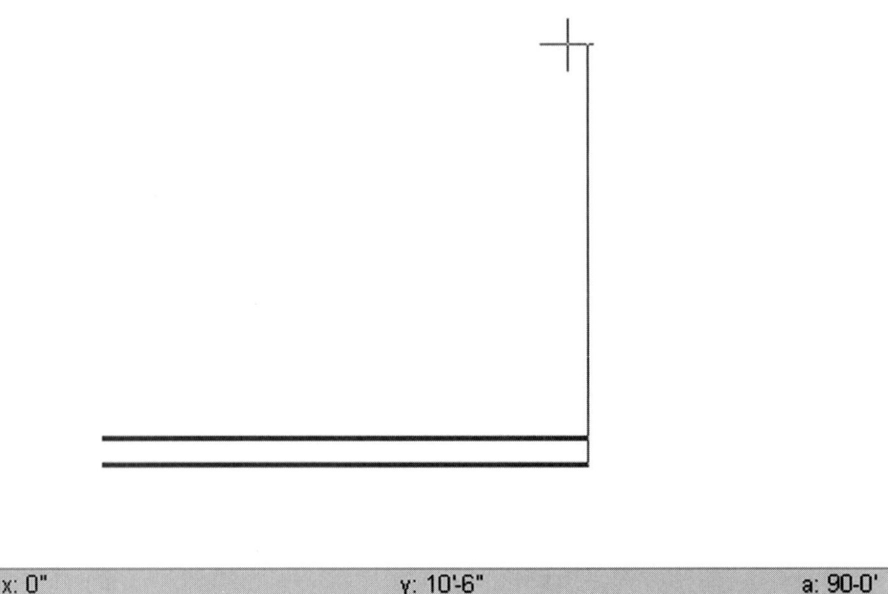

menu, then T-intersections will automatically be cleaned as well, as shown in Figure 11-15.

Notice also that in Step 9 you were not prompted to select the inside of the wall this time. That's because you already set the inside wall location when you drew the first wall. Since these two walls are contiguous, the inside wall line of the second wall can only exist on the same side as the previous wall.

At any time during the freeform wall process, even right in the middle of it, you could have pressed the **Oo** key to stop drawing in Ortho mode. This would enable you to draw a wall at any angle, but getting both the angle and the distance correct this way is difficult because of the sensitivity of the mouse. Let's say you needed a 30-degree wall with a length of 20′-0″ [6096]. Without having to *right-click* to stop the freeform process, you could have pressed the **Space** bar to enable you to type in the desired distance and angle. That's the next method of drawing walls that we will describe.

Walls by Distance and Angle

Walls can also be drawn by typing in their exact distances and angles. This can be further broken down into the two primary input methods of *Relative*

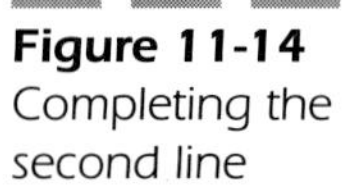

Figure 11-14 Completing the second line

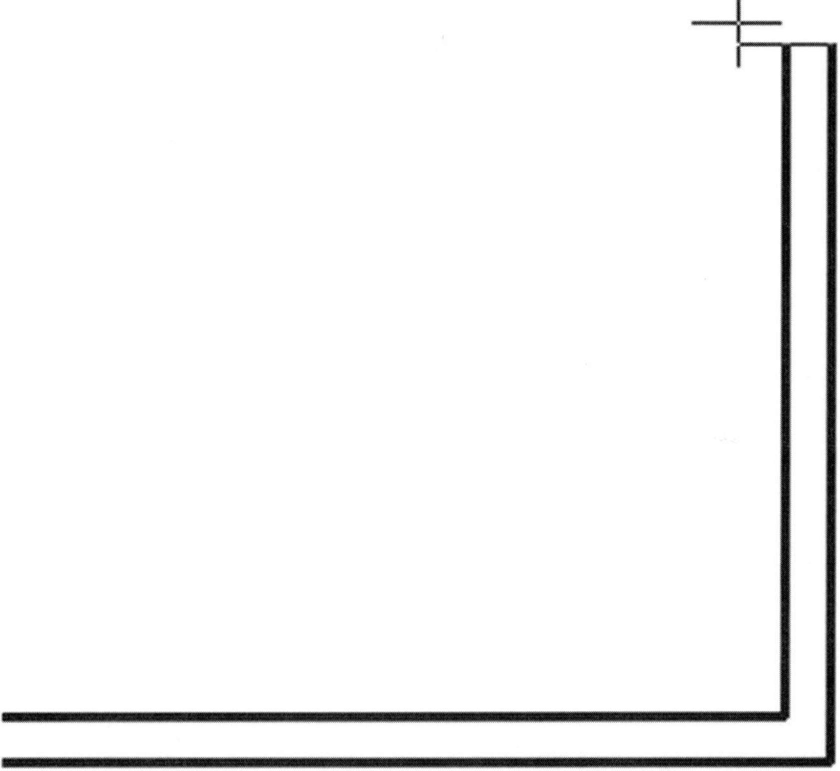

Polar and *Relative Cartesian*. Both methods are nearly identical, but we'll demonstrate both anyway.

Relative Polar

In case you've forgotten, Relative Polar means that walls and lines are input by typing in a distance and an angle (0 to 359 degrees). In the following example, we draw two walls, one to the left at 180 degrees, and one at an angle of 150 degrees:

1. Make sure the Walls setting is on (either press the equals (=) key or select **Edit/Architct/Walls**). The W in SWOTHLUD should be an upper case W.
2. Set the wall Width (after turning Walls on with the equals (=) key or with **Edit/Architct/Width**).

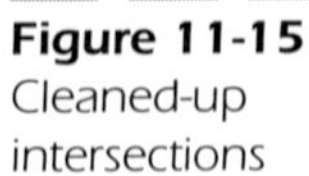

Figure 11-15 Cleaned-up intersections

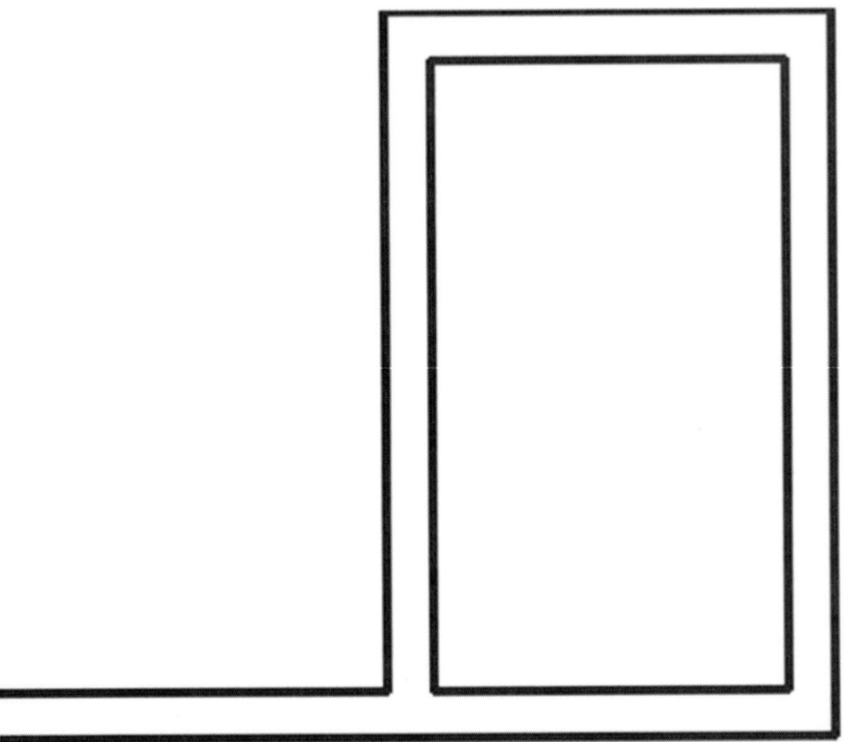

3. *Click* anywhere in the Drawing Window to set the start point of the new wall. The Coordinate Readout line will reset all values to 0.
4. To create a wall that is 12′-6″ [3810] long going to the left (180 degrees), we need to type those values in. To do this, press the **Space** bar. This is always the method that DataCAD uses to enter distances and angles. You will be prompted to *"Enter relative distance: 0.0."*
5. Type in **12.6** (DataCAD shorthand for 12′-6″ [3810]) and press **Enter** to accept the value.
6. Now you will be prompted to *"Enter relative angle: 0.0.0."* To draw a horizontal line to the left, type in **180** and press **Enter** to accept the value.
7. You will be prompted to *"Select a point to define the Inside of the wall." Click* to the side of the line where you want the inside line of the wall to be drawn. DataCAD will draw the second line of the wall, and you

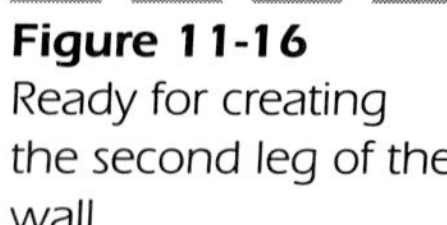

Figure 11-16
Ready for creating the second leg of the wall

will be prompted to *"Select next end point of the line / wall"* (see Figure 11-16).

8. If you do not want to draw any more walls, then *right-click* to finish. We want to continue, however, so continue on to the next step. Drawing the second, 150-degree wall is just like drawing the first one, except that the angle will be different. Let's draw this second wall with a length of 8′-5″ [2565].
9. Press the **Space** bar to enter a distance and angle. You will be prompted to *"Enter relative distance: 0.0."*
10. Type in **8.5** (DataCAD shorthand for 8′-5″ [2565]) and press **Enter** to accept the value.
11. Now you will be prompted to *"Enter relative angle: 0.0.0."* To draw a horizontal line to the left and up at an angle of 150 degrees, type in **150** and press **Enter** to accept the value. Both the outside and inside wall lines will be drawn, as shown in Figure 11-17.
12. You can continue to draw more walls or *right-click* to finish.

Walls by Offset

The final method for creating walls is to simply draw single lines representing one side of the wall and then use the **Offset** command to create the other side of the walls (unlike some other CAD programs, DataCAD will not allow you to offset polylines to create walls, only 2D lines and arcs). This

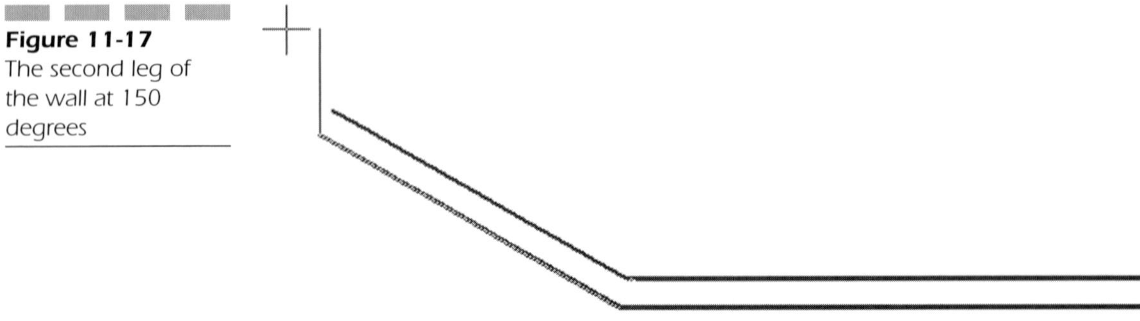

Figure 11-17
The second leg of the wall at 150 degrees

Figure 11-18
Three lines that will be offset to form walls

method does not make use of any of the Walls features, like automatic cleanups and end caps, since the **Offset** command is part of the **Geometry** menu, and not part of the Walls function. To utilize the **Offset** command, follow these steps:

1. Draw the 2D lines and arcs that will form the inside or outside face of the walls. You can use any of the previously mentioned methods: Rectangle, Freeform, or Distance and Angle (see Figure 11-18).

2. Select **Utility/Geometry/Offset**.
   a. If the **Dynamic** toggle is on, you will be prompted to *"Select object to offset."* You can use this option to dynamically drag a copy of a selected line into position. This is usually not the option you want to use since it is difficult to get the correct offset distance due to the sensitivity of the mouse.
   b. If the **Dynamic** toggle is off, a **PerpDist** (perpendicular distance) button will appear in the menu, and you will be prompted to *"Select first point of the offset distance."* This is the option you will want to use.
3. Select **PerpDist** (if you don't see that option, turn **Dynamic** off). You will be prompted to *"Enter perpendicular distance to offset: 0.4."* Select the width of the walls you desire. In this example, we will type **.8** for an 8″ [203] wall.
4. You are again prompted to *"Select object to offset."* Click on one of the 2D lines you drew. It will change to a dashed line to show that it was selected, and you will be asked, *"Offset to which side?,"* as shown in Figure 11-19.

Figure 11-19
Selecting an object to be offset

Figure 11-20 Offsetting to the right side

5. Place the cursor anywhere on the side of the line where you want the offset line to be created and *click* the mouse. In this example, we clicked on the right side of the line to offset the line to that side (see Figure 11-20).
6. Do the same for the other walls (if you want some of the walls to be of a different width, select the **NewDist** menu option to enter a new offset distance). See Figure 11-21.
7. Notice that the walls were not cleaned up at the corners. To clean them up, you will need to use the **2LnTrim** option in the **Edit/Cleanup** menu (refer to Chapter 9, "Editing"). After cleaning up the corners, you can also draw a new 2D line across the ends of the walls if you want to cap them. The final product would look like Figure 11-22.

Curved Walls

With only a few exceptions, drawing curved walls is no different than drawing straight walls, except, of course, that you can't draw a rectangle with a curve. You can create walls with most of the curve options in the **Curves** menu, except for **Polyline**, **Bezier**, and **B-Spline**. You also cannot draw an

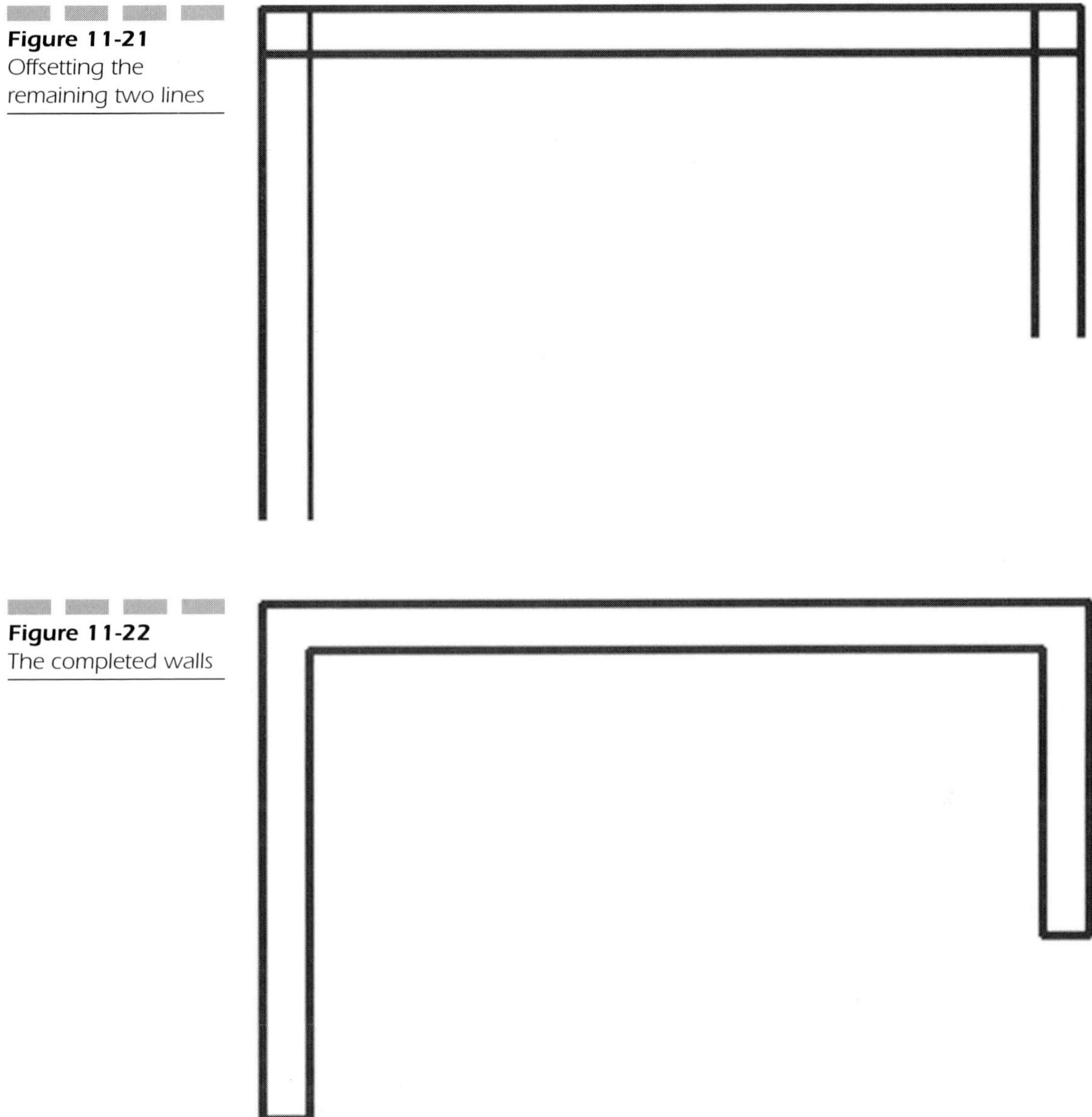

Figure 11-21 Offsetting the remaining two lines

Figure 11-22 The completed walls

elliptical wall by simply turning on the Walls option. Instead, you must first draw the **Ellipse** and then use the **Offset** method described earlier to create the other side of the wall.

Unfortunately, you cannot do certain things with curved walls after you have created them:

- Although a curved wall is still a wall to DataCAD, some of the automatic drawing features in DataCAD, such as the insertion of doors and windows, will only work in straight walls.
- The automatic wall cleanup functions (**WeldLine**, **WeldWall**, **TIntsct**, **LIntsct**, and **XIntsct**) do not work on curved walls.

Figure 11-23 shows some walls drawn with curves.

Wall Attributes

The previous descriptions of drawing walls assumed that you were only drawing simple 2-line walls, all of the same linetype and the same color. But the Walls option enables many other options for drawing walls and even allows you to save all those settings so that you can reuse them later without having to reset all the options. Keeping in mind the four methods for drawing walls, the following are the various attributes that can be set for 2-line, 3-line or 4-line wall types:

Walls

Toggle this button on to draw walls. Toggle it off to draw single lines. The shortcut key for toggling this option on and off is the equals (=) sign on your keyboard.

Figure 11-23 Some curved wall examples

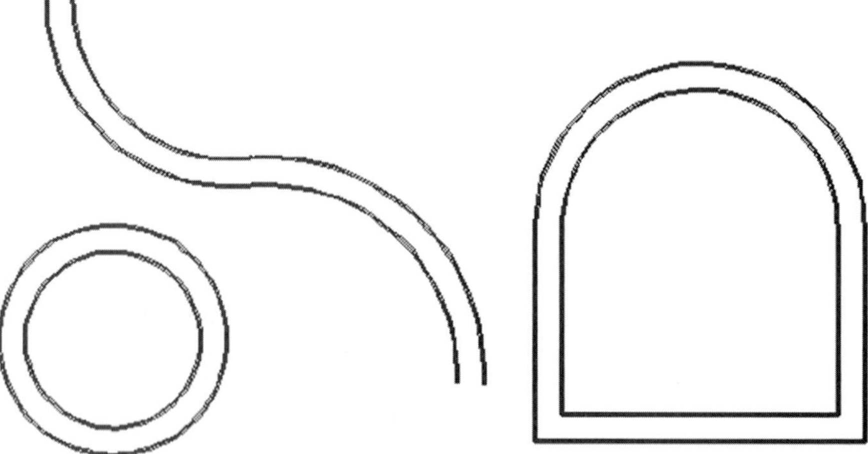

Width

Select this option to set the current width for walls to be drawn at. Select a dimension from the menu or type it into the command line.

Hilite

Use this option to set the attributes of the **Outside** or **Inside** wall line. Only one of those two lines can be highlighted at one time, so the Outside and Inside options are mutually exclusive. With **Hilite** on, the selected wall will be highlighted. If it is off, then no walls will be highlighted.

Outside If toggled on, the Outside wall line will be highlighted with the current highlight color, line weight, and line type. No option exists for linetype spacing, so the current spacing is used.

Inside If toggled on, the Outside wall line will be highlighted with the current highlight color, line weight, and line type. No option exists for linetype spacing, so the current spacing is used.

Color The option selects the color for the highlighted wall.

LineWgt The option selects the line weight for the highlighted wall.

LineType The option selects the linetype for the highlighted wall. No option exists for linetype spacing, so the current spacing is used.

Outside

When toggled on, all walls will be defined by drawing the outside wall first. **Outside**, **Inside**, **CntrWall**, and **CntrCvty** are mutually exclusive; only one can be selected at one time. Unless overridden by the **Hilite/LineType** option, the current linetype is used to draw the outside line.

Inside

When toggled on, all walls will be defined by drawing the Inside wall first. **Outside**, **Inside**, **CntrWall**, and **CntrCvty** are mutually exclusive; only one can be selected at one time. Unless overridden by the **Hilite/LineType** option, the current linetype is used to draw the inside line.

CntrWall

When toggled on, all walls will be defined by drawing the centerline of the wall. If **Hilite** is turned off, the two sides of the walls will be automatically drawn on both sides of that imaginary centerline without any further prompting. If **Hilite** is turned on, then you must select the Outside or Inside of the wall. **Outside**, **Inside**, **CntrWall**, and **CntrCvty** are mutually exclusive; only one can be selected at one time.

CntrCvty

When toggled on, all walls will be defined by drawing the center of the cavity of a 4-line wall. If the 2-line or 3-line wall options are selected, then the **CntrCvty** option acts just like the **CntrWall** option, drawing the walls by the centerline. If **Hilite** is turned off, the two sides of the walls will be automatically drawn on both sides of the cavity without any further prompting. If **Hilite** is turned on, then you must select the Outside or Inside of the wall. **Outside**, **Inside**, **CntrWall**, and **CntrCvty** are mutually exclusive; only one can be selected at one time.

Clean

With this option turned on, all T-intersections will be cleaned up as you draw the walls. This option will only work on straight wall segments, not curves. See Chapter 9 for how 2-line, 3-line and 4-line walls are cleaned at T-intersections. **Clean** and **Cap** are mutually exclusive; only one can be selected at one time.

Cap

With this option turned on, all walls will be capped at their ends. **Clean** and **Cap** are mutually exclusive; only one can be selected at one time (see Figure 11-24).

Multi-Line Walls

The following descriptions pertain to the options available in the menus on the side of the screen, in the **Edit/Architct** menu. However, you can input these same options in a more Windows-like fashion with the new **Wall style manager** dialog box, accessed from within the **Architct** menu by pressing the **WallStyl** button. Figure 11-25 shows what the dialog box looks like.

We'll describe this dialog box in more detail in the following section.

Walls in drawings are often made up of more than just two lines. For that reason, DataCAD enables you to draw 3-line and 4-line walls as well as simple 2-line walls. In the **Architct** menu, different menu options appear, depending on whether you select the **3LnWalls** or **4LnWalls** options, as described in this section.

3-Line Walls (3LnWalls)

A 3-line wall consists of an outside line, an inside line, and a centerline. Unless overridden by the **Hilite/LineType** option, the current linetype is used to draw both the inside and outside lines. The centerline is always drawn down the center of the wall and cannot be offset to one side or the other.

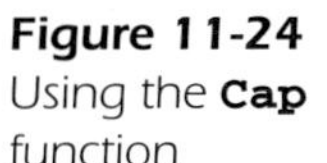

Figure 11-24 Using the **Cap** function

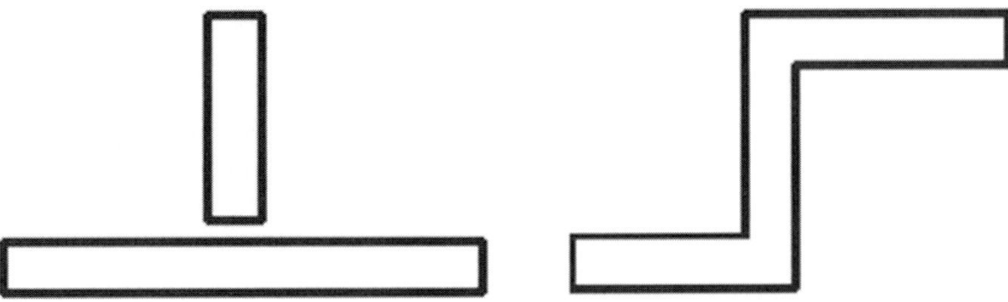

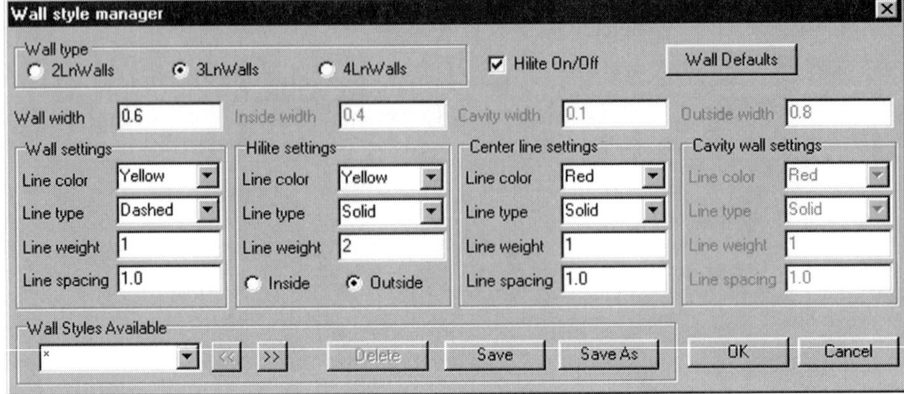

Figure 11-25 The Wall style manager

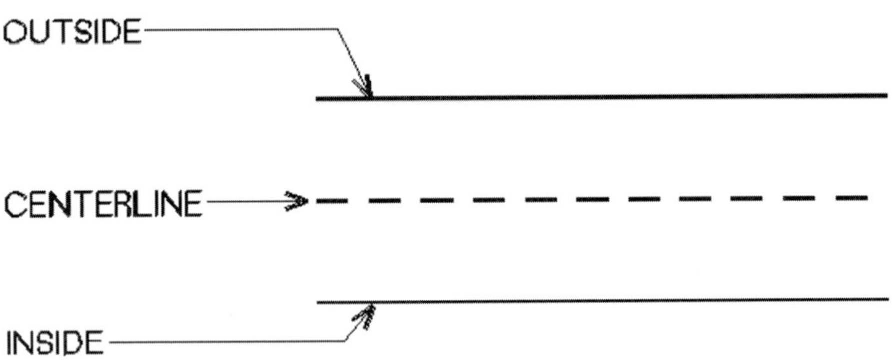

Figure 11-26 The components of a 3-line wall

CntrLine Besides the options described previously, when you select **3LnWalls**, a **CntrLine** option will appear to enable you to set the attributes of the centerline (see Figure 11-26). It contains the following options:

- **Color** selects the color for the centerline of the wall.
- **LineWgt** selects the line weight for the centerline of the wall.
- **LineType** selects the linetype for the centerline of the wall.
- **Spacing** selects the spacing of the selected linetype for the centerline of the wall.

Figures 11-27 and 11-28 shows a few examples of walls drawn with **3LnWalls**, and the settings used to create them.

4-Line Walls (4LnWalls)

A 4-line wall consists of an outside line, an inside line, and two interior lines forming two cavities. Unless overridden by the **Hilite/LineType** option, the current linetype is used to draw both the inside and outside lines (see Figure 11-29).

When you select the **4LnWalls** feature, three new options appear. Notice that the **Width** option disappears from the menu, since all wall and cavity widths for 4-line walls are controlled by the three new menu options:

- **Exterior** sets the distance from the Outside wall line to the adjacent cavity line.
- **Interior** sets the distance from the Inside wall line to the adjacent cavity line.
- **Cavity** sets the attributes of the cavities and the cavity lines. It includes the following options:
  - **Width** sets the width of the Interior cavity from one Interior line to the other.
  - **Color** selects the color for both Interior cavity wall lines.
  - **LineWgt** selects the line weight for both Interior cavity wall lines.
  - **LineType** selects the linetype for both Interior cavity wall lines.
  - **Spacing** selects the spacing of the selected linetype for both Interior cavity wall lines.

Figures 11-30 and 11-31 show a few examples of walls drawn with **4LnWalls** and the settings used to create them.

Wall Styles: Saving and Loading

OK, you've spent time creating the perfect 2-, 3-, or 4-line wall, getting all the settings and attributes just right. It would be a shame to have to trash all those settings just to create another wall type. Fortunately, DataCAD enables you to save your different wall types and then recall them later. All this is done through the **WallStyl** option, and the **Wall style manager dialog**, in the **Architct** menu (see Figure 11-25).

Note that you can also use this dialog to input all of your 2, 3 and 4-line wall options, rather than using the menu buttons on the side of the screen, as described in the previous section.

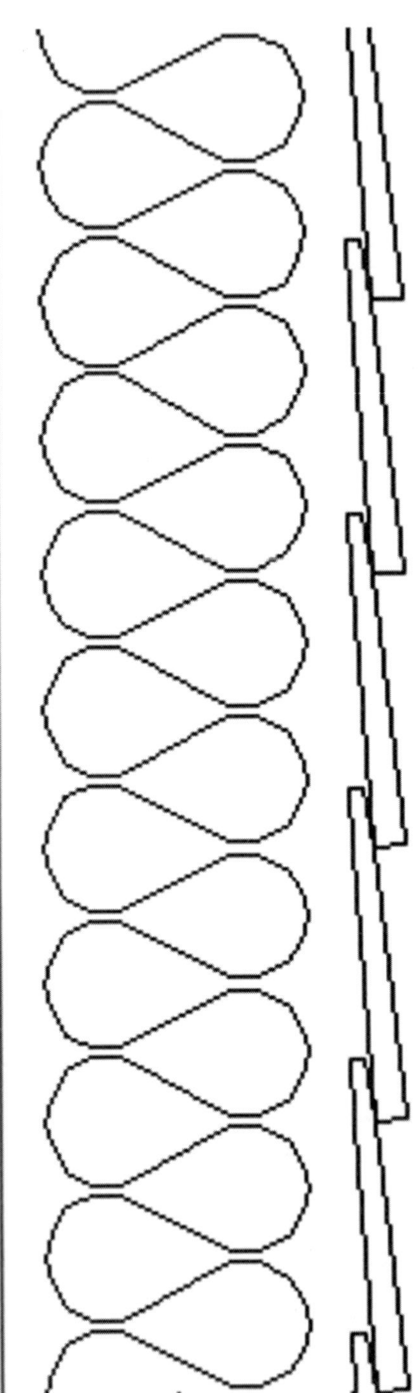

Figure 11-27
A 3-line wall example

Current Linetype = Solid
3LnWalls = on
Width = 1′-0″
CntrLine LineWgt = 1
LineType = Insul1
Spacing = 4½″
Hilite = on
Inside = on
LineWgt = 1
LineType = DShngl_R
Outside = on
Cap = off

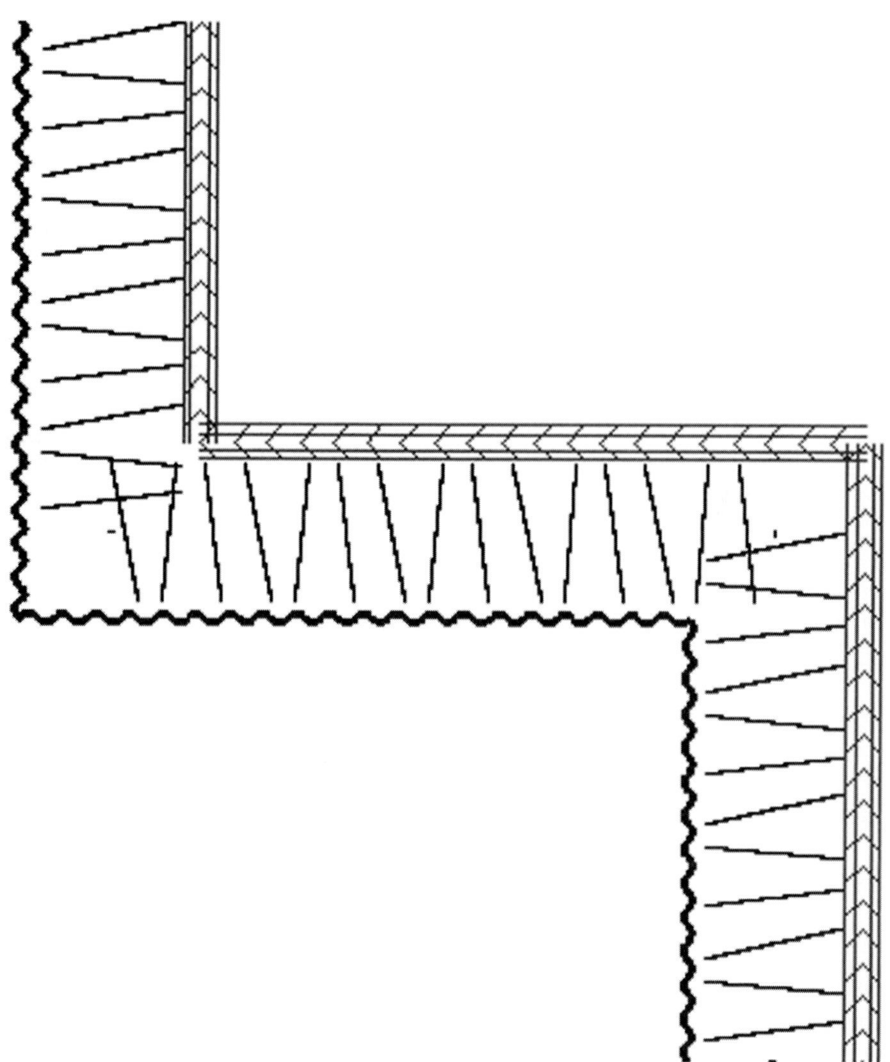

Figure 11-28 Another 3-line wall example

Current Linetype = NewPlywd
Current Line Spacing = $1^1/_2''$
3LnWalls = on
Width = 8″
CntrLine LineWgt = 2
LineType = RndmLite
Spacing = 8″

Hilite = on
Outside = on
LineWgt = 2
LineType = Undulate
Outside = on
Clean = on

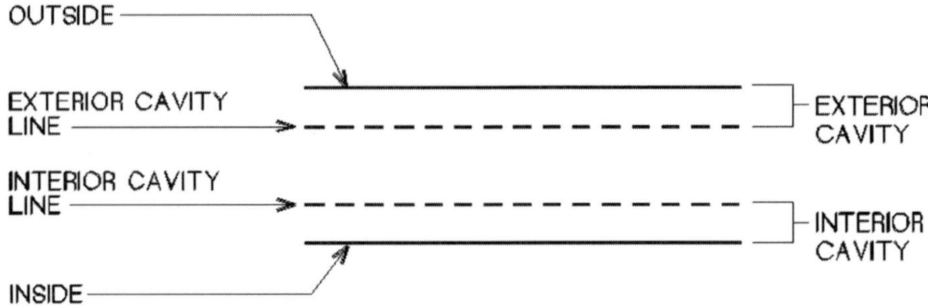

Figure 11-29 The components of a 4-line wall

Figure 11-30 A 4-line wall example

Current Linetype = Solid	LineType = NewPlywd
4LnWalls = on	Spacing = 1″
Exterior = 2″	Hilite = off
Interior = 4″	Outside = on
Cavity Width = 5″	Clean = on
LineWgt = 1	

To save a wall style, perform the following steps:

1. Set all your wall attributes, either with the menu buttons on the side, or by selecting **Architct/WallStyl**, and usign the **Wall style manager dialog**.
2. In the **Wall style manager** dialog select **Save As** to save the new wall settings with a new name (names can only be 9 characters long).

   If there is a wall style displayed in the **Wall Styles Available** box, selecting the **Save** option will save the settings to that wall style.

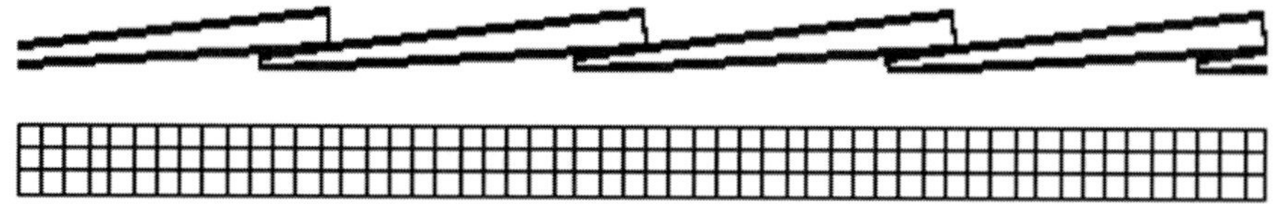

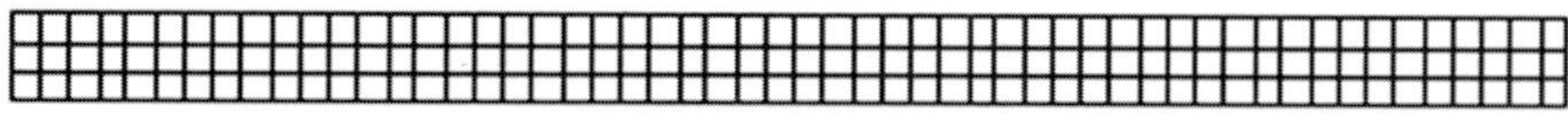

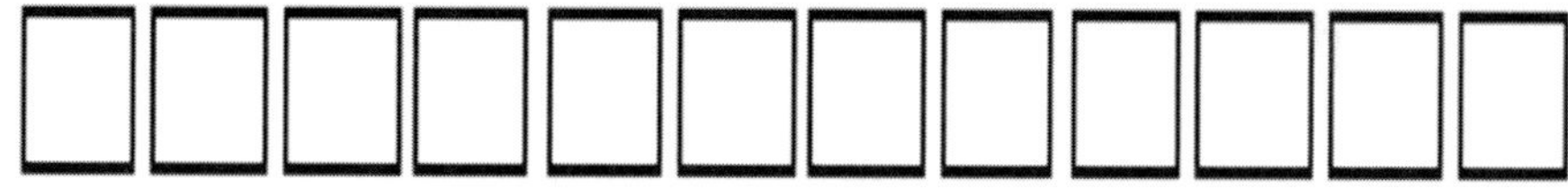

Figure 11-31 Another 4-line wall example

Current Linetype = DShngl_R
Current Line Weight = 2
Current Linetype Spacing = 6″
4LnWalls = on
Exterior = 2″
Interior = 5″
Cavity
Width = 6″
LineWgt = 1

LineType = RigdIns1
Spacing = $1^{1}/_{2}$″
Hilite = ON
Inside = on
LineWgt = 2
LineType = Brick
Outside = on
Clean = on

3. Select **OK** to accept all of the current settings, or **Cancel** to exit the dialog without saving the settings.

NOTE: *Wall styles are saved in an ASCII text file called WALL.STL, located in the DataCAD \SUP directory. If you open that file with a text editor you can see how all the settings are saved. Don't change anything, though, or your saved wall style will be affected.*

To recall a saved wall styles:

1. Select **Architct/WallStyl**. The **Wall style manager** dialog appears.
2. Pick an existing wall style by
   - a. Selecting the down-arrow to the right of the **Wall Styles Available** box

   or

   - b. Selecting the << or >> scroll buttons to the right of the **Wall Styles Available** box
3. All the settings in the dialog box will change to match the selected wall style.
4. Click on **OK** to exit the dialog while retaining the current style settings, or click on **Cancel** to retain the settings you had before accessing the **Wall style manager** dialog.
5. If the **Walls** function was turned off, loading a saved wall and existing with the **OK** button will turn the **Walls** function on.

To delete a saved wall style, perform the following steps:

1. Select **Architct/WallStyl**. The **Wall style manager** dialog appears.
2. Pick an existing wall style by:
   - a. Selecting the down-arrow to the right of the **Wall Styles Available** box

   or

   - b. Selecting the << or >> scroll buttons to the right of the **Wall Styles Available** box
3. Click on the **Delete** button to remove the selected wall style.
4. Click on **OK** to exit the dialog while retaining the current style settings, or click on **Cancel** to retain the settings you had before accessing the **Wall style manager** dialog.

To view saved wall style settings, do the following:

1. Select **Architct/WallStyl**. The **Wall style manager** dialog appears.
2. Pick an existing wall style by
   - a. Selecting the down-arrow to the right of the **Wall Styles Available** box

   or

   - b. Selecting the << or >> scroll buttons to the right of the **Wall Styles Available** box.

3. All the settings in the dialog box will change to match the selected wall style.
4. Click on **OK** to exit the dialog while retaining the current style settings, or click on **Cancel** to retain the settings you ahd before accessing the **Wall style manager** dialog.

Wall Defaults

This option in the Wall style manager dialog is used to return the options in the dialog to the DataCAD defaults (2 line wall: width = $4^1/_2''$, color = white, linetype = solid, line weight = 1, and line spacing = 1).

Walls That Are Not Walls

Walls in DataCAD are simply parallel lines whether they are straight lines or arcs. With that in mind, think of all the other items that you draw that are just parallel lines, or parallel lines with caps at both ends: sidewalks, countertops, wood studs, and so on. The examples in Figure 11-32 were drawn with the Walls feature. Note that some were drawn with **Cap** turned on and some with it off.

Cut Wall

At the bottom of the **Architct** menu is the **CutWall** tool, which is used to create clear openings in walls. In terms of where it belongs in this chapter, the **CutWall** feature sort of belongs with the description for drawing walls but also with the description for drawing doors. So, of course, that's where we have placed it. Figure 11-33 is a wall before and after using the **CutWall** feature.

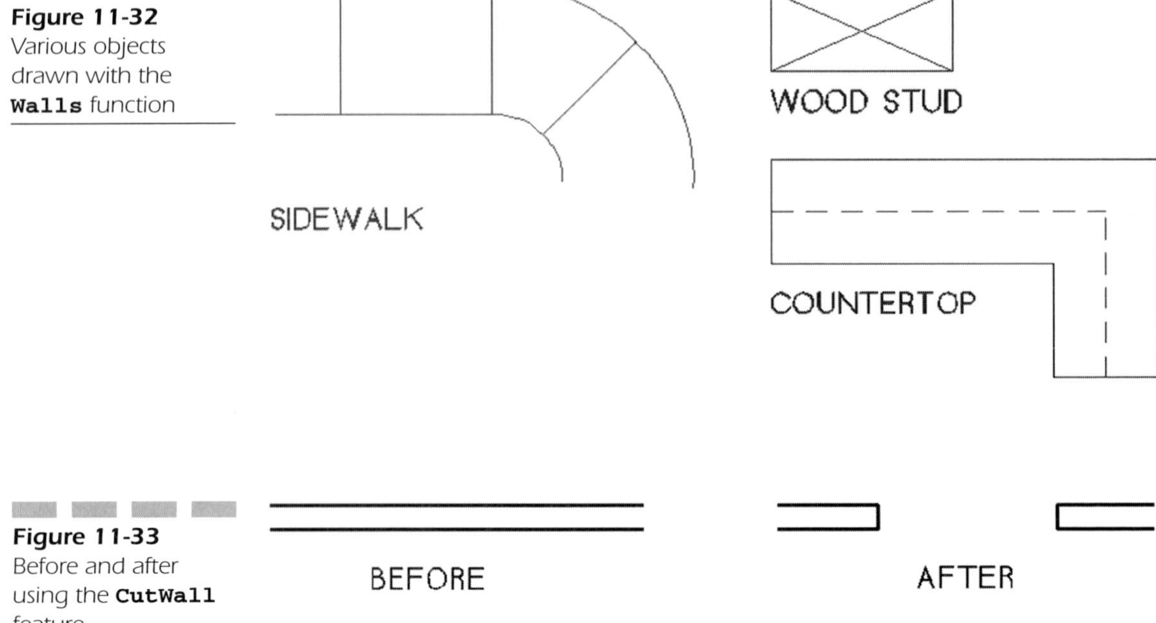

Figure 11-32 Various objects drawn with the **`Walls`** function

Figure 11-33 Before and after using the **`CutWall`** feature

The first thing you must be aware of when cutting a wall is which layer the wall is on. You can cut a wall that is on the active layer only, or you can tell DataCAD to search (via the **LyrSrch** option) for the wall on a specific layer. Using the **LyrSrch** option, you can only cut a wall that is on the specific layer you tell DataCAD to search on.

To cut a wall, follow these steps:

1. Go to the **Edit/Architct** menu and press the **CutWall** button.
2. Turn **LyrSrch** on or off as required:
   a. With **LyrSrch** off, only walls on the active layer will be cut.
   b. When you turn **LyrSrch** on, the current list of layers will be displayed and you will be prompted to *"Select the layer to search for walls:."* Select the layer that the wall resides on or use the **Match** function.

NOTE: *The **Match** function is found throughout DataCAD. Instead of selecting a layer from the layer list, you can select **Match** and then click on an entity in the Drawing Window. Whatever layer the entity is on is the layer that will be selected. This makes it easy to select the wall to be cut, even if you don't know the name of the layer it is on. You should get to know this feature.*

3. Turn **DrwMarks** on or off.

  a. With **DrwMarks** on, a small X will be placed at each of the two pick locations. This is useful for confirming that you have selected or snapped to the correct points.

  b. With **DrwMarks** off, no X's will be displayed.

4. Select the first side of the cut. You can use snapping and/or the reference point function (`) to accurately locate the first point.

5. Select the second side of the cut. You have several options for selecting the second point:

  a. Move the cursor and *click* to select the second side.

  b. Use object snapping to snap to the point of the second side.

  c. Use the reference point function (`) to locate the second point in reference to another point.

  d. Press the **Space** bar and type in the exact distance and angle to the second point.

6. The wall will be cut after selecting the second side of the opening.

How DataCAD Finds the Wall to be Cut

To determine which lines to cut, DataCAD looks first for the 2D line nearest the center of the cursor. Then it looks for the next closest parallel line to the first line. The wall cut is made within those parallel walls. For that reason, you cannot use **CutWall** to cut an opening in two non-parallel lines.

If **LyrSrch** is on, then DataCAD also checks to see if the closest lines to the cursor are on the current search layer. If not, then the walls will not be cut. If **LyrSrch** is off, then DataCAD also checks to see if the closest lines to the cursor are on the active layer. If not, then the walls will not be cut.

If the two lines closest to the cursor are not on the correct layer or are not parallel, but two other parallel lines on the correct layer are nearby, then DataCAD may select them instead. But DataCAD is also smart enough to know that these lines might not be the ones you want to cut because of how far away they are from your cursor. In this case, DataCAD will cut those lines, but it will also prompt you, *"Is this correct? (Yes/No)."* If you select **Yes**, then the wall will remain cut. If you answer **No**, then the wall will not be cut and DataCAD will tell you *"No walls found to cut. Please check search layer."*

Removing Cut Walls

You can remove a cut wall at any time, but just as **LyrSrch** has rules for creating cut walls, it has similar rules for removing them. To remove a cut wall, follow these steps:

1. In the **CutWall** menu, select **Remove**.
2. Turn **LyrSrch** on or off as required:
   a. With **LyrSrch** off, only walls on the active layer will be cut.
   b. If you turn **LyrSrch** on, then all layers that are currently on will be searched.
3. Draw an area bounding box completely around the wall cut. The cut will be removed and the wall will be "healed" (see Figure 11-34).

You can remove more than one wall cut at one time, and the wall cut does not require capped wall ends (see Figure 11-35).

Only lines that are aligned directly across from each other will be healed (see Figure 11-36).

Doors

Walls without openings hardly make useful buildings, so you are probably going to want to place some doors in your floor plans. DataCAD provides

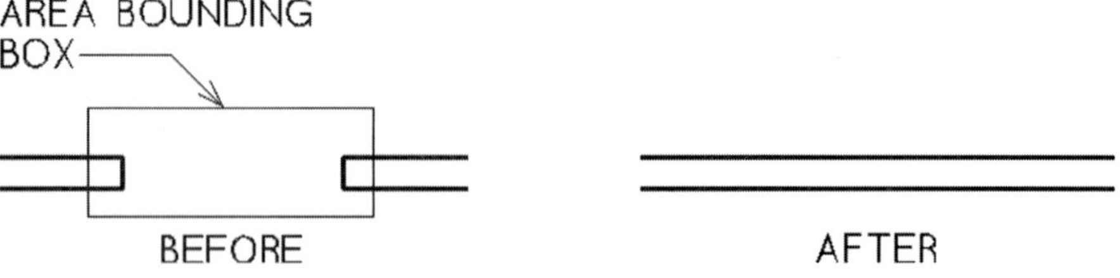

Figure 11-34
Healing a cut wall

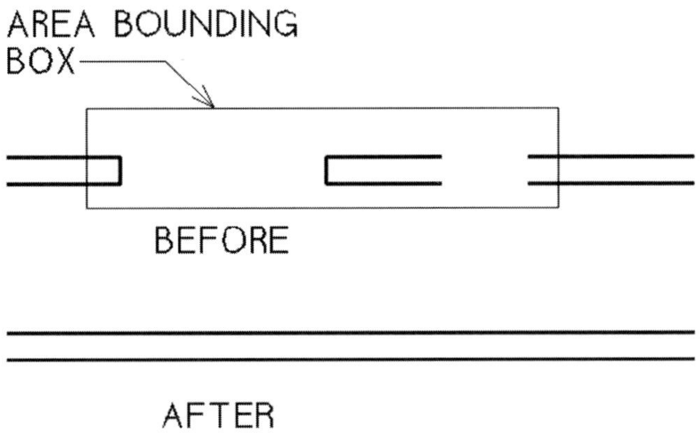

Figure 11-35 Removing more than one wall cut

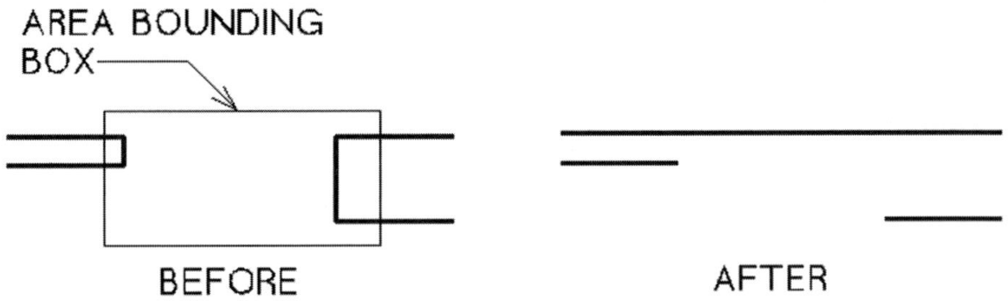

Figure 11-36 Repairs can only be made to lines directly across from one another.

you with the tools to automatically draw doors in your floor plan and to automatically remove them as well. The 2D door features are found in the **Edit/Architct** menu under **DoorSwng**. Figure 11-37 displays the six styles of doors that can be automatically drawn.

Adding Doors

Each of the six door types has a number of options that control how they will be drawn, but the basic operation of inserting doors is mostly the same for each of the door types. The rules for how DataCAD determines what lines constitute a valid wall for door insertion are the same as for wall cuts, described earlier, so you should read that section if you haven't already.

Before we go into all the options for the display and creation of doors, let's take a look at how a typical door might be inserted into a wall. Don't worry too much about which options are selected at this point, just note the basic steps. Let's add a single swinging door, open to 90 degrees, with a curve to

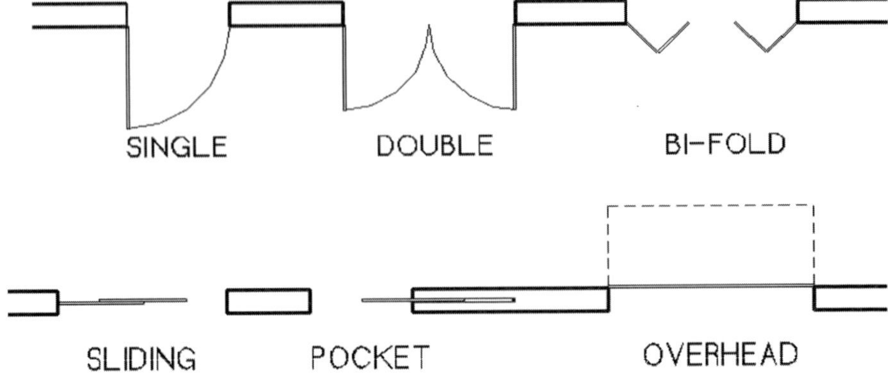

Figure 11-37
The six door styles

display the door swing. Doors are drawn in the current layer color, so make sure you set the color before you begin to draw your doors.

1. Go to the **Edit/Architct** menu and select **DoorSwng**.
2. Select **DoorStyl/Single** to create a single swinging door.
3. Select **SwngStyl/Arc** and then turn on **MtchDoor**. *Right-click* back to the **DoorSwng** menu.
4. Select **Angle** and then select or type **90** degrees. *Right-click*.
5. Select **Thicknss** and then type in **.1.3/4** (DataCAD shorthand for 1 $^{3}/_{4}$″ [45]).
6. In the **DoorSwng** menu, set these buttons as follows:

   Sides = on
   CntrPnt = on
   Cutout = on
   InWall = on
   DrawJamb = on
   JambIn = off
   MtchWall = on

7. If it is not already on, press **LyrSrch** to turn it on. You will be presented with a list of all the layers in your drawing. You need to tell DataCAD which layer contains the walls that the doors will be cut into. In this example, the wall is on the **Walls** layer, so that's the layer we will select.
8. The Prompt Line reads, *"Select hinge side of door."* Place the cursor at the point where you want to insert the door and then *click* the mouse. You do not have to actually touch the wall; you just need to be close to it.

9. You will then be prompted to *"Select the strike side of door."* You can do it one of two ways:
   a. Drag the cursor to the right or left until the Coordinate Readout line displays the width of the door, such as 3′-0″ [914]. Then *click* the mouse to select that distance.
   b. Press the **Space** bar to enter the door size by typing. If you are using the relative polar method, then you will be prompted to enter a distance (width) for the door. After typing the distance, press **Enter** to accept the typed distance. Now you will be prompted to enter an angle to the second point. After typing the angle, press **Enter** to accept it.
10. Now the Command Line will read, *"Select direction of door swing."* Pick a point anywhere on the side of the wall that the door is to swing to and *click* the mouse. The wall will be cut, but the door will not be drawn yet.
11. The Prompt Line will read *"Select any point on the outside of the wall."* Pick a point anywhere on the side of the wall that the door is to swing to and *click* the mouse. The door is inserted.

NOTE: *You may be asking yourself why DataCAD asks you twice to pick a point on the side of the wall that the door is to swing to. It's because if you turn the* ***CntrPnt*** *option on, DataCAD will place the snapping dot on the outside of the new door, and the outside of the door is not necessarily the same side as the door swing. That's what the second pick in Step 11 is indicating.*

In Figure 11-38, we opted to draw our door with jambs and with a center snapping point, but if you don't want either of these, then you would simply make sure those options were turned off prior to creating the door.

NOTE: *In Step 8, we said to place the cursor at the point where you want to insert the door. In most cases, you will want to be more accurate than just getting the door close. Accurately locating each side of the door is the same as accurately locating anything else in DataCAD. You can use the Reference Point (`) key to locate the door from another fixed point, or you can draw a perpendicular guideline across the wall and then snap to the guideline / wall intersection.*

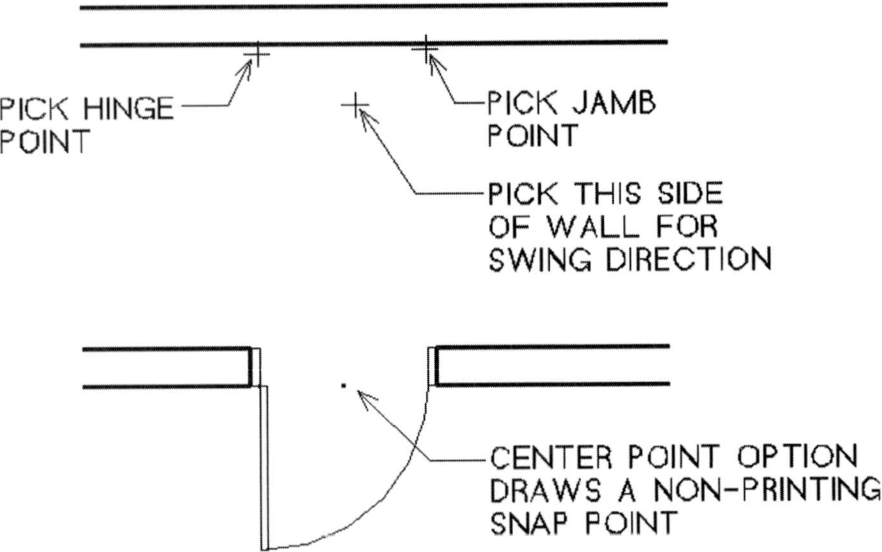

Figure 11-38 Creating a door with jambs and a center snapping point

Removing Doors

Now let's see how to quickly remove a door and heal the wall that was cut. We'll use the door we just inserted as an example:

1. In the **Edit/Architct/DoorSwng** menu, select **Remove**. The Prompt Line will read *"Select first point of box around door or window to remove."*
2. Pick two points to place a bounding box around the door and both jambs. Be sure to have all the endpoints of the wall cuts and all the entities that make up the door within the bounding box. Also be sure that no extraneous endpoints or entities are within the bounding box (see Figure 11-39).
3. The door will be erased and the wall is healed. *Right-click* back to the **DoorSwng** menu.

Inserting the Other Door Types

The steps for drawing and removing the other door types are essentially the same as for a single swinging door, although some of the Command Line prompts will vary slightly, depending on which door type you select.

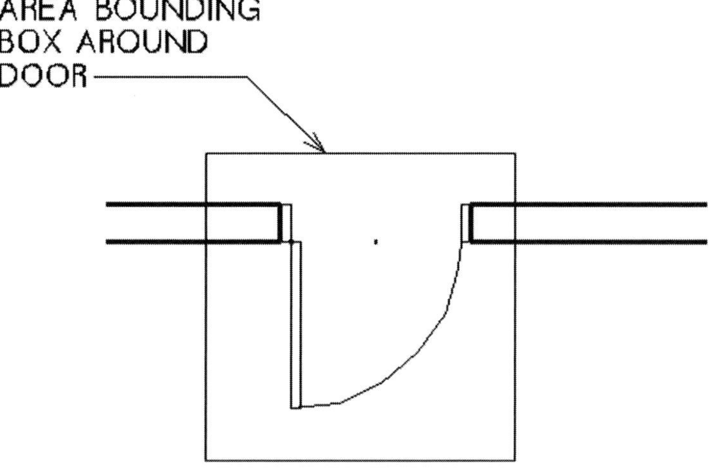

Figure 11-39
Placing a bounding box around the door and jambs to be removed

Figure 11-40
The healed wall with the door removed

Door Options

Here is what each of the options in the **DoorSwng** menu does.

Sides With this option on, doors are selected by their jambs. With this option off, doors are defined by their centers and one jamb (the strike side if it is a door type with a strike).

CntrPnt With this option on, a non-printing snap point is placed at the center point between the two door jambs. This point is useful if you dimension your doors to their centers rather than their jambs (see Figure 11-38).

Cutout With this option on, the wall will be cut when the door is drawn. If it is turned off, then the wall will not be cut when the door is drawn (see Figure 11-41). Doors drawn with this option on can be removed with the **Remove** option. Doors that are drawn with this option off cannot be

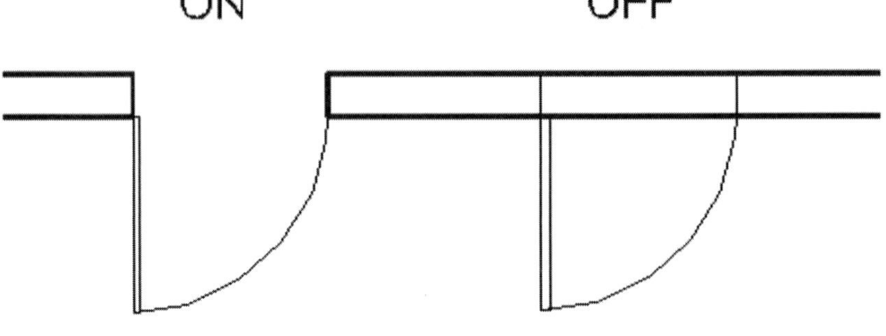

Figure 11-41
`Cutout` on and off

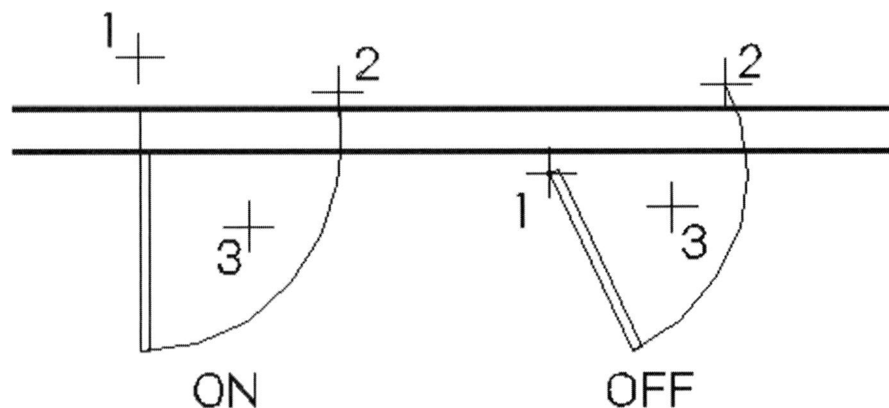

Figure 11-42
`InWall` examples

removed with the **Remove** command. Use the **Erase/Area** command instead.

InWall With this option on, doors will snap to the wall or jamb. With this option off, all walls are ignored and the doors are drawn exactly between the two points selected with the cursor even if those two points are at an angle and even if they are not within the wall opening (see Figure 11-42). The **InWall** option cannot be turned off when the **Cutout** option is turned on.

LyrSrch The **LyrSrch** option only appears when **InWall** is turned on. With **LyrSrch** off, only walls on the active layer will be cut. With **LyrSrch** on, the current list of layers will be displayed and you will be prompted to

"Select the layer to search for walls." Select the layer that the wall resides on or use the **Match** function.

NOTE: *This function is found throughout DataCAD. Instead of selecting a layer from the layer list, you can select **Match** and then click on an entity in the Drawing Window. Whichever layer the entity is on is the layer that will be selected. This makes it easy to select the wall to be cut, even if you don't know the name of the layer it is on. You should get to know this feature.*

LyrSrch is the option that tends to confound people when they are trying to insert doors. It is intended to make sure that you only cut the walls that you really intended to cut. If you are drawing doors on the first floor, you may be cutting walls on the 1-WALLS layer. If you then want to cut walls on the second floor on layer 2-WALLS, you must select the **LyrSrch** button twice; once to turn it off and then again to turn it back on after which you can then select the 2-WALLS layer to be the new layer in which the door cutouts will be made.

DrawJamb With this option turned on, doors will be drawn with jambs. If it is off, no jambs will be drawn. If **Cutout** is also on, then the cut walls will be capped in addition to the jambs being drawn (see Figure 11-43).

JambIn This option will only appear when **DrawJamb** is turned on. With **JambIn** turned on, the door will be defined by its rough opening (outside of jamb to outside of jamb). With this option turned off, the door will be

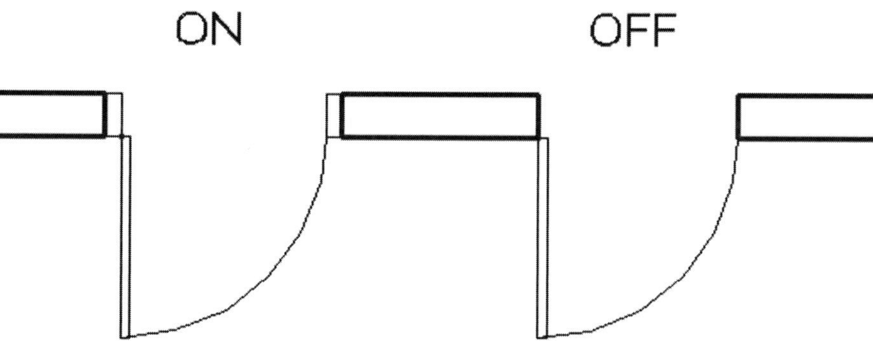

Figure 11-43 Doors with and without jambs

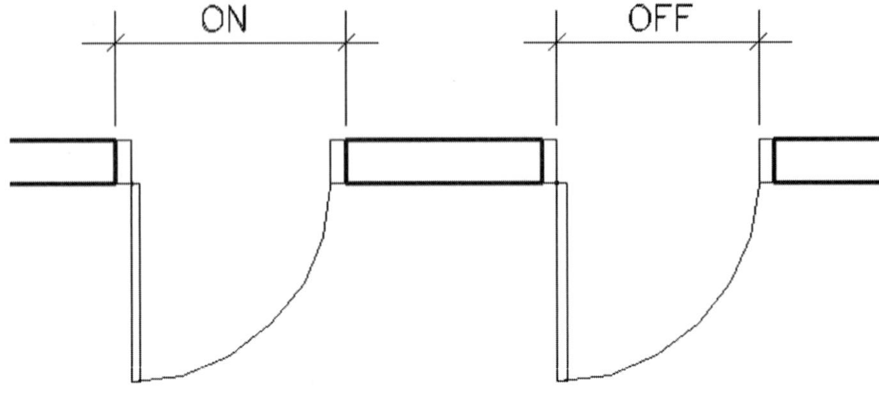

Figure 11-44
The width of the door with **`JambIn`** on and off

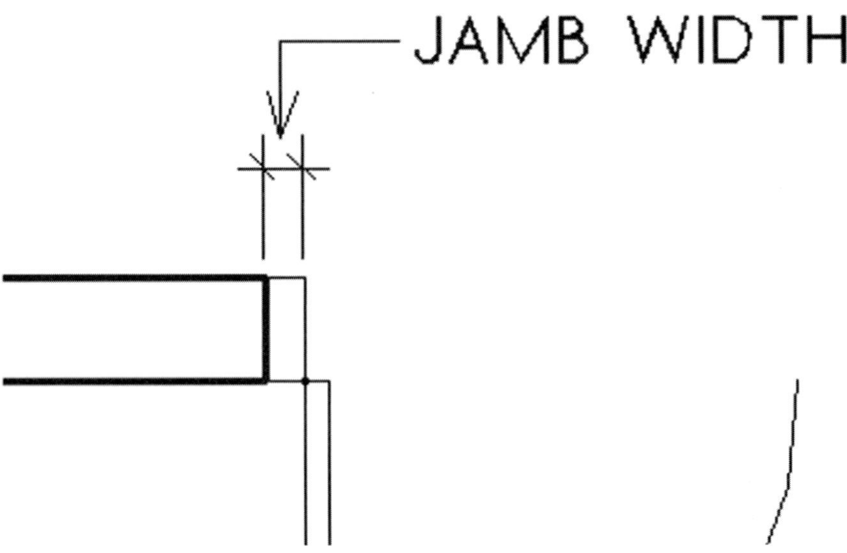

Figure 11-45
Use **`JambWdth`** to set jamb widths.

defined by its clear opening (inside of jamb to inside of jamb), as shown in Figure 11-44.

JambWdth This option will only appear when **DrawJamb** is turned on. Select this option to set the width of jambs when they are drawn (see Figure 11-45).

MtchWall This option will only appear when **DrawJamb** is turned on. With **DrawJamb** on, the depth of the jamb will match the depth of the wall. With this option turned off, the **MtchWall** option will be displayed, allowing you to set the depth of the jamb to something other than the width of the wall.

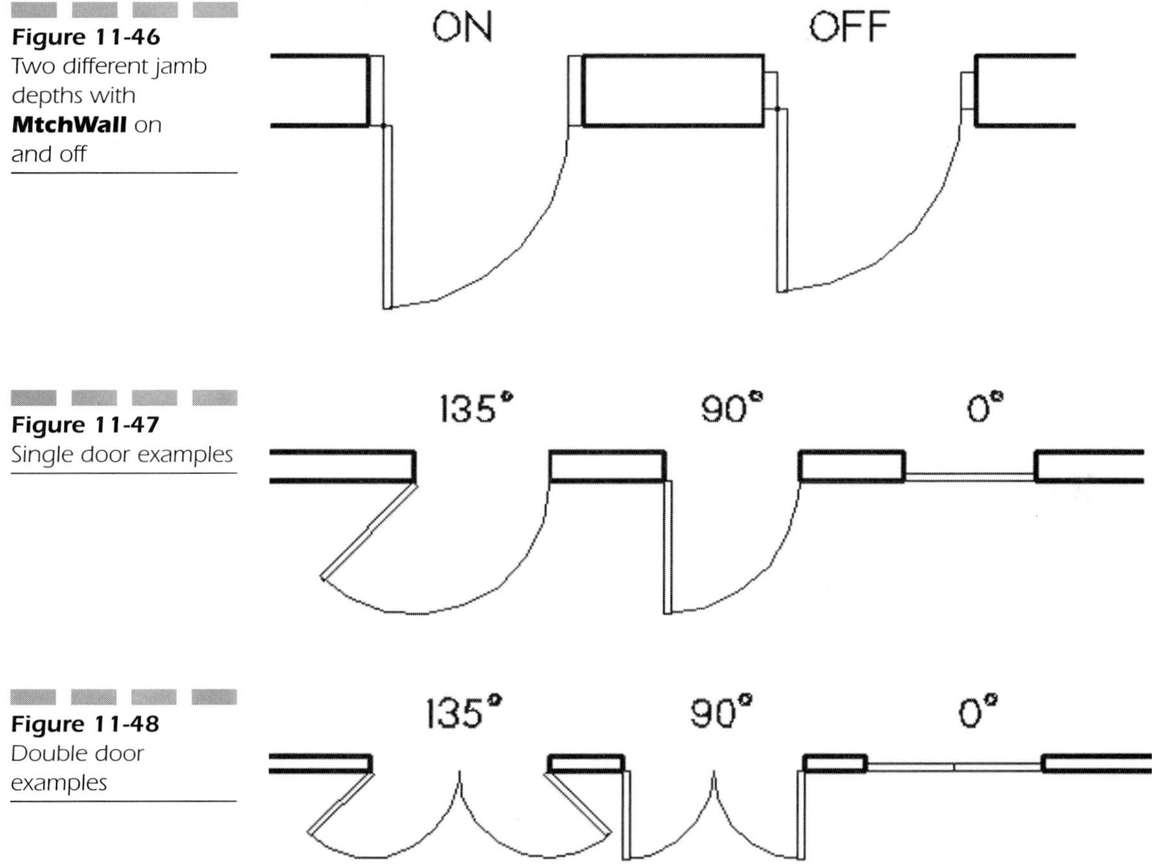

Figure 11-46 Two different jamb depths with **MtchWall** on and off

Figure 11-47 Single door examples

Figure 11-48 Double door examples

JambDpth This option will only appear if the **MtchWall** option is turned off. Use **JambDpth** to select a new jamb depth. In Figure 11-46, the wall is 12″ [305] thick, while the jamb depth of the door on the right is 6″ [152].

DoorStyl Use this option to select one of the six door styles: Single, Double, Bi-fold, Sliding, Pocket, and Overhead.

Angle Use this option to select the degree of openness of each door type. The word angle obviously refers to a swinging door, and although sliding, pocket, and overhead doors do not swing open, the **Angle** option is still used to determine the degree of openness of those door types as well. In general, you can enter any angle from 0 to 180 degrees, but as you will see in Figures 11-47 through 11-53, some door types don't support more than 90 degrees, lest you get some odd results.

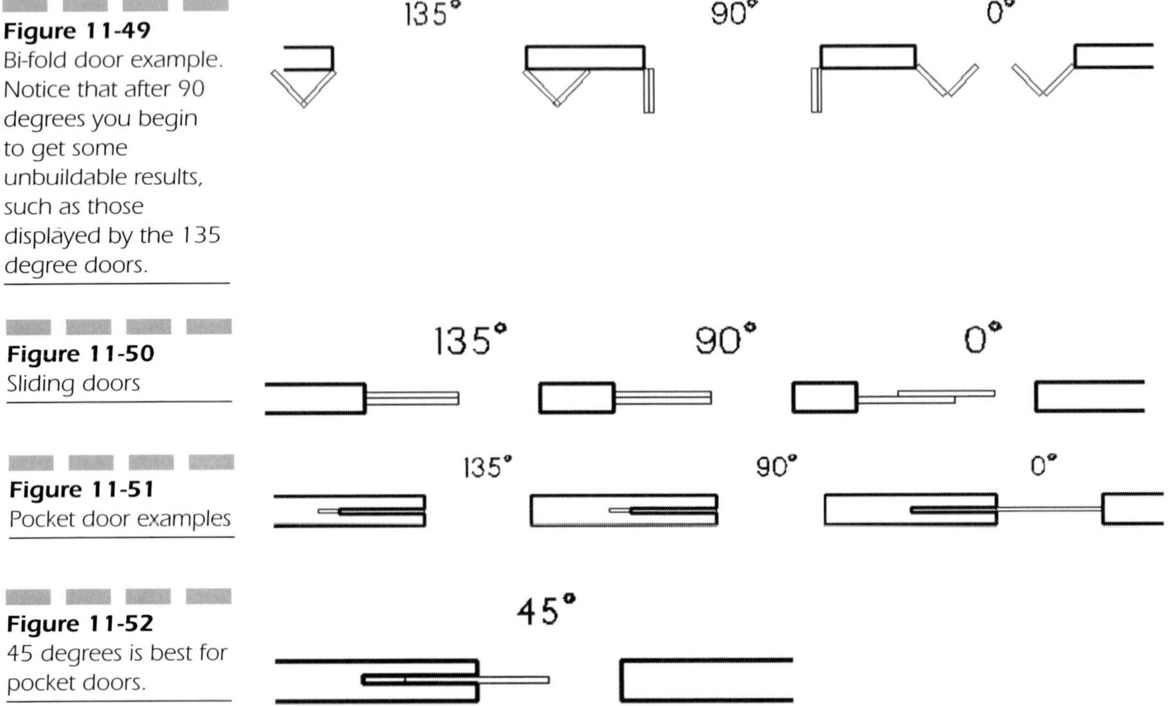

Figure 11-49 Bi-fold door example. Notice that after 90 degrees you begin to get some unbuildable results, such as those displayed by the 135 degree doors.

Figure 11-50 Sliding doors

Figure 11-51 Pocket door examples

Figure 11-52 45 degrees is best for pocket doors.

Sliding doors cannot use an angle of more than 90 degrees. Doing so will still yield results that look exactly like a door specified at 90 degrees (see Figure 11-50).

Pocket doors are drawn a bit strangely in DataCAD. You will get different degrees of openness depending on how wide the door opening is drawn (Figure 11-51). This is probably a bug in the program, so don't be surprised if this gets fixed. In Figure 11-52, you can see that an angle of 45 degrees seems to work better for this door type.

Pocket doors cannot be automatically removed with the **Remove** command. Instead you have to use **Erase/Area** and then weld the walls back together (see Figure 11-52).

With overhead doors, the depth of the dashed line (representing a partially raised door overhead) is drawn according to the dimension specified for the **HeadHgt** (see Figure 11-53). Notice that the **Angle** specified has no impact on how an overhead door is drawn.

Thicknss **Thicknss** controls the thickness of the door itself. A setting of 0 will draw a door of only one line with no thickness (see Figure 11-54).

HeadHgt This controls the height of the door in $2^{1}/_{2}$ D. Since DataCAD is capable of extruding 2D lines to display a height, you can set the head

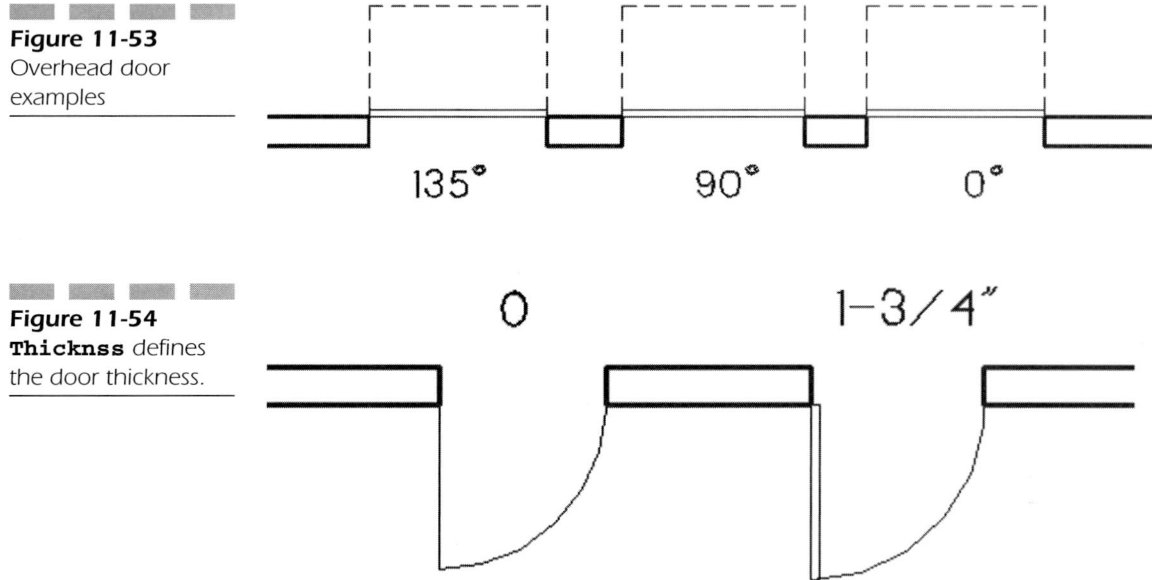

Figure 11-53
Overhead door examples

Figure 11-54
`Thickns` defines the door thickness.

height if you want to get a rough 3D idea of what the doors will look like. The wall header above the door is not filled in, so you won't get a true 3D wall with a true 3D door; that you would have to do with the 3D door tools in the AEC_MODL macro in the **Toolbox**. In Figure 11-55, the walls are 8′-0″ [2438] high, and the door height (**HeadHgt**) is 6′-8″ [2032].

SwngStyl This option determines how the line showing the swing of the door will appear (see Figure 11-56).

MtchDoor This will only appear when **Line** or **Arc** is selected. Selecting this option will draw the door swing in the same color as the door.

SwngColr This will only appear when **MtchDoor** is turned off. When you select this option, the standard list of DataCAD colors appears. Select a color for all your door swings.

Remove Use this option to automatically remove a door. You can remove more than one door at a time if they exist within the same wall. This function does not work on pocket doors. See the example at the beginning of the "Doors" section for a description of how to remove a door. When removing a door, if you have some stray entities within the area box, then DataCAD may become confused about which lines are doors, which are walls, and which are something else, so you may need to turn off a layer or two prior

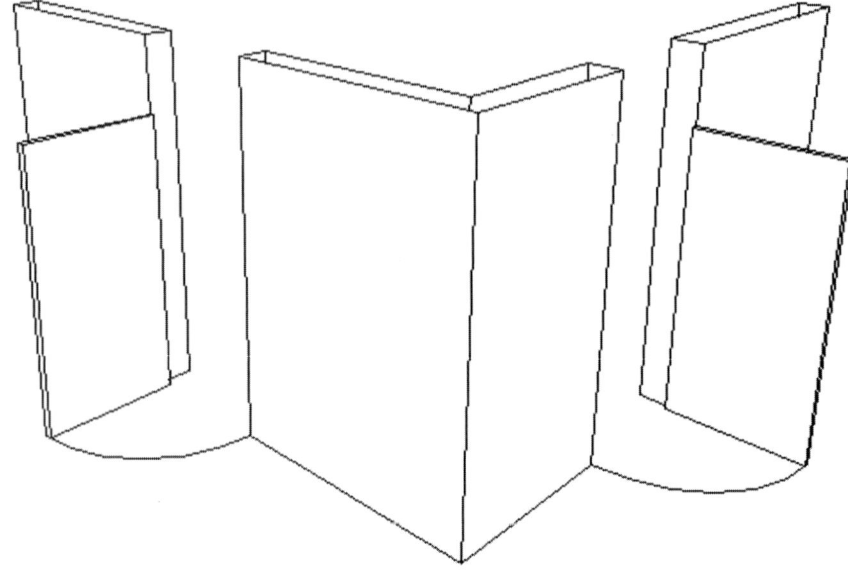

Figure 11-55 A $2^1/_2$ D version of some doors

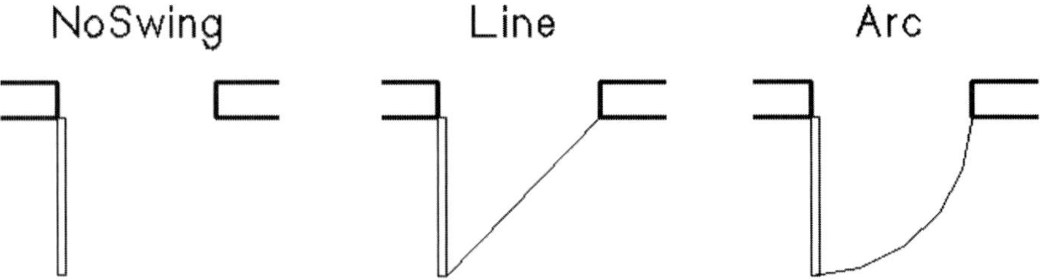

Figure 11-56 The three available swing styles

to trying to remove doors just so there aren't a lot of stray entities around the door.

DrwMarks With **DrwMarks** on, a small X will be placed at each of the cursor pick locations. This is useful for confirming that you have selected or snapped to the correct points. With **DrwMarks** off, no X's will be displayed.

Windows

Windows are drawn very much like doors and follow the same general rules. DataCAD provides you with the tools to automatically draw windows in your floor plan and to automatically remove them as well. The 2D window features are found in the **Edit/Architct** menu under **Windows**. Unlike doors, there are no styles of windows since most window types in a plan (sliding, double hung, casement, and so on) are generally drawn the same way. However, you do have control over sills and glass, as well as the heights of the sill and head, for display in $2^1/_2$ D. Figure 11-57 demonstrates what a window might look like in a plan and Figure 11-58 shows the same window in $2^1/_2$ D.

The rules for how DataCAD determines which lines constitute a valid wall for window insertion are the same as for wall cuts, described earlier, so you should read that section if you haven't already.

Before we go into all the options for the display and creation of windows, let's take a look at how a typical window might be inserted into a wall. Don't worry too much about what options are selected at this point; just note the basic steps. We will insert a window with an interior and exterior sill and a

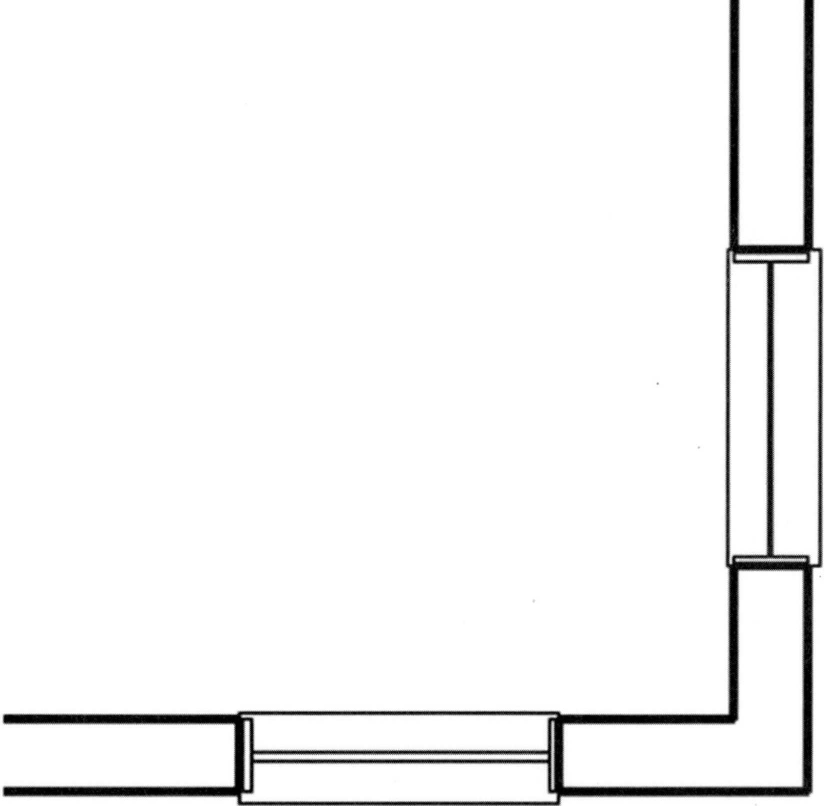

Figure 11-57
A couple of windows in a floor plan

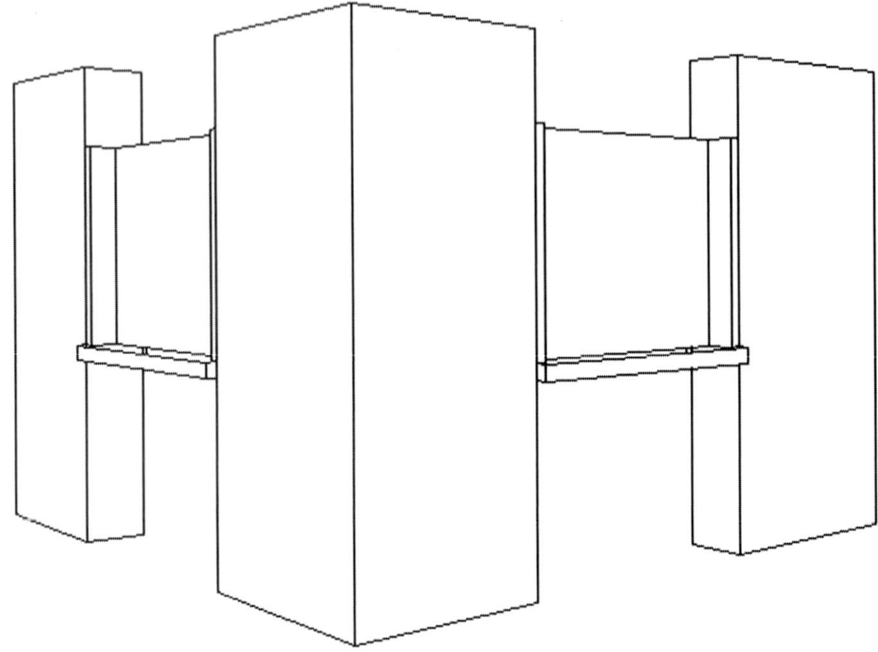

Figure 11-58 The same windows in $2^1/_2$ D

1″ [25.4] thick insulating glass with a sill height of 3′-4″ [1016] and a head height of 6′-8″ [2032]

Windows are drawn in the current layer color, so make sure you set the color before you begin to draw your doors:

1. Go to the **Edit/Architct** menu and select **Windows**.
2. In the **Windows** menu, set these buttons as follows:

   Sides = on
   CntrPnt = off
   Cutout = on
   InWall = on
   DrawJamb = on
   JambIn = off
   MtchWall = on

3. Select **JambWdth** and type **.2** (draws a jamb of 2″ [51] wide) and then press **Enter** to accept it.
4. Select **OutSill** and type **.2** (draws an outside sill of 2″ [51]) and then press **Enter** to accept it.

5. Select **InSill** and type **.1** (draws an inside sill of 1″ [25.4]) and then press **Enter** to accept it.
6. Select **GlassThk** and type **..1/2** (DCAD shorthand for a $^{1}/_{2}$″ [12.7]) and then press **Enter** to accept it.
7. Select **SillHgt** and type **3.4** (3′-4″ [1016]) and then press **Enter** to accept it.
8. Select **HeadHgt** and type **6.8** (6′-8″ [2032]) and then press **Enter** to accept it.
9. If it is not already on, press **LyrSrch** to turn it on. You will be presented with a list of all the layers in your drawing. You need to tell DataCAD which layer contains the walls that the windows will be cut into. In this example, the wall is on the **Walls** layer, so that's the layer we will select.
10. The Prompt Line reads, *"Select one jamb of window."* Place the cursor at the point where you want to insert the window and then *click* the mouse. You do not have to actually touch the wall; you just need to be close to it.
11. You will then be prompted to *"Select second jamb of window."* You can do it one of two ways:
    - **a.** Drag the cursor to the right or left until the Coordinate Readout line displays the width of the window, such as 3′-0″ [914]. Then *click* the mouse to select that distance.
    - **b.** Press the **Space** bar to enter the window size by typing. If you are using the Relative Polar method, then you will be prompted to enter a distance (width) of the window. After typing the distance, press **Enter** to accept the typed distance. Now you will be prompted to enter an angle to the second point. After typing the angle, press **Enter** to accept it.
12. Now the Command Line will read, *"Select any point on the outside of the wall."* Pick a point anywhere on the outside of the wall and *click* the mouse. The wall will be cut and the window will be added (see Figure 11-59).

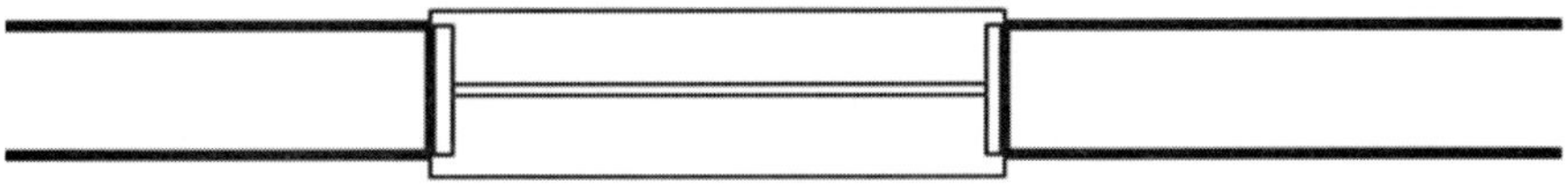

Figure 11-59
Adding a window

NOTE: In Step 10 we said to place the cursor at the point where you want to insert the window. In most cases you will want to be more accurate than just getting the window close. Accurately locating each side of the window is the same as accurately locating anything else in DataCAD. You can use the Reference Point (`) key to locate the window from another fixed point, or you can draw a perpendicular guideline across the wall then snap to the guideline / wall intersection.

Removing Windows

Now let's see how to quickly remove a window and heal the wall that was cut. We'll use the window we just inserted as an example:

1. In the **Edit/Architct/Windows** menu, select **Remove**. The Prompt Line will read, *"Select first point of box around door or window to remove."*
2. Pick two points to place a bounding box around the window and both jambs (see Figure 11-60). Be sure to have all the endpoints of the wall cuts and all the entities that make up the window within the bounding box. Also be sure that no extraneous endpoints or entities are within the bounding box.
3. The window will be erased and the wall is healed. *Right-click* back to the **Windows** menu.

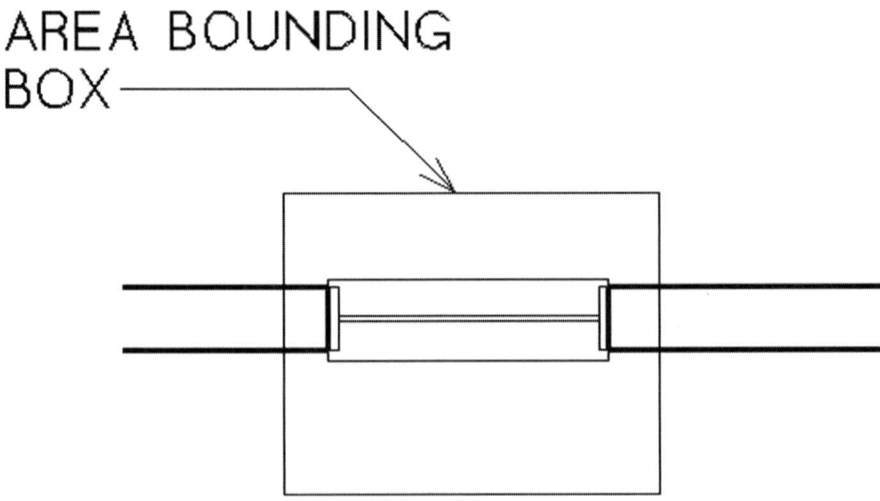

Figure 11-60 Remove a window by drawing a bounding box around it.

Window Options

Here is what each of the options in the Windows menu does. The first five options in the menu are the same as those in the **DoorSwng** menu, so if you are familiar with creating doors, then you will be familiar with the basics of creating windows.

Sides With this option on, windows are selected by their jambs (sides). With this option off, windows are defined by their centers and one jamb.

CntrPnt With this option on, a non-printing snap point is placed at the center point between the two window jambs (see Figure 11-61). This point is useful if you dimension your windows to their centers rather than their jambs.

Cutout With this option on, the wall will be cut when the window is drawn. If it is turned off, then the wall will not be cut when the window is drawn (see Figure 11-62). Windows drawn with this option on can be removed with the **Remove** option. Windows that are drawn with this option off cannot be removed with the **Remove** command. Use the **Erase/Area** command instead.

InWall With this option on, a window will snap to the wall or jamb. With this option off, all walls are ignored and the windows are drawn exactly between the two points selected with the cursor, even if those two points

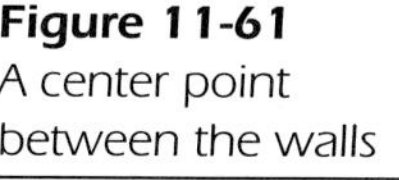

Figure 11-61
A center point between the walls

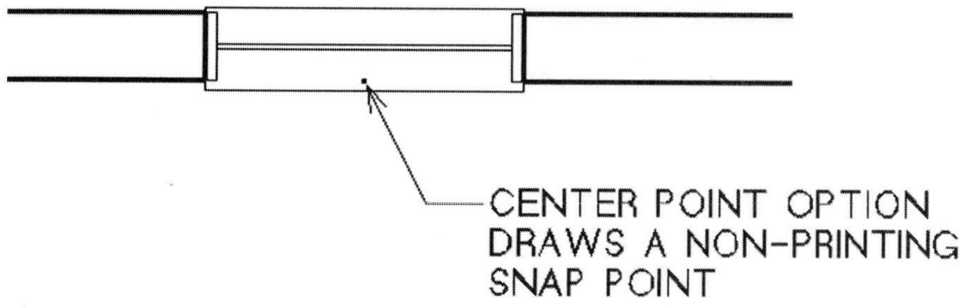

Figure 11-62
Examples of `Cutout` on and off

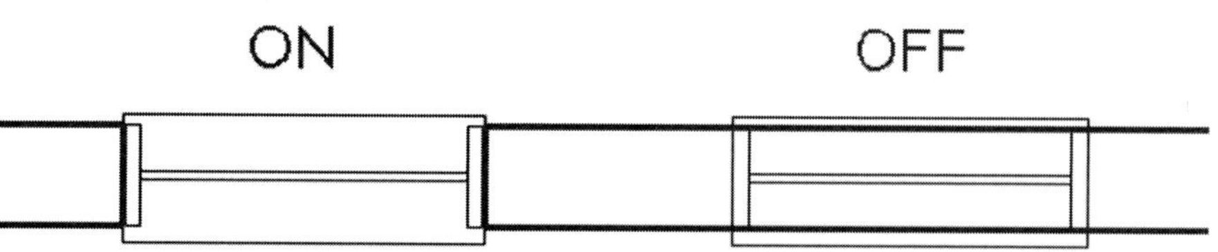

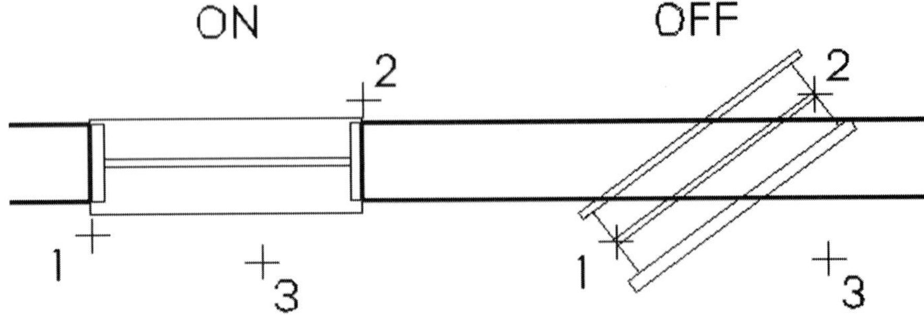

Figure 11-63 Examples of `InWall` on and off

are at an angle, and even if they are not within the wall opening (see Figure 11-63). The **InWall** option cannot be turned off when the **Cutout** option is turned on.

LyrSrch The **LyrSrch** option only appears when **InWall** is turned on. With **LyrSrch** off, only walls on the active layer will be cut. With **LyrSrch** on, the current list of layers will be displayed and you will be prompted to *"Select the layer to search for walls:"* Select the layer that the wall resides on or use the **Match** function.

NOTE: *This function is found throughout DataCAD. Instead of selecting a layer from the layer list, you can select* ***Match*** *and then click on an entity in the Drawing Window. Whatever layer the entity is on is the layer that will be selected. This makes it easy to select the wall to be cut, even if you don't know the name of the layer it is on. You should get to know this feature.*

LyrSrch is the option that tends to confound people when they are trying to insert windows. It is intended to make sure that you only cut the walls that you really intended to cut. If you are drawing windows on the first floor, you may be cutting walls on the 1-WALLS layer. If you then want to cut walls on the second floor, on layer 2-WALLS, you must select the **LyrSrch** button twice; once to turn it off and then again to turn it back on after which you can then and select the 2-WALLS layer to be the new layer in which the window cutouts will be made.

DrawJamb With this option turned on, windows will be drawn with jambs. If it is off, no jambs will be drawn. If **Cutout** is also on, then the cut walls will be capped in addition to the jambs being drawn (see Figure 11-64).

JambIn This option will only appear when **DrawJamb** is turned on. With **JambIn** turned on, the window will be defined by its rough opening (outside of jamb to outside of jamb). With this option turned off, the window will be defined by its clear opening (inside of jamb to inside of jamb), as shown in Figure 11-65.

JambWdth This option will only appear when **DrawJamb** is turned on. Select this option to set the width of jambs when they are drawn (see Figure 11-66).

MtchWall This option will only appear when **DrawJamb** is turned on. With this option turned on, the depth of the jamb will match the depth of the wall. With this option turned off, the **MtchWall** option will be displayed, allowing you to set the depth of the jamb to something other than the width of the wall.

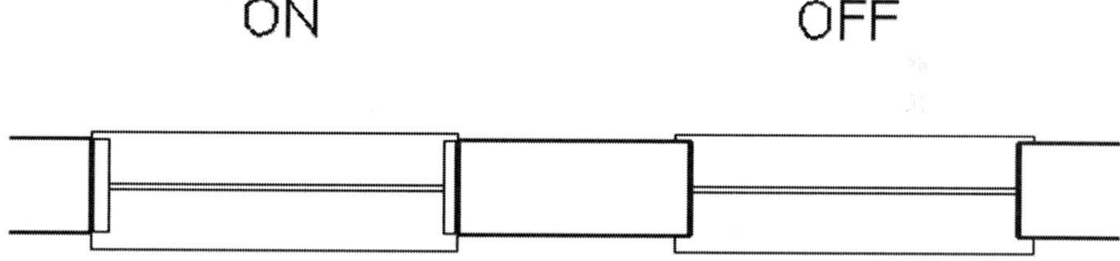

Figure 11-64 The difference between `DrawJamb` on and off

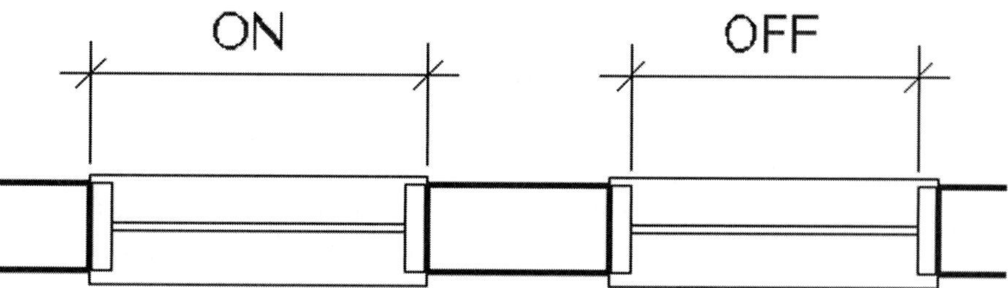

Figure 11-65 `JambIn` on and off

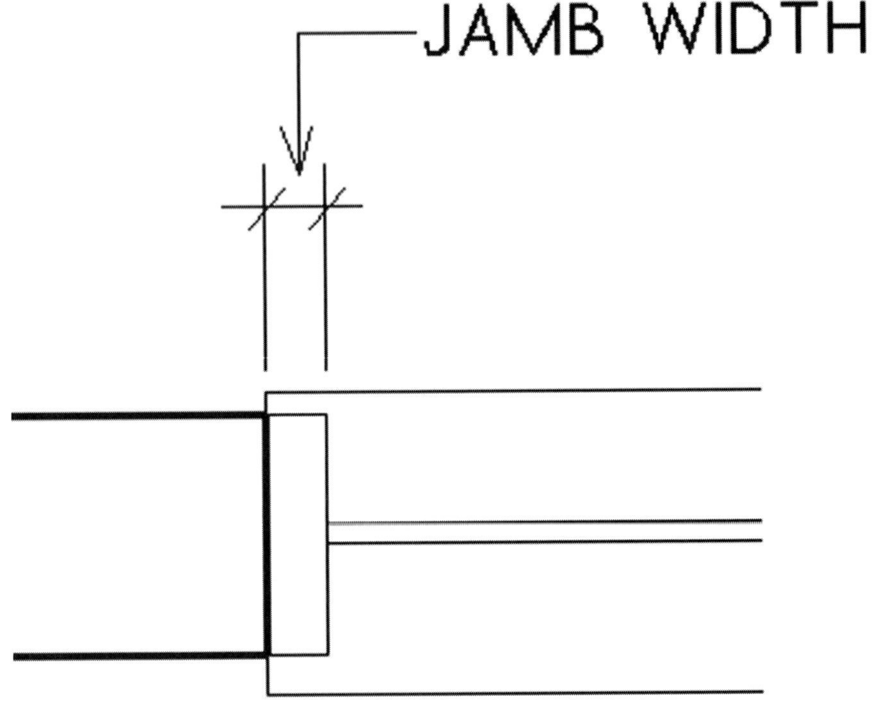

Figure 11-66
JambWdth to set the jamb width

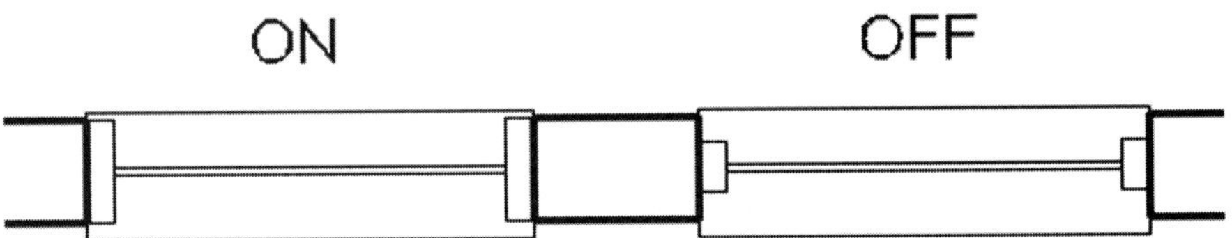

Figure 11-67
JambDpth sets jamb depth.

JambDpth This option will only appear if the **MtchWall** option is turned off. Use this option to select a new jamb depth. In Figure 11-67, the wall is 12″ [305] thick while the jamb depth of the window on the right is 6″ [152].

OutSill This option sets the depth of the sill extension on the outside of the wall (see Figure 11-68). A setting of 0 will cause a sill extension to not be drawn.

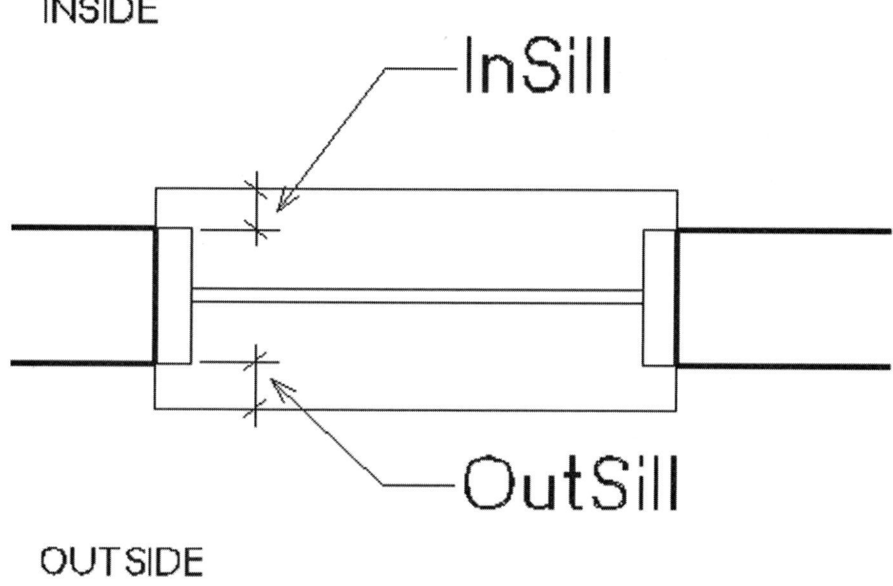

Figure 11-68 `InSill` and `OutSill` examples

InSill This option sets the depth of the sill extension on the inside of the wall (see Figure 11-68). A setting of 0 will cause a sill extension to not be drawn.

GlassThk This option sets the thickness of the glass in the window. A setting of 0 will draw a window of only one line.

SillHgt This controls the height of the bottom of the window in $2^1/_2$ D. Since DataCAD is capable of extruding 2D lines to display a height, you can set the sill height if you want to get a rough 3D idea of what the windows will look like in a space, as in Figure 11-69. Notice that the wall sill under the window is not filled in, so you won't get a true 3D wall with a true 3D window. You would have to do that with the 3D window tools in the AEC_MODL macro in the **Toolbox**. In this example, the walls are 8′-0″ [2438] high, and the window sill height (**SillHgt**) is 3′-4″ [1016].

HeadHgt This controls the height of the top of the window in $2^1/_2$ D. Since DataCAD is capable of extruding 2D lines to display a height, you can set the head height if you want to get a rough 3D idea of what the windows will look like in a space, as in Figure 11-69. Just as the wall sill is not filled

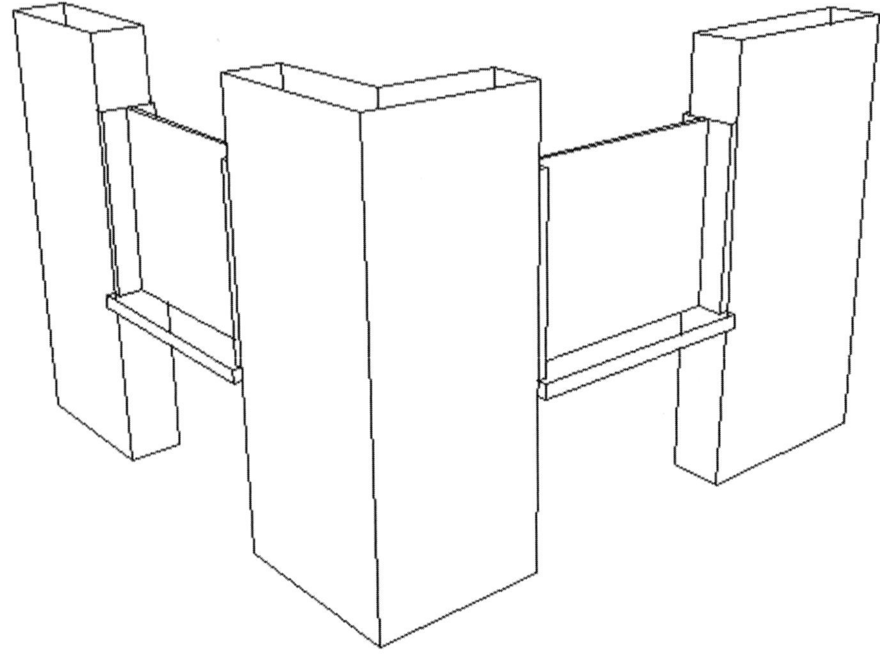

Figure 11-69 `SillHgt` and `HeadHgt` enables you to see a rough 3D version of the window.

in under the window, the wall header above the window is not filled in either, so you won't get a true 3D wall with a true 3D window; that you would have to do with the 3D window tools in the AEC_MODL macro in the **Toolbox**. In Figure 11-69, the walls are 8′-0″ [2438] high, and the window head height (**HeadHgt**) is 6′-8″ [2032].

Remove Use this option to automatically remove windows. You can remove more than one window at one time if they exist within the same wall. This function does not work on pocket windows. See the example at the beginning of the Windows section for a description of how to remove a window. When removing a window, if you have some stray entities within the area box, then DataCAD may become confused about which lines are windows, which are walls, and which are something else, so you may need to turn off a layer or two prior to trying to remove windows, just so there aren't a lot of stray entities around the windows.

DrwMarks With **DrwMarks** on, a small X will be placed at each of the cursor pick locations. This is useful for confirming that you have selected or snapped to the correct points. With **DrwMarks** off, no X's will be displayed.

More window options are available for strings of windows, including curved windows and commercial storefront types, in the DCAD_AEC macro in the macro **Toolbox**.

Elevations

For drawing interior or exterior elevations, you should become familiar with at least three methods for doing so. There is the time-honored method of extending guidelines from elements in your floor plans, there is a free **Toolbox** macro called *Window Master* (**Wdwmastr**), and there is the $2^1/_2$ D method.

Guidelines

If you are drawing both your floor plans and elevations in the same drawing file, or using the new XREF feature (Chapter 13) then the process is fairly simple:

1. Turn on only the floor plan layers that you need (walls, windows, doors, and so on).
2. Create new layers for your elevation elements.
3. Snap to your floor plan elements and draw guidelines out to where your elevations will be drawn.
4. Draw your floor, roof, or ceiling lines at the appropriate heights.
5. Draw your doors and windows using the guidelines, or use door and window symbols that you have created (see Figure 11-70).

After cleaning up the guidelines, you'll have something like Figure 11-71.

Window Master

To help with the creation of windows in elevations, DataCAD has made Window Master (by Patrick McConnell) available for free from the DataCAD LLC Web site as well as on the CD-ROM that comes with this book. You can create nine different window types, and you can control dozens of

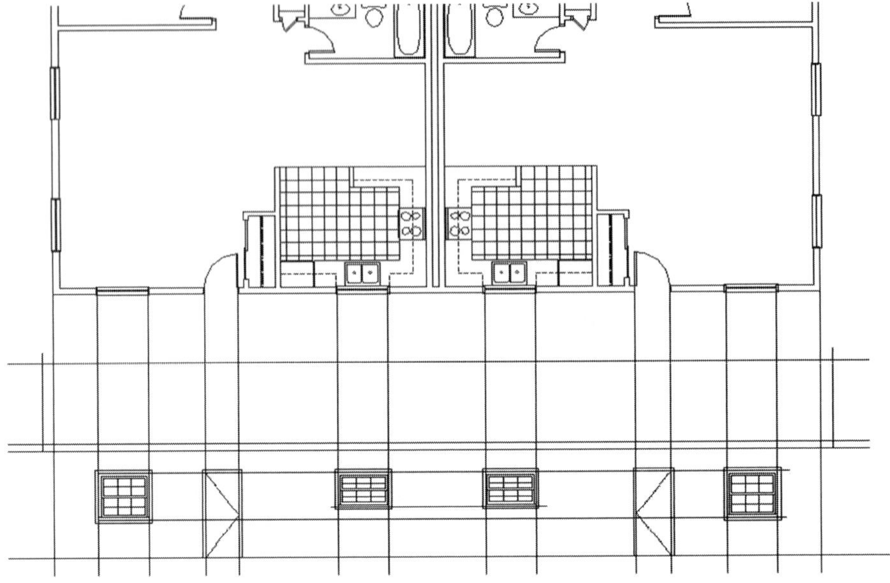

Figure 11-70
An elevation drawn from guidelines from the plan

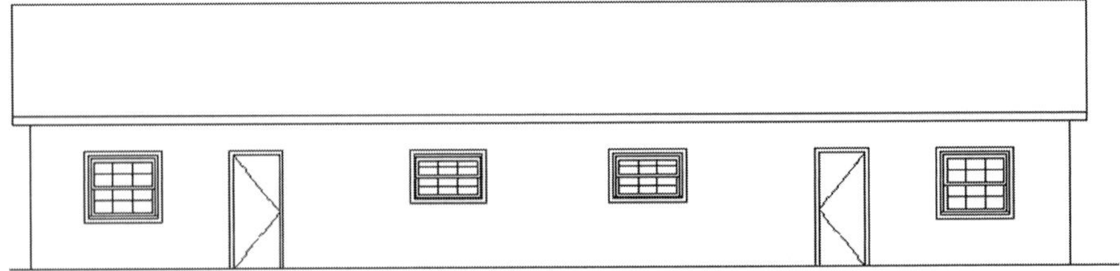

Figure 11-71
Cleaned up elevation

different settings. One of its great features is that you can save all the settings you come up with so that you can instantly retrieve any window that you have created in the past.

Figure 11-72 shows the nine window types you can create.

You can even create banks of windows, as shown in Figure 11-73.

With the macro is a ReadMe.txt file that describes the operation of each of the settings of the macro. Following is a list of the basics.

SetStyle

This option sets the type of window being created: Awning, Casement, Picture, Slider, Double Hung, Single Hung, Half Round, Quarter Round, and Circle.

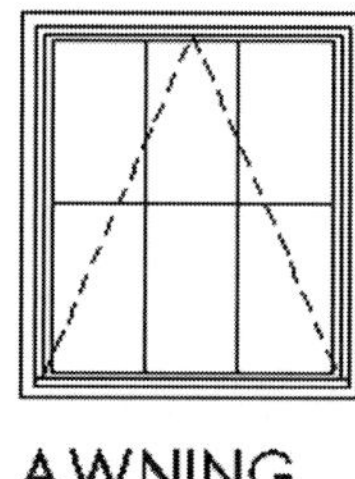

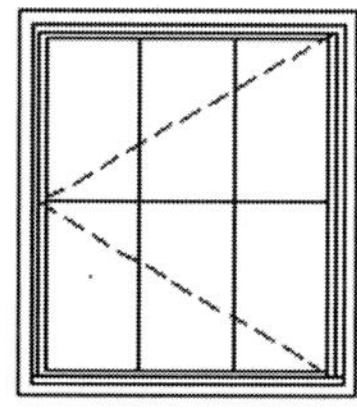

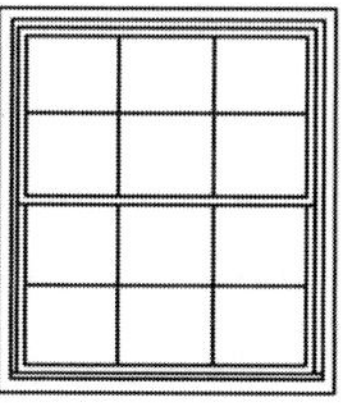

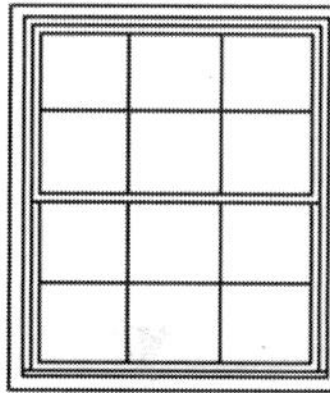

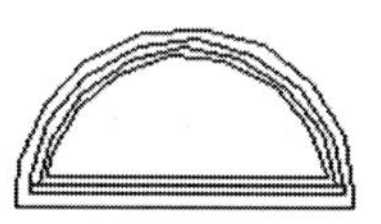

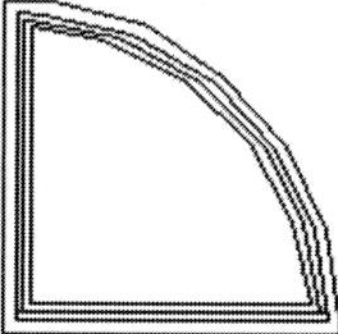

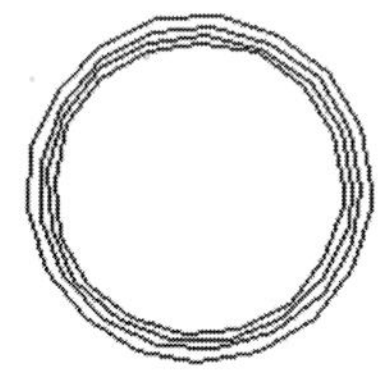

Figure 11-72
The nine window types of Window Master

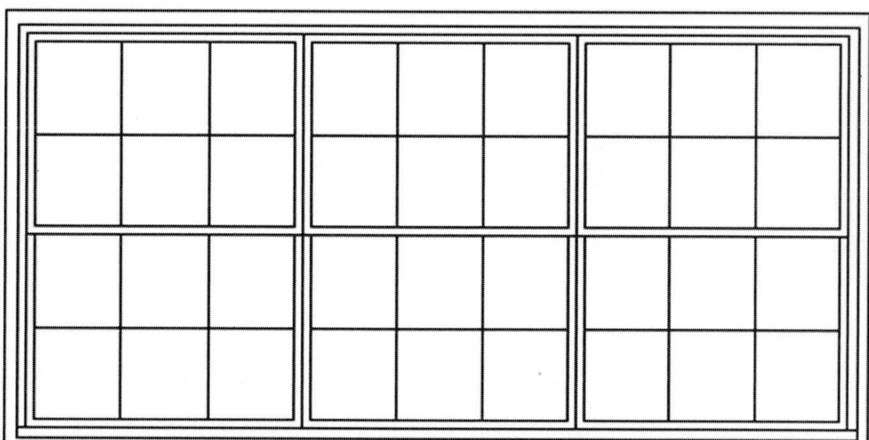

Figure 11-73
A bank of windows drawn with Window Master

No.Units

This option sets the number of window units to be drawn at one time.

Settings

The Settings menu contains the following options:

- **SetStyle** The same as main menu, this option enables you to choose from the nine window types.
- **Width** This option sets the width of the window. This is the total width of all units. If you want four casements at 2′-0″ [610] each, for instance, set the width to 8′-0″ [2440]. This setting is only used in one point input mode.
- **Height** This option sets the height of the window unit and is only used in two- or one-point input mode.
- **HeadHite** and **SillHite** These options are provided if you prefer using them instead of the Width and Height settings. These settings will be updated by changing the Width and Height or you can manually set the head and sill heights.
- **Outline** This option sets all the parameters for drawing the outline of the window.
- **Frame** The window frame is optional. If it is toggled on by selecting DoFrame, you can set the parameters used to draw the frame. If extended sill is selected, the bottom line of the frame will be extended to the full window width.
- **Sash** This option sets all the parameters for drawing the sash of the window.
- **Muntins** This option sets all the parameters for drawing the muntins of the window. The muntins of the upper or left sash are Muntins 1 and the lower or right sash are set by Muntins 2. If a window only has one sash (for example, a picture or awning window), then the Muntins 2 settings are ignored. Muntins are not currently supported in curved windows.
- **Trim** This option sets all the parameters for drawing the window trim. The trim is wrapped around the perimeter of the window.

- **Header** This option sets all the parameters for drawing the window header. The header may extend beyond the width of the window. Set this extension as desired.
- **Sill** This option sets all the parameters for drawing the window sill. The sill may extend beyond the width of the window. Set this extension as desired. This sill is placed below the outline of the window and is in addition to the extended sill that may be drawn under the frame settings.
- **Shutters** This option sets all the parameters for drawing the window shutters. Shutters can have any number of panels in them. Set the width of both the frame and the entire shutter; the panel is what is left over. Specify the desired number of panels.
- **Mullions** This option toggles on the drawing of mullions on windows of more than one unit.
- **MullWidth** This option sets the width of the joining mullions.

InptMode

Select the input mode and it will offer the following options:

- **ByHead** Windows can be inserted by their head or from a given offset distance from the head. This option can be set with the offset setting.
- **Offset** This option sets the distance from your cursor where the window head will be drawn. This is useful for tracing window locations from plans while drawing a new elevation.
- **3 point** This is the fully dynamic input mode. Select two points to describe the window width and then a third point to indicate the unit height. None of the width and height settings currently set in the menus are used.
- **2 point** This mode is dynamic only in the specification of the width of the window. Two points are selected to specify the window width and the height is taken from the current settings as set in the settings menu.
- **1 point** This mode operates similar to placing a symbol. The window width and height are taken from the settings menu and you are placing the window at the point selected. You may specify that the point chosen is either the left, center, or right side of the window head.

*NOTE: When in **1 point** mode and drawing quarter round windows, you can set whether the window is right- or left-handed via a toggle in the main menu. This toggle only appears when necessary.*

*When using **1 point** mode, only the window size itself will be indicated until it is placed. Shutters, headers, and so on will not be shown until a point is selected.*

FileMenu

Access the file menu. The file menu is used to load or save your window settings. The macro allows for a lot of customization, so it is wise to save your settings once you are happy with the results. You can save and load an unlimited number of settings files. These settings files will be saved as standard text files and will have the extension *.WM1*. Do not manually alter these settings files or the macro may not be able to load them. **FileMenu** offers the following options:

- **LoadWdw** This option loads a previously saved window style and all its settings.
- **SaveWdw** This option saves the current window style and settings.
- **Delete** This option deletes an existing .WM1 window style.
- **Rename** This option renames an existing .WM1 window style.
- **Copy** This option copies an existing window style to a new location.

$2^1/_2$ D Method

The final method is similar to what you will see in the 3D chapters of this book, except that in this case you use DataCAD's $2^1/_2$ D entities to create 2D elevations. It's not perfect, but it can give you a head start. To make it work, you need to take the extra time to set the Z-heights of everything in the floor plan. You can do this after the fact, but it's generally more efficient to set them during the process of creating the floor plan entities. Let's use the example of the building used previously in this section. You will see later how you can put a 3D roof on a 2D building, so let's show a roof on this building. In this example, all the walls are 8′-0″ [2438] high. The windows are inserted with the sills at 3′-4″ [1016] and the heads at 6′-8″ [2032] above the floor. Figure 11-74 shows it in plan.

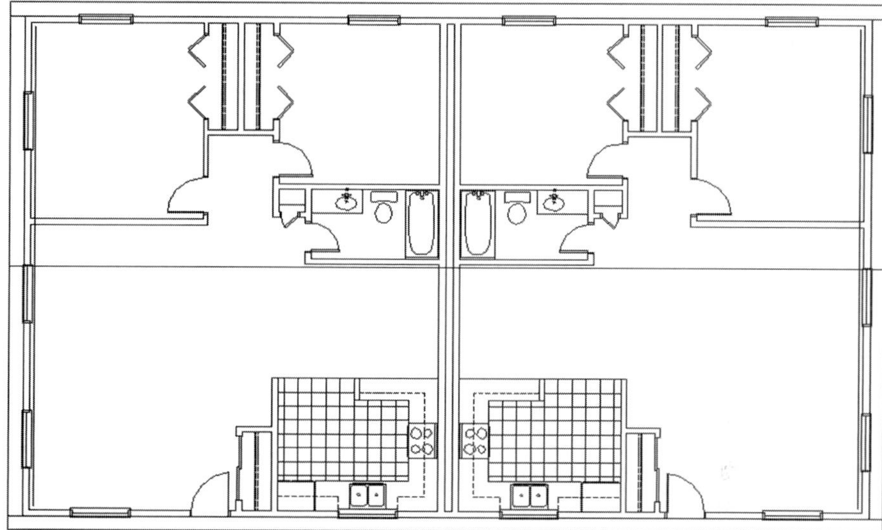

Figure 11-74
The current floor plan

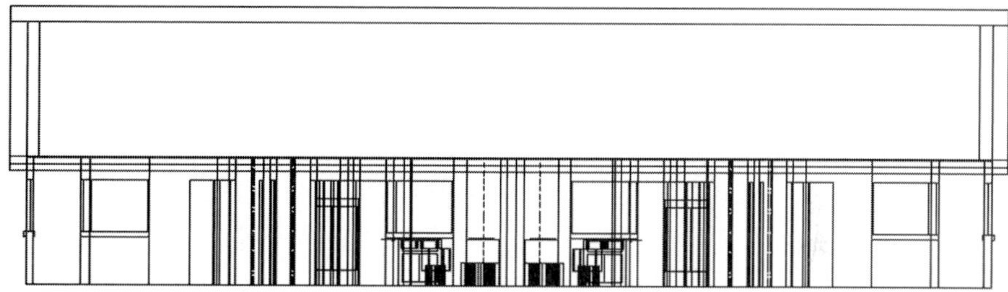

Figure 11-75
A wireframe view of the front of the building

By pressing the **Front** button in the Projection Pad, we get a wireframe view of the front of the building (see Figure 11-75).

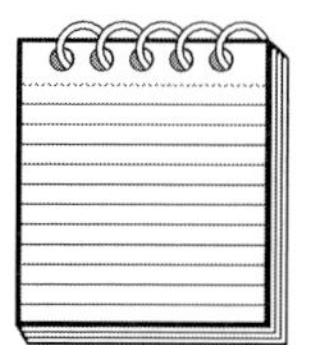

NOTE: *Figure 11-76 shows how DataCAD determines what is Front, Back, Left, and Right in the Projection Pad.*

Because it is a wireframe view, you see every entity from the front of the building to the back, but since you only need the entities at the front face of

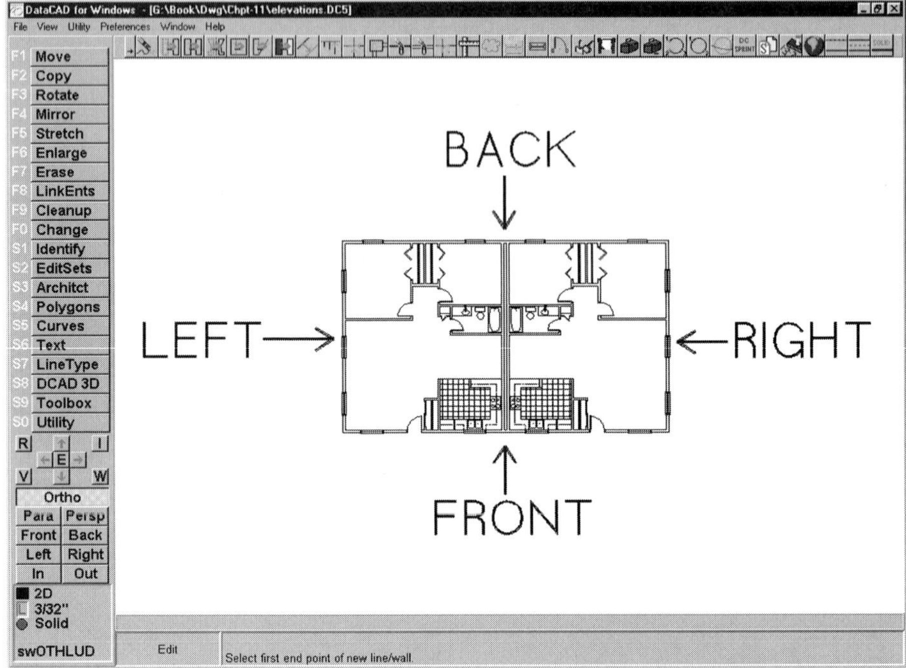

Figure 11-76 Determining projections

the building, you need to hide the rest of the entities behind the front wall. To do this, you will need to go to the **DCAD 3D** menu:

1. Select **Edit/DCAD 3D**.
2. Select the **Hide** button. Several options will appear. Set them as follows:
   a. **SaveImg** = on
   b. **CropImg** = on
   c. **HLRpart** = off
   d. **Pierce** = on
   e. **ClipBox** = off
   f. **All** = on
3. Select the **Options** button. Several options will appear. Set them as follows:
   a. **Join** = on
   b. **DelDoubl** = on

c. **IgnorClr** = on
d. **DrawHidn** = off
e. **BackFace** = on
f. **CloArc2** = off
g. **ClosElip** = off
h. **NoEdge** = off

4. *Right-click* back to the main **Hide** menu. Go to the top of the menu and select **Begin**. DataCAD will run through some calculations and then finally show you a hidden line image of the front face of the building (see Figure 11-77).
5. At this point, DataCAD has only generated a hidden line image, but it is not saved anywhere. You need to save this image to another layer so that you can work on it. To do this, select the **NewLyr** option that should have appeared as soon as the hidden line image was created.
6. You are prompted to *"Enter name of new layer to be added."* Type in a new layer name. This will send the image you see on your screen to that new layer.
7. Press **Esc** and you will see that the original wireframe image is still there. To begin working on the saved 2D image, *right-click* out of the **Hide** menu and then select the **DCAD 2D** button (or press the **;** key) to get back to the 2D menus.
8. Select the **Ortho** button from the Projection Pad to get back to a plan view.
9. Now go to the **Layers** menu. Turn off the plan layers and turn on the layer you saved the hidden line image to.

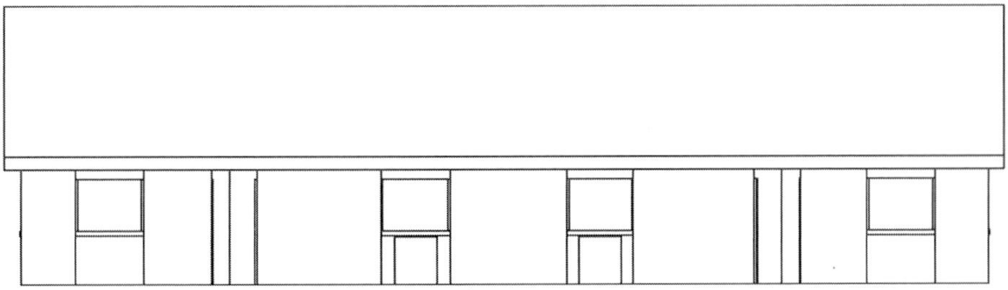

Figure 11-77
A hidden line image of the building

10. Press the **R** (Recalc) button in the Navigation Pad. You should see your saved 2D image filling the screen.
11. Now with a little cleaning up you will have the start of your elevation, as shown in Figure 11-78.

Reflected Ceiling Plans

A *reflected ceiling plan* (RCP) is much like a floor plan, but it shows what is on the ceiling of a space, rather than what is on the floor. The key to creating good RCPs is using the appropriate floor plan layers while adding additional layers unique to the ceiling entities. All of this takes appropriate prior planning in order to make it work. But if you don't take the time to plan, and if you don't take the time to create standards for layer naming, then you might as well go back to drawing with pencil on paper, since you're not going to get any efficiency out of your CAD program. To create RCP's, you can use one of two methods:

- Keep all the floor plan and ceiling plan entities in one drawing file and draw your RCP over the floor plan layers

or

- Use the new *External File Referencing* (XREF) feature to draw your RCP over the floor plan layers in a second drawing file.

The methodology of both these options is exactly the same. The only difference is in where the various layers reside. If you are not familiar with the new External File Referencing feature, you should read Chapter 13. Here we will describe the first method, where all the layers exist within one drawing file. After that, we'll describe how the process differs for the second method of External File Referencing. This reflected ceiling plan example (RCP.DC5) can be found on the accompanying CD-ROM.

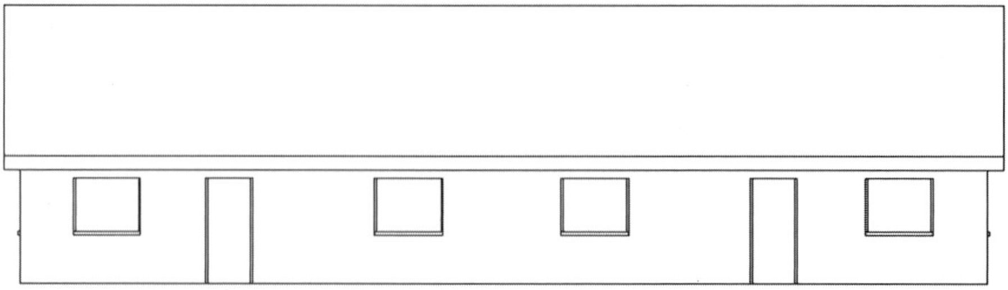

Figure 11-78 The cleaned up elevation

RCPs in One Drawing File

In the following section, we have several basic floor and ceiling plan layers, all in one .DC5 drawing file. The key to making this work is to set up your layers and your entities so that you can reuse as much as possible between both the floor plan and the reflected ceiling plan. Here are the layers in this drawing:

1F-WALLS	First floor walls
1F-LWWAL	First floor low walls (like half-height walls)
1F-DRWIN	First floor doors and windows
1F-FIXTR	First floor fixtures
1C-GRID	First floor ceiling grid
1C-FIXTR	First floor ceiling fixtures
1C-HEADR	First floor door and window headers

To display your floor plan, you would turn on all of the 1F (floor plan) layers (see Figure 11-79).

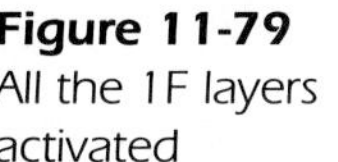

Figure 11-79
All the 1F layers activated

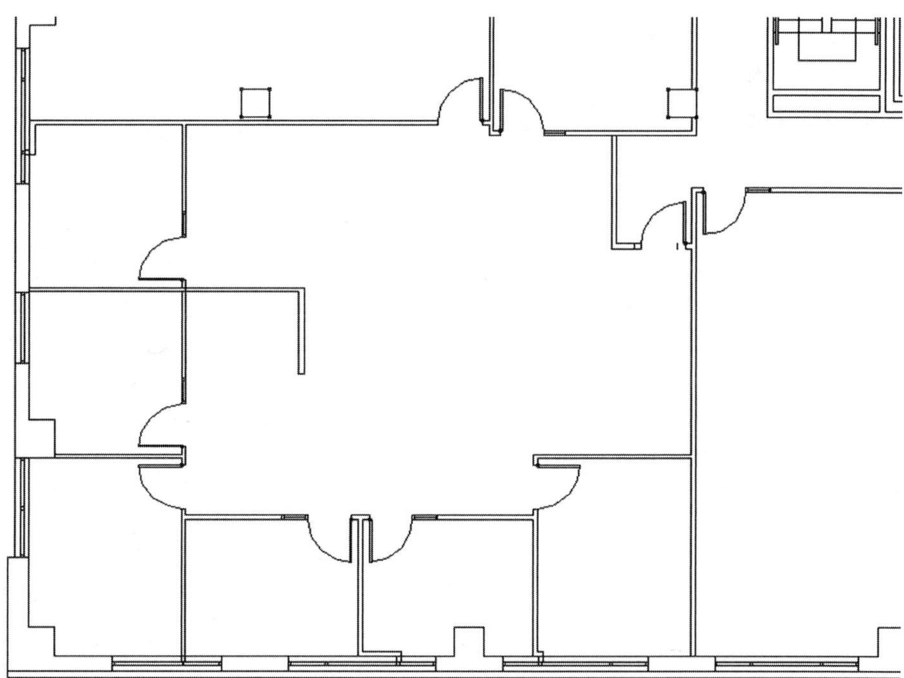

In a reflected ceiling plan, you need to turn off the layers that contain entities that are not displayed in the reflected ceiling plan, like the low walls, doors, and plumbing fixtures. You then turn on only the layers that are needed for the ceiling plan:

1. In this example, we have made sure that all half-height walls and stairs are on the 1F-LWWAL layer, all doors and windows are on the 1F-DRWIN layer, and all plumbing fixtures and counters are on the 1F-FIXTR layer.
2. The suspended acoustic ceiling tile grids are located on the 1C-GRID layer, the light fixtures are on the 1C-FIXTR layer, and the lines that make up the headers of the doors and windows are on the 1C-HEADR layer.

To display the reflected ceiling plan, turn on only the following layers (see Figure 11-80):

1F-WALLS	First floor walls
1C-GRID	First floor ceiling grid
1C-FIXTR	First floor ceiling fixtures
1C-HEADR	First floor door and window headers

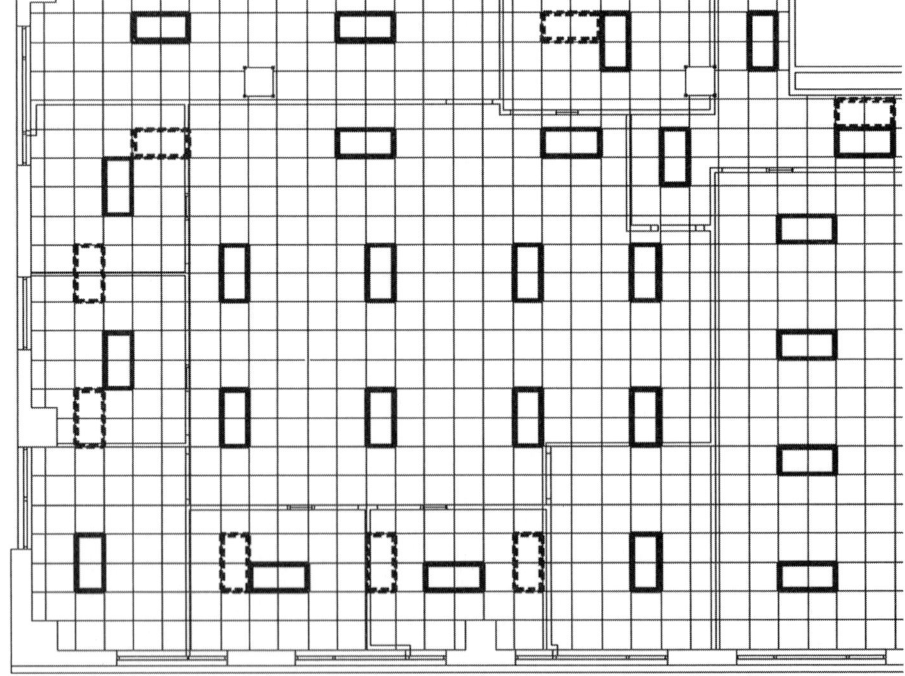

Figure 11-80 The plan with the 1F-WALLS, 1C-GRID, 1C-FIXTR, and 1C-HEADR

One of the keys to efficiency in CAD is to never draw anything twice. To that end, in the previous figure, we used the floor plan's 1F-WALLS layer in the RCP, rather than copying the walls to another layer for use only by the RCP. The advantage of this should be immediately clear: if you change the walls in the floor plan, the reflected ceiling plan is instantly updated as well.

RCPs with File Referencing

The process of creating a reflected ceiling plan using the new File Referencing feature is exactly the same as earlier, except that the floor plan layers would be in one drawing file, like FLOORPLAN.DC5, while the reflected ceiling plan layers would be in a second drawing file, like RCP.DC5. The process is as follows:

1. Create all the 1F layers in the FLOORPLAN.DC5 file and draw your floor plan.
2. Open a second drawing file called RCP.DC5 and create all the 1C layers.
3. While still in the RCP.DC5 file, use the new External Reference File function in the **Insert/Reference File** pull-down menu to insert, or reference, the FLOORPLAN.DC5 drawing file. Now you can see the floor plan layers (but you can't edit them).
4. Using the Reference File Manager (**Insert/Reference File Management/Manager**), turn on only the 1F-WALLS layer, as we did in the figure previously.
5. In the RCP.DC5 drawing, turn on all the 1C layers and then begin drawing your RCP entities on those layers.

You now have a reflected ceiling plan that used the first-floor plan walls, just as in Figure 11-80, but you did not have to keep all the drawing entities in one drawing file.

NOTE: *This is by far the most important new feature in DataCAD 9. If you haven't quite got the concept down, or if you don't yet see its power, it is worth checking out Chapter 13 again. If you can master file referencing, then you will find yourself on a whole new plane of CAD utility and efficiency.*

Text and Notes

Once you have some plans and elevations drawn, you will need to add notes and dimensions. Chapter 5, "Basic Drawing," goes into great depth about text, so we won't say too much about it here, other than to say that you should become very familiar with the **TxtScale** function in the **Text** menu. Here is a recap.

Without **TxtScale**, offices used to come up with elaborate charts to remind everyone what the text size was supposed to be set to for different scale drawings and details. A $^1/_4$" [1:50] scale drawing may have used text that was, at a $^1/_4$" [1:50] scale, defined as 6" [152] high for the text to plot out at $^1/_8$" high (as measured on the paper output). What **TxtScale** does is figure out all those conversions for you on the fly. You tell **TxtScale** what the scale of the detail/drawing will be (with **Edit/Text/PltScale** or **Utility/Plotter/Scale**) when you plot it (key concept here) and how big you want the text to be (like $^1/_8$" high) when it is printed out on paper. Then **TxtScale** automatically figures out how big the text should be drawn in the Drawing Window. Best of all, **TxtScale** carries over into DataCAD's dimensioning.

NOTE: *If you do not see a **PltScale** button in your **Text** menu (at the **S3** position, between **Arrows** and **Justify**), then you need to make a few changes in DataCAD. See the Plot Scale section in Chapter 5 for instructions.*

It is also important to remember that when using the **Arrows** function DataCAD sets the size of the arrowheads relative to the current plot scale and text size. Just as you should be sure to have the correct plot scale selected prior to adding text, so should you make sure it is properly set prior to adding arrows.

You must learn to use Text Scale if you really want to get the most out of text, arrows, and dimensions in DataCAD. If you haven't already, do yourself a favor and try the Text Scale mini-tutorial in Chapter 5.

Importing Text

One of the new features of DataCAD 9 is that you can now paste text data from other Windows applications into DataCAD. This will allow you to copy text from a spreadsheet or a word processing document and paste it into DataCAD while generally retaining its formatting.

Cutting and pasting text from a word processor into DataCAD is much like it is between two word processing documents. Use **Edit/Copy** or **Edit/Cut** in the word processor to save the text to the Windows clipboard. Then use the **Edit/Paste** or **Edit/Paste Special** menu option in DataCAD to insert the text.

When text is pasted into a drawing file, DataCAD will try to maintain formatting as much as possible. How the text is formatted in DataCAD depends to some extent on how it was created in the word processing document. It also depends on the current font in DataCAD at the time the text is pasted. To copy text from a word processor to DataCAD, perform these steps:

1. Use **Edit/Copy** or **Edit/Cut** in the word processor to save the text to the Windows clipboard.
2. Switch to DataCAD and select the **Edit/Paste** or **Edit/Paste Special** menu option from the drop-down menu.
   a. If you use **Edit/Paste** or **Edit/Paste Special**, then your cursor will display a bounding box that represents the extents of the text to be pasted.
   b. If you use **Edit/Paste Special/Text**, then only the standard DataCAD text cursor will be displayed. Where you click is where the pasted text will begin from.
3. Alternatively, if you only want to use the standard **Paste** option, then you can use the Windows shortcut of **Ctrl+V** to paste the text.
4. The only font setting that has any relevance is **FontName**, so you can change that now if you want to. None of the other **Text** menu settings will have any impact on the text to be pasted.
5. Select **Flip** or **Enlarge** as desired.
6. *Click* the mouse to place the text.

Edit/Paste is sufficient for most text imports, but some formatting may not come through as you would expect. Here are some rules to keep in mind:

- The size of the pasted text in DataCAD depends on the size of the text in the word processing document.
- Tabs will cause a string of text to be broken at that point. Each tab will cause a new break in the string of text.

- Spaces are not interpreted as breaks in the text string, so you might want to use spaces instead of tabs in the word processing document.
- Formatting such as italics and bold will also cause a string of text to be broken at that point.
- Outline-formatted text will be broken between the paragraph number/text and the rest of that line of text.
- The bullets in a bullet list will be converted to sequential numbers like this: 1), 2), 3), and so on

Paste Text Options

After you select one of the Paste options, you will see three new options in the menu. Select them prior to placing the text in the Drawing Window.

Flip Selecting this option will cause the lines of text to be displayed in their opposite order. On the left is the normal text, and on the right is the text after selecting the **Flip** option, as shown in Figure 11-81.

Enlarge Select this option to paste the text at a larger size. Choose an enlargement factor prior to placing the text.

Figure 11-81 Normal text and text inverted with the **`Flip`** option

```
1) Outline Formatting
   a) Sub-text
      i) Sub-sub text
      ii) Sub-sub text
   b) Sub-text
      i) Sub-sub text
      ii) Sub-sub text
2) Outline Formatting
   a) Sub-text
      i) Sub-sub text
```

```
      i) Sub-sub text
   a) Sub-text
2) Outline Formatting
      ii) Sub-sub text
      i) Sub-sub text
   b) Sub-text
      ii) Sub-sub text
      i) Sub-sub text
   a) Sub-text
1) Outline Formatting
```

FontName Select this option to change the current font name. If you don't select a new font, then the current font will be used.

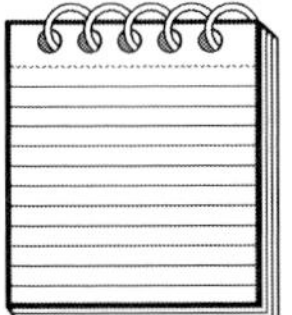

NOTE: Some DataCAD fonts do not support lower case letters, only upper case. If you try to paste lower case text after selecting one of these fonts, then you will get some interesting results, such as Figure 11-82.

Importing Tables

This has been another long-awaited feature. Importing tables will enables you to copy a spreadsheet or a vector graphic table (not a bitmap graphic, however) and paste it into DataCAD as either text or as standard drawing entities while generally retaining its formatting.

Cutting and pasting a table from a spreadsheet (like Excel or Lotus) into DataCAD is much like it is between two spreadsheets. Use **Edit/Copy** or **Edit/Cut** in the spreadsheet to save the table to the Windows clipboard. Then use the **Edit/Paste** or **Edit/Paste Special** menu option in DataCAD to insert the table and text.

The font used in the pasted table depends on the current font in DataCAD at the time the text is pasted. To copy a table from a spreadsheet to DataCAD, follow these steps:

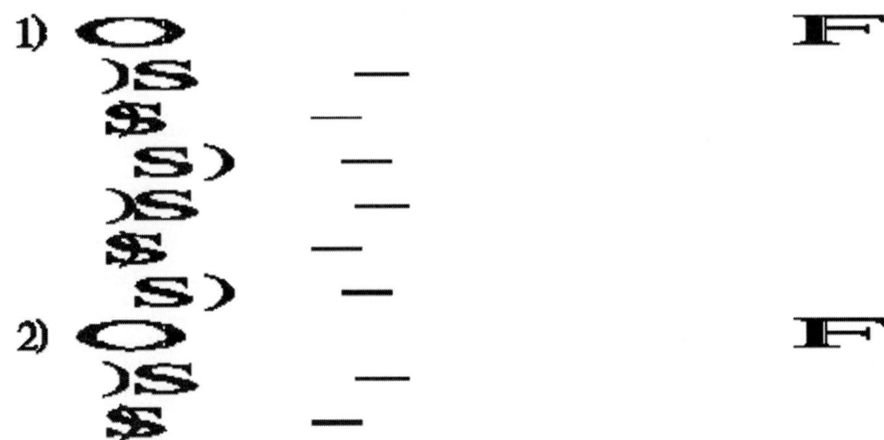

Figure 11-82 Some strange results from a **Paste** option

1. Use **Edit/Copy** or **Edit/Cut** in the spreadsheet to save the table to the Windows clipboard.
2. Switch to DataCAD and select the **Edit/Paste** or **Edit/Paste Special** menu option from the drop-down menu.
   a. **Edit/Paste** (or **Ctrl+V**) will place the full table and text into the Drawing Window.
   b. **Edit/Paste Special/Text** will place only the text from a table into the Drawing Window. However, you are limited to a total length of 80 characters, so your text may wrap around to the next line if it is longer than that.
3. The only font setting that has any relevance is **FontName**, so you can change that now if you want to. None of the other **Text** menu settings will have any impact on the text to be pasted.
4. Select **Flip** or **Enlarge** as desired.
5. *Click* the mouse to place the table.

Figure 11-83 is an example of an original spreadsheet.

Figure 11-84 is that same spreadsheet pasted into DataCAD with **Edit/Paste** (**Ctrl+V**).

Notice that if the spreadsheet table has visible borders and lines, then those lines will also be pasted into DataCAD. However, special lines, like thick or double lines, do not translate. Figure 11-85 is the same spreadsheet, but with all the lines as double lines in the original spreadsheet pasted into DataCAD.

If you use **Edit/Paste Special/Text**, then only the text from a table will be pasted into the Drawing Window. However, you are limited to a total length of 80 characters, so your text may wrap around to the next line if it is longer than that (see Figure 11-86).

Figure 11-83 The original spreadsheet

	A	B	C	D	E	F
1	DS$ = "move	drawing	" +	SYMPATH$ +	"\sym\00-grphc\drawing	"
2	SHELL DS$					
3	DS$ = "move	b3ibrff1	" +	SYMPATH$ +	"\sym\12-furn\bedroom\b3ibrff1	"
4	SHELL DS$					
5	DS$ = "move	b3ibrff2	" +	SYMPATH$ +	"\sym\12-furn\bedroom\b3ibrff2	"
6	SHELL DS$					
7	DS$ = "move	b3ibrfr1	" +	SYMPATH$ +	"\sym\12-furn\bedroom\b3ibrfr1	"
8	SHELL DS$					
9	DS$ = "move	b3ibrmt1	" +	SYMPATH$ +	"\sym\12-furn\bedroom\b3ibrmt1	"
10	SHELL DS$					
11	DS$ = "move	b3ibrmt2	" +	SYMPATH$ +	"\sym\12-furn\bedroom\b3ibrmt2	"
12	SHELL DS$					
13	DS$ = "move	b3ibrpb1	" +	SYMPATH$ +	"\sym\12-furn\bedroom\b3ibrpb1	"
14	SHELL DS$					
15	DS$ = "move	b3ibrpb2	" +	SYMPATH$ +	"\sym\12-furn\bedroom\b3ibrpb2	"
16	SHELL DS$					

Figure 11-84 The spreadsheet pasted in DataCAD

DS$ = "move	drawing	" +	SYMPATH$ +	"\sym\OO-grphc\drawing	"
SHELL DS$					
DS$ = "move	b3ibrffl	" +	SYMPATH$ +	"\sym\12-furn\bedroom\b3ibrffl	"
SHELL DS$					
DS$ = "move	b3ibrff2	" +	SYMPATH$ +	"\sym\12-furn\bedroom\b3ibrff2	"
SHELL DS$					
DS$ = "move	b3ibrfrl	" +	SYMPATH$ +	"\sym\12-furn\bedroom\b3ibrfrl	"
SHELL DS$					
DS$ = "move	b3ibrmtl	" +	SYMPATH$ +	"\sym\12-furn\bedroom\b3ibrmtl	"
SHELL DS$					
DS$ = "move	b3ibrmt2	" +	SYMPATH$ +	"\sym\12-furn\bedroom\b3ibrmt2	"
SHELL DS$					
DS$ = "move	b3ibrpbl	" +	SYMPATH$ +	"\sym\12-furn\bedroom\b3ibrpbl	"
SHELL DS$					
DS$ = "move	b3ibrpb2	" +	SYMPATH$ +	"\sym\12-furn\bedroom\b3ibrpb2	"
SHELL DS$					
DS$ = "move	b3ibrwfl	" +	SYMPATH$ +	"\sym\12-furn\bedroom\b3ibrwfl	"
SHELL DS$					

Figure 11-85 Special lines do not translate into DataCAD

```
DS$ = "move    drawing    " +    SYMPATH$ +    "\sym\OO-grphc\drawing                   "
SHELL DS$
DS$ = "move    b3ibrffl   " +    SYMPATH$ +    "\sym\12-furn\bedroom\b3ibrffl           "
SHELL DS$
DS$ = "move    b3ibrff2   " +    SYMPATH$ +    "\sym\12-furn\bedroom\b3ibrff2           "
SHELL DS$
DS$ = "move    b3ibrfrl   " +    SYMPATH$ +    "\sym\12-furn\bedroom\b3ibrfrl           "
SHELL DS$
DS$ = "move    b3ibrmtl   " +    SYMPATH$ +    "\sym\12-furn\bedroom\b3ibrmtl           "
SHELL DS$
DS$ = "move    b3ibrmt2   " +    SYMPATH$ +    "\sym\12-furn\bedroom\b3ibrmt2           "
SHELL DS$
DS$ = "move    b3ibrpbl   " +    SYMPATH$ +    "\sym\12-furn\bedroom\b3ibrpbl           "
SHELL DS$
DS$ = "move    b3ibrpb2   " +    SYMPATH$ +    "\sym\12-furn\bedroom\b3ibrpb2           "
SHELL DS$
```

Figure 11-86 Spreadsheet information wrapped around

```
 DS$ = "move                         drawing     " +   SYMPATH$ +
"\sym\OO-grphc\drawing               "
 SHELL DS$
 DS$ = "move                         b3ibrffl    " +   SYMPATH$ +
"\sym\12-furn\bedroom\b3ibrffl       "
 SHELL DS$
 DS$ = "move                         b3ibrff2    " +   SYMPATH$ +
"\sym\12-furn\bedroom\b3ibrff2       "
 SHELL DS$
 DS$ = "move                         b3ibrfrl    " +   SYMPATH$ +
"\sym\12-furn\bedroom\b3ibrfrl       "
 SHELL DS$
 DS$ = "move                         b3ibrmtl    " +   SYMPATH$ +
"\sym\12-furn\bedroom\b3ibrmtl       "
 SHELL DS$
 DS$ = "move                         b3ibrmt2    " +   SYMPATH$ +
"\sym\12-furn\bedroom\b3ibrmt2       "
 SHELL DS$
 DS$ = "move                         b3ibrpbl    " +   SYMPATH$ +
"\sym\12-furn\bedroom\b3ibrpbl       "
 SHELL DS$
 DS$ = "move                         b3ibrpb2    " +   SYMPATH$ +
"\sym\12-furn\bedroom\b3ibrpb2       "
 SHELL DS$
```

TIP: *If you want a table without lines, but you need to get around the 80-character length, remember that a spreadsheet table made up of double lines will not display the lines when pasted into DataCAD, as in Figure 11-85. Notice in that figure that the text was not wrapped to the next line.*

Paste Table Options

After you select one of the Paste options, you will see three new options in the menu. Select them prior to placing the text in the Drawing Window. The options are as follows:

- **Flip** Selecting this option will cause the lines and text of a table to be displayed in their opposite order
- **Enlarge** Select this option to paste the table at a larger size. Choose an enlargement factor prior to placing the table.
- **FontName** Select this option to change the current font name. If you don't select a new font, then the current font will be used.

Exporting Text

Just as you can import text into DataCAD, so too can you export text to the Windows clipboard or directly to any Windows program that accepts text, like a word processor. Formatting in DataCAD is quite basic, so the output you get will likewise be very basic. To export text, follow these steps:

1. Select **Edit/Clipboard Select** (**Ctrl+E**).
2. Pick the text by one of the EGAFS methods. The selected text will turn gray. You can continue to make multiple selections.
3. Once you are done making all your selections and while they are still gray, pick **Edit/Cut** (**Ctrl+X**) or **Edit/Copy** (**Ctrl+C**). You will be prompted to *"Select reference point for Clipboard Copy."*
4. Pick a reference point. The text will be copied to the Windows Clipboard.
5. You can now paste the selected text back into DataCAD with **Edit/Paste** (**Ctrl+V**), or you can paste it into another Windows program.

The first thing you should be aware of is that when you select text, if you also select other drawing entities like lines, arcs, and so on, they too will be saved to the Windows Clipboard. If you paste that information back into DataCAD, then all of the entities will be pasted. If you use a simple **Paste** command in another Windows program, then only the text will be inserted. If you instead use the **Paste Special** command, then you can paste a .WMF image of the text and entities into any Windows program that accepts metafiles, like a word processor. The difference is that this is only a graphic image and not editable text.

In Step 4, you were prompted for a reference point. This point has no function outside of DataCAD, but within DataCAD that point will be attached to your cursor for precise placement.

If you look at the Prompt Line while you are selecting text, you may notice that you are being prompted to select entities to "<add to selection set>". If you don't know what a selection set is, read about them in Chapter 9. Then you will understand the final two options in the selection menu:

- **DelFrom** Use this option to delete text or entities from the active selection set.
- **Mask** Use this option to search for particular types of entities by masking by Entity, Color, LineType, or Weight.

Dimensions

What would a construction drawing be without dimensions! DataCAD has some very powerful dimensioning options, which are found in the **Utility/Dmension** menu. If you have not done so already, you should read the section regarding the Text Scale feature in Chapter 5, since it is integral to the dimensioning methodology that we present here.

Types of Dimensions

When you select the **Utility/Dmension** button, you will see four basic options: **Linear**, **Angular**, **Diameter** and **Radius**. Figure 11-87 shows an example of each.

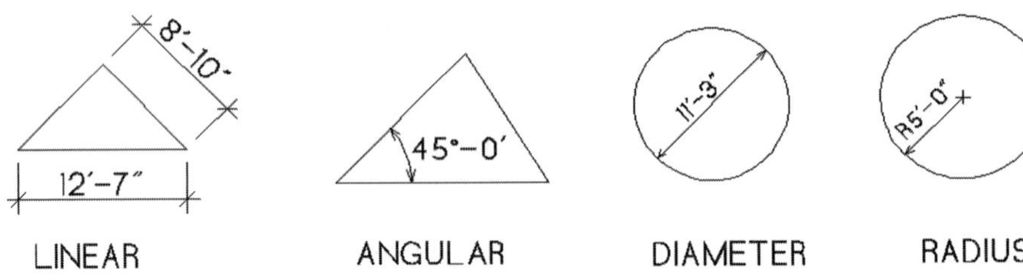

Figure 11-87 The four types of dimensions

Linear Dimensions

Use this option to create a straight-line dimension between two points. This is the most common type of dimension that you will use. If you are using the default drawing that we outlined in Chapter 2, then most of the dimension settings you will need will already be set, so you won't have to reset them. But here are the typical steps for creating the most common associative linear dimension settings. You only have to make these settings one time in each drawing file. Once set, you generally won't need to set them again.

1. Select **Utility/Dmension/Linear**.
2. Select the appropriate type of linear dimension. For this example, select the **Horizntl** dimension type.
3. Make sure **Assoc** (associative dimensioning) is on.
4. Select **TextStyl** and set the following options:
   a. **TextSize** = 1/8" (..1/8) [3.2]
   b. **Aspect** = 1.25
   c. **TxtScale** = on
   d. **OutHorz** = off
   e. **Above** = on
   f. **Auto** = off
   g. **Color** = [select the desired text color]
   h. **FontName** = [select the desired font name]
5. *Right-click* back to the main **Linear** menu.
6. Select **DimStyl** and set the following options:
   a. **Line1** and **Line2** = on
   b. **Offset** = 1/8" (..1/8) [3.2]

c. **Overlap** = $^1/_8$″ (..1/8) [3.2]
d. **Overrun** = $^1/_8$″ (..1/8) [3.2]
e. **FixdDist** = off
f. **DINStd** = off
g. **Rounding/RoundIt** = off
h. **Limits** and **Tolrance** = off

7. *Right-click* back to the main **Linear** menu.
8. Select **ArroStyl** and set the following options:
   a. **TickMrks** = on
   b. **Size** = .5
   c. **Color** = [select the desired text color]
9. *Right-click* back to the main **Linear** menu.

With the **TxtScale** setting turned on, you must make sure you change the current plot scale (see Chapter 5 for a full explanation of the Text Scale feature) prior to creating your dimension strings.

NOTE: *Just remember that all dimension settings are made relative to the current plot scale. So if you change your plot scale from a* $^1/_4$″ *[1:50] to an* $^1/_8$″ *[1:100], all your text, offsets, arrowheads, and so on will still be displayed and plotted at the correct proportions for the new scale.*

You can set the plot scale with **Utility/Plotter/Scale** or **Edit/Text/PltScale**. Let's use the previous settings to dimension this simple walled box (see Figure 11-88).

The currently active DataCAD color is the color that will be used to draw the dimension lines and the leader lines, so the first step in dimensioning is to make sure you have the correct color set prior to creating your dimension lines:

1. Be sure that you have the correct plot scale selected. For this example, we are using a $^1/_4$″ [1:50] scale.
2. Go to **Utility/Dmension/Linear**. You will be prompted to *"Select first point of distance to dimension."*
3. Dimensioning one of the outside lines can be accomplished with one of two methods:

Figure 11-88 A walled box to be dimensioned

a. Select **Entity**. *Click* on the line you want to dimension, such as the top line of the box. Dimension leaders will extend from the endpoints of the selected line.

or

b. *Snap* to the first endpoint of the line to be dimensioned and then *snap* to the second point.

4. With both of these methods, the horizontal line of your cursor represents the horizontal line of the dimension string. Move the cursor to where you want the dimension string to be located and then click the mouse.
5. Your cursor will now be attached to the dimension text. Move the cursor to where you want the text to be located and then *click* the mouse.

   The final result should look something like Figure 11-89.

Linear Dimension Settings

The first four settings (**Horizntl**, **Vertical**, **Aligned**, and **Rotated**) are mutually exclusive. Only one of the four can be selected at one time.

Horizntl This dimensions the horizontal distance between two points, even if the two points are not in the same horizontal plane. Figure 11-90 shows

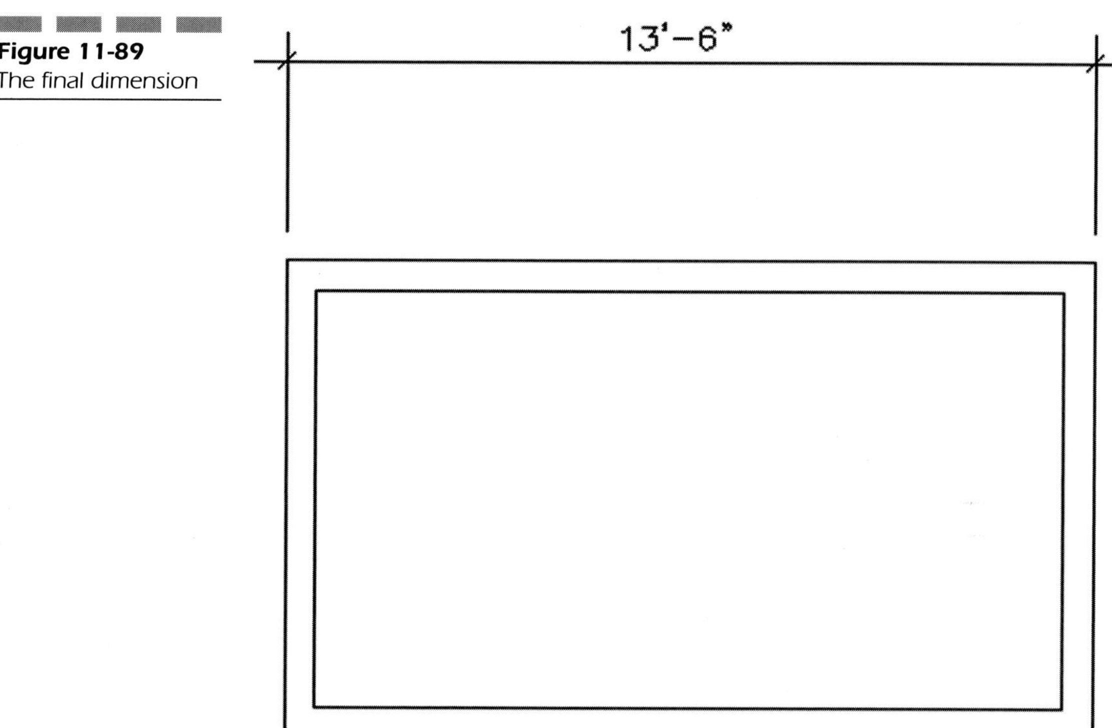

Figure 11-89
The final dimension

two examples. Notice that in both cases it is the horizontal distance that is being dimensioned, not necessarily the length of the line itself.

Vertical This dimensions the vertical distance between two points, even if the two points are not in the same vertical plane. Figure 11-91 shows two examples. Notice that in both cases it is the vertical distance that is being dimensioned, not necessarily the length of the line itself.

Aligned This dimensions the actual distance between two points, regardless of the angle between them. The dimension string is drawn parallel to the two points selected (see Figure 11-92).

Rotated This dimensions the distance between two points at a specified angle, even if the two points are not in the same plane. Notice in Figure 11-93 that the dimension is different depending on the angle that you input.

Assoc This is the most important of all the dimension options. It stands for Associative, meaning that the dimension is smart enough to change the

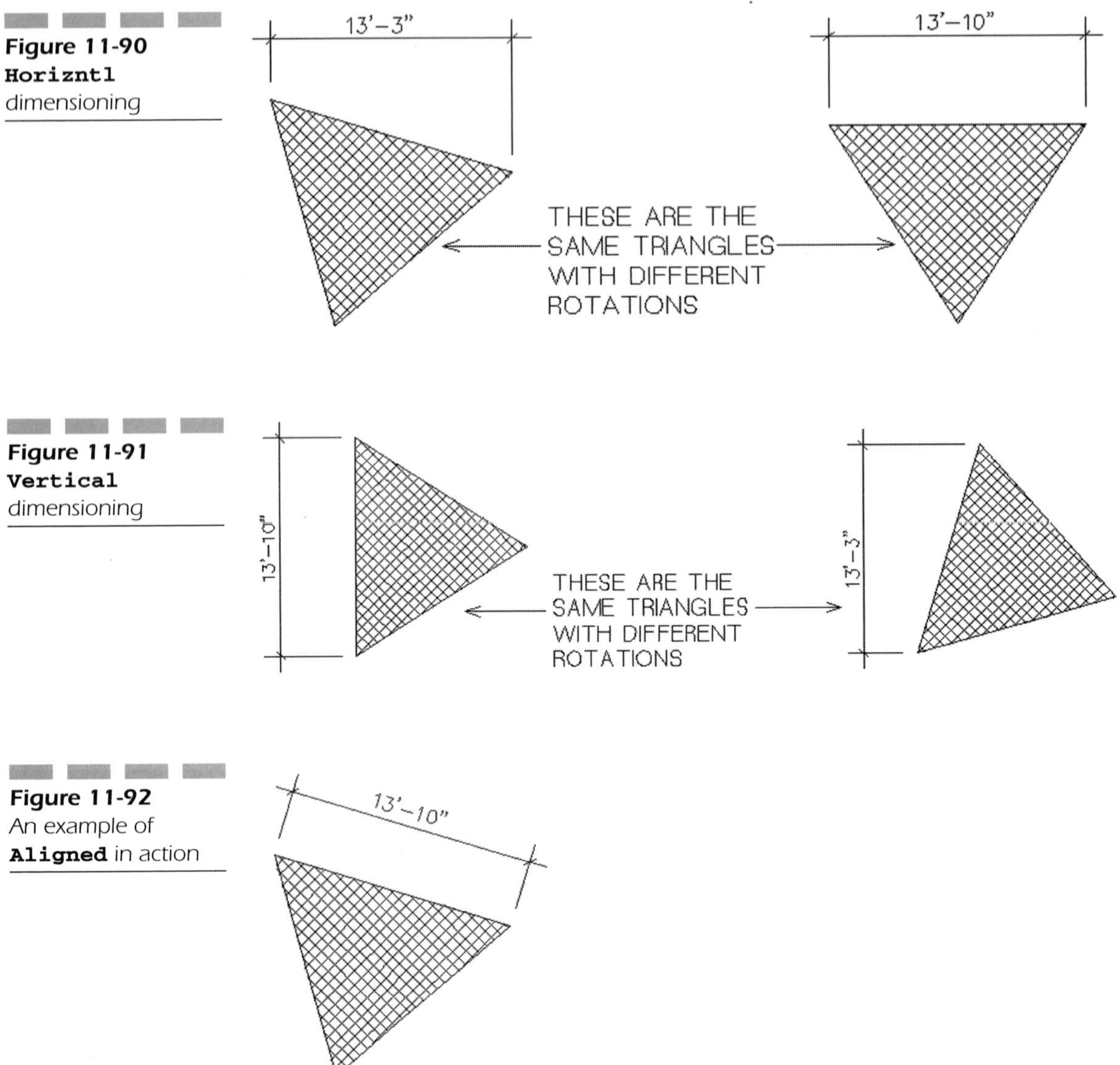

Figure 11-90 **Horizntl** dimensioning

Figure 11-91 **Vertical** dimensioning

Figure 11-92 An example of **Aligned** in action

way it displays itself when it is stretched or moved. If a horizontal associative dimension displays a dimension of 4′-10″ [1473] and you stretch it 1′-6″ [457] to the right, the dimension will then stretch the leader and the arrowhead, and the numeric dimension display will update itself to show that the dimension is now 6′-4″ [1931], as in Figure 11-94.

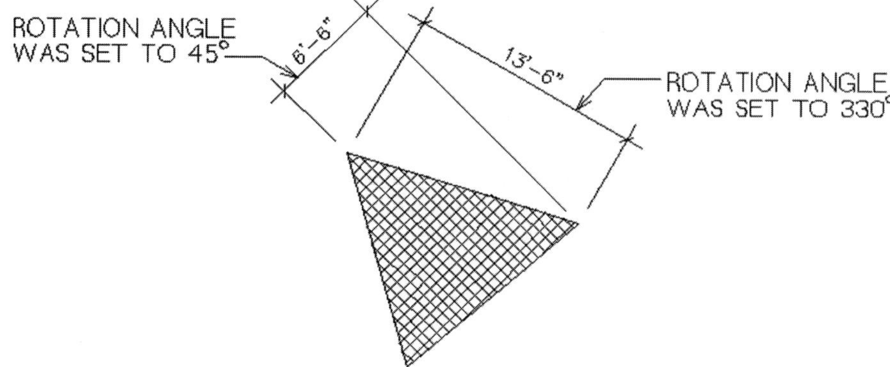

Figure 11-93 The **Rotated** option measures distances along a specific angle.

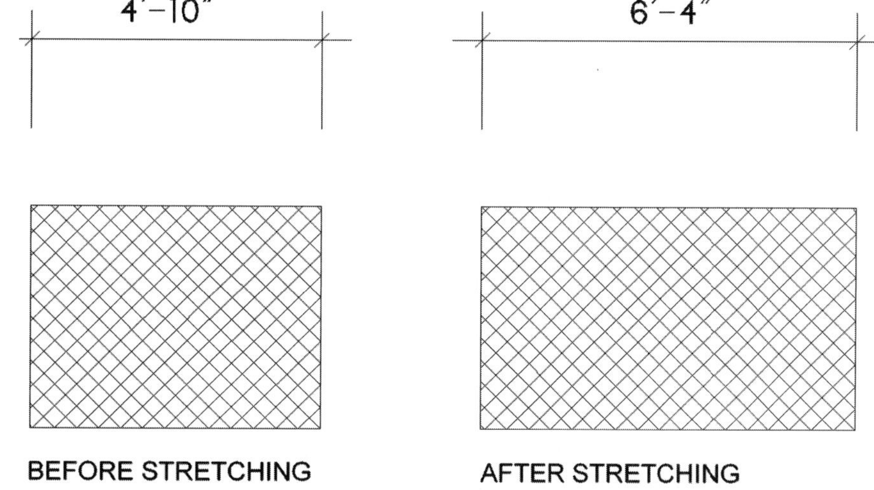

Figure 11-94 **Assoc** enables the dimension to be updated when stretching.

So if you stretch a room in your building or stretch the entire building, as long as their dimensions are turned on when you stretch the building, then those dimensions will automatically be updated.

And if you move an associative dimension, the entire dimension (leaders, arrowheads, text, and so on) will move together as a group.

Entity Use this option to select an entity to dimension, rather than snapping to two points to define the points to be dimensioned. You can select lines, circles, or arcs. The current dimension type (**Horizntl**, **Vertical**, **Aligned**, or **Rotated**) will be used to create the dimension.

AutoDim This is a whole feature unto itself. With it, DataCAD will attempt to automatically dimension all the relevant points between two user-selected points. The two key words here are *attempt* and *relevant*, since DataCAD can only guess at what entities you really intend to dimension based on the current **ObjSnap** (Object Snap) settings in the **Utility** menu and the settings in the **Dmension/Linear/AutoStyl** menu. For instance, you may not want to dimension the door and window jambs, but only the window openings themselves. But since a jamb creates two snappable intersections (at either side of the jamb), DataCAD will dimension the width of the jambs as well as the width of the window opening between the jambs. Keep this in mind before running the **AutoDim** feature.

When you select **AutoDim**, you will be prompted to "Select first endpoint of line to dimension along." Figure 11-95 shows one wall of a simple floor plan that was dimensioned with **AutoDim** by simply selecting the two points shown. DataCAD then automatically dimensions everything between those points as long as they fit into the **ObjSnap** and **AutoStyl** settings. In this example, the windows were created with snapping points at the center of the windows, and the doors were created with jambs, so DataCAD dimensions to those points as well.

AutoStyl The settings in this menu control what is drawn when you use the **AutoDim** feature. (We are covering this option out of order from its appearance in the menus, since it affects only the previously described **AutoDim** feature). The **AutoStyl** settings are as follows:

- **Baseline** Selecting this option will enable you to create a series of dimensions that all share the same reference point. As you pick each subsequent point, a new dimension string is drawn from the first point

Figure 11-95 Automatic dimensioning of a wall

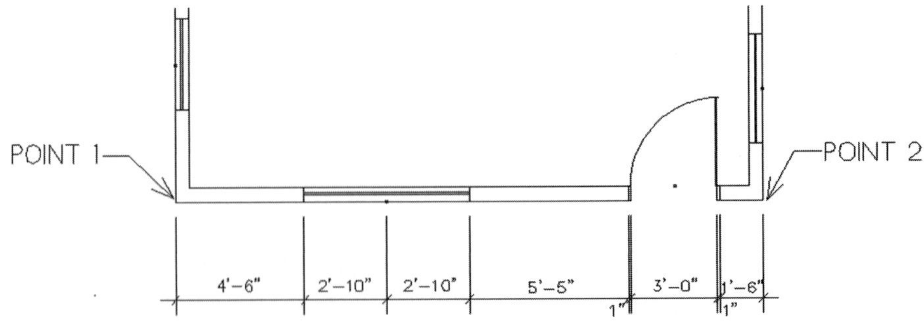

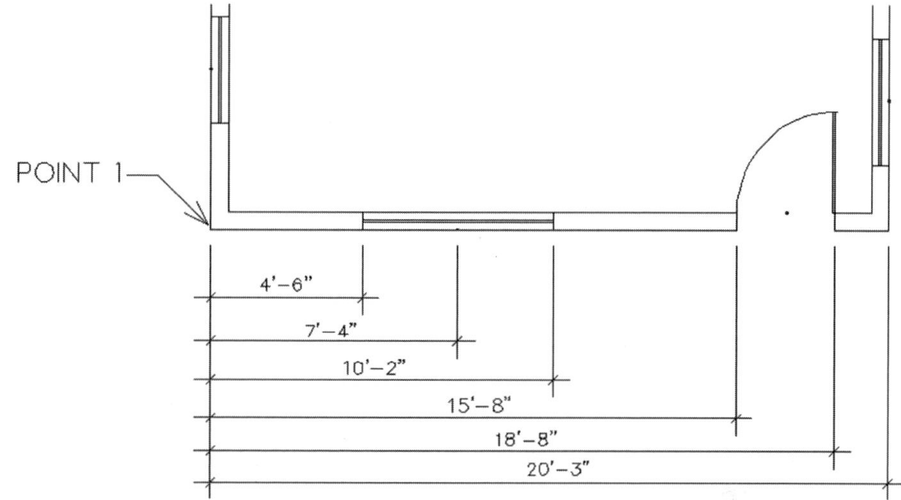

Figure 11-96 Automatic dimensioning with the **Baseline** option

selected (see Figure 11-96). **Baseline** and **StrngLin** are mutually exclusive settings.

- **StrngLin** This is the default setting and will cause all of your dimensions to be drawn in a continuous string by simply selecting the start point and then each subsequent point to dimension to without having to re-select the previous point of each dimension (see Figure 11-97). **Baseline** and **StrngLin** are mutually exclusive settings.
- **Overall** This option will appear only after a complete string line is created. It will create a new dimension string that gives the overall dimension from the first point to the final point in the string line.
- **OnlyPnts** This option will cause DataCAD to dimension only to the centerpoint marks of objects such as doors and windows (if they were created with center marks), which is how many offices like to dimension doors and windows. **OnlyPnts** and **NoPnts** are mutually exclusive settings.
- **NoPnts** This option will cause DataCAD to dimension only between line endpoints that are at a distance greater than or equal to the distance set with the **MinDist** setting. **OnlyPnts** and **NoPnts** are mutually exclusive settings.
- **MinDist** Use this option to set the minimum dimension distance between line endpoints when using the **NoPnts** option. Setting this to zero will result in all line endpoints being dimensioned. Setting it to something like 4″ [10.16] would be a good way to avoid dimensioning door and window jambs that might be only 2″ to 4″ widths.

- **MissDist** Use this option to set the search distance perpendicular to the dimension line. All line endpoints within the Miss Distance will be dimensioned.
- **LyrSrch** With this option turned on, all line endpoints on all layers will be dimensioned (as long as they are within the miss distance).
- **DirDim** With this option turned on, **AutoDim** is the default dimensioning mode and will be accessed automatically whenever you enter the **Dmension** menu.

Baseline This option only appears after you have created the first dimension. Selecting this option will enable you to create a series of dimensions that all share the same reference point. As you pick each subsequent point, a new dimension string is drawn from the first point selected. The results will look something like Figure 11-96.

StrngLin This option only appears after you have created the first dimension. Selecting this option will cause all of your dimensions to be drawn in a continuous string by simply selecting the start point and then each subsequent point to dimension to without having to reselect the previous point of each dimension. In this example, we picked Point 1, Point 2, and then selected **StrngLin**. After that, we only had to snap once to Points 3 through 7 without ever having to go back to snap to the previous point first (see Figure 11-97).

Overall This option will appear only after a complete string line is created. It creates a new dimension string that gives the overall dimension from the first point to the final point in the string line.

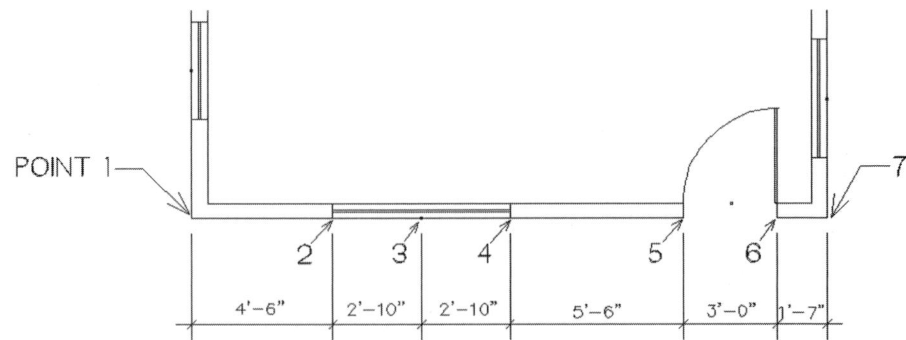

Figure 11-97 **StrngLin** draws all your dimensions in a continuous string by simply selecting the start point and then each subsequent point to dimension to.

TextStyl Use this to set all the text options for the dimension text. The options are as follows:

- **TextSize** This option sets the size of the dimension text.
- **Weight** This option sets the line weight (width) of the dimension text.
- **Slant** This option sets the slant, or angle, of the letters of the dimension text.
- **Aspect** This option changes the width-to-height aspect ratio of the dimension text.
- **TxtScale** This option sets the size of the text relative to the current plot scale (see Chapter 5 for a full description of Text Scale).
- **InHorz** This setting controls only dimension text that is drawn within the dimension line string (**Above** must be off). With **InHorz** turned off, all dimension text will be parallel to the dimension line. With this setting turned on, all dimension text will be drawn horizontally (see Figure 11-98).
- **OutHorz** This setting controls only dimension text that is drawn outside the dimension line string (**Above** must be on). Like the

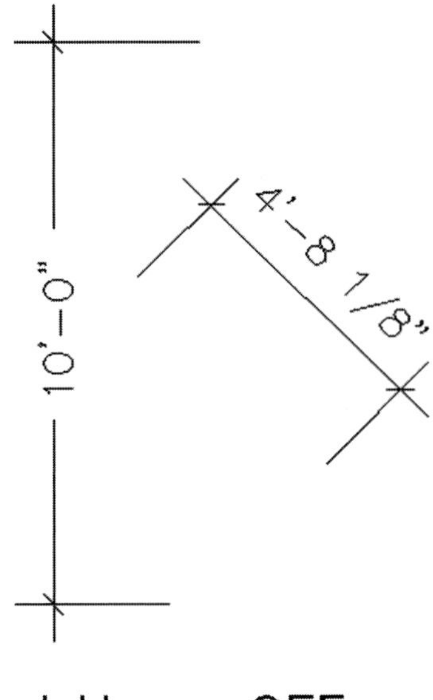

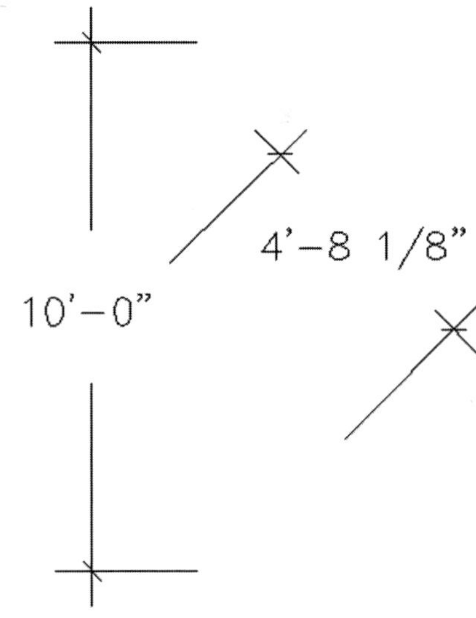

Figure 11-98
InHorz on and off

previous option, with **OutHorz** turned off, all dimension text will be parallel to the dimension line. With this setting turned on, all dimension text will be drawn horizontally.

- **Above** With this setting turned on, all dimension text will be drawn above the dimension line string. With this setting off, all dimension text will be drawn within the dimension line string (see Figure 11-99).
- **Offset** This option sets the distance of the dimension text above the dimension line string. This setting is only applicable when **Above** and **Auto** are both turned on (see option E in Figure 11-101).
- **Auto** With this setting turned on, all dimension text will automatically be placed without user intervention. Text is placed according to the current settings for **InHorz**, **OutHorz**, **Above**, and **Offset**. With this setting turned off, you will be prompted for the new text location each time you create a dimension. A bounding box representing the dimension text will be attached to your cursor. Locate the text wherever you want it and then click the mouse to place it. Before placing the text, you can select the **Rotate** or **DrwLeadr** options, or you can change the current Text Style.
  - **Rotate** Select this option if you want to rotate the dimension text prior to placing it.
  - **DrwLeadr** Select this option prior to placing the text if you want to draw a leader from the dimension text to the dimension line (see Figure 11-100).
  - **TextStyl** This is a trap door to the standard Dimension **TextStyl** menu.
- **Color** Use this option to select the color of the dimension text.
- **FontName** Use this option to select the font name for the dimension text.

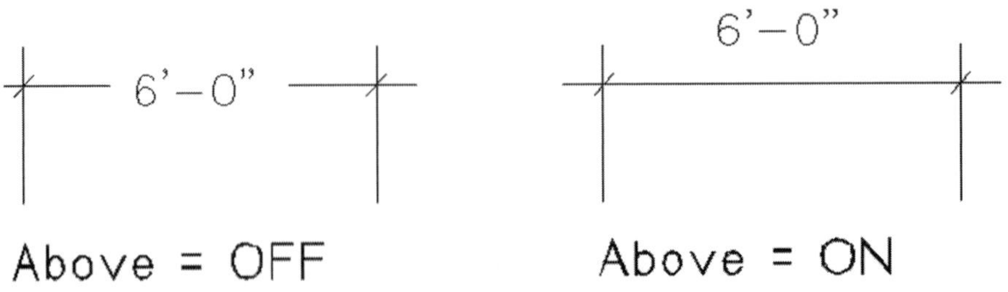

Figure 11-99
Above off and on

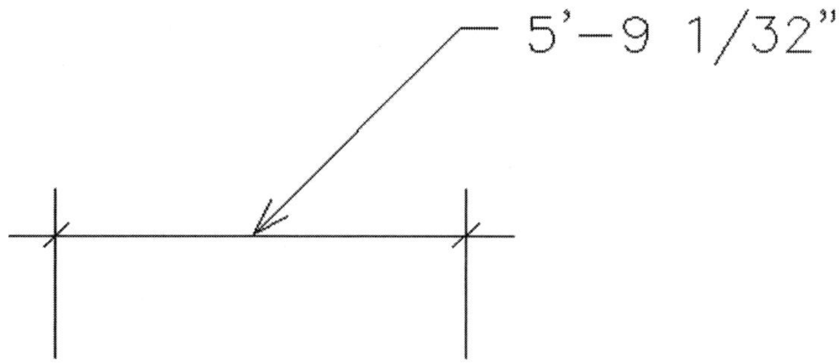

Figure 11-100
A **DrwLeadr** example

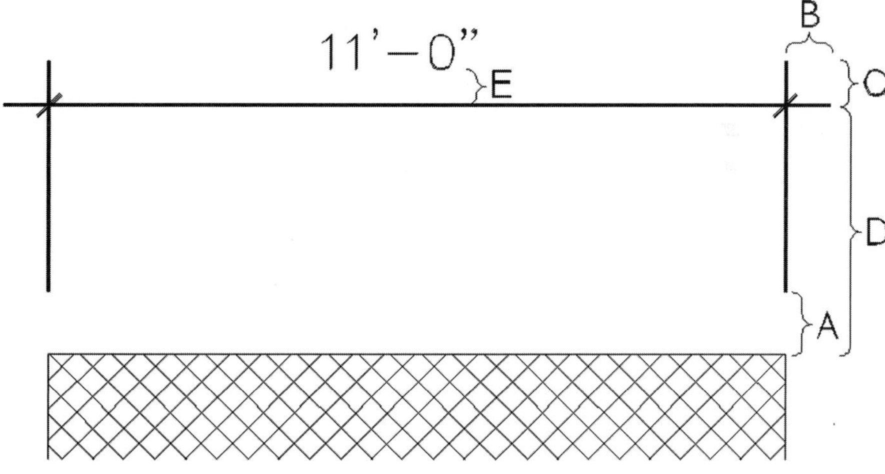

Figure 11-101
DimStyl sets all the options for the dimension lines.

A = Offset
B = Overrun
C = Overlap
D = Increment (with FixedDist on)
E = Text offset

DimStyl Use this option to set all the options for the dimension lines. For the first six options, refer to Figure 11-101.

It is important to note that if you have **TxtScale** turned on (which is highly recommended), then all the distance settings will be relative to the current plot scale. For example, with **TxtScale** turned on, the **Incrment** distance is figured relative to the current plot scale. In Figure 11-101, the real-world distance of D might be 3′-6″ [1066], but relative to the current plot scale (a $^1/_4$″ [1:50] in our case), the **Incrment** distance is $^7/_8$″ [22.2] ($^7/_8$

= 3.5/4 ; hence, the 3′-6″ [1066] real-world dimension). The same is true for distances A through E. The dimension line options are as follows:

- **Line1** Turn this option on to display the first extension line of the dimension.
- **Line2** Turn this option on to display the second extension line of the dimension.
- **Offset** Use this option to set the distance between the extension lines and the object being dimensioned (see A in Figure 11-101).
- **Overlap** Use this option to set the distance of the extension line overlap (see C in Figure 11-101).
- **Incrment** This setting applies only to two other options: **Baseline** and **FixdDist**. When using **Baseline**, **Incrment** sets the standard incremental distance between each dimension line, each time a new one is drawn. When **DimStyl/Incrment** is turned on, **Incrment** sets the distance of the dimension line from the object being dimensioned (see D in Figure 11-101). **Incrment** has no effect when **StrngLin** is on, unless it is set to zero, in which case the first tick mark is dropped and you get a running dimension.
- **Overrun** Use this option to set the distance of the dimension line past the extension lines (see B in Figure 11-101).
- **FixdDist** This setting applies only to two other options: **Baseline** and **Incrment**. Use this option with the **Baseline** mode to increment the distance between each subsequent dimension line. If you are not using **Baseline**, with **FixdDist** turned on all your dimension lines will automatically be drawn at a fixed distance from the object being dimensioned (see D in Figure 11-101). Use the **Incrment** option to set the distance.
- **DINStd** Use this option to create dimensions conforming to the *Deutsches Institut fur Normung* (DIN Standard or the German Institute for Standardization). This specifies the Offset distance. When you select the DIN standard (**DINStd**), the **Offset** option will therefore disappear from the menu.
- **Rounding** Choose this option, followed by **RoundIt**, to round your dimension text to a certain precision. This is a global setting that affects all associative dimensions currently in the drawing file, not just the next dimension you create. This means that you cannot selectively round only certain dimensions in the drawing; it is an all-or-nothing option.

- **RoundUp** Fractions and decimals will be rounded up to the next value specified with the **Precsion** option.
- **Bankers** Fractions and decimals will be rounded up to the next value specified with the **Precsion** option if the fraction is greater than half the **Precsion** dimension. Fractions and decimals will be rounded down to the next value specified with the **Precsion** option if the fraction is less than half the **Precsion** dimension.
- **RoundDn** Fractions and decimals will be rounded down to the next value specified with the **Precsion** option.
- **Precsion** Use this option to set the precision by which all dimensions will be rounded.

- **Limits** With this option, you can have two dimensions displayed in the dimension string: one for an upper limit as specified by **UpperLim**, and one for a lower limit as specified by **LowerLim**. For instance, if you have upper and lower limits of 1″ [25.4] and you enter a dimension whose true length is 3′-6″ [1066] the dimension will look like Figure 11-102. **Limits** and **Tolrance** are mutually exclusive options.
- **Tolrance** Use this option to display a dimension along with a positive and negative tolerance displayed after it. If the + and the − tolerance

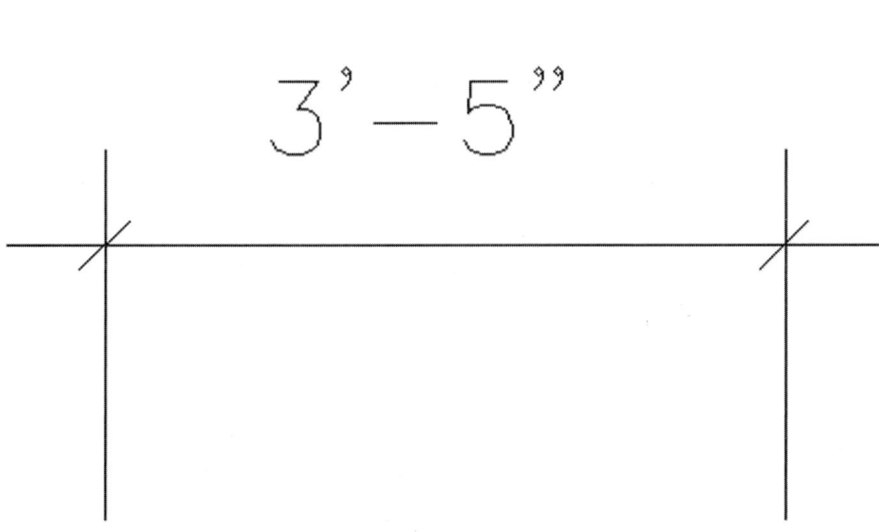

Figure 11-102 A dimension with upper and lower `Limits`

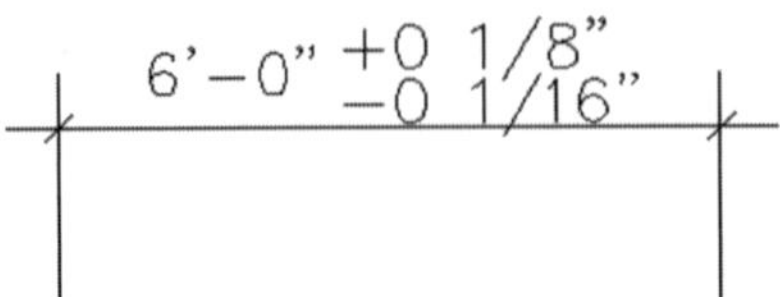

Figure 11-103
Tolrance displays a dimension along with a positive and negative tolerance.

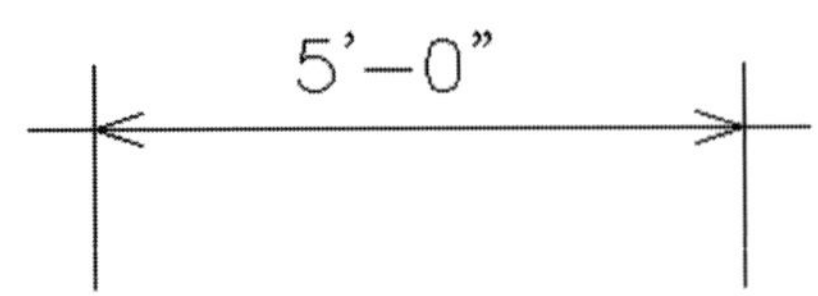

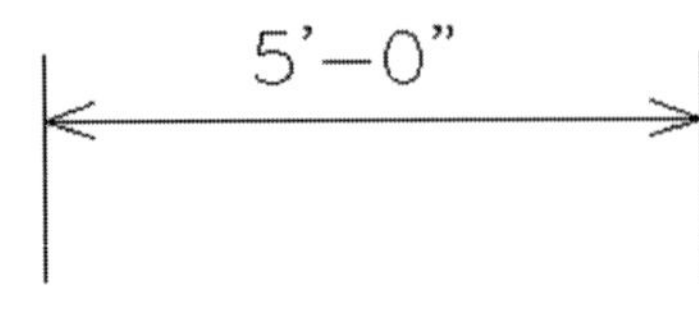

WITH Overrun NO Overrun

Figure 11-104
Arrowheads with **Overrun** on and off

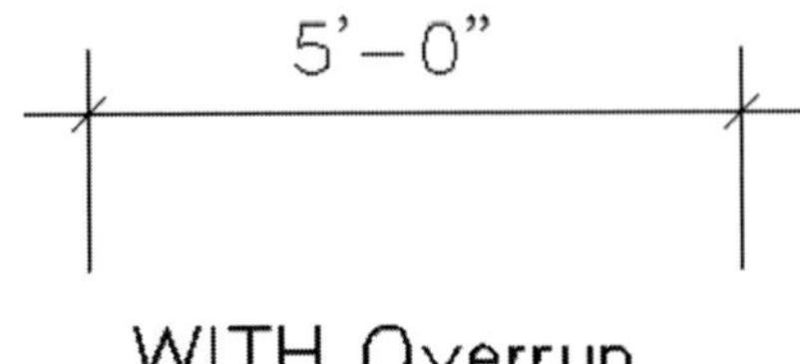

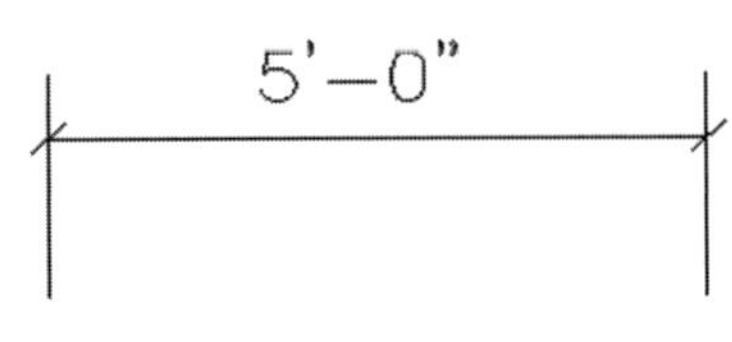

WITH Overrun NO Overrun

Figure 11-105
Tick marks with **Overrun** on and off

numbers are the same, then only one tolerance number is shown, as in the second example in Figure 11-103. **Limits** and **Tolrance** are mutually exclusive options.

- **+Tolrnce** Enter a positive tolerance value.
- **–Tolrnce** Enter a negative tolerance value.

- **UpperLim** This option will only be displayed once **Limits** is toggled on.
- **LowerLim** This option will only be displayed once **Limits** is toggled on.

ArroStyl Here you can set the visual properties of the arrowheads (a generic term, since not all are arrows) of your dimensions. Three basic types of arrowheads exist: **Arrows**, **Tick Marks**, and **Dots**.

- **Arrows** creates open style arrowheads. They can be modified with the **Size**, **Aspect**, and **Color** options (see Figure 11-104).
- **TickMrks** creates tick marks as arrowheads. They can be modified with the **Size**, **Weight**, and **Color** options (see Figure 11-105).

Figure 11-106 Dots with **Overrun** on and off

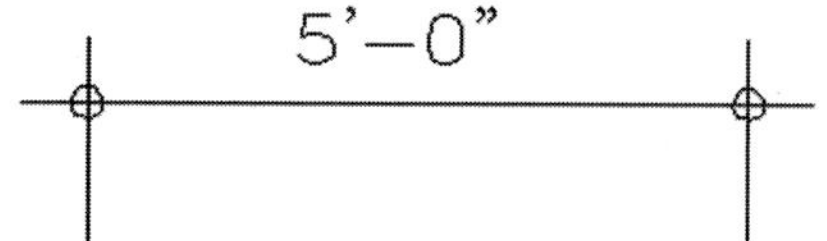

Figure 11-107 Tick marks with two different **Size** settings

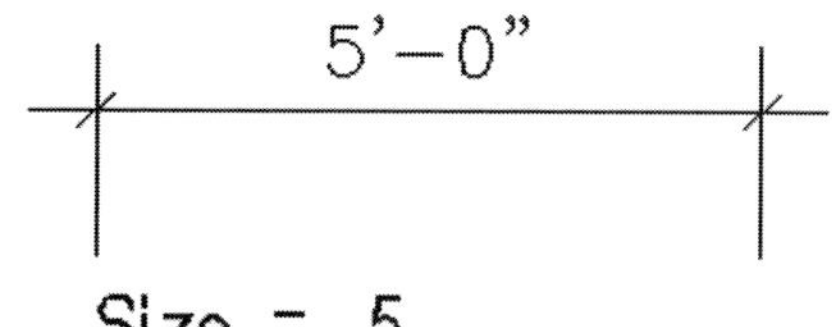

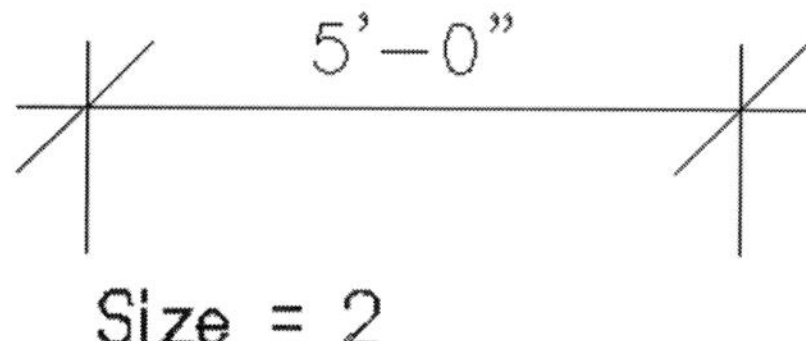

Figure 11-108 Examples of how **Size** and **Aspect** affect arrowheads

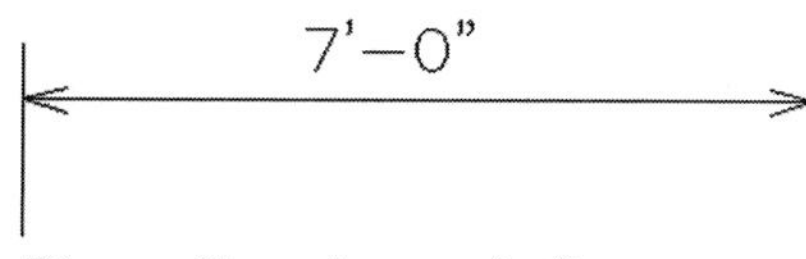

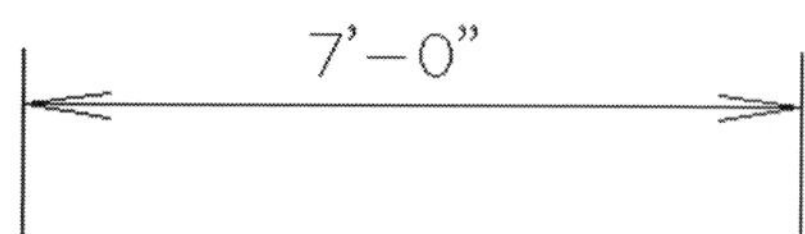

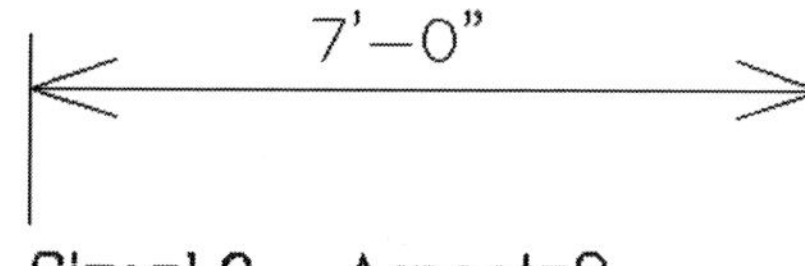

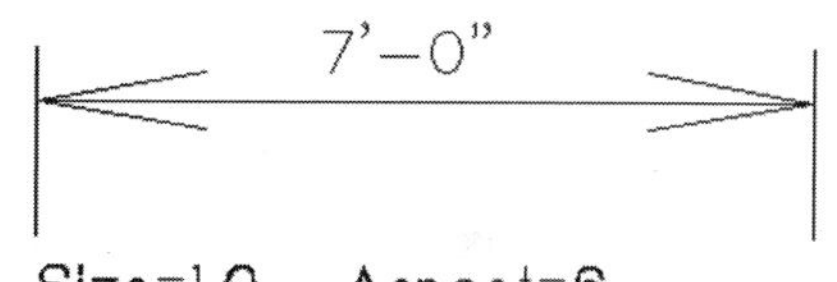

- **Dots** creates a circle as arrowheads. Because they are simple circles, they can only be modified with the **Size** and **Color** options (see Figure 11-106).

ArroStyl also offers some options for customizing the arrowheads:

- **Size** Use this option to adjust the arrowhead size relative to the current plot scale (see Figure 11-107). Choose a factor from the menu window or type in a value. The most common arrowhead factor is .50. Experiment to see what works for you. For more examples, see the **Aspect** description.
- **Aspect** This option is only available when **Arrows** is selected. The aspect is defined as the length of the arrowhead divided by its height (length/height). In the examples shown in Figure 11-108, notice how the sizes of the arrowheads decrease as the aspect values decrease

while the angles of the arrows increases. To keep the arrowheads larger, you would need to increase the **Size** value as you decrease the **Aspect** value. The ANSI standard arrowhead has an aspect ratio of 6 to 1. A popular aspect value is 3.0.

- **Weight** Use this option to set the weight, or thickness, of the individual strokes of the arrowheads.
- **Color** Use this option to set the color of the arrowheads. The color of the dimensions will not be affected, only the color of the arrowheads. If you are mapping your line colors to your pen thicknesses, then this option is how you control how heavy the arrowheads are drawn.

Explode Use this option to make existing associative dimensions non-associative. Each part of the dimension reverts to simple lines, arcs, and text with no association to one another. Stretching the exploded dimension will therefore not change the text to match the distance of the stretch, and entity properties can no longer be changed with the **Dmension/Linear/Change** commands.

Change Use this option to change the settings of any existing associative dimensions in the drawing. You can select one or multiple settings to change all at once; you do not have to change one setting at a time. For instance, you can select changes to **TextSize**, **FontName**, **Offset**, **Limits**, **TickMrks**, and **Color** all at once. Then all you have to do is select each existing associative dimension and all of the new settings will be applied at once.

To make any changes, you must be aware of one fact. After selecting all the changes to be made, you will be back in the **Change** menu, but you cannot apply the changes while you are still in this menu. You must *right-click* one more time back to the main **Linear** menu. Then you can select the entities to be changed, including the following:

- **TextStyl** Use this option to change dimension text settings. Current selections will remain green. Pick all the attributes that you want to change and then *right-click* back to the **Change** menu. Select more changes or right-click back to the **Linear** menu to apply the changes.
- **DimStyl** Use this option to change extension line settings. Current selections will remain green. Pick all the attributes that you want to change and then *right-click* back to the **Change** menu. Select more changes or right-click back to the **Linear** menu to apply the changes.
- **ArroStyl** Use this option to change dimension arrowhead settings. Current selections will remain green. Pick all the attributes that you

want to change and then *right-click* back to the **Change** menu. Select more changes or right-click back to the **Linear** menu to apply the changes.

- **TxtPostn** This is a very important option even though it's kind of hidden way down at the bottom of this menu. Select this option to reposition the text in an existing associative dimension. After selecting **TxtPostn**, click on an associative dimension. The text will turn to a rectangular bounding box attached to your cursor. Locate the text in its new position and then click the mouse to place it (see Figure 11-109).

Display Dimension Control Points

One helpful display option for working with associative dimensions is activated by turning on **Utility/Display/DimPoint**. Turning this on will display the control points (the small dots below the leaders in the graphic below) for an existing associative dimension (see Figure 11-110). Displaying these control points makes it much easier to select the ends of dimension leaders in order to stretch the dimension line.

Associative dimensions can be stretched only by selecting the actual points from which the dimensions were created (represented by the control points). If you do not have your dimension leaders set with an offset from the object you are dimensioning, then the dimension points, the ends of the

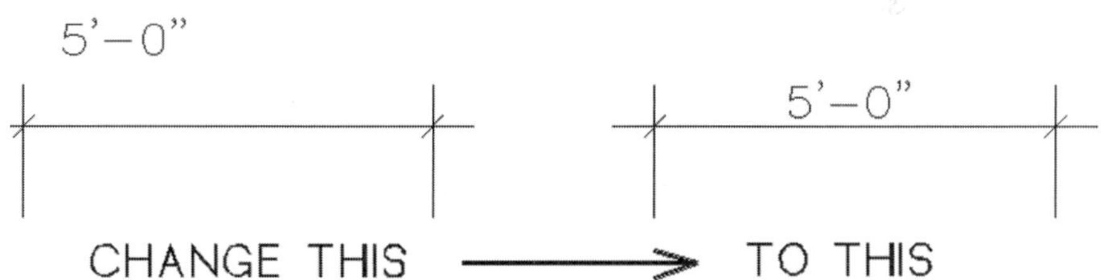

Figure 11-109
TxtPostn example

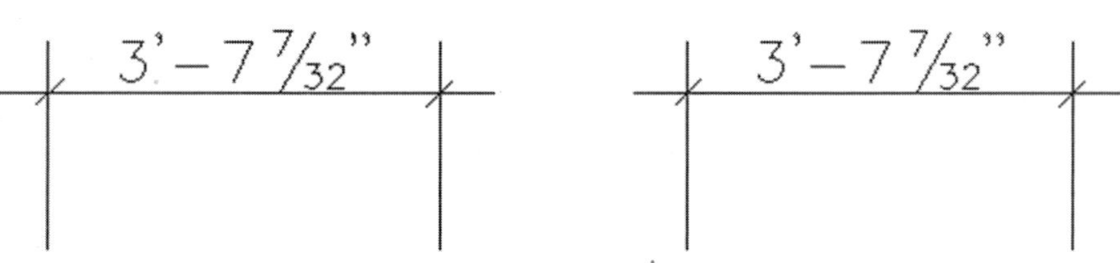

Figure 11-110
Turning the control points on

dimension leaders, and the object you are dimensioning all coincide with one another. If your dimension leaders are offset from the object you are dimensioning, however, then the ends of the leaders do not coincide with the dimension points, as in Figure 11-110.

Stacked Fractions

Another setting that you should be aware of if you are using the **Arch** (Architect) scale type is **StakFrac** (stacked fractions). I like this option because it makes the dimension text considerably more compact. Figure 11-111 displays examples of dimensions with **StakFrac** off and on.

To enable this option, select **Settings/ScaleTyp/Arch**, after which the **StakFrac** option will appear, and you can turn it on. Be aware that this is a global setting, meaning that when it is turned on, all current and future associative dimensions will be affected. It is an all-or-nothing option.

Another thing to be aware of is that when you **Explode** a dimension with stacked fractions, the dimension text will be broken into several pieces. In Figure 11-111, the **5′-1** would be one piece of text, the **17** would be another, the **/** is another, the **32** is another, and the **″** is yet another. This can be a bit of a nightmare later on if you have to move or edit the text.

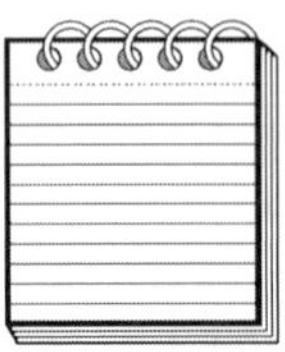

NOTE: *Nearly all DataCAD fonts will work with* ***StakFrac****, but a very few will not. If you find that your stacked fractions are being displayed over the top of some of the other numbers in the dimension, then you will need to switch to a different font for the dimension text.*

Identify/Set All

Perhaps the most useful trick with regards to associative dimensions is that the **Identify/Set All** command will reset all of the dimension settings to

Figure 11-111 **StakFrac** off and on

5’-1 17/32”

StakFrac = OFF

5’-1 17/32”

StakFrac = ON

match a selected associative dimension. Think about that one. The font, the size of the text, all the colors, the type and size of the arrowheads, and so on will all be reset to match an existing dimension.

Just remember that all dimension settings are made relative to the current plot scale, so change the current Plot Scale before you use **Identify/ SetAll**. Your dimension settings will then be changed to match those of the selected dimension.

Angular Dimensions

Use this option to show an angle designator between two lines, as in Figure 11-112.

Angular dimensions are not associative, so the lines, arcs, and text are all individual entities.

LyrSrch

With **LyrSrch** on, all layers will be searched for lines to dimension between.

TextStyl

Use this option to set all the text options for the dimension text:

- **TextSize** sets the size of the dimension text.
- **Weight** sets the line weight (width) of the dimension text.
- **Slant** sets the slant, or angle, of the letters of the dimension text.
- **Aspect** changes the width-to-height aspect ratio of the dimension text.

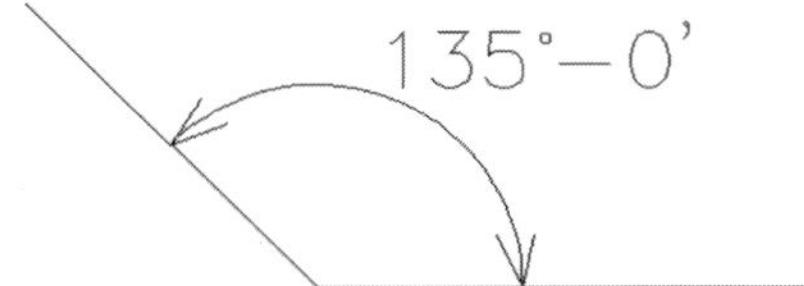

Figure 11-112 The angular dimensions option at work

- **TxtScale** sets the size of the text relative to the current plot scale. See Chapter 5 for a full description of Text Scale.
- **Color** selects the color of the dimension text.
- **FontName** is used to select the font name for the dimension text.

DimStyl

Use this option to set the properties of the arc and of the limits and tolerances of the dimension (see Figure 11-113).

DimStyl offers the following options:

- **Offset** When the dimension arc is outside of a line being dimensioned, an extension line is displayed, as in Figure 11-113. Use this option to set the distance between the origin of the extension line and the object being dimensioned.
- **Overlap** When the dimension arc is outside a line being dimensioned, an extension line is displayed, as in Figure 11-113. Use this option to set the distance between the dimension arc and the end of the extension line.
- **Limits** With this option you can have two dimensions displayed: one for an upper limit, as specified by **UpperLim**, and one for a lower limit, as specified by **LowerLim**. For instance, if you have upper and lower limits of 5 degrees, and you create a dimension whose true angle is 135 degrees, the dimension will look like Figure 11-114. **Limits** and **Tolrance** are mutually exclusive options.

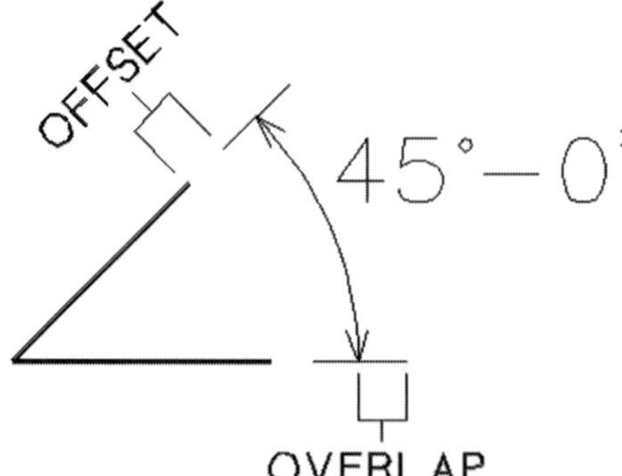

Figure 11-113 Angular **Offset** and **Overlap**

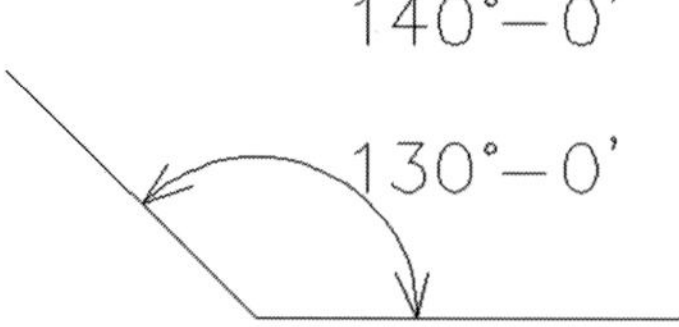

Figure 11-114 Setting the upper and lower limits

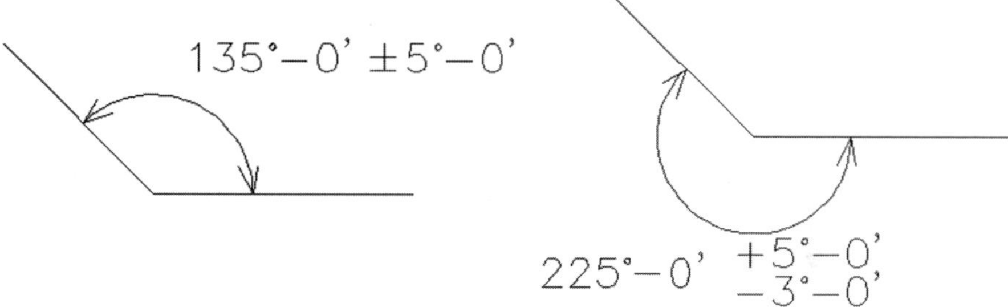

Figure 11-115 Two examples of dimensions with various tolerances

- **Tolrance** Use this option to display a dimension along with a positive and negative tolerance displayed after it. If the + and the − tolerance numbers are the same, then only one tolerance number is shown, as in the second example in Figure 11-115. **Limits** and **Tolrance** are mutually exclusive options.
  - **+Tolrnce** Here you enter a positive tolerance value.
  - **−Tolrnce** Here you enter a negative tolerance value.
- **UpperLim** This option will only be displayed once **Limits** is toggled on.
- **LowerLim** This option will only be displayed once **Limits** is toggled on.

ArroStyl

See the descriptions of these options in the Linear Dimension section.

Working with Rotated Plans

One very common question from new DataCAD users is how to handle floor plans that are not orthogonal in orientation; that is, the plans are rotated

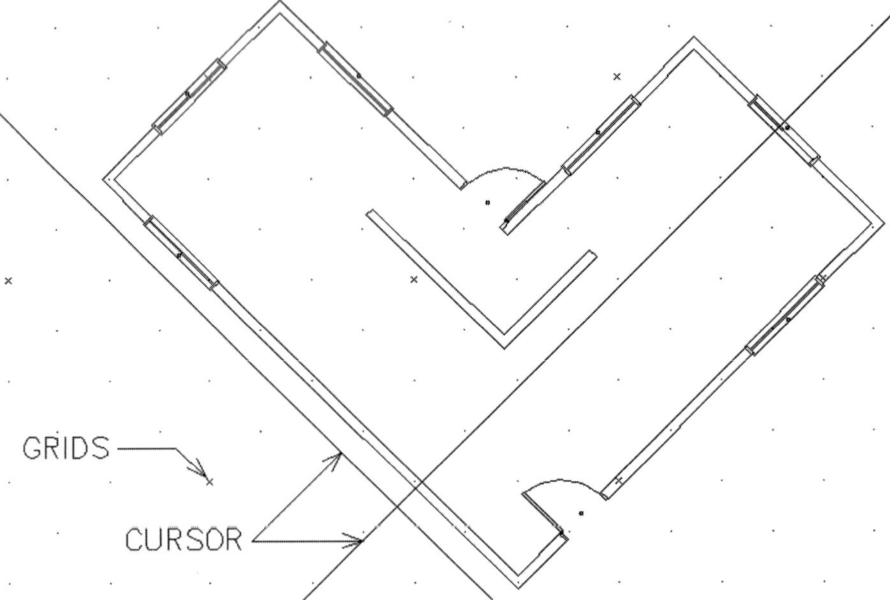

Figure 11-116
A rotated plan with several helpful options displayed

off of the horizontal/vertical. DataCAD has several tools that will help. Refer to Figure 11-116 for the following section.

Grids

In the **Utility** menu is the **Grids** feature. *Click* on this and you will see a number of options. Chapter 3 has an in-depth explanation of all these options, but let's just recap the ones that will really help you when working with rotated plans:

- **DspGrid1 (Display Grid 1)** Turn this option on to display a uniform grid of reference dots in the Drawing Window. You set the grid distance with the **GridSize/SetGrid1**.
- **DspGrid2 (Display Grid 2)** Turn this option on to display a uniform grid of reference dots in the Drawing Window. You set the grid distance with the **GridSize/SetGrid2**.
- **Angle (Grid Angle)** Use this option to rotate all three grids (**SnapGrid**, **DspGrid1**, and **DspGrid2**) simultaneously on the current layer. Choose or type an angle value from 0 to 90 degrees and press **Enter**. To match the grid angle to an existing entity in the drawing,

choose the **Match** option in the **Value** menu and select an existing line or a line defined by any two points.

Notice that the cursor crosshairs rotate along with the grids. This is really the key to working with rotated entities. But also note that your X and Y axes do not rotate along with the grids. X remains along the horizontal and Y remains along the vertical axis no matter which grid angle you choose.

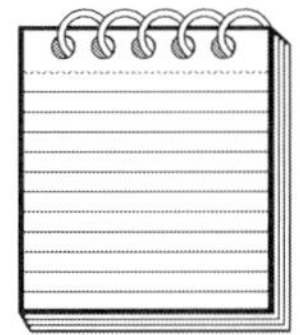

NOTE: *Although your cursor is rotated, DataCAD will not rotate any rectangular bounding boxes, such as those for selecting by* ***Area*** *for* ***Cleanup*** *functions or when drawing a polygon or rectangle.*

Rotated Cursor (Tangents)

In the Grids section, you learned that the cursor automatically rotates when you activate a rotated grid. Whether you have a rotated grid/cursor or not, however, you can independently rotate the cursor itself with the **Utility/Geometry/Tangents** option (the shortcut is the **Bb** key). To use it, you need to have an entity already drawn so that the **Tangents** option can match the angle of that entity.

To use this feature, follow these steps:

1. Select **Utility/Geometry/Tangents** (or **Bb**). You are prompted to *"Select the line to draw the tangent to."*
2. *Click* on the line in the Drawing Window whose angle you want to match.

The cursor will be rotated, while any grid settings you may have set will remain unaffected. But whether or not the X and Y coordinate readouts are also rotated with the cursor depends on a setting in the **Utility/Settings** menu. If **DistSync** is turned on, then the X and Y coordinates will be rotated to match the rotated cursor. If **DistSync** is turned off, then the X and Y coordinates will not be rotated to match the rotated cursor and will remain horizontal and vertical with the screen.

To turn off the Tangents feature, you must do the following:

1. Select **Utility/Geometry/Tangents** (or **Bb**).
2. Select **Cancel**. The cursor will be returned to its original orientation.

*NOTE: Be aware that Ortho mode does not function when Tangents is turned on. If you want to draw orthogonally horizontal or vertical lines, then you must temporarily turn **Tangents** mode off.*

*If you believe that you are back to the standard Ortho mode after using **Tangents** mode, and you can't get the **Ortho (Oo)** option to toggle on / off, then you are probably still in **Tangents** mode without realizing it, even if your cursor is perfectly horizontal and vertical. This can happen if you accidentally select a horizontal or vertical line to draw tangent to. Even though this will return the cursor to a horizontal / vertical orientation, you will still be in **Tangents** mode where the **Ortho** toggle doesn't work. This is a common mistake when using the **Tangents** mode.*

Bi-sect

Select this option to rotate the cursor to an angle that bisects the angle between two existing lines.

1. Select **Utility/Geometry/Tangents** (or **Bb**) and then **Bi-sect**.
2. Select the first line and then the second one. The cursor will be rotated to match the angle between the two selected lines (see Figure 11-117).

TanDivs

This setting is similar to the **Utility/Grids/SnapAng** setting (refer to Chapter 3). **TanDivs** sets the number of cursor snap segments (like slices

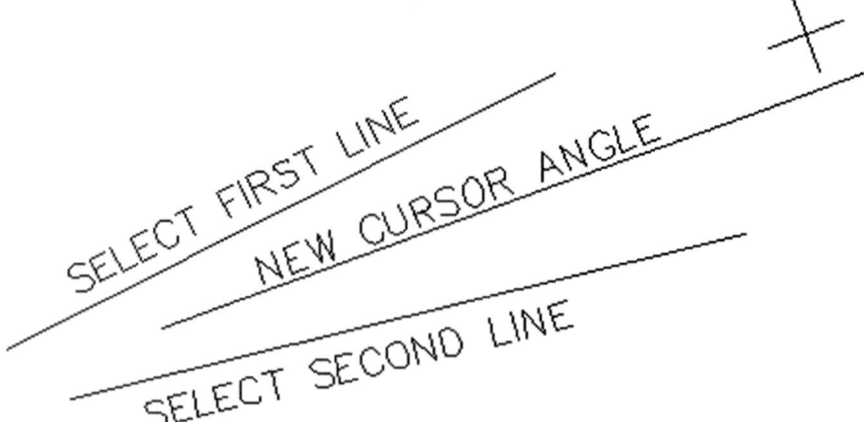

Figure 11-117 Bisecting the angle between two lines

of a pie) to be used when one of the **Tangents** options is active. This enables you to move your cursor and draw entities at exact angles by simply moving your mouse and without having to type any coordinates or angles. By default, DataCAD uses 45-degree segments for a total of eight segments (45 degrees × 8 segments = 360 degrees).

Site Plans

If you need to enter site plan information into DataCAD, two features will help:

- The **CurvData** feature is used to input surveyor data. You can read about it in Chapter 5.
- If you select **Utility/Settings/AngleTyp**, then you can select the **Bearings** option to enable you to input meets and bounds data, such as N 48-14′-23°E.

If you need to create a 3D site model, read about the **DropMesh** feature in the DataCAD Online Help.

Toolbox Macros

Several DataCAD **Toolbox** macros can help you automate the process of creating basic plan and elevation drawing elements.

DCAD_AEC Macro

The **DCAD-AEC** macro is a small but very useful program found in your **Toolbox**. It includes many common construction drawing functions that will make drafting a much simpler and quicker task. Here is a list of some of the things it can do. The Online Help has a good description of each of these options:

- The creation of drawing sheets and titleblocks
- The creation of column grids
- The creation of strings of windows, including curved windows and commercial storefront types

- The creation of individual elevators or banks of them
- The creation of double-back stairs in a plan
- The creation of room and door labels with options for placing them in boxes or capsules
- A square footage calculator that calculates and builds square footage tables for tenant layouts
- The creation of suspended ceiling grids

Door and Window Labels

Once you have inserted doors and windows in your floor plan, you will most likely need to label them with numbers or letters for coordination with a door or window schedule. Four labeling tools can help you do this. Two of them come with DataCAD, while the third one is part of another third-party macro.

Room Labels

One of the macros that comes with DataCAD is called **DCAD_AEC**, located in the **Toolbox**. Within that macro is another macro called **RmLabels**. But its name belies what else it can do for you. It can also automate the process of creating alphanumeric labels for your doors and windows. Around the text, you can choose to draw a box, a capsule, or no box at all. You can set the font and box attributes, including color. Best of all, alphanumeric labels can be automatically increased with each placement in the drawing. If you place room number 100 in the drawing, the next label will automatically be numbered 101, then 102, and so on (see Figure 11-118).

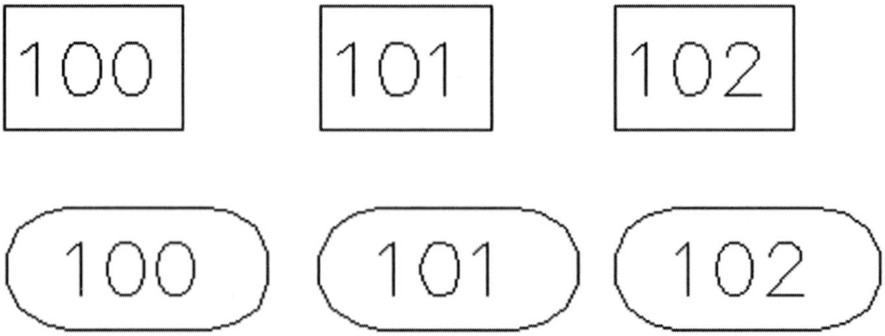

Figure 11-118 Automatic room labels

100 101 102

100 101 102

Door Label and Window Label Macros

The second and third useful macros that come with DataCAD are called **DoorLabl** and **WndwLabl**. Not only do they create labels, or tags, for doors and windows, but they will maintain a database of information regarding those tags, allowing you to then automatically create a door or window schedule.

DCSprint/Labels

In Chapter 24, "Third-Party Macros," you can read about a third-party macro called *DC Sprint*. Within it is another macro called **Labels**, which acts much like the **RmLabels** macro described earlier, but with a few more options.

CHAPTER 12

Advanced Construction Drawings

Multi-Scale Plotting (MSP)

DataCAD 7 introduced *Multi-Scale Plotting* (MSP) to the base program, and with it the program took a giant leap forward. Before MSP, you had two basic choices if you wanted to create a drawing sheet with details of more than one scale. The first option was to use the **Enlarge** function to resize the entities in the details, but then the details would no longer reflect true dimensions. Associative dimensions would not work, and editing the details later would be problematic since you would have to remember by what factor you enlarged each detail. The second option was to lay out all details of one scale, plot them out, reload that same printed sheet into the plotter, turn on all the details of another scale, making sure to lay them out where you hoped they would not interfere with the previously plotted details, and then plot those details to the plot sheet. You can imagine the coordination headaches here!

The basic premise of MSP is to enable users to continue to draw all of their plans and details at true scale, and then to arrange them on a drawing sheet at whatever drawing scale is required. MSP enables you to change the plotted scale of each detail on the fly simply by using the **PgUp** and **PgDn** keys.

Think of the basic steps this way:

1. Decide at what scale you want to plot a particular detail (this is important only for text size).
2. Draw your detail, complete with notes.
3. Enter the Multi Layout (**MltLyout**) menu and locate the detail on the drawing sheet in the MSP window.
4. Repeat Steps 1 through 3 for subsequent details.

You will not find a menu or an option called *Multi-Scale Plotting* (MSP). Instead, you access this feature by selecting **Utility/Plotter/MltLyout**.

How It Works

If you use and understand 3D GotoViews (refer to Chapter 10, "Creating Efficiency and Order"), you will easily grasp how MSP works. Just keep in mind that 3D *GotoViews* (GTVs) are not directly related to MSP and are not required to be used with MSP. 3D GTVs simply make it easier for you to make subsequent edits to each detail. If you don't yet have a grasp on 3D GTVs, I would highly recommend that you review 3D GTVs in Chapter 10 prior to continuing with this chapter.

As an example, if you turn layers 1, 2, 3, and 4 on and then go to MSP, DataCAD uses the X and Y extents of the entities in the activated layers to define the boundaries of the "detail." A detail is defined by DataCAD as "which layers are turned on, and what are the X and Y extents of their outermost entities," very much like 3D GTVs. When you go into the **MltLyout** menu and choose **Layout**, you will see a bounding box that represents the X and Y extents of the "detail."

The following questions and answers will help illustrate the point. Notice the similarities with GTVs. Assume that you have a number of details already laid out in MSP and then you want to make some changes to one or more of them:

Q: So what happens when you add/delete entities to a detail in MSP?

A: As long as they are on the same layers as originally defined when you placed your detail, they will show up properly in MSP.

Q: What happens if you delete a layer from the drawing file, and it was a layer that used to be in a detail laid out in MSP?

A: Since the layers do not exist in the drawing file any more, then DataCAD can obviously not display them.

Q: What happens if you add new layers to your detail?

A: They will only show up in the MSP window after you **Update** the detail in MSP since it determines which layers to display in each detail based on which layers were turned on at the time you originally placed the detail. So if you add new layers to a detail, you need to use **Update** to tell MSP to include them.

Q: What happens if you turn layers off in your detail?

A: MSP will still show the off layers. Just as you must update a detail in MSP when new layers are added, you must also update the detail if you turn layers off.

NOTE: *For the examples in this chapter you will need a printer driver that supports large paper sizes like 24″ × 36″ [610 × 914] and 30″ × 42″ [762 × 1066]. If you don't have one, try the OCE 9800 driver, available for free at www.oce.com.*

Part I: Basic MSP Concepts

Quick Steps

Here are the basic steps for adding detail to an MSP layout.

1. Get the detail to be laid out in MSP in your Drawing Window. Make note of the scale that you want the detail to be plotted at.
2. Press **ReCalc** to check the X/Y extents of the detail and to make sure that the entities to be placed are the only entities that are on (this step is not required but is useful for cross-checking for stray entities or layers).
3. Select **Utility/Plotter/MltLyout/Layout**.
4. Use **PgUp** or **PgDn** to change to the correct scale for the detail. The scale is shown in the Coordinate Readout line.
5. Locate the detail in the MSP window and *click* the mouse to place it.
6. Name the detail and press **Enter** (if you don't name it, then the detail will not be saved).
7. *Right-click* out of the **MltLyout** menus.

The Basics

We'll use a sample drawing from the CD to illustrate the use of MSP. Copy the BAR.DC5 file from the CD to your computer's hard drive, then open the file in DataCAD. This file contains a series of details that we want to plot at various scales on three drawing sheets. When you first open up the file, you will be presented with a 24″ × 36″ [610 × 914] drawing sheet with a title block. It is important to note that the title block measures a true 36 inches wide by 24 inches high. The L-shaped entities in the four corners of the title block represent the corners of a 24″ × 36″ [610 × 914] sheet of paper.

1. Go to the **Layers** (**Ll**) menu and turn on layer **A3-TEXT**. Only two layers should be on now: **TTLBLK** and **A3-TEXT**. *Right-click* out of the **Layers** menu.
2. **ReCalc** the Drawing Window to be sure you have the correct extents of the detail.

   Note that everything we place in MSP, even the title block, is called a detail. This word is simply a convention. An MSP detail can be a single drawing detail, a group of details, a titleblock, border, or even an entire drawing sheet full of details of the same scale.
3. First, you must set up your paper size (see Chapter 7, "Printing and Plotting," for help in selecting the correct printer drivers and settings).

a. Go to **Plotter/Setup**.
b. Pick a printer driver that supports a 24″ × 36″ [610 × 914] paper size.
c. A great many printer drivers will try to leave a border around the paper, refusing to let you have the full paper area. If you are having this sort of trouble, then try using the **Properties** button to set a custom-sized paper size in your driver that would eliminate the border, or try the custom paper size options in DataCAD's own **Setup** menu.
d. Select **OK** to exit the Setup dialog.

DataCAD places details in the MSP window based on which layers are turned on when you open the **MltLyout** menu and select **Layout**. Since the layers **TTLBLK** and **A3-TEXT** are currently the only layers that are on, those are the ones that will be placed in the MSP drawing window.

4. Go to MSP via **Utility/Plotter/MltLyout/Layout**.

NOTE: *Two basic menus can be found in MSP: the* ***Sheet*** *menu and the* ***MltLyout*** *menu.*

Use the ***Sheet*** *menu to edit the properties of MSP sheets. For instance, you will use the sheet menu when you want to view the current drawing sheet, change the current drawing sheet or rename the current drawing sheet.*

The ***MltLyout*** *menu is where you access and edit the individual details on the current MSP drawing sheet. This includes the layout of new details and the editing of current detail layers and parameters. It also includes a button to take you to the Sheet menu. Unless you just want to view the current MSP drawing sheet, you'll most often want to pick the* ***MltLyout*** *menu.*

5. By default, you will be on **Sheet1**. You will see four dashed lines that define the edges of your 24″ × 36″ [610 × 914] sheet and two more dashed lines crossing in the center of your sheet. These are Layout Division lines and can be modified via the **LyoutSet/LyoutDiv** option to enable you to define how many dashed lines will be displayed horizontally and vertically on the sheet. They are simply guidelines to help you lay out your details. They do not plot and cannot be snapped to.
6. Move your cursor into the drawing area. The message area will say, *"Locate detail on sheet of paper. "Enter" to end."* You will see a small

bounding box at the intersection of the cursor crosshairs. The bounding box defines the extents of the entities that make up the title block. Notice that the title block detail is much smaller than the dashed lines that represent the outside edges of your paper. We need to change the scale of the title block to make it fill the entire sheet of paper.

7. Press **PgDn** once. The bounding box gets a little bigger. The Message Line will read *"Scale for this detail is now 3/8″."* Continue to press **PgDn** (think of it as moving down closer to the drawing) until the title block detail fills the dashed, rectangular paper extents in the MSP window. The Message Line will read, *"Scale for this detail is now 12″."*

   The scale is 12″ (full scale) because the title block is drawn at a true 24″ by 36″ size. This is an important concept. Without MSP, if you wanted to create three different drawing sheets, each with different scale details on them, you would have to copy the title block two more times, and then use the **Enlarge** function to scale each copy of the title block up or down to match the scale of the details to be placed on them. More on this later, after we place another detail.

8. Align the outside edges of the detail's bounding box with the outside four dashed lines in the MSP window. To place the title block detail, *click* the mouse or hit **Enter**. The message area will say, *"Enter name of new detail: Detail1."*

9. Now we must give the detail a name. Type in **TTLBLK** and press **Enter**. You must name the detail after placing it in the MSP window. If you do not name it before *right-clicking* or exiting the layout, then the detail will not be saved. It will appear that it has been saved, but when you reenter the MSP layout window, the detail will not be there (try it and see). After successfully typing in a name, your screen should look something like Figure 12-1.

NOTE: *An important tip for laying out details: MSP uses the current snap grid of your drawing file, so prior to entering MSP, you can change your snap grid and then make sure Snap Grid is toggled on (**Utility/Grids/SnapGrid**). This will help when you need to align details with one another.*

10. Now that the first detail has a name, let's name the drawing sheet as well. While still in the Layout menu (the one with your new detail

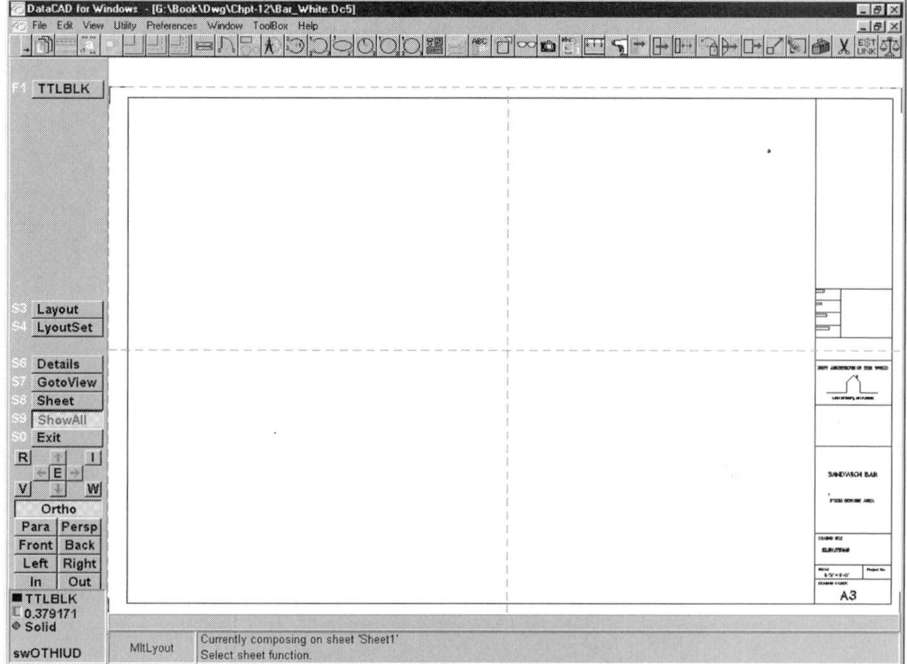

Figure 12-1 The first detail is the titleblock and border.

name **TTBLK** at the top), press **Sheet** to go to the **Sheet** menu. You will see a list of default drawing sheet names. Each name represents a separate drawing sheet. **Sheet1** is on and is therefore highlighted. By default, we laid out our first detail on **Sheet1**, but as part of our drawing set, this sheet should be called A3, so let's do that.

11. Press **Rename**. The Message Line says, *"Select sheet to rename."* Select **Sheet1**. Now the Message Line reads, *"Enter new name: Sheet1."* Type in **A3** and press **Enter**. Our MSP sheet has now been renamed.
12. Since we know that we have two more sheets to lay out, A4 and A5, let's go ahead and name them now. With **Rename** still selected, select **Sheet2**, type **A4**, and then press **Enter**. Select **Sheet3**, type **A5**, and then press **Enter**.
13. *Right-click* several times to exit the **MltLyout** menu and get back to the 2D menu.

We will look at a couple of methods of placing the rest of your details in MSP. First, we'll do it the long way and then we'll learn the shortcuts.

*NOTE: In the **MltLyout** menu, be sure that the **ShowAll** button is turned on. If not, then you won't see your details when you return to the **MltLyout** menu.*

The Long Method of Placing Details

This sample drawing file already has several 3D GTVs (refer to Chapter 10) saved for all the details and drawing sheets. These are not required in order to use MSP, but finding and editing details outside of MSP is much easier with them, so we've saved you some time by setting them up. View them by pressing the **V** in the Navigation Pad and then selecting **GotoView** or by selecting **Utility/GotoView**. You will see the following views:

- **ELEVNS**
- **SECTNS**
- **DTL-1**
- **DTL-2**
- **DTL-3**
- **DTL-4**
- **DTL-5-6**
- **DTL-7**
- **DTL-8**
- **DTL-8**
- **A3**
- **A4**
- **A5**

The next set of details we will lay out are the bar elevations.

1. Select the **ELVNS** 3D GTV from the list and then *right-click* back to the 2D menus.
2. To verify the extents of the elevations, **ReCalc** the Drawing Window. This will show you that only the expected elevation entities are on.
3. Go to the **Plotter** menu and press **MltLyout/Layout**. Notice that the MSP window with the current A3 drawing sheet appears. The MSP window will always display whichever sheet was last accessed.

4. Move the cursor into the MSP window. You probably won't see any bounding box representing your new detail. That is because we are still at 12″ (full) scale from the last detail we laid out, so we just can't see it. Press **PgUp** until the Message Line reads, *"Scale for this detail is now* $^1/_2$*″,"* which is the scale we want these details to plot at. Your screen should look like Figure 12-2.

   Since all of these elevations are to be displayed at the same scale on the drawing sheet, all the elevations are grouped together as one detail. We could just as well have separated each elevation as different details by using Clip Cubes (more on this later) or by separating each elevation's entities into separate groups of layers, but there was really no need.

5. *Click* to place the detail on the drawing sheet. Remember we have to save the detail by giving it a name. Type in **Elevns** and then press **Enter**.

6. Now let's say you want to relocate a detail to another location on the drawing sheet. All you have to do is *click* on the detail name in the **MltLyout** menu. The detail's entities will disappear from view in the MSP window and you will see the bounding box again. Try it now by pressing the **Elevns** button. Move the bounding box to a new location and *click* the mouse.

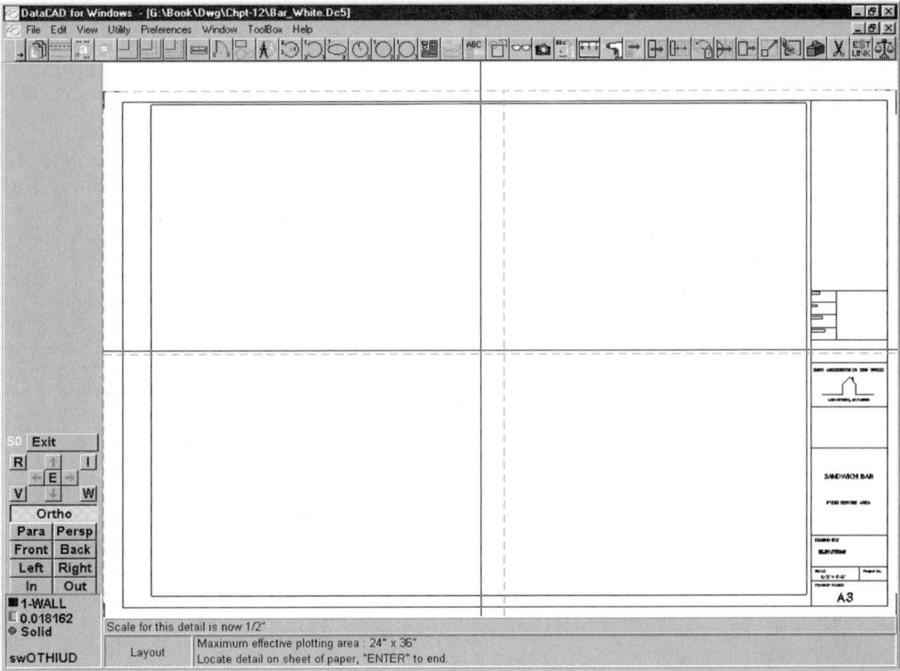

Figure 12-2 Adding the bar elevations, displayed as a bounding box

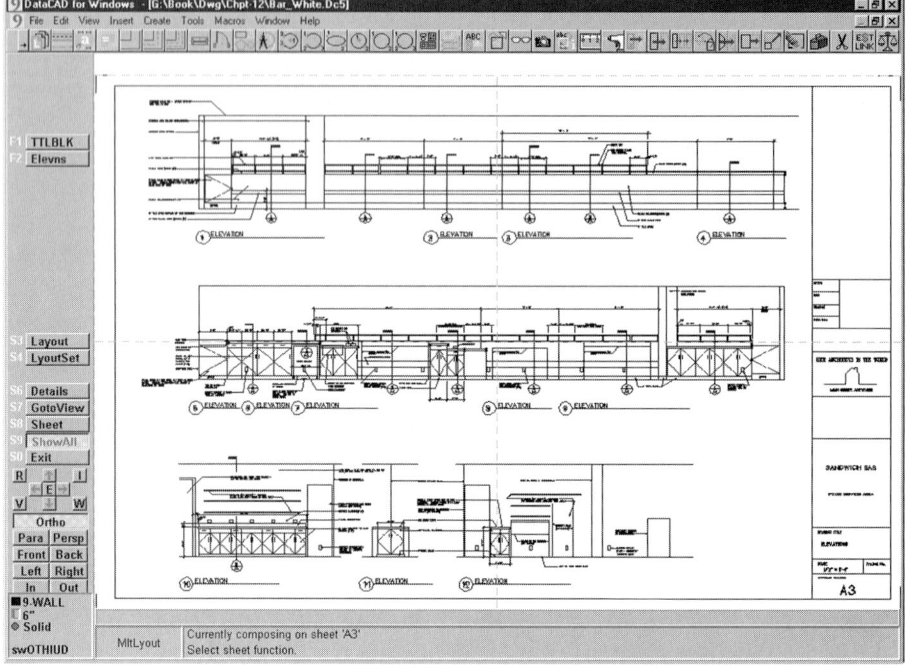

Figure 12-2A Correct location of the **Elevns** button

7. Select the **Elevns** button again and move the bounding box around again, but now *right-click* the mouse instead. The detail will revert back to its original location. This is handy if you change your mind or if you accidentally select the wrong detail.
8. Select the **Elevns** button again and make sure it is placed properly in the MSP window. Your sheet should look like Figure 12-2A.

Placing Details with GotoViews

The previous exercise is the long method of placing a new detail on an MSP sheet. Now let's look at a quicker method. We're done with sheet A3, so let's change to MSP sheet A4:

1. Go to the **MltLyout** menu, select **Sheet**, and then pick the **A4** button that we created earlier. A new set of dashed lines representing the extents of the new A4 drawing sheet appears.
2. *Right-click* once, back to the **MltLyout** menu, and then select **GotoView**. The list of 3D GTVs that we saw earlier will appear. Now

we can select a ready-made detail to place. The first one we need to place is our title block for A4.

3. Press **ScrlFwrd** and then select **A4**. Press **PgDn** until the scale is 12″ [1:1] and the detail fits the drawing sheet. Place the detail, *click* the mouse, and then name it **TTLBLK**. Notice that you are still in the 3D GTVs menu.
4. *Right-click* back to the **MltLyout** menu, where you will now see the TTLBLK detail name that you just placed.
5. Select **GotoView** again. **ScrlBack** if you have to, and select **SECTNS**. You won't see the bounding box because the scale is still set to 12″ from the last detail we placed. We want these details to be displayed at $1^1/_2$″ [1:10] scale. Press **PgUp** until the Message Window tells you that the scale is **$1^1/_2$″** [1:10]. Place the detail and name it **Sectns**.

Creating Additional Drawing Sheets

We're done with sheet A4, so let's start sheet A5 and lay out some more details there.

1. *Right-click* once, back to the **MlyLyout** menu. Then select **Sheet/A5**.
2. To place the sheet border, *right-click* once and select **GotoView/ScrlFwrd/A5**.
3. Press **PgDn** until the scale reads 12″ [1:1] and the detail fills the MSP window. Place the detail and name it **TTLBLK**.
4. To place a detail, *right-click* once and then select **GotoView/ScrlBack/DTL-1**. Press **PgUp** until the scale reads **3″** [1:5]. Place the detail in the upper-left corner of the sheet and name it **Detail-1**. *Right-click* once to see the list of details on the drawing sheet (see Figure 12-3).
5. Now press **GotoView/DTL-2**. Place the detail to the right of Detail-1 and name it **Detail-2**. It's at the same scale, so you don't have to use **PgUp** or **PgDn**. *Right-click* once, back to the **MltLyout** menu (see Figure 12-4).

   The previous two details were displayed at 3″ scale, but the next detail needs to be displayed at 6″ [1:2] scale.
6. Press **GotoView/DTL-3**. Now press **PgDn** once until the scale reads 6″ [1:2]. Place this detail to the right of Detail-2. Don't worry if the new detail's bounding box overlaps Detail-2. We can move it later. Name it **Detail-3**. Notice in Figure 12-5 that Detail-2 and Detail-3 overlap one another. We need to move them around a bit to make them fit. *Right-click* once, back to the **MltLyout** menu (see Figure 12-5).

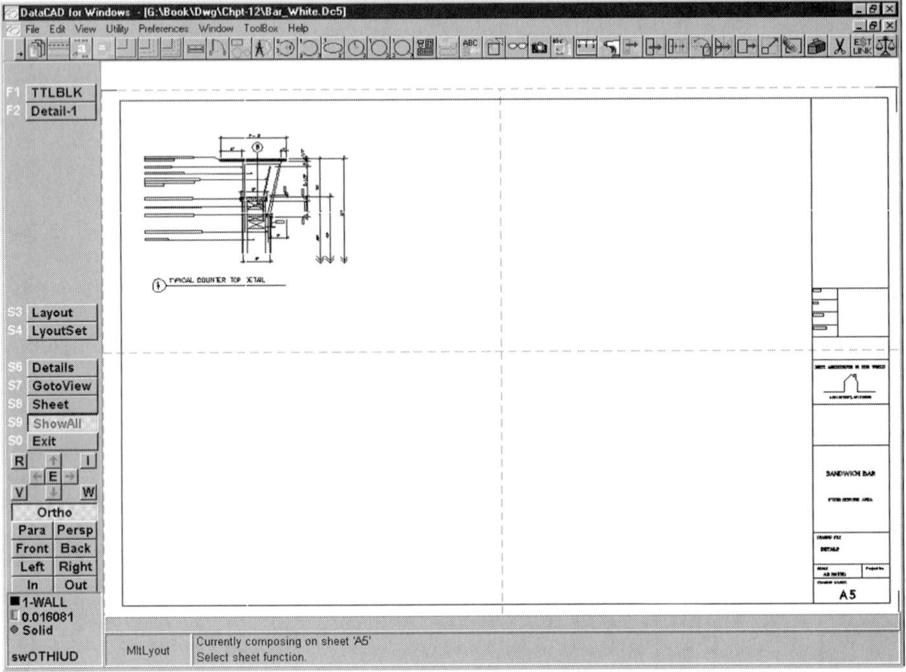

Figure 12-3
Placing a detail

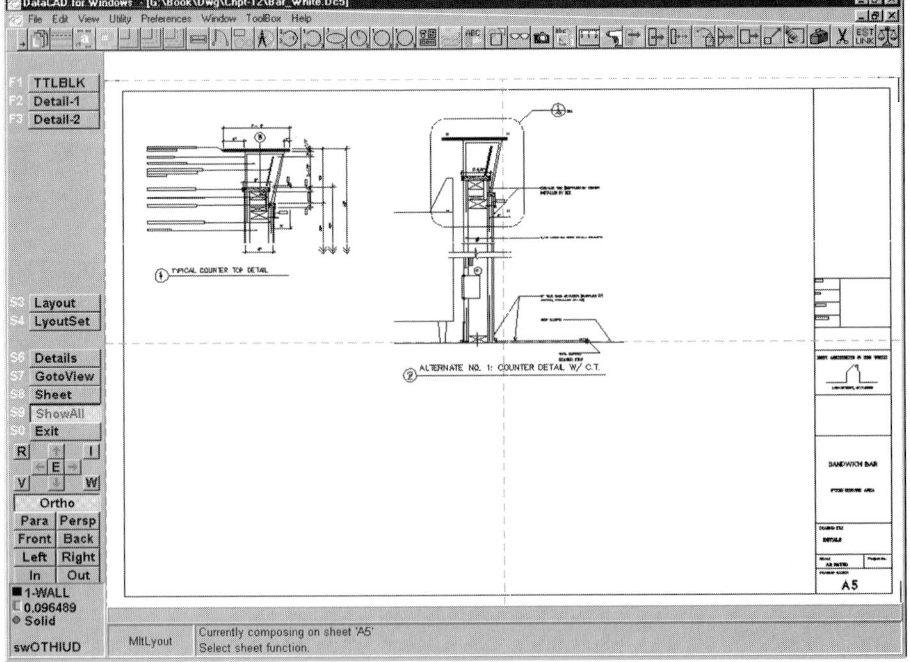

Figure 12-4
Adding another detail

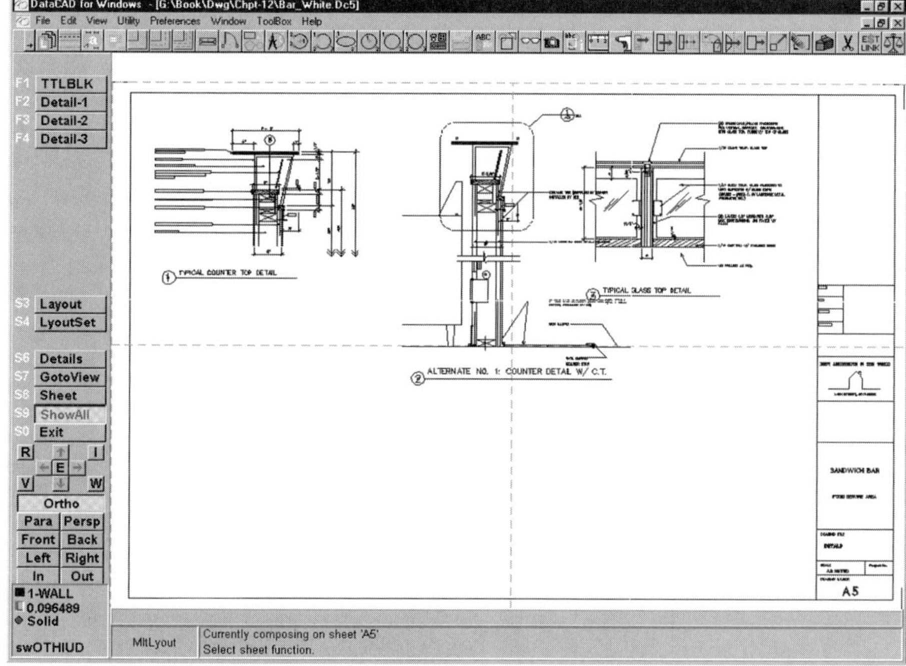

Figure 12-5 Adding a third detail that overlaps with the second one

7. Press the **Detail-2** button and move the detail to the left as necessary. *Click* the mouse to place the detail in its new location. Now do the same for **Detail-3** as required. Your sheet should look something like Figure 12-6.

Changing Scales after Placing Details

The previous detail was displayed at 6″ [1:2] scale, while the next detail is to be displayed at 3″ [1:5] scale. This time we will change the scale *after* we place the detail.

Select **GotoView/DTL-4**. Place the detail at the bottom-left corner of the sheet and name it **Detail-4**. Now notice that the detail appears too large because the scale is too large. We want the scale to be 3″ [1:5]. *Right-click* back to the **MltLyout** menu. To change the scale of Detail-4, simply press the **Detail-4** button, press **PgUp** once until the scale reads 3″ [1:5], and then place the detail where you want it, somewhere in the lower-left corner.

Notice that the title below Detail-4 does not have any text below it to indicate the scale of the detail. Let's see how we would add the scale text below the title of the detail and have it show up in this MSP sheet.

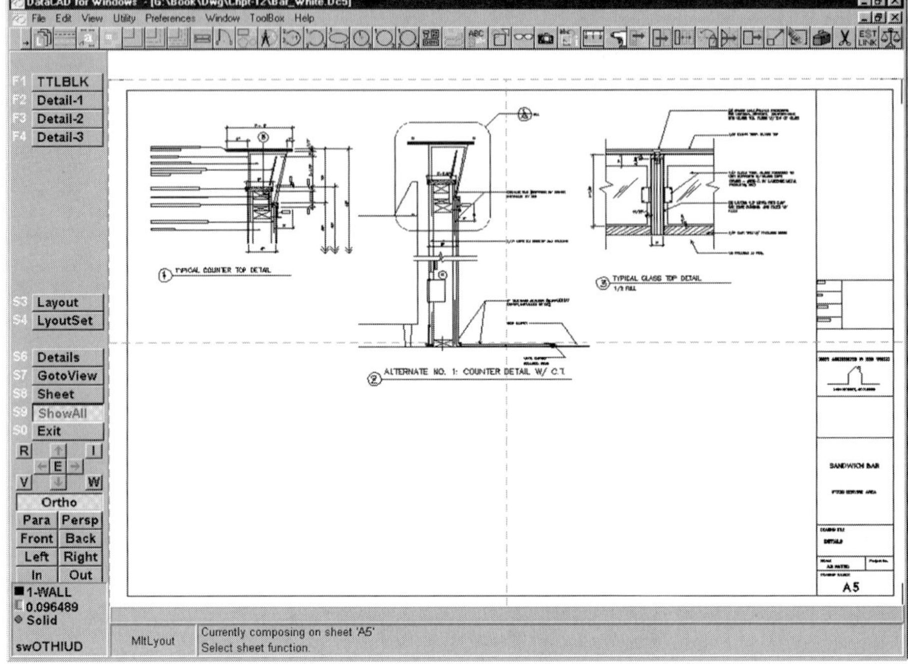

Figure 12-6 Correcting the overlap of the two details

Editing Entities of MSP Details

Details cannot be edited in the MSP window. In order to do so, you must exit from the **MltLyout** menu, change your details, and then return to MSP to see the changes. You cannot Window In to the details or otherwise move around within the MSP window, either (though you can do so with the **Plotter/Preview** option). Let's see how we would use MSP to view and edit Detail-4:

1. In the **MltLyout** menu, press **Details/MakeCurr**. This will allow you to choose which of the four details you want to edit by making it the current detail in the Drawing Window.
2. The Message Line will read, *"Select the detail you wish to use as the current view."* Press **Detail-4**, since that is the one we want to edit. Your screen will now look like Figure 12-7.

   What DataCAD has done is to exit MSP and turn on only the layers that make up the detail you have chosen. DataCAD is doing this in exactly the same manner as it does for 3D GTVs; it remembers and displays the layers that make up the detail and the outermost extents of the entities on those layers. But remember that 3D GTVs and the

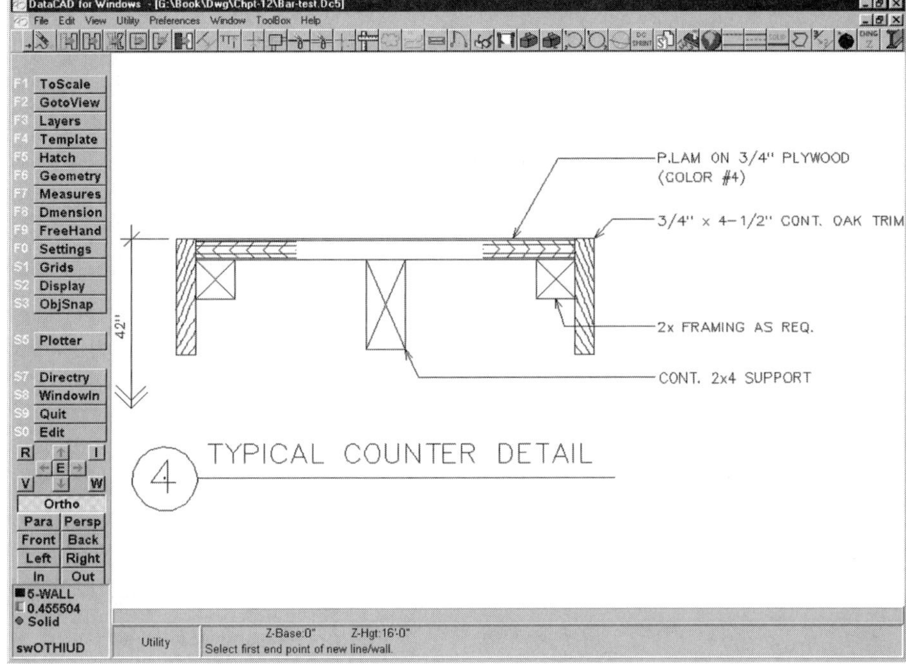

Figure 12-7 Selecting the detail to edit

details in MSP are independent of one another, so you do not have to have one to have the other.

3. Now go to the **Text** menu (**Edit/Text**) and then place the cursor in the Drawing Window.

   Your text cursor is probably displaying too big or too small for your needs. Also notice that the text cursor is Light Cyan, and the current layer is **5-WALL**. We want to change all of those settings to match the text in the detail's title. The easy way to do this is to use the **Identify** command.

4. Press the **I** button (for Identify) in the Navigation Pad. Now locate your cursor on some part of the text and *click* the mouse. (How you select the text depends on whether or not you are running Display List. Refer to Chapter 3 or 5 for an explanation). The text will become highlighted. If it doesn't, try it again. Now select **SetAll** from the menu. All your settings (layer, font name and size, and color) have now changed to match those of the detail's text. *Right-click* once to go back to the **Text** menu.

5. Locate the cursor below the horizontal line in the detail's title, *click* the mouse, and type in **Scale: 3″=1′-0″**. Then *right-click* or press **Enter** to accept it.

Now go back to MSP to see how our detail has changed.

6. *Right-click* out of the Text menu and then go to **Utility/Plotter/MltLyout/Sheet** to display the sheet again.

   Notice that Detail-4 now displays the text you just typed in. This is a key concept. The details you see in MSP are not "copies" of the details. They are the actual details themselves, just displayed in one common location (the MSP window). So whatever changes are made to the layers that make up your detail will be displayed in real time in the MSP window.

7. *Right-click* to the **MltLyout** menu and then use the **GotoView** button to select **DTL-5-6**. Use **PgUp** or **PgDn** to change the scale of the detail to 12″ [1:1] (full) scale. Place it in the lower-right corner of the sheet and label it **Dtl-5-6**.

   Remember that you can change the scale before or after placing the detail. I very often don't worry about what scale the detail is supposed to be displayed at. Rather, I place the detail, then select the detail from the **Layout** menu, and use **PgUp** or **PgDn** to change the scale to whatever is required. Your drawing should look something like Figure 12-8.

8. Place **DTL-7** as you did in Step 7. This detail is to be displayed at a $1\frac{1}{2}''$ [1:10] scale. Label it **Detail-7**. Don't worry if you are running out

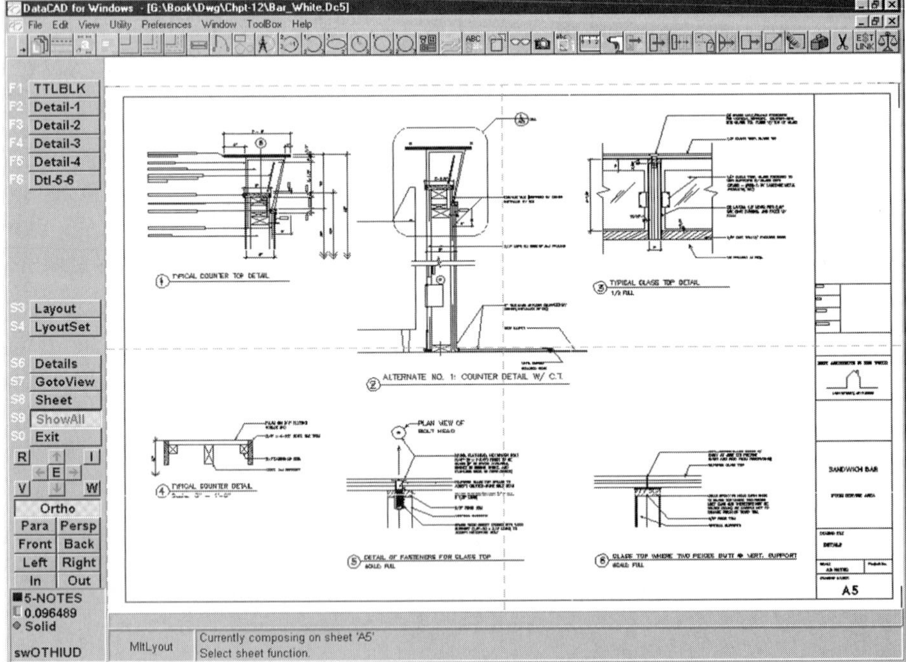

Figure 12-8
Placing details 5 and 6, together

of room for the detail. We'll rearrange the details after placing one final detail.

9. Place **DTL-8**, naming it **Detail-8**. This detail is to be displayed at 6″ [1:2] (half-full) scale.
10. After placing the detail, *right-click* back to the **Layout** menu. Pick and move each detail, starting with **Detail-3**. When you are done moving details around, you might have something like Figure 12-9

Helpful Hints for Layouts

It is sometimes difficult to relocate your details in MSP, especially in small increments, so here are few helpful tricks:

- In MSP, pick **Layout** and place the detail in MSP by clicking the mouse. Do not give it a name. If you don't like where it's located, simply leave the image on the screen (remember it has not been

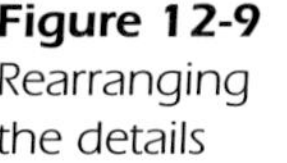

Figure 12-9 *Rearranging the details*

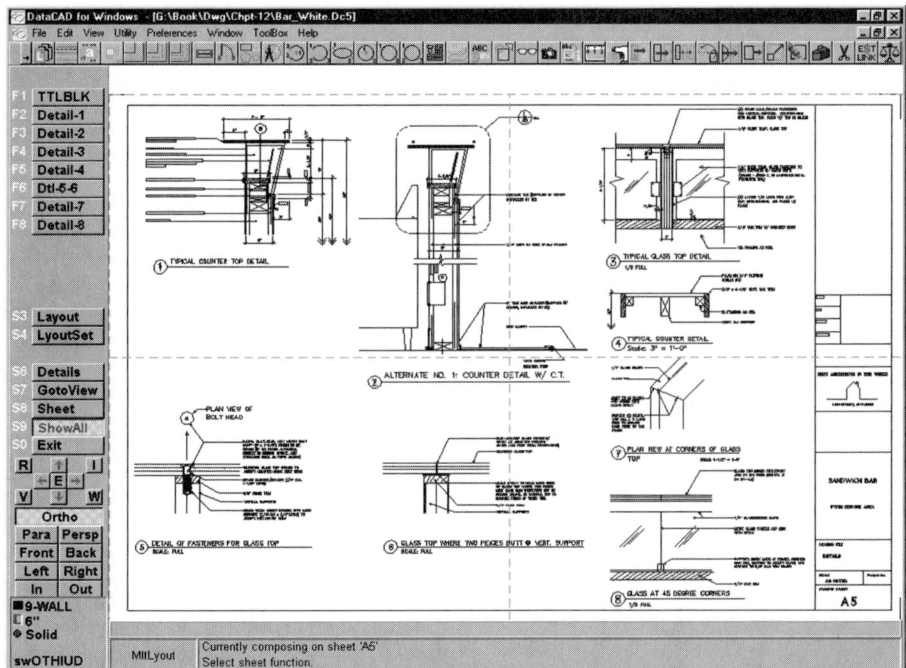

saved), *right-click* once to back up, and then click on **Layout** again. Place the detail where you want, using the old image (which is still on the screen) as a guide. You can do this an infinite number of times until you get it where you want it. Once you do, give the detail a name and hit **Enter**. Only the final location is saved, not all the other "ghosts" still on the screen.

- Similar to the previous method, once your detail is placed and named, pick **Layout** again and place the detail using the previous detail as a guide. You can now do just as you did in the previous method, trying new locations until you get it placed right. Once you do, give it a different name, and then go back and delete the original detail.

Layout Options

In the MSP **MltLyout** menu is a button labeled **LyoutSet**. Press that button and you will see two more options: **LyoutDiv** and **Extents**.

1. Press the **LyoutDiv** button. You get a menu starting at **1** and ending at **80**. The Message Line reads, *"Enter X-divisions for layout sheet: 2."* You can pick a number from the menu or type one in. First, let's pick one from the menu.
2. Select the number **5** from the menu. Then *right-click* or press **Enter**.
3. Now the Message Line reads, *"Enter Y-divisions for layout sheet: 2."* This time type in the number **4** and then press **Enter**.
4. *Right-click* once to get back to the **MltLyout** menu. Notice that now a new set of dashed lines is shown in the MSP window. You now have five horizontal (X) divisions and four vertical (Y) divisions. These lines do not print; they are just guidelines to help you lay out your details (see Figure 12-10).
5. Now press **LyoutSet** and toggle on **Extents** by pressing it. *Right-click* once. Now all of your details are shown only as bounding boxes representing the outermost extents of each detail. This enables the MSP window to refresh much faster each time you open it. Note that when you add a new detail, the new detail's entities will be shown in the MSP window. But after placing and naming the detail, exiting MSP, and returning, that detail too will be represented only by a bounding box (see Figure 12-11).
6. Now turn **Extents** off by selecting **LyoutSet/Extents** and *right-clicking*. Your details are all displayed in their entirety once again.

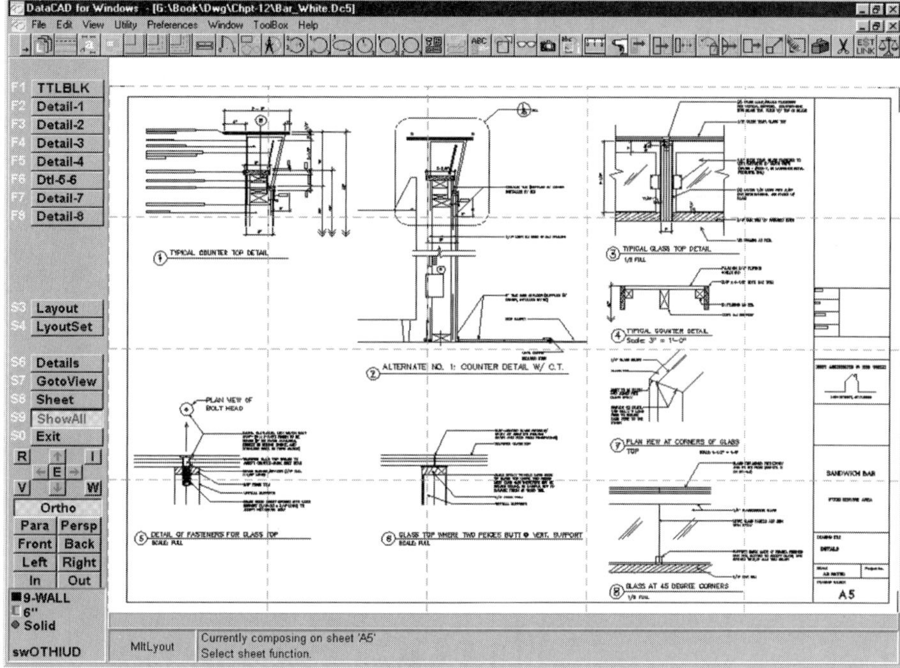

Figure 12-10 Adding horizontal and vertical guidelines

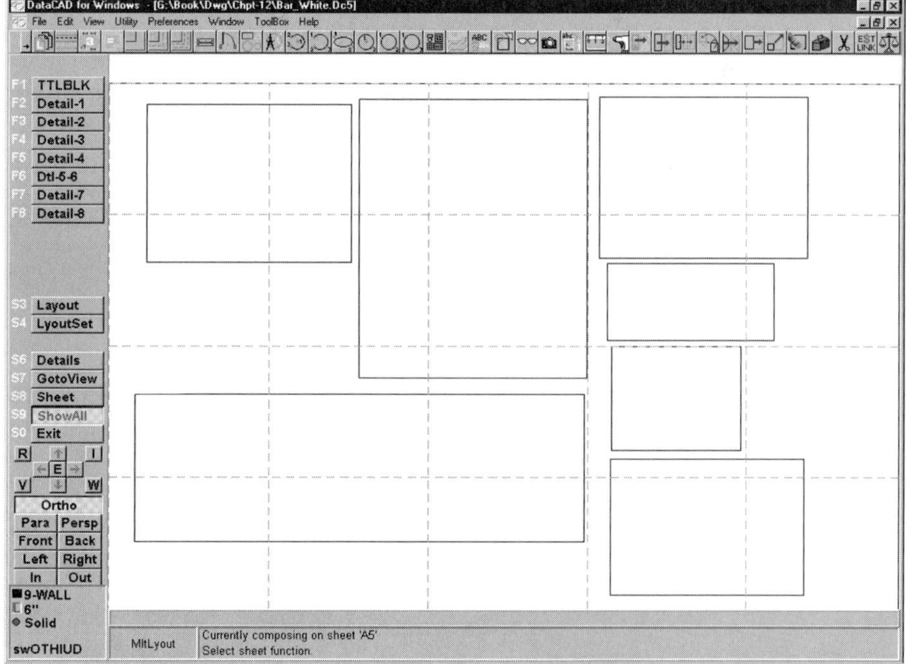

Figure 12-11 **Extents** enables you to see the details as only bounding boxes.

Deleting Details

Just as surely as you will want to add details, you will also want to delete details from an MSP sheet. Let's add another detail to the MSP sheet and then delete it:

1. Use the **GotoView** option in the **MltLyout** menu to place another copy of **Detail-8** in the MSP window. Place it anywhere, since you'll be deleting it anyway. Name it **Temp-8** and then *right-click* once back to the MltLyout menu.
2. Select **Details/Delete**. The Message Line reads, *"Select detail to delete from list."*
3. Select **Temp-8** from the menu. The detail will be deleted from the MSP window.
4. *Right-click* twice to go back to the **MltLyout** menu.

Turning Layers On and Off

There will be times when, after placing a detail in MSP, you will need to delete an entire layer from a detail or add new layers to a detail. DataCAD enables you to do this directly from within the **MltLyout** menu:

1. From the **MltLyout** menu, select **Details/LyrTogl**. The Message Line reads, *"Select the detail to toggle layers on and off."*
2. Select **Detail-1**. The standard DataCAD Layer list is displayed. Scroll forward (**ScrlFwrd**) once until you see the layers that make up Detail-1. Notice that the detail number doesn't have to correspond to the layer names.
3. Press **3-NOTES**. That layer is now shut off and will not be displayed in the MSP window or your plotted output. To prove that the change is permanent, *right-click* back to the Drawing Window.
4. Now go back to the **MltLyout** menu and display the drawing sheet again by pressing **Sheet**. Detail-1 still has the **3-NOTES** layer turned off.
5. *Right-click* once back to the **MltLyout** menu and select **Details/LyrTogl/Detail-1/ScrlFwrd/3-NOTES**. That layer is now turned back on again.

   Now let's see what happens if we turn on a layer that is not part of the detail we have drawn.

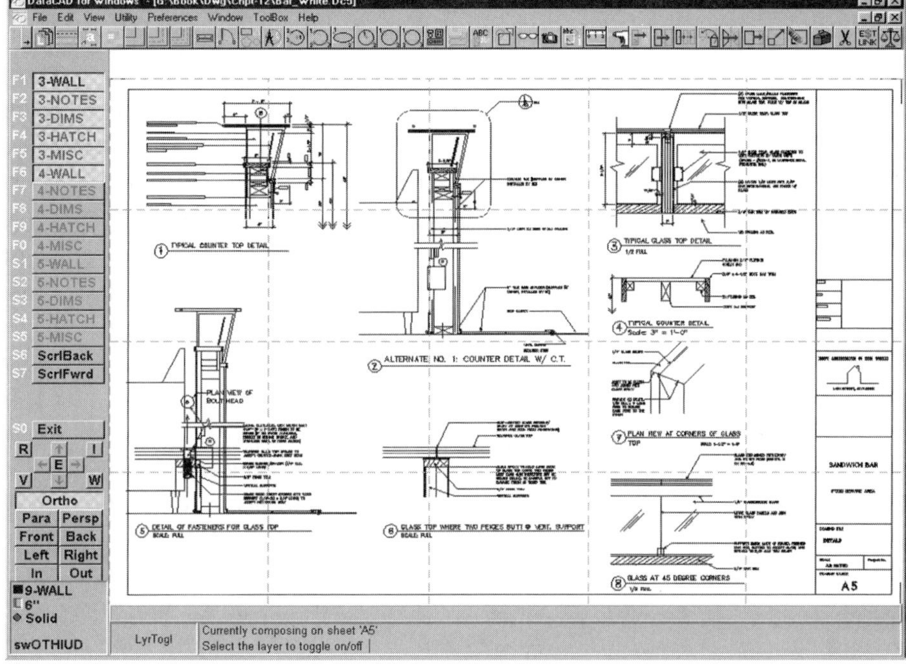

Figure 12-12
Turning on a layer not part of the detail we have drawn (look at the bottom left corner)

6. While still in the Layer menu, select the layer named **4-WALL**. Notice that the entities on layer 4-WALL appear in the MSP window in the lower-left corner, overlapping Detail-5 (see Figure 12-12).

Editing MSP Details

You may be wondering why those entities show up where they do and how you are supposed to know where entities will show up when you turn on a layer. Well, let's see. We'll use the **MakeCurr** feature to see this. The **MakeCurr** feature allows you to display and edit any of the details that you have in the MSP window. When you select **MakeCurr**, DataCAD takes note of which layers are currently turned on for that detail and determines what are their outermost extents. DataCAD then exits out of MSP and places you in the Drawing Window with the detail displayed at its extents. To edit MSP details, follow these steps:

1. *Right-click* twice back to the **Details** menu and then select **MakeCurr/Detail-1**. All of the current entities of the MSP detail called Detail-1 are displayed.

Notice that the entities of layer **4-WALL** are also displayed. That's because we used the **LyrTogl** command in MSP to make the entities on layer 4-WALL part of Detail-1.

2. Go back to the MSP Details menu (**Plotter/MltLyout/Sheet/Exit**). Turn off layer 4-WALL by selecting **Details/LyrTogl/Detail-1/ScrlFwrd/4-WALL**. Layer 4-WALL is removed from Detail-1.
3. To prove this, and to see how the MakeCurr option handles the change, *right-click* twice back to the **Details** menu. Select **MakeCurr/Detail-1**. You can see that DataCAD displays the current entities of Detail-1 and displays them at their extents.

NOTE: *If you don't have a fast computer, MSP details can sometimes take awhile to redraw themselves every time you re-enter the Sheet menu. You can skip the redraw of individual details in the MSP window by pressing the* ***Del*** *or* ***End*** *keys as an MSP detail begins to redraw itself. Pressing* ***End*** *will stop the redraw of the current detail. The next detail will begin to redraw instead. You can press* ***End*** *again to stop the redraw of this detail as well. This is useful if you have several details and only want to work on the last one. You can stop the redraw of the first set of details in order to more quickly get to the final one.*

Pressing ***Del*** *will stop redrawing any further details. This is useful if you accidentally press the* ***Sheet*** *option and want to stop the lengthy redraw process.*

4. Exit and **Save** the drawing. We'll move onto another drawing for the next demonstrations.

Recalculating the Extents of Details

One final note before proceeding. Let's say you have a detail that you have already laid out in the MSP window. Then you add, delete, or move entities in your detail, which changes its outermost extents. Now when you go to MSP and select that detail in order to move it to a new location, the bounding box that defines the extents of the detail will probably not reflect the new extents. DataCAD does not automatically know that it is supposed to update the detail's extents.

However, an undocumented feature in MSP will do a **ReCalc** of the extents of a detail for you. Any time the detail is attached to your cursor and displayed as a simple bounding box, just press the **F1** key on your keyboard and you will get an instant **ReCalc** of the extents of that detail.

Summary

Those are the basics of MSP. Before we go on to the more advance methods of using MSP, we must first go over how to use 3D *Clip Cubes* (CCs). Using CCs in concert with MSP is a very powerful feature of DataCAD. If you have ever used AutoCAD's Viewports, then you will find similarities with DataCAD's ClipCubes.

Part II: Advanced Multi-Scale Plotting Concepts

The next section of this chapter deals with a more advanced use of MSP. In doing so, we will also learn to use 3D CCs to allow us to place and view portions of details, thereby avoiding the need to copy various parts of drawings in order to show them at a larger scale.

We'll use another sample drawing from the CD for this section. Copy the **WALLSECT.DC5** file from the Chapter 12 directory to your computer's hard drive. Open up the drawing in DataCAD. This file contains a 30″ × 42″ [762 × 1066] drawing sheet with a title block and a wall section from a three-story wood framed building. In this section, we will place this detail on our drawing sheet at a $^3/_4$″ [1:20] scale using MSP. Then we will use these very same drawing entities to display three enlarged details with a new set of dimensions and notes at an enlarged scale of $1^1/_2$″ [1:10].

When you first open up the file, you will be presented with a 30″ × 42″ [762 × 1066] drawing sheet and a three-story wall section with notes and dimensions. It is again important to note that the title block truly measures 42 inches wide by 30 inches high (see Figure 12-13).

Let's first get our drawing sheet set up:

1. First, you must set up your paper size (refer to Chapter 7, "Printing and Plotting," for help in selecting the correct printer drivers and settings):
   a. Go to **Plotter/Setup**.
   b. Pick a printer drive that supports a 30″ × 42″ [762 × 1066] paper size.

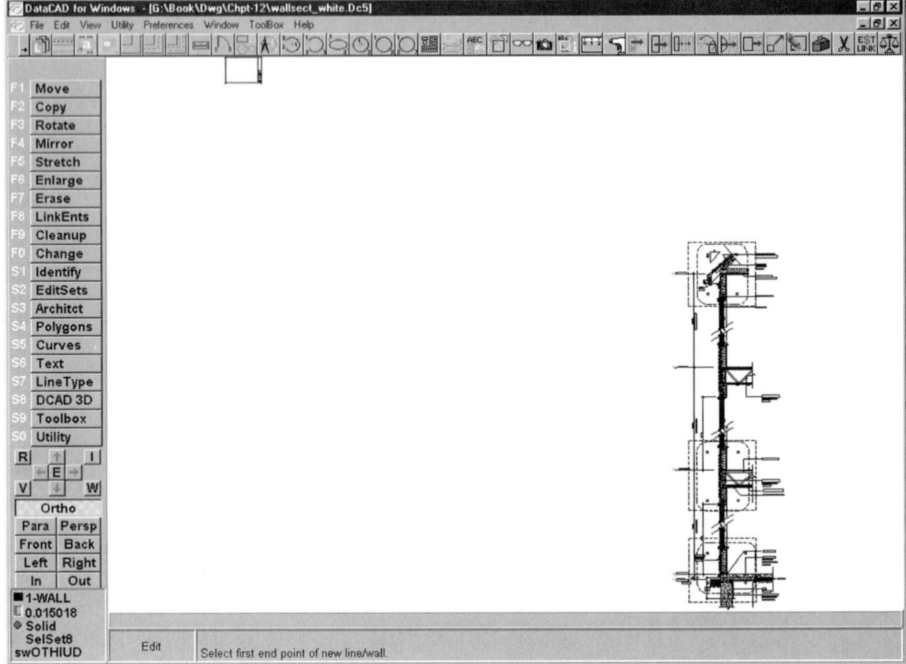

Figure 12-13 What you see when you first open the drawing

c. A great many printer drivers will try to leave a border around the paper, refusing to let you have the full paper area. If you are having this sort of trouble, then try using the **Properties** button to create a custom-sized paper in your driver, or try the custom paper size options in DataCAD's own **Setup** menu.

d. Select **OK** to exit the Setup dialog.

2. Press **Ll** to go to the **Layers** menu.
3. Select **ActvOnly/TTLBLK**, *right-click* twice, and then **R**eCalc the screen. The drawing sheet will fill the DataCAD Drawing Window.
4. Lay out this TTLBLK in MSP:

a. Select **Sheet/Rename/Sheet1** and enter **A-10.** *Right-click* twice, back to the **Plotter** menu.

b. Select **MltLyout/Layout**. Press **PgDn** until the scale is shown as a 12″ [1:1] (full) scale.

c. Locate the TTLBLK detail in the MSP window, taking care to center it from edge to edge. *Click* to place the detail and then name it **TTLBLK**.

d. *Right-click* back to the main DataCAD Drawing Window.

Now to work on the wall section detail.

5. Press **Ll** to go to the **Layers** menu.
6. Select **On/Off** and then turn on all the layers except **CC** and **TTLBLK** (**1-WALL, 1-HATCH, 1-NOTES, 1-DIMS**, and **1-WINDOR**).
7. *Right-click* and then press **R**eCalc to center the wall section in the Drawing Window. Check out the drawing yourself to see what's there. WindowIn to various parts of the detail. When you are done, press **R**eCalc again (see Figure 12-14).
8. Lay out this wall section detail in MSP (see Figure 12-14):
   a. Select **Plotter/MltLyout/Layout**. Press **PgUp** until the scale is shown as $^3/_4''$ [1:20].
   b. Locate the wall section detail in the MSP window near the right side of the drawing sheet. *Click* to place the detail and then name it **WALLSECT**.
   c. *Right-click* back to the DataCAD Drawing Window.

   Let's make a 3D GTV of this view so we can quickly get back to it later. This is not a requirement, just a convenience.
9. Select **Plotter/3DViews/GotoView/AddView**. Name it **1-2SECT**.
10. *Right-click* back to the **Utility** menu.

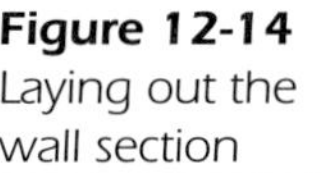

Figure 12-14
Laying out the wall section

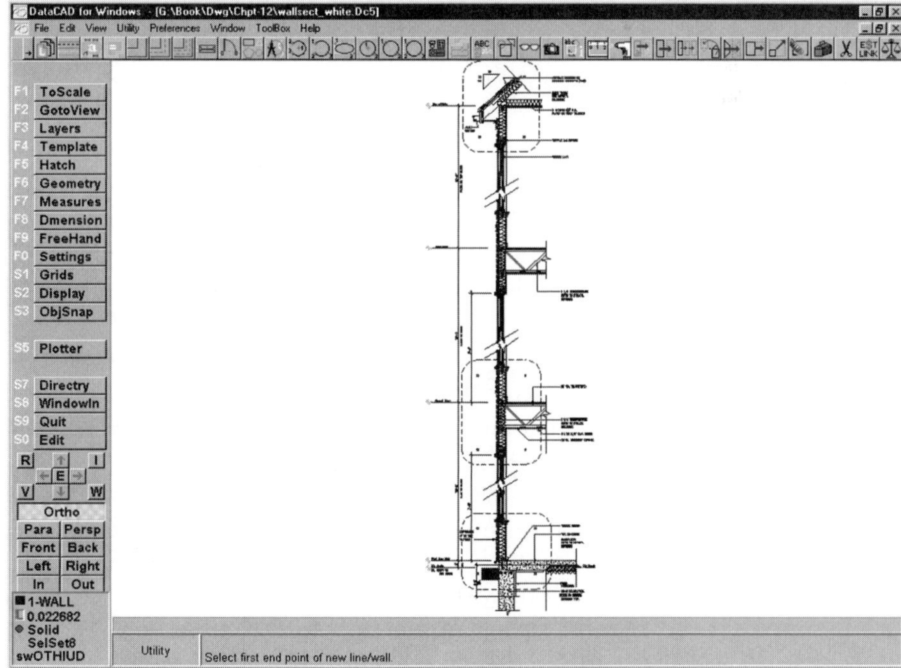

Notice the three sets of dashed green detail bubbles. These are the three details that we want to display on the drawing sheet at a $1^1/_2''$ [1:10] scale. One method to do this would be to copy each of these three areas, tidy up the copies, and then place them on the drawing sheet in the MSP window. The disadvantage of this method is that each time you make changes to the $^3/_4''$ [1:20] details, you have to make the same changes to the $1^1/_2''$ [1:10] scale copies, and vice-versa. This method is inefficient and inevitably leads to errors in coordination between the variously scaled details. It is no better than drawing all those details with a pencil. With CAD, there is a better way.

It makes more sense to avoid the process of copying the wall section entities by simply using those same entities, but at a larger scale. But how do we "localize" and display only selected portions of the drawing? The answer is DataCAD's 3D CCs. These act as masks by displaying only what is inside of them, both 2D and 3D. Once this is done, each clip-cubed detail can be placed at any scale in MSP. Each of these details can also be saved as 3D GTVs, since 3D GTVs will also save the CCs.

Using Clip Cubes (CCs)

By using the 3D CC function, you can further extend the usability and time-saving possibilities of DataCAD and MSP. It takes a little getting used to, a little more layer management, and a little more forethought, but once you get the hang of it, it's pretty slick and can save a lot of redrawing time and coordination mistakes. If you have ever used AutoCAD's Viewports, then you will recognize the same concepts in CCs.

3D CCs enable you to define a specific viewing area by "clipping" everything outside the cube from your view. All the entities outside the cube are still there; they are just obscured from view. MSP can then use the defined CCs as details, showing only those entities visible inside the cube. The outermost X/Y extents of the detail as displayed in MSP are the outermost X/Y extents of the CC.

By using CCs, you can take your $^1/_8''$ [1:100] floor plan, make a CC of the stair, and show an enlarged $^1/_2''$ [1:20] scale detail of that stair plan without having to copy the detail for use at a larger scale, and you won't have to go back and make changes to two separate details. Because the CC is just another view of something you have already drawn, any changes made to your $^1/_8''$ [1:100] plan are immediately shown in the $^1/_2''$ [1:20] scale CC detail of the stair. Read this paragraph again and consider the possibilities for saving time and enhancing coordination!

The reason that these clipping boxes are called 3D cubes is because of DataCAD's capability to create $2^1/_2$ D extrusions and full 3D objects. If the CC were only a flat, 2D box, many entities would be cut off because they are outside of the 2D drawing plane. They are above it unless you squashed all your entities to a height of zero. In this chapter, we are dealing with 2D construction drawings, but CCs can and should be used for 3D work as well. More on that later.

A ClipCube is an invisible box. Because the edges of a CC cannot be seen while you are working on your drawing, it is very helpful to draw some guidelines to define the X/Y extents of your CC. In this wall section detail, we have already drawn some for you. (These guidelines should become unnecessary, perhaps by the time you read this. In a future update it appears that CC's will automatically show their extents with non-printing dashed lines.) To see them, follow these steps:

1. Press **Ll** for the **Layers** menu.
2. Turn on the **CC** layer. A set of dashed white boxes will appear around the areas of our three proposed details.
3. While still in the **Layers** menu, turn off the **1-NOTES** and **1-DIMS** layers. *Right-click* twice, back to the **Utility** menu.
4. WindowIn to the lowest detail, down at the foundation (see Figure 12-15).
5. To get to the ClipCube menu, select **Plotter/3DViews/ClipCube**. Set the **Z-Max** to a height that will ensure all your entities are in the cube

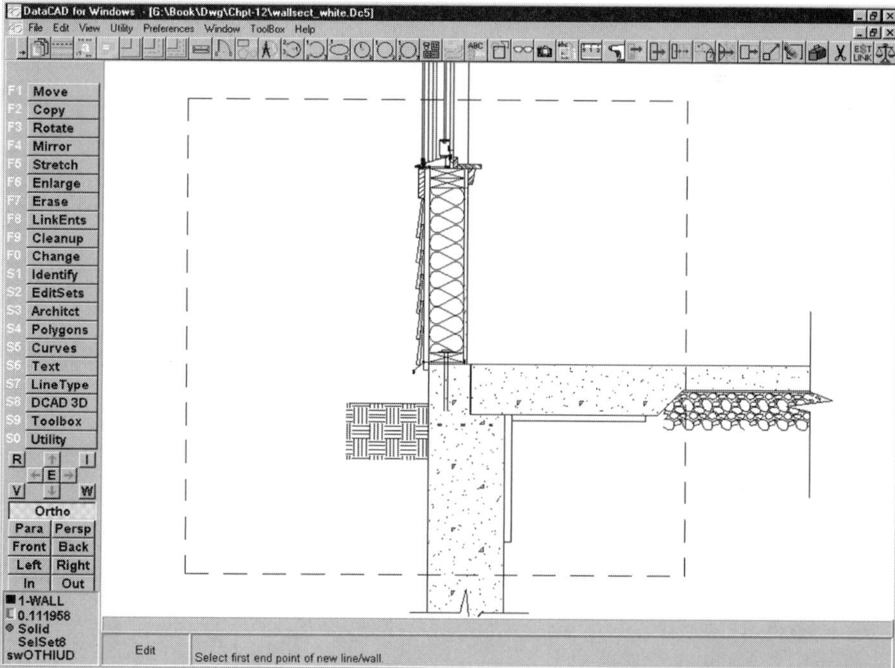

Figure 12-15 The dashed lines of the CC layer

(like 10 or 20 feet). Make sure your Z-Min also includes all your entities. We will use a **Z-Max** of 10′ high.

Select the **Z-Max** button in the menu, type in **10**, and then press **Enter**.

NOTE: *Remember that entities in DataCAD default to a base elevation (**Z-Base**) of 0′-0″ and a height (**Z-Hgt**) of 8′-0″, although you can change this at any time by selecting **Edit/Change/Z-Base** or **Z-Hgt**. When selecting the Z-Base (called **Z-Min**) or Z-Height (called **Z-Max**) in the ClipCube menu, you must make sure that your drawing entities are within the CC (although, like the English language, for every rule, there is a set of exceptions, as we will see).*

*The entities in our current wall section example were all drawn with both their Z-Bases and Z-Heights at 0′-0″, so any CC Z-Height (**Z-Max**) from 0′-0″ and higher is OK. In Figure 12-16, you can see two graphic depictions of CCs at different heights, both of which will work. In Figure 12-17, one of the CCs has been turned on. Notice that everything within the CC is displayed, but everything outside is clipped from view.*

6. While still in the **ClipCube** menu, pick **NewCube**. Define a new CC box by clicking and dragging a bounding box just outside the dashed guidelines.
7. *Right-click* to accept the new CC and then select the **ClipOn** button to turn on the CC. You will see all the entities outside the CC disappear

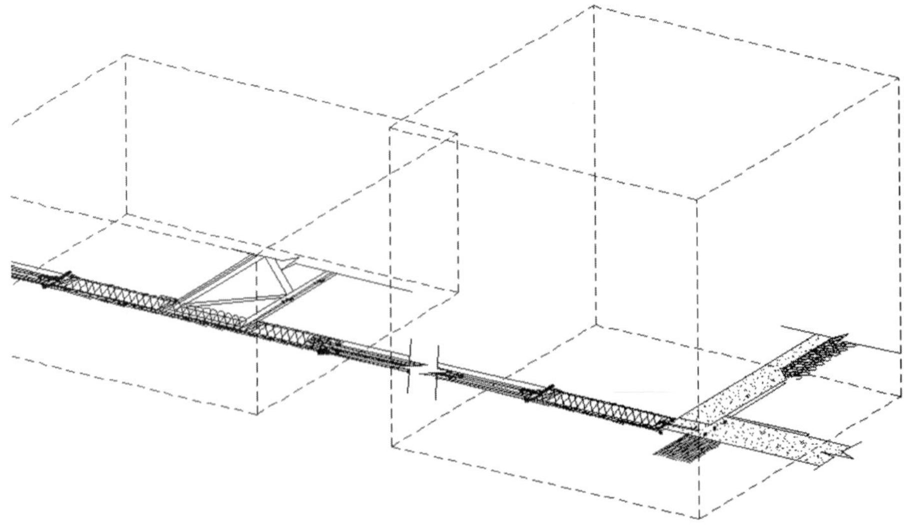

Figure 12-16
Two CCs at different heights

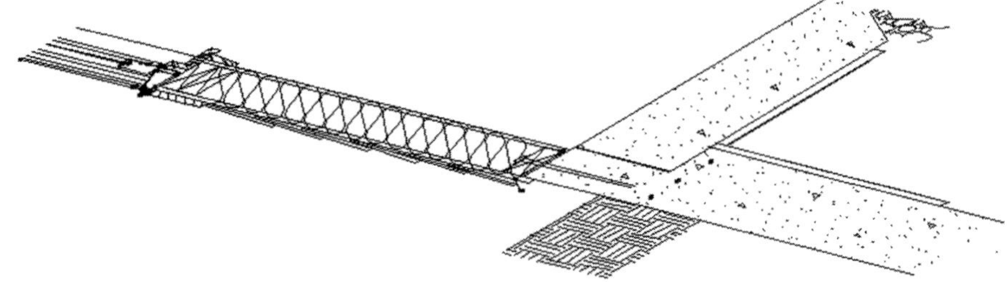

Figure 12-17
Turning a CC on clips the view of entities outside

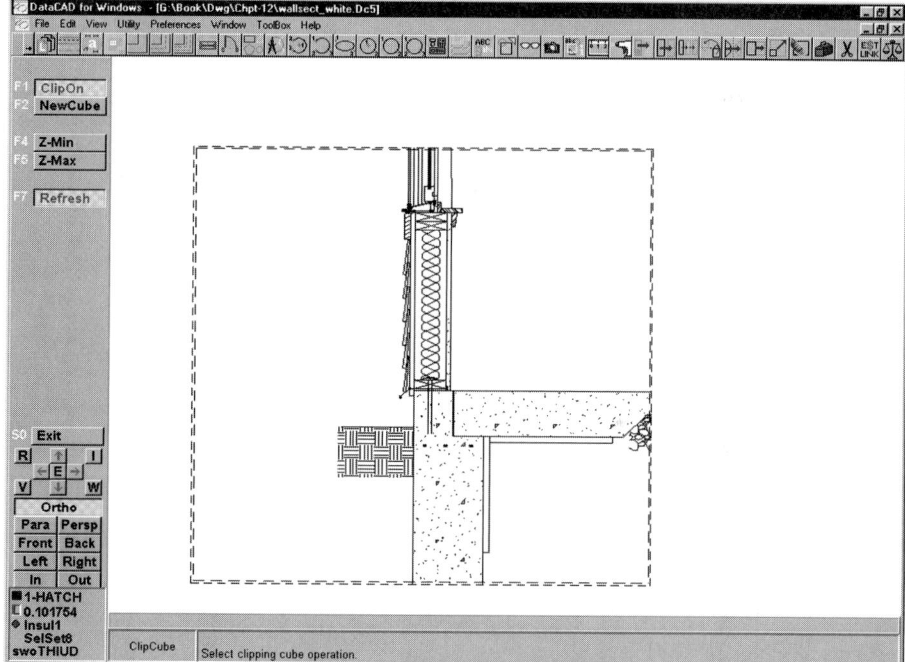

Figure 12-18
Viewing the CC with **ClipOn** selected

from view. They are still there, but not visable, and they will not be part of the MSP detail that we are going to create (see Figure 12-18).

8. Notice that two dashed sets of lines form around the detail now. One set is, of course, the guidelines that we drew on the **CC** layer. The other is a temporary set of dashed lines that show the extents of the current CC. They will disappear from view when you exit the **ClipCube** menu. *Right-click* to exit the CC menu now and then press the **;** key to get back to the 2D menus.

9. Before we add this detail to our drawing sheet, let's make a 3D GTV of it. Remember that this is not required for MSP, but doing so makes it easy to later return to this detail for editing. Select **Utility/GotoView/AddView**. Type in **DTL-2** for a name and then *right-click* back to the 2D menus.

NOTE: It is important to note that CCs are saved along with the entities in 3D GotoViews. This is a great feature and is another good reason to save your details via 3D GotoViews.

10. You don't want the dashed guidelines on the **CC** layer to be part of your detail in MSP, so go to the **Layers** menu and turn off the **CC** layer. You can see that without the dashed guidelines, it is difficult to see exactly where the outside edges of the CC are.
11. You do not need to do a **R**eCalc in order for DataCAD to find the outside extents of the CC. DataCAD is smart enough to know that the extents of your CC are what should be placed in the MSP window. Doing a **R**eCalc won't hurt anything, but it won't help either. Let's add this detail to the MSP drawing sheet we created earlier.
12. Go to the MSP window (**Utility/Plotter/MltLyout**) and pick **Layout**. You will see a bounding box that defines the extents of your CC. Use **PgUp** or **PgDn** to change the scale of the new detail to 1$^1/_2$″ [1:10]. Place the detail at the bottom of the page, as in Figure 12-19, and name it **DTL-2**.
13. Now you have an enlarged detail with no extra drawing involved, and that will be updated at both scales every time you change it. What? You say you want proof? OK, then try this:
    a. *Right-click* back to the Drawing Window. Draw a couple of diagonal lines across the clipped detail. If your linetype is set to Dashed, you might want to change it to Solid first.
    b. Now go back to the **MltLyout** menu and select **Sheet** to view your drawing sheet.
    c. Notice that the lines you drew show up in both the $^3/_4$″ [1:20] wall section and the 1$^1/_2$″ [1:10] detail. Pretty cool, huh?
    d. Go back to the Drawing Window and delete the extra lines you drew.

To add the next detail to the MSP sheet, you need to create another CC around the next area of the wall section:

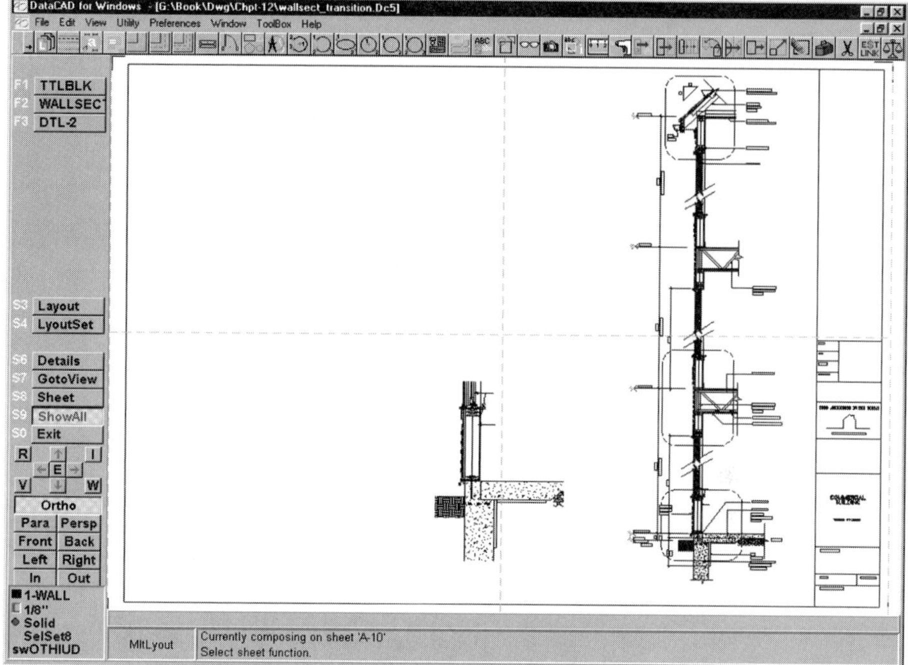

Figure 12-19
Adding an enlarged detail to the bottom of the page

1. Go to the **Layers** menu and turn the **CC** layer back on.
2. Now go to the **ClipCube** menu (**Plotter/3Dviews/ClipCube**) and turn the current CC off by pressing the **ClipOn** button so that it's not highlighted anymore.
3. In the Drawing Window, move up to the next detail area, the one at the floor joist.
4. In the **ClipCube** menu, select **NewCube**.
5. Draw a bounding box just outside the dashed guidelines. *Right-click* to accept the new CC and then select ClipOn to turn it on.
6. Save this detail as a new 3D GTV (**Utility/GotoView/AddView**). Call it **DTL-3**.
7. Do not turn off the **CC** layer this time. Leave it turned on.
8. Go to MSP (**Utility/Plotter/MltLyout**) and pick **Layout**. Add the new detail above the first one (see Figure 12-20).
9. This brings up one of MSP's shortcomings. It is difficult to get two related details to align perfectly in the MSP window. You can do a few things to help yourself. The first method is the one you should use now:

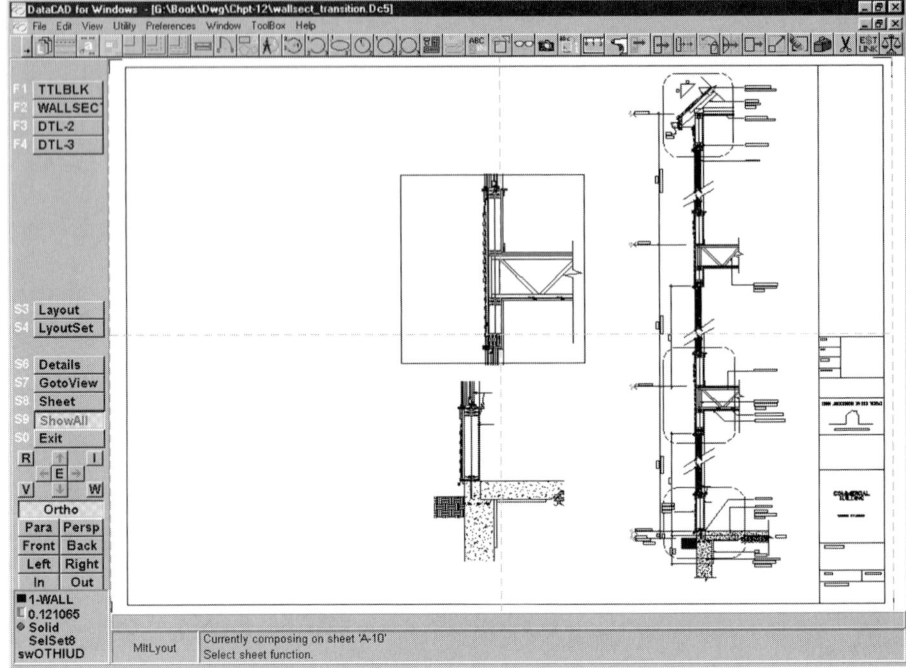

Figure 12-20 Adding another enlarged detail

a. In the **Layout** menu, select **Details/LyrTogl/DTL-2** and select the **CC** layer to turn it on. *Right-click* back to the **Layout** menu. Select **DTL-3**. Align the detail bounding box with the detail below it. Click to place it. This, of course, only works if you draw all your CC guidelines at the same width ahead of time.

b. As mentioned earlier in the chapter, you can also set the snap grid to something like 4″. The details will then snap around the MSP window at that grid size.

c. In the **Utility/Settings** menu, you can select **DrwMarks**. When you go back to the MSP window, you will find that details are located by the centerpoint of the detail. The mark at the center point of each detail remains on the screen even when you pick a detail to move it. This can help you to place the detail by allowing you to see where the detail needs to be moved in relation to the last mark or centerpoint.

10. For the final detail of the roof, make a new CC and 3D GTV, name it **DTL-4**, and place it in the MSP window. It should look something like Figure 12-21.

11. Lastly, let's turn off the dashed **CC** layers from each of the three enlarged details in the MSP window. From the MSP **MltLyout/Layout**

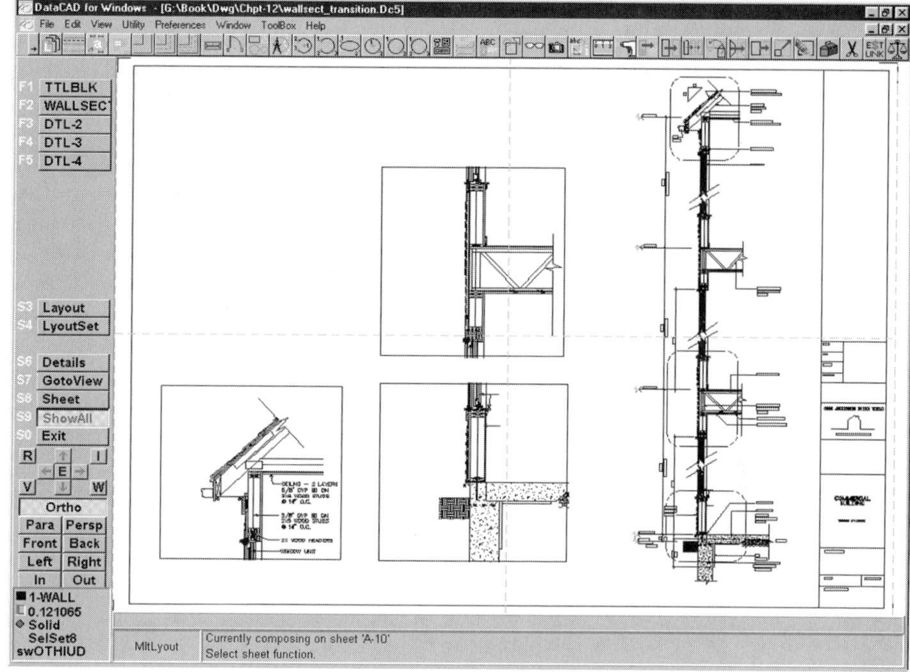

Figure 12-21
Adding the third enlarged CC detail

menu, select **Details/LyrTogl/DTL-2/CC** and then *right-click*. Do the same for DTL-3 and DTL-4. Now your MSP sheet will look like Figure 12-22.

Adding Notes to Enlarged Details

Before laying out the 1-$^1/_2$″ [1:10] details, we turned the text layers (**1-NOTES** and **1-DIMS**) off. The 1-$^1/_2$″ [1:10] details could potentially use the same notes from the $^3/_4$″ [1:20] details, but that would not be very useful for the following reasons:

- The scale of the text would be twice as large. If the text is $^1/_8$″ [3.2] of an inch high in the $^3/_4$″ [1:20] scale detail, that same text would be $^1/_a$″ [6.4] high in the 1-$^1/_2$″ [1:10] details.
- The purpose of larger scale details is to convey a greater level of detail, both in materials and notes. Therefore, your 1-$^1/_2$″ [1:10] details should have more notes than could be reasonably described in the $^3/_4$″ [1:20] detail.

Drawing." To change the current plot scale, select **Plotter/Scale** and pick $^3/_4$" [1:20].

5. Go back to the **Text** menu. Press **Size**, type in **..1/8** , and then press Enter. This will set the text size as $^1/_8$" [3.2] tall. The text cursor will be active now. Select **FontName** and chose **Romans.chr**.
6. Place the text cursor within the dashed guidelines and add some notes to the right side of the detail. Some sample notes to add might include the following:
   - $^5/_8$" GYP BD on 2 × 6 wood studs @ 16" O.C.
   - Window unit
   - 2x wood headers
   - Ceiling: 2 layers $^5/_8$" GYP BD on wood roof trusses
7. Add leaders with arrowheads to point from the notes to the materials:
   a. From the **Text** menu, select **Arrows**. The arrow settings should be something like these (or whatever suits you):
      - **Size** = 0.5
      - **Style** = Open
      - **Aspect** = 3.0
      - **Weight** = 1
      - **Color** = Green

   b. Start the leader just to the left of the text by placing the cursor there and *clicking*. Drag the cursor to the material to point to and *click* again. *Right-click* to end the leader and automatically place the arrowhead. Continue to do this for all the notes. When finished, it should look something like Figure 12-23.

   Now go to the MSP **MltLyout** menu where we have to update the detail to include the two new layers. It would be more efficient to do this later when all three details have had their notes and dimensions added, but let's do this first detail now so we can see the results.
8. Select **Plotter/MltLyout/Sheet** and then *right-click* back to the **MltLyout** menu.
9. Select **Details/LyrTogl/DTL-4** and select both **1-2NOTES** and **1-2DIMS**. You will see your notes appear in the roof detail (see Figure 12-24).
10. *Right-click* back to the DataCAD Drawing Window. Add some notes to the other two details just as you did for DTL-4.
11. Go to the 3D **GotoView** menu and select **DTL-3** to switch to that detail. *Right-click* back to the 2D menus.

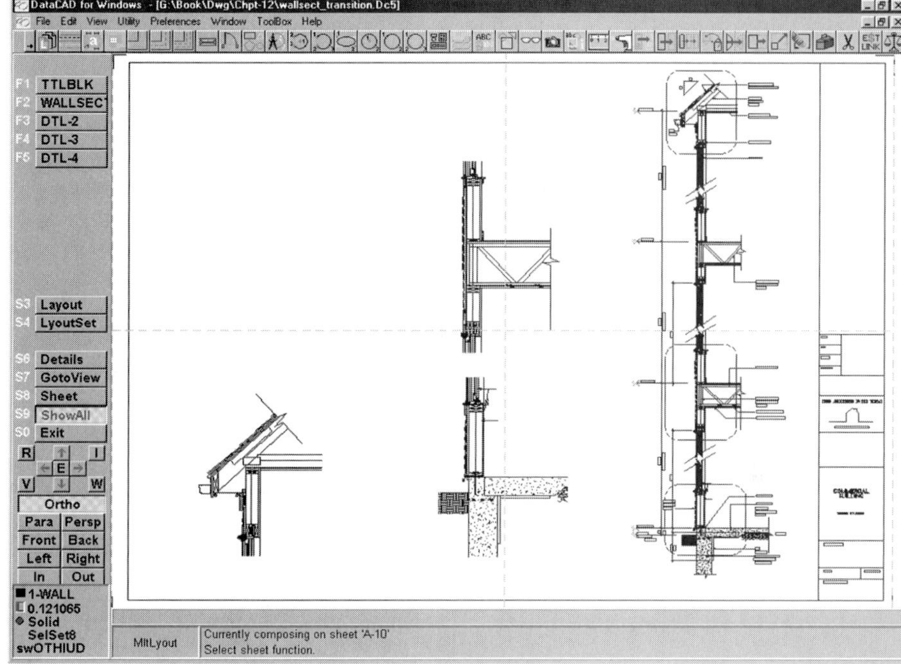

Figure 12-22 Turning off the dashed CC layers

For these reasons, we therefore need to create a new set of layers for the notes and dimensions of the $^3/_4''$ [1:20] details to be displayed:

1. Go to the **Layers** menu (**Ll**) and create two new layers called **1-2NOTES** and **1-2DIMS**.
2. Assuming the roof detail (DTL-4) is the current detail in your Drawing Window, make sure the two new layers are on. Then update the 3D GTV of DTL-4.
   a. Select **Plotter/3DViews/GotoView/Update/DTL-4**. This will add the two new layers to the 3D GTV for easy editing later (for more information, refer to Chapter 10).
   b. *Right-click* back to the 2D menus.
3. Press **Tab** until the active layer is **1-2NOTES**. Go to the **Text** menu and make sure **TxtScale** is on.
4. Set your current plot scale to $^3/_4''$ [1:20]. This is very important; otherwise, your text will not be sized properly for this scale detail. If you need to, review the Text Scale section of Chapter 5, "Basic

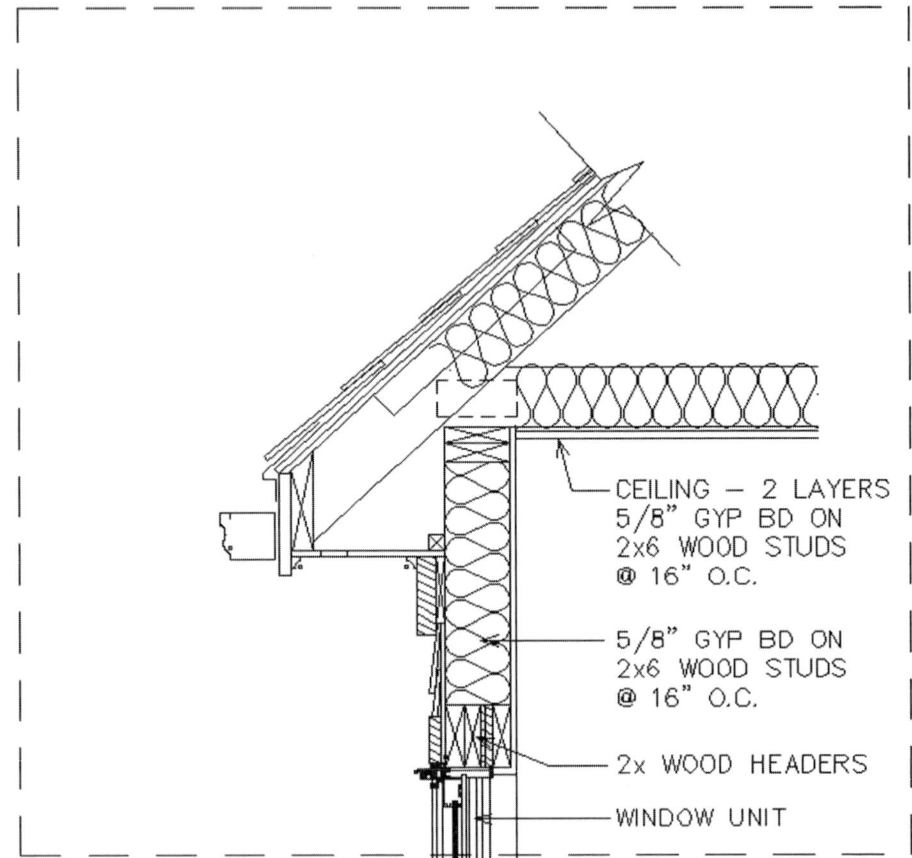

Figure 12-23 Adding notes and arrows

12. Go to the **Layers** menu (**Ll**) and turn on layers **1-2NOTES** and **1-2DIMS**. Return to the 3D **GotoView** menu and select **Update/DTL-3** to add the new layers to this 3D GTV. Now go back to the 2D menu.
13. Press **Tab** until the layer **1-2NOTES** is the active layer.
14. Go to the **Text** menu and add some notes and leaders to the right side of the detail. Perhaps you might add the following:
    - Window unit
    - PTD wood sill
    - $^5/_8$″ GYP BD on 2 × 6 wd studs @ 16″ O.C.
    - PTD wood base
    - $^1/_2$″ plywd floor'g on $^3/_4$″ sheathing
    - $^5/_8$″ GYP BD on resilient chnls on wood joists

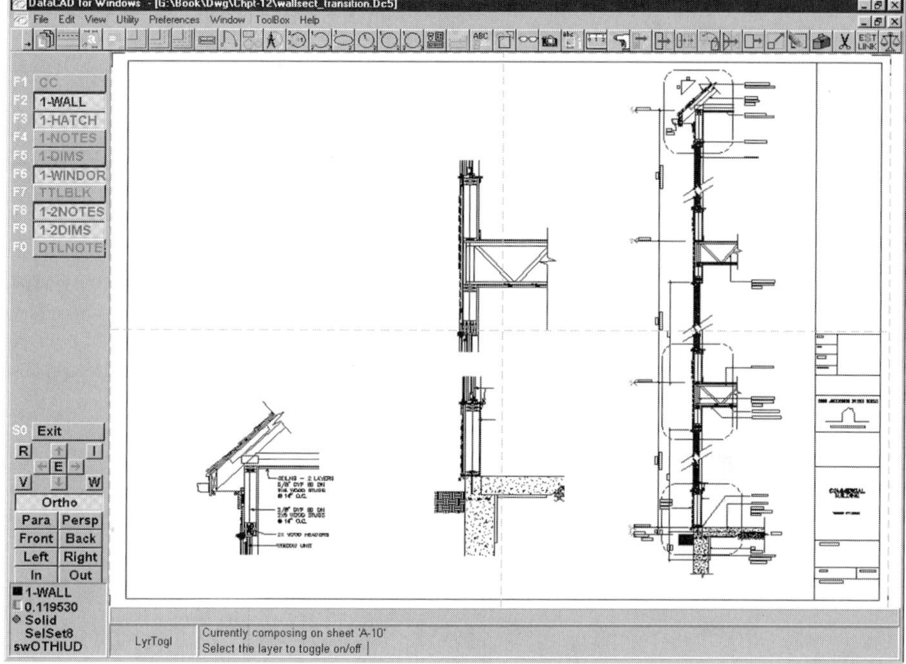

Figure 12-24 Updating the detail to include the notes layers (see bottom left corner)

- 2x wood header
- Window unit

15. Go to the MSP **MltLyout** window and use the **LyrTogl** option to add the two new layers to **DTL-3**. Your notes will appear in the MSP detail.

Repeat the process again for DTL-2. Make sure you update the 3D GTV and be sure to add the new text to the correct layer.

Spanning Dimensions between Enlarged Details

Adding dimensions to the three 1-$^1/_2$″ [1:10] details is, for the most part, just as straightforward as adding text. However, we need to get around the fact that in some cases you may want to run dimension strings that span between two enlarged details. Because the CCs will keep parts of the dimension lines from being displayed, you need to keep this in mind when placing dimensions and dimension text.

Let's add a dimension string that defines the window opening between **DTL-2** and **DTL-3** (see Figure 12-25):

1. Go to the 3D **GotoView** menu and turn on **DTL-2**.
2. Go to the **ClipCube** menu and deselect **ClipOn**. Use **PgUp** and **WindowIn** to get the window centered in the Drawing Window.
3. Keep the dashed guidelines on the **CC** layer turned on.
4. Press **Tab** until **1-2DIMS** is the active layer.
5. Go to the Dimension menu (**Utility/Dmension/Linear/Vertical**).

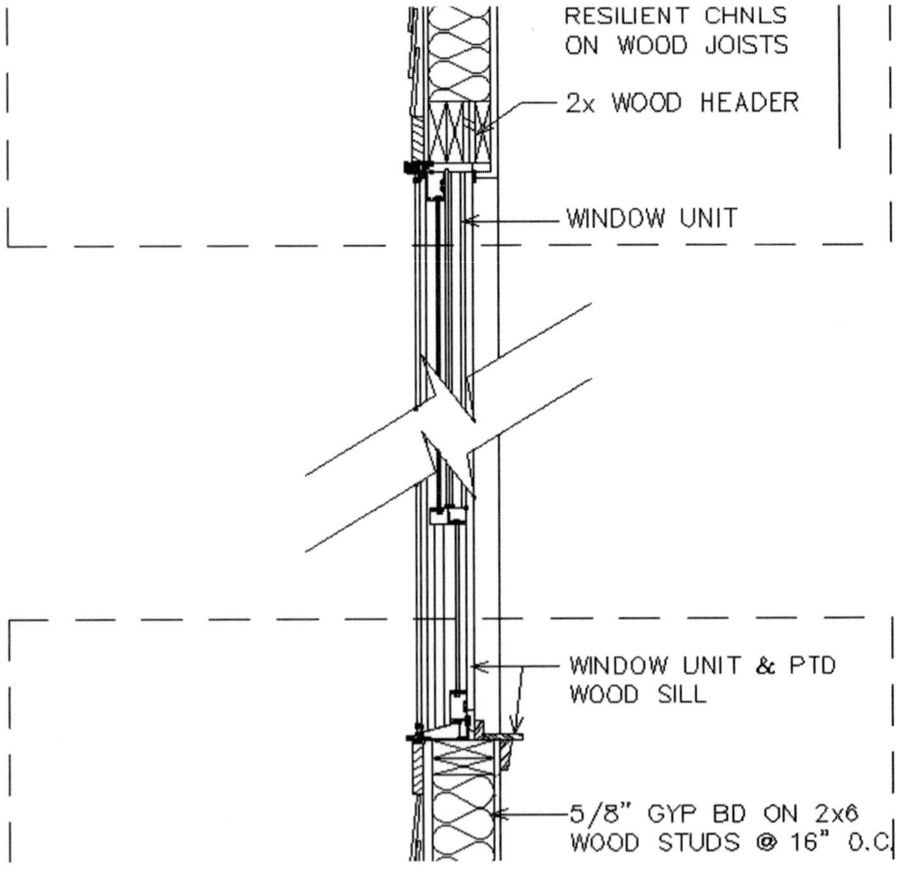

Figure 12-25 Viewing the window to be dimensioned, with the bottom and top of two clip cube areas displayed

a. Turn **Assoc** off.
b. In the **TextStyl** menu, make sure **TxtScale** and **Above** are on, **Auto** is off, and set **TxtSize** to $^1/_8$″ [3.2].
c. In the **DimStyl** menu, make sure **Offset**, **Overlap**, and **Overrun** are set to $^1/_8$″ [3.2].

6. Add a dimension string to define the rough opening of the window (from the 2x wood header to the 2x wood sill). If you need help on adding dimensions, see Chapter 11, "Basic Construction Drawings."
   a. The trick here is the placement of the dimension text. This is where the dashed guidelines come in handy.
   b. After selecting the two points between which you want to dimension, drag the dimension line to the left and select where you want to place it. Enter a dimension text of **4′-10″** [1473].
   c. Place the text within the dashed lines around the lower detail, making sure that the whole dimension text is within the guidelines (see Figure 12-26).
7. To see how this displays in the MSP window, go there now. Although this may not be the ideal way to display a dimension line, it serves the purpose (see Figure 12-27).

Adding Detail Numbers and Text to Enlarged Details

The combination of CCs and MSP makes it a little more difficult to place text, such as detail numbers and titles, in the MSP window. How do you label a detail when nothing will display outside of the CC? There are at least two methods that I have used. The first is to keep the text as part of each detail, and the second is to place each detail number by using a CC, placing it in the MSP window like any other detail. We'll try both.

Method #1:

1. Go to the **Plotter** menu and make sure the scale is set to **1-$^1/_2$″** [1:10].
2. Activate the 3D GTV **DTL-2** from the 3D **GotoView** menu.
3. Press the **Tab** key until **1-2NOTES** is the active layer.
4. Create a circle at the lower-left corner of the detail within the dashed guidelines. Go to **Edit/Curves/DiaCirc** and make the circle 7″ [178] in diameter.

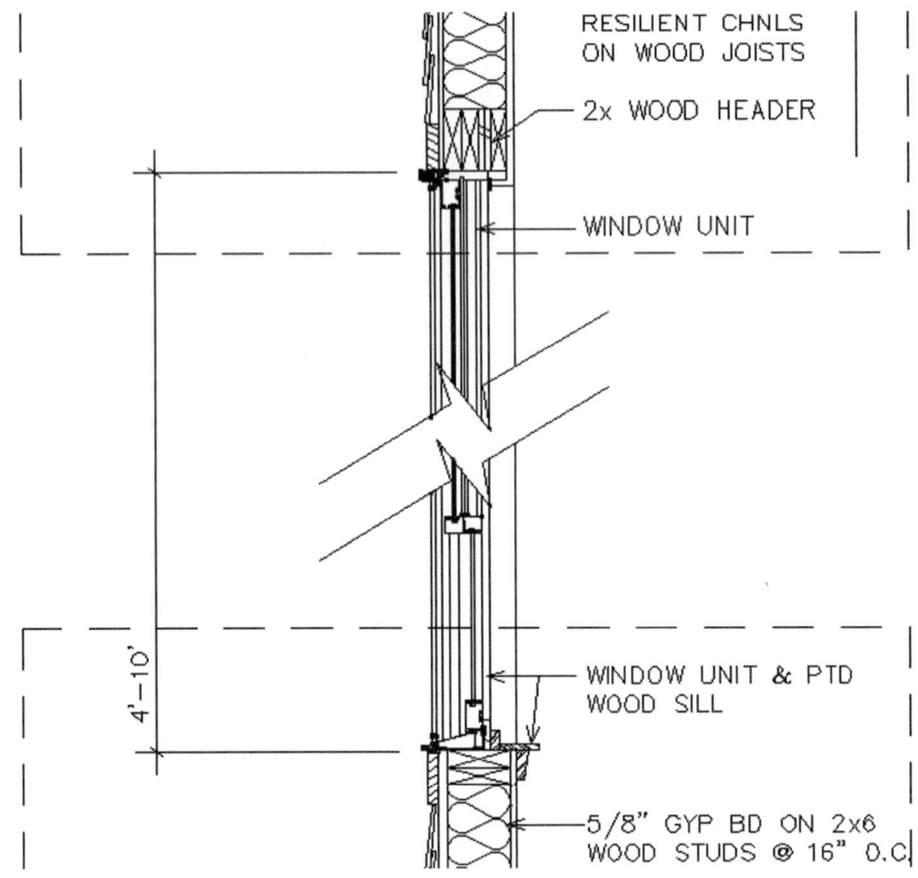

Figure 12-26 Placing the dimension text within the dashed lines

5. Draw a horizontal line from the right side of the circle, extending to the foundation wall.
6. Now go to the **Text** menu and change the **Size** to $^3/_8$ " [9.5].
7. Above the line you just drew, add the text "**FNDN DETAIL.**"
8. Change the text size to a $^1/_4$" [6.4] and add the text "**1-$^1/_2$"=1′-0"**" under the line.
9. Change the text size once again to a $^1/_2$" [12.7] and add the number **2** inside the circle. The completed note should look like Figure 12-28.
10. Go to the MSP window and look at your handiwork.

The drawback to this method is that you have to leave enough room to the right or left of your detail to add the note. If you don't have room, you're out of luck, since the use of CCs prevents you from adding the note below

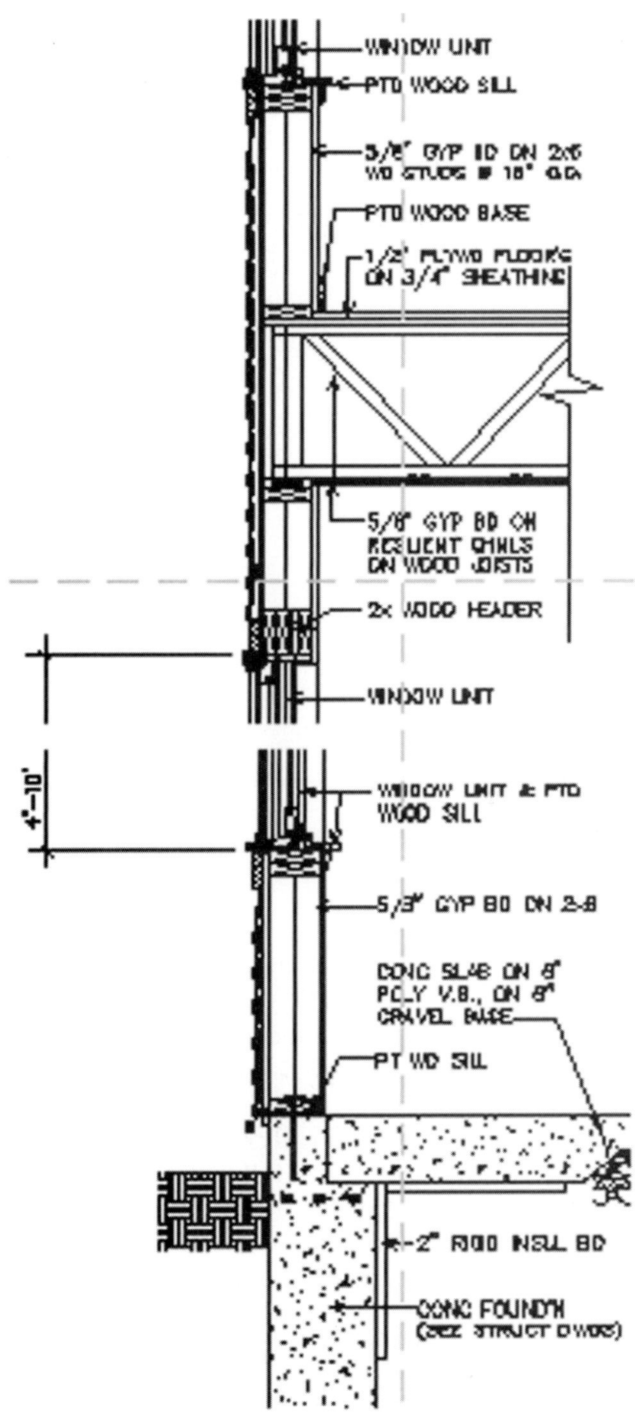

Figure 12-27
The dimension in the MSP window

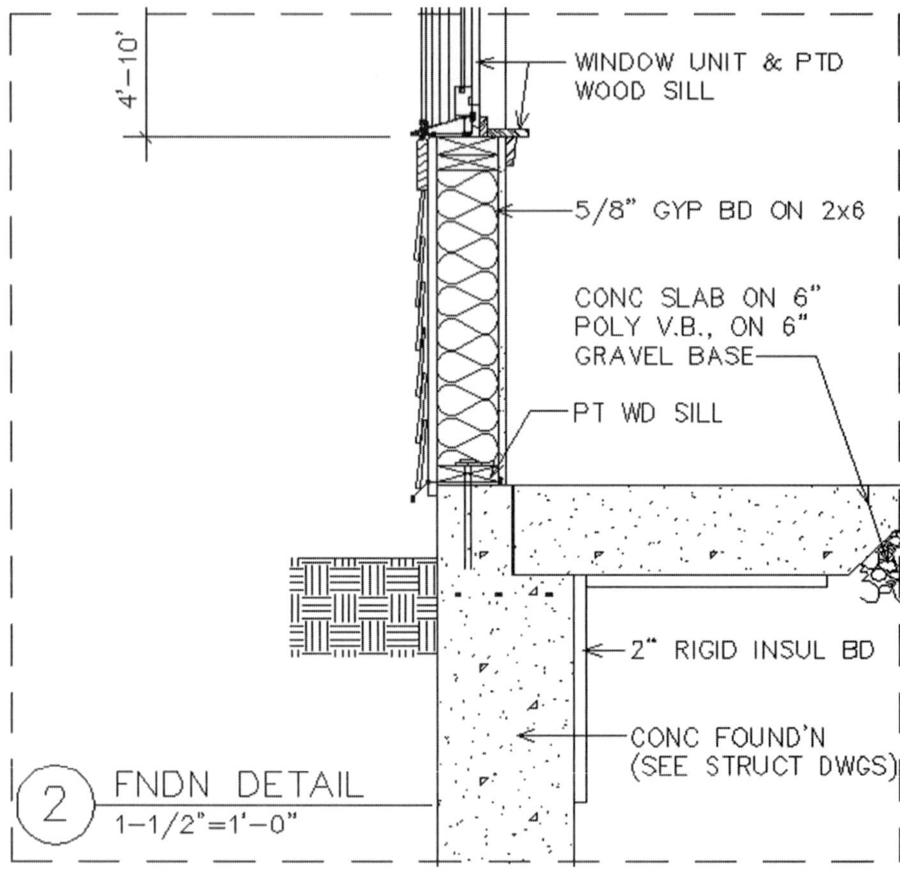

Figure 12-28 The completed note

the detail. You can make this method very workable if you leave enough room to do so and if you don't mind the detail label being located next to the detail instead of under it.

The alternative is to use a CC for each detail label. You will be better off creating a new layer for these notes. Since these details will be created and placed independently of the details they will describe, we can make them at a scale of 12″ [1:1] (full scale). This is a benefit, since the circle and line you will draw can be drawn with "real" dimensions. In the previous exercise, you had to draw a circle of 7″ [178] to approximate the size you wanted on your drawing. In this example, you will not have to do that.

Method #2:

1. Go to the **Plotter** menu and make sure the current scale is 12″.
2. Go to the **Layers** menu and add a new layer. Call it **DTLNOTES**.

3. Now pick **AllOn** to turn all the layers on. This is simply so that you can see where there is some open real estate in the Drawing Window to draw the new labels.
4. Turn off the current CC in the **ClipCube** menu.
5. Do a **R**eCalc of your drawing. Pick an area away from your detail entities, such as the area near your Titleblock. Then **WindowIn** to that area.
6. Go to the **Layers** menu and make the **DTLNOTES** layer the only layer that is on (**ActvOnly/DTLNOTES**).
7. Now go to the **Edit/Curves** menu and make a $^5/_8$″ [16] diameter circle. Draw a line 5″ [127] long, extending from the right of the circle.
8. Go to the **Text** menu and make the text size $^3/_8$″ [9.5]. Now add the number **3** inside the circle.
9. Change the text size to a $^1/_4$″ [6.4] and add the text "**WALL DETAIL AT SECOND FLOOR**" above the line you drew.
10. Change the text size to $^3/_{16}$″ [4.8] and add the text "**1-$^1/_2$″=1′-0″**" below the line.
11. Make another detail note just as you did previously. This time the text will read **4** , **ROOF DETAIL AT EAVE,** "**1-$^1/_2$″=1′-0″**" (see Figure 12-29).
12. Go to the **ClipCube** menu and select **NewCube**. The **Z-Min** and **Z-Max** settings are not important here, so you don't need to reset them.
13. Drag a bounding box around the whole detail label number **3**.
14. Select **ClipOn** to turn it on. Only note number 3 will be visible now (if the text is not visible, then you may need to make the CC a little bigger).
15. Make a 3D GTV of this detail for easy editing. Call it **NOTE-3**.

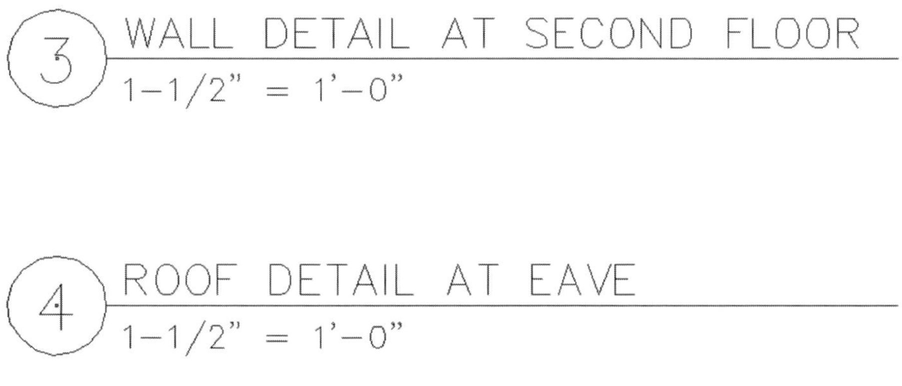

Figure 12-29
Creating two more detail notes

16. Go back to the **ClipCube** menu and select **NewCube**. Make a similar bounding box around note number **4**. If the **ClipOn** button is not already on, select it to turn it on.
17. Make a 3D GTV of this detail as well. Call it **NOTE-4**. Now let's add these two detail notes to our MSP drawing sheet.
18. Go to the MSP **MltLyout** window, select **Sheet**, and then *right-click*. If you need to make some room under the second floor detail, select **DTL-3** and move it up.
19. In the **MltLyout** menu select **GotoView** then select **NOTE-3**. If the note doesn't look like the correct size, use **PgUp/PgDn** to make sure your scale is set to **12″** [1:1].
20. Locate the detail note under the second floor detail and click to place it. Name it **NOTE-3**.
21. Locate the next 3D GTV detail (**NOTE-4**) below the roof eave detail. Name it **NOTE-4**.
22. If you don't like where you've placed any of these notes, you can move them (see Figure 12-30). That's one of the benefits of this system.

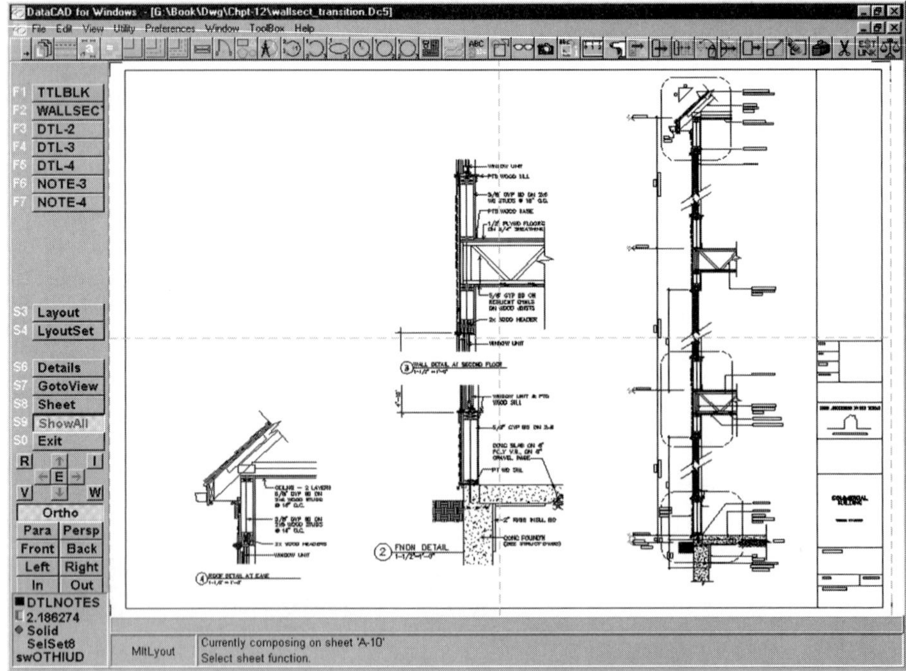

Figure 12-30
The completed MSP sheet

If you have access to a plotter you might want to print out this sheet to take a look at it all. Or you could use the Check Plot feature (Chapter 7) to plot a mini-version of the sheet.

Part III: Printing and Plotting

In this section, we'll examine the processes involved in printing and plotting the plans we've created.

Hatching

In the previous exercise, we created two layers, 1-2NOTES and 1-2DIMS, so that you could place notes and dimensions that would only be displayed in the 1-$^1/_2$″ [1:10] scale details. In our office, we usually also create an additional layer for each detail for larger scale hatching as well. Like notes and dimensions, a different level of hatching and poché is needed for large-scale drawings. Too much hatching in the smaller scale drawings will cause the details to bleed together and look muddy.

Pen Settings

Keep in mind that pen width settings (**Plotter/PenTable**) will appear differently at different scales. If you plan your pen colors to plot wide, dark lines for your 1-$^1/_2$″ [1:10] scale enlarged details, those same lines may appear too thick when plotting the $^3/_4$″ [1:20] details. If you plan your pen colors to plot thinner and more legibly for the $^3/_4$″ [1:20] details, they may appear too thin when plotting your 1-$^1/_2$″ [1:10] details. You will have to experiment to see what works for you. It will always be a compromise.

Printing and Plotting MSP Sheets

Printing and plotting MSP sheets is covered near the end of Chapter 7. Essentially, the steps are as follows:

1. Make sure you have the correct **Printer/Name**, **Paper/Size**, and other appropriate settings selected in the **Plotter/Setup** menu.
2. Make sure you have the correct .PEN file loaded from the **PenTable** dialog box.
3. Make sure the current MSP sheet is the one you want to print.
4. Press the **Plot** button in the **Plotter** menu.
5. If you want to save the output file to your hard disk instead of directly to paper, pick **ToFile** instead of **Plot**.

Check Plots of MSP Sheets

If you want to print a small version of your full MSP sheet, perhaps on 8-$^1/_2$″ × 11″ paper so that you can check your layouts, you can do this via the **Plotter/Setup/Check Plot** option. You can read about this in more detail in Chapter 7, but here are the essential steps, all of which take place in the Printer/Plotter Settings dialog box, accessed via the **Plotter/Setup** option:

1. Make sure the plotter driver in the **Printer/Name:** box contains the full size plotter driver that supports the large paper size you arranged your MSP details on. This is not the place to select the smaller check plot printer.
2. Make sure the **Paper/Size:** box contains the full paper size you arranged your MSP details on 24″ × 36″ [610 × 914]. This is not the place to select your check plot paper size.
3. Go to the **Check Plot** area and select the **Fit to:** option.
4. In the **To this printer:** box, select the printer that you want to print your scaled down check plot to.
5. In the **Fit to:** box, select the paper size you want to print your scaled down check plot on. Note that you must first select the correct driver in the **To this printer:** box or you won't see the correct paper sizes in the **Fit to:** box.
6. If you want to print more than one copy of your Check Plot, select the **Copies** option and change it to the value you want.
7. Select **OK**.
8. Select **Plot**. The small check plot will be sent to the printer selected in the **To this printer:** box.

   This **Check Plot** feature works for the **QwkLyout** option as well.

Partial Printing and Plotting of MSP Sheets

Unfortunately, there is currently no way to print out a partial area of an MSP sheet in DataCAD. You can make whole or partial prints of individual details within the DataCAD Drawing Window, but not of anything within the MSP window.

Transferring MSP Sheets to Other CAD Programs

One of the issues with MSP is how to transfer each of your MSP sheets intact to another CAD application (earlier DataCAD versions, AutoCAD, and so on). Unfortunately, it is not currently possible to do this from DataCAD or most other CADD programs. If you create a .DXF or .DWG file of your DataCAD .DC5 file, all the drawing entities will be saved intact (see Chapter 25, "Converting File Formats," for more on conversion topics), but your MSP sheet layouts will not. That's because every CAD program has its own way of printing multiple scale details on one drawing sheet. Right now there is no way to transfer any of those layout methods from one program to another. Of course, you could get around this by creating a new drawing file, enlarging or shrinking your details to fit, and then arranging them on a sheet just like in the old days. MSP sheets do transfer completely between DataCAD 8 and 9 applications with the exceptions that will be noted later.

Because we deal with a lot of AutoCAD users, we have learned to use their multiple scale plotting method, called *Paper Space*. After drawing our project in DataCAD, we can then recreate our MSP layouts in AutoCAD. See Chapter 25 for more information.

Part IV: How to Stay out of Trouble

The title of this section is more ominous than it sounds. Now that you have learned about the power of MSP, you need to know some of its problems in order to keep yourself out of trouble. Here are a few tips about MSP to keep in mind:

- The display of entities in the MSP window is dependent on the resolution that your computer's graphics card is set at (800 × 600, 1024 × 768, and so on). This alone does not present a problem, but if drawings with MSP details in them are edited between computers with differing resolutions, the MSP details will display differently on each computer.

For instance, assume computer A has a resolution of 800 × 600, and computer B has a resolution of 1280 × 1024. If computer B opens an MSP drawing created on computer A, the details will all be much smaller in the MSP window of computer B. If computer B edits those MSP details and then gives the file back to computer A, all the MSP details will be larger than the MSP window of computer A. The only way to fix this problem is to layout all the details from scratch! And that's a good way to ruin somebody's week. However, simply opening the drawing file or even working on the entities in the file will not harm anything, but computer B should not do any editing in MSP.

Also note that the problem is more acute with laptop computers. Even if your desktop system and your laptop are both at an 800 × 600 resolution, you will more than likely have the same MSP problems as if you had two different resolutions.

- You must be careful when using the **LyrUtil** (Layer Utility) macro. When you select **LyrSave** in the macro, you will see two highlighted options: **3D Views** and **Details**. The **Details** option relates directly to MSP. If this option remains highlighted, then all the MSP details will remain intact when you later open a new file and select **LyrUtil/LyrLoad**. If the **Details** option is toggled off, then your drawing entities will remain intact, but none of the MSP sheets or detail layouts will be retained.

  It is critical to understand that MSP sheets and details from two different drawing files cannot be combined. If you try to merge two drawing files that both contain MSP information, the outcome will be corrupted MSP layouts. Selecting MSP sheets or details in a drawing with corrupted MSP information will nearly always crash DataCAD and make your file irretrievable. So if you plan to merge two or more drawings that contain MSP layouts, make sure you turn the **Details** option off in all the saved drawings or in all but one of those files. As always, you should also get in the habit of making a backup of a critical file before making major modifications, like merging two drawings.

- Currently, you have no way to rename the details in the MSP Layout menu, so choose them carefully the first time.

- There is currently no way to rearrange the order of the details in the MSP Layout menu. Keeping details in some kind of order makes things easier to find but does not affect the operation of MSP.

- Although MSP sheets can be renamed, their order cannot be rearranged.
- DataCAD 9 for Windows files are backward-compatible with previous Windows and DOS DataCAD versions back to and including Version 5. MSP sheets created solely in Version 8 or 9 for Windows are not backward-compatible with pre-Version 8 DataCAD programs. However, MSP sheets created in any 7.x version are forward-compatible with DataCAD 8 and 9 for Windows.

Here are a few pointers about CCs to keep in mind:

- CCs cannot be moved, stretched, copied, mirrored, or otherwise edited. So you cannot, for instance, copy a CC from one detail to another. Instead you have to create a new CC just for that detail.
- You can, however, place a CC around a detail, and then place it rotated in the Multi-Scale Plot Window:
  - Before placing a detail in the MSP window, select **Plotter/Rotate** and then pick the angle of rotation.
  - Select **MltLyout/Layout** and place the detail in the MSP window. It will be inserted rotated.

WARNING: *If you insert a rotated CC in the MSP window, then use the **Details/MakeCurr** option. A serious bug in the program may cause DataCAD to crash and will certainly garble the entities in your drawing file to the point of making parts of the drawing file unusable.*

- Text must be completely enclosed in a CC for the text to be visible. If even one small bit of a string of text is outside the CC, then none of that string of text will be displayed.
- Very large radius curves made of custom DataCAD linetypes may show up faceted instead of smoothly curved when located in a CC. You may have to experiment to find out what does and does not work. Ellipses are not affected, so you may be able to use an ellipse with equal X and Y components to form a circle or a pseudo arc.
- It is very easy to forget that you have a CC turned on. Remember that unless you change to a new 3D GTV or manually turn the CC off, it will stay on. So if you've been working with CCs and suddenly find that some or all of your drawing has disappeared, try going to the CC menu and making sure that **ClipOn** is turned off.

- Because CCs are 3D elements, it is important to always know the Z-Hgt and Z-Base of your entities as well as the Z-Min and Z-Max of the CC.
- If you try to print a detail with a CC around it by using the **Rotate** option in the **Plotter** menu, or the **Rotated** option in the **Setup** menu, the entities of the detail will rotate, but the CC will not. This can yield some very unexpected printing results.

SUMMARY

In this chapter, we have seen how the combination of MSP, CCs, and Text Scale can come together to save you time and better coordinate your details. MSP has its quirks. It may not do everything you want it to the way you want to do it, but it is still a very powerful and useful feature of DataCAD. It is well worth spending some time to learn and will certainly change the way you do business.

Although the final enlarged details example was of a wall section, keep in mind that this combined method of CCs and MSP has endless possibilities. Floor plans are a good example. After drawing your floor plans, you can use this method to create enlarged plan-details of stairs, elevators, bathrooms, kitchens, and so on. Every time your $^1/_8$″ [1:100] floor plans are changed, the enlarged details in the CCs are immediately updated as well. You end up having to draw half as much while gaining twice the coordination.

CHAPTER 13

External Reference Files (XREFs)

External Reference Files (XREFs) area brand new feature in DataCAD, one that has been very anxiously awaited, and with good reason. What it does is enable you to view, or reference, another drawing file within the file you are working on. As a reference drawing, it cannot be edited, but you can see it, snap to it, draw on top of it, and turn its various layers on and off. Because the referenced drawing file is actually another drawing file outside of, or external to, the main drawing file, it is called an External Reference File, or XREF for short.

The XREF file is not a different kind of drawing file. It is a standard .DC5 drawing file. The only thing that makes it an XREF is the fact that other drawings use it as a reference. Any drawing file can be an XREF, you can XREF more than one drawing file at one time, and you can even nest multiple XREFs inside of one another.

Before continuing, let's establish two key definitions:

- *Master file* The active .DC5 drawing file in which you are drawing and editing
- *XREF file* The externally referenced .DC5 drawing file

Why is this such an important feature? For the following reasons:

- It keeps the size of the master file smaller and faster, since the XREF file is only referenced, and is not actually a part of the master file.
- For drawings like floor plans, it enables one person to work on the main floor plan, while someone else uses it as an XREF for the reflected ceiling plan, while another person uses it as an XREF for enlarged floor plan details, while still another person uses it as an XREF for laying out the roof plan.
- In all the cases described in item 2, if the person working on the floor plan changes the plan layout, everyone else working with that plan as an XREF will have instant access to the changes being made.

Learning to use DataCAD's new XREF feature should be considered an essential task for any office, even the sole practitioner.

NOTE: *You should download the free version 9.01 patch from the DataCAD Web site (*`www.datacad.com`*) before continuing with this chapter. Several XREF improvements are included.*

XREF Example

Before going over each feature of DataCAD's XREFs, let's look at a sample file from the CD to get an idea of how a project can use XREFs. Copy the contents of the SAMPLE folder to your hard drive from the Chapter 13 section of the CD (TTLBLK.DC5, 5-PLAN.DC5, A1.DC5, A2.DC5, and A3.DC5). When you copy the files to your hard drive, they most likely will not be located in the exact file folders on your hard drive as they were on mine.

In order to reference a drawing file, DataCAD needs to know exactly where on your computer that drawing file is located. The exact path and file name is stored with the master file drawing. If you move the XREF file to a new location on your computer, then DataCAD won't be able to find it. The XREF file has become an *orphan*. To use the XREF file, you need to tell DataCAD its new location:

1. Open the A1.DC5 file. You will most likely get the error message shown in Figure 13-1 for the reasons described earlier.
2. Click **OK**. All you will see on the screen is some lonely text (press **R**eCalc if you don't see it). There is supposed to be an XREF of a sheet border and title block, but since DataCAD is looking in the path that I originally created on my computer for that XREF, you need to tell DataCAD where to find it on your computer.
3. From the drop-down menus, select **Insert/Reference File Management/Orphans**. You'll see a dialog box like the one in Figure 13-2.
4. Highlight the first path (the one with the TTLBLK.DC5 file) and then select **Redefine Path**.
5. A **Browse for Folder** dialog will appear. Find and click on the folder where the TTLBLK.DC5 file is located and then click on **OK** (by default, DataCAD will begin searching in the current drawing file directory).

Figure 13-1
The orphan warning message

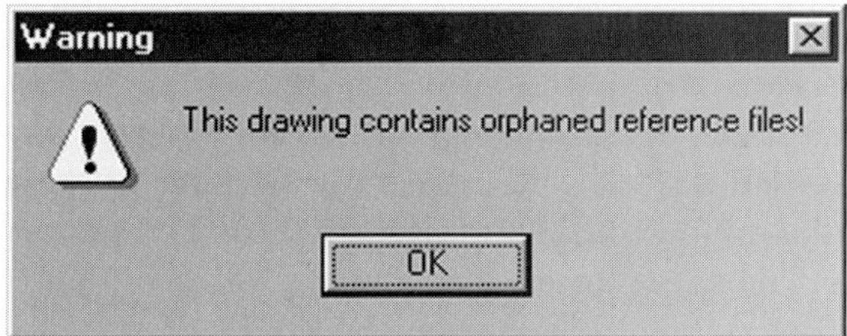

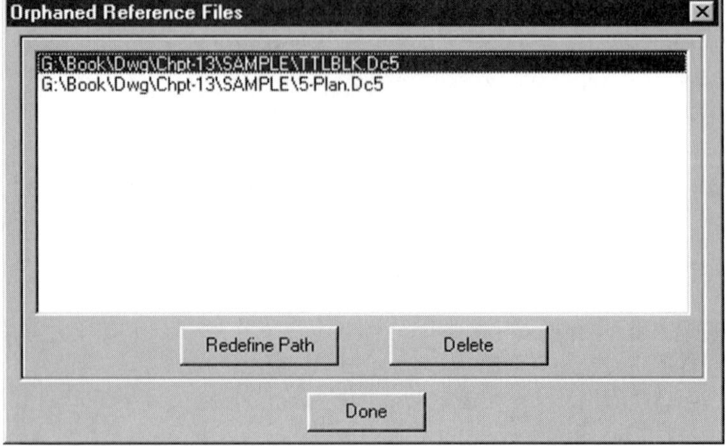

Figure 13-2 All the orphaned XREFs are displayed.

The path with the TTLBLK.DC5 file in it will disappear from the Orphaned Reference Files dialog, indicating that a new path has been selected, and the correct drawing has been found there. You should also see the sheet border and title block appear in the Drawing Window.

Now notice that we need to take care of one more orphaned XREF file. Follow Steps 4 and 5 to locate the 5-PLAN.DC5 file. Alternately you could have highlighted both files to take care of them both at once, as long as both files are in the same directory.

6. *Click* on **Done** to exit the **Orphans** dialog box. You should see a title block and border on the screen. Select **Utility/Plotter/MltLyout/Sheet** to see the entire layout of sheet A1 (it was laid out on a 24″ × 36″ [610 × 914] sheet).
7. *Right-click* back to the main Drawing Window.
8. Use **Edit/Identify** (or the **I** in the Navigation Pad) and select one of the lines that make up the sheet border.

   Notice how the entire border and title block, except for the drawing title, scale, and drawing number text, are highlighted. At the very top of the Menu Window (adjacent to F1), you can see that the object is being identified as an **XREF**. The next menu button down (adjacent to F2) tells you that the object is on the **X-TTLBLK** layer. So, the entities that make up the border and title block are not actually part of this drawing file. They are part of a drawing file being referenced, and that referenced file is located on the **X-TTLBLK** layer.

9. *Right-click* back to the main Drawing Window and then press **Ll** to go to the **Layers** menu. Select **On/Off** and then turn off layer **A1-TEXT** by selecting it. The drawing title, scale, and drawing number text disappear from view. The entities on that layer are native to the current master file and are not XREFed.
10. Turn on only layers **5E-DEMO** and **X-PLAN** and then press **R**eCalc. You will see a floor plan with walls and doors, along with dashed lines that represent walls and doors to be demolished.
11. Go back to the **Layers** menu and turn **5E-DEMO** off. Only the red dashed lines disappear from view. The entities on that layer are native to the current master file and are not XREFed.

    The rest of the entities on layer **X-PLAN** are part of another referenced file. The referenced file has several layers to it, but just because the XREF file is located on the **X-PLAN** layer doesn't mean that you can't control those layers.
12. From the drop-down menu, select **Insert/Reference File Management/Manager**. The **Reference File Manager** (RFM) dialog box will appear, as shown in Figure 13-3.

Figure 13-3 The **Reference File Manager** (RFM) dialog box

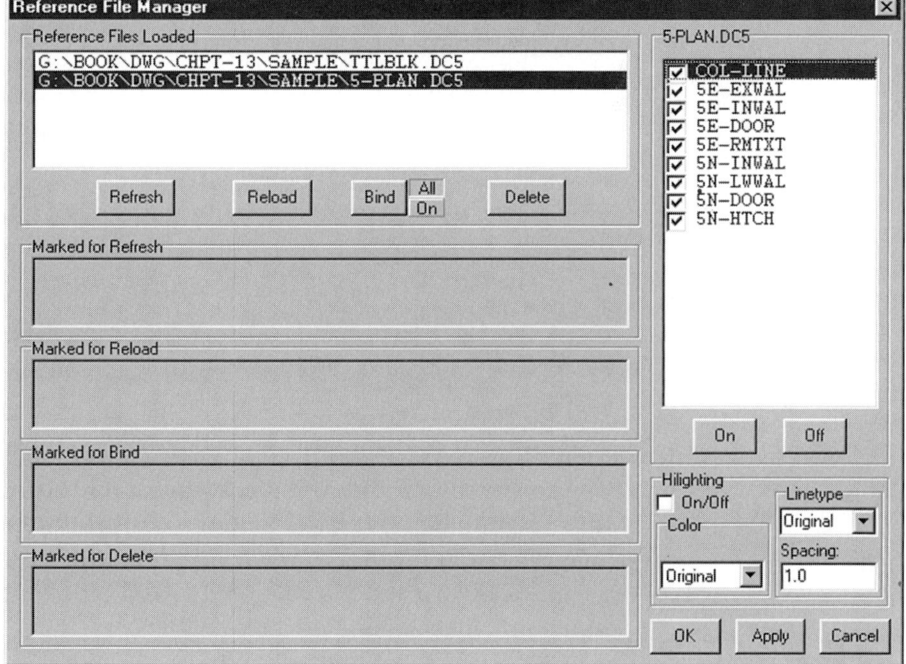

13. Highlight the path containing the 5-PLAN.DC5 file. The layers in that drawing file will be displayed in the Layer List at the right side of the dialog box. All of the boxes should be checked. A box with a check mark in it means that that layer will be displayed in the Drawing Window. Turn the **COL-LINE** layer off by *clicking* on the check mark or by highlighting the layer name and pressing the **Off** button.
14. *Click* on **OK** to close the dialog box and return to the Drawing Window. The column lines, column bubbles, and the dimensions (all of which were on that layer) are turned off.
15. Go back to the RFM and turn the **COL-LINE** layer back on.
16. While still in the RFM, go to the **Highlighting** section of the dialog and set the following:
    a. Check the **On/Off** box.
    b. Select **Linetype/Dot-Dash**.
    c. Type a spacing of **3.0**.
    d. Select **Color/Brown**.
17. Select **OK**. All the layers of the XREF files have been changed to match the settings in the Highlighting section of the RFM. This can be very useful for visually determining which entities are part of the XREF. If you print this drawing now, DataCAD will print exactly what you seen on the screen. Highlighting affects each selected XREF independently, without affecting the others (v.9.01).
18. To return the XREF layers to their original state, go back to the RFM and then deselect the **Highlighting/On/Off** option. *Click* on OK and the XREFs will be returned to its original state.

If you are asking yourself, "why bother with all this," let's look at another drawing file that uses the same XREF files (TTLBLK.DC5 and 5-PLAN.DC5):

1. Open the A2.DC5 file (follow Steps 2 through 5 in the previous example if you get an orphans warning message).
2. Use **Edit/Identify** (or the **I** in the Navigation Pad) and select one of the lines that make up the sheet border.

   The entire border and title block, except for the drawing title, scale, and drawing number text, are highlighted. At the very top of the Menu Window (adjacent to F1), you can see that the object is being identified as an **XREF**. This is the very same XREF file that we were referencing in the previous example: TTLBLK.DC5.

3. Go to the **Layers** menu and turn on only **5N-NOTES**, **5N-DIMS**, and **X-PLAN**. **R**eCalc the Drawing Window to display the entire plan.

   The X-PLAN layer contains the very same XREF file that we were referencing in the previous example: 5-PLAN.DC5.

4. Select **Utility/Plotter/MltLyout/Sheet** to see the entire layout of sheet A2 (it was laid out on a 24″ × 36″ [610 × 914] sheet)
5. *Right-click* back to the Drawing Window.
6. Go to the RFM and look in the Reference Files Loaded box. The file names are exactly the same as those in the previous example of drawing A1.

Now let's see how a change in an XREF File, such as the 5-PLAN.DC5 file, impacts the other drawing files. This is a key concept, so it's really worth trying out:

1. With the A2.DC5 file still open, also open the 5-PLAN.DC5 file.
2. Draw two diagonal lines across the screen to form a big X across the floor plan. Do not save the drawing, but don't close it either.
3. Switch back to the A2.DC5 drawing (select **Window** from the drop-down menu and then pick the A2.DC5 file). Make sure that only layers **5N-NOTES**, **5N-DIMS**, and **X-PLAN** are on, and that you can see the floor plan in the Drawing Window.
4. Go to the RFM, highlight the 5-PLAN.DC5 XREF, and then *click* on the **Refresh** button. *Click* **OK** to exit the RFM.

   No change will be made to the floor plan on your screen. That's because when we drew the X in drawing 5-PLAN.DC5 we did not save the drawing. Changes to the XREF can only be registered after the XREF drawing is saved, since the XREF information is being read directly from the .DC5 file and not from the screen of the computer working on that XREF file.

5. Switch back to the 5-PLAN.DC5 drawing (which should still have a big X across it). Now save the drawing.
6. Switch back to the A2.DC5 drawing. Go to the RFM, highlight the 5-PLAN.DC5 XREF, and then *click* on the **Refresh** button. *Click* **OK** to exit the RFM.

   Now you should see the big X across the floor plan. This is one of the greatest features of XREFs. If someone is working on the floor plan (adding doors, changing the plan, and so on), each time they save the drawing, all those changes become available to anyone who has an XREF of that plan in their drawing file.

If you want one more example, a reflected ceiling plan can be found in drawing file A3.DC5 that references the same two files as the previous two examples.

What's It All Mean?

Taking all three of the previous examples into account, let's look at how four people on a computer network could work on this project together:

- Drafter #1 works on the base plan (5-PLAN.DC5), which has no XREFs in it.
- Drafter #2 works on the demolition plan (A1.DC5), which references the 5-PLAN.DC5 drawing that Drafter #1 is working on.
- Drafter #3 works on the 5th floor plan (A2.DC5), which references the 5-PLAN.DC5 drawing that Drafter #1 is working on.
- Drafter #4 works on the 5th floor reflected ceiling plan (A3.DC5), which references the 5-PLAN.DC5 drawing that Drafter #1 is working on.

Since drawings A1 through A3 all reference the same TTLBLK.DC5 file, if a change is made to the title block (perhaps the name of the project changed), then that change will be reflected in all three drawings at once.

Insert a Reference File

To reference an external DataCAD drawing file within the current drawing file, perform the following steps:

1. XREFs are inserted into the currently active drawing layer. So, in most cases, you will want to create a new layer just for the XREF, though you can move it to a layer later. Make that layer the active layer (**Layers/SetActiv**) before going onto the next step.
2. Select **Insert/Reference File**. The Insert Reference File dialog box will appear, as shown in Figure 13-4.
3. *Double-click* on the file to be XREFed, or highlight the file and select **Open**. You can only select and insert one drawing file at a time.

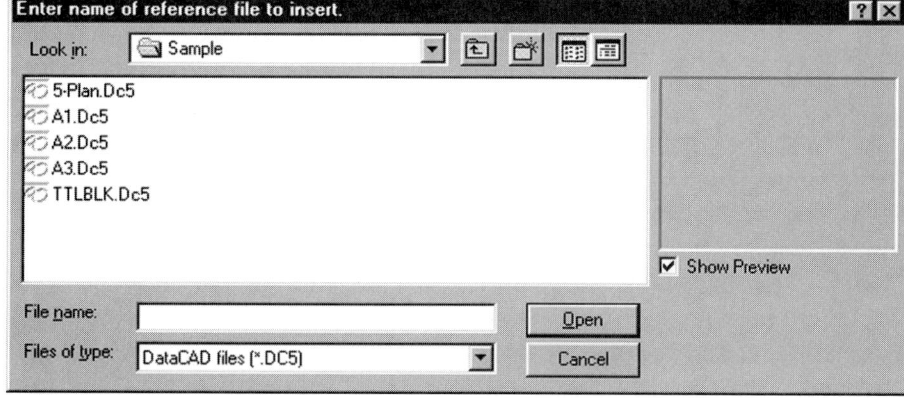

Figure 13-4 The **Insert Reference File** dialog box

a. By default your cursor will be attached to the absolute zero point of the .DC5 file that is being referenced, and the extents of the entities of that file will be represented by a rectangular box. Unless the entities of the XREFed drawing were drawn on top of the absolute zero point of the drawing, the cursor will probably be offset from the bounding box as shown in Figure 13-4A.

   Often the absolute zero point may be dozens or even hundreds of feet away from the entities that make up the XREF, making placement of the XREF difficult. In that case you may want to use the **ByCenter** option.

b. When you select the **ByCenter** option, the absolute zero points are ignored and the cursor will be located in the center of the extents of the XREF entities, making placement of the XREF easier and more precise (see Figure 13-5).

4. To place the XREFed file in the current drawing:

   a. Locate the bounding box where you want the XREF entities to be located. Click the mouse to place the XREF in the drawing at that location.

   or

   b. Click on the **AbsZero** button. This will force the absolute zero point of the Master File and the XREF File to be overlaid directly on top of one another.

5. To insert more XREFs, follow Steps 1 through 4 for each file to be added.

6. Because the entities of an XREF look just like any other entities in the drawing file, it's easy to accidentally **Erase**, **Move**, **Copy** or otherwise

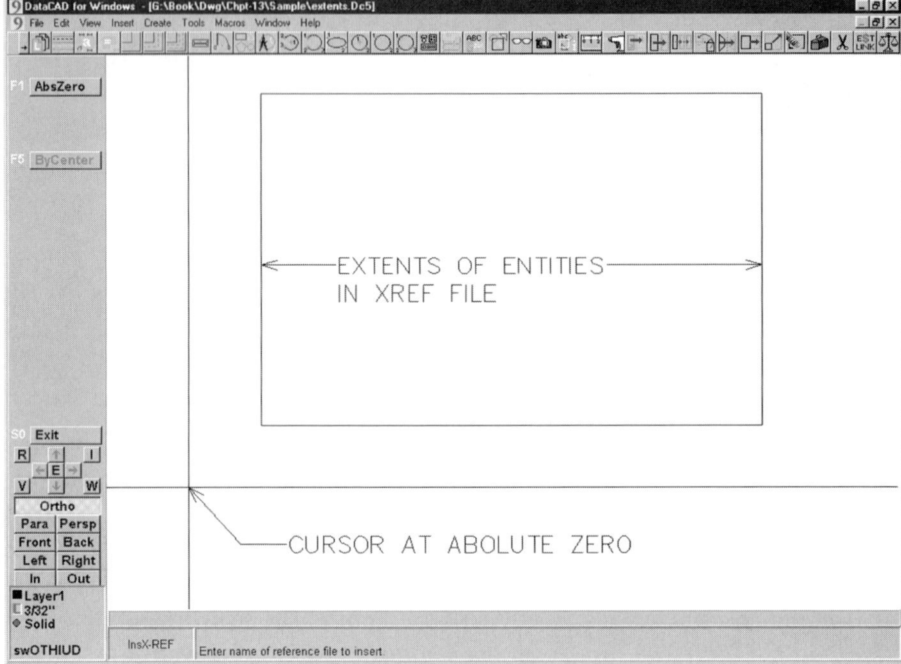

Figure 13-4A Placing an XREF with the curser at absolute zero

edit an XREF. For that reason, after you insert an XREF you might want to use **Layer/Lock** to lock the layer, since a locked layer cannot be edited. But be aware that you cannot change settings in the Reference File Manager while the layer is still locked.

NOTE: *XREFs are inserted into the currently active drawing layer, but you cannot insert an XREF into the active layer if that layer is locked. Either switch to an unlocked layer or unlock the current layer prior to inserting the XREF.*

Insert by Absolute Zero

All entities in a CAD drawing are referenced off of the absolute zero point of the file. You may not see this happening, but the CAD program certainly does.

As noted previously, placing an XREF with the **AbsZero** button will force the absolute zero point of the master file and the XREF file to be overlaid directly on top of one another. In most cases, this isn't necessary. The

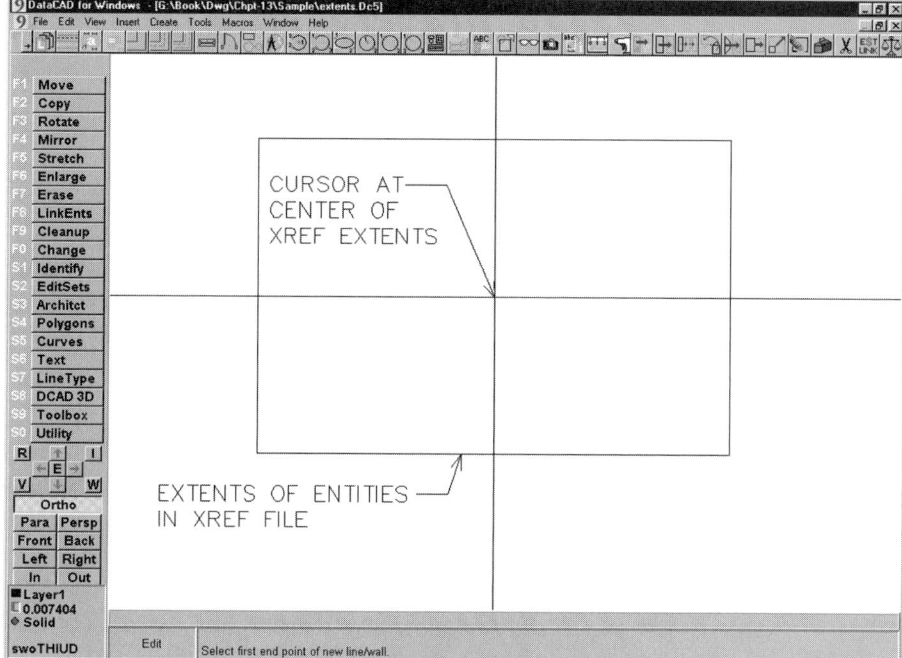

Figure 13-5 Placing an XREF with the **ByCenter** option

exception is when you have entities in both files that are drawn in the same relative location from the absolute zero point of each drawing file, and you need to keep them aligned when the XREF is inserted.

Let's say that you have created a floor plan, and you then send it to your mechanical engineer. Your engineer does his work and you want to overlay his work with yours to make sure there are no conflicts. As long as the engineer did not move the plan you sent away from its original location, when you **Insert** his drawing as an XREF into your drawing using the **AbsZero** option, the two plans will display perfectly aligned. If you had manually placed the engineer's XREF into the drawing, then you would have to **Move** the XREF by snapping to identical points between the two plans. This takes more time and has fewer guarantees of accuracy.

Once an XREF file is inserted into a master file, if you then move the entities in the XREF file to a new location, then those entities will no longer be aligned with the entities in the master file, since they are in different locations relative to the absolute zero point. For this reason, it is very important to make sure that when you are using XREFs, users don't, for instance, move the floor plan over a few feet in the XREF file (like the 5-PLAN.DC5 example). If they do, then the plan will be misaligned in every drawing that referenced it, and that could ruin your day.

Manage Reference Files

Once an XREF File is inserted into the master file, management of that XREF is done from the *Reference File Manager* (RFM). To access it, go to the drop-down menu and select **Insert/Reference File Management/ Manager**.

NOTE: *Because the RFM is such an important aspect of using XREFs, I have made my own shortcut to it by modifying the DCADWIN.MNU file. Chapter 23 has more info on how to do this, but here it is:*

1. *Make sure DataCAD is not running.*
2. *In the DataCAD /SUP directory, open the DCADWIN.MNU file with a text editor like Notepad.*
3. *Find the line that says "&Manager". Change it to read* "&Manager|R" *(the | character is called a pipe and is located on your keyboard above the \ character).*
4. *Save the file.*

*Now when you open DataCAD you can access the RFM by pressing **Ctrl+R**.*

Refresh

Selecting this option will cause all the highlighted XREFs to be refreshed. DataCAD goes to each highlighted XREF file, reads the last saved information from it, and then updates the current drawing. **Refresh** will reread the XREFed file to update the entities while maintaining the layer on/off settings that are currently set for that XREF in the RFM.

Reload

Selecting this option will cause all the highlighted XREFs to be reloaded. DataCAD goes to each highlighted XREF file, reads the last saved information from it, including layer settings, and then updates the current drawing. Reload will reread the XREFed file to update the entities but will use the layer on/off settings that were last set in that XREFed file, rather than using the settings that were previously set in the RFM. This option differs from the Refresh option because it changes the layer settings.

Bind

When you select an XREF file to be bound, all the layers in the referenced file will be added to the current drawing file, and the XREF will be deleted. These layers and all their entities are now part of the current drawing file, increasing the file size by however many entities were in the XREF drawing file. You can bind all the layers of the XREF or only the layers that were on.

You will want to **Bind** your reference files prior to exporting a DXF or DWG file. If you do not first bind an XREF, the XREF entities will be exported, but all the entities in that XREF will be placed on the single layer that the XREF is located on in the master file. Another instance would be if you are sending a DataCAD file to another DataCAD user that does not have version 9. When binding files, you have two options to select from:

- **All** Use this command to bind all the layers of the highlighted XREF files into the current drawing file, regardless of whether those XREF layers are currently turned on or off. Select **All** before selecting **Bind**.
- **On** Use this command to bind only the layers that are turned on in the highlighted XREF files. In the process, the XREF will be deleted along with any XREF layers that were turned off when the XREF was bound. Select **On** before selecting **Bind**.

Both the **All** and the **On** options only affect the XREF files that are highlighted in the RFM. To bind an XREF file, perform the following steps:

1. From the drop-down menu, select **Insert/Reference File Management/Manager**.
2. Select each XREF and turn on/off layers as appropriate.
3. Press **Apply** so that the new layer settings are saved. If you do not do this *prior* to binding the drawing, then all the layers of all the selected XREFs will be bound, regardless of which ones you turned on and off.
4. Highlight each XREF to be bound. You can select one or multiple XREFs. To select more than one, you can use the standard Windows selection methods of Alt+click, Ctrl+click, or Shift+click.
5. Select **All** (the default option) or **On**, as appropriate.

6. Click on **Bind**. The XREF will be moved to the **Marked for Bind** window.
7. Select **Apply** or **OK** to **Bind** the selected XREFs.

If a layer name in the XREF File is identical to the layer name in the master file, no duplicate layer names will be created in the **Bind** process. The XREF entities on that layer will be added to the identically named layer in the master file.

Delete

To delete an XREF from the current master file, highlight all the XREF files that you want to delete and then select **Delete**. After selecting **Apply**, or exiting the RFM with the **OK** button, all those XREFS will be deleted from the master file. If you mistakenly delete an XREF, use the Undo feature (**Ctrl+Z**) to bring it back.

Layer Control

Use the Layer Control box at the right of the RFM to control which layers are on or off in each XREF. Highlight one of the XREFs and its layers will be displayed in the box. Layers will be turned on when a box is checked and will be turned off when a box is unchecked. These settings will not take effect until you press **Apply**, or **OK** to exit the RFM.

These layer settings are not "retained" by *GotoViews* (GTVs) or MSP details. See the following section regarding XREFs, GTVs, and MSP details.

Highlighting

These options will display all the entities of all XREFs with a specific line-type and/or color. This makes it easy to identify which entities belong to XREFs and which are native to the master file. It is also great for creating background plans, over which electrical, plumbing and HVAC entities can be added. You can select from these highlighting options:

- **On/Off** Check this box to enable the Highlighting feature.
- **Color** Use this option to display all the XREF entities in one particular color. To return the display of the XREF entities to their original colors, select the **Original** option from the top of the drop-down list.
- **Linetype** Use this option to display all the XREF entities in one particular linetype. All the linetypes in the current DCADWIN.LIN file are displayed in the drop-down list. To return the display of the XREF entities to their original linetypes, select the **Original** option from the top of the drop-down list.
- **Spacing** Use this option in combination with the **Linetype** option to set the linetype spacing. Type in a value.

DataCAD's printing is WYSIWYG (what you see is what you get), so with the **Highlighting** option turned on, DataCAD will print exactly what you see on the screen. If your **Highlighting** linetype is set to **Dashed** and the color to **Red**, then DataCAD will print all the entities of the XREF as dashed red lines.

Orphans

In order to reference a drawing file, DataCAD needs to know exactly where on your computer that drawing file is located. The exact path and file name is stored with the master file drawing. If you move the XREF File to a new location on your computer, then DataCAD won't be able to find it. The XREF file has become an *orphan*. You will know if you have an orphaned file if you open a master file and get the message shown in Figure 13-6.

Click on **OK** to close the error message. To continue using the XREF file, you need to tell DataCAD its new location with the Orphans dialog. You cannot use the Orphans dialog to locate the XREF if you have renamed the XREF, only if you have relocated it. So be sure not to rename XREFs if you

Figure 13-6 The orphan warning message

want to continue to use them. To access the Orphans dialog, go to the drop-down menu and select **Insert/Reference File Management/Orphans**. The **Orphaned Reference Files** dialog box will appear, as shown in Figure 13-7.

The main window of the dialog will display all the currently orphaned XREFs along with their original locations.

Redefine Path

Use this option to tell DataCAD where the XREF Files have moved to:

1. First, make sure you know the file folder that the XREF is now located in. Then highlight one or more of the XREF files in the main window of the **Orphans** dialog.
2. *Click* on **Redefine Path**. A **Browse for Folder** dialog will appear. Find and *click* on the folder where the XREF file is now located (by default, DataCAD will begin searching in the current drawing file directory).
3. *Click* on **OK**. If you have selected the correct new location, the old path with the XREF file in it will disappear from the **Orphaned Reference Files** dialog, indicating that a new path has been successfully selected, and the correct drawing has been found there.
4. *Click* on **Done** to exit the **Orphans** dialog.

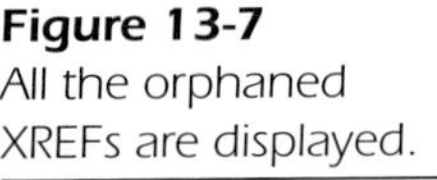

Figure 13-7
All the orphaned XREFs are displayed.

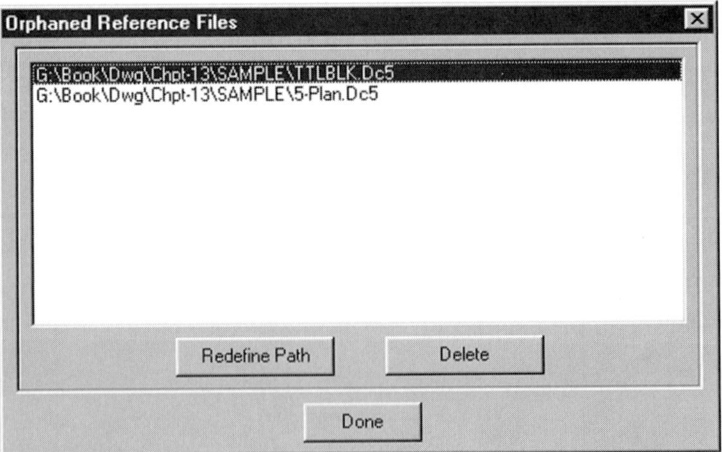

Delete

Use this to remove the reference to an orphan. Since an orphaned XREF is not actually a part of the current drawing file, no entities are deleted. It is only the reference to that XREF that is deleted from the drawing file.

XREFs, GTVs, and MSP Details

As mentioned earlier, GTVs and *Multi-Scale Plot* (MSP) details do not "retain" the XREF layers that were on and off when the GTV or MSP detail was created. Instead, XREFs always display the current XREF layer settings. This presents a few major problems:

- GTVs become far less useful, since their whole purpose is to remember layers.
- If you use an XREF in several MSP details, but you want different XREF layers turned on in each detail, you just cannot do it, since XREFs only display the current layer on/off settings, regardless of what was selected when the MSP detail was created.

But have no fear, there is a way to get around these problems. You can have more than one instance of any XREF in your drawing file, and each of those XREFs can have its own layer settings. So what you do is this:

1. Create a new layer for each XREF to be inserted.
2. **Insert** one instance of the XREF on the first layer.
3. Use the **Copy** command to copy the XREF to the next new layer.
4. Continue to do this for as many instances of the XREF as you need, placing each one on its own layer. Each copy will display in the RFM with a number at the end (see Figure 13-8).
5. For each detail that needs to have its own XREF layer settings, turn on the layer that has the desired XREF on it. If you forgot which XREF copy this is, use the Identify command and *click* on the XREF in the Drawing Window. The full XREF name (including the number at the end) will be displayed in the Coordinate Readout line. Use that XREF for your GTV and/or MSP detail.
6. Turn on and off the desired XREF layers. You can save the GTV or MSP before or after you fo this.

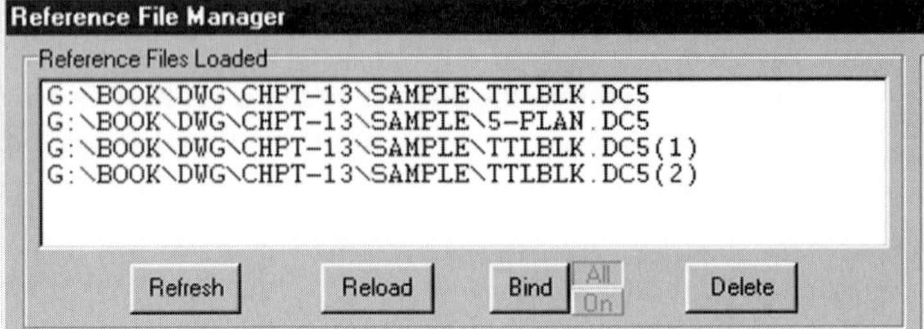

Figure 13-8 Multiple instances of an XREF (TTLBLK.DC5)

GTV Cheap Trick

Another way to gain some of the functionality of GTVs when using XREFs is to use the **Reload** command in the RFM, since that function will reload the XREF file using the current on/off layer settings in that .DC5 file. So, assuming that you already have an XREF in the master file:

1. Keep both the master file and the XREF file open.
2. Make all the 3D GTVs that you want in the XREF file.
3. Select the 3D GTV that you want to use in the XREF file.
4. Save the XREF drawing. If you don't, the master file cannot read the updated layer settings.
5. Now switch to the master file and open the RFM.
6. Highlight the XREF file, then select **Reload**, and then **Apply**.

The XREF in the master file will be updated with the current on/off layer settings in the XREF file. Convoluted? Yes. Inexpensive work-around? Absolutely. In the future, I expect that DataCAD's GTVs and MSP details may be capable of retaining XREF layer settings.

Manipulating Reference Files

Here are some general notes regarding the way XREFs are handled in DataCAD:

- XREFs are inserted on the current, active layer, but once inserted, you can move the XREF to another layer if you want to.
- The entities of the XREF are not part of the database of the master file, so the size of the master file only increases by one entity with the addition of an XREF.

- XREFs are loaded into RAM, so you can insert however many XREFs your system's RAM can handle. No DataCAD limit exists for the number of XREFs that can be inserted.
- XREFs can be nested, meaning that you can have an XREF inside of an XREF, inside of an XREF, and so on. In fact, you can have circular XREFs, where a .DC5 file that already references Drawing-1 can be inserted back into Drawing-1. For instance, if Drawing-2 has an XREF of Drawing-1 in it, Drawing-2 can then be XREFed into Drawing-1. Drawing-1 therefore contains a reference to itself.
- Unfortunately, there is currently no way to turn layers on/off in a nested XREF. To do that, you must return to the nested file, change the layers to on or off there, and save the file.
- XREFs are treated by DataCAD more or less like a symbol, except that you can turn the individual layers of the XREF on or off.
- Like a symbol, you can edit the position of the XREF with commands like **Move** (including **Move/ToLayer**), **Copy**, **Rotate**, and **Erase**, but . . .
- You cannot edit any of the entities in the XREF. To do that, you have to open the .DC5 file that is being referenced and make the changes there.
- Like a symbol, you can enlarge an XREF.
- The XREF will adhere to your current object snap settings, so you can snap to the entities, intersections, and endpoints in the XREF, just like with any other entities.
- Because an XREF is treated as a single entity, you cannot use the lines in it for trimming. For instance, you cannot select 1-Line Trim and then pick a line in the XREF to trim to.
- XREFs will **Hide** and **GLShade** just like native drawing entities.

Here are some notes on how you can use XREFs:

1. Because XREFs are inserted into a layer, I usually make a separate layer for each XREF. For instance, the layer with the XREF of the floor plan might be called X-PLAN. Once there, I can simply turn that layer off if I want to quickly turn off the display of that XREF. If I left the XREF on some other drawing layer, then I could only turn the whole XREF off by going into the RFM, turning off all the layers, and then exiting the RFM.
2. To keep from inadvertently erasing, moving, copying, or otherwise editing an XREF, after you insert an XREF then you may want to use **Layer/Lock** to lock the layer, since a locked layer cannot be edited.

3. If you haven't read through the examples at the beginning of the chapter, you should do so. It is the best way of describing how you can use XREFs to enable multiple users to work on related drawing files at the same time.
4. Because the **Highlighting** feature in the RFM is WYSIWYG (what you see is what you get), it is useful for creating background images. For instance, engineers often take the architectural floor plans and change the color of all the entities to a light gray color so that they print lightly in the background, allowing the engineering entities to read darkly in the foreground. All the engineer would have to do now (assuming they are using DataCAD) is change the **Highlighting** color to a color that will print lightly on their plotter.
5. If you were working on a big project that encompassed several city blocks, you could divide the project into smaller .DC5 drawing files in order to keep the file size down and the drawings faster, bringing them all together into a seamless single project with XREFs.
6. You can do likewise with 3D models, which can get quite complex, thereby increasing the file size. Using XREFs would enable you to break the project down into smaller parts. For instance, you could have a complex 3D entry canopy in one .DC5 file, the main building in another .DC5 file, and the highly detailed interior entities in another. All could then be brought together via XREFs.
7. Remember that it is very important to make sure that when you are using XREFs, users don't, for instance, move the floor plan over a few feet in the XREF File (like the 5-PLAN.DC5 example). If they do, then the plan will be misaligned in every drawing that referenced it.
8. As with anything in DataCAD, keeping your drawing entities, including the inserted XREFs, close to the absolute zero point of the master file will keep things more accurate.

Transferring Reference Files

XREFs can be transferred from one drawing file to another using the **LyrUtil** (Layer Utility) macro, using Sticky Backs created with the third-party DC Sprint macro, using cut and paste operations, or as part of a symbol.

Layer Utility

As long as the XREF to be saved is located on one of the layers to be saved with **LyrUtil/LyrSave**, then the XREF will be saved and transferred with the Layer Utility macro.

Sticky Backs

In the third-party macro called **DC Sprint** (see Chapter 24, "Third-Party Macros") is a sub-macro called **StkyBack** (Sticky Back). XREFS that are selected with the **CutStky** option will be copied to the drawing that the Sticky Back is pasted into.

Cut and Paste

Like the Sticky Back macro, an XREF that is copied to the Windows Clipboard with the **Cut** or **Copy** commands will be copied to the drawing that the XREF is pasted into.

Symbols

Since symbols cannot save layers, when you create a symbol that contains an XREF, the XREF becomes part of the symbol, just as if the XREF had first been bound with the **Bind** command in the RFM. However, what is saved is WYSIWYG, so only the XREF layers that are currently turned on will be saved in the symbol.

Because DataCAD erases entities from the drawing file when they are made into a symbol, the XREF will be deleted from the current drawing file, even if the symbol is made from only a few of the XREF layers. If you don't need the XREF anymore, then just continue on, but if you do need it select **Undo**. The symbol will remain in the template, but the XREF will be returned to the Drawing Window and the drawing database.

XREFs and AutoCAD Drawing Files

AutoCAD also has XREFs, but for now DataCAD's XREF feature will not directly import XREFs from AutoCAD, nor will it directly export XREFs to AutoCAD. However, the implementation of XREFs in AutoCAD is essentially the same as in DataCAD, so if you plan to trade drawing files with an AutoCAD office, what you send to them will be structured just like the DataCAD XREFs. For instance, if you were to send the sample building project (from the beginning of this chapter) to an ACAD office, you would send them .DWG versions of the same files. They would therefore get these drawings:

- TTLBLK.DWG
- 5-PLAN.DWG
- A1.DWG
- A2.DWG
- A3.DWG

You would then give them the same instructions that you would if these were DataCAD drawing files:

1. 5-PLAN.DWG is the base floor plan to be XREFed into A1.DWG, A2.DWG, and A3.DWG.
2. TTLBLK.DWG is the base drawing of the sheet border and title block to be XREFed into A1.DWG, A2.DWG, and A3.DWG.
3. Open A1.DWG and Attach the TTLBLK and 5-PLAN drawings (the equivalent of the **Insert** command is called **Attach** in ACAD). Insert them by 0,0,0 (this aligns the absolute zero points of the XREFs and the master file drawing).
4. Do the same for A2.DWG and A3.DWG.

Because these steps align the absolute zero points of all the drawings, all the entities will be aligned just as they were in DataCAD.

Like DataCAD, if you update the floor plan and send it to the ACAD office, remember to do the following:

- Make sure that you never move the floor plan or it will be misaligned in all drawings that used it as an XREF.
- Make sure that the name of the file remains exactly the same as the original file.

CHAPTER 14

Templates and Symbols

What Is a Symbol?

A symbol is an entity or a group of entities that have been saved as a single object, which can then be placed repeatedly in any drawing. Creating symbols allows you to draw complex objects only once, such as 3D objects that are often time-consuming to draw and repetitive 2D drawing objects such as wood studs, masonry units, steel sections, furniture, and notational drawing symbols. Figure 14-1 shows the DataCAD drawing screen with a group of symbols displayed in a series of template windows.

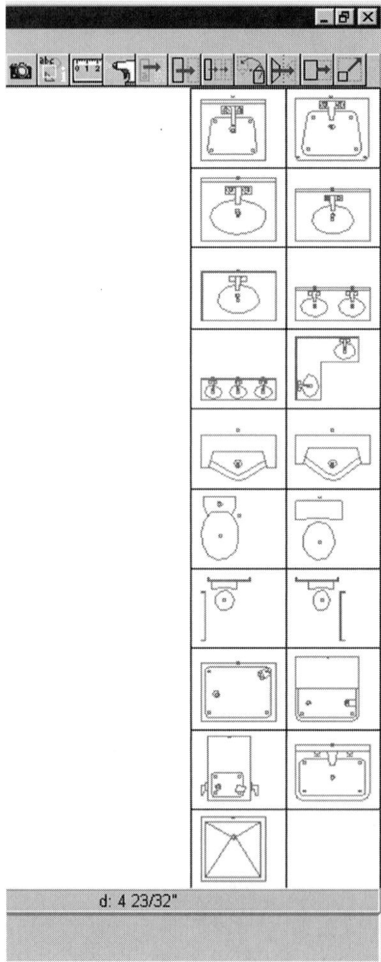

Figure 14-1 Typical template windows

DataCAD symbols exhibit the following characteristics:

- Symbols can include any DataCAD entities, 2D or 3D.
- In DataCAD, symbols do not have layers; they exist on any one layer, just as any other entity does.
- A symbol is treated as one single entity, no matter how many entities are in it. However, if 20 entities are in a symbol, then 20 entities will be added to the drawing database, but only for the first instance of that symbol. Each additional instance of that symbol placed in that drawing will add only one more entity to the drawing database. This is referred to as *symbol instancing*.
- If a symbol is placed in a drawing "exploded", or is exploded later on, then it ceases to be a symbol. All the entities in that symbol revert to being separate entities once again but are all placed on the currently active layer.
- The only way to change the colors or other properties of entities in a symbol is to explode the symbol, make the changes, and then resave the symbol.
- Once placed in the Drawing Window, you can use the same snapping options (endpoint, midpoint, quadrant, and so on) on any entities in the symbol, just as you can with any other entities in the drawing. This is a new feature of DataCAD 9.

Templates versus Symbols

Templates and symbols in DataCAD are symbiotic in nature. A symbol has an .SM3 file extension and is the actual image that will be placed in your drawing. A symbol is a mini-DataCAD drawing with only one layer. Symbols are stored in the \SYM directory (either in the root DataCAD directory or elsewhere, like your office network server). A template has a .TPL file extension and is a simple text file that tells DataCAD where to find the symbols, just as the table of contents in a book tells you where to find a chapter. Templates are stored in the \TPL directory (either in the root DataCAD directory or elsewhere, like your office network server). In Figure 14-2, we have opened the 3D-BED1.TPL template file.

In Figure 14-2, the text next to each symbol was added to show you the location of that symbol on the computer. The whole process works like this:

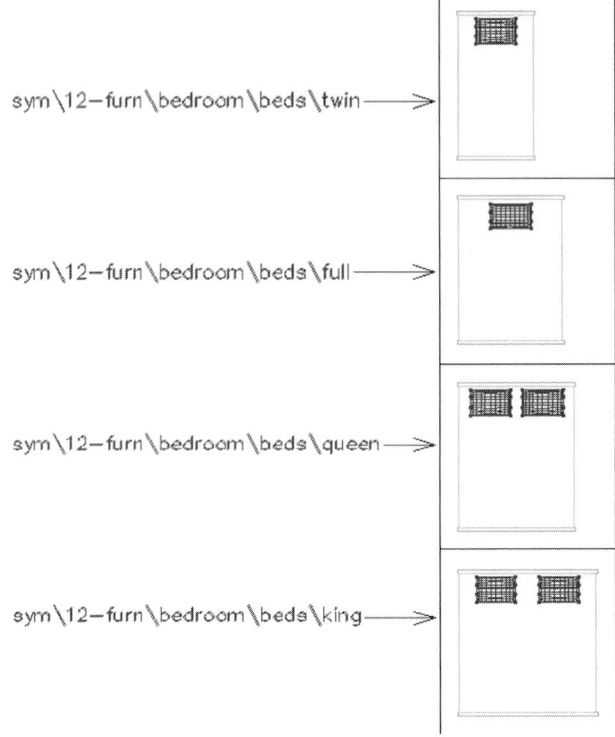

Figure 14-2 The 3D-BED1.TPL template file

1. When you call up a template file, the rectangular windows are displayed. The template file contains instructions that tell DataCAD how many horizontal and vertical windows to display.
2. Each window has a computer path associated with it (this information is saved in the template file when you save a symbol), so DataCAD next follows the path for each window, finds the symbol, and then displays it in the template window.

Here is the text from that 3D-BED1.TPL template file. You can see this yourself by opening the file with a text editor like Windows Notepad. Notice how the paths match the image in Figure 14-2.

```
DataCAD template file. version 01.10.
1
4
*
sym\12-furn\bedroom\beds\twin
sym\12-furn\bedroom\beds\full
sym\12-furn\bedroom\beds\queen
sym\12-furn\bedroom\beds\king
```

The “DataCAD . . . 01.10” line is the header. It is important to maintain this header format without any changes or the template will not work. The 1 tells DataCAD to display one horizontal row of template windows, and the 4 tells DataCAD to display four vertical columns of windows when you turn on the template in DataCAD. The * is a required character that separates the row and column numbers from the next group of lines. Those lines tell DataCAD where to find the symbols. The actual name of the symbol is the final name after the last \ character. The .SM3 extension is not written. The name of the first symbol is therefore TWIN.SM3, and it can be found in the \SYM directory in the \12-furn\bedroom\beds\ subdirectory.

Notice that the full path of the symbol does not need to be written.

NOTE: *The total number of template windows, as defined by the number of rows and columns, does not have to equal the number of symbols referenced by the template. You can have 10 template windows while having only six symbols. The extra windows will simply appear empty. You can also have fewer windows than you have symbols, so you could have six template windows and 10 symbols. Without template windows to display them, however, you will not be able to see or access those additional four symbols. You should be careful of this as you add symbols to a template. If you need more template windows, you can use the* ***Division*** *option in the* ***Template*** *menu to add more windows.*

This template file example will also work if your symbols are located in either one of these actual locations. Notice that the entire path is shown.

- **c:\cadstuff\dcadwin\sym\12-furn\bedroom\beds\twin**
- **c:\dcadwin\sym\12-furn\bedroom\beds\twin**

Writing template files with only the SYM\ directory shown first and not the full path to the symbols is called the *Relative Path method*. The other method is called the *Absolute Path method*. The previous example, written in this later method, would look something like this:

```
DataCAD template file. version 01.10.
1
4
*
c:\dcadwin\sym\12-furn\bedroom\beds\twin
c:\dcadwin\sym\12-furn\bedroom\beds\full
c:\dcadwin\sym\12-furn\bedroom\beds\queen
c:\dcadwin\sym\12-furn\bedroom\beds\king
```

The difference is that the Relative Path method only identifies the general or relative location of the symbols as being in the \SYM directory, which is assumed to be located within the main DataCAD directory. The Absolute Path method identifies the exact location of the symbols as being on the C:\ drive (in the previous example), in the dcadwin\SYM directory. Both methods have some distinct advantages and disadvantages and how they behave depends partly on whether you have your symbols located on a local hard drive or a network drive. More on these two methods later.

The symbols within a template file do not all have to be stored in the same folder locations, or even on the same computer or hard drive. They can be located anywhere, as in this example from part of a template file:

```
c:\dcadwin\sym\00-grphc\drawing\north1
d:\symbols\02-site\bsbnc2\bsbn2-1
s:\sym\11-equip\kitchen\bikc1\bikc1-1
s:\sym\16-elec\elec\rcptcl\1recout
```

Placing Symbols

Placing symbols is as simple as pick-and-click, but numerous options are available to extend their usefulness. Once placed in the drawing, even more options are available that you'll want to become very familiar with.

Before we talk about placing symbols, you need to know what a *symbol insertion point* is. When a symbol is created, you must define the point to which the cursor is automatically attached when the symbol is later selected from a template. An insertion point is the point around which the entire symbol is defined. This becomes even more important when you use symbol editing functions like **Replace** and **Redefine**. Symbol insertion points should therefore be chosen to make future positioning and editing easy. For instance, the outside corner of a piece of furniture would make sense.

An insertion point is displayed as a small X in the symbol, but it is only displayed after the symbol is placed in the drawing, as shown in Figure 14-3.

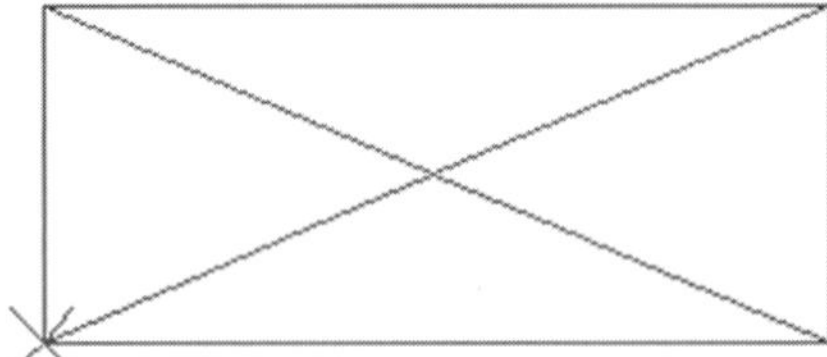

Figure 14-3
An insertion point is displayed as an X after the symbol is placed in the drawing.

An insertion point does not have to be located on or near an entity within the symbol. It doesn't even have to be located anywhere near the symbol's entities. It could be a mile away if you so desire, but in most situations you will want the insertion point to be located within or near the symbol's entities.

Picking and Snapping

To place a symbol, follow these steps:

1. Place your cursor in the template window containing the symbol and then *click* the mouse or press **Enter**. The symbol will be attached to the cursor at the insertion point that was defined when the symbol was created. As you move the cursor around the drawing, rather than displaying all the entities that make up the symbol, a bounding box will be displayed at the outer extents of the symbol's entities, like the image on the left in Figure 14-4.
2. To place the symbol in the drawing, just click the mouse. The symbol's insertion point will be placed at the location of the cursor. To place a symbol more precisely, you can use object snapping (*middle-snap* or **Nn**) to place the symbol's insertion point at any of the currently active object snap points (**Utility/ObjSnap**).

Placement Options

Before selecting a symbol in a template window, you can select several options that will affect how the symbol is placed in the drawing.

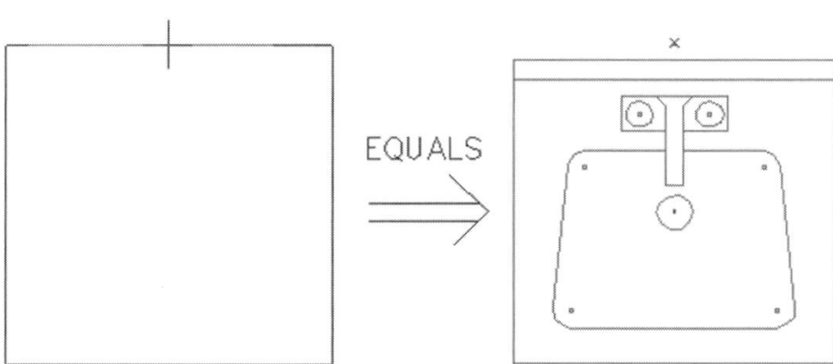

Figure 14-4 A bounding box is displayed at the outer extents of the symbol's entities.

DynamRot (Dynamic Rotate) With the **DynamRot** option turned off, DataCAD defaults to inserting symbols at the current default rotational angle. In most cases, you will want to leave this angle at 0 so that symbols are inserted just as you see them in the template window. To change the default insertion angle, perform the following steps:

1. Turn **DynamRot** on by clicking on the button and then *click* on it a second time to turn it off. You will be prompted to *"Enter angle of rotation from keyboard."* The current default value will be shown.
2. Type in a new angle or select an angle from the menu. To match the rotation angle to an existing drawing entity, select the **Match** function. Pick an existing entity and DataCAD will enter the angle (you will also see several other options that are described later). The other option for selecting an angle under the **Match** function is to select **2Points** and define the angle by selecting two points in the Drawing Window. *Right-click* to accept the new angle.
3. *Right-click* or press **Enter** to accept the new angle. You will be sent back to the main **Template** menu. All symbols will be entered at the new angle until you change the setting again.

When selecting the **Match** function, you will see several other options:

- **Original** This option reverts back to the originally selected angle. This is useful if you change the angle using **Invert**, **Complmnt** or **Suplemnt**, and then want to return to the original angle that you started with.
- **Invert** Use this option to pick the inverse of the selected angle. The inverse is 360 degrees minus the selected angle. For instance, if the current angle is 35 degrees, the inverse angle will be 325 degrees (360 − 35 = 325).
- **Complmnt** Use this option to select the complimentary angle of the currently displayed angle. Complementary angles are two angles whose measures combine to equal 90 degrees. For instance, if the current angle is 35 degrees, the complimentary angle will be 55 degrees (35 + 65 = 90).
- **Suplemnt** Use this option to select the supplementary angle of the currently displayed angle. Supplementary angles are two angles whose measures combine to equal 180 degrees. For instance, if the current angle is 35 degrees, the supplementary angle will be 145 degrees (35 + 145 = 180).

With **DynamRot** turned on, you can place a symbol and rotate it dynamically to the position you desire:

1. Turn **DynamRot** on.
2. Pick a symbol from the template window.
3. Place the symbol where you want it in the Drawing Window.
4. As you move the cursor away from the symbol, a handle will extend out from the symbol's insertion point. Use this handle to rotate the symbol about the insertion point. Press **Oo** to turn the **Ortho** setting on or off while rotating the symbol.
5. Once the symbol is turned to the angle you want, *click* the mouse to place the symbol.

Enlarge Use this option to place symbols at a larger or smaller size than they were created. You can enlarge symbols in any one axis, a combination of the three axes (X, Y, or Z), or in all three axis. Each axis can have its own enlargement factor. For instance, you could enlarge the X axis by 1.5 times, the Y axis by 2.7 times, and the Z axis by .75 times all at the same time. To enlarge something equally in all axes, select the **SetAll** option, which saves you the time and trouble of having to enter the same factor for each axis individually.

You can also enlarge the spacing or line factor of linetypes within a symbol with the **LineFact** option. By selecting this option and entering a factor, all the linetypes in the symbols will be enlarged by the factor entered:

1. Select **Enlarge** and pick the enlargement axis (X, Y, Z, or **SetAll**). In most cases, you will select **SetAll** to enlarge equally in all three axis.
2. Enter an enlargement factor and then *right-click* or press **Enter** to accept it. For example, a factor of 2 will enlarge the symbol by two times.
3. To enlarge the spacing or line factor (**LineFact**) of all linetypes within a symbol, select the **LineFact** option, enter a factor, and then *right-click* or press **Enter** to accept it. **SetAll** will affect this setting as well as the X, Y, and Z factors.
4. *Right-click* or press **Exit** to exit the **Enlarge** menu to accept all the settings.
5. Select a symbol and place it in the drawing.

If you enter a negative number for an enlargement factor, then the symbol will be placed in the drawing mirrored, or "flipped over." This can sometimes be useful to you, but I have on occasion done this accidentally and wondered why my symbols were being mirrored. Selecting a negative X enlargement will mirror the symbol along the Y-axis. Selecting a negative Y enlargement will mirror the symbol along the X-axis. Selecting a negative Z enlargement will mirror the symbol so that it is upside down.

Once you set the enlargement factors, they will remain with those settings until changed to something else, so don't forget to check the current enlargement factors before placing future symbols. To check the current enlargement factors, simply select the **Enlarge** option and DataCAD will show you the current settings at the bottom of the screen. However, only the X, Y, and Z factors are shown. The line factor is not displayed, so there's no visual way to tell what that current setting is.

Z-Offset Select this option to insert symbols at a selected vertical height above the current Z-Base. For instance, if the current Z-Base is 12′-0″ [3657], selecting a Z-Offset of 3′-0″ [914] will place the symbols at a total Z-height of 15′-0″ [4571]. This setting will remain in effect until it is changed back to 0 or to some other dimension.

Explode Selecting this option will cause all the entities in the symbol to revert to being separate entities once again. These entities are no longer databased as a symbol. All the exploded entities are placed on the currently active layer. Generally, the only time you would place symbols exploded is if you need to edit any of the entities in the symbol.

Direct Symbol Selection

You can also select a symbol directly without having to open a template at all. This method assumes that you know where the symbol is located and what its name is. Its real purpose is to allow the direct selection of a symbol with an icon toolbar button, but you can also use it more directly within DataCAD. Follow these steps to do so:

1. Select the quote (″) key, or select **Insert/Symbol** from the drop-down menus. A dialog box with the current symbol files path will be displayed (see Figure 14-5).
   a. Note that the dialog box may not be showing a directory with symbols in it. That's because the dialog box that opens is determined by the path currently shown in **Tools/Program Preferences/**

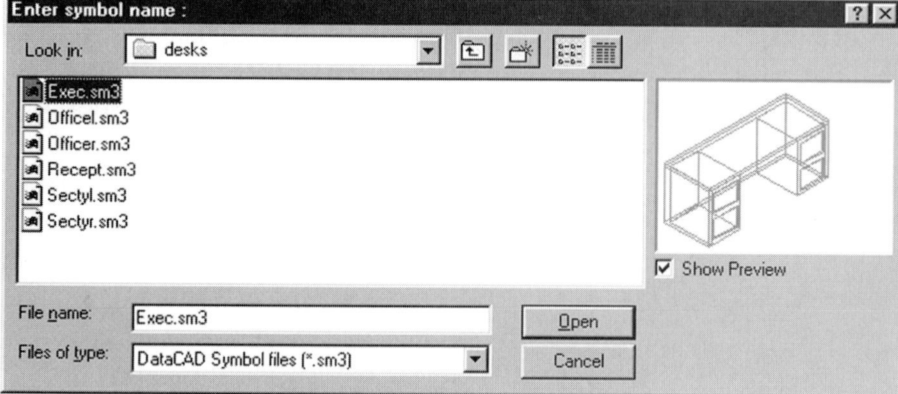

Figure 14-5 Directly selecting a symbol without a template

Pathnames, which often is just the path to your main DataCAD directory, such as C:\DCADWin\.

2. Navigate to the symbol directory containing the symbol you are looking for.
3. Click once on the symbol to see it displayed in the preview window to the right (as long as "Show Preview" is checked, as in Figure 14-5).
4. Once you find the symbol you want to place, make sure it is highlighted and then press **Enter** or select **Open**. The symbol is attached to your cursor, ready for placement in the drawing. Note that your only option is to place the symbol by its insertion point, where the cursor is attached.
5. The Menu Window will display several more options (explained earlier). Select any of these options prior to placing the symbol.
6. Place the symbol in your drawing.

In the previous example, note the options in the dialog box:

- **Files of type** The default option is for DataCAD Symbol files (*.sm3). By pressing the down arrow, you can also select DataCADplus Symbol files (*.sm+). DataCAD Plus is a new program from DataCAD LLC that is lacking some of the features of DataCAD 9, but adds others, such as parametric drawing objects.
- **Show Preview** Checking this box will display an image of the selected symbol when you click once on the symbol in the main selection window. The preview image in the example is displayed in

an isometric view. You have the option to display all symbols this way, even 2D symbols. To turn this option on or off, go to **Tools/Program Preferences/Misc/Symbol Preview**. Select **Orthographic View** for a 2D plan view of symbols. Select **Isometric View** for a 3D isometric view.

After selecting a symbol you will see five familiar options: **XEnlgmt**, **YEnlgmt**, **ZEnlgmt**, **DynamRot**, and **Explode**. For an explanation of these options, see the descriptions earlier in this chapter.

Editing Options

Once placed in the drawing, you may later need to edit the symbol, or you might want to replace it with another symbol. DataCAD provides several options to accomplish these tasks.

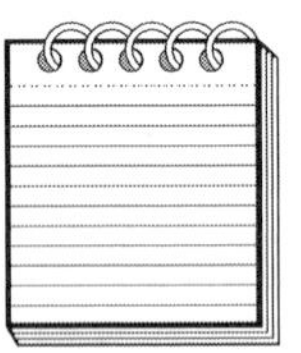

NOTE: *If you are accessing symbols directly from a CD-ROM, then you cannot edit or make changes to symbols and templates since CD-ROMs are a read-only media, and you cannot save the changes back to the CD-ROM.*

DelSymb (Delete Symbol) Use this option to permanently remove a symbol from the template and/or from the computer. When you select **Delete**, you will be asked by DataCAD if you really want to do this. If you answer **Yes**, the symbol will be deleted from the template file, though it will still be visible in the template window. Next, you will be asked if you want to delete the actual .SM3 symbol as well. If you answer **Yes**, then the symbol will be permanently removed from the computer. If you select **No**, then the symbol will remain on the computer, but the template will still no longer show the symbol. If no other template files on your computer refer to that symbol (remember that more than one template file can reference the same symbol), then you will have "orphaned" the symbol, since there is now no way to find it via a template. It exists on the computer but cannot be accessed by a template.

SymName (Identify Symbol Name) To identify the full path and name of a symbol in a template window, select the **SymName** option and click on a symbol in one of the windows. As always, the Coordinate Readout line will display the Item Name of the symbol, as it was entered when the symbol was created. The **SymName** option, however, will additionally display the full path of the symbol in the Message Line at the bottom of the screen with the .SM3 symbol name at the end of the path. Even if the .TPL template file references the symbol via the relative path method, the full path will be shown using the **SymName** command.

Replace This option is one of the truly great features of DataCAD's symbols. It will enable you to replace one or more instances of a symbol already in the drawing file with a different symbol. Let's say you had 80 symbols of square fluorescent light fixtures in your ceiling plan, but you later decide that you want them all to be symbols of round incandescent fixtures. Instead of having to erase the square fixtures and then add the round fixtures in their place, you can tell DataCAD to **Replace** all 80 square fixtures with the round ones.

You must be very aware of one thing, however. This is where the location of the symbol reference points becomes very important. When a symbol is replaced, DataCAD identifies the insertion point of each symbol in the drawing file and then places the insertion point of the new symbol where the old one was. So if the insertion point of the square fixture was in the upper-left corner, and the insertion point of the round fixture were in the center, you will end up with round fixtures that are not centered like the square fixtures were.

To use **Replace**, follow these steps:

1. Turn on all the layers that have the symbol that you want to replace. DataCAD will only replace symbols that are in layers that are currently on.
2. Select **Replace**. You will be prompted to *"Point to the symbol to replace."*
3. *Click* on any one instance of the symbol in the Drawing Window. You will then be prompted to *"Point to the symbol to replace with."*
4. *Click* on the new symbol in the template window that you want (you are not confined to picking symbols from a template. You can pick any symbol that is already in the drawing file as well).
5. Turn **LyrSrch** on or off as appropriate.
6. Select all the symbols to be replaced by using one of the **EGAFS** selection methods.
7. Use the **All** option to replace every instance of the symbol in the drawing file. Remember that DataCAD will only replace symbols that are in layers that are currently on.

When you first select the **Replace** function, the **AnySymb** option will be displayed. This can be a useful function but can also be dangerous if misused, since it will literally replace any symbol selected with **EGAFS** or **All**. For instance, if you select **Replace/AnySymb** and then choose to replace by **Area**, every symbol within that area box will be changed to the new symbol, even if that includes toilets, sinks, and furniture symbols.

To use **AnySymb**, perform the following steps:

1. Select **Replace/AnySymb**. You will be prompted to *"Point to the symbol to replace with."*
2. *Click* on the new symbol in the template window that you want.
3. Turn **LyrSrch** on or off as appropriate.
4. Select all the symbols to be replaced by using one of the **EGAFS** selection methods. Any symbol you select, no matter what it is, will be replaced.
5. Use the **All** option to replace every instance of every symbol in the drawing file. 99.9 percent of the time you will never want to select this, since it will change every symbol on every activated layer in the drawing file, whether that includes toilets, sinks, furniture or any other symbols. Remember that DataCAD will only replace symbols in layers that are currently on.

Redefine Use this option to update or change an existing symbol in a template. As always, be mindful of the insertion points you define. If you are updating an existing symbol, then you will probably want to make sure the new symbol's insertion point matches the old one.

If you have a particular symbol already placed in the drawing file, when you **Redefine** the symbol in the template, all those symbols in the drawing file will be updated immediately. If you open another drawing file with the same symbol in it, however, you will notice that both the template and all the instances of that symbol in the drawing file are not displayed with the updated information. This may not seem to make sense at first, but it is actually very sensible. This feature was developed so that redefining a symbol in one drawing would not adversely affect previous drawings with that same symbol in it. Here's what's going on.

Every time a symbol is selected from a template and placed in the drawing file, that particular version of the symbol is added to the drawing database, and it remains there even if the symbol is later deleted from the drawing (see the **Purge** and **Re-load** commands). If you then open a different drawing file, place that same symbol in the drawing, and then decide to **Redefine** it in the template file, this redefinition will not affect the symbol in the original drawing since that drawing has the original symbol definition stored in its database. If you decide that the old drawings should receive the redefined symbol, then you would use the **Re-Load** command, described later.

To redefine a symbol, follow these steps:

1. Go to the **Template** menu and open the template containing the symbol you want to redefine.
2. Select the **Explode** option.
3. Pick the symbol and place it in the drawing file. It will be inserted as exploded entities.

NOTE: *Be sure to make note of the current insertion point of the symbol before placing it in the drawing. You must make sure to define the same insertion point for the new symbol as for the old one. Otherwise, all the previously placed symbols will be offset from the old insertion point after the symbol is redefined.*

If you have trouble finding or remembering the insertion point, here is a workaround that I use often. Before picking the symbol, draw a single line. Make sure ***Explode*** *is on. Place the symbol in the drawing file, snapping to the endpoint of the line. Continue with Step 4. The end of line you drew will be at the old insertion point so that you can snap to it when you have to define the insertion point of the new symbol.*

4. Make your changes to the symbol entities.
5. Select **Redefine**. You will be prompted to *"Point to symbol in template window to redefine."*
6. *Click* on the symbol in the template window.
7. Turn **LyrSrch** on or off as appropriate. Use the **EGAFS** selection methods to select all the entities to be made into the new symbol.
8. Select **Continue** if you have finished selecting all the correct entities.
9. You will be prompted to *"Select reference point for symbol."* Pick or snap to the new reference point. The new symbol will be displayed in the template window in place of the old one. Without doing anything further, all instances of that symbol in the current drawing file will also be updated automatically.

PurgeSym (Purge Symbol) To delete all references to old, deleted symbols from the current drawing file, use the **PurgeSym** option. If you have not done so already, read about the **Redefine** command for an explanation of how DataCAD adds all selected symbols to the drawing database. If you erase all instances of a particular symbol from a drawing, it will still remain in the database. Not only does this take up space, but it also means that if you have redefined a symbol and try to place the new symbol in this drawing, the old symbol will be displayed in the template window and be placed in the drawing file. Using the **PurgeSym** command will fix all this so that DataCAD will no longer reference the old symbol file. A **Purge Symbols** command in the **Utility/Directry** dialog does the exact same thing as this one.

Re-Load This option is closely related to both the **Redefine** and **Purge** commands. It will search all the symbols in the current drawing's database, search the original paths to those symbols, and then update the symbols in the drawing with the latest symbol definitions.

The only way DataCAD knows how to update a symbol is if the path to the new symbol is exactly the same as the path to the old symbol. If you have moved the new symbol to a new path or renamed it, then DataCAD will not update the symbol in the drawing file, since it won't be able to locate the symbol.

You can reload one symbol at a time or all of the symbols throughout the entire drawing file, even if the layers they are on are turned off.

To reload one symbol, perform these steps:

1. Open the template containing the symbol to be reloaded.
2. Select **Re-Load**. You will be prompted to *"Point to symbol in template window to reload from disk."*
3. *Click* on the symbol in the template window to be reloaded.
4. The full symbol path and name will be displayed at the bottom of the screen, and a **Yes** and **No** option is displayed. Select **Yes** to reload the symbol. Select **No** to go back to the previous menu without reloading.

To reload all symbols, perform the following steps:

1. Select **Re-Load**. The **All** option will be displayed.
2. Select **All**. You will be prompted to *"Reload all symbols in drawing?"*
3. Select **Yes** to reload all the symbols throughout the drawing file. Select **No** to go back to the previous menu without reloading.

Although the **Purge** command will purge all unused symbols from a drawing file, the **Re-Load** command will only update all the symbols it can find along the original symbol paths. If any paths or names have changed, not only will those symbols not be reloaded, but they will also still remain in the drawing file database. So even if you use the Re-Load command, it is still a good idea to also use the Purge command.

WARNING: *When you purge symbols, the undo buffer will be purged. The "undo" buffer is the database that DataCAD keeps of all your past actions. DataCAD uses this database to remember what you did in a drawing session so that you can use the Undo and Redo functions. If you think that you will need to use Undo/Redo then do not purge your symbols at this time. Wait until you are sure there is nothing that you need to Undo/Redo, then purge them. Once the buffer is cleared a new one will be started.*

EditFlds Use this option to edit the information that was entered or omitted from each of the available fields when the symbol was created. These fields include the following:

- Item Name
- Manufacturer
- Model Number
- Remark 1
- Remark 2
- Cost

For a description of what can be entered in each of these fields, see the section about creating symbols later in this chapter.

To add or change information in these fields, perform the following steps:

1. Go to the **Template** menu and open the template containing the symbol you want to redefine.
2. Select **EditFlds**. You will be prompted to *"Select Symbol to edit."*
3. *Click* on the symbol to be edited in the template window. You will be prompted to *"Select Field to modify."* The six fields will be displayed in the side menu bar.
4. *Click* on one of the six fields. The button will be depressed and the Command Line will prompt you with the current information for that field.

5. Add to or change the information in the field. Press **Enter** to accept the changes. *Right-click* or press **Exit** if you do not want to save the changes.
6. Repeat Steps 4 and 5 if you want to edit other fields.
7. *Right-click* or press **Exit** to stop editing fields.

Picking Symbols for Editing

Certain guidelines must be kept in mind when you are picking symbols, especially if you have the **Display List** option enabled (**Tools/Program Preferences/Misc/Display List**).

Entity and Group With **Display List** turned on, the **Entity** and **Group** commands will enable a symbol to be selected by any visual entity within it. With it turned off, the **Entity** and **Group** commands can only select symbols by their insertion points.

Area With the **Area** command, a symbol must be completely enclosed within the area bounding box to be selected. This is true even when **Display List** is on.

Area/Crossing With Display List turned on, the **Area/Crossing** bounding box in Figure 14-6 will select the symbol even though the bounding box does not encompass the entire symbol, and even though the symbol's inser-

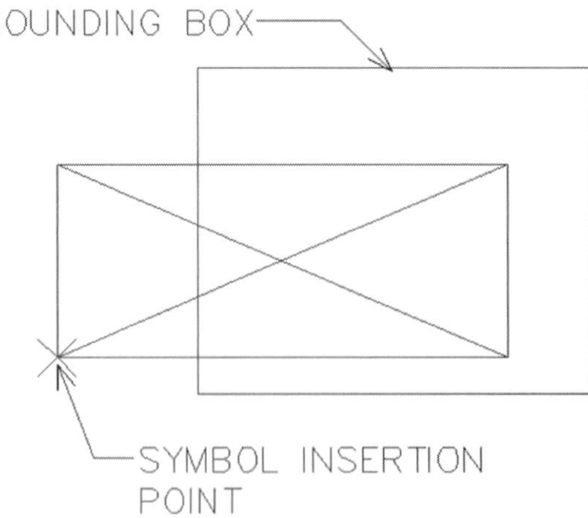

Figure 14-6 The **Area/Crossing** bounding box

tion point (the small X in the bottom-left corner) is not enclosed by the bounding box. With **Display List** turned off, the **Area/Crossing** bounding box will not select the symbol at all, since the entire symbol must be within the bounding box, even with the **Crossing** option enabled.

Some longtime DataCAD users may have used the inability to choose symbols with **Area/Crossing**, except on their insertion point, as a feature and incorporated it into their editing practices. Be advised that without turning display list off, you will no longer be able to work this way because selecting anywhere on a complex entity, such as text, a complex linetype, or a symbol, is considered valid as far as **Display List** is concerned.

Fence With the **Fence** command, a symbol must be completely enclosed within the fence to be selected. This is true even when **Display List** is on.

The SymExp (Symbol Explode) Macro

In the macro **Toolbox** is a macro called **SymExp** (Symbol Explode), which is used to explode a symbol that is already placed in the drawing file. This is exactly like the **Template/Explode** option, except that this macro enables you to explode symbols after they are already in the drawing. Just as when you place a symbol with the **Explode** option turned on in the **Template** menu, the entities in the symbol will cease to be a symbol. All the entities in that symbol revert to being separate entities once again but are moved from the layer the symbol was on to the *currently active layer*.

Mirrored Symbols

DataCAD has one nasty little quirk that you should be aware of when exploding symbols that have been mirrored. First, a little lesson on how DataCAD handles 2D and 3D entities within symbols.

In DataCAD, a 2D entity can only exist in the 2D X/Y plane, but you can rotate or mirror a symbol with 2D entities about the X or Y axis, thereby placing those 2D entities into the Z plane. However, when you explode a symbol, those 2D entities cannot exist or be displayed in the third dimension. To rectify this, DataCAD attempts to revert all symbols back to their original orientation when exploded. Unfortunately, DataCAD often mirrors the symbols back along the wrong axis, so that the symbol may not actually wind up in the proper orientation after all.

A third-party macro called *Global Edit* has a better symbol explode routine. It's not foolproof, but it does a much better job than the standard Symbol Explode macro.

Nested Symbols

You know the old trick of placing a gift inside a box, then placing that box in a second box, then that one into a third, and so on? The concept of symbol nesting is very much the same. You can create one symbol and then create a second symbol with the first one inside of it. DataCAD supports nested symbols up to eight levels deep. Each time you explode a nested symbol only the last symbol level is exploded. So if you explode a symbol that is eight levels deep, only the symbol you last created is exploded, while the other seven symbol levels remain intact as symbols. How can this help you? We're glad you asked.

Symbol Nesting 101

You can use a group of previously saved symbols to create a new symbol, rather than first exploding them or redrawing the whole thing. In Figure 14-7, every element of the plan is a symbol. The counter, stove, sink, and refrigerator symbols are then saved together to form a new symbol of the complete kitchen plan. The individual symbols are therefore nested in the new symbol.

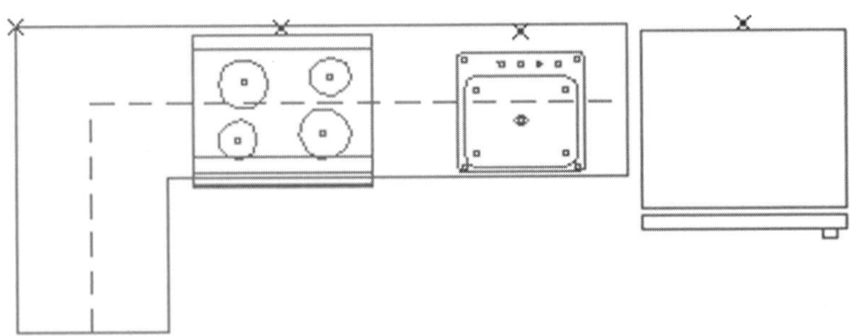

Figure 14-7
Nesting symbols in a kitchen plan

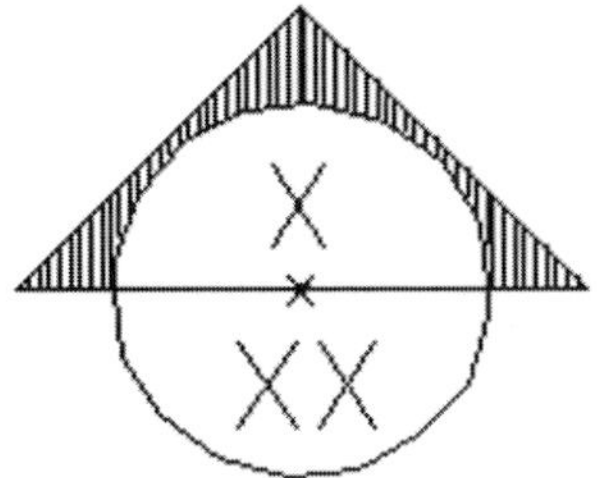

Figure 14-8
A nested symbol with text

You might also want to create a symbol of your entire floor plan that could be brought into another drawing file to use as a reference. Because symbols can be nested, you can do this even if the floor plan already contains symbols.

Symbol Nesting 201

Figure 14-8 was created by first making a symbol of the detail bubble without any text in it. With that symbol placed in the drawing unexploded, we added the text (the X's are simply placeholders for text that will be changed later). Then we made a new symbol that included both the first symbol and the text. Now here's the really cool part.

If you turn the **Explode** option on in the **Template** menu and then place the new symbol, only the last saved level of the symbol will be exploded. This means that the text is exploded and is no longer part of the symbol, but the detail bubble remains a symbol. By using **Change/Text/Contents**, you can now change the text in the detail bubble. This ensures that you always have the correct font name, size, aspect, color, and so on when placing the detail bubble.

Symbols in DataCAD do not have layers, but a third-party macro called *Bundle* (part of the DC Sprint macro available from DataCAD LLC) enables you to save symbols with layers. When you explode the symbols, the layers are created in the current drawing (if they don't already exist) and the entities are exploded to their respective layers.

Symbol Instancing

One of the hallmarks of symbols in DataCAD is their capability to be "instanced" in the drawing database. This means that the definition of a symbol is saved to the drawing database only once, no matter how many instances of the symbol are subsequently placed in the drawing. So even if you place 1,000 Sink.SM3 symbols in the drawing, the drawing database only contains the definition for one of those symbols. DataCAD simply references that information for the remaining 999 instances of that symbol without making your drawing database larger. This is very helpful if you are placing lots of 3D trees composed of 500 entities into a plan. The 500 entities are added to the database only once. All the rest are simply instanced. (Actually, each instance of the symbol would add one entity to the database, which is still far less than the 500 entities.)

The definition of a symbol remains in the drawing database even if every instance of that symbol is erased from the drawing. This can eventually bloat your drawing file if too many of these are left behind. To fix this, you can run the purge command (**Template/PurgeSym** or **Utility/Directry/Purge Symbols**), which will empty the database of all unused symbol definitions.

Relative and Absolute Paths

We mentioned this topic at the beginning of the chapter. Now that you know how symbols are used and edited in DataCAD, you need to know more about how symbols are created and saved. To recap, the Relative Path method only identifies the general or relative location of the symbols as being in the \SYM directory, which is assumed to be located within the main DataCAD directory. The Absolute Path method identifies the exact location of the symbols. Here are some examples:

Relative Path method: sym\12-furn\bedroom\beds\twin

Absolute Path method: c:\dcadwin\sym\12-furn\bedroom\beds\twin

Relative Path Method

All the original DataCAD template files that come with the program are written with the Relative Path method. Presumably, part of the reason behind this is so that the installation program does not have to deal with writing new paths to all the template files, based on what drive and path the user might want their symbols in. Using this method, the TWIN.SM3 file, mentioned earlier, would get saved like this:

\sym\12-furn\bedroom\beds\twin

NOTE: *One thing about the DOS version of DataCAD. As long as your symbols are in a subdirectory called \SYM, located in the root DataCAD directory, DataCAD will find your symbols regardless of whether they are written via the Relative Path or Absolute Path methods. Once you place the SYM directory outside of the root DataCAD directory, however, DataCAD gets very touchy about how its template files are written.*

What's **good** about using the Relative Path method?

- If done properly, it enables you to freely relocate your \SYM directory at will. For instance, you could move your symbols from the C:\DCAD\SYM directory to the D:\DCADWIN\SYM directory and they would still work.
- It makes it easy to transfer and copy TPL and SYM directories to other users and computers and have them work without rewriting any of the .TPL (template) file paths.
- As noted previously, as long as your symbols are in a subdirectory called \SYM, located in the root DataCAD directory, Relative Path and Absolute Path (explained next) definitions will coexist peacefully.

What's **bad** about using the Relative Path method?

- If your symbols are not located in the \SYM directory in the root DataCAD directory (perhaps they are on a network drive or server like S:\SYM), mixed Relative Path and Absolute Path definitions will cause your templates to go horribly wrong. This is because when a new symbol is saved, that path of that symbol is appended to the symbol path setting in the **Tools/Program Preferences/Pathnames** menus (or the CONFIG program in the DOS version). This is the path used by the AUTOPATH option in DataCAD (see the end of this document for an explanation). After that happens, DataCAD will be looking in the wrong location for symbols.
- If your symbols are located in a different directory than the root DataCAD\SYM directory, keeping your Relative Path templates working properly takes a lot of user intervention. This is because DataCAD does not enable you to save new symbols via the Relative Path method, so you will have to open the .TPL file every time you create new symbols in order to manually delete the full path to the symbol. More on this later.

Curiously, one difference exists between the way the DOS and Windows versions of DataCAD handle the path settings in the **Tools/Program Preferences** (Windows) or Configuration (DOS) menus for Relative Path templates. In DataCAD DOS, as long as the \SYM directory is in the DataCAD root directory, DataCAD will always find that \SYM directory, even if the path in the Configuration program points to a different drive or directory. So in the DOS version, none of these path settings would keep DataCAD from finding the \SYM directory (assuming DataCAD is in C:\DCADWIN):

- Symbol files path = C:\DCADWIN\SYM\
- Symbol files path = C:\
- Symbol files path = D:\
- Symbol files path = C:\WINDOWS\

However, the same is not true for the Windows version of DataCAD. It is very sensitive to the path settings for the \SYM directory. In the previous \sym\12-furn\bedroom\beds\twin example, only the following will work (in the **Tools/Program Preferences/Pathnames** menu):

```
Symbol files path = C:\DCADWIN\
```

If your symbols are located on drive S:\ in the SYM directory (S:\SYM), then you must have your path set to

```
Symbol files path = S:\
```

So if you ever have the problem in which your template windows are displayed without symbols in them, check your path settings and your template paths in the .TPL file.

IMPORTANT: *Note that there is no way in DataCAD to save new template files via the Relative Path method. The only way to do it is to first save the symbols in DataCAD. Next, you would have to open that .TPL file in an plain text editor (like Windows Notepad) and delete all Absolute Path references, like this:*

Original .TPL line: t:\sym\80-detls\test
Revised .TPL line: sym\80-detls\test

Absolute Path Method

The other way to save symbol locations in a template file is via the Absolute Path method. This is the most consistent and foolproof method of creating and saving symbols, and only becomes tedious if you need to move symbols to another path on your computer or network or if you want to share symbols with another DataCAD user. Since those situations are the exception rather than the rule, it is recommended that you use the Absolute Path methods. Using this method, the TWIN.SM3 file might get saved in the template file like this:

```
c:\dcadwin\sym\12-furn\bedroom\beds\twin
```

This means DataCAD will only search for the TWIN.SM3 file in that specific location and nowhere else. If you decide to change your DataCAD directory name from c:\dcad to c:\dcadwin (as in the previous example), then you will have to rewrite all of your template files to start with c:\dcadwin instead of c:\dcad.

This becomes a long and tedious process to rewrite a hundred or more template files, but Windows utilities help eliminate much of this tedium. You might ask, "When would I ever want to rename or move my template or symbol directories?" Perhaps for the following reasons

1. If you start a network in your office, you will want the template and symbol directories on their own drive partitions (like T:\ and/or S:\).
2. If you want to share your symbols with consultants
3. If you need to move or copy symbols to a computer with different template and symbol paths

Recommendations

Here are some recommendations for using the Absolute Path method:

- If you plan to keep all your symbols and templates in the root DataCAD directory (such as C:\DCADWIN\SYM) and you are not going to put them on a network server, then leave the Relative Paths that are created when you install DataCAD.
- If you are installing your templates and symbols in directories other than the root DataCAD directory (such as D:\SYM, S:\SYM, or

T:\TPL), then you should change all the symbol paths in the .TPL files to Absolute Paths, pointing to the proper location. To do this, use a program like the Ultra Edit program (a shareware program included on the CD-ROM) or something similar to change them all at once. By doing this, it will be even easier in the future to make mass changes to your .TPL files, since replacing by C:\SYM\ . . . is more accurate than just SYM\, since many templates may contain names like SYM\TABL-SYM\TABLE-1, which would confuse a Search/Replace function in a text editor.

Accessing Symbols from a CD-ROM

You can access symbols directly from a CD-ROM drive without having to first save the templates and symbols to your hard drive. This is useful for accessing the symbols on the DataCAD CD or the Cheap Tricks Ware CD (see Chapter 27, "DataCAD Resources"). To do this, you need to change the path settings for both the templates and the symbols:

1. Select **Tools/Program Preferences/Pathnames**.
2. Leave the **Template Files** path alone. There is no need to change anything here.
3. *Click* on the file folder button to the far right of the **Symbol Files** window and navigate to the CD-ROM drive. *Click* on the drive icon and select **OK**. The path should now say something like **E:**, assuming your CD-ROM is the E: drive. *It should not have the \SYM directory in its path*. This is the key to making this work.
4. Select **OK** to close the Program Preferences dialog box.
5. Now go to **Utility/Template/NewFile** and navigate to the template folder on the CD-ROM. If you are accessing the DataCAD CD, then this will be either the **Tpl** or **CSI-tpl** folders.
6. Select the template file you want to access and press **Open**. The symbols on the CD will appear in the template windows.

The reason all this works is that the template files on the CD all point to the symbols in a folder called SYM. With the path set to **E:** in the **Symbol Files** window in Step 3, DataCAD will add the two paths together and look for the symbols at E:\SYM. If you entered E:\SYM in Step 3 instead, then DataCAD would look for symbols in a folder called E:\SYM\SYM, which doesn't exist.

Locations of \TPL and \SYM Directories

If you are a sole practitioner running DataCAD on only one computer, then you will most certainly want to leave your templates and symbols in the local \TPL and \SYM directories within the main DataCAD directory, like this:

```
C:\DCADWIN\TPL
```

and

```
C:\DCADWIN\SYM
```

But if you run DataCAD from two or more computers in the same office, you will certainly want to network them together (see Chapter 26, "Networking") and then locate all the templates and symbols on only one computer. If you don't, then every computer in the office will be placing and saving only the templates and symbols that they have on their own hard drives. As each user creates new templates and symbols on his or her own computer, they will not be accessible to anyone else. To make them available to everyone, each day you will have to manually copy all the new templates and symbols from every computer to a floppy disk and then copy them to everyone else's computer, making sure to get all the paths correct. This is tedious, inefficient, and highly prone to errors.

On the other hand, networking the templates and symbols creates some very important and distinct advantages:

- Networking is the only way to make sure that every user is using the same and most current symbols.
- Everyone on the network will have immediate access to every template and symbol in the office.
- Every time a new template or symbol is added, it is immediately available to everyone else on the network.

If you have two or more DataCAD workstations in the same office, there is really no reason that you should not have them networked together and set up to share not only templates and symbols, but drawing files, plot files, and so on. See Chapter 26 for a full explanation of networking your computers, files, templates, and symbols.

CSI Format or Not?

CSI is an acronym for the Construction Specifications Institute, which has developed the 16 material and product classifications that construction industry professionals are familiar with. For instance, Section 5 deals with metals, and Section 6 deals with wood and plastics. Since most everyone is familiar with these sections, this makes specifying materials easier for all involved.

When you install DataCAD, you have the option of installing all the templates via the CSI classifications system or via a more generic classification. We heartily recommend the CSI format because of its generally universal acceptance and its more intuitive breakdown.

Installing from the DataCAD CD

We mentioned earlier that the templates are installed in CSI format. We did not make an error in omitting the word "symbols" along with it, since the DataCAD installation does not install the symbols in CSI format. The symbols are installed in a more generic format. Don't ask me why this is, but it is certainly an omission in my eyes. Here are the folders created when you choose to install the template and symbol files in CSI format:

Template Directories	Symbol Directories
00-GRPHC	Concrete
01-GENRL	Conveync
02-SITE	Door&win
03-CONC	Electrcl
04-MSNRY	Equpmnt
05-METAL	Finishes
06-WD&PL	Furnitur
07-TH&MP	General
08-DR&WN-	Job01
09-FINSH-	Masonry

Template Directories	Symbol Directories
10-SPEC	Mech
11-EQUIP	Metals
12-FURN	Moisture
13-SP_CO	Site
14-CONVE	Sp_const
15-MECH	Specity
16-ELEC	Symbols
80-DETLS	Walltype
90-JOBS	Wood&plas

The final two template sections, Sections 80 and 90, are not part of the standard CSI categories. They are empty directories provided so that you will have a ready location to begin saving your own templates that don't fall into one of the other categories.

To fix the situation noted earlier, we have included a "Template and Symbol Reorganizer" program on the CD-ROM. It's a DOS program I wrote a few years ago to reinstall the DataCAD templates and symbols so that both the template and symbols are in CSI format. Even if you don't choose to use this program, you ought to consider creating CSI-formatted symbol directories for storing all your future symbols, as shown in the following table:

Template Directories	Symbol Directories
00-GRPHC	00-GRPHC
01-GENRL	01-GENRL
02-SITE	02-SITE
03-CONC	03-CONC
04-MSNRY	04-MSNRY
05-METAL	05-METAL

Template Directories	Symbol Directories
06-WD&PL	06-WD&PL
07-TH&MP	07-TH&MP
08-DR&WN-	08-DR&WN-
09-FINSH-	09-FINSH-
10-SPEC	10-SPEC
11-EQUIP	11-EQUIP
12-FURN	12-FURN
13-SP_CO	13-SP_CO
14-CONVE	14-CONVE
15-MECH	15-MECH
16-ELEC	16-ELEC
80-DETLS	80-DETLS
90-JOBS	90-JOBS

Using Ultra Edit to Edit Multiple .TPL Files at Once

The CD-ROM contains a shareware utility called Ultra Edit that will enable you to open all of your .TPL files at one time, change all the paths in them in one fell swoop, and save them with one click. With this utility, it took me 10 minutes to change the paths of a little over 200 .TPL files.

Let's assume you want to change all your Relative Path template files to absolute paths. First, back up your TPL directory just in case. Then follow these steps:

1. Open Ultra Edit.
2. Change one configuration setting:
   a. Go to **Advanced/Configuration/Backup**.
   b. Check **No Backup**.
   c. If you don't do this, then all the files you change will have .BAK files saved as well, which you will have to manually clean up later.
   d. If you want the backup file, then skip all of Step 2 and go to Step 3.

3. Press **Apply** and then **OK**.
4. Select **File/Open** and Navigate to your template (\TPL) directory.
5. Go into each folder and open every .TPL file. You can pick multiple files in the Windows dialog box by dragging a bounding box around all the files and then pressing Open. This will save a lot of time. Be sure to open all your .TPL files at once. This is what will save you a great deal of time.
6. Choose **Search/Replace**.
7. Assuming your templates currently say —> SYM\ . . . , do the following:
   a. **Find What** = SYM\.
   b. **Replace With** = S:\SYM\ (or whatever your path is).
8. Check **All Open Files** and then select **Replace All**. All your files will be changed.
9. Press **Cancel** when finished.
10. Select **File/SaveAll**.
11. Select **File/Close All Files**.

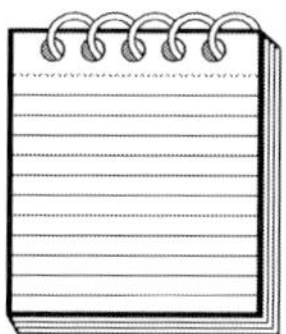

NOTE: *You must be careful of one thing. If some of your templates already have absolute paths, such as*

```
C:\DCAD\SYM\
```

and some of your templates have relative paths, changing all your files at once, as in the previous example, will result in some text being duplicated like this:

```
C:\DCAD\S:\SYM\
```

I got around this by first replacing all C:\DCAD\SYM\ text with SYM\. Then I did the replace just as in the previous instructions.

Also, if a path says something like

```
SYM\NEW-SYM\SINKS
```

then you will get something that looks like

```
S:\SYM\NEW-S:\SYM\SINKS
```

You probably won't run into too many of these, but if you do, you will just have to rewrite the path correctly in a text editor.

Creating Templates and Symbols

Once you have all your symbols in CSI or whatever format you desire, you will want to ensure that all the new symbols you create are maintained in that order. By default, DataCAD tries to place all the symbols in the \SYM directory where you last saved a symbol, which is almost never the correct location. Thus, you must "force" DataCAD to place them in the correct directories.

This turns out to be both easy and difficult. It's easy to do, but difficult to get users to do properly and consistently. The trick is to make sure you write down the full and complete template path when you create it so that you can make an identical path in the symbol directory.

Let's assume our templates are located on the T:\ drive and the symbols are located on the S:\ drive. We want to create a template by the name of TEST.TPL, and we want to place a symbol in it, called FIRSTSYM.SM3. When we are done, it will look like this:

```
T:\TPL\80-DETLS\TEST.TPL
```

and

```
S:\SYM\80-DETLS\TEST\FIRSTSYM.SM3
```

Creating a Template

To create a template, follow these steps:

1. Select **Utility/Template**. A dialog box appears (assuming there is not already a template being displayed).
2. Navigate to the directory you want to save the new template file in. Let's use this as an example: **c:\dcadwin\tpl\04-msnry**.
3. In the **File name:** box, type in a name for the template, such as **CMU**. Write down the exact path to the new template, such as **c:\dcadwin\tpl\04-msnry\CMU**. You will need this later when you create the first symbol.
4. Press **Open** to accept the name.
5. You will be prompted to ***"Create new file?"*** Select **Yes**.
6. You will be prompted for a ***"Field name:"***. DataCAD has six standard fields that are part of every symbol (for saving the symbol's model number, cost, and so on). If you want to add more custom fields, you can do this when you are prompted for a ***"Field name:"***.

a. Type one in if you want to (explained later).
b. If you don't want anything entered, press Enter to skip it and go onto Step 7.
c. *Right-clicking* will skip you ahead to Step 9.

7. You will be prompted to *"Select Field type."* Select **Text**, **Number** of **Dollar**.
   a. Type in the appropriate text if you want it (explained later).
   b. If you don't want anything entered, press **Enter** or *right-click* to skip it.
8. *Right-click* when done. The **Template** menu reappears, along with a series of template windows at the right side of your screen.
9. To change the number of template windows, select **Divisions**.
10. You will be prompted ***"Number of X-divisions in template:"***. Type or select a number and then press **Enter**.
11. You will be prompted ***"Number of Y-divisions in template:"***. Type or select a number and then press **Enter**.
12. *Right-click* to exit the **Template** menu.

Custom Template Fields

In Steps 6 and 7 earlier, you have the option of adding additional fields to a symbol. This information can later be used to print out a spreadsheet (called a *symbol report*) of symbol information and calculations. DataCAD has six standard fields that are part of every symbol, even if you don't fill them in when creating the symbol:

Field	Value Type
Item Name	Text
Manufact.	Text
Model No.	Text
Remark 1	Text
Remark 2	Text
Cost	Dollar

If you want to add more custom fields, you can do this when you are prompted for a ***Field name***, as described previously in Step 6. You cannot add, delete, or change field names after you create the template, so you must define the new fields while creating the new template. A template file that has added fields will look something like this:

```
DataCAD template file. version 01.10.

3
12
SSymbol
SType
SManufacturer
SModel Number
SBulb
$Wholesale Cost
$Retail Cost
SSpecification
#Installation Time
#Delivery Time
SMisc.
*
c:\dcadwin\sym\16-elec\elec\fl_light\incds1
c:\dcadwin\sym\16-elec\elec\fl_light\incds2
```

Note the additional text between the column and row information (the 3 and the 12), and the symbol paths. The S before the text means that the field is a text field. The $ before the text means that the field is a dollar/cost field. The # before the text means that the field is a numerical field. The text after each of those characters is the text that describes the additional fields beyond the standard six. This is the text that will be printed when you create a symbol report. You can view these additional fields with the **EditFlds** option, but only the first eight characters (not including the S, $, and #) can be displayed on the **EditFlds** buttons in DataCAD. However, all of the characters can be printed to the symbol report. The following section will describe how to create a symbol and how to use these additional fields.

Creating a Symbol

Here is how you would create a symbol in DataCAD and have it saved in a symbol directory that matches the template directory (CSI or otherwise). For this example, let's assume you've already created the template file as in the previous example. Now we want to add some *Concrete Masonry Unit* (CMU) symbols. CMU is located in the CSI directory **04-MSNRY**.

1. Select **Utility/Template**. A dialog box appears (assuming there is not already a template being displayed).

2. Make sure **AutoPath** is turned off.
3. Navigate to the template that the symbol is to be located in. Using the previous example, that would be *c:\dcadwin\tpl\04-msnry\CMU*.
4. Pick **SaveSymb**.
5. The path of the last symbols saved will automatically pop up. For instance, you might see ***"Path S:\sym\90-jobs\20003\cmu\ does not exist. Create it?"***
6. Select **NO**. This is very important, since DataCAD almost never gets the correct path.
7. A dialog box appears, displaying symbol directories. Navigate to the symbol directory with the identical directory name as the template file. Using the previous template example, that would be *c:\dcadwin\sym\04-msnry*. Notice that there is no CMU directory to match the CMU directory created in the previous template example, so we need to create one.
8. Place your cursor inside the main window of the symbol dialog box and *right-click* the mouse. Navigate down the resulting dialog and pick **New/Folder**, as shown in Figure 14-9.
9. Type in the new symbol directory. In this case, this is CMU. Then press **Enter**.

NOTE: *Stop for a moment and think about what we are trying to do here. Most of this is not a requirement for saving symbols in DataCAD, but it is imperative to understand and follow Steps 5 through 9 if you want to keep your symbols orderly. Developing the good habit now of matching your template and symbol directories will serve you long into the future.*

From experience, I can tell you that the most difficult aspect of keeping orderly symbol files is getting your employees to buy into the idea and to follow these steps religiously, but a little effort goes a long way.

10. *Double-click* on the new folder to open it.
11. In the **File name:** box, type in a name for the symbol, such as **8inCMU**. Press **Save** or **Enter**.
12. Select **Entity**, **Group**, **Area**, **Fence**, or **SelSet**. Turn **LyrSrch** on or off as appropriate.
13. Let's assume you are going to use the **Area** command. In this case, you will be prompted with the following: ***"Select first corner of area to <CREATE SYMBOL>."***

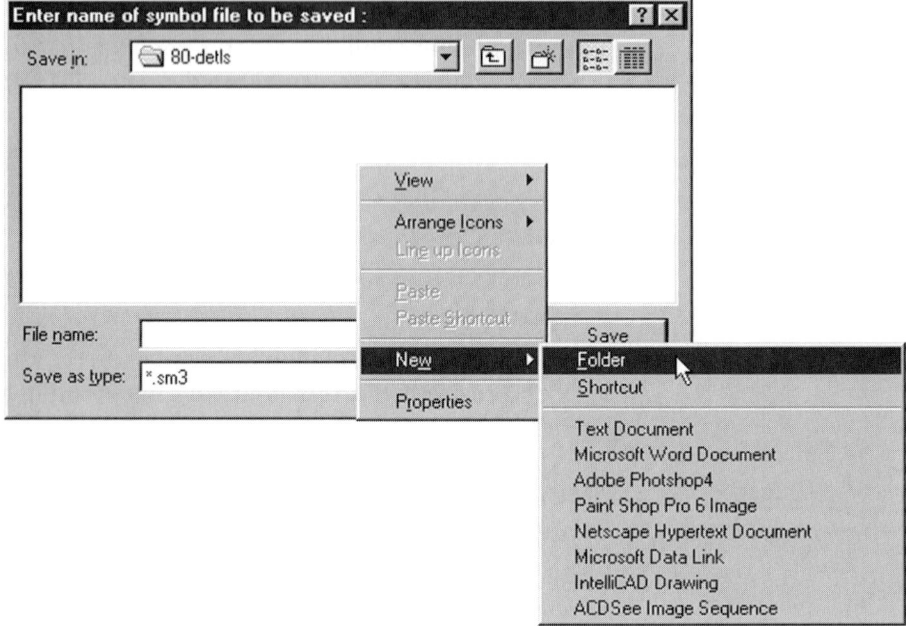

Figure 14-9 Creating a new folder

14. Draw a bounding box around the entities that you want to create a symbol of.

WARNING: *When DataCAD creates a symbol, it destroys the original entities in the Drawing Window. So if you want those entities to remain in your drawing, copy them first and then make the symbol from the copy. Alternatively, you can save the symbol and then select* ***Undo****. The symbol will still exist, but the entities will return to the Drawing Window.*

15. You will next be prompted to ***"Select reference point for symbol."*** Pick a reference point that makes sense for placing the symbol (explained in more detail later). Use the middle mouse button to snap to that point.
16. You will be prompted to enter an ***"Item Name."*** Type in a name like "8 inch CMU" and then press **Enter**.
    a. This will be the text that will appear at the bottom of the screen when you pass your cursor over the symbol in the template and will also be used for symbol reports.
    b. Unlike most things in DataCAD, this name can be up to 30 characters long and can use spaces and special characters.

  c. If you forget to enter the text here or want to add or change the text later, you can do this with the **EditFlds** option in the Template menu.

17. You will be prompted for a **"Manufact.:"** (manufacturer). If entered, this information can be used later for symbol reports:
    a. It can be up to 30 characters in length.
    b. If you don't want anything entered, press **Enter**.
    c. If you *right-click* to skip it, then you will skip forward to Step 23.
18. You will be prompted for a **"Model No:."** If entered, this information can be used later for symbol reports:
    a. It can be up to 30 characters in length.
    b. If you don't want anything entered, press **Enter**.
    c. If you *right-click* to skip it then you will skip forward to Step 23.
19. You will be prompted for **"Remark 1:."** If entered, this information can be used later for symbol reports.
    a. It can be up to 30 characters in length.
    b. If you don't want anything entered, press **Enter**.
    c. If you *right-click* to skip it then you will skip forward to Step 23.
20. You will be prompted for **"Remark 2:."** If entered, this information can be used later for symbol reports.
    a. It can be up to 30 characters in length.
    b. If you don't want anything entered, press **Enter**.
    c. If you *right-click* to skip it then you will skip forward to Step 23.
21. You will be prompted for **"Cost:."** If entered, this information can be used later for symbol reports.
    a. It can be up to 11 digits in length.
    b. If you don't want anything entered, press **Enter**.
    c. If you *right-click* to skip it, then you will skip forward to Step 23.
22. If you created additional fields when you created the symbol template file, you will be prompted for those additional fields. If entered, this information can be used later for symbol reports. You can define any number of additional fields.
    a. If you don't want anything entered, press **Enter**.
    b. If you *right-click* to skip it, then you will skip forward to Step 23.
23. When you are done with all your inputs, the symbol dialog box will appear. You can now go back to Step 11 to save another symbol or click on **Cancel** to stop.

Text fields can accept letters or numbers that cannot be used in symbol report calculations. Dollar fields can only accept real numbers with two fixed decimal places. Number fields accept integers only. You can define any number of additional fields.

Automatic Symbol Numbering

Whenever you save a symbol that has a number at the end of it, such as CMU1 or 8inCMU3, the next time you try to save a symbol, DataCAD will automatically add the next sequential name at the prompt line, such as the 8inCMU4 example in Figure 14-10, just to save you some time. If you want to accept it, just press **Enter**. If not, then type something else.

Picking an Insertion Point

An insertion point is the point around which a symbol is defined. At its simplest level, it is the default point on or within the symbol that your cursor will attach itself to when selected from the template window. At its more complex level, it is the point that all future changes and redefinitions are based on.

The only way to create an insertion point is to pick it when the symbol is first created. You cannot change an existing insertion point without exploding the symbol and resaving the symbol all over again. It is important to pick an insertion point that makes sense, especially if alignment will be important when placing the symbol. For instance, if you made a symbol of

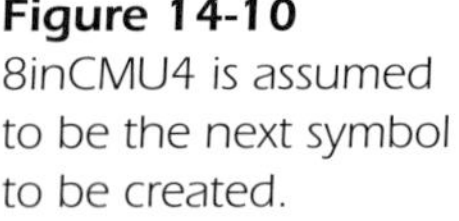

Figure 14-10 8inCMU4 is assumed to be the next symbol to be created.

a wood stud, you would want to pick an insertion point that would enable you to place the stud in alignment with other materials. This stud has its insertion point at one corner of the stud, which makes it easy to snap to the edge of the slab, as shown in Figure 14-11.

These options will be explained later, but if the insertion point (shown by the X at the lower left corner of the stud in Figure 14-11) is not displayed, follow these steps:

1. Check to make sure the **Explode** option is turned off in the **Template** menu prior to placing the symbol.
2. Go to **Utility/Display** and make sure **ShowIns** is turned on.

The insertion point is also critical to using the **Replace** and **Redefine** commands, as you will read about shortly.

Automatic Symbol Pathing

The **AutoPath** function in DataCAD is a good concept gone awry. The DataCAD manual says it "appends a subdirectory, with the same name as the template file, to the symbol path that you set in **Tools/Program Preferences/Pathnames**," but its implementation falls far short of that seemingly simple description.

The concept was that you shouldn't have to keep track of where the symbols are saved, so the **AutoPath** setting would automatically save them to a logical location based on the name of the current template file. It does this

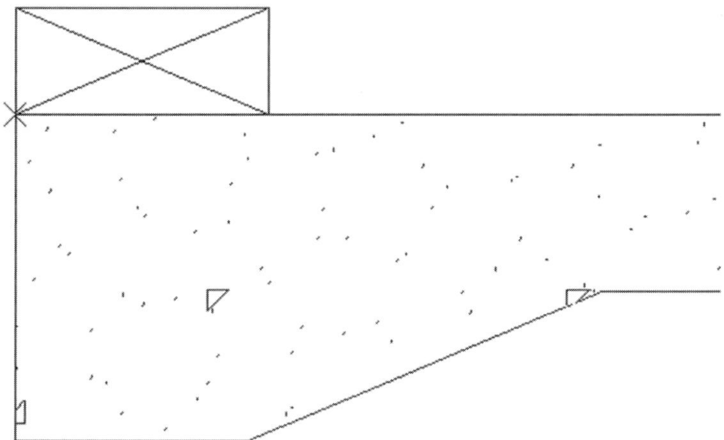

Figure 14-11
A wood stud with its insertion point at one corner

by creating a new path with the name of the template file and appends it to the current symbol path. The idea was that the symbol location would always be in the DataCAD\SYM directory, so that all the symbols would be saved to a location with the current template's name. If the template file is called SINKS, then the symbols would be saved to \SYM\SINKS. What really happens, however, is that because the current symbol path is constantly changing, AutoPath is actually storing the symbol files in an overly appended path, such as SYM\SINKS\SINKS\SYM\SINKS.

We recommend that you do not use the AutoPath feature, but feel free to experiment yourself to see how it all works (or doesn't work). Just be careful that you watch and pay attention to what is happening. In DataCAD for Windows, you will find **AutoPath** in the **Template** menu. Make sure the button is turned off. In DataCAD DOS, you will see the AutoPath prompt when you try to save a new symbol.

Rearranging Symbols in Templates

As noted at the beginning of this chapter, the symbols in any one template file do not all have to be stored in the same symbol folder locations or even on the same computer or hard drive. They can be located anywhere, as in this example from part of a template file:

```
c:\dcadwin\sym\00-grphc\drawing\north1
d:\symbols\02-site\bsbnc2\bsbn2-1
s:\sym\11-equip\kitchen\bikc1\bikc1-1
s:\sym\16-elec\elec\rcptcl\1recout
```

At some time or other, you will probably find that you want to create a template that contains a bunch of previously saved symbols. You might have CMU detail symbols in one template file, brick details in another, and window sections in yet another. But maybe you want to create a new template file for a specific project that would contain all or some of those far-flung symbols in one convenient location. This is actually very easy to do.

Remember that a template file is just a table of contents that tells DataCAD where to find the symbols. Symbols do not have to be exclusive to one template file. You can have one symbol that is accessed from several template files. So in the previous scenario, all you have to do is create a new template file with a text editor, or alter an existing one, adding in the correct paths to all the symbols you want to use. The key concept here is that you do not have to copy the symbols anywhere, and you do not have to move them. All you do is create a new template file and add in the text that describes the existing paths to the existing symbols.

Because the proper header information and format of a template file is critical, it is usually best to copy an existing template file, keeping the header information intact, and then substitute the new symbol locations for the previous ones. Make sure you locate the new template file in the proper directory.

Symbol Redirection

DataCAD has the capability to display one thing in a template window while accessing and placing a different symbol entirely. In Figure 14-12, all of the sinks are displayed in the template window as text, but when we select the 26 × 22 wall hung sink and place it in the drawing, we get the symbol of the actual sink.

This bit of display trickery is called *symbol redirection*. Unfortunately, DataCAD won't do this automatically. To accomplish it, you will need to create two symbols (the one to be displayed and the one to be placed) and then do a little manual editing of the .TPL file. The displayed and actual symbols

Figure 14-12 *The wall hung sink menu displays text instead of the symbol.*

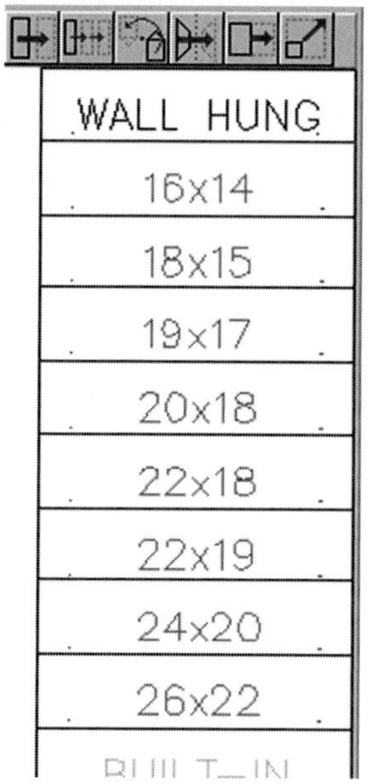

can be text, 2D entities, or 3D entities, anything that DataCAD can create. Here are the basic steps:

1. Create the two objects for the template: the text or object to be displayed in the window, and the actual symbol to be placed in your drawings.
2. Create a new template file. Be sure to write down the full path so you won't forget it when saving the symbols.
3. Create a symbol of the first object, the one that will be displayed in the template window. Be sure to save the symbol to a path that mirrors the template path (see the section about creating symbols if you need help). This symbol will be displayed in one of the template windows.
4. Create a symbol of the second object, the actual symbol to be placed in a drawing. You must save it with a different name than the first symbol. Make sure you save it to the same path as the first symbol (this is not a requirement, but it keeps your symbols much more orderly). This symbol will also be displayed in the next available template window, but we want to change this.
5. If you want to save more symbols, continue repeating Steps 3 and 4. Once completed, move onto Step 6.
6. Close the template file and minimize DataCAD.
7. Open the newly created .TPL file in an ASCII text editor such as Windows Notepad. If you created only two symbols, it might look something like the following. Note that the first path is the first symbol we created, the one to be displayed in the template window. The second path is the actual symbol to be placed in the drawing:

   ```
   C:\dcadwin\sym\15-mech\plumb\sinks\16x14
   C:\dcadwin\sym\15-mech\plumb\sinks\16x14wh
   ```

8. At the end of the first symbol path, add a pipe (|) character (that's the one you get by pressing **Shift+**).
9. Using cut/paste, move the second symbol path directly after the pipe. Make sure the full path remains intact. Now it will look something like this:

   ```
   C:\dcadwin\sym\15-mech\plumb\sinks\16x14|C:\dcadwin\sym\15-
   mech\plumb\sinks\16x14wh
   ```

   This has told DataCAD to display the symbol described in the first path, but to actually place only the symbol in the second path after the pipe. Pretty cool.

10. Save the template file, maximize DataCAD, and then reopen the new template file.
11. Pick the new symbol and place it in the Drawing Window to make sure it works.

Now let's take a look at the first example of the \TPL\15-MECH\PLUMB\SINKS.TPL template to check out a couple of other little tricks. Here is the beginning of the SINKS.TPL file with the header information omitted:

```
c:\dcadwin\sym\15-mech\plumb\sinks\wallhung
c:\dcadwin\sym\15-mech\plumb\sinks\labels\16x14|c:\dcadwin\sym\15-
mech\plumb\sinks\16x14wh
c:\dcadwin\sym\15-mech\plumb\sinks\labels\18x15|c:\dcadwin\sym\15-
mech\plumb\sinks\18x15wh
```

Figure 14-13 shows the first three windows of the template file that correspond to the symbols in the previous three lines.

Notice that the first window has a symbol that is just a text header for the symbols in the windows below. That text was saved as a symbol but has no redirected symbol associated with it since it simply acts as a header and is not meant to be placed in a drawing.

Notice also that the text symbols for the sinks are located in a subfolder called *labels*. This is by no means a requirement, but some symbol authors find it useful for organizing and separating the displayed symbols (that act as labels) from the actual symbols. If you want to do this, just create the subfolder and move the appropriate symbols there.

In this example, we created a hidden-line isometric view of the furniture and saved it as a 2D image. This became the symbol saved for viewing in the template window, while the actual symbol for placement was the original 3D piece of furniture in plan view (see Figure 14-14).

Here is the .TPL file for this group of symbols:

```
DataCAD template file. version 01.10.
1
```

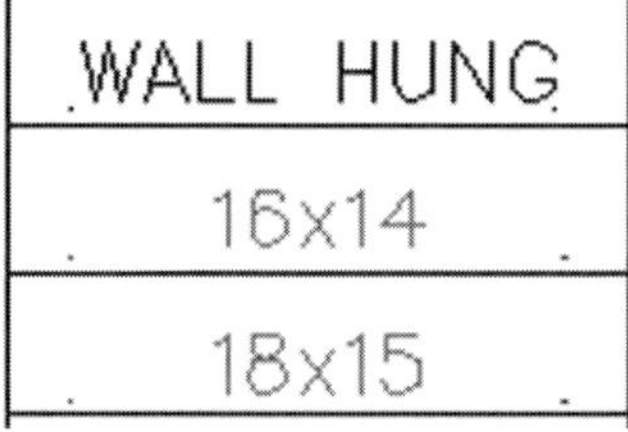

Figure 14-13 The first three windows of the template file

Figure 14-14
The furniture symbols displayed in 3D

```
4
*
G:\DCADWIN\SYM\80-DETLS\CHAIRS\ISOCHR-1|G:\DCADWIN\SYM\80-DETLS\
CHAIRS\CHAIR-1
G:\DCADWIN\SYM\80-DETLS\CHAIRS\ISOCHR-2|G:\DCADWIN\SYM\80-DETLS\
CHAIRS\CHAIR-2
G:\DCADWIN\SYM\80-DETLS\CHAIRS\ISOCHR-3|G:\DCADWIN\SYM\80-DETLS\
CHAIRS\CHAIR-3
G:\DCADWIN\SYM\80-DETLS\CHAIRS\ISOCHR-4|G:\DCADWIN\SYM\80-DETLS\
CHAIRS\CHAIR-4
```

For another example of the usefulness of symbol redirection, take a look at the \SYM\05-METAL\COLUMNS template. Also be sure to check out the "Troubleshooting" section at the end of this chapter for some information about a problem where symbols cannot be placed in the drawing because of excessively long .TPL paths, which is very easy to do with the long paths created with symbol redirection.

Building Details with Symbols

Symbols can be used like a child's building blocks to build a detail, rather than drawing the same detail line by line. In Figure 14-15a, a detail has

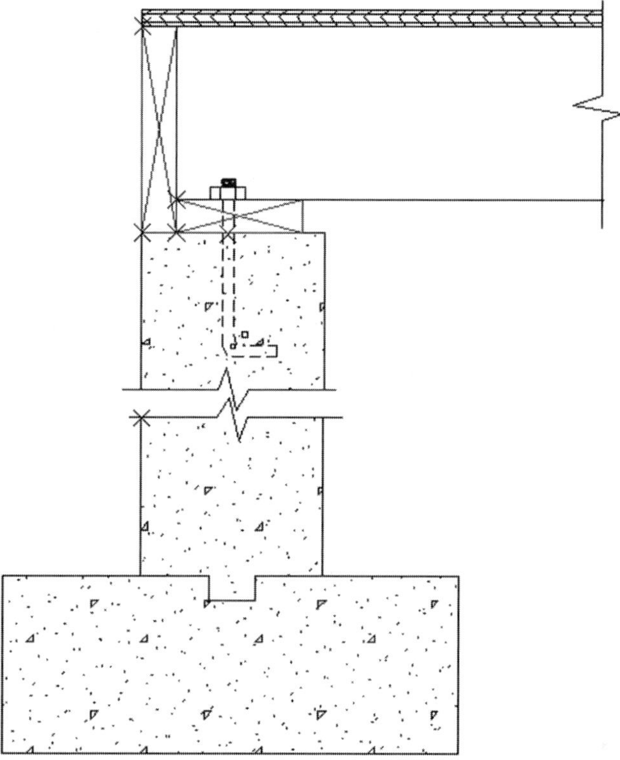

Figure 14-15a
The completed detail

been created solely by using symbols (see Figure 14-15b) and took only about 30 seconds to create.

This is an excellent example of how symbols can greatly increase your speed and productivity. Symbols do not have to be limited to furniture and detail bubbles.

To make the best use of this concept, you should spend some time creating templates and symbols of the most common detail elements that you use in your office. These symbols can be simple or complex. They could consist of individual elements, such as wood studs in section, or could even be an entire wall section. Here are some examples of repetitive symbols that our office has created over the years:

- Wood studs
- Metal studs (tracks and sections)
- Partition types, with all notes and ratings indicated
- Drawing symbols

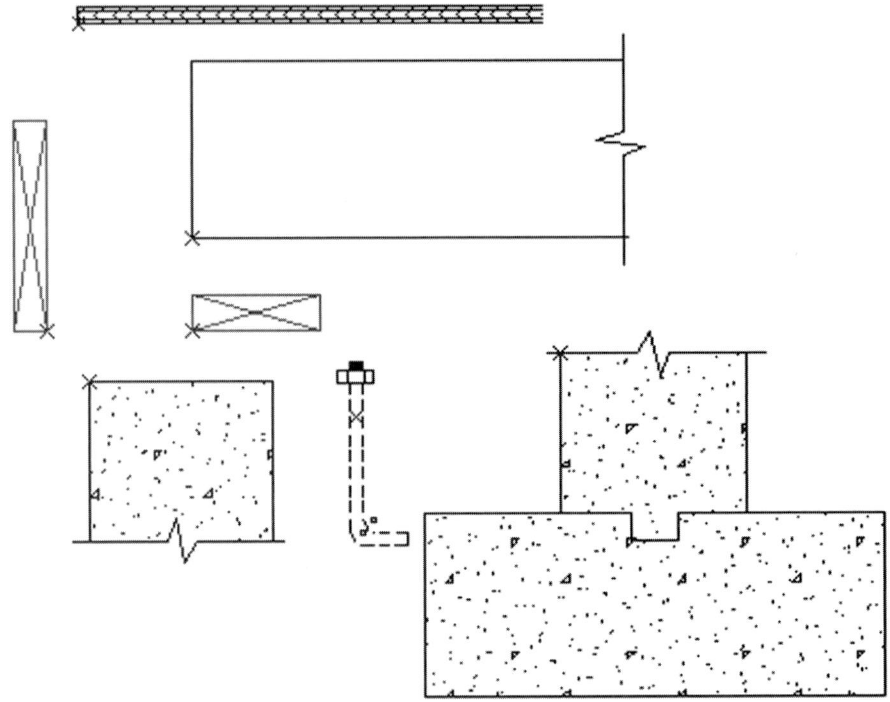

Figure 14-15b The symbols that make up the detail

- Masonry units (bricks and blocks)
- Doors, windows, and frames in elevation

Symbol Reports

This is a feature that is little used by most people because it's not particularly user-friendly or intuitive, though it does prove useful to those who need its capabilities. It's also actually very simple to use. A symbol report in DataCAD is sort of a poor man's spreadsheet for symbol information, allowing you to quantify the symbols in a drawing file, and to report total costs and other data. The format and the methodology is based on the old DOS version of DataCAD, which explains its lack of a Windows-type interface.

All forms are located in the DataCAD \FRM directory and have a .FRM file extension. Forms are simple ASCII text files and can be opened in any text editor like Windows' Notepad or Write programs. To create your own forms, you should save one of the existing forms with a new name and then edit that form.

DataCAD comes with five predefined symbol reports, called *forms*:

- *Dcadcost.frm* A generic form to report all symbol data, quantities, and costs
- *Dcadlist.frm* A generic form to list basic symbol data
- *Dcadqty.frm* A generic form to list basic symbol data and report quantities
- *Elecls.frm* A basic lighting fixture schedule including quantities and totals
- *Elecskd.frm* A more detailed lighting fixture schedule including extended data fields, installation times, cost, pricing, tax, overhead, and totals. This shows basic math routines as they apply to symbol reports.

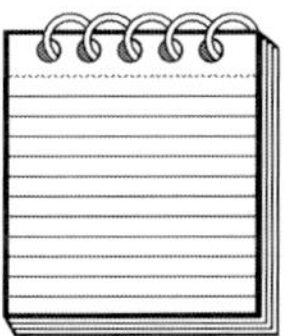

NOTE: *Remember that DataCAD symbols have six standard data fields per symbol. If you want more data fields than that, such as Type, Model, and Rough Opening, you need to define them when the template file is created and the symbols are saved. They cannot be added later.*

Here is the Elecskd.frm file to output a lighting fixture schedule:

```
@ZZZZZZZZZZZZZZZZZZ

 @AAAAAA @BBBBBBBBB @CCCCCC @DDDDDD @EEEEEEEE @FFFF  @GGG  @HHHHHHH
 @IIIIII @JJJJJJJJJ @KKKKKK @LLLLLL @MMMMMMMM @NNNN        @OOOOOOO
 ______________________________________________________________________
*
   @hh   @iiiiiiiii @jjjjjj @kkkkk  @llllllllll @mmmm  @nnn  @ooooo
*
 ______________________________________________________________________

                                        @QQQQQQQQQQQQQQQQQ  @qqqqq
                                        @RRRRRRRRRRRRRRRRR  @rrrrr
                                        @SSSSSSSSSSSSSSSSS  @sssss
                                        @TTTTTTTTTTTTTTTT   @ttttt
 ______________________________________________________________________
```

```
Z= \L 'LIGHTING FIXTURES'
A= \L 'Symbol'
B= \L 'Manufact.'
C= \L 'Model'
D= \L 'Install'
E= \L 'Delivery'
F= \L 'Unit'
G= \L 'Qty.'
H= \L 'Extended'
I= \L 'Type'
J= \L 'Name'
K= \L 'Number'
L= \L 'Time'
M= \L 'Time'
N= \L 'Cost'
O= \L 'Price'
Q= \R 'SubTotal Cost:'
R= \R 'Tax 4%:'
S= \R 'Overhead 10%:'
T= \R 'Total Cost:'
h= <Item name>
i= <Manufact.>
j= <Model No.>
k= <Installation>
l= <Delivery Tim>
m= <Cost>
n= <#>
o= m * n
q= [o]
r= q * 0.04
s= 0.1 * (q + r)
t= q + r + s
```

Here is a sample of the report form output generated from this form:

```
LIGHTING FIXTURES

Symbol  Manufact.  Model   Install  Delivery  Unit    Qty. Extended
Type    Name       Number  Time     Time      Cost         Price
--------------------------------------------------------------------
A       Visa       342789  1 hr     45 days   205.00  12   2460.00
B       Lightolir  249002  .5 hr    30 days   120.00   5    600.00
C       Premier    682147  1.5 hr   35 days   110.00  20   2200.00
--------------------------------------------------------------------
                                 SubTotal Cost:            5260.00
                                 Tax 4%:                    210.40
                                 Overhead 10%:              547.04
                                 Total Cost:               6017.44
```

Note that the Quantity and Unit Cost fields are automatically right-justified. Let's walk through this example to see how it all works. If you are familiar at all with creating spreadsheets, then this will all seem very familiar to you.

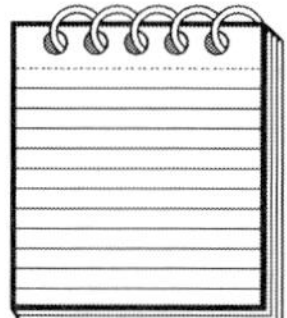

NOTE: *If you want to create your own form from scratch, you should type a sample of the final output, as in the previous example, and then use the overtype option in your text editor to type the @ symbols and letter definitions (A, B, a, b, and so on) over the formatted text.*

Definitions for Reports

The following is a list of the various components of a report:

- **Sequence Section** The top section of the form above the long dashed line. It defines the format of the final output.
- **Definition Section** The bottom section below the long dashed line. It defines the variables used in the top section. For instance, Z= \L 'LIGHTING FIXTURES' tells DataCAD to use the LIGHTING FIXTURES text in place of the @ZZZZZZZZZZZZZZZZZ text. The \L tells DataCAD to left-justify the text. The @ symbol and the number of Z's determines the maximum number of characters to be substituted. For instance, if the variable definition were only @ZZZZ (the @ symbol plus four Z's), then the output would only appear as LIGHT, since only five characters were defined.
- **Form Separator** A hyphen (-) in the first position of a line marks the separation between the sequence section and the definition section. Any hyphens after the first one are optional but are useful for visually separating the two sections. The hyphen is one of only two characters that is placed in the form without a space in front of it.
- **Field** Each group of @+letters is called a field.
- **Field Title** The heading of each column is called a field title.
- **Sequence Line** Each line of the symbol report is called a *sequence line*, except for the lines that start with an asterisk (*). Sequence lines cannot begin with an asterisk or a hyphen.
- **Report Definition** The line that tallies and reports the information about each symbol is called the *report definition*. To define the report definition section, an asterisk (*) must be placed before and after the report definition. The asterisk is one of only two characters that is placed in the form without a space in front of it. Each report definition line can report on only one symbol.

- **Variable Definition** Each variable in the definition section is called a *variable definition*. Each variable definition must start with a letter followed by an equals (=) sign. The second part of a variable definition can contain formatting information, but it does not have to. The possible letters are as follows:
  - **\R** Here output is aligned to the right edge of the field (right-justified). This is the default format for the number and dollar fields.
  - **\L** Here output is aligned to the left edge of the field (left-justified).
  - **\C** Here output is aligned to the center of the field (center-justified).

  The final part of a variable definition can contain any of the following:
  - A text string enclosed in single quotes. The text inside the quotes is what will be placed in each field in the sequence section. An example would be 'SubTotal'.
  - A field name enclosed with less-than and greater-than symbols. The name must correspond exactly to the field name defined in the template and symbol. An example would be <Item name>.
  - A number sign enclosed with less-than and greater-than symbols. This will count up the total number of occurrences of the symbol in the drawing file. An example would be <#>.
  - An arithmetic operator (see the examples later in this section). An example would be 0.1 * (m + n).
- **Reserved Characters** These are characters that are used for specific symbol report operations and can therefore not be used for any other operation in the sequence section. These include ( ) ! @ # * + - / { } [ ].
- **Comment** The ! character can be used to add non-printing characters to the .FRM. These are useful to remind you what you were doing at a particular point in the form. Everything after the ! is ignored when the report is printed. An example would be "! This is a comment." If you are going to place a comment at the beginning of a line, you must place a blank space in the first position of a line, followed by the !.

Rules for Reports

Here are some guidelines to follow when working with reports:

- The number of characters in a field title is independent of the number of characters in the fields below it, so you can have a field title that is six characters in length with a field under it of only two characters.

- A template report cannot display more than 24 individual symbols on the screen at one time.
- Although a symbol report reports information about symbols placed in the drawing file, each report can only report on symbols located in one particular template (.TPL) file. Any symbols that you want to group together in a report must all be grouped together in a common template file.
- Reports cannot be wider than 80 characters. The five DataCAD .FRM samples are all 70 characters wide, so you can increase them by 10 more characters if you need to.
- Reports can only contain a maximum of 132 horizontal rows.
- Every line of a report must start with a blank space, except as noted.
- Before you perform arithmetic operations, you must first define all variables.
- You can insert blank lines for clarity in the .FRM file, but those blank lines will be ignored when the report is printed out. If you want the output to display those blank lines, try adding a space and then a period at the beginning of the line. The period will print but is small enough to be barely noticeable.
- Since symbol reports are ASCII text files, you can add them to the current drawing, send them to a printer, or save them as a text file.

Math Operations in Symbol Reports

Using the following math operators you can perform math operations on fields in the sequence section containing numeric or cost data. All of the following math operators are available:

+	Addition
−	Subtraction
*	Multiplication
/	Division
( )	Order of evaluation (Mathematical operations inside the parenthesis will be operated on before operations outside the parentheses.)
[ ]	Reports the sum of a field with the report definition line (more on this later)

You can use any valid mathematical formulas involving these math operators, such as

$$\times\ (m + n)$$

$$q \times 0.04$$

$$q + r + s$$

DataCAD ignores spaces between mathematical operations, numbers, and letters. Adding spaces can make the mathematical formulas in your forms easier to read. For instance, all of the following are acceptable:

$$0.1 \times (m + n)$$

$$\times\ (m + n)$$

$$0.1 \times (m + n)$$

All variables must be defined before a mathematical formula can operate on it. In the following example, the line "s=" is valid because the variables q and r were defined before the mathematical expression in line s was performed:

$$n = \ < \# >$$

$$o = m \times n$$

$$q = [o]$$

$$r = q \times 0.04$$

$$s = 0.1 \times (q + r)$$

$$t = q + r + s$$

Line "s=" is not valid in the following example, since the t in $(t + r)$ is defined in the line after line s:

$$n = \ < \# >$$

$$o = m \times n$$

$$q = [o]$$

$$r = q \times 0.04$$

$$s = 0.1 \times (t + r)$$

$$t = q + r + s$$

Summing Fields with Brackets

Placing a field variable inside the left and right brackets will cause DataCAD to add up the total numeric or cost data for a common field between each of the reported symbols. Here is part of the output from the previous lighting schedule:

```
A   Visa        342789   1 hr     45 days   205.00   12   2460.00
B   Lightolir   249002   .5 hr    30 days   120.00    5    600.00
C   Premier     682147   1.5 hr   35 days   110.00   20   2200.00
-------------------------------------------------------------------
                                        SubTotal Cost:    5260.00
```

Field "*o*" is the last column, all the way to the right. Field "*q*" is the SubTotal Cost. The formula for field q is $q = [o]$. This tells DataCAD to add up all the cost values for field o for all three reported symbols (A, B, and C), so we have the following:

$$\begin{aligned} q &= [o] \\ &= 2460.00 + 600.00 + 2200.00 \\ &= 5260.00 \end{aligned}$$

More Forms

Two additional forms are provided on the CD-ROM that comes with this book. One is for a window schedule, the other for a door schedule.

To generate a symbol report, follow these steps:

1. The report will be generated based on the currently displayed template, so open the template file containing the symbols that you want to report on.
2. Select **Reports** from the menu. A dialog box will appear, displaying all the form files in the \FRM directory.
3. Select the form you want to use, and then press **Enter** or **Open**.

4. The drawing and template windows will disappear, and the symbol report will be displayed. You will be prompted to *"Press any key to continue."* After doing so, the drawing and template windows will return, and you will see three new menu buttons. Nothing will be displayed if no symbols from the current template are located in the drawing file.
5. Select **Exit** or *right-click* if you don't want to save the symbol report. Otherwise, select one of the three options: **ToDrwing**, **ToFile**, or **ToPrntr**.

ToDrwing Selecting this option will place the completed symbol report in the drawing. To activate **ToDrwing**, follow these steps:

1. Select **ToDrwing**. The drawing cursor will revert to a text cursor and seven font options will appear in the menu. The current font settings will be used unless you change them.
2. Place the cursor where you want the report text to start and then *click* the mouse or press **Enter**. The report will be printed on the current layer.
3. *Right-click* or press **Exit** to back out of the **Reports** menu.

ToFile Selecting this option will enable you to save the completed symbol report to a text (.TXT) file. By default, the text file will be saved to the path currently defined for Text Output Files (**Tools/Program Preferences/Pathnames**), or you can navigate to a new location. To save the completed symbol report to a .TXT file, follow these steps:

1. Select **ToFile**. A dialog box will appear. If you are not already there, navigate to the directory where you want to save the symbol report.
2. Type a name for the symbol report. Note that the default file type is an ASCII .TXT file. Press **Enter** or **Save** to save the file.
3. *Right-click* or press **Exit** to back out of the **Reports** menu.

ToPrntr Selecting this option will send the completed symbol report directly to the printer. To do so, follow these steps:

1. Select **ToPrntr**.
2. The symbol report will be sent directly to the current Windows default printer.
3. *Right-click* or press **Exit** to back out of the **Reports** menu.

Two Ways to Put Symbols in Drawings for Reporting

Two basic methods can be used to include symbols in reports. Take the previous example using window symbols. Either the windows themselves are placed in the drawing as symbols, or you can create symbols of the window-type designators. Figure 14-16 shows two examples. In the one on the left, the window itself is the symbol that is reported on. In the example on the right, the "B" in the hexagon is the symbol that is reported on. In both cases, the symbol fields, such as cost, rough opening, and manufacturer, are the same. It is just the symbol that is different.

Symbol Shortcomings

Throughout this chapter we have mentioned some of the shortcomings of DataCAD's symbols. Although those shortcomings are few, you should be aware of them. We'll summarize them here.

Changing Entities in Symbols

Once a symbol is saved, you cannot alter any aspect of the entities within it (except linetype spacing) without first exploding the symbol, making the changes, and then using the **Redefine** command. You can therefore not change colors, fonts, linetypes, and so on unless you first explode the symbol.

A third-party macro called Template Manager is available through Cheap Tricks Ware. It enables you to alter certain aspects of entities within

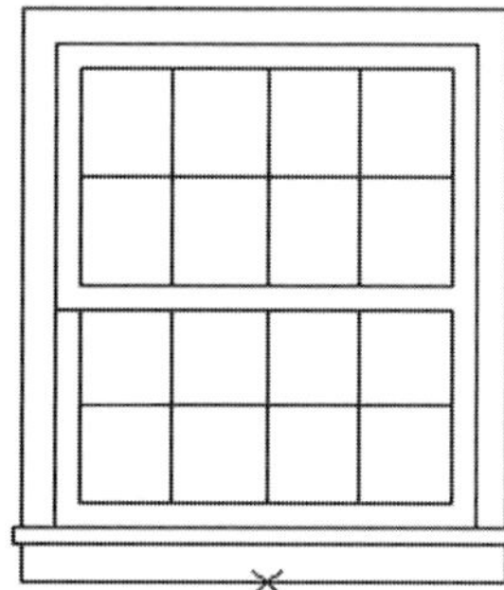

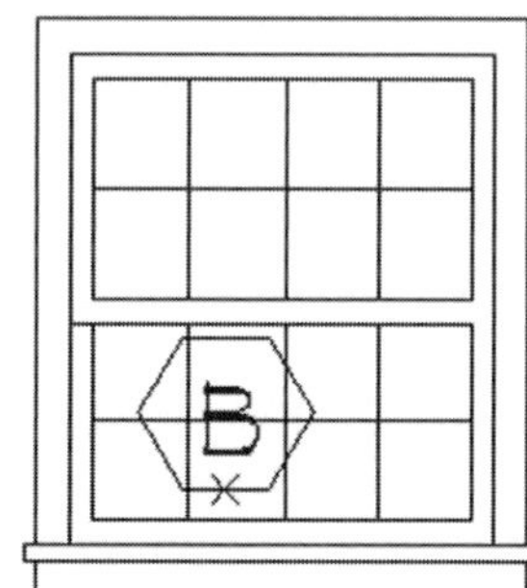

Figure 14-16 Two different window symbol examples

a symbol by automating the process of exploding, changing, and redefining the symbol. It can change colors, linetypes, lineweights, line spacing, and symbol attributes.

No Layering in Symbols

Symbols are singular entities without layers. Even if the symbol was created from entities contained on a dozen layers, the symbol is reduced to only one. Symbols therefore only exist on the layer on which they are placed. Even when a symbol is exploded, the entities are placed only on the currently active layer.

A third-party macro called Bundle that is part of the DC Sprint macro available from DataCAD LLC does allow you to save symbols with layers. When you explode the symbols, the layers are created in the current drawing (if they don't already exist) and the entities are exploded to their respective layers.

Exploding Symbols

Two problems can arise when you explode a symbol:

- When a symbol is exploded with the standard **SymExp** macro, the entities are placed on the currently active layer. They do not remain on the layer on which they were placed. You can get around this with a new free macro called Symbol Explode 2. You can find it on the CD-ROM or on the DataCAD Web site, in the DDN section (`www.datacad.com/dcal/source.htm`).
- When a symbol is mirrored and then exploded with the standard SymExp macro, DataCAD attempts to revert all symbols back to their original orientation. Unfortunately, DataCAD often mirrors the symbols back along the wrong axis, so that the symbol may not actually wind up in the proper orientation after all. A third-party macro called Global Edit has a better symbol explode routine. It's not foolproof, but it does a much better job than the standard Symbol Explode macro.

Template Bugs

It is important to be aware of a couple of bugs with regards to templates. If you save a symbol with a particular name and then want to change it later, you can use the **Template/EditFlds/ItemNam** function to rename it. However, the new name will not be displayed until you close the template and then reopen it.

Also, DataCAD can only read a line of text of up to 64 characters in the template file (not including the final backslash and symbol name). Most paths will never be this long, but one notable exception is when you use symbol redirection, which will essentially double the number of characters in each line of a template file. If this happens, you will see the symbol image in the template window, but you won't be able to place it in a drawing. You'll have to find a way to shorten the path so that it has 64 characters or less.

Troubleshooting

This section will offer some solutions to common problems with DataCAD symbols, specifically with empty windows and cursors.

Empty Windows

If, when you open a template file, the template boxes appear, but one or more symbols are missing, or no symbols exist in any of them, there can be one of several problems.

1. The most likely cause is that your paths to the symbols do not match the paths for the templates. Correct the .TPL file and reset the symbol path to rectify the problem.
   a. Open the offending .TPL file in a text editor like Windows Notepad.
   b. Make sure all the paths are either relative (c:\dcadwin\sym\ . . .) or absolute (sym\). We suggest they all be absolute.
   c. Make sure the correct path is shown. It might have old path references.
2. If the offending .TPL file has relative paths in it, then the probable cause of the problem is that the Symbol Files path in the **Tools/Program Preferences/Pathnames** menu (the Configuration program in the DOS version) now has an absolute path in it. That's fine if you have only absolute paths, but this confuses templates with relative paths. If you want to keep the offending symbols in a relative path in the .TPL file, go to the **Tools/Program Preferences/**

Pathnames menu and change the path back to relative (just type in the path of your DataCAD directory, like this: c:\dcadwin\). Otherwise, just change the path in the .TPL file to absolute. Note that in DCAD DOS you have to quit DataCAD, change the path in the CONFIG.EXE program, and then restart DCAD for the changes to take effect.

Empty Cursor

If you can see all the symbols in the template windows, but when you try to place one in your drawing, the symbol is not attached to your cursor and can't be placed in the drawing, the most likely culprit is that the paths in the template file have become too long for DataCAD to read. This is caused by the bug mentioned earlier, where DataCAD can only read a line of text of up to 64 characters in the template file (not including the final backslash and symbol name). Most paths will never be this long, but one notable exception is when you use symbol redirection, which will essentially double the number of characters in each line of a template file, such as the following:

```
C:\dcadwin\sym\05-metal\columns\labels\04x04|C:\dcadwin\sym\
05-metal\columns\04x04s
```

In this case, what is happening is that DataCAD will read and display the first symbol whose path ends at the pipe (|), but the second symbol can't be read and selected by DataCAD because the line of text is too long.

To fix the problem, you will need to shorten the path to the symbols. This means you will have to physically move the symbols to new folders where the path won't be so long, like this:

```
C:\dcadwin\sym\05-metal\labels\04x04|C:\dcadwin\sym\05-metal\04x04s
```

Lastly, rewrite all the lines in the template file to show the new path and save the file.

CHAPTER 15

3D Basics

Introduction

There are many reasons to learn how to create 3D models in DataCAD. In the not-so-distant past, firms concentrated their CAD work on two-dimensional production drawings in the form of plans, details, and sections, as these are the devices we use to construct buildings. Today, many practices are discovering that investing time in building a three-dimensional CAD model is well justified. If modeled accurately, the CAD model is an invaluable resource for both construction documents and presentation material. The 3D model can be used to generate sections and elevations. With the aid of rendering software, it can also produce illustrations to convey what the built work will look like to those who don't understand how to read construction drawings. There is no question that the layman will have a much easier time understanding a designer's proposal looking at a perspective drawing or rendering rather than a dimensioned plan drawing.

Additionally, three-dimensional work is a marketable skill for the architectural practice. The firm that can effectively communicate its ideas to a client is more likely to secure commissions. Showing a project three-dimensionally minimizes misunderstandings about what the project will look like once built, and it aids the designers as a pre-flight check for structural and mechanical conflicts.

With the abundance of three-dimensional animation in the advertising and entertainment industries, many clients simply expect their architect to take them on a virtual tour of the building as part of the design process. Many clients also think erroneously that these digital environments are available to the designer at the push of a button. While that is certainly not the case, varying degrees of representation can be achieved relative to the time and cost warranted. Educating the client about the benefits of investing in some three-dimensional CAD work as part of the design services provided is a good way to introduce this type of work to your practice.

Lastly, CAD models offer a wealth of material for use in portfolios and marketing materials. Digital illustrations convey to the prospective client that you are on the cutting edge of technology. A well-executed CAD model may be used over and over again to render an infinite number of views.

DataCAD includes a versatile and comprehensive set of 3D modeling functions designed expressly for architectural work. This section will teach you the essential concepts of 3D modeling in DataCAD. Once you master the basics, you will be able to create everything from quick-study models to presentation models with stunning complexity. You will find that once you understand how to work two-dimensionally, it is a short jump to working in the third dimension.

You only need to master the six 3D concepts and five modeling methods described to begin building your own virtual buildings. We will start with some basic study exercises, and then move on to increasingly more complex constructions. In the advanced 3D-workshop section, complex 3D models are built step by step. The CD includes all the files illustrated, so that you may follow along in the process, or take the model apart to see how it was created. A gallery illustrates some of the possibilities with DataCAD.

Overview

DataCAD's modeler is a *polygon surface primitives* type. This means that it can only create the surfaces of shapes, as opposed to solid objects. Think of the difference between making a box by gluing together six pieces of cardboard as opposed to cutting it out of a solid block of wood. The surfaces created to form the volumes are called polygons. The primitives part refers to the fact that DataCAD can use these surfaces to build elemental geometric volumes like cubes, cylinders, spheres, cones and tori.

There are many different modeling engines used in CAD programs today. Technologically, DataCAD's is not one of the most advanced. It does not support advanced operations like subtracting one volume from another or finding the difference between two intersecting volumes. Despite these limitations, you will find that you can do almost anything with a little creative thinking and the techniques illustrated here.

Surface modeling offers many advantages over more advanced systems. It is extremely versatile, easy to learn, and very efficient in its data storage requirements. All of the sample models included on the CD are less than 1 MB in file size.

Before You Begin

The 3D section of this book assumes that you are comfortable drawing 2D plans and the like. If you are still getting acquainted with the 2D CAD drawing fundamentals covered in the first half of this book, you may wish to review before going further. We will be using all that you have learned about 2D drawing to launch into the third dimension. There are only a few basic concepts that you need to understand to start modeling on your own. Read through them, then pick an advanced exercise to experience applying the concepts to an actual project.

Ground Zero: The DataCAD 3D Menus

DataCAD's 2D portion contains two primary menus: **Edit** and **Utility**. When you select **DCAD 3D** from the **2D Edit** menu, you enter the 3D portion of the program, which also has two primary menus: **3D Edit** and **3D Entity** (see Figure 15-1). Through these two menus all of the necessary 3D functions and entities can be accessed.

The **3D Entity** menu enables the creation of the basic 3D geometric primitives used for modeling, including the following: 3D lines, polygons, blocks, slabs, 3D arcs, cylinders, cones, spheres, tori, contours, mesh surfaces, revolved surfaces and 3D markers.

The **3D Edit** menu contains functions for manipulating 3D entities. These include **Move**, **Copy**, **Rotate**, **Mirror**, **Stretch**, **Enlarge**, **Erase**, **Change**, **Explode**, and so on. For the most part, the 3D editing functions work identically to their 2D counterparts. Additional 3D-specific functions include **Hidden Line Removal** and the **Shader**.

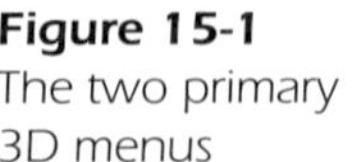

Figure 15-1 The two primary 3D menus

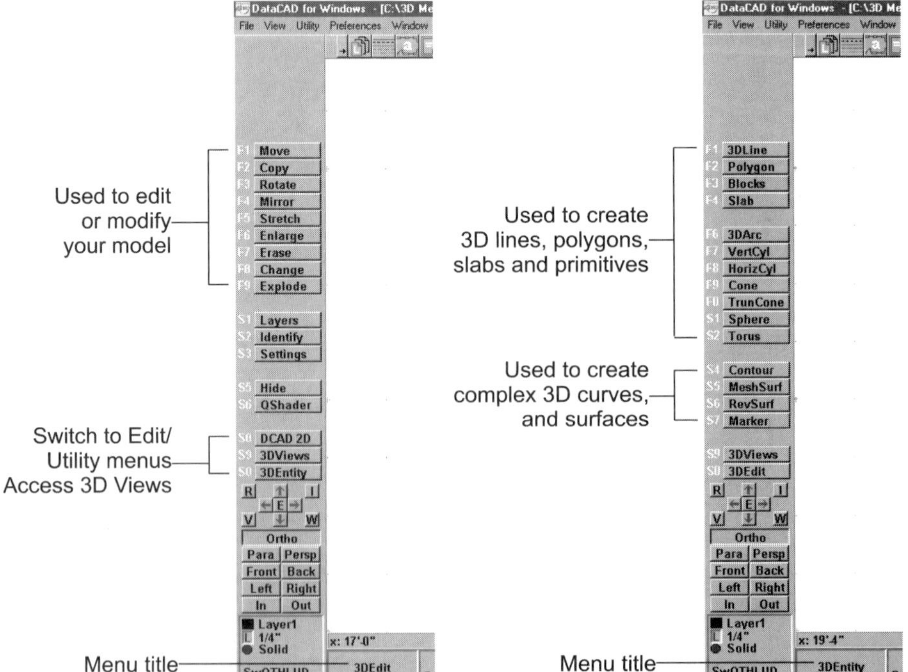

Concept 1: Z-Base and Z-Height

You may already be familiar with the concept of extruded Z-height, as DataCAD allows 2D lines to carry a Z-base and Z-height value. DataCAD creates 3D entities in much the same way, by beginning at the current Z-base setting and extending up through space to the Z-height. In addition to a base and height, some 3D entities may also have a thickness associated with them, as with slabs, or a radius, as with cylinders. By combining the variables of base, height and thickness, you are able to model an endless variety of shapes. DataCAD determines Z-heights by measuring from an imaginary plane in space where Z = 0; also called the World Zero Plane. As measured from the zero plane, positive Z values extend toward you, out of the screen, and negative Z values project downward, into the screen. Think of the World Zero Plane as the table on which you build your model. Using the tabletop as a reference, we can specify whether an element is to be created above, below, or extending through this zero plane.

The world zero plane is fixed in space—it never moves. This means that while looking down on your model from a plan view, you can confidently set your **Z-Base** value to 9′-0″ and your **Z-Height** value to 10′-0″, knowing that the next slab you model will be properly positioned in space to make a second floor. Then, when it is time to model the basement floor slab, you can change these values to −9′-0″ and –10′-0″, respectively.

Setting the Z values is easy. Just press the Z key on your keyboard, enter a value for the base, press enter, type in a value for the height, and press enter again. Whenever you press the **Z key**, DataCAD will prompt you to accept the existing value for Z-base and Z-height or enter a new one. After you set the Z values, they will apply to every object you create thereafter until you change it (see Figure 15-2).

Z-User 1 and **Z-User 2** In addition to **Z-base** and **Z-height**, you have the ability to set two auxiliary heights, **Z-User 1** and **Z-User 2**. This gives you the ability to store two other commonly used heights. For example, you might set the **Z-base** and **Z-height** values to first story settings, and **Z-User 1** and **Z-User 2** to second story settings.

Concept 2: Screen Coordinates versus World Coordinates

As we have observed so far, the World Zero Plane is fixed, which allows us to easily relate to things like floor slabs and window sill heights. Whenever

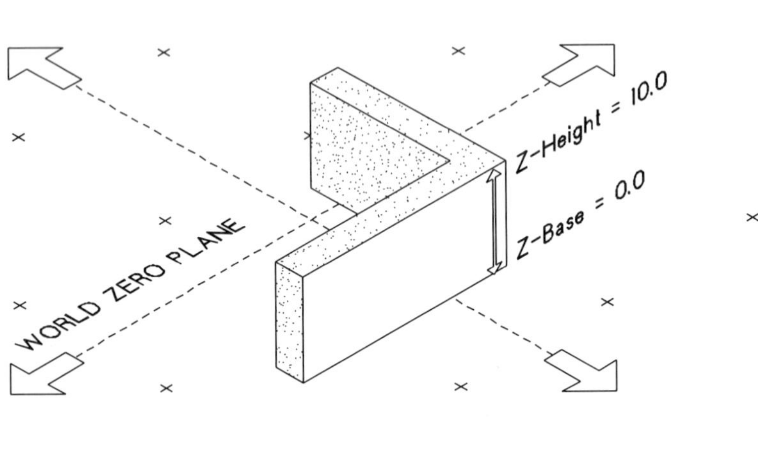

Figure 15-2 Isometric view of the World Zero Plane, and a slab created from Z-base 0.0 to Z-height 10.0

you are viewing your model in Ortho (plan) view, you can think in terms of the fixed world Z-plane.

However, there are many occasions where this is inconvenient. For example, we may wish to model a piece of window trim while viewing the model in elevation. To do this, we wish to think of the trim thickness as the **Z-base** and **Z-height** values, not its top and bottom edges. Fortunately, when you change your viewpoint from Ortho to Parallel (Elevation or Isometric) mode, the Z-plane stays relative to the screen. Think of your monitor frame as a picture frame that allows you to work in any view, confidently knowing that it defines how the Z-base and Z-heights are set. The glass in the frame is absolute zero. Positive values extend out of the screen, and negative values project into it. So, when we define an elevation view to model window trim, as long as we set the elevation point in the plane of the wall, we can set our Z values to the thickness of the trim and know that it will come out right.

Screen coordinates allow you to input distances and Z settings relative to the screen and current view, while world coordinates rely on a stationary point in space that never changes, no matter how you view the model. We will be using screen coordinates for all of our modeling exercises.

3D Viewing

As you create your model, you will change your viewpoint constantly to evaluate your work and watch things take shape. The majority of modeling

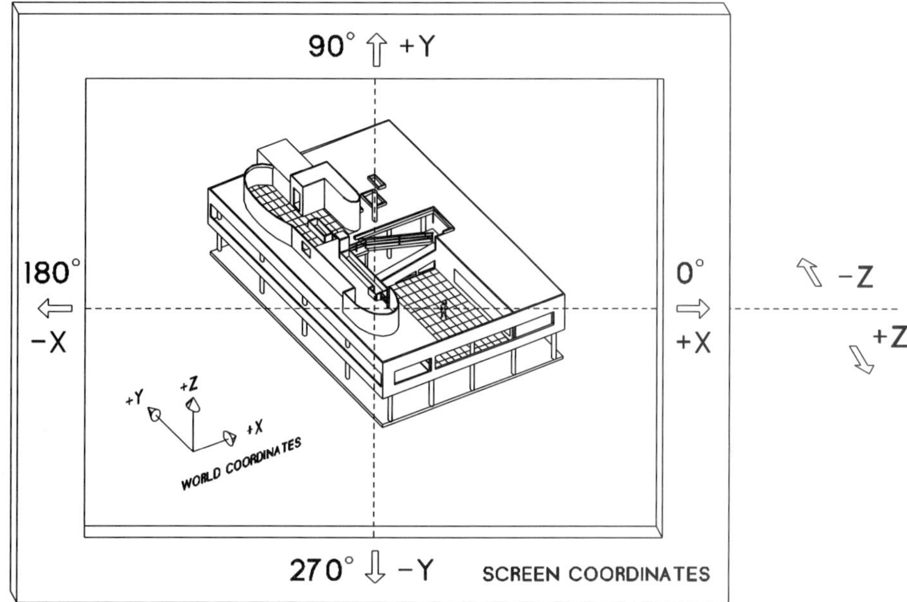

Figure 15-3 Picture Frame Analogy of the coordinate systems: World axes never change, while screen coordinates are always relative to your current view. Z values extend out of or into the screen.

operations are performed from a plan or elevation view. Isometric views are useful to verify that the elements you modeled are positioned properly, while perspectives allow you to experience the view as if you were looking through a camera in a built environment (see Figure 15-4).

Ortho

DataCAD supports several different view types, with orthographic, or plan, view being the one you'll do most of your work in. To view your model in plan, choose **Orthographic** from the **View** menu in the Menu Bar, or **Ortho** from the **3D Views** menu in the Menu Window.

Parallel

Parallel restores the last parallel view you generated, typically an elevation or isometric view. Although elevations and isometrics share the same view projection, two separate buttons allow you to access them independently, as follows.

Elevation Views Elevations are a type of parallel view. When you view an elevation in plan, the front of the model faces the bottom of the screen.

Figure 15-4
DataCAD supports many different view projections in addition to plan and elevation.

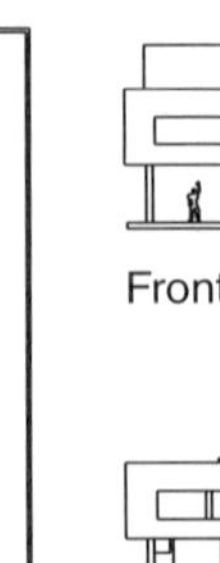

Front (Parallel View)

Left (Parallel View)

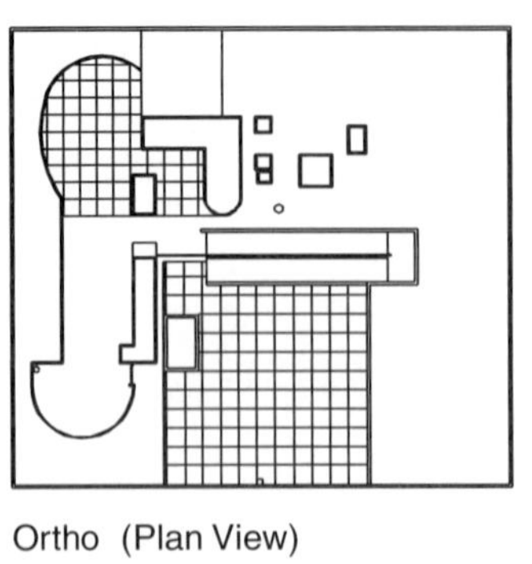

Ortho (Plan View)

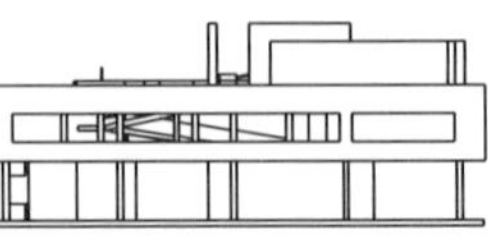

Back (Parallel View)

Right (Parallel View)

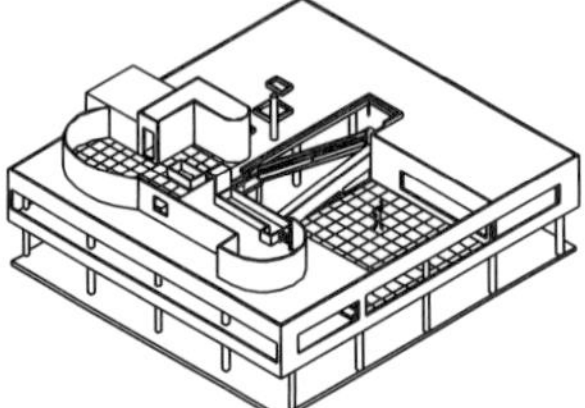

Parallel (Isometric View)

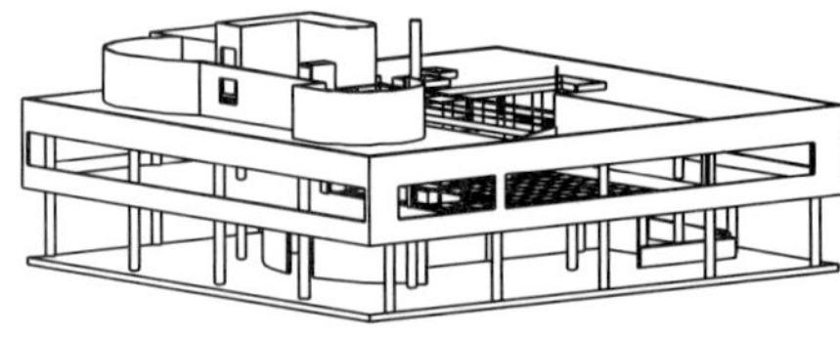

Parallel (Isometric View)

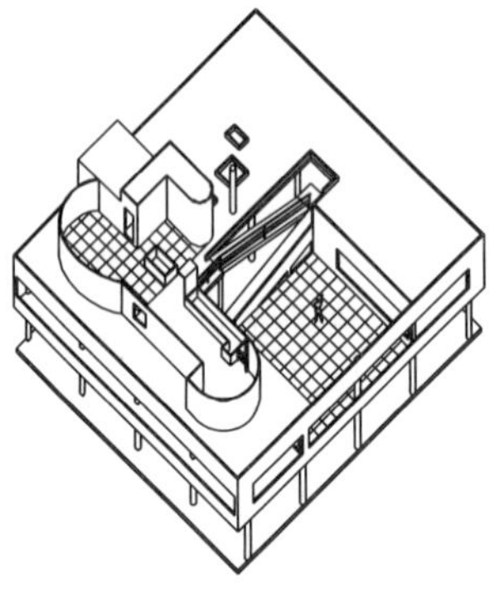

Plan Oblique (Axonometric View)

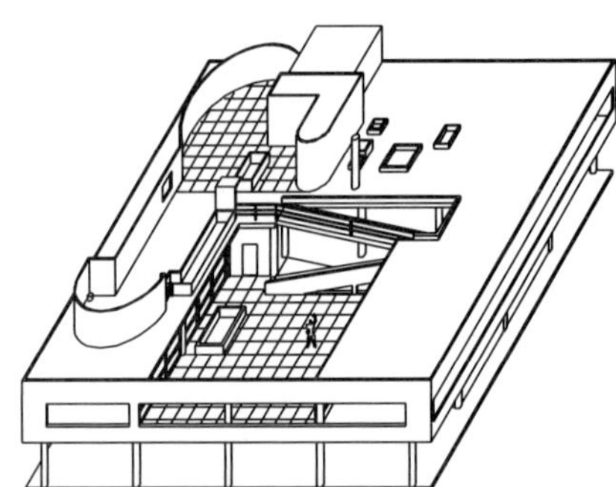

Elevation Oblique

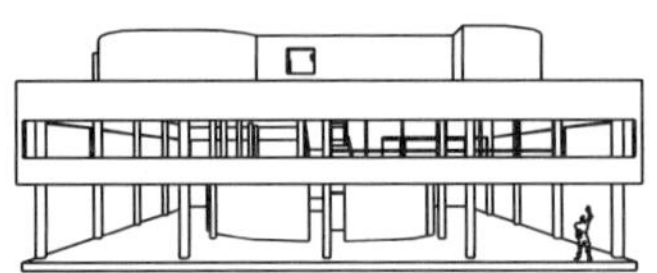

Perspective (One point)

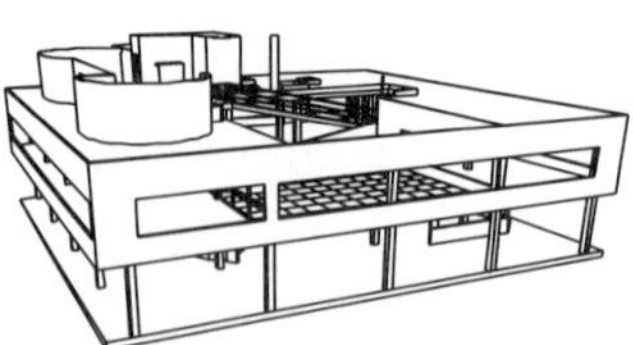

Perspective (Three point)

Perspective (Two point)

Use **Front** to view the front elevation of the model. Use **Back** to view the back elevation of the model. Use **Right** to view the right elevation of the model. Use **Left** to view the left elevation of the model.

In addition to the four preset elevations—Front, Back, Left, and Right—you can create orthogonal and non-orthogonal elevations from any place in the model with the **New Elevation** command.

Isometric Isometric views are indispensable for 3D modeling, as they allow you to quickly view your model from different vantage points. To create an isometric view, simply choose **Isometric** from the window menu. When you choose **Isometric**, the view globe appears. Use the view globe to view your model from several directions and verify that it is coming together properly.

View Globe

The globe display is used to reorient your eye point in a parallel view; it is automatically displayed on screen whenever you choose a parallel view. Selecting a point inside the globe display indicates the new observer position for viewing the model.

In a parallel view, the globe sets the observer's eye point above or below the ground plane at which the observer stands. Click in the center circle to define viewing points above the model. Generally, clicking in the inner quadrants works best. To view the model from below, click within the ring defined by the inner and outer circles.

NOTE: *If your model spins off the screen when you click in the View Globe, you need to reset the view center. Choose **3D Views, Controls, View Center**, and middle-button snap anywhere close to the center of the model with the 3D Cursor.*

The globe also appears in perspective views. Here, the globe changes the elevation of the observer's eye point and the direction from which you observe the model (north, south, east, or west), but it does not change the view center, the distance to the object, or the cone of vision.

Globe Perspective

To visualize how the globe operates, imagine that the model sits inside a large sphere, as in Figure 15-5. Half of the sphere is above the ground

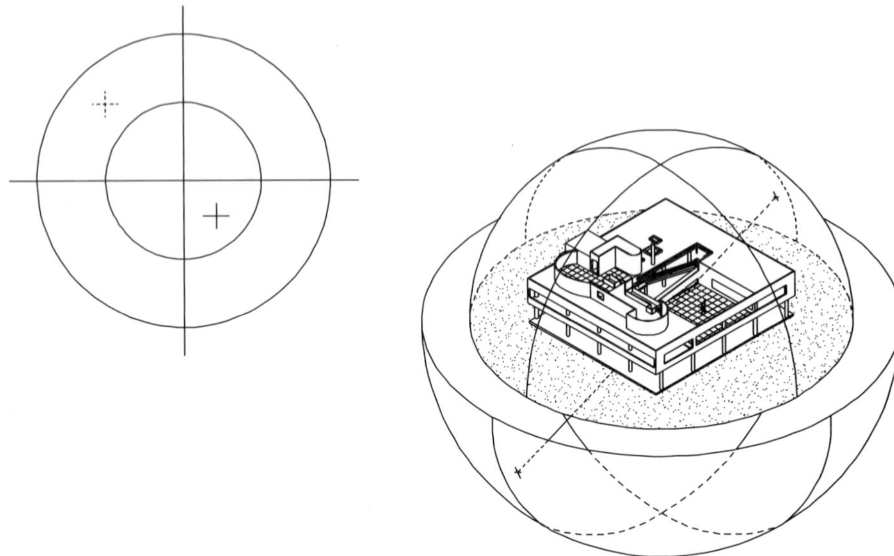

Figure 15-5
Globe Display and diagrammatic representation of the inner and outer circle

plane, and half of it is below, with the ground plane cutting the sphere at the equator. The equator of this sphere is represented as the inner circle on the globe. All points within the inner circle define views from above the model, while points within the outer ring define views from below.

Viewing Your Model in Parallel View, Step-by-Step

1. Select **3D Views**, **Parallel**
2. The globe appears, and the last parallel view is displayed. If you can't see any of your model, or only part of it, select **Window In**, **Extents**.
3. To view your model from the major vantage points, click successively within each of the four quadrants in the inner circle.

   If the model spins off screen, perform a **Window In**, **Extents**; then choose **3D View**, **Controls**, **View Center**, and middle-button snap to an entity vertex anywhere near the center of the model. Click the quadrants again.
4. To **Shade** or **Hide** the current parallel view, select **3D Edit**, **Shader**, **Shade**, or **3D Edit**, **Hide**, **Begin**.

Moving the Globe Display

While the Globe display is in the upper-left corner of the screen by default, you can use the Globe option in the 3DViews menu to reposition the globe on the screen. You can move the globe to a different location on the screen so that you can continue to work on a part of the model without interference, or change its size. A larger globe gives you finer control over the view rotation. To move the Globe display,

1. Choose **Globe** from the **3DViews** menu. The **Globe** menu is displayed:

   Default — Returns the globe to its customary position in the upper-left corner of the Drawing Window

2. Choose a new center for the globe. Move your mouse to dynamically size the Globe display. Choose a point on the perimeter of the new globe. The globe moves to the selected position.

Perspective View

Choose **Perspective** from the **View** pull-down menu or **Perspective** from the **3D Views** menu in the Menu Window to restore the last perspective view you generated. If this is your first perspective for this drawing, the **Set Perspective** options are displayed to set the options for the perspective view (see Figure 15-6).

Figure 15-6
The **Set Perspective** options

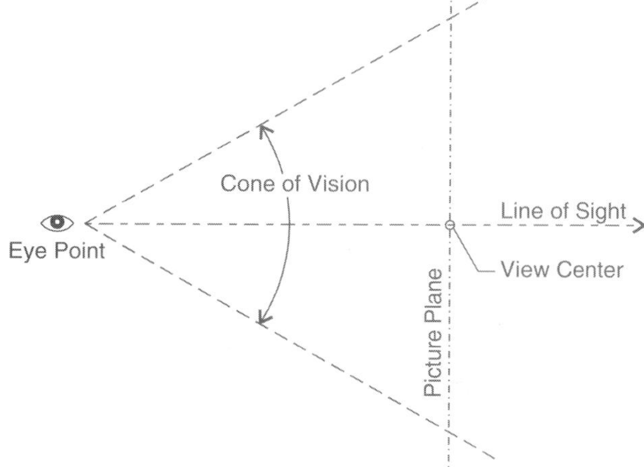

To set a perspective view at any time choose **Set Perspective** from the **3D Views** menu.

1. Choose **Set Perspective** from the **3D Views** menu. DataCAD automatically reverts to a plan view of your drawing.
2. Select **Eye Point Z** if you wish to change the Z-height of your eye point.
3. Select **Center Point Z** if you wish to change the Z-height of the center of the view.
4. Select **Reset** to return to a perspective view using the current settings in this menu.
5. Use **Hither Clip** to reduce "fisheye" distortion that can appear in a close-up perspective. You can also increase the hither clip distance to partially cut your model shown in perspective view, and then create a shaded "sectional perspective" with the **Shader** menu. The hither clip plane is displayed as a crosshair on the cone of vision (see Figure 15-7).
6. Toggle **Fixed Cone** ON to fix the width of the cone of vision. The width of the cone of vision in a fixed cone is set using either **Cone Angle** or **Camera**.
7. To set the width using **Cone Angle**, select it from the menu and select or type a value from 1°–179°, then press **Enter**.
8. To set the width using **Camera**, select it from the menu and choose or type a value, then press **Enter**.

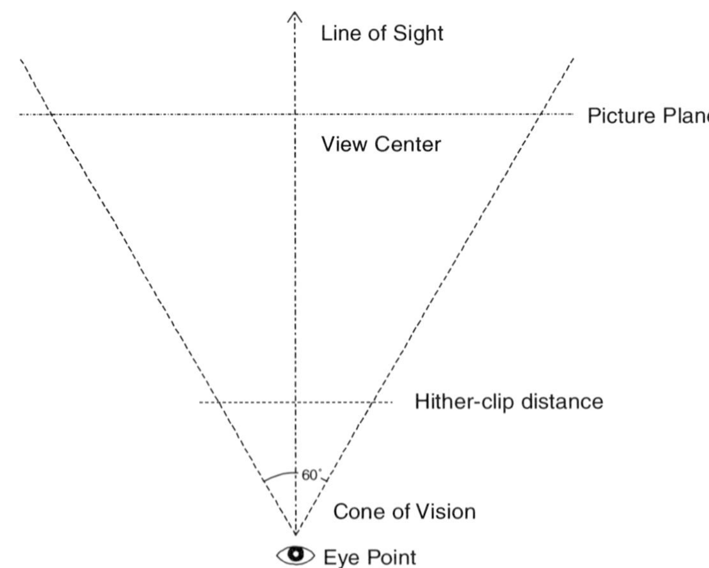

Figure 15-7
The hither clip plane, displayed as a crosshair on the cone of vision

Oblique View

Oblique views include the plan oblique (axonometric view), commonly used for architectural representation, and the elevation oblique. Oblique views are underutilized in CAD because they are less flexible than isometrics. Plan obliques and elevation obliques are excellent for presentation drawings where you need to show clear relationships to plans, as the projected view remains scalable.

Choose **Oblique** to restore the last oblique view you generated. If this is the first oblique view you've accessed for this drawing, the **Set Oblique** option appears, allowing you to set the parameters for an oblique view.

1. Choose **Set Oblique** from the **3D Views** menu.
2. Toggle **Plan Oblique** to create a plan oblique (architectural axonometric), and set your angle and factor.
3. Toggle **Elevation Oblique** to create an elevation oblique, set a new elevation, and then set your angle and factor.
4. Isometric creates a true isometric view. Isometric views are a special parallel view where the X- and Y-axes of the model form a 30° angle from the horizontal. To make sure that the angle is precise, choose **Isometric** from the **3D Views** menu in the Menu Window or choose **Isometric** from the **View** menu in the Menu Bar at the top of the screen.
5. Isometric views are always generated from the left side of the model. To create an isometric view from some other vantage point, use the **Controls** menu to rotate the view by 90° increments around the Z-axis in world coordinates.

Concept 3: Setting the Working Plane with New Elevation

Setting a New Elevation allows you to work on any face of your model, and next to Z-base and Z-height, it is the most important concept to understand in 3D modeling. Once you relate your screen to the zero plane it is quite easy to grasp.

The zero plane is the plane in the model where Z = 0. In plan view, the zero plane is fixed, and you cannot move it. In parallel views, the zero plane is relative to the screen. You can think of your screen as a picture frame containing the zero plane from which DataCAD measures the positive and negative Z values.

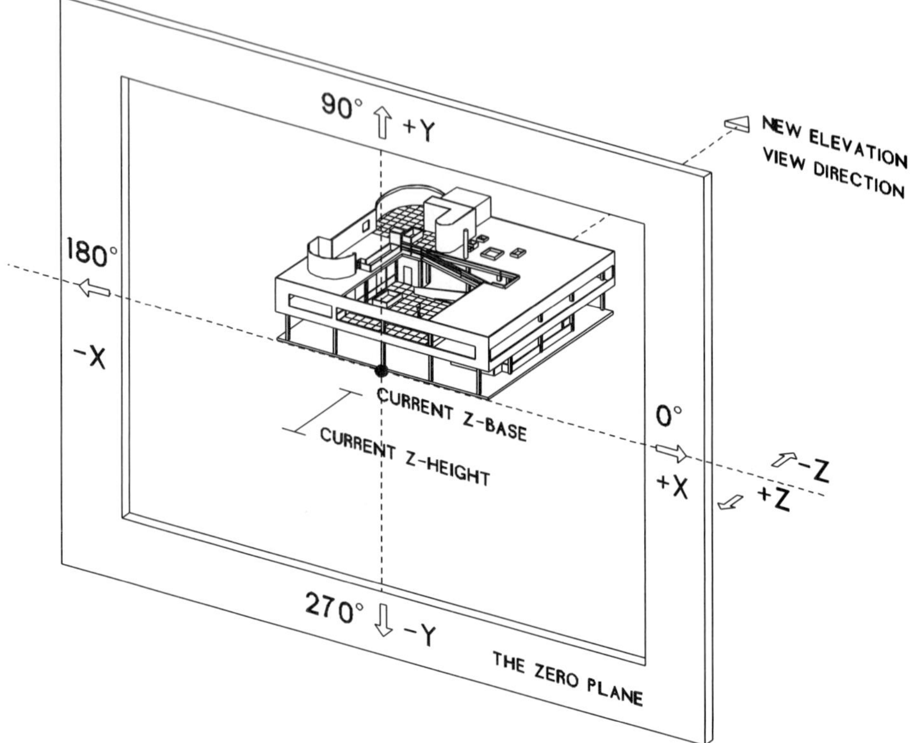

Figure 15-8 Z-base and Z-height will move relative to the screen

Think of it this way: As you look down on your model in plan view, Z-base and Z-height relate directly to elements like floor slabs. If we choose a wall façade as our next working view, the Z-base and Z-height will move relative to the screen, and will more approproiately relate to trim and glass panes (see Figure 15-8). So, when you change your view, you usually reset your Z-values as well.

Defining a New Elevation Step-by-Step When working in 3D, it is often necessary to create an entity on a vertical plane or an inclined plane; for example, to add a trim piece as a slab to an existing window opening in a vertical slab,

1. From ortho (plan) view, choose **3D Views** from the **3D Edit** menu.
2. Choose **Elevation** and then **New Elevation**.
3. DataCAD will prompt you to pick a point on the viewing plane. Snap with the middle mouse button to a corner of the slab surface you want to work on, then point in the direction you want to view.

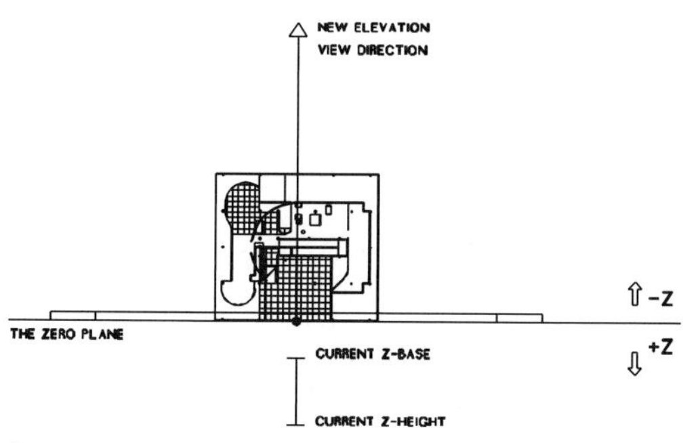

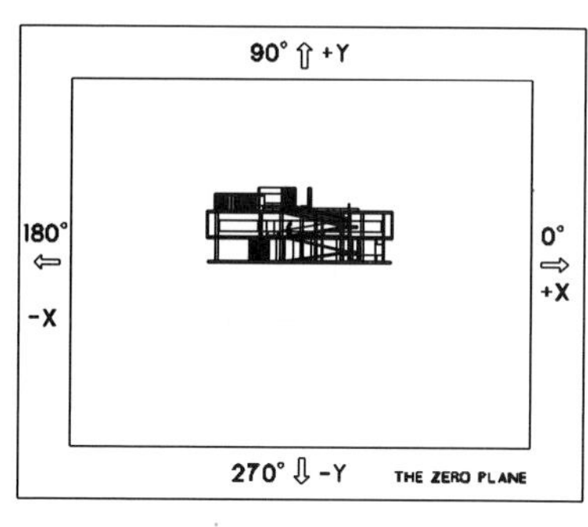

Figure 15-9 Defining a New Elevation: 1. The zero plane is placed in plan view and the direction is established. 2. The resulting screen view

4. The elevation appears perpendicular to the screen. Z-base zero is on the viewing plane defined in ortho view. Set Z-height to desired material thickness, and create trim entities with rectangular slabs. See Figure 15-9.

Exercise: Cantilevered Construction Using this method of construction, you will use **New Elevation** to change the working plane so that you can edit entities in 3D. By setting a new elevation you can change the orientation of the working plane to edit entities that appear on different planes.

1. Set DataCAD to the ortho (plan) view and begin by creating a vertical slab from Z-base 1.0 to Z-height 10.0.
2. Next, create a rectangular slab from Z-base 0.0 to Z-height -1.0 that encompasses the vertical slab. Changing to an isometric view reveals that the vertical slab rests on top of the rectangular one.
3. To change the working plane in order to draw on the vertical slab, go to the **3D Views** menu and select **Elevation**, then choose **New Elevation**.
4. Make sure **Ortho line mode** is toggled on. DataCAD prompts you to *"select the first point of the elevation line."* Snap to the lower left corner of the vertical slab.
5. DataCAD prompts you to *"indicate the direction of the elevation view."* Move the mouse in the direction you wish to view (towards the top of the screen) and click the left mouse button. On the screen, the tail of the arrow represents the Z-height in the new view.

   Note that the elevation line displayed in plan view will become your zero plane in elevation after you left-click to indicate the direction.

6. DataCAD presents the new view. Now you can model on the surface of this plane by setting a new Z-base and Z-height relative to it.
7. Create another vertical slab, this time from Z-base 0.0 to Z-height 5.0. Draw the slab in the center of the existing one, which is now horizontal to the screen.
8. Set your Z-base to -7.0 and your Z-height to 7.0; draw one final vertical slab.
9. Set an **Isometric** view of your model and shade the image, by selecting **3D Edit**, **Shader**, **Shade.**
10. Experiment with different **New Elevation**, Z-base/height, and slab combinations. Try to place multiple elements on the faces of the various slabs. Note how the **New Elevation** command provides you with the ability to create a working plane anywhere in the model, and then create an entity relative to it. See Figure 15-10.

Plane Snap Use Plane Snap in the **3D Views** menu to specify the zero plane by snapping to three points in your model. This is often the easiest way to add elements to sloping or inclined surfaces.

Use the **View** menu in the Menu Bar or the **3D Views** menu in the Menu Window to generate a parallel view that provides easy access to the three points that lie on the plane to which you want to align the zero plane.

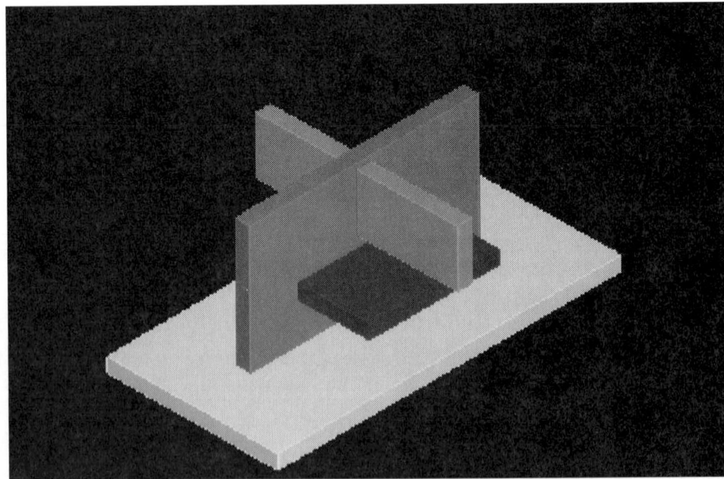

Figure 15-10 View of New Elevation model exercise

1. Choose **Plane Snap** from the **3D Views** menu.
2. Use the 3D world snapping cursor to snap to the three points. Keep the following in mind when using the 3D world snapping cursor:
   a. Snap to the three points in a clockwise direction
   b. The first two points snap to align with the bottom of the screen in the new view; choose these first two points in such a way that the model makes sense to you when the new view generates

For example, Figure 15-11 shows a building with a roof where you want to place a skylight. To do this, you need to make that part of the roof match up with the zero plane. Because the angle of the roof is a difficult one to match, use **Plane Snap** to align the zero plane.

Snap to the first two points of the plane along the bottom edge of the roof rather than along the side. Snapping to the first two points along the side of the roof would align those two points along the bottom of the screen, turning the entire building onto its side.

Plane Snapping Step-by-Step The **Reset** option in the **Plane Snap** menu changes the current view to a parallel view that matches the last orthographic view. This brings the model back into view if it moves off-screen. To add the current parallel view to the list of saved views,

1. Choose **Add View** from the **Plane Snap** menu.
2. Type the name of the new view, up to eight characters, and press **Enter**. The view is added to the list of saved views.

Methods to Define the Zero Plane: Setting a New Elevation is the most convenient, but not the only, way to establish a non-plan working plane. You can orient the zero plane in any of the following five ways:

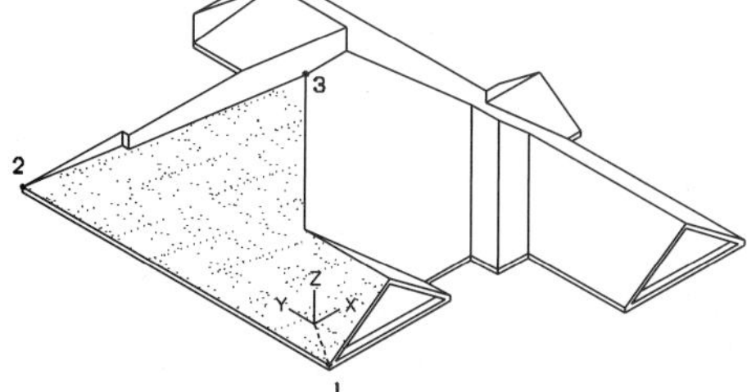

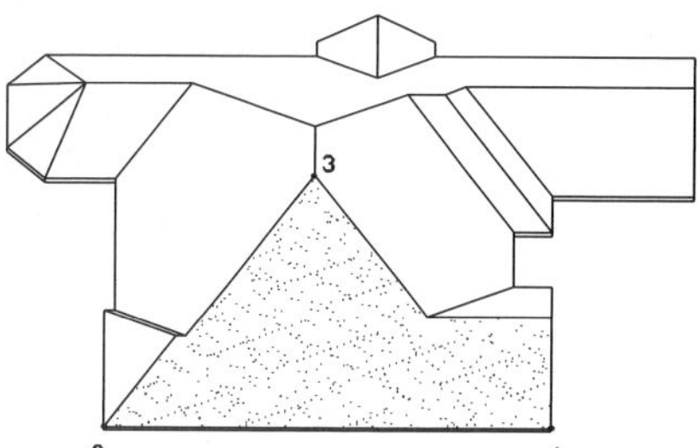

Figure 15-11 Using the **Plane Snap** command to reorient the zero plane to align with an inclined surface

- *Choose **Ortho** view* Orthographic has a fixed zero plane, also referred to as the ground plane. Use orthographic view to set the zero plane only when what you are modeling can be drawn from plan view.
- *Use **New Elevation*** Use this method when you can cut the new zero plane through the model from plan view. This always forces the view back into a plan view so that the elevation line cuts through the model or aligns with an elevation plane.
- *Use **Edit Plane*** Use this method when you can cut the new zero plane through the model from any parallel view. This method works like **New Elevation** except the view is not forced into an orthographic view first.
- *Use **Plane Snap*** Use this method when the model is in some isometric-like parallel view, and you can snap to three points, like the corners of a slab. The points that you snap to define the new zero plane.
- *Use **Controls*** The **Controls** menu can orient a model in any parallel view. When you define a view in this way, the position of the zero plane is established so that it passes through the view center.

Hidden Line Removal

This function allows us to remove lines from view that would not normally be seen in a real-world view. Looking at our model in perspective view, we can see lines which represent the opposite side of the model. Performing a Hidden Line Removal on the current view will create a 2D line drawing that can be saved to a new layer and then edited. This allows you to utilize your 3D models to create elevation or section drawings that can be used as drawings for construction documents.

To access the **Hide** function, select it from the **3D Edit** menu. The option **Save Image** appears, along with other settings. Toggle this on if you wish to save the resulting image to a new layer. Select **Begin** to start the process of removing hidden lines. As this is a math-intensive process, it may take a few minutes for DataCAD to complete the task. Once finished, you have the option to save the image to a new layer, name that layer, or turn it on or off. After you save the image, you can view it as another layer in the drawing.

To view the hidden line image, change your view back to **Ortho**. Select **Active Only** from the **Layers** menu. Pick the new layer from the layer list. The image is displayed. Choose **Window In**, **Extents**, if you can not see all or part of the Hidden Line image. As this image is composed of 2D lines, you may continue to edit it and add notation after you save it.

The HLR Process Step-by-step

1. Set up a perspective, parallel, or other 3D view that you wish to hide and save.
2. Choose **Hide** from the **3D Edit** menu.
3. Make sure the following options are toggled on:

Save Image	Gives you the option to save the image when it is complete, or discard it by right-clicking
Crop Image	Crops the image to the current screen view. Toggle off if you wish to calculate the image beyond the screen. This may produce distorted lines with close-up views using wide camera angles.
HLR Partial	Speeds up the HLR process by breaking the scene down into panels.
All	Hides all layers currently turned on

4. Next, choose **Options** and toggle on:

Join	Joins lines together in the final saved image to make editing easier
Delete Double	Deletes lines that are one on top of the other in the final image
Ignore Color	Ignores the color of the lines for the above two operations
Back Face	Speeds calculation by removing back-faced polygons from the HLR calculation
No Edge	Removes facets from cylinders, and only draws their outer edges
Divisons	Enter **36**. (This is the maximum allowed; this number may be lowered if smooth curves aren't required.)

5. Exit the **Options** menu and choose **Begin**.
6. When hidden line removal is complete, DataCAD will prompt you to choose a destination for the saved line drawing. Note: If you *right-click* at this point, you lose the hide! Only right-click if you don't care to save the image.
7. Choose **New Layer**, to save the hide to a new layer in the file or **Layer File** to save it to a layer outside the file, which can be loaded into a different file later.
8. DataCAD will prompt you to type in a layer name. Type a name and press **Enter**.

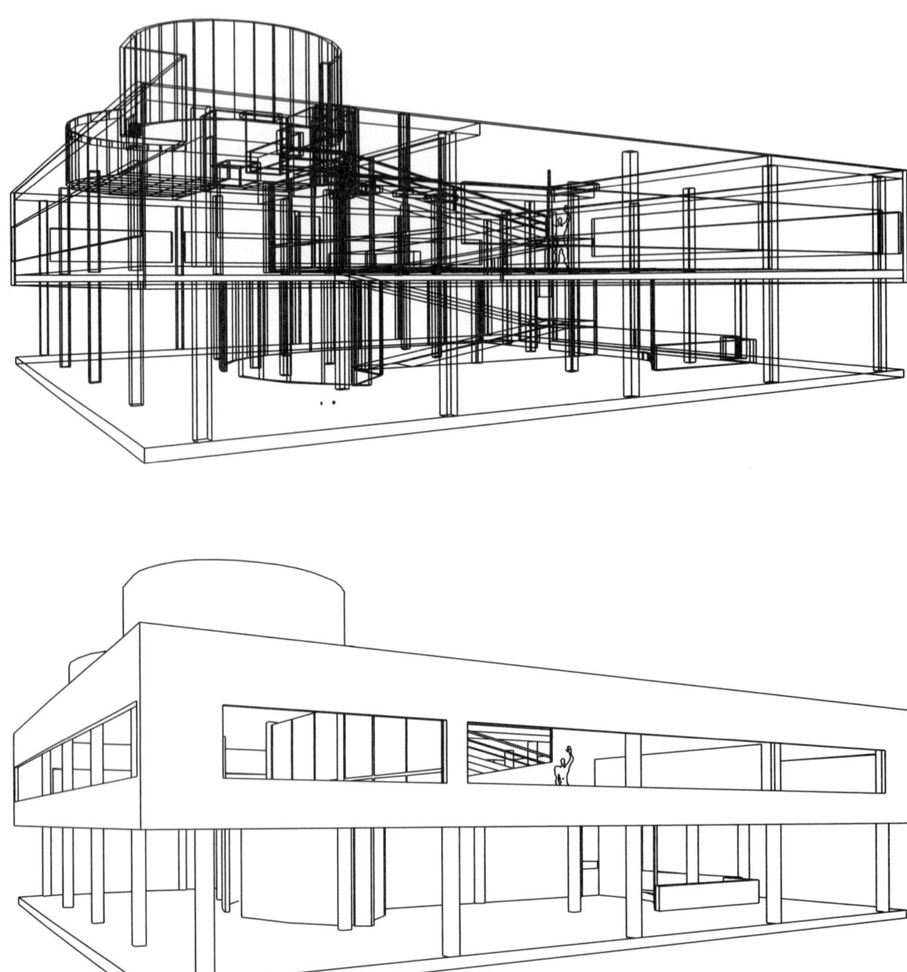

Figure 15-12 Model before and after Hidden Line Removal process

9. If a new layer is being saved to your file you will be prompted to have it **On** or **Off**. Choose **Off**.
10. To view the hide, change to ortho view and set the new hidden line layer to **Active Only**. You may now edit and notate the image just like a drafted 2D drawing. See Figure 15-20.
11. To print your Hidden line Drawing, proceed to the Layout and printing process, with this layer displayed as **Active Only**.

GL Shader

The **Shader** is an invaluable resource for 3D modeling. With it you can intermittently create shaded views of your model to judge how things are coming together. More often than not, this will be an isometric or parallel view, and you will discard the image without saving it. As the **Shader** does not support textures, shadow, or complex lighting, its utility derives from its ability to quickly convey a solid impression of the model. Shaded views are also good for design study presentations, where it is more desirable to communicate the design idea rather than a photo-realistic representation. In this regard, the unsophisticated quality of the **Shader** is an asset.

The **Shader** supports multiple light sources. You can define and place up to seven light sources in your model and use ambient light as well. Ambient light refers to the overall lightness or darkness of the scene. These light sources are specific points that radiate light out in all directions and can be toggled on or off to create different lighting effects. Each model surface is illuminated independently of the others, so one surface will not cast a shadow on its neighbor. The only factor that affects the light on a surface is the distance of the light source from it.

The **Shader** supports all 256 colors in the DataCAD palette as well as a rendering palette of 16.8 million colors.

To set the **Shader** background color, choose **Program Preferences** from the **Tools** pull-down menu and click on the **Misc** tab. Click in the colored rectangle under the **Shader Background Color** heading to open the **Color** dialog box; select a color, and click OK. This background color setting is independent from the background color for the Drawing Window.

Shader Settings Dialog Box

To create a shaded image of your model,

1. From the **Edit** menu in the Menu Window, choose **DCAD 3D**. The **3D Edit** menu is displayed.
2. Choose **Shader** from the **3D Edit** menu. The **Shader** menu is displayed.
3. Choose **Settings** from the **Shader** menu. The **Shader Settings** dialog box is displayed (see Figure 15-13). The settings in this dialog box are also available from the **Shader** menu in the Menu Window. You can use either the dialog box or the menu to change settings; all changes will be reflected both in the dialog box and the menu options.

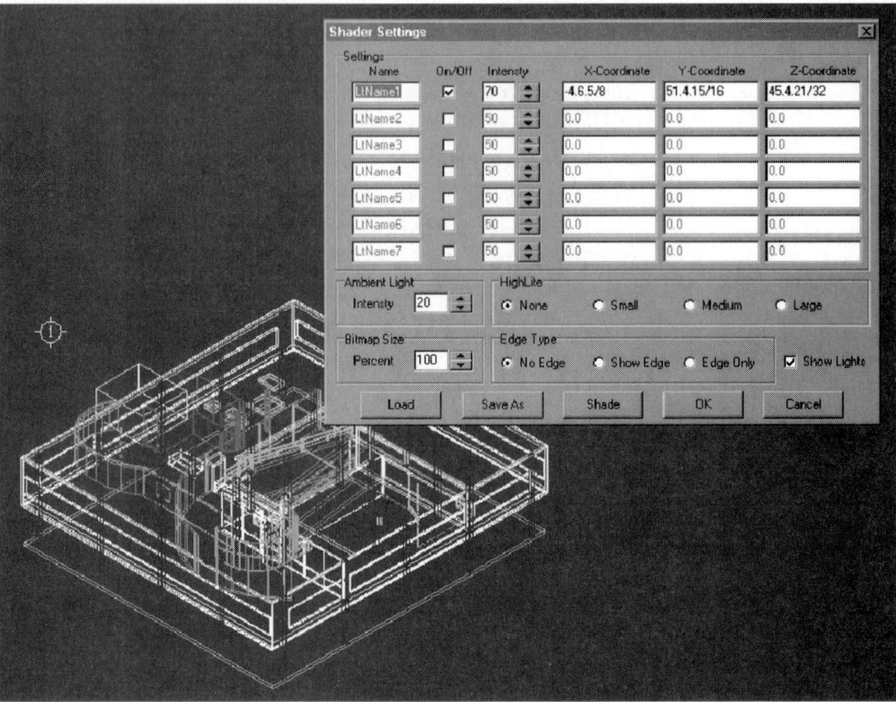

Figure 15-13 The **Shader Settings** dialog box

4. The light sources are named **LtName1**, **LtName2**, etc., by default. To rename a light source, click in the input box and type a new name.
5. To toggle a light source on or off, click in the checkbox in the **On/Off** column next to the light source name.
6. The light source's intensity is the percentage of white, with values ranging from 0–100. To set the intensity of a light source, use the up and down arrows to increase or decrease the percent value.
7. Each light source can be positioned either by entering X-, Y-, and Z-coordinates or visually by using your cursor.

   To enter specific coordinates for a light source, click in the **X-Coordinate**, **Y-Coordinate**, and **Z-Coordinate** input boxes and type new coordinates.

   To visually place the light source, toggle on **Show Lights** in the Menu Window, then click on the light source name, e.g. **Light Source 1**. Click **On/Off** in the menu if the light source has not yet been toggled on, and then choose **Position** from the menu. A 3D cursor is displayed in the Drawing Window; note the cursor's coordinates displayed in the Message Window. Click to place the light source. A light symbol indicator displays the position of the light (see Figure 15-14).

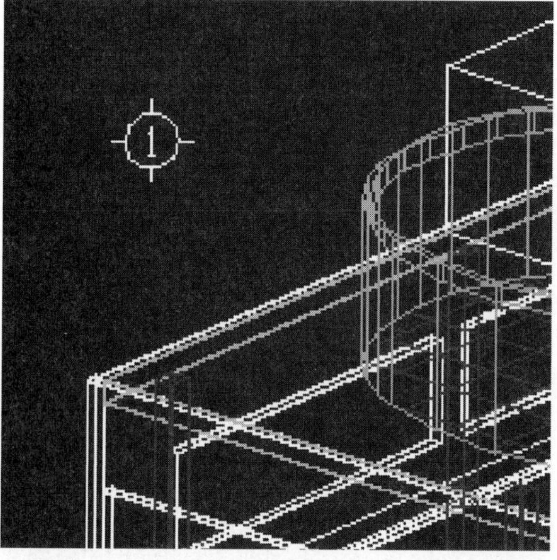

Figure 15-14 Visually placing a light source

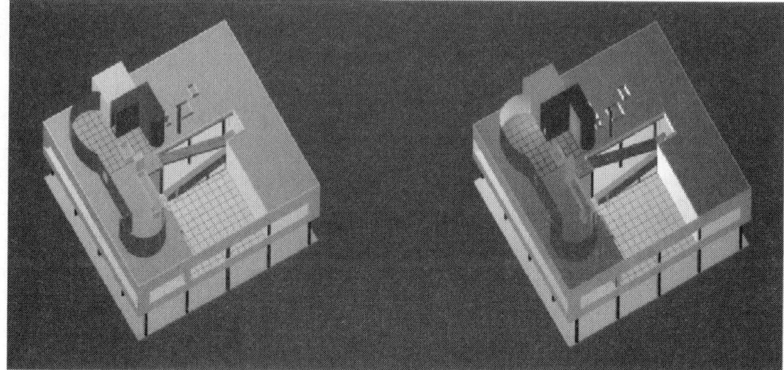

Figure 15-15 Using the **Highlight** setting

8. The intensity of ambient light can be adjusted just as light source intensity was in Step 6. To set the intensity of ambient light, use the up and down arrows to increase or decrease the percent value.
9. The shaded image can be saved as a bitmap (.BMP) file. You can adjust the size of this bitmap by using the up and down arrows to increase or decrease the percent value, with 100% being the full size of the Drawing Window. You can go as high as 1,000%, depending on how much system memory you have available.
10. The **Highlight** setting produces a visible area of light around a light source, much like a lamp throws on a wall and ceiling (see Figure 15-15). This setting is off by default, but you can set it to three intensities, with **Large** being the most intense.

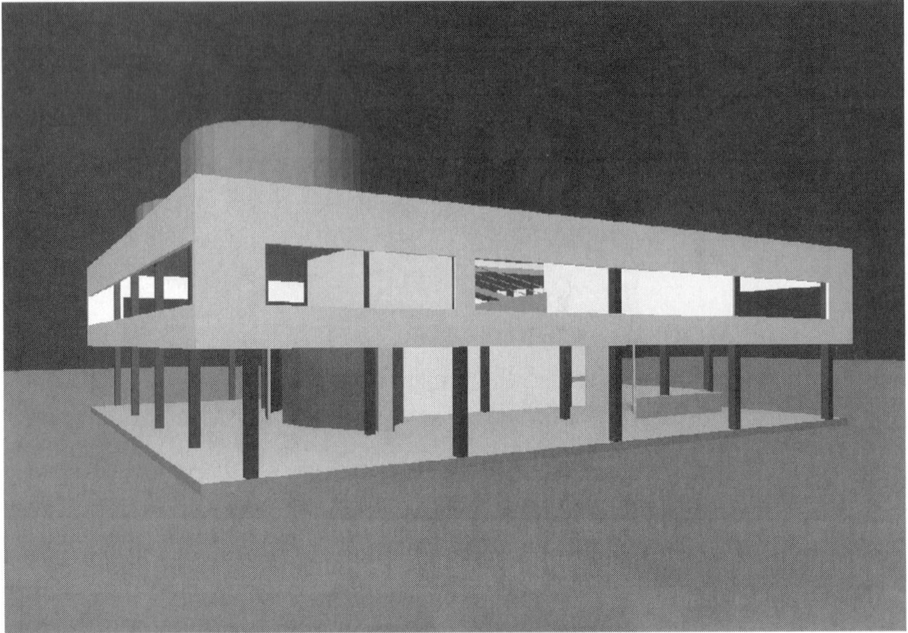

Figure 15-16 Model shaded using DataCAD's **Open GL Shader**

11. You can choose to shade only the surfaces of your model, both the edges and surfaces, or only the edges, which is similar to a hidden line removal with lights. The default setting is **No Edge**.
12. The **Show Lights** toggle is checked **on** by default. This option displays the light sources as small symbols around your model, letting you see exactly where they are positioned. These symbols do not print; they are only displayed for visual reference.
13. Click **OK** to save your settings changes and close the dialog box.
14. Choose **Shade** in the Menu Window. Your model is shaded and the bitmap image is displayed in the Drawing Window (see Figure 15-16). You can now save the shaded image as a bitmap file or return to your drawing file.

To quickly shade your model without changing any **Shader** settings, choose **Shade** from the **View** pull-down menu. Your model is shaded.

To save your shaded image,

1. Choose **Save** or **Save As** from the **File** menu, or right-click in the bitmap window to display a pop-up menu and choose **Save bitmap** or **Save bitmap as**. If you chose **Save bitmap**, the bitmap is saved in

Figure 15-17 Image converted to sepia tone in Photopaint

the **BMP** folder in your DataCAD folder.

2. If you chose **Save bitmap as** in Step 1, type a name for the bitmap file, and click **Save**. The bitmap is saved in the **BMP** folder in your DataCAD folder.

To close your shaded image without saving it and return to the drawing,

1. Right-click on the shaded image; a dialog box appears.
2. Select **Close**, then **No** to discard the image.

To print your shaded image:

1. Open the saved image in an image editing program like Microsoft Paint, Adobe PhotoShop, or Photopaint (see Figure 15-17).
2. Select **File**, **Print**.

CHAPTER 16

3D Modeling

Introduction to the Geometric Primitives

Modeling in DataCAD involves creating, and then editing 3D entities in such a way that they create a representation of a design idea. The process is very similar to physical construction, or building a study model whereby various pieces of cardboard are cut and glued together. In DataCAD, the raw material for our models exists in the form of geometric shapes that have particular properties we may coax to assume a particular shape. These 3D forms, called primitives, are defined by the view you create them in, and the active variables you have set, such as Z values, thickness, radius, divisions, etc.

A major difference between 2D entities, whose properties you are already accustomed to, and 3D entities, are the object snapping methods you may use on them. 2D entities support snapping of virtual points, such as midpoint, intersection and perpendicular. 3D entities only support object snap at their corners, known as *vertices*. The following is an inventory and short description of the 3D entities we have at our disposal in 3D modeling:

3D Line

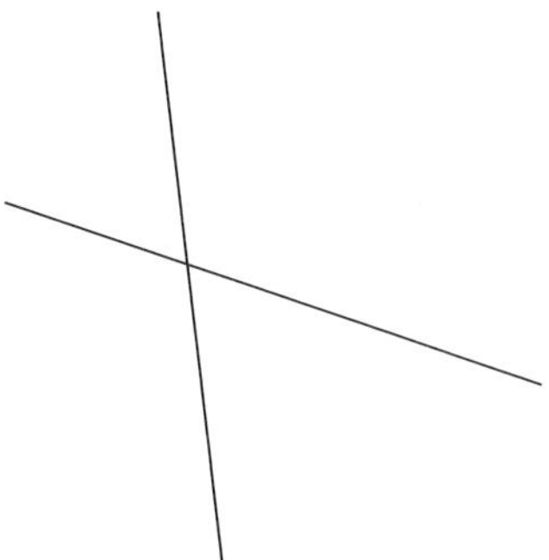

3D lines are the simplest 3D entity. It is simply a line that connects two points in space. 3D lines are useful for indicating a joint or pattern on a surface, or creating a **Sweep** path. When creating 3D lines, you may draw them by **Z-base** only, **Z-height** only, or from b-**base** to **Z-height**.

1. 3D lines are like wires or threads in space. Unlike 2D lines, 3D lines have no Z-height. Furthermore, 3D lines are not constrained to lie in the XY plane; they can connect any two points in 3D space.
2. When you draw 3D lines from **Z-base** to **Z-height** and you select two points at exactly the same place on the screen, you create a line that is perpendicular to the screen. Such a line looks like a dot from the current view (a line viewed end-on). Although this method works, it is usually conceptually easier to draw such lines flat across the screen in an appropriate elevation view.
3. You can use the 3D cursor to connect to the entities themselves rather than connecting to the projections of those entities. Drawing with 3D cursor is the only time you can draw in DataCAD without concern about the location of **Z-base** and **Z-height.**

Polygon

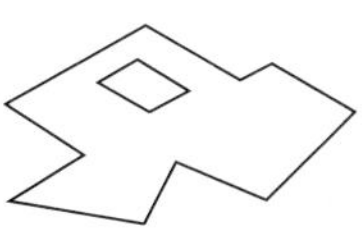

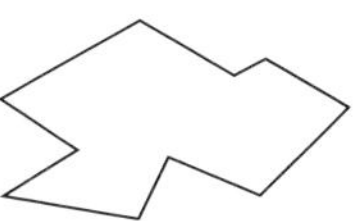

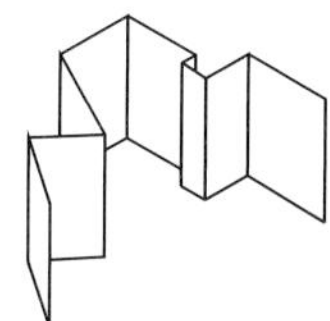

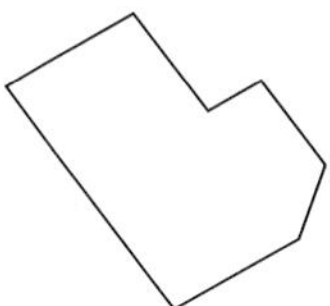

Polygons are the simplest 3D surface you can create, and the most versatile of all the 3D entities. Experienced 3D models utilize 3D polygons for the majority of their modeling for this reason. Polygons can have up to 36 sides. All of these points must lie within the same plane. Polygons also support voids and hatch pattern fills.

1. Select **3D Entity, Polygon** to create 3D polygons. Do not confuse 3D polygons with their 2D counterparts. A 3D polygon is not a group of lines like a 2D polygon; it is a single entity that describes a surface. 3D polygons are opaque to the hidden line removal process.
2. Polygons are like flat sheets of paper. They can be either horizontal or vertical. For horizontal polygons, you can select points along the Z-Base plane up to a maximum of 36 sides. *Right-click* your mouse to close the shape. When drawing vertical polygons, they are constructed from Z-Base to Z-Height; with the **Chain** option toggled on, they will be drawn connected, as with 2D lines. This method is often used in the construction of walls.
3. Polygons support voids. To create a void in a polygon, two polygons are drawn, one within the other. The encompassing polygon is the Master polygon, and the enclosed polygon is subtracted from it to make a void. Select **3D Entity, Polygon, Horizontal, Void, Add Void.** Choose the master polygon first, then the polgons within it that are to become voids.
4. You can edit the shape of a **Polygon** after you have created it. Use the **3D Entity, Polygon, Partial, Add/Delete/Move, Vertex** function to add up to 36 sides, delete sides to a minimum of three, or move existing vertices.

Slab

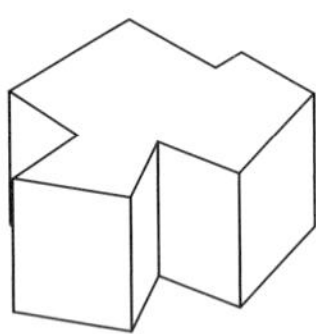
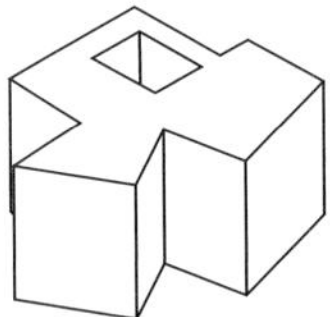
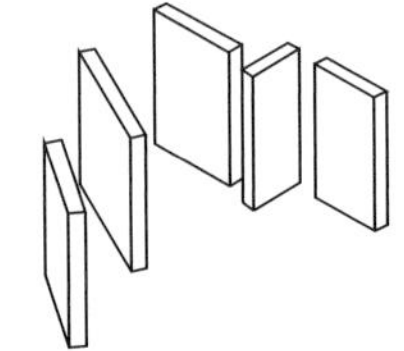
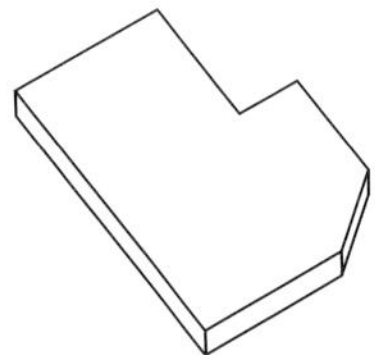

Slabs are the second-most common entity type used in a 3D model. **Slabs** share similar properties with polygons, but they have the added thickness dimension. Common uses for slabs are floor planes and masonry walls.

1. The one important distinction between slabs and polygons is that slabs possess a reference face that controls the direction in which you can pass a void. Every slab has a special face called the *reference face* that orients the slab in the DataCAD database. The reference face is important when you move or add vertices or voids to a slab.
2. You can edit the shape of a **Slab** after you have created it. Use the **3D Entity, Slab, Partial, Add/Delete/Move, Vertex** function to add up to 36 sides, delete sides to a minimum of three, or move existing vertices. You may only add or delete points from the Reference Face of the slab.
3. Depending also on the creation order and the kind of slab you create, the reference face location varies:

Vertical Slabs

When you draw a slab from the top of the screen to the bottom, the reference face falls on the left. When you draw a slab from the bottom of the screen to the top, the reference face falls on the right.

Inclined Slabs

The reference face is always on the underside of the slab.

Use Slab References to display markers that indicate the reference face for a slab as well as the direction of extrusion. The reference face of the slab is determined by the points selected on the screen. The extrusion is determined by the relative Z-base and Z-height (or Z-User1 or Z-User2) settings you have entered.

All that you need to remember about reference faces is that when you intend to make a void in a slab, you should create the master slab and the

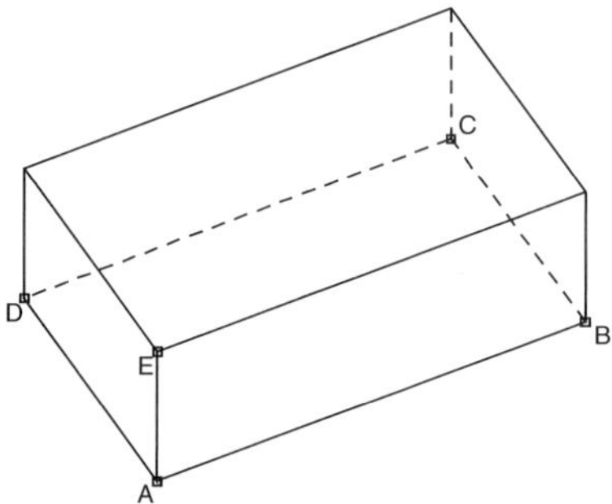

Reference Face = ABCD
Extrusion Value = AE

Z base = A
Z height = E

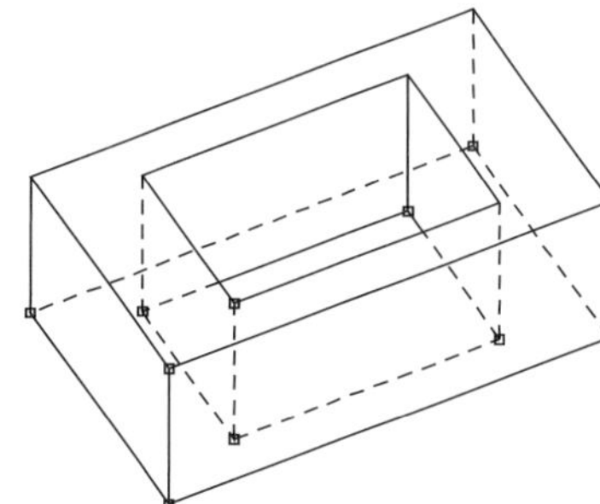

When creating voids, the reference faces of the master and secondary slabs must be in the same plane, and their extrusion values must be equal.

void slab in the same view, and in the same direction. This will ensure that DataCAD creates the void properly (If it doesn't you will notice a strange, inside out, M.C. Escher effect when you perform a hide). For example, in the case of a vertical slab, if you create the master in plan view from left to right by center, you should do the same for the void. In the case of horizontal slabs, if you create the master in elevation view counter-clockwise, do the same for the void.

3D Arc

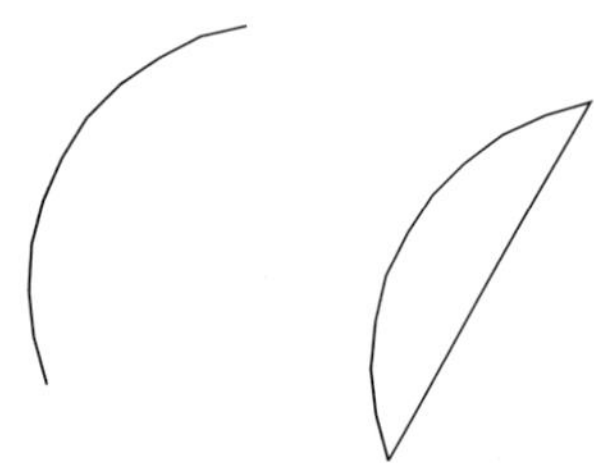

3D Arcs are seldom used, but useful to create circular surfaces. 3D arcs are essentially the same as their 2D counterparts, with the following exceptions:

- 3D arcs can exist at any orientation in space.
- 3D arcs can be specified as open or closed.
- 3D arcs are approximations of arcs consisting of many short line segments called *divisions*.

The first eight options on the **3DArc** menu are quite common and are seen wherever there is a curved 3D entity. Creating this 3D arc footprint is often the first step in creating a curved 3D entity. This footprint defines the **Sweep** angles for the entity.

To make a 3D arc into a circular surface, create the entity with **Closed** toggled on, then **Explode, To Polygons**.

Vertical Cylinder

Vertical cylinders are frequently used as columns or posts in 3D modeling. They are created standing on end. The column diameter is dynamically specified by drawing the footprint of the cylinder using one of the standard 3D arcs. The column height is controlled by setting the **Z-base**, **Z-height**,

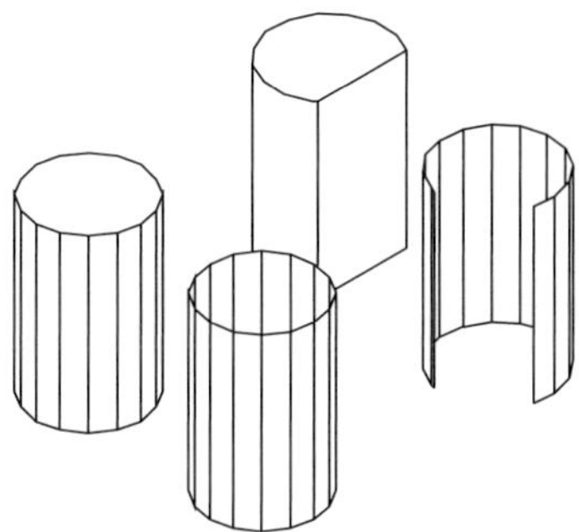

or **thickness**. The faceting of the cylinder is controled by its **divisions** setting. When performing a hidden line routine, you have the option not to display the cylinder facets.

Horizontal Cylinder

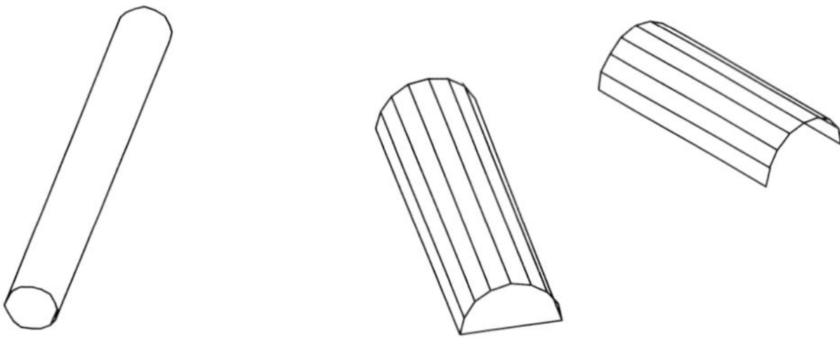

Horizontal cylinders are commonly used for handrails, piping and the like. They are drawn flat across the screen. You must set the radius before you create them. They are then entered at Z-base, Z-height or from base to height. Note that Vertical and Horizontal cylinders may not be stretched after you create them. To increase or decrease a cylinder's length, you must use the **Enlarge** command, or erase and re-create it.

Cone

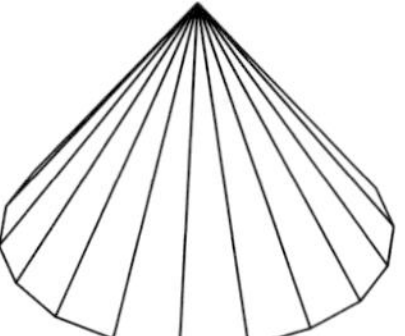

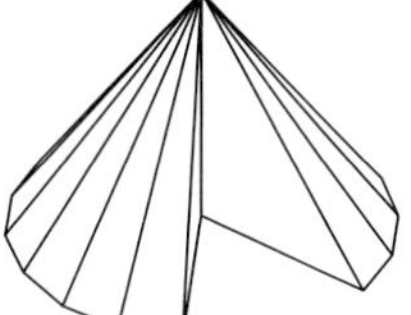

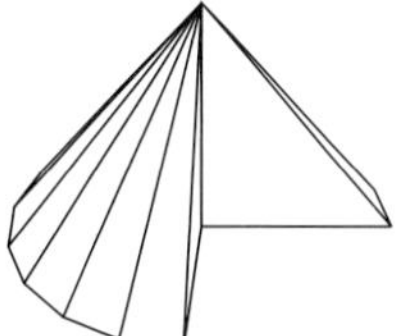

Cones are entities that are not used often in architectural modeling, but they are sometimes seen in spires and pylons. You can create cones by drawing the base of the cone (at Z-base) using the standard **3DArc** menu and placing the vertex of the cone at Z-height. DataCAD automatically places the vertex, thus creating a right cylinder, or you can place the vertex yourself, creating a skewed cylinder.

Truncated Cone

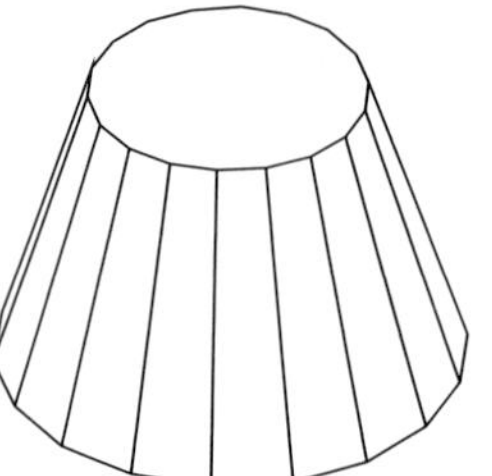
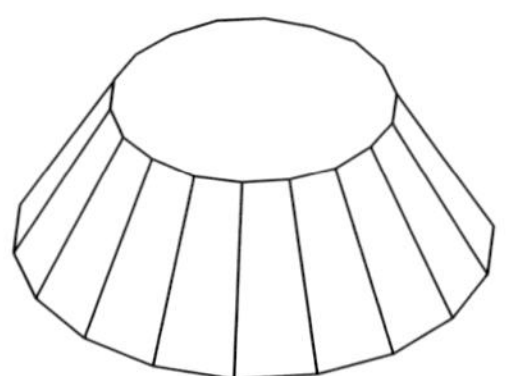
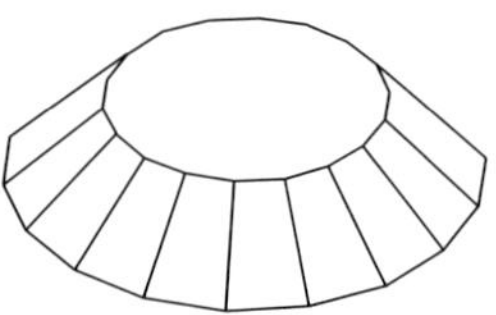

Truncated cones have sheared-off tops. Creating a **truncated cone** is similar to creating a regular cone, except you must also specify the diameter of the top part of the truncated cone after setting the base. If you set the base and height to zero or any equal value, the result is a flat, circular plane.

Sphere

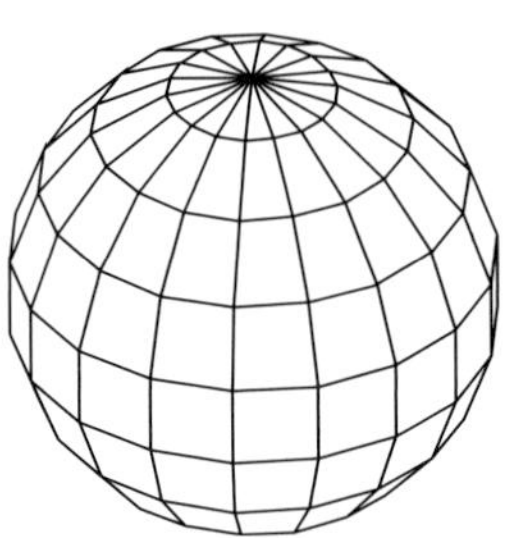
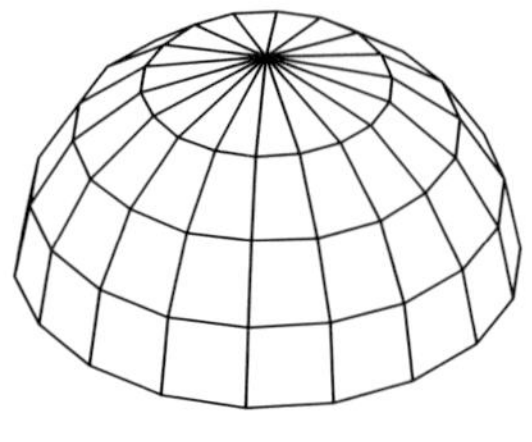
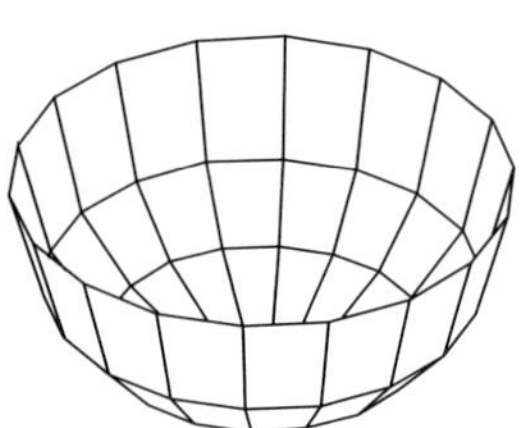

Spheres can be used to create domes, or even a sky-like back-drop for a 3D model. Spheres can be a full sphere, or section of a sphere. The menu combines features from both the vertical and horizontal cylinder menus as well as the **3DArcs** menu.

Spheres may be drawn a number of different ways: full or half, by radius circle, diameter circle or 3 point circle. You can control the number of facets that appear by changing the **Primary** and **Secondary** divisions.

Torus

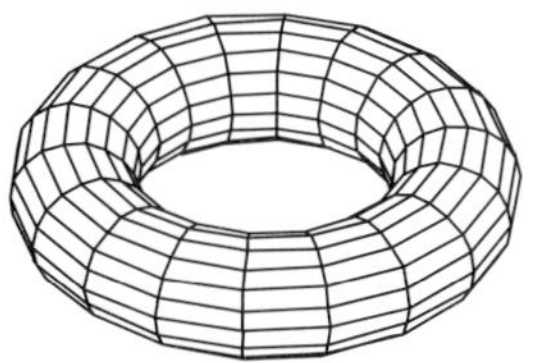
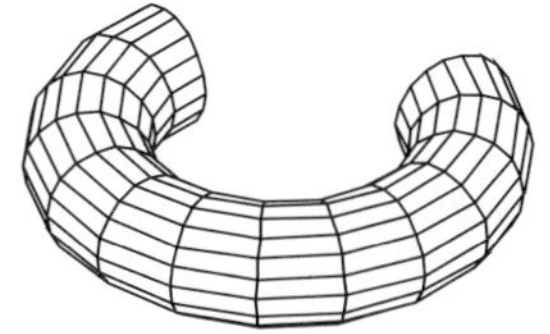
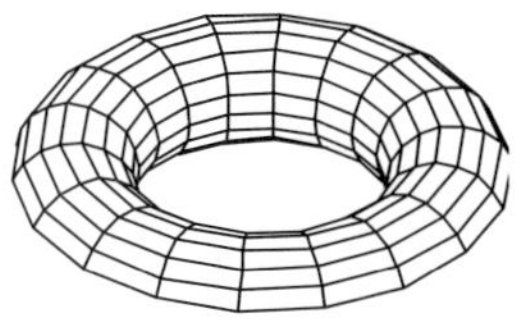

A **torus** is a donut-shaped entity. Full torii are obviously very rare in architectural modeling, but sections of tori are fairly common. Typical applications include turns in railings, elbow connectors in pipes, etc. Torii are entered at the selected Z setting, at the entered radius. You may create a full torus, or any portion thereof.

Revolved Surface

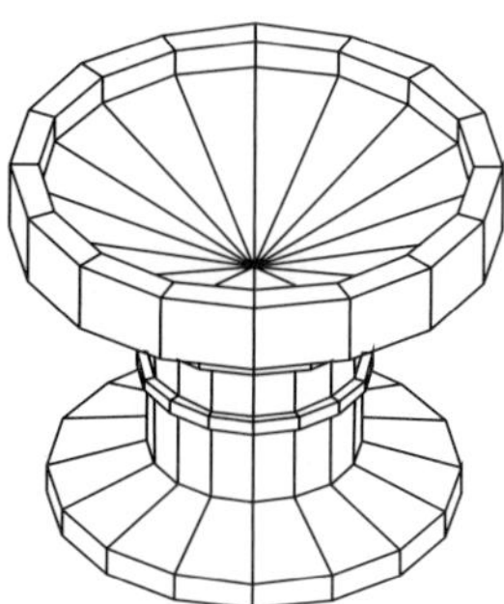
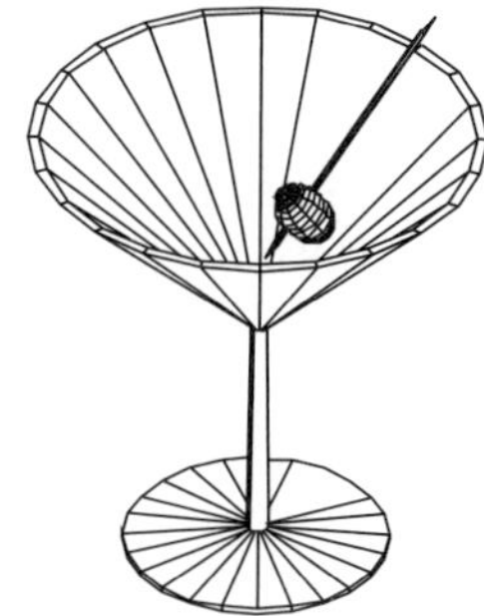

Use **Revolved Surface** to create 3D solids with radial symmetry. Columns, lampshades, fountains, and canopies are all examples of objects with radial symmetry that you can model using **RevSurf**.

Surfaces of revolution are so named because of the way you construct them: a polyline profile revolved around an axis of revolution. As the profile revolves around a given axis, it "sweeps out" a surface in its wake. This surface is the surface of revolution. To create any surface of revolution, you must determine what profile and axis combination generates the surface you want.

Surfaces of revolution are single entities. To modify or edit a surface of revolution, you can explode it and stretch it, or enlarge it to create complex shapes. If you're trying to model a shape that doesn't have radial symmetry, but is close, try starting with a surface of revolution. Then through enlargements, explosions, and stretches, shape the solid into the form you need.

Revolved surfaces may not be stretched after you create them. To change a Revolved Surface's size, you must use the **Enlarge** command, or erase and re-create it.

Contour

Contours are a curved version of 3D lines. Like 3D lines, contours have no thickness, and they don't need to be straight. Use contours to create topographical maps.

Creating contours is similar to creating Bezier or B-spline curves. You place a number of control points that define the curve. DataCAD accepts up to 36 control points. Unlike Bezier or B-spline curves, the contour curve that DataCAD generates *passes through* the points that you draw. Contour points may be stretched to new positions after you have drawn them, but you can't add or subtract points. To edit contours further, install the dhContour macro from the bonus CD.

Mesh Surface

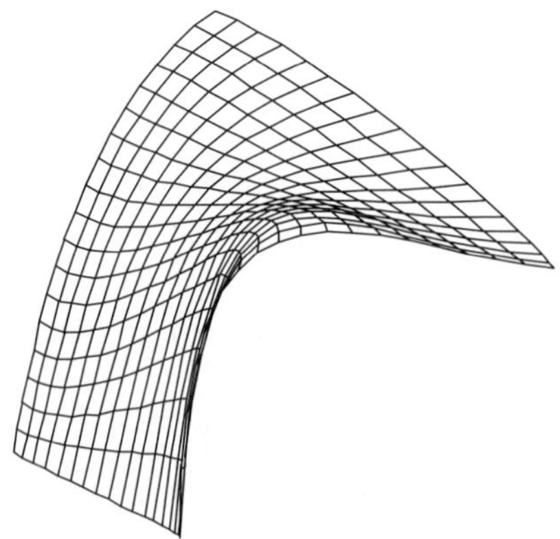

A **mesh surface** is a smooth patch that you can mold into a variety of useful shapes including free-form built structures, tents, and canopies. Mesh surfaces are the surface equivalents of Bezier curves.

Like contours, mesh surfaces are created through a series of control points that you specify. The greater the variation in the elevation of these control points, the more dramatic the curves in the surface.

There are always exactly 16 control points to specify for a mesh surface, and these 16 control points always form a 4×4 grid of points, although the spacing of these points may not be uniform.

Marker

A marker is a 3D snapping point. Use markers to mark create snapping points in space to guide the creation of other entities. Markers may be entered at Z-base, Z-height, or dynamicaly with the 3D Cursor.

The Sampler.dc5 file included with DataCAD 9 includes examples of each 3D Entity type:

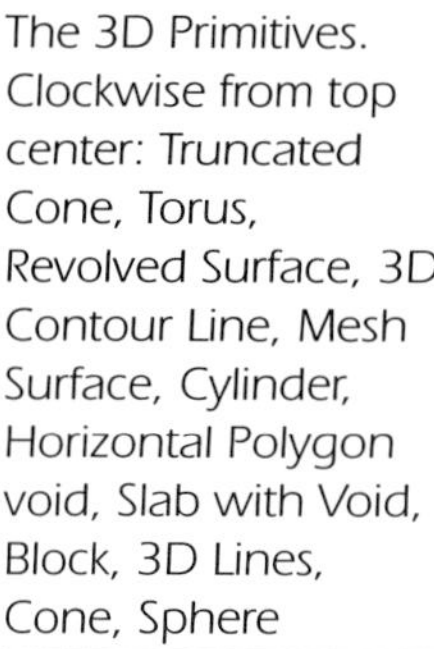

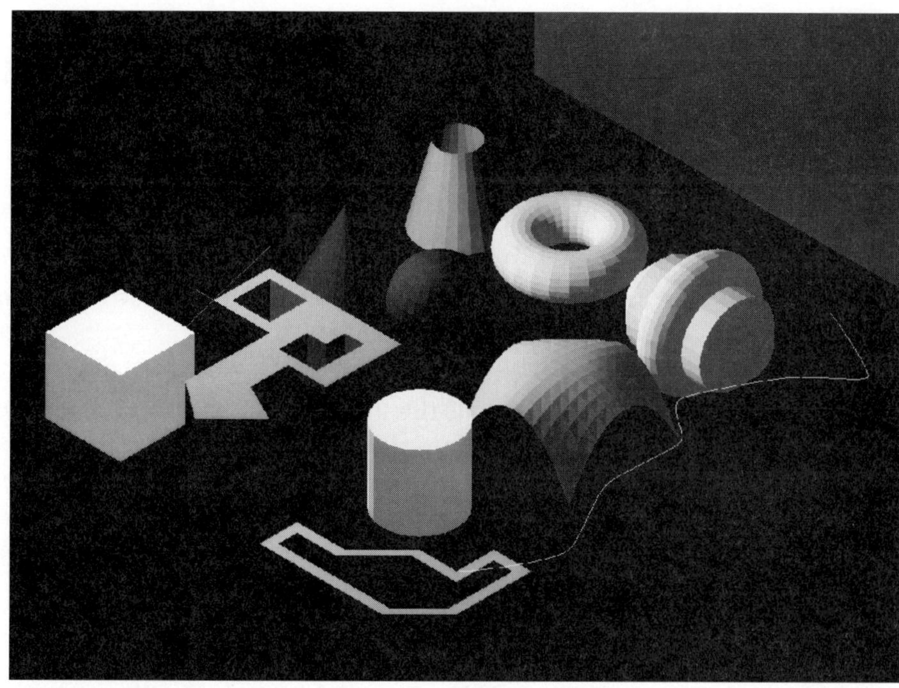

The 3D Primitives. Clockwise from top center: Truncated Cone, Torus, Revolved Surface, 3D Contour Line, Mesh Surface, Cylinder, Horizontal Polygon void, Slab with Void, Block, 3D Lines, Cone, Sphere

Creating Specialized Shapes Using the Primitives

Inclined polygons can be used to create roof elements. Roofs that you create with this menu do not have a thickness. To create roofs with thickness, use the **Inclines** menu in the **3Dentity,Slab** menu.

NOTE: *DataCAD needs **Z-base** and **Z-height** information about the roofs that you are going to create—information that is easily calculated but not immediately at hand. Calculate the heights and breadths of roofs before using the **Inclines** menu.*

To create roof elements using inclined polygons:

1. Use the **3D Views** option from the **Polygon** menu to place the model in an orthographic view.
2. Set **Z-Base** to the height of the lowest point on the roof.
3. Set **Z-Height** to the height of the ridge line (peak) of the roof.
4. Choose **Inclines** from the **Polygon** menu.
5. Continue with the following sections to draw the three- or four-edged inclined polygon you need.

To create roof elements using inclined slabs:

1. Place the model into orthographic (plan) view.
2. Choose **Inclines** from the **Slab** menu.
3. Choose **Thickness** to set the depth of the roof element; choose or type a value and press (Enter).
4. Choose whether to add the thickness perpendicularly or vertically. **Vertical** adds the thickness of the slab to the roof element vertically, or perpendicular to the ground. This produces a plumb cut on the edge of the roof element. **Perpendicular** adds the thickness of the slab to the roof perpendicularly to the roof slope.
5. Continue with the following instructions to draw the three- or four-edged inclined slab you need.

To draw gable, A-frame, and shed roofs:

1. Choose **4 Edge Parallel** from the **Inclines** menu.
2. Select the first corner point at the bottom edge of the slope.

3. Select the second corner point at the bottom edge of the slope.
4. Select the first corner point on the ridge of the roof. Enter this third point opposite the first point, as shown next. Locate this point by snapping or with a reference point shortcut and coordinate entry. The fourth corner point is automatically derived.

Four-edged parallelograms

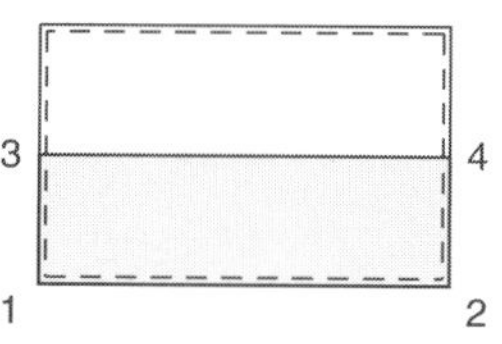

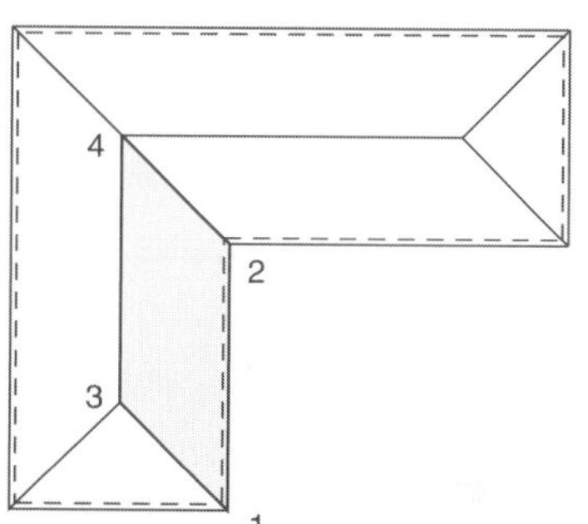

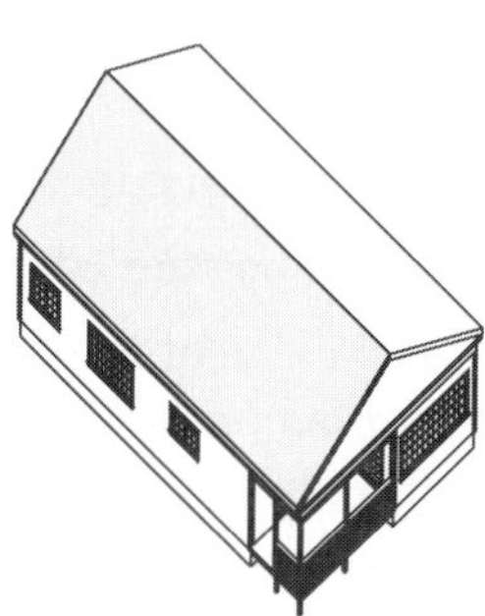

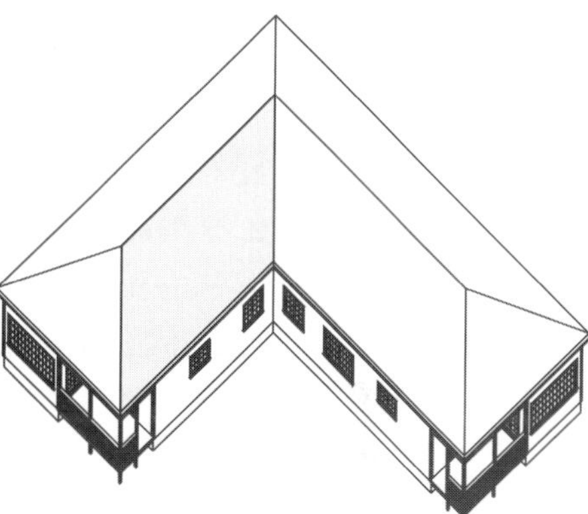

To draw hip and mansard roofs

1. Choose **4 Edge Trapezoid** from the **Inclines** menu.
2. Select the first corner point at the bottom edge of the slope.
3. Select the second corner point at the bottom edge of the slope.
4. Select the first corner point on the ridge of the roof. Enter this third point opposite the first point, as shown next. Locate this point by snapping or with a reference point shortcut and coordinate entry. The fourth corner point is automatically derived.

Four-edged trapezoids

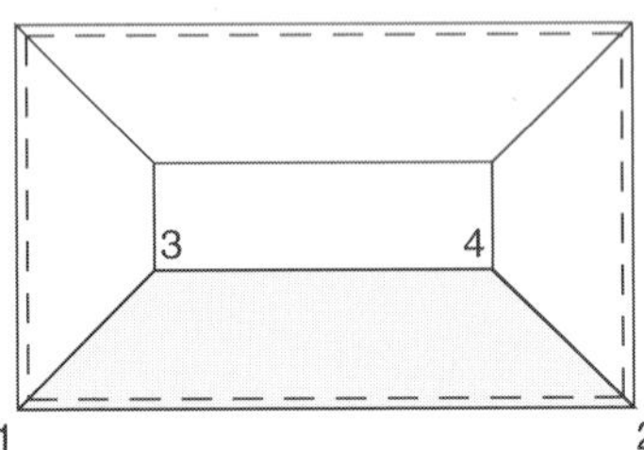

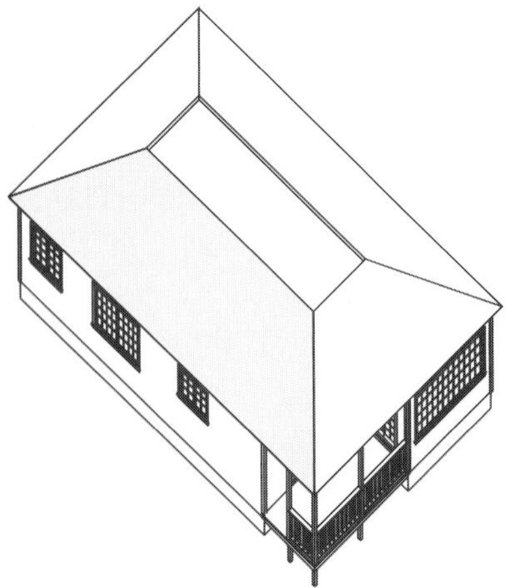

When a portion of a roof has four edges, but the opposite edges are not parallel due to unusual intersections or valleys with other roofing elements, use the four-edged general inclined polygon or slab option.

To draw dormers and other unusual roofs:

1. Choose **4EdgGen** from the **Inclines** menu.
2. Select the first corner point at the bottom edge of the slope.
3. Select the second corner point at the bottom edge of the slope.
4. Select the first corner point on the ridge of the roof. Enter this third point opposite the first point, as shown next. Locate this point by snapping or with a reference point shortcut and coordinate entry.
5. Select the last corner point on the ridge of the roof.

Four-edged general incline

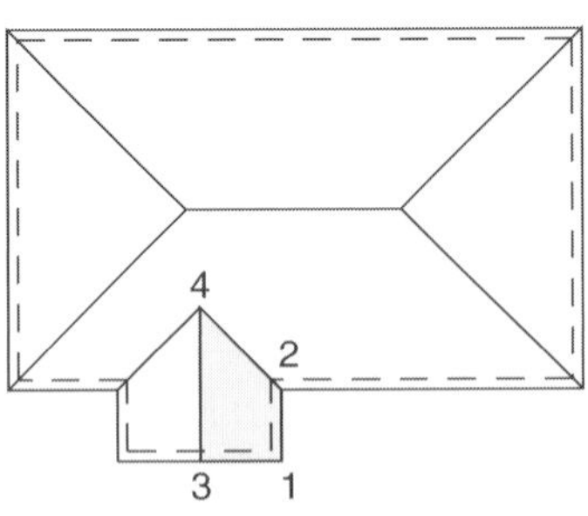

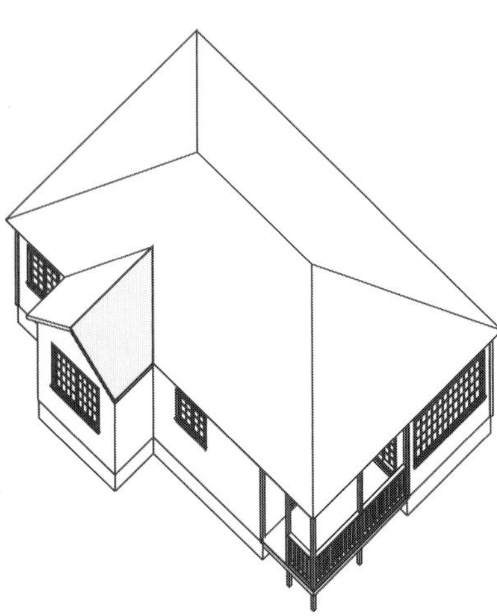

To draw the triangular pieces of hip roofs:

1. Choose **3 Edge Bottom** from the Inclines menu.
2. Select the first point at the bottom edge of the slope (at **Z-base**).
3. Select the second corner point at the bottom edge of the slope (also at **Z-base**).
4. Select the corner point on the ridge of the roof (at **Z-height**). Locate this point by snapping or with a reference point shortcut and coordinate entry.

Three-edged bottom incline

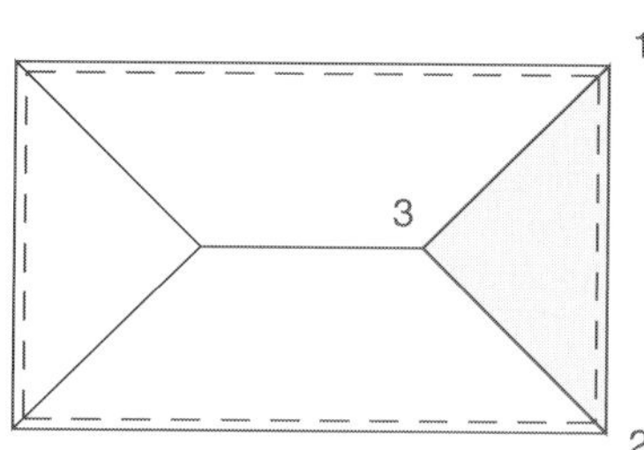

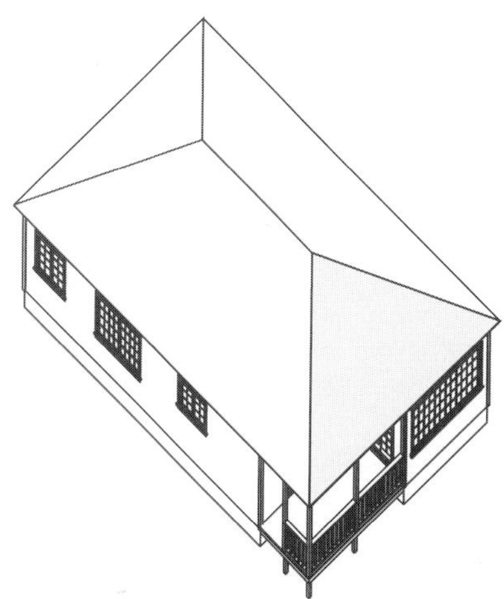

To draw simple dormers

1. Choose **3 Edge Top** from the **Inclines** menu.
2. Select the first corner point on the ridge of the roof (at **Z-height**).
3. Select the second corner point on the ridge of the roof (also at **Z-height**). Locate this point by snapping or with a reference point shortcut and coordinate entry.
4. Select the corner point at the bottom edge of the slope (at **Z-base**).

The four-edged vertical incline is created edge on, very much like the vertical polygon or slab, except that the incline has an independent thickness associated with it that controls the depth of the beam.

Three-edged top incline

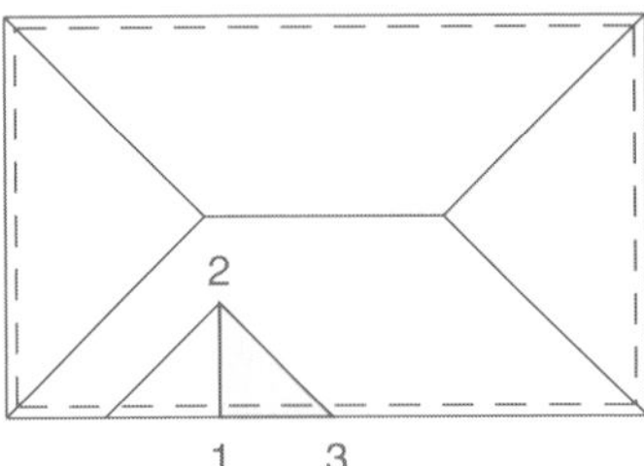

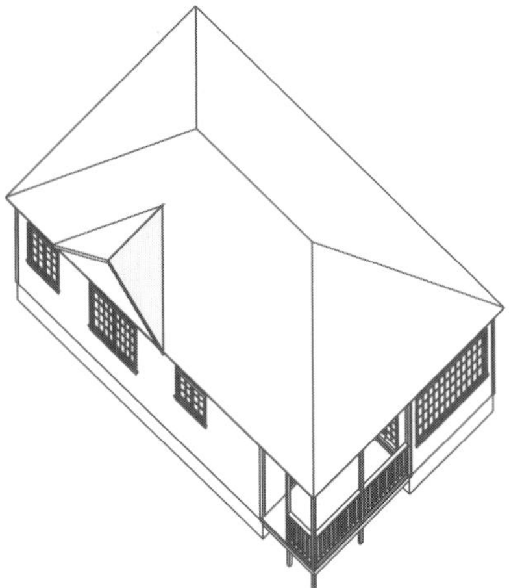

To draw rafters using vertical inclined polygons or slabs

1. Choose **4 Edge Vertical** from the **Inclines** menu.
2. Choose **Thickness** to set the depth of the polygon or slab (rafter). A list of values appears.
3. Choose or type a value and press (Enter).
4. Select the location for the lower edge of the polygon or slab, or the base of the incline, including the thickness of the polygon or slab.
5. Select the location for the upper edge of the polygon or slab, or the top of the incline. The polygon or slab ends at the **Z-height**, including the thickness of the polygon or slab.

To create three-edged vertical, inclined polygons or slabs

1. Choose **3 Edge Polygon** from the **Inclines** menu.
2. If you're drawing a slab, set the thickness of the slab using the **Thickness** option.
3. Snap to the first corner point on the incline.
4. Snap to the second corner point on the incline.
5. Snap to the third corner point on the incline. The polygon or slab is drawn.

NOTE: *Use the 3D cursor to snap to the corners of roof elements already in place. You don't have to select the points in any particular order.*

Dropping a Polygon Mesh

Drop Mesh is used to create a regular, triangulated, polygon mesh that lays over selected geometry or points, which can be created from a survey data file. This mesh can then be shaded or processed through a hidden line removal to provide a model of the building site. The **Drop Mesh** function is accessed from the **3D Entity**, **Polygon** menu.

A drop mesh acts like a cloth dropped over the geometry. The way it drapes over the geometry can be adjusted with the **Div-X, Div-Y, Stiffness, Smooth,** and **Smooth Pass** options. The mesh is made up of a grid of polygon divisions, with lines running in both the X and Y directions. These divisions are set with the **Div-X** and **Div-Y** options, and increasing these values adds flexibility to the mesh. **Smooth** adds points between the geometry to flatten the surface. **SmthPass** increases the number of smoothing passes. **Stiffnss** controls how closely the mesh coincides with the actual geometry; in other words, the lower the stiffness setting the more tightly it fits to the geometry.

To drop a mesh

1. Choose **Drop Mesh** from the **3D Entity**, **Polygon** menu. Notice that the current settings are displayed in the Message Window.
2. Choose **Div-X** and **Div-Y** to set the number of X and Y divisions in the mesh grid.

NOTE: *A drop mesh is a group of triangulated polygons, not a single entity; thus, it can increase your drawing file size substantially. A mesh with 100 X divisions and 100 Y divisions will add 20,000 polygons to your drawing (X divisions × Y divisions × 2 polygons per grid division, or 100 × 100 × 2), increasing your drawing file by 3MB. Additionally, since entities are not deleted from the drawing database even after you delete them from the drawing itself, your drawing file size can increase rapidly if you add a drop mesh, delete it, add another one, and so on. To remove unused entities from the drawing database, use Layer Utility in the* ***Edit/Toolbox*** *menu. It is recommended that you only use as many divisions as is necessary for a reasonably accurate model.*

3. Choose **Stiffness** to set the stiffness of the mesh. Enter a number from 1 through 10, where 1 is the stiffest, and press (Enter). The default setting is 5.

4. Choose **Smooth** to toggle on smoothing. This setting will have no effect on points created from an imported survey data file; it only works with vector geometry. Use of this option increases the time it takes DataCAD to generate the mesh.
5. If **Smooth** is toggled on, the **Smooth Pass** option becomes available on the **Drop Mesh** menu. Choose **Smooth Pass** to set the number of smoothing passes. Enter a number from 1 through 10, where 1 is the minimum number of passes. 1 is the default setting.
6. Toggle **Drop Edge** on to model the site with an edge base. When **Drop Edge** is toggled on, **Edge Height** becomes available on the menu. Choose **Edge Height** to set the distance below the lowest, calculated point of the site model. The default edge height is 10 feet.

NOTE: *You can enter negative values for the edge height, but DataCAD will use the absolute value. For instance, if you entered –10, DataCAD would use 10 as the setting.*

7. Toggle **Show Diagonals** on if you want to display the diagonal polygon edges of the mesh.
8. A mesh can be dropped over either selected geometry or imported, survey points. To drop a mesh over existing geometry, use the selection menu at the top of the **Drop Mesh** menu to select geometry in your drawing. You can select all of the geometry or only a portion of it. When selection is complete, continue with Step 10. You can choose **Clear** at any time to cancel your selections. To use survey data instead, skip to Step 9.

NOTE: *You can drop a mesh over any geometry in the drawing except geometric primitives, such as spheres, cones, or cylinders. To drop a mesh over geometric primitives, you must first explode the entities.*

9. To use survey data, it must be formatted in columns in a text file in one of four data formats, including X-Y-Z, Y-X-Z, Angle-Distance-Height, and Distance-Angle-Height. To import one of these files, first choose **Import Scale** from the **Drop Mesh** menu to set the scale for the data of the imported file. Then choose **Import** from the **Drop Mesh** menu,

select a data format from the Files of Type drop-down box, and select a file to import and click Open. Points based on the survey data are displayed in the Drawing Window.

10. Choose Begin from the **Drop Mesh** menu to begin drawing the drop mesh. Press **End** to cancel the mesh at any time.
11. If you do not like the result, erase by group, change the settings and try it again.

The following series of figures demonstrate the effect that the divisions, stiffness, and smoothing options have on a drop mesh.

Site plan drawn using contours

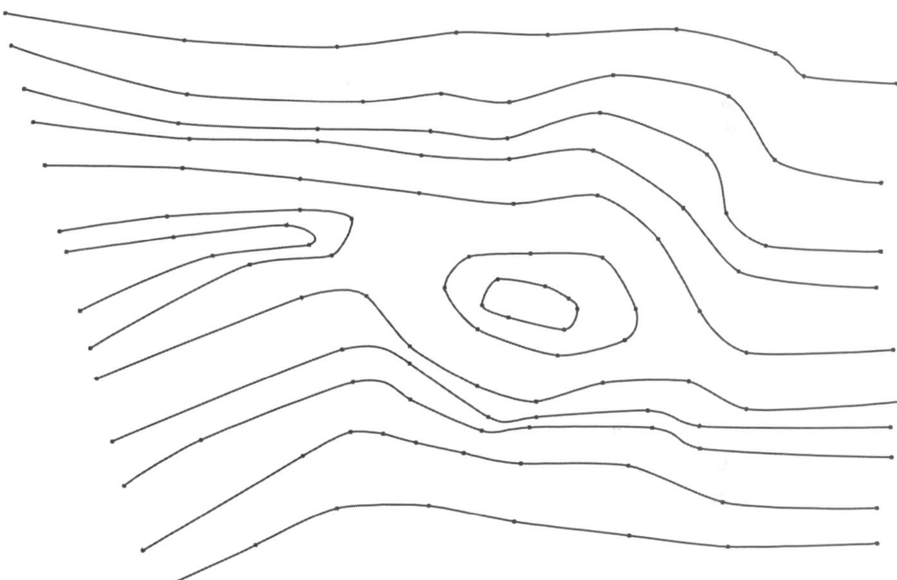

Isometric view of the contour line drawing

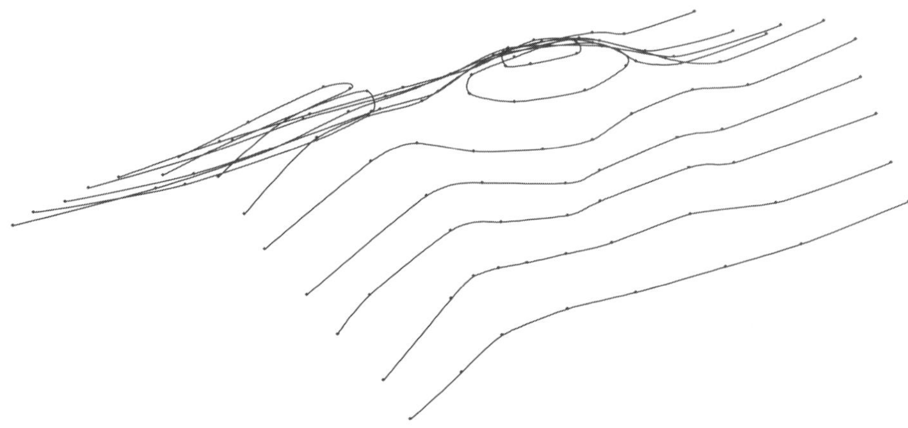

Model using 16 × 30 divisions, a stiffness setting of 5, and **Smooth** toggled off. The isometric view to the right shows the results of these drop mesh settings. As you can see previously, the mesh (represented by the gray line) does not always meet the points on the contour lines (represented by the darker line and dots). The result is a rougher approximation of the contour geometry.

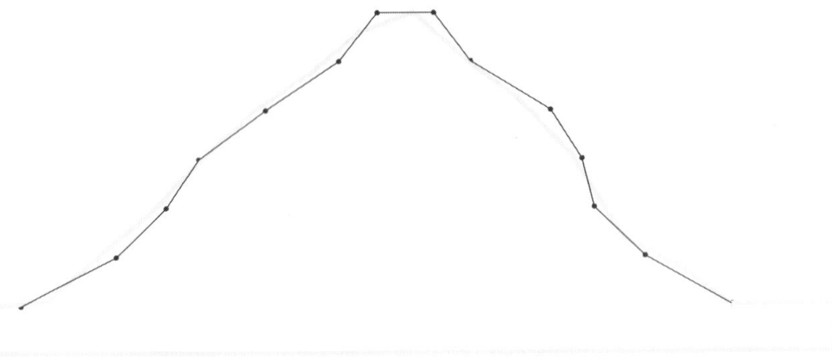

Divisions are increased to 32 × 60, while the stiffness and smooth settings remain the same. A higher resolution mesh produces a stair-step effect with this stiffness setting.

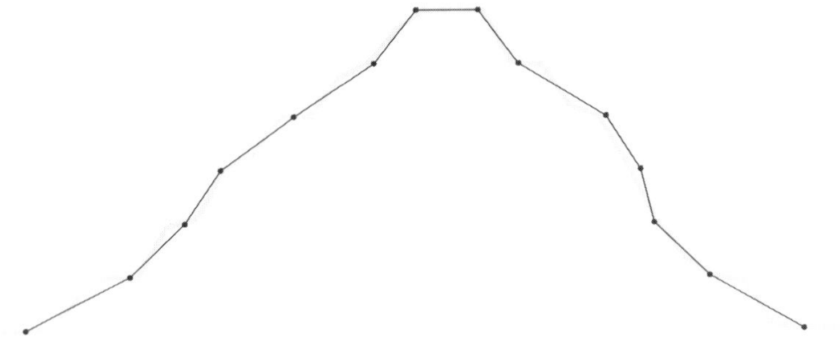

Isometric view of model using 16 × 30 divisions with stiffness set to 5 and **Smooth** toggled off

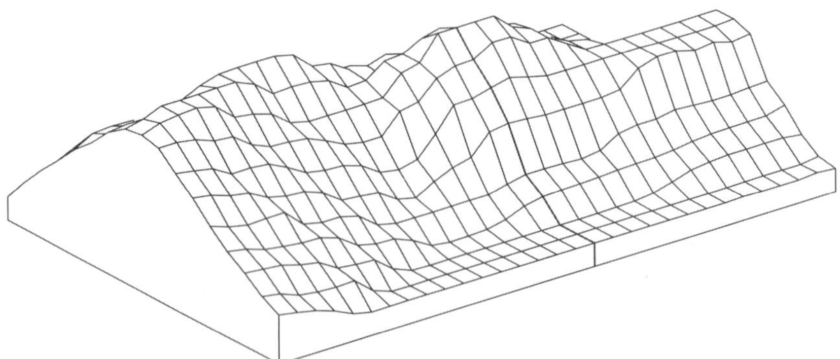

Isometric view of model with divisions increased to 32 × 60. Note stair-step effect.

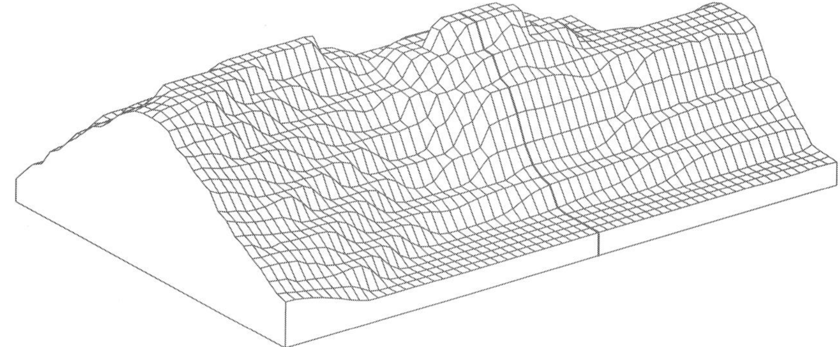

Divisions remain at 32 × 60 with **`Smooth`** toggled off, but stiffness is increased to 9. The stiffness is adjusted to counter the stair-stepping effect, resulting in a closer relationship to the contour points.

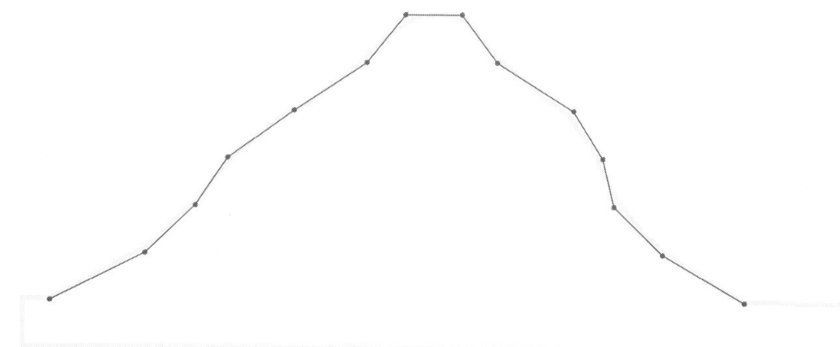

Another way to smooth the drop mesh is to use the **`Smooth`** option. Here the divisions and stiffness have been reduced to their original values of 16 × 30 and 5, respectively. However, Smooth has been toggled on and **`SmthPass`** has been set to 5. The interpolation of the smoothing operation creates smoother transitions between contour points.

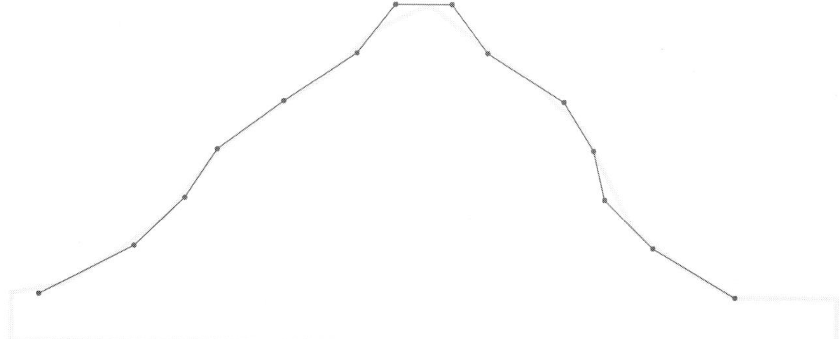

Isometric view of model with Divisions at 32 × 60, stiffness increased to 9.

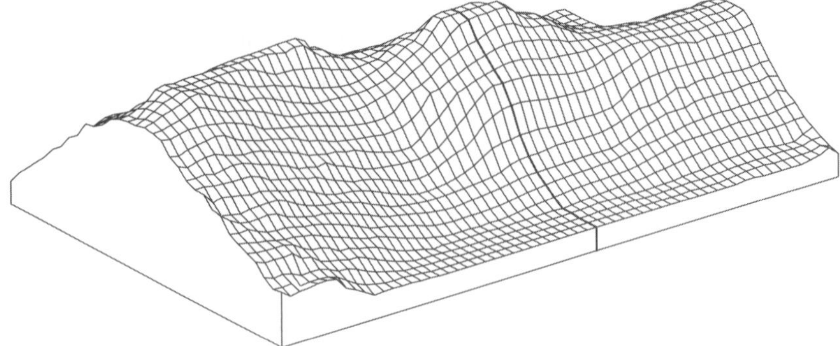

Isometric view of model with Divisions at 16 × 30, **`Smooth`** toggled on and set to 5.

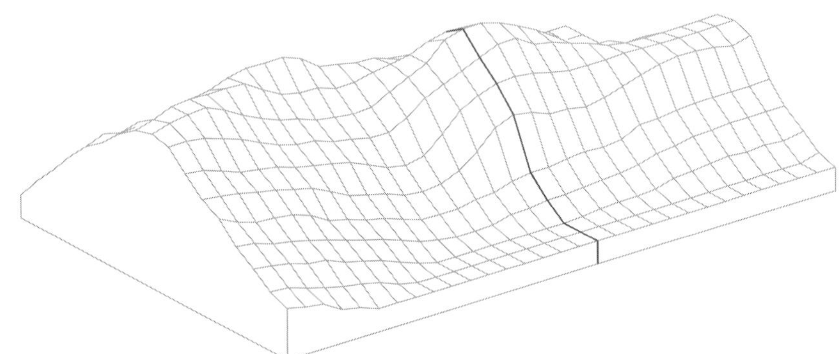

Concept 4: Creating Voids

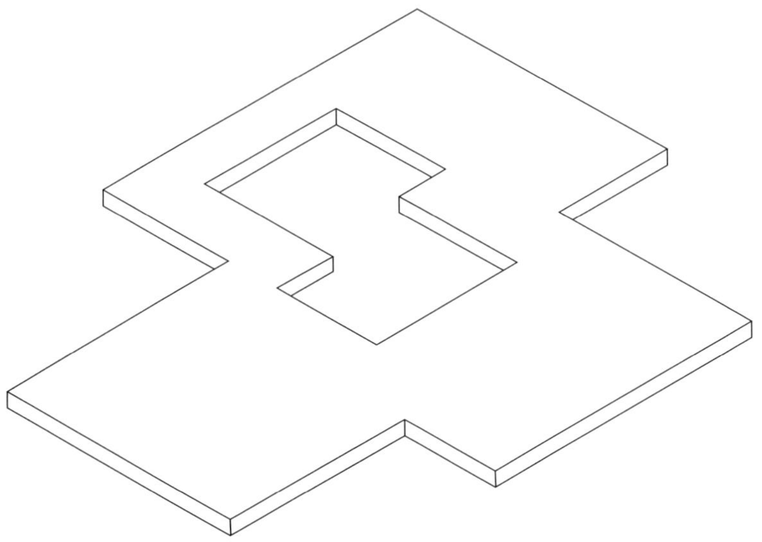

DataCAD supports the creation of voids in slabs and polygons to allow placement of doors and windows in your walls. When creating a void, it is important to draw both the master polygon (or slab) and the void in the same direction, and in the same working plane. As an example, draw a horizontal slab from Z-Base 0 to Z-Height 1. Now draw another horizontal slab in the same plane, whose boundaries are totally within the shape of the first slab. From the slab menu, select **Voids**. DataCAD prompts you to select the master slab to process voids. Select the larger slab. DataCAD prompts you to select the entity to add void. Select the smaller slab. DataCAD tells you that 1 entity has been selected. Exit the menu, the view the slab in an isometric view and shade. You will see a hole cut through the original slab.

The same can be done for vertical slabs. Create a vertical slab from Z-Base 1 to Z-Height 10. Draw another vertical slab within this slab from Z-Base 3 to Z-Height 9, drawing in the same direction as the original slab. Change to an isometric view, and repeat the process of adding the void. Shade the model. You will see what appears to be a wall with an opening for a window.

Void created with proper alignment of master faces (left) and reversed faces, (right)

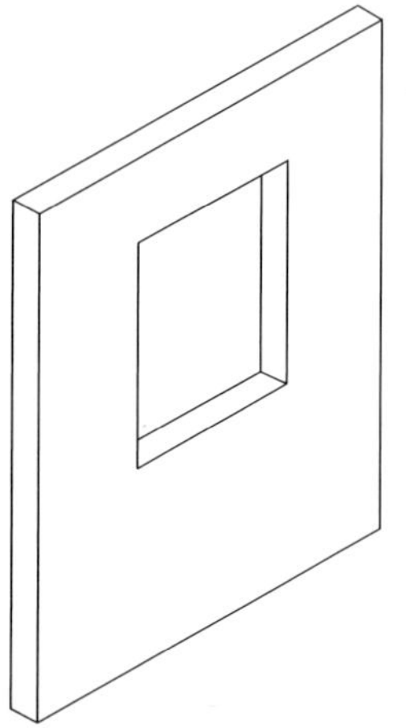

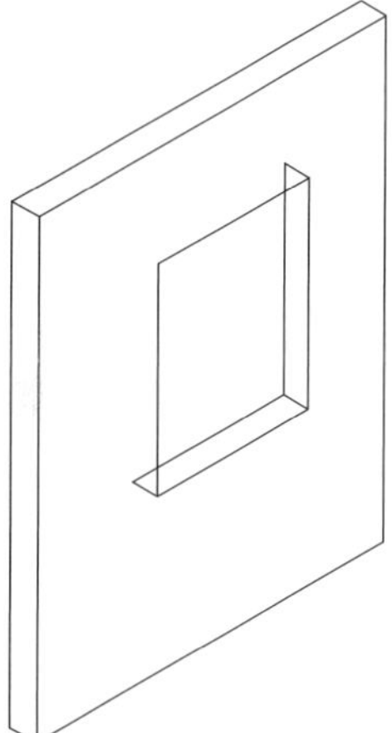

As you can see, voids are created by drawing a slab within a slab and then converting the interior entity into a void. The exclusive slab is called the "master." The slab to be converted to a void should be entirely contained within the master slab, even if only by a 1/32 of an inch. Additionally, all slabs have a "reference face," or primary surface. The reference faces of the master slab and the void must be in the same plane for the void to be created properly. To ensure this, create the master slab and void slab(s) in the same view and in the same direction, i.e. left to right, top to bottom or counter-clockwise. Voids may be moved and stretched, but not copied.

To Create a Void

1. Create the master slab.
2. Create the interior slab to be made in to a void. Remember to create it with the same orientation and thickness of the master slab so that the reference faces align.
3. Choose **3D entity/Slab/Voids** from the 3D Entity menu.
4. Choose **Add Void**.
5. Choose the master slab by pointing to a slab edge with the left mouse button.
6. Choose the entity/entities to be converted to voids within the master slab.
7. This menu also contains functions for erasing and converting voids back to slabs.

Exercise: Wall Openings

1. Select **3D Entity**, **Slab**, **Vertical**, **Center**.
2. Set the **Z-base** to 0.0 and the **Z-height** to 20.0.
3. Create a vertical slab 30′-0″ long or so.
4. Change your **Z-base** and **height**, and place multiple slabs of different proportions within the first.
5. Define a New Elevation to view the façade of the slab.
6. Continue to create additional sub-slabs using **3D Edit**, **Copy**. Change their proportions with **3D Edit**, **Stretch**.

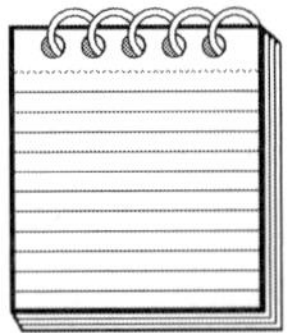

NOTE: *If you invoke the 2D shortcut for **Stretch** while in a parallel view, your view will change to **Ortho**, as this shortcut calls the 2D version of **Stretch**. To quickly get back to your parallel view, select **3D Views**, **Parallel**, or use the shortcut **y**, **F2**. Then choose **3D Edit**, **Stretch** from the main menu.*

7. Select **3D Entity**, **Slab**, **Void**. Pick the first slab you created as the Master Slab. Choose **Add Void**, set your selection method to **Area**, and

define a selection box that encloses the sub-slabs to make them into voids.

8. Check your work with parallel and perspective views. Perform **Hides** and **Shades** to examine your design.

Vertical wall with multiple voids of differing proportions

Concept 5: The Right Hand Rule

Positive axis of rotation

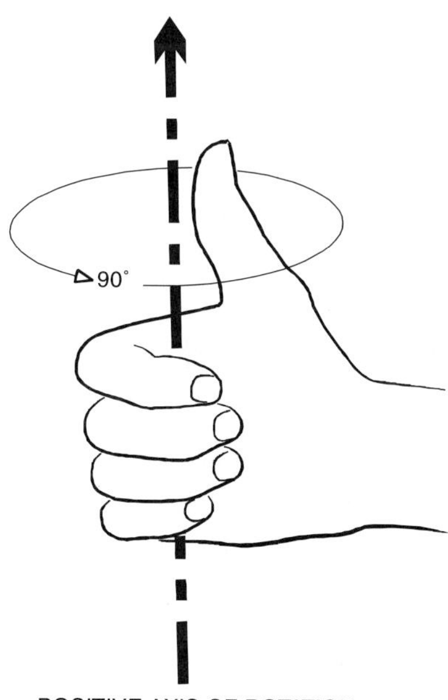

In 2D, positive rotation angles go counterclockwise; in 3D, the concept is a bit more complex. This rotation angle of –90° flips the model to the left. DataCAD uses the right-hand rule to indicate the direction in which positive angles run. To understand the right hand rule, point the thumb of your right hand in the direction of the axis in question. Your fingers curl in the direction of positive rotation. The Y-axis goes up, so when you point the thumb of your right hand upwards, your fingers naturally curl toward you in a counter-clockwise direction. Negative angles, therefore, run in a clockwise direction.

The right hand rule as it relates to the three primary axes

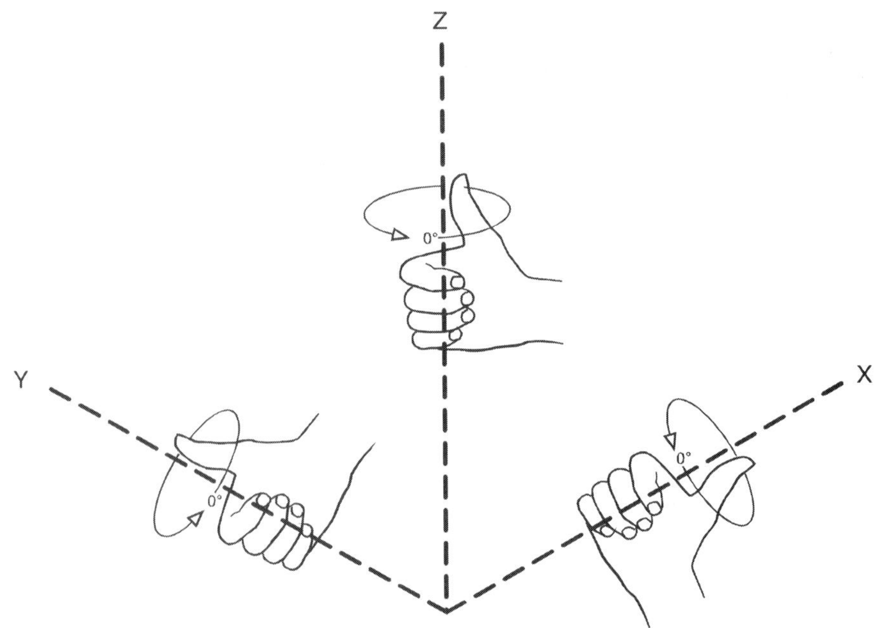

Applying the Right Hand Rule:

1. Create a vertical shape as if lying on the ground. For this example we will model a Tuscan Column.

   A Revolved Surface will be utilized to make the column. First, the profile is drawn and the center axis established. **Z-base** and **Height** settings are set to zero. Open the file Concept 5_RHR.Dc5 from the bonus CD to follow along.

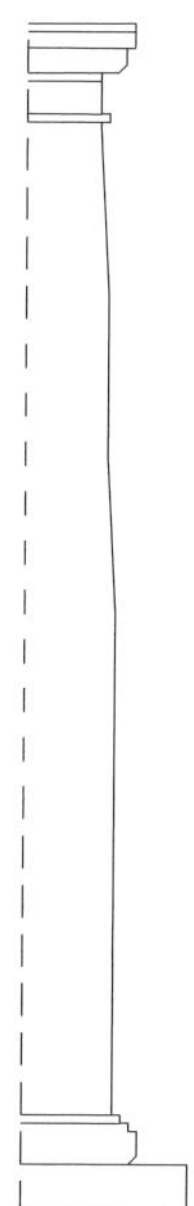

2. To create the **Revolved Surface**, select **3D Entity**, **Revolved Surface**, **Open**. Choose **Primary Divisions** and enter **4**. Trace the base shape from points A through C. When C is reached, right click to finish. Snap to points on the centerline to define the center axis of the **Revolved Surface**. The surface is created.

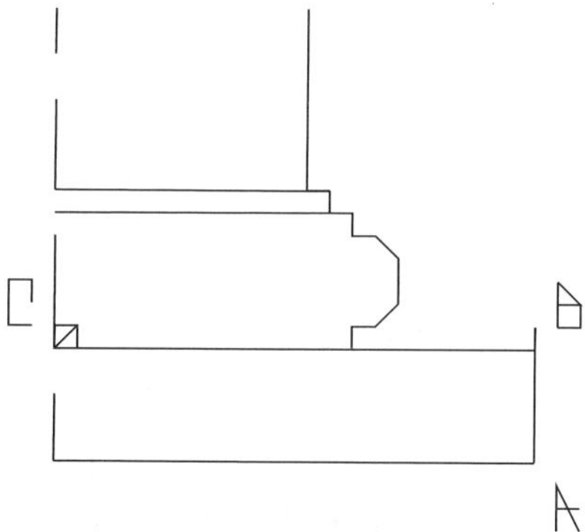

3. Change Primary Divisions to **18** and repeat for the body of the column. Note that the curved portions of the profile are broken down into line segments.

Note that you can set up an action, i.e. rotate, in one view; and then change your viewpoint to complete it.

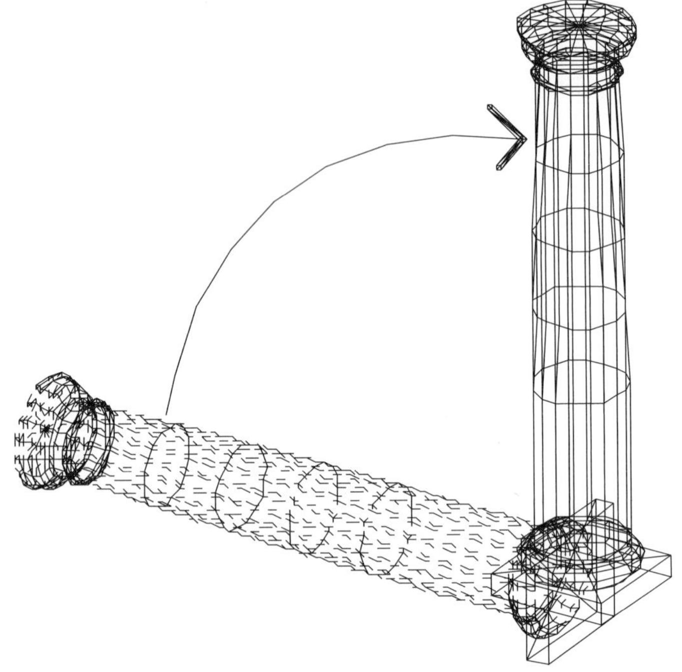

4. The column is now complete, but must be rotated vertically. This is where the **Right Hand Rule** applies. First, the axis of rotation must be determined. In the case of the column, the hinge of rotation is along the X-axis. Next, the rotation angle must be determined: point the thumb of your right hand in the positive direction of the X-axis, which is to the right. Your fingers curl towards you, indicating a rotation angle of 90 degrees.
5. Select **3D Edit, Rotate.** You are prompted to "Select the center of rotation." Middle-button snap to the bottom center of the column. Now choose **X-axis, X-angle,** and enter **90** degrees. Now choose **3D Views, Isometric**. Right-click once. You are prompted to pick the entity to rotate. Pick the column. You see it rotate vertically in an isometric view.

Concept 6: Curves Don't Exist in 3D

Unlike their 2D counterparts, all curved 3D objects are approximations of true curves. The surfaces are approximated by replacing the true curve with a number of short line segments that create facets. The exact number of these line segments varies from 4 to 36. We have seen this already in the exercise for the column, when curved sections of the profile were broken down into segments, and the round column itself was represented as a facet shape with 18 sides.

Change this setting with **Primary Divisions** and **Secondary Divisions** in the **3D Edit**, **Settings** menu. You can change an entity's divisions after you have created it by selecting **3D Edit, Change Primary/Secondary Divisions**. You can also toggle a shape **Open** or **Closed** from the **Change** menu.

Primary Divisions

Some objects like cylinders, 3D arcs, cones, and truncated cones only curve in one direction, around the entity's center point or central axis. For these entities, you only need to make one division setting, the primary division. Sometimes primary divisions are referred to simply as "divisions."

Secondary Divisions

Entities that curve in more than one direction at a time, like spheres, torii, surfaces of revolution, and mesh surfaces, have both primary and secondary divisions. Primary divisions control the number of curve divisions in the plane of the screen (Sweep angle) and secondary divisions control the number of divisions for curves that bulge into or out of the plane of the screen (Roll or Rise angles).

One notable exception to this is the horizontal cylinder. Even though a horizontal cylinder curves into and out of the screen, its divisions are controlled by primary divisions, as with vertical cylinders, which are the same entity type.

Setting the Number of Divisions

When the number of divisions is set too high, causing a large number of facets in each curved entity, the resulting model is dense and difficult to read. When the number of divisions is set too low, the approximation results in non-circular renditions. Each curved entity has its own division values. Change these with **3D Change** or override them with **Global Divisions** in the **3D Settings** menu. Temporarily lowering (or overriding) the number of divisions of complex curved objects can significantly reduce the amount of time needed for hidden line removal screen regenerations.

Primary and secondary divisions relate to the way a sphere, torus, revolved surface or mesh surface is divided.

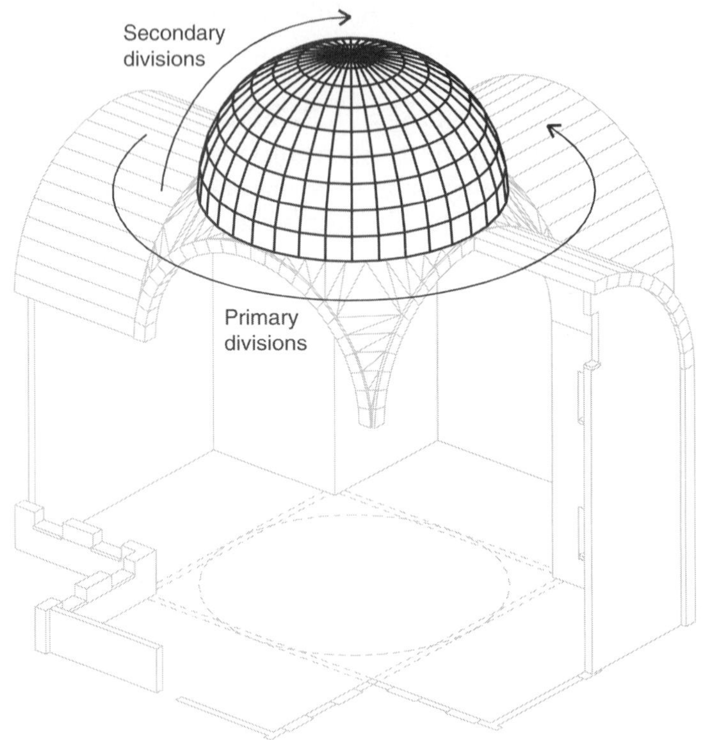

Exercise: Creating Curves in 3D

1. Open the sample file on the Bonus CD for Concept 6. Create four new layers, one for the base, top, spandrels and glass.
2. Model the curvilinear curtain wall: Use the 2D markers on the plan layer to guide the creation of horizontal slabs for the base and top pieces. You will need to use multiple slabs, so only snap 16 points per side.

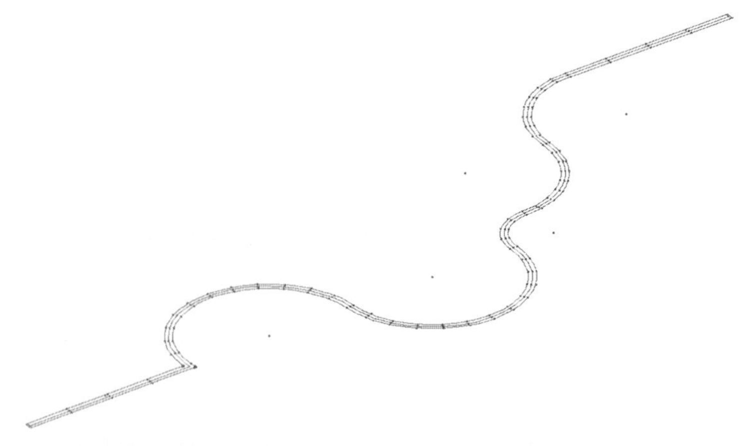

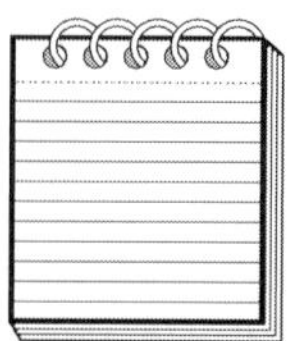

NOTE: *Once you create the base you can use* ***Move, And Copy****, to create the Top. Use* ***Vertical Slabs*** *for the spandrels, drawn by* ***Center*** *with a thickness of 4″, and vertical polygons for the glass.*

Vary the Z-heights for each major element as follows:

Base: 0.0—0.6″

Spandrels: 0.6″—19′-6″

Glass: 0.6″—19′-6″

Top: 19′-6″—20′-0″

3. When complete, turn on the 3D layers and explore the design using **3D Views**, **Shade** and **Hide**.

Curves don't exist in 3D. You must break curved elements into straight segments.

Concept 7: The Clip Cube

The **Clip Cube** allows you to define a cube in space beyond which nothing is viewable. It is useful for isolating a section of a model to work on, or limiting the plotting area for a plan fragment such as an enlarged plan. The **Z-base** and **height** may be set so as to exclude elements above and below the cube. Note that the **Clip Cube** restricts what is visible, but does not alter the drawing in any way. There are three independent **Clip Cubes**, one for **Ortho** view, one for **Parallel** views including elevations, and one for **Oblique** views.

When modeling, use the **Clip Cube** to restrict your view of the model, so that it is easier to edit entities within a restricted area, and not affect others unintentionally. Only the entities inside the rectangular clipping region are visible. If an entity is only partially inside the clip cube, DataCAD clips it at the point where it passes outside the cube. This reduces the visual clutter of a complex model as it develops. You cannot edit entities outside of the clip cube. An entity that is partially within a **Clip Cube** is editable at it visible points only.

The three **Clip Cubes** are independent of one another. This means that the **Clip Cube** you define for a plan view will not display when you change to a parallel view and vice-versa. When you begin the **Clip Cube** operation from plan view, DataCAD uses the orthographic **Clip Cube**. When you use **Clip Cube** while in a parallel view, the parallel **Clip Cube** is used.

The active **Clip Cube** settings are stored with the **3D Views, Go to View, Add View** function. Restoring a saved view also restores the **Clip Cube** that was in use when you saved the view. This allows you to save multiple **Clip Cubes** of various configurations for later recall.

The three independent **Clip Cubes** and their respective views

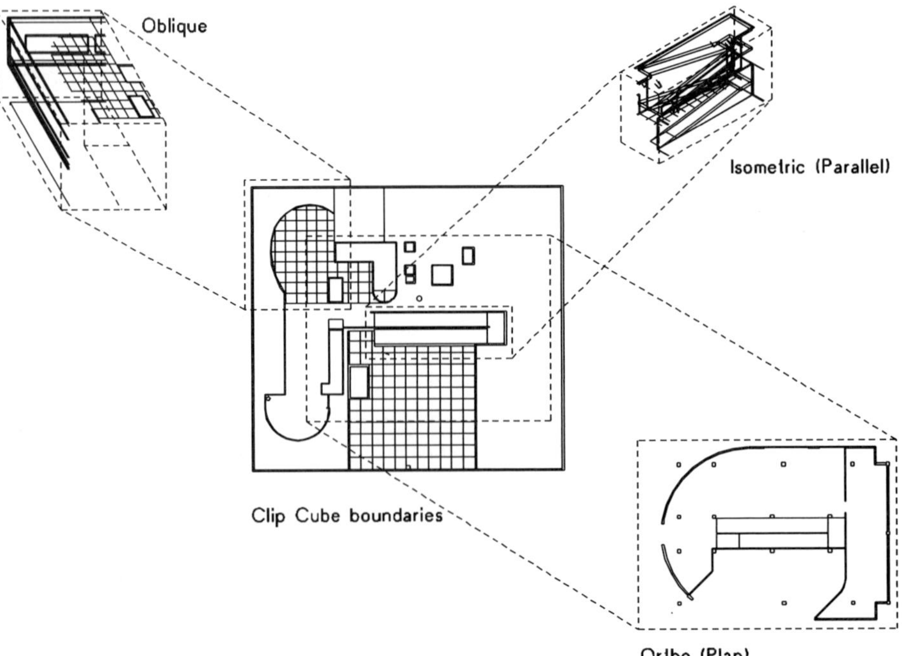

When calculating a Hidden line Removal, enabling **Clip Box** will only hide the entities displayed within the current **Clip Cube**.

Defining a **Clip Cube** entails three steps. First, choose the view projection: **Ortho**, **Parallel**, or **Oblique**. There is an independent **Clip Cube** for each of these view types. Second, define the boundary of the **Clip Cube**, as well as its **Z-base** and **height**. Third, turn the **Clip Cube** On.

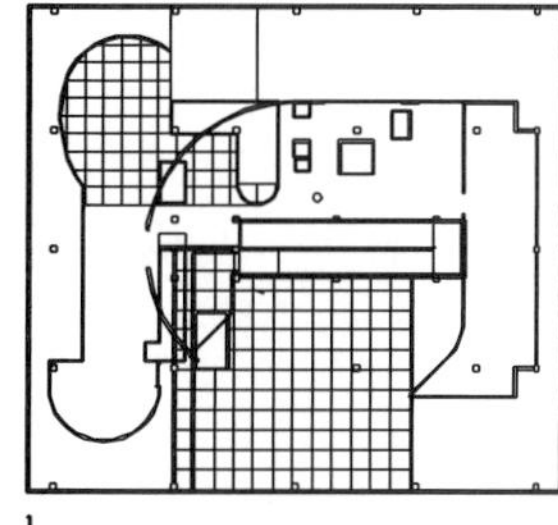

1.

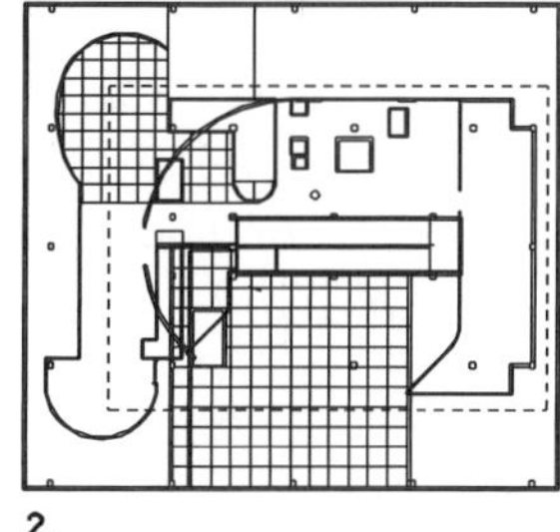

2.

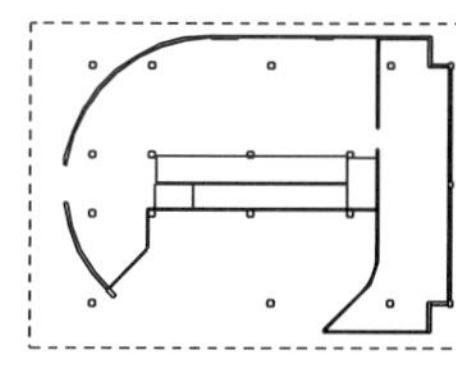

3.

1. **Clip Cube** defined for a Plan (Ortho) view. 2. **Clip Cube** boundary is defined. 3. **Clip Cube** is activated. Note how the Z-settings for the **Clip Cube** affect the entities that are displayed when it is activated.

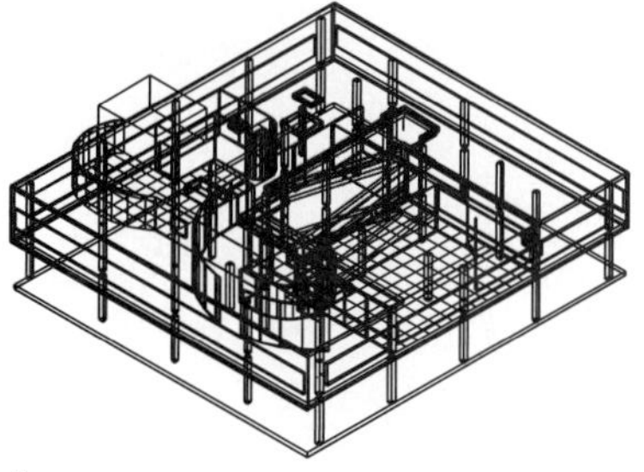

1.

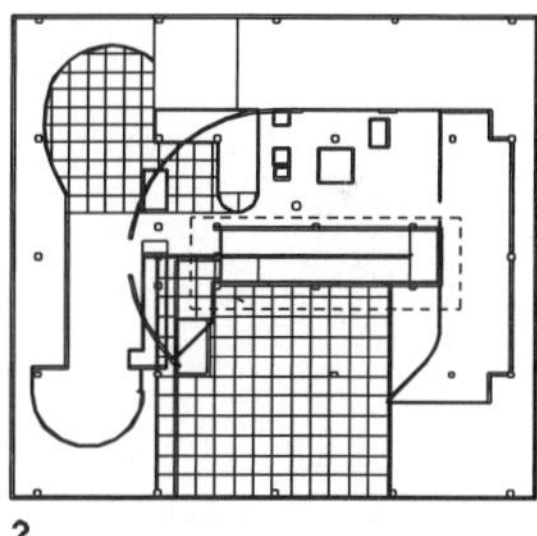

2.

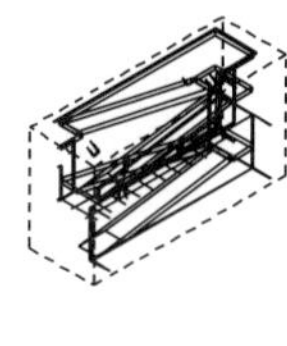

3.

1. **Clip Cube** defined for Parallel (Isometric) view 2. Selecting **New Cube** calls a plan view to define the **Clip Cube** boundary. 3. Resulting parallel view with the **Clip Cube** turned on

Defining a New Clip Cube

1. To access the **Clip Cube**, go to the **3D Views** menu and select your desired view—**Ortho**, **Parallel**, **Isometric**, etc. Note: **Clip Cubes** are supported by all view types except **Perspective**.
2. Choose **Clip Cube**.
3. Choose **New Cube**.
4. If you are not in **Ortho** view, DataCAD will temporarily return you to plan view.

5. Set desired **Z-min** and **Z-max**. Select the area to define the **Clip Cube** with a rubberband box.
6. Select refresh then **Clip On**.

   DataCAD returns to your view as in Step 1, and displays only what is within the cube extents. Editing operations will only effect what is in view. Don't forget to disable the cube when you are through.

NOTE: *Saving a* ***Go to View*** *with the* ***Clip Cube*** *on and another one with it off is a convenient method to quickly call the view when you need it, and dismiss it when you don't. Turning on a* ***Clip Cube*** *that encloses a blank area of the drawing will cause your model to apparently vanish!*

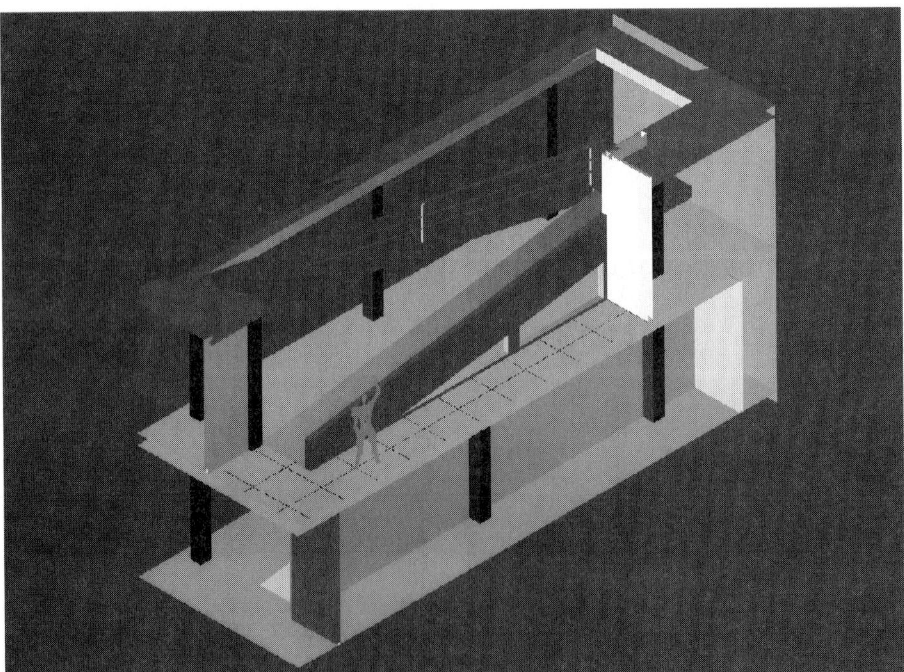

The **Shader** supports the ability to shade only what is contained by the **Clip Cube** set for the current view.

Exercise: Mastering the Clip Cube

1. Open the sample file for Concept 7.
2. Select **DCAD 3D, 3D Views.**
3. Toggle between the **Otho**, **Parallel**, **Isometric**, **Oblique** and **Elevation** views.
4. For each view, turn the **Clip Cube** on and off, then on again.
5. Perform a **Shade**, and **Hide** with **Clip Box** enabled for each view. Note the results.
6. Define new **Clip Cubes** for each view. Vary the **Z-settings** for each.
7. Save each one as a **Go to View**.
8. Change the **Clip Cube** boundary for each of the **Ortho**, **Isometric**, and **Oblique** views.
9. Save each of these as additional **Go to View**s. Save several different views of each.
10. Recall each **Go to View** in succession. Note how the **Clip Cube** status and settings are retained for each view.

CHAPTER 17

3D Modeling Methods

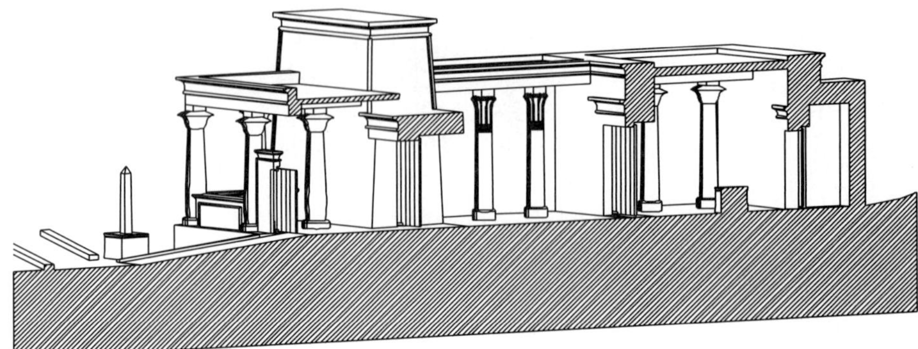

Over the years, 3D CAD users have developed various ways to create a model, usually guided by a 2D plan. Initially, modeling was achieved by simply placing 3D primitives in different arrangements and Z-heights to create different forms. The desire for greater detail and compatibility with rendering programs demands new methods and tools to accomplish the modeling task. This section will describe some of the techniques that you can use to model your project. Ultimately, you will combine the different techniques illustrated here to satisfy your own needs. Once you master the basics, you can adapt these methods to model just about any form you may require.

We will examine five different methods to construct a 3D model using the DataCAD 3D primitives and the **AEC_Model** macro to complete the windows and doors. Refer to the chapter on 3D Toolbox Macros for more detail on the AEC Modeler macro.

Each of these approaches has different strengths and weaknesses, as well as a level of complexity. We will start with the simpler methods first, and then progress to the more complex ones. You are encouraged to open one of your 2D projects and experiment with the methods described here as you read.

The 2D Base Plan

The base plan provides an underlayment upon which to construct the 3D model, and to a large degree, may also be used to derive the building envelope as in method four. 2D lines also serve as a means to create surrogate geometry to aid in placing and editing 3D entities. For example, since you can't snap to the mid-point of a slab, you can draw temporary 2D lines over the slab to facilitate the operation.

The 2D component of DataCAD allows a **Z-height** extrusion. Though not truly a 3D model, a plan composed of extruded 2D lines will convey the walls and a sense of enclosure in a 3D view. The utility of the base plan lies primarily in its edit-ability. 2D lines may be trimmed and altered with the cleanup functions whereas 3D elements may not. Hence, the 2D drawing provides for efficient generation of design plans, and serves to guide creation of the 3D model as well.

All of the methods described in this section presume you have created a basic 2D plan to model from. At the very least, you should draw a 2D floor plan delineating the walls and openings you wish to model, and a roof plan. If you are only concerned with creating a model for exterior views, there is no need to draw the interior elements. The 2D entities used to draw the plan should be entered at a Z-base and height of zero, so that the 3D geometry you place over them is easily distinguished when the plan and model are displayed simultaneously in a 3D view. It is also helpful, though not usually critical, to have a 2D section drawing completed as well.

The 3D Model

The DataCAD 3D model consists of three-dimensional primitives such as polygons, slabs, spheres, cones, etc. These shapes, when combined in various relationships to one another, create building forms of walls, windows, doors, roofs and so on. They are most easily created and placed in space with the aid of the 2D plan previously described.

The model represents the 3D construction of the 2D plans, and provides you with the opportunity to build virtually from your own plans. This aspect alone makes 3D modeling rewarding and informative. Relationships and dependencies in the structure are perceived in an almost tactile way, which frequently contribute to design decisions or indicate unforeseen aesthetic or even structural conflicts.

While 3D models are expected to provide renderings and other visualization-oriented material, they may also contribute a significant amount of data to the preparation of 2D construction documents. A benefit of taking the time to build a 3D model is that not only you can derive an unlimited number of perspective views from it, but it may also yield Elevations, Sections, Axonometrics, as well as the raw information necessary for construction details.

Coordinating 2D and 3D Information

The 2D drawings and the 3D model are not dynamically linked together in DataCAD. This dictates the need for coordination between the 2D documents and 3D model throughout the life of the project. Once a model is created, changes made to the 2D plan must be performed together with the model, and then carried through to the extracted 2D elevations, sections and details as required. Conversely, changes made to the model require adjustments to the affected 2D documents. If the model is being used only for design purposes this is less critical, but if the model is used to continuously to generate elevations and sections, some amount of strategy is required to manage design changes. Multiple drawing files with External File Referencing or a single project file are typical methods employed to ensure that parity is maintained between the model and the construction drawings.

Modeling Method 1: Slab on Slab

This is the simplest method to create three-dimensional walls with a thickness that support voids. Use your completed 2D plan drawings as a base, and place slabs of the appropriate thickness over them to create the walls. The AEC_Model macro can be used to insert 3D doors and windows. The 2D drawing guides the placement of the fenestration as well as the walls. The slabs can be butted together or overlapped at the corners, creating one or two visible lines. This is a side effect of this modeling technique. The main advantage of slab modeling is that it is very quick. It is less flexible for rendering, as the interior and exterior wall surfaces are not easily separated. If a slab model is used to produce hidden line drawings, the joints created by the slabs at corners can be a problem. In a simple model you can easily erase them, but if the model has a lot of corners, it will require a lot of editing.

To create a Slab model:

1. To enter the vertical slabs, Set the z base and height required for the walls.
2. Next, create a new layer for your 3D walls, set it active, and turn off all other layers except for the 2D guide-plan. Select **3D Entity, Slab, Vertical** to create vertical slabs.
3. Trace the exterior walls using vertical slabs. You may enter a vertical slab by its left or right edge, or its centerline. These options are toggled

The 2D/3D hierarchy (bottom to top): 2D lines, 2D lines with extruded height, 3D entities

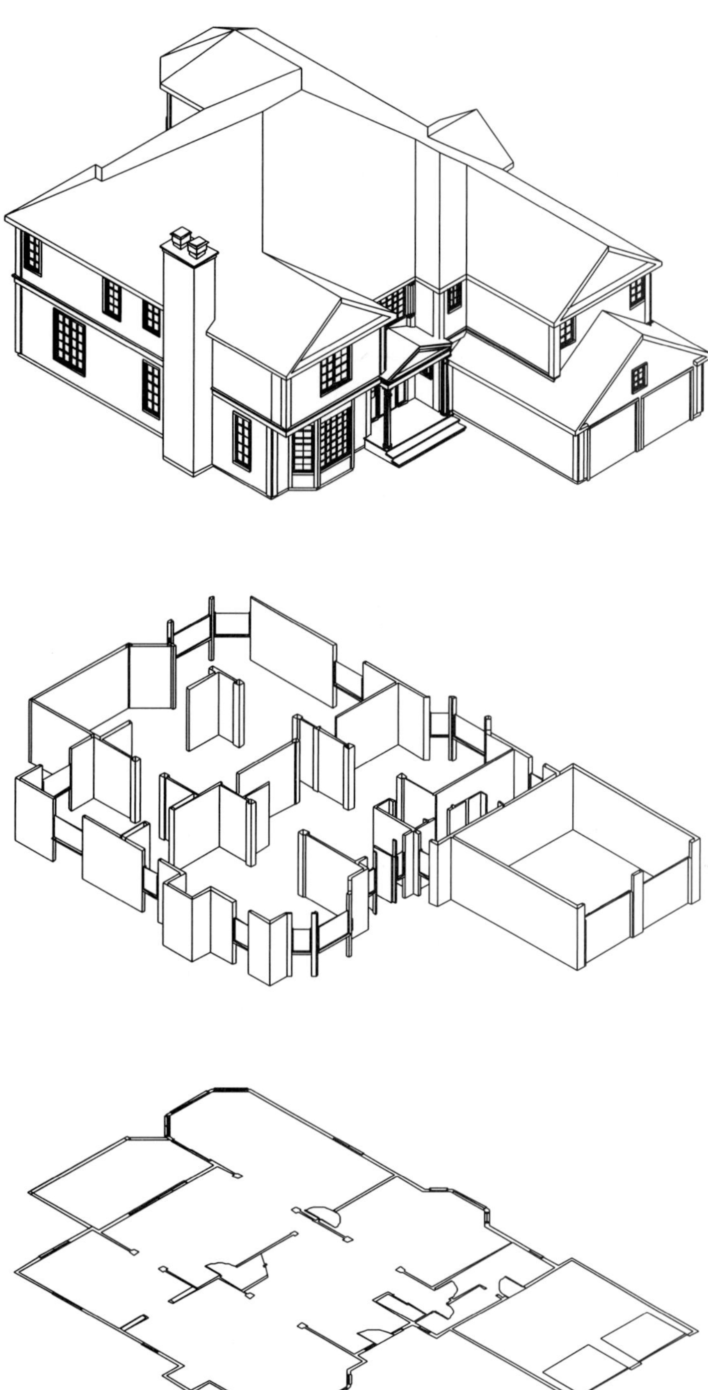

from the main menu. Use the overlap or butt joint technique at the wall corners.

4. Create a new layer for the 3D windows, and set it active.
5. Open the **AEC_Model** macro in the Toolbox and select Windows. Make sure **Cut Wall** and **Layer Search** is toggled on.
6. Select **Unit Type**: Double-hung, for example.
7. Set the **sill** and **head** heights: i.e., 3 feet for the sill and 7 feet for the head height.

At this point, you should have the 2D walls, 3D walls and 2D window layers turned on. Insert your 3D windows based on their position on the 2D-window layer. The openings in the 3D walls are cut automatically, and DataCAD constructs the 3D windows based on your settings. Refer to the chapter on 3D Toolbox macros for details on the **AEC_Model**.

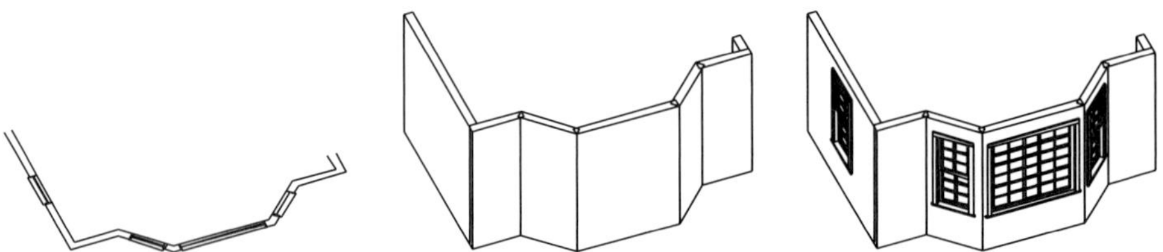

Slab model method: 2D Plan (left) is overlayed with 3D Slabs (middle), and completed with AEC_Model macro (right).

Modeling Method 2: Plane by Plane

This method allows you to use your completed 2D plan drawings as a base, and model with the more flexible 3D polygon entities for the walls. As with the slab method, the AEC_Model macro is used to insert the 3D doors and windows. Using your 2D draft plans as a guide, you snap in vertical polygons to create the walls. You will need two polygons for each wall: one for the inner surface, and one for the outer surface. This method has advantages in rendering, where you typically want to map different textures for the inside and outside surfaces. As each wall is composed of inner and outer polygons, you can place them on separate layers so that the render treats them as individual objects. Another advantage over the slab method is that the corner conditions are resolved properly, as shown in the following illustration.

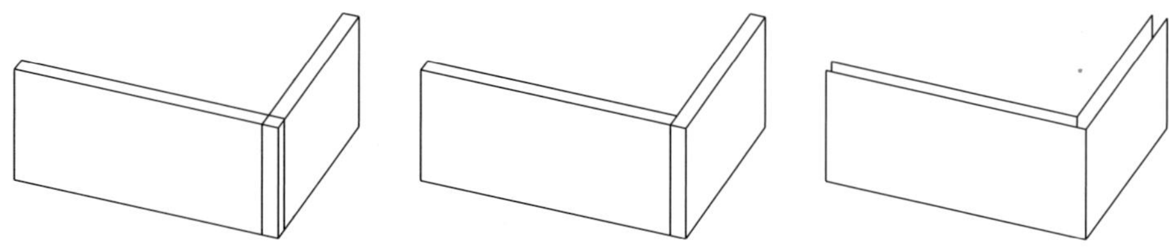

The Corner Dilemma: Slabs intersecting or abutting at corners creates an undesirable joint (left and center). Vertical polygons (right) provide for a clean resolution of the corner condition.

To create a model Plane by Plane:

1. To enter the vertical polygons, first set the z base and height required for the walls.
2. Next, create a new layer called for the 3D walls and set it active. Turn off all other layers except for the 2D plan.
3. Trace the exterior wall using vertical polygons with chain toggled on. Chain allows you to create polygons connected end to end as when drawing 2D lines.
4. Create a new layer for the 3D windows, and set it active.
5. Open the **AEC_Model** macro in the Toolbox and select **Windows**. Make sure **Cut Wall** and **Layer Search** are toggled on.
6. Select **Unit Type**: Casement, for example.
7. Set the **sill** and **head** heights: i.e., 3 feet for the sill and 7 feet for the head height.

At this point, you should have the 2D walls, 3D walls and 2D window layers turned on. Insert your 3D windows based on their position on the 2D-window layer. The openings in the 3D walls are cut automatically, and DataCAD constructs the 3D windows based on your settings. Refer to the chapter on 3D Toolbox macros for details on the **AEC_Model**.

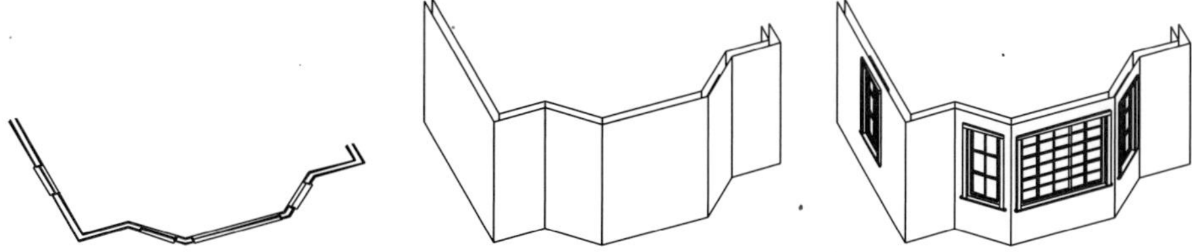

Polygon model method: 2D Plan (left) is overlayed with 3D polygons (center), and completed with AEC_Model (right).

Modeling Method 3: Tilt-Up Slab

In the tilt-up slab method, façade models are created with slabs and voids in plan view, rotated vertically, and moved into position using the 2D plan as a reference. This technique allows you to utilize the 2D elevations as a guide in addition to the plan. The Tilt-Up Slab method is most appropriate for situations where the 2D plan and elevation drawings are complete.

Tilt-Up Slab modeling revolutionized the way models were produced in the early days of CAD, because users tended to have many 2D documents that they wanted to make into models. With this method, it was possible to use that information to produce a 3D model relatively quickly. Today, it is more likely that the 3D model will be used to generate the elevation drawings in the first place. Even so, this technique is very versatile, and is utilized for many aspects of the exercises shown illustrated in the Advanced Modeling Exercises.

Creating a model with the Tilt-Up Slab method:

1. Utilize the 2D-plan and elevation layers to guide the model. Create new layers for the various 3D façades.
2. Trace the façade outlines with slabs of the proper wall thickness. Set wall thickness via Z-base and Z-height. Example: for a 6″ wall, set z-base to zero, z-height to 6″.
3. Trace the fenestration outlines with similar slabs and convert to voids.
4. Add additional 3D elements to the facade slabs to create detail and relief, such as window frames, trim etc. The **AEC_Model** macro can be used to model 3D windows in plan view, though it will not cut the voids automatically.

Tilt the slabs vertically using the 3D Rotate command. Rotation angle follows the Right-hand Rule. In tilt-up slab modeling, rotation is typically about the Z-axis at 90 degrees.

Once rotated to a standing position, the façades must be moved into place in plan.

Side and rear elevation models will have to be rotated about the Z-axis before they can be positioned properly.

Some adjustment is required to account for the wall width where two façade slabs meet at a corner. Typically one slab is stretched back to create a lapped joint.

Tilt-Up Slab Method: 2D Elevation drawing (left) is used as a guide for a slab model (center) that is rotated 90 degrees when complete (right).

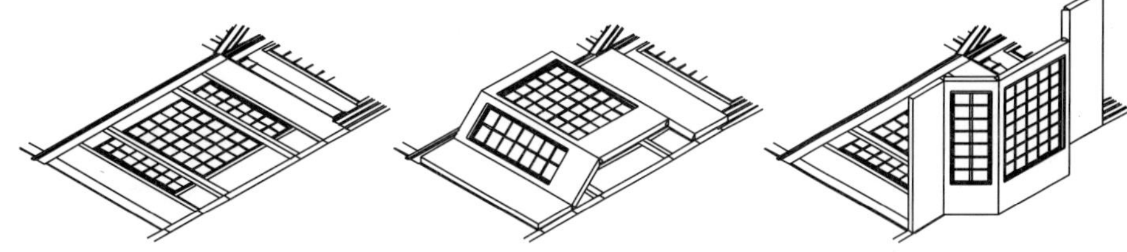

Tilt-Up Slab in detail:

The 2D Elevation is displayed to guide the creation of a 3D façade model.

3D slabs are modeled over the 2D elevation at varying Z-heights.

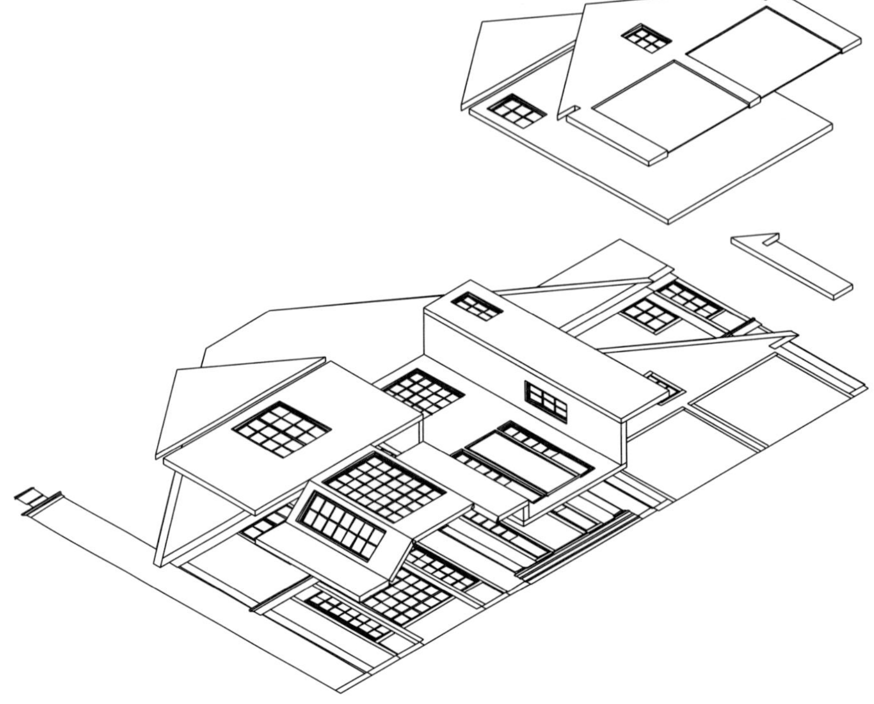

When the façades are complete, they are rotated 90 degrees about the X-axis, and moved into position on the 2D plan.

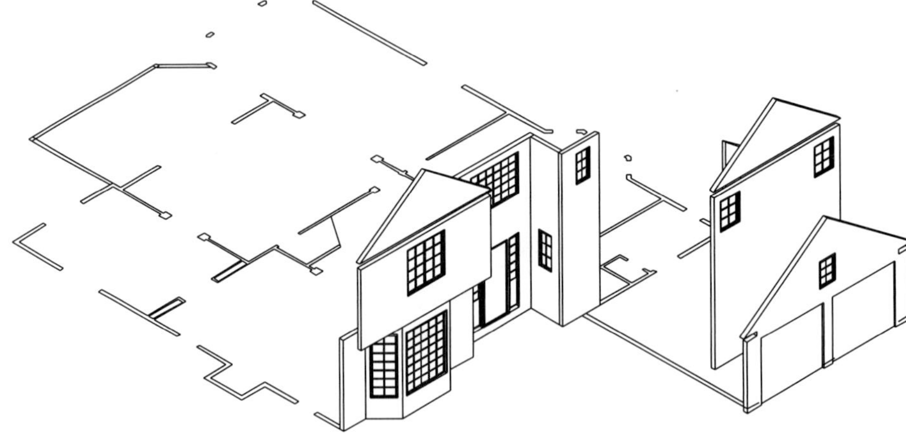

Once vertical, the façades must be moved to align with the 2D plan. Side and rear elevations are rotated again, this time about the Z-axis, and are also moved into place.

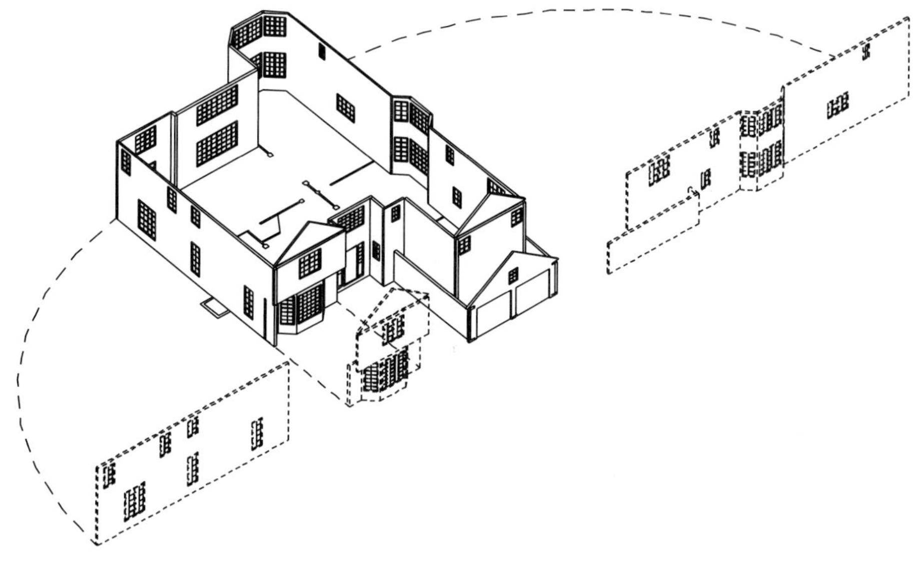

Roofs and other elements are added to complete the Tilt-Up Slab model. Note the seams created by slabs at the corners.

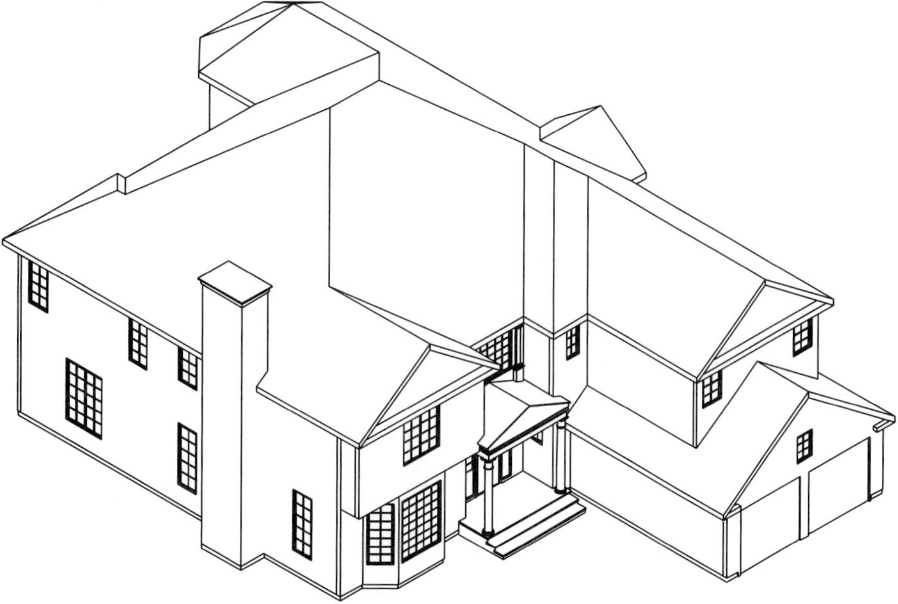

Modeling Method 4: 2D to 3D Conversion

After taking the time to draw a 2D plan, only to model over it with 3D entities, you may begin to tire of continuously tracing and then re-tracing the same points with different entity types. With the Line to Polygon conversion method, you can take advantage of DataCAD's ability to convert entities from one type to another. The 2D lines from a floor plan can be converted to 3D polygons to create a model.

The conversion process involves copying the 2D plan to a new layer, removing the wall openings, assigning a Z-height to the lines, "exploding" them to polygons, and finishing the model by inserting 3D windows and doors with the AEC_Model macro. The advantage of this method lies in its speed and recycling of existing data, and that the 2D lines yield 3D surfaces that can receive voids.

Converting a 2D Plan to a 3D Envelope Model:

1. Create a **New Layer** for the 3D wall polygons
2. Set the 2D walls layer active only, and copy the 2D lines to the 3D wall layer with the **Copy**, **To Layer** command. Copying by **Area** is the most direct method.
3. Set the 3D walls layer with the copied 2D entities active only. Change the color of these lines to a new one with the **Change**, **Color** command so that they will be distinguishable from the 2D plan.
4. Remove all door and window openings with the **Architect, Window, Remove** function. If the model does not require interior partitions, remove these as well, and heal any gaps left behind in the exterior walls.
5. Change Z-base and height to appropriate top of foundation and top of wall plate heights.

NOTE: *Performing the changes from isometric view allows you to easily create differing Z-heights.*

6. Go to the **3D Edit** menu and choose **Explode**, **To Polygons** then select the plan by **Area**. The 2D plan entities are now converted to true 3D polygons.
7. Insert the windows and doors with the **AEC_Model** macro, model the roof, and add other required 3D elements.

Line to Polygon Conversion: 2D lines (left) are assigned a Z-height and exploded to polygons (center). The resulting polygon model can be completed with the AEC_Model macro (right).

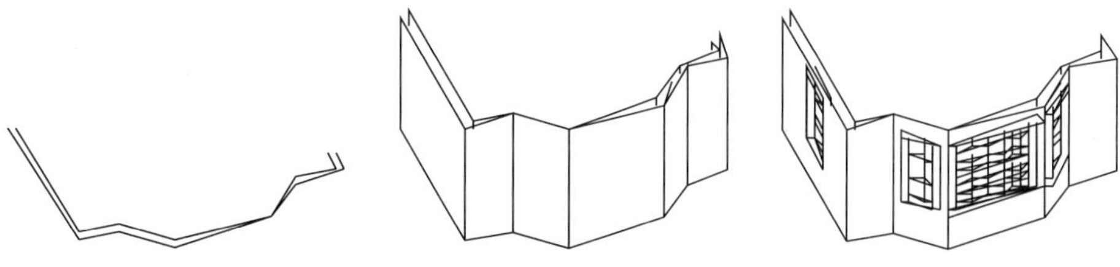

Line to Polygon Conversion in detail:

The 2D line to 3D model conversion process: 1. The 2D plan. 2. The plan is stripped of openings and interior walls, and the lines are given a Z-height. 3. The walls are modified and windows are cut through using the AEC_Model macro. 3. Roofs and details are added to complete the model.

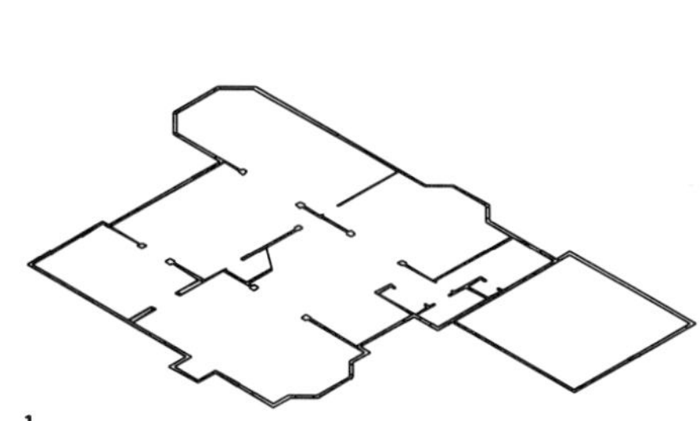

1.

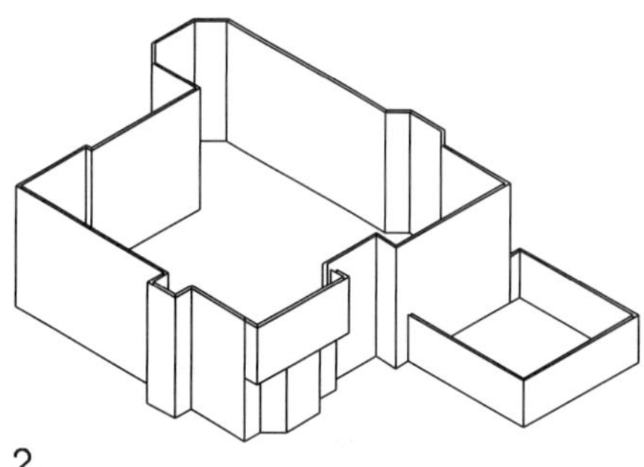

2.

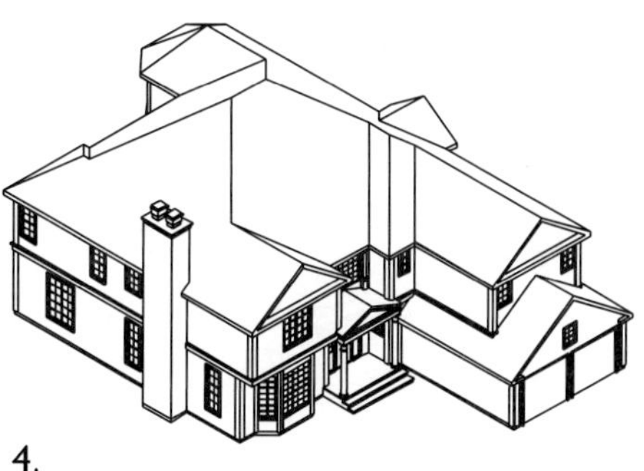

4.

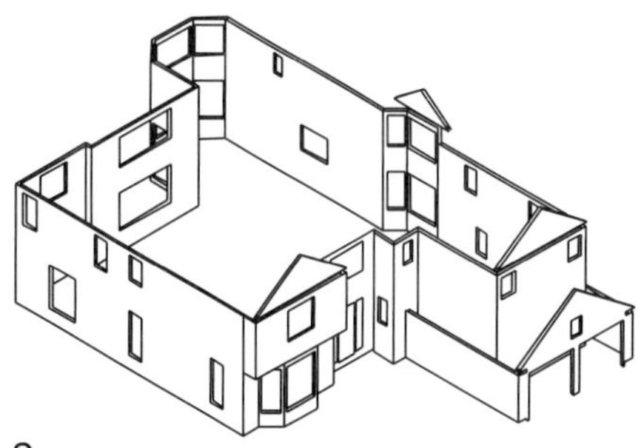

3.

Modeling Method 5: Sweeps

Sweep is part of the 3D Power Tools suite of macros. With it, you can produce complex forms very quickly. A Sweep consists of a cross-section and a path. As the cross section moves along the path it "sweeps" out a form in its wake. A Surface of revolution is an example of a radial sweep. The **Sweep** function provides you with the ability to apply this concept to irregular boundaries.

Revolved Surfaces are examples of radial sweeps.

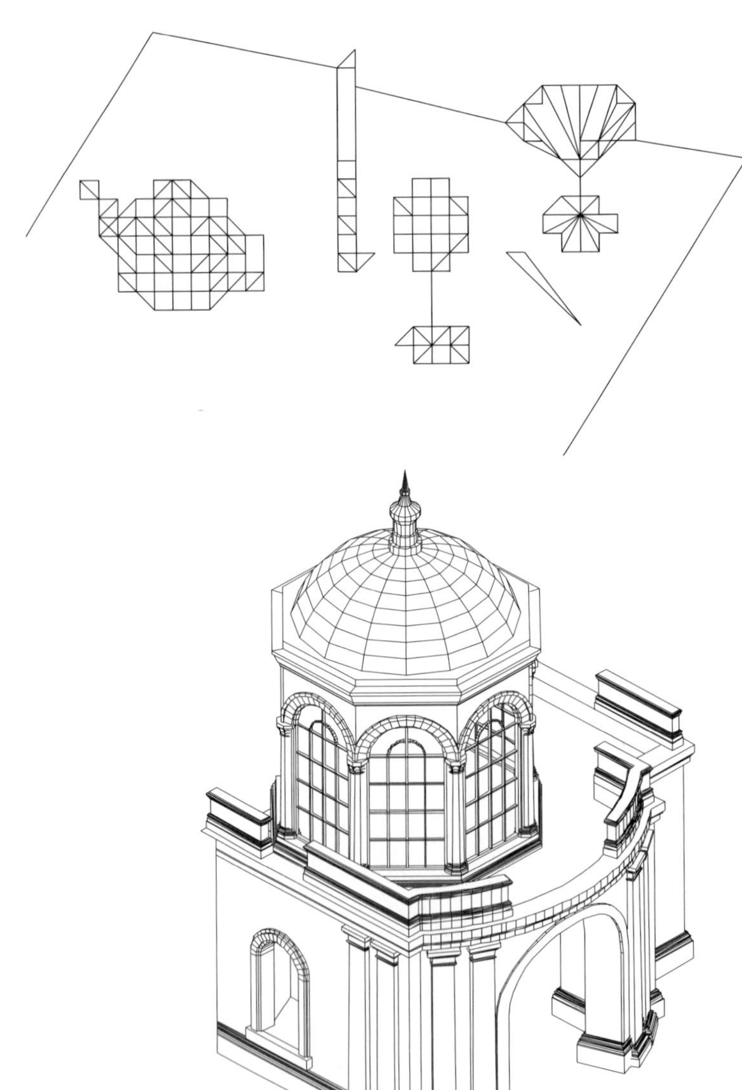

The Sweep cross section never changes, but the path may vary or even curve. It is also possible to scale or twist the cross section along the path. Sweeps are capable of producing complex model with convoluted profiles very rapidly.

Sweeps are most typically used in 3D modeling to produce building envelopes, cornice lines, trim-work and roof gutters.

Exercise: Instant Building

In this example, a small building is generated in one step from a polygon cross-section. A polyline outline defines the outer wall of the building. Next to it, a polygon that represents the cross section of is drawn. The **Sweep** function of the 3D Tools macro, is used to map the cross section around the polyline path to create a polygon model. The model is edited and enhanced further with the AEC_Model macro to insert 3D windows and doors.

Open the file Method-5.dc5 from the Bonus CD to try this technique.

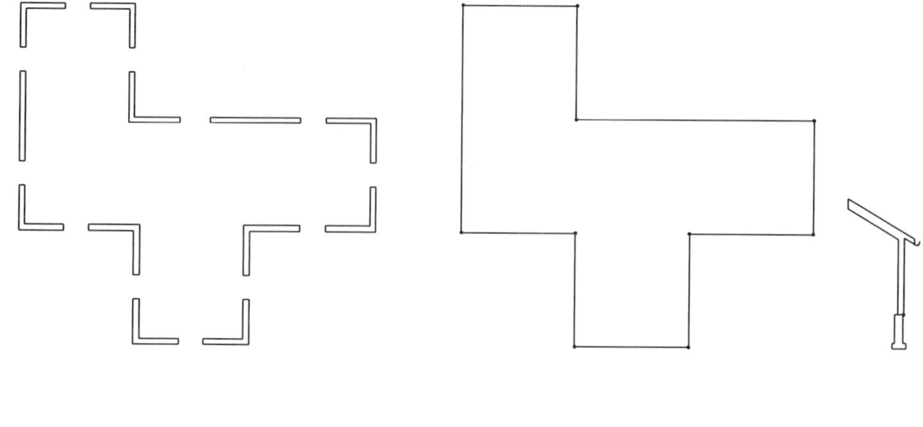

*A polyline (center) is traced around a plan (left) to define the **Sweep** path. Polygons in the shape of the building section (right) are created for the **Sweep** profile. An alignment point on the polygon indicates where it aligns with the **Sweep** path. The polygons are then applied to the path to create the **Sweep**.*

1. Define the **Sweep** Path: The first step is to draw the path that the sweep cross-section will take. In this example, a closed polyline defines the building footprint. **Sweep** paths may also take the shapes of 2D lines, 3D lines or 3D contour lines.

Create the **Sweep** cross-section

2. Create the cross section polygon: Press the **Z**-key on your keyboard, and set your **Z-base** and **Z-height** to 0′-0″. Draw the shape of the cross-section you wish to use for the **Sweep**. In this example, two polygons describe the shape of the wall section and the foundation. Give consideration to where the shapes will align with the **Sweep** path. Here it is the lower outside corner of the exterior wall. Note that the point does not have to be a point directly on the polygon.

3. Execute the **Sweep**: Select **Toolbox, 3D Tools, Sweep**. You are prompted to "Select polygon entity that defines sweep cross-section." Pick the polygon that you created in Step 2. Next, you are prompted to "Enter a point for sweep path origin." Snap to the alignment point on the polygon cross-section.

Additional menu options appear. Select **Settings**. Set the following options:

Toggle off **Vertical** (applies cross section perpendicular to **Sweep** path).

Toggle on **Show Latitude** (shows horizontal divisions between segments).

Toggle on **Show Longitude** (shows vertical divisions between segments).

Toggle on **Z-Base** (aligns cross-section to Z-base of path).

Exit the **Settings** menu by right clicking your mouse or choosing **Exit** from the menu.

4. You are prompted to "Select entity to use for sweep path." Select the polygon for the wall section. The **Sweep** is created.
5. Select **New Section**, pick the foundation polygon, and snap to the same origin point. Pick the polyline path again. The next **Sweep** is created, and the model is complete.

The resulting model generated by the **Sweep** function

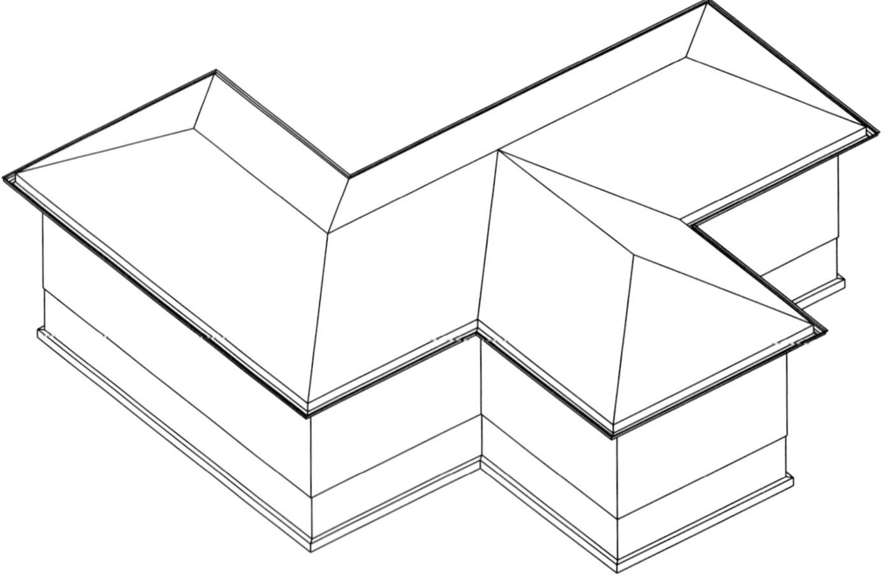

6. The **Sweep** function generates parallel polygons for the walls. 3D doors and windows may be cut into these with the **AEC_Model** macro.

Sweep model completed with **AEC_Model** and **Touch Up**.

CHAPTER 18

3D Toolbox Macros

Figure 18-1
Model by Gregory Dysart, Architect

The **Toolbox** contains several macros that assist with modeling particular architectural components such as windows, doors, stairs and roofs (see Figure 18-1). In addition to the standard macros provided with the program, DataCAD LLC offers a supplemental collection of 2D and 3D utilities called the Macro Tabouret. These macros are highly recommended for 3D modelers. **3D Power Tools** alone will significantly enhance your ability to model complex forms with the **Sweep** and **Knife** functions. Also included is **DC Sprint**, which has a multitude of 2D- and 3D-assisting functions, and **Touch-Up**, which allows you to apply hatch patterns three-dimensionally. We have already documented the importance of **3D Power Tools**, as many of the exercises outlined in this book rely on its functionality.

A third-party programmer by the name of Bill D'Amico authored the collection of macros offered in the Tabouret. Mr. D'Amico has attained legendary status in DataCAD circles for his Merlin-like ability to craft incredibly elegant and functional macros for the program. Many a user's plea for enhanced functionality has been satisfied with one of Bill's creations, myself included.

Macros offer an important window of enhanced functionality to the DataCAD user. They can be programmed to perform simple repetitive functions automatically, such as drawing a 2D stair, or they can bring previously unknown functionality to the program, as is the case with **Sweep** and **Knife**. DataCAD's programming language, DCAL, ensures that users can extend and customize the program to meet their needs. In this section we will review the macros that are pertinent to 3D modeling. You can find information on the other macros provided with DataCAD in the 2D portion of this book.

AEC Modeler

AEC Modeler is a standard DataCAD macro, but architect Patrick McConnell has enhanced it to perform more reliably across layers. The modified version of this macro is provided on the bonus CD. To install it, simply copy the **AEC_Modl.DCX** file to your **DATACAD\DCX** directory.

The **AEC Modeler** macro focuses on 3D door and window modeling, offering you simplified creation of 3D model components, while reducing creation time. Specifically, **AEC Modeler** allows you to generate 3D windows and doors based on your entered parameters. It will also create different window types, such as fixed, awning, double-hung and casement. To use the **AEC Modeler**, choose **Toolbox** from the **Edit** menu and then choose **AEC_MODL** from the **Toolbox** dialog box.

There are two creation options available with this macro: plan and elevation. It is most advantageous to use the plan option, as this is the only view which supports automatic insertion of voids into the slab or polygon wall.

Using the **In Plan** option, you model windows and doors looking down on them in plan view. The entry points you assign specify the width of the window and wall thickness. The height of the component is determined by the **Sill Height** and **Head Height** options. If **Cut Wall** is toggled on, the unit thickness is determined automatically.

Using the **In Elevation** option, you can create entities in an **Elevation** view and view a component as if you were looking at it from the side. When creating the window or door, the entry points you assign specify the width and height, while you assign wall thickness (depth) in a menu. You must take care of your **Z** settings when modeling in this view to assure that the unit is placed properly within the wall. Use the **New Elevation** technique to define a working plane that is coincident with the surface of the wall. Placing the window void ahead of time provides reference points for defining the unit size.

Additionally, the following four options appear in many of the **Windows** and **Doors** menus:

- *Color* Sets or resets the representative color for any modeling component; see “Color Menus” in “The Drawing Board” chapter for more information

- *Line Weight* Sets the line width (weight) for any modeling component that appears on screen or on printed or plotted output; choose **LineWgt**, then choose or type a value and press **Enter**
- *Line Type* Sets the representative line to solid, dotted, dashed, or dot-dashed
- *Spacing* Sets the line spacing between dots, dashes, and so on; choose **Spacing**, then choose or type a value and press **Enter**

In addition to the **Windows** and **Doors** menus, **AEC Modeler** has several other options.

Cutouts creates an opening in a wall according to head height and sill height settings. When you choose **Cutouts** from the **AEC_MODL** menu, the following options appear:

- *Head Height* Sets the head height; choose or type a value and press **Enter**
- *Sill Height* Sets the sill height; choose or type a value and press **Enter**
- *Layer Search* Places cutouts in walls on layers other than the active layer

You can use **Remove** to remove a window, door, or cutout from a wall. If **Cut Wall** was used when creating the window, door or cutout, **Remove** also removes the void in the wall.

The **Hide**, **3DViews**, and **3D Entity** options are shortcuts to those menus.

Modeling Windows

The **Windows** menu in **AEC_MODEL** serves as the master level for accessing all three-dimensional window component groups. You can access window data, load and save previously designed windows, and select or enter window components and data.

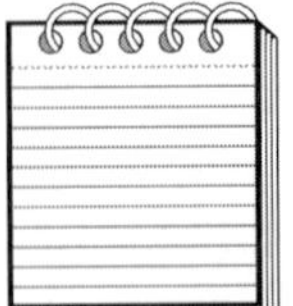

NOTE: *All windows open outward, away from you.*

Cutting Walls **Cut Wall** is a toggle used to automatically create voids in slabs for windows. It is available in the **Windows** menu when **In Plan** is toggled on. When **Cut Wall** is toggled on, a **Layer Search** toggle also becomes available.

Window Files The **Window File** option is a convenient way to expand on initial template designs, tailoring the product to your needs. The files then become additional resources for quick recall. The **Window File** menu options include the following:

- *Load Window* Loads window settings from saved files; select or type a window filename and press **Enter**; the window appears with all previously saved settings
- *Save Window* Saves window settings to a file; type the filename and press **Enter**
- *Delete Window* Deletes saved window files from the hard disk. To help prevent accidental deletions, the system prompts you for the name of the file you want to delete. Select the file you want to delete and press **Enter**. Choose **Yes** to delete the window file; otherwise choose **No**.
- *Rename Window* Renames existing files; select a file to rename, type a new filename of up to eight characters long, and press **Enter**
- *Copy Window* Copies the contents of an existing window file to a new file; select the file to copy from and press **Enter**, then type the file to copy to and press **Enter**
- *Form* Displays a window data form

Window Forms With the **Window Form** option, you can view the settings for the current window. The design form offers a comprehensive view of the window parameters on one screen. These window parameters are divided into ten major groups.

NOTE: *To save **Form** settings, press **Esc**.*

Saving Windows as Symbols Use the **Template** option to create and save windows as symbols, developing a library of window symbols that you can add quickly to new projects.

Window Height and Wall Thickness You can set the head height and sill height of a window with the **Head Height** and **Sill Height** options, as well as use the **Wall Thk** option to change the wall thickness. The **Head Height** and **Sill Height** options only appear when **In Plan** is toggled on, while **Wall Thickness** is only available when **In Elevation** is toggled on. When you choose **Head Height**, **Sill Height**, or **Wall Thickness**, a list of values is displayed. Choose or type a new value and press **Enter**.

*NOTE: The **Head Height** and **Sill Height** settings are relative to **Z-base**.*

Window Types The **Unit Type** menu displays available window types. Hinged window types open outward, away from the designer. Sliding windows, when viewed from the interior design position, have a fixed sash on the right and a movable sash on the left.

Choose from the following window types:

- *Fixed* Windows with fixed panes of glass
- *Casement* Casement windows
- *Awning* Awning windows, hinged at the top
- *Hopper* Hopper windows, hinged at the bottom
- *Double Hung* Double-hung windows
- *Sliding* Sliding windows
- *% open* Determines the amount that the window appears open in the model, with 0% being fully closed and 100% being fully open

Window Casings The **Casing** option controls the casing or exterior window group (see Figure 18-2). You can set any of the following window casing options:

- *At Head* Models casing elements at the window header
- *Width* Sets the casing width at the window header, the window jamb, and the sill across the window plane; choose the **Width** option under **At Head**, **At Jamb**, or **At Sill,** and then choose or type a width and press **Enter**

Figure 18-2
Window casing

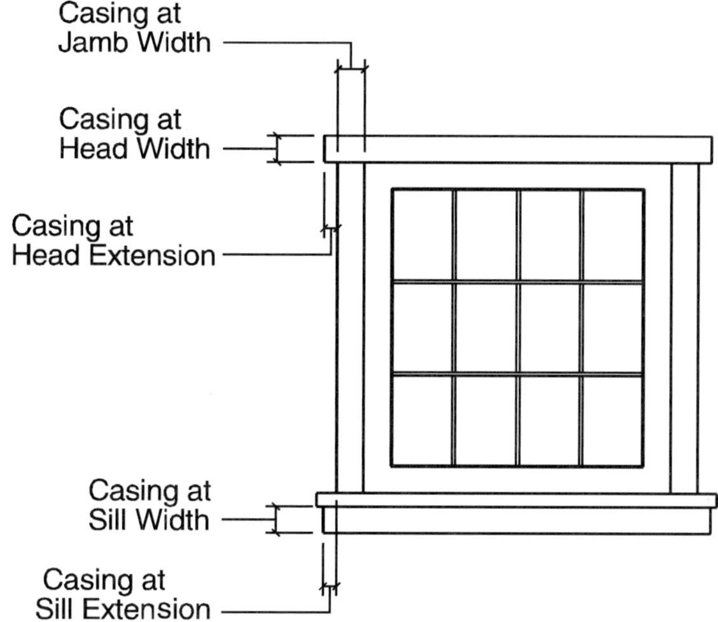

- *Thickness* Sets the casing thickness at the window header, the window jamb, and below the sill; choose or type a thickness and press **Enter**
- *Extension* Sets the extension length past the outside edge of the vertical casing components (jamb casing) and past the outside edge of the vertical jamb casing component to the sill. Choose or type an extension length and press **Enter**. For any extension, the measurement is always **0** when the outside edge of the vertical casing and the extension end are aligned. Positive numbers indicate an extension; negative numbers indicate a setback.
- *At Jamb* Models the casing element at the jamb (**Width** and **Thickness** appear again below **At Jamb**)
- *At Sill* Models the casing elements below the sill. (**Width**, **Thickness** and **Extension** appear again below **At Sill**)

Window Trim Use **Trim** to control the trim or interior window component group (see Figure 18-3). You can vary elements, such as the trim apron, according to aesthetic preference or design requirements.

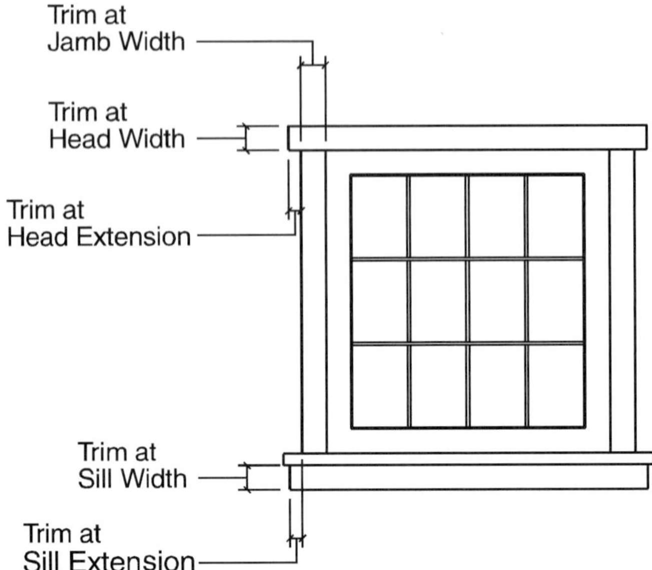

Figure 18-3
Window trim

- *At Head* Models trim elements at the window header
- *Width* Sets the trim width at the window jamb, across the window jamb; choose or type a value and press **Enter**
- *Thickness* Sets the trim thickness at the window jamb and the window sill; choose or type a value and press **Enter**
- *Extension* Sets the trim extension length past the outside edge of the vertical jamb trim components (jamb casing) or past the outside edge of the vertical jamb casing component of the sill; choose or type a value and press **Enter**. For any extension, the measurement is always **0** when the outside edge of the vertical casing and the extension end are aligned. Positive numbers indicate an extension; negative numbers indicate a setback.
- *At Jamb* Models trim elements at the jamb (**Width** and **Thickness** appear again below **At Jamb**)
- *At Sill* Models trim elements at the sill (**Width**, **Thickness** and **Extension** appear again below **At Sill**)

Window Headers The **Head** option lets you define the window header, or top of the frame (see Figure 18-4). This area can become exceedingly detailed in a model rendering, with the inclusion of its structural and aesthetic elements in one master group. You can individually set the following components to provide a clearer view:

Figure 18-4
Window header

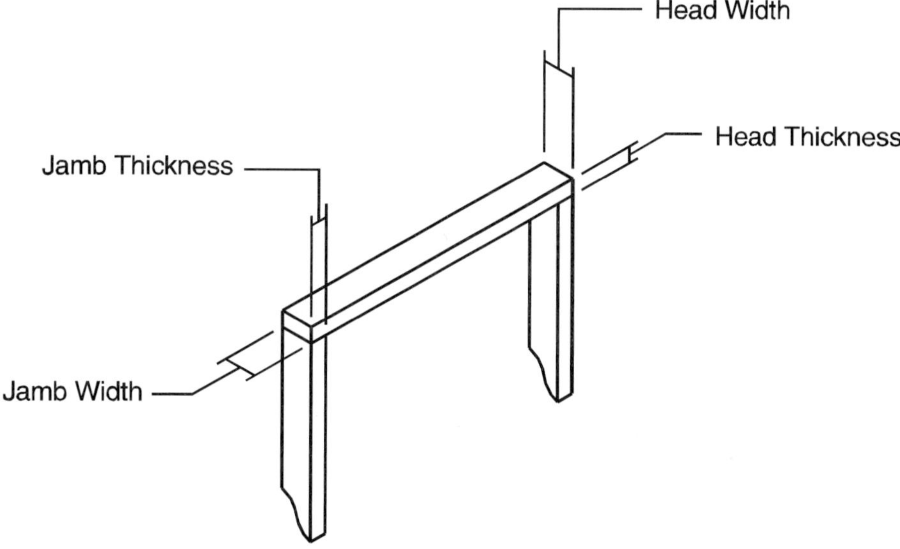

- *Do Head* Models the header component
- *Head Width* Sets the header width. Measure through the wall, from the inner surface of the wall, outward. This option appears when **Wall Width** is off. When you choose **Head Width**, a list of header width values appears. Choose or type a header width and press **Enter**.
- *Head Thickness* Sets the window header thickness. Measure from top, the wall opening, to bottom. When you choose **Head Thickness**, a list of header thickness values appears. Choose or type a header thickness and press **Enter**.
- *Wall Width* Calculates the header width equal to the thickness of the wall. When **Wall Width** is off, the width of the head is controlled by the header width setting.

Window Jambs Use the following Jamb options to define the window jamb (see Figure 18-5):

- *Do Jamb* Models jamb components
- *Jamb Width* Sets the jamb width. Measure through the wall from the inner surface of the wall, outward. This option appears when **Wall Width** is off. Choose **Jamb Width**, then choose or type a jamb width and press **Enter**.
- *Jamb Thickness* Sets the jamb thickness. Measure from the void edge, or wall opening, inward toward the window center along the wall plane.

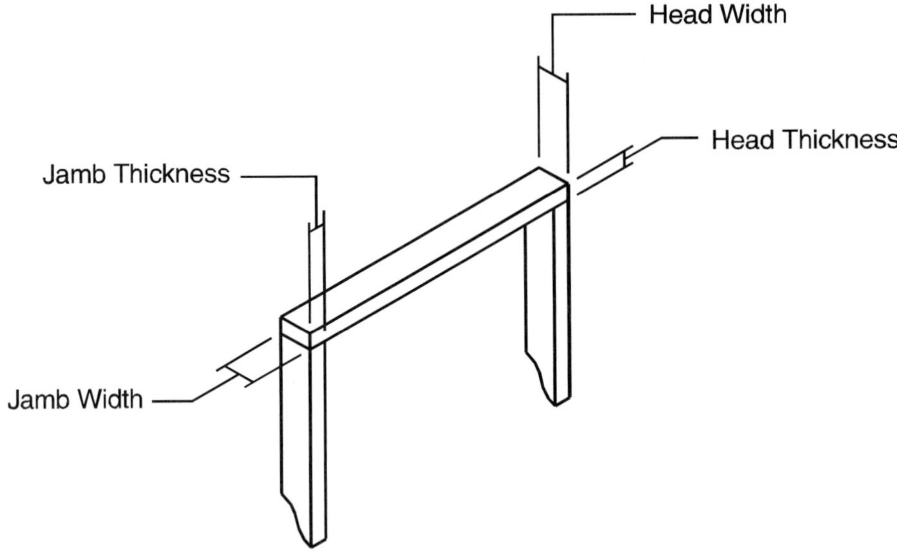

Figure 18-5
Window jamb

Choose **Jamb Thk**, and then choose or type a jamb thickness and press **Enter**.

- *Wall Width* Sets the wall width. When this toggle is off, the width of the jamb is controlled by **Jamb Width**.

Window Sills **Sill** lets you define the window sill. The sill is not broken down into stool (interior) and finish (exterior) sill components. Also, the sill horns are proportionate and do not carry different lengths on the interior and exterior extensions.

- *Do Sill* Models the sill
- *Thickness* Sets the sill thickness from the bottom of the window opening, upward; choose **Thickness**, and then choose or type a sill thickness and press **Enter**
- *In Extension* Sets the inside sill extension away from the plane of the wall toward the center of the room; choose **In Extension**, and then choose or type an inside sill thickness and press **Enter**
- *Out Extension* Sets the outside sill extension away from the plane of the wall outward; choose **Out Extension**, and then choose or type an outside sill extension and press **Enter**
- *Side Extension* Sets the inside and outside (sill horn) extensions from the sides of the window opening, outward, along the plane of the wall;

Figure 18-6
Window sill

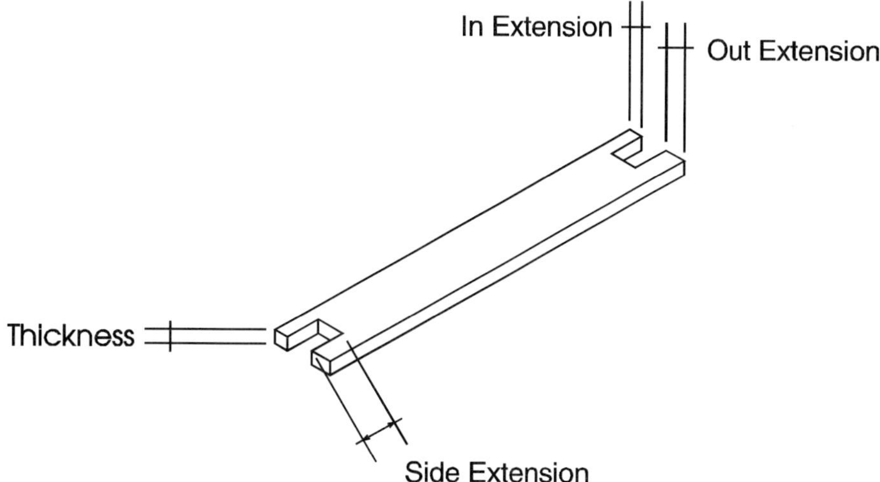

choose **Side Extension**, and then choose or type a side extension and press **Enter**

Window Sashes **Sash** defines the window sash assembly. The sash is the frame that holds the muntins and glass; without it, the muntins and glass would be suspended in 3D space. Single pane or double thermopane windows, while not having muntins, must have **Centered** toggled on to function within the confines of the wall plane.

- *Do Sash* Models the sash components; when **Do Sash** is off, muntins and glass are not added to the drawing even if **Do Muntin** and **Do Glass** are on
- *Sash Width* Sets the sash width for all components (top and bottom rails and stiles); choose **Sash Width**, then choose or type a sash width and press **Enter**
- *Sash Thickness* Sets the sash thickness for all sash components; choose **Sash Thickness**, and then choose or type a sash thickness and press **Enter**
- *Centered* Centers a sash within the window frame. **Centered** and **Offset** are mutually exclusive; only one can be toggled on at any given time
- *Offset* Determines the sash offset placement from the inside of the window jamb. The distance is from the interior surface of the wall to the interior surface of the sash; choose **Offset**, then choose or type a sash offset and press **Enter**.

Muntins Use **Muntins** to define and place window pane dividers within the sash assemblies. You can also set the number of window panes and their placement.

- *Do Muntin* Models muntins, or window pane dividers
- *Muntin Width* Sets the muntin width along the plane of the wall; the width applies equally to vertical and horizontal muntins; choose **Muntin Width**, then choose or type a muntin width and press **Enter**
- *Muntin Thickness* Sets the muntin thickness, which is measured through the window; choose **Muntin Thickness**, then choose or type a muntin thickness and press **Enter**
- *Pane Horizontal* Sets the number of panes per window horizontally; choose **Pane Horizontal**, then choose or type a value and press **Enter**
- *Pane Vertical* Sets the number of panes per window vertically; choose **Pane Vertical**, then choose or type a value and press **Enter**

NOTE: ***Pane Horizontal*** *and* ***Pane Vertical*** *are also available in the* ***Glass*** *menu. They are the same settings, so when you change either option in the* ***Muntins*** *menu, the new values are also displayed for the* ***Pane Horizontal*** *and* ***Pane Vertical*** *options in the* ***Glass*** *menu.*

- *Centered* Centers a muntin within the sash; **Centered** and **Offset** are mutually exclusive: only one can be toggled on at any given time
- *Offset* Sets the offset placement of the muntins from the interior surface of the sash to the interior surface of the muntins; choose **Offset**, then choose or type a muntin offset and press **Enter**; **Centered** and **Offset** are mutually exclusive: only one can be toggled on at any given time
- *Full Edge* Sets a full-edge muntin at the window sash
- *Half Edge* Sets a half-edge muntin at the window sash
- *No Edge* Sets a no-edge muntin at the window sash
- *Slabs* Constructs muntins as slabs; when **Slabs** is toggled off, muntins are constructed as 3D lines

Glass Use **Glass** to define the number of glass panes horizontally and vertically. It also sets the glass thickness for the window panes in the model. Since glass is represented as a slab or solid object, it is often left out of a model so that a hidden line removal can be performed as if the glass were transparent.

- *Do Glass* Models window glass in the window panes
- *Glass Thickness* Sets the glass thickness; choose Glass **Thickness**, and then choose or type a glass thickness and press **Enter**
- *Pane Horizontal* Sets the number of horizontal panes per window; choose **Pane Horizontal**, and then choose or type a value and press **Enter**
- *Pane Vertical* Sets the number of vertical panes per window; choose **Pane Vertical**, and then choose or type a value and press **Enter**

***NOTE:* *Pane Horizontal* and *Pane Vertical* are also available in the *Muntins* menu. They are the same settings, so when you change either option in the *Glass* menu, the new values are also displayed for the *Pane Horizontal* and *Pane Vertical* options in the *Muntins* menu.**

- *Centered* Centers a window pane glass in the muntin, through the window; **Centered** and **Offset** are mutually exclusive: only one can be toggled on at any given time
- *Offset* Sets the offset of the glass panes from the interior surface of the muntin to the interior surface of the glass; choose **Offset**, then choose or type a glass pane offset and press **Enter**; **Centered** and **Offset** are mutually exclusive: only one can be toggled on at any given time

Modeling Doors

The **Doors** menu lets you access door data, load or save previously designed doors, and select or enter door components and data.

Cutting Walls **Cut Wall** is a toggle used to automatically create voids in slabs for doors. It is available in the menu when **In Plan** is toggled on. When **Cut Wall** is toggled on, a **Layer Search** toggle also becomes available.

Door File Use the **Door File** option to load, save, delete, rename, or copy door files. Choose from the following **Door File** options:

- *Load Door* Loads door parameters from saved files; choose or type a door filename and press **Enter**
- *Save Door* Saves door parameters to a file; type a filename and press **Enter**

- *Delete Door* Deletes saved door files from the hard disk. To help prevent accidental deletions, the system prompts for the filename requested. To delete a door file, select a filename and press **Enter**. Choose **Yes** to delete the file; otherwise choose **No**.
- *Rename Door* Renames existing files; select a file to rename, type a filename of up to eight characters long, and press **Enter**
- *Copy Door* Copies the contents of an existing door file to a new file; type the name of the file to copy from and press **Enter**, then type the filename to copy to and press **Enter**
- *Form* Displays a door data form

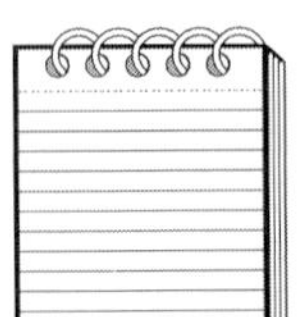

NOTE: *All doors are created to open inward, toward you.*

Door Forms The **Door Form** menu lets you view the settings for the door on which you are currently working. The design form offers a comprehensive view of the door parameters on one screen.

Saving 3D Doors as Symbols Use the **Template** option to create and save doors as symbols, developing a library of door symbols that you can quickly insert into new projects. For more information, see "Templates and Symbols" in the "Drawing Elements" chapter.

Door Heights and Wall Thickness You can set the head height and sill height of a door with the **Head Height** and **Sill Height** options, as well as use the **Wall Thickness** option to change the wall thickness. The **Head Height** and **Sill Height** options are available only when **In Plan** is toggled on, while **Wall Thickness** is only available when **In Elev** is toggled on. When you choose one of these options, a list of values is displayed in the menu window. Choose or type a new value and press **Enter**.

NOTE: *The **Head Height** and **Sill Height** settings are relative to **Z-base**.*

Door Types **Unit Type** displays available door types. Doors swing inward (toward you), so you should build designs from the interior of a structure.

Choose a door type from the following options:

- *Single* Single door, hinged
- *Double* Double doors, hinged
- *Bi-fold* Bi-fold door
- *2xBifold* Double bi-fold doors
- *Sliding* Sliding door
- *Pocket* Pocket doors
- *% open* Determines the percentage that a door appears open in the model, with 0 percent being fully closed and 100 percent being fully open

Door Casings Use **Casing** to control the casing or exterior door component group (see Figure 18-7). You can set any of the following door casing options:

- *At Head* Models casing elements in the door header model
- *Width* Sets the casing width at the door header, across the door plane, and at the door jamb; choose or type a value and press **Enter**
- *Thickness* Sets the casing thickness at the door head and at the door jamb; choose or type a value and press **Enter**
- *Extension* Sets the extension length past the outside edge of the vertical casing components (jamb casing). Choose or type a value and press **Enter**. For any extension, this measurement is always 0 when the outside edge of the vertical casing and the extension end are aligned. Positive numbers indicate an extension; negative numbers indicate a setback.
- *At Jamb* Models casing elements in the door jamb model (**Width** and **Thickness** appear again after this option)

Door Trim Use **Trim** to control the trim, or interior door component group (see Figure 18-8). You can vary such elements as the trim apron, according to aesthetic preference or design requirements.

- *At Head* Models trim elements at the door header model
- *Width* Sets the trim width at the door header and across the door jamb

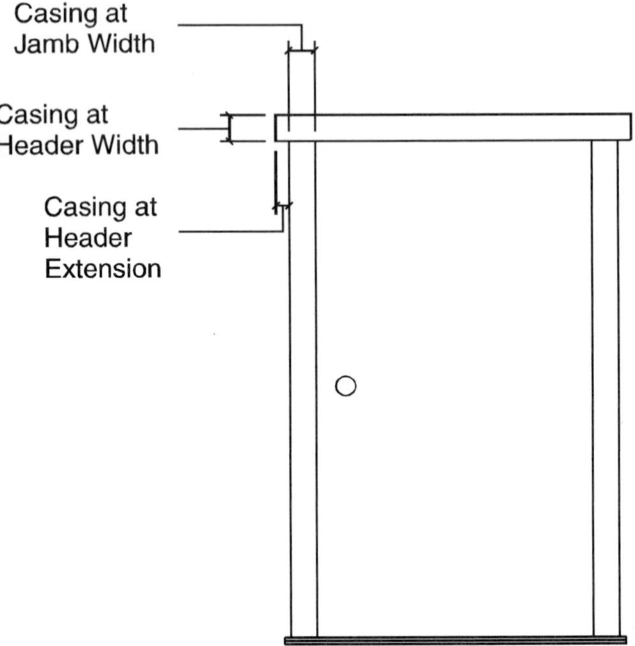

Figure 18-7
Door casing

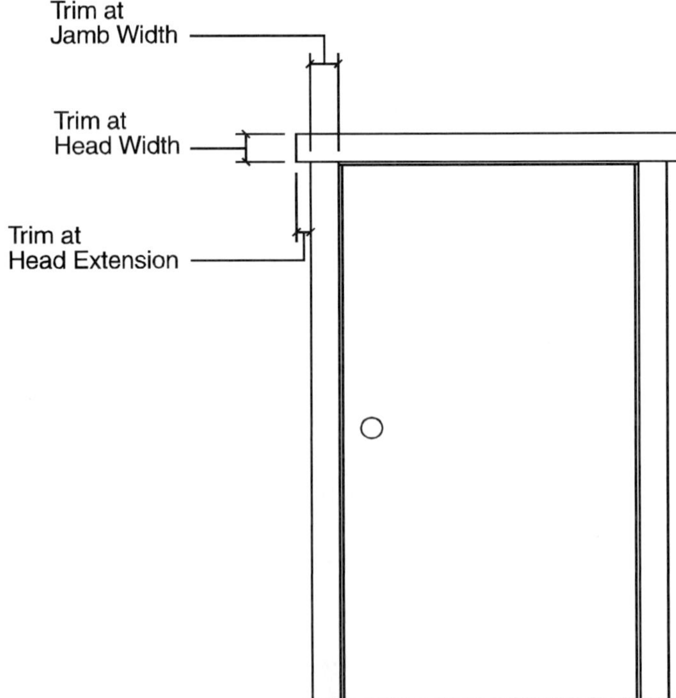

Figure 18-8
Door trim

- *Thickness* Sets the trim thickness at the door header, through the door, and at the door jamb
- *Extension* Sets the trim extension length past the outside edge of the vertical jamb trim components
- *At Jamb* Models trim elements in the door jamb model (**Width** and **Thickness** appear again after **At Jamb**)

Door Headers Use **Header** to define the door header. This area can become exceedingly detailed in a model rendering, including its structural and aesthetic elements in one master group. With **Header** you can break down the various components to provide a clearer view (see Figure 18-9).

- *Do Head* Models the door header
- *Head Width* Sets the door header width. Measure through the door, from the inner surface of the door outward. This option appears when **Wall Width** is off; choose **Head Width**, and then choose or type a header width and press **Enter**
- *Head Thickness* Sets the door header thickness. Measure from top (wall opening) to bottom. When you choose **Head Thickness**, a list of thickness values appears. Choose or type a header thickness and press **Enter**.

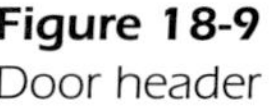

Figure 18-9
Door header

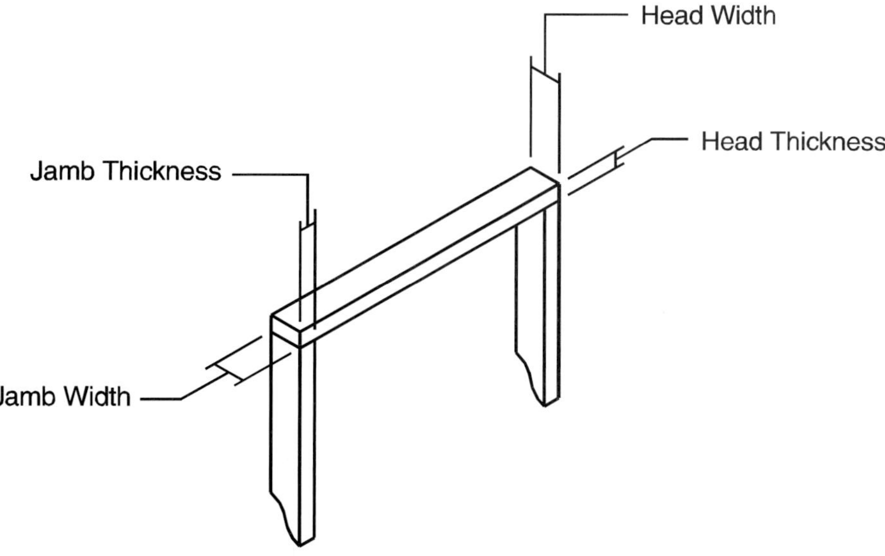

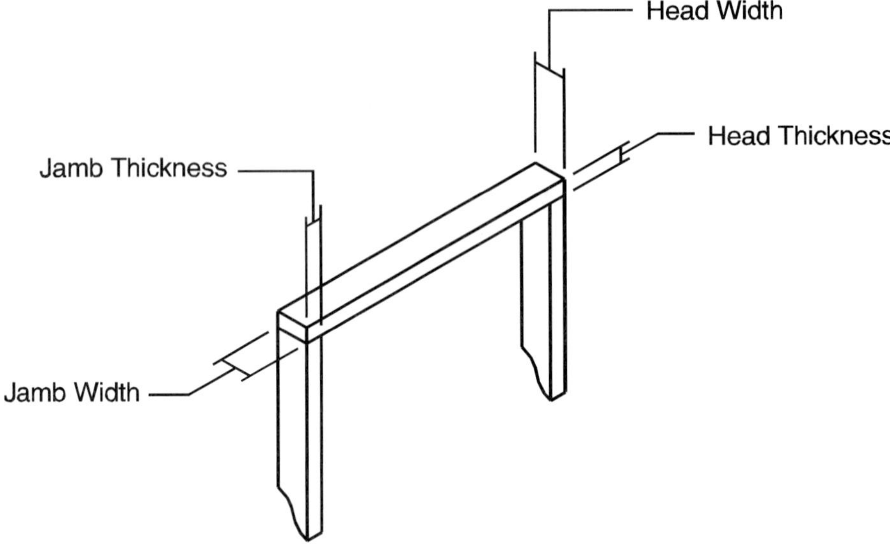

Figure 18-10
Door jamb

It calculates a header width equal to the wall thickness; when **Wall Width** is off, the width of the header is controlled by the header width (**Head Width**) setting.

Door Jambs The **Jamb** menu defines the door jamb (see Figure 18-10), using the following options:

- *Do Jamb* Models jamb components
- *Jamb Width* Sets the door jamb width (measure from the inner surface of the door, outward through the door); available only when **Wall Width** is off: choose **Jamb Width**, and then choose or type a jamb width and press **Enter**
- *Thickness* Sets the door jamb thickness (measure from the void edge, or wall opening, inward toward the door center along the wall plane); choose **Thickness**, then choose or type a jamb thickness and press **Enter**
- *Wall Width* Sets the jamb width equal to the wall width; when **Wall Width** is off, the width of the jamb is controlled by **Jamb Width**: choose **Wall Width**, then choose or type a jamb width and press **Enter**

Door Sills The **Sill** menu defines the door sill (see Figure 18-11), using the following options:

Figure 18-11
Sill thickness

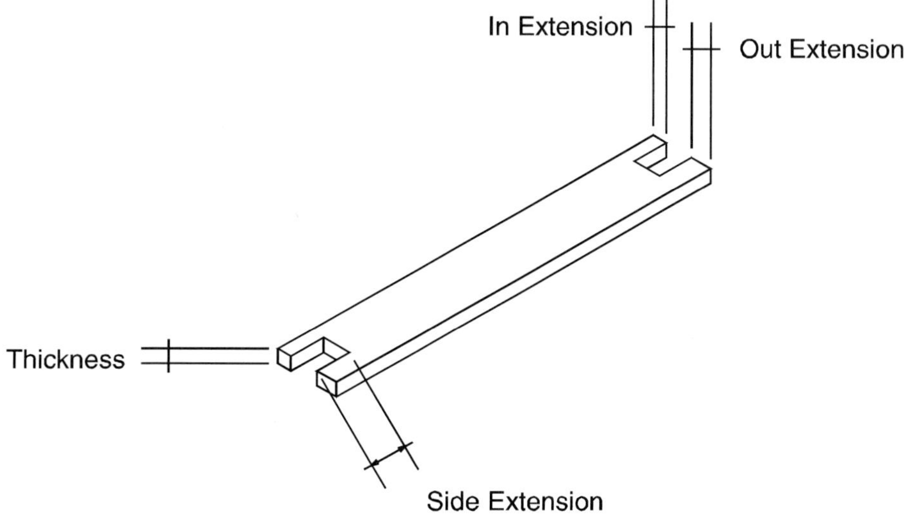

- *Do Sill* Models the sill
- *Thickness* Sets the sill thickness from the bottom of the door opening, upward at the sill middle; choose **Thickness**, then choose or type a sill thickness and press **Enter**
- *Edge Thickness* Sets the thickness of the tapered edge; choose **Edge Thickness**, then choose or type an edge thickness and press **Enter**
- *In Extension* Sets the inside sill extension away from the plane of the door toward the center of the room; choose **In Extension**, then choose or type an extension and press **Enter**
- *Out Extension* Sets the exterior sill extension across the door plane from the exterior wall outward; choose **Out Extension**, then choose or type an extension and press **Enter**
- *Side Extension* Sets the exterior sill extension along the wall plane from the outside edge of the trim outward (also known as the "sill horns"); choose **Side Extension**, then choose or type an extension and press **Enter**

Door Stops Use **Stop** to place and size the door stops. The stop is drawn to the exterior of the door, whether the door is flush or offset. Door stops are not created for sliding or pocket doors, regardless of whether **Do Stop** is toggled on.

- *Do Stop* Models the door stop
- *Stop Width* Sets the stop width for all components (measure the stop width in the same direction as the jamb width and across the wall

plane); choose **Stop Width**, then choose or type a stop width and press **Enter**

- *Stop Thickness* Sets the stop thickness dimension (measure the stop thickness in the same direction as the jamb thickness and across the wall plane); choose **Stop Thickness**, then choose or type a stop thickness and press **Enter**

Door Options The **Door** menu sizes, positions, and hinges a door, using the following options:

- *Do Door* Models the door
- *Door Thickness* Sets the door thickness; choose Door **Thickness**, and then choose or type a door thickness and press **Enter**
- *Hinge Right* Hinges doors on the right; determine the right side of the door by standing inside the structure and looking at the door
- *Hinge Left* Hinges doors on the left; determine the left side of the door by standing inside the structure and looking at the door
- *Flush* Sets the door flush with the outside wall
- *Offset* Sets the offset placement of the door from the interior wall surface across the wall plane; choose Offset, and then choose or type a value and press **Enter**

Door Knobs Use **Knob** to place, size, and identify the door knob; only round door knobs are available. When **Do Door** is off, you can't create a knob, regardless of the **Do Knob** setting.

- *Do Knob* Models the knob
- *Diameter* Sets the diameter of the door knob; choose **Diameter**, then choose or type a door knob diameter and press **Enter**
- *Extension* Sets the knob extension out from the door; choose **Extension**, then choose or type a knob extension and press **Enter**
- *Knob Height* Sets the height of the knob from the bottom of the door; choose **Knob Height**, then choose or type a knob height and press **Enter**
- *Offset* Sets the knob offset from the door edge; choose **Offset**, then choose or type a knob offset and press **Enter**
- *Inside* Models a knob on the door interior
- *Outside* Models the knob on the door exterior; **Outside** and **Inside** can be toggled individually or together

3D Stairs

The **3DStairs** macro provides everything you need to create a set of three-dimensional stairs. **3DStairs** creates real stairs, not stacked slabs, making this representation of stairs more accurate than ever before. The **3DStairs** macro has the following features:

- *Seven predefined stair types* Single-run, straight-run, double-back, L-shaped, open-well, curved, and spiral
- *Stair calculator* Makes it easy to calculate the values of stair settings and readjust the **Settings** menu values

With the stair forms and other settings, you can customize the shape and placement of stair components. There are three ways to define the settings for the type of stairs you want to model:

- Use the **3DStairs/Settings** menu to select options and enter individual values for the various stair components.
- Use the **Stair Form** to display the settings on the screen for reference and/or customization.
- Save your stair settings to a file through the **StrFile** menu; when you want to model that specific stair, type it again and you can recall the file.

NOTE: *Because of the interactive nature of stair components, changing one option can cause other options to change automatically. For example, when you close stringers, you can't have tread extensions.*

The sections that follow describe how to create each stair type. You can only use the **Stair Calculator** for single-run stairs; see "The Stair Calculator" later in this section for more information.

To draw 3D stairs, do the following:

1. Choose **Toolbox** from the **Edit** menu.
2. Choose **3DStairs** from the **Toolbox** dialog box.
3. Choose a stair type (see Figure 18-12).
4. Set the options for the stair style you want. You can use either the **Settings** menu, the **Stair Form**, or you can load a stair file. When you model single-run stairs, you can use the **Stair Calculator** to

Figure 18-12
Stair types

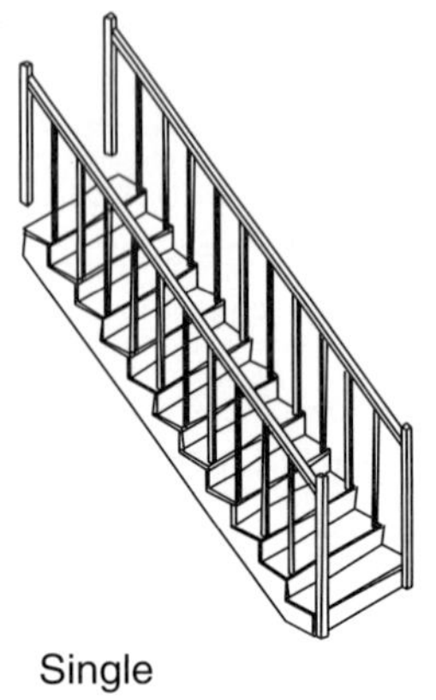

Single

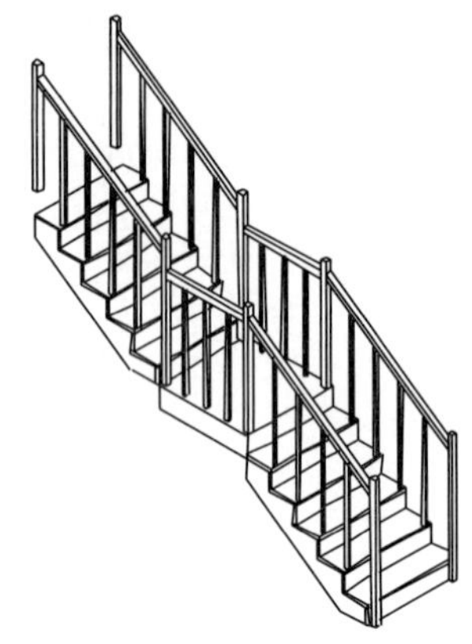

Straight

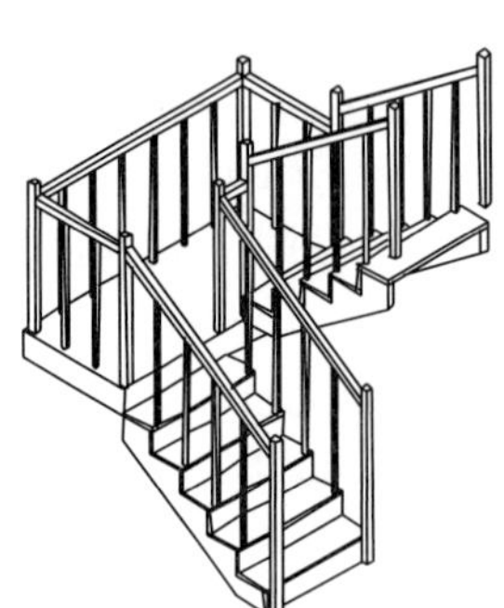

Double Back

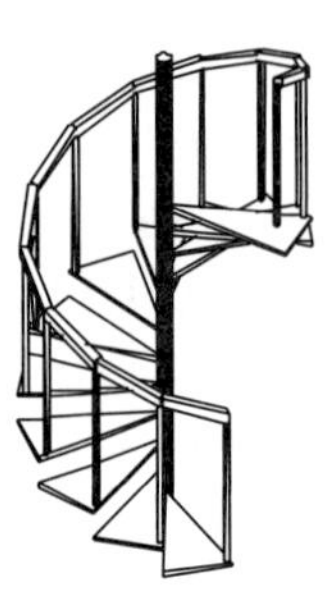

Spiral

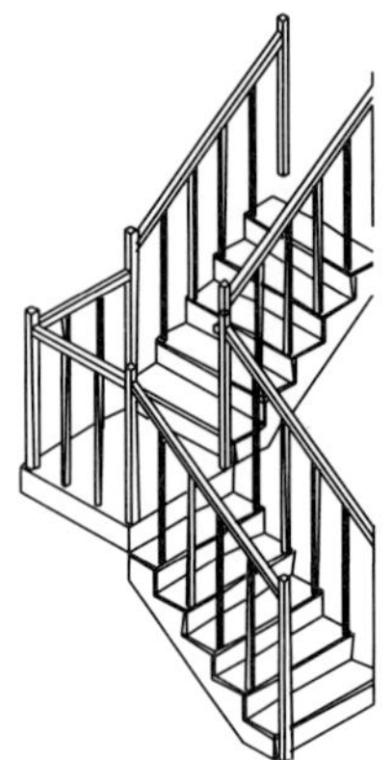

L-Shaped

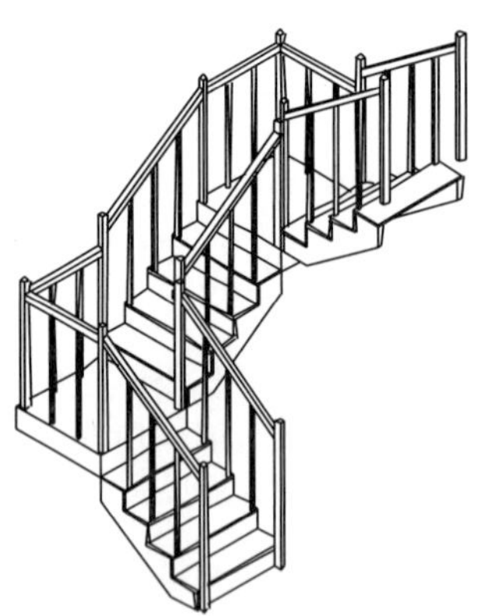

Open Well

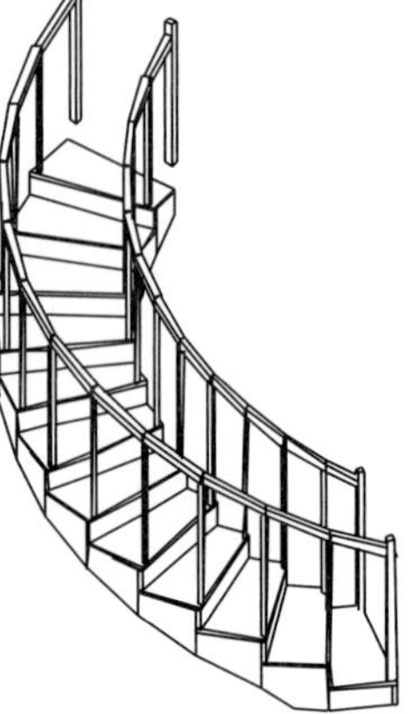

Curved

determine the settings. See "Stair Settings" later in this section for more information.

5. Choose **Begin** to place the stairs. The tool automatically changes to a plan view and a boundary box appears, representing the run of stairs.

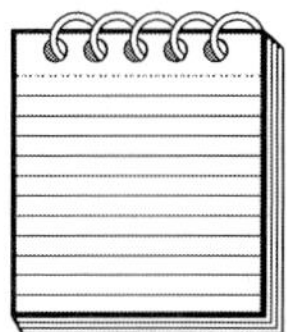

NOTE: *When **Ortho** mode is on, the boundary box does not track with the cursor. Press **0** to toggle **Ortho** mode off.*

6. Drag the stairs by the lower outside corner of the first run to the new stair position. The boundary box rotates as you move your cursor.
7. Rotate the stairs to the angle you want. The box rotates as you move the cursor.
8. Click to place the stairs. The stair model appears in the selected location with the settings you chose.
9. To change the results, erase the last group by pressing **Shift + a** and repeat Steps 3 through 7.

Stair Settings

Use the **Settings** menu to change the settings for **3DStairs**. Different options appear depending on the stair type you selected from the **3DStairs** menu.

The options in the top half of the menu are global settings that apply to all stair types. The bottom half of the menu are stair component submenus. Each of these submenus contains the various settings for that particular component.

When you choose any of the **Settings** options, a list of values appears. Choose a value or press **Enter** to accept the given value.

Base Elevation Use **BaseElev** to change the base elevation of a flight of stairs. Base elevation is the height, or elevation, at which the flight of stairs begins (see Figure 18-13).

Treads Use **TrdLngth** to change the tread length. The tread length is the horizontal distance from one side of the tread to the other side. It does not include tread extension values. Any tread extensions are added to the tread length and only apply to stairs modeled with open stringers.

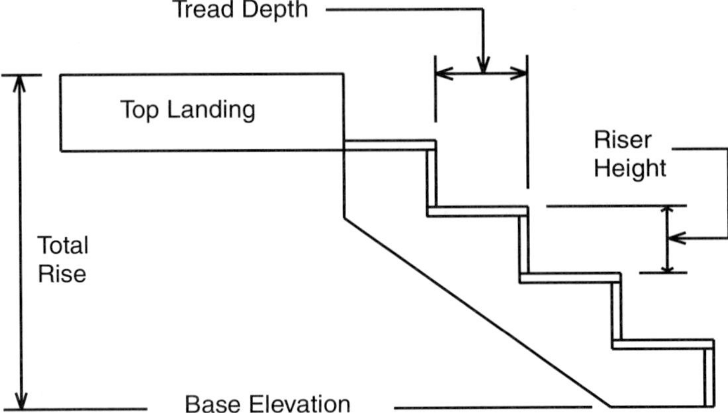

Figure 18-13
Base Elevation

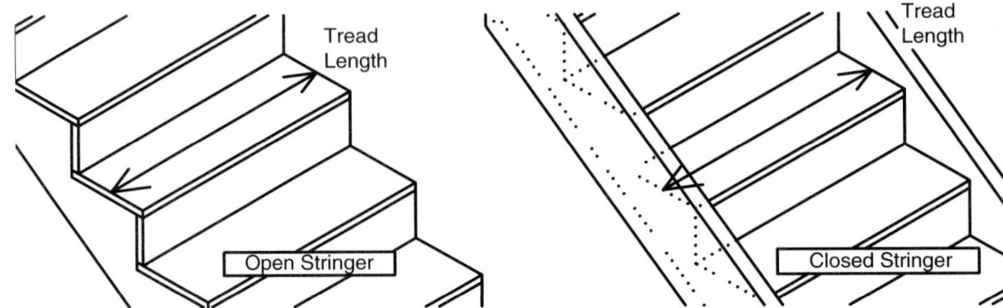

Figure 18-14
Tread length

Use **Trd/Run1** to change the number of treads in the first run of stairs. This applies to all stair types. The number of treads does not include the landing at the top of the run and is equal to one less than the number of risers.

Use **Trd/Run2** to change the number of treads in the second run of stairs. **Trd/Run2** is only available for straight, double-back, open-well, and L-shaped stairs.

Use **Trd/Run3** to change the number of treads in the third run of stairs. **Trd/Run3** is only available for open-well stairs.

Radius of Curved Stairs Use **Radius** to change the inner radius of curved and spiral stairs. Measure **Radius** from the exact center point to the inside face of the stringer for curved stairs and the radius of the center pole for spiral stairs (see Figure 18-15).

Clockwise The **Clockwise** option applies to all stair types except single and straight. You can place the stairs clockwise or counterclockwise depending on your needs.

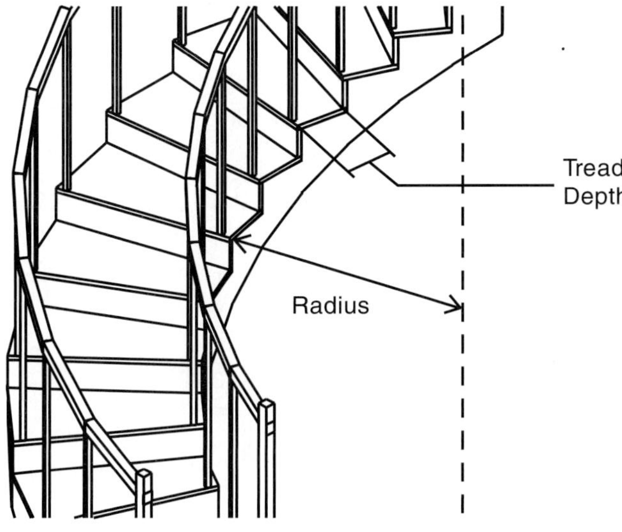

Figure 18-15
Radius on a curved flight of stairs

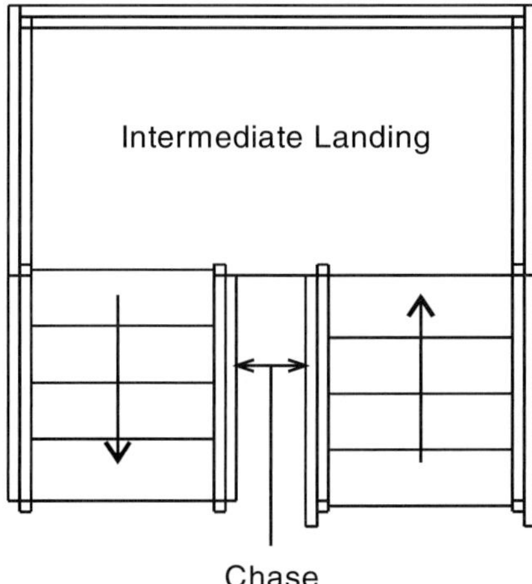

Figure 18-16
Chase

Chases Use **Chase** to control the horizontal distance between the runs of a double-back flight of stairs (see Figure 18-16). This setting affects the width of the intermediate landing and the tread length.

Landings The top, bottom, and in some cases, intermediate landings all have depth and thickness settings. You can change the specific settings for each landing you create using the **Landing** menu in the **3DStairs/**

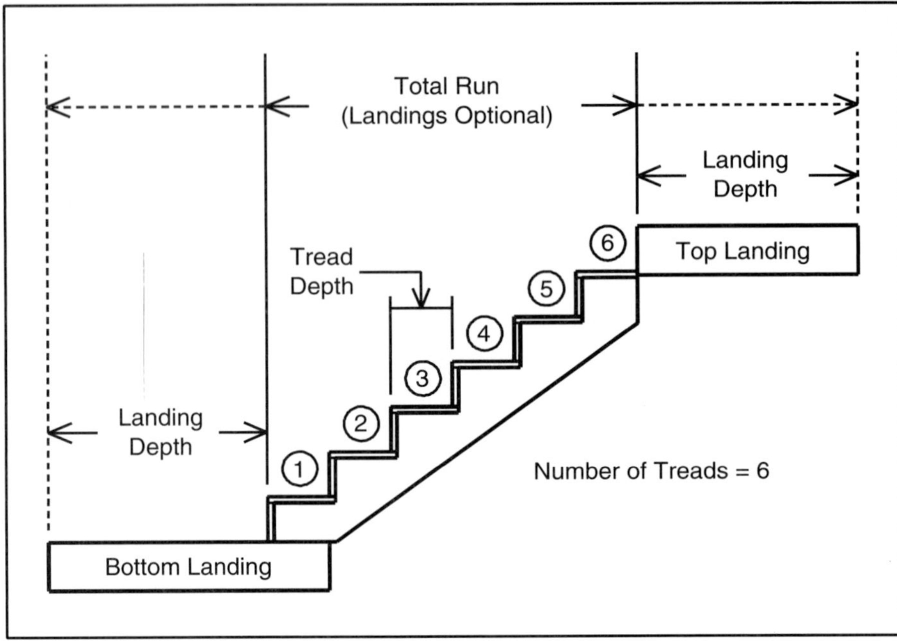

Figure 18-17
Landing depth

Settings menu. (Changing the settings on the top landing does not affect the settings of the bottom landing, and vice versa.)

Toggle **DoLndTp** on to model a landing at the top of the flight of stairs; toggle **DoLndBt** on to model a landing at the bottom of a flight of stairs.

The **Depth** option sets the landing depth. The *landing depth* is the horizontal distance from the front of the landing to the rear of the landing (see Figure 18-17). When you model L-shaped or open-well stairs, you cannot set a landing depth on the intermediate landing. The landing depth equals the width of the run of stairs.

Thickness shows the current thickness of a landing. Measure the landing thickness vertically from the underside to the top side. **Thickness** sets the top, intermediate, and bottom landing thicknesses.

For stair types that have intermediate landings, **IntrLand** is always on. It is available for straight, double-back, open-well, or L-shaped stairs only. For rectangular or curved intermediate landings, use the **Rectangl/Curved** option; this is available for L-shaped, double-back, or open-well stairs only (see Figure 18-18).

Use **Color** to set the landing color. All landings in a single flight of stairs are modeled using the same color. See "Color Menus" in "The Drawing Board" chapter for more information about how to choose a color.

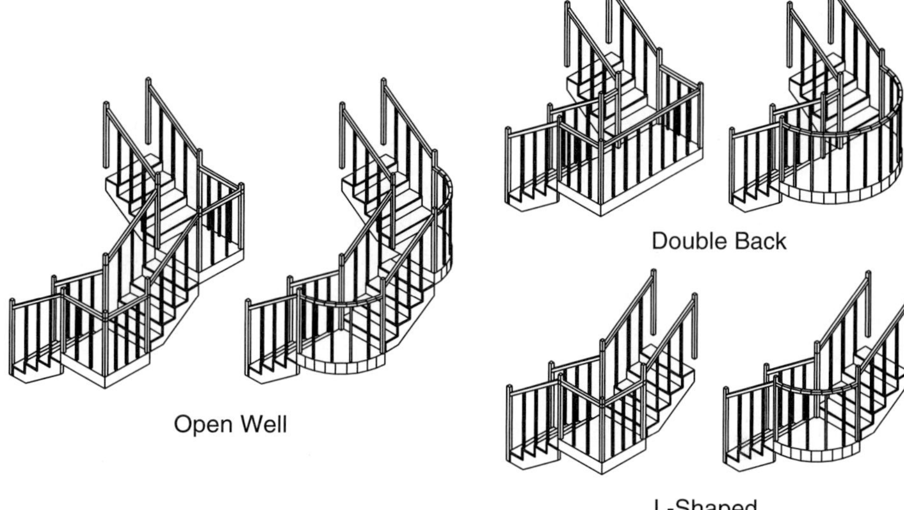

Figure 18-18
Rectangular/curved intermediate landings

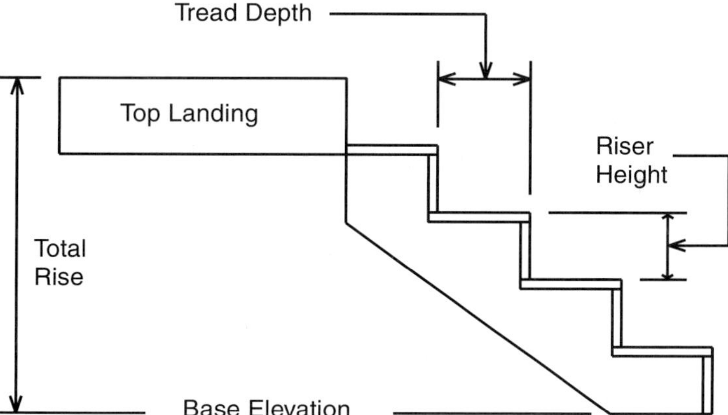

Figure 18-19
Riser height, total rise, and tread depth

Risers Use the **Riser** menu to set options that create risers and determine riser height, thickness, and color. Toggle **DoRiser** on to model risers on your flight of stairs.

RiserHgt displays the current riser height in the Message Window and lets you reset it. The riser height is the vertical distance from the top of one tread to the top of the next (see Figure 18-19).

The **Thickness** option displays the current riser thickness in the Message Window and lets you reset it. The *riser thickness* is the horizontal distance from the front side of the riser to the backside of the riser when you choose **Thickness**.

Use **Color** to set the landing color. All landings in a single flight of stairs are modeled using the same color. See "Color Menus" in "The Drawing Board" chapter for more information about how to choose a color.

Treads Use **Tread** to set options that model treads and determine the depth, thickness, and nosing of the treads. Toggle **DoTread** on to automatically calculate and model treads on a flight of stairs.

Depth displays the current tread depth in the Message Window and lets you reset it. The tread depth is the horizontal distance from the front of one tread to the front of the next tread (see Figure 18-20). When you select a spiral stair, **TredDegs** (tread degrees) is displayed in the menu in place of depth. The **TredDegs** Message Window displays the inclusive angle of the treads on spiral stairs and lets you reset it.

Thickness displays the current tread thickness in the Message Window and lets you reset it. The tread thickness is the vertical distance from the top side of the tread to the bottom side of the tread.

LeftExtn and **RghtExtn** display the current left and right extension, respectively, of the tread in the Message Window and let you reset it. Left extension includes tread length, the distance from one stringer to the next (see Figure 18-21). **LeftExtn** controls the distance that the tread extends beyond the outside of the stringer. **LeftExtn** is available only for stairs with open stringers.

When **Nosing** is toggled on, you can set the nosing length and type for treads. **3DStairs** prompts for an overhang value when you toggle nosing on. Nosing controls the distance that the front of the tread extends beyond the front of the riser.

Figure 18-20
Tread depth

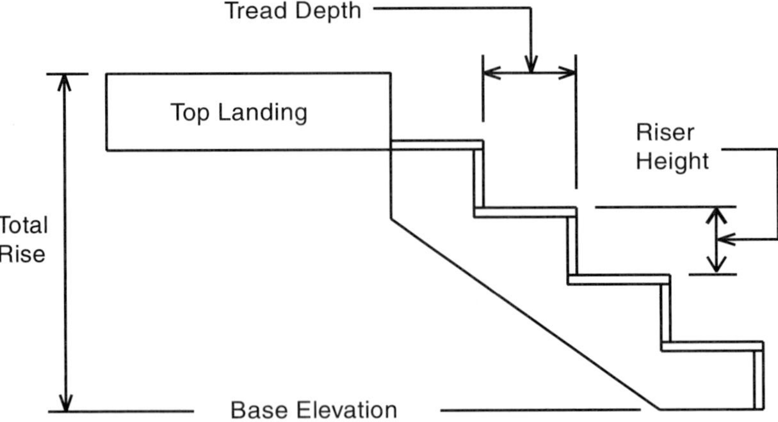

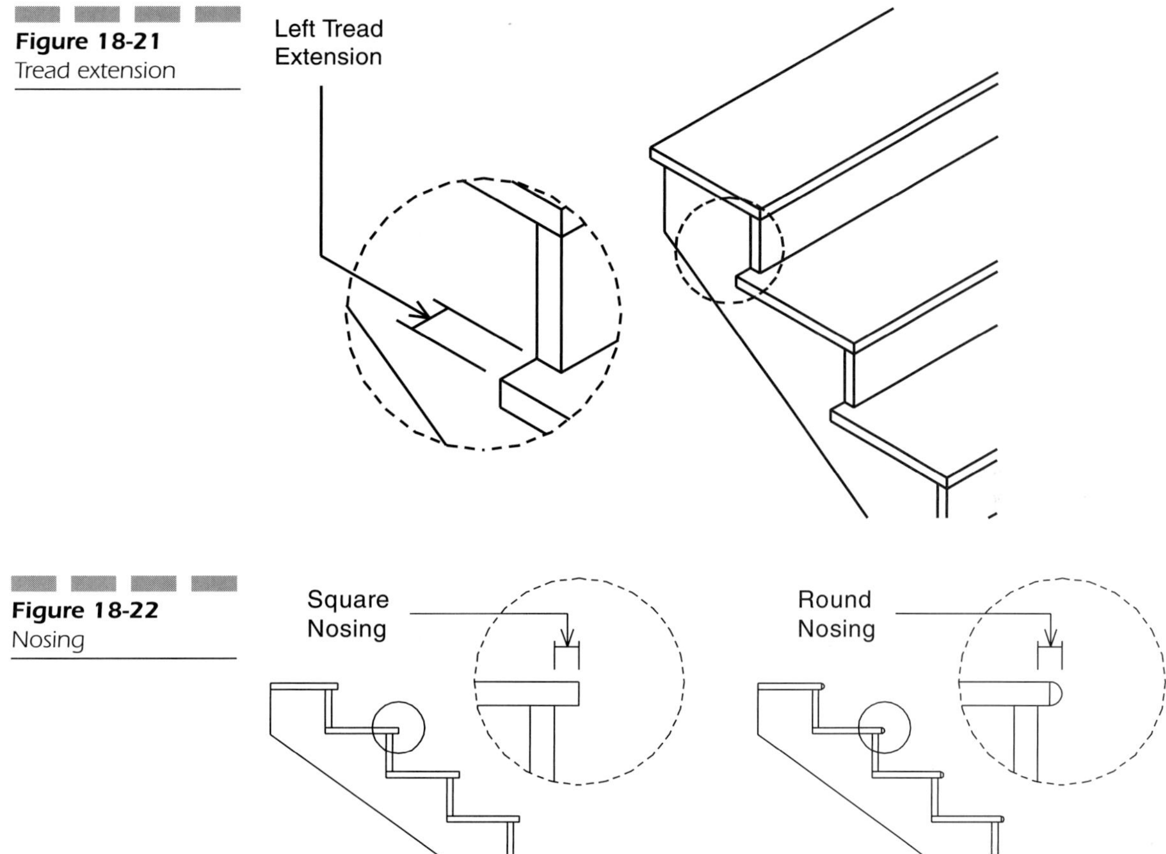

Figure 18-21
Tread extension

Figure 18-22
Nosing

To set the nosing type to square or round, click on the **RndNosng/ SqrNosng** toggle. This **Nosing** option changes between round and square just by clicking on it. If **RndNosng** is displayed in the **Nosing** menu, round nosing will be drawn on the stair treads; if **SqrNosng** is displayed in the **Nosing** menu, square nosing will be drawn on the stair treads (see Figure 18-22).

Use **Color** to set the color for the treads. All treads in a flight of stairs are the same color. See "Color Menus" in "The Drawing Board" chapter for more information about how to choose a color.

Stringers **Stringer** allows you to set options that create stringers and determine where and what kind you want to use. Toggle **DoLeft** on to model a stringer on the left side of the run of stairs, toggle **DoCtr** on to model a stringer in the center of a run of stairs, or toggle **DoRght** on to model a stringer on the right side of the stairs.

NOTE: *When using* ***DoCtr****, the stringer is always modeled as an open stringer.*

Use the **ClosdStr/OpenStr** toggle to set the stringer type to open or closed. When **ClosdStr** is displayed in the **Stringer** menu, stringers will be modeled as closed; when **OpenStr** is displayed in the **Stringer** menu, stringers will be modeled as open. Choose **ClosdStr** or **OpenStr** to toggle between the two options.

Width displays the current width of the stringer in the Message Window and lets you reset it. Measure the width from the top edge of the stringer to the bottom edge of the stringer (see Figure 18-23).

The stringer thickness is the horizontal distance from the inside of the stringer to the outside of the same stringer (see Figure 18-24). Choose **Thickness** from the **Stringer** menu, then choose or type a new thickness value and press **Enter**.

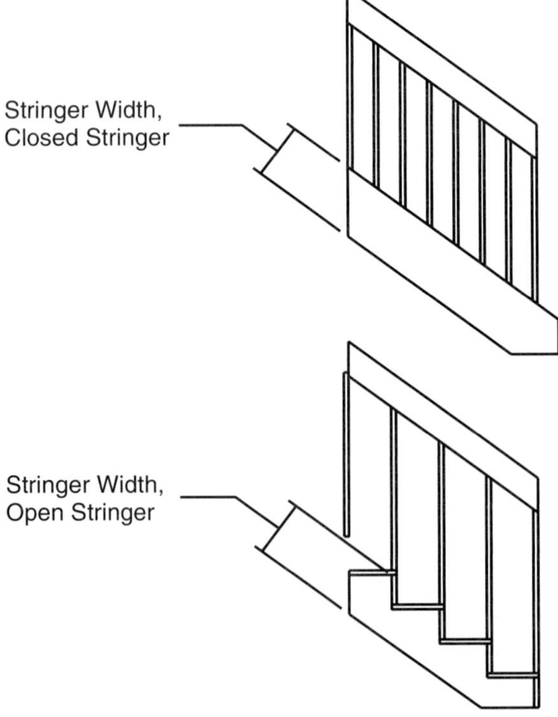

Figure 18-23 Stringer width (open or closed)

Figure 18-24
Stringer thickness

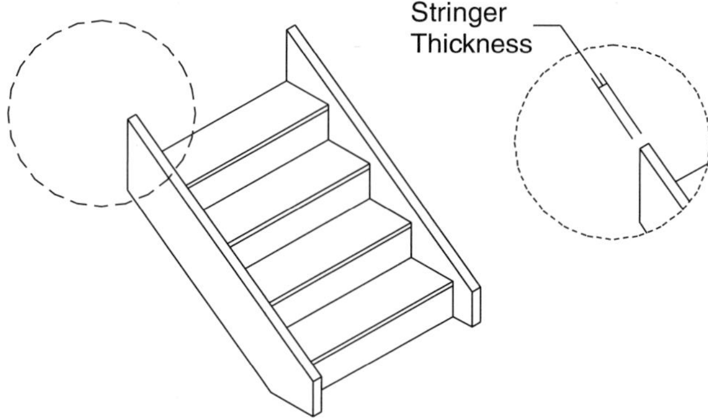

Use **Color** to set the color for the stringer. All stringers in a flight of stairs are modeled using the same color. See "Color Menus" in "The Drawing Board" chapter for more information about how to choose a color.

Handrails Toggle **DoLeft** on to model a handrail on the left side of the flight of stairs. Toggle **DoRght** on to model a handrail on the right side of a flight of stairs. When either of these options are toggled off, **3DStairs** does not model the handrails on that side of the stairs.

The **Rectangl/Cylinder** toggle lets you choose between modeling a cylindrical or rectangular handrail. When you select **Cylinder**, the **Radius** and **Division** options appear for you to enter settings. When you select **Rectangl**, the **Width** and **Depth** options appear for you to set.

The **Radius/Width** toggle displays the current radius or width of the handrail in the Message Window, depending on whether you selected cylindrical or rectangular handrails. When you model a cylindrical handrail, you can adjust the radius of the cylinder. When you model a rectangular handrail, you can adjust the width of the rectangle. Measure width from the top of the handrail to the bottom of the handrail (see Figure 18-25).

The **Depth/Division** toggle displays the current depth or number of divisions in the handrail in the Message Window, depending on whether you select cylindrical or rectangular handrails. When you model a cylindrical handrail, you can adjust the division of the cylinder, or the number of sides in the cylinder. When you model a rectangular handrail, you can adjust the depth of the rectangle. Measure depth from one side of the handrail to the other side. You can use up to 36 divisions, but the more divisions you have, the longer it takes DataCAD to redraw.

Use **Height** to see or reset the current height of the handrail. Measure handrail height from the top surface of the handrail to the top of the tread (see Figure 18-26).

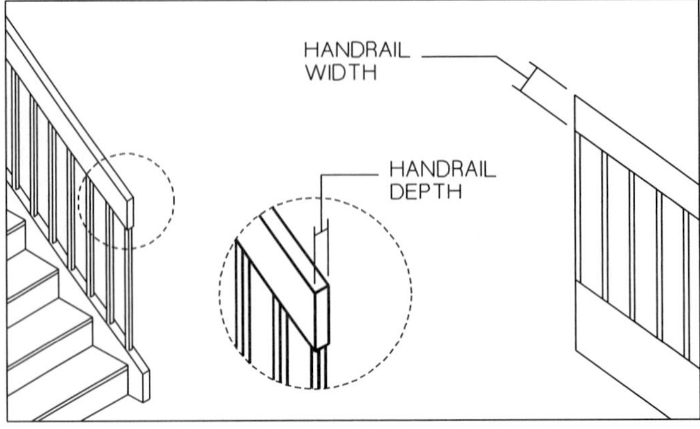

Figure 18-25
Handrail width and depth

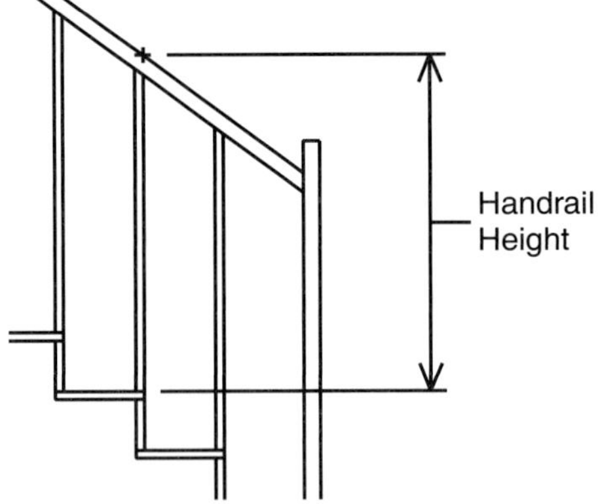

Figure 18-26
Handrail height

Use **Color** to set the color for the handrails. All handrails in a flight of stairs are modeled using the same color. See "Color Menus" in "The Drawing Board" chapter for more information about how to choose a color.

Balusters Use the **Baluster** menu in the **3DStairs/Settings** menu to choose what kind of balusters, if any, to use for your stairs. Toggle on **DoBalus** in the **Baluster** menu to model balusters on your flight of stairs.

The **Baluster** menu has the following three options for baluster placement (see Figure 18-27):

- **OnSide** models balusters on the sides of the stringers.
- **OnTread** models balusters on the treads.
- **OnStrng** models balusters on the center of the stringer. **OnStrng** is available only when you model a closed stringer.

Figure 18-27
Baluster placement

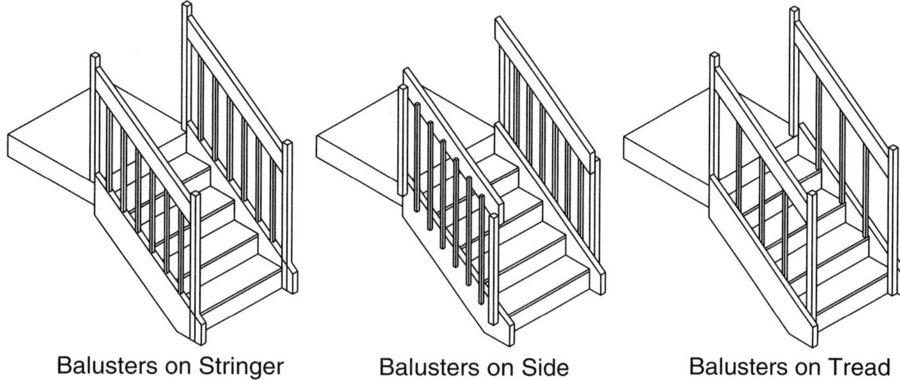

Depending on whether you choose **OnSide**, **OnTread**, or **OnStrngr** for baluster placement, either the **Bal/Tred** or **Bal/Run1** option is displayed in the **Baluster** menu. The **Bal/Tred** option displays the number of balusters per tread in the Message Window and lets you change the value. **Bal/Tred** is only available when you use **OnTread**. The **Bal/Run1** option displays the current number of balusters in the first run of stairs in the Message Window and lets you change that value. **Bal/Run1** is available only when you place the balusters on the side of the stringer (**OnSide**) or on the stringer (**OnStrng**).

The **Bal/Run2** option displays the current number of balusters in the second run of stairs in the Message Window and lets you change that value. **Bal/Run2** is available only when you place the balusters on the side of the stringer (**OnSide**) or on the stringer (**OnStrng**) on straight, double-back, open-well, and L-shaped stairs.

The **Bal/Run3** option displays the current number of balusters in the third run of stairs in the Message Window and lets you change that value. **Bal/Run3** is available only when you place the balusters on the side of the stringer (**OnSide**) or on the stringer (**OnStrng**) on open-well stairs.

Use **Cylinder/Rectangl** to choose between modeling a cylindrical or rectangular baluster. When you select **Cylinder**, the **Radius** and **Division** options appear. When you select **Rectangl**, the **Width** and **Depth** options appear.

The **Radius/Width** toggle displays the current radius or width of the baluster in the Message Window, depending on whether you selected cylindrical or rectangular balusters. When you model a cylindrical baluster, you can adjust the radius of the cylinder. When you model a rectangular baluster, you can adjust the width of the rectangle. Measure width from one side to the other of one baluster.

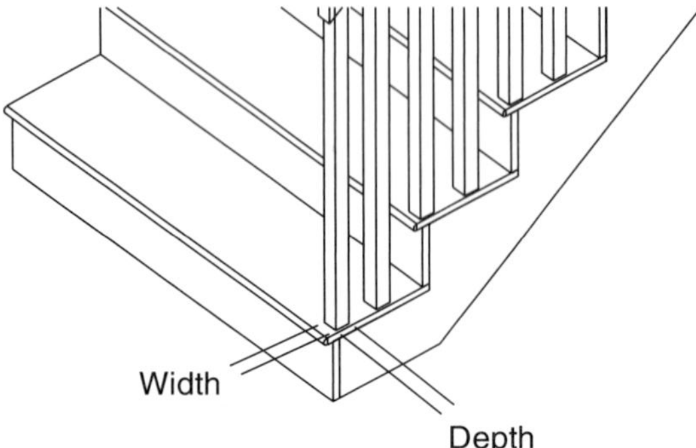

Figure 18-28 Baluster width and depth

The **Depth/Division** toggle displays the current depth or number of divisions in the baluster in the Message Window, depending on whether you selected cylindrical or rectangular balusters. When you model a cylindrical baluster, you can adjust the division of the cylinder, or the number of sides in the cylinder. When you model a rectangular baluster, you can adjust the depth of the rectangle. Measure depth from the front to the back of the baluster (see Figure 18-28).

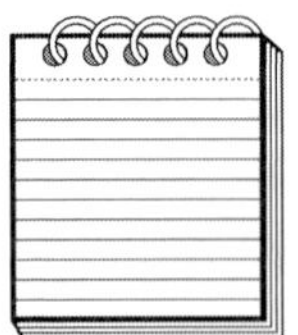

NOTE: *You can use up to 36 divisions, but the more divisions you have, the longer it takes DataCAD to redraw.*

Use **Symbol** to model balusters as symbols. When balusters are symbols, you can replace the balusters produced by **3DStairs** with different, more detailed baluster symbols that you create.

To use custom balusters in stairs, do the following:

1. Create your detailed baluster in the DataCAD drawing window and save it as a symbol. See "Saving and Deleting Symbols" in the "Templates and Symbols" section of the "Drawing Elements" chapter.
2. Choose **3DStairs** from the **Toolbox** dialog box.
3. Set the baluster options as necessary.
4. Toggle **Symbol** on to draw the balusters as symbols.
5. Create your run of stairs.
6. Replace the existing balusters with the symbol you created in Step 1 using the **Replace** option in the **Template** menu. See "Replacing a Symbol" in the "Templates and Symbols" section of the "Drawing Elements" chapter.

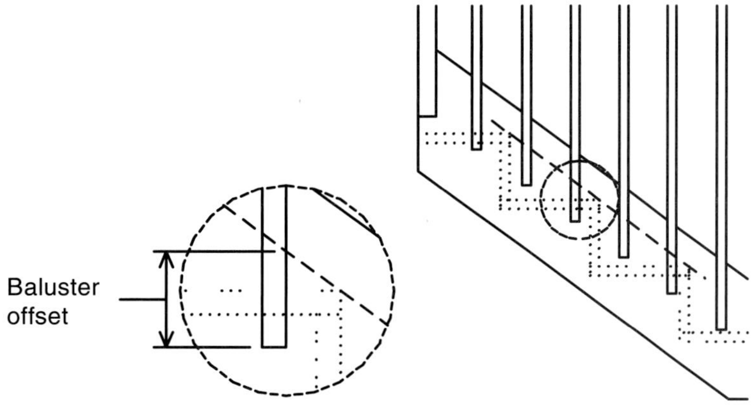

Figure 18-29
Baluster vertical offset

Baluster offset

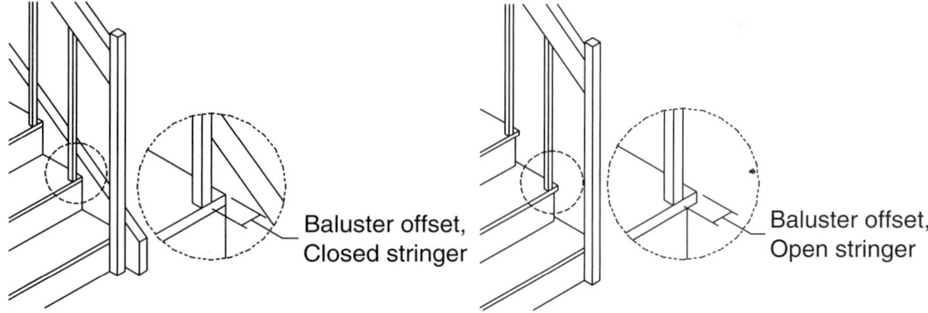

Figure 18-30
Baluster horizontal offset

VertOfst shows the current vertical offset of the baluster. Vertical is the distance that the bottom end of the balusters extends below the treads (see Figure 18-29). Vertical offset appears only when you select **OnSide**.

HorzOfst shows the current horizontal offset of the baluster. When you model balusters on the treads, measure the offset from the side edge of the tread to the middle of the baluster (see Figure 18-30). **HorzOfst** only appears when you select **OnTread**.

Use **Color** to set the color for the balusters. All balusters in a flight of stairs are modeled using the same color. See "Color Menus" in "The Drawing Board" chapter for more information about how to choose a color.

Newel Posts Use **NewelPst** to model newel posts on the stairs. Toggle **DoNlPst** on to model newel posts on a flight of stairs. When **DoNlPst** is toggled off, **3DStairs** models balusters in place of newel posts (see Figure 18-31).

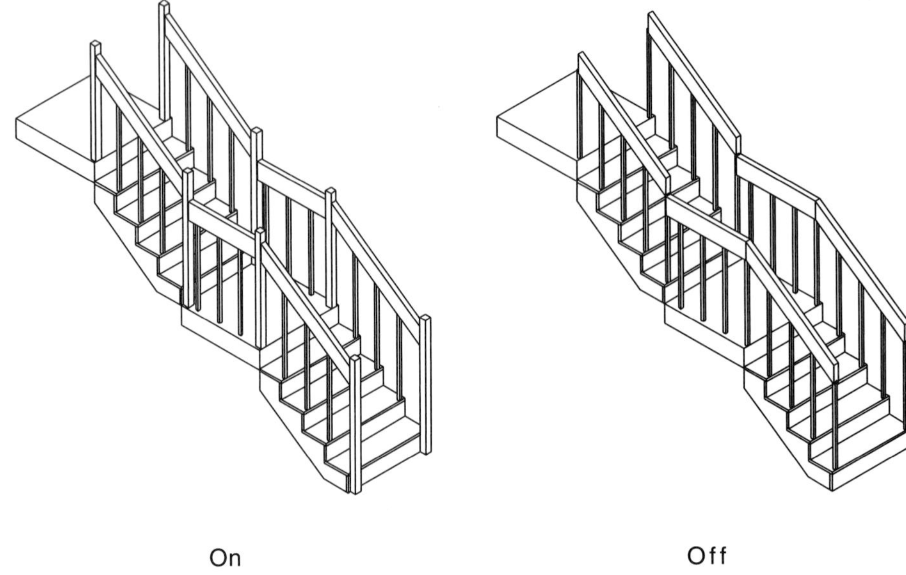

Figure 18-31
Newel posts on/off

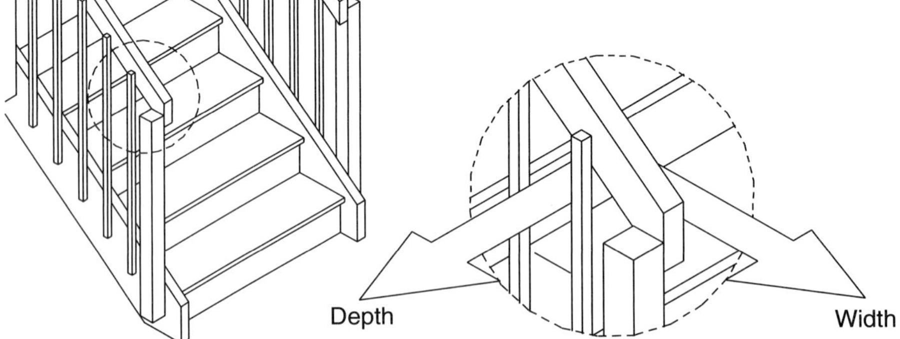

Figure 18-32
Newel post width and depth

Use **Cylinder/Rectangl** to choose between modeling a cylindrical or rectangular newel post. When you select **Cylinder**, the options **Radius** and **Division** are displayed. When you select **Rectangl**, the **Width** and **Depth** options are displayed.

The **Radius/Width** toggle displays the current radius or width of the newel post, depending on whether you selected cylindrical or rectangular newel posts, in the Message Window. When you model a cylindrical newel post, you can adjust the radius of the cylinder. When you model a rectangular newel post, you can adjust the width of the rectangle. Measure width from one edge, near a baluster, to the other edge, facing out, of the same newel post (see Figure 18-32).

The **Depth/Division** toggle displays the current depth or number of divisions in the newel post, depending on which of the previous options you selected, in the Message Window. When you model a cylindrical newel post, you can adjust the division of the cylinder, or the number of sides in the cylinder. When you model a rectangular newel post, you can adjust the depth of the rectangle. Measure depth from the edge of the newel post near the handrail to the open edge.

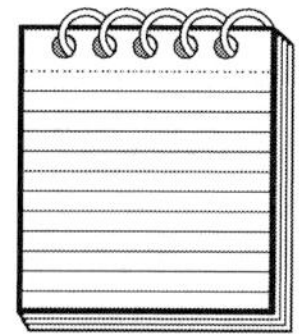

NOTE: *You can use up to 36 divisions, but the more divisions you have, the longer it takes DataCAD to redraw.*

Use **Symbol** to model newel posts as symbols. When newel posts are symbols, you can replace the newel posts produced by **3DStairs** with custom newel post symbols that you create.

To use custom newel posts in stairs, do the following:

1. Create your detailed newel post in the DataCAD drawing window and save it as a symbol. See "Saving and Deleting Symbols" in the "Templates and Symbols" section of the "Drawing Elements" chapter.
2. Choose **3DStairs** from the **Toolbox** dialog box.
3. Set the newel post options as necessary.
4. Toggle **Symbol** on to draw the newel posts as symbols.
5. Create your run of stairs.
6. Replace the existing newel posts with the symbol you created in Step 1 using the **Replace** option in the **Template** menu. See "Replacing a Symbol" in the "Templates and Symbols" section of the "Drawing Elements" chapter.

Use **Color** to set the color for the newel posts. All newel posts in a flight of stairs are modeled using the same color. See "Color Menus" in "The Drawing Board" chapter for more information about how to choose a color.

Stair Form

To display values and settings for all stair types, do the following:

1. Choose **StrForm** from the **3DStairs** menu.
2. Press **Tab** to move forward or **Shift + Tab** to move backward through each line of the stair form.

3. Scroll through the two-page form by selecting **ScrlFwrd** or **ScrlBack** from the menu.
4. Press **Esc** to save settings and exit the **StrForm** menu, or choose **Exit** to leave the **StrForm** menu without saving any changes. For settings with options, you can scroll through them using **Spacebar**.

Stair File

Use the **StrFile** option in the **3DStairs** menu to save stair settings for future use, as well as load, save, delete, and rename stair setting files.

LoadStr	Loads a stair settings file; choose **LoadStr** from the **StrFile** menu and type the name of the file to load
SaveStr	Saves your current stair settings to a file; choose **SaveStr** from the **StrFile** menu and type the name of the file to save
DelStr	Deletes a stair settings file; choose **DelStr** from the **StrFile** menu and type the name of the file to delete
Rename	Renames an existing stair file; choose **Rename** from the **StrFile** menu, choose the file to rename, and type a new name
StrForm	Shortcut to the **Stair Form** menu; see "Stair Form" for more information

The Stair Calculator

The **Stair Calculator** (**Calculat**) is available only when you create single-run stairs. Use it to calculate one of six possible values: total rise, number of risers, riser height, total run, number of treads, or tread depth (see Figures 18-33 and 18-34).

Use the **Stair Calculator** as you would use a desktop calculator; prompts are displayed in the Message Window at the bottom of the screen. Save the answers to automatically update the **3DStairs/Settings** menu for single-run stairs.

1. Choose **Calculat** from the **3DStairs** menu.
2. Select an option from the following **Calculat** menu to calculate and modify:

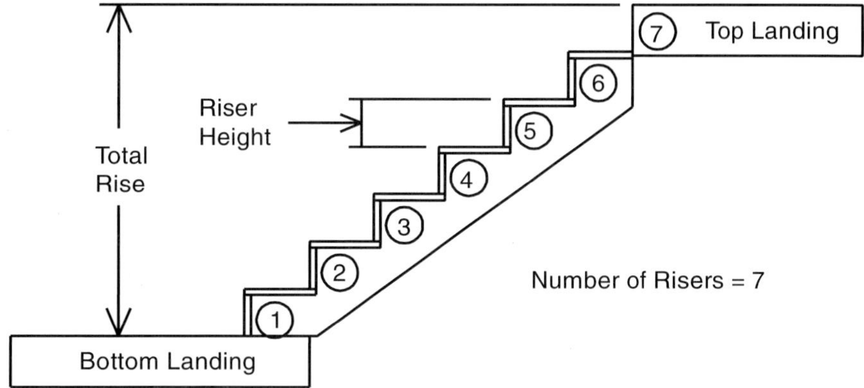

Figure 18-33 Total rise, riser height, and number of risers

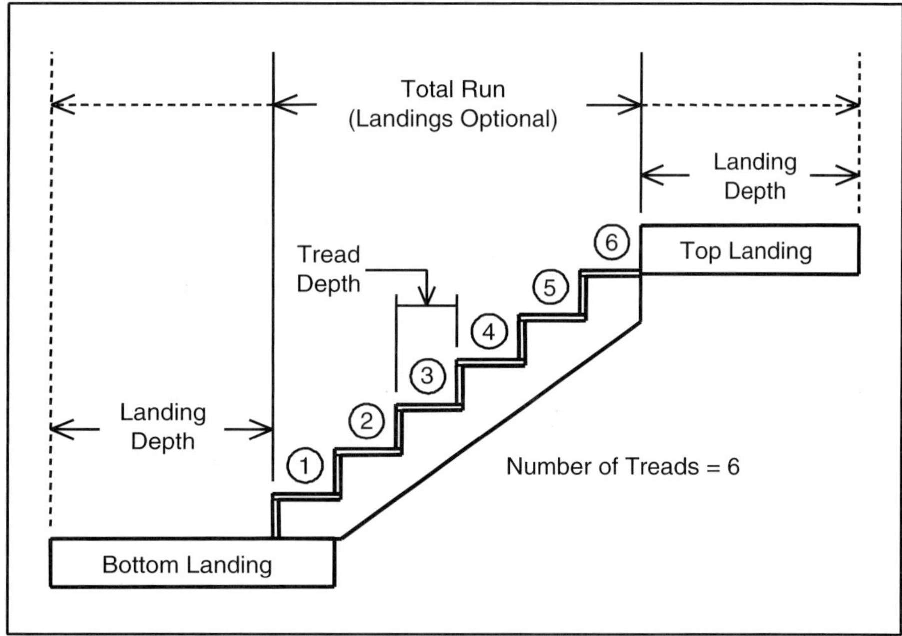

Figure 18-34 Total run, tread depth, and number of treads

TotlRise The number of risers times the riser height; same value regardless of whether **LandTop** is on or off

#Risers The total rise divided by the riser height, including the rise of the top landing; changing the number of risers automatically changes the number of treads and vice versa

RiserHgt The vertical distance from the top of one tread to the top of the next tread; the total rise divided by the number of risers

TotalRun	The horizontal distance from the top of the run of stairs to the bottom of the stairs; the number of treads times the tread depth (the distance can include landing depth for the top and bottom landings when those toggles are turned on)
#Treads	The number of treads in the stair run; changing the number of treads automatically changes the number of risers and vice versa (you can set **TreadDth** from this menu to recalculate the number of treads)
TreadDth	The horizontal distance from the face of one riser to the face of the next riser; the total run divided by the number of treads (you can change # Treads in this menu)
LandBtm	When **LandBtm** is on, the total run value includes the bottom landing depth
LandTop	When **LandTop** is on, the total run value includes the top landing depth
Reset	Resets your calculator settings to the default values
Update	Overwrite the old settings in the **Setting** menu with new settings; saves your calculated values
Begin	Use **Begin** to start modeling stairs. Follow the Steps in the "Using Stairs" section for information on using **Begin**.
Template	Shortcut to the **Template** menu; select a template from the menu, or type a template name, and press **Enter**
Hide	Perform hidden line removals; see "Hide" in the "3D Editing" chapter for more information
3DViews	Shortcut to the **3DViews** menu; see the "3D Viewing" chapter for more information

3. With some **Calculat** menu options, an additional submenu is displayed. If this occurs, choose an option to modify; if not, skip to Step 5.
4. Choose a new value from the menu or type a different value. As you change the values of a setting, the results are automatically calculated and displayed.
5. To modify other options, repeat Steps 3 and 4.
6. Choose **Update** to update the settings in the **3DStairs/Settings** menu.

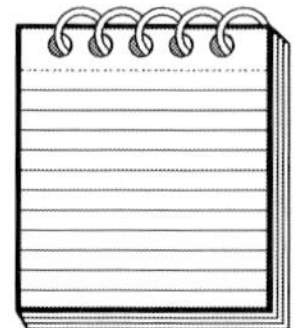

NOTE: *The stair calculator only accepts a whole number of treads or risers as a result. If the values you enter in a calculation result in a fractional solution, the stair calculator modifies the values.*

For example, to calculate the number of risers through changing **RiserHgt**, use the following table:

	Initial Settings	Modified Settings	Stair Calculator Results
TotlRise	120	120	123
RiserHgt	6″	7″	7″
#Risers	24	20.57	21

Because **#Risers** is a fractional number, the **Stair Calculator** automatically adjusts the **TotlRise** so that the results are whole numbers. If you change **TotlRise**, **Stair Calculator** adjusts RiserHgt to find a whole number solution.

3D Tools

3D Power Tools contains eight functions that create and manipulate custom architectural elements. You can shape 3D slabs and polygons, convert polyline shapes to polygons, edit planes in **Ortho** view while keeping the pitches and angles, cut holes in roof planes, and easily create handrails, ramps, moldings, fluted columns, and more. When you choose **3DTools** from the **Macros** menu, the following options appear:

Sweep

Knife

3DPlane

Pln2Pgon

Triangle

SnapShot

MultiPOF

Kaboom!

Sweep

Use **Sweep** to create complex geometric forms. You can create a section in **Ortho** and apply it to any extrusion path defined by 2D lines, arcs, polylines, 3D lines, or contours (see Figure 18-35). You can apply a taper to the section independently in X and Y over the length of the sweep, and then apply a twist of any number of revolutions to create complex geometric forms.

Use **Sweep** to create extruded shapes such as moldings, gutters, handrails and ramps. Use **Twist** and **Taper** to create twisted balusters, fluted columns, and other forms.

1. Create a horizontal 3D polygon to be used as a cross-section.
2. Choose **Sweep** from the **3DTools** menu.
3. Select the 3D polygon entity that defines the sweep cross-section. The 3D polygon should be drawn in **Ortho** (flat on the X, Y plane).
4. Select a point on the polygon as the origin of the sweep path. The following options appear:

   NewSect

   Reverse

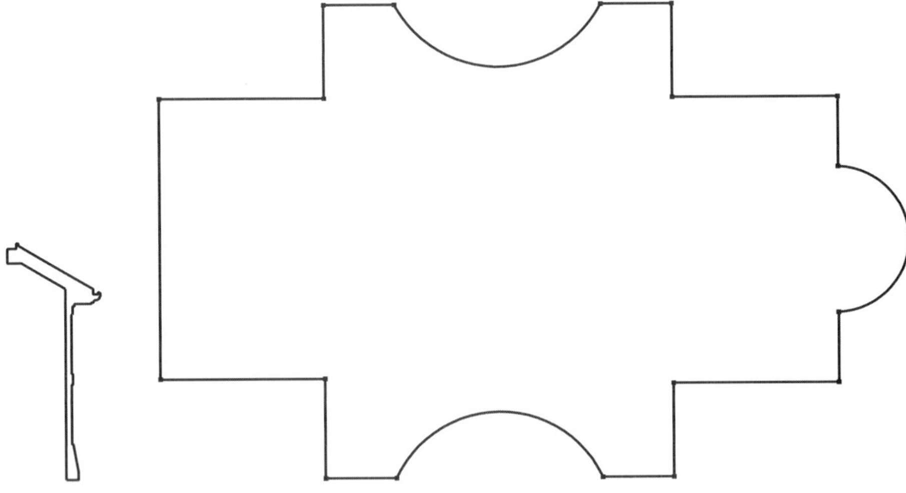

Figure 18-35 When performing a **Sweep** operation, a section shape polygon (left) is applied to a polyline path (right).

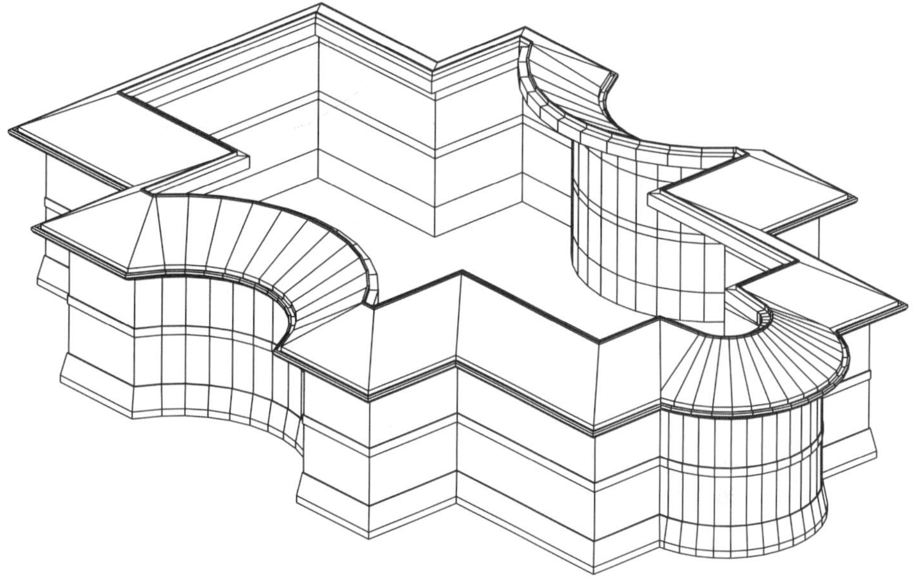

Figure 18-36
The resulting model generated by the **Sweep** function

CapEnds

Taper

Twist

Settings

5. Using the **Selection** menu, select the entity or entities to use for the sweep path (see Figure 18-36).

New Section (NewSect) Use **NewSect** to select a new polygon to extrude.

Reverse Use **Reverse** to create the extruded entity mirrored on the opposite side of the sweep path.

Cap Entity Ends (CapEnds) Use **CapEnds** to cap the ends of the extruded entity, making the sweep appear solid; otherwise, the ends are open.

Taper Use **Taper** to set the X and Y range for tapering the extruded entity along the sweep path. A taper of 1.0 is equal to no taper. Choose or type the X and Y values and press **Enter**.

Twist Use **Twist** to set the number of rotations to twist a section during the sweep. The extruded entity rotates the selected number of times around the point selected as the origin of the sweep path. Choose or type the number of twists and press **Enter**.

Sweep Settings (Settings) **Settings** accesses options that let you further define the sweep.

CirDivs	Sets the number of divisions (segments) for circular entities (circles and arcs)
TwstDvs	Sets the number of divisions for each full twist
Vrtical	Places the cross-section vertical to the X-Y plane when toggled on; when toggled off, the cross-section is perpendicular to the sweep path
ShowLat	Shows or hides the latitudinal polygon lines generated by the sweep
ShowLon	Shows or hides the longitudinal polygon lines generated by the sweep
Z-Base	Sweeps the extruded entity along the Z-base (bottom) of the sweep path
Z-Hite	Sweeps the extruded entity along the Z-height (top) of the sweep path
Base/Ht	Starts the extruded entity sweep at the Z-base of the sweep path and ends it at the Z-height of the sweep path.

NOTE: *If **Base/Ht** is toggled on and you are selecting a group of 2D lines or polylines that close on themselves, the entity sweeps only at Z-base, not from Z-base to Z-height.*

Knife

Use **Knife** to cut away 3D slabs and polygons. Use this one-line trimming interface to slice through slabs at custom angles to create mitered shapes.

1. Choose **Knife** from the **3DTools** menu.
2. Select the first point to define the starting point of the knife line.

3. Use the rubberband line to select the second point on the knife line. The knife line is an infinite cutting line.
4. Select on the side of the knife line to keep. When the entity or entities are cut, the geometry on the keep side stays, while the geometry on the opposite side is deleted.
5. Use the **Selection** menu to select the section to cut.
6. To change the cutting line, select **NewLine** and repeat Steps 2 through 5.

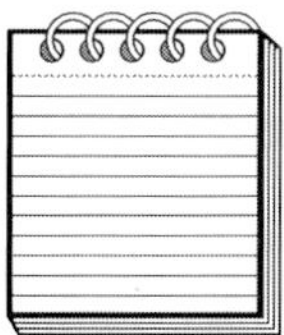

NOTE: ***Knife*** *will not cut polygons or slabs with voids.*

3D Plane (3DPlane)

Use **3DPlane** to edit 3D entities in **Ortho** view while maintaining specific pitches and angles. This is useful when editing angular or inclined polygons or slabs because it maintains the pitch and angle. Use the **SmrtVoid** option to cut holes of any shape through the roof planes for stove pipes, skylights, chimneys, and so on.

1. Choose **3DPlane** from the **3DTools** menu. The following options appear:

   Polygons

   Slabs

   PitchSlb
2. Choose **Polygons**, **Slabs**, or **PitchSlb**.
3. Choose the editing options you want to edit the 3D entities.

Polygons To define the polygon you want to edit in a plane, use the following options. See the *DataCAD 8 Reference Manual* for more information on the standard DataCAD **Polygon** menu options: **Horizntl**, **Vertical**, **Rectngle**, **Inclined**, and **Voids**.

Horizntl

Vertical

Rectngle

Inclined

Voids

SmartPrt

SmrtVoid

Smart Partial (SmartPrt) Smart partial editing aligns the vertices to the plane of the polygon. As you move or add vertices to a polygon, the reference face of the polygon is maintained. You can add, move, or delete vertices.

In addition, when you use **SmartPrt**, you can identify vertices and place snapping points using the following functions:

IdentVrt	Identifies the **Z** value of the vertex end nearest the selected point
SnapPt	Places a snapping point on the same plane as the polygon

NOTE: *A polygon may have no more than 36 vertices. When adding vertices, make sure your current editing plane is parallel to the polygon you are editing.*

Smart Void (SmrtVoid) **SmrtVoid** automatically snaps the void you are creating to the correct reference face on the master polygon.

Slabs Use the following options to define the slab you want to edit in a plane. See the *DataCAD 8 Reference Manual* for more information on the standard DataCAD **Slab** menu options: **Horiztl**, **Vertical**, **Rectngle**, **Inclined**, and **Voids**.

Horizntl

Vertical

Rectngle

Inclined

Voids

SmartPrt

SmrtVoid

Smart Partial (SmartPrt) Smart partial editing aligns the vertices to the plane of the slab. As you move or add vertices to a slab, the reference face of the slab is maintained. You can add, move, or delete vertices.

In addition, when you use **SmartPrt**, you can identify vertices and place snapping points using the following functions:

IdentVrt Identifies the **Z** value of the vertex end nearest the selected point

SnapPt Places a snapping point on the same plane as the slab

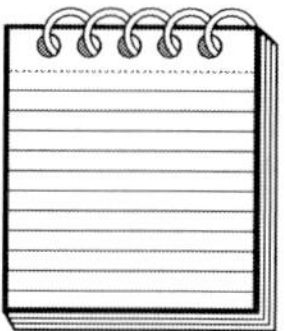

NOTE: *A slab may have no more than 36 vertices. When adding vertices, make sure your current editing plane is parallel to the slab you are editing.*

Smart Void (SmrtVoid) **SmrtVoid** automatically snaps the void you are creating to the correct reference face on the master slab.

Pitched Slabs (PitchSlb) Use these options to define the pitch and thickness of the slab you want to create or edit in the plane (see Figures 18-37 and 18-38). When you choose **PitchSlb** from the **3DPlane** menu, the following options appear:

Pitch

Thickness

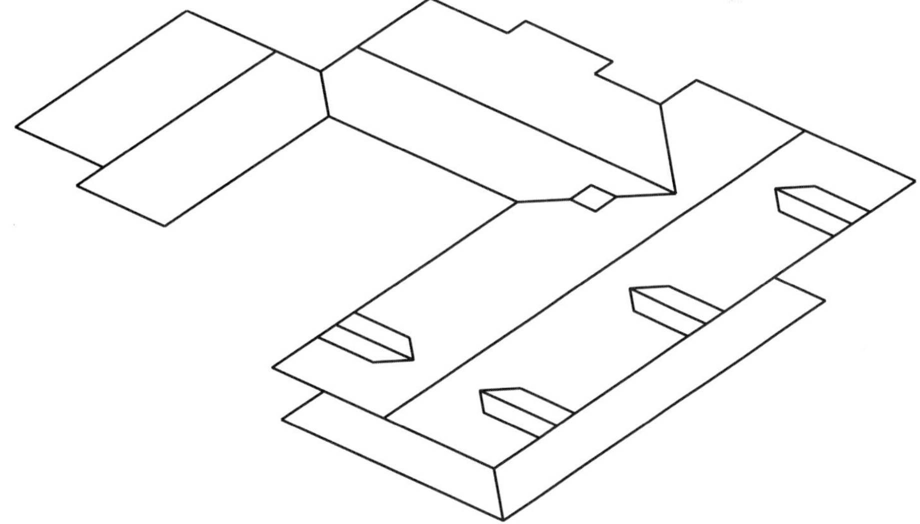

Figure 18-37
To create a roof model, begin with a 2D roof plan drawing.

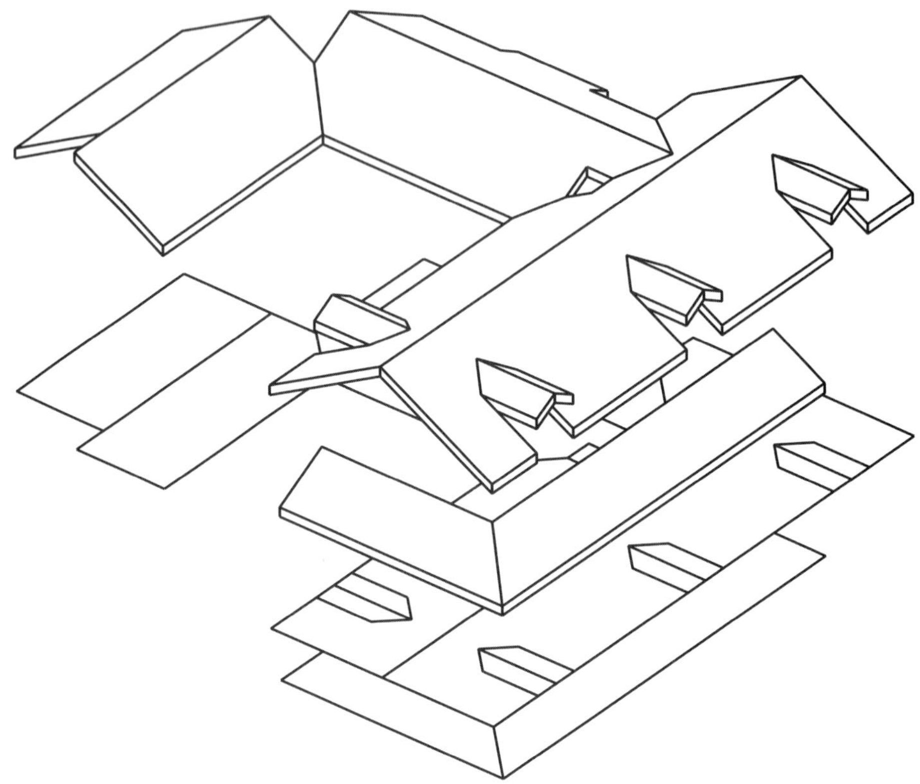

Figure 18-38 Using the **PitchSlb** function, trace over the 2D roof plan at the desired Z-height.

1. Choose **Pitch** to set the slab pitch.
2. Choose a pitch setting from the list, or choose **Custom** to create a user-defined slab pitch.
3. Choose **Thickness** to set the slab thickness.
4. Choose or type in the thickness of the slab and press **Enter**.
5. Set the Z-base value equal to the plate height of the wall.
6. Define the inside plate of the wall by snapping to the inside of the wall.
7. Select a point in the direction that indicates the top of the pitch.
8. Define the roof slab by selecting the first point where the base of the roof with overhang starts.
9. Define the width of the roof by selecting a point parallel to the wall where the roof ends.
10. Define any other points along the top of the pitch. The roof slab is created (see Figure 18-39).

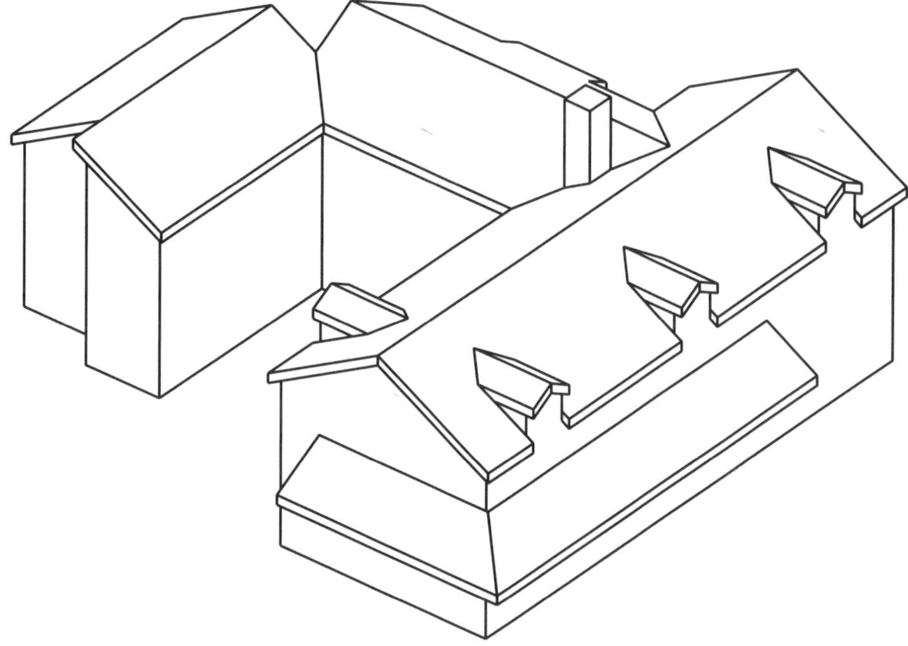

Figure 18-39
The completed model

NOTE: *A roof can be created with three or more edges, with a maximum of 36 vertices.*

Polyline to Polygon (Pln2Pgon)

Use **Pln2Pgon** to convert polyline shapes to polygons. **Pln2Pgon** supports chained polylines with an infinite number of sides by automatically triangulating polygons that exceed 36 sides.

1. Choose **Pln2Pgon** from the **3DTools** menu. The following options appear:

   Top

   Sides

   Bottom

   ShowEdg

2. Since polylines can have Z-base and Z-height values, they can appear to have a thickness; therefore you can toggle **Top** and **Bottom** on, and **Pln2Pgon** creates these parts.

3. Toggle **Sides** on to have **Pln2Pgon** convert these parts to polygons.
4. Toggle **ShowEdg** on or off to show/hide abutting polygon edges.
5. Use the selection menu to select the polyline entity or entities to convert.

NOTE: *When the polylines are converted they are temporarily stored on a layer called* ***_TRIHID*** *in the* ***Layer*** *menu. This layer is available for the* ***Undo*** *function; do not delete it.*

Triangle

Use **Triangle** to make your 3D entities compatible for **.DXF** transfer and to make your DataCAD models consistent with third-party rendering programs. Once the triangulation process is complete, you can write **.DXF** files (see Figures 18-40 and 18-41).

1. Choose **Triangle** from the **3DTools** menu. The following options appear:

   Begin

   ShowEdg

   File I/O
2. Toggle **ShowEdg** on to show abutting triangle edges; otherwise, they remain hidden.
3. Choose **Begin** to start triangulating polygons and slabs with more than four sides or those that contain voids. You can use **Undo** to undo any triangulation of entities during the current session.
4. Choose **File I/O** to access DataCAD's **File I/O** menu to output the file to **.DXF**.

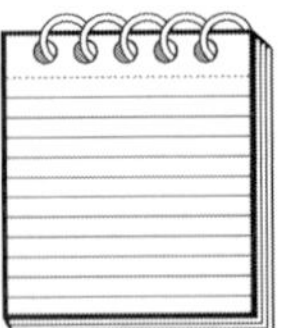

NOTE: *When the entities are triangulated, they are temporarily stored on a layer called* ***_TRIHID*** *in the* ***Layer*** *menu. This layer is available for the* ***Undo*** *function; do not delete it.*

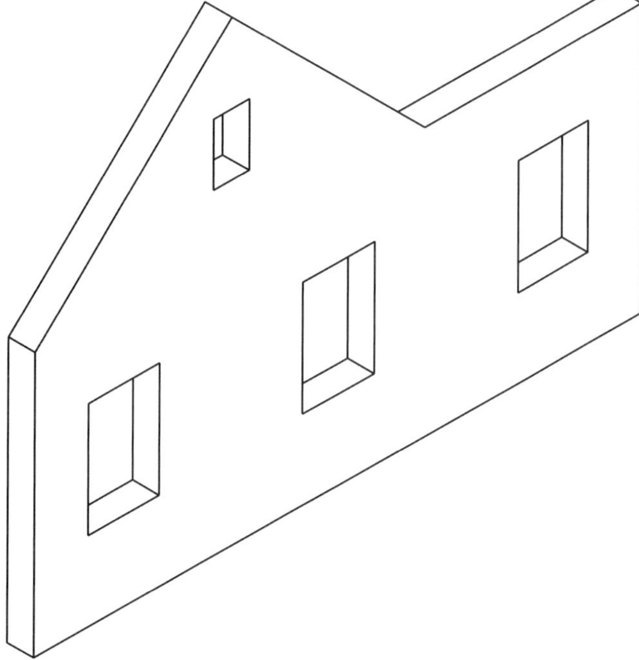

Figure 18-40 Slab before triangulation

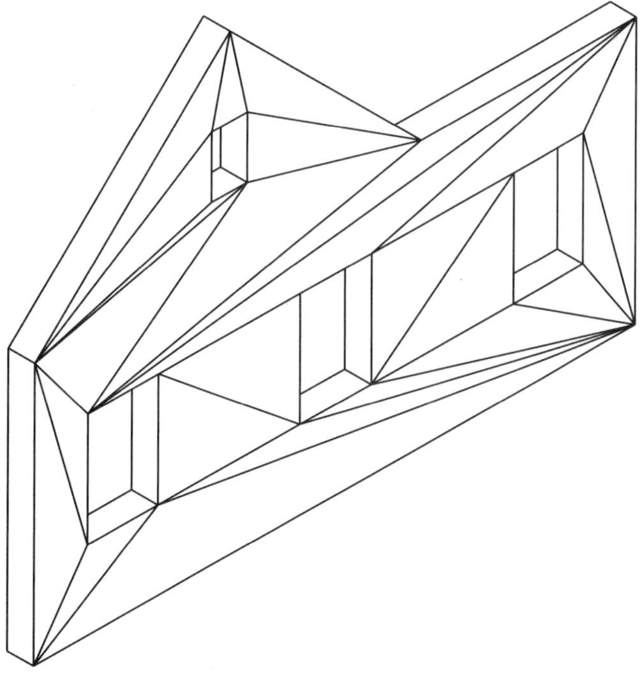

Figure 18-41 Slab after triangulation, **ShowEdg** option on

Snap Shot (SnapShot)

Use **SnapShot** to generate elevation, isometric, and perspective views from 3D models. **SnapShot** displays each selected view, performs hidden line removals, and saves the results on new layers.

1. Choose **SnapShot** from the **3DTools** menu. The following options appear:

ELFRONT	Front elevation
ELREAR	Back elevation
ELLEFT	Left elevation
ELRIGHT	Right elevation
ISO	Isometric
PRSPECT	Perspective
Begin	Generate snap shots

2. Choose a view option: **ELFRONT**, **ELREAR**, **ELLEFT**, **ELRIGHT**, **ISO**, or **PRSPECT**.
3. Choose **Begin**. Choose **Yes** or **No** to respond to the prompt "Ready to begin generating snapshots?" **SnapShot** displays the created view layers and turns off all other layers (see Figure 18-42).

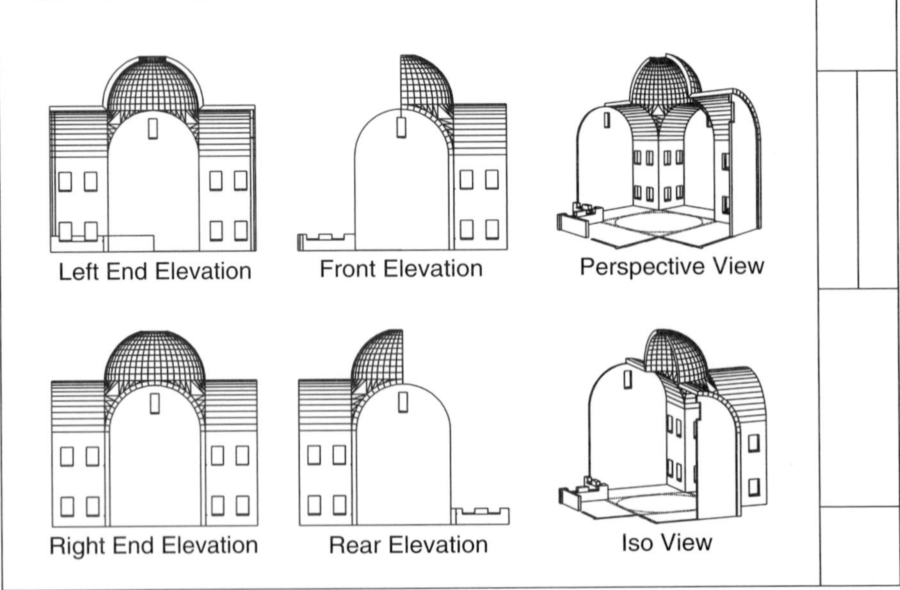

Figure 18-42 Use **SnapShot** to create a layout of views from a 3D model.

Kaboom!

Use **Kaboom!** to separate entities in a drawing for detailing, create organic shapes such as trees, or extract individual forms from a complex model (see Figures 18-43 and 18-44).

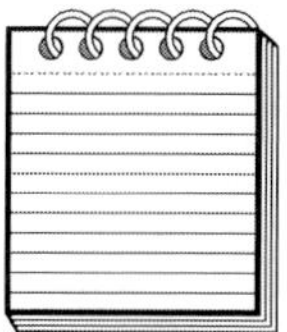

NOTE: *Once you've exploded a drawing, it cannot be undone. Make a backup of your file before using* ***Kaboom!***

1. Choose **Kaboom!** from the **3DTools** menu. The following options appear:

   Severity

   NumSteps

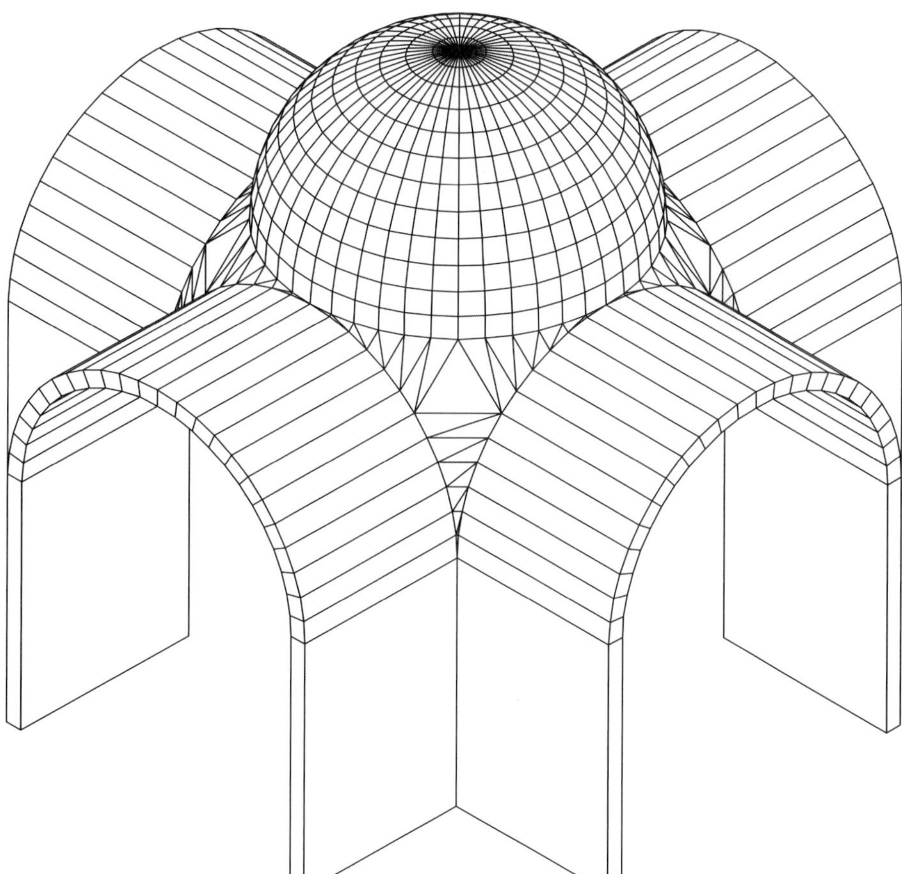

Figure 18-43
Model before **Kaboom!** operation

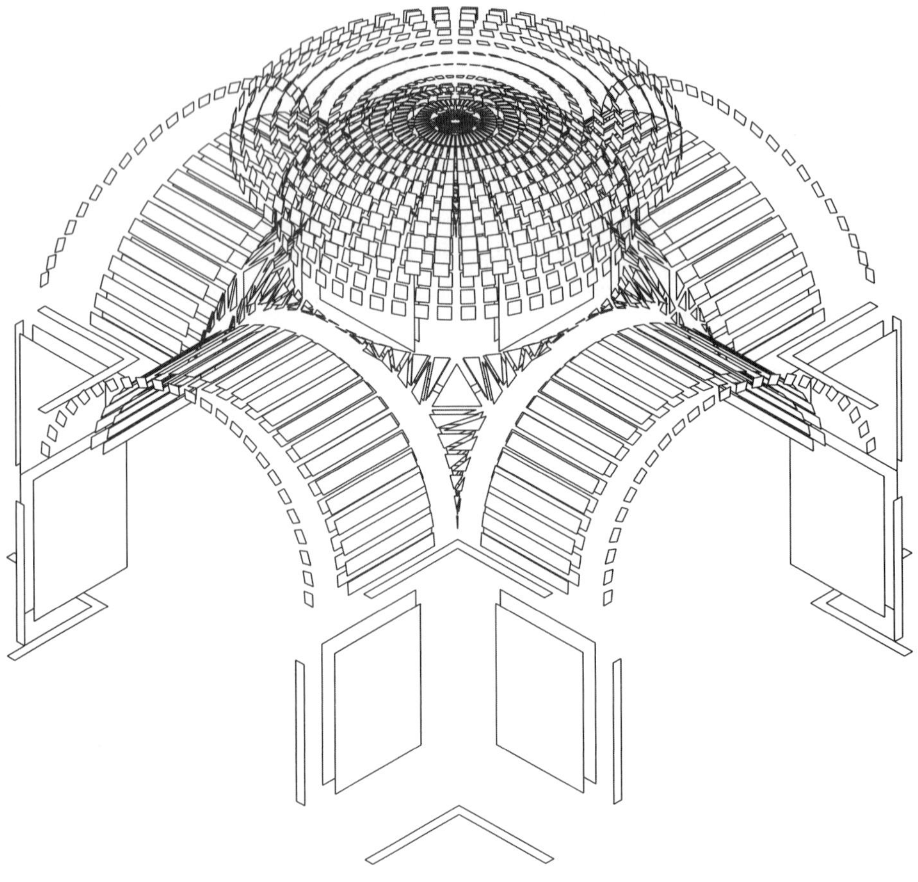

Figure 18-44 Model after **Kaboom!** operation with 20% severity, bomb placed at model's center

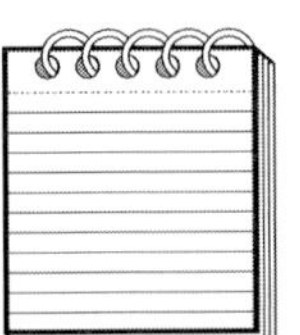

NOTE: *These options determine how far the entities are separated from the explosion point.*

2. Choose **Severity** to set the severity of the explosion from 1 (least severe) to 100 (most severe). Choose or type in the severity of the explosion and press **Enter**.
3. Choose **NumSteps** to set the number of steps for the explosion. Choose or type in the number of steps and press **Enter**.

4. Select the explosion point. When you move the cursor over the drawing, the cursor changes into the image of a bomb.
5. Choose **Yes** or **No** to respond to the prompt, "Are you sure you want to explode this drawing?"

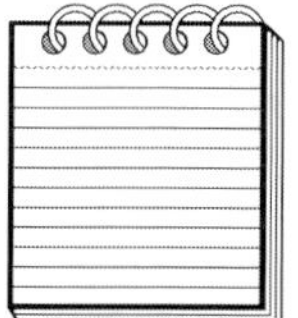

NOTE: *All entities on all visible layers move radially out from the point where the bomb was dropped.*

Touchup

Touchup allows you to produce 3D hatching on DataCAD entities to visually enhance your 3D models for production of presentation-quality images. Unlike rendering programs, **Touchup** uses vector entities (3D lines) that can be scaled up or down without loss of clarity and can be outputted directly to a printer or plotter.

Unlike DataCAD's 2D hatching function, which requires the definition of a hatch boundary, selecting a surface to be hatched in **Touchup** is as simple as clicking on it with the mouse. Hatchable entities include 2D lines (if Z-base and Z-height are not equal), polygons (with or without voids), slabs (with or without voids), and blocks.

These 3D hatch patterns are truly on the entity face in the form of the DataCAD 3D line entity; the model may be viewed from any point and in any viewing projection, and the pattern will be correct. A subsequent hidden line removal will result in the displaying of the pattern on the model face, giving it a much more realistic and textured appearance than the typical plain, empty faces of ordinary DataCAD models.

NOTE: *As with 2D hatching in DataCAD, the more complex the pattern, the quicker the drawing size grows. Therefore, the amount of hatching to be applied to a model should be weighed against the negative impact of increased drawing size and longer hidden line removal times.*

To produce 3D hatching with **Touchup**, first select a pattern (along with the desired scale and rotation of the pattern), then select the face of an entity to be hatched; the pattern is generated.

Selecting Entities for 3D Hatching

1. Choose **Touchup** from the **Toolbox** menu. You are prompted as follows: "Choose menu option, or select the entity to Touchup."
2. Click on an entity to select it. If there is more than one entity near the point you chose (as is often the case in a 3D view when looking through many entities, or when the entity selected shares edges with other nearby entities), **Touchup** will highlight the first entity found. You are prompted as follows: "Is this the right face? (Y or N)."
3. Verify that the flashing entity is the one to hatch by selecting **Yes** from the menu. Selecting **No** causes **Touchup** to highlight the next nearest entity and again ask for verification. Once the correct face is verified, pattern generation will begin. The selected hatch pattern is generated in the current color, on the current layer, and with the current linetype, spacing, and so on; the **Touchup** of that surface is complete (see Figures 18-45 and 18-46).

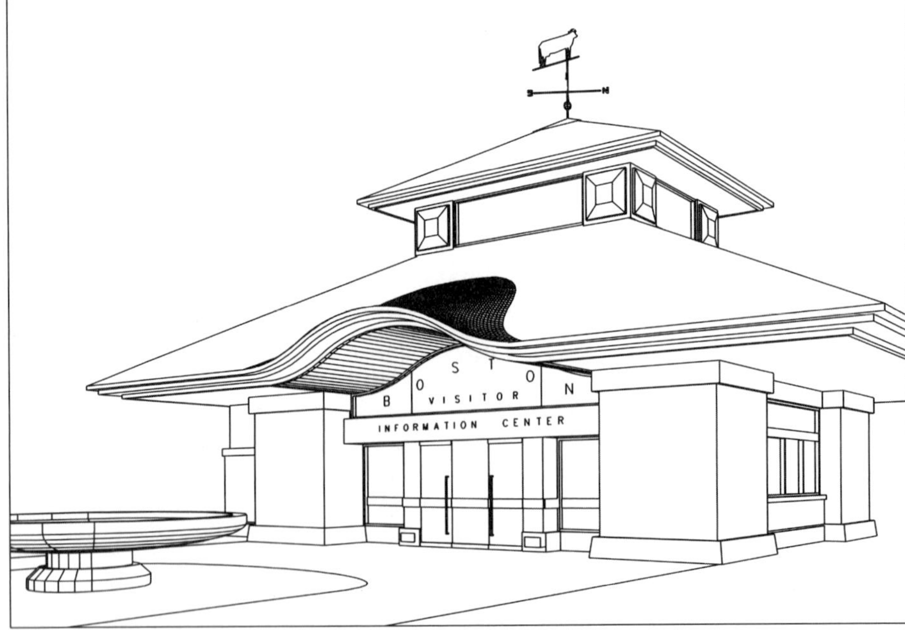

Figure 18-45 3D model before applying hatch patterns with **Touchup**

Figure 18-46 3D model after applying hatch patterns with **Touchup**

Touchup's Menu Options

The **Touchup** main menu contains the following options:

Pattern

Scale

Angle

Layer Search

SideBrds

DoVoids

SBWidth

Sides

NmbrSides

3DViews

Hide

Polygons

Slabs

Exit

Touchup Hatch Pattern Select a 3D hatching pattern from this menu. **Touchup** can read your DataCAD 2D hatch pattern file (**DCAD.PAT**) and will therefore be able to produce in 3D any pattern that is available in 2D. This includes any pattern the user has created, if it is in that file.

Touchup comes preloaded with DataCAD's standard 2D hatch patterns in its memory. If your hatch pattern file includes other patterns you would like to use, you must load those patterns into the **Touchup** memory.

1. Choose **Pattern** from the **Touchup** menu to view the currently loaded patterns.
2. Choose **Rescan** at the bottom of the **Pattern** menu. You are asked if you want to re-read your hatch pattern file.
3. Select **Yes** to load the patterns in your 2D hatch pattern file into the **Touchup** memory. This process may take some time.

NOTE: *Should you make any subsequent changes to DataCAD's hatch pattern file, you will have to choose **Pattern, Rescan** in **Touchup** to load those changes.*

Touchup Hatch Scale Scale sets the enlargement scale for the pattern, similar to DataCAD's hatch scale factor.

The most commonly used pattern is **Line**, and a scale factor of 1.0 translates into lines spaced $^1/_{32}$″ apart. Therefore, to get a 6″ **Line** pattern on a wall to indicate 6″ clapboards, you would set the **Spacing** to **192** (6 × 32). The default is **200.0**.

Touchup Hatch Pattern Angle **Angle** sets the hatch pattern angle. An angle for a 3D pattern is made relative to the face that is being hatched. For the simple case of hatching a face that is parallel to the X-Y plane (such as a slab or polygon representing a floor, flat roof, ceiling, or tabletop), a zero angle is the same as it is for 2D hatching, that is, to the right. For any other face, a zero hatch angle is parallel to the X-Y plane (a zero angle when viewed in elevation), and its angle in plan view is equal to the intersection line of the face and the X-Y plane. Thus a zero hatch angle (the default) will produce horizontal clapboards when viewed in elevation. Likewise for roofs, a zero angle will show shingles aligned in rows parallel to the ground (X-Y plane).

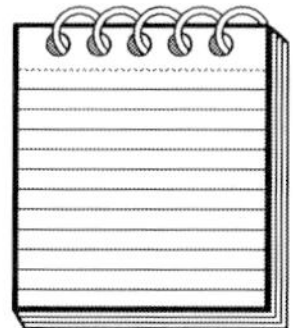

NOTE: Any time a nonzero hatch angle is specified by the user, the specified angle is added counter-clockwise to the zero direction, as previously described.

Layer Search in Touchup This is the DataCAD **LyrSrch** toggle. When on, it allows the user to select the entity to be hatched no matter what layer the entity is on. When off, only entities on the current layer may be selected.

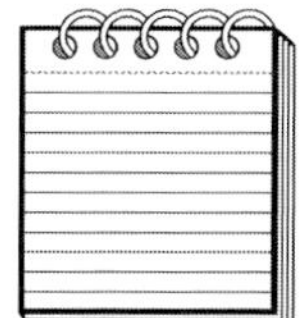

NOTE: Remember that the pattern is always generated on the active layer, regardless of the layer that the selected entity is on.

Sideboards This toggle, when on, allows the user a variety of choices in creating sideboards offset from entity edges by a given amount. This gives the end effect of having, for instance, 6″ clapboards along a wall that terminate 4″ short of the corner, where a vertical board is drawn. When **SideBrd** is off, the face will not be offset and the pattern will extend to all edges of the selected face.

Do Voids If a polygon or slab that is being hatched contains voids (as with windows and doors), the user is allowed the option of having sideboards automatically drawn offset from those voids, giving the visual effect of trim around the void. If **DoVoids** is on, sideboards will be drawn having the width specified in **SBWidth**. If **DoVoids** is off, no offsetting of voids will be performed.

Sideboard Width Use **SBWidth** to set the sideboard size. Sideboards will be offset automatically on the sides specified using the **Sides** option.

Side Selection When selected, **Sides** will bring up a list of toggles for the various side numbers that are to have sideboards. Turning the toggle off for a side will cause hatching to be done right up to the edge of the selected face. When the side is to contain an offset sideboard, its toggle must be on.

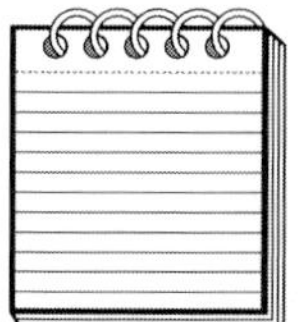

*NOTE: To determine side numbering for any given face of a model, select the face with **NmbrSides** on.*

Number Sides Use **NmbrSides** to number the sides of the selected face of an entity.

1. Choose **NmbrSides** from the **Touchup** menu.
2. Select the entity face to be numbered. The sides of the selected face are numbered.

You can now specify which sides will contain automatically generated offset sideboards.

3DViews The **3DViews** menu helps you generate views for plotting or presentations. By orienting the model during the drawing process, you can add new entities to the right place on the model. For complete information about the **3DViews** menu, see the *DataCAD 8 Reference Manual*.

Hide Use **Hide** to perform hidden line removals in DataCAD. The hidden line removal system recognizes only two basic types of data: closed polygons (which may have voids) and lines. DataCAD's hidden line removal system is a two-pass system with an additional initialization step. The initialization step performs a few simple and quick operations: allocating memory, calculating the limits of your model for a specified view, and initializing the temporary hidden line removal database. For complete information about the **Hide** menu, see the *DataCAD 8 Reference Manual*.

Polygons Use **Polygon** to create 3D polygons. Do not confuse 3D polygons with their 2D counterparts. A 3D polygon is not a group of lines as a 2D polygon is; it is a single entity. 3D polygons are opaque to the hidden line removal process. For complete information about the **Polygons** menu, see the *DataCAD 8 Reference Manual*.

Slabs Slabs are usually the most common entity type in a model. Slabs are similar to polygons; you might, in fact, think of slabs as polygons with thickness. The one important distinction between slabs and polygons is that slabs possess a reference face that controls the direction in which you can pass a void. For complete information about the **Slab** menu, see the *DataCAD 8 Reference Manual*.

Exit Touchup This command will exit the **Touchup** macro. The current settings for pattern, scale, angle, and sideboards will be remembered by **Touchup** the next time the macro is run.

Keyboard Shortcuts You can use DataCAD's primary keyboard shortcuts (listed in "The DataCAD Drawing Board" chapter of your *DataCAD 8 Reference Manual*) while in **Touchup**.

Additionally when at the main **Touchup** menu, the user may issue any keyboard command, such as pressing **M** for move (except those that use **Alt** keys). Once the menu operation is complete and the user chooses **Exit**, control is transferred back to the **Touchup** main menu.

This is a one-level function: for instance, pressing **M** for move and **T** for template, without first exiting the **Move** menu, will in fact take the user to the **Template** menu and entirely out of the **Touchup** macro.

Tips for Erasing Touchup Entities If you generate a hatch pattern with the wrong scale, angle, color, and so on, you can immediately rehatch it. Because the 3D hatch pattern is a group and if it is the most recent group added to the drawing, it can be quickly removed using the **Erase Last Group** keyboard shortcut, <. Patterns can also be removed quickly as a group by using the **Erase** menu and using the **Group** option.

CHAPTER 19

Advanced Modeling Workshop

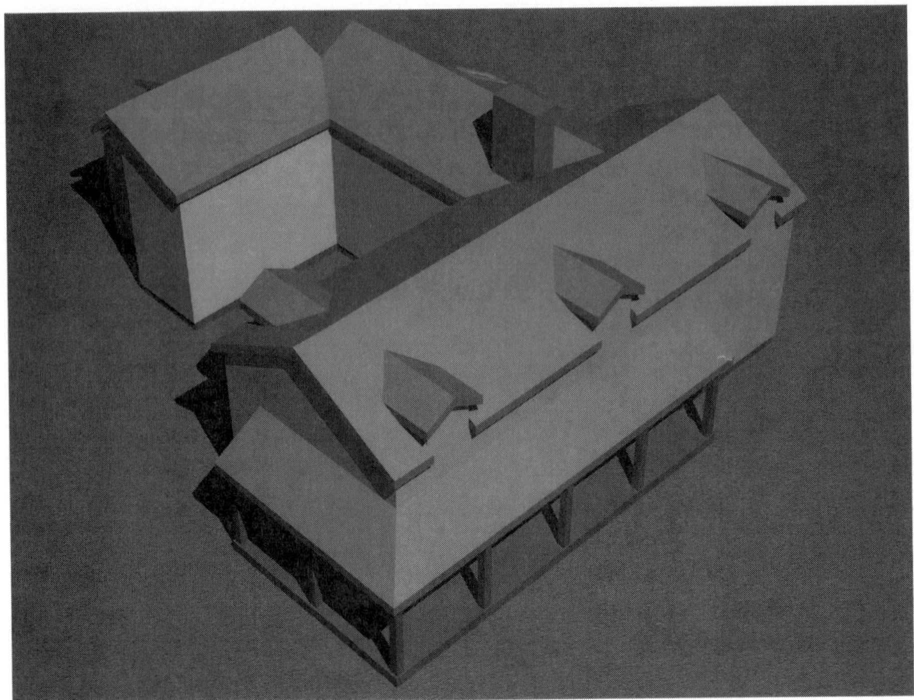

Exercise 1: Complex Roof

Roofs are as challenging to model as they are to build in the real world. The modeler must be adept at dealing with complicated intersections between inclined planes of varying slopes and eave heights. In this example, we will build a sample project that contains many of the complications encountered in typical roof projects: dormers of varying types, varying eave heights, skylights, and chimneys.

Open the file **Exercise_1.DC5** if you wish to follow along in the model file. You will also need to have the DataCAD add-on macro **3D Power Tools** installed to complete this exercise.

Step 1

The first step is to draw a roof plan with 2D lines at a Z-base and Z-height of zero. The dashed lines indicate the line of the walls below the roof. The wall thickness should be indicated if you wish to insert windows and doors into your model. In this example, we will only construct the roof and the walls below it.

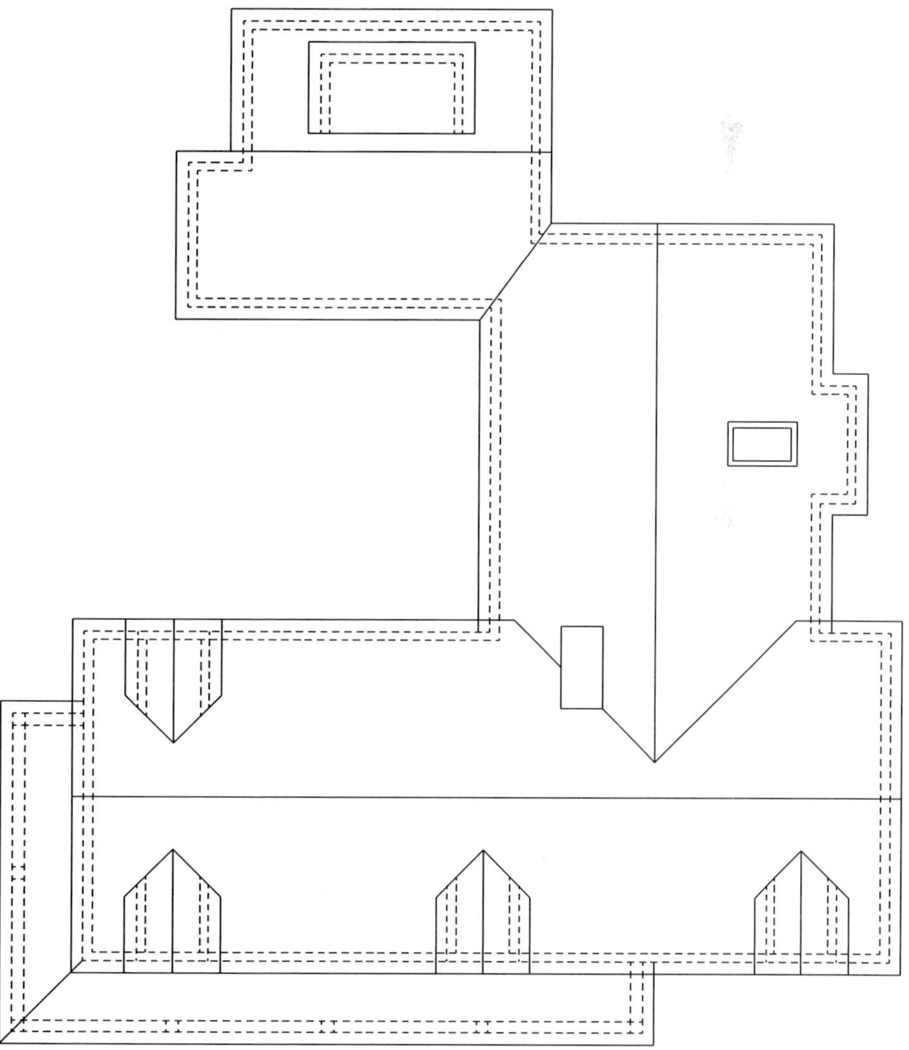

Step 2

We will begin by modeling the roof planes over the porch. We will ignore the overhang of the roof for now, and model the roof as if it aligns exactly with the wall below. This is important, as the roof planes will provide us with the points we need to easily model the walls later. After the walls are complete, the roof will be extended beyond the walls to create the overhang.

2.1 Create a new layer for the porch roof and make it active.

2.2 Press the **Z** key on your keyboard, and set the Z-base and Z-height to a value of 8′-0″. Make sure **Ortho** line mode is on.

2.3 Select **Toolbox**, **3DTools**, **<Open>**, **3D Plane**, **Pitched Slab**.

Select **Pitch** and select the option or 7:12.

Select **Thickness** and enter a value of 10″.

2.4 You are prompted to "Enter first point on horizontal roof line (Z-base)." Middle-button snap to point A in the plan.

You are prompted to "Enter second point on horizontal roof line (Z-base)." Drag anywhere toward point B and left-click. You are prompted to "Enter point on rising side of roof." Left-click anywhere on the plan that is higher than the line of the Z-base. The macro just needs to know to which side of the line you described is the rising side.

2.5 Begin tracing your roof shape. Middle-button click to point A, then points B, C, and D. After snapping to point D, click the right mouse button to finish. The first roof plane is modeled. If you make a mistake or mis-snap while tracing, you can press **F7** to **Backup**.

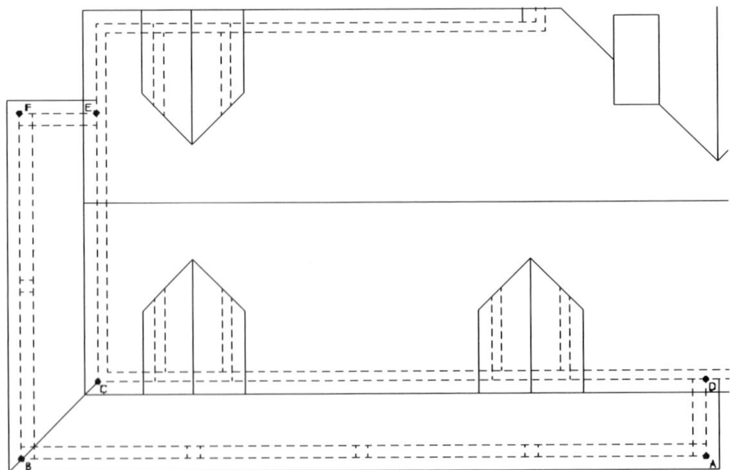

2.6 Repeat Step 2.5 with points B, C, E, and F. The porch roof is complete! Exit the macro by pressing the right mouse button three times. Here is how it looks in isometric view:

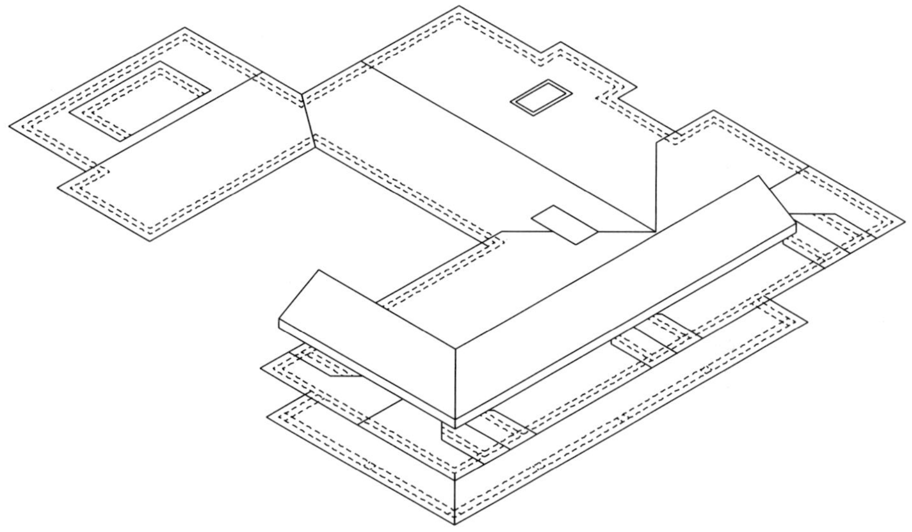

Step 3

Model the main roof. Create a new layer for the main roof and set it active.

3.1 Press the **Z** key on your keyboard, and set the Z-base and Z-height to a value of 8′-0″. Make sure **Ortho** line mode is on.

3.2 Select **Toolbox**, **3DTools**, **<Open>**, **3D Plane**, **Pitched Slab**.

Select **Pitch** and select the option or 7:12.

Select **Thickness** and enter a value of 10″.

3.3 You are prompted to "Enter first point on horizontal roof line (Z-base)." Middle-button snap to point G in the plan.

You are prompted to "Enter second point on horizontal roof line (Z-base)." Drag anywhere toward point W and left-click. You are prompted to "Enter point on rising side of roof." Left-click anywhere on the plan that is higher up the roof than the line you described previously as the Z-base.

3.4 Begin tracing the roof shape. Middle-button click to point G, then points H through Y. Note the notched areas created for the dormers. After snapping to point Y, click the right mouse button to finish. The first main roof plane is modeled.

3.5 Repeat for the opposite side, noting the snap points 1–15 shown in the plan. Also take note of the notch made for the chimney, and the adjustment for the lack of overhang at the gable intersection. You must plot the point where the line of the outside wall would meet the roof valley. Here is a view of the project so far:

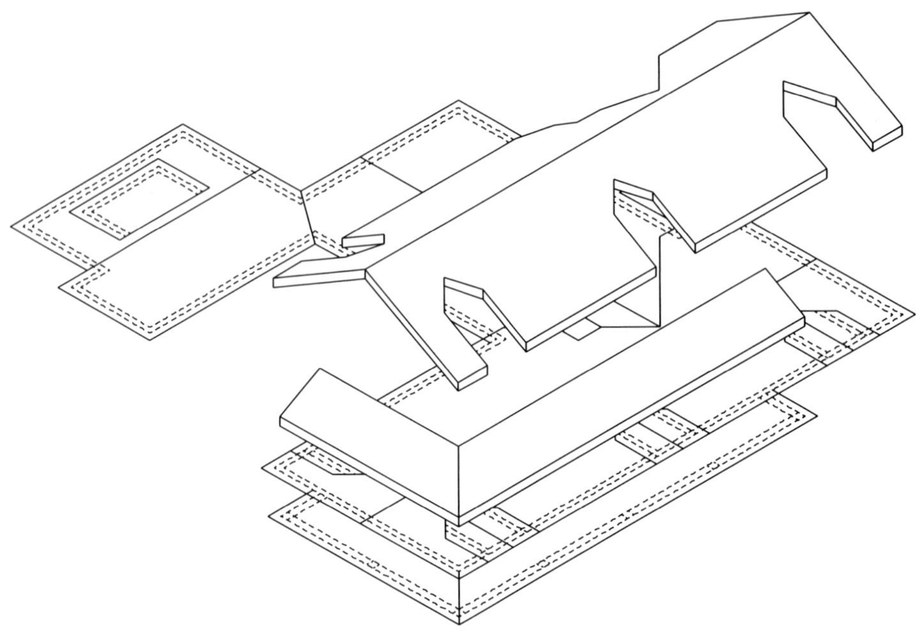

Step 4

Model the north wing roof. Create a new layer for the north roof entities, and set it active.

Press the **Z** key on the keyboard, and set the Z-base and Z-height to 14′-10″. Using the roof plane technique described in Step 3, model the north wing roof. Refer to the following illustration for the snap points. Note that we will skip the angled portion for now. Use points A–F and 2–3 for the base lines of the roof planes.

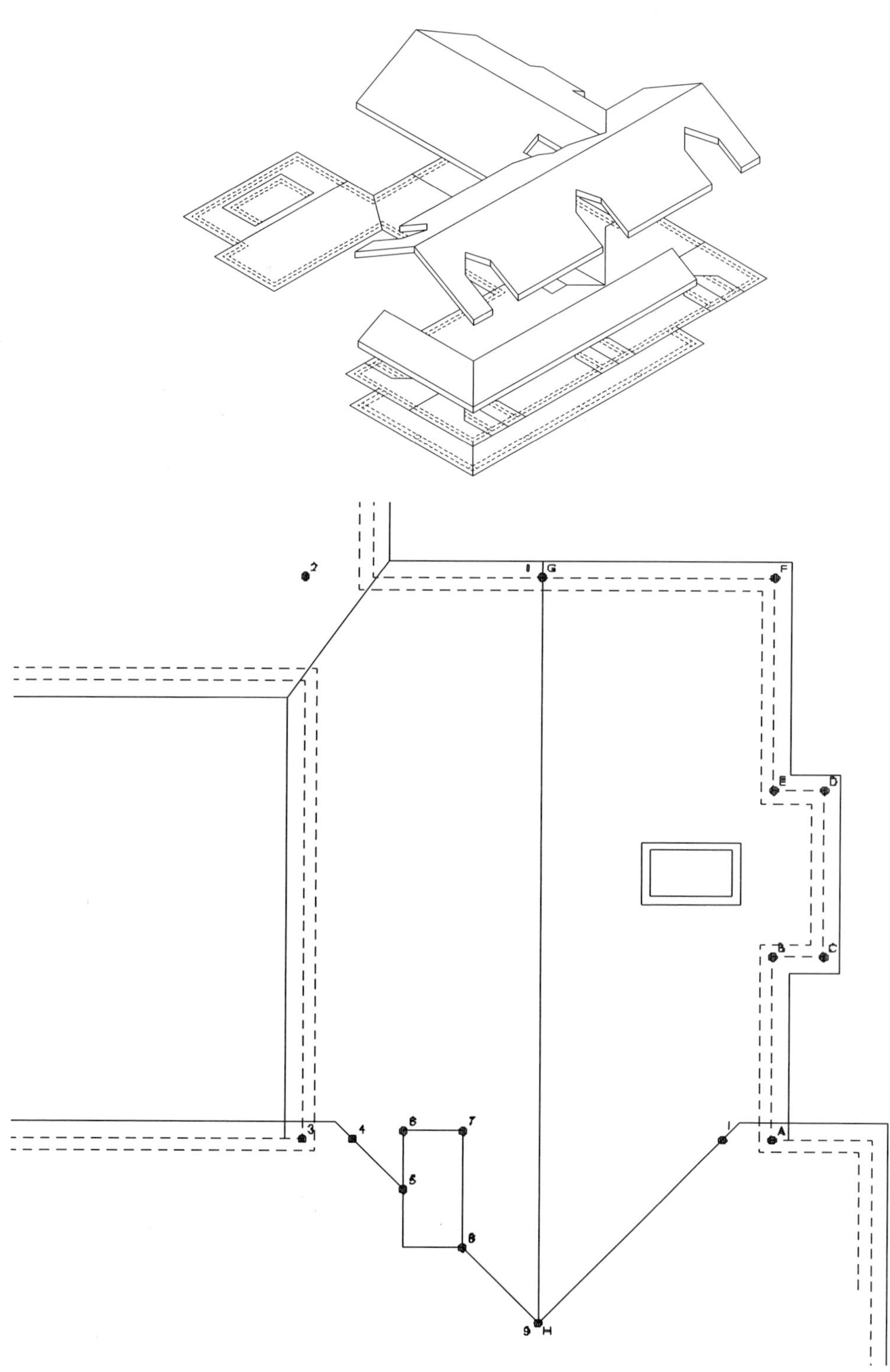
G
F
E
D
C
A
H

Step 5

Trim the north wing roof.

5.1 Select **Toolbox**, **3DTools**, **<Open>**, **Knife**.

5.2 You are prompted to "Enter first point of Knife line." Using the 2D Plan as a guide, middle-button snap to the end points of the angled line defining the valley joining the north and west wings.

5.3 The macro prompts you to "Enter point on the 'keep' side of knife." Left-click anywhere on the side of the roof slab that is not to be trimmed away. The macro then prompts you to choose the entity, group or area to trim. Change your selection mode to entity and choose the slab on the left side of the north wing. The slab is trimmed to the knife line.

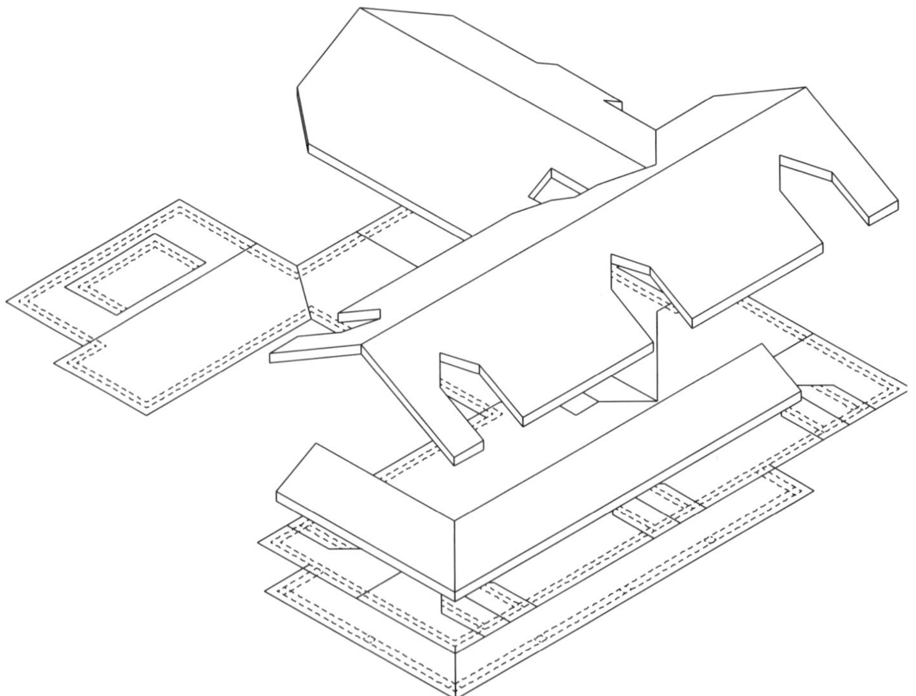

Step 6

Model the west wing roof. The valley joining the north and west wings presents a challenge, because the oblique angle makes its slope unknown. We will use some clever 3D techniques instead of trigonometry to solve the problem. We know that the top of wall is at 14′-10″, so we will start working from there. Create a new layer for the west wing roof, and set it active.

6.1 Set the Z-base and Z-height to 14′-10″. A minimum of 3 known points are required to define a plane. Two points are defined by the clip in the north roof, and the third lies at the far west corner.

6.2 Select **3D Entity**, **Marker**, and middle-button snap to point C in the following plan. This places a visible marker in space at the third point. Now, to create the roof plane.

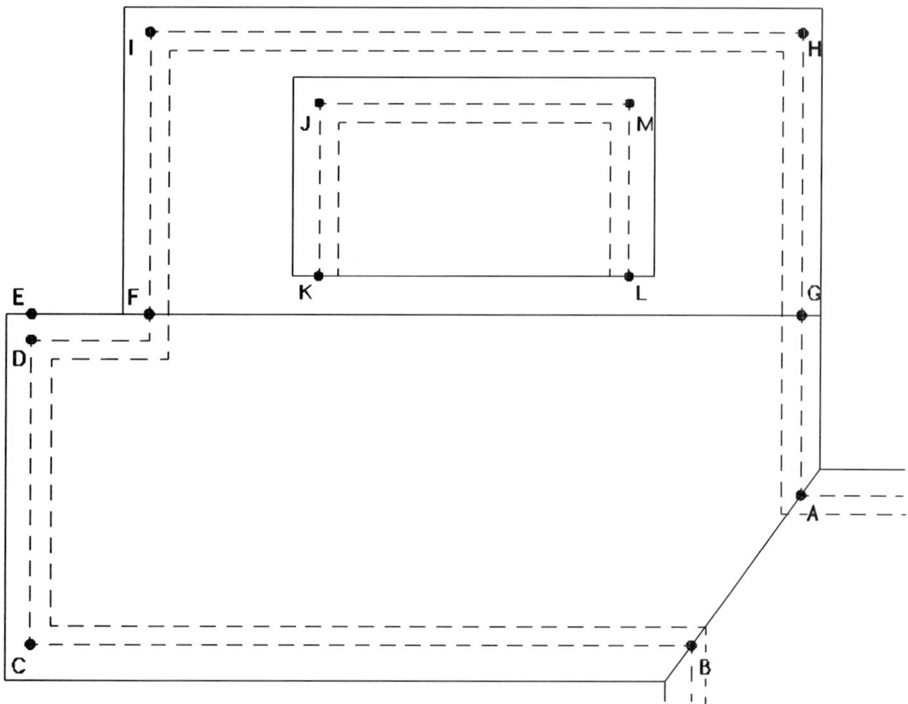

6.3 Change to an isometric view. Adjust the view if necessary so that you can clearly see all three points.

Select **3D Entity**, **Slab**, **Inclines**, **Partial**, **3 EdgePolygon**. Toggle on **Vertical** to create a slab with a plumb fascia. Select **Thickness**, and enter a value of 10″. You are presented with a 3D cursor showing you the X-, Y-, and Z-axis of your current view, and prompted to "Select first point of triangle (world snap enabled)." Middle-button snap to the marker in space that you entered previously at point C. You are then prompted to select the second point. Middle-button snap to point B at the underside of the slab, then to point A to select the third point and close the shape. Here is what the model should look like at this stage:

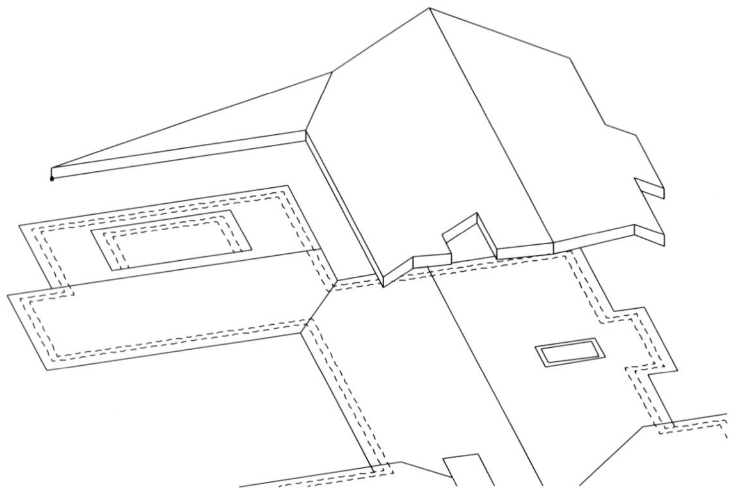

6.4 Return to **Ortho** view. Next we will use this 3-sided slab to create the rest of the roof at the proper slope. Select **Toolbox**, **3D Tools**, **<Open>**, **3D Plane**, **Slabs, Smart Partial**. This function allows you to edit inclined slabs while maintaining the slope. Select **Add Vertex**. Left-click anywhere along the angled edge of the 3-edge slab. The edge becomes elastic. Middle-button snap to point D. Repeat again, and middle-button snap to points E and G. The roof plane is complete. Here is an isometric view:

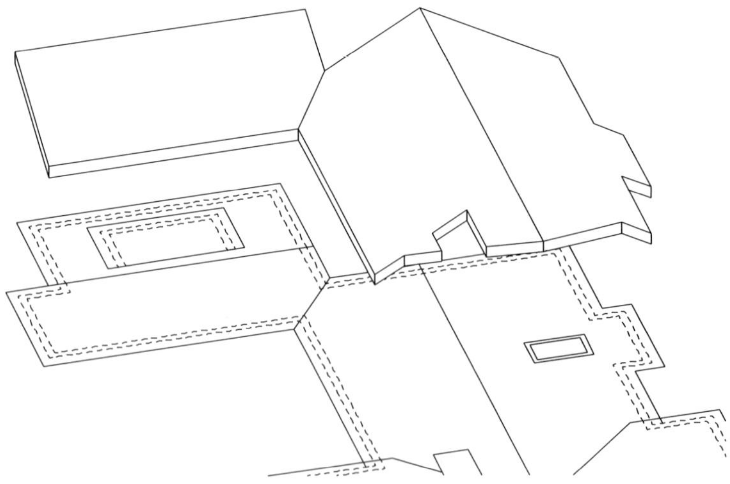

Step 7

Model the opposite half of the west wing roof. This roof plane has a steeper pitch of 10:12.

7.1 Make sure that **Ortho** line mode is on. Set your Z-base to 12′-0″, a temporary value. After modeling the roof plane, we will move it to meet the ridge. Select **Toolbox**, **3D Tools**, **3D Plane**, **Pitched Slab**. Select **Pitch**, and set to **10:12**. Select **Thickness**, and set to **10″**. You are prompted to "Enter first point on horizontal roof line (Z-base)." Middle-button snap to point H. Drag toward point I and left-click. You are prompted to "Enter point on rising side of roof." Drag toward the bottom of the screen and left-click. You are prompted to "Select first point of the roof slab." Trace out points F, G, H, I by middle-button snapping. Right-click after snapping to point I. The roof plane is created.

7.2 Exit the **3D Tools Macro**. **Select DCAD 3D**, **3D Views**, **Elevation**, **New Elevation**. Middle-button snap to point I. Drag toward point H and left-click. Your model is displayed in Elevation as shown.

7.3 Select **3D Edit**, **Move**. Make sure that **And Copy** is toggled off. Middle-button snap to point A, shown in the following diagram, to define the first point of the distance to move, then to point B for the second point. Choose **Entity** as your selection method and pick the uppermost slab. The slab is moved to align with the ridge of the roof plane modeled in Step 6.

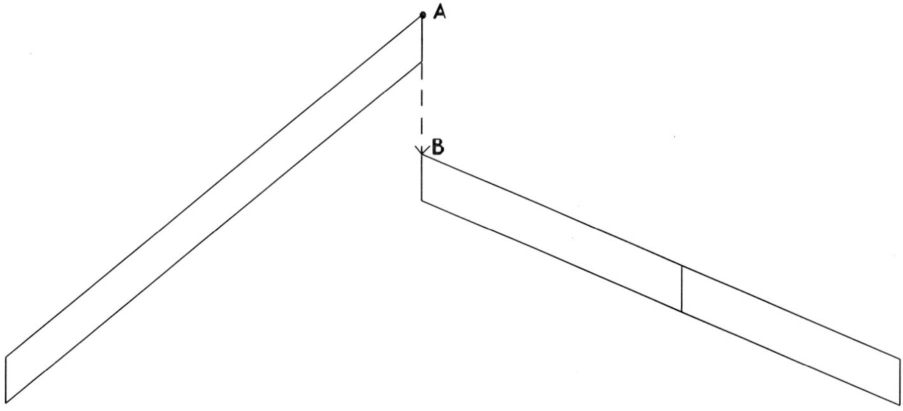

Step 8

Model the shed dormer roof. Create a new layer for the dormer entities, and set it active.

Change your working view to **Ortho**. Make sure that **Ortho** line mode is on. Set your Z-base to 18′-0″, a temporary value. After modeling the roof plane and creating an opening in the primary roof, we will move it into position. Select **Toolbox**, **3D Tools**, **3D Plane**, **Pitched Slab**. Select **Pitch**, and set to **4:12**. Select **Thickness**, and set to **10″**. You are prompted to "Enter first point on horizontal roof line (Z-base)." Middle-button snap to point L. Drag toward point I and left-click. You are prompted to "Enter point on rising side of roof." Drag toward the bottom of the screen and left-click. You are prompted to "Select first point of the roof slab." Trace out points J, K, L, M by middle-button snapping. Right-click after snapping to point M. The roof plane is created.

Step 9

Create the roof openings for the shed dormer and skylight. We will use the Smart Void function in 3D Tools to project openings through the roof slabs.

9.1 Exit the **3D Tools** macro. Press the **Z** key on your keyboard, and set your Z-base to 0′-0″ and your Z-height to 1′-0″. Select **3D Entity**, **Slab**, **Rectangle**. Trace over the 2D plan to define the openings by middle-button snapping to points A, B, C, and D, as shown in the following figure.

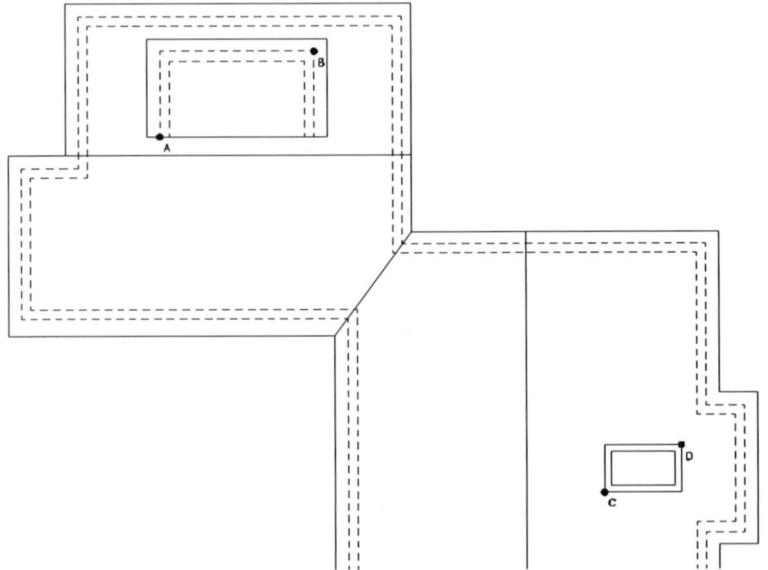

9.2 Change your view to isometric. Click various points in the view globe until you can easily distinguish between the slabs you just created and the roof planes they should project through:

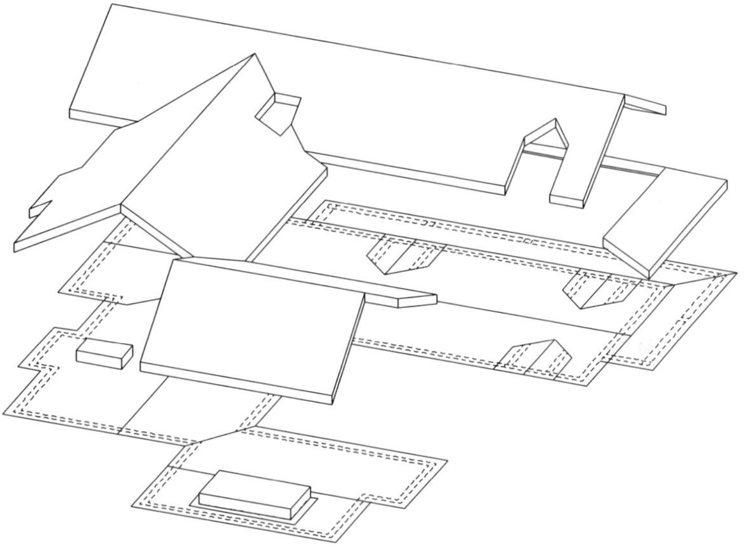

9.3 Select **Toolbox**, **3D Tools**, **3D Plane**, **Slabs**, **Smart Void**. You are prompted to choose the master slab to receive voids. Pick the roof slab for the shed dormer. Next you are prompted to choose the entity to make into a void. Select the following rectangular slab. The slab is projected through the roof plane to create a void! Right-click to finish processing this roof slab. Select the roof slab to receive the skylight as the next master slab. Choose the rectangular slab defining the skylight to make into a void. Right-click until you exit the **3D Tools** macro. The roof voids are complete.

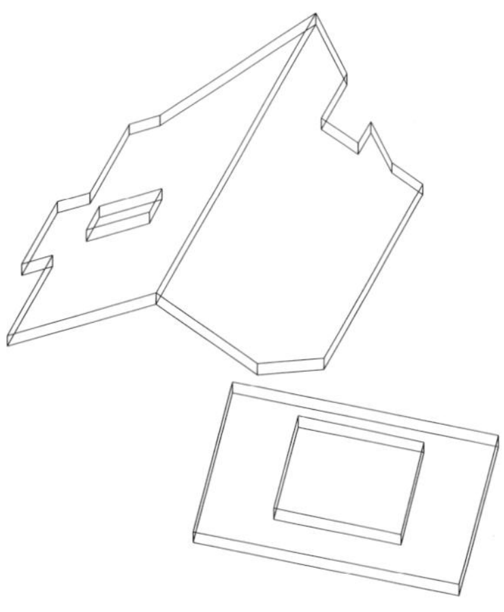

Step 10

Move the shed dormer into position.

10.1 Select **DCAD 3D**, **3D Views**, **Elevation**, **New Elevation**. Middle-button snap to point H. Drag toward point G and left-click. Your model is displayed in Elevation as shown.

10.2 Select **3D Edit**, **Move**. Make sure that **And Copy** is toggled off. Middle-button snap to point A, shown in the following diagram, to define the first point of the distance to move, then to point B for the second point. Choose **Entity** as your selection method and pick the shed dormer. The slab is moved to align with the roof opening created in Step 9.

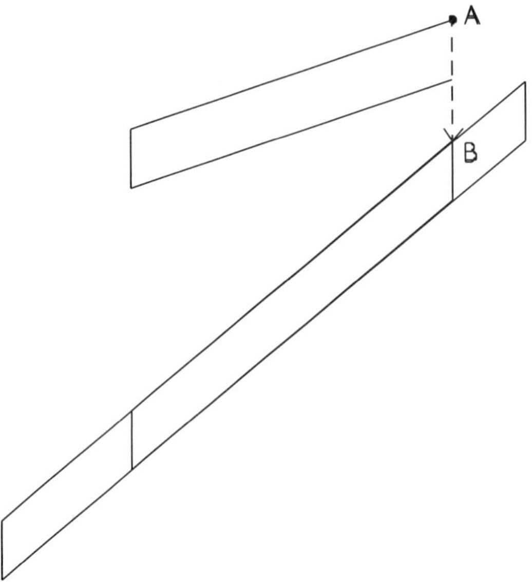

Step 11

Model the dormers on the main roof.

11.1 Change your view to **Ortho**. Turn on the layer for the 2D plan to use as a guide, and set the layer you created for the dormers as active. Make sure that **Ortho** line mode is on. Set your Z-base to 20′-0″, a temporary value. After modeling the roof planes, we will move them to align with the dormer opening. Select **Toolbox**, **3D Tools**, **3D**

Plane, Pitched Slab. Select **Pitch**, and set to **7:12**. Select **Thickness**, and set to 10″. You are prompted to "Enter first point on horizontal roof line (Z-base)." Middle-button snap to point A. Drag toward point D and left-click. You are prompted to "Enter point on rising side of roof." Drag toward the right and left-click. You are prompted to "Select first point of the roof slab." Trace out points A, B, C, and D by middle-button snapping. Right-click after snapping to point D. The dormer roof plane is created.

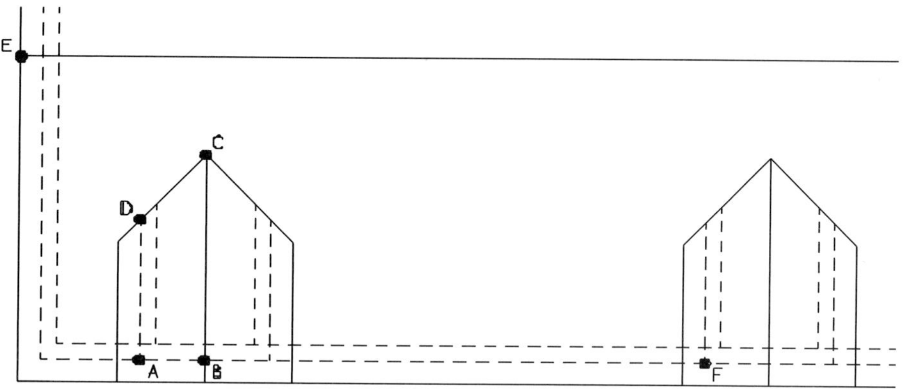

11.2 Exit the **3D Tools** macro. Select **3D_Edit, Mirror**. Middle-button snap to points B, C to define the line of reflection. Make sure **And Copy** is on. Select the dormer slab to **Mirror And Copy**.

11.3 Define a new parallel view as earlier in Step 7. Repeating the method used in Step 7, move the dormer slabs downward to align with the cutout in the main roof. The dormer is complete.

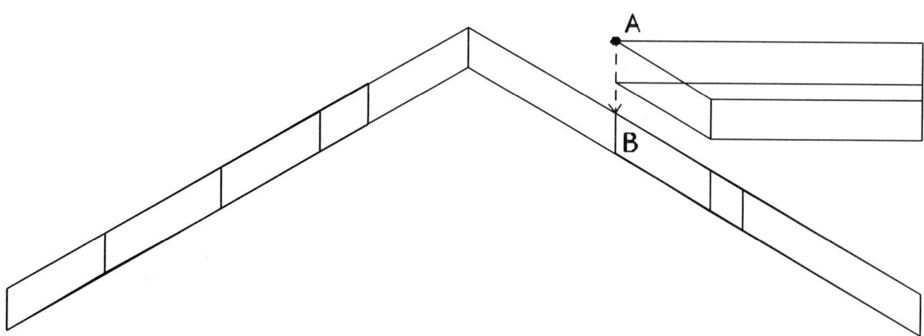

Step 12

Create the remaining dormers.

12.1 Mirror and copy the two dormer slabs horizontally from point E to create the dormer on the opposite side.

12.2 Copy the two dormer slabs from point A to point F, and similarly for the remaining dormer. All the roof planes are now complete!

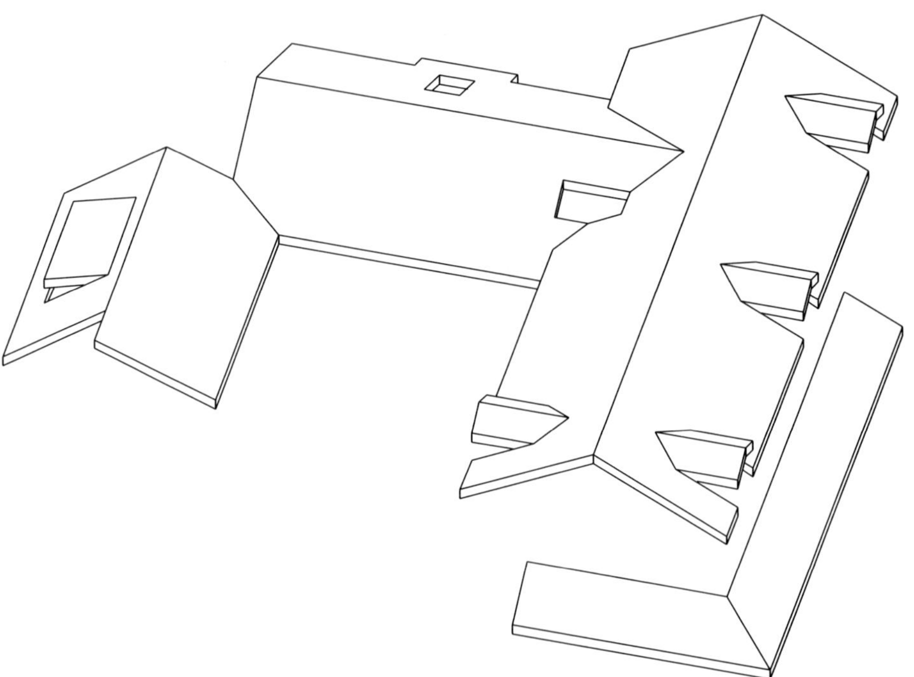

Step 13

Create the walls below the roofs. Here, we will use a technique to convert 2D entities to 3D polygons to avoid having to retrace the plan. Create a new layer for the 3D wall polygons.

13.1 Set the layer that contains the 2D plan as **Active Only**. Copy the lines of the building footprint and porch columns to the 3D wall layer. Set the 3D wall layer as **Active Only**.

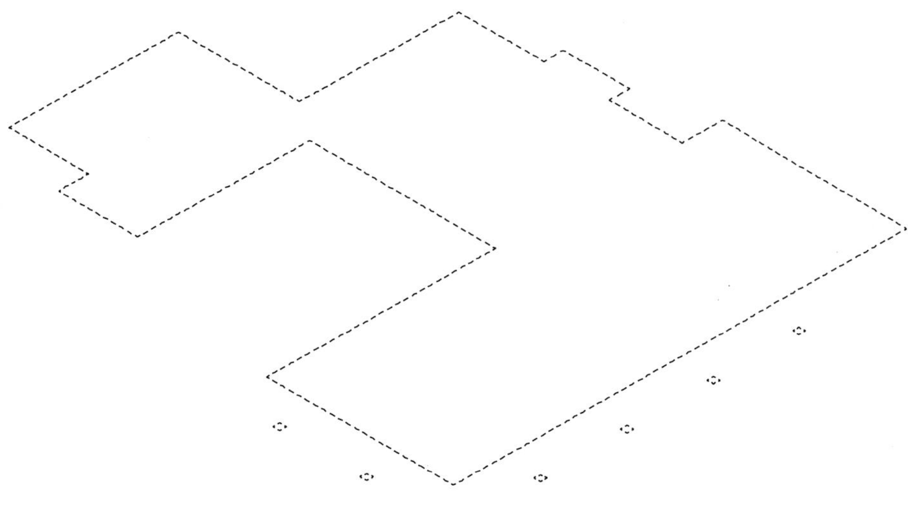

13.2 Select **2D Edit**, **Change Linetype**, and choose solid. Next, select **Z-height**, and enter 16′-0″. Select **Z-base** and enter 0′-0″. Set your selection method to **Area**, and define a selection box encompassing all of the 2D lines. The 2D lines change to match the new parameters. Now, select **Z-height** again, and enter 7′-0″ as the value to change to. This time, only select the porch columns. Here is the result:

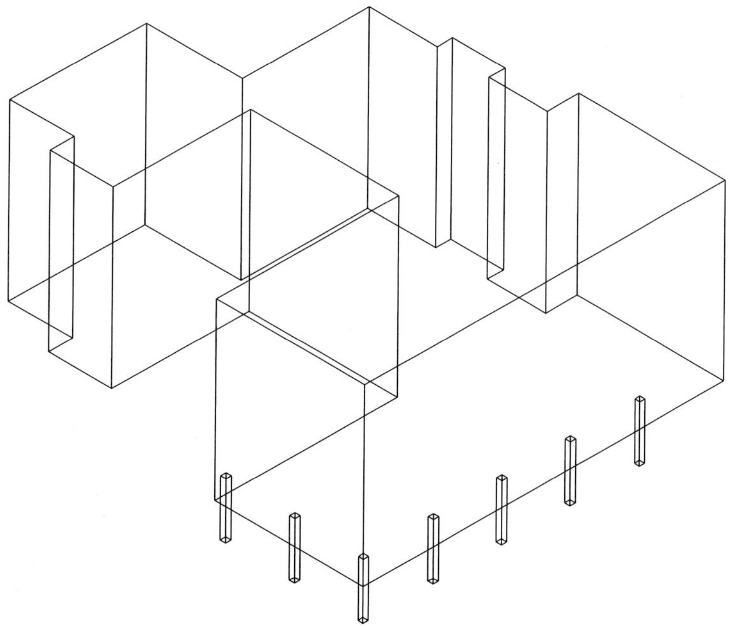

13.3 Select **3D_Edit**, **Explode**, **To Polygons**, and area-select all the 2D lines. The 2D entities are converted from 2D lines to three-dimensional polygon surfaces. This allows us to modify their shapes and place voids within them.

Step 14

Modify the wall polygons to align to the roof.

14.1 Define a **New Elevation** perpendicular to the wall line at point E. Select **3D Views**, **Clip Cube**, **New Cube**. When prompted to define clip cube boundary, click to opposing points to create a box as shown in the following illustration. Pick **Z-Min** and set to −1′-0″, then **Z-Max** and set to 25′-0″. Right-click to exit, then choose **Clip On**. The view of your model is confined to the limits of the clip cube, allowing you to easily select the appropriate polygon to modify. Exit the **Clip Cube** menu. Turn **Layer Search** off so that you won't risk picking entities that are not on the wall polygon layer.

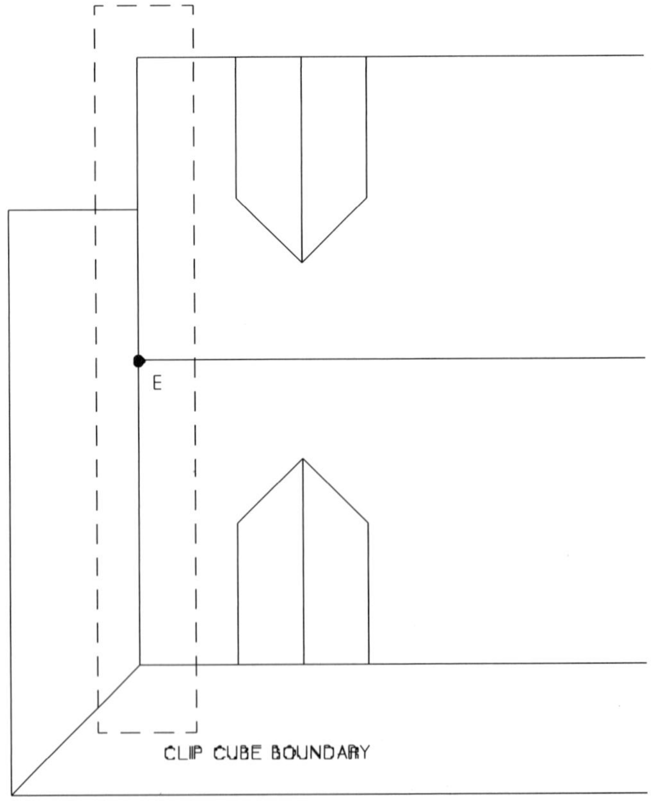

14.2 Select **3D_Edit**, **Polygon**, **Partial**, **Add Vertex**. Pick the wall polygon edge anywhere near point A shown in the following diagram. The polygon boundary becomes elastic. Middle-button snap to the underside of the gable ridge at point B. Exit back to the **3D_Edit** menu. Select **3D Views**, **Clip Cube**, **New Cube**, and define a similar clip cube at the opposite gable. Exit and choose **Clip On**. Repeat the **Add Vertex** on this wall polygon.

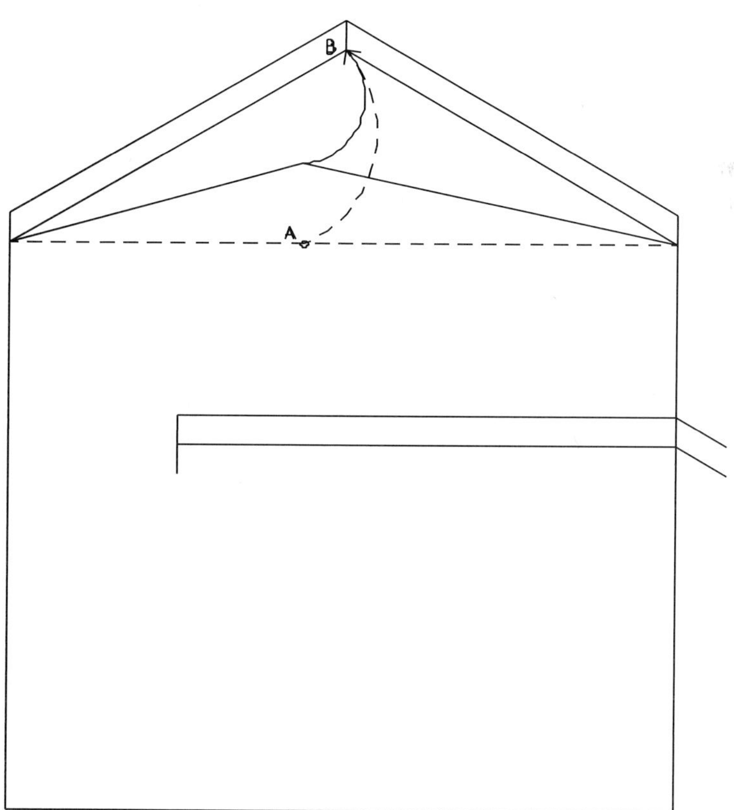

14.3 Define a new elevation perpendicular to the front façade of the main building. Define a clip cube isolating this wall as well. Use the **Polygon**, **Partial**, **Add Vertex** function to match the polygon profile to the roof and dormers as illustrated. This entails middle-button snapping to points A, B, C, D, E, F, G, H, I, J, K, L, M, N, and O.

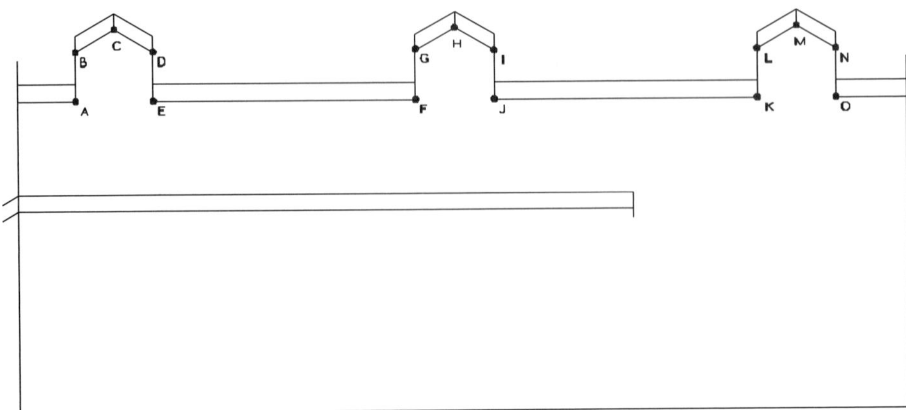

14.3 Repeat for the single dormer on the opposite side. Select **3D Views**, **Clip Cube**, **Clip Off**. Here is a view of the model thus far:

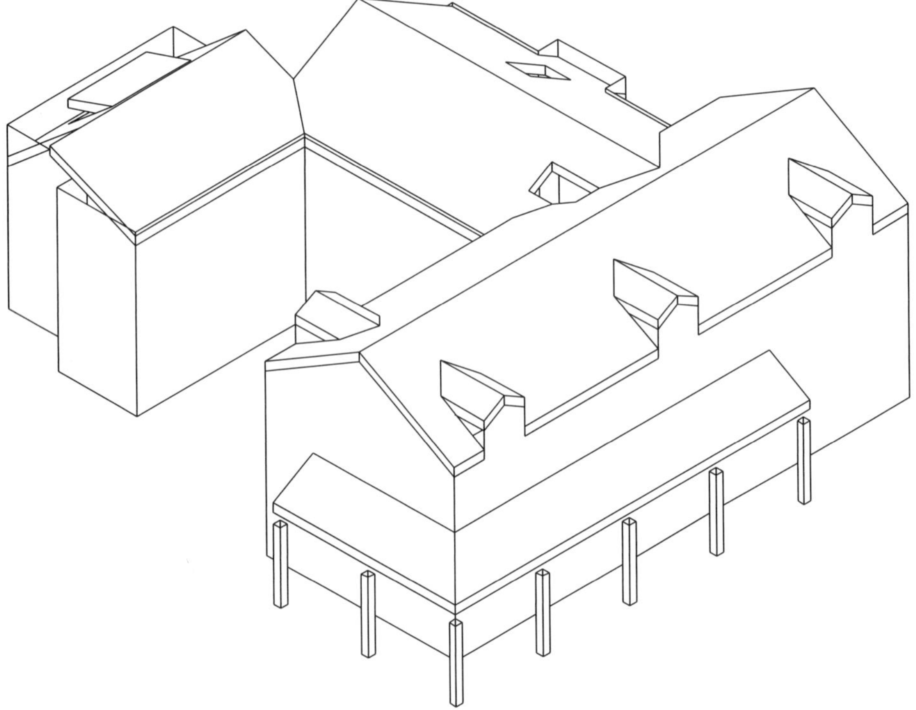

14.4 Create a new layer to further separate the 3D walls. Move the wall polygons for the north and west wings to this layer as shown. Set this layer active and make sure that **Layer Search** is off. Turn on the layer for the 3D roof planes.

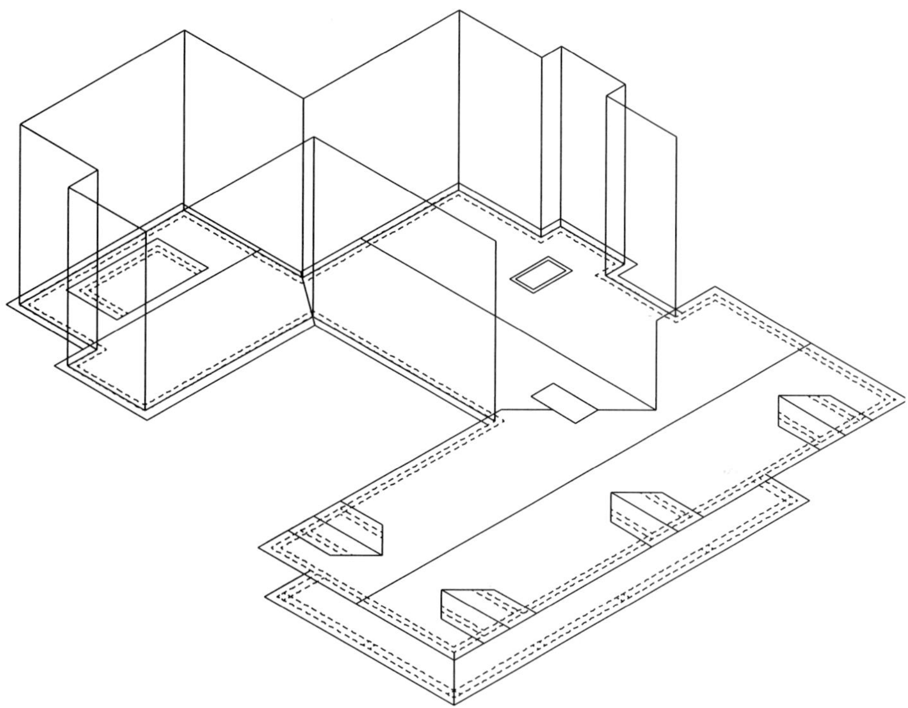

14.5 Define a **New Elevation** facing east. Select **3D Edit**, **Stretch**, **by Area**, and stretch the existing top of wall (line A–C) from point A to B. Next, stretch point C to point D.

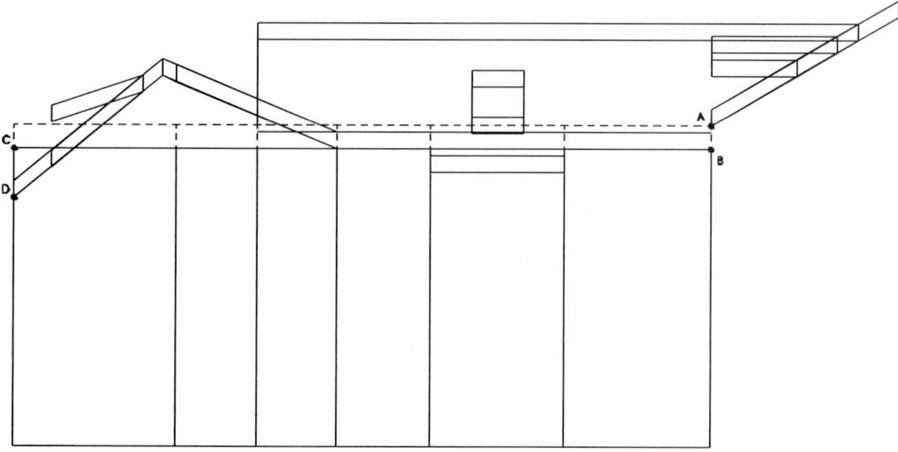

14.6 Stretch point B to point A.

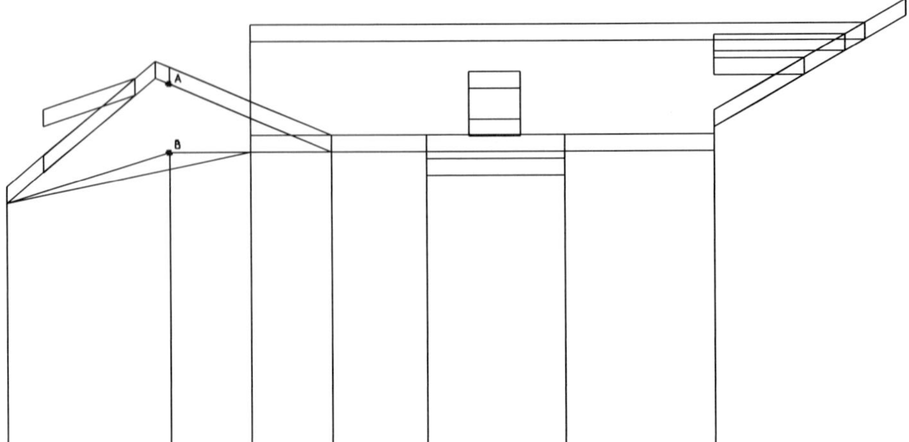

14.7 Select **3DEntity**, **Polygon**, **Partial**, **Add Vertex**. Add a vertex to the polygon edge at point B. Drag and middle-snap to define its position at point A. Repeat for the polygon edge at point C. Add another vertex to the resulting edge at point D, then middle-button snap to place the vertex at point E.

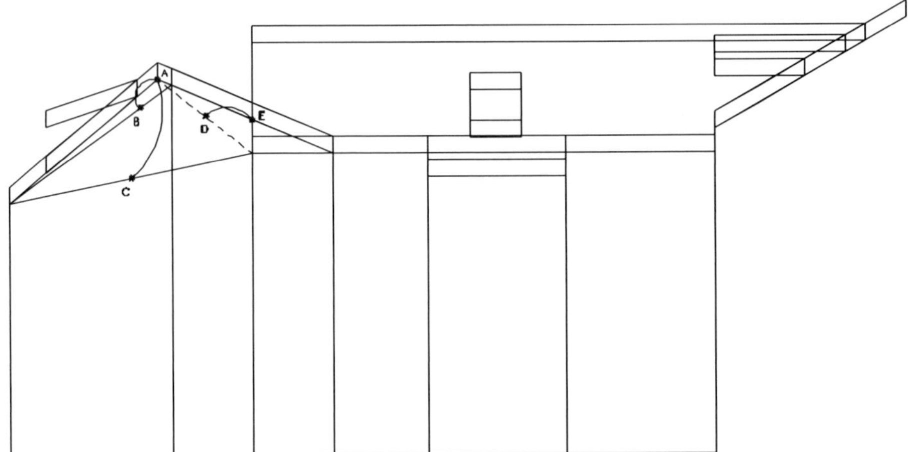

14.8 Define a **New Elevation** looking south. Refer to the following illustration. Stretch point A to B. Select **3D Entity**, **Polygon**, **Partial**, **Add Vertex**. Pick the polygon edge at point C to add a vertex, then middle-button snap to point D to place it. Repeat with the resulting edge at point E, placing it at point F.

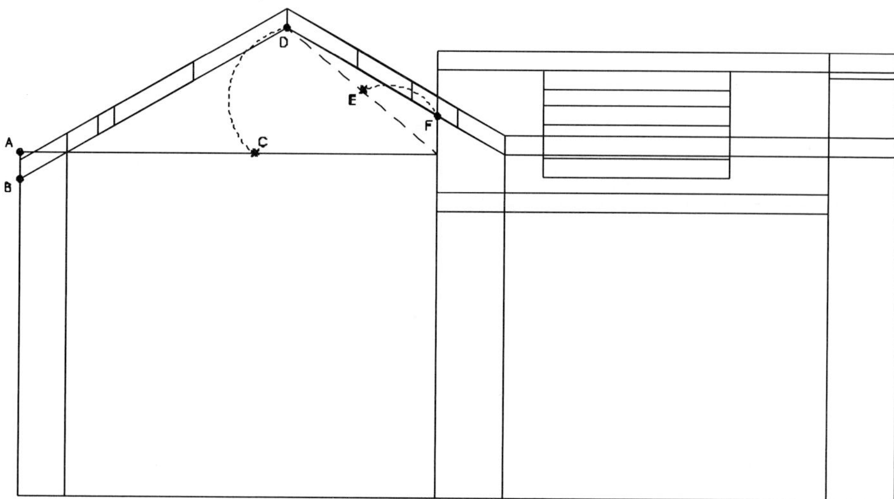

The exterior walls are complete!

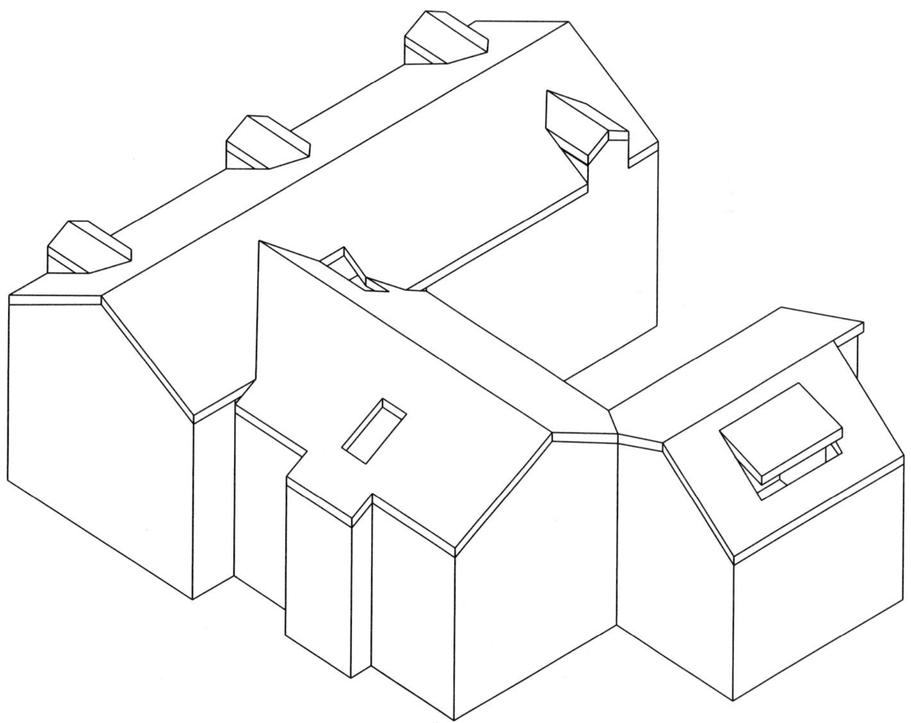

Step 15

Complete the details with dormer cheek walls.

Turn on the layers containing the dormer walls and roofs. Change to an isometric view similar to the following diagram. Select **3D Entity**, **Polygon**, **Inclines**, **3 Edge Polygon**. You are presented with a three-dimensional cursor. Middle-button snap to points A, B, and C to create the cheek wall. Repeat for the opposite side. Repeat for remaining dormers and shed dormer on west wing.

> Hint: You can use **Mirror**, **And Copy**, and **Move**, **And Copy** to complete the remaining gabled dormers.

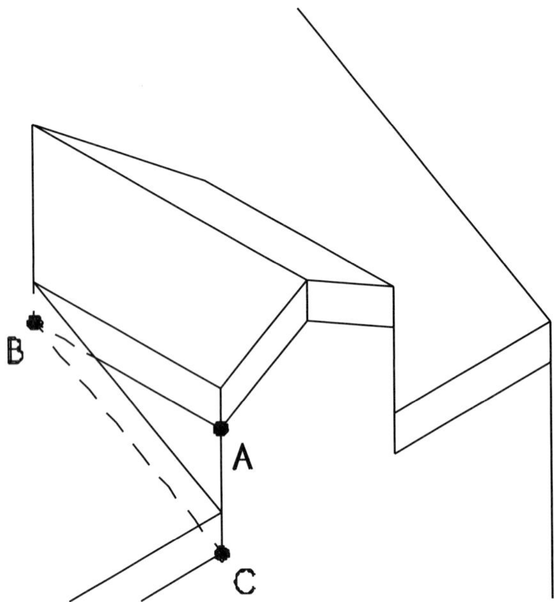

Step 16

Add a chimney and complete the porch.

16.1 Change to **Ortho** view. Turn on the layers associated with the main and north wing roofs. Create a new layer for the chimney and set it to active. Press the **Z** key on your keyboard; set the Z-base to 0′-0″ and the Z-height to 26′-0″. Select **3D Entity**, **Slab**, **Rectangular**. Using the roof layer as a guide, middle-button snap to two opposite corner points to define the chimney block.

16.2 Turn on the layer for the 2D Roof plan. Set the Z-base to 7′- 0″ and the Z-height to 8′-0″. Using the plan as a guide, trace off points A, B, C, D, E, F, G, H, I, and J.

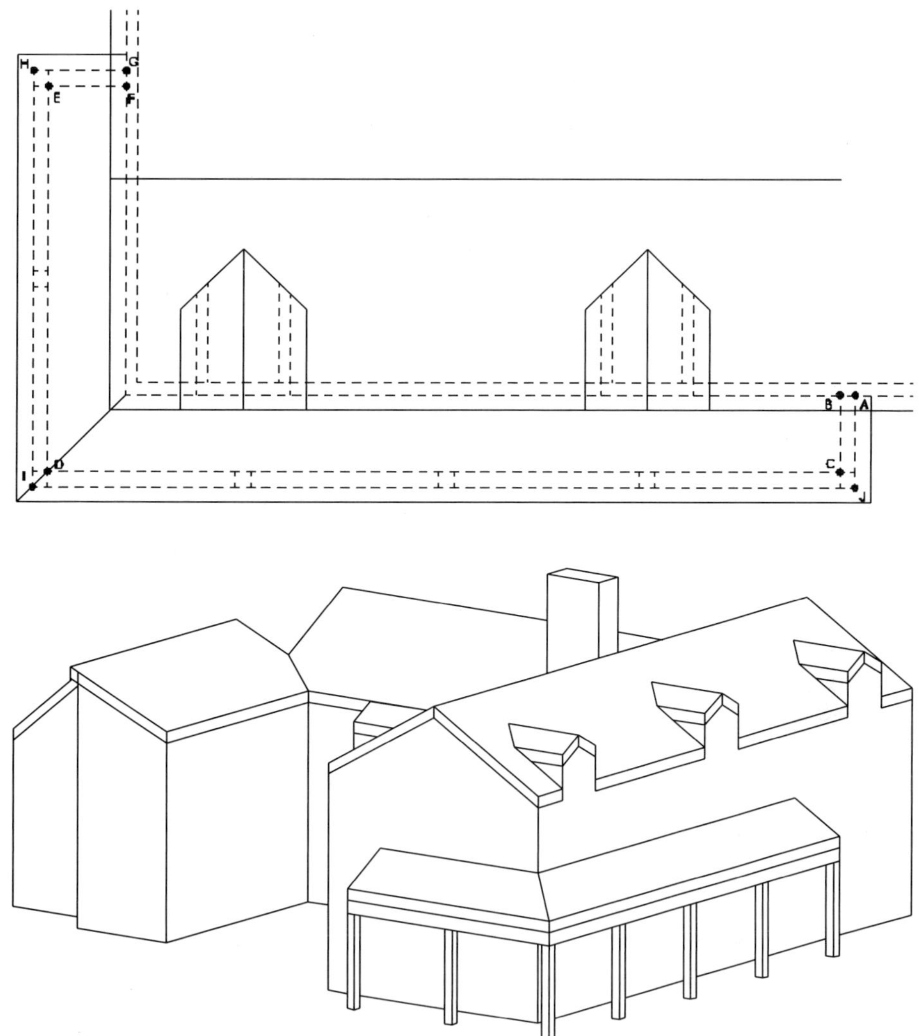

Step 17

Add the overhang.

The last step to complete the model is to adjust the roof planes for the overhang. As you have seen, modeling the roof planes without the overhang initially makes creating the walls much easier, as the roof provides the necessary snap points. We will use the **Smart Partial** function of the **3D**

Power Tools macro to move the roof line outward while preserving the slope.

17.1 Set the layer for the 2D Roof Plan as **Active Only**. Turn on the layer for the porch roof. Turn **Layer Search** on. Select **Toolbox, 3D Tools, 3D Plane, Slabs, Smart Partial, Move Vertex**. Refer to the following diagram. Left-click the slab vertex at point A and move it to point B, middle-button snapping to place it. Next, select the vertex at point C and move it to point D. Repeat similarly for the remaining points shown. Note that two edges meet at points E and M, so you will perform the operation twice at this point. After the porch roof is complete, turn it off and turn on the layer for the next section. This will prevent you from mis-picking vertices at shared roof plane edges.

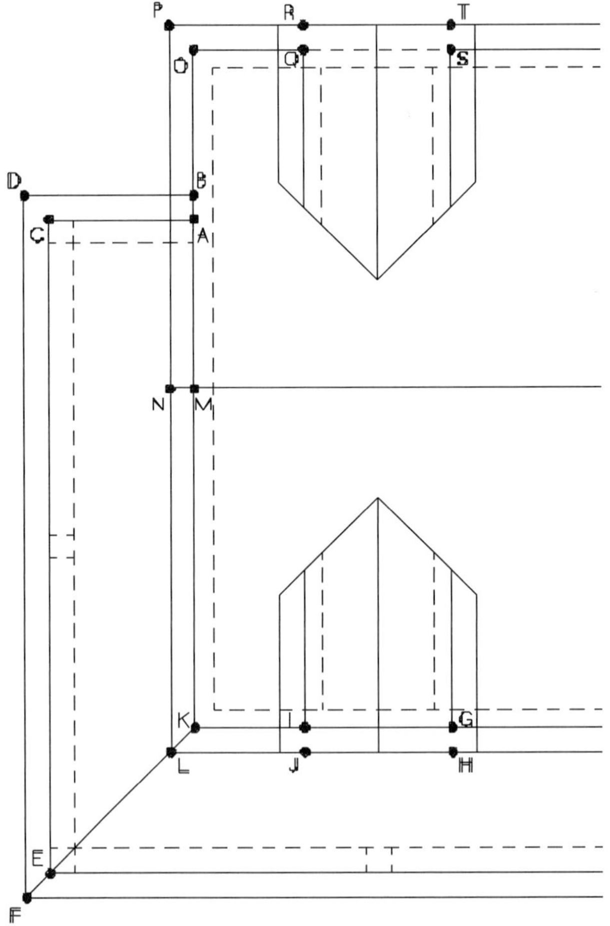

Hint: You can place a snapping marker at the point where any two lines would intersect with the **2D Utility**, **Geometry**, **Intersect** command. This is useful for determining the proper snap points at the dormer edge.

17.2 Using the preceding process, continue adjusting the remaining roof vertices, moving each roof corner out to meet the line delineating the roof overhang on the 2D roof plan layer.

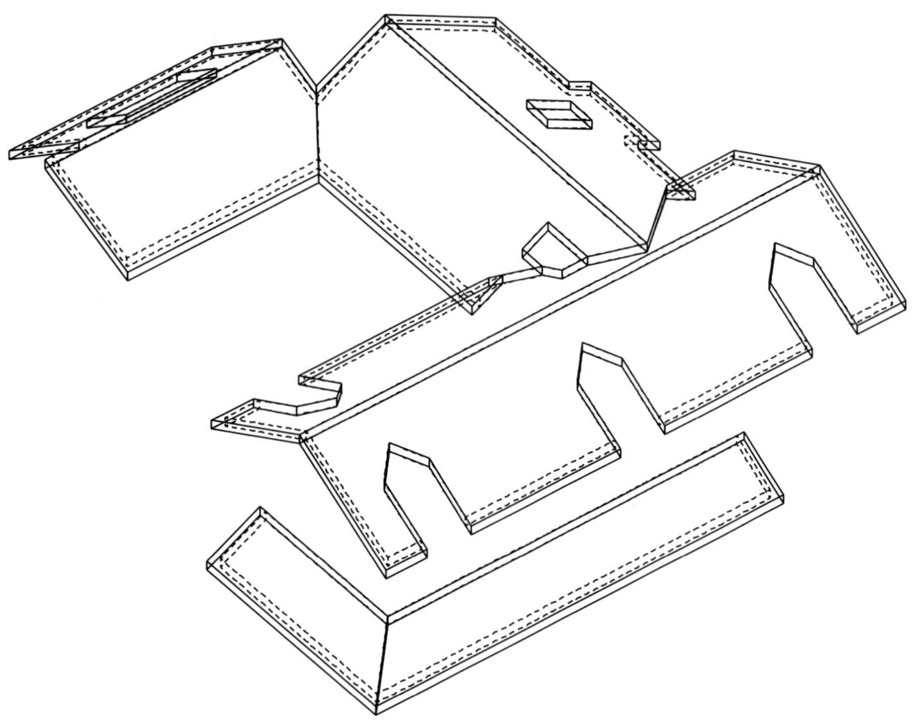

17.3 Turn off the 3D roof layers. Turn on the layer for the 3D dormers. Apply the same process to the dormers. Note again that wherever two vertices meet, as at point E, you will move each one in turn.

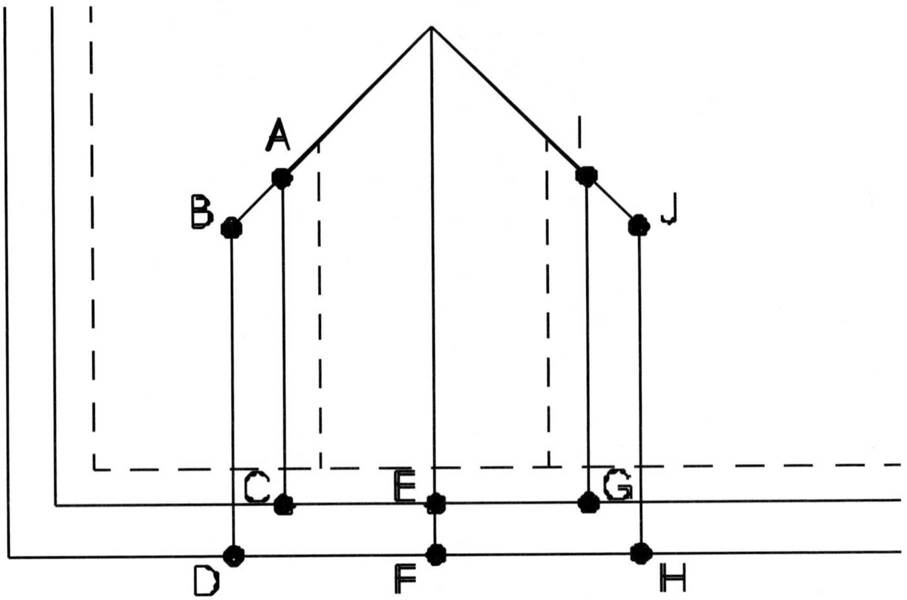

The model is complete!

Exercise 2—Pendentive Dome

In this example we will model a small building with a pendentive dome. We will face the same dilemma as the builders of antiquity, who were challenged by the space where the dome meets the vault.

Open the file **Exercise_2.DC5** if you wish to follow along in the model file. The enhanced **AEC_Modl** macro located on the bonus CD should be installed before undertaking this exercise.

Step 1

The fist step is to draw the floor plan with 2D lines at a Z-base and Z-height of zero. The dashed lines indicate the perimeter of the dome and the vaults that support it. The symmetry of the plan allows us to model one quarter of the structure and then use **Mirror**, **And Copy** to complete it.

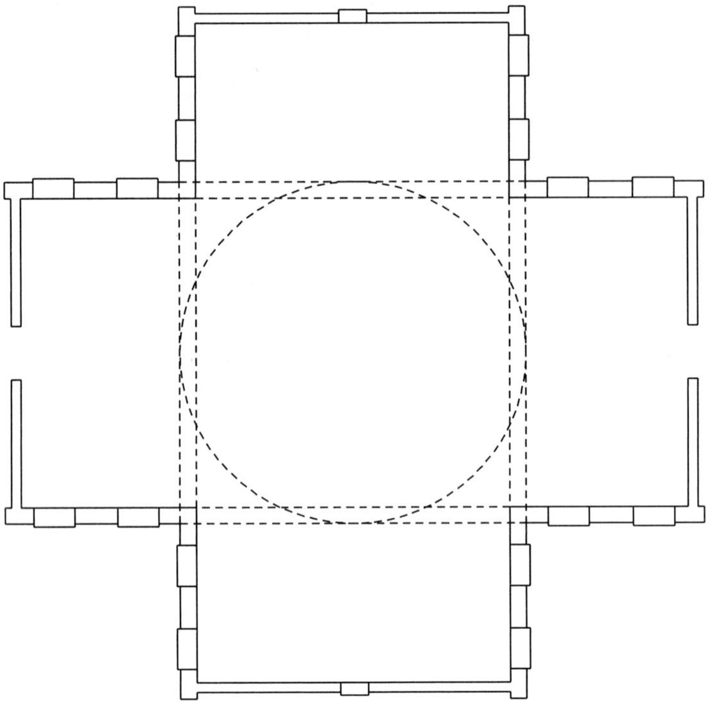

Step 2

We will begin with the vaults.

2.1 Create a new layer for the exterior surface of the first barrel vault and make it active.

2.2 Set the Z-base to a value of 50′-0″.

2.3 Select **3D Entity**, **Horizontal Cylinder**, **Top Half**, **Z base**. Make sure that **Closed** is toggled off. Select **Divisions** and enter a value of **24**. Select **Radius**, and enter 26′-6″.

2.4 You are prompted to "Select center point of first end of Cylinder." Middle-button snap to the center of the top wing. Drag downward and middle-button snap to the center of the first dashed line, where the dome is tangent to the vault.

2.5 Repeat with the right wing as well to create a total of two horizontal cylinders.

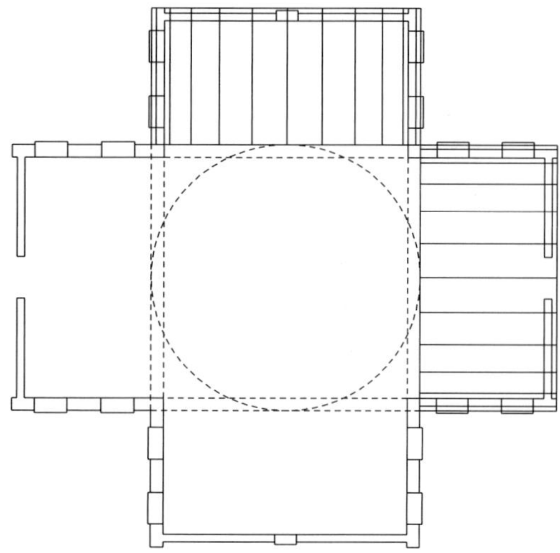

Step 3

Model the exterior dome surface.

3.1 Create a new layer for the exterior dome and make it active.

3.2 Select **3D Entity**, **Sphere**, **Radius Circle**, **Top Half**, **Z Base**. Press the **Z** key on your keyboard, and set the **Z Base** to a value of 76′-6″, which is the height of the barrel vault.

3.3 You are prompted to "Select center point of Sphere." Middle-button snap to the center of the circle on the 2D plan layer, and then middle-button snap again to the top vault at its midpoint. The result should look like this:

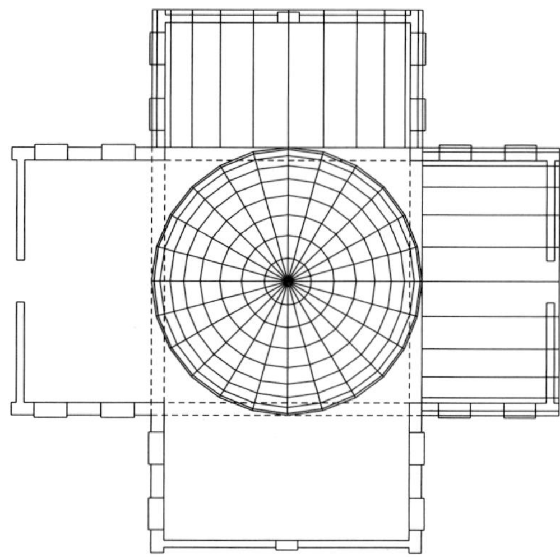

Step 4

Model the first pendentive.

4.1 Create a new layer for the pendentive and make it active.

4.2 View the model in isometric as shown:

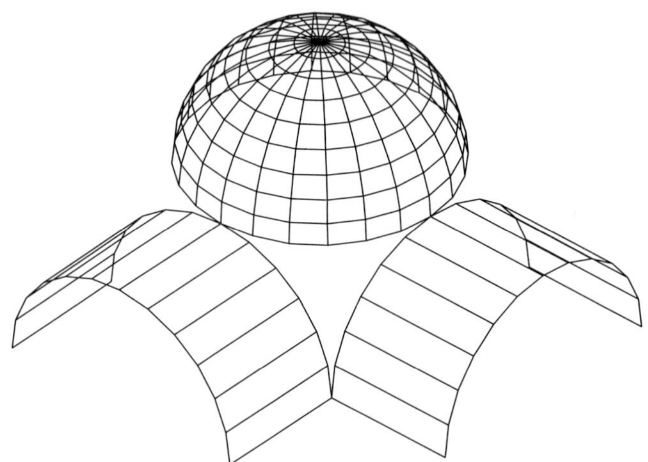

4.3 Select **3D Entity**, **Polygon**, **Inclines**, **3 Edge Polygon**. You are presented with a three-dimensional cursor, showing you the X-, Y-, and Z-axis of your current parallel view. A three-edge polygon is defined by selecting three points. Middle-button snap to the center vertex of each vault and then the dome. A triangular shape should then occupy the void like this:

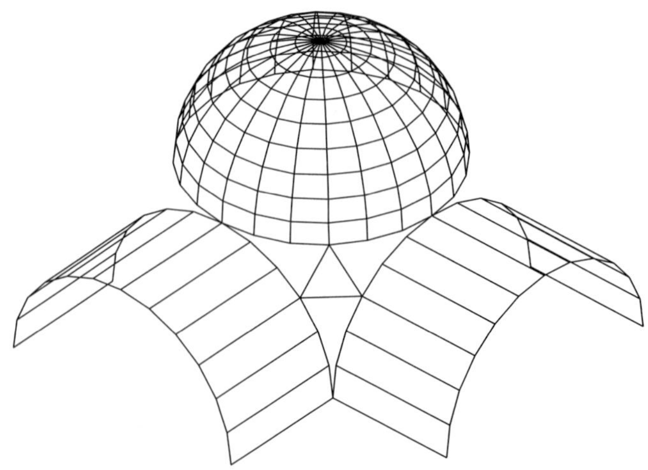

4.4 Continue creating 3-edge polygons until you have "stitched" the void closed to form a pendentive. A total of 16 triangular polygons comprise the shape.

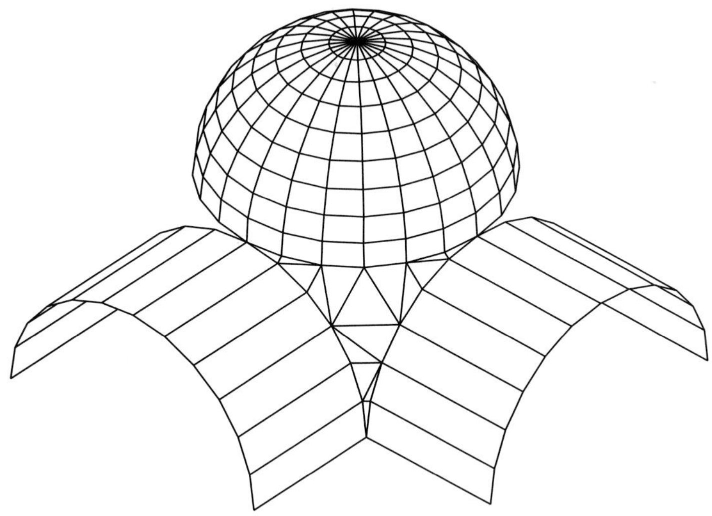

Step 5

Create the interior vaults.

5.1 Return to **Ortho** view, and make the 2D plan layer **Active only**.

5.2 Create a new layer for the interior surface of the first barrel vault and set it as the active layer.

5.3 Set the Z-base to a value of 50′-0″.

5.4 Select **3D Entity**, **Horizontal Cylinder**, **Top Half**, **Z base**. Make sure that **Closed** is toggled off. Select **Divisions** and enter a value of **24**. Select **Radius**, and enter 24′-0″.

5.5 You are prompted to "Select center point of first end of Cylinder." Middle-button snap to the center of the top wing. Drag downward and middle-button snap to center of the second dashed line, just beyond the line of the dome.

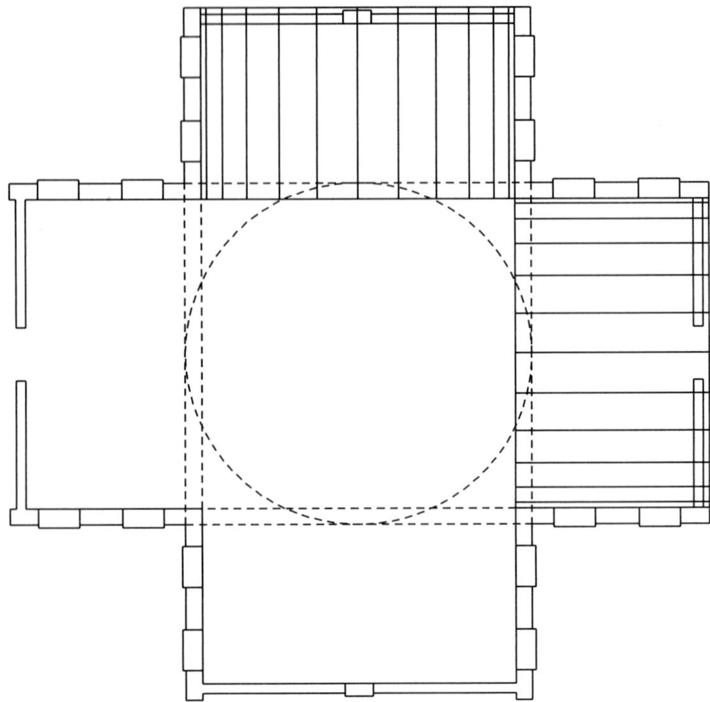

5.6 Repeat with the right wing as well to create a total of two horizontal cylinders.

Step 6

Model the interior dome ring. The ring allows the bases of the interior and exterior domes to align properly.

6.1 Create a new layer for the interior of the dome and its ring, and set it active.

6.2 **3D Entity**, **Vertical Cylinder**, **Radius Circle**, **Base/Height**. Make sure **Closed** is toggled off, select **Divisions** and set to **24**. Press the **Z** key on your keyboard, and enter 74′-0″ for the Z base, and 76′-6″ for the Z-height.

6.3 You are prompted to "Select center point of Cylinder." Middle-button snap to the center of the circle in the 2D plan, and then to the midpoint of the interior vault. Here is the result when viewed in isometric:

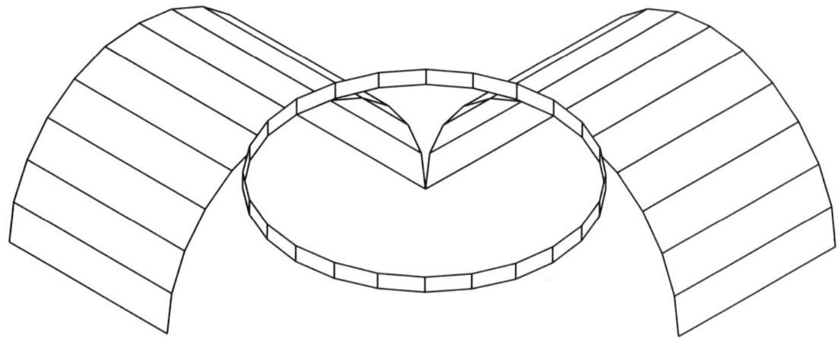

Step 7

Model the interior dome surface.

7.1 Create a new layer for the interior of the dome.

7.2 Select **3D Entity**, **Sphere**, **Radius Circle**, **Top Half**, **Z Base**. Press the **Z** key on your keyboard, and set the Z-base to a value of 76′-6″, the height at the top of the ring.

7.3 You are prompted to "Select center point of Sphere." Middle-button snap to the center of the circle on the 2D plan layer, and then middle-button snap again to the interior vault at its midpoint. Here is the result showing the dome, ring and vaults:

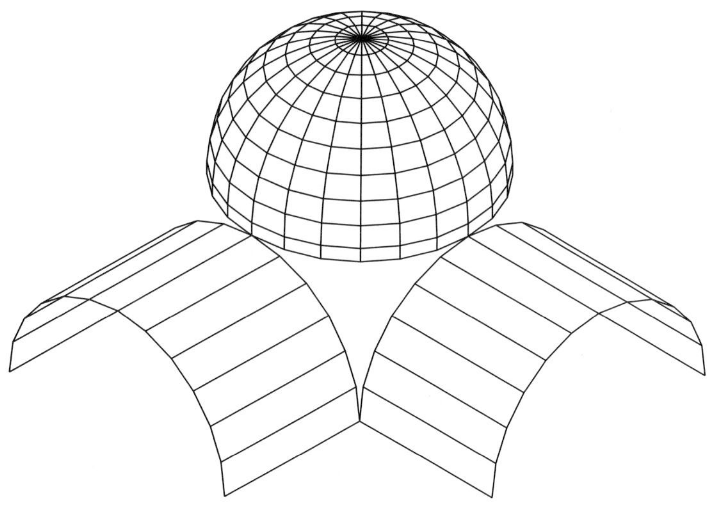

Step 8

Repeat Step 4 to model the interior pendentive:

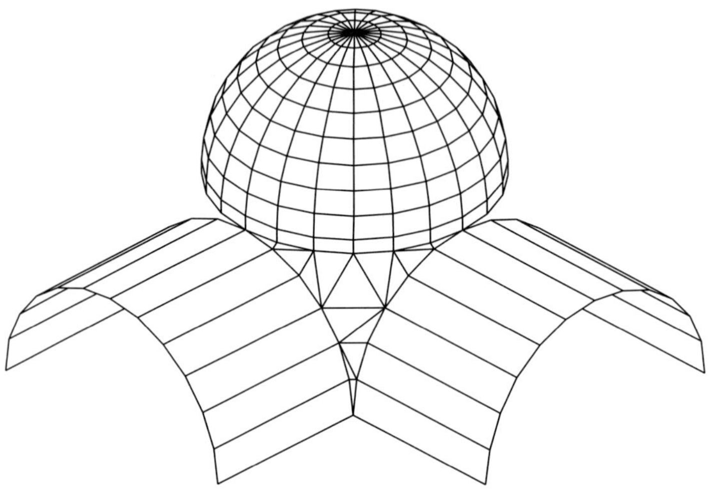

Step 9

This view shows the open space between the two vault surfaces. Next, the outside ends of the vaults will be capped.

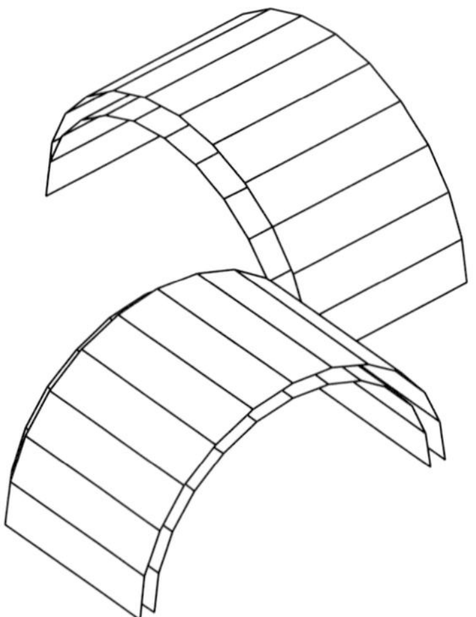

We will employ a 3D trick to create a perfect closure for the two cylinders using a truncated cone.

Normally, a truncated cone looks like the figure on the left. However, if we set the Z-base and Z-height to equal values, we can create a shape like the one on the right.

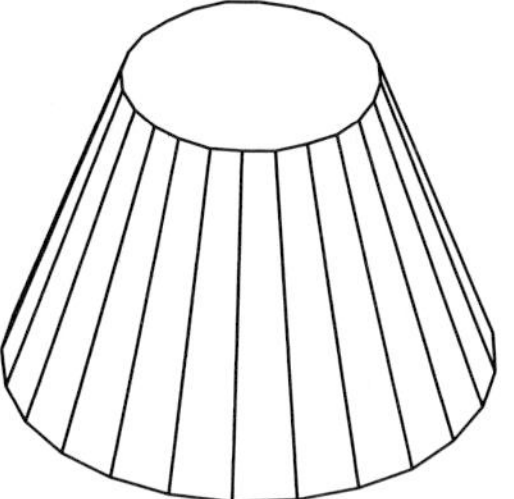

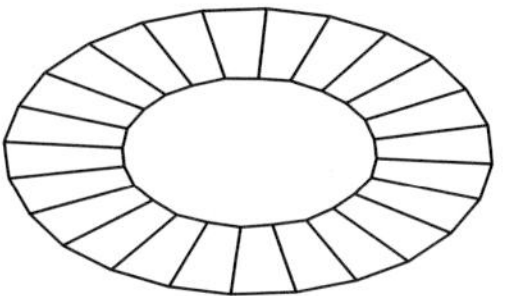

9.1 Turn off all displayed layers except the two for the interior and exterior vaults. Create a new layer for the vault caps, and set this layer active. Make sure **Ortho** mode is toggled on.

9.2 Set the Z-base and Z-height values to zero.

9.3 We will start with the vault on the right. Select **3D Views, Elevation, New Elevation**.

9.4 You are prompted to "Select first point of Elevation Line."

Middle-button snap to the center of the vault as shown then drag to the left and left-click. You are presented with an end-on view of the vault.

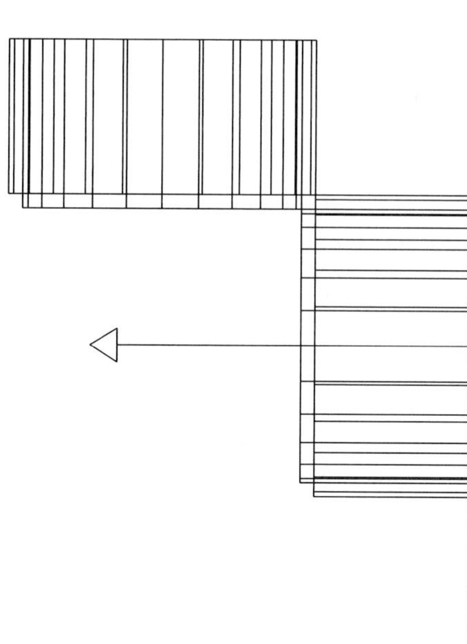

9.5 Select **3D Entity**, **Truncated Cone**, **Radius Circle**. Make sure **Closed** is off and **Auto Height** is on. Select **Divisions**, and set to **24**.

9.6 You are prompted to "Select center point of Truncated Cone." Middle-button snap to the center marker of the vault cylinder, then drag to the lower edge of the outside vault, and middle-button snap again. This defines the extreme edge of the cone. You are now prompted to "Indicate radius of top of truncated cone." Middle-button snap to the lower edge of the inside vault. The result should look like the figure on the left. On the right is an isometric view.

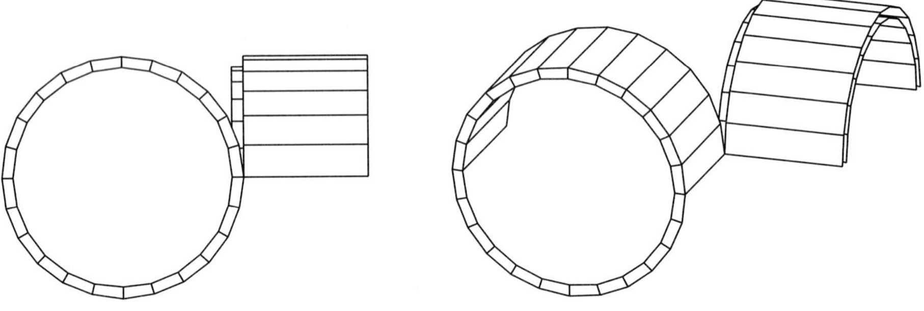

Step 10

Now we will eliminate the lower half of the flat ring. To do so, we must "explode" it first so that it consists of individual pieces instead of one entity.

10.1 Select **3D Edit**, **Explode**, **To Polygons**. Set your selection mode to **Entity**, and left-click anywhere on the lower half of the truncated cone. DataCAD returns the message "Exploded entity is a Truncated Cone." The cone is now composed of 24 polygons, and we may erase the 12 in the lower half.

10.2 Select **3D Entity**, **Polygons**, **Partial**, **Mark All**. Change your selection mode to **Group**, and left-click anywhere on the lower half of the exploded truncated cone. This will force DataCAD to display all of the edges of the individual polygons, even if two edges share the same line. This makes editing easier.

10.3 Now select **3D Edit**, **Erase**. Change your selection mode to **Entity**, and erase all the polygons that comprise the lower half of the flat ring. The result should look like this:

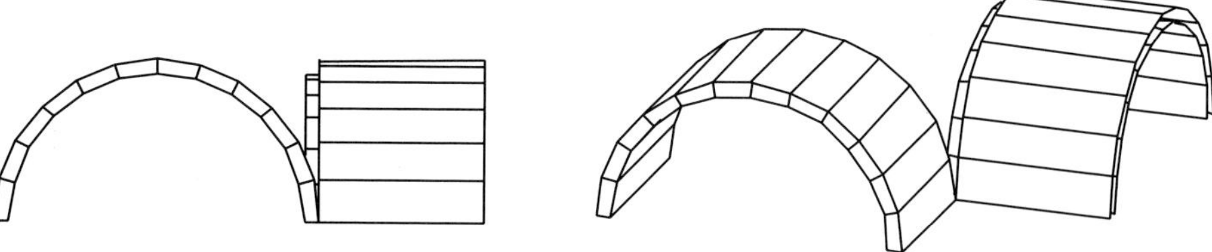

Step 11

Repeat Steps 9 through 10 on the other remaining vault to cap its end as well.

Step 12

Model the walls below the vaults.

12.1 Create a new layer for the 3D walls, and set it to **Active Only**. Set view to **Ortho**. Turn on the layer for the 2D plan.

12.2 Press the **Z** key on the keyboard, and set the Z-base to 0′-0″ and the Z-height to 50′-0″.

12.3 Select **3D Entity**, **Polygon**, **Vertical**. Middle-button snap to the lines that correspond to the "L" shape that is formed by the two vaults. This will require 6 polygons, as shown in the following isometric view. Note that we are only modeling the exposed surfaces. The open tops will be closed off by the cylinders above.

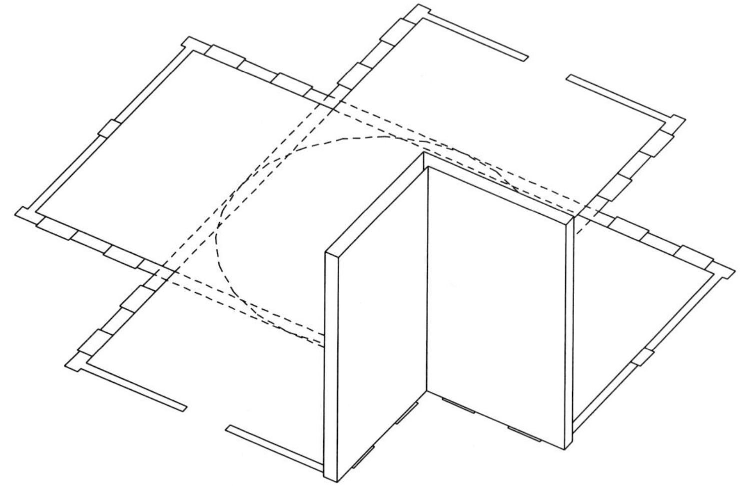

Step 13

Model the infill end walls.

13.1 Define a **New Elevation** vault-end view as in Step 9.3, but middle-button snap to the outside line of the infill wall instead of the vault.

13.2 Turn on the layer that contains the vault-end caps modeled in Steps 10 and 11.

13.3 Select **3D Entity**, **Polygon**, **Rectangle**. Press the **Z** key on the keyboard and set the Z-base and Z-height to 0′-0″.

13.4 You are prompted to "Select first corner of Rectangular polygon." Middle-button snap to the inside corner of the vault cap at the top left and then again to the lower corner of the inside wall at the lower right, as shown in the following diagram.

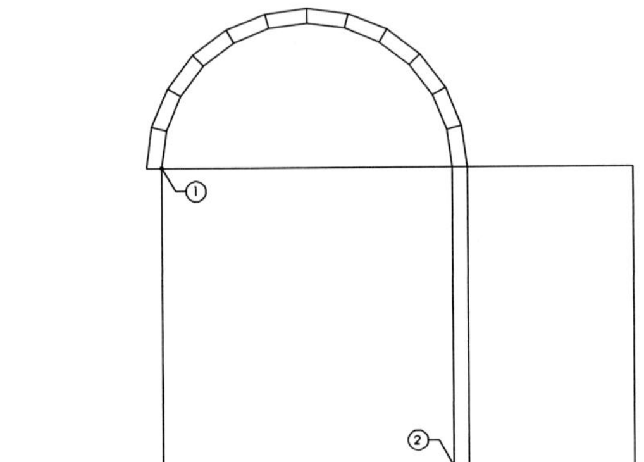

13.5 Select **3D Entity**, **Polygon**, **Partial**, **Add Vertex**. Left-click anywhere on the top edge of the rectangular polygon that you just created. Its edge becomes a flexible boundary. Middle-button snap to the inside center of the vault.

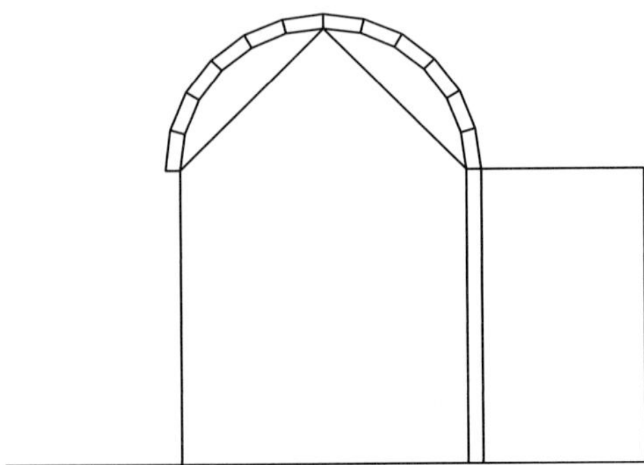

13.6 Continue to add vertices until you have filled in the vault.

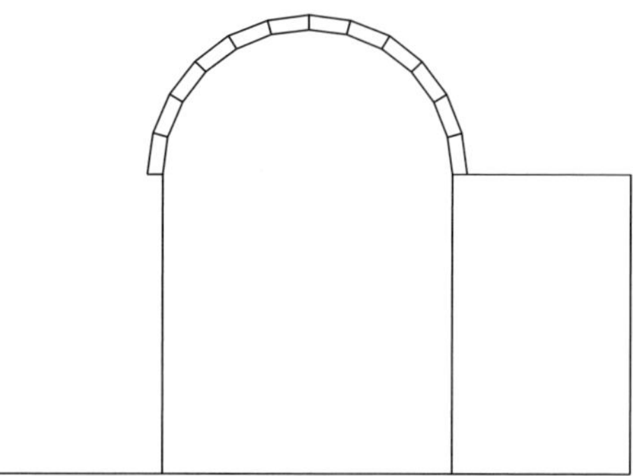

13.7 Select **3D Edit**, **Copy**, **Z-Only**, **Z Distance**, and enter −1′-6″. Pick the polygon anywhere on its left edge. This will copy it downward 1′-6″ to create the other side of the wall. Here is the result when viewed in isometric:

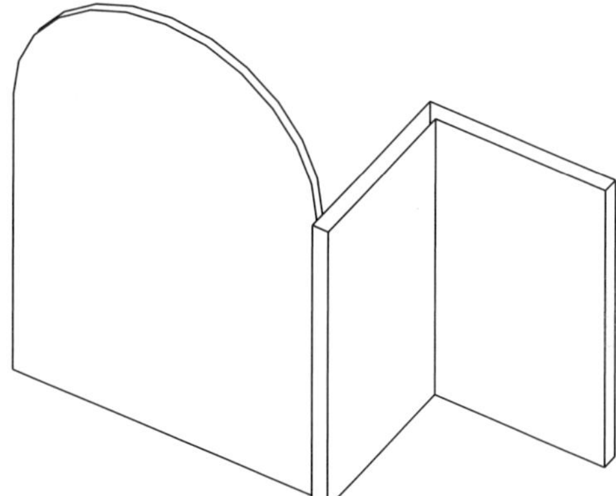

Step 14

Model the window cutouts and sills using the **AEC Model** macro. An improved version of this macro is included with the bonus CD provided with this book, and should be installed before proceeding further.

14.1 Make the layer for the 3D walls **Active Only**, then turn on the layer for the 2D plan.

14.2 Select **2D Edit**, **Toolbox**, **AEC_Modl**, **Open**. Next pick **Windows**. Make sure **In Plan** and **Cut Wall** are toggled on. Select the following buttons and enter these values:

Head Height = 15.6

Sill Height = 5.6

Unit Type = **Fixed**

Casing = **At Head**, **At Jamb** and **At Sill** toggled off

Trim = **At Head**, **At Jamb** and **At Sill** toggled off

Head = **Do Head**; **Wall Width** toggled on, select **Head thickness** and enter 1″

Jamb = **Do Jamb**; and **Wall Width** toggled on, select **Jamb thickness** and enter 1″

Sill = **Do Sill** toggled on, **Thickness** to 4″, **In Extension** to zero, **Out Extension** to 6″, **Side Extension** to zero.

Sash = **Do Sash** off

Muntins = **Do Muntin** off

Glass = **Do glass** off

This will model a cased opening with a sill only.

14.3 You are prompted to "Enter a point on first inside corner of window." Refer to the following diagram, and middle-button snap to the inside corner of the first window in the wall at point A, then again at point B. Repeat this procedure for the next window, at points C and D.

14.4 Select **Sill Height**, and change it to 31′-6″, select **Head height** and change it to 41′-6″.

14.5 Repeat Step 14.3 at points A, B, C, and D to model the windows above.

14.6 Repeat the window modeling process in Steps 14.0–14.5 on points E, F, G, and H.

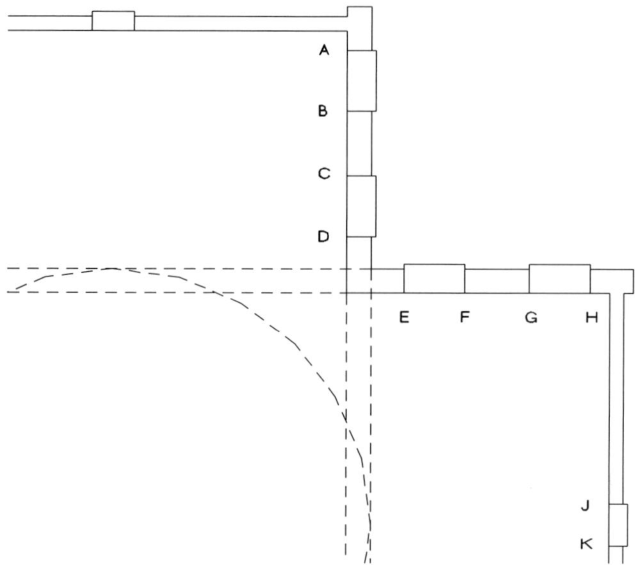

14.7 Select **Sill Height** and change it to 58′-6″, select **Head height** and change it to 68′-6″.

14.8 At the prompt, "Enter a point on first inside corner of window." Middle-button snap to the inside corner of the first window in the wall at point J, then again at point K.

You have now completed the windows; here is how it looks in isometric view:

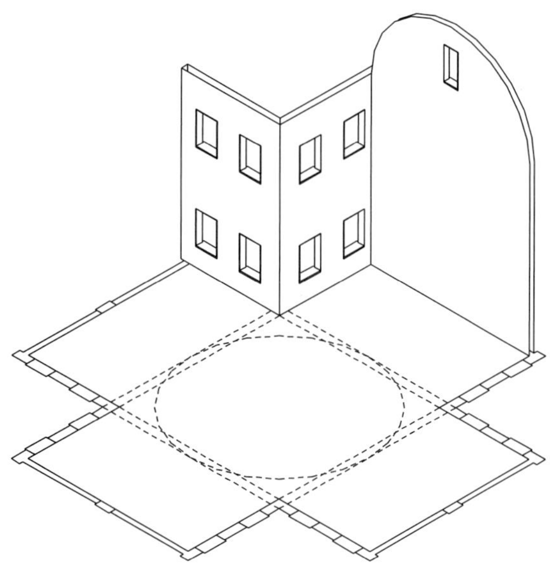

Step 15

Complete the building.

15.1 Select **2D Utility**, **Layers**, **On-Off**. The layers for the 2D plan and 3D walls should already be on. In addition to these, turn on the layers for the interior and exterior pendentives. Exit. Select **Lock**, and pick the 2D plan layer.

15.2 Select **2D Edit**, **Copy**, **Circular Array**. You are prompted to "Select center point of array." Middle-button snap to the center of the 2D circle in plan. You are next prompted to "Select center of object." Middle-button snap to the center of the 2D circle in plan again. At the "Select angle between items" prompt, type 90.0 and press **<Enter>**. DataCAD prompts you to "Enter number of objects in array." Type **4** and press **<Enter>**. Make sure that **Layer Search** is toggled on. Set your selection mode to **Area**, and define an area box that encompasses the entire model. The model is rotated and copied about the center point 3 times. Note that when copying by array, the existing elements constitute the first instance of the array.

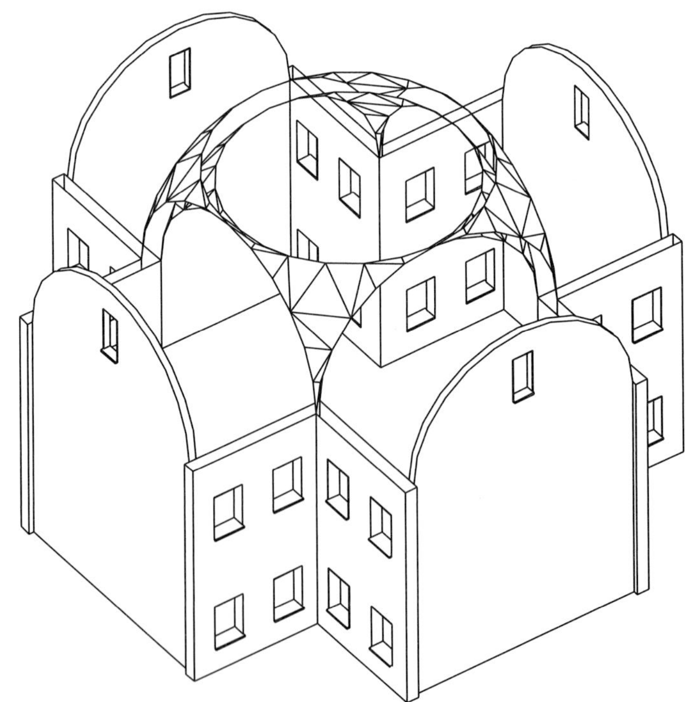

Step 16

Complete the vaults.

16.1 Select **2D Utility**, **Layers**, **Active Only** and pick the locked 2D plan layer. Select **Layers**, **On-Off**, and turn on the layers that contain the interior and exterior barrel vaults, and the vault caps.

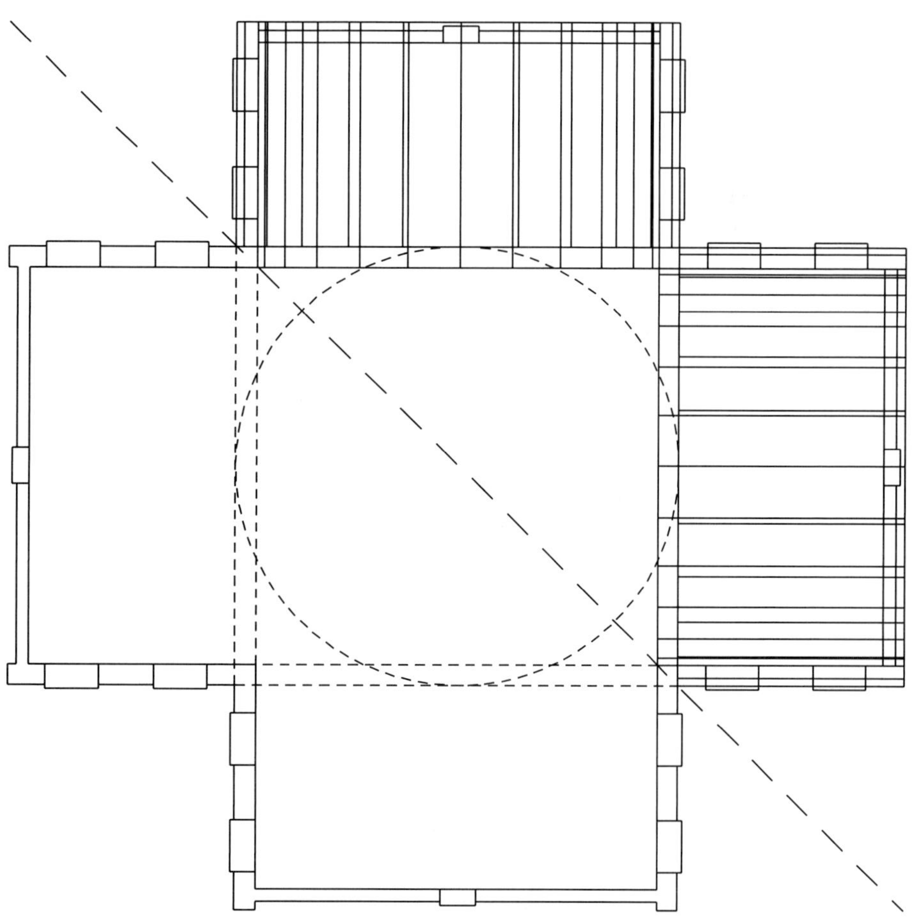

16.2 Select **2D Utility, Mirror**. You are prompted to "Select first point along the line of reflection." Make sure that **Ortho** mode is toggled on. Middle-button snap to the center of the circle in the 2D plan. Drag down to the lower right at a 45-degree angle and left-click. A dashed line indicating the line of reflection is drawn. Make sure that **Area**, **Layer Search** and **And Copy** are toggled on. You are prompted to "Select first corner of area to Mirror." Define an area box that encompasses the entire model. The vaults and end caps are mirrored and copied to the other side, completing the model.

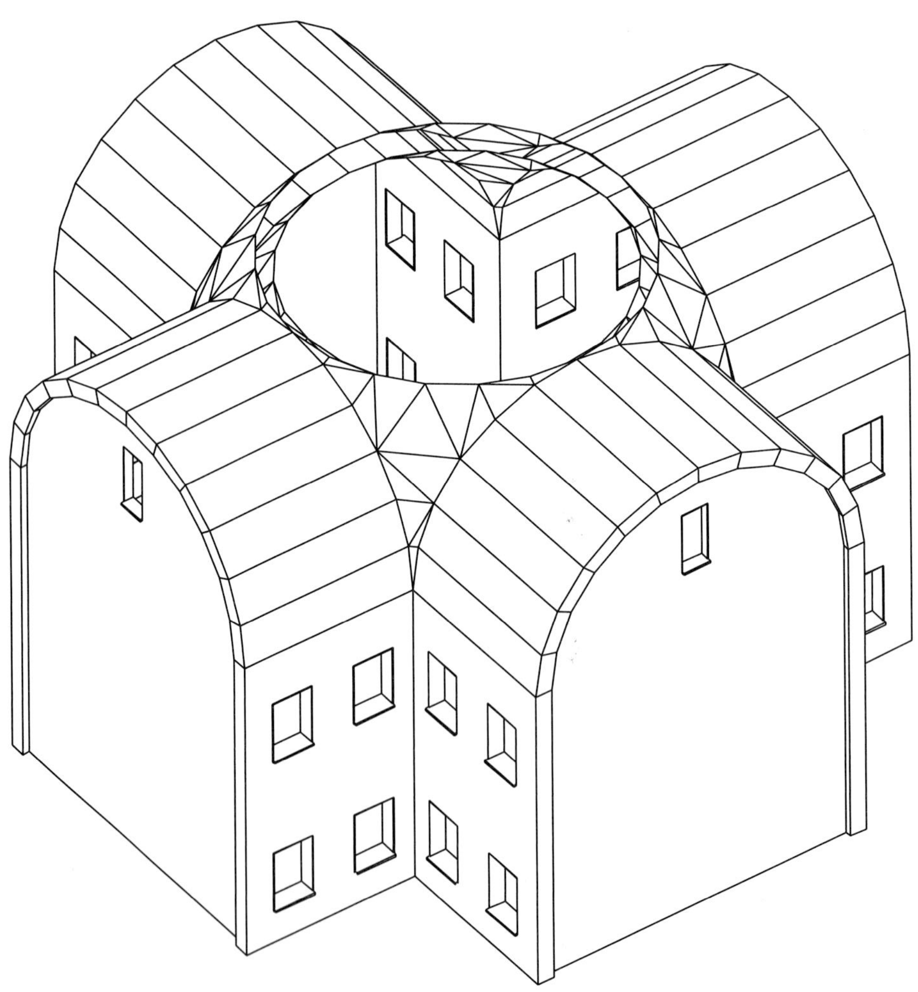

Step 17

Creating the oculus.

17.1 Select **2D Utility**, **Layers**, **Active Only**, and choose the layer that contains the exterior dome. Select **3D Edit**, **Explode**, **To Polygons**. Change your selection mode to **Entity** and left-click anywhere on the dome. The dome is exploded into individual polygons.

17.2 Select **3D Entity**, **Polygons**, **Partial**, **Mark All**. Change your selection mode to **Group** and left-click anywhere on the lower half of the exploded dome.

17.3 Change your view to **Ortho**.

17.4 Select **3D Edit**, **Erase** and choose **Fence**. Define a fence that encloses the inner circle at the top of the dome to erase only those polygons:

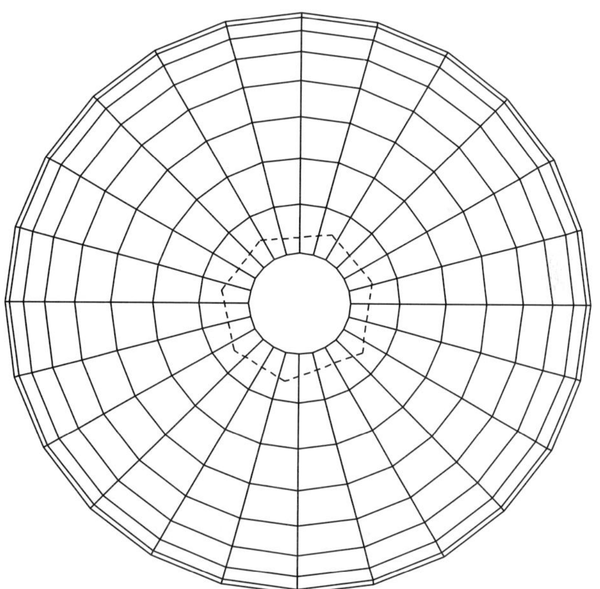

Step 18

Repeat Steps 17.1–17.4 on the layer containing the interior dome.

Step 19

Model the rim of the oculus.

19.1 Turn on the layers for the interior and exterior dome. Notice that the two openings created at the top do not quite align. We will use an open, inverted, truncated cone to close the gap.

19.2 First, we will use a trick to set the Z-heights: Select **3D Entity**, **3D Line**, **3D Cursor**. Middle-button snap to points A and B, as illustrated in the following diagram. The line created spans between the two domes. Select **Identify**, **Set All**, and pick the 3D line with your left mouse button. This sets all properties to match the line. Now erase the line.

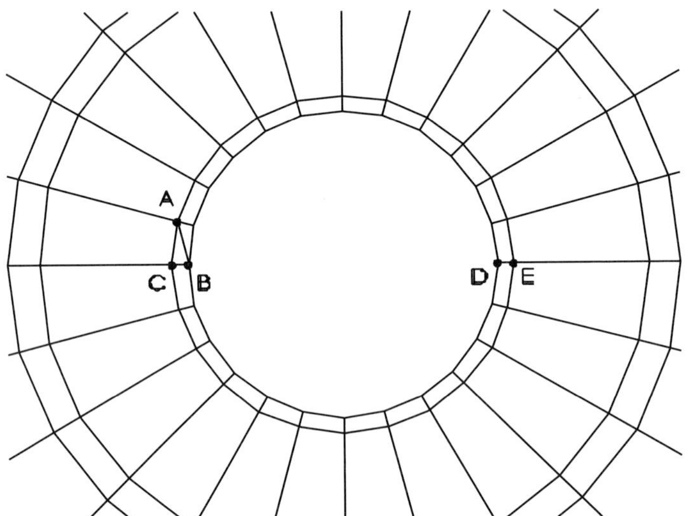

19.3 Select **3D Entity**, **Truncated Cone**, **Diameter Circle**. Make sure **Closed** is off and **Auto Height** is on. Select **Divisions** and set to **24**. You are prompted to "Select first point of diameter for Truncated Cone." Middle-button snap to point C then to point E. You are now prompted to "Indicate radius of top of truncated cone." Middle-button snap to point D. The truncated cone is created, and the model is complete!

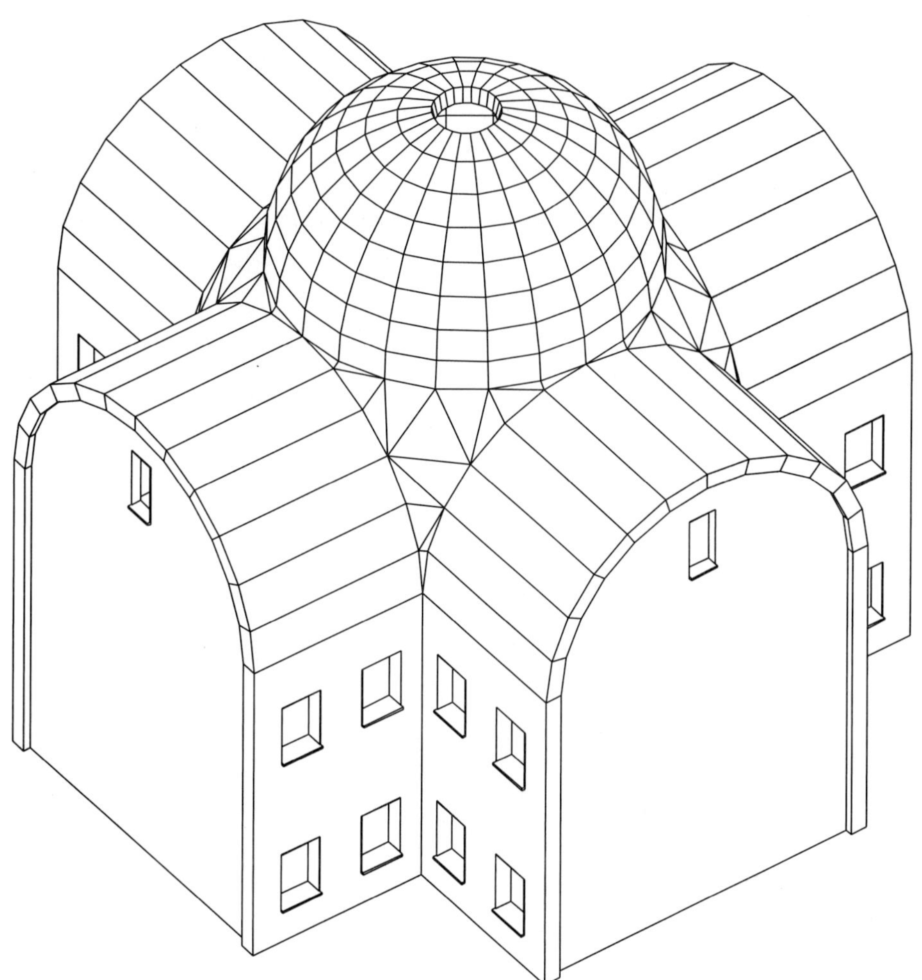

Exercise 3—The Classic Sweep: A Stair Handrail

This exercise introduces the sweep concept in 3D modeling. A sweep consists of a cross-section and a path which it is applied to. The result is a collection of polygons that reflects the cross-section as it is swept along the path. The cross-section remains constant, while the path is variable. The illustration of the handrail clearly shows the cross-section and path relationship. This exercise will take you through the process of performing this type of sweep.

Before you begin, you will need to install **the 3D Power Tools** macro, available separately from DataCAD LLC. Open the file **Exercise_3.DC5** to follow along in the model file.

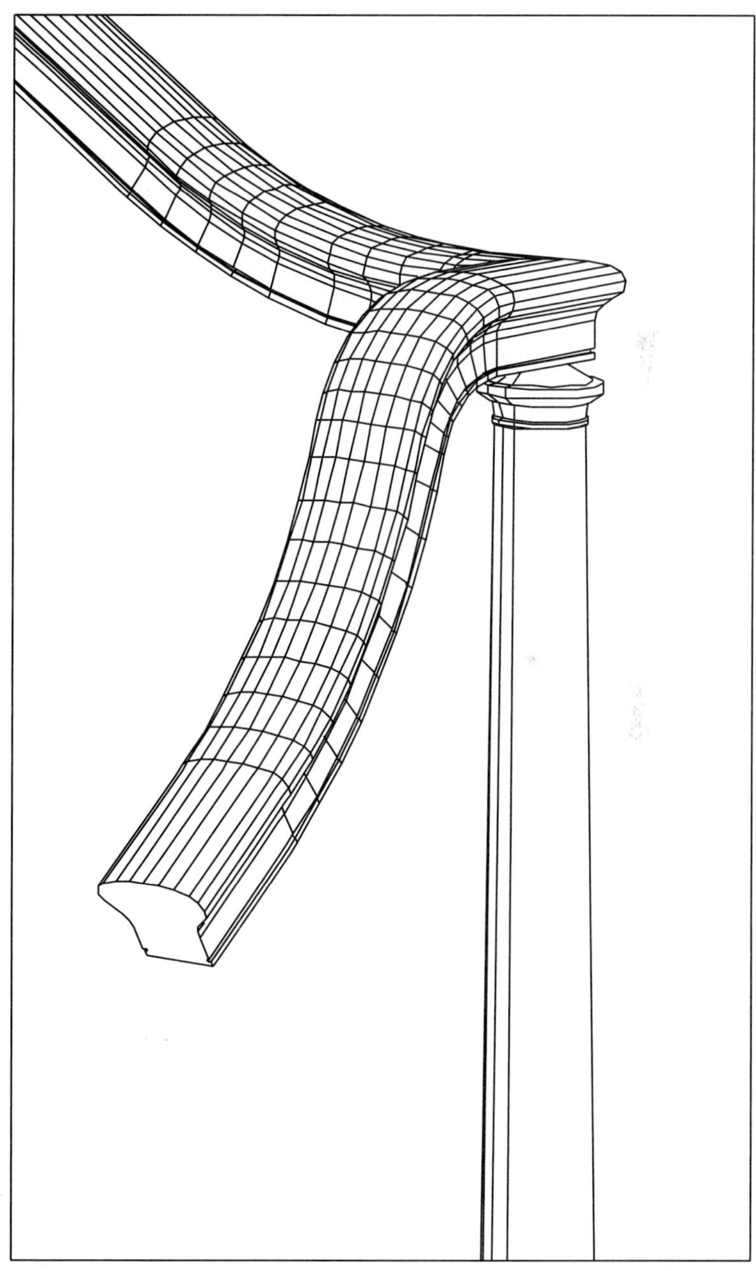

Step 1

The first step is to draw the path that the sweep cross-section will take. In the example, 3D lines were traced out from parallel New Elevation views. Sweep paths may also take the shapes of open or closed polylines (most common), 2D lines, 3D lines or 3D contour lines. To create the 3D path shown, make **Active Only** the layer for the guide points. Change to an isometric view that allows you to clearly distinguish the order of the marker points. Select **3D Entity**, **3D line**. Toggle **Chain** on. Middle-button snap from the first to the last point of the rail path, from top to bottom. It is important to create the path in one continuous chain so that it will sweep correctly from entity to entity. Here is the result:

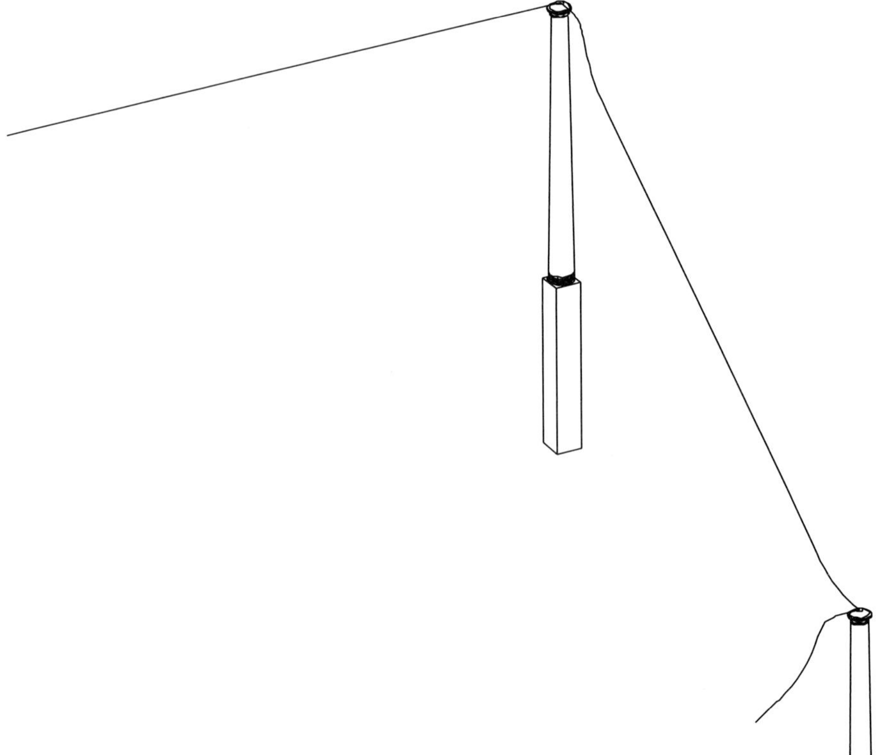

Step 2

Create the sweep cross-section.

2.1 Create a layer for 2D guide lines. Press the **Z** key on your keyboard, and set your **Z-base** and **Z-height** to 0′-0″. Draw the shape of the cross-section you wish to use for the sweep. Give consideration to where it will align with the sweep path.

2.2 Change to **Ortho** view and change your active layer to that of the sweep path. Select **3D entity**, **Polygon**, **Horizontal**. Toggle on **Z-base** to create the polygon at 0′-0″ in elevation. Trace the shape of the cross-section from the 2D guide line layer. It is helpful to place a vertex at the point of alignment with the sweep path, shown here as point A.

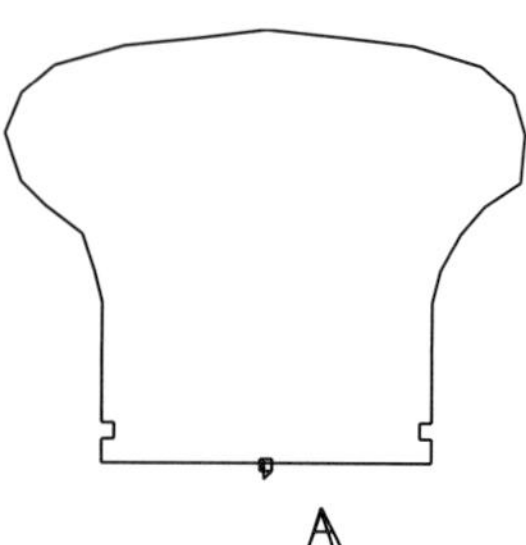

Step 3

Execute the sweep.

3.1 Change your view to **Ortho**. Check to see that **Layer Search** is toggled on. Create a **New Layer** for the 3D handrail, and make it **Active Only**. Turn on the layer for the cross-section polygon and the 3D path. Make sure that all the elements are visible on your screen. Select **Toolbox**, **3D Tools**, **Sweep**. You are prompted to "Select polygon entity that defines sweep cross-section." Pick the polygon that you created in Step 2. Next, you are prompted to "Enter a point for sweep path origin." Snap to point A on the polygon cross-section. Additional menu options appear. Select **Settings**. Set the following options:

Toggle off **Vertical** (applies cross-section perpendicular to sweep path)

Toggle on **Show Latitude** (shows horizontal divisions between segments)

Toggle on **Show Longitude** (Shows vertical divisions between segments)

Toggle on **Z-Base** (aligns cross-section to Z-base of path)

3.2 Exit the **Settings** menu by right-clicking your mouse or choosing **Exit** from the menu.

3.3 You are prompted to "Select entity to use for sweep path." Toggle on **Group** as your selection method. Toggle on the option for **Cap Ends**. This will close the openings at each end of the sweep.

3.4 Left-click on the 3D path to choose it. The sweep is created. Here is the result viewed in isometric:

Turn on the other layers associated with the 3D Stair to view the completed model:

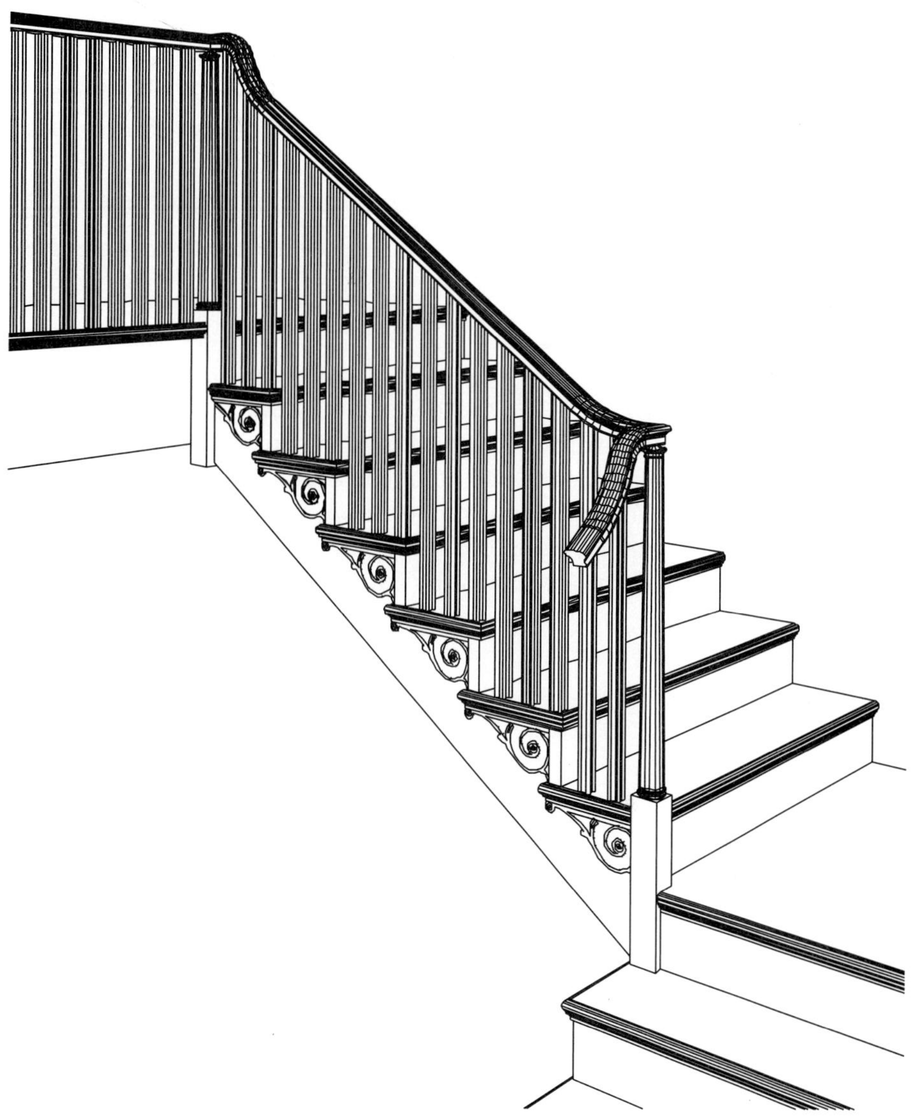

Knife Cuts and Mitres

When working with swept or extruded forms, it is common to need to create an angled or mitred joint. Using the previous example as a point of departure, here is how to create a 45-degree turn with an extruded shape.

Step 1

Model the shape with slabs, polygons or the **Sweep** function. If you opted to use slabs, explode them to polygons before proceeding to the next step.

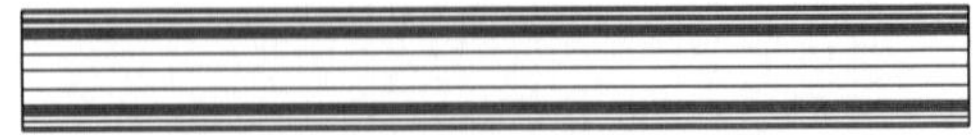

Step 2

Draw a 2D guideline indicating the mitre. Select **Toolbox**, **3D Tools**, **Knife**. You are prompted to "Enter first point of knife line." Middle-button snap to the end point of the 2D guide line, then again to opposite end point. You are next prompted to "Enter a point on the keep side of the knife." Click any point on the side of the line that is not to be cut away. Finally select the entities to be cut by **Entity**, **Area**, **Group**, **Fence**, or **Selection Set**.

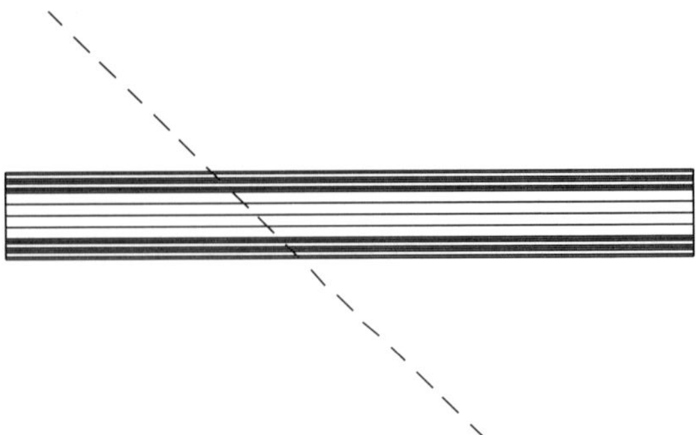

The polygons are cut back to the knife line.

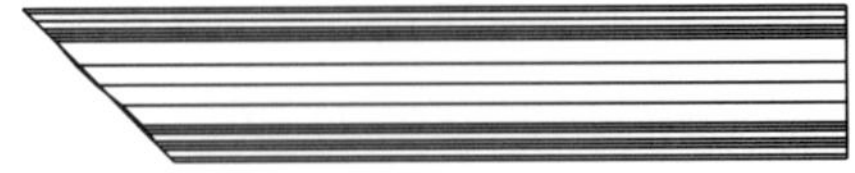

Step 3

Select **2D Edit**, **Mirror**, And **Copy**. Mirror the entities about the mitre line.

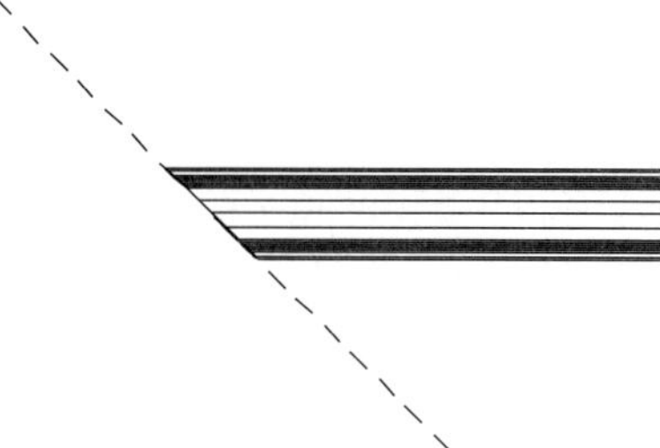

The polygons are mitred to create a 45-degree mitred joint.

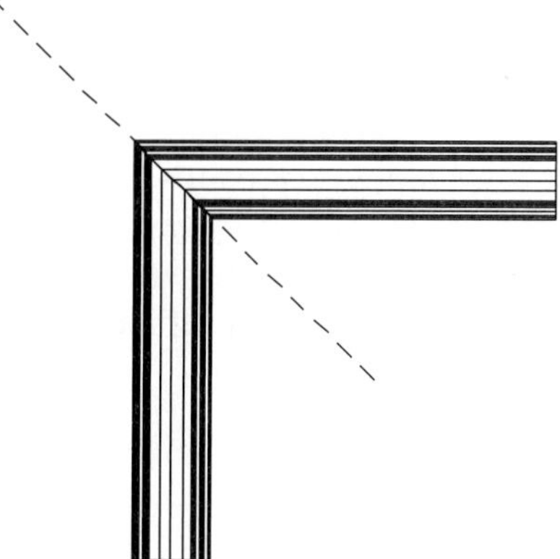

The completed miter is shown.

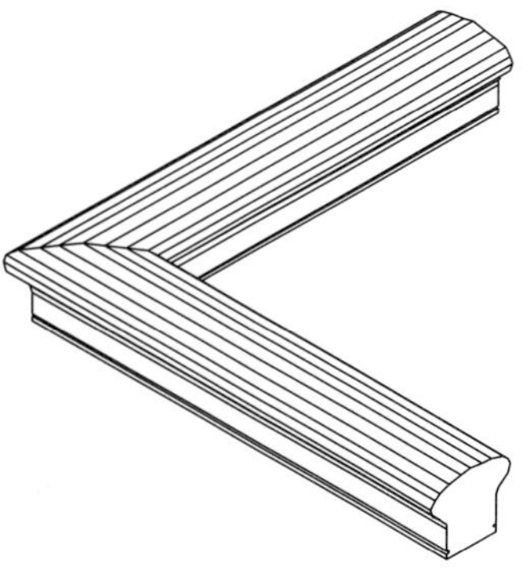

A Note about the Model The file used in this example contains a complete 3D model of a complex stair. We have only focused on how the handrail was created, but you may be interested to know how the rest of the model was accomplished. The risers are simple rectangular slabs. The stringers were created by tracing their shapes after defining a **New Elevation** at their edge. The treads and landings were created with the **RevSurf** $\times$ $\sqrt{\mathbf{2}}$

technique outlined in Chapter 19, Exercise 8. The balusters are horizontal slabs, some of which were exploded to polygons and then cut-to-fit with the **Knife** tool. The detailed scroll work on the risers was traced from a digital photo using a shareware program called **Win Topo**. The trace was pasted into DataCAD, then traced over again with a horizontal slab. **The Tilt-Up Slab** technique was used to rotate the slab vertically, then it was positioned on the stringer and copied to each tread.

Variations

Sweeps are incredibly versatile and useful to the 3D modeler. You can achieve different results with a sweep by enabling the options for **Twist** and **Taper**. This allows you to specify the number of revolutions the cross-section will take as it is applied to the path, or change its proportion as it progresses from start to finish. Illustrated are some of the possibilities:

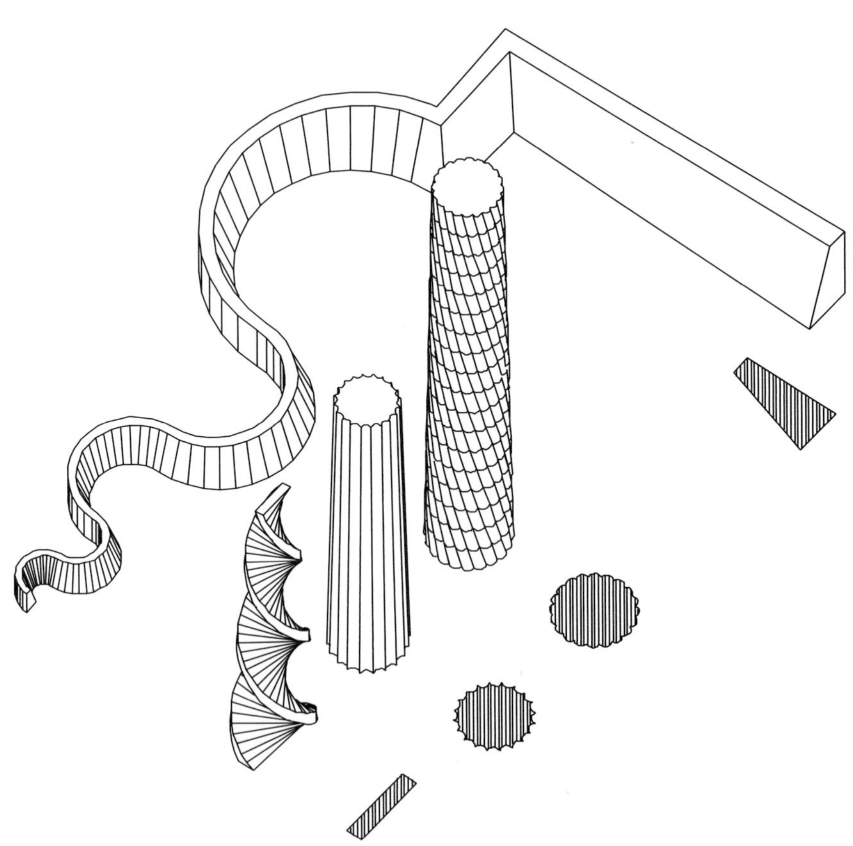

Exercise 4—Land Forms: Four Methods

Before you begin, you will need to install the **PolyMaster**, **dhContour**, and **dhTIN** macros* found on the bonus CD. If you wish to complete all of the exercises shown, you will need to have the **3D Tools** and **DC Sprint** macros installed as well. These macros, available separately from DataCAD LLC, are highly recommended, as they make certain tasks feasible that would otherwise be difficult or impossible to perform. Open the file **Exercise_4.DC5** if you wish to follow along in the model file.

These examples provide an overview of site modeling in DataCAD. Four methods of varying complexity are described, listing the advantages and disadvantages of each. Terrain modeling itself is a complicated subject. There are a multitude of programs dedicated to this one task. Though the

*The **PolyMaster** macro was contributed by Patrick McConnell, an architect in Valatie, New York, USA. The **dhTIN** and **dhContour** macros were written by David Henderson of dhSoftware, Springwood, Australia.

methods shown here are quite simplified, they encompass the typical routines used by terrain modeling software. The goal here is to provide you with some proven techniques to create a landscape for your architectural models. Choose the method that best suits your project and the complexity of your site conditions.

All terrain modeling functions have one thing in common: They generate a multitude of triangular polygons to form a surface passing through the points you have delineated, to represent the site topography. Triangles are used because they are the simplest form of a surface. Any array of points can be connected with triangles to create a contiguous surface.

We will examine four techniques to create a terrain model.

- *Method One* Drop mesh
  - *Difficulty Level* Very easy
  - *Data required* 3D contour site drawing or point field
  - *Advantages* Extremely easy and fast to produce, once you have entered or imported the data. Renders acceptably.
  - *Disadvantages* Does not convey the terrain at all when viewed in plan, though somewhat in isometric and perspective. Extremely uniform appearance. Difficult to control the results.
  - *Suitability* Best for use in rendering, or elevation studies where you need to establish the ground line. Good for all types of sites, although it works best with large changes in elevation.
- *Method Two* Slab on slab
  - *Difficulty Level* Easy to moderate
  - *Data required* 2D contour site drawing
  - *Advantages* Easy to produce from contour plan, conveys topography in a familiar way. Contours are discernable. Only method that makes it easy to cut out portions to receive your model.
  - *Disadvantages* Can be time-consuming and tedious to complete. Does not render realistically.
  - *Suitability* Best for design studies. Hides and renders well, though not realistically. Good for all types of sites.
- Method Three Contour rail sweep
  - *Difficulty Level* Moderate
  - *Data required* 3D contour site drawing
  - *Advantages* Conveys topography well in all views. Contour lines are discernable. Renders well.

- *Disadvantages* Produces a very dense model, and hence does not hide well. Can be difficult and tedious to execute on complex sites.
- *Suitability* Best for use in rendering, and on sites with gentle inclines and little complexity.

- *Method Four* *Triangular Irregular Network* (TIN) modeling
  - *Difficulty Level* Easy
  - *Data required* Point field
  - *Advantages* Easy to produce once you have entered or imported the the data. Renders extremely well. Produces an interesting and attractive model.
  - *Disadvantages* Can take hours for the computer to process the points. Contours are not discernable.
  - *Suitability* Best method for rendering. Hides well. Good for all types of sites.

Entering the Site Data

The sample file contains the different types of site data used on individual layers so that you can try each method yourself.

The first step is to draw the site plan showing the contour elevation lines. In some cases, this information is available from the surveyor. As all the terrain modeling methods require three-dimensional data, the contour lines should be entered at their relative Z-base. You can use 3D contour lines, 3D lines, or 2D lines set to equal Z-base and Z-height. Check the modeling method you wish to use for what it requires.

NOTE: *When entering topographic lines hundreds of feet above sea level, set the lowest line at 0'-0" instead of its actual height. This will make the increments easier to follow as you draw, and will keep the model reasonably close to the zero plane.*

Option 1: Polylines Polylines offer many advantages for site drawings. They support Z-base and Z-height values, can be open (start and end at different points) or closed (continuous loop), and are easily modified. When used in conjunction with the **3D Tools** macro, polylines can generate a slab site model very quickly. They are also effective for drop mesh models.

The following site drawing was created with open 2D polylines set to corresponding Z-base and Z-height values.

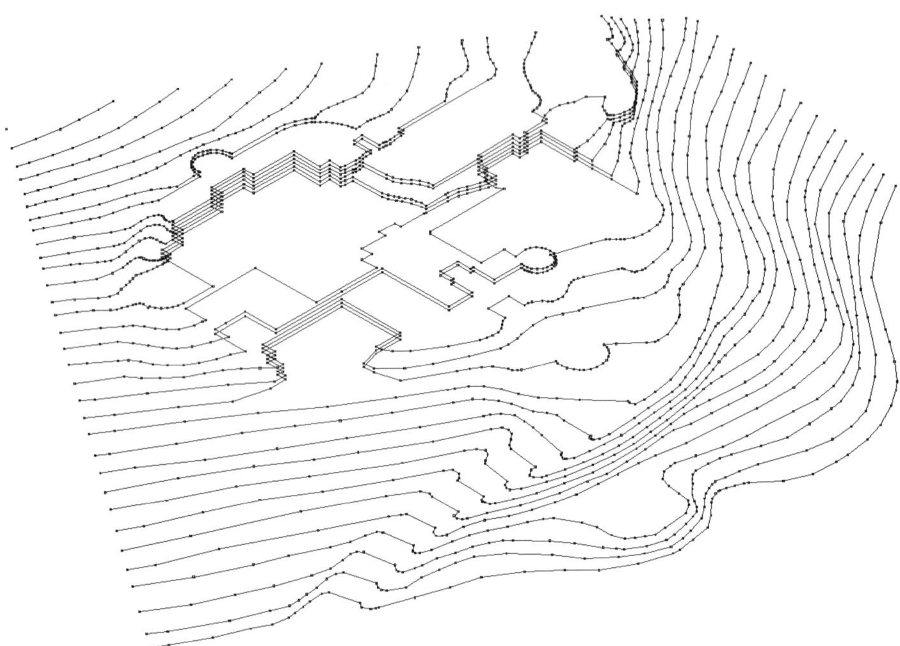

To create **polylines**, perform the following steps:

1. Set **Z-base** and **Z-height** accordingly, such as 0–10, 10–20, 30–40, etc.
2. Select **2D Edit Curves**, **Polyline**, **Open** or **Closed**.
3. Trace a circuit for each elevation contour of the site plan. Change your **Z-base** and **Z-height** settings between polylines.

NOTE: *To change or modify a polyline after it has been created, use the* ***Polyline*** *macro listed in your* ***Toolbox****.*

Option 2: 3D Contours 3D Contours are similar to Bezier curves, in that they are defined and controlled from their input points. When entering contour lines you are allowed to specify up to 36 points, and the contour line is plotted between them. You can modify the **Stiffness** setting to control the degree of curvature between points, and the **Divisions** setting for the apparent smoothness.

Contour lines make an attractive site drawing with smooth fluid lines, but they don't support linetypes well. The pattern gets expanded and compressed as it is applied to the line. They can also be a bit tricky to get just right. It takes some practice to get a feel for how many control points to place along a line. Study the sample drawing and try creating a few on your own. Contour lines are effective for 3D rail sweeps and drop mesh models. The following is a topography drawing with 3D contour lines:

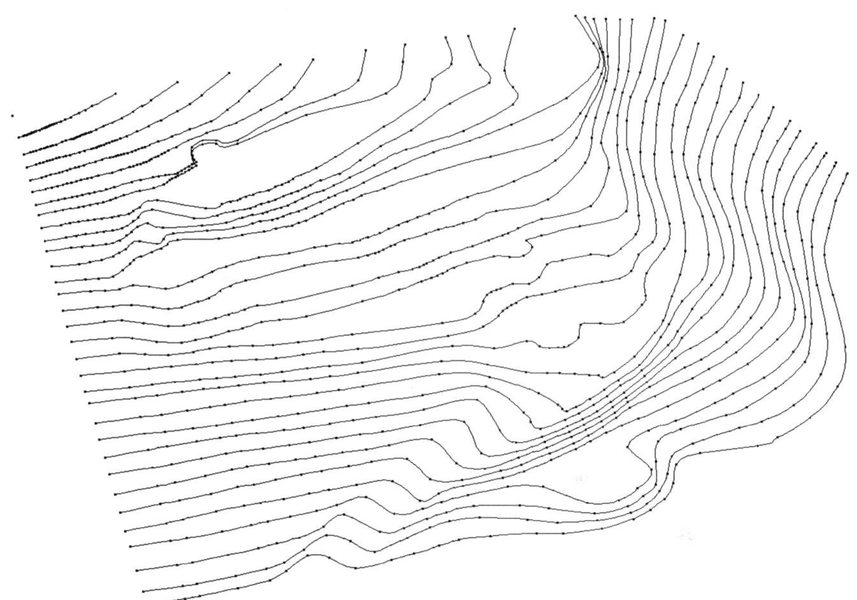

To enter **3D contours**, perform the following steps:

1. Press the **Z** key and set your **Z-base** to the elevation of the first topographic line.
2. Select **3D Entity**, **Contour**, **Natural**. Toggle on **Z-base**. Set **Divisions** between 4 and 10, and **Stiffness** to **.5**. Begin placing points along the topographic lines. You can enter a maximum of 36 points. If you run out of points before you reach the end, begin a new contour line from the previous one.
3. Change your Z-settings and start the next contour line.

NOTE: *You can set the **Add Increment** value to your elevation interval to easily add to the current **Z-base** setting.*

Editing Contour Lines One of the drawbacks to using contour lines is that once they are created, you cannot add or delete points. You can, however, stretch the control points around. On the bonus CD is a macro called **dhContour**. This remarkable tool allows you to add, remove or relocate points on a contour line. Simply launch the macro and select the contour to edit.

Option 3: 3D Line Segments 3D Lines are a primitive but effective way to create site data for use in drop mesh models. Simply **Trace** the site with chained 3D lines at the appropriate Z-height. The disadvantage is that if you need to edit them later, you will have to erase and redraw many line segments. The following is a topography drawing with 3D line segments.

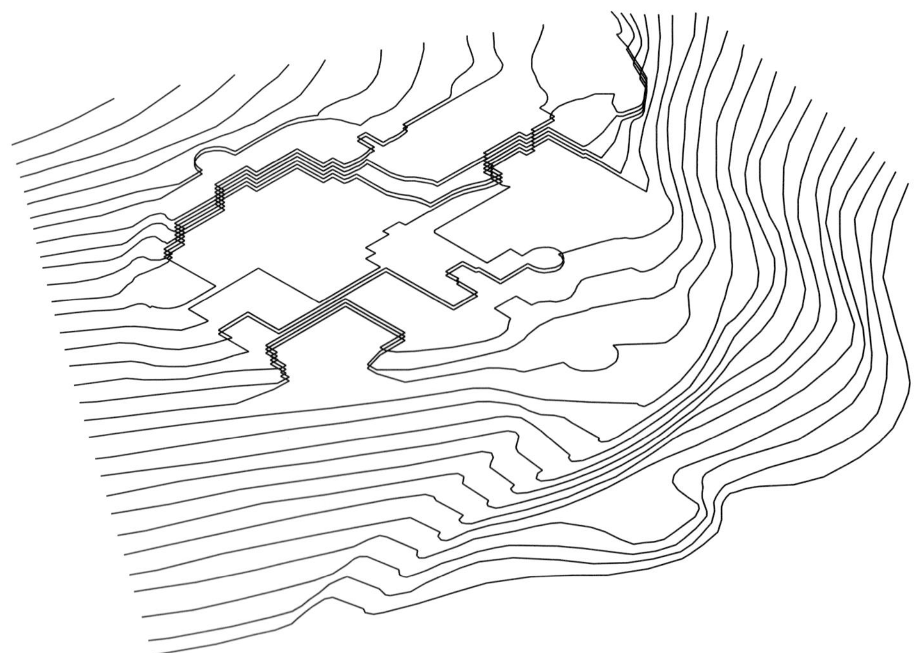

To enter 3D lines, perform the following steps:

1. Press the **Z** key and set your **Z-base** and **Z-height** to the elevation of the first topographic line.
2. Select **3D Entity**, **3D Line**. Toggle on **Z-base** and **Chain**.
3. Begin placing points along the topographic lines. When you reach the end, change your Z-settings and start the next line.

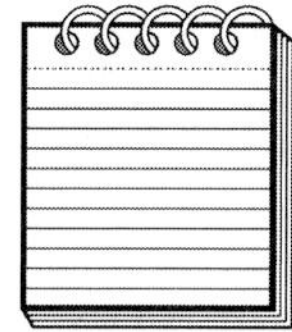

NOTE: *You can also quickly create a 3D line drawing by tracing the site with 3D contours first, and then exploding them to 3D lines. Select* ***3D Edit, Explode, To Lines****, and pick the contours.*

Option 4: Point Field Data Point data is sometimes available from the land surveyor, and can be imported directly via a **.DXF**, **.DWG**, or a **.XYZ** text file. You can also create a point field yourself from a 3D contour drawing: The objective is to achieve a sufficient number of points without too much density, as shown in the following point field drawing with 3D markers applied randomly along contour lines. Point field drawings are good for creating a drop mesh and essential for TIN models.

To insert **3D Markers**, perform the following steps:

1. Select **3D Entity**, **Marker**, and toggle on **3D cursor**.
2. Middle-button snap to place markers at semi-regular intervals along the contours. Areas of greater intricacy require more points.

NOTE: *You can explode the contours to 3D lines to generate entities between the contour control points. This allows you to enter many more points. Copy the contours to a new layer so that you don't lose them.*

Site Method One: Drop Mesh

The **Drop Mesh** function uses grid-point averaging to create a surface constructed of three-sided polygons. The diagonal edges of the polygons may be hidden, which produces a model with a grid appearance. Though rather bland in appearance, it is easy to execute and good for quick site renditions. Any source data will do, as long as it is three-dimensional. The **Drop Mesh** function samples points at regular intervals, hence the resulting grid. The process can produce undesirable results if you have high concentrations of data in one place, as the sampling will average all of the points in that area together. You can also use this effect to your advantage. For example, you might add additional 3D markers around the building perimeter to give it greater computational "weight" to push or pull the terrain toward it.

To create a drop mesh model, perform the following steps:

1. Enter the site topography data using 2D or 3D lines, contour lines, or markers at the appropriate Z-base and Z-height.
2. Create a **New Layer** for the site model and set it Active. Select **3D Entity**, **Polygon**, **Drop Mesh**.
3. Check the settings: Select **Divisions**, **X and Y** and enter a value relative to the proportion of the site. For a site that is 800×400 meters, you might start with 40×80. Choose **Stiffness** and enter **5**. This affects the degree to which the polygons sag between data points. A stiffness setting between 5 and 10 works well with most sites. Optionally, toggle on **Smooth** to perform an additional smoothing pass on the mesh. This will soften its appearance somewhat. When **Smooth** is toggled on, an option for **Smoothing Passes** appears. Select it and enter a value between 1 and 3. Toggle on **Drop Edge** if you wish to create a base around the perimeter of the model. Pick **Edge Height** to set the base depth.

4. **Turn Layer Search** on and choose **Area** as your selection mode, and define a selection box that encompasses all of the site data. Select **Begin**. The complete drop mesh model is created.

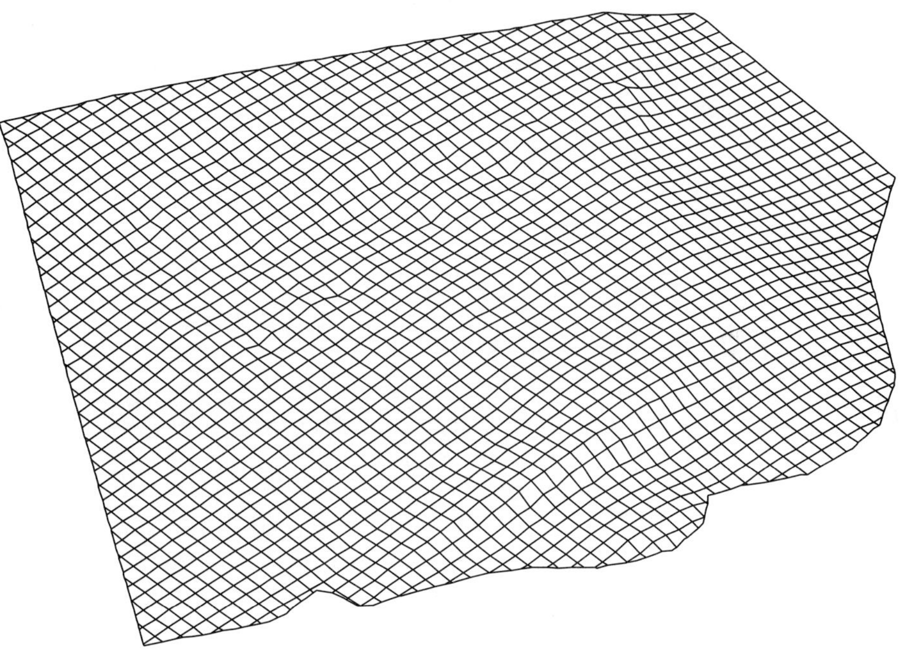

NOTE: *The* ***Drop Mesh*** *function always generalizes your site to its nearest rectangular shape. In the example illustrated, the* ***Knife*** *function in the* ***3D Tools*** *macro was used to create the irregular boundary.*

To automatically generate a drop mesh model from a point file provided by a surveyor, perform the following steps:

1. Copy the point file to your **DATACAD\XFER** directory. The file can be in any text format with spaces between the coordinate values. A sample file called **DropMesh.XYZ** is provided on the bonus CD.
2. Open a file for the model. Create a new layer for the drop mesh model and set it Active.
3. Select **3D Entity**, **Polygon**, **Drop Mesh**. Check your settings for **Divisions**, **Stiffness**, and so on, as previously described.
4. **Choose Import**. A file dialog appears. Navigate to the **XFER** directory and pick the point data file. Under the "Files of Type" drop down box,

set the file's data order: **XYZ**, **YXZ**, **Angle/Distance/Height** or **Distance/Angle/Height**. If you do not know what format the file is in, ask the provider. The sample file provided on the CD is in **XYZ** format.

5. Choose **Open**. The file is processed and the model is created. The following figure is the result of XYZ data imported into **Drop Mesh**. The **Drop Edge** setting is toggled on.

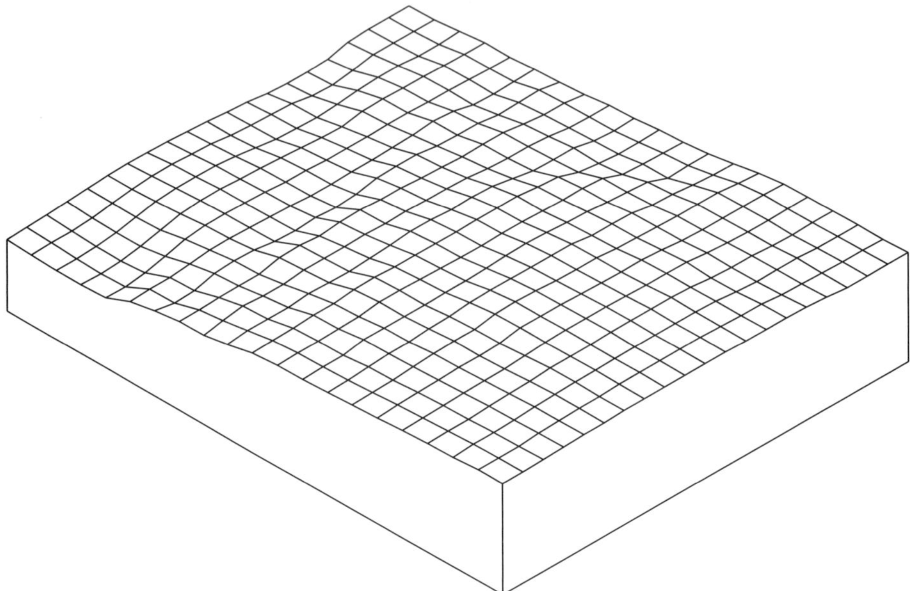

Site Method Two: Slab Model

The slab model method creates a result similar to physical architectural models fabricated with paperboard or cork. When compared with the previous drop mesh, this model type has a far greater sense of elevation height and visual interest. Additionally, this method is the only one which allows you to easily create cuts and recesses that clearly show the building footprint, which facilitates coordination with a 3D building model.

To create a slab-on-slab site model manually using **3D Slabs**, perform the following steps:

1. Enter the site data with 2D lines at a Z-base and Z-height of 0′-0″.
2. Create a **New Layer** for the site model and set it Active.
3. Press the **Z** key on your keyboard and enter the **Z-base** and **height** for the contour elevation, such as 0–10, 10–20, 30–40, and so on. Toggle off

Ortho line mode. Select **3D Entity**, **Slab**, **Horizontal**. Toggle on **Base/Height**. Begin tracing a circuit to form the first level. As slabs are limited to a maximum of 36 sides, you will most likely run out of points before make a complete loop. To overcome this limitation, use multiple slabs to make the complete shape. The mating edges between the slabs can be hidden with the **Mark Visible** function. To hide a slab edge, select **3D Entity**, **Slab**, **Partial**, **Mark Visible**. Click on the edge you wish to hide. If two slabs meet at the edge you wish to hide, you will have to click twice, once for each edge.

4. Once a level is complete, change your **Z-base** and **Z-height** settings and proceed to the next level. Eventually, you will reach the top. The result hides and renders nicely. Here is the completed slab model:

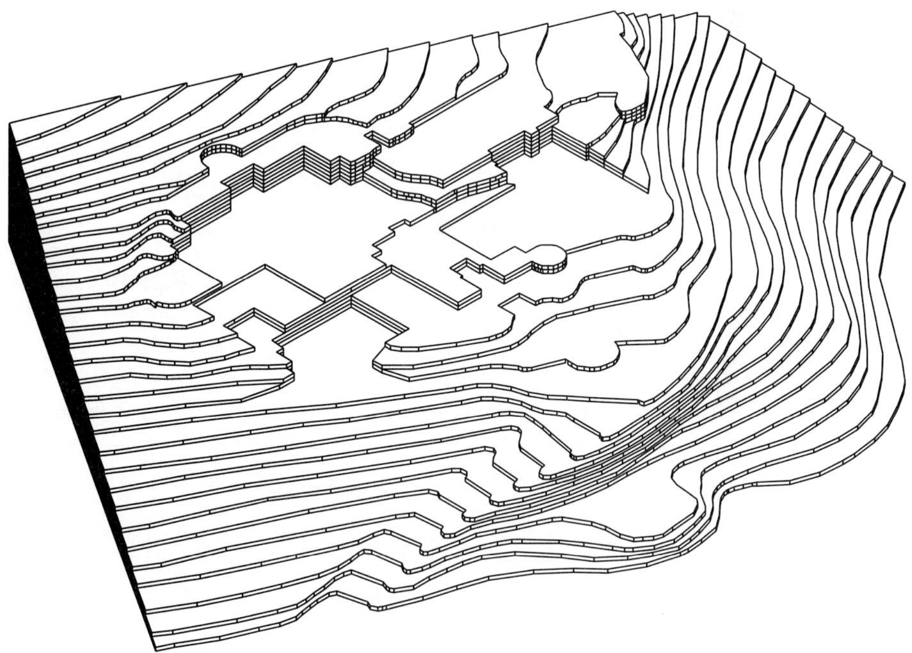

To create a slab-on-slab site model automatically from polylines, use the following method. It is far easier to execute, but requires the **3D Tools** macro.

1. Trace the topographical contours with **2D Polylines** at their respective Z-base and Z-height. Select **2D Edit**, **Curves**, **Polyline**. The polylines must form closed loops around the site, so make sure **Closed** is toggled on. There is no limit to the number of points you can use, but strive for economy to keep the polygon count as low as possible.

NOTE: *It is also possible to convert 2D lines to polylines with the* ***DC Sprint*** *macro.*

2. Select **Toolbox**, **3D Tools**, **Polyline to polygon**. Toggle on **Top** and **Sides**. Toggle off **Show Edge**. Turn **Layer Search** on and choose **Area** as your selection mode, and define a selection box that encompasses all of the site data. The polylines are converted to polygons, and the model is complete. Note that the converted polylines are copied to a new layer called **_TRIHID** for later use. The following figure is a site drawing created with 2D polylines.

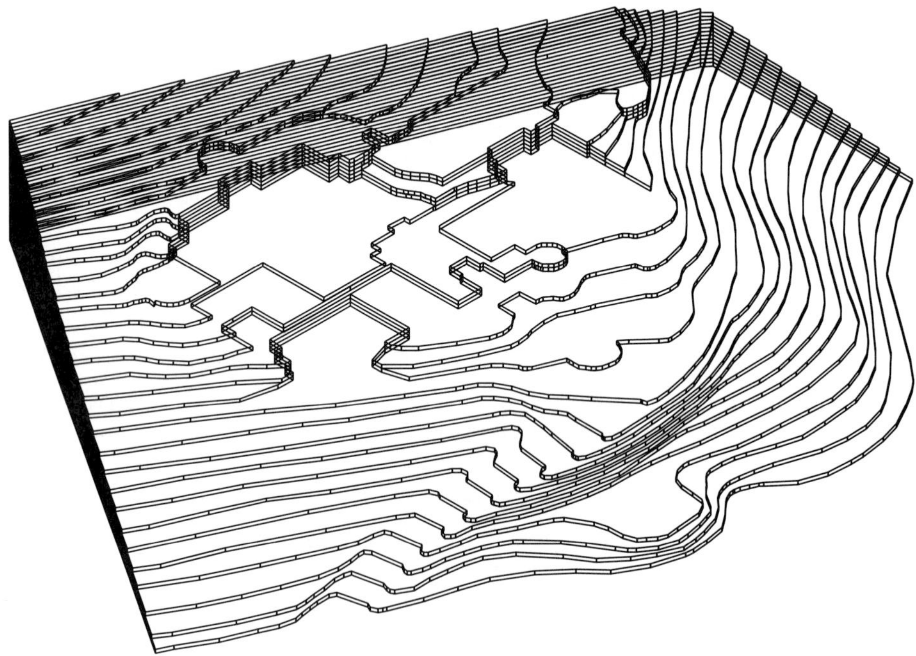

The completed polyline-to-polygon conversion model is shown. It looks identical to the slab model created manually, but it is composed of many individual polygon entities. This model contains some 3,156 polygons!

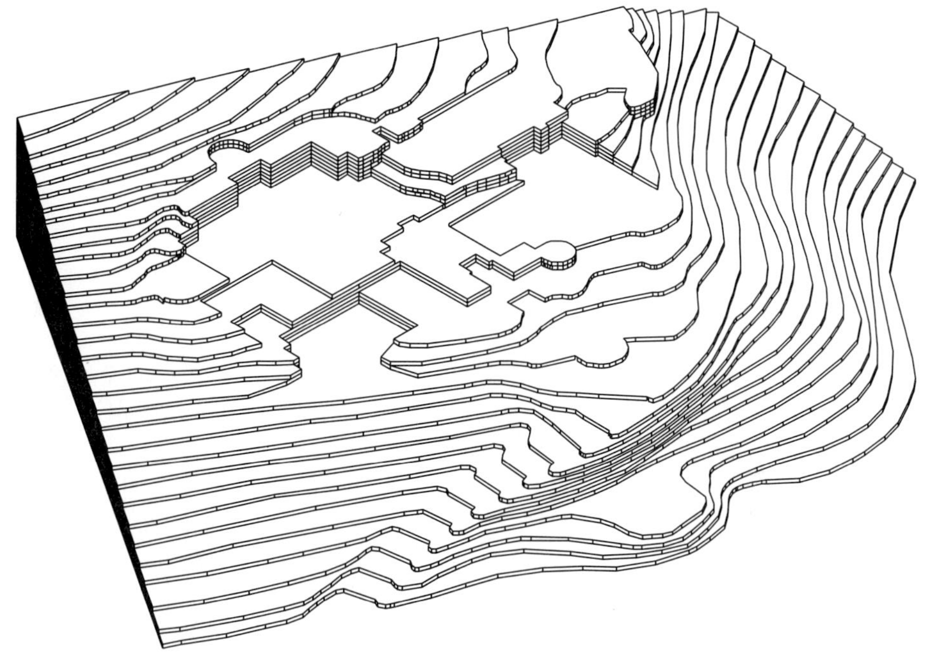

Site Method Three: Contour Rail Sweep

This modeling method is useful for creating surfaces between irregular boundaries. It can be used to model terrain or any object for which you can define contoured outlines at its key points. To create this model type, you must install the **PolyMaster** macro included on the bonus CD.

The rails referred to are two contour lines separated in space. The sweep occurs when polygons are stitched between these contour lines to create a surface. The example shown pushes the limits of its practicality, as it generates a very dense array of polygons. This modeling method is the most literal translation of contours to a surface, and they remain discernible in the completed model.

To create a contour rail sweep surface model, perform the following steps:

1. Trace the site topographical drawing using 3D contour lines at their respective Z-heights. This method works best if you place an equal number of control points on each line as shown in the following illustration. You must take care when tracing over curves. If one contour has more points than the other, or points too far ahead of the

other, the end result will produce undesirably distorted or overlapping polygons. Experiment with different configurations to see what works best before starting a new model. It is best to use the lowest **Divisions** setting of **4**. Also set the **Stiffness** low, between **0.0** and **0.3**. The following figure is the topography drawing with 3D contour lines used for the rail sweep.

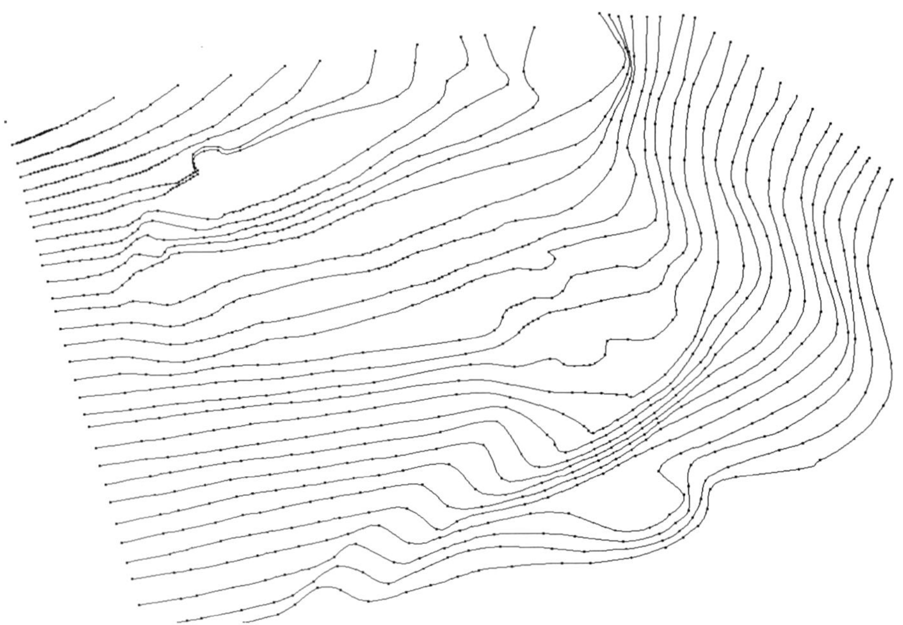

2. Once the contour lines are defined and the control point locations are worked out to their best effect, the rail sweep is easy. Create a **New Layer** for the site model and set it Active. Select **Toolbox, PolyMaster**. Choose **Site Mode**. Toggle on **Show Diagonals** and pick the lowest contour. **PolyMaster** counts the points and reports them at the command line, then prompts you to pick the next contour. Pick the one above the first. The rail sweep is created. Pick the next contour in sequence to sweep again. Keep picking contours until you reach the top.
3. To create the last triangular piece shown in the sample file, when you reach the last contour, select **Point to Contour** from the menu. Choose the marker point at the top corner of the site, then the contour closest to it. **PolyMaster** generates radial polygons between the marker and the contour.

The finished Rail Sweep model renders well, and provides a good sense of the surface flow between contours. Note that keeping the number of points consistent from contour to contour produces very dense areas at the top of the model where the contour lines are short. Also note the "polygon acceleration effect" created in the twists and turns as the contour control points change relative to one another between contour lines. The following figure shows the completed Contour Rail Sweep model, weighing in at 8,298 polygons!

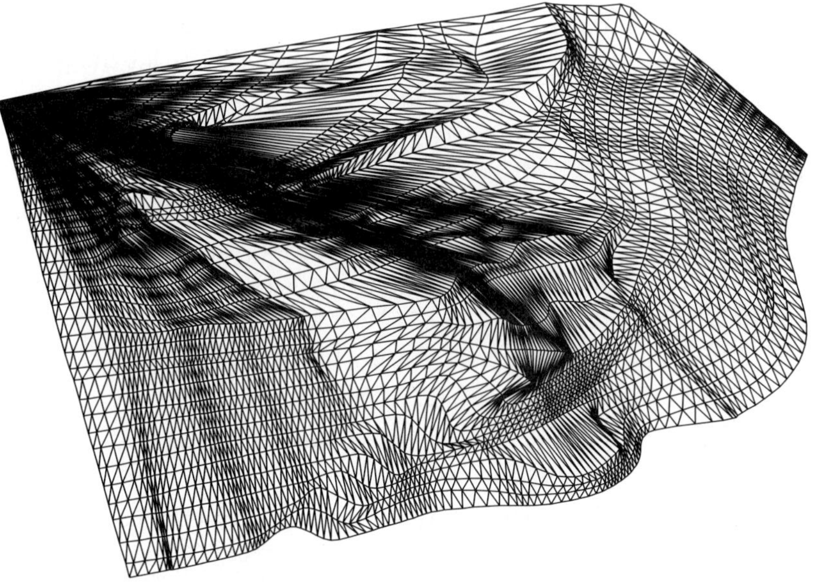

NOTE: *If you edit or change the contours that define the model, you will have to erase and re-sweep those sections. Use the* ***Group*** *selection mode when erasing these sections to remove them in one pick. You can also use* ***Polymaster*** *to* ***Rail Sweep*** *between* ***Cyclic Contours*** *and individual points.*

Site Method Four: TIN model

This exercise introduces you to the *Triangular Irregular Network* (TIN). A routine is used to analyze and connect a random array of points with triangular polygons. It is the most efficient and effective method to represent site data accurately with a minimum of polygons. This makes it ideally suited to models that will be rendered. The rendering process blends the edges together, and the variation of the points lends good shadow and depth.

The TIN model is interesting to look at and communicates massing and elevation effectively, but it is not indicative of the contours used to produce it. It is much better at reproducing site features. Compare the driveway in the following model to the others from the previous exercises.

You cannot control how the routine will connect the points, as any three that are connected by a surface are considered valid. Sometimes this produces awkward results, which you can erase and replace manualy.

The primary disadvantage to the TIN method in DataCAD is that it is computation-intensive. It can take many hours to process a significant number of points. It is suggested that you test this method on a limited number of points in the sample file, or let the process run over a night or weekend. You are forewarned! The macro is a demo version and is limited to a maximum of 1000 points.

To create a TIN model, perform the following steps:

1. Enter the site topographic data with **3D Markers**. These can be imported via a **.DXF** or **.DWG** file provided by the surveyor, or entered manually by snapping marker points over an existing 3D contour drawing.
2. Create a **New Layer** for the TIN model and set it Active. Select **Toolbox**, **dhTIN**.
3. Change your selection mode to **Area** and select the marker points to be processed.

   The macro analyzes and connects the points with triangular surfaces. The completed TIN model is shown.

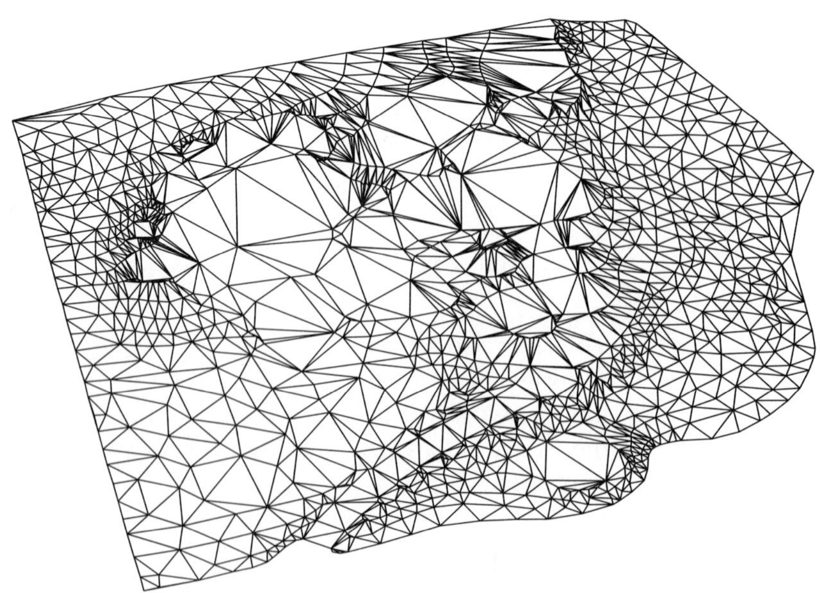

Exercise 5—Ornate Wrought Iron Railing

In this example we will model an ornate railing design. We will make repetitive use of simple geometric primitives. This example is also a good introduction to the technique of tilt-up slab modeling, whereby the model is created flat on the ground plane and rotated to vertical when complete.

The railing looks complicated, but is actually quite simple to execute. Using three-dimensional shapes for the curving design provides the necessary depth and realism in the final result.

Open the file **Exercise_5.DC5** if you wish to follow along in the model file.

Step 1

The first step is to draw the railing elevation with 2D lines at a Z-base and Z-height of zero. Three-point arcs spaced $^1/_2''$ apart are used for the curved

portions of the design. Line segments designate where each arc begins and ends. The railing will be modeled as if lying on the ground and then rotated to vertical when complete.

Step 2

Upon closer inspection, it is apparent that the symmetry of the design permits modeling portions of the design and then using **Mirror**, **And Copy** to complete it.

Copy the 2D design to a new layer, and remove all but the critical elements to model. Draw guide lines indicating the lines of symmetry.

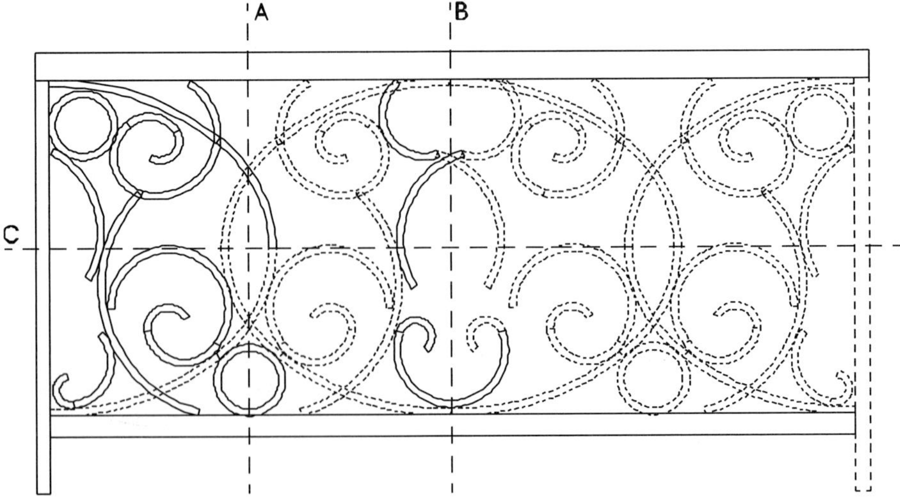

Step 3

The following numbered diagram will be used to refer to the different elements of the model. A scroll composed of multiple curves will be referred to as an assembly. Fourteen different components comprise the basis of the model.

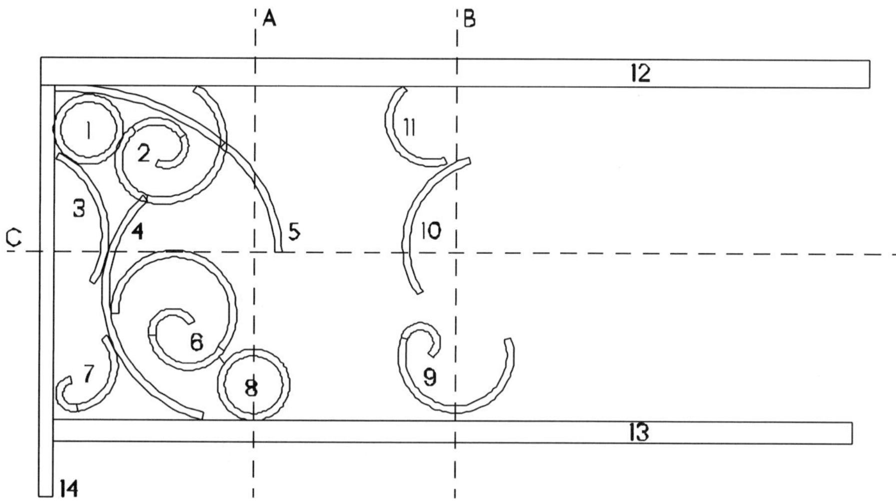

Enable **Quadrant** under **2D Utility**, **Object Snap**. Using **Geometry**, **Divide**, **Divisions = 2**, middle-button snap from arc to arc to place a marker at the midpoint between each curved piece.

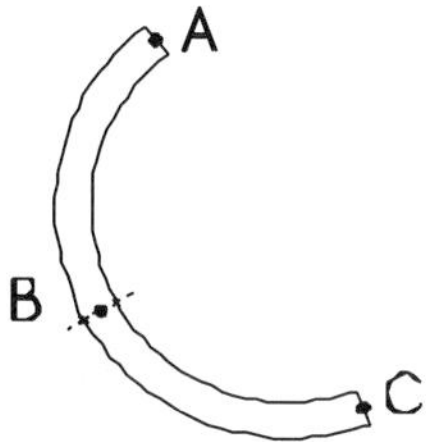

Step 4

Construct shapes 3, 4, 5, and 10.

Create a new layer for the 3D portion of the rail design, and set it active. Select **3D Entity**, **Torus**, **3 Point Arc**. Toggle on **Z-base**. Select **Primary Divisions** and set it to **36**. Select **Secondary Divisions** and set it to **5**.

Select **Radius** and enter $^1/_4''$. For each arc, middle-button snap to point A and C, then to the midpoint marker at point B. The completed tori are shown in the following figure.

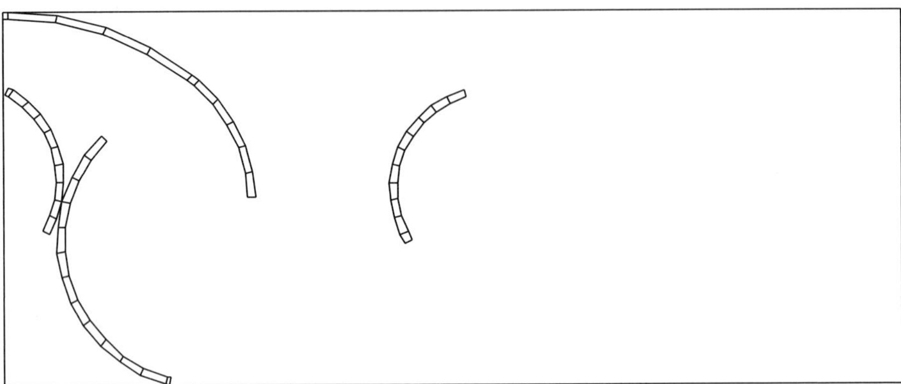

Step 5

Model the major arcs from assemblies 2, 6, and 9.

Select **3D Entity, Torus, 3 Point Arc**. Toggle on **Z-base**. Select **Primary Divisions** and set it to **24**. Select **Secondary Divisions** and set it to **5**. Select **Radius** and enter $^1/_4''$. For each arc, middle-button snap to point A and C, then to the midpoint marker at point B. Here is the result:

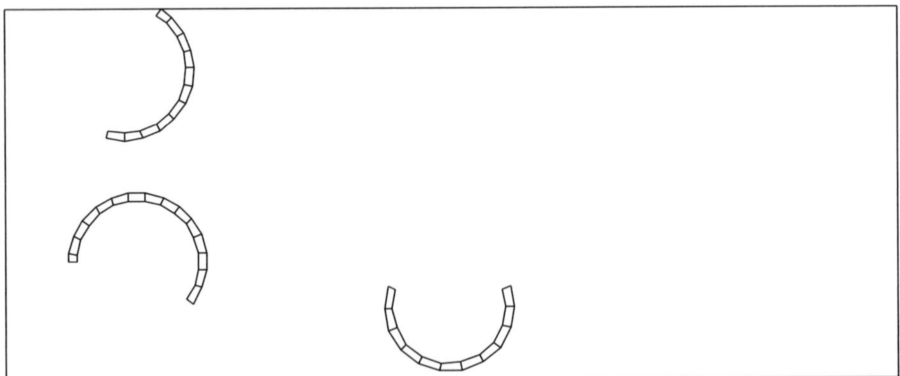

Step 6

Model the secondary arcs from assemblies 1, 2, 6,7, 8, and 11.

Select **3D Entity**, **Torus**, **3 Point Arc**. Toggle on **Z-base**. Select **Primary Divisions** and set it to **18**. Select **Secondary Divisions** and set it to **5**. Select **Radius** and enter $^1/_4$″. For each arc, middle-button snap to point A and C, then to the midpoint marker at point B. The completed tori are shown in the following figure.

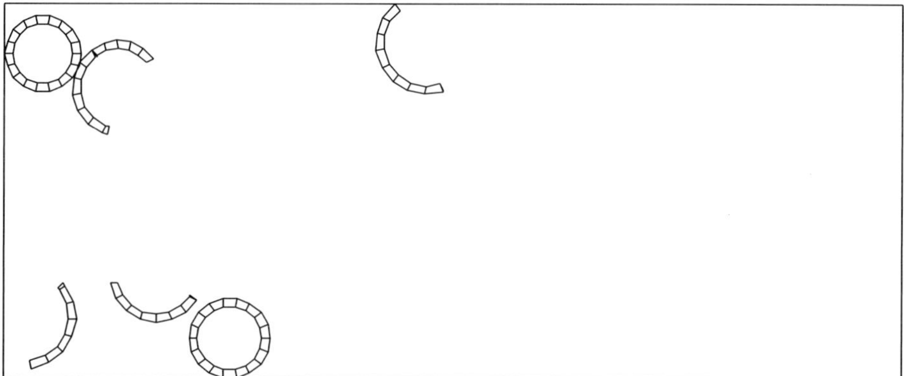

Step 7

Model the tertiary arcs from assemblies 2, 6, 7, and 9.

Select **3D Entity**, **Torus**, **3 Point Arc**. Toggle on **Z-base**. Select **Primary Divisions** and set it to **12**. Select **Secondary Divisions** and set it to **5**. Select **Radius** and enter $^1/_4$″. For each arc, middle-button snap to point A and C, then to the midpoint marker at point B. The completed tori are shown in the following figure.

Step 8

Model the top of the rail frame.

Create a new layer for the 3D frame, and set it active. Press the **Z-key** on your keyboard. Set the **Z-base** to −1″, and the **Z-height** to 1″. Select **3D Entity**, **Slab**, **Rectangle**. Middle-button snap to opposing corners to create the top rail.

Step 9

Model the rest of the frame.

Press the **Z** key on your keyboard. Set the **Z-base** to −½″ and the **Z-height** to ½″. Select **3D Entity**, **Slab**, **Rectangle**. Middle-button snap to opposing corners to create the left side and bottom rail.

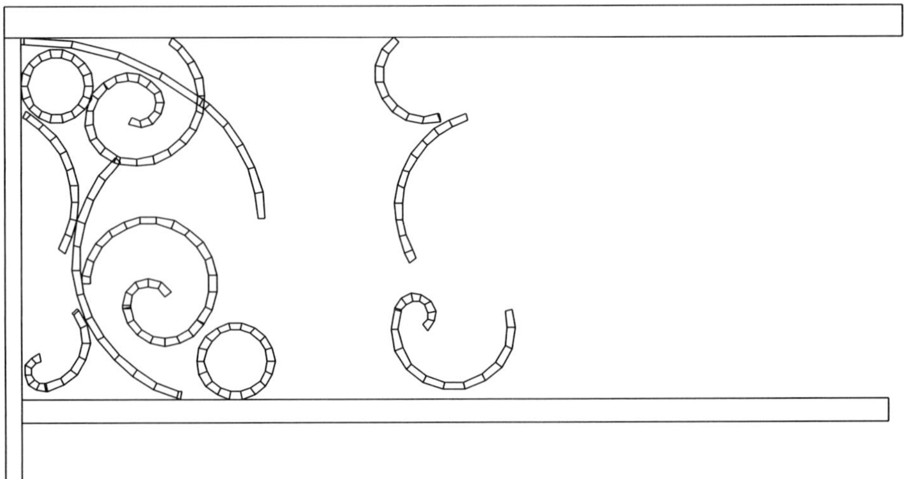

Step 10

Mirror the horizontal elements.

Turn on the layer with the mirroring guide lines. Make sure **Ortho** line mode is toggled on. Select **2D Edit**, **Mirror**, **And Copy**. Middle-button snap to **Mirror** line C. Drag the line of reflection to the right and click. Pick assembly 5 to **Mirror**.

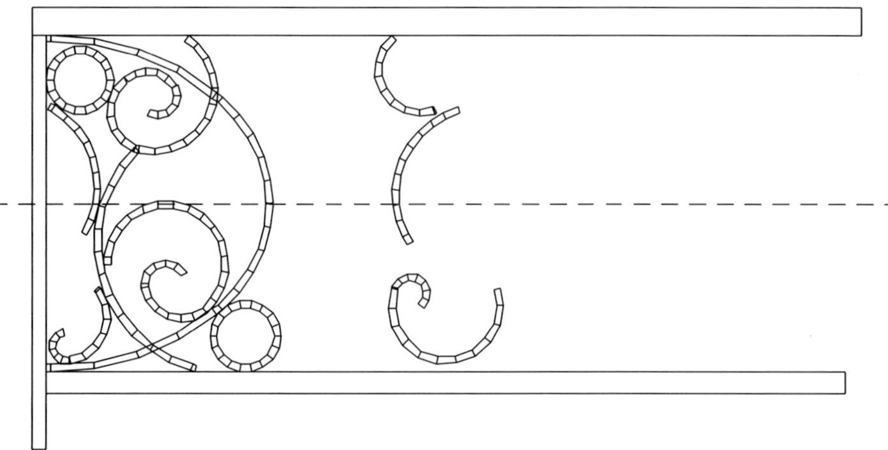

Step 11

Mirror the first of the vertical elements.

Select **2D Edit**, **Mirror**, **And Copy**. Middle-button snap to **Mirror** line A. Drag the line of reflection towards the bottom of the screen and click. Pick assemblies 2, 4, 5, 6 to mirror.

Step 12

Mirror the last of the vertical elements

Select **2D Edit**, **Mirror**, **And Copy**. Middle-button snap to **Mirror** line B. Drag the line of reflection towards the bottom of the screen and click. Pick assemblies 1, 2, 3, 4, 5, 6, 7, 8, 10, 11, and 14 to mirror. Lastly, choose only the smallest segment of assembly to mirror. The 3D portions are complete.

Step 13

Complete the model.

Return to **Ortho** view. Turn off the 2D layers, and turn on all the 3D layers associated with the railing. Select **3D Edit**, **Rotate**, **X-axis**, **X-angle** and enter 90 degrees. Select **New Center** and middle-button snap to the lower left corner of the vertical post. Turn on **Layer Search** and select the entire model by area. The railing rotates to a vertical position. The model is complete!

Here is the completed railing in context:

Exercise 6—Corinthian Column

In this example we will model a Corinthian column. As this style of column is heavily ornamented, the goal is to create a convincing stylized representation that will render well, yet have a low polygon count. Open the file **Exercise_6.DC5** if you wish to follow along in the model file.

Step 1

The first step is to draw the column elevation with 2D lines at a Z-base and Z-height of zero. The curved profiles of the base and neck below the capital are broken down into line segments. The column will be modeled as if lying on the ground and then rotated to vertical when complete.

Step 2

Create a new layer for the midsection of the column. Using the 2D drawing as a guide, the main body of the column is modeled using a revolved surface.

2.1 Select **3D Entity**, **RevSurf**.

2.2 Select **PrimDivs** and enter **12** as the number of primary divisions for the revolved surface.

2.3 Select **Open** to create a revolved surface with open ends.

2.4 You are prompted to "Select first point on <Surface of Revolution>."

2.5 Begin tracing the right profile of the column, from bottom to top, as shown in the following figure at left.

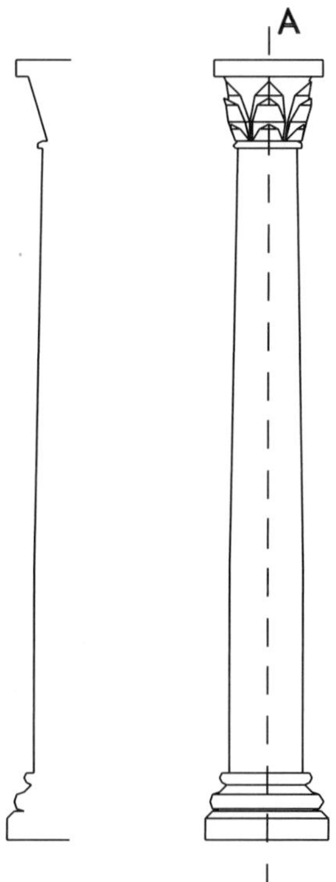

2.6 When the last point is reached, press the right mouse button to finish.

2.7 You are prompted to "Select first point of surface center axis." Make sure that **Ortho** mode is on, and middle-button snap to the midpoint (line A) of the line at the top of the column. Drag the center axis line down towards the column base and left-button click. The revolved surface is created.

The result should look like this when viewed in isometric:

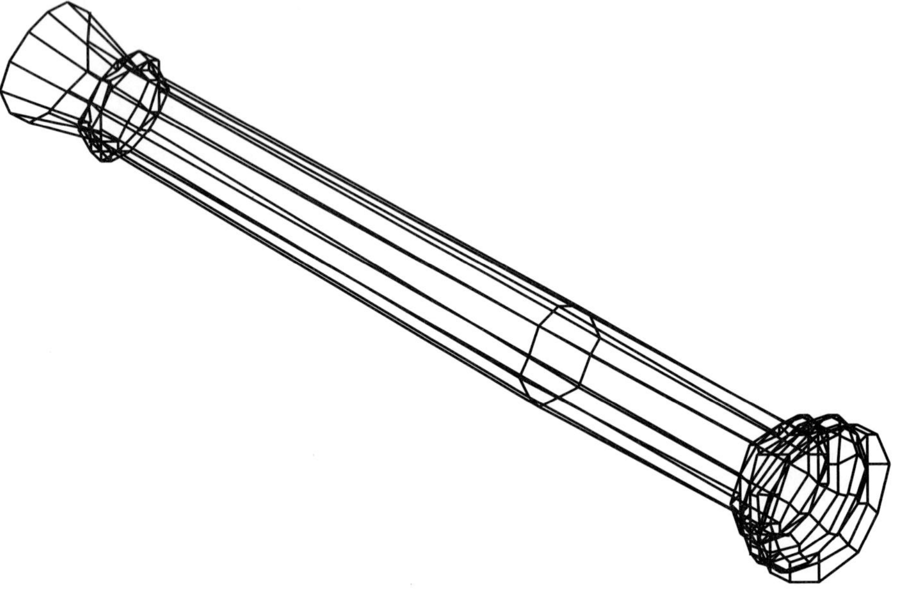

Step 3

Model the square column base.

3.1 Create a new layer for the base. The base measures 1′-3$^{1}/_{4}$″ on a side. Set the Z-base to −7$^{5}/_{8}$″ and the Z-height to 7$^{5}/_{8}$″.

3.2 Select **3D Entity**, **Slab**, **Rectangle**, and trace over the rectangular base by middle-button snapping to two opposite corners.

Step 4

Model the square portion of the column capital. This piece will not be square in the finished model, as we will modify its shape later.

4.1 The column top, or abacus, measures 1′-1$^{1}/_{2}$″ square and 2″ thick. Set the Z-base to −6$^{3}/_{4}$″ and the Z-height to 6$^{3}/_{4}$″.

4.2 Select **3D Entity**, **Slab**, **Vertical**, **Right**, and set the thickness to 2″.

4.3 Create the abacus by tracing its lower edge from right to left.

The completed column base, body and abacus should look like this when viewed in isometric and the hidden lines removed:

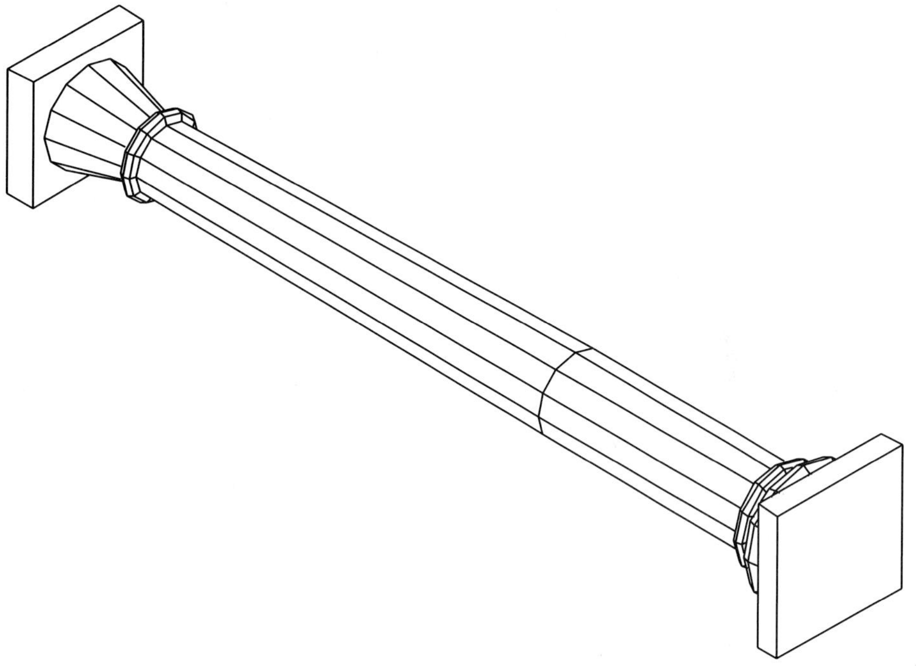

Step 5

Model the abstracted acanthus leaves.

5.1 Create a new layer named Acanthus. Set the Z-base to $-2^1/_4''$ and the Z-height to $2^1/_4''$.

5.2 Select **3D Entity**, **Polygon**, **Vertical**; using the 2D drawing as a guide, trace the vertical edge of the three nested leaves. Make sure that **Chain** is on, so that you can create them in three groups.

The result looks like this:

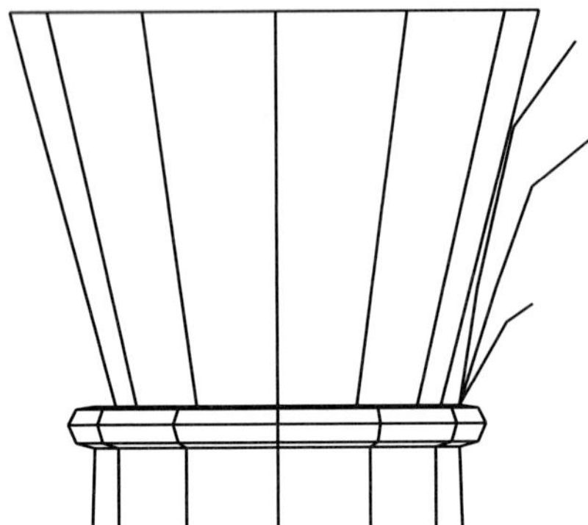

Step 6

Next, the "leaves" will be rotated 90 degrees so they are facing the screen flat on, and then stretched to shape.

6.1 Set Z-base and Z-height to zero, then select **3D Edit**, **Rotate**.

6.2 Select **Y-axis**, then **Y-angle**, and set to 270 degrees. (Remember the right-hand rule?)

6.3 Select **New Center**. You are prompted to "Select center of rotation." Middle-button snap to the top center of the column.

6.4 You are now prompted to select the entities to rotate. Make sure that **And Copy** is off, choose **Group** as the selection method, and select each group of leaves to rotate.

Here is the result:

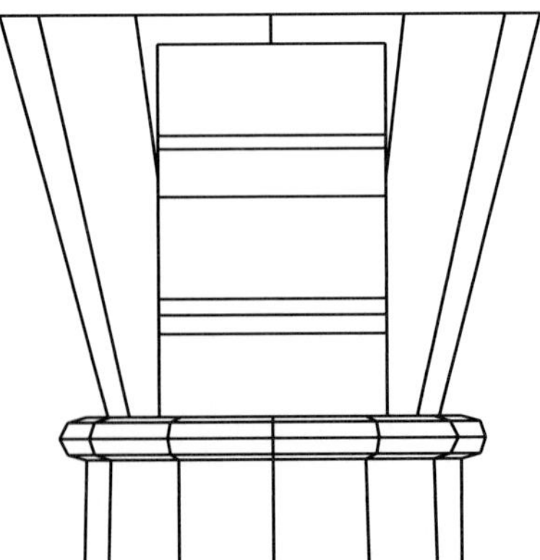

Step 7

Select **3D Entity**, **Polygon**, **Partial**, **DelVrtex**, and delete the top left vertex from each group. The top of each group is a three-sided polygon, so this additional point is not needed.

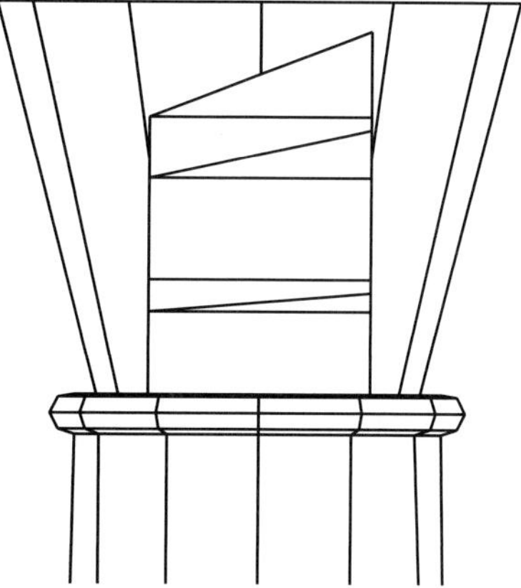

Step 8

Turn **Layer Search** off, and set the Acanthus layer as active. Using the 2D drawing as a guide, stretch the vertices of the polygons to match the shape of the abstracted acanthus leaves:

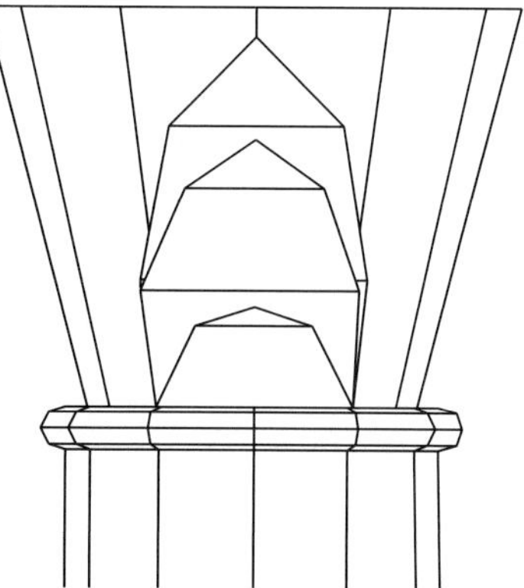

Step 9

Next, the leaves are rotated and copied around the capital.

9.1 Select **3D Views**, **Elevation**, **Back**, to view the column from the "top." **Zoom**, **Extents** if necessary.

9.2 Select **3D Edit**, **Rotate**, **Z-Axis**, **Z-angle**, and enter **60 degrees**.

9.3 Select **New Center**. You are prompted to "Select center of Rotation." Middle-button snap to the center of the column.

9.4 Make sure **And Copy** is toggled on and that your selection mode is set to **Group**. Select the outer-most group. Keep picking this group until you have completed one circuit around the column.

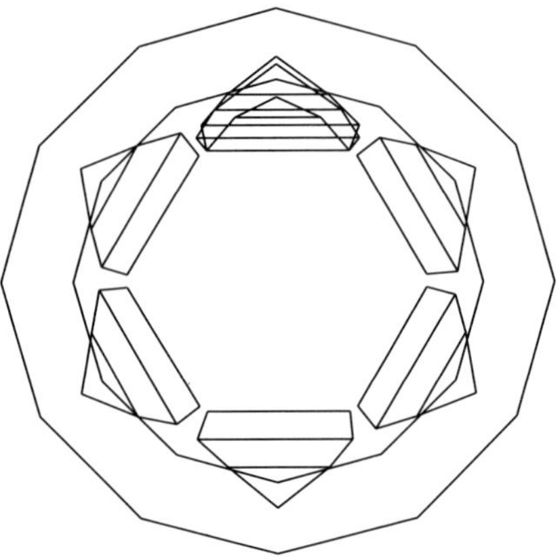

9.5 Repeat the previous step with the next two groups.

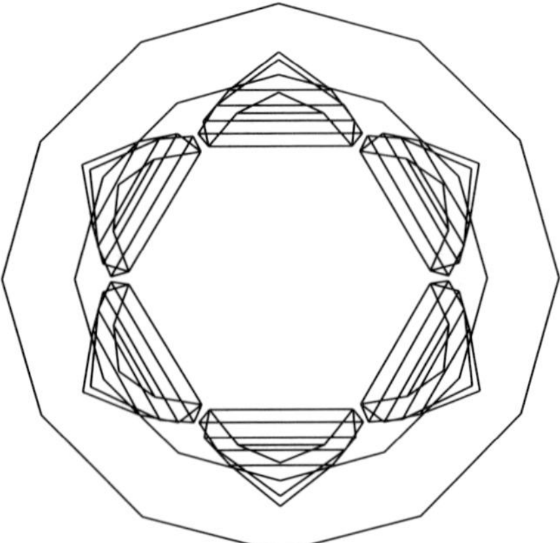

Step 10

Now select **Z-angle** and enter **30 degrees**. Select **by Entity**, and choose the outer-most triangular polygon. Select **Z-angle** again, and enter **60**

degrees. Continue selecting this polygon until you have completed one circuit around the column capital. Here is the result:

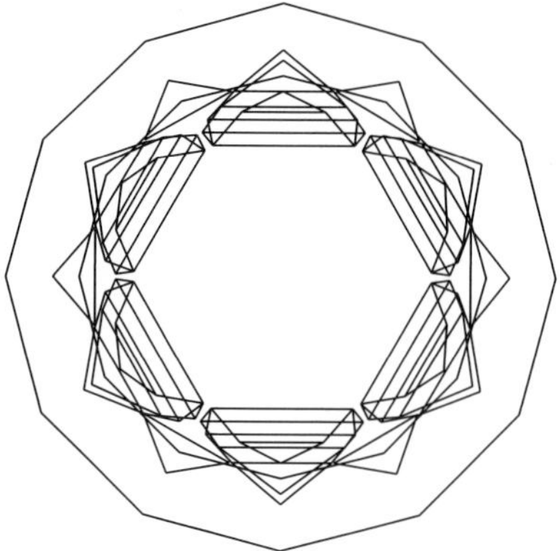

Step 11

Rotate the column to a vertical position.

Return to **Ortho** view, and turn on all the 3D layers associated with the column. Press the **Z** key on your keyboard, and check to see that the Z-base and Z-height values are set to 0′-0′. Select **3D Edit**, **Rotate**, **X-axis**, **X-angle** and enter 90 degrees. Select **New Center** and middle-button snap to the lower left corner at the very bottom of the column base. Make sure **And Copy** is toggled off. Turn on **Layer Search** and select the entire column by **Area**. The column rotates to a vertical position.

Step 12

Turn on the 2D drawing and the Abacus layers. Turn **Layer Search** off and set the 2D drawing layer as active. Trace the outline of the curved abacus with 2D lines and curves, using the 3D abacus slab as a guide. Use **Geometry**, **Divide, Divisions**, **5**, to place markers along the curves as a guide.

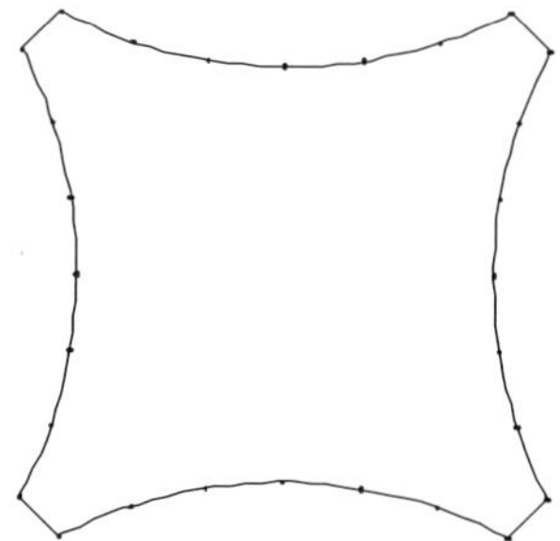

Step 13

Set the Abacus layer as active. Select **3D Entity**, **Slab**, **partial**, **Add Vertex**, and modify the abacus slab to align with the 2D guide.

The column is complete!

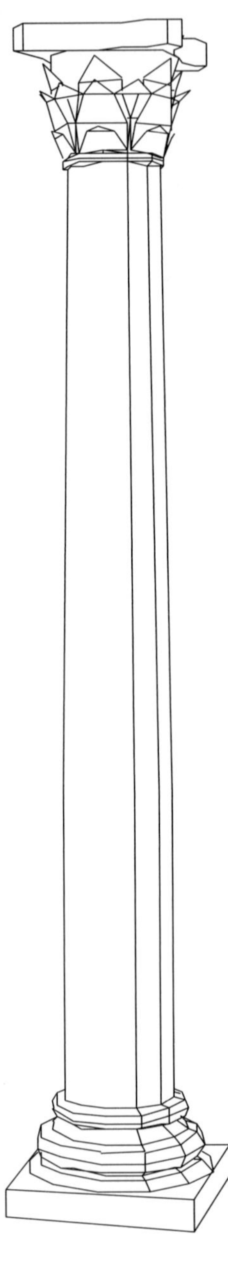

Exercise 7—Arched Colonnade

In this exercise we will take the Corinthian column from the previous example and incorporate it into a colonnade. The primary focus will be on the arched opening. The sweep function will also be used to create the cornice and base profiles. Open the file **Exercise_7.DC5** if you wish to follow along in the model file.

Step 1

The first step is to draw the colonnade elevation with 2D lines at a Z-base and Z-height of zero. This drawing will be used as a guide to create the 3D model. The curved arches between each set of columns are broken down into line segments that approximate an elliptical opening. Care is taken to consider where the repetitive design elements will meet. Dashed lines indicate two identical bays flanked by two ending bays. The design is symmetrical, so only half needs to be modeled.

The arcade will be modeled as if lying on the ground and then rotated to vertical when complete. On a separate layer, the profile of the wall cornice and base is also drawn and a polyline is defined for the sweep path. This path represents the footprint of the wall in plan.

Step 2

Import the column from the previous example via **Cut** and **Paste**. Using **3D Rotate** about the **X-axis** at a value of 270 degrees, flip the column so that it lies flat on the ground plane. Position the column within the 2D drawing of the arcade. The center of the column should located at a **Z-base** of 0′-0″.

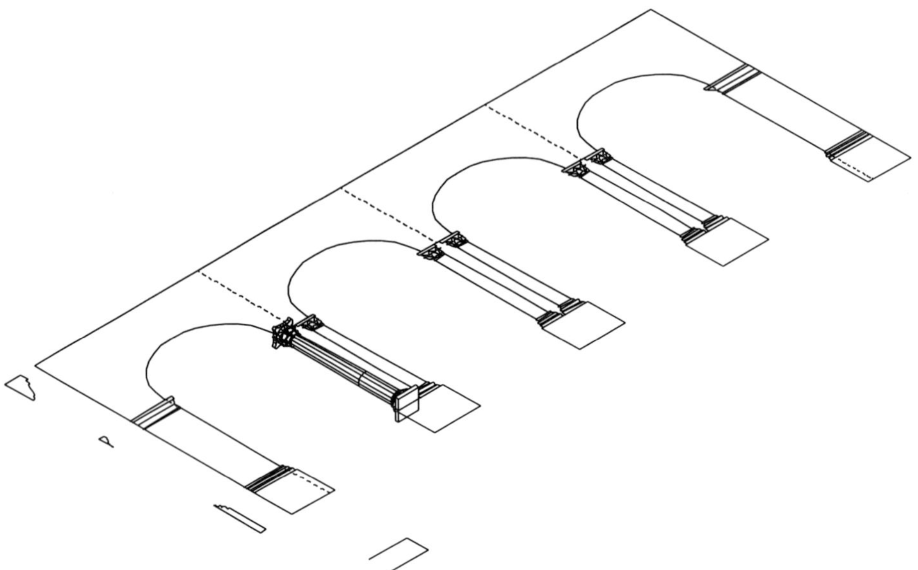

Step 3

Create the column pair.

Change your view to **Ortho**. Select **Copy**, and pick a point on the screen. Press the space bar, and enter a value of 1′-4$^{3}/_{4}$″ at 0 degrees. Area-select the 3D column to copy. You now have two columns side by side.

Step 4

Model the column pedestal.

4.1 Press the **Z** Key on your keyboard; set the **Z-base** to −8″ and the **Z-height** to 8″. This will create a slab that is a total of 1′-4″ thick, centered within the plane of the 2D drawing.

4.2 Select **3D Entity**, **Slab**, **Rectangle**. Middle-button snap to two opposing corners to create the pedestal.

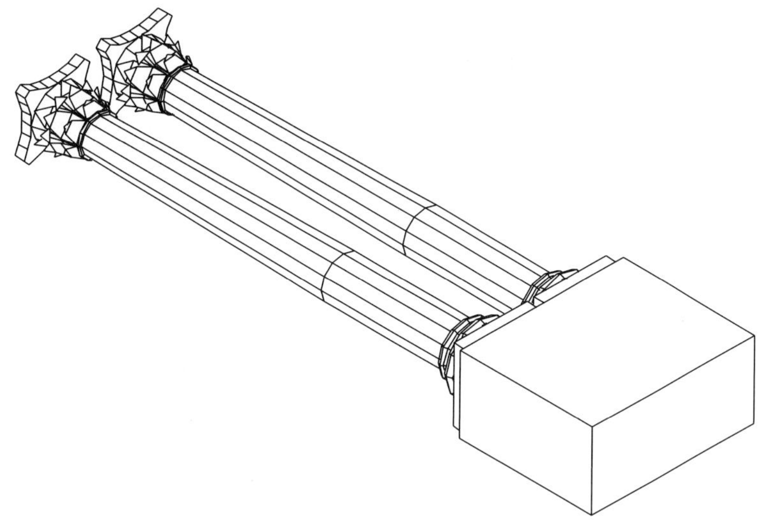

Step 5

Model the arched wall openings.

5.1 Create a new layer for the 3D walls and **Set Active**. Press the **Z** key, and set **Z-base** to −4″ and **Z-height** to 4″, to create a wall 8″ thick. Toggle **Ortho** line mode off. Select **3D Entity**, **Slab**, **Horizontal**. Make sure that **base/height** is toggled on. Using middle-button snaps, trace out the central arch shape as shown:

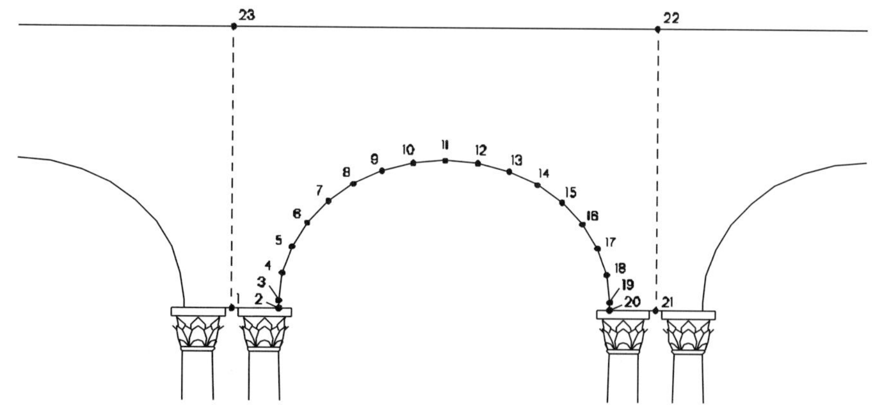

5.2 Repeat for the next arch as shown in the following figure.

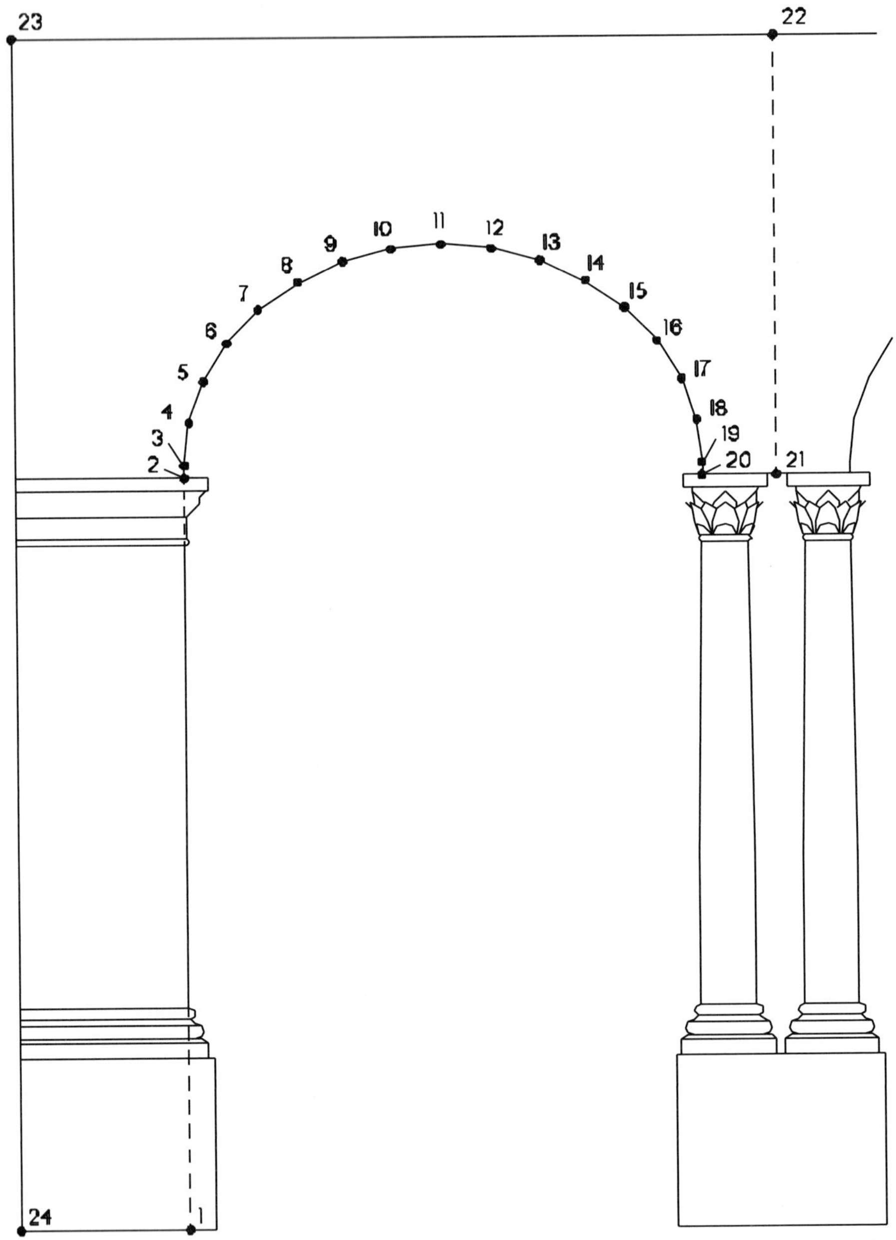

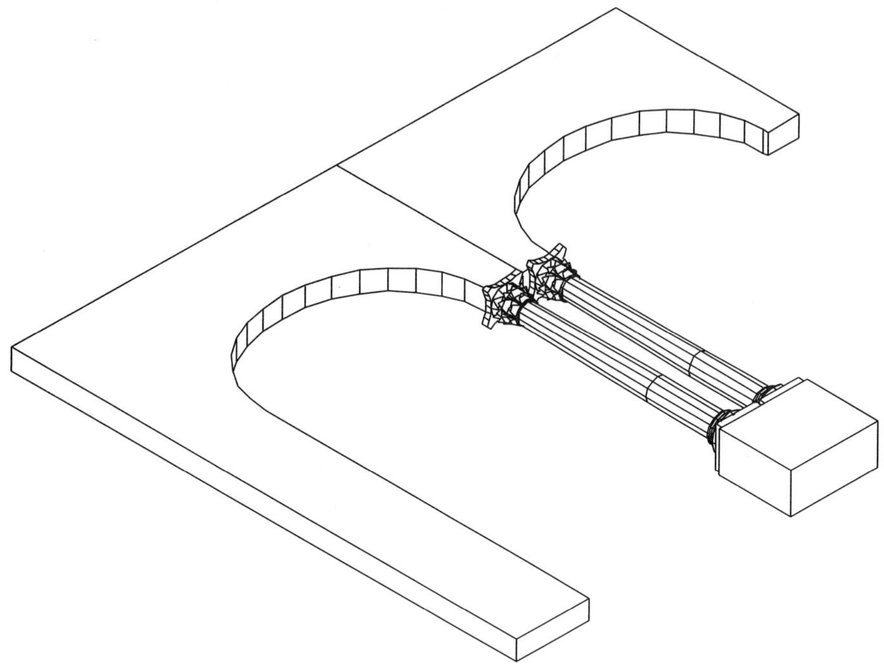

Step 6

Hide the edges of the mating pieces.

A polygon or slab is limited to a maximum of 36 vertices. To overcome this limitation, you can break down complex shapes into smaller pieces. The **Mark Visible** function allows you to hide the edges where the pieces join.

Select **3D Entity**, **Slab**, **Partial Mark Visible**. Pick the slab edges where they join at points A, B, and C, as shown in the following figure. You will have to click, refresh the screen, **<esc>**, and click again at points A/B where two slabs share the same edge. Here is the result:

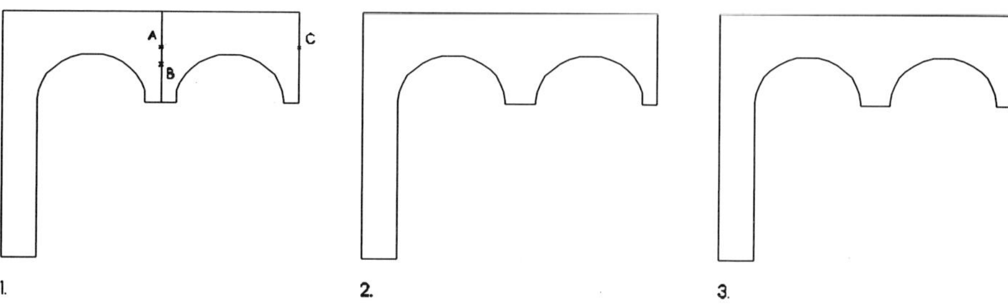

Step 7

Rotate the model to the vertical position.

Turn on all the 3D layers associated with the 3D colonnade modeled thus far. Set Z-base and Z-height values to 0′-0″. Select **3D Edit**, **Rotate**, **X-axis**, **X-angle** and enter 90 degrees. Select **New Center**, and middle-button snap to the lower left corner at the very bottom of the wall. Turn on **Layer Search** and select the entire model by area. The model rotates upright.

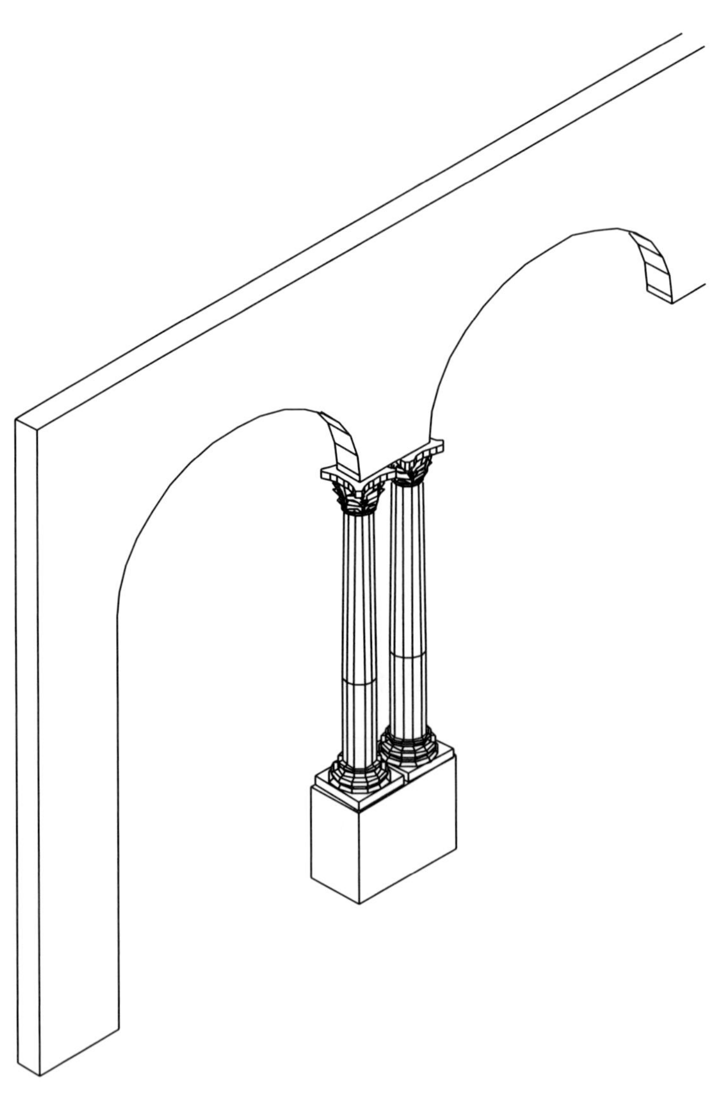

Step 8

Turn on the layer associated with the 2D polyline path and profile shapes. Using the **2D Edit**, **Move** command, move the previously drawn polyline to align with the model.

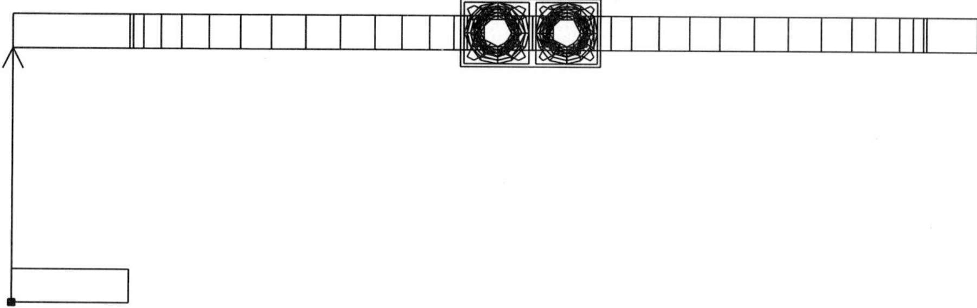

Step 9

Create the sweep profiles.

9.1 Set the layer with the 2D drawings as **Active Only**. Turn on the layer with the sweep path and set it as the active layer. Press the **Z** key and set **Z-base** and **Z-height** to 0′-0″. Using the 2D outlines as a guide, trace over the profile shapes with horizontal 3D polygons. This will produce three polygons. These shapes will be applied to the polyline path to produce the sweep. Turn off the layer for the 2D guide.

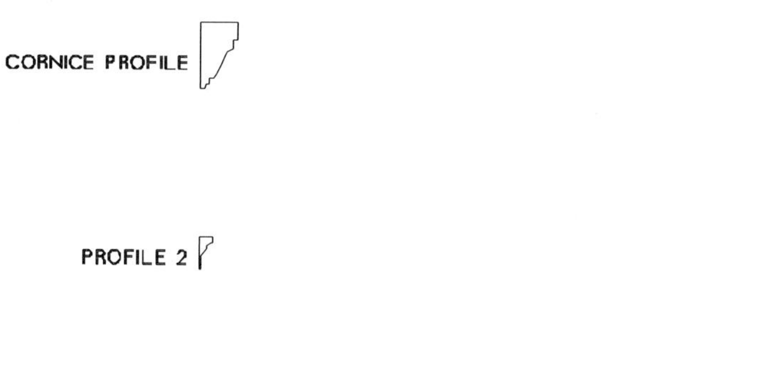

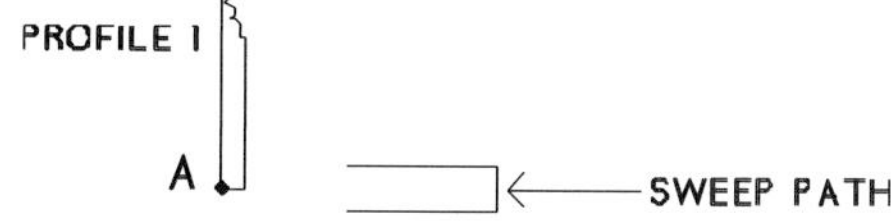

9.2 Create a new layer for the sweep entities and set it as the active layer. Select **Toolbox**, **3D Tools**, **Sweep**. You are prompted to "Select polygon entity that defines the sweep cross section." Pick the polygon for profile 1. Next you are prompted to "Enter a point for sweep path origin." Middle-button snap to point A as indicated in the illustration. Pick **Entity** as your selection mode. Toggle on the option for **Cap Ends**. Select **Settings**. Toggle on the following options: **Vertical** (applies the polygon section vertically) **Show Latitude**, **Show Longitude** (shows all polygon edges) **Z-base** (uses Z-base as the basis for the sweep)

Pick the polyline that represents the sweep path. The sweep is created.

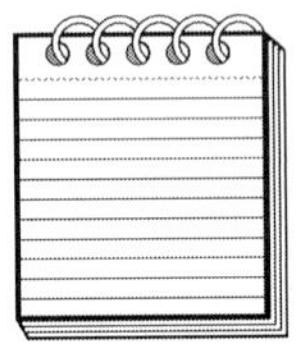

NOTE: *Verify that the sweep was applied to the* outside *of the path (that is, the resulting shape should be larger than the path). If it swept the wrong side, pick* ***Undo****, toggle on* ***Reverse****, and pick the polyline path again.*

9.3 Select **New Section**. Pick profile 2. When prompted for the sweep path origin, again snap to point A. Check your settings for **Cap Ends** and **Reverse** (if necessary) and pick the polyline path again. The next sweep is created above the first. Exit the macro and check your work in isometric view.

9.4 On the polyline path layer, create a 2D line at **Z-base** 0′-0″ that spans the entire length of the façade. This will be used for the cornice path. Create a new layer for the cornice sweep and set it active. Restart the **3D Tools** macro. Choose the cornice polygon as the sweep cross section. Snap to point A to define the sweep path origin. Check the macro settings for **Cap Ends** and **Reverse**. Pick the 2D line to use as the sweep path. The sweep is created. Verify that the sweep was applied to the proper side of the line. The following figure is an exploded view of the sweep elements.

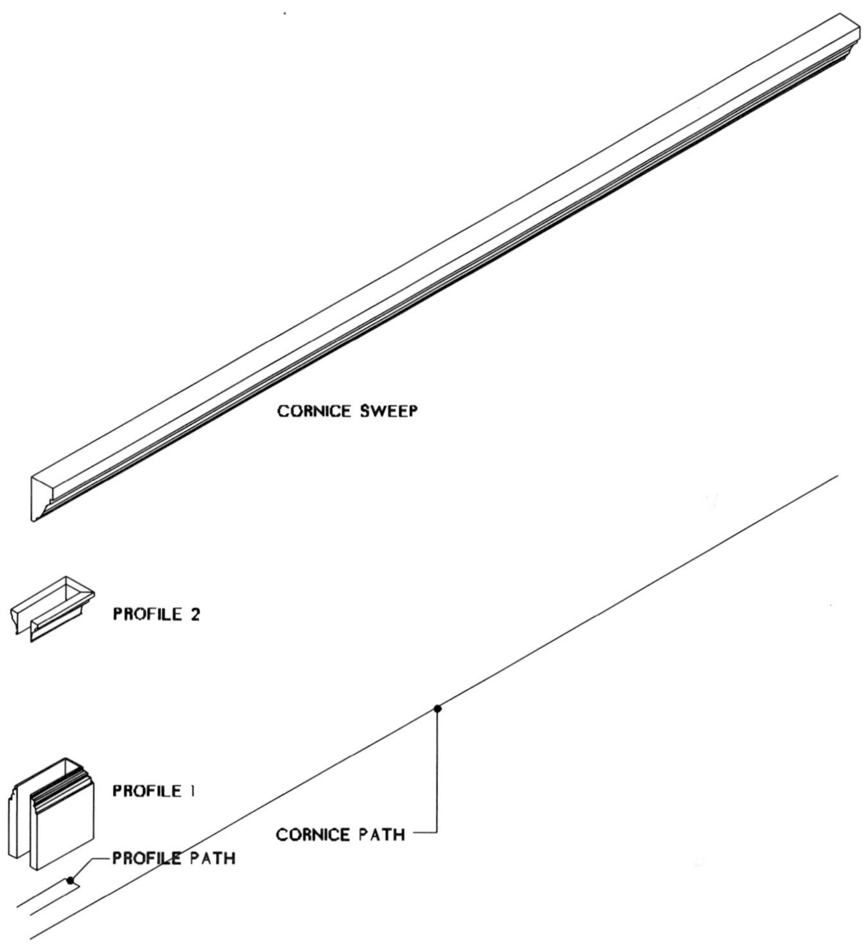

Step 10

Complete the model.

10.1 Change your view to **Ortho**. Set the 3D Columns layer as **Active Only**. Turn on the layers for the 3D wall and sweep profiles 1 and 2. Use **Mirror**, **And Copy** to create the opposite half of the model.

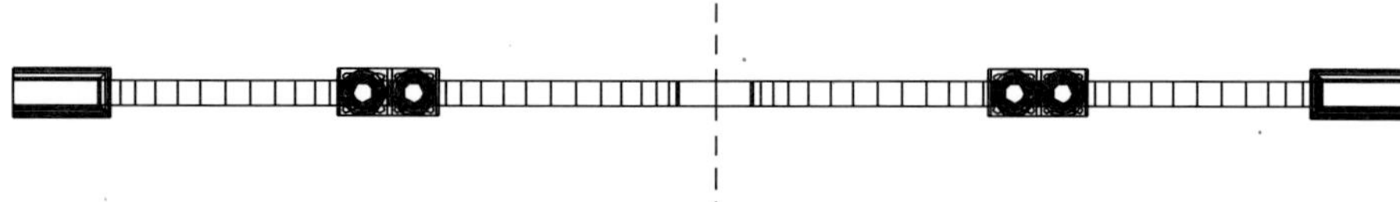

10.2 Set the column layer as **Active Only**. **Copy** the column and pedestal assembly 9′-0″ to the right to the middle bay. Turn on the layer for the cornice sweep. The model is complete.

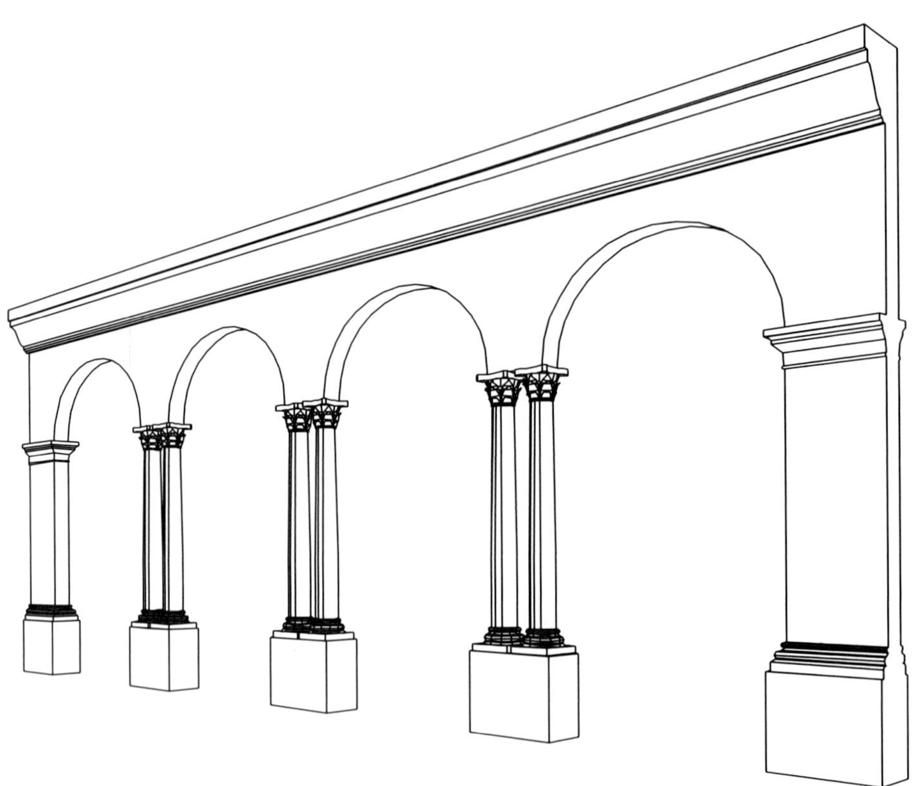

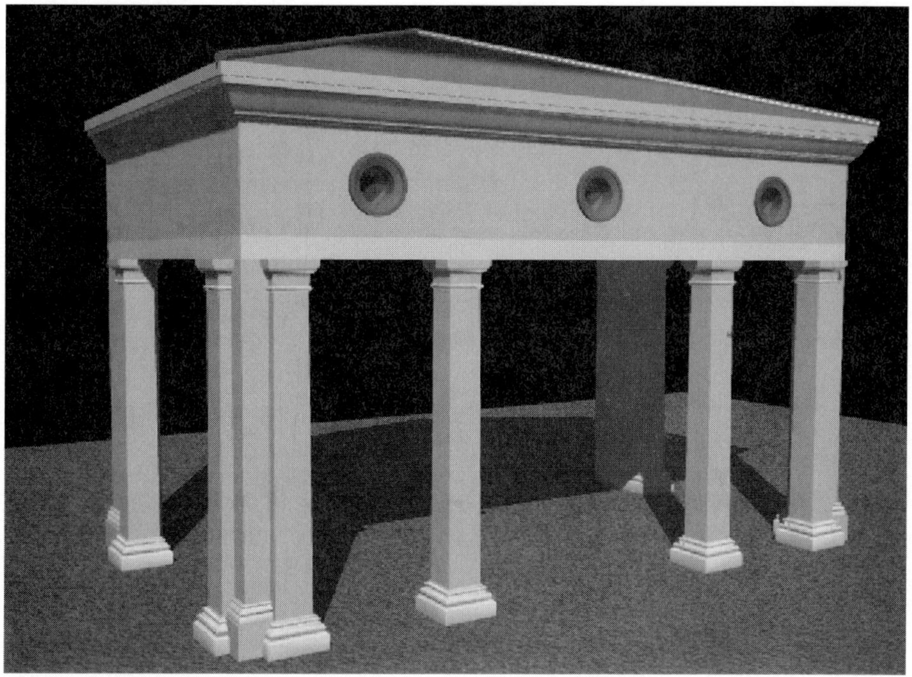

Exercise 8—Porte-cochere: RevSurf × $\sqrt{2}$

This example employs a useful technique to quickly create four-sided shapes with complex profiles. It is a variation of the sweep method, but much faster. It is also an excellent example of economy of effort in modeling: three primary shapes create the entire structure. Before you begin, you will need to install the **PolyMaster** macro* found on the bonus CD. The file **Exercise_8.DC5** contains the step-by-step procedure.

Step 1

The first step, as always, is to establish the critical guiding 2D elements at a Z-base and Z-height of zero. The curved profiles of the pilasters and cornice line are broken down into line segments. The entire model can be

* The **PolyMaster** macro was contributed by Patrick McConnell, Architect, Valatie, New York, USA.

generated from three pieces: The roof/entablature assembly, the wall section, and the pilasters. These profiles are indicated in the following figure as items 1, 2 and 3. Also note the axis lines for mirroring and trimming, labled as A, B, C, D, E, and F. The various elements will be modeled as if lying on the elevation, then rotated vertically and re-distributed on the 2D plan to create the finished model.

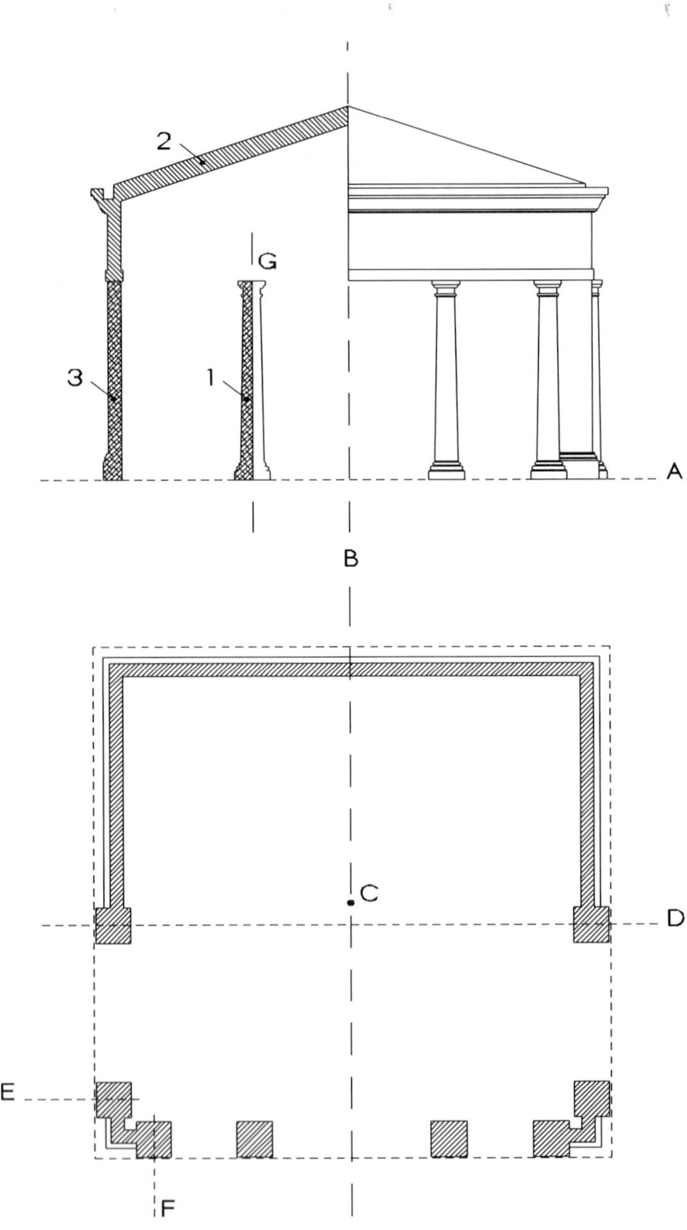

Step 2

Create the pilaster.

2.1 Create a **New Layer** for the 3D pilaster and set it as active. Press the **Z** key on your keyboard, and set the Z-base and Z-height values to 0′-0″.

With the 2D elevation drawing as a guide, the pilaster is modeled using a revolved surface.

2.2 Select **3D Entity**, **RevSurf**.

2.3 Select **PrimDivs** and enter **4** as the number of primary divisions for the revolved surface.

2.4 Select **Closed** to create a revolved surface with open ends.

2.5 You are prompted to "Select first point on <Surface of Revolution>."

2.6 Begin tracing the profile of the column, from bottom to top as shown in the following figure at left.

2.7 When the last point is reached, press the right mouse button to finish.

2.8 You are prompted to "Select first point of surface center axis." Make sure that **Ortho** mode is on, and middle-button snap to the midpoint at line G at the top of the column. Drag the center axis line down towards the column base and left-click. The revolved surface is created.

If you were to view the pilaster end-on, it would look like the diamond shape shown at right. This is because the profile of the revolved surface defines the edge of the surface along a 45-degree axis, in the case of a four-sided surface of revolution.

Confused? All you need to know is that to make it look right, we need to rotate it 45 degrees about its center axis, and enlarge it by the square root of two. Fortunately, we have a tool to accomplish this with little effort. Here is the concept again, illustrated graphically in the following figure.

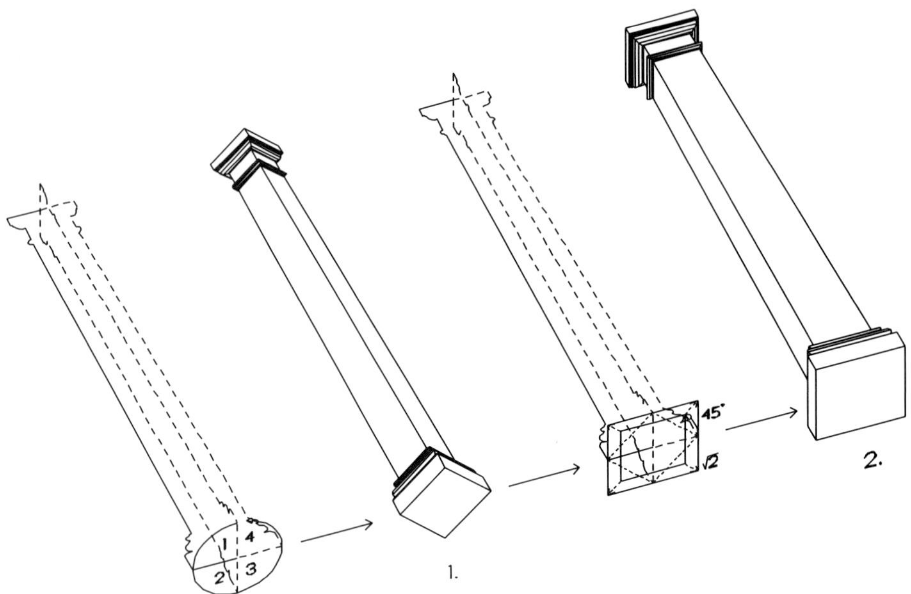

From left to right: The profile is traced and revolved four times to produce shape 1. To correct the shape, it must be rotated 45 degrees and enlarged by $\sqrt{2}$ to produce shape 2.

Step 3

RevSurf $\times\sqrt{\mathbf{2}}$ to correct the shape.

Now for the fun part. Fun, because the macro does all the work.

3.1 Select **Toolbox**, **Polymaster** (which you installed before starting this exercise).

3.2 Choose **Adjust**. Toggle on **Y-axis** (we are "adjusting" an element that is lying flat on the ground, about its vertical, or Y dimension).

3.3 Change your selection mode to **Entity**, and pick the revolved surface to modify. The revolved surface flips 45 degrees and enlarges appropriately in one step. Done!

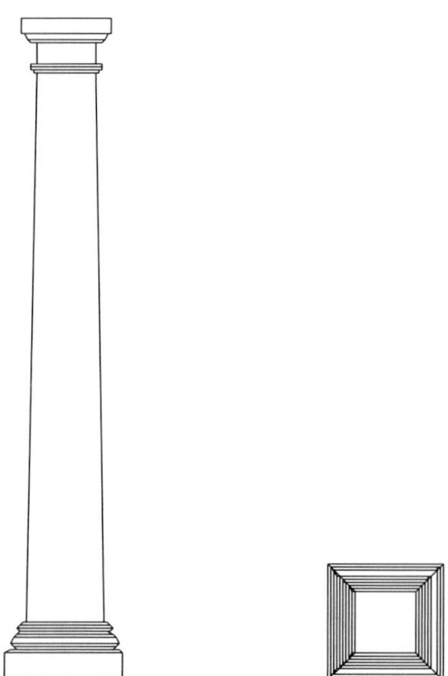

If you were to look at the pilaster end-on again, it would now look like the square figure at right.

Step 4

Repeat Steps 2 and 3 on profile 2 to create the roof and entablature.

4.1 Create a **New Layer** for the roof and set it as active. Press the **Z** key on your keyboard, and set the Z-base and Z-height values to 0′-0″.

Using the 2D elevation drawing as a guide, the roof and entablature is modeled using a revolved surface.

4.2 Select **3D Entity**, **RevSurf**.

4.3 Select **PrimDivs** and enter **4** as the number of primary divisions for the revolved surface.

4.4 Select **Open** to create a revolved surface with open ends.

4.5 You are prompted to "Select first point on <Surface of Revolution>."

4.6 Begin tracing the profile of the column, from bottom to top as shown at left in the following figure.

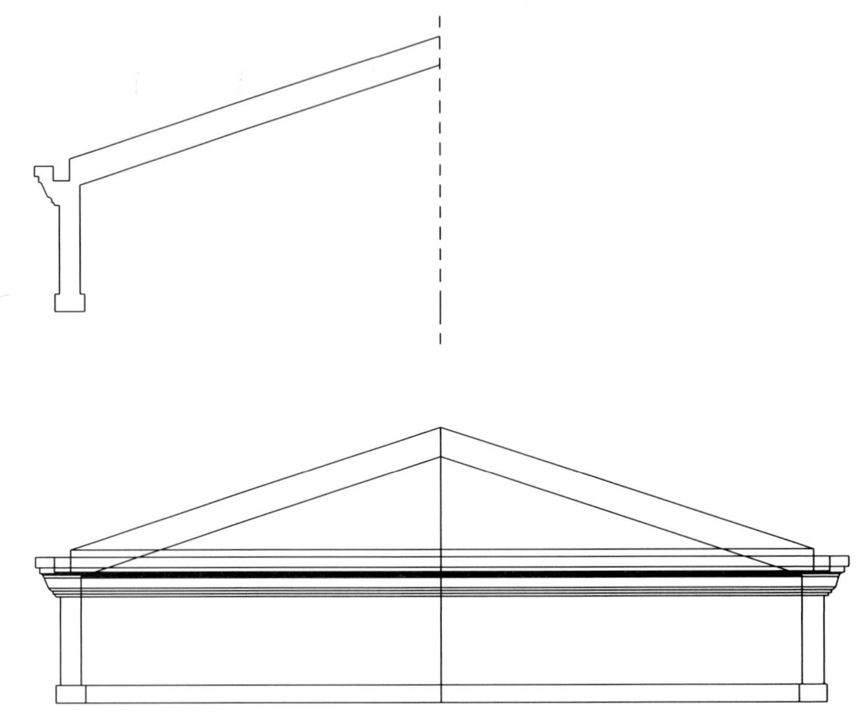

4.7 When the last point is reached, right-click to finish.

4.8 You are prompted to "Select first point of surface center axis." Make sure that **Ortho** mode is on, and middle-button snap to the midpoint at line B at the top of the roof peak. Drag the center axis line down towards the column base and left-click. The revolved surface is created.

Step 5

Correct the roof shape.

5.1 Select **Toolbox, Polymaster**.

5.2 Choose **Adjust**. Toggle on **Y-axis**.

5.3 Change your selection mode to **Entity**, and pick the revolved surface to modify. The revolved surface flips 45 degrees and enlarges by a factor of 1.41 in the X and Z directions.

Step 6

Repeat Steps 2 and 3 on profile 3 to create the walls.

6.1 Create a **New Layer** for the 3D walls and set it as active. Press the **Z** key on your keyboard and set the Z-base and Z-height values to 0′-0″.

Using the 2D elevation drawing as a guide, the wall section is modeled using a revolved surface.

6.2 Select **3D Entity**, **RevSurf**.

6.3 Select **PrimDivs** and enter **4** as the number of primary divisions for the revolved surface.

6.4 Select **Open** to create a revolved surface with open ends.

6.5 You are prompted to "Select first point on <Surface of Revolution>."

6.6 Begin tracing the profile of the column, from bottom to top, as shown at left in the following figure.

6.7 When the last point is reached, right-click to finish.

6.8 You are prompted to "Select first point of surface center axis." Make sure that **Ortho** mode is on, and middle-button snap to the midpoint at line B at the top of the roof peak. Drag the center axis line down towards the column base and left-click. The revolved surface is created.

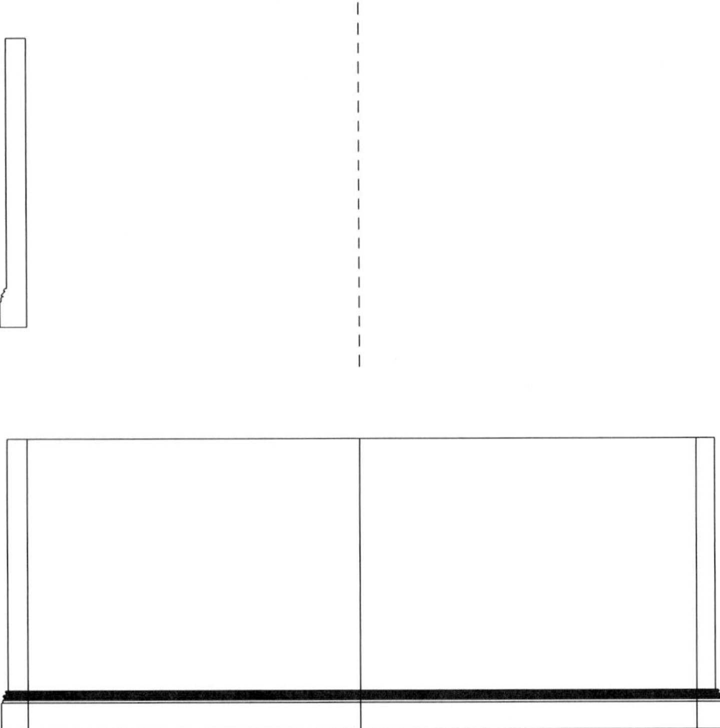

Step 7

Correct the wall shape.

7.1 Select **Toolbox**, **Polymaster**.

7.2 Choose **Adjust**. Toggle on **Y-axis**.

7.3 Change your selection mode to **Entity** and pick the revolved surface to modify.

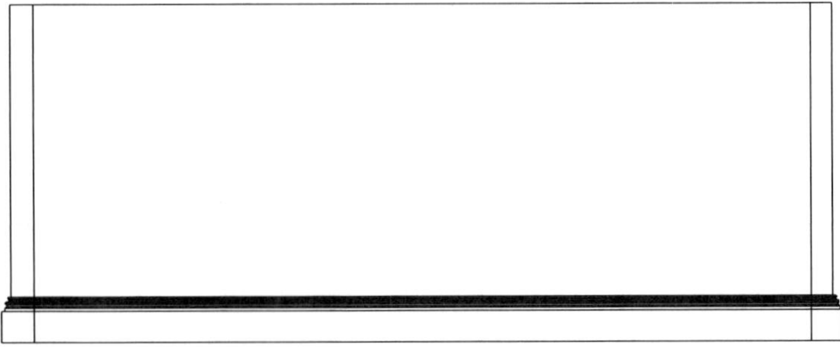

Step 8

Rotating to the vertical.

Now the modeled pieces will be rotated to their proper orientations in relation to **Ortho** view.

8.1 Turn on the layers for the 3D pilaster, roof and walls.

8.2 Press the **Z** key on your keyboard, and check to see that the Z-base and Z-height values are set to 0′-0″.

8.3 Select **3D Edit**, **Rotate**, **X-axis**, **X-angle** and enter **90 degrees**.

8.4 Select **New Center** and middle-button snap to the bottom left wall base along line A. Make sure **And Copy** is toggled off.

8.5 Turn on **Layer Search** and select the entire model by **Area**. The model rotates to a vertical position.

Here is an isometric view:

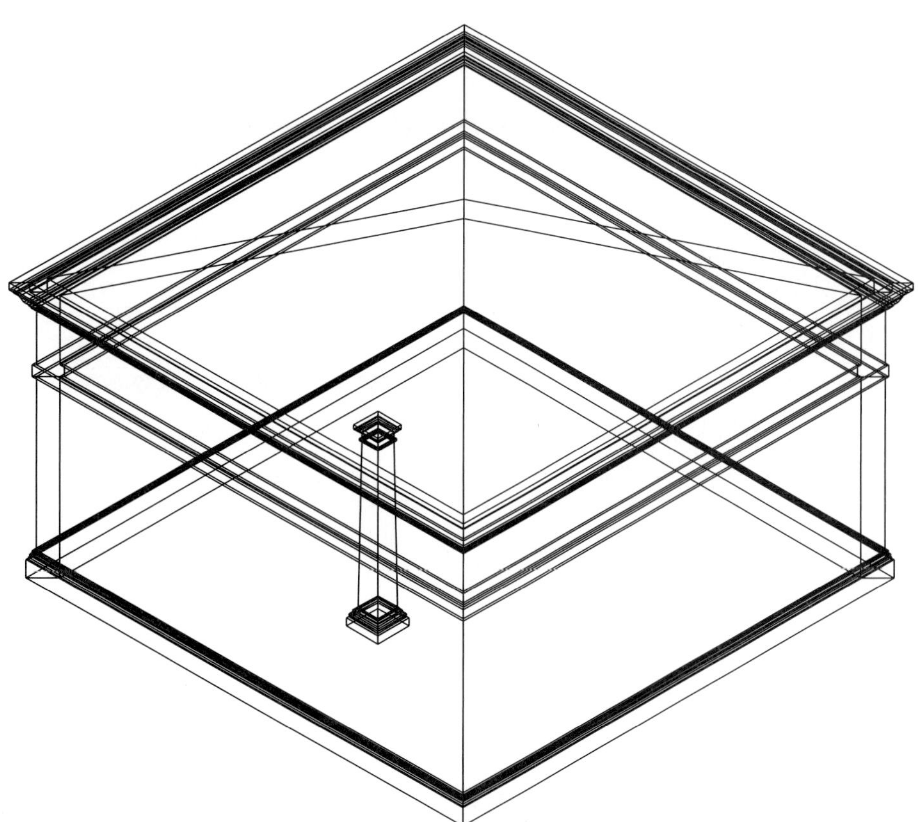

Step 9

Move the elements into position.

The pilaster, roof, and wall are not properly aligned to the plan. Select **3D Edit**, **Move** and reposition the modeled pieces on the plan. Make sure **And Copy** is off and **Layer Search** is on. Middle-button snap for each point selected. To place the roof and walls, snap from the center of ridge to point C on the plan. To position the pilaster, snap from the corner of the 3D element to the respective corner of its position in plan.

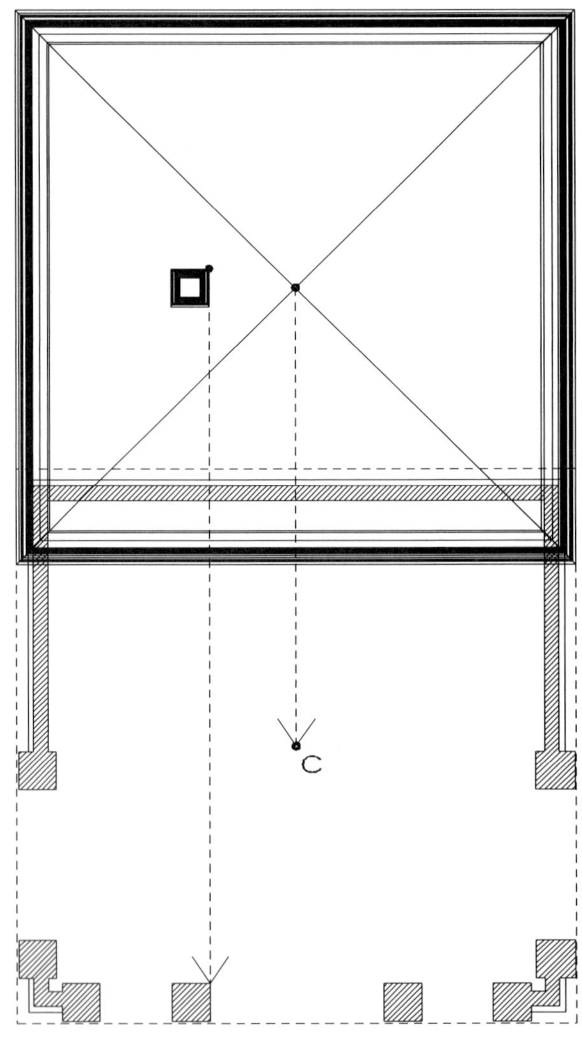

Step 10

Cutting out the openings.

10.1 Set the layer for the 3D walls as **Active Only**. Select **3D Edit, Explode, To Polygons**, and pick the wall element. The revolved surface comes apart into individual 3D polygons, which may be edited further.

10.2 Trim the walls: Select **Toolbox**, **3DTools**, **Knife**.

10.3 You are prompted to "Enter first point of Knife line." Using the 2D Plan as a guide, middle-button snap to the end points of line D as shown in the key plan in Step 1.

10.4 The macro prompts you to "Enter point on the 'keep' side of knife." Left-click anywhere above the knife line. The macro then prompts you to choose the entity, group or area to trim. Change your selection mode to **group** and pick the wall polygons. The polygons are trimmed to the knife line as shown in the following figure.

10.5 To create the corner piece, **Mirror, And Copy** the walls by **group** horizontally from point C. Then, using the knife technique described previously, cut the walls back along lines E, and F as shown in the following figure.

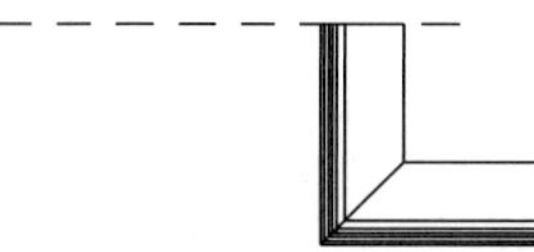

10.6 To create the opposite corner, **Mirror**, **And Copy** vertically from point C.

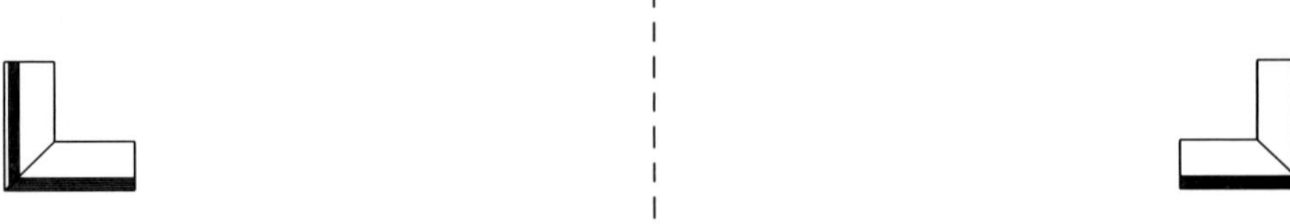

Step 11

Copy the pilasters about the plan.

Select **3D Edit**, **Move**, **And Copy**. Copy the pilaster to the remaining locations on the plan.

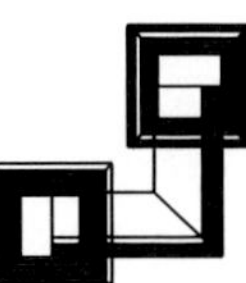

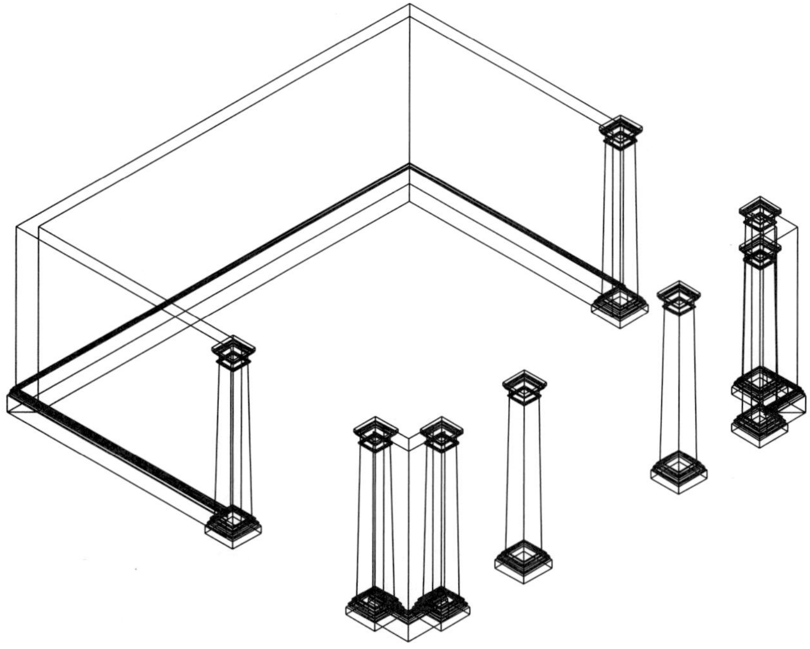

Step 12

Turn on the various layers for the 3D components, and change your view to isometric or perspective. The model is complete!

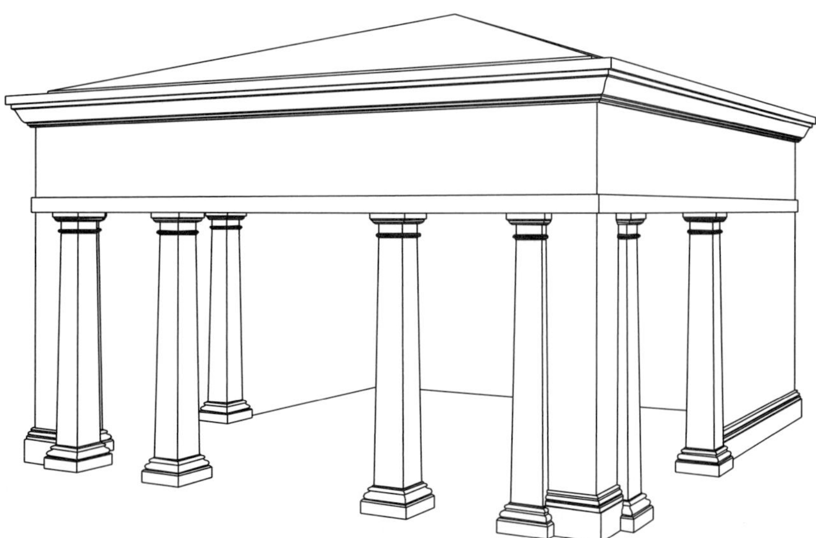

Variation

You can create a variation of the model by selecting **3D Edit**, **Change Primary Divisions**. Enter **18** and select the two supporting pilasters to change them to 18-sided columns.

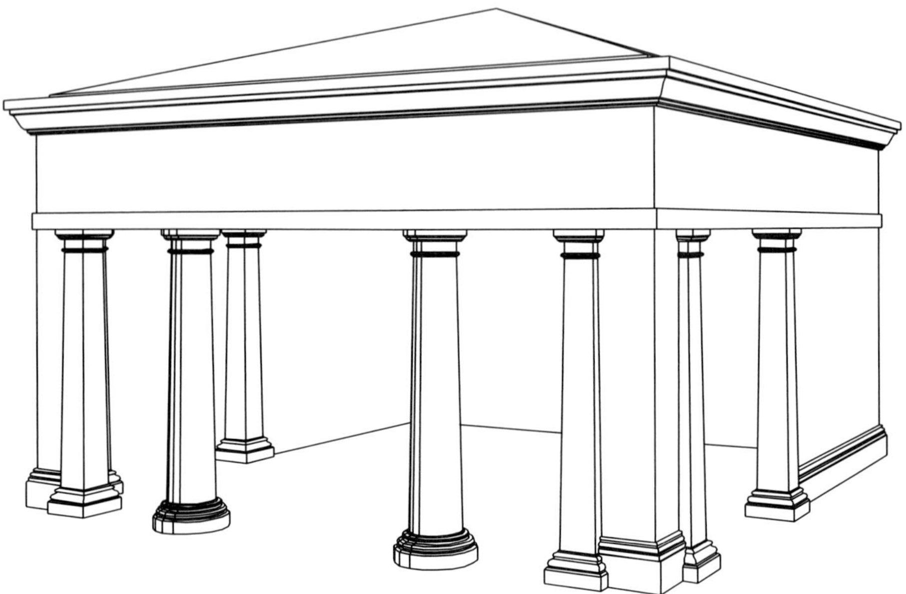

CHAPTER 20

3D Rendering

Advanced 3D Workshop 10: Notes on Rendering

Before you can render a model, you must, of course, create the model in DataCAD. While this is a complete subject unto itself, a couple of points should be noted. First, it is generally easiest to render a model if it is organized by layer and material in DataCAD. That is, for each material you plan to render, model that element on a single layer in DataCAD: You might have one layer for brick, another for glass, yet another for the metal frames containing the glass, and so on. The materials could be further separated by their location to facilitate rendering only what is within view, so you might create four layers for your brick material for the north, south, east, and west elevations. Secondly, when the polygons and slabs are created in the model, you will want to keep the face orientation for each plane in mind. This is the 'normal' face that is rendered when imported to **Renderize**. See the section on polygon normals for further details.

Once your model is finished and organized by layer and material in DataCAD, you are ready to export it to **Renderize**.

1. Save the model as a .DXF file to your DATACAD\XFER directory. **Renderize** requires .DXF format to read the file.
2. Start **Renderize Live** by choosing it from the **DataCAD 8** menu under the **Start/Programs** menu on your Windows desktop. When **Renderize** opens, a new project is automatically created.
3. From the **File** menu, choose **Load Object**.
4. At the **Load Object** dialog, change to your DATACAD\XFER directory, choose the file you exported as .DXF, and click **Open**.
5. In the **Create Materials From** dialog, check **Layer**.
6. In the **Create Objects From** dialog, check **Layer**.
7. Change **Set by Level** from **0** to **1**.
8. Uncheck all four options at the very bottom of the dialog box and click **Read**.
9. As the file reads, you will receive a prompt telling you that the imported file includes "children" and will ask if these should be included. Choose **Include**. If you do not receive this message, you did not properly configure the **Load Object** options, and you should quit and start over.

   Your .DXF file is now loaded as a **Renderize** project.

Setting Object Property Options

Before you start rendering, there are a few key elements to configure.

1. Choose the **Object Resource** button from the left menu.
2. Scroll through the object list, and locate the one that appears as **Node**; the **Node** is the parent object, or master folder, for the project. Any changes or settings made to the parent will affect all the dependent, or child, objects as well and relieve you of having to set all these options individually.
3. Drag and drop it onto the **Edit** well, located in the lower-left corner of the screen. The **Edit Object** dialog appears.
4. Check **Pass to Children**, **Render Backface**, **Cast Shadows** and **Receive Shadows**. Then click on **Save**, and, when prompted, click **Replace**.

NOTE: *When you are more familiar with the polygon normal issue and modeling, you will uncheck the **Render Backface** option. While enabling this function allows you to ignore normals for the time being, it will conflict with some types of texture and reflection mapping.*

5. Minimize the Object Edit menu to display the application window. As a rule, you should minimize the dialog whenever you are finished with a setting. Move the minimized window to the right of the screen if it is more convenient. Minimizing these windows instead of closing them will conserve system resources.

Establishing the View to Render

1. Click on the **Move** icon in the upper menu bar.

2. Click on the **4 Window View** button in the upper menu bar. Four windows appear. The window in the lower left shows the camera view, which is the view that will be rendered.

3. Click on the **Move Around Target** button.

4. In the **Plan View** window in the upper left, click on the location indicating the target for the camera.

5. Click on the **Move Camera** icon and move the camera into position. Check the camera view window to verify the desired view. You may access either elevation by clicking in the window to activate it, and then moving the camera or target.

6. Other move options may be selected which constrain the camera movement to left/right, up/down, or in/out.
7. Additionally, you may move around the target or camera in various X-, Y-, and Z-axis positions by selecting these icons.

8. Once you are satisfied with the view, choose the **Camera View** icon to return to full-screen camera view.

Creating a Test Render

1. Click on the **Render** icon to display the **Render** menu.

2. From the **Render** menu options, click the first icon for a full screen rendering.

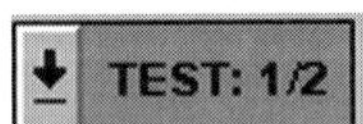

You can adjust the resolution of the test render image in this drop-down dialog. A value of **1** will render at the highest screen display, 640×480. You may final render at any resolution. Note the lighting and background in the model.

Loading a Background Image

1. Click on the **Image Resource** icon.
2. Choose **Load Image** from the **File** menu.
3. Locate an image to load; several are provided in the **VREAL/TUTOR** directory.
4. Choose **SKY.TIF** and click **Open**. The image appears in the resource window.

5. To use this image as a background image, click the **View Resource** icon, and drag and drop the **Default_1** view to the **Edit** well.

6. Click on the **Image Resource** icon again, and drag and drop the **Sky** image to the background well displayed in the **Edit View** dialog box.
7. Click **Save** and then **Replace**, and minimize the dialog box. The **Sky** background will now appear in your rendering. Perform a test render.

Positioning and Adjusting Lights

1. Click on the **Light Resource** icon.
2. Drag **Default_1** light to the **Move** well located in the upper left of the screen.
3. Click on the **Move** icon in the upper menu bar.
4. Click on the **4 Window View** button in the upper menu bar.
5. Using the target and move controls as previously described, position the light relative to the model. To simulate sunlight in an architectural rendering, the most realistic effect will be achieved by positioning the light a great distance from the model.
6. Drag the **Default_1** light to the edit well.
7. In the **Edit Light** dialog box, check **Cast Shadows**.
8. Set the self shadow bias to a value between 0.001 and 0.01. The self shadow bias is the distance a shadow falls relative to its object. While you want it to be close, if it is too close, the object will appear striped when rendered. Trial and error will determine the best setting.
9. Set the shadow map size to 2048 or higher. The shadow map size is the pixel density of an object's shadow. Higher values will produce sharper shadows, but at the expense of slower rendering times. Start out reasonably low, and change to a higher setting for the final render.
10. Click **Save** and then **Replace**.
11. Edit objects that do not need to cast shadows by going to the **Object Resource** menu, dragging the object to the **Edit** well, and disabling **Cast Shadows**. An example for this would be the ground plane, which needs to receive shadows but usually does not cast them. You may also selectively enable the **Smooth** option on objects that are curved to produce a smooth contour when rendered. Don't forget to **Save** and **Replace** the settings for each change you make.
12. Additional light sources may now be added, or a light below the ground plane might be added to simulate reflected light, as **Renderize** does not calculate reflected light.

13. To add a light, click on the **Light Resource** icon, and drag **Default_1** to the edit well. Change its name to **Default_2**, and click **Save**. You have now cloned an additional light and may place it and change its settings accordingly. The process for setting multiple views is done in the same manner.

14. Click on the **Render** icon to display the **Render** menu.

15. In the **Options** box, toggle the first two buttons on: **Calculate Shadows** and **Anti-alias**.
16. Perform a test render, and make changes if desired.

Setting Material Properties

1. Click on the **Materials** icon.
2. A list of material colors and textures is displayed for each object in the model. To change or adjust one, scroll through the list to locate the object, and drag and drop it onto the **Edit** well.
3. Adjust the materials properties as necessary, using the following parameters:

- *Constant* Constant color; no shade, transparency or highlights
- *Matte* Flat color with transparency, but no highlights
- *Shiny* Glossy color with transparency and highlights
- *General* Transparency of the object: 1=opaque, 0=transparent
- *Edge* Transparency at edge: 1=opaque, 0=transparent. Like a soap bubble, it should appear nearly opaque at its edge, so the value would approach 1.
- *Highlight* Transparency of highlight: 1=opaque, 0=transparent. Like a glass bowl, the highlight spot on the bowl would appear nearly solid, so it should have a value approaching 1.
- *HiLite Size* Relative size of the highlight

Ignore **Reflect** settings for now.

4. After setting your options, don't forget to click **Save** and **Replace**.

To adjust the color of the object:

1. Choose the **Color Wheel** icon.
2. Click in the **Matte** well, then adjust the color as desired by sliding the **HSV** or **RGB** buttons.
3. When you achieve the proper color, click **Save** and **Replace**.
4. Test render and adjust as necessary.

To map a texture, first load the texture as an image via the **Image Resource** menu, as you did previously for the sky background. Once loaded, it can be dragged and dropped to the **Color Texture** well shown previously in the **Edit Material** dialog box. The texture will now appear in your rendered image.

To scale a texture, click the **Object Resource** icon, and drag and drop the object receiving the texture to the **Edit** well.

You will notice an outline defining the texture map superimposed on the object. Adjust its size with the **Move** and **Enlarge** tools below the window.

Next, choose a mapping type. **Ortho** or **Planar** are best to try first.

Save, **Replace**, minimize the window and test render.

Creating the Final Rendering

When you are satisfied with the results of your rendering setup, you can generate a final rendered image at a higher resolution.

1. From the **Render** menu, enter an X and Y resolution in the **Render to File** box.

2. You may adjust the aspect if desired by clicking on the drop-down arrow.
3. Click the checkmark.
4. Choose a directory to save the image in, enter a filename, choose a graphic format, and click **Save**. The image is rendered to your hard disk. Complex scenes may take a fair amount of rendering time.

Saving Your Work

1. Choose **Save** from the **File** menu to save your project.
2. Choose **New** from the **File** menu to close this project and start a new one.

   If you have loaded new objects to your scene during your session, you must choose **Save As** from the **File** menu and overwrite your existing file to update it with the new objects. Choosing **Save** from the **File** menu only updates your views and materials.

NOTE: *Your project is saved as an* ***.EYE*** *file. The* ***.EYE*** *file is simply a text file that maps the objects, saved in the directory as* ***.GED*** *files, to their corresponding textures, which may be in other directories. As such, if you move an* ***.EYE*** *file, or relocate your texture maps, when you reload the* ***.EYE*** *file, it may not be able to locate its referenced objects and textures. You may edit the* ***.EYE*** *file in any text editor to change texture and object paths if necessary.*

CHAPTER 21

Techno-Files

Although the title of this chapter may tempt all but the most technology-obsessed geeks (like me) to run for cover, there is some very important information that you will get from it, especially the first two sections about file recovery and CAD files.

File Recovery

Computer data is simply a bunch of digital zeros and ones that occasionally realign themselves into digital rubbish. Electricity is at best a fickle mistress that tends to desert you without so much as a warning, and with the simple slip of a finger that benign little rodent that is your mouse might accidentally delete eight hours of work. For these and other potentially catastrophic events, we shall delve into the wonderful world of file recovery.

In order to understand how to recover drawing files, you first need to know about the various files created by DataCAD every time you open or save your drawing file. Besides the standard .DC5 file, four other files are created by DataCAD. These files are the .$$$ (semaphore), .BAK (backup), .ASV (autosave), and .SWP (swap) files. Every time you open a drawing file, DataCAD immediately creates the .$$$ file and the .SWP file. The .BAK file is created or updated every time you save the drawing, and the .ASV file is created or updated every time DataCAD does an automatic file save.

The .SWP file is the only file that is modified continuously during a user's drawing session. The .SWP file is created by DataCAD by simply copying the .DC5 file at the time you open the .DC5 file. The DC5 file then sits there completely untouched by DataCAD while the user works on this .SWP file. All commands executed by the user result in modifications being made directly to the .SWP file, and only the .SWP file. Only upon saving the drawing file (**File/Save**, **Ctrl+S,** or **F**) is the entire .SWP file copied back to the .DC5 file. The original .DC5 file is copied to the .BAK file just prior to that.

For all of the following explanations, we will assume that you are working with a DataCAD drawing file called MYFILE.DC5.

The Semaphore (.$$$)File

This file is created as soon as you open an existing drawing file or create a new one. It will be located in the same directory as the main .DC5 file and will have the same name as the main .DC5 file. Thus, the MYFILE.DC5 file would have a corresponding MYFILE.$$$ file. If you *right-click* on the file in Windows Explorer and check its properties, you will see that it has a file size of zero kilobytes. This is because it actually contains no information. It's what is known as a *semaphore file*, because it signals to DataCAD that the .DC5 file with the same name is currently open. As soon as you exit or close the drawing file, the .$$$ file is removed from the computer, but if the drawing is not exited normally, the .$$$ file will remain.

Why is this important? For two reasons:

- If you are working on the MYFILE drawing and another user on your computer network tries to open the file, this semaphore file will cause DataCAD on the other computer to display the message shown in Figure 24-1.

  The other user will know that someone else is working on the file and hopefully not open it. The .$$$ file only signals DataCAD to display this message. It does not keep someone else from opening the file as well. If you try to open a file on your own computer that you already have open, you will get the message shown in Figure 24-2, and you won't be allowed to open the second copy of the file.

  Why shouldn't someone else open MYFILE if you are already working on it? Because if two users have the same file open at the same time, both copies of the file can be edited and saved by both users, but with dire consequences. If both of you are working on the file, every time you

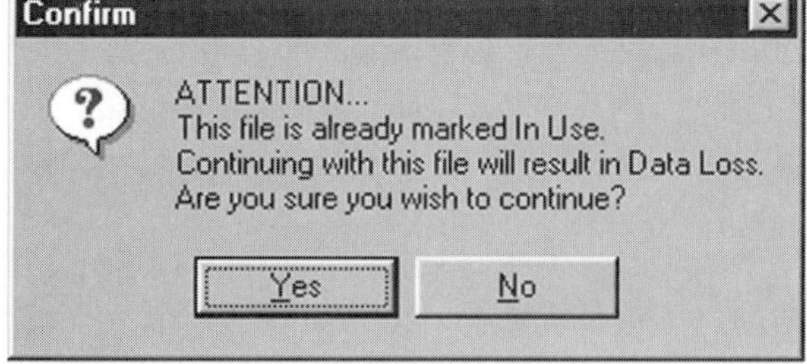

Figure 24-1
A message that the file to be opened is being used by another computer

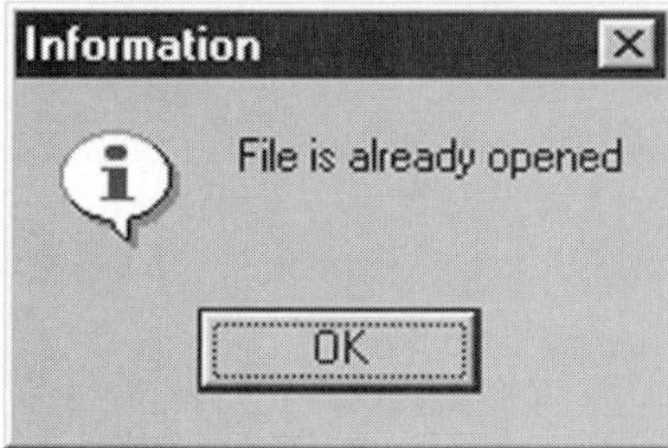

Figure 24-2
What happens when you try to open a file that's already open

save the drawing, all of your work is saved as MYFILE.DC5; all of the other user's work is overwritten, but she doesn't know it. Every time the other user saves, all of your work is overwritten, but you won't know it. And whoever saves last wins; the other user loses, since none of their changes will be saved!

- If you are working on MYFILE and DataCAD closes abnormally (such as a computer crash or a DataCAD crash), the .$$$ will still be on your computer. When you open MYFILE again, you will get the same Attention message displayed earlier, telling you the file is already in use. If an .ASV file is on the computer as well, and you choose to continue, another message will follow asking if you want to rename the .ASV file, as shown in Figure 24-3 (more on this in the .ASV description).

  If you select **No**, then the drawing file will not be opened and the .ASV file will be deleted. You will then have to select a new drawing file to open via the **File** drop-down menu. If you select **Yes** instead, then you will see the message shown in Figure 24-4.

  If you select **No**, then the original drawing MYFILE.DC5 will be opened. If you select **Yes**, then you will be prompted to rename the .ASV drawing to a new name to avoid writing over the original file. If the original file was called MYFILE.DC5, then perhaps you name the new file MYFILE2.DC5. DataCAD will leave the original MYFILE.DC5

Figure 24-3 The .ASV warning box

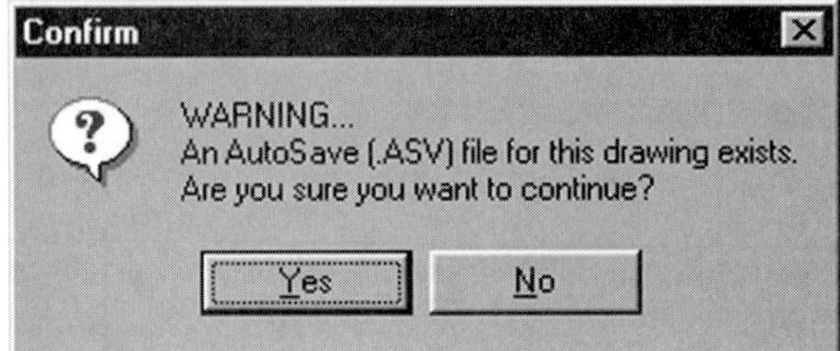

Figure 24-4 The prompt to rename the .ASV file

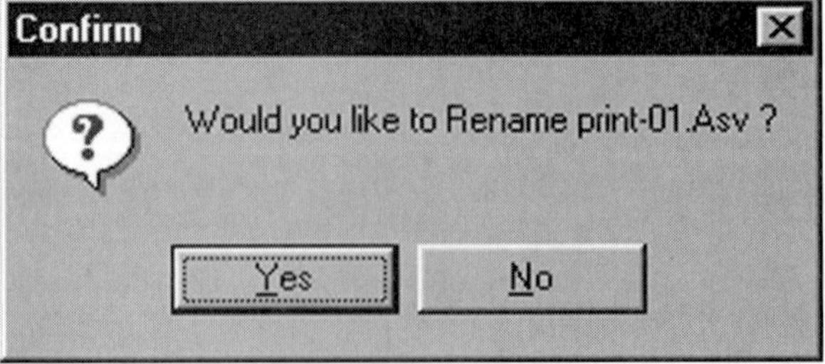

drawing alone and will instead start the new drawing MYFILE2.DC5. In this way, you will retain the original MYFILE.DC5 while examining and editing the MYFILE2.DC5 drawing.

WARNING: *If User #1 has MYFILE.DC5 open, User #2 opens it anyway, and then User #2 exits the drawing without saving (as he should), User #2 will not receive any future warnings when he tries to open MYFILE.DC5. Once User #1 exits MYFILE.DC5, when she or someone else opens it again later, User #2 will again see the warning when trying to open the file.*

The Backup (.BAK) File

Like most computer programs these days, whenever a file is opened, a backup file is created at the same time in the same directory as the main .DC5 file. It contains the same data as the original file so that if something unrecoverable happens to the main MYFILE file, then you can rename the .BAK file extension with a .DC5 extension (through Windows Explorer, not through DataCAD), open the file, and begin working again.

The .BAK file is updated every time you first open a file, every time you save the file (using the **File/Save** command **Ctrl+S,** or the **F** key), and every time you exit the file and choose to save. Therefore, if you open a file, work on it for 10 minutes, and then the computer or DataCAD crashes, the .BAK file will not have all the changes you made in the last 10 minutes. It will only contain the data that was saved when the file was first opened. But if you save the drawing, then the .BAK file will be updated to include the current drawing information. If the computer or DataCAD crashes now, you will not have lost anything. You can rename the MYFILE.BAK file to MYFILE.DC5 and continue working.

The .BAK file is persistent, meaning it remains on your computer and is not deleted when the file is closed or if the computer or DataCAD crashes. It remains as a permanent backup file in case something happens to the main .DC5 file.

The Autosave (.ASV) File

This file is created every time DataCAD does an automatic save. The number of minutes between auto saves is set in DataCAD under **Utility/Settings/SaveDlay**. The .ASV file is like the .BAK file except that DataCAD saves the file after a certain number of minutes, even if you don't. While an auto save is going on, DataCAD will suspend all operations until the save is complete, so you may notice short or long pauses, depending on how large your file is.

Like the semaphore file, the .ASV file is deleted from your computer if the drawing file is exited normally. The file remains, however, if the drawing file is exited abnormally. When DataCAD updates the .ASV file, the .BAK file is updated as well.

The .ASV file may or may not be the most recent backup of your drawing file. If your .BAK file was the last one saved, then that is the most recent file. If DataCAD did an auto save anytime after you saved (**Save** or **F**), then the .ASV file is the most recent one. To find out, you can check the Properties of both the .ASV and .BAK files in Windows. Whichever one has the most recent date and time is your best candidate for recovering the most recent information.

Perhaps the best way to recover the information in an .ASV file is to not open DataCAD after a crash. Instead, use Windows Explorer to rename the .ASV file with a new name and a .DC5 file extension. Then you can open the new drawing in DataCAD just like any other drawing file. The location of the .ASV file is set in **Tools/Program Preferences/Pathnames/Autosave Files**.

The Swap (.SWP) File

This is perhaps the most important ancillary file related to your main .DC5 file. As you work on your drawing file, DataCAD is constantly updating the .SWP file. In fact, the .SWP file is the only file that is modified continuously during a user's drawing session. The .SWP file is created by DataCAD by simply copying the .DC5 file at the time you open the .DC5 file. The .DC5 file then sits there completely untouched by DataCAD while the user works on this .SWP file. All commands executed by the user result in modifications being made directly to the .SWP file, and only the .SWP file. Only upon saving the drawing file (**File/Save Ctrl+S,** or **F**) is the entire .SWP file copied back to the .DC5 file so that the information in both files is the same again.

Because the .SWP file is constantly being updated by DataCAD, it is most likely the most up-to-date file. But like the .ASV file, if you exit your drawing file abnormally and a .SWP file is present on your computer, reopening your original drawing file will cause the .SWP file to be deleted, thereby losing any chance of recovering the information in the .SWP file.

Instead of risking the deletion of this file, rename it with a .DC5 file extension before trying to reopen the main .DC5 file. The location of the .SWP file is set in **Tools/Program Preferences/Pathnames/Autosave Files**.

WARNING: *Do not delete the MYFILE.SWP file while you are actively working in the MYFILE.DC5 drawing file. If you do, you may damage the .DC5 file and cause DataCAD to lock up or crash.*

Recovering from Catastrophe

Now that you know a bit about these other files, let's see how to use them. The following chart should help you figure out the best recourse should you need to recover from file crashes, errors, and operator-induced mistakes.

What Happened?	How to Recover (Assume the file is called MYFILE.DC5)
DataCAD crashed and I had to restart the program. OR Windows or the computer crashed and I had to restart my computer.	1. Do not open DataCAD or MYFILE.DC5. 2. In Windows, *right-click* on the file and choose Properties to check the date and time of each of these files: a. MYFILE.DC5 b. MYFILE.SWP c. MYFILE.BAK d. MYFILE.ASV 3. If the original MYFILE.DC5 file is the most up-to-date, then open DataCAD and open MYFILE.DC5. If it is not the most up-to-date then . . . 4. Pick the file with the most recent date and time and rename it to a new, unique name with a .DC5 file extension, such as MYFILE-2.DC5. 5. Move the file to the same directory as the original MYFILE.DC5 file. 6. Start DataCAD and open the newly named file (MYFILE-2.DC5). This should now be the most up-to-date drawing file. Assuming you elect to keep the file, then . . . 7. Use **SaveAs** to save the file under the original name (MYFILE.DC5) in DataCAD. When asked if you want to replace the original file, select **Yes**.

What Happened?	How to Recover (Assume the file is called MYFILE.DC5)
I made a major drawing change that I can't recover from (assuming that **Undo** did not work because the undo buffer had already cleared). I need to get the most up-to-date drawing from before my changes.	1. Do not save the drawing or exit DataCAD. 2. Minimize DataCAD (click on the dash in the upper-right corner of the program). 3. In Windows, *right-click* on the file and choose Properties to check the date and time of each of these files: a. MYFILE.SWP b. MYFILE.BAK c. MYFILE.ASV 4. Pick the file with the most recent date and time, and then rename it to a new, unique name with a .DC5 file extension, such as MYFILE-2.DC5. 5. Move the file to the same directory as the original MYFILE.DC5 file. 6. Maximize DataCAD and then select **New**. When asked if you want to save the current drawing, select **Yes**. (Even though you may not want to keep the current file, it may end up being better than the .SWP, .BAK, or .ASV file. Better safe than sorry.) 7. Open the newly named file (MYFILE-2.DC5). This should now be the most up-to-date drawing file. Assuming you elect to keep the new file . . . 8. In DataCAD use **SaveAs** to save the file under the original name (MYFILE.DC5). When asked if you want to replace the original file, select **Yes**.

Understanding CAD Files

The following section will focus on examining CAD files and will discuss drawing and saving files.

Drawing Files as Scrolls

You may have noticed that even if you erase all the entities in a drawing file and then save it, the file does not get any smaller! What's going on?! Think of a CAD file as a kind of paper scroll. If you spend a few hours writing text across that scroll, it may be a few feet long by the time you're done. If you go back and erase a few paragraphs or even a lot of paragraphs, certainly less text is on the scroll, but the scroll is still the same length. As you write new text on the scroll, it keeps getting longer. The same concept is true for a CAD file. The data that makes up a file is linear; each piece of data is written sequentially, one after the other. Although you may erase a little or a lot of information, the length or size of the file doesn't get any smaller, at least not without a little user intervention, and as you write more data to the file, the file size continues to grow.

"Scrunching" Files with Layer Utility

Now for the user intervention we just spoke of. Ideally, the process we will describe here will be done on the fly whenever a file is saved, but currently that does not happen in DataCAD, so you have to do it manually via the *Layer Utility* (LyrUtil) macro in the **Toolbox**. Incidentally, this process is also used to fix files that are corrupted, crashing, or behaving oddly.

The process is affectionately referred to as file scrunching, consolidating the data into a new drawing file to make it smaller. What it does is take all the information on the scroll (the drawing file); remove all the blank areas left after erasing, moving, and altering the data; and then consolidate all that information onto a new scroll (drawing file), whose size will be smaller than the original.

Saving All the Drawing Layers to a New File

To see how to use the Layer Utility macro, we will use an imaginary file called **House1.DC5** for this example. You can try it yourself by opening any DataCAD drawing file. To see how large the file currently is, go to **Utility/Directry** and read the number after *"Current drawing size =."* In our example, it is 4704K, or 4.7 MB, a pretty large drawing file. Follow these steps:

1. Select **Edit/Toolbox** and then open the **Lyrutil** macro. The menu will look like Figure 24-5.
2. Select **LyrSave**. The menu will look like Figure 24-6.

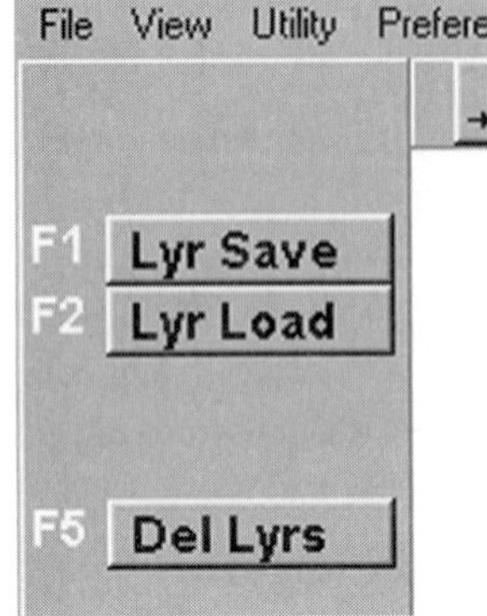

Figure 24-5
The **Lyrutil** menu

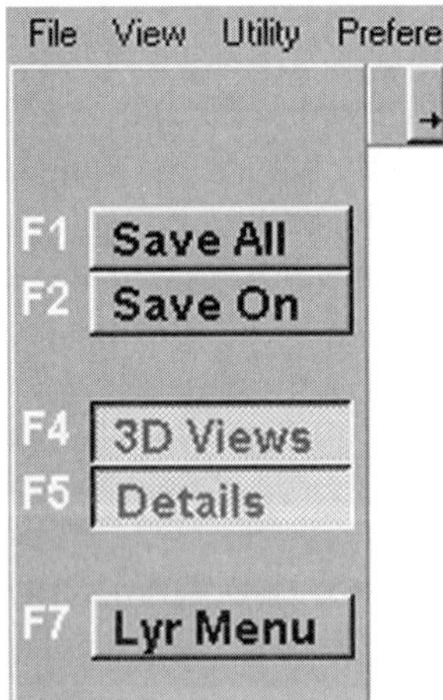

Figure 24-6
The **LyrSave** menu

3. The **3D Views** button and **Details** buttons will always default to being on. You must decide to leave one or both of them on or turn them off prior to using either the **SaveAll** or **SaveOn** options. Here is what each one does:
   a. **3D Views** will save all the currently saved 3D *GotoViews* (GTVs) (see Chapter 10) so that when the file is scrunched, the new file will retain those 3D GTVs. Leave this option on if you want to save the 3D GTVs; turn it off if you do not need to save them.
   b. **Details** will save all the currently saved *Multi-Scale Plotting* (MSP) details so that when the file is scrunched, the new file will retain those MSP details. Leave this option on if you want to save the MSP details; turn it off if you do not need to save them.

WARNING: *You cannot merge together two drawings with 3D GTVs and/or MSP details in them. If you intend to use* ***LyrUtil*** *to save the layers of File1, then read those saved layers into File2, you should toggle off the* ***3D Views*** *and* ***Details*** *buttons when saving File1 with* ***LyrUtil****. If you do not, the resulting merged file will be VERY messed up, and will probably become corrupted in short order. Even if you save only one layer with* ***LyrUtil****, all of the 3D GTVs and MSP details are saved in the process.*

4. Select **SaveAll**. This will save all the layers in the drawing file, even if they are empty. A dialog box will pop up. Note that the dialog box defaults to the **\LYR** directory that is created when DataCAD is installed (see Figure 24-7).
5. In this example, the drawing called **House1.DC5** was previously saved with the LyrUtil macro. This created a layer folder called **House1** and a layer file called **House1.lys**. All the individual layers are saved in the House1 folder. The House1.lys file tells DataCAD where to find those layers; it points to the House1 folder. The **.lys** file is important because this is the one you will select in the next phase of using the LyrUtil macro.
6. To see what's in the House1 folder, I can minimize DataCAD and then navigate to the House1 folder and open it. It will look something like Figure 24-8.

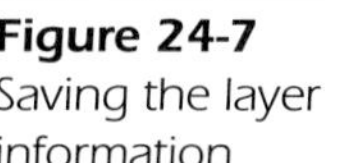

Figure 24-7
Saving the layer information

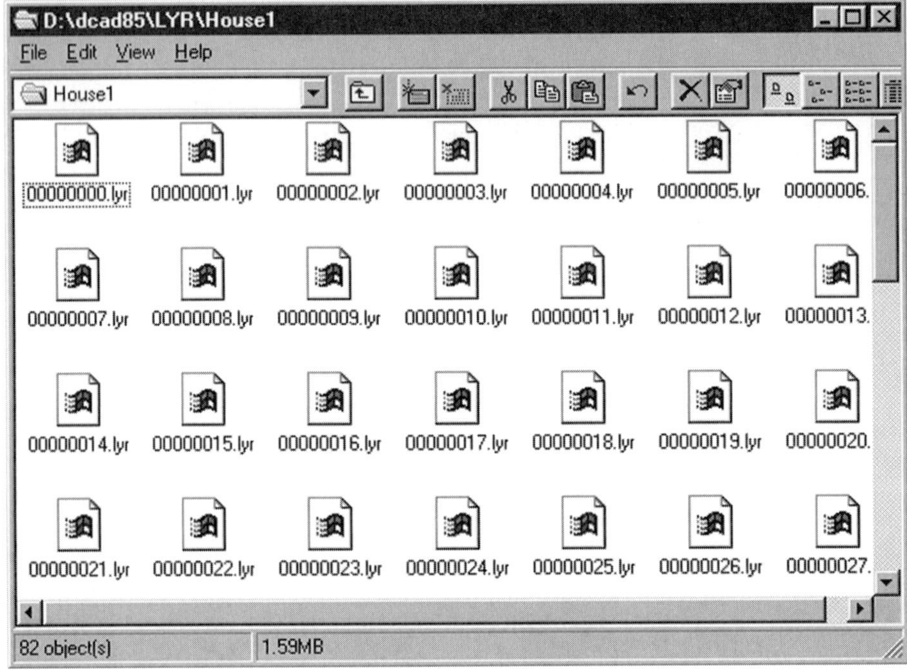

Figure 24-8
The contents of the House1 folder

7. All the layers from the drawing file are stored here in numerical order, starting with layer **00000000.lyr**. Note that the file extension here is **.lyr**. It's not terribly important that you know all this, but you should know that you should not delete or rename any of these **.lys** or **.lyr** files.
8. Exit the **LyrUtil** macro by *right-clicking*.

Reading All the Drawing Layers into a New File

Now you need to open a new blank .DC5 drawing file, and then read the scrunched layers back into this clean drawing:

1. To start a new drawing file, pick **File/New**, type a new name in the **Filename:** box, then press **Create**. In this case we named our new drawing **House2**.
   a. Whichever default drawing you pick should be clean, meaning there are no entities in it. The process will function perfectly well if you use a default drawing with entities in it, but the purpose of scrunching the drawing file is to get rid of excess data, so opening a default drawing with stuff in it would be counterproductive.

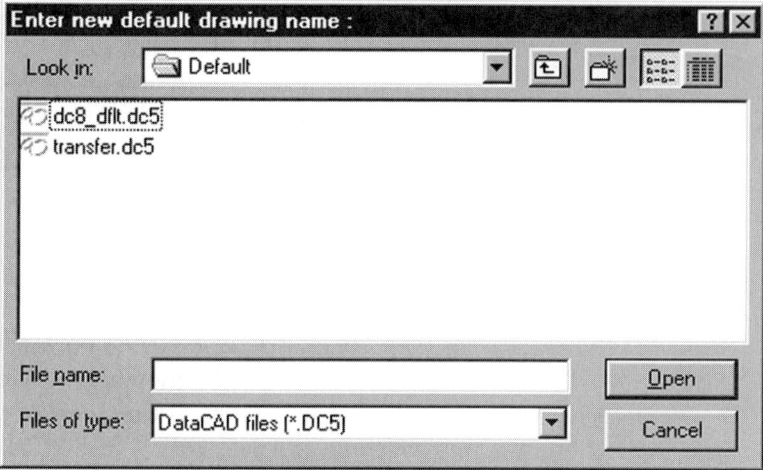

Figure 24-9 The current default drawing (**Tools/ Program Preferences/Misc/ Default Drawing File**) is used.

2. Now you need to read the layers you saved earlier, back into this clean drawing file. To do this, you will run the **LyrUtil** macro once again.
3. Select **Edit/Toolbox** and open the **Lyrutil.dcx** macro.
4. Select **LyrLoad**. The same **\Lyr** dialog box you saw earlier appears again.
5. Pick the **.lys** file that you want to read in (see Figure 24-7). In this example, we select the **House1.lys** file and press **Open**. DataCAD reads the layers into the current drawing.

The command line will now read, *"Would you like to delete this session's control and layer files?"* If you select **Yes**, then the **House1.lys** file is deleted from your hard drive to save space. If you select **No,** then it will remain on your hard drive. If you will need, or think you will need, to read these layers in again, then select **No**. Otherwise, select **Yes** to keep the **\Lyr** directory clean.

If you select **Yes**, then you may notice a flaw in the program. Only the **House1.lys** file is deleted. The **House1** folder with all the **.lyr** files in it has not been deleted. This is a bug and not a feature. To get rid of this folder, you will have to manually delete it in Windows.

To see how much smaller this new scrunched file is, select **Utility/ Directry** and read the number after *"Current drawing size ="*. Previously in our example, it was 4.7 MB. Now it is only 1416K or 1.4 MB. That's a 335 percent decrease in size.

The smaller size of the new file is important for two major reasons:

- The speed of the drawing display, entity selection, and object snapping will all be greatly increased by the smaller size.
- A smaller, cleaner drawing is less likely to produce crashes or errors.

For these reasons, running the LyrUtil macro on all your active drawing files should become at least a weekly habit.

WARNING: *Using **LyrLoad** will reset the pens (colors mapped to pen numbers) of the current file to those of the incoming saved layers. For instance, let's say your current drawing, called File1, uses a Pen Table called **PEN-1.PEN**. You then run the **LyrSave** routine on File2, which uses a Pen Table called **PEN-2.PEN**. While in File1, your run **LyrLoad** to read in the layers saved from File2. **Your PenTable** settings will now have changed to match the settings from the File2 **PEN-2.PEN** table. To get your original pen settings back, go to **Plotter/PenTable** and **Load** your original **PEN-1.PEN** table.*

Program Bugs

No computer software and no storage media is perfect. On occasion you may find a problem that causes something in DataCAD not to work or, worse, may cause the program to crash. Such problems are affectionately referred to as *program bugs*. DataCAD LLC tends to be a very responsive company when it comes to correcting software problems, so the first thing you should do is send an e-mail to the company to let them know about the problem.

Reporting Bugs

Before you send a bug report, try running the **LyrUtil** macro on the file, loading the saved layers back into a clean default drawing (refer to the section on file scrunching earlier in this chapter if you don't know how to do this). Open the new file and see if the problem persists. If it does, follow these steps:

1. Write down exactly what happened, including any keys you pressed, and any error messages you received. Specifics are important.
2. Note whether or not you can recreate the problem a second time.

3. Send an e-mail to the DataCAD LLC technical support address (`techsupport@datacad.com`) with all the specifics.
4. If your drawing file keeps crashing, send a copy of the problem file along with your e-mail. As a courtesy to you and DataCAD LLC, compress the file with ZIP compression software to make the file smaller.

Getting around Bugs

Sometimes a program bug is well known to a few or a large number of users. Many of these users may have found a suitable workaround to the problem, a way of getting around the bug so you can still get your work done. How do you find out about these workarounds? If you have Internet and e-mail access, the best source is the *DataCAD Boston User Group* (DBUG) listserv forum. Here you will find hundreds of fellow DataCAD users who will be more than happy to let you in on their tried and true DataCAD hints and workarounds. Check out the DBUG Mailing List section of Chapter 27, "DataCAD Resources," for more information.

Known Bugs in DataCAD 9

A few bugs are already known about in the latest release of DataCAD. Many of them only affect a small number of users and are the result of specific hardware/software conflicts. The remainder are bugs that DataCAD LLC knows about and is trying to correct. Check out the bug fix list in the ReadMe.win file that is installed in the main DataCAD directory and in any software update you may have downloaded from the DataCAD LLC Web site.

Error Messages

On the CD-ROM accompanying this book is a file called **errormsg.txt.** It has a list of error messages that you might get from DataCAD if you experience a problem with the program, a macro, or a file. Unfortunately, this error message file was compiled when DataCAD was still a DOS program and has not been updated since, so many of the messages don't apply to the Windows version of the program. However, if you get an error message from DataCAD, you can read this file to help you figure out what the problem might be.

Besides error messages, this file also contains some common fixes to problems usually caused by inadvertently pressing a key or turning something off. Here are a couple of examples:

Problem. The status line with X,Y,A,D readouts has disappeared.

Solution. Go to Settings/DistDlay and set it to 3.

Problem. The Import DWG file and Ortho snap angle is set to 72 degrees.

Solution. Go to Grids/SnapAngl and set it to 8.

One other caveat about this error message file should be mentioned. You will rarely get an error message that exactly matches one in this file. Instead, you will get something close, at least enough to narrow down the problem. So think of the error message file simply as something to help you narrow down the problem.

Speeding Up DataCAD

If you are running DataCAD on a very fast computer, you may be quite happy with the program's speed. If you are looking for even faster performance, however, this section offers some hints.

Computer Settings

The following are the settings that would enable a faster performance:

- Use a Pentium II or III computer.
- In Windows, go to **My Computer\Control Panel\System\Performance**.
- Under the Performance Status heading status, File System and Virtual Memory should be **32-bit**.
- Select **File System**. Under "Typical role of this computer:," select **Network server** (even if you are not on one, because it increases the cache settings). Make sure "Read-ahead optimization:" is set to **Full**.
- Select **Graphics** and set the slider to **Full**.

- Select **Virtual Memory**. Check the option "Let me specify my own virtual memory settings." Set the **Minimum** and **Maximum** boxes (the permanent swap file size) to 2.5 to three times the memory you have on your computer. For instance, if you have 32 MB of RAM on your computer, set the minimum and maximum to between **80** and **96**. You will get a warning message. Ignore it and continue on. Restart your computer for the settings to take effect.

DataCAD Settings

All of the following settings are explained in greater detail in Chapter 3, so refer to that chapter for more information. In this section, we will simply tell you that you can speed up DataCAD by enabling some or all of these settings. They are all related to how DataCAD searches for and displays entities onscreen.

Display List (D/L) Enabling *Display List* (D/L) is the greatest enhancement you can make and will greatly increase the speed of DataCAD with few if any drawbacks. You can choose to enable the standard D/L or to enable the Maximum Acceleration setting for even faster operation. To enable D/L, go to **Tools/Program Preferences/Misc/Display List**. Check the box labeled **On/Off**. To enable Maximum Acceleration, check the box labeled **Maximum Acceleration**.

Utility/Display Settings Try some of these settings:

- **Small Text** = 4. You can try various numbers to see what works for you.
- **ShowWgt** = Off.
- **SmallSym** = 5. You can try various numbers to see what works for you.
- **ArcFactr** = 1.

Of Mice and . . . Well, Mice

The primary input device for DataCAD is the mouse, but not all mice are created equal. This section offers a list of mice that other DataCAD users have had success with, along with descriptions of custom setting some of them are using. By far, the most popular mice for DataCAD are those made by Logitech, but the most important thing to look for in a mouse for DataCAD is that it must have at least three primary buttons, since you will want the middle button for object snapping in DataCAD.

Logitech Cordless C-RA1

Manufacturer Logitech

Web Page `www.logitech.com`

Number of buttons Two buttons plus center wheel/button

Middle button Set to "Middle Button" to enable middle-snapping for DataCAD. This setting will still enable the center wheel to scroll in Windows programs but will not double-click in Windows.

Logitech Trackman Marble FX

This is a departure from a traditional mouse. Instead of your hand moving the mouse, the mouse stays put and your fingers move a track ball sitting on top of the mouse. Many people like this because it reduces or eliminates repetitive stress injuries to their wrists. Most users report it takes one to three days to get used to the new arrangement.

Web Page `www.logitech.com`

Number of buttons Four buttons. The first, second, and fourth buttons are for the right thumb, and the third button in used by the right ring finger or pinkie.

Programmability The mouse buttons are programmable for middle-snapping or other tasks.

The middle button can be assigned to do both Windows double-clicking and DataCAD snapping, or it can be assigned to do only one or the other. Some people set the third button for Office 97's Autoscroll/Middle button. It works fine in DataCAD and will scroll in Internet browsers. The fourth button could then be set to double-click. The Office 97 setting must be selected in the Autoscroll options.

Logitech Mouseman Plus

Web Page www.logitech.com

Number of buttons Two buttons plus center wheel/button

Settings Set the thumb button to function as the middle-snap button in DataCAD. The right and left buttons operate in a standard manner. The wheel doesn't do anything for us in DataCAD, but it will scroll up/down in many other windows applications.

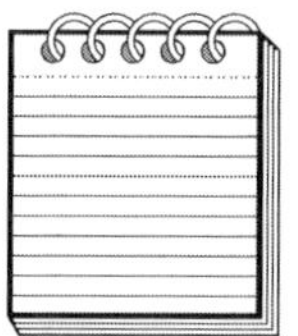

NOTE: *In Windows NT 4, service pack 4, some users report that the wheel sometimes locks. While using the wheel to scroll, the screen will suddenly stop scrolling and from then on the wheel won't work until the computer is restarted.*

Mouse-Trak Model B-5XX9P

Manufacturer Itac Systems, Garland, Texas. Phone: 972-494-9073

Web Page www.mousetrak.com

Number of buttons Three buttons plus one for speed adjustments.

Programmability In DataCAD, the middle button is used for snapping and the other buttons can be individually programmed. A fourth button adjusts the speed of the cursor across the screen. If you needed very small movements, you could push this fourth button and the cursor would move half as fast for greater accuracy.

Kensington Thinking Mouse (Programmable)

Manufacturer Kensington Technology Group

Web Page www.kensington.com

Number of buttons Four buttons ($^2/_2$)

Programmable Settings For standard Windows, the setting are as follows:

- Upper-left: *Double-click*
- Upper-right: *Middle-click*

Here is an example of one user's additional settings that take effect when DataCAD is opened:

- Ctrl and *left-click*: Identify, Set All
- Ctrl and *right-click*: Text
- Ctrl and *top-left-click*: Change Text
- Ctrl and *top-right-click*: Arrows
- Shift and *left-click*: Measure Point-point
- Shift and *right-click*: Measure, Area
- Shift and *top-left-click*: Tangents
- Shift and *top-right-click*: Geometry, Divide, Set Divisions

Buttons can also be programmed for Ctrl and Shift, Ctrl and Alt, and Alt and Shift. Programming for the Alt button alone doesn't work well in DataCAD. It tries to invoke the menu bar when pressed.

Microsoft Intellimouse

Manufacturer Microsoft Corporation, Inc.

Web Page `www.microsoft.com`

Number of buttons Two plus the center wheel/button

Windows Center Wheel/Button is set to Windows double-click.

DataCAD Set Center Wheel/Button is set to middle-snap.

It is very important for the proper operation of your mouse to have the latest software driver. The previous descriptions have the Web pages listed. After installing your mouse, you should check its Web page to be sure you have the latest driver. If you have a mouse that is more than a few months old, you should definitely check for a new driver.

DataCAD Applications Language (DCAL) Programming

The *DataCAD Applications Language* (DCAL) is the language that all of DataCAD's previous and current toolbox macros are written in. It is a "proprietary" language, meaning it was developed solely for DataCAD. Because of this, you will not find any DCAL programming references on your local computer or bookstore shelves. DataCAD LLC, however, has provided full documentation of the language on your DataCAD installation CD-ROM, and other online sources for information and help are available if you are interested in programming your own toolbox macros. If you did not install DCAL or have since erased it from your drive, you can install it from your original DataCAD CD-ROM.

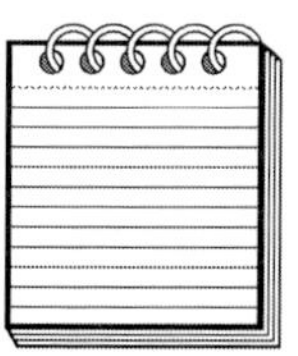

NOTE: *DataCAD itself is not written in DCAL. The original DOS version of DataCAD was written in MS Pascal. The Windows version is written in the Delphi visual programming language.*

DCAL is a block-structured programming language. It is based on Pascal and is nearly identical to Pascal in its syntax. If you are familiar with languages such as Turbo Pascal or Delphi, then you should feel right at home with DCAL. DCAL supports recursive procedures and functions, as well as nested routines, local variables and constants, reference and value parameters, and conformant arrays.

If you are a novice at programming but want to learn to program in DCAL, I would suggest getting a beginner's book on programming in Pascal.

With a modular programming language, you can create separate blocks of code (called *modules*) and compile them separately. These modules can be used by different macros. This is one of many powerful features of DCAL that you should learn and take advantage of. For example, let's say you create a piece of DCAL code that draws a circle on the screen in a macro to draw a door symbol. If this piece of code is in its own module, it can also be shared with another macro that uses this circle for a completely unrelated purpose. This is the preferred method for sharing code, but any procedure that you create can be shared.

Because of this modular approach, DCAL requires both compilation and linking of the different modules to produce the final macro. This will be explained further a little later. In actuality, DCAL is made up of two major pieces: the DCAL compiler (DCC.EXE) and the DCAL linker (DCL.EXE). It also contains various help files and sample code listings, all of which reside in your DCAL subdirectory.

DCAL has many hooks (ways of tying into) into low-level procedures of DataCAD itself. A DCAL macro can appear as an extension of DataCAD itself or it can take control of the DataCAD interface and use it in new and different ways. The possibilities are only limited by your skills and imagination.

As of the writing of this book, DataCAD LLC announced that new programming commands will be added to DCAL when DataCAD 9 is released. Those additions did not make it to the DCAL material on the CD that comes with this book, so check out the DataCAD and DCAL Web pages for the most up-to-date information.

Compiling Macros

DCAL has two executables, or programs, that are used to compile the macro. A compiler is a program that converts the plain text programming language that we humans can understand into a machine language that the computer can understand. More precisely, DCAL macros are compiled and linked into code that can be understood by DataCAD. All DCAL macros are written in a text format (the source code) that you as the programmer will come to understand. However, DataCAD doesn't understand the source code that you write. In order for this source code to be understood by DataCAD, you must translate it into the instructions that DataCAD will be more familiar with. The process goes something like this:

1. First, you write out your source code in a text editor. The source code will be plain text files.
2. Next, you compile and link any optional modules to create the compiled macro. The macro will have a .DCX file extension.
3. This compiled macro can be understood and executed by DataCAD. Place the new macro in your DataCAD **Toolbox**, open DataCAD, and then select it. If all is well, the new macro should run.

Now if you were to open up and look at the code that DataCAD understands (the DCX file), you would not see much that made sense to you. It's all in computer gibberish. This is due to the compilation process. When you compile the macro, it eliminates all the unnecessary information from your source code and then converts the source into binary code that can quickly be processed by DataCAD.

Setting Up Your System for DCAL

You will need a few things in order to write DCAL macros. First, you will need the DCAL compiler and linker. If you have a default installation of DataCAD, you should have a subdirectory under your main DataCAD directory named *DCAL*. If you do not have this directory, you can install DCAL by running the DataCAD installation program and selecting "Install DCAL" as an option. Since the procedure for installation is different for both DOS and Windows, we will not explain these procedures here. It is not necessary to install anything else from the DataCAD installation while installing DCAL, so avoid reinstalling the main DataCAD program as this may write over your custom settings. Make a note of the directory you install DCAL into.

You will need to add the path to the compiler and linker to your path. The compiler and linker will be located in the directory you specified during the install (typically C:\DCADWIN\DCAL). You should also add the path to your *include files* (more on these later). The include files are typically located in a subdirectory of the main DCAL directory named \INC (C:\DCADWIN\DCAL\INC). If you do not add the path to your DCAL files within your path, you will not be able to compile your source code easily from other directories. You can modify your path by editing your AUTOEXEC.BAT file in either DOS or Windows. In Windows NT, you can change the path via the system control panel applet. Consult your DOS or Windows help files for more info on changing your path.

You can also create a batch file to modify the path only when needed. Such a batch file might look like this:

```
PATH=%path%;C:\DCADWIN\DCAL;C:\DCADWIN\DCAL\INC
```

This will add the path to the compiler and linker and include files to your current path. Modify these paths as needed for your system. By putting this in a batch file, you will only have these items in your path when you run the batch files. This is most useful in Windows. Be aware that adding too many directories to your path can slow down Windows. Simply call this path from a DOS session when you need to work in DCAL. This way, when you exit this DOS session, the path is reset to the original path. This method is only helpful for Windows users.

Next, you will need a text editor. This is one area where you will have some freedom. You can use any editor capable of generating unformatted text. Note that you can use your favorite word processor, but you should avoid this temptation. The reason not to use a word processor is that it will default to adding formatting when you attempt to save your source code. The compiler will not understand your source code if your editor adds formatting codes such as fonts, margins, and so on. You can use Windows Notepad, which ships with Windows, but we recommend getting an editor designed for programming as these tools have powerful functions that can aid you. Check out the *DataCAD Developer Network* (DNN) Internet links page at `www.datacad.com/DCAL/DDN.html` for two editors we recommend. Select any editor that you are comfortable using as long as it saves plain, unformatted text files. (The Ultra Edit shareware that comes on the CD in this book is an excellent choice.)

The last thing needed is a DOS session. If you are working in DOS, you will simply use your standard command line. In Windows, you will need a DOS session. You can use your default DOS window or full-screen session, but as you build more complex macros, you may benefit from increasing the memory available to the DOS session by changing the parameters for the session. Check your windows help files for information on modifying a DOS session. Once these things are configured, you are ready to test your setup.

Testing Your Setup

Once you have DCAL installed and have selected an editor, you will want to check your installation to be sure that it is working correctly. Open up your editor and type in the following code. Pay close attention to punctuation; it's important:

```
PROGRAM test;

BEGIN
        wrterr ('Hello World!');
        pause (5.0);
    END test.
```

For now, don't worry about what this code means. We will look at that later.

Save this text file as TEST.DCS and make a note of where you save it. A good practice is to create a subdirectory in either the DCAL directory or another directory of your choice to store all your source code. When you get into more involved macros, you will want to create separate subdirectories for each macro's source code. Let's assume you put the TEST.DCS source code in the directory C:\DCADWIN\DCAL\SOURCE\.

Next, open up a DOS session or if you are working in pure DOS, close your editor and return to the command line. From this point on, we will assume you are working in Windows. If you have a choice of DOS or Windows (or OS/2 and so on), it is helpful to work in a multitasking operating system (such as Windows or OS/2) while writing DCAL macros. The reason for this is that you will be able to quickly switch between your editor, compiler, and DataCAD. As you will see, you will be switching back and forth quite a bit in the debugging process.

Once you are at your DOS prompt, switch to the directory where you saved your TEST.DCS source code. In our example, this is C:\DCADWIN\DCAL\SOURCE\ Once you are at the correct location, type the following:

```
DCC TEST
```

If all goes well, you will see something like:

```
C:\DCADWIN\DCAL\Source>dcc test
DCAL (tm) Compiler Version 5.00
(C) Copyright 1986-1993 CADKEY, INC.
Pass One.

Usage of dcc1 : dcc1 [-v45 -v50] [-f]filename
     Input parameter list : test
     Input file [.dcs] : test
     Input options [45/50] : 50

     test...

     Done.

     DCAL (tm) Compiler Version 5.00
     (C) Copyright 1986-1993 Cadkey, Inc.
     Pass Two.

     Input file [.dci]: test

     32 bytes written.
     No errors detected.
```

The important part here is the last line "No errors detected." If you see anything else, your code contains errors. We will get to that issue in a minute, but what if you don't even see this text? If you receive an error such as,

```
The name specified is not recognized as an internal or external
command, operable program, or batch file.
```

then you have not correctly set your paths and the operating system is unable to locate the compiler. Go back and check your settings to make sure the compiler (DCC.EXE) is in a directory that appears in your path. For more information, review the Setup section. The message may vary based on your use of DOS or the different flavors of Windows. If you receive any other system errors, consult your operating system for more information.

Another example would be that your compiler ran, but you saw something like this:

```
DCAL (tm) Compiler Version 5.00
(C) Copyright 1986-1993 CADKEY, INC.
Pass One.

Usage of dcc1 : dcc1 [-v45 -v50] [-f]filename
Input parameter list : test
Input file [.dcs] : test
Input options [45/50] : 50

test...
4: wrterr (Hello World!');
^
undefined symbol
4: wrterr (Hello World!');
^
right paren expected
4: wrterr (Hello World!');
^
a semicolon must end a statement
5: pause (5.0);
^
undefined symbol
5: pause (5.0);
^
a semicolon must end a statement

Compilation Errors.
```

The first thing to notice here is the last line "Compilation Errors." This is the compiler telling you that errors exist in your source code. To generate this error, we intentionally modified the code. Here is the line we changed:

```
wrterr (Hello World!');
```

Notice that the Hello World! string is not enclosed in single quotes (') as in the original listing. Your error may vary, but we are interested in what the compiler tells us more than the nature of the error.

The process of compiling your source code, reviewing the compiler messages, and adjusting your source code is called *debugging*. It is critical that you are comfortable debugging your code. Your macros will almost never compile on the first try. It is a fundamental law of programming that you will make mistakes. Fortunately, the compiler will catch your mistakes and try to explain them to you. Think of the compiler as your translator and the debugging process as your proofreader.

Studying the following lines will help us find the error:

```
test...
4: wrterr (Hello World!');
^
undefined symbol
4: wrterr (Hello World!');
^
right paren expected
4: wrterr (Hello World!');
^
a semicolon must end a statement
5: pause (5.0);
^
undefined symbol
5: pause (5.0);
^
a semicolon must end a statement
```

The first line of this piece of code tells you which module the compiler is working on. In our example, there is only one module (TEST.DCS), so naturally the only one the compiler bothers with is TEST.

The next line is the first error line. The first number on this line is the line in your source code where this error occurs. The line under this line has a caret (^) symbol that will attempt to point out the specific location on the line above where the error occurs. Notice that all we did was remove one single quote from the code to trigger five separate error messages. Now you can begin to see how a simple omission or error can lead to troublesome compiler errors.

The compiler will try to find the errors and point them out to you, but it is not perfect. Look at the compiler errors making a note of the lines where the error occurs and then return to your source code. Often the compiler will point out an error on one line when the problem lies on another. For this reason, you will need to review your source code rather than the code pieces shown in the compiler errors. As you can see in the example, nothing is wrong with line 5, but the compiler flags it since the error on line 4 is making problems for the rest of the code. Often the compiler will not even show the line where the actual error occurs. Always work with the source code. Relying on the lines of code the compiler flags to spot the errors will not always work and you will waste a lot of time looking. Always try to start at the first line it points out as incorrect (in our case, line 4) and then work up or down until you spot the error.

The caret (^) character will try to point out the problem but may not be able to narrow it down. In our case since we have confused the compiler, it is unable to point out exactly where the error occurs and it simply points to the beginning of the line. Since the compiler will read through your code from top to bottom, it will often flag the line below the error. We will cover debugging in more detail in future lessons. For now, use the hints the compiler gives to spot the error and fix it. After fixing the error, recompile the source. Repeat this step as often as needed.

Once you have gotten the code to compile correctly, you will need to link it. If your macro is composed of more than one module, you would want to join all the code together to form one macro. In our example, only one module exists (TEST.DCS).

If you were to look in the directory where you are compiling your code, you should see a file called TEST.DCO. This file is an object file and was generated by the compiler. The object file is a compiled file but must be linked with other optional modules before it can be understood by DataCAD.

To link our sample macro, type

```
DCL TEST;
```

Notice the line ends with a semi-colon. You should see an output such as

```
C:\DCADWIN\DCAL\Source>dcl test;

    DataCAD Macro Linker Version 4.01
    (C) Copyright 1986-1990 CADKEY, INC.

    Closing files
```

Since no errors occurred and this is such a small program, you will not see much more of interest. If you output looks like this

```
DataCAD Macro Linker Version 4.01
(C) Copyright 1986-1990 CADKEY, INC.

    Object modules:
```

you did not put the semi-colon at the end of the line. In this case, the linker is waiting for you to tell it the name of the next module to link. If you are done (as you are in our example), simply enter the semi-colon and press Enter. We will cover the linker more in future lessons.

A directory listing in our source director should now show something like this:

```
Directory of C:\DCADWIN\DCAL\Source

    12/16/98 10:26 4,096 TEST.DCO
    12/16/98 10:26 79 TEST.DCS
    12/16/98 10:30 4,096 TEST.DCX
```

By now, you should be getting familiar with these file types. The TEST.DCO file is the compiled unlinked object file. The TEST.DCS file is your original source code. The TEST.DCX is the compiled macro. This last file is the one that DataCAD can run as a macro.

Now you can copy this file to your macro directory (normally C:\DCADWIN\DCX) or you can change directories while in the toolbox menu to point to your source directory in order to run this macro.

Open up DataCAD and go into the TOOLBOX menu. Depending on whether you copied the TEST.DCX file to your default macro directory or not, you may need to change the path to locate your new macro. If are running DataCAD for DOS, you will not see the file extension (.DCX); rather, you will only see the file name (TEST). Select this macro and DataCAD will execute it.

If you look towards the bottom of the screen where the Z-Base and Z-Hite are normally displayed, you should see the message "Hello World!" You will not be able to do anything in DataCAD until the macro is done running since we put in a pause statement in order for you to see your message. Your macro should show your message for five seconds and then exit.

Congratulations, you have just written your first DCAL macro! Of course, this macro doesn't do much, but the concepts you learned here will carry over into every macro you write from this point forward.

Try changing the message or the pause delay. Save the file with different names, and so on. It is critical that you understand these basic steps before you try to move on to more advanced lessons.

If you are interested in more challenging exercises, look in your SAMPLES directory for sample macros that are part of your DCAL kit. Try to compile some of these macros. Start with the very simple ARROW sample and see if you can get it to run. See if you can debug any errors you receive. Don't worry if you can't do much with many of the samples yet. We will cover more advanced compilation and linking issues in later lessons.

Basics

Fully describing the DCAL language is far beyond the scope of this book, but we will describe a few things about the language here and point to some resources if you are interested in learning more about it. This information comes mostly from the documentation on the DataCAD CD-ROM.

Angles

All angles in DCAL (and therefore internal to DataCAD) are expressed in radians starting at zero to the right (east). Angles increase counterclockwise.

NOTE: The internal representation has no effect on how DataCAD displays angles to the user.

Distances

Internally, DataCAD stores all distances and coordinates as one unit equal to $^1/_{32}$″ [0.8]. This is also true for metric units. When converting distances to strings or strings to distances, the current scaling type (feet-inches fractions, meters, decimal feet, and so on) determines the format of distance strings. This does not, however, affect the way that DataCAD maintains distances internally.

Entities

Every primitive graphic shape that DataCAD understands is called an entity. Currently supported entities include the following:

3D arc	ellipse	3D line
line	arc	point
associative	polygon	dimension
bezier	skew box	
slab	surface	bspline
symbol	circle	text string
cone	torus	cylinder
truncated cone	dome	

Each entity has certain information associated with it, such as which layer it is on, its color, line type, line spacing, line weight, overshoot factor, and an attribute (or integer). Each entity also has an address associated with it that is a handle to the entity in the drawing file. The address of an entity does not change either during an editing session or between editing sessions.

Keyboard

A DataCAD macro perceives the keyboard differently from other programs. To simplify writing macros, all keys are returned from the keyboard as integers, not as characters. This enables DataCAD to represent the entire 256 extended ASCII character set as well as any special keys that return escape codes.

A special key is any key that has a macro-interpretable function. For instance, the function keys are special keys, as are the arrow keys (Ins) and (Del). Some keys are special keys only some of the time. For instance, (q) returns an escape code under certain circumstances. This causes DataCAD to increment the current line type and at other times (while reading a string) returns a normal key code.

Several of the DCAL routines return a result that tells you how they were exited. These result codes are *res_normal* for a normal finish and *res_escape* for an interruption because an escape code was read. A result of res_escape typically means that the user pressed a function key while the macro was requesting some type of input. The escape code that was pressed returns from these routines as an integer.

Layers

Each entity is on a layer. Each layer can be scanned during any search of DataCAD's database. When a layer is not searched, every entity on that layer is ignored. Ignoring certain layers (such as layers that are off) or confining a search to a specific layer or set of layers improves the search speed.

Selection Sets

DataCAD supports 16 selection sets. A selection set is a collection of entities. Every entity can be a member of some, none, or all of the selection sets. Selection sets are convenient ways of grouping entities so you can act upon them together. Selection sets also provide an alternative to using layers for grouping entities. The first eight selection sets are available to you and DCAL, while the other eight are reserved for DataCAD's internal use.

Selection sets exist provides many uses in a drawing file. That is, you can set up and manipulate a selection set. If the drawing is exited and then reentered, the selection set still exists. In this respect, selection sets are similar to layers as a way to group entities.

Local Variables

Routines can declare local variables. These variables can only be used inside this routine or any routine local to it. They are pushed on the stack and exist only during the lifetime of the routine.

Symbols

A symbol is similar to any other DataCAD entity, but a symbol is a reference to a group of other entities that are drawn where the symbol is inserted. A symbol, as created at the DataCAD template menu, is referenced by its file name (including the path). With a symbol defined by a macro, however, all you need is a unique symbol name. Symbols are instanced; that is, their geometry exists only once in the database. They can be globally updated but no longer individually edited.

White Space

White space is ignored by the compiler. It consists of spaces, new lines, and tabs that are not in a string or character literal.

Program Layout

In this section, we'll briefly examine certain aspects of DCAL's layout, including macro headers, subprograms, procedures, and functions.

Macro Header

Two types of macro headers exist, depending on whether the .dcs file contains a program or a module:

- *Programs* As the sample macro spiral shows, DCAL program macros begin with the keyword program followed by the name of the macro, which is followed by a semicolon.

- *Modules* DCAL modules begin with the keyword module followed by the name of the module, which is followed by a semicolon. Modules have no body and thus no corresponding begin keyword. However, they do have an end followed by the module name and a period. The sample module wrtutl.dcs is an example of a module.

Subprograms

A subprogram, or routine, is a procedure or function. A function is identical to a procedure except that it returns a value and can be used in an expression if its type is appropriate for the expression. In this manual, the terms subprogram and routine mean either procedure or function.

Procedures

Procedures are declared by the keyword procedure followed by the procedure name, which is followed by an optional parameter list. After that, a procedure can declare local constants, types, variables, and other routines. A procedure ends with control falling through to the final end statement or by using the return statement.

Functions

Functions are similar to procedures, except they return a value of a certain type. Functions are declared with the keyword function followed by the function name, an optional parameter list, a colon, and the type of variable the function returns. A function may declare local constants, types, variables, and other routines. In the body of the function, the return value is returned in the return statement. The returned value follows the keyword return. If control falls through to the final end in a function without encountering a return statement, a fatal runtime error occurs. Functions may only return scalar types.

NOTE: *In the DCAL manual, the function or procedure is listed with a description followed by the parameters with descriptions of the parameters concluding the section.*

Sample Macros

In the DCAL directory is the source code for a number of sample macros. Opening and dissecting these macros is by far one of the best ways to learn more about DCAL. The ARROW macro even has a great deal of explanatory text in the source code to explain what is going on in each part of the macro.

Each macro is in a separate subdirectory. Many macros include a .lnk or a .bat file. The .lnk file contains the linker statement used by the dcl.exe linker. When more than one object file exists, the .lnk file simplifies the linking process. Take, for example, the CONCRETE macro. To link it, use the command

```
dcl <concrete.lnk
```

The .bat files are like MAKE files in that they recompile all modules of the macro and then relink the macro. To recompile and relink the entire CONCRETE macro, type

```
CONCRETE
```

The batch and linker files assume that each sample macro is in a separate subdirectory and that the system includes files are in a subdirectory at the same directory level. As well, macros that use the WRTUTL library also assume that the WRTUTL subdirectory is at the same directory level, like this:

```
DCAD —|— DCX —|— INC    System includes files
 |     |— WRTUTL   Message utilities
 |     |— ARROW   Arrow macro
 |     |— CONCRETE  Concrete macro
 |       .
 |       .    etc.
 |       .
 |— DWG
 |— SUP
 |— SYM
 |— TPL
```

In each macro and in each .lnk file, the pathnames for system includes files and object files are designed with the previous directory structure in mind. All pathnames are relative, such that the name of the DCX subdirectory is unimportant, except for the location of the INC and WRTUTL directories relative to the macro source code.

DCAL Resources

The best DCAL resource is part of the DataCAD LLC Web site. This is the *DataCAD Developer Network* (DDN). Although this page is part of the DataCAD LLC site, it is moderated and maintained by a third party, currently Patrick McConnell. Most of the text in this DCAL section was written by Patrick for the DDN site and is printed here with his permission. You can go directly to the DDN page at `www.datacad.com/DCAL/DDN.html`. Here you will find an excellent DCAL tutorial, sample macros with full source code, various programming references, bug lists, and a wealth of other information about non-DCAL related features like custom linetypes, hatches, and fonts.

One of the best downloads available on the site is the formatted Word for Windows version of the DCAL manual. Although the same text is available on your DataCAD CD, the formatting is not all that good. The downloadable version is much better and is worth the time to get it. We suggest you print out all 200 pages and place them in a three-ring binder for easy reference.

CHAPTER 22

Frequently Asked Questions

The intent of this chapter is to try to answer some of the questions that we have seen new and seasoned users ask about. Some questions can be answered outright, while others may require you to look for the answer elsewhere in this book. Even if you don't have any particular questions, you will probably learn a few new things by reading through the questions and answers anyway.

A. How Do I Draw Things?

A1: How do I draw a line?

In DataCAD, that's easy. The answer is you don't have to do anything, since the program defaults to drawing lines! Well, you don't have to select any menu options anyway. Just place the cursor in the drawing window and *click* the left mouse button. This is the point from which the line will begin. Now move the cursor where you want the end of the line to be and *click* the mouse again. If you continue to move and click, you will continue to create lines. To stop drawing lines, press the right mouse button or press the **Exit** button in the menu. Drawing lines in DataCAD can be done in many other ways, but this is the most basic method.

A2: How do I draw a wall?

In DataCAD, a wall is essentially two or more lines running parallel to one another. So drawing a wall is just like drawing a line (see the previous FAQ answer), but you first need to turn on the **Walls** function and set the width of the wall. To do this, go to **Edit/Architct** and select the **Walls** option (or press the equals (=) key to get there with one key press). Select the **Width** option and you will be prompted to *"Select the wall width."* Type in a width and press the **Enter** key. Now you can draw walls just like you draw lines. In the **Architct** menu, you should also note that you can define your walls by the **Outside** line of the wall, by the **Inside** line, by the **CntrWall** (centerline), or by the **CntrCvty** (center of the cavity). See Chapters 5 and 11 for more info.

A3: Can I draw walls with more than one line, like for a brick cavity wall?

Yes, you can draw walls of up to three or four parallel lines. Follow the previous directions for drawing a two-line wall, but after selecting **Edit/**

Architct, select either **3LnWalls** or **4LnWalls**. As you select each one, a different set of option buttons will appear to enable you to select the color and linetype of each line. Once you get something you like, you can save those settings for future use by using the **WallStyl/Save As** option and giving your new wall style a name.

A4: How do I quickly draw a simple rectangle?

Select **Edit/Polygons/RectAngl** (if you leave the **CntrPnt** option on, a snappable dot is placed in the center of the rectangle).

A5: How do I work with plans at an angle?

There is more than one way to do this, but the simplest is to leave the plan at an angle while rotating the CURSOR instead. To do this, select **Utility/**
Geometry/Tangents (or press the **B** or **b** key to get there in one step). The message line at the bottom of your screen will say, *"Select the line to draw tangent to."* Pick any line in the rotated plan and the cursor will automatically rotate to match the angle of that line. Now you can draw everything off of that angle. To rotate the cursor back to its original orientation, go back to the **Tangents** menu and select **Cancel**.

A6: How do I draw a property boundary using surveyor's data?

Switch to **Relative Polar** input mode by pressing the **Insert** key repeatedly, and then changing to Bearing angle mode by going to **Utility/Settings/AngleTyp/Bearings**. *Click* on the drawing area to locate the first point of the property line and then press the **Space** bar. Supply the necessary distance and angle for the succeeding points.

If you need to supply irregular angles such as N 35 degrees, 48 minutes E, type in N35.48E. Note that the dot here is being used as a delimiter between degrees and minutes. (Note: For those using the Imperial/English System, usage of the dot as a delimiter in distance input depends on the current scale type.)

A7: How do I draw text aligned to the angle of a line?

Use **Edit/Text/Angle/Match** and then select the line you want to align the text to. *Right-click* to accept the angle displayed. The current text angle will be changed to match the line you selected. You also have the option of clicking at two points to create the angle to match.

B. How Do I Edit and Move Things?

B1: How do I erase only part of a line?

Go to **Edit/Erase/Partial** and then select the line you want to partially erase. Select or snap to the first point, and then select or snap to the second point. The area of the line between those two selected points will be erased. You can do this in the middle of a line or at the end of a line.

B2: How do I erase only part of an arc or circle?

The process is nearly the same for an arc or circle as it is for a straight line. Go to **Edit/Erase/Partial** and then select the arc or circle you want to partially erase. Select or snap to the first point, and then select or snap to the second point. To complete the process, you must tell DataCAD which part of the circle or arc is to be removed by selecting the section to be erased. The area of the circle or arc between those two selected points will be erased. You can do this in the middle of an arc or at the end of an arc. You can do this anywhere around a circle.

B3: How do I copy something multiple times without having to constantly pick it and select new distances?

Use **Move/Drag/AndCopy**. Select the item you want to copy. Pick a reference point to move from, then locate the copy where you want it, and *click* the mouse button. Continue to copy more instances of the selected item by placing the cursor and clicking. You can continue to do this until you press the right mouse button or press **Exit** in the menu. If you want to select more than one item to be moved at one time, select **Move/Drag/AndCopy/Multi**. Choose all the entities you want to copy and press **Begin**. Pick a reference point to move from, then locate the copy where you want it, and *click* the left mouse button.

B4: How do I change or copy only green arcs of a Dot-Dash line-type (or other such specific entity combinations)?

You can use edit sets and masking to accomplish this. Refer to Chapter 9, "Editing."

C. Dimensions

C1: Sometimes the dimensioning process only places a tick mark or arrowhead at one end of the dimension line, rather than both ends of the line. How do I fix this?

Try going to the **Dmension** menu and then to **Linear/DimStyl/Incrment**. If that is set to **0**, you will only get one tick or arrow. Set it to anything but 0 and that should fix the problem.

C2: How do I add text like +/−, V.I.F, O.C., R.O., and so on to my dimensions?

Some CADD programs enable you to add text to associative dimensions. Currently, DataCAD does not. If you need to add text like +/−, V.I.F, O.C., R.O., and so on, you have two basic options:

- Create your associative dimensions as you normally would. Then go to the standard **Text** menu and add the extra text after or below the dimension. This text is not part of the dimension string and must be edited separately, but it works.
- Use non-associative dimensions instead. When prompted for the dimension string to input, move your cursor to the end of the dimension and manually type in the additional text you want. Press **Enter** and the dimension and text will be entered. The dimension text is now just simple "text," rather than an associative dimension, so you can edit the dimension and text with the **Edit/Change/Text/Contents** option.

D. Customizing Things

D1: How do I customize my keyboard shortcuts?

The process is not difficult, but it does take some explanation. See Chapter 23, "Customizing DataCAD," for a thorough explanation.

D2: How do I customize my icon toolbar?

The process is very much like customizing keyboard shortcuts, but it's a little more involved. See Chapter 23 for a thorough explanation.

D3: How do I customize the DataCAD screen and menus?

In the drop-down menu options, select **Tools/Program Preferences/ Interface Settings**. You will get a graphic image that looks like your DataCAD screen. Move your cursor over the item you want to change (menu buttons, icon toolbar, drawing screen, or message area) and *click* on it to change the current settings (see Chapter 23 for more information). Then select OK to accept the changes.

D4: How do I move my menu buttons from the left side of the screen to the right, or from the right to the left?

In the drop-down menu options, select **Tools/Program Preferences/ Interface Settings**. You will get a graphic image that looks like your DataCAD screen. Place your cursor over the image of the square area below the menu buttons. The DataCAD message line will read, *"Click here to change the menu position." Click* once and the menu buttons will switch sides. Select **OK** to accept the changes.

D5: How do I make my drawing cursor bigger or smaller?

Press the **+** key (this is easy to remember since the plus sign looks like your cursor crosshairs).

E. Printing and Plotting

E1: Can I plot out a half-size drawing set without composing these sheets all over again in different scales? For example, my original layouts are on 24 × 36 sheets. I want to print out a half-size check set at 12 × 18. Can I do this?

Yes. Note that Step 1 is the most important step of the process:

1. In the **Plotter/Setup** dialog, you must leave the **Printer/Name** and **Paper/Size** settings set to the full size driver and paper size. So if your MSP (or **QwkLyout**) layout was made with your HP plotter driver and a 24″ × 36″ sheet size, you must leave those settings alone.

2. Now go to the **Check Plot** section and check the **Fit to:** circle.
3. Below the **Fit to:** box, select the printer that you want to print the half-size drawing to. So if you want to print to your Panasonic printer, select that driver.
4. To the right of the **Fit to:** box, select the paper size you want to print to, like 12″ × 18″, or $8^1/_2$″ × 11″, or 11″ × 17″. The **Use this** factor: box will automatically change to the percent reduction required to fit your drawing on the paper size selected. Sometimes DataCAD rounds this number up, causing the drawing to not quite fit. If this happens, just decrease the percentage by one.
5. If you want to print a half-size set to your large format plotter (the same device you would print your full size sheets to), just select that plotter below the **Fit to:** box and then select **50%** in the **Use this factor:** box.
6. Select **OK** to accept the settings and press **Plot**.

See Chapter 7, "Printing and Plotting," for more information.

E2: How do I send plot files to an outside plotting service?

Set up everything the way you normally would to send the plots to your own plotter, except that you must pick the plotter driver used by your plotting service. Then use the **Plotter/ToFile** option to save a .PLT file. See Chapter 7 for more information.

E3: We cannot get the printed image to rotate properly on our plotter. The plotter uses a 24-inch-wide roll feed and, regardless of DataCAD settings, the 18-inch dimension of the drawing sheet plots along the 24-inch dimension of the paper.

The answer comes from Neil Blanchard. For some reason, the **Rotate** option doesn't work in Windows on some Hewlett Packard plotters due to the software driver. If it has an Auto Rotate setting, then that might work, but you have to be aware of the printable margins and the Plot Size setting. The Software option often prevents Auto Rotate from working because it "thinks" it needs the whole 24-inch dimension of the sheet. Its five-millimeter nonprintable margins on the edges of the roll reduce the 24 inches to roughly $23^5/_8$ inches. Setting the Plot Size to Inked Area is best, but you need to design your border to center itself. The plot will always go right up against the lower-right corner of the printable margins. The nonprintable margins on the cut edges of the sheet are 17 millimeter (approximately $^3/_4$ of an inch) on each end. So, if the 18 × 24 border has a $^3/_4$-inch margin at the top and the bottom, and the right end needs to be bigger than five millimeters ($^3/_{16}$ of an inch), you will need to put a small, light line on the lower-right corner of the sheet to push the inked area to the desired distance from the edge of the paper. You must also allow enough room on the left (stapled) edge to fit within the 23 $^5/_8$ inches or so. Otherwise, the plot will not automatically rotate, assuming that it does rotate at all.

E4: The "Dots" hatch pattern will not print on my laser printer, but it prints fine on the plotter.

Some printers, especially Hewlett Packard printers, have a couple of options in the software driver control panel that say something like Use Vector Graphics and Use Raster Graphics. Set it to Use Raster Graphics and your dots should print. If not, try setting the Pen weight to something higher (like .12mm).

E5: It takes 15 minutes to print to a file. How can I speed things up?

One solution is as follows:

1. Go to the Windows **Control Panel/Printer** folder.
2. *Right-click* on your printer, then go to **Details/Spool Settings**, and make sure the following are checked:
   a. "Spool print jobs so program finishes printing faster"
   b. "Start printing after first page is spooled"
3. *Click* on **OK**.
4. Now go to the adjacent **Port Settings** and make sure the following are checked:
   a. "Spool MS DOS print jobs"
   b. "Check port state before printing"
5. *Click* on **OK**.

F: Hatching and Linetypes

F1: How do I change hatch patterns or settings without redefining the hatch boundary all over again?

The most important advice is to use associative hatching (**Edit/Hatch/Associat**). Associative means that all the lines in the hatch pattern are grouped together as one entity. Non-associative hatching is just a collection of dozens or hundreds of individual entities. If you use associative hatching, here is all you have to do to change an existing hatch pattern in a drawing: Set the new hatch settings and then select the existing associative hatch by **Entity**, **Group**, **Area**, or **Fence**. Now select **Begin** and the hatch pattern will change to reflect the new settings.

F2: How do I change the boundary of an associative hatch, especially if two hatches share some of the same vertices?

Use the **Polyline.dcx** macro in the **Toolbox**. It will enable you to add, delete, and move the vertices of any individual hatch polyline. See Chapter 5, "Basic Drawing," for more information.

G: Text and Fonts

G1: How do I import text into DataCAD?

You can use the Cut/Copy/Paste options in DataCAD, just like any other Windows program. See Chapters 8 and 11, "Basic Construction Drawings, for a thorough explanation.

G2: How do I export text from DataCAD?

You can use the Cut/Copy/Paste options in DataCAD, just like any other Windows program. See Chapters 8 and 11 for a thorough explanation.

G3: When editing text, if I hit the space bar too long, the computer beeps and the keyboard freezes up until the system catches up. Can I change this behavior?

1. Go to the Windows **Control Panel** and run the **Keyboard** function.
2. Select **Speed**.
3. Adjust the following settings (*right-click* on them for a description of what they do):
   a. **Repeat Delay**: Try setting it all the way to the right at Short.
   b. **Repeat Rate**: Try setting it directly in the middle.

H. Toolbox Macros and Add-Ons

H1: If one of my toolbox macros stops working or behaves oddly, what can I do?

Usually, this is an indication that the macro has been corrupted somehow. Reinstalling the macro (.DCX) file should fix the problem.

H2: What is the DataCAD Productivity Pack and how do I install it on my system?

The Productivity Pack features, which include Program Files, Linetypes, and Keyboard Macros, is a series of six linked icon toolbars that enable you to work in a task-oriented way. Each toolbar has links to the other toolbars for easy switching between them. Additionally, a series of new linetypes is added to your DCADWIN.LIN file, and a new set of keyboard macros (**Alt+Letter**) is added to your \SUP directory. See the end of Chapter 23 for a complete description of this option.

J. Multi-Scale Plotting (MSP) and Clip Cubes (CCs)

J1: When using CCs, entire lines of text will sometimes disappear from the screen and will not be printed either.

Most fonts must be completely within the CC boundaries in order for the text to be displayed. To fix the problem, either enlarge the boundaries of your CC or decrease the length of the text line so it fits within the existing CC boundaries. Some fonts, such as the Arch font, will display correctly even if part of the font is outside the CC boundaries, but 99 percent of the others will not.

K. Templates and Symbols

K1: Why are all my symbols being inserted in my drawing backwards (mirrored)?

Open the **Template** menu and select the **Enlarge** option. *Click* on each of the settings to check the current enlargement factor. If one of the numbers is a negative number, that will cause the symbols to be inserted mirrored. Make sure all the settings are positive numbers.

K2: Some of my templates don't display the symbols in the windows. How do I get them back?

This probably means that the template file has a mix of both absolute and relative paths. You need to reset the paths in the template file to have either one or the other type of path. See Chapter 14, "Templates and Symbols," for a complete description of how to do this.

K3: I can see my symbols in the template windows, but I can't place them in the drawing.

The most likely culprit is that the paths in the template file have become too long for DataCAD to read. This is caused by a program bug, whereby DataCAD can only read a line of text of up to 64 characters in the template file. Most paths will never be this long, but one notable exception is when you use symbol redirection, which will essentially double the number of characters in each line of a template file. See the end of Chapter 14 for some ways to fix this.

L. Importing and Exporting DWG and DXF Files

L1: We received an AutoCAD file for a project and imported it into DataCAD, but then it had different scaling in the X and Y axes. Also, the crosshair in DataCAD was very jerky around the screen.

DataCAD currently has what is called *single-point precision*, meaning that when the drawing gets very far (thousands of meters or many kilometers away) from the absolute zero point of the drawing file, DataCAD's accuracy begins to suffer due to the calculations involved in the long distances. The answer that follows refers to physically moving the entire drawing close to or on top of the absolute zero point of the drawing file in order to increase the accuracy of the entities in the drawing.

Moving the imported information to DataCAD's absolute zero solves the jerky crosshair movement but not the scaling problem about the two axes. In AutoCAD, when we identified an entity, its distance from the AutoCAD origin was very far away. We moved the drawing closer to the origin in AutoCAD, resaved the drawing, and then imported the new drawing into DataCAD. The cursor moved properly, and all the dimensions in the X and Y axes were correct.

To avoid this problem yourself, especially if you don't have AutoCAD in your office, you should ask the person sending the file to make sure that their drawings are closer to the AutoCAD origin before sending you the file. The person sending the file will probably even find that they will have fewer problems themselves once they do.

M. Techno Files

M1: How can I open old .DC3 files that are dated 1987?

The answer comes from Paul Nida. When Microtecture (a previous owner of the DataCAD program) came out with version 3.5, they changed the drawing database format, and you were required to convert older files. Version 3.5b shipped with a file called CONVERT.EXE to convert those files. The command was *convert filename.dc3*. You could also convert multiple files using *convert *.dc3*. If you still have a copy of this file, you could try it to see if this works, but it might be best to try it on a copy just in case.

M2: I saved a DataCAD drawing with a password, but I forgot the password. How can I open the file?

There is no straightforward way to do it, but there is a way. Open the drawing file with a Hexidecimal editor (found in PCTools or Norton Utilities, among others). You should find the password in plain text somewhere in the file. While in the drawing file, don't change anything or you could really damage the file.

M3: I copied some DataCAD macros from a CD-ROM to my hard drive, but they don't work.

When files are copied from a CD-ROM disk, they still have a Read-only attribute associated with them, which will keep a DataCAD macro from running. To fix this, just *right-click* on the macro on your hard drive, *click* on **Properties**, and then uncheck the Read-only box.

M4: What does it mean when I get a "List index out of bounds" error message?

A list index is a programming term. It refers to the number of items in an array. An array is simply a list of items. Here is an example. Let's say we put the days of the week into an array (or list.) List index 1 points to Sunday and list index 7 points to Saturday. If we try to access list index 8, we will get the list index out of bounds error. In other words, you are trying to access something that is beyond the number of items in the list.

In the case of third-party macros, authors should check that they are using the Display List-enabled DCAL functions and make sure that when an item is erased or added that the display list is updated correctly.

I am afraid there isn't much that you as a user can do except be aware of situations that cause these anomalies and try to work around them. This may not help much, but perhaps it helps to understand the error messages.

N. Miscellaneous Stuff

N1: How do I find out which layer something is on, which linetype an entity is made of, what it's z-height is, or any other information about a specific entity?

Use the **Identify** command. Either select **Edit/Identify**, select the letter **I** in the little gray square at the bottom of your menus, or press **Shift+I** (upper case **I**). Pick the entity you want to identify. DataCAD will display most of the information about the entity in the side menu. Further information is displayed in the message window at the bottom of your screen.

N2: How do I get a bitmap or scanned image into DataCAD?

This is one of the most asked questions of all time, and unfortunately there is no easy answer. The short answer is that you can't, since DataCAD cannot import bitmap images. The longer answer is that you can use various other third-party programs to first convert a bitmap image to a vector image (DXF, DWG, or WMF), which can then be imported into DataCAD. See Chapter 8 for a more complete description of the various methods.

N3: How can I speed up DataCAD?

The best way is to enable the Display List option found in the **Tools/Program Preferences/Misc** tab. Chapter 2, "Getting Started," describes some other little Windows software tricks for speeding up DataCAD.

N4: How do I track the time I spend on a drawing file?

Go to the **Utility** menu and select **Directry**. The time to the right of Total is the total time that the drawing file has been open. The time to the right of This Session is the time that the drawing has been open during the current session.

N5: How do I increase or decrease the time between auto-saves?

Select **Utility/Settings/SaveDlay** and enter a new value in minutes.

N6: What is the difference between vector and raster files? Which of the two does DataCAD create?

Len Nasman replies to the question. Basically, vector files store mathematical rules for creating stuff while raster files store all of the dots required to make a display. Vector graphics are used mostly for line drawings, while raster (or bit-mapped) graphics are used for shaded images. OK, that might be too simple. It is possible to make a mathematical rule for saving gradient fills (smoothly changing colors) or pattern fills, and it is possible to save compressed bit-mapped graphics by having the system look ahead a pixel and only save when there is a color change, resulting in not having to save every dot on the display.

One big difference between the two types is noticed when you enlarge a graphic. In the case of a vector graphic, the mathematical rules are used to generate a new image that will have a resolution that is as good as your output device (your display screen or plotter). In the case of a bit-mapped graphic, the dots (or pixels) making the image get enlarged, resulting in a jagged or blocky appearance to lines or the edges of stuff in images.

DataCAD and other CAD software store drawing files as vector images. So, your printed drawings will be as good as the resolution of your plotter or printer, not your computer display. Computer paint programs typically store bit-mapped images, but if you want to save a DataCAD GLShader image, you will be saving the image as a bit-mapped graphic. If you want to use this image in a large printed display (like a poster or billboard), you should set your computer display to the highest resolution possible with your system before capturing the screen. To get beyond this, you can export the DataCAD model to something like Renderize or Caligari True Space for rendering where it is possible to create bit-mapped image files that have a higher resolution than the computer display. Then a service bureau can be used to print large-format, bit-mapped images. The problem with high-resolution bit-mapped images is that they get very large (frequently too big for a floppy disk). If you are capturing shaded images for documents that are $8^1/_2$ by 11 in the United States (or A4 elsewhere), you will probably find that an 800 by 600 display resolution will be adequate.

N7: Is there an easy way to paste a GLShader image into a word processing document?

Once the GLShader image is created, just *right-click* the mouse and select **Save bitmap** (this saves the image to the \BMP directory in DataCAD), or select **Save bitmap as** . . . to save the image to another directory of your choice. Open your word processor, select **Insert/Picture** (or whatever command is used for inserting images), and select the bitmap image that you saved.

N8: How do I move DataCAD from its present location on my computer to another location, such as a different hard drive partition?

Move all your DataCAD files to the new drive partition (let's say you moved it from the C:\ drive to the D:\ drive). Then go to the DCADWIN.INI file in the Windows directory and change all the C: paths to D: (the new drive letter). If any other paths have changed (perhaps you changed from C:\DCAD to D:\DCADWIN), make sure you change them too.

One caveat must be mentioned though. If your template (.TPL) files have absolute paths (like C:\DCADWIN\SYM\ . . .), then you will also need to edit all of those paths to reflect the new symbol locations. To get around this, you could leave the current templates and symbols where they are on your C:\ drive, and just move everything else. If your templates and symbols are already on a network server, then nothing needs to be changed, since they will not have moved.

P. When Things Don't Act Like They Should

P1: When I type in the coordinates for entities, they instead seem to be drawing themselves in random locations around the screen.

More than likely, you accidentally pressed the **Insert** key on the keyboard. This will change the Current Input Mode (Relative Polar, Absolute Polar, Relative Cartesian, or Absolute Cartesian; see Chapter 5 for an explanation of each of these). Since only four options are available, press the Insert key until you get back to the input method you want to draw with (we use Relative Polar, but some people swear by Relative Cartesian).

P2: A little box suddenly appeared around the intersection of my drawing cursor. What is it and how do I get rid of it?

The box represents the *miss distance*. DataCAD searches within this square for things to identify and to snap to. When the box is turned off, DataCAD still searches within that box, but it is just not displayed. To turn if off, select **Utility/ObjSnap** and turn off the **Aperture** option.

P3: My drawing files are getting bigger and slower. How do I fix that?

Use the Layer Utility (LyrUtil.dcx) macro in the **Edit/Toolbox** to scrunch the file down to a smaller and faster size. See Chapter 21, "Techno-Files," for a thorough explanation.

P4: My drawing keeps crashing. What do I do?

Occasionally, any computer file can become corrupted with bad data. CAD programs appear to be more susceptible than other programs. Use the Layer Utility (LyrUtil.dcx) macro in the **Edit/Toolbox** to clean up the file. See Chapter 21 for a thorough explanation.

P5: A few times every hour DataCAD seems to lock up and won't let me do anything for a few seconds. What's going on?

This is the AutosSave function that automatically saves the current drawing (as a .ASV file) every few minutes, causing DataCAD to pause until the save is completed. If DataCAD crashes or closes abnormally, the .ASV file can be renamed to .DC5 to get most of your drawing back. The Autosave function will save at specific intervals as defined by the number of minutes set in **Utility/Settings/SaveDlay**. See Chapter 23 for more information.

P6: Why can't I access my Toolbox macros either through the menu, by keyboard shortcuts, or through my icon toolbar?

Check to see if the active layer is Locked. You cannot access the **Toolbox** if it is.

P7: I just installed DataCAD and have observed two anomalies: 1) The cursor doesn't seem to appear within the drawing area, although the Windows arrow is visible and usable to pick menu items, and 2) the images of all DataCAD drawings I bring into the program are inverted and mirror-imaged from the normal!

The answer comes from Mark Humienny. The problem may be with an old graphics card and computer display settings. After experimenting with adjustments to the number of colors and the monitor resolution, you may be able to get everything back to normal. In my case, I had an older two-MB Diamond Stealth card and I had to turn the colors down to 16-bit, high color with the resolution at 1024 × 768. Since upgrading to the Viper 770 graphics card, I can turn the colors back to 32-bit.

P8: When we type in text, it comes in reversed like we're seeing it in a mirror. How can I get rid of this?

Your text **Aspect** is set to a negative instead of a positive number. Change it to a positive number and your text will be entered normally.

P9: I'm drawing entities and placing symbols, and they keep disappearing!

Check to see if you have a *ClipCube* (CC) turned on, since nothing outside of a CC will be visible. Go to **Utility/Plotter/3DViews/ClipCube** and make sure the **ClipOn** setting is turned off.

P10: I am having trouble with drawing access over our network. It crashes with an access error, and if I can get the drawing open later, it opens with portions of the drawing missing. Plots that are sent across the network to the plotter are coming up incomplete and/or garbled, while plotting locally works fine.

The problem may be a low-cost ethernet adapter with one of the Realtek 8029 chipsets. There is a known corruption problem with the Rtl8029.sys driver included with Windows 98, as documented at `http://support.microsoft.com/support/kb/articles/q189/7/78.asp`.

For some users, getting an updated driver as recommended may clear things up, but other users have continued to experience other problems like random system lockups. If you continue to have problems, you should replace the network card with one from a different manufacturer.

P11: While using DataCAD, many of the keyboard actions have become very slow. Some actions take up to five or six seconds to respond.

If you have recently installed or updated an anti-virus software program, it may be that the software is checking every keyboard action before allowing it to happen. Check your anti-virus software and set it to exclude DataCAD and all the commonly used keyboard macros.

P12: In some drawings, the cursor keeps snapping to imaginary points, no matter what my *ObjSnap* settings are. How do I stop this?

It's a minor corruption of the drawing file that can usually be corrected by running the **LyrUtil** (Layer Utility) macro. See Chapter 21 for more information on this macro.

Q: External Reference Files (XREFs)

Q1: I made several *Gotoviews* (GTVs) and *Multi-Scale Plotting* (MSP) details using an XREF, but the XREF layers I had turned on and off are now different than when I saved them.

Currently, the on/off states of XREF layers are not saved with GTVs or MSP. Instead, they will display the layer settings as they are currently set, but there is a solution to this problem. You can have multiple instances of each XREF in the same drawing file. So if you have two different GTVs that use the same XREF file but have different layer requirements, simply create a new layer and copy a second instance of the same XREF to that layer. Now you can use the *Reference File Manager* (RFM) to independently set the on/off layers of each of the two XREFs. See Chapter 13, "External Reference Files (XREFs)," for more information.

Q2: The XREF floor plan in my drawing has suddenly moved to a new location on the screen, but I'm sure that I didn't move it. How did it move on its own?

It appears that someone moved the floor plan in the XREF file (the .DC5 file that is being referenced), causing the entities to also move in master file. It is very important that once a .DC5 drawing file is referenced by other drawings, the locations of the entities in that file are not moved. To fix it, you need to determine the distance that the XREF file entities were moved and then move them all back to where they were originally located.

Q3: In the RFM, I turned XREF layers on and off and then used the *Bind/On* command to bind only those layers that were on. After I bound the XREF, the new layer settings were not reflected in the drawing file. Why didn't the layer settings take?

After making any layer changes in the RFM, you need to press the **Apply** button first in order for the changes to be recognized. Then you can run the **Bind/On** command.

R: 3-Dimensional Questions

R1: Move, Copy, Mirror, and other editing functions don't seem to be working like they should, or they don't function at all.

Most likely, you are trying to use 2D editing commands on your 3D entities. Although some 2D editing commands may work on some 3D entities, in most cases you must make sure that you use the 3D editing commands. So when editing 3D entities, be sure that you are in the 3D menus, and that you aren't using 2D keyboard shortcut commands.

R2: All my curved 3D surfaces appear with faceted surfaces. Can I get rid of them?

In DataCAD, no true curved surfaces exist, only surfaces that approximate a curve with multiple facets. To smooth a curve when you create it, you can increase the number of facets of each curved surface, up to 36 facets, with the **3DEdit/Settings/PrimDivs** and the **3DEdit/Settings/ScndDivs** settings. If you save your 3D drawing as a 2D image (with the **3DEdit/Hide** command), then you can manually erase the lines that make up the facets.

R3: When I use the *Roofit* macro, how come my roof is upside down?

You must define the roof in a clockwise direction with the Roofit macro. If you define it in a counterclockwise direction, then you will get an upside down roof.

R4: I want to create a 3D site, but I can't find the DropMesh feature. Where is it?

Because this feature creates a 3D mesh surface made of multiple triangular polygons, it is located in the **3DEntity/Polygon** menu.

R5: Now that I have successfully constructed my 3D site using the Drop-Mesh function, how do I create drives, parking lots, and other paving so that it will appear slightly above the ground surface, making it possible to shade and render?

One way is to place a thin horizontal slab for each car park. Using these slabs, rather than polygons, will mean that if a few gaps occur they will be closed by the slab thickness. You are likely to find individual areas of the drop-mesh too high for the slab mechanism to accommodate. If that is the case, it may be worthwhile to explode the mesh and trim the polygons to the edge of the parking spaces. You can explode the mesh with the **3DEdit/Explode/ToPgons** option or the third-party 3D Power Tools macro.

R6: What is the easiest way to model a site that contains retaining walls? A simple drop-mesh doesn't build correctly on either side of the retaining wall.

Model the retaining walls and the like separately. Then trim the mesh up to them as discussed in the previous FAQ answer. Curbs require similar treatment. Using separate layers and colors for every different element will help make this task easier. Often you will need to fill a few gaps by adding three-edge, triangular polygons with the **3DEntity/Polygon/Inclines/3EdgPoly** feature.

CHAPTER 23

Customizing DataCAD

This chapter is all about customizing DataCAD to work the way you want it to. We will cover how to make DataCAD look different, how to more quickly access your most commonly used commands, how to create your own linetypes and hatches, and how to convert AutoCAD fonts for use in DataCAD.

Customizing the Graphical User Interface (GUI)

You might first ask, "what the heck is a GUI?" It's pronounced "gooey," and it's short for Graphical User Interface. That's a fancy phrase that refers to the whole graphical DataCAD screen: the Drawing Window, menu buttons, icons, message lines, and so on. DataCAD enables you to change how many of those GUI features look, including their colors, fonts, and sizes, among other things. Some of these can be adjusted inside DataCAD, while others can be adjusted by various other means. Let's look at each of these.

DataCAD Preferences

While DataCAD is running and a drawing file is open, go to the Windows drop-down menus and select **Tools/Program Preferences**. A dialog box with five tabs appears. Each tab has a series of options to enable you to determine how DataCAD operates and how it looks.

Pathnames The Pathnames option determines the paths to various files on your computer or network (see Figure 23-1). All of these paths are automatically filled in when you first install DataCAD, but often you will later want to change some of them. *Click* on the file folder buttons at the far right of each path box to display a Browse dialog to search for and insert the correct paths.

Most of these settings will also change automatically as you select paths within DataCAD. For instance, if, while working in DataCAD, you select a different symbol template, the new path will automatically be inserted in the Template Files box.

Notice that in many cases you may not have to have the full path displayed. If the directory is located under the main DataCAD directory, only the directory will be displayed. For instance, if all your fonts are stored in

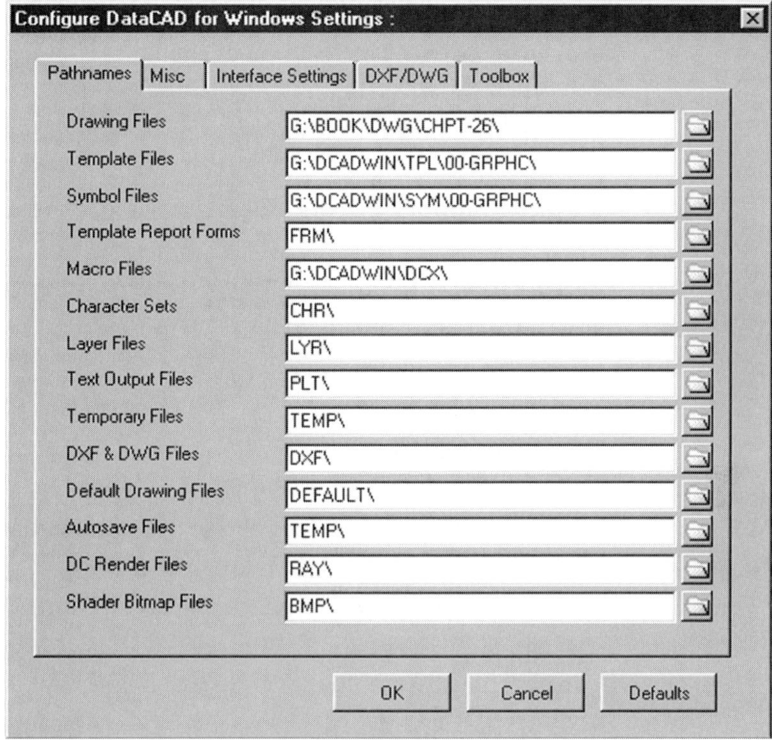

Figure 23-1
The Pathnames tab

the C:\DCADWIN\CHR folder, then the Pathname will simply show CHR\ as the path, since DataCAD knows that this path is within the main DCADWIN directory.

WARNING: *Remember two important things about entering paths:*

- *Typing the wrong paths can sometimes keep DataCAD from working properly, or at all, so make sure they are filled in correctly.*
- *If you type in a path that does not exist, DataCAD will not let you do anything until you type in one that does. So if you type an incorrect path, you can't even get the cursor out of the box to press the Cancel button. If you get stuck and can't remember a valid pathname, either use the Browse button or just type in* ***C:*** *and then move on. You can go back later to fill in the correct path.*

The **Pathnames** tab contains the following settings:

- **Drawing Files** The location of your DataCAD .DC5 drawing files
- **Template Files** The location of the last template file you selected while in DataCAD
- **Symbol Files** The location of the last symbol file you selected while in DataCAD
- **Template Report Forms** The location of the forms where template reports are stored
- **Macro Files** The location of the **Toolbox** macros
- **Character Sets** The location of your DataCAD fonts
- **Layer Files** The location of the .LYR files created when you save individual layers or run the LyrUtil (Layer Utility) macro
- **Text Output Files** The location of saved plot (.PLT) files and the location of text files saved from or saved into the Text menu
- **Temporary Files** The location of temporary files, such as the swap (.SWP) file, saved by DataCAD while a drawing file is open
- **DXF & DWG Files** The location of .DWG and .DXF drawing files exported from or imported into DataCAD
- **Default Drawing Files** The location of all your default DataCAD drawings (.DC5s)
- **Autosave Files** The location of the autosave (.ASV) file created by DataCAD while a drawing file is open
- **DC Render Files** The location of the 3D model (.RAY) files that DataCAD creates when you want to render a model in the third-party DC Render program
- **Shader Bitmap Files** The location of any bitmap (.BMP) files that you save after running the GLShader in the 3D section of DataCAD

Misc As the name suggests, the Misc tab has a number of miscellaneous options within it (see Figure 23-2). As with the paths on the previous tab, these options are automatically filled in when you first install DataCAD, but later you may want to change some of them. Figure 23-2 is actually the tag as revised in version 9.01.

Default Drawing File The .DC5 drawing file shown in this window will be the default drawing that DataCAD uses every time you start a new drawing file. These default drawings are found in the \Default directory defined in the Pathnames tab, described earlier. When you *click* on the down arrow

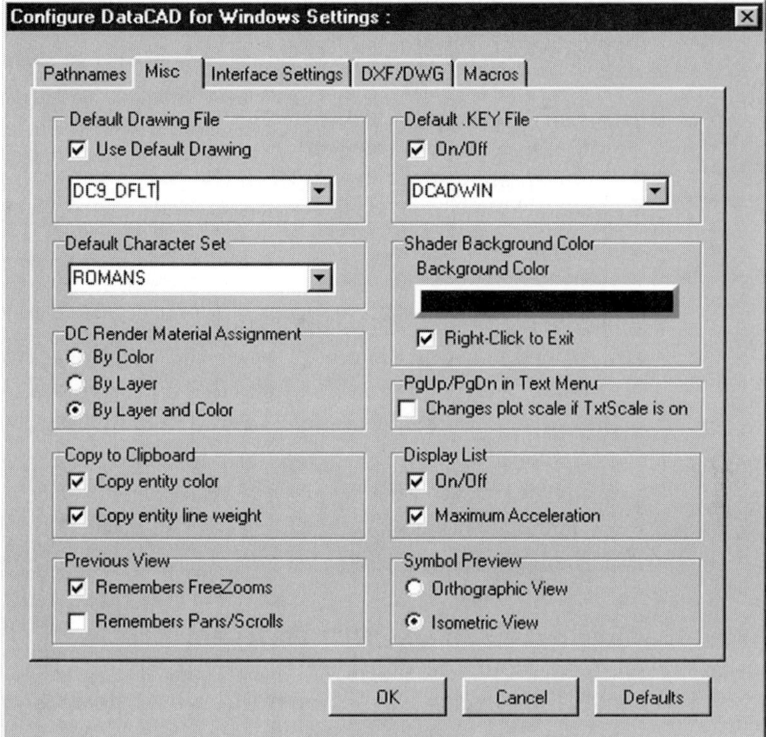

Figure 23-2
The Misc tab

to the right of the white box, the name of every .DC5 file in that default directory is shown. To select a new default drawing, press the down arrow, highlight the appropriate .DC5 file, and then *click* on it.

Leaving the **Use Default Dwg** box checked will force DataCAD to default to using the drawing shown in Figure 23-2. If you uncheck the box, then DataCAD will use its own default settings to create new drawing files and will not use any of the default drawings in the \Default directory.

Default .KEY File The icon toolbars across the top of the DataCAD screen are controlled by files with a .KEY extension. These files tell DataCAD which icons to display and what DataCAD should do when each one is clicked on. These files are always located in the \SUP\MENUPOF folder in your DataCAD directory. DataCAD comes with two pre-made icon bars: DCADWIN (for 2D work) and DCADWIN3 (for 3D work). Later in this chapter you will find out how to create your own icon toolbars. You can select the icon toolbar to be displayed by clicking on the down arrow and clicking on one of the files shown. If you do not want an icon toolbar to be displayed, simply uncheck the **On/Off** box.

Default Character Set This option sets the default font name (.CHR) that DataCAD will use in the **Text** menu whenever you start a new drawing file. By default, all of these fonts are stored in the \CHR folder in your DataCAD directory. To change the default font, *click* on the down arrow and *click* on one of the fonts shown.

Shader Background Color This is the color that DataCAD will use for the background behind your 3D model when the **GLShader** option is clicked in the 3D menu. To change the color, just *click* on the colored horizontal bar and the standard Windows color dialog box will appear, as shown in Figure 23-3. Pick a new color and then select **OK**. The horizontal bar in DataCAD will change to match that color. With **Right-Click to Exit** checked, right-clicking will exit the GLShader without saving the image. With it un-checked, right-clicking will display several options for saving the shaded image.

DC Render Material Assignment There is another third-party 3D rendering program made to work specifically with DataCAD, called "DC Render." You can directly export a DC Render file (with **File/Export/RAY**) for use in that program. The options you see here allow you to determine how the materials in DC Render will be applied to the layers and colors in your DataCAD drawing when exported. The default is **By Layer and Color**. Selecting **By Color** will assign materials based solely on the color of the

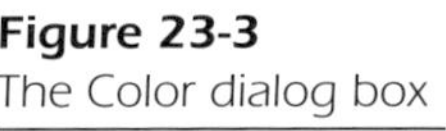

Figure 23-3
The Color dialog box

entities (all RED entities will be displayed as brick). Selecting **By Layer** will assign materials based solely on the layer the entities are on (all entities on the WALLS layer will be displayed as brick).

PgUp/PgDn in Text Menu This option was added in the free 9.01 upgrade. When this option is checked, **PgUp** and **PgDn** will change the current plotting scale in the **Text** and **Dimension/TxtStyle** menus. When un-checked, pressing PgUp and PgDn while in the Text menu will cause the Drawing Windows to zoom in and out. We recommend you leave it un-checked.

Copy to Clipboard These settings determine how entities are copied to the Windows clipboardwhen you use the Cut and Copy options. Both options are selected by default. If you un-select **Copy entity color**, then all entities will be copied to the clipboard as black. If you un-select **Copy entity line weight**, then all entities will be copied to the clipboard with the same, thin line weight (thickness).

Display List This determines whether the Display List feature is on or off and how it will function. Turning Display List on (by checking the box) will greatly speed up DataCAD's operations. For a full explanation of what Display List is and how it works, refer to Chapter 3, "Settings and Display Options." That chapter will also explain what the **On/Off** and **Maximum Acceleration** options do.

Previous View With both of these options un-checked pressing the **Pp** key will return the previous view to the Drawing Window, but only if **WindowIn/FreeZoom** is turned OFF. That is, the extents of what you were viewing before you moved to a new view will be redisplayed. With only **Remembers FreeZooms** checked, up to the previous 25 views will be returned when **WindowIn/FreeZoom** is turned ON; but only views which were created by using **WindowIn**, **PgUp**, or **PgDn**. Previous pans (up, down, left and right) are not returned. With **Remembers Pans/Scrolls** checked, up to the previous 25 views will be returned; but only views that were created by panning with the arrow keys or the red arrows in the Navigation Pad. With both options checked, all views will be returned, regardless of how they were created, even with **WindowIn/FreeZoom** turned ON.

Symbol Preview This option controls the preview window when selecting symbols directly. Select **Orthographic View** for a 2D plan view of symbols, or **Isometric View** for a 3D isometric view. See "Direct Symbol Selection" in Chapter 14 for more information about this option.

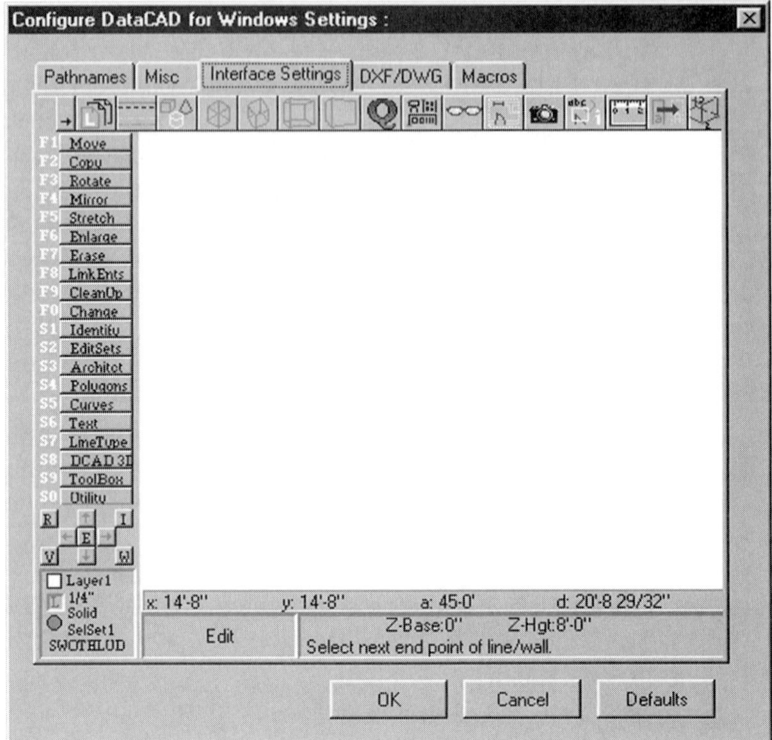

Figure 23-4
The Interface Settings tab

Interface Settings This dialog is actually a small graphic picture of a typical DataCAD screen. It is not an actual picture of your own screen. By clicking on the different parts of this screen, you will be presented with various options for customizing how it is displayed (see Figure 23-4).

Menu Buttons *Click* on the vertical menu buttons to change how the text on your menu buttons is displayed. The fonts that are displayed are all the fonts available to your Windows system, most of which are True Type fonts. You can also choose the font style, size, and script (see Figure 23-5).

WARNING: *If you select font settings that create large text, some text may not be fully displayed in some menu buttons and the Projection Pad and Navigation Pad buttons may disappear due to a lack of room to display them, as shown in Figure 23-6.*

If this happens, you have two options:

Figure 23-5
The Font dialog box

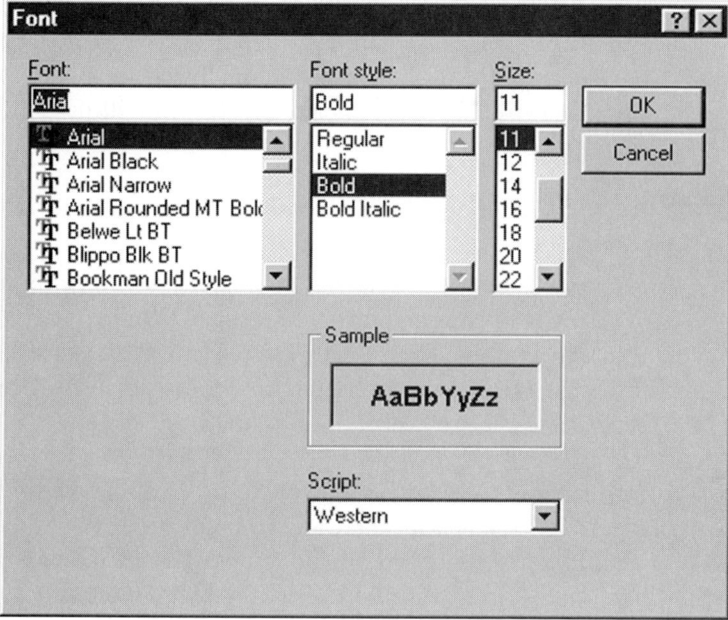

Figure 23-6
The Projection and Navigation Pads do not appear if text is set too large.

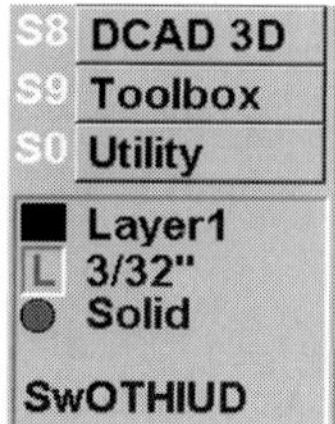

- *Reduce the size of the text, select a different font, and/or change the font style.*
- *Edit the BTN_WIDTH setting in the DCADWIN.INI file. There will be more information about editing this file later in the chapter.*

Icon Bar *Click* on the icon toolbar across the top to turn the toolbar off. If no icon toolbar is displayed, *click* at the top of the screen where it would be and the toolbar will be turned on.

Drawing Window *Click* on the part of the graphic that looks like the Drawing Window. The standard Windows color dialog box will appear, as shown previously in Figure 23-3. Pick a new color and then select **OK**. The color of the graphic Drawing Window will change to match the color you selected. By default, DataCAD starts with a black Drawing Window (my preference), but some people prefer a white background. Be aware that if you change your Drawing Window color to white you will have to adjust how the other colors in DataCAD are displayed, since some colors, most notably yellow, don't display well on a white background. More about this will be discussed later in the chapter.

Message Window *Click* on the bottom of the graphic where the Message Window is to change how the Message Window text is displayed. The fonts that are displayed are all the fonts available to your Windows system, most of which are True Type fonts. You can also choose the font style, size, and script (see Figure 23-5).

Status Area *Click* on the bottom of the graphic where the status area is (the square with the SWOTHLUD in it) in order to change which side the vertical menu buttons are displayed on. Each time you *click* on the graphic of the status area the menu buttons will switch sides, as shown in Figures 23-7 and 23-8.

DXF/DWG These settings determine how DataCAD imports and exports .DXF and .DWG drawing files (see Figure 23-9). You can learn more about these options in Chapter 25, "Converting File Formats."

Macros Use this dialog area to add and remove macros from the drop-down **Macros** menu. Select the macros you want from the list on the left, and then *click* **Add** to add them to the **Macros in Menu** area on the right. To remove macros from the area on the right, just highlight them and *click* **Remove**. If you have more than one Toolbox directory, or you need to navigate to the correct directory somewhere else, *click* on the square with the yellow file folder in the upper-left corner to browse for the correct directory. You can add macros from more than one directory to the **Macros in Menu** area to the right (see Figure 23-10). The macros will be displayed in the order you add them, so be mindful of the order in which you add them.

Macros Drop-Down Menu

In the previous section, we described how to add macros to the **Macros** drop-down menu at the top of your screen. Clicking the **Macros** option will display the current list of macros (see Figure 23-11). Just *click* on a macro to run it.

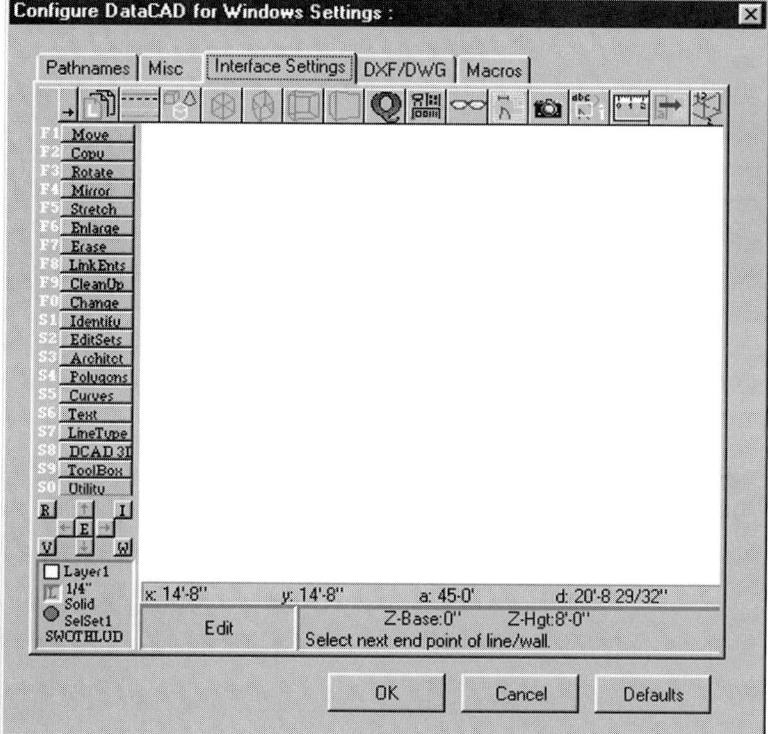

Figure 23-7
The Status Area of the image enables you to change the menu buttons from one side . . .

Click on **Configure** to bring up the **Macros** tab from the **Tools/Program Preferences** menu. Remember that the macros will be displayed in the order you add them, so be mindful of the order in which you add them.

Customizing DataCAD with the DCADWIN.INI File

An .INI file, such as the DCADWIN.INI file, is an initialization file, which means that it is only read by DataCAD at the initial startup of the program. DataCAD does not read this file again until the program is restarted, but DataCAD does write information to this file while it is running.

If you are running Windows 95 or 98, you will find the DCADWIN.INI file in your C:\WINDOWS folder. If you are running Windows NT, you will find it in your C:\WINNT folder. It is a simple text file that DataCAD looks at every time you start the program, telling it about your current settings and

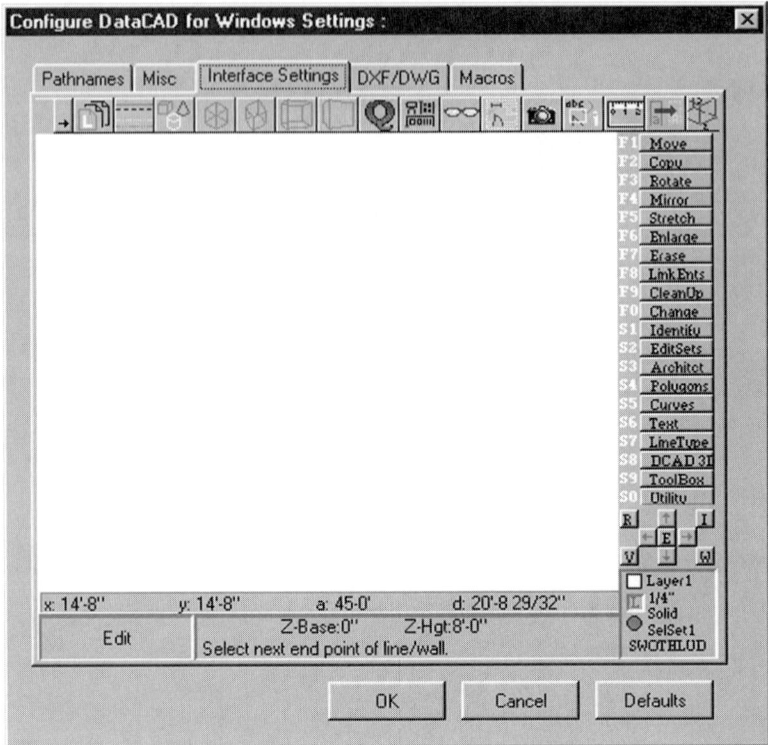

Figure 23-8
. . . to the other.

paths. Most of the settings within the DCADWIN.INI can be changed within DataCAD. The DCADWIN.INI file is not really meant to be edited by the average user, but some options and features can only be activated or customized by doing so. The order of the items in the file does not matter as long as they are in the correct sections, and the options are not case-sensitive.

WARNING: *Improperly editing the DCADWIN.INI file can cause DataCAD to run improperly, or not at all. For this reason, you should always save a backup copy of the file prior to attempting any editing.*

Although many of your settings will be different, let's use the DCADWIN. INI file from my computer to illustrate what's in the file and what you can customize. The *italic* text in the following section is our description of what that particular line in the file does and is not part of the .INI file. Some of these descriptions will give you information that you can use to customize

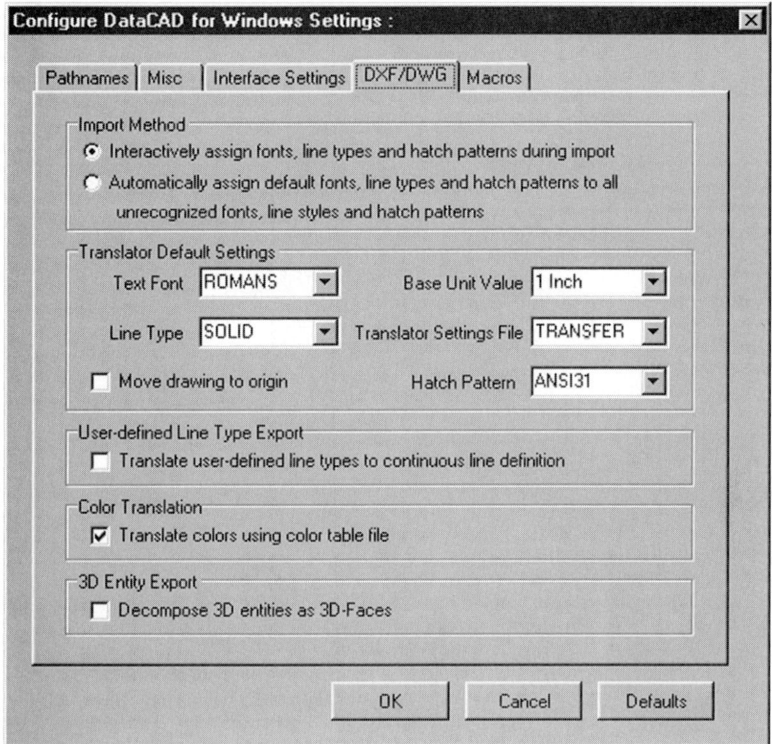

Figure 23-9 The DXF/DWG tab

your DataCAD interface, while other descriptions will tell you how you can manually fix some problems (by typing in the correct text) that might arise if DataCAD is not functioning properly. Again, notice that this is simply a text file:

```
[General]
FONTNAME=ROMANS
   This is the default DataCAD font (.CHR) file used in the Text
   menu. This option is also available within DataCAD via
   Tools/Program Preferences/Misc/Default Font Name. If DataCAD ever
   won't start because it says it can't find a particular font, open
   the DCADWIN.INI file and change the font shown to Romans. Save
   the file and restart DataCAD and it should work.
DEFAULT_DRAWING_NAME=DEFAULT.DC5
USE_DFLT_DWG=TRUE
   This determines whether new drawings in DataCAD will be created
   by using the default drawing listed in the line above (TRUE), or
   whether DataCAD will ignore all the default drawing files and use
   its own default settings (FALSE). This option is also available
   within DataCAD via Tools/Program Preferences/Interface Settings.
```

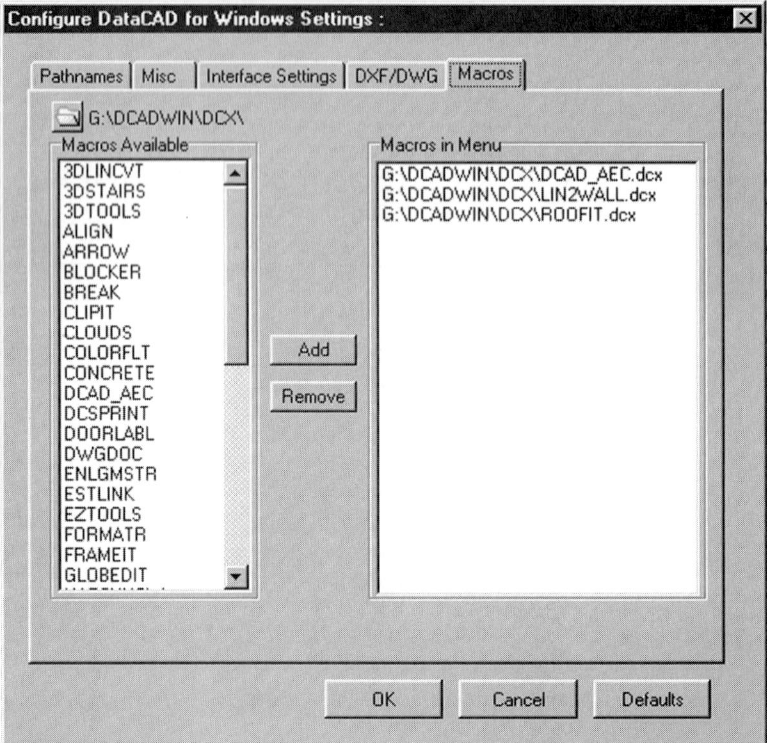

Figure 23-10
The Macros tab

Figure 23-11
The current list of macros in the Macros menu

```
MENU_WINDOW=left
   This determines whether the menu buttons are located on the right
   or left of your screen. This option is also available within
   DataCAD via Tools/Program Preferences/Misc.
CAVWALL=false
   When this option is changed to TRUE, the insertion of doors and
   windows into cavity (4-line) walls trims the outer jamb lines to
   the cavity lines.
CHECK_SYM=false
   This addresses a bug in the Symbol/Purge function of previous
   DataCAD versions. If TRUE, it tells DataCAD to scan all symbol
   instances (upon loading a drawing) to check if valid symbol
   definitions are in the drawing file headers. This will clean
```

files previously corrupted by the PurgeSym command. The downside is that it slows drawing file loading somewhat.

[Paths]

All the PATH lines below tell DataCAD where to find various directories. Most are self-explanatory, but we'll describe a few of the most important ones. Note that many of these paths can be changed via the Tools/Program Preferences/Pathnames dialog.

INSTALL_DIR=C:\DCADWIN\

This is the location of your DataCAD program. It is the directory that the DCADWIN.EXE program resides. Generally, you'll have no reason to change this unless you rename or relocate your DataCAD directory.

PATH_ASV=C:\DCADWIN\

This is the location for the autosave (.ASV) file that DataCAD creates whenever a drawing is opened and worked on.

PATH_SWAP=C:\DCADWIN\TEMP\

This is the location for the swap (.SWP) file that DataCAD creates and uses whenever a drawing is opened and worked on. Refer to Chapter 21, "Techno-Files," for an explanation of the very important swap file.

[KeyFile]

KEYFILENAME=DCADWIN

This is the name of the currently displayed .KEY file from the SUP\MENUPOF directory (notice that the .KEY file extension is not used here). .KEY files are the files that control the icon toolbar at the top of the DataCAD screen. If there is no file shown after the = sign, then DataCAD would not display any icon toolbar.

[PaletteFile]

RGBFile=DEFAULT.RGB

This is the name of the current .RGB (Red/Green/Blue) color palette. .RGB color palettes are used to define the 256 colors available for use in DataCAD and affect all entities in the drawing file. Refer to Chapter 3 for more information about color palettes and .RGB files.

[MenuTextColor]

These settings determine how the DataCAD menu buttons look. Three sets of numbers can be found after each equals (=) sign. This is known as a color triplet and describes the Red, Green, and Blue color mixtures that make up all the available colors on your computer. Each number can range from 0 to 255. By changing the values of each triplet, you can change the colors that DataCAD displays. See the section below about finding and picking these triplets.

HighLight=0 255 0

This is the color of the highlight around a menu button when the cursor is passed over it.

FunctionKey=255 255 138

This is the color of the F# and S# text next to each menu button.

NormalButton=0 0 0

This is the color of the standard text on all the menu buttons. 0 0 0 is black.

ToggleButtonOn=0 128 0

This is the color of the text on a menu button when the button is toggled ON. 0 128 0 is green.

ToggleButtonOff=255 0 0

```
   This is the color of the text on a menu button when the button is
   toggled off. 255 0 0 is red.
LayerButtonActive=0 0 183
   This is the color of the text on a layer name button when it is
   the currently active layer.
NavPadButton=0 0 0
   This is the color of the text on the Navigation Pad buttons.
StatusArea=0 0 0
   This is the color of the text on the Status Area buttons.
MessageArea=0 0 0
   This is the color of the text in the Message Area.

[BackGroundColor]
QuickShaderColor=0 168 168
   This is the color that DataCAD will use for the background behind
   your 3D model when the GLShader feature is run from the 3D menu.
DrawingAreaColor=255 255 255
   This is the color that DataCAD displays in the main Drawing
   Window. 255 255 255 is pure white.

[MenuBtn]
   These are the settings that DataCAD uses for the text on the menu
   buttons. These options are also available within DataCAD via
   Tools/Program Preferences/Interface Settings.
FONTNAME=Arial
POINT_SIZE=12
STYLE=BOLD
BTN_SPACING=0
   This is the distance between each menu button. Spacing is
   measured in pixels and the default is 0.
BTN_WIDTH=0
   This enables you to add pixels to the right side of all your menu
   buttons to make them wider. This should take care of the problem
   where some long text is not shown at the right side of some
   buttons. The default is 0.

[MessageArea]
   These are the settings that DataCAD uses for the text in the
   Message Area. These options are also available within DataCAD via
   Tools/Program Preferences/Interface Settings.
FONTNAME=Arial
POINT_SIZE=10
STYLE=

[TemplateWindow]
Width=210
   DataCAD defaults to displaying the template window width at 200
   pixels. You can increase or decrease that number by typing in a
   new number (like the 210 shown). If the option is left blank, the
   default of 200 is used.

[Recent File List]
   This is where DataCAD stores the last four drawing files you
   accessed. The names you see here are the names displayed at the
   bottom of the File drop-down menu.
FILE1=C:\Book\Dwg\Chpt-26\CustLine.dc5
FILE2=C:\Book\Dwg\Chpt-26\custhatch.dc5
FILE3=
FILE4=
```

```
[DXF_DWG]
   This is where DataCAD saves the .DWG and .DXF import and export
   settings that are selected in the Tools/Program
   Preferences/DXF/DWG dialog. You can learn more about these
   options in Chapter 25.

[SHADER]
SMOOTH=3
   This corresponds to the highlight setting under
   GLShader/Settings. Valid options are 0, 1, 2, or 3. The higher
   the value, the denser the tessellation of polygons is for
   lighting calculations, and the more apparent the display of
   highlights becomes.

[DropMesh]
MaxDivisions=100
   This number sets the maximum number of X and Y divisions in the
   3D Drop Mesh function (DCAD_3D/3DEntity/Polygon/DropMesh). The
   default value of 100 is a number meant to keep new users from
   inadvertently setting the X or Y values too high, thereby slowing
   down the program considerably. With the value set to 100, even if
   you enter an X or Y value higher than 100, DataCAD will not
   accept the number. You will have to enter a number at or below
   100. But before you go changing this value, keep this in mind:
   One X/Y unit yields a square, which is further divided into two
   triangles or polygons. That makes a mesh of 200 by 200 polygons,
   which would yield a total polygon count of 40,000. That should be
   far more than any model would require. And such a large polygon
   count in your drawing file would really slow things down. This is
   a setting that can only be changed here in the INI file.

[Clipboard]
   The following three options control aspects of the Cut and Copy
   features in DataCAD as they apply to the Windows Clipboard. These
   options are available in the Tools/Program Preferences menu.

[ToolBox]
   These are the macros under the Macros option in the drop-down
   menu. You can rearrange the order of the macros by rearranging
   the file names here.
Macro_1=G:\DCADWIN\DCX\DCAD_AEC.dcx
Macro_2=G:\DCADWIN\DCX\LIN2WALL.dcx
Macro_3=G:\DCADWIN\DCX\ROOFIT.dcx
```

Menu Text Colors in DCADWIN.INI

In the previous DCADWIN.INI file, all the settings under the [MenuTextColor] field describe the colors to be used for various menu settings. The settings do this via three numbers separated by spaces called *color triplets*, such as these two examples:

```
HighLight=0 255 0
FunctionKey=255 255 104
```

These triplets describe the Red, Green, and Blue color mixtures that make up all the available colors on your computer. Each number can range from 0 to 255. You could try to select new triplet colors for the menus by trial and error, but this is very difficult. The easiest way to do it is to use the standard Windows color picker, which is accessible within DataCAD. Once you visually find the color you want, just write down the three numbers that make up that color. Now you can type that number in as a color triplet.

Let's try an example. Any menu that offers you color options in DataCAD can be used to bring up the Windows color picker. We'll access it through the **Tools** menu option at the top of your screen:

1. Select **Tools/Program Preferences/Interface Settings**.
2. *Click* on the graphic of the main Drawing Window. The Windows color picker will be displayed (see Figure 23-12).
3. Pick a color in one of three ways:
   a. Pick a color from the grid of **Basic colors** or **Custom colors**.
   b. Use the color window and slider at the right of the dialog box.
   c. Manually type in values for **Hue**, **Sat**uration, **Lum**inance, **Red**, **Green**, or **Blue**.
4. Once you find the color you like, look at the numbers in the **Red**, **Green**, and **Blue** boxes and write them down. Be sure to write them in

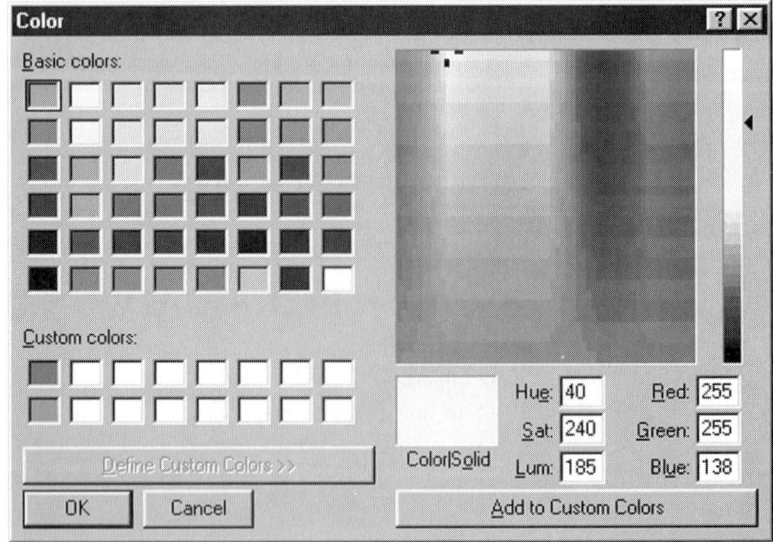

Figure 23-12 The Windows color picker

order from the top down. In Figure 23-12, the RGB values are 255, 255, and 138.

5. Cancel out of the color picker and Program Preferences dialogs.
6. Exit DataCAD. The new color settings in the .INI file are only read by DataCAD when the program is initially started, so you must exit DataCAD before making the changes.
7. Open the DCADWIN.INI file in a text editor, such as Windows Notepad, and type in the new triplet next to whichever setting you want to change. So if you wanted to change the text for the FunctionKey setting you would type the following, making sure you leave a space between each number:

```
FunctionKey=255 255 138
```

8. Save the .INI file (make sure you save it as a plain text file with no formatting).
9. Run DataCAD and your new color setting is in place.

DCADWIN.INI and Foreign Character Sets

If you have the foreign language version of Windows 95 or NT installed, you can get DataCAD to use those character sets in the menu window, status area, navigation pad, and Message Window. Under the [General] heading of the DCADWIN.INI file, you can add the following:

```
[General]
CHAR_SET=0
```

The default value is 0 (DEFAULT_CHARSET). The valid values range from 0 to 18, but you must have the foreign language version of Windows 95 or NT installed for foreign character sets to appear. The corresponding character sets are listed here.

For Windows 95/98/2000 and NT, the foreign character sets are as follows:

0: DEFAULT_CHARSET;

1: ANSI_CHARSET;

2: SYMBOL_CHARSET;

3: SHIFTJIS_CHARSET;

4: GB2312_CHARSET;

5: HANGEUL_CHARSET;
6: CHINESEBIG5_CHARSET;
7: OEM_CHARSET;
8: HANGEUL_CHARSET;

For Windows 95 only, the foreign character sets are as follows:

9: JOHAB_CHARSET;
10: HEBREW_CHARSET;
11: ARABIC_CHARSET;
12: GREEK_CHARSET;
13: TURKISH_CHARSET;
14: THAI_CHARSET;
15: EASTEUROPE_CHARSET;
16: RUSSIAN_CHARSET;
17: MAC_CHARSET;
18: BALTIC_CHARSET

Color Adjustments

Besides the color adjustments to DataCAD's GUI, as described earlier, you can also adjust the colors displayed in the Drawing Window. Here are several reasons why you might want to do so:

- To improve the differentiation between blues, greens, reds, and so on (such as Blue and LtBlue, and Green and LtGreen). People who are red/green color blind especially like this feature.
- To fine-tune the colors for your particular monitor and video card. Place two different computer displays next to each other and you will see that no two displays look the same.
- When changing the Drawing Window from black to white, DataCAD's colors need to be adjusted for better contrast. Most colors, especially yellow, do not show up well on a white screen.

DataCAD uses a special file with a .RGB file extension to determine the way the colors of all screen entities are displayed. RGB color palettes are used to define the 256 colors available for use in DataCAD and affect all entities in the drawing file. The default DataCAD file is called, appropriately enough, DEFAULT.RGB.

You can adjust the color definition of one or all of the 256 colors on the fly, but unless the changes are saved to the existing or a new .RGB file, they will not remain in effect after the drawing file is reopened.

When a .DC5 drawing file is saved, the .RGB file name is stored as an attribute with the file, so it is possible to have a unique pallet for each drawing. If an associated RGB file is not found, DataCAD will use the RGB file as defined in the DCADWIN.INI file.

If you open it with a text editor like Notepad, you will see that it, like many other files in DCAD, is just a simple text file. Here are the first several lines of the DEFAULT.RGB file:

```
0 0 0
255 255 255
255 0 0
0 255 0
0 0 255
0 170 170
170 0 170
250 120 0
170 170 170
```

Each line has three sets of numbers, which are called *triplets*. They describe a particular color for DataCAD to display. Refer to the previous section on "Menu Text Colors in DCADWIN.INI" for an explanation of how this works.

Each .RGB file contains 256 lines. Each line of the file corresponds to one of the 256 colors available in DataCAD, and the triplet on each line describes it. The color triplet on that line, therefore, tells DataCAD which color to display that element in. For instance, the very first line, 0 0 0, defines the color black. The second line, 255 255 255, defines the color white, which is the first color available in all of DataCAD's **Color** menus. The next 14 lines of the .RGB file define the next 14 colors in DataCAD's standard **Color** menus: Red, Green, Blue, and so on.

Changing RGB Screen Entities

To change the RGB screen items, select **Utility/Display/Palettes**. When selected, two additional options will appear:

- **LoadRGB** Use this option to load an existing .RGB file from the \SUP directory. Once loaded, that RGB file will be associated with the current drawing file.

- **SaveRGB** Use this option to save the current color settings if you have made any changes. You can save the settings to an existing file or name a new file.

You can create your own custom RGB palettes (described later), but DataCAD comes with several that are ready for you to use. However, all of them, except for the Default.rgb file, were created to only affect colors 16 through 256. That's because these RGB files were created primarily to affect how the Quick Shader images display, without affecting the 16 main DataCAD colors (Quick Shader was the original shading utility prior to the new GLShader).

Here are the RGB files that come with DataCAD, located in the **\SUP** directory:

- **Default.rgb** This is the standard default RGB file.
- **DCViewer.rgb** The colors are optimized for use with the DC Viewer program module.
- **Destijl.rgb** The colors are reduced to their simplest, primary colors, excluding more complex colors.
- **Greyscal.rgb** A monochromatic grey RGB file
- **Sepia.rgb** A monochromatic cream or sepia tone RGB file
- **Transfer.rgb** The colors optimized for use in DWG/DXF transfers. Colors are supposed to have a 1:1 relationship with AutoCAD's fixed color palette.
- **White.rgb** The colors optimized for use with white Drawing Windows

To use one of the RGB files, go to **Utility/Display/Palettes/LoadRGB**. In the /SUP directory, select one of the available .RGB files. Entities in the Drawing Window will now be displayed with the colors defined in the .RGB file.

To see the true effects of using one of these .RGB files, open the 256CUBE.DC5 from the Samples directory (on the CD or on your hard drive) and then load one of the .RGB files. Refresh the screen and you will see all of the 256 colors as they will be applied in DataCAD.

Adjusting Individual Colors

To adjust any of the 256 colors available to DataCAD, you must change the color of each one and then save the new settings. You can do this within

DataCAD, one color at a time, through any menu that has a **Color** option. Four of those locations are the following:

- **LineType**
- **Layers**
- **Change**
- **Text/Arrows**

To change the current colors, go to one of the above menus and follow these steps:

1. Select **Color/Adjust**.
2. DataCAD prompts you to "Select the color to adjust."
3. Pick one of the standard 15 DataCAD colors, or pick **Custom** to select any of the other 240 available colors by typing in a number from 16 to 255.
4. The Windows color picker appears.
5. Pick a new color and then select OK (if you need help using the color picker, refer to the previous section on "Menu Text Colors in DCADWIN.INI").
6. Repeat the process for all the colors you want to adjust.

To save the current color settings to an RGB file, select **Palettes/SaveRGB**. Save the file with a new name, or select one of the existing RGB files if you want the current changes to be added to it. New RGB files are saved in the **/SUP** directory.

THE INI-OUTI Editor

A couple of years ago I wrote a Windows program to enable users to more easily customize many of the features mentioned so far. Since the program accesses and alters the DCADWIN.INI file, I called it the INI-OUTI Editor. We have included this utility on the CD-ROM.

The most difficult things to change in DataCAD are the interface colors, since you have to use a lot of trial and error to get the triplet color codes for the colors you want. Figures 23-13, 23-14, and 23-15 show the three main pages of the program:

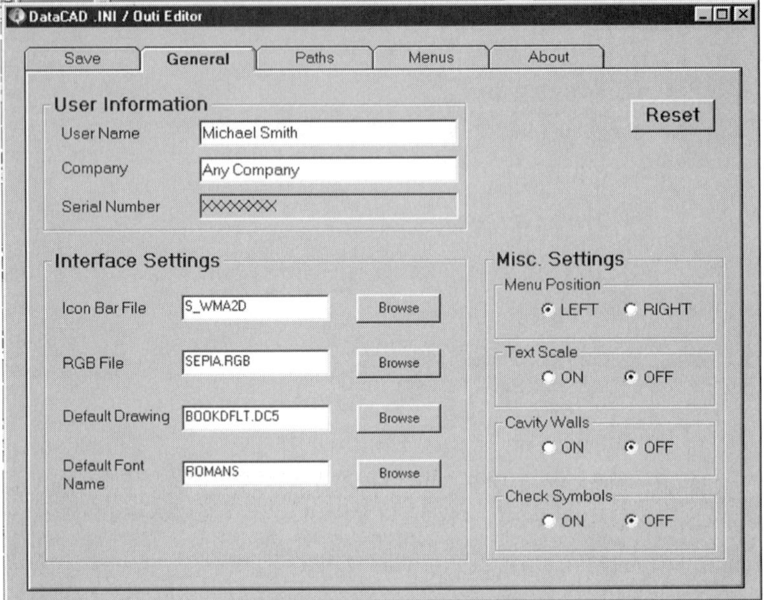

Figure 23-13
The General tab

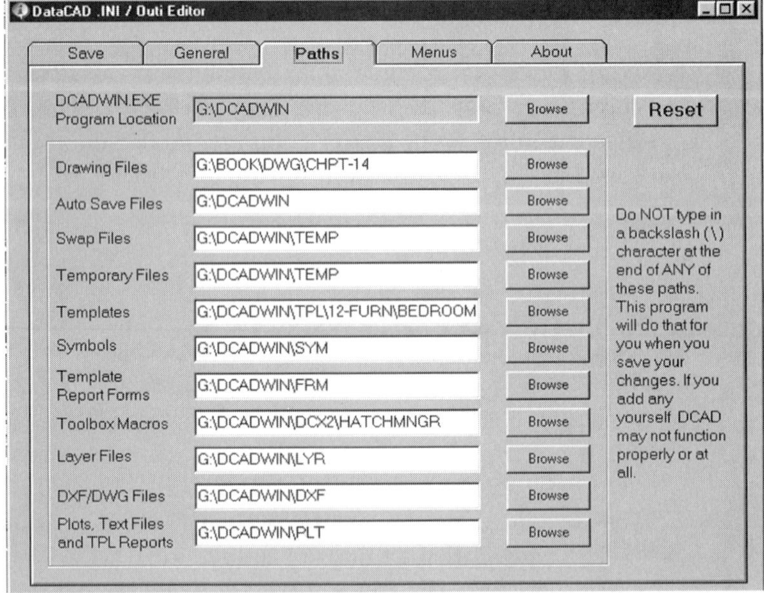

Figure 23-14
The Paths tab

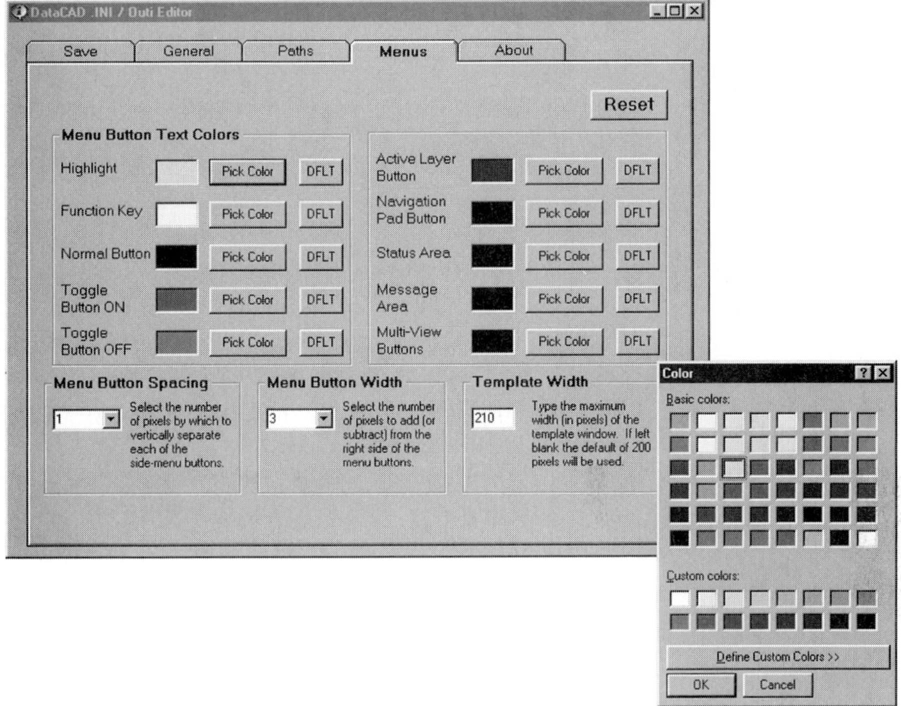

Figure 23-15
The Menus tab

We won't go into any detail about this utility here, but you can read more about it in the README file that is installed with the program.

Customizing the MSG File

Much of the information here was derived from an article written by Evan Shu for his *Cheap Tricks* DataCAD newsletter.

In the \SUP directory is an American Standard Code for Information Interchange (ASCII) text file called DCADWIN.MSG. It contains all the text messages that DataCAD uses to prompt users and display error messages, but two of the lines in this file, lines 140 and 141, assign menu functions to the keyboard interrupts. For instance, it assigns the lowercase **m** to the Move command, and the upper case **M** to the macro Toolbox. Maybe you want the lowercase **m** and the upper case **M** to both access the Move command. By editing the DCADWIN.MSG file, you can force DataCAD to do this.

WARNING: *Editing this particular file is not for the faint of heart, as an improperly edited MSG file can cause DataCAD to act erratically, or not at all. If you choose to edit, be very sure to have made a copy of the file prior to making any changes to it in case you make a mistake.*

For a full Excel spreadsheet of all the available functions that can be edited in these two lines, check out the CD-ROM that comes with this book.

Let's use the **Mm**/Move function as an example. Open the DCADWIN. MSG file in a text editor. You cannot use Windows Notepad for this file because the file is too large, so you will have to use some other text editor like Windows Write, but be very sure when you save the file to save it as plain text, not as formatted text.

The first things you will notice are the two sets of three-digit numbers, starting from 001 and going up to 999. Scroll down to the second set of digits until you see the following two lines:

```
140| npozuds,<.i\+qQ=f/xXh`~|-_wWjJkKmcregalv;:t!@#$%^&*()b{?[]
   '"SAyYVFIM}
141| NPOZUDs,<>i\+qQ=f/xXH`~|-_wWjJkKmCREGaLv;:T!@#$%^&*()B{?[]
   '"SAyYVFIM}
```

Each character in each line has a very precise location, and each character has a very definite meaning and importance, so it's important to change only characters that you are sure about. You did make a backup copy of the file, right?

To change the capital **M** so that it accesses the Move command, go to the first lower case **m** (reading left to right) in line 140 and change it to a capital **M**, like this:

```
140| npozuds,<.i\+qQ=f/xXh`~|-_wWjJkKMcregalv;:t!@#$%^&*()b{?[]
   '"SAyYVFIM}
```

Save the file and you're done, but not all changes are this simple. DataCAD reads each line from left to right and sequentially assigns each function to a key. If a letter or character has already been used, DataCAD will skip any duplicate characters it runs into after reading the first one. For that reason, in order to change or disable a function, you have to substitute a previously unassigned character. Here is an example.

In the standard MSG file, the lower case **i** is assigned to the Digitizer, but few of us use a Digitizer, so let's change things so that the **i** will access the

Identify command, just as the upper case **I** already does. Go to the next-to-last digit of line 140 and change the upper case **I** to a lower case **i**. If you look near the beginning of line 140, you will see that there is already a lower case **i** that is assigned to the Digitizer. Because DataCAD reads this lower case **i** before reading the one you just changed at the end of the line, you need to change this first lower case **i** to some other unused character. In this example, we have changed that first lower case **i** to a left bracket, **{**, as shown here:

```
140| npozuds,<.{\+qQ=f/xXh`~|-_wWjJkKmcregalv;:t!@#$%^&*()b{?[]
  '"SAyYVFiM}
```

Keyboard Shortcuts

Keyboard shortcuts are one- and two-key keyboard commands used to quickly access often-used commands within DataCAD. Instead of digging through menus and submenus with a mouse, you can go directly to the function you want by using the keyboard. Refer to Chapter 5, "Basic Drawing," for more information regarding how these shortcuts work and the various types of shortcuts available. One of the most powerful features of DataCAD is its capability to be customized by the average user, not just by computer programmers.

This section of the book describes how to create your own custom keyboard shortcuts. The two types we are concerned with in this section are those you access by holding down the **Alt** or **Ctrl** key while simultaneously pressing a letter key. In the Appendix, and on the CD that comes with this book, you will find a chart outlining the standard DataCAD keyboard shortcuts along with some custom shortcuts that we use or have developed over the years.

Alt-Key Shortcuts

Any keystrokes you can make while in DataCAD can be recreated for an Alt-key keyboard shortcut. This makes DataCAD's keyboard shortcuts very powerful.

The default keyboard macro file is called DCADWIN.MCR, and is located in your main DataCAD directory in the \SUP folder. Open this file with any ASCII text editor. I usually just use the Windows Notepad program, but any text editor will do. Just make sure that you always save the file as a plain text file with no formatting. Do not use a word processor like Word for Windows, which will save a formatted document rather than one with plain text.

WARNING: *Always make a backup copy of any DCAD files prior to editing them in case you make mistakes or want to reinstall the default files. Call the original DCADWIN.MCR file something like DCADWIN.MC1.*

Here is roughly what you should see when you open your DCADWIN. MCR file. Note that your file will not look exactly the same as this one since it is my own highly customized file. The standard DataCAD .MCR file is pretty boring, and you wouldn't learn as much from it, so that's why we're using our .MCR file here:

```
A^:^F3^F7^F2^S0^S0^S6^S2^
B^:^S5^F4^
C^;^S8^S9^S7^
D^:^S5^S1^S8^S0^
E^:^F3^F1^S8^$^S0^S0^;^
F^:^S3^F8^;^
G^;^S8^S9^S5
H^;^S2^F4^F8^F5^F0^F1^F8^S0^F2^F8^F3^F0^F2^S9^
I^;^S1^$^S8^S0^
J^;^S8^F9^F3^F8
K^;^F8^
L^;^S7
M^:^S5^S1^S3^
N^
O^:^F8^F1^S7^F3^
P^:^S5^
Q^:^F2^S6^F1^S0^
R^;^F9^F4^F1^
S^;^F5^
T^;^S6
U^M^UNLOCK$^F0^S0^S0^
V^;^F0^S6^S7^S6^
W^;^F9^F7^
X^;^F0^S8^F1^S2^$^
Y^:^F3^F9^S0^F6^S7^S7^S7^S7^S7^S5^S4^S3^S2^S1^F0^F9^F8^F7^F6^F5^F4^
F3^F2^F1
Z^;^S8^S9^S5^S5
```

The first capital letter of each line corresponds to the letter you will press (in combination with the Alt key) to accomplish a specific action (since only 26 letters exist in the English alphabet, you can only have a maximum of 26

Alt-key shortcuts). Everything that follows the letter tells DataCAD what do when these keys are pressed. The following section outlines are what some of these symbols mean.

Action Codes

The following is a list of symbols and commands to be used in each line of an Alt+key keyboard shortcut (these are also the same codes used in the Action column when creating custom toolbar icons, which will be covered later in this chapter):

- **; (colon)** Goes to the DataCAD EDIT menu
- **; (semicolon)** Goes to the DataCAD UTILITY menu
- **^ (caret)** A required separator between each command
- **Fx (where x is a number)** This corresponds to the appropriate Function key to press if you were walking through the DataCAD menus to execute a desired DataCAD function.
- **Sx (where x is a number)** This corresponds to the appropriate Shift-Function key to press if you were walking through the DataCAD menus to execute a desired function.
- **$** The same as pressing the **Enter** key
- The blank space between carets (like this ^ ^) is DataCAD's Immediate Mode Command for a keyed in coordinate (for inputting distances or angles); the same as if you pressed the **Space** bar while in DataCAD.
- **ALT+#** By using the combination of the Alt key plus a two- or three-digit number, you can get DataCAD to access a few more menu options and do a few more tricks as well, as described later in this section.
- **M** Use this to accesses a DataCAD Toolbox macro. Place a caret after the M, then type the name of the macro, with a $ on the end of it (see the previous "U" example). If the macro is located in the default \DCX macro directory then you don't have to type a complete path to the macro, but if you have a second toolbox directory, then you need to type the complete path to the macro after the caret.

Let's use one of the shortcuts shown in the .MCR file at the beginning of this section as an example to see how it all works. This keyboard macro will turn Perpendicular Object Snapping on or off:

```
F^:^S3^F8^S0^
```

The **F** means you would press the Alt and F keys together on the keyboard to invoke this macro.

The **:** tells DataCAD to go to the UTILITY menu (you can put a caret in front of it or not; it works either way).

The **S3** means that while in the DataCAD Utility menu you would press the **Shift** and **F3** keys together (**ObjSnap**).

The **F8** means that while in the ObjSnap menu you would press the **F8** key (**Perpend**).

The **S0** means that while in the Perpend menu you would press the **Shift** and **F10** keys together (**Exit** the current menu).

To see how it works, go into DataCAD and press the same keys as those shown in the macro (**Utility** menu, **S3**, **F8**, and **S0**).

You could alter this macro to turn on or off Object Snapping by End Point instead of by Perpendicular. All you would have to do is to substitute the **F8** with **F2** and then save the DCADWIN.MCR file.

You can edit the DCADWIN.MCR file while you are still running DataCAD. Just make the changes in your text editor, save it, and then run the macro in DataCAD to see if your changes to the DCADWIN.MCR file worked properly. This makes troubleshooting your macros much easier than having to exit and reenter DataCAD.

You can make up your own keyboard macros by just copying down the sequence key that DataCAD requires to run something. The trick is to get the correct keys in the correct sequence and not forget simple things like starting with the correct DataCAD menu (**:** or **;**), using the **$** for **Enter**, or adding a space (or a keyed-in coordinate where required).

Here's a final example of a more complicated keyboard macro (see Figure 23-16). It comes from Dale Weiss and was published as one of the "Best of Cheap Tricks" in the January 1997 *Cheap Tricks* newsletter. What it does is automatically zoom you to your cursor location at a scale that is good for entering text. Place your cursor at the area you want to note and then execute the macro (in this case, Alt-A). When you finish with your text entry, hit **P** (previous view) and you're back to your original, overall view.

Note that this macro assumes a default coordinate system of Relative Cartesian. If your default coordinate system is something else, such as Relative Polar, you would have to modify it to work with that type of input.

To see how it all works, go into DataCAD and press each key exactly as you see it in Figure 23-16, starting with the Window In (/) key. Make sure you first press the **Insert** key until your coordinate system is set to Relative Cartesian.

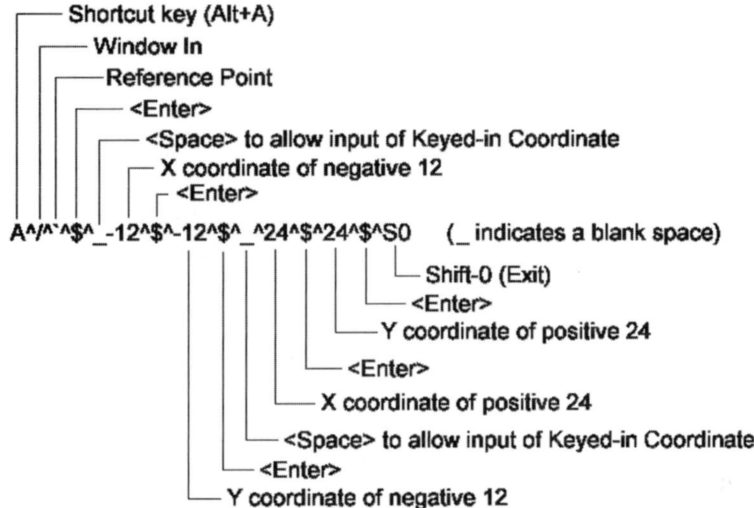

Figure 23-16 A more complicated Alt+key macro

Alt + # Combinations

This feature was defined in the Action Code description earlier in this section, but here is how it fits into keyboard shortcuts. This feature is a throwback to the DOS days of DataCAD but can still be useful. Back when computers had 286 or 386 processors, it took a lot longer for a computer to process information. Graphics cards were also quite a bit slower, so it took time for computers to display screen information. For both of these reasons, DataCAD enabled features like the Alt+key combinations that allowed DataCAD to shortcut some of the processing and screen display chores, thereby making the program run faster. With DataCAD 9, many new combinations have been added that do far more than just speed things up.

These Alt+# sequences are separated by carets, just like all the other keyboard sequences. To input these key sequences in the DCADWIN.MCR file, turn **NumLock** off at your keyboard and then while holding down the Alt key, press the two- or three-digit number on the number pad of your keyboard. It will not work if you use the numbers at the top of the keyboard. Release the Alt key and a funny looking character will be entered in the macro file, such as one of the following:

```
· Í _ º „ ! ú û
```

These are character representations of the Alt+# combination pressed and are different for each number value used. You will also see a different character depending on which text editing program you use. For instance, the symbols look different in Word for Windows than they do in Notepad, or in the DOS Edit program. Be aware this can make troubleshooting difficult.

NOTE: *Windows text editors interpret ASCII key sequences differently than do their DOS counterparts. In most Windows editors, you must prefix the ASCII code with a zero (**0**). In other words, to enter ASCII character 221 you would hold down the **Alt** key and enter **0221**.*

NOTE: *If you are going to use the Alt-key sequences, then you may want to do all your editing in a DOS editor, such as the EDIT.EXE program. Windows has an annoying habit of transposing ASCII values in the system fonts installed on your machine, so the numbers you enter may not work. Using a DOS program to edit the .MCR file will keep this from happening, ensuring that you get the intended results.*

The following Alt+key characters are part of the standard ASCII 437 English Character Set, but because they were developed for DOS computer programs, some key combinations do not work with Windows (or with the Windows version of DataCAD). If you want to use them, you'll just have to experiment a little to see which combinations do and don't work.

The following values with an asterisk (*) denote new functions that have been added to DataCAD 9:

13	[ENTER]	**201**	Turn display of messages off
18	ESC to refresh display list/database	**202**	Turn display of messages on
		203	Set Z offset for symbols
27	[ESC], also Alt-028)	**204***	Turn AndCopy on (all 2D functions)
31	WindowIn/Extents		
32	[Spacebar]	**205***	Turn AndCopy off (all 2D functions)
33	Goto 3D View 1		

35	Goto 3D View 3	**206***	Turn aperture on
36	[ENTER]	**207***	Turn aperture off
37	Goto 3D View 5	**208**	NOT USED
38	Goto 3D View 7	**209**	Arrow left
40	Goto 3D View 9	**210**	Arrow right
41	Goto 3D View 10	**211**	Arrow up
64	Goto 3D View 2	**212**	Arrow down
176	Set selection to entity	**213***	Turn parallel clip cube on
177	Set selection to group	**214***	Turn parallel clip cube off
178	Set selection to area	**215***	Turn layer search on
179	Set input mode to abs. cartesian	**216***	Turn layer search off
180	Set input mode to rel. cartesian	**217**	NOT USED
181	Set input mode to abs. polar	**218**	Page up
182	Set input mode to rel. polar	**219**	Page down
183	Turn display of menus off	**220***	Set window and door input to sides
184	Turn display of menus on		
185	Turn walls on	**221***	Set window and door input to center
186	Turn walls off		
187	Enter walls by centerline	**222***	Turn associative hatch on
188	Enter walls by sides	**223***	Turn associative hatch off
189	NOT USED	**224***	Turn associative dimension on
190*	Turn symbol explode on		
191*	Turn symbol explode off	**225***	Turn associative dimension off

192* Turn symbol dynamic rotate on

193* Turn symbol dynamic rotate off

194* Turn template off

195* Turn geometry/offset on

196* Turn geometry/offset off

197* Turn dynamic on (2D curves and off polygons)

198* Turn dynamic off (2D curves and polygons)

199 NOT USED

200 Turn display of messages off

226* Set 2D polygon = polygon

227* Set 2D polygon = rectangle

228* Turn wall cap on

229* Turn wall clean on

230* Turn wall cap/clean

231* Turn draw jambs on

232* Turn draw jambs off

233* Turn 3D dynamic on

234* Turn 3D dynamic off

Although many of the Alt+Key sequences will access commands that can be accessed in other ways, some commands cannot be accessed in other ways, such as numbers 183, 184, 200, 201, 202, and 209 through 219. Here are two examples of how these can be useful.

Example 1: Message Line Display As DataCAD works its way through an action sequence in a keyboard shortcut, things like the menu buttons and the message line at the bottom of the screen will all be displayed just as if you were selecting individual commands outside of the keyboard macro. All of this extraneous display happens so fast that you only see them quickly flash on your screen. Although it is too fast for you to read, displaying them on your screen does take time. Granted, it may not be long with today's faster computers, but it does take time nonetheless. Although this issue was a big deal in the DOS days, in most cases nowadays considering it is up to you.

To speed things up, DataCAD came up with these Alt+# options:

- **(Alt+183)^(Alt+200)^** Turns DataCAD's Menus & Messages off. Note that each Alt+# sequence is separated by a caret
- **(Alt+184)^(Alt+201)^** Turns DataCAD's Menus & Messages back on. Note that each Alt+# sequence is separated by a caret

You would place the first sequence at the beginning of a keyboard shortcut to tell DataCAD not to display the menus and messages during the shortcut's execution. Then you would place the second sequence at the very end of the shortcut to tell DataCAD to resume displaying menus and messages. Remember do not to type the information. Insert it by pressing the Alt Key and the number pad numbers.

Example 2: PgUp As you know, in DataCAD you can zoom in and out to your drawing by pressing the **PgUp** (Page Up) and **PgDn** (Page Down) keys. You cannot do this inside the DataCAD menus, so if you want to use one or both of these zooming functions during a keyboard macro, you need to use the Alt+218 and Alt+219 sequences. Here's an example:

```
Q^:^ú^F2^S6^F1^S0^
```

Notice the funny symbol (**ú**) after the colon. This is the symbol that represents the Alt+218 function. Remember that depending on which text editor you use, these symbols may appear differently.

In this macro example, when the **Alt+Q** button is pressed, DataCAD will do the following:

1. **:** Access the **Utility** menu.
2. **ú** **PgUp** (zoom out). Hold down the Alt key and press 218 on the keyboard number pad.
3. **F2** GotoView
4. **S6** Update
5. **F1** Selects 3D View #1
6. **S0** Exit

This macro is useful for quickly saving your current screen view as 2D **GotoView** number 1, so that you can work somewhere else in DataCAD and then quickly return to the same view (including the correct layers, etc.) that you were working on. The reason for the **PgUp** command is just to get a better overview of your location. You may not want to add this to your own macro, but we used it here mainly to illustrate the Page Up function.

This particular keyboard macro will only work if you first create a Goto View, since if there is no Goto View saved, then there is nothing to "Update."

Hardwired DataCAD Keyboard Shortcuts

The following is a list of the standard DataCAD keyboard shortcuts. These shortcuts are considered to be "hardwired" into the program since they cannot be altered by the user. Note that we have not listed any Alt+Key shortcuts because those can be, and frequently are, altered by the user. If you want a list of the standard DataCAD Alt+Key shortcuts, as they are installed by DataCAD, you can find it in the Appendix and on the CD-ROM disk. If both an upper and lower case letter are shown together, it means that the command is not case-sensitive and can be invoked with either case letter.

<table>
<tr><td>TAB</td><td>Change active layer</td><td>m</td><td>Move menu</td></tr>
<tr><td>A</td><td>Append Selection Set</td><td>N n</td><td>Object snap</td></tr>
<tr><td>a</td><td>Architect menu</td><td>X</td><td>Object snap menu</td></tr>
<tr><td>+</td><td>Big/small cursor</td><td>Ctrl+F1</td><td>Online help</td></tr>
<tr><td>|</td><td>Cap Walls toggle</td><td>O o</td><td>Ortho mode toggle</td></tr>
<tr><td>\</td><td>Clean T toggle</td><td>(Arrows)</td><td>Pan (up, down, left, right)</td></tr>
<tr><td>K k</td><td>Change color</td><td>Home</td><td>Pan to cursor</td></tr>
<tr><td>?</td><td>Coordinate identification</td><td>P p</td><td>Previous view</td></tr>
<tr><td>C c</td><td>Copy menu</td><td>`</td><td>Reference point</td></tr>
<tr><td>D d</td><td>Dimension menu</td><td>ESC</td><td>Refresh screen</td></tr>
<tr><td>Ctrl+F12</td><td>2D Goto Views</td><td>U u</td><td>Regenerate Display List</td></tr>
<tr><td>F</td><td>Forced save (use this often)</td><td>F12</td><td>Right elevation</td></tr>
<tr><td>;</td><td>Edit menu</td><td>R r</td><td>Rotate menu</td></tr>
<tr><td>S</td><td>Edit Sets menu</td><td>_</td><td>Set overshoot</td></tr>
<tr><td>I</td><td>Identify entity</td><td>s</td><td>Snap grid</td></tr>
<tr><td>,</td><td>Erase last entity</td><td>x</td><td>Snap grid toggle</td></tr>
<tr><td><</td><td>Erase last group</td><td>≅</td><td>Snap point</td></tr>
<tr><td>E e</td><td>Erase menu</td><td>End</td><td>Stop layer refresh</td></tr>
<tr><td>[</td><td>Grid 1 display</td><td>Del</td><td>Stop refresh</td></tr>
<tr><td>]</td><td>Grid 2 display</td><td>B b</td><td>Tangents menu</td></tr>
</table>

G g	Grid commands	T t	Template menu
H h	Hatch menu	.	Undo last entity erase
Insert	Input menu	>	Undo last group erase
Spacebar	Keyed-in coordinate	:	Utility menu
‘	Layer search toggle	=	Walls toggle
L l	Layers menu	/	Window In
F11	Left elevation	Z z	Z Base and Height
-	Line overshoot toggle	Pg Dn	Zoom in
f	Line spacing	Pg Up	Zoom out
Q q	Line type toggle	j	3D Edit menu
W w	Line weight	J	3D Entity menu
“	Load symbol	Y	3D Hide menu
M	Macros menu	V, y	3D Views menu

Customizing the CTRL Key Shortcuts

In the days of DataCAD for DOS, no shortcuts used the **Ctrl** key on the keyboard. In the Windows version, the Ctrl key has been enlisted to provide additional shortcuts to various commands, some of which are standard Windows **Ctrl+Key** shortcuts. Here is a list of the standard DataCAD **Ctrl+ Key** shortcuts (there could be some others by the time the final version of DataCAD 9 is released). The italic text are commands that are unique to DataCAD and not one of the Windows standards.

Ctrl + B	*Batch Plot*
Ctrl + C	Copy
Ctrl + E	*Select (use before **Ctrl**+X or **Ctrl+C**)*
Ctrl + N	New drawing
Ctrl + O	Open drawing
Ctrl + S	Save
Ctrl + P	Printer menu

Ctrl + V	Paste
Ctrl + W	*Turn the multi-view windows ON or OFF*
Ctrl + X	Cut
Ctrl + Y	Redo
Ctrl + Z	Undo
Ctrl + F4	*Exit the current drawing*

Any of the previous shortcuts can be customized by the user, but you should refrain from editing the standard Windows shortcuts, lest you really confuse someone else using your computer. Here's how it works. In the \SUP folder within your main DataCAD directory is a file called DCADWIN.MNU. This, like many other support files in DataCAD, is a simple text file. If you open the file with a plain text editor like Windows Notepad, the beginning of the file will look like the following:

```
File
&New|N
&Open|O
&Close
Clos&e All
&Save|S
Save &As...
Sa&ve All
&Import
&Export
```

All of these are Windows standards. Some of the lines without the ampersand (**&**), like **File**, are the titles of the drop-down menu options at the top of the DataCAD screen. All the other lines, with or without the ampersands, represent the various menu options that you can *click* on. The lines with the ampersands in them, either at the beginning of the word or within it, are options that can be accessed with the keyboard as well as the mouse. The lines with the vertical pipes (**|**) represent commands that can be accessed with **Ctrl+key** combinations. So the piece of the .MNU file shown previously will display the part of the **File** menu shown in Figure 23-17.

Let's dissect the **&Save|S** line to see how it works. In the drop-down DataCAD menus at the top of the screen, the ampersand (**&**) tells DataCAD to place an underscore under the letter that follows it. As you can see in the previous image, the **S** is therefore underlined. Whenever the **Save** option is displayed, the underscore tells DataCAD that if the **S** key is pressed, the Save command should be initiated.

When a menu option is not visible, the **|** (pipe) character (on the same key as the backslash, \) with a letter after it tells DataCAD which letter

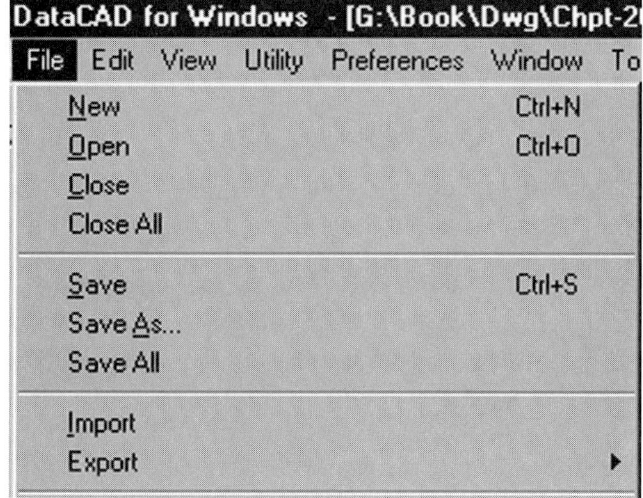

Figure 23-17 Part of the actual File menu

will activate that command when used in combination with the **Ctrl** key. So in the Save example, the **|S** means that when you simultaneously press the **Ctrl** key with the letter **S**, DataCAD will save the current drawing. Likewise, pressing **Ctrl+O** will open a new drawing.

NOTE: *The letter pressed with the **Ctrl** key is not case-sensitive. The function will work whether you press **Ctrl+S** or **Ctrl+s**.*

Now let's customize one of these existing options. The DataCAD shortcut of **Ctrl+W** will open and close the Multi-View Windows (see the **Window** drop-down option). So if we change the W to an M in the **&Multi-View|W** line, it will then read **&Multi-View|M**. Now pressing **Ctrl+M** will open and close the Multi-View Windows.

Now let's create a new shortcut (you can only create shortcuts of the menu options that are already in the .MNU file). Near the bottom of the DCADWIN.MNU file is a line that reads:

```
DataCAD &Help
```

There is no | symbol and no letter at the end, so there is no shortcut to the help command. To change this so that pressing **Ctrl+H** will activate the DataCAD Help feature, change the line to read

```
DataCAD &Help|H
```

The |H added to the end of the line is what makes it work. The &H tells DataCAD to put an underscore under the letter H, but this actually has no bearing on the **Ctrl+H** function. You could just as well have put the ampersand before the D in DataCAD to underline the D, in which case **Ctrl+H** would still activate the Help option. Also notice that in the drop-down menu, DataCAD has automatically added the text **Ctrl+H** to the right of the Help command (see Figure 23-18).

Like the Alt+key shortcuts, because only 26 letters exist in the alphabet, you can make only 26 shortcuts, but try to keep your shortcuts pneumonic. For instance, **O** is for open, **S** is for save, and so on. It will make your shortcuts far more useful if they are easy to remember. Remember that you can only have one of each letter in the DCADWIN.MNU file and can only customize menu text that is part of the original DCADWIN.MNU file—so you cannot add your own.

Custom Icon Toolbars

If you learned how to create custom keyboard shortcuts, then that will make the explanation of icon bars much easier, because it works on the same principle. It is worth reading that section now if you haven't already. DataCAD 6 (for DOS) introduced user-definable icon toolbars. Instead of being limited to only 26 keyboard shortcuts, you could now use graphic icons across the top of your DataCAD screen to invoke whatever commands or macros you would like. DataCAD comes with two standard icon toolbars, one for 2D work and one for 3D.

Figure 23-18 Adding the `Ctrl+H` shortcut for the Help menu

The real power of these toolbars is their capability to be customized like crazy. You can even have a toolbar icon bring up an entirely new icon toolbar, so that you can have a limitless number of icons if you so choose. You can also create your own graphic icons and then define what they will do. The process of defining what each button will do is nearly identical to the keyboard shortcut process described earlier. So if you can do that, then you can do this.

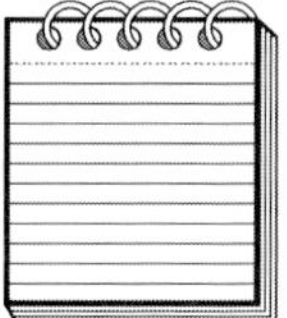

NOTE: *The Productivity Pack of custom toolbar icons and .KEY files (those are the files that tell DataCAD what to do when each icon is pressed) is one of the options in the DataCAD installation. Hundreds of additional custom files can be found in the productivity pack, all of which can be cut and pasted to create your own toolbars. So once you get the hang of creating your own icon toolbars, check out the Productivity Pack. If you see something you want to change, by all means, go ahead!*

Our office uses two primary icon toolbars, one for 2D and one for 3D work. We only make one toolbar for each because we feel that too much clicking through multiple menu bars defeats the purpose of having the menu bars in the first place. One exception to this rule might be to have a third icon bar (or more) to access certain standard office defaults. For instance, you might make an icon bar with options for standard text and dimension settings. Each icon button could be set for text or dimensions at scales of $^1/_8$, $^1/_4$, $^1/_2$, or 1″. Each one would be set up to make your entities of the correct size, color, orientation, and all other optional settings. You will have to be your own judge of how many toolbars are enough.

Each icon of the icon bar is a .BMP (Windows bitmap) image with a common size. The standard size is 24 pixels by 24 pixels, but you can make them larger or smaller as long as they are all the same size. If you create more icons than will fit across the top of one screen, scroll arrows will appear to allow you to access the icons off the screen. The number of icons that fit across your screen without having to scroll over to see more of them depends on the resolution of your computer's display. The total maximum number of icons allowed in any one icon bar is 50. If you try to use more than this, then DataCAD will not start.

The icon toolbars consist of two parts: a text file with a .KEY file extension that tells DataCAD what to do when an icon is pressed, and a bitmap-format (.BMP) graphic icon. If you have not already done so, you should read the previous section in this chapter regarding the creation of keyboard shortcuts, since the commands in the .KEY file are nearly identical.

.KEY Files

A .KEY file is a simple text file that tells DataCAD what to do when a toolbar icon is clicked. It is located in the DataCAD \SUP\MENUPOF directory. One .KEY file exists for each icon toolbar, so you will probably have two or more .KEY files in the MENUPOF directory. DataCAD comes with two primary .KEY files named DCADWIN.KEY (for 2D drafting icons) and DCADWIN3.KEY (for 3D construction icons). You can either modify these existing files with new icons and commands, or you can create new ones. We would suggest you copy these files and alter them to create new ones. That way you will always have the originals to fall back on.

The Anatomy of a .KEY File

They say that the best way to make a small fortune is to start with a large one, and likewise, the best way to make a new .KEY file is to start with an existing one. This is especially true since there is a very precise format to a .KEY file. I started with the custom .KEY file and icons from the *Productivity Pack* and from the *Cheap Tricks Ware Icon Bar Collection* by Rick Ferrara. I then substituted some of the icons from those files with some others that I liked better and created new ones using Paint Shop Pro (more on this later).

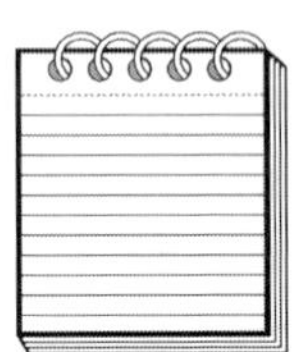

NOTE: *As always, before making any changes to important files like the icon toolbar .KEY files, be sure to make a backup copy of the files before editing them.*

I will use our 2D icon bar as an example. Unlike the keyboard macro file, you can name your icon bar files anything you want, as long as they end with the file extension of .KEY. However, the format for the .KEY file is so exact that you cannot deviate from it by even one space or it will not work. All of the pipes (|) are important and must align, which is a good way to verify that your spacings are correct. Don't forget the last pipe at the end of the Message column.

Opened with the Windows Notepad program, an entire .KEY file would look like Figure 23-19 if you opened it full-screen. Notice that all the pipes are perfectly aligned.

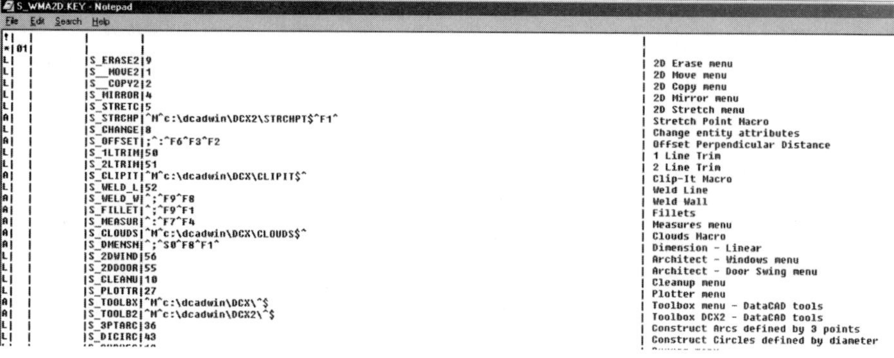

```
S_WMA2D.KEY - Notepad
File Edit Search Help
!|  |        |        |                                       |
*|01|        |        |                                       |
L|  |        |S_ERASE2|9                                      | 2D Erase menu
L|  |        |S__MOVE2|1                                      | 2D Move menu
L|  |        |S__COPY2|2                                      | 2D Copy menu
L|  |        |S_MIRROR|4                                      | 2D Mirror menu
L|  |        |S_STRETC|5                                      | 2D Stretch menu
A|  |        |S_STRCHP|^M^c:\dcadwin\DCX2\STRCHPT$^F1^        | Stretch Point Macro
L|  |        |S_CHANGE|8                                      | Change entity attributes
A|  |        |S_OFFSET|;^:^F6^F3^F2                           | Offset Perpendicular Distance
L|  |        |S_1LTRIM|50                                     | 1 Line Trim
L|  |        |S_2LTRIM|51                                     | 2 Line Trim
A|  |        |S_CLIPIT|^M^c:\dcadwin\DCX\CLIPIT$^             | Clip-It Macro
L|  |        |S_WELD_L|52                                     | Weld Line
A|  |        |S_WELD_W|^;^F9^F8                               | Weld Wall
A|  |        |S_FILLET|^;^F9^F1                               | Fillets
A|  |        |S_MEASUR|^:^F7^F4                               | Measures menu
A|  |        |S_CLOUDS|^M^c:\dcadwin\DCX\CLOUDS$^             | Clouds Macro
A|  |        |S_DMENSN|^;^S0^F8^F1^                           | Dimension - Linear
L|  |        |S_2DWIND|56                                     | Architect - Windows menu
L|  |        |S_2DDOOR|55                                     | Architect - Door Swing menu
L|  |        |S_CLEANU|10                                     | Cleanup menu
L|  |        |S_PLOTTR|27                                     | Plotter menu
A|  |        |S_TOOLBX|^M^c:\dcadwin\DCX\^$                   | Toolbox menu - DataCAD tools
A|  |        |S_TOOLB2|^M^c:\dcadwin\DCX2\^$                  | Toolbox DCX2 - DataCAD tools
L|  |        |S_3PTARC|36                                     | Construct Arcs defined by 3 points
L|  |        |S_DICIRC|43                                     | Construct Circles defined by diameter
```

Figure 23-19 An entire .KEY file

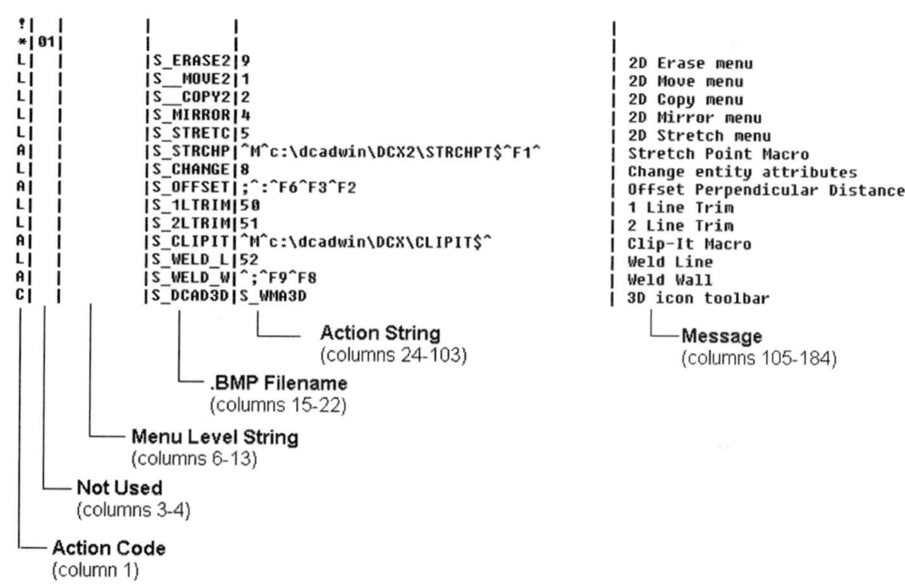

```
!|  |        |        |                                       |
*|01|        |        |                                       |
L|  |        |S_ERASE2|9                                      | 2D Erase menu
L|  |        |S__MOVE2|1                                      | 2D Move menu
L|  |        |S__COPY2|2                                      | 2D Copy menu
L|  |        |S_MIRROR|4                                      | 2D Mirror menu
L|  |        |S_STRETC|5                                      | 2D Stretch menu
A|  |        |S_STRCHP|^M^c:\dcadwin\DCX2\STRCHPT$^F1^        | Stretch Point Macro
L|  |        |S_CHANGE|8                                      | Change entity attributes
A|  |        |S_OFFSET|;^:^F6^F3^F2                           | Offset Perpendicular Distance
L|  |        |S_1LTRIM|50                                     | 1 Line Trim
L|  |        |S_2LTRIM|51                                     | 2 Line Trim
A|  |        |S_CLIPIT|^M^c:\dcadwin\DCX\CLIPIT$^             | Clip-It Macro
L|  |        |S_WELD_L|52                                     | Weld Line
A|  |        |S_WELD_W|^;^F9^F8                               | Weld Wall
C|  |        |S_DCAD3D|S_WMA3D                                | 3D icon toolbar
```


Figure 23-20 The components of a .KEY file

Figure 23-20 shows a few lines from my own 2D .KEY file (called WMA-2D. KEY). We have shortened the number of blank spaces in the Action String column to fit it across the page. Notice that the first required line of a .KEY file is blank except for the exclamation (!) and the pipes (|), which must be located in exactly the correct columns. The second required line is similar to the first, but it has an asterisk at the beginning and the number 01 between the first two pipes. Always make sure that these are the first two lines in a .KEY file. After those two lines, you can start to write your own custom lines.

NOTE: *Within the .KEY files, you cannot use **Tabs**. Every blank space must be an individual blank space entered with the keyboard **Space** bar.*

Because each line of a .KEY file is 185 columns in length, we can't fit a whole line in its entirety across the page of this book. For that reason, the columns are shown compressed, so that you can see the columns and what's in them (see Figure 23-20).

The following information is from the DataCAD Reference Manual with a few additional comments, and it refers to the columns shown in the .KEY file shown earlier.

Column	Name	Function
1	Action Code	This contains an Action Code that identifies the function or program to access. It tells DataCAD what to expect to see in the Action String column. Both of these columns must agree with one another or the icon function will not work.
3–4		Not currently used. Leave this blank.
6–13	Menu Level String	This contains up to an eight-character description for the Message Window. Any characters are valid and you can enter whatever you want as long as it fits.
15–22	.BMP filename	This contains the .BMP file name (without the .BMP file extension). This is the graphic picture that makes up an icon. It can only be eight characters in length.
24–103	Action String	This contains the action to perform, such as the menu or macro to access or the keyboard sequence to perform. This is the heart of the .KEY file.
105–184	Message String	Whatever is typed in here will be displayed in two locations when the cursor moves over the icon: a yellow tag under the icon and the coordinate readout line. You can use up to 79 characters, but you probably want to keep it short and readable.

Just as when you were editing the DCADWIN.MCR file for keyboard macros, you can edit and troubleshoot the .KEY files while you are still in DataCAD by using Windows Notepad or another plain text editor.

Action Code Column

Action Codes are the single characters that go in column 1 of the icon .KEY file. You must make sure that this code matches the action you will be performing in the Action Column (columns 24 through 103). Failure to coordinate these two columns is the most common cause of failure for a macro not to execute. If you *click* on an icon in DataCAD and nothing at all happens, check to make sure these two columns agree with one another. Again, remember there is an Action Code column and an Action Column. Here are all the Action Codes that DataCAD recognizes:

! This provides a method for you to place comments or to leave a note for yourself within the key file. DataCAD takes no action on the comments after this symbol. Here's an example:

```
! The previous line will execute the 2D erase function.
```

A This executes a macro key sequence, just like a keyboard macro. A key sequence, such as ^;^F0^S8^F1^S2^$^, calls function keys, or keystrokes, in this order. When making a key sequence to access a toolbox macro, enter the name of the macro as it appears in the DCX directory (without the .DCX extension), followed by a dollar sign (**$**). Here are two examples:

```
A|  |          |CHGTXT   |^;^F0^S8^F1^S2^$^      | Change text
   contents
A|  |          |ROOFIT   |^M^c:\dcadwin\dcx2\ROOFIT$^ | Roof It
   macro
```

C This is used to load a new .KEY file for a new toolbar. It can be up to eight characters in length, and must match the name of the .KEY file it is to access. This is a great feature because it lets you easily display a new icon bar by pressing an icon on your present toolbar, rather than having to dig through DataCAD's menus (**Tools/Program Preferences/Misc**) to change to a new icon bar.

```
C|  |          |3D_KEY   |DCADWIN3                | 3D icon toolbar
```

(DCADWIN3 [.KEY] is the name of the key file that you want to load when the 3D_KEY[.BMP] icon is pressed.)

K This executes a hardwired internal DataCAD keyboard shortcut by clicking on an icon. Using this, you can assign one of nearly any of DataCAD's keyboard shortcuts to an icon. For example, in DataCAD, the lower case **a** keyboard shortcut takes you to the **Architect** menu. By entering the letter a in the Action String column (not an upper case **A**; these characters are case-sensitive) of the .KEY file, you would access the **Architect** menu, like this:

```
K|  |          |ARCHTCT  |a                  | Architect menu
```

With the K action code, you cannot use the Alt, Tab, or arrow keys in the Action String column.

L This provides access to an internal DataCAD function, such as 2D Erase. In the Action String column, you will simply type in a number from 1 to 117 from the Internal Function Codes list shown later in this section. Here is an example:

```
L|  |          |ERASE    |9                         | 2D Erase
```

M This executes a Toolbox macro located in the standard \DCX macro directory. In the Action String column, you will type the name of the macro exactly as it appears in the DCX directory, but without the .DCX file extension:

```
M|  |          |ROOFIT   |Roofit                  | Roof It macro
```

See the more in-depth discussion of this feature later in this section, which discusses instances when you might want to use the **A** Action Code method instead of the **M** Action Code.

T This provides access to a template file. In the Action String column, you would type in the letters **TPL**, followed by a backslash, and then the location and template file filename. A comma is required between the path and filename. Note that the filename must include the file extension, **.TPL**. In the following first example, the \TPL path exists within the DataCAD root directory, so you do not have to type the full path. In the second example, the \TPL path exists on a network server with the drive letter T:, so you have to type the full path. More on this later in this section.

```
T|  |   |CONC    |TPL\03-CONC,CONCRETE.TPL          | Concrete
   template
T|  |   |CONC    |TPL\T:\TPL\03-CONC,CONCRETE.TPL    | Concrete
   template
```

Menu-Level String Column

This column contains up to an eight-character description for the Message Window. Any characters are valid and you can enter whatever you want. This description appears in the Message Window at the bottom of your screen and will appear just in front of the message that you place in the Message columns (columns 105 to 184). Both messages display when you place your cursor over the icon on the menu bar.

.BMP Filename Column

This column contains a .BMP file name of up to eight characters (with no .BMP file extension). This is the graphic picture that makes up an icon. An explanation of how to create them is covered later in the chapter.

Action String Column

The Action String is the heart of the .KEY file, specifying exactly what DataCAD is to do when an icon is pressed. Its format is nearly identical to the format of a keyboard shortcut, covered earlier in this chapter. One of the shortcomings of the icon .KEY files is that the total number of characters allowed in the Action String column is limited to 80. That may sound like a lot, but plenty of macro instructions could fill all of that space and still require more.

Like keyboard macros, any keystrokes possible while in DataCAD can be recreated in the Action String column of a .KEY file, so this is where a working knowledge of keyboard macros comes in. All the symbols and commands are the same, with a few additional tricks added for good measure. The Alt+# Key sequences can be used in the Action String Column as well. See the lengthy list earlier in the chapter, in the Keyboard Macro Section. The next section discusses what can go into the Action String.

Action String Codes

When using the **A** Action Code in column 1, you will use the following symbols and commands in the Action String column. You cannot use these commands in conjunction with the other Action Code letters. Note that these are precisely the same codes used for writing keyboard shortcuts, covered earlier in this chapter.

- **;** (colon) Goes to the DataCAD Edit menu
- **:** (semicolon) Goes to the DataCAD Utility menu
- **^** (caret) This is a required separator between each command.
- **Fx** (where x is a number) This corresponds to the appropriate Function key to press if you were walking through the DataCAD menus to execute a desired DataCAD function.

- **Sx** (where x is a number) This corresponds to the appropriate Shift-Function key to press if you were walking through the DataCAD menus to execute a desired function.
- **$** The same as pressing the **Enter** key.
- M Use this to accesses a DataCAD Toolbox macro. Place a caret after the M, then type the name of the macro, with a $ on the end of it. If the macro is located in the default |DCX macro directory then you don't have to type a complete path to the macro, but if you have a second toolbox directory like I do, then you need to type the complete path to the macro after the caret.

A blank space between carets (like this ^ ^) is DataCAD's Immediate Mode command for a keyed-in coordinate (for inputting distances or angles), which is the same as if you pressed the **Space** bar while in DataCAD.

Weld Wall and ClipIt Macro are two examples of .KEY file lines using action codes. We have shortened the number of blank spaces in each line so that it will fit across the page. If you're not sure what's going on in each file, go back and review the section of this chapter regarding keyboard shortcuts.

Weld Wall This example runs the Weld Wall function located in the **CleanUp** menu.

```
A|  |WeldWall|WELD_WL |^;^F9^F8                    | Weld Wall
```

ClipIt Macro This example accesses the **ClipIt** macro located in the macro toolbox (in the **\DCX2** directory) and picks the **Area** selection method.

```
A|  |ClipIt   |CLIPIT   |^M^c:\dcadwin\DCX2\ClipIt$^F1^    | Clip
   It Macro
```

You can also have the toolbar insert common text strings or extended characters into your drawings by placing the text in the Action String area. This can be useful for inserting common and repetitive notes in your drawings with just the *click* of a button.

To do so, start by typing a caret ^ symbol in the Action String portion of the file and then type the text you want to insert. You can use an optional **$** character to insert linefeeds between text (see the following example).

To use this feature, place the cursor where you want the text to be located while you are in Text mode and *click* the mouse as you normally would to begin typing text. Now go up to the icon toolbar and press the toolbar button where you put your text string. The text will be added to the drawing. In this example, if you put this string into the .KEY file,

```
^TEST$TEST$TEST$
```

you will get the following result:

```
TEST
TEST
TEST
```

Notice that the text is on three separate lines due to the **$** characters forcing a line feed. You are limited in that the Action String can only be 80 characters, depending on your use of line feeds. Of course, you can also use multiple buttons to input longer strings of text.

Internal Function Codes

When using the **L** Action Code in column 1, you must use one of the following numbers in the Action String column. You cannot use these commands in conjunction with the other Action Code letters, and you cannot use more than one code in each line of the .KEY file. Each of these internal function codes accesses a specific internal DataCAD function, such as MOVE. In the following list, 2DMOVE would be Internal Menu Function Number **1**. 3DMOVE would be number **98**.

1 - 2d move	41 - ellipse	81 - sphere
2 - 2d copy	42 - rad_circle	82 - torus
3 - 2d rotate	43 - dia_circle	83 - mesh surface
4 - 2d mirror	44 - 3pt_circ	84 - rev_surface
5 - 2d stretch	45 - 2d move drag	85 - marker (3D)
6 - 2d enlarge	46 - 2d rect array	86 - 3d polygon
7 - identity	47 - 2d circ array	87 - slab
8 - 2d change	48 - fillets	88 - horizontal polygon
9 - 2d erase	49 - chamfer	89 - horizontal slab
10 - cleanup	50 - 1 line trim	90 - rectangular polygon
11 - architect	51 - 2 line trim	91 - rectangular slab
12 - 2d polygons	52 - weldline	92 - vertical polygon
13 - curves	53 - T intersection	93 - vertical slab

14 - text	54 - L intersection	94 - inclined polygon
15 - linetype	55 - architect door	95 - inclined slab
16 - hatch	56 - architect window	96 - voids
17 - 3d edit	57 - cut wall	97 - partial
18 - window in	58 - divide	98 - 3d move
19 - to_scale	59 - intersection	99 - 3d copy
20 - geometry	60 - tangents	100 - 3d rotate
21 - 2d gotoview	61 - selection sets	101 - 3d enlarge
22 - grids	62 - 3d viewer	102 - 3d erase
23 - layers	63 - elevation	103 - 3d stretch
24 - template	64 - view controls	104 - 3d change
25 - 2d settings	65 - edit plane	105 - 3d explode
26 - measure	66 - plane snap	106 - 3d edit
27 - plotter	67 - set perspective	107 - 3d entity
28 - dimension	68 - walk thru	108 - quick shader
29 - display	69 - set oblique	109 - linear dimension
30 - object snap	70 - 3d gotoview	110 - layer on/off
31 - freehand	71 - save image	111 - layer name
32 - link entities	72 - clip cube	112 - layer delete
33 - file I/O	73 - hide	113 - DXF read
34 - directory	74 - 3d line	114 - DXF write
35 - 2pt_arc	75 - 3d settings	115 - pixel out
36 - 3pt_arc	76 - blocks	116 - toolbox
37 - cent_ang_arc	77 - vert cylinder	119 - undo
38 - cent_arc	78 - horz cylinder	120 - redo
39 - cent_chord	79 - cone	
40 - arc_tangent	80 - trunc cone	

Hundreds of DataCAD commands exist, but Internal Function Codes only access 118 of them. If you don't see a function here in this list, then you will have to access it through a series of Action Codes, covered earlier in this section.

Here is an example of a .KEY file line using an internal function code. We have shortened the number of blank spaces in each line so that it will fit across the page. If you're not sure what's going on in each file, go back and review the section of this chapter regarding keyboard shortcuts.

Text This example will take you directly to the **Text** menu.

```
L|  |Text    |TEXT    |14                        | Text menu
```

M is for Macro

As described earlier, you can use an **M** in the Action column to execute a toolbox macro located in the standard \DCX macro directory:

```
Ex: M|  |        |ROOFIT  |Roofit                | Roof It macro
```

The advantage of using the **M** action code rather than using the **A** action code method is that it lets you temporarily interrupt what you are currently doing to run the macro. Upon exiting the macro, you are returned to the exact state you were in when you clicked on the macro icon. For instance, if you were in the middle of entering text from the **Text** menu and you then clicked on an icon to run a macro, upon exiting the macro you would be returned back to the **Text** menu. With the **A** action code sequence, you would be returned to the main **Edit** menu instead.

However, you may sometimes want to use the **A** Action Code method instead of the **M** Action Code at least two reasons. First, if you use the **M** Action Code to access an icon, then you must exit the macro by right-clicking the mouse or pressing the **S0 Exit** menu button. If you don't, then you can sometimes temporarily lose the use of all the icon bar icons. The **A** Action Code method does not have this drawback.

The second reason is that, if you have more than one macro toolbox, you will need to use the **A** Action Code method so that you can type the entire path to the macro directory. For instance, I keep my most often used macros in the main \DCX directory, but I keep the lesser used ones in another directory called \DCX2. If I use the **M** Action Code, then DataCAD will not be able to find the second macro directory. If the **Roof It** macro were in my \DCX2 directory, then I would use this line instead:

```
A|  |        |ROOFIT  |^M^c:\dcadwin\dcx2\ROOFIT$^     | Roof It
  macro
```

.KEY File Examples

In this section, we'll be accessing "hardwired" DataCAD commands in a .KEY File. Refer to the following example from the beginning of a single line in the earlier .KEY file:

```
L|  |           |ERASE        |9
```

The letter L is the Action Code. It indicates that what is in the Action column is a code to access an internal DataCAD function, in this case the 2D Erase command. The ERASE text is the name of the .BMP image of the icon toolbar button to be displayed. The number 9 is DataCAD's own internally hardwired code for 2D erase.

Accessing Other DataCAD Commands in a .KEY File

Refer to the following example from the beginning of a single line in the earlier .KEY file:

```
A|  |           |OFFSET      |^:^F6^F3^F2
```

The letter A is the Action Code. It indicates that what is in the Action column is a series of Action Codes (refer to the earlier list) to access various DataCAD functions. Refer to the earlier section in this chapter regarding keyboard shortcuts if you're not sure how to do this. The OFFSET text is the name of the .BMP image of the icon toolbar button to be displayed.

Accessing Toolbox Macros in a .KEY File

Refer to the following example from the beginning of a single line in the earlier .KEY file:

```
A|  |           |3dtools  |^M^3DTOOLS$^F3^F2^F5^
```

The letter A is the Action Code. It indicates that what is in the Action column is a series of Action Codes (refer to the earlier list) to access various DataCAD functions. The first 3Dtools text is the name of the .BMP image of the icon toolbar button to be displayed. The letter M is another Action Code that tells DataCAD to access the Toolbox macro after the next carret, which in this case is the 3DTOOLS macro. Just as with keyboard macros, you must place the dollar sign ($) directly after the name of the macro.

However, you don't have to stop there. The next three commands, F3, F2, and F7, correspond to the menu buttons found in the macro when DataCAD runs it. In this example, DataCAD runs the 3DTOOLS macro and then executes the commands represented by the F3 (3DPlane), F2 (Slabs), and F5 (Voids) menu buttons.

The previous example assumes that the Toolbox macro you are accessing is in the \DCX directory within your main DataCAD directory. If you want to access a macro in a different directory, then you need to include the path to the directory before the name of the Toolbox macro. In this example, the CLOUDS macro is located in a directory called \DCX2.

```
A|   |          |Clouds    |^M^C:\DATACAD\DCX2\CLOUDS$^F0^F9^
```

Now you might ask why you would place macros in a location other than the standard \DCX directory. For me, it's simply a matter of tidiness. I have so many macros that it gets difficult to search through them all when I access the toolbox. So I place my less often used macros in a directory called \DCX2. Then I keep two toolbox icons on my icon toolbar, one for my primary toolbox, and one for the secondary (see Figure 23-21).

Templates from a .KEY File

You can also access DataCAD templates from the icon toolbar. Just like accessing Toolbox macros, there are slightly different ways to access templates—depending on whether they are located in the default DataCAD\TPL directory or if they are somewhere else, such as on your network server.

To access templates that are located in the default template directory (the \TPL subdirectory in your DataCAD directory), use this format:

```
T|   |          |CONC      |TPL\03-CONC,CONCRETE.TPL
```

Figure 23-21 The two toolbox icons (the second one has a "2" over it)

The letter T is the Action Code. It indicates that what is in the Action String column is a template (.TPL) file. The CONC text is the name of the .BMP image of the icon toolbar button to be displayed. Notice that the format of the commands in the Action String column are a bit different than the ones used to access Toolbox macros. Start the Action String column with the letters TPL, which again tell DataCAD you are going to access a template. Now add the path of the directory containing the template file. In the example, the CONCRETE.TPL template is being accessed, and it is located in the DataCAD\TPL\03-CONC directory. Next comes a comma followed by the name of the template to be opened, including the .TPL extension at the end.

If that same CONCRETE.TPL template were located somewhere other than the main DataCAD\TPL directory, then you would need to give the full path to the template file. In this example, the CONCRETE.TPL file is located on a network server whose drive letter is T: in a directory called TPL\03-CONC. Note that you must still start the Action String column with the letters TPL. This has nothing to do with the path to the template. It is simply part of the format that DataCAD requires to access the template.

```
T|  |        |CONC    |TPL\T:\TPL\03-CONC,CONCRETE.TPL
```

Symbols from a .KEY File

To access an individual symbol rather than a whole template, you need to use the **A** Action Code method with a specific format in the Action String column. Also, similar to the template access method described earlier, accessing symbols can be done in different ways, depending on whether they are located in the default DataCAD\SYM directory or if they are somewhere else.

To access symbols that are located in the default symbol directory (the \SYM subdirectory in your DataCAD directory), use this format:

```
A|  |        |IntElev |^"^SYM\DWGSYMB\INTELEV$^
```

The letter A is the Action Code. It indicates that what is to come in the Action String column is a symbol (.SM3). The IntElev text is the name of the .BMP image of the icon toolbar button to be displayed. Start the Action String column with a carret, followed by quotation marks ("), followed by another carret. This tells DataCAD that you are going to access an individual symbol. Now add the path of the directory containing the template file. In the example, the INTELEV [.SM3] symbol is being accessed, and it is located in the DataCAD\SM3\DWGSYMB directory. Be sure to add the dollar sign to the end of the symbol name.

If that same INTELEV symbol were located somewhere other than the main DataCAD \SYM directory, then you would need to give the full path to the symbol. In this example, the INTELEV symbol is located on a network server whose drive letter is S: in a directory called SYM\DWGSYMB:

```
A|  |         |IntElev |^"^S:\SYM\DWGSYMB\INTELEV$^
```

Graphic Toolbar Icons

Now it's time to make the graphic icons that make up the icon toolbars. The filenames of these icons are what go in the .BMP columns 15 to 22. These icon names can only be up to eight characters, and the .BMP extension is not included in these columns.

To create icons, all you need is a graphics program that can show you each pixel enlarged so that you can edit each one individually. It must also be able to allow you to easily pick one of the standard 255 colors supported by DataCAD. I have found *Paint Shop Pro* (PSP) to be a good one. It is a shareware program and is included on the CD that comes with this book.

I find the best way to make a new icon is to start with an existing one and then edit it as you see fit. Find an existing icon in the DataCAD \SUP\MENUPOF directory that is similar to the new one you want to create. Then start the PSPprogram and open the .BMP icon file. Make sure you use the **Save Copy As** function and give your new icon a new name.

Zoom into the icon a couple of times until you can see the individual pixels (squares) that make up the image (see Figure 23-22). Now you can edit each individual pixel, zooming back out to see what the icon will look like at its actual size. If you want to make a new icon of a different size, open a new file in PSP and specify the size in pixels.

Standard icons are 24 x 24 pixels in size, but you can make them a little smaller or a little larger. Smaller icons will enable you to fit more icons across the top of your screen, while larger icons will allow fewer icons. The actual number of icons that will fit across the screen will also vary, depending on the resolution of your computer screen. Keep in mind, however, that smaller icons have fewer pixels, so it may be more difficult to fit all your graphics in them, and most important, all the icons in a toolbar must be the exact same size. If they are not, then some of the icons will display as blank,

Figure 23-22
Zooming into an icon in PSP

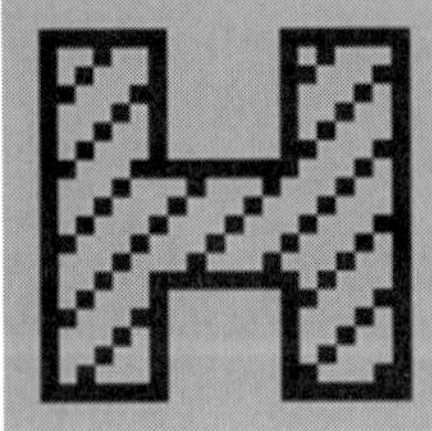

Figure 23-23
An icon example without text

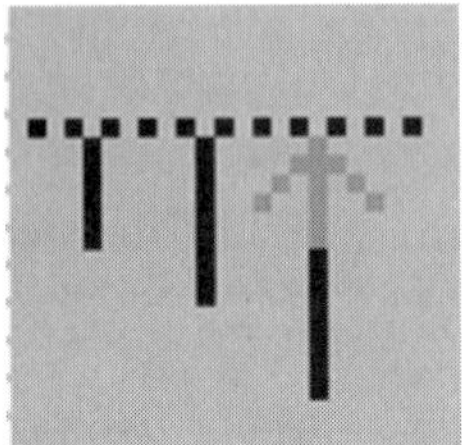

Figure 23-24
An icon example without text

gray buttons. You can test this out yourself by changing the size of the first icon to something a little smaller or larger than the rest of them.

Your icons do not have to have pictures on them. You can simply draw text. Often a picture is worth a thousand words, but sometimes it's just easier to use text. Figures 23-23 through 23-26 are some icons that I have created over the years.

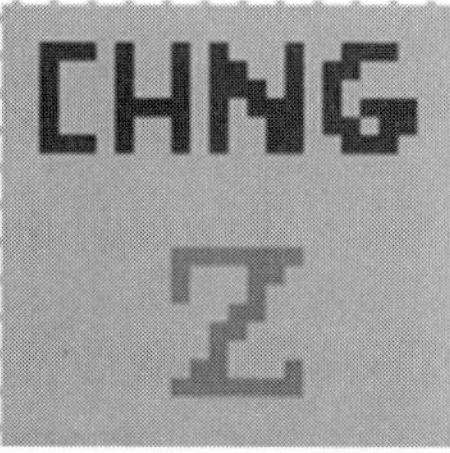

Figure 23-25 An icon example with text

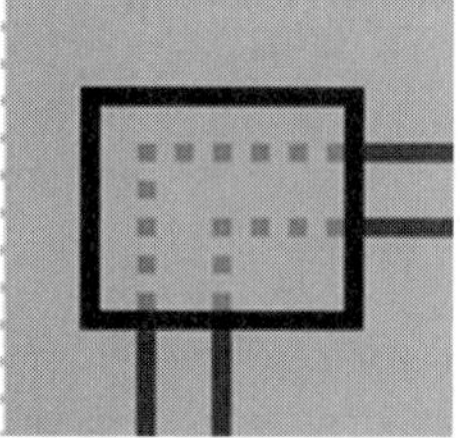

Figure 23-26 An icon example without text

Custom Linetypes

Creating custom linetypes is not terribly difficult, but it does take some patience and perseverance. Linetypes all reside in the DCADWIN.LIN file, normally found in the DataCAD \SUP directory. This file is a simple text file (an ASCII text file) and can be edited using any plain text editor like Windows Notepad. You can use Windows Write or even your word processor as long as you remember to save the file in an ASCII or text-only format.

Simply put, each linetype definition is a listing of graphic instructions. To visualize how they are created, think of your hand, a pencil, and paper. Each linetype definition tells DataCAD to do the following:

1. Place the point of the pencil on the paper.
2. Move the pencil at angle X and distance Y. A line is drawn on the paper.
3. Lift the pencil off the paper.
4. Move the pencil at angle X and distance Y. A line is not drawn, but the pencil is moved to a new location.
5. Place the point of the pencil on the paper.
6. Move the pencil at angle X and distance Y. A line is drawn on the paper, and so on.

If you have ever seen an old pen-plotter work, then you will understand the concept right away.

Because the commands are described by linear movements in either the X or the Y axis, only straight-line segments are possible. To create curves, you must draw a series of small straight lines that approximate a curve. Try zooming into the INSUL linetype in the DCAD Drawing Window to see a good example of this.

Each linetype is defined within a square box of 100 by 100 units. Note that the units are not specifically inches, millimeters, or any other convention. Like the pencil and paper example, you will need to write instructions of direction and distance within this 100 x 100 box. The instructions can arbitrarily begin anywhere within the box, but they must return to their starting point at the end of the sequence, since the linetype pattern will repeat itself over various distances. Use the same angle direction that you use in DataCAD; namely, due East is 0 degrees, North is 90, West is 180, and South is 270.

NOTE: *As mentioned earlier, a linetype definition must end where it began so that a repeating pattern will be formed. However, since a linetype is a repeating pattern, the end of a linetype definition can end on the opposite side of the 100 x 100 square. To see an example of this, see the figure of the complex linetype later in this section.*

To show how this works, let's first see how the CentrLin linetype is created. Then we'll create our own more complex hatching linetype.

Here is the CentrLin linetype definition you will find in the DCADWIN.LIN file:

>CentrLin	The linetype name (up to eight characters, preceded by the > character)
8.0	The scaling factor (optional)
0 42.5	0 is the angle; 42.5 means draw a line 42.5 units long
0 − 5	0 is the angle; − 5 means move five units *without* drawing a line
0 5	0 is the angle; 5 means draw a line five units long
0 − 5	0 is the angle; − 5 means move five units *without* drawing a line
0 42.5	0 is the angle; 42.5 means draw a line 42.5 units long

The rules for linetype definitions are as follows:

- The first line of a linetype definition must start with either a greater than symbol (>) or an asterisk (*). All of the standard DataCAD linetypes start with a greater than symbol, which causes DataCAD to repeat the pattern as many times as possible to fill the specified distance. It then stretches the pattern slightly to fill in any remaining space. If an asterisk is used, then DataCAD will repeat the pattern as many times as possible to fill in the specified distance. It then draws a short, solid line, called a *tail*, to fill any remaining space (this is how the old DCAD linetypes were displayed, but it fell out of favor since the visual tails at the ends of the lines drove people a little nuts).
- The name of the linetype (as displayed in DataCAD) follows the greater than or asterisk symbols and can only be up to eight characters in length.
- Notice that there is a space between the angle and distance. You can have any number of spaces, but you must have at least one so DataCAD knows that they are two numbers.
- A linetype can consist of only 20 lines of instructions after the name of the linetype. This limit includes the optional scaling factor that, if used, is inserted on the line directly below the name of the linetype. So if your custom linetype needs 21 or more lines of direction/distance code to define it, then you will have to redesign your linetype to be less complex.
- Each angle/distance definition must be on its own line.
- By default, any linetype definition will start at the left edge of the 100 × 100 box halfway up (50 units). So if you want to define the linetype along the top or bottom of the line (or anything other than the center), then you must start your definition at the center-left, move the pen up or down to the preferred start point, and then define a negative line length (which will move the pen without drawing a line).
- A positive distance means DataCAD should draw a visible line in the Drawing Window.
- A negative distance means DataCAD should move to a new location but should not draw a visible line over that distance.
- A positive angle moves the pen counterclockwise. A negative angle moves it in a clockwise direction (an angle of 270 degrees is the same as an angle of − 90 degrees). The angles are as follows:

  0 = Right

  90 = Up

180 = Left

270 = Down

Figure 23-27 shows what this linetype looks like when used in DataCAD.

Figure 23-28 is how the CentrLin linetype is described in the linetype definition step by step. The square on the outside is the imaginary 100 x 100 box for reference. They are not lines to be identified or drawn as part of the linetype.

The definition here can be broken down as follows:

1. The scaling factor of 8.0 tells DataCAD that when the Spacing setting in the LineType menu is set to 1.0, the linetype will be drawn 8′-0″ (when using feet/inches) long and will repeat at 8′-0″ intervals (more on this later).
2. Start at point 1 (by default).
3. The first line after the scaling factor of 8.0 is 0 42.5. The 0 is the angle at which the first line will be drawn (horizontally to the right). The

Figure 23-27
The CentrLin linetype

Figure 23-28
A step-by-step definition of the CentrLin linetype

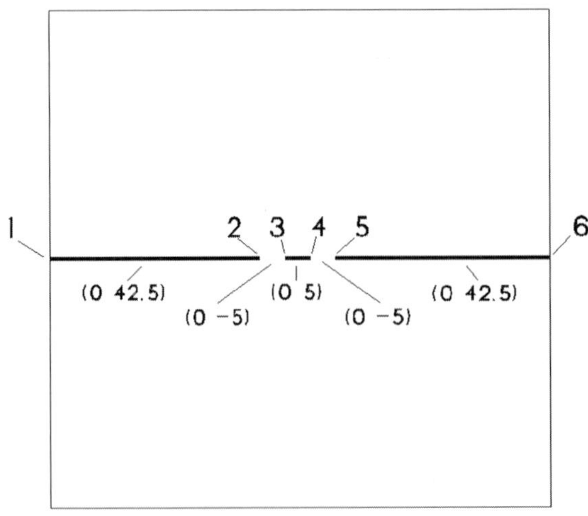

(positive) 42.5 units is the distance of the line that will be drawn from point 1 to point 2.

4. The space from point 2 to point 3 is defined by the next line of the definition: 0 − 5. Remember that a negative number means DataCAD will move over that many units, but *without* drawing a line. So DCAD moves over from point 2 to point 3 at 0 degrees and at a distance of 5 units, but without drawing a line.
5. The next line of the definition is 0 5. Thus, a line is drawn (because the 5 is a positive number) at 0 degrees and at a distance of 5 units.
6. The next line of the definition is 0 − 5. Like the space from point 2 to point 3, the space from point 4 to point 5 is created when DataCAD moves over at 0 degrees at a distance of 5 units, but *without* drawing a line (because the 5 is a negative number).
7. The final line of the definition is 0 42.5. Like the first line from point 1 to point 2, this final line is drawn at 0 degrees at a distance of 42.5 units.

So what happens if you use this linetype to draw a line in DataCAD that is longer than 100 units? The answer is that the linetype definition just keeps repeating. So if you draw a line that is 200 units long, DCAD would draw a line from point 1 to point 6. Then, starting at point 6, the full linetype would be drawn again, as shown in Figure 23-29.

A More Complex Linetype

In this example, we will create a more complex crosshatch pattern. To start off, we will disregard the scaling factor, adding this in at the end of the process instead. The best way to figure out all the angles and distances is to let DataCAD do it for you, and then write down the information for each vector. To do this, we will set a decimal scale type, draw a 100 × 100 unit box for reference, and then draw the new linetype pattern within the box. Once that is done, you can use the **Utility/Measures/PntToPnt** function in DataCAD to measure the distance of each line of the linetype definition and

Figure 23-29 How the linetype repeats

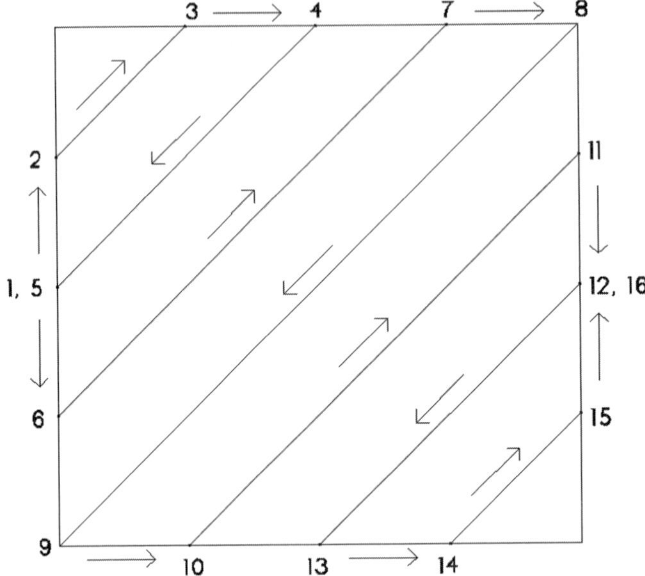

Figure 23-30
The linetype pattern

the **Identify** function to find the angles of each line. Figure 23-30 is the pattern we will create now.

1. Start a new drawing in DataCAD.
2. Go to **Utility/Settings/ScaleTyp**. Pick either the Imperial **Decimal** option or the **Metric** option. Both will give you a whole number followed by a decimal number. This is important. If you select the Decimal option, pick **DoFloat** as well.
3. In the Drawing Window, draw a reference box of 100 units by 100 units.
4. Draw the pattern you want to create. It must fit completely within the reference box, and the right and left edges of the linetype must work together to form a repeating pattern.
5. The start point for any DataCAD linetype is the center left of the box, represented by point 1.
6. Although the first line to be drawn is from point 2 to point 3, we first have to tell DataCAD to move the pen from start point 1 to point 2. The first line of this linetype definition does this by telling DataCAD to move at an angle of 90 degrees (up) at a distance of −25 units. Now that the pen is at point 2, that is where the first line will start from (see Figure 23-30).

Note that the square outline is just a guide and is not part of the linetype, so we don't figure those lines into the definition. Here is the linetype definition:

```
>Hatching
90 -25
45 35.355
0 -25
225 70.711
-90 -25
45 106.066
0 -25
225 141.421
0 -25
45 106.066
-90 -25
225 70.711
0 -25
45 35.355
90 -25
```

Figure 23-31 is what the completed linetype will look like when drawn in DataCAD.

To figure the lengths and angles, for standalone line segments like point 2 to point 3 or point 4 to point 5, the **Identify** command (the **I** button in the Navigation Pad) will give you the length and angle of the line.

NOTE: *When you identify a line in DataCAD, you will get one of two angles. Since an angle is measured from its start point to its end point, you will get a different angle readout depending on which point was the start point when it was drawn. To see how this works, draw a vertical line from top to bottom and then draw one from bottom to top. Now use the* ***Identify*** *command to identify each one. The one drawn top to bottom will identify as 270 degrees. The one drawn from bottom to top will identify as 90 degrees, the 180-degree opposite of 270 degrees.*

When you identify a line in DataCAD, you must decide whether the angle displayed is correct or whether you need to add or subtract 180 degrees.

Figure 23-31 The completed linetype

For lines that do not stand alone, such as the four lines that make up the reference box, use the **Utility/Measures/PntToPnt** function to identify the distance of each movement of the pen (remember the analogy at the beginning of this section). For instance, if you use **PntToPnt** to identify the first move of the pen (from point 1 to point 2), DataCAD will tell you that the distance is 25 units.

Unfortunately, in this case the **PntToPnt** function will not give you the angle of each line, but the **Identify** function will. *Click* on the **I** button in the Navigation Pad and then *click* on the line. Decide if the angle is correct or if you need to add or subtract 180 degrees (see the previous note about identifying angles).

Here is the final linetype definition. See if you can follow along. Note that negative angles can be used as well as positive angles. For instance, an angle of −90 degrees is the same as an angle of 270 degrees; they both describe an angle that is 270 degrees in a counterclockwise direction or 90 degrees in a clockwise direction. Using either one is acceptable. We use negative numbers here simply because they are easier for us to quickly comprehend.

Notice that the final movement of the pen is from point 15 to point 16. No line is drawn between these two points (hence the negative distance), but it is critical to make this final move in order to get the pen back to the point from which the linetype definition can start repeating itself. In other words, point 16 now becomes point 1 for the next repetition of the linetype. If you're still not clear on this, try leaving that last line out of your linetype definition and see what happens; it's not pretty.

Now open up the DCADWIN.LIN file (in the \SUP directory) with a text editor like Notepad. Go to the very end of the file and add the angles and distances that you wrote down for each segment of the linetype definition. Add this new linetype definition on the line directly below the last listed linetype, then save the file.

If you are still currently in a drawing file in DataCAD, the new linetype will not appear at the end of the linetype list. DataCAD still has the old one (without the new linetype in it) stored in memory. To access your new linetype, simply exit out of the drawing you are in and then reopen it or open any other drawing. Now select **Edit/LineType** and **ScrlFwrd** until you see the Hatching linetype listed at the end. Notice that when you place your cursor over the hatching linetype button a preview image is automatically displayed in the preview window.

For now, just select the **Hatching** linetype. Make sure the **Spacing** is set to **1.0**. Draw a line in the Drawing Window. Does it look like the image below? If not, go back and see if you made a mistake in the definition in the

Figure 23-32
The correct linetype outcome

.LIN file. Chances are you'll get something wrong the first time through (I did the first time I created this linetype). Just correct any potential problems, save the .LIN file, go back to DCAD, save and reopen the drawing, and look at the line you drew. Keep doing this until the line looks correct, as in Figure 23-32.

Scaling Factor

The scaling factor is an optional part of the linetype definition. If a linetype definition does not have a scaling factor, as in our new Hatching linetype example, then the scaling factor is assumed to be 1.0. If you do use it, then it gets inserted as the very first line after the name of the linetype, such as the 0.6667 in the following example:

```
>Brick
0.6667
0 -2.605
270 45.3125
```

As you now know, a linetype is defined within a 100 × 100 unit box. What the scaling factor does is scale the box up or down according to the factor used. A scaling factor of 2.0 will enlarge the box by two times, making a linetype that is 2′-0″ long (if using feet and inches). A scaling factor of 0.50 will scale the box down by half, making a linetype that is 6″ long. Thus, the scaling factor's purpose is to force a linetype to be of a certain length or width. This is especially useful for linetypes that need to be dimensionally accurate to represent dimensional materials like brick or wood siding. We'll step through an example of this later in this section.

Let's say you plan to use this Hatching linetype to represent the hatching of 8″ concrete blocks. You may want the width of the hatching linetype to always be 8″ as long as the **Spacing** is set to **1.0** in the **LineType** menu. Perhaps you want the linetype to be 1″ wide with the **Spacing** set to **1.0**. This way you could set a **Spacing** of **1.0** to hatch walls that are 1″ thick, or you could set a **Spacing** of **5.0** to hatch walls that are 5″ wide. DataCAD determines this characteristic of a linetype through the *optional* scaling factor at the beginning of the linetype definition. You can see that this scaling factor you use will have a profound effect on how the linetype is used and displayed.

The unit generally used in imperial units is feet, which is further divided into 12 inches. Let's assume that in the **LineType** menu, with a linetype **Spacing** of **1.0**, we want the line to be 1″ wide. The scaling factor would be calculated by dividing 1″ by 12″, which is equal to .083334. If, with a linetype **Spacing** of **1.0**, we want the line to be 4″ wide, then the scaling factor would be calculated by dividing 4″ by 12″, which is equal to .33334. For 8″, the scaling factor would be 8″ divided by 12″, or .66667. For 12″, the scaling factor would be 12″ divided by 12″, or 1.0. For 4′-0″ (48 inches), the scaling factor would be 48″ divided by 12″, or 4.0.

But what if you want the linetype **Spacing** to more directly determine the dimensions of the linetype? Perhaps you want a **Spacing** of **0.4** to represent a 4″-wide line, a **Spacing** of **0.8** to represent an eight-inch-wide line, or a **Spacing** of **3.0** to represent a 3′-0″-long line. This seems to make a lot of sense since you would only have to develop one linetype, allowing the linetype menu's **Spacing** option to control the line's dimensions in a one-to-one relationship.

All you have to do is not use a scaling factor, in which case the 100 × 100 unit box is equal to 12″ × 12″. When you select a **Spacing** of **0.8** in the **LineType** menu, the line will be drawn at 8″ long and will be repeated every 8″ thereafter. A **Spacing** of **2.0** would draw and repeat the line every 2′-0″. Take a look at the **Hatching** linetype we created earlier. Since we did not use a scaling factor, this is how this linetype works.

Keep in mind that only one scaling factor is possible per linetype, which means that the factor will be applied equally to both the X and Y axes. To create a linetype that represents a material with both a specific thickness and a specific length (like the brick coursing or wood siding mentioned previously), you need to do the following:

1. Scale the 100 × 100 box up or down to match either the required X dimension of the linetype or the Y dimension of the linetype (in the vast majority of cases it will be the X axis). This is done with the scaling factor of the linetype definition.
2. Figure out the unit lengths of all the lines in your linetype, and then multiply them by the scaling factor. The resulting numbers are the ones you type into the linetype definition.

Clear as mud? Hopefully, this is starting to make sense. If it's still not clear, you might need to read through this information a few times and try to work through the following Brick linetype example for an illustration of these steps.

Dimensional Accuracy

In the previous two examples, we created linetypes that did not have to be created at any specific scale, but if you want to create a linetype to represent a dimensionally correct building material like brick or shingles, then you just need to do a little extra math. Let's look at part of the DataCAD Brick linetype definition to see how the 0.6667 scaling factor is used.

The following code is the definition for DataCAD's standard Brick linetype (see Figure 23-33). First, you should notice that it is defined off of the centerline and to one side of it. Thus, the Brick linetype is drawn from the outside face of the brick rather than down its center. Try drawing a line with the Brick linetype in DataCAD and you'll see what we mean.

```
>Brick
0.6667
0 -2.605
```

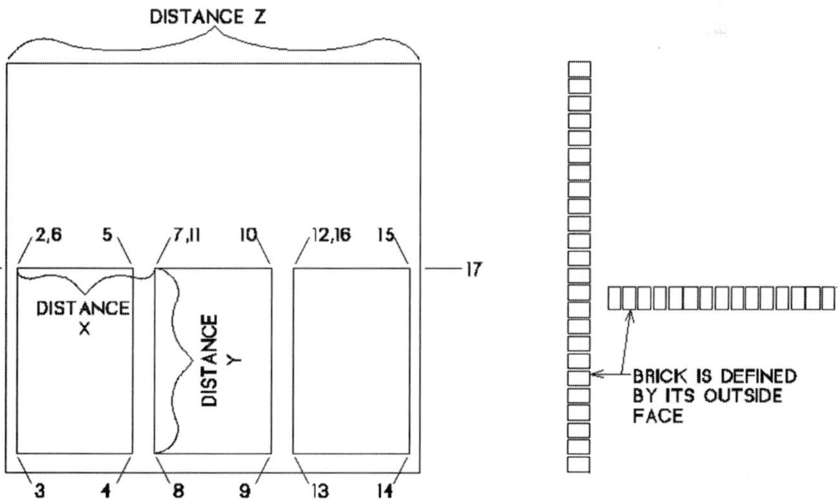

Figure 23-33 The linetype in the unit box

```
270 45.3125
0 28.125
90 45.3125
180 28.125
0 -33.333
270 45.3125
0 28.125
90 45.3125
180 28.125
0 -33.333
270 45.3125
0 28.125
90 45.3125
180 28.125
0 -2.604
```

To understand the three bricks drawn in the box in Figure 23-33 you need to know that here in the United States a single course of brick (standard brick with one joint) is $2^2/_3$″ (all you metric users stop laughing!), which equals a full CMU course of 8″. $2^2/_3$″ is $^1/_3$ of 8″, or 33.333. We therefore drew three equal brick courses in the 100 × 100 box, so that the distance between point 6 and point 7 is 33.333 units.

For this linetype, we need distance Z to be 8″, meaning that the 100-unit box will be equal to 8″ on each side, which will cause three bricks plus three joints to be displayed as 8″ high when the **LineType/Spacing** is set to **1.0.** The 0.6667 scaling factor (derived by dividing 8″ by 12″) properly scales the brick coursing in the X axis.

The true scaling factor for the Brick pattern is 0.666666 repeating. The 7 at the end of the factor was just to round the number off. But rounding can give you slightly inaccurate results. For that reason when you are making your calculations you should always use as many decimal places as possible. So this linetype definition would have been better off using a factor of 0.6666666 instead of 0.6667.

NOTE: *What if the scaling factor is not used in this linetype? With this scaling factor, and the **LineType/Spacing** set to **1.0**, drawing a line 8″ long will draw 3 dimensionally correct courses of brick. If you draw a line 16″ long, then 6 courses of brick are drawn. If the scaling factor were omitted from the linetype definition (both in the second line and in the calculations to follow), then you would have to set the **LineType/Spacing** to 0.8 (8″) before drawing those same 8″ or 16″ lines. This would get very confusing since there is no way to warn the user that to get dimensionally accurate brick courses they have to first set a **Spacing** of **0.8**.*

Figure 23-33 is the linetype as it is described inside the 100 × 100 unit box. As always, the definition begins at the center left of the box at point 1.

The brick dimension from point 7 to point 8 (Distance Y) is to be $3^5/_8$″; the standard brick depth in the United States. That distance is described by the line: 270 45.3125 unit dimension was derived like this:

- $^5/_8$″ is 5 divided by 8 = 0.625″
- The entire brick is therefore 3″ + 0.625″ = 3.625″
- 3.625″ divided by 12″ = 0.3020833 of a foot
- 0.3020833 × 100 units = 30.20833 units
- 30.20833 divided by the scaling factor of .6666666 = 45.312499
- Round to 45.3125 units

I would not have rounded the final number, but that's what was done with this linetype.

Metric Scale Factors

The scale factor is related to the standard dimensional unit you are using. For dimensionally accurate linetypes, that factor will be different depending on whether you use imperial or metric units simply because the materials that the linetypes are describing (like brick or plywood) are different dimensions depending on whether they are metric or imperial. We'll take a look at both, but let's first look at our new Hatching linetype in imperial units.

The unit generally worked with in metric units is the meter, which is further divided into 100 centimeters. Let's assume that with a linetype **Spacing** of **1.0** we want the line to be one centimeter wide. The scaling factor would be calculated by dividing one centimeter by 100 centimeters, which is equal to .01. If, with a linetype **Spacing** of **1.0**, we want the line to be 25 centimeters wide, then the scaling factor would be calculated by dividing 25 by 100, which is equal to .25. For 50 centimeters, the scaling factor would be 50 centimeters divided by 100 centimeters, or .50. For 100 centimeters (one meter), the scaling factor would be 100 divided by 100, or 1.0. For four meters (400 centimeters) the scaling factor would be 400 centimeters divided by 100 centimeters, or 4.0.

So what if you want to define your linetype in millimeters, instead? The metric system makes that easy. The scaling factors will be exactly the same as the centimeter example as long as you use identical units. For example, let's assume that with a linetype **Spacing** of **1.0** we want the line to be 10 millimeters wide. The scaling factor would be calculated by dividing 10 millimeters by 1,000 millimeters, which is equal to .01, which is the same factor derived in the previous example.

Linetype Caveats

The following is a list of things to watch out for when using linetypes:

- You have 175 new slots available in the DCADWIN.LIN file after the mandatory four of Solid, Dotted, Dashed, and Dot-Dash, so you can have a total of 179 total linetypes.
- Also, you should note that DataCAD uses linetypes in a drawing file not by name, but by order, so that if you are continually switching your linetypes, your old drawings may come up looking very weird. For instance, the program will use the fifteenth linetype in you current DCADWIN.LIN file, not what used to be the fifteenth linetype at the time you made the file. This factor also makes it difficult to swap drawings between different DataCAD users if they have radically different linetype files. So you should modify your DCADWIN.LIN file very carefully and thoughtfully with the long range view in mind.
- Although the linetype **Spacing** function in the **LineType** menu gives you the flexibility to resize any linetype, if you need a precise width or length of the linetype pattern, you will first need to look at the scaling factor of the linetype definition so that you can accurately decide how much larger or smaller the spacing needs to be changed. A **Spacing** of **2.0**, for instance, will double the scaling factor so that a linetype designed to create a 4″-wide pattern will now create an 8″-wide pattern (2.0 × 4″ = 8″). Or you can find out by trial and error.
- The best way to learn the nuances of linetypes is to pick an existing one and then disect it until you understand how and why it works.

Certainly keep copies of all your previous versions of DCADWIN.LIN under a different name, particularly the file that came with a major release (save it with something like DCAD9.LIN.) You may need it at later time. Even with these drawbacks in mind, judicious use of customized linetypes is a very powerful tool indeed.

Linetype Preview Images

Once you have created a custom linetype in the DCADWIN.LIN file, it will be displayed in the **LineType** menu and is immediately accessible for use. A preview image of the new linetype will also be immediately available for display in the linetype preview window as you pass your cursor over each linetype. DataCAD uses the linetype definition in the .LIN file to display the linetype, based on the current linetype settings in the drawing file.

Custom Hatches

Some of the information here was derived from an article written by Evan Shu for the *Cheap Tricks* DataCAD newsletter

Writing custom hatch patterns is more complicated and daunting than writing custom linetypes, so creating new patterns may not be for everyone. As with linetypes, the best way to learn about hatch patterns is to select an existing one and pick it apart until you understand it.

Like the custom linetype file (DCADWIN.LIN), all the hatch patterns are kept in a simple ASCII (plain text) file called DCADWIN.PAT found in the \SUP directory under your main DataCAD directory. If you pull this file up in your word processor, you will see each individual hatch pattern starts with an asterisk and the hatch name and then a series of lines that are made up of numbers and commas. Some of you might recognize this file format as a "comma-delimited" database. Basically, that means that each line is one record or instruction and each number separated by a comma represents a field or category of that record/instruction. Once you understand what each of these numbers represents, you'll understand custom hatch patterns.

Figure 23-34 is the **ansi31** hatch pattern found at the beginning of your list of DataCAD hatches.

In a smaller, more close-up form, it looks like Figure 23-35.

Really only one line is used in this hatch pattern, the dark one shown in Figure 23-36.

Once that line is defined, it is then repeated as many times as it takes to fill the area you are hatching. Here is the line of code from the standard DCADWIN.PAT file that describes this hatch pattern:

Figure 23-34
The ansi31 hatch pattern

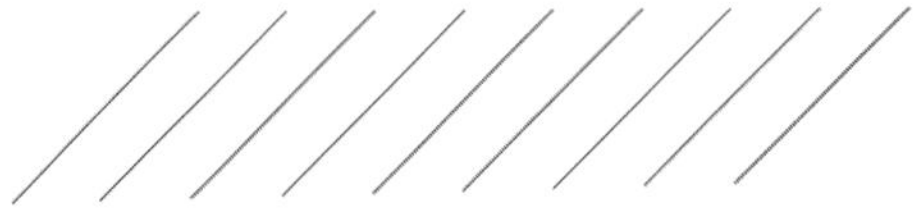

Figure 23-35
A close-up of the asi31 hatch pattern

Figure 23-36
The hatch is defined by only one line.

```
*ansi31
0, 0, 45, 0, 0, 0, 1
```

The format of this line is called *comma-delimited*. Each digit is separated by a comma and, as such, DataCAD only reads the digits and the commas, disregarding any spaces. Many of the hatch patterns in the DCADWIN.PAT file therefore have additional spaces in each line just to make each line more readable when working on the pattern. So the ansi31 pattern could also look like this if you wanted it to:

```
*ansi31
0,      0,      45,      0,      0,      0,      1
```

The ansi31 hatch pattern is such a simple pattern that it can be described in only one line. Most other hatch patterns aren't this simple, but its simplicity makes it the best pattern to use to begin to describe how a hatch pattern works. Here is what each number in the ansi31 hatch definition does:

1. The first number stands for line weight, which is always 0 in DataCAD because DataCAD controls line weight separately.
2. The second number is line color, which again is always 0 in DataCAD. (We keep saying "in DataCAD" because AutoCAD also utilizes the exact same hatch file format, but they do utilize the first two numbers.)
3. The third number, as you might guess, stands for Angle, in this case 45 degrees.
4. The fourth number stands for the X-axis offset distance from the origin to the start of the line to be drawn. If a line starts from the origin point, then this number will be 0.
5. The fourth number stands for the Y-axis offset distance from the origin to the start of the line to be drawn. If a line starts from the origin point, then this number will be 0.
6. The sixth number stands for the offset distance along the X axis to where the line will repeat (for solid lines, this number will always be 0.) If the line to be described is at any angle other than 0 (horizontal), then you must mentally rotate the line to be horizontal (zero degrees) in order to determine this X offset.
7. The seventh number stands for the offset distance along the Y axis to where the line will repeat, which is, more often than not, your basic spacing distance. If the line to be described is at any angle other than 0 (horizontal), then you must mentally rotate the line to be horizontal (0 degrees) in order to determine this Y offset.
8. The final numbers (not used in the ansi31 definition) describe the actual line to be drawn. This description can add up to six more numbers to the line, but the values must alternate between positive and negative numbers. A positive number means "pen down" and negative number means "pen up," so you can have up to 3 different line segments before the pattern repeats.

NOTE: *If none of the additional numbers are used as described in line 8, then DataCAD will default to drawing a solid line based on lines 1 through 7. Such is the case with the ansi31 pattern.*

This format only allows for straight-line segments, so curved arcs are not possible. If you want curves, then you will have to simulate a curve via a series of segmented lines.

Here is a more complex hatch pattern that makes use of the numbers in Step 8:

```
*Squares
0, 0, 0, 0, 0, 0, 8, 4, -4
0, 0, 90, 0, 0, 0, 8, 4, -4
0, 0, 180, 0, 0, 0, 8, 4, -4
0, 0, 270, 0, 0, 0, 8, 4, -4
```

To help define what each line does, let's break down the last line and put a label on each digit for future reference:

1	2	3	4	5	6	7	8	9	10	11	12	13
Wt	Clr	Ang	Xor	Yor	Xoff	Yoff	Dis1	Dis2	Dis3	Dis4	Dis5	Dis6
0,	0,	270,	0,	0,	0,	8	4	– 4				

Wt = Line weight   Xor = From X origin   Yoff = Y offset

Clr = Color   Yor = From Y origin   DisX = Line segment distances

Ang = Angle   Xoff = X offset

While custom linetypes are defined within a 100 × 100 unit box, hatch patterns have no such limitations. In describing the distances in a hatch pattern, the basic unit of distance is 1 unit = $^1/_{32}$″. For hatch patterns that don't describe dimensionally accurate materials (ceramic tile or acoustic ceiling tiles), it is advisable to use small, easy-to-figure units, such as 1 to 10. It's not important to use a specific scale since the size of the hatch can be enlarged or reduced with the Scale option in the Hatch menu. But as you design your non-dimensional hatch pattern, try to create your lines with as many whole numbers as possible just to make your life a little easier.

Let's see how the ansi31 hatch pattern works. Here it is again:

```
*ansi31
0, 0, 45, 0, 0, 0, 1
```

The first line of any hatch pattern starts with an asterisk (*), followed by the hatch pattern name of up to eight characters. The first two digits of every line of your hatch pattern will generally always be 0, 0 for the reasons mentioned previously, although we'll see later that even this does not have to be so.

The 45 tells DataCAD to draw a line at a 45 degree angle. The angle is defined from the starting point of the line, as defined by the next two numbers. The next two digits are 0, 0 because the line is drawn directly from the origin without any X or Y offset from it.

The sixth digit is also 0, but this is where things get tricky. Any hatch line that is at an angle other than 0, such as the 45 degree line in the ansi31 hatch pattern, must be mentally rotated to 0 degrees (horizontal) before determining the offset. If we rotate the ansi31 hatch pattern about the origin point, it would look like Figure 23-37.

With the line rotated, you can see that there is no need to define a horizontal offset, since the line is already being drawn horizontally. Because of this, the X offset is 0.

With the hatch line still mentally rotated, you can now determine the Y offset as well. In the case of the ansi31 pattern, each successive line is drawn 1 unit above the first line, so the Y offset is 1 (see Figure 23-38).

No further numbers are required in this hatch definition since it is only a single solid line that is being drawn. DataCAD will always default to drawing a solid line if you don't tell it to do otherwise.

Now for a more complex pattern! If you understand the concepts of the ansi31 hatch pattern, then this one is just a bit longer and requires a little more math. But just as we used DataCAD to help figure out the angles and line lengths required for a linetype definition, so we will make use of those functions for figuring out our more complex hatch.

Figure 23-37 Rotating the ansi31 hatch pattern

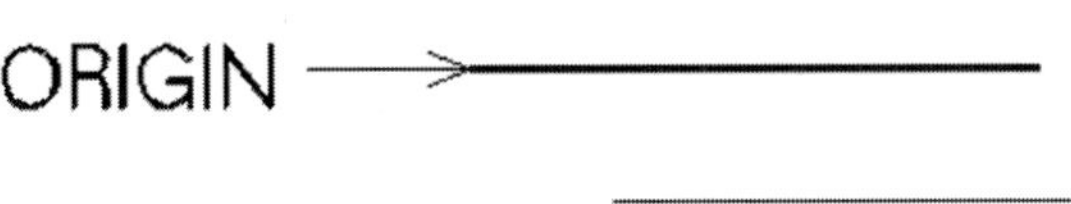

Figure 23-38 Determining the Y offset

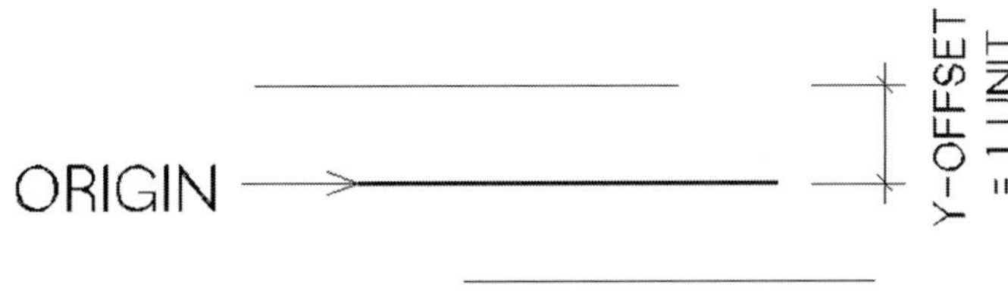

Complex Tile Pattern

Let's step up our custom hatch study by moving up one level in complexity by creating the tile hatch pattern shown in Figure 23-39.

For this pattern, we have chosen to create the large tiles with a dimension of 1″ × 1″ [25.4 × 25.4]. Because 1/32 of an inch in DataCAD equals one hatch unit, in order to get a one-inch tile dimension, we need to multiply 1″ by 32 units for a total hatch dimension of 32 units × 32 units. You might ask why we are doing this, since earlier we said it was easier to use small units between 1 and 10. We do this so that when a hatch scale of 0.1 (DataCAD shorthand for 1″) is selected, then the large tiles will be displayed as 1″ squares. If a scale of 0.8 (DataCAD shorthand for 8″) is selected, then the large tiles will be displayed as 8″ squares. To me, this makes a lot of sense and makes it easy for the CAD operator to figure out which scale factor he has to use in order to get his tile at the proper scale in the drawing.

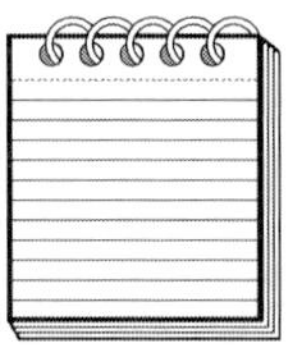

NOTE: *Be sure to use all the decimal places you can when defining your hatch patterns. If you don't use enough decimal places or you round them off, then your hatch lines may not line up or intersect the way you intended.*

In the end, we want the hatch pattern to work this way: when a hatch **Scale** of **0.1** is selected, then the large tiles will be displayed as 1″ squares. And if a **Scale** of **0.8** is selected then the large tiles will be displayed a 8″ squares. To me this makes a lot of sense, and makes it easy for the CAD

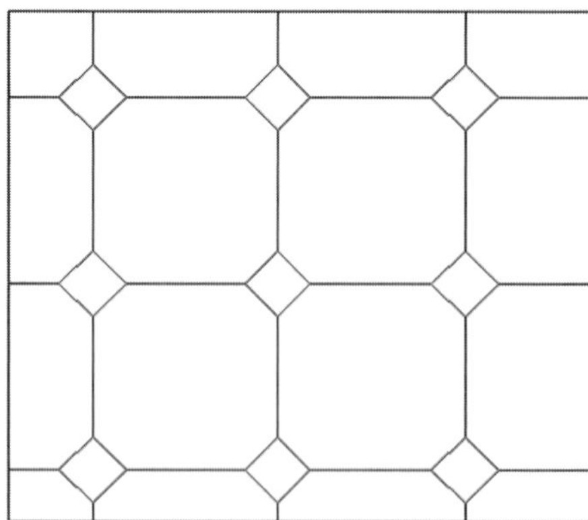

Figure 23-39
A complex hatch pattern

operator to figure out what scale factor he has to use in order to get his tile at the proper scale in the drawing.

When you decide to design a custom hatch you need to determine the repeating horizontal lines, the repeating vertical lines, and the repeating angled lines. To determine those angles and dimensions we are going to draw the pattern with 2D lies in the DataCAD drawing window, as described below, and then let DataCAD tell us all the pertinent dimensions and angles that will define the hatch pattern. You also need to pick an origin point that makes your life as easy as possible—namely a point where a lot of your lines intersect, so that they all can use 0.0 as their origin point.

Next, you need to determine all the X and Y offsets. As mentioned above, all of this work can be made simpler if you first draw your intended pattern in DataCAD. That way you can use DataCAD's **Measures** utility, and other drafting aids, so figure out all your angles and distances.

Drawing the Pattern in DataCAD

For this diamond tile pattern example we drew the large square tile at a dimension of 1″-0″ × 1″-0″ [305 × 305]. W could have drawn it at 1″ × 1″ [25.4 × 25.4] since that is how we want to define our final tile size, but it's easier to pan and zoom around the Drawing Window with one foot units rather than one inch. It didn't matter too much in this case, as long as we drew the pattern at a size of 1 unit by 1 unit.

Setting the Scale Type

To get the proper decimal values for the hatch pattern, we will eventually need to set the **ScaleTyp** in DataCAD to either Decimal or Metric. If you need to draw imperially dimensioned materials (such as brick or other materials involving fractions of an inch), then you should initially set the **ScaleTyp** to **Arch** to draw the hatch pattern. After the pattern is drawn, then change the scale type to a decimal format as follows.

NOTE: *Unfortunately, DataCAD's accuracy for decimal dimensions only extends out to three decimal places. If the true length of a line is 1.64234, it will be rounded to and displayed as 1.642, thereby losing some of its accuracy. This is generally not a problem for hatches, since an accuracy of greater than three digits is visually imperceptible. Just be aware that after creating a hatch pattern, if you zoom into it really close (I mean really close), you may see that some lines don't intersect quite properly.*

Figuring Out the Pattern

Once the pattern is drawn, you can use the **Utility/Measures/PntToPnt** function in DataCAD to measure the distance of each line of the linetype definition and the **Identify** function to find the angles of each line, but first you need to have a decimal **ScaleTyp** enabled. Go to **Utility/Settings/ScaleTyp**. Pick either the imperial **Decimal** option or the **Metric** option. Both will give you a whole number followed by a decimal number. If you select the **Decimal** option, pick **DoFloat** as well to get more decimal places.

Our tile pattern actually has only six repeating lines: one set of horizontals, one set of verticals, and four sets of diagonal repeats, as displayed by the bold lines with the labels on them in Figure 23-40.

We have selected the point where three lines intersect as our origin, one horizontal and two diagonal lines, as indicated in the figure. Here is what the final hatch definition looks like, with the identifying letter for each line at the beginning:

```
   *DmndTile
A  0, 0, 0, 0, 0, 0, 1, .646, −.354
B  0, 0, 90, −.177, .177, 0, 1, .646, −.354
C  0, 0, 135, 0, 0, −.707, .707, .25, −1.164
```

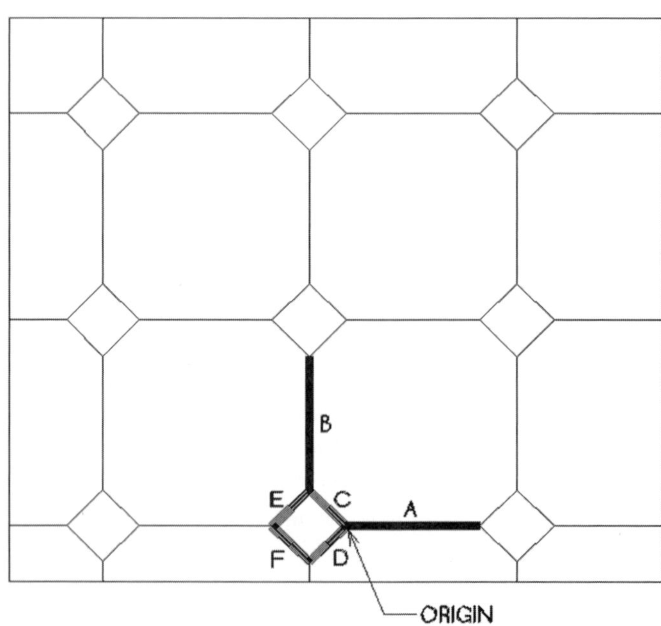

Figure 23-40 Only six repeating lines are used for this pattern.

D 0, 0, 45, −.354, 0, .707, .707, .25, −1.164
E 0, 0, 315, −.354, 0, .707, .707, .25, −1.164
F 0, 0, 225, 0, 0, .707, .707, .25, −1.164

Hatch Editor

In this section, we will pick apart each line of this hatch definition to see how it all works, but first a word about editing. You can use any simple text editor, such as Windows' Notepad, to edit the DCADWIN.PAT file containing your hatch pattern definitions. Just make sure you always save the file as plain text with no formatting. While working on a new hatch pattern, you can switch between the text editor and DataCAD as you work. As long as you first save the DCADWIN.PAT file, DataCAD will read and recognize the changes the next time you select the hatch pattern from the list. Being able to switch between open programs in Windows makes writing custom hatches much easier. You can write your hatch into the DCADWIN.PAT file and then switch back into DataCAD to test the hatch. If you've made a mistake, switch back to Notepad, make corrections, and then switch back to DataCAD to test your change.

Hatch Name

The first line of any hatch pattern is the hatch pattern name. It begins with an asterisk and has a hatch name of up to eight characters. In our case, it is * DmndTile (short for Diamond Tile).

Line A

1. The first two digits are always 0, 0 in DataCAD.
2. From the beginning of line A, it is drawn at 0 degrees, so the next number is 0.
3. Since the start point of line A starts at the hatch origin, the offset to the start of the line is 0 in both the X and Y directions. Therefore, the next two numbers are 0, 0 (see Figure 23-41).
4. The next two numbers define the X and Y distances to the repeated line. The dashed line Ao (A offset) is where we will define the repeated line. Think of how you would tell DataCAD to copy line A to the location of line Ao. In this case, you don't have to copy line A at any X

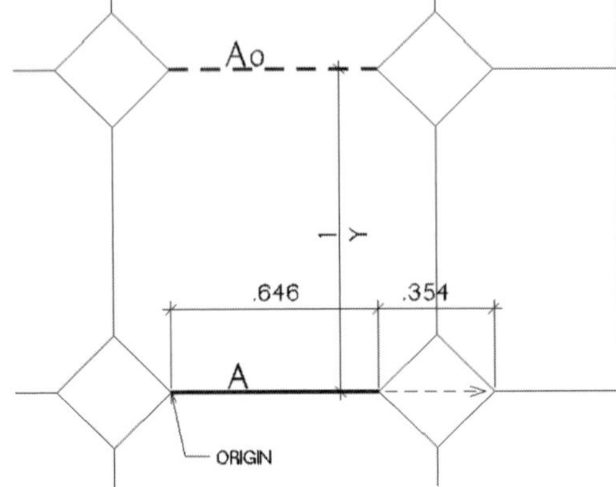

Figure 23-41
Line A

distance to the right or left. Therefore, the X offset is 0. Line A has to be copied 1 unit in the Y direction to get to Ao, so the Y offset is 1.

5. Notice that so far we have not defined line A itself, only its angle and offsets. Now we need to describe the actual line itself. Beginning from the start point for line A, it is drawn as a solid line at 0.646 units long, so we enter .646 as the next number. Remember that a positive number equals a solid line (pen down).
6. Any line of a hatch pattern needs to be described completely from one origin to the next, so we still need to define the movement of our pen back to the new origin. From the far right end of line A, we need to lift the pen up so that a line is not drawn and then move the pen 0.354 units to the right, which will put us at the next origin. To do that, we enter a minus sign (pen up) so that the final number is −.354.

Line B

The first two digits are always 0, 0 in DataCAD.

7. From the beginning of line B, it is drawn at 90 degrees, so the next number is 90. The beginning of line B does not start at the hatch origin, so we need to describe where the start of the line is in relationship to the hatch origin. The start of the line is .177 units to the left. Numbers that go to the left are negative numbers, so the next number in the sequence (the X offset) is 2 .177 (see Figure 23-42).
8. The start of line B is .177 units up from the hatch origin. Numbers that go up are positive numbers, so the next number in the sequence (the Y offset) is .177.

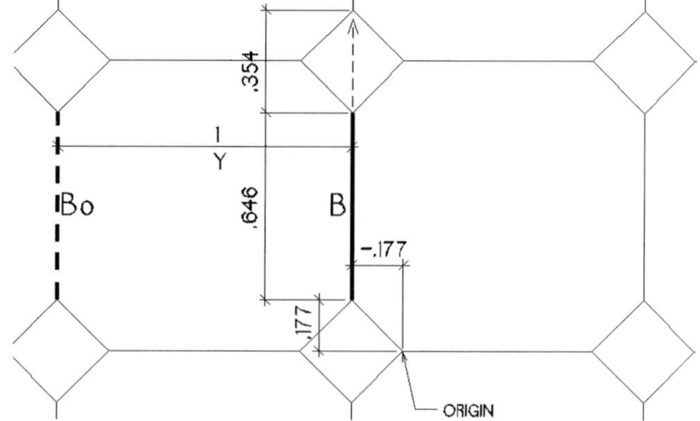

Figure 23-42
Line B

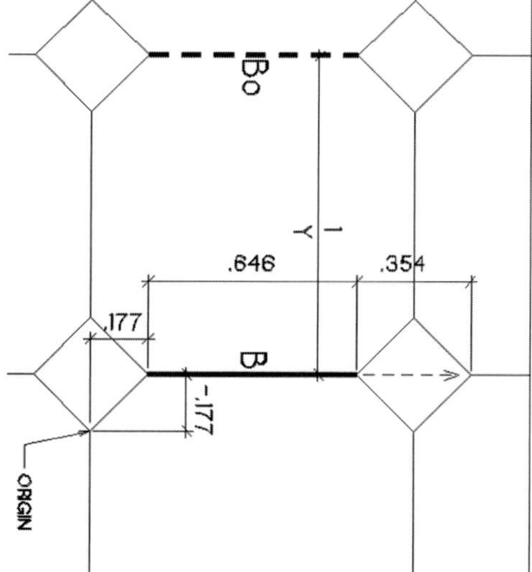

Figure 23-43
Rotation of the hatch pattern

9. The next two numbers define the X and Y distances to the repeated line. The dashed line Bo (B offset) is where we will define the repeated line, but the trick here is that the X and Y distances must be described with the line in a horizontal orientation. To do this, you need to mentally rotate the hatch pattern, as shown in Figure 23-43.

   Think of how you would tell DataCAD to copy line B to the location of line Bo. In this case, you don't have to copy line B at any X distance to the right or left. Therefore, the X offset is 0. Line B has to be copied 1 unit in the Y direction to get to Bo, so the Y offset is 1.

10. Again, notice that so far we have not defined line B itself, only its angle and offsets. Now we need to describe the actual line itself. Beginning from the start point for line B, it is drawn as a solid line at 0.646 units long, so we enter .646 as the next number. Remember that a positive number equals a solid line (pen down).
11. Any line of a hatch pattern needs to be described from one origin to the next, so we need to define the movement of our pen back to the new origin. From the far right end of line B, we need to lift the pen up so that a line is not drawn. Then move the pen 0.354 units to the right, which will put us at the next origin. To do that, we enter a minus sign (pen up) so that the final number is − .354.

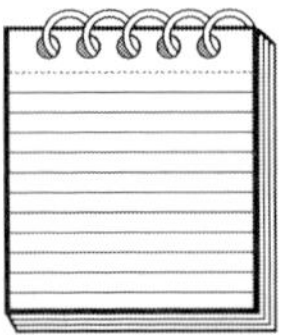

NOTE: *A long-standing bug can be found in the hatch routine in DataCAD. Programmers have tried to track it down, but as we went to press, it still has not been corrected.*

In the previous example of line B, we selected the line to the left of B to be line Bo. We could have selected the line to the right of Bo, and if we had, the definition for that line would have been

```
0, 0, 90, -.177, .177, 0, -1, .646, -.354
```

The only difference is the **–1** *instead of a positive 1. This should work, but it doesn't because of a bug with this X/Y offset. If you run into this problem, the work-around is to simply select a different line for the offset line (Bo in this example).*

Line C

1. The first two digits are always 0, 0 in DataCAD.
2. From the beginning of line C (at the hatch origin), it is drawn at 135 degrees, so the next number is 135.
3. Since the start point of line C starts at the hatch origin, the offset to the start of the line is 0 in both the X and Y directions. Therefore, the next two numbers are 0, 0 (see Figure 23-44).
4. The next two numbers define the X and Y distances to the repeated line. The dashed line Co (C offset) is where we will define the repeated line, but the trick again is that the X and Y distances must be described with the line in a horizontal orientation. To do this, you need to mentally rotate the hatch pattern, as shown in Figure 23-45.

   Think of how you would tell DataCAD to copy line C to the location of line Co. In this case, you have to copy line C to the left (negative

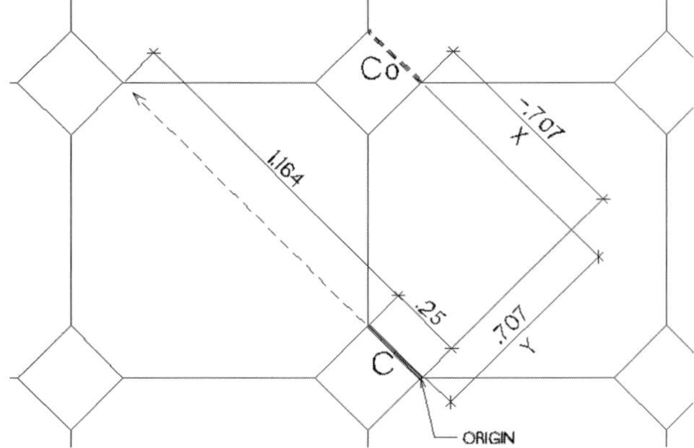

Figure 23-44
Line C

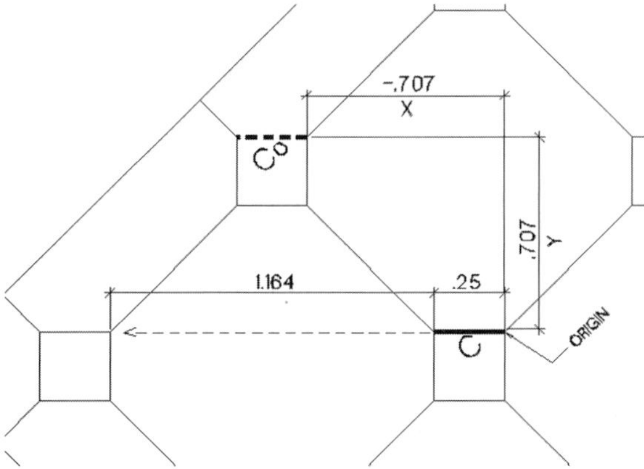

Figure 23-45
Rotating the hatch pattern

direction) at a distance of .707. Therefore, the X offset is −.707. Likewise, line C has to be copied upward (positive Y direction) a distance of .707 to get to Co, so the Y offset is .707.

5. Again notice that we have not yet defined line C itself, only its angle and offsets. To describe the actual line itself, begin from the start point for line C. It is drawn as a solid line at 0.25 units long, so we enter .25 as the next number. Remember that a positive number equals a solid line (pen down).
6. Any line of a hatch pattern needs to be described from one origin to the next, so we need to define the movement of our pen back to the new origin. From the far left end of line C, we need to lift the pen up so that a line is not drawn. Then move the pen 1.164 units to the left, which will put us at the next origin. To do that, we enter a minus sign (pen up) so that the final number is −1.164.

Line D

1. The first two digits are always 0, 0 in DataCAD.
2. From the beginning of line D (left of the hatch origin), it is drawn at 45 degrees, so the next number is 45.
3. The beginning of line D does not start at the hatch origin, so we need to describe where the start of the line is in relationship to the hatch origin. The start of the line is .354 units to the left. Numbers that go to the left are negative numbers, so the next number in the sequence (the X offset) is −.354 (see Figure 23-46).
4. The next number in the sequence (the Y offset) is 0, since the start of line D is not offset up or down from the hatch origin.
5. The next two numbers define the X and Y distances to the repeated line. The dashed line Do (D offset) is where we will define the repeated line, but the trick again is that the X and Y distances must be described with the line in a horizontal orientation. To do this, you need to mentally rotate the hatch pattern, as shown in Figure 23-47.

   Think of how you would tell DataCAD to copy line D to the location of line Do. In this case, you have to copy line D to the right (positive direction) at a distance of .707. Therefore, the X offset is .707. Likewise, line D has to be copied upward (positive Y direction) a distance of .707 to get to Do, so the Y offset is .707.
6. We have not yet defined line D itself, only its angle and offsets. To describe the actual line itself, begin from the start point for line D. It is

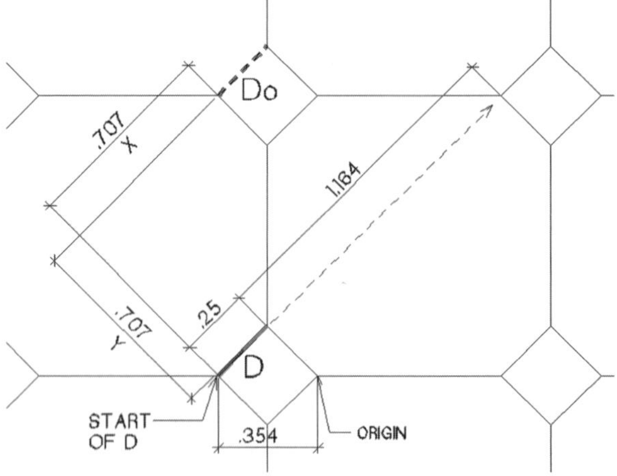

Figure 23-46
Line D

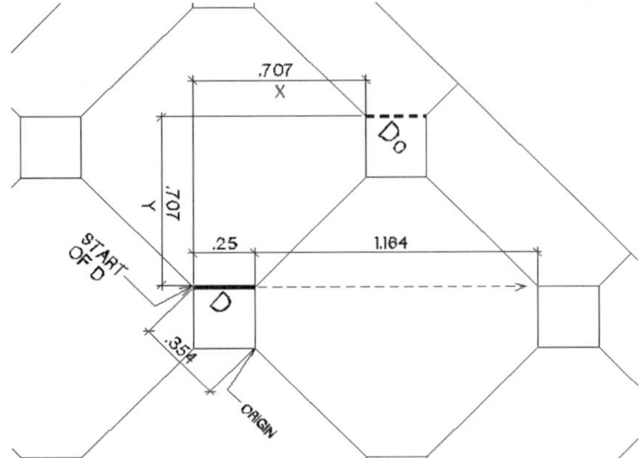

Figure 23-47 Rotating the hatch pattern

drawn as a solid line at 0.25 units long, so we enter .25 as the next number. Remember that a positive number equals a solid line (pen down).

7. Any line of a hatch pattern needs to be described from one origin to the next, so we need to define the movement of our pen back to the new origin. From the far right end of line D, we need to lift the pen up so that a line is not drawn. We then move the pen 1.164 units to the right, which will put us at the next origin. To do that, we enter a minus sign (pen up) so that the final number is −1.164.

Line E

1. The first two digits are always 0, 0 in DataCAD.
2. From the beginning of line E (left of the hatch origin), it is drawn at 315 degrees, so the next number is 315.
3. The beginning of line E does not start at the hatch origin, so we need to describe where the start of the line is in relationship to the hatch origin. The start of the line is .354 units to the left. Numbers that go to the left are negative numbers, so the next number in the sequence (the X offset) is −.354 (see Figure 23-48).
4. The next number in the sequence (the Y offset) is 0, since the start of line E is not offset up or down from the hatch origin.
5. The next two numbers define the X and Y distances to the repeated line. The dashed line Eo (E offset) is where we will define the repeated line, but the trick again is that the X and Y distances must be described

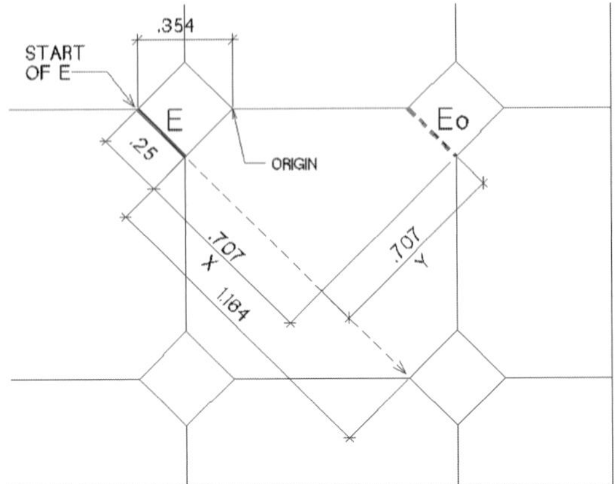

Figure 23-48
Line E

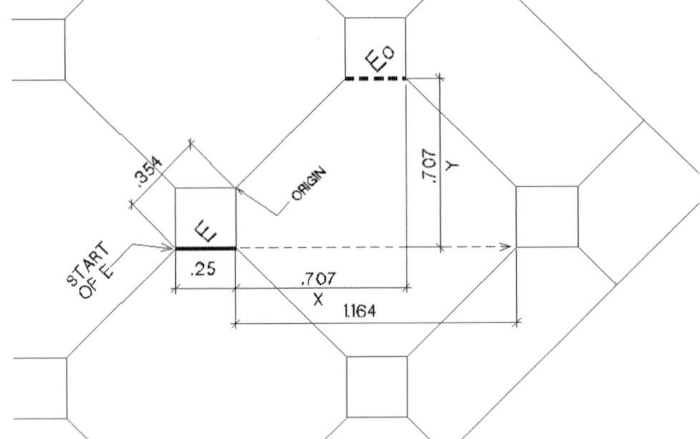

Figure 23-49
Rotating the hatch pattern

with the line in a horizontal orientation. To do this, you need to mentally rotate the hatch pattern, as shown in Figure 23-49.

Think of how you would tell DataCAD to copy line E to the location of line Eo. In this case, you have to copy line E to the right (positive direction) at a distance of .707. Therefore, the X offset is .707. Likewise, line E has to be copied upward (positive Y direction) a distance of .707 to get to Eo, so the Y offset is .707.

6. We have not yet defined line E itself, only its angle and offsets. To describe the actual line itself, begin from the start point for line E. It is drawn as a solid line at 0.25 units long, so we enter .25 as the next number. Remember that a positive number equals a solid line (pen down).
7. Any line of a hatch pattern needs to be described from one origin to the next, so we need to define the movement of our pen back to the new

origin. From the far right end of line E, we need to lift the pen up so that a line is not drawn and then move the pen 1.164 units to the right, which will put us at the next origin. To do that, we enter a minus sign (pen up) so that the final number is −1.164.

Line F

1. The first two digits are always 0, 0 in DataCAD.
2. From the beginning of line F (at the hatch origin), it is drawn at 225 degrees, so the next number is 225.
3. Since the start point of line F starts at the hatch origin, the offset to the start of the line is 0 in both the X and Y directions. Therefore, the next two numbers are 0, 0 (see Figure 23-50).
4. The next two numbers define the X and Y distances to the repeated line. The dashed line Fo (F offset) is where we will define the repeated line, but the trick again is that the X and Y distances must be described with the line in a horizontal orientation. To do this, you need to mentally rotate the hatch pattern, as shown in Figure 23-51.

   Think of how you would tell DataCAD to copy line F to the location of line Fo. In this case, you have to copy line F to the right (positive direction) at a distance of .707. Therefore, the X offset is .707. Likewise, line F has to be copied upward (positive Y direction) a distance of .707 to get to Fo, so the Y offset is .707.
5. We have not yet defined line F itself, only its angle and offsets. To describe the actual line itself, begin from the left end of line F (notice

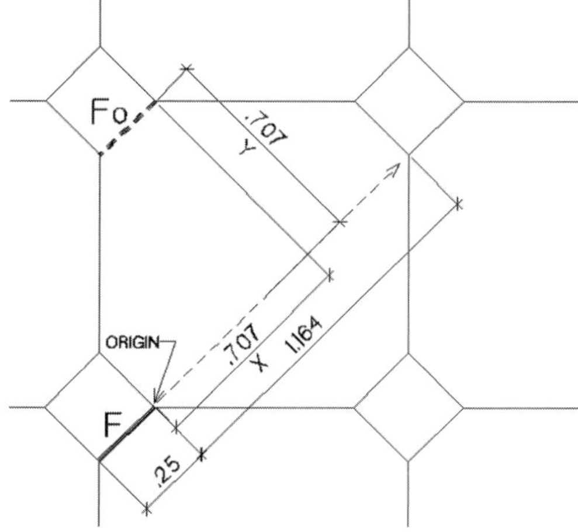

Figure 23-50
Line F

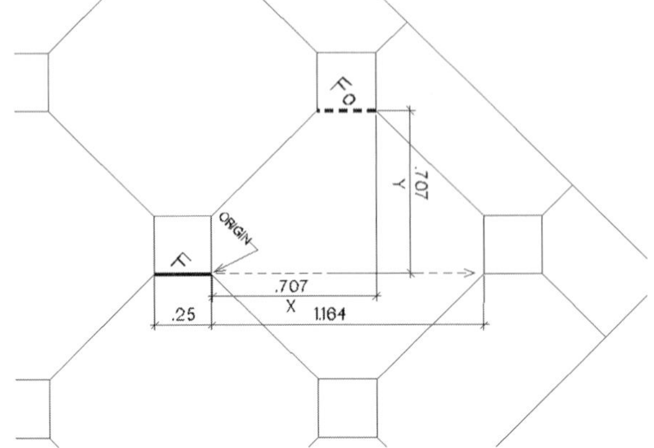

Figure 23-51 Rotating the hatch pattern

that this is not the start point of line F as defined in Step 3; it doesn't have to be). It is drawn as a solid line at 0.25 units long, so we enter .25 as the next number. Remember that a positive number equals a solid line (pen down).

6. Any line of a hatch pattern needs to be described from one origin to the next, so we need to define the movement of our pen back to the new origin. From the far left end of line F, we need to lift the pen up so that a line is not drawn and then move the pen 1.164 units to the left, which will put us at the next origin. To do that, we enter a minus sign (pen up) so that the final number is −1.164.

Completing the Pattern

Our final result looks like this:

```
*DmndTile
0, 0, 0, 0, 0, 0, 1, .646, −.354
0, 0, 90, −.177, .177, 0, 1, .646, −.354
0, 0, 135, 0, 0, −.707, .707, .25, −1.164
0, 0, 45, −.354, 0, .707, .707, .25, −1.164
0, 0, 315, −.354, 0, .707, .707, .25, −1.164
0, 0, 225, 0, 0, .707, .707, .25, −1.164
```

We have now created a pattern where the large tiles are 1 unit × 1 unit in size. But as we noted in the beginning of this example, we want to the size of the large tiles to be 1″ [25.4] when a **Scale** of **0.1** (1″) is selected. To do this we just have to multiply all of the dimensions (not the angles) in the pattern by a factor of 384 (12″ × $^1/_{32}$nd of an inch). When we do so, the final result looks like this:

*DmndTile

0, 0, 0, 0, 0, 0, 384, 248.064, −135.936

0, 0, 90, −67.968, 67.968, 0, 384, 248.064, −135.936

0, 0, 135, 0, 0, −271.488, 271.488, 96, −446.976

0, 0, 45, −135.936, 0, 271.488, 271.488, 96, −446.976

0, 0, 315, −135.936, 0, 271.488, 271.488, 96, −446.976

0, 0, 225, 0, 0, 271.488, 271.488, 96, −446.976

Once you finish defining the pattern, the final definition can be inserted into your DCADWIN.PAT file in any order, and the hatches will be listed in the Hatch menu in the order in which they occur in the file.

NOTE: *Because of decreasing accuracy as you move farther from the 0,0 origin of the drawing file, if a hatch pattern is located very far from that origin, the pattern may not display properly, with some of the hatch lines seeming to display all over the place.*

When a hatch pattern is drawn in the Drawing Window, its initial hatching origin is at the drawing file's 0,0 origin. By selecting ***Hatch/ Origin*** *and picking a new origin within the hatched area, the hatch origin will move from the drawing's 0,0 origin to the newly selected point. That should fix any incorrectly displayed hatch patterns. Keep this in mind when creating new hatch patterns, too.*

We would be remiss if we didn't tell you more about the third-party Hatch Manager macro, available from *Cheap Tricks Ware* (see Chapter 27, "Data-CAD Resources") by John Fornaro and Design/Program Associates. It won't scan the pattern you have drawn and then automatically create the hatch definition for you, but if you understand how a hatch pattern is created, it will take you line by line through the creation of a pattern, measure your line lengths and angles for you, and add the new definitions to your .PAT file. Thus, you don't have to worry about your geometry lessons as much. The other great feature of Hatch Manager is that it gives you control of the first two parameters of line width and line color. You can use it to actually design hatches that have a multiplicity of line weights and line colors, which can be accessed through the Hatch Manager macro (and only through the macro).

We must, however, mention one shortcoming of this macro, which exists as we went to press. The macro was originally written for the DOS version of DataCAD, which used a .PAT file called DCAD.PAT. In order to use the macro, you will need to copy your DCADWIN.PAT file and rename it DCAD.PAT. After creating your final hatch pattern, you will then have to rename the DCAD.PAT file to DCADWIN.PAT. Remember to always keep a backup of the original DCADWIN.PAT file in case you make a mistake and need to go back. Perhaps the macro will be updated by the time you read this.

Hatch Pattern Preview Images

Once you have created a custom hatch pattern in the DCADWIN.PAT file, it will be displayed in the **Hatch/Pattern** menu and is immediately accessible for use. A preview image of the new hatch pattern will also be immediately available for display in the hatch preview window as you pass your cursor over each hatch pattern. DataCAD uses the hatch definition in the .PAT file to display the hatch, based on the current hatch settings in the drawing file.

Custom Scales, Angles, and Distances

You can use the **Settings/EditDefs** option to customize the default scales, angles, and distances that DataCAD uses in many of its menus, and when pressing the space bar to input distances and angles. The settings are saved in the DataCAD |SUP directory as DCANWIN.SCL (scale), SCADWIN.ANG (angle), and DCADWIN.DIS (distance).

Scales

To edit the values displayed in the **Utility/Plotter/Scale** menu, follow these directions:

1. Select **Utility/Settings/EditDefs/Scales**.
2. If the **Scale** list is already full with 18 scales, then you can only change existing scales. Select **Change**, then pick the scale to be changed.

3. Enter the new scale text, up to 8 characters, to be displayed on the scale menu button.
4. Enter the new value of the new scale (see the following directions). It must be expressed in decimal form.
5. Saving new scale settings:
   a. The new angle will appear in the menu as long as you remain in the current drawing file, but if you do not save the change, then you will lose all your changes upon exiting the drawing.
   b. You can save the settings to the DCADWIN.SCL file, which is the default file that DataCAD opens each time a drawing is opened.
   c. Select the **SaveFile** option, select the **DCADWIN.SCL** file, then select **Save**. The new settings will now be in effect every time you open a DataCAD drawing.
   d. You can also save the new settings to a different file, that can be called up with the **LoadFile** option. To do this, select **SaveFile**, then type a new name, like **WMA.SCL**.

The DataCAD manual gives you the scaling value for the most common scales. To calculate the scaling value for architectural scales:

- scale / 12

  Ex: (1/4″ scale would be fi (1/4)/12 5 .2083333

  Or → .25/12 = .02083333

  Ex: $1^1/_2$″ scale would be → (1.5)/12 5 .125

Calculating the scaling values for imperial engineering scales is mostly the same as for architectural scales, requiring division by 12:

- scale / 12

  Ex: 1:20 scale would be fi (1/20)/12 5 .00416666

  To calculate the scaling value for metric scales:

- 1 / scale

  Ex: 1:20 scale would be fi 1/20 5 .05

  Caveats:

- If you use the **LoadFile** option to load a custom scale, angle, or distance file (like WMA.SCL), then save your drawing, DataCAD will save those settings with the drawing file, and will use those settings every time the drawing is opened, even if it is opened on another computer.

- Whenever you start a new drawing, or open an existing drawing that did not have custom scale settings, DataCAD always uses the DCADWIN.SCL file, the DCANWIN.ANG file, or the DCADWIN.DIS file.
- The maximum number of values is 18; the number of buttons that fit vertically in one menu. So if there are already 18 scales in a menu you will have to get rid of one existing scale for every new scale you add.
- Scales, angles or distances are always shown in ascending order; you can not alter that order. If you have a keyboard or icon macro that uses the scales in the scale menu, make sure that changing the values has not affected your macro.
- The previous caveat also can have one other serious repercussion: It can wreak havoc with previous MSP layouts, since the Scale settings of MSP details are dependent on the order of the values in the Scale menu. That is, the scales are assigned by the order they appear in the list of scales. So if you use custom scales, or alter the DCADWIN.SCL file with new scale values, you will probably need to check and re-lay out MSP layouts in previously created drawing files. Ditto if another DCAD user receives your drawings.

Angles

This is easier than editing scales. All you do is add or edit the angle values directly:

1. Select **Utility/Settings/EditDefs/Angles.**
2. To add a new angle, select **Add** and then type in a new angle, such as **75** for 75 degrees. Done!
3. Saving new angle settings:
   a. The new angle will appear in the menu as long as you remain in the current drawing file, but if you do not save the change, then you will lose all your changes upon exiting the drawing.
   b. You can save the settings to the DCADWIN.ANG file, which is the default file that DataCAD opens each time a drawing is opened.
   c. Select the **SaveFile** option, select the **DCADWIN.ANG** file, and then select **Save**. The new settings will now be in effect every time you open a DataCAD drawing.
   d. You can also save the new settings to a different file that can be called up with the **LoadFile** option. To do this, select **SaveFile**, and then type a new name such as **WMA.ANG**.

Caveats:

- If you use the **LoadFile** option to load a custom angle file (like WMA.ANG) and then save your drawing, DataCAD will save those settings with the drawing file and will use those settings every time the drawing is opened, even if it is opened on another computer.
- Whenever you start a new drawing, or open an existing drawing that did not have custom angle settings, DataCAD always uses the DCADWIN.ANG file.
- The maximum nuber of values is 18; the number of buttons that vertically in one menu. So if there are already 18 angles in a menu you will have to get rid of one existing angle for every new angle you add.
- Angles are always shown in ascending order; you can not alter that order. If you have a keyboard or icon macro that uses the angles in the angle menu, make sure that changing the values has not affected your macro.

Distances

Creating custom distances is much the same as for angles. All you do is add or edit the distance values directly:

1. Select **Utility/Settings/EditDefs/Distnces**.
2. To add a new distance, select Add, and then type in a new distance such as **3.6** for 3′-6″ or **.3.5/8** for 3 ⁵⁄₈″.
3. Save the new distance settings:
   a. The new distance will appear in the menu as long as you remain in the current drawing file, but if you do not save the change, you will lose all your changes upon exiting the drawing.
   b. You can save the settings to the DCADWIN.DIS file, which is the default file that DataCAD opens each time a drawing is opened.
   c. Select the **SaveFile** option, select the **DCADWIN.DIS** file, and then select **Save**. The new settings will now be in effect every time you open a DataCAD drawing.
   d. You can also save the new settings to a different file that can be called up with the **LoadFile** option. To do this, select **SaveFile** and type a new name such as **WMA.DIS**.

When customizing angles, scales, and distances, certain caveats should be kept in mind:

- If you use the **LoadFile** option to load a custom scale, angle, or distance file and then save your drawing, DataCAD will save those settings with the drawing file, and will use those settings every time the drawing is opened, even if it is opened on another computer.
- Whenever you start a new drawing or open an existing drawing that does not have custom distance settings, DataCAD always uses the DCADWIN.DIS file for distances, the DCADWIN.ANG for angles, and the DCADWIN.SCL file for scale files.
- The maximum number of values is 18, the number of buttons that fit vertically in one menu. So if 18 scales, angles, or distances already exist in a menu, you will have to get rid of one for every new one you add.
- Scales, angles, and distances are always shown in ascending order, which cannot be altered. If you have a keyboard or icon macro that uses the scales, angles, or distances in the respective menu, make sure that changing the values has not affected your macro.
- With regard to Scales, the previous caveat can also have one other serious repercussion. It can wreak havoc with previous MSP layouts since the Scale settings of MSP details are dependent on the order of the values in the Scale menu. That is, the scales are assigned by the order in which they appear in the list of scales. So if you use custom scales or alter the DCADWIN.SCL file with new scale values, you will probably need to check and re-lay out MSP layouts in previously created drawing files. You'll also need to do so if another DCAD user receives your drawings.

Converting AutoCAD Fonts to DataCAD

You might sometimes see a font in AutoCAD that you would like to use in DataCAD, or you may need to share drawing files between yourself and an AutoCAD user. To do this, you need to have the identical fonts on each system, but AutoCAD and DataCAD fonts cannot be shared directly, since they use totally different formats. However, some AutoCAD fonts can be converted to DataCAD format.

AutoCAD fonts have a .SHX file extension, while DataCAD fonts have a .CHR extension, but you can't convert a SHX file to a CHR file directly. You first need to convert the SHX file to a shape (.SHP) file format. To do this, you need a piece of free software that is available from the Internet called SHX2SHP.EXE. Luckily for you, we have included it on the CD-ROM so that you won't have to go searching for it. When you install DataCAD, a DOS program called SHP2CHR.EXE is installed in your \CHR directory. This little program will convert a SHP file to DataCAD's CHR format.

Earlier we said that only some AutoCAD fonts could be converted to DataCAD format. That's because AutoCAD uses some shape descriptions that are not supported in DataCAD. The only way you will find out which fonts work and which don't is to convert them and try them out in DataCAD.

First, The Easy Way

O.K., you have two incredibly easy ways to get AutoCAD/DataCAD-compatible fonts:

- Download a whole slew of ACAD and DCAD fonts from the DataCAD Web site.

or

- Get those very same fonts from the CD-ROM included with this book.

Pretty simple, huh? And while you're there, you can look for all the international fonts as well. DataCAD LLC does a pretty good job of updating these files on their Web site, so if you don't see something you like on the CD, keep checking the DataCAD Web site.

Next, The Long Way

The more involved method of converting AutoCAD fonts to DataCAD format uses the method described earlier: convert the AutoCAD .SHX font file to a .SHP file with the freeware SHX2SHP.EXE program, and convert the resulting .SHP file to a DataCAD .CHR font file with the SHP2CHR.EXE program. Then follow these steps:

1. Create a new file folder somewhere convenient and copy the SHX2SHP.EXE program to it. Then move all the .SHX fonts that you want to convert to that same file folder.
2. Look in the Conversion section of the CD-ROM that comes with this book for a DOS batch file called SHX2SHP.BAT. Copy this file to the directory created in Step 1. This is a very simple mini-program that

will convert any number of SHX files with the *click* of a button. If you choose not to use this batch program, you can still convert the SHX files, but you will have to do them one at a time.

a. Open the SHX2SHP.BAT file with a text editor like Notepad. It will look like this:

```
COPY *.shx *.
FOR %%N IN (C:\DCADWIN\CHR\CONVERT\*.) DO shx2shp %%N %%N
del *.
```

b. Change the path within the parentheses to match the path to the files created in Step 1 (be sure to leave the ***** just before the right parentheses). This will tell the program where to look for all the SHX fonts and where to find the SHX2SHP.EXE program. A batch file is a DOS program, so in order for it to work, your path names must conform to DOS standards (no long file names over eight characters, no spaces, and no special characters). Then save the .BAT file.

3. *Double-click* on the SHX2SHP.BAT file. A DOS window will appear and all the SHX fonts will automatically be converted to SHP files with identical names.

4. Look again in the font section of the CD-ROM that comes with this book for another DOS batch file called SHP2CHR.BAT. Copy this file to the directory created in Step 1. This program will convert any number of SHP files with the *click* of a button. If you choose not to use this batch program, you can still convert the SHP files, but you will have to do them one at a time.

a. Open the SHP2CHR.BAT file with a text editor like Notepad. It will look like this:

```
COPY *.shp *.
FOR %%N IN (C:\DCADWIN\CHR\CONVERT\*.) DO shp2chr %%N
del *.
```

b. Change the path within the parentheses to match the path to the files created in Step 1 (be sure to leave the ***** just before the right parentheses) This will tell the program where to look for all the SHP fonts and where to find the SHX2SHP.EXE program. If you are an astute reader, you might notice that line 2 in the SHP2CHR.BAT file has only one %%N at the end of the line, although the SHX2SHP.BAT file had two. This is correct and is not a typo. Save the .BAT file.

5. Double-*click* on the SHP2CHR.BAT file. A DOS window will appear and all the SHP fonts will automatically be converted to CHR files with identical names.

6. Copy all the new .CHR fonts to the DataCAD \CHR directory.
7. Test the new fonts in DCAD to see if they work.

This may seem like a lot of steps, but Steps 2 and 4 only have to be done once. After that, you won't have to ever do it again.

Creating Custom Fonts

DataCAD's fonts are proprietary, which means that only DataCAD uses them, and they have to be in a format that DataCAD understands. All DataCAD fonts have a .CHR file extension (short for "character"), but to create a font in DataCAD, you must first create a shape file to describe each line that makes up each letter and character of a particular font style. This shape file has a .SHP file extension and is a simple ASCII text file. This file is then compiled into a .CHR file that DataCAD can use. Although a SHP file is an editable text file, the compiled .CHR file is not. Your DATACAD\CHR directory contains some sample SHP files to explore. Here is the shape file code for a capital letter A:

```
*65,0, capital A
1,8,(42,100),1,8,(42,-100),2,8,(-71,31),1,8,(58,0),2,8,(37,-31),0
```

The first line is the ASCII character # and the font title. The second line is the code definition for the character "A."

If you read through the instructions in this chapter about how to create custom linetypes, then you will begin to recognize the similarities here. Much like a linetype definition, the shape code describes coordinates for each stroke, or vector, of the letter. Working from a predefined grid of points, you map each vector as a pen down or pen up stroke in relative x, relative y fashion.

NOTE: *If you have ever worked with AutoCAD's shape files and proprietary fonts, then you will quickly recognize the similarities between DataCAD's and AutoCAD's shape files. The two are, in fact, nearly identical. That's why, in the previous section of this chapter, you can convert some AutoCAD shape files to DataCAD fonts.*

The two previous SHP file lines show the beginning of the entire SHP file associated with the capital letter A. The first few lines in the header describe the font name, set the grid variables, and define the distance (line feed) between carriage returns. Next comes the code for a space and then all the characters themselves. The carriage return code is optional but is used in all AutoCAD fonts. All the characters on your keyboard have a numerical ASCII value that the operating system uses to display the character on your screen. For instance, the number 65 designates the ASCII assignment for the letter A. Here is a list of all the standard English ASCII character codes:

value	name	value	char	value	char	value	char
0	NUL	32	space	64	@	96	`
1	SOH	33	!	65	A	97	a
2	STX	34	"	66	B	98	b
3	ETX	35	#	67	C	99	c
4	EOT	36	$	68	D	100	d
5	ENQ	37	%	69	E	101	e
6	ACK	38	&	70	F	102	f
7	BEL	39	'	71	G	103	g
8	BS	40	(	72	H	104	h
9	HT	41	)	73	I	105	i
10	LF	42	*	74	J	106	j
11	VT	43	+	75	K	107	k
12	FF	44	'	76	L	108	l
13	CR	45	-	77	M	109	m
14	SO	46	.	78	N	110	n
15	SI	47	/	79	O	111	o
16	DLE	48	0	80	P	112	p
17	DC1	49	1	81	Q	113	q
18	DC2	50	2	82	R	114	r
19	DC3	51	3	83	S	115	s
20	DC4	52	4	84	T	116	t
21	NAK	53	5	85	U	117	u
22	SYN	54	6	86	V	118	v
23	ETB	55	7	87	W	119	w
24	CAN	56	8	88	X	120	x

25	EM	57	9	89	Y	121	y
26	SUB	58	:	90	Z	122	z
27	ESC	59	;	91	[	123	{
28	FS	60	<	92	\	124	|
29	GS	61	=	93	]	125	}
30	RS	62	>	94	^	126	≅
31	US	63	?	95	_	127	del

Character codes 1 through 31 are reserved for the operating system and cannot be used. That is why your first font definitions will generally start with character 32, a space. The first 127 characters are the most often used, but 128 other extended characters are available for total of 255. Some of these are relevant to us because they include fractions and other special characters, as in the following mini-table. The full list of extended English language characters can be found in the Appendix.

value	char
171	½
172	¼
237	∅
241	±
248	°

When the DataCAD user wants to use one of these special, extended characters (assuming that you have created the corresponding font character), they hold down the **Alt** key while pressing the numbers on the numeric key pad. Upon letting up on the **Alt** key, the character will appear. These characters will *not* work if you press the regular numbers across the top of your keypad.

However, you do not have to recreate the corresponding extended character that you see in the chart. You can make up your own special characters (for instance, architectural symbols like centerline, angle, or diameter) and assign them to a specific **Alt+#** key sequence. You will have to be sure to let the user know that these characters are part of the font set, however, and tell them the **Alt+#** key sequence to access it. From our experience, if you don't provide written documentation of these facts, then the special characters tend to be forgotten.

Here is a sample of one font set with characters *33 through *64 omitted:

Code	Description
Font	Font name
*0,4, HLV-850 Latin I 100,100,0,0	Grid size X, Y
*10,0, LINE FEED 2,8,(0,140),0	Line feed drop distance
*13,0, CARRIAGE RETURN 6,0	Carriage return code (optional)
*32,0, SPACE 2,8,(84,0),0	Spacebar Code definition for a space
*65,0, capital A 1,8,(42,100),1,8,(42,-100),2,8,(-71,31), 1,8,(58,0),2,8,(37,-31),0	ASCII Character # and Title Code definition for character A

Following the code sequence for the capital A, we see that the letter begins with a pen down command, designated through code as 1,8. The pen stroke, or vector, goes 42 units in the relative x direction and 100 units in the relative y direction to produce an angled line. Commas separate the pen commands and vector values, and brackets help to distinguish the pen commands from the vector values. These values must be whole units that lie within the grid specified in the header. Next, the pen down command is given again to signal the next vector, and it continues 42 units in the x direction and this time 100 units downward in the y direction. Now a pen up command is signaled with the code 2,8, and the pen moves 71 units to the right and 31 units upward. The pen is brought down again with a 1,8 command and continues 58 units horizontally to the right and zero units vertically. Finally, the pen is parked in the up position 37 units to the right and downwards 31 units. This leaves a blank space, or kerning, of 24 units between the leg of the A and the start of the next character. Zero designates the end of the code sequence for this character. Figure 23-52 shows the code sequence illustrated graphically.

Although creating a font is fairly simple, the complexity comes from the scores of vector definitions that might make up a complicated letter. Further complexity can be introduced by defining octant arcs, introducing scaling factors, and making the code more efficient by chaining multiple pen commands together. For most fonts, however, straight-line vector geometry is sufficient. Curves can be achieved by breaking the arc down into line seg-

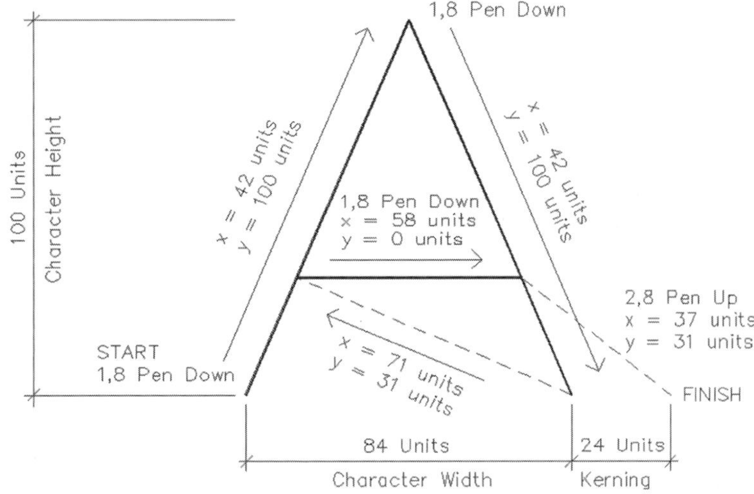

Figure 23-52
An illustration of the code sequence

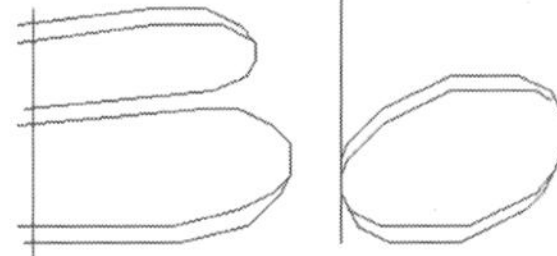

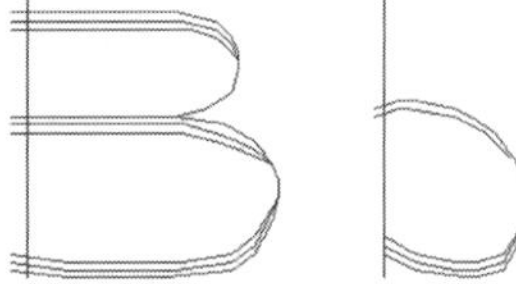

Figure 23-53
The vectors that make up the fonts.

Figure 23-54
Part of a font set

ments. This is how most DataCAD fonts are made. If you zoom in very close to a font in DataCAD, it will looke like Figure 23-53.

Creating a Font Set

The first step in creating an entire font set is to draft out all of the characters, as shown in Figure 23-54 (the numbers underneath the letters are the ASCII codes for that letter).

For an English-only font, the entire font set would consist of ASCII characters 32 through 126, about 89 characters including punctuation. You can also add some of the custom characters like fractions and architectural symbols. It is important to keep your linework on a predefined grid so the coordinate values are whole numbers. For a high-resolution font, we like to use a 100-unit grid. This adds a lot of extra work and code but provides for smoother looking arcs. A high-resolution font for all 256 characters can take well over a thousand lines of code. Most of the fonts supplied with DataCAD are based on a grid of 27 units and provide acceptable resolution. You must also determine the kerning distance between each character and the amount of line feed distance required. For this, you must take descenders into account, which are the strokes that go below the base line, as with the small letters g and y. Foreign characters like the Scandinavian Å require extra space above the character as well.

DATACAD LLC has made numerous foreign font sets available for free download from the DataCAD Web site. Some of the currently available fonts include Slavic, Baltic, and Cyrillic languages among others.

Font Preview Images

Once you have created a custom font set and placed it in the \CHR directory, it will be displayed in the **Text/FontName** dialog box and is immediately accessible for use. A preview image of the new font will also be immediately available for display in the preview window as you *click* once to highlight the font because DataCAD uses the .CHR file to display the text preview.

The Fontmaster Macro

As with the creation of custom hatches, we feel it is appropriate to mention a third-party DataCAD macro called *Fontmaster* (available from Cheap Tricks Ware) that will help you create new fonts within DataCAD. After you have drawn out your entire character set in the DataCAD Drawing Window, the macro enables you to trace each letter. Then it records the relative x and y coordinates for each stroke and marks it as a pen up or pen down sequence. This eliminates all the math work involved in figuring out the vectors and ensures that each line of code is written in the correct format. Here's how it works.

Once your font is drafted out, you start the Fontmaster macro and begin tracing. The macro will prompt you for a font name, the grid size, the line feed value, and the distance in units for a blank space. A default value of 1 unit = 1′ is assumed, so setting your grid snap to one foot is helpful. Next, you simply trace each letter, and then when finished, enter its ASCII code and name. The macro will enable you to choose colors for pen up and down vectors so it is easy to monitor your progress, and it even has a built-in utility for determining the proper ASCII code for each character. The macro will also optimize the code for multiple pen down commands to make the code more efficient. Another option enables you to see the code onscreen as you work. All that remains is to proceed through each letter in your font.

As this process may take several sessions and can be limited by one's endurance, the macro allows you to open previously created .SHP files and add to them. Once complete, or to check your progress along the way, the macro has a function to compile the final .SHP file into DataCAD's .CHR font format.

The Productivity Pack

One set of options that you are presented with when installing DataCAD is the inclusion of the Productivity Pack features, which include program files, linetypes, and keyboard macros. In a nutshell, the Productivity Pack is a series of six linked icon toolbars that enable you to work in a task-oriented way. Each toolbar has links to the other toolbars for easy switching between them. Additionally, a series of new linetypes is added to your DCADWIN. LIN file, and a new set of keyboard macros (**Alt+Letter**) is added to your \SUP directory. Here is what each of the toolbars look like and what they do.

2D Toolbar

This toolbar provides access to the most often used 2D drafting commands and features, such as walls, windows, cleanup functions, and *GotoViews* (GTVs) (see Figures 23-55a and 23-55b). It also has icons to access all of the other five Productivity Pack toolbars.

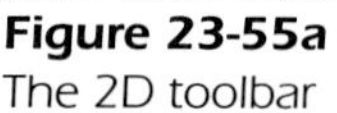

Figure 23-55a
The 2D toolbar

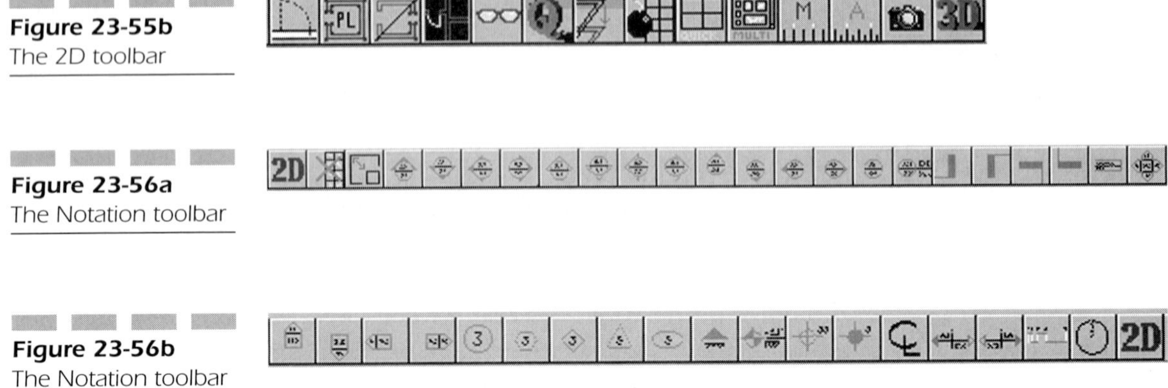

Figure 23-55b
The 2D toolbar

Figure 23-56a
The Notation toolbar

Figure 23-56b
The Notation toolbar

Figure 23-57a
The Dimensioning toolbar

Notation Toolbar

This toolbar provides access to the most commonly used detail symbols and notes, including symbols for wall types, door and window types, section cuts, and elevations (see Figures 23-56a and 23-56b). One great aspect of using these symbols is that after placing a symbol you will automatically be prompted to enter new text (detail number, sheet number, and so on). This is done by using *nested symbols*, where the underlying symbol remains as an intact symbol and the text is exploded so that you can change it. You can read more about this in Chapter 14, "Templates and Symbols."

Dimensioning Toolbar

This provides access to the most commonly used text, dimension, and arrow functions at various sizes to match the scale of the detail you are working on. Make sure that your current plot scale is set to the scale of the detail you are working on. Clicking on the appropriate icon will then enable you to create text, dimensions, and arrows of the correct size to match the detail you are working on (see Figures 23-57a and 23-57b).

Figure 23-57b
The Dimensioning toolbar

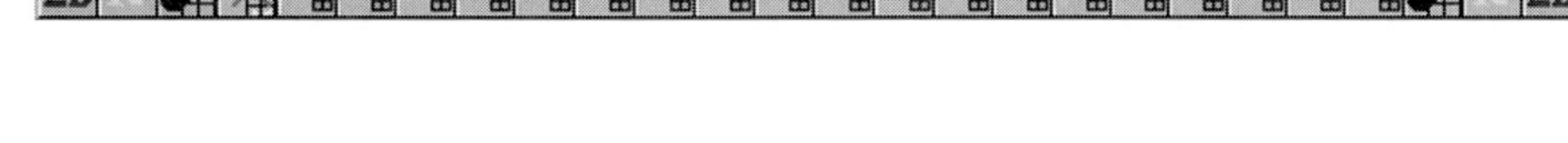

Figure 23-58
The CSI Template toolbar

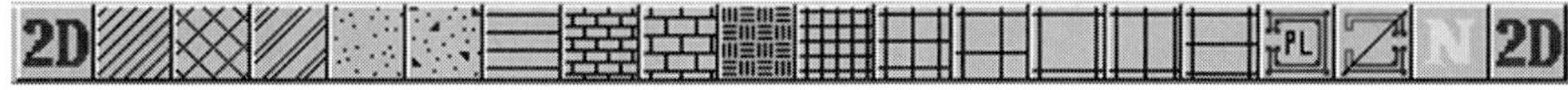

Figure 23-59
The Hatch toolbar

CSI Template Toolbar

This toolbar provides access to each of the 16 CSI-formatted template files, along with a miscellaneous Details and Jobs file. Two other icons enable you to place exploded symbols and to turn off the display of the template windows (see Figure 23-58).

Hatch Toolbar

This toolbar provides access to 15 of the most commonly used standard DataCAD hatch patterns as well as polyline creation and editing options (see Figure 23-59).

3D Toolbar

This toolbar provides access to 39 common 3D commands, some of which create primitive shapes, slabs, and polygons or provide access to perspective views, parallel views, the GLShader, and the Hide command (see Figures 23-60a and 23-60b). A few icons access features in the 3D Power Tools and DC Sprint macros, which are third-party macros that do not come with DataCAD but can be purchased from DataCAD LLC as part of the Macro Tabouret CD-ROM.

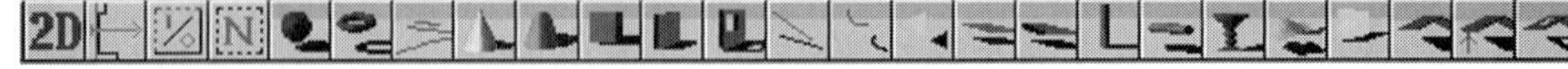

Figure 23-60a
The 3D toolbar

Figure 23-60b
The 3D toolbar

Installing the Productivity Pack Options

You can install the Productivity Pack options when you first install DataCAD, or you can install them later. If you install them later, just run the SETUP.EXE program on the CD-ROM. When you get to the Setup Type dialog box, choose the Custom option, after which you will see the Select Components dialog box. Scroll down until you see the Productivity Pack options. Make sure all the options are unchecked, except for the three Productivity Pack options. Select **Next** and follow the directions. The following sections outline what is installed by each of these options.

Program Files This option accomplishes the following tasks:

- Adds seven new icon toolbar .KEY files to the \SUP\MENUPOF directory:
  2Dtl_win.key
  3Dtl_win.key
  Ardm_win.key
  Htch_win.key
  Note_win.key
  Sscl_win.key
  Tpl-win.key
- Adds 181 new .BMP icon toolbar buttons to the \SUP\MENUPOF directory
- Adds 19 new template (.TPL) files to the \TMP directory. All the new template files are placed in CSI-formatted template directories. If you did not originally install your DataCAD templates and symbols in CSI directories, they will be created for you now.

- Adds 92 new symbol (.SM3) files to the \SYM directory. The new symbol files are placed in miscellaneous directories, generally by material names.
- Adds new Help File information about the Productivity Pack additions to the \HELP directory

Linetypes This option creates a new DCADWIN.LIN file with 60 new linetypes in addition to the standard linetypes that are part of the original .LIN file when DataCAD is first installed.

WARNING: *Make sure you copy your old DCADWIN.LIN file* before *installing the Productivity Pack, since the original .LIN file will be replaced by the new one without any backup being created first. Since DataCAD accesses its linetypes by their order in the .LIN file and not by their names, you would be well advised to copy the new linetypes from the ProPack .LIN file onto the end of your original .LIN file, rather than using the ProPack .LIN file itself.*

The Linetypes option also adds 60 .BMP preview images to the \SUP\LINEPOF directory. The new linetypes are as follows:

Arrow	Bar	Boxes	Brikcors
Bubble	Clngtile	Corugate	Criscros
-dash-	Dblarrow	Diamond	Dotdash2
Dshngl_l	Dshngl_r	Hedge1	Hedge2
Hexagon	i-line	Insul1	Insul2
Insul3	Insul4	Mapline	Marble
Msnryblk	Mtldeck	Mtlstud	Newbrick
Newplywd	Newsectn	Parkln_l	Parkln_r
Pshngl_l	Pshngl_r	Rigdins1	Rigdins2
Rndfence	Rndmhevy	Rndmlite	Shiplapl
Shiplapr	Sqrfence	Sqr-post	Stipple
Stone	T&g-sid	Traintrk	Treeline
Undulate	Waterlin	Wiggle	Woodgrn1
Woodgrn2	Woodgrn3	Woodstud	Woodwall
-x-	x-fence	Zigzag	z-line

Keyboard Macros This option creates a new DCADWIN.MCR keyboard macro file in the \SUP directory.

WARNING: *Make sure you copy your old DCADWIN.MCR file* before *installing the Productivity Pack, since the original .MCR file will be replaced by the new one without any backup being created first.*

Here is what each of the new keyboard shortcuts (**Alt+Letter**) does:

A	Curves menu	N	Not used in Windows version
B	1-Line trim	O	One-line trim, select outside of line
C	Change menu	P	Plotter menu
D	Divide entity into two divisions	Q	Quit
E	Enlarge menu	R	Polygons menu
F	Fillet, enter radius	S	Stretch menu
G	Offset perpendicular, enter distance	T	Text menu
H	Selection Set, mask by color	U	T-intersection cleanup
I	Identify/Set All	V	Change all settings to match current
J	Two-line trim	W	Weld Line menu
K	Link entities	X	Measures menu
L	L-intersection cleanup	Y	Create new layer and prompt for name
M	Mirror menu	Z	Change text contents

Although you may choose not to use some of these Productivity Pack options, they will become extremely useful for creating your own custom icon toolbars and keyboard macros, as described earlier in this chapter. You can use or modify the .BMP icons for your own toolbars, and the underlying .KEY and .MCR coding is well worth exploring, understanding, and editing for you own custom uses. And for the history books, Richard Morse, co-author of this book, was the person who created the entire Productivity Pack.

CHAPTER 24

Third-Party Macros

No matter how good a software program is, others will always step in to make it even better. Using DataCAD's own DCAL programming language, many users have created new **Toolbox** macros to help you get even more out of the program. The following are some of those macros. The first set of macros is available for free from the DataCAD LLC Web site. The second set is some of my favorite macros available for nominal amounts from a DataCAD software distributor called Cheap Tricks Ware. The final group of macros is available on a single CD-ROM from DataCAD LLC.

DataCAD macros reside in the main DataCAD directory in the \DCX folder. All macros have a .DCX file extension. Once placed in the \DCX folder, all you have to do to run the macro is *click* on the **Edit/Toolbox** menu button, pick the macro you want to run, and then press **Enter** or **Open**. The macro will then run. That's all there is to it.

Free Macros from DataCAD LLC

These macros and documents are all currently available for free from the DataCAD Web site (`www.datacad.com/support/downloadpg.htm`). The list of macros is always growing, so consider this as only a partial list, and check their Web site often for new additions. To save you some computer-time we have put them all on the CD-ROM in the back of this book.

3D Line Conversion

Although it has other uses, the time you will use this macro most is after importing a .DWG drawing file from AutoCAD or other CAD programs. Many of the lines and arcs created by these programs are 3D entities. Since 3D lines cannot be edited with DataCAD's 2D editing functions, this macro will quickly convert all 3D lines to 2D lines to facilitate editing. This macro will only convert 3D lines that lie flat in the X/Y plane. It will not convert 3D lines that have different Z-heights at both ends. For those types of lines, use the **SubLine** macro.

SubLine

This macro is similar to the 3D Line Conversion macro, but it converts 3D lines to 2D lines using the 3D lines' X and Y coordinates. This macro differs from the 3D Line Conversion macro because it will automatically flatten 3D lines that have different Z-heights at both ends.

Annotate

Annotate is a collection of macros that help you to quickly and efficiently place annotation symbols on the drawing. Figure 24-1 shows some examples.

ArcArrow

DataCAD's standard text arrows can only be drawn with straight-line leaders. This macro enables you to draw arcs with either open or closed arrowheads at the end (see Figure 24-2). If the arrowhead is closed, it can be filled with crosshatching. You can also turn off the arrowhead and daisy-chain as many arcs together as you want. Then you can turn on the arrowhead to be drawn at the end of the last arc (great for electrical plans).

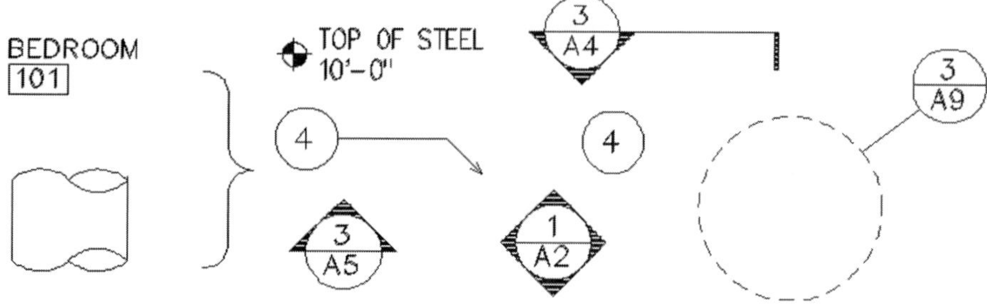

Figure 24-1 A variety of annotation symbols

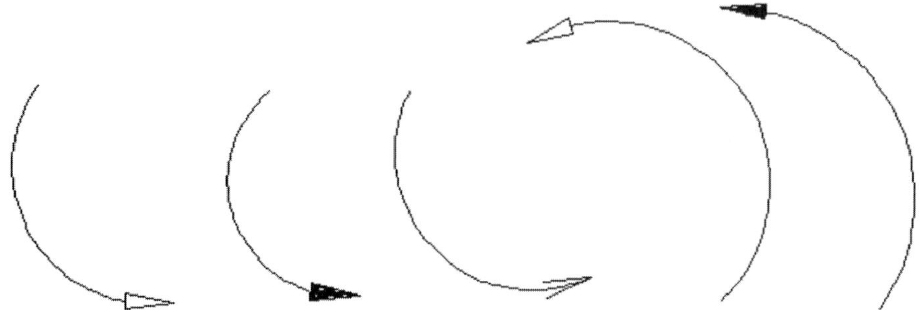

Figure 24-2 Drawing arcs with **`ArcArrow`**

ASCIIMax

This is a macro for processing ASCII text files into a schedule form on your drawings. Figure 24-3 displays a simple text file created in Windows Notepad.

Figure 24-4 shows the schedule created by the ASCIIMax macro from the previoustext file.

Break

This macro breaks an existing line at a given point, leaving a space in the line. The distance of the space can be defined with the macro. In addition, you can also have an arc drawn at the break location to indicate one line "jumping" over another (see Figure 24-5).

CountSym

Use this macro to count symbols on your drawing and place a description of the symbols, the number found, and the cost into an ASCII text file that can

Figure 24-3
A Notepad text file

schedule.txt - Notepad
File Edit Search Help

D O O R S C H E D U L E

No.	Size	Material	Frame Type	Finish
01	3'-0"x7'-0"	S.C. Wood	H.M.	Paint
02	2'-6"x7'-0"	S.C. Wood	H.M.	Paint
03	1'-6"x7'-0"	H.C. Wood	Wood	Clear Finish
04	(2)2'-0"x7'-0"	S.C. Wood	Wood	Clear Finish
05	3'-6"x6'-8"	S.C. Wood	H.M.	Paint
06	3'-0"x7'-0"	S.C. Wood	H.M.	Paint

Figure 24-4
The resulting ASCIIMax macro

D O O R S C H E D U L E

No.	Size	Material	Frame Type	Finish
01	3'-0"x7'-0"	S.C. Wood	H.M.	Paint
02	2'-6"x7'-0"	S.C. Wood	H.M.	Paint
03	1'-6"x7'-0"	H.C. Wood	Wood	Clear Finish
04	(2)2'-0"x7'-0"	S.C. Wood	Wood	Clear Finish
05	3'-6"x6'-8"	S.C. Wood	H.M.	Paint
06	3'-0"x7'-0"	S.C. Wood	H.M.	Paint

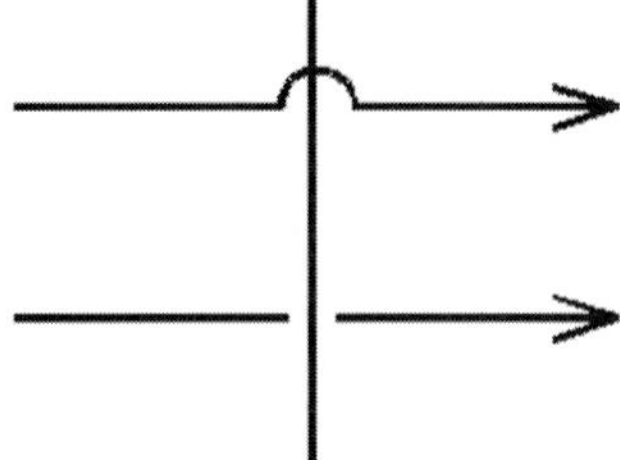

Figure 24-5 Two Break options

then be printed. The ASCII text file can also be appended to add new symbol count information.

Delete Smalls Macro

This macro enables you to delete or place all 2D and 3D lines under a user-specified length in a Selection Set. This is useful for cleaning up after a hidden line removal where there may be hundreds of little extraneous entities cluttering up your drawing and making the file size quite large. After running this macro, you may also want to run the LyrUtil macro to decrease the overall file size after the small entities have been deleted.

Dynamic On/Off Toggles

This macro toggles the **Dynamic** setting on or off as desired. Simply add a reference to this macro in your keyboard or toolbar macro sequence and you can have the **Dynamic** toggle set exactly the way you want.

DuctPro

DuctPRO is used to insert HVAC ductwork in your drawings. By simply selecting a width and depth (for rectangular ductwork) or a diameter (for round ductwork) and then locating the points on the duct run, the user can create a full 2D or 3D ductwork drawing. Choices for the duct type (rectangular or round), sizes, Z elevation, elbows, and branches are all available from the menu. In addition, the user can set DuctPRO to draw ductwork to the left, center, or right of given points as well as flat-on-bottom, flat-on-top,

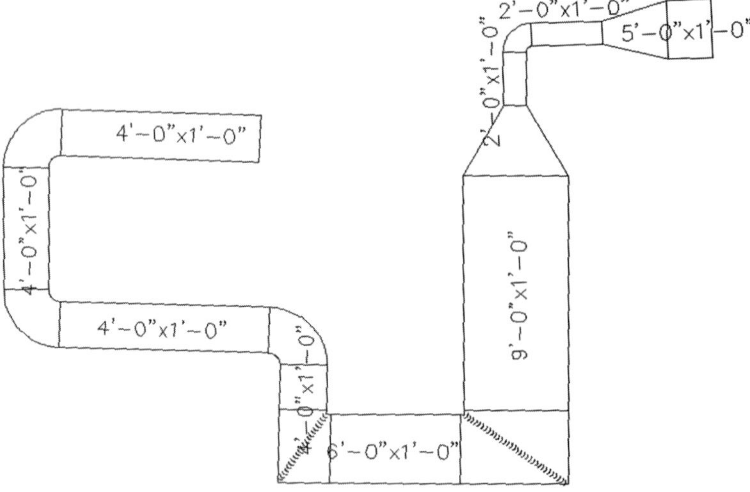

Figure 24-6
Ducts and text created with DuctPro

or centered on the Z elevation. The ductwork elbows, reductions, branches, vanes, and text are all drawn as you select points (see Figure 24-6).

GenOff (Generate Offset)

This macro is designed for constructing (generating) multiple offset lines. GenOff eliminates the need for offsetting multiple lines and then trimming them, and also allows for multiple repetitions of offset lines. Offset distances can be changed at any line segment during the macro to allow for unequal offset lines.

Gridlock

Running this macro sets every layer's grid settings to match the current active layer's grid settings. This is a great way to reset all the grid settings in all the layers in your drawing file in one shot.

LgtCalc

This macro calculates the number of light fixtures required for a selected space by setting the lumens, number of lamps, foot candles, and so on.

Macro Manuals

This is not a macro, but you can download and print out Adobe .PDF files with text and pictures regarding the use of these popular third-party macros: 3D Power Tools, DC Sprint, Blocker, Wavy, Template Librarian, and Touchup.

Macro Online Help

This is not a macro, but when this file is downloaded and installed, online help is added to the DataCAD **Help** menu for these third-party macros: 3D Power Tools, DC Sprint, Blocker, Wavy, Template Librarian, and Touchup. If you already own these macros, now you can get the answers to your macro questions with a *click* of the mouse. While in DataCAD, just choose **Add-ons** from the **Help** menu and then choose the **Macro_Tabouret** help file.

RampMaster

RampMaster enables the dynamic creation of both 2D and 3D ramps (see Figure 24-7). Ramps can include curbs and railings. They can be modeled as concrete masses or have a thickness to emulate other construction methods, such as wood. 2D and 3D ramps can be created at one time using Layer Memory to track the layers for each 2D and 3D component. Layer Memory also automatically places the components on the layers you specify. All settings are user-configurable and can be saved to text files for reuse.

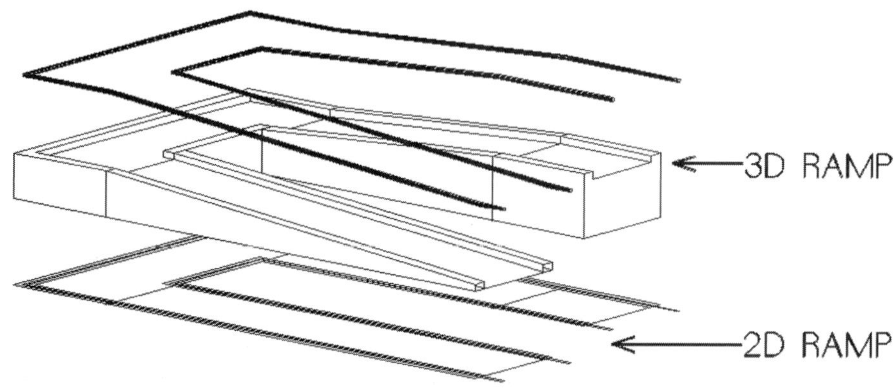

Figure 24-7 Creating 2D and 3D ramps

SteelPro

SteelPro will automatically draw virtually all the steel shapes currently listed in the AISC Manual of Steel Construction (see Figure 24-8). All wide flanges, tees, channels, angles, tubes, and pipes are available. It even draws open web steel joists. All steel sections can be input into your drawing in either top, side, or section views. Use SteelPro to draw all of your steel shapes and know for certain that your details are drawn accurately. This macro is one of my personal favorites.

SubText

This Substitute Text editing macro is used to search and replace (substitute) characters and/or words within a line or multiple lines of text using DataCAD's selection menu. For instance, all at once you can change all instances of "vinyl siding" to "wood siding" by running the macro and selecting the word "vinyl" to be replaced with the word "wood." SubText also enables the user to substitute text with text strings that are longer than the DataCAD normal maximum of 80 characters.

TxtLine (Text Line)

This macro virtually eliminates the need for creating special linetypes with text in them and allows you to change the text size and text-to-text spacing.

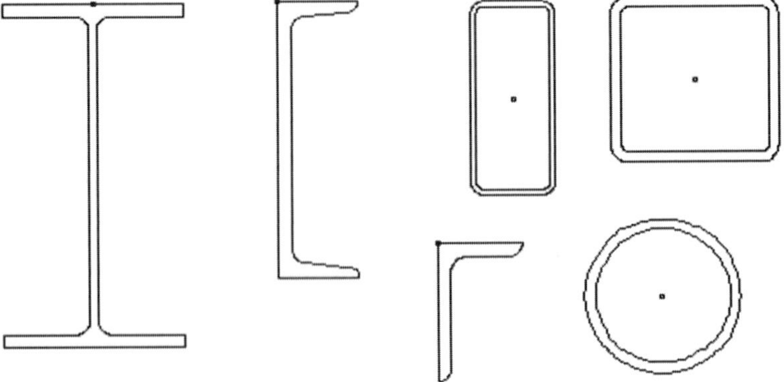

Figure 24-8 Some of the shapes available through SteelPro

You can also select any of DataCAD's linetypes for the line portion of TxtLine. Another feature, not allowable with linetypes, enables you to select different colors for your text and lines. Lines go in as individual (grouped) entities, and text pieces that cross one another can be moved by using Move or Stretch. This macro is especially useful for site and plot plan drawings where various utilities must be shown. Figure 24-9 shows some example of lines created with this macro.

Window Master

Window Master draws windows in elevation. It can draw awnings, casements, and pictures, as well as double-hung, slider, quarter-round, half-round, and full-circle window types (see Figure 24-10). You can set all the settings of the windows to match your office standards. From the color to the width of the sashes, it is all customizable. Optionally, you can insert sills, headers, trim, and shutters. All the settings can be saved to and loaded from settings files, allowing you to develop a library of standard window types. Window Master

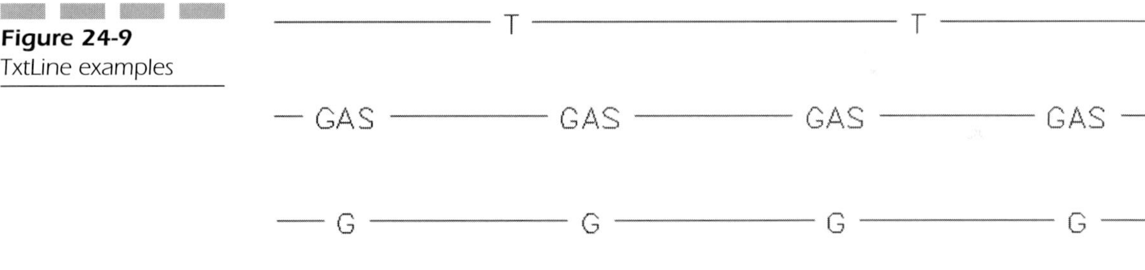

Figure 24-9 TxtLine examples

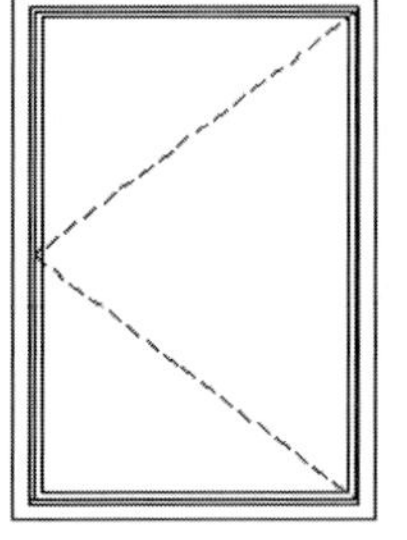

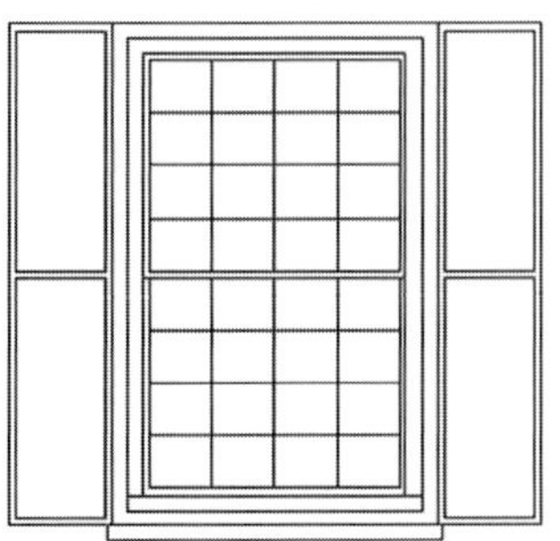

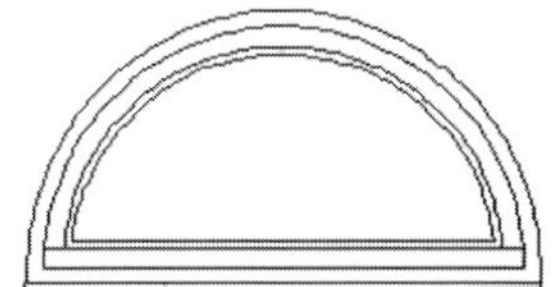

Figure 24-10 Some windows available in Window Master

features multiple input modes to enable you to insert the window the way you like. Up to 12 units can be mulled together at any one time.

Z Master

Z Master lets you save and recall different Z-base/Z-height combinations. In addition, these combinations may be linked to a specific layer so that when the setting is used, the Z-base, Z-height, and layer are all set to the user's specifications. Groups of these settings may be saved and loaded for future use.

Third-Party Macros

The following macros are all available from Cheap Tricks Ware and were created by numerous authors. Besides macros, they have templates and symbols, sample drawing files, hatch patterns, linetypes, and lots of other related programs and utilities. These items cost anywhere from $5 to $35. See Chapter 27, "DataCAD Resources," for their Web and contact information.

Stretch Point Macro

This DataCAD macro will selectively stretch lines with a common end point. It previews the results of each stretch one at a time, with a confirmation before performing the stretch. Figure 24-11 shows before and after verions of a line that **Stretch Point** was used on.

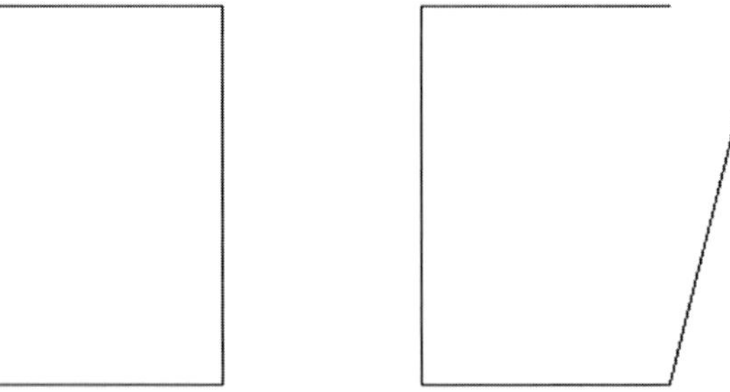

Figure 24-11 Before and after using **`Stretch Point`**

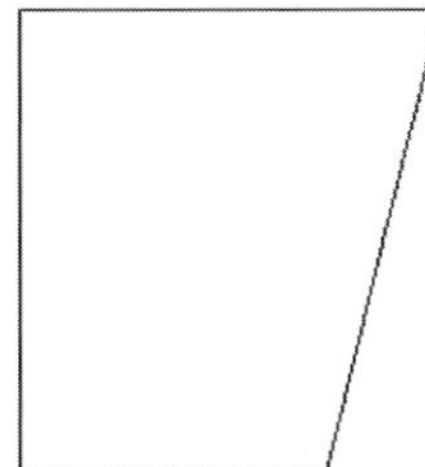

Figure 24-12 Using the stock DataCAD **`Stretch`** command

If we had used the standard **Stretch** command in DataCAD, the result would have looked like Figure 24-12.

Global Edit Macro

This macro is designed to supplement the regular DataCAD **Change** and **Text/Change** menus to enable the reassignment of multiple colors and linetypes, to explode entities, to provide text searches and replacement, to manipulate layers (change to all uppercase), and much more all from one location. The preferences can also be saved. Two of our favorites include a more accurate symbol explode routine and a routine to delete all empty layers in drawing the file.

This macro actually has two menus, but there is a little bug in this program that keeps you from accessing the second menu of options until after you select an option from the first menu. To work around this, select any option in the first menu, such as **SymProp**, and then turn that option off again. Now you will see a **Do It** button at the bottom of the menu. Press it to go to the second menu.

Clouds Macro

This macro enables you to use the mouse to draw revision clouds as three-point arcs with only two mouse picks (which is 50 percent faster than the three-point arc method). You can adjust the Bulge Factor to suit your own tastes. It draws clouds in PermaColor, ensuring color consistency, and draws in PermaLayer as well to automatically draw clouds on the correct layer, returning you then to your original layer (see Figure 24-13). It also has a trap door to the **Templates** menu for revision symbols, and an option to draw clouds as polylines. This is one of my absolute favorite macros.

Figure 24-13 A product of the Clouds macro

Figure 24-14 The Drawing Documentation Macro

```
 -------------------------------------------------------------------------
|                    DATACAD DRAWING FILE INFORMATION                      |
|                                                                          |
 -------------------------------------------------------------------------

DRAWING NAME: G:\BOOK\DWG\CHPT-12\BAR-FINL.DC3

 -------------------------------------------------------------------------
|                          GENERAL INFORMATION                             |
 -------------------------------------------------------------------------
Description    :
               :

 -------------------------------------------------------------------------
|                           LAYER INFORMATION                              |
 -------------------------------------------------------------------------
      LAYER                                                        # OF
      NAME       DESCRIPTION                           COLOR     ENTITIES
---------------------------------------------------------------------------
   6 1-WALL                                            Lt Red        894
   7 1-NOTES                                           Green         576
   8 1-DIMS                                            Yellow        236
   9 1-HATCH                                           Lt Cyan         0
  10 1-MISC                                            Lt Blue         0
```

Drawing Documentation Macro

This is a great aid in documenting your drawing layer information for transfers to consultants or even for in-house management of drawings. The macro will load up existing layer names with their on/off status and current layer colors noted. You can also add or edit remark information to explain the use of each layer (see Figure 24-14). Note that you can get all of this information from the main DataCAD **Utility/Directry** screen as well.

Curved Slab Macro

Unfortunately, DataCAD does not create curved or round 3D slabs on its own, but fortunately this macro enables you to dynamically draw curved 3D slabs in either an arc or circle format, and even puts snap points on the center of the curve. By using division control, you can even use this macro to draw polygonal slabs (see Figure 24-15).

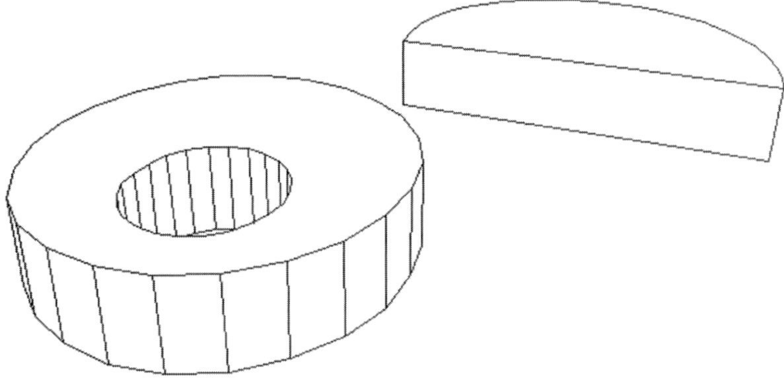

Figure 24-15 The Curved Slab macro at work

Macro Tabouret

The Macro Tabouret is a CD-ROM containing six different macros that do specialized tasks not available with DataCAD itself. Some are 3D-related, but most are 2D tools. The Adobe Acrobat documentation files for these macros comes on the Tabouret CD, or you can download them from the DataCAD LLC Web page. You can also get them from the CD that comes with this book.

3D Power Tools Macro

This single macro has eight functions that facilitate creating and manipulating custom architectural elements even more efficiently than you can with DataCAD's built-in features:

- **Knife:** This function slices through slabs and polygons like, well, a knife. You can elect to keep the entities on one side only or on both sides. You can model some complex shapes that could otherwise not be created with DataCAD.
- **Sweep:** This function will enable you to create a cross-section of any shape and then extrude it along a simple or complex path in 3D. This is great for creating 3D mouldings, ramps (including curved ramps), stair rails, and even roofs.
- **3DPlane:** This function will enable you to create more complex polygons and slabs than with DataCAD alone.
- **Pln2Pgon:** Use this function to convert 2D polylines to 3D polygons.

- **Triangle:** Use this function to triangulate polygons and slabs. This is often necessary for importing 3D models to other 3D programs.
- **SnapShot:** This function will automatically perform a hidden-line routine on your current 3D model. It also produces 2D elevations, an isometric, and a perspective view.
- **MultiPOF:** Use this function to generate multiple GLShaded views of your 3D model based on some or all of your saved 3D *GotoViews* (GTVs).
- **Kaboom!:** This one is kind of fun, but make sure you save a copy of your drawing file before doing it. You set the center point of your explosion and then the macro will literally explode your 3D model into all its individual 3D entities and faces, blowing them outward in an ever-expanding explosion. You can stop the explosion at any distance to wind up with an exploded view (no pun intended) of your model.

Blocker Macro

The macro is for space planning and reporting. It provides dynamic control over the portioning of space in your designs. The Blocker macro quantifies square feet and room dimensions as you draw and make changes, and then produces reports of square footage take-offs and room sizes. As you outline and name rooms with polylines, square footage and room dimensions are automatically calculated and shown on your drawing. The Blocker macro also includes special tools for automatically numbering rooms and converting outline diagrams to walls for production drawings.

DC Sprint Macro

This is a macro containing 17 editing and time-saving functions. The DC Sprint Macro contains the following options:

- **MovToLyr:** Use this option to move entities to another layer while simultaneously changing their colors to match the color of the layer the entities are moved to.
- **MCopy:** This macro adds features that the standard **Copy** command doesn't have. It will enable you to make multiple entities in any direction, display copy distances in Cartesian or polar coordinates,

control the grouping of entities (including retaining the original grouping), determine the color of the copied entities, and avoid copying entities on top of one another on a layer.

- **StkyBack:** This macro acts much like DataCAD's templates and symbols, but all the selected entities are saved on their current layers. When the stickybacks are placed in another drawing file, if those layers don't already exist, then they will be created. Unlike symbols, when stickybacks are created, the original entities are not destroyed. This is a great way to cut and paste details between projects and drawing files. Stickybacks are saved in their own directory and can be called up at any time like symbols.
- **Bundle:** This macro enables you to save symbols with their layers intact. Upon placing the symbols, the entities will be located on their original layers. More steps are involved here than in DataCAD's standard symbols' processes, but if you need this functionality, it works like a charm.
- **Roundoff:** Use this option to round off non-associative dimensions. You can round up, down, or to the nearest value. The precision of the round-off is user-definable.
- **SymUpdat:** This option will automatically update all symbols in the current drawing file with the latest version of each of those symbols.
- **3DLinCvt:** This option does the same thing as the 3Dline Conversion macro mentioned earlier in this chapter.
- **ExpldSym:** Use this option to explode symbols back to individual entities. It has extra features that enable you to predetermine the number of nested symbol levels to explode, to explode the symbol to the current or original layers, to keep or delete the original symbol, and to draw a point where the symbol's insertion point used to be. It also has a toggle that will automatically "unbundle" symbols that were saved with the **Bundle** macro.
- **MkTemplt:** This option will create a new template containing all the symbols in the current drawing file. This is great for saving all the symbols in a drawing sent to you by a third party.
- **Captaliz:** This option enables you to change selected text so that all the letters are capitals, only the first letters of each word are capitalized, or all the text is changed to lower case letters.
- **ScaleTxt:** Here you can scale text up or down by a factor. It works like the **Enlarge** command, but you do not have to choose a new center of

enlargement for every piece of text. Instead the enlargement will automatically be made from the text left, right, or center.

- **ScaleSym:** Here you scale symbols up or down by a factor. It works like the **Enlarge** command, but you do not have to choose a new center of enlargement for every symbol.
- **LayrSort:** This option enables you to sort your current drawing layers alphanumerically. You can select sorting with case-sensitivity and in ascending or descending order. Be aware that your GTVs and *Multi-Scale Plot* (MSP) layouts will be irretrievably jumbled because they are created by layer order, not by name.
- **DwgDoc:** This option works the same as the Drawing Documentation Macro mentioned earlier in this chapter.
- **Labels:** This option automatically creates labels for things like doors, windows, room numbers, wall types, and so on. You can choose from various box shapes and font attributes, as well as underlined text. Alphanumeric labels can be automatically increased with each placement in the drawing (if you place room number 100 in the drawing, the next label will automatically be number 101, then 102, and so on).
- **Lin2Pln:** This option will convert any contiguous 2D lines and arcs into a single open or closed polyline. This is great for hatching things like walls in a plan. Copy all your walls to another layer and then run the Lin2Pln macro to convert the copy of the walls into a series of polylines. Since associative hatching is just a filled-in polyline, now you can use the standard **Hatch** function to convert the new polylines to a hatch pattern. Now you have hatching inside all your walls.
- **SetPens:** Use this option to instantly view the pen numbers assigned to each of DataCAD's standard 15 colors. This has a **SetAll** option to change them all to one pen number.

Template Librarian Macro

Use this macro to create custom template catalogs. The Template Librarian automatically lays out 2D-wireframe or 3D-hidden-line catalog-page views of any DataCAD template you choose. It also automatically hides 3D isometric views of each symbol and labels each one. Use your printer to create catalogs for easy reference.

Touchup Macro

This is an interactive, point-and-shoot 3D hatching tool that works completely within DataCAD. Using the Touchup macro, you can apply any 2D DataCAD hatch pattern onto any 3D surface with a few mouse clicks. Enhance your DataCAD models by adding 3D hatching and create presentation quality images without resorting to time-consuming, full-color rendering programs. You can view models from any vantage point and the pattern will be right. It supports 2D walls, 3D polygons, slabs, blocks, and voids (see Figure 24-16).

Wavy Macro

This macro enables you to give a hard-line CADD drawing a hand-drawn look. The Wavy macro randomizes the spacing and overshoot values of

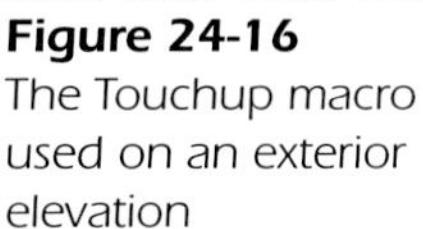

Figure 24-16 The Touchup macro used on an exterior elevation

selected lines in your DataCAD drawing to produce a hand-drawn, friendlier look to your presentation drawings. This macro also includes the famous Minnesota Wiggle linetype. Figure 24-17a and b shows before and after versions of the Wavy macro in action.

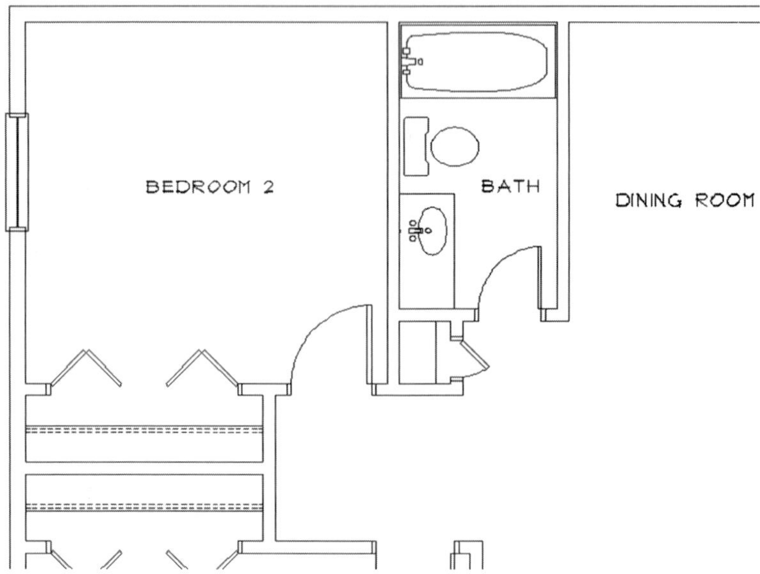

Figure 24-17a A standard CADD drawing

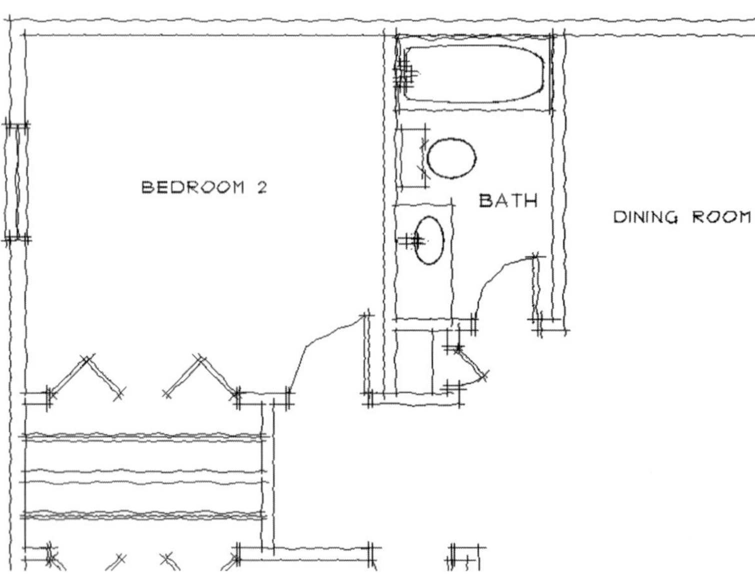

Figure 24-17b The same drawing converted with the Wavy macro

CHAPTER 25

Converting File Formats

Living in an AutoCAD World

Like it or not, the most pervasive CAD program used by architects and construction industry engineers is AutoDesk's AutoCAD. I used AutoCAD before trying DataCAD, and I continue to use AutoCAD insofar as my company needs to transfer AutoCAD drawings between us, our engineers, and our clients. I won't get into what I really think of AutoCAD, but suffice it to say that I enjoy going to the dentist far more than I enjoy working with AutoCAD. Once I began using DataCAD, my eyes were opened to the fact that here was a CAD program that was actually easy to learn, easy to use, was architect-friendly, and had the most efficient menus and features that I had ever run across.

With that said, most of our clients and engineers use AutoCAD and we must be able to trade files back and forth with them. DataCAD 9 makes this task pretty simple: just select **File/Export/DWG** and save the drawing (the .DWG file extension is AutoCAD's native file format, just the way that the .DC5 file extension is DataCAD's native drawing file format). Importing a .DWG file is likewise very simple to do with **File/Import**, but you will still have to deal with a number of issues:

- AutoDESK does not publish AutoCAD's file format, so it becomes very difficult for other software products, like DataCAD, to create 100 percent-compatible drawing files.
- Some entities in AutoCAD are not supported in DataCAD. Likewise, some entities in DataCAD are not supported in AutoCAD.
- DataCAD has *Multi-Scale Plotting* (MSP) to enable you to lay out multiple-scale details on one sheet. AutoCAD has a similar feature called *Paper Space* (PS). Unfortunately, although they both operate in a very similar manner, they are fundamentally different enough that there is currently no way to maintain those layouts between DataCAD and AutoCAD. This again is due mainly to the fact that AutoDESK does not publish the .DWG format.
- AutoCAD users tend to use any of 255 available colors in their drawing files. These extended colors are only identified by their color number, rather than a name. DataCAD, on the other hand, is capable of using all those same 255 colors, but the 15 basic colors cover the spectrum pretty well and have easily identifiable names. When importing a drawing from AutoCAD to DataCAD, it is often necessary to convert some of those 255 colors to the standard 15 DataCAD colors just to make the resulting file easier to work with.

- The way AutoCAD handles symbol insertion points will sometimes cause those symbols to be displayed dozens or hundreds of feet away from their original locations.
- Although AutoCAD recently received associative hatches, they are currently not well supported when imported into DataCAD. When exported from DataCAD to AutoCAD, associative hatches translate only as simple lines. When imported into DataCAD, AutoCAD's associative hatches translate as symbols.
- Many of AutoCAD's entities are 3-D by default and translate that way into DataCAD, but DataCAD's 2-D editing functions can only operate on 2-D entities, so you must first change the imported 3-D entities to 2-D ones. This is very simple to do with the DataCAD 3DLine Convert macro, but it is something to keep in mind.
- AutoCAD symbols (called *blocks* in AutoCAD) are capable of having layers, while DataCAD's symbols are not. Thus, an imported AutoCAD symbol that has a number of layers will be imported into DataCAD with all its entities on only one layer. This can be a problem in DataCAD if you don't want to display all those entities. This can also be a problem if you then export the drawing back to .DWG format, and the person on the receiving end needs those symbols to maintain their layers in AutoCAD.

For a complete list of how DataCAD and AutoCAD entities translate back and forth between the two programs, consult DataCAD's Online **Help** resource. Look under **DWG Translator**, and view the **Importing** and **Exporting** subjects, or see the Appendix at the back of this book.

Exporting DataCAD Drawings to the AutoCAD Format

The actual conversion process for exporting DataCAD drawings to AutoCAD is very simple:

1. Run the DataCAD drawing through the *Layer Utility* (LyrUtil) macro (see Chapter 21 for more information). This will make the file smaller and more efficient. You don't *have* to do this step, but the results are always better if you do.

2. If you have any XREFs in the drawing file, you first need to use the **Bind/All** command in the Reference File Manager to make them a native part of the drawing file. If you don't, the XREF entities will be exported, but all the entities in that XREF will be placed on the single layer that the XREF is located on in the master file.
3. Select **File/Export/Dwg/All Layers** to export all the layers in the drawing file or select **File/Export/Dwg/On Layers** to export only the layers that are currently turned on in the drawing file. A dialog box will appear (see Figure 25-1).
4. Navigate to the folder where you want to save the new drawing. By default, DataCAD selects the \XFER directory. In the example above it's the \DXF directory.
5. In the **Save as type:** box, *click* on the down arrow to select which .DWG version you want to save to. You can select from R12 through R2000.
6. In the **File name:** box, type the name of the file to be saved. You do not need to type the .DWG file extension. By default, DataCAD uses the name of the current .DC5 drawing file.

NOTE: *When naming a .DWG file, keep in mind that not all programs can read long filenames (long being more than eight characters) or special characters like spaces, periods, and slashes. For instance, AutoCAD 13 and earlier cannot read long file names. Your best bet is to keep the name simple and to a maximum of eight characters.*

Figure 25-1
The DWG/DXF export dialog box

8. *Click* on **Save**. DataCAD will create the .DWG drawing file. In most cases this will happen fairly quickly, but if you have a large drawing with a lot of individual entities, or one with a large amount of hatching, this process could take a couple of minutes.

Colors and .TBL Files

The following information is derived primarily from the DataCAD Online **Help** file.

Both DataCAD and AutoCAD define colors in terms of their place in each program's internal color list. For instance, in DataCAD's code, Color 1 is white, Color 2 is red, and Color 3 is green. Although we see white, red, and green colors onscreen, DataCAD recognizes these only as Color 1, Color 2, and Color 3, respectively. AutoCAD works similarly, except AutoCAD displays Color 1, Color 2, and Color 3 as red, yellow, and green, respectively. Correcting these differences is the purpose of the **Color Translation** option in the **DWG/DXF** tab of the **Tools/Program Preferences** dialog.

Without the Color Translation option checked, when you export a DataCAD file, the resulting DWG/DXF file will only associate Color 1, Color 2, and so on with each entity. When the file is displayed in AutoCAD, the program recognizes Color 1 and displays the entity as red, instead of white as it was drawn in DataCAD. With the Color Translation toggle on, a .TBL table file (in the \SUP directory) corrects for this, so that white entities in DataCAD are displayed as white when exported to a DWG/DXF file and vice versa.

The Color Translation toggle, which is on by default, creates a table (.TBL) file so that DataCAD colors translate properly when exported to or imported from a DWG/DXF file. This file is created automatically during translation and requires no input from you. It is given the same filename as the currently selected RGB file and is saved to the \SUP subdirectory. A .TBL file is a simple text file. Within it are two tables, one for exporting and one for importing. Here is part of the Default.TBL file:

```
[Export]
D_Color_1=7
D_Color_2=1
D_Color_3=3
D_Color_4=5
D_Color_5=132
D_Color_6=212

[Import]
```

```
A_Color_1=2
A_Color_2=15
A_Color_3=3
A_Color_4=13
A_Color_5=4
A_Color_6=14
```

The number after D_Color is the DataCAD color with the equivalent AutoCAD color after the equals sign. The number after A_Color is the AutoCAD color with the equivalent DataCAD color after the equals sign. There are a total of 255 colors, so a total of 510 lines are in each .TBL file. In AutoCAD, if you use a color value of 256, the entity will display as white but will not print.

I have found that the DEFAULT.TBL file that is installed with DataCAD has some inaccuracies in it. Some of those inaccuracies are highlighted in the previous example. The following is what the values should be. I have verified only the values shown here. In order to get proper imports and exports, you should modify your file with these values:

```
[Export]
D_Color_1=7
D_Color_2=1
D_Color_3=3
D_Color_4=5
D_Color_5=4
D_Color_6=6
D_Color_7=30
D_Color_8=9
D_Color_9=8
D_Color_10=11
D_Color_11=91
D_Color_12=171
D_Color_13=131
D_Color_14=211
D_Color_15=2

[Import]
A_Color_1=2
A_Color_2=15
A_Color_3=3
A_Color_4=5
A_Color_5=4
A_Color_6=6
A_Color_7=1
A_Color_8=9
A_Color_9=8
A_Color_10=2
A_Color_11=10
A_Color_30=7
A_Color_91=11
A_Color_131=13
A_Color_171=12
A_Color_211=14
```

The full, modified DEFAULT.TBL can be found on the CD-ROM at the back of this book, in the "Custom" section.

Custom Linetypes

DataCAD's capability to use complex, custom linetypes to represent various materials (plywood, batt insulation, and so on) is one of the hallmarks of the program. Fortunately, you have the means to maintain those custom linetypes in AutoCAD. When a DataCAD drawing that contains custom linetypes is exported, an AutoCAD compliant .SHX file with the same name as the drawing is also created. For instance, if you export a DataCAD drawing named MYDRAWING.DC5, the translator will create two files: MYDRAWING.DWG and MYDRAWING.SHX.

During export, the line definitions and shape codes for DataCAD linetypes are stored in a .SHX file. In order for the linetypes in the DWG file to display properly, the .SHX file must be either be placed in the \SUPPORT subdirectory in the AutoCAD directory or in the same file as the .DWG drawing. However, if you choose to export a DataCAD file to a Release 12 DWG file, and the Export Linetypes toggle is off, all user-defined linetypes in DataCAD will be displayed as Continuous (equivalent to DataCAD's Solid linetype) in the DWG/DXF file, since AutoCAD 12 doesn't support complex line shapes. These DataCAD linetypes, however, are not permanently changed to the Continuous line shape. Instead, they retain the DataCAD linetype name and merely display as Continuous in AutoCAD. When you import the file back into DataCAD, the proper linetypes will once again be displayed.

If the Export Linetypes toggle is on all user-defined linetypes in DataCAD will still be displayed as Continuous, but they will not retain their linetype names when re-imported into DataCAD. The first four linetypes in DataCAD (Solid, Dotted, Dashed, and Dot-Dash) are already directly supported in AutoCAD, so those four linetypes will not be part of the .SHX linetype file.

Fonts

Both DataCAD's and AutoCAD's fonts are proprietary, meaning that they can only be used in their respective programs. Thus, when a DataCAD font is exported in a .DWG file, if the same font name cannot be found on the AutoCAD system then AutoCAD will use a default font, usually Romans. The same is incidentally true in reverse when an AutoCAD drawing is imported into DataCAD. At least two ways exist for getting around this limitation:

- DataCAD LLC has made various DataCAD and AutoCAD fonts available on their Web site. They are also included on the CD-ROM at the back of this book.
- You can trick AutoCAD into using a specific font by copying and renaming an existing DataCAD font.

The first method is very simple. The DataCAD fonts are copies of standard AutoCAD fonts, so just use the AutoCAD-compatible .CHR fonts in your DataCAD drawing. Then when you export it to a .DWG file, the AutoCAD user should see the same fonts that you did. A few of the DataCAD fonts are not standard AutoCAD fonts, so you would simply send along the matching AutoCAD .SHX fonts for the AutoCAD user to add to their current list of fonts.

The second method involves a bit of trickery, and the results may not be exact. Neither DataCAD nor AutoCAD care what the name of a font is. They just display whatever is in the font. If you know that you have a DataCAD font that looks like an AutoCAD font, just copy the DataCAD font and give it the same name as the AutoCAD font. When the DataCAD drawing is exported to AutoCAD, the name of the font will go with it. AutoCAD will then display the font that matches that name. Here's an example:

1. Your Archx.CHR DataCAD font looks like the Eng.SHX AutoCAD font. Make a copy of the Archx.CHR font, and name it Eng.CHR.
2. Export the drawing to a .DWG file.
3. When AutoCAD opens the file, it will display the Eng.SHX file as long as it exists on the AutoCAD user's system.

One thing that you must keep in mind is that different fonts have different aspect ratios. In other words, if you use one font in DataCAD and another in AutoCAD a string of text may be displayed longer or shorter in AutoCAD than it was in DataCAD. Figure 25-2 is a string of text in DataCAD, and Figure 25-3 is what is displayed in AutoCAD.

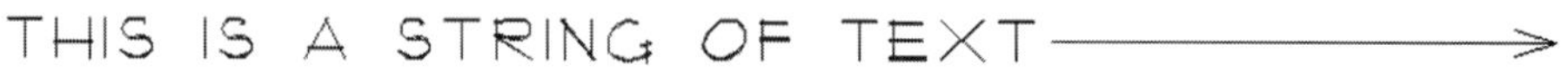

Figure 25-2
DataCAD text

Figure 25-3
That text translated in AutoCAD

3-D Entities

Any 3-D surfaces in a DataCAD file will be translated as a single entity, polyface mesh when exported to a DXF or DWG file. A polygon with five or more vertices and/or voids will be exported as multiple, triangulated 3-D faces. A 3-D line is translated to a Line; and a 3-D arc is translated to an Arc. See the Appendix of this book for a more complete list of 3-D entities and how they translate between DataCAD and AutoCAD.

However, by checking the "Decompose 3-D Entities as 3-D Face" option (in the **DXF/DWG** tab of the **Tools/Program Preferences** dialog), each facet of a 3-D entity will be turned into an individual polyface. So a single sphere in DataCAD can be translated into many separate entities arranged in a sphere shape in a DWG file. By default this option is unchecked (OFF). Checking the Decompose option can make translation faster and the translated entities will look the same after a shade or hidden line removal. However, editing may be more difficult since these translated 3-D entities are made up of many smaller and separate entities.

Troubleshooting

The following section will cover some common troubleshooting problems regarding color map tables, exploding symbols, and working with title blocks.

Color Map Table If you get an error message in the Message Line saying something like, "DESTIJL.TBL—color map table file not found. Translation aborted," here is what is going on: if any .RGB file except Default. RGB is being used in the current DataCAD drawing file, DataCAD will look for a color table with the same name when importing the drawing. In this example, the current .RGB file in the DataCAD drawing is called DESTIJL. RGB, so DataCAD is looking for a .TBL color table file with the same name. It doesn't exist, so the error message is displayed.

To rectify the problem, you have two different options to choose from. The first is to *click* on **Utility/Display/Palettes/LoadRGB** and select either the **Default.rgb** file or an .RGB file with a name that matches a previously saved .TBL file. Then *right-click* back to the 2-D menus.

The second option would be chosen if you want to maintain the current .RGB file in your drawing. Perform the following steps:

1. Go to the \SUP directory and copy the Default.TBL file.
2. Rename the copy with the same name as your .RGB file (for instance, if your .RGB file is called MYCOLOR.RGB, then name the table file to

MYCOLOR.TBL). A custom .RGB file only affects the way the colors are displayed on your screen and printed, not how they are translated to another file format, so you don't actually have to create a new .TBL file specifically for your .RGB file.

Now when you run the **File/Import** option, the file should read properly. DataCAD comes with five pairs of matching .RGB and .TBL files:

- DCViewer.RGB / DCViewer.TBL
- Default.RGB / Default.TBL
- Greyscal.RGB / Greyscal.TBL
- Sepia.RGB / Sepia.TBL
- White.RGB / White.TBL

Explode Symbols You generally do not need to explode the symbols in your drawing, but if you are having trouble with the way some of your symbols are exporting, you may want to explode those symbols prior to trying the export process again. Make sure you do this on a *copy* of the original file.

Exploding symbols means you lose the benefits of having symbols: smaller drawing files, the interchangeability of symbols, uniformity, and so on. If the people you are sending the drawings to do not really need those benefits (engineers who only want a background floor plan for their own work), then you don't lose much.

The **Toolbox** macro called SYMEXP.DCX is not always a good option, since symbols you mirrored during the drawing process will sometimes be unmirrored in strange ways (along a different axis) when you run that macro. Therefore, I highly recommend using the Cheap Tricks Ware macro called GLOBAL EDIT to explode all your symbols (it has a lot of other nice features besides exploding symbols). Global Edit will explode nearly any symbol and in most cases its entities will remain in place, as they should.

To use Global Edit, run the macro from the **Toolbox**, select **Symbol/Do It**, and pick one of the EGAFS methods. All selected symbols will be exploded. If nested symbols (symbols inside of symbols) are being used, you may have to run the macro more than once to pick them all up.

Title Blocks Often, a lot of variables are used in a title block. When your title block is exported to .DWG format, you will often find that your finely tuned graphics and multiple fonts have taken on a new and unwanted look. We use a little trick in our office to make sure that everything in the title block appears in the .DWG file just as it does in the DataCAD .DC5 file.

We explode all the entities in the title block, text included, into individual entities. This increases the file size, but it ensures the integrity of the title block. To perform this explosion, follow these steps:

1. Turn on only the layer(s) associated with the title block.
2. Select **DCAD_3D/Explode/ToLines**.
3. Pick the whole title block area by **Area** with **LyrSrch** on.
4. All the selected entities will be converted to hundreds of 3-D lines. For instance, text is no longer text, but a series of 3-D entities that just look like text.
5. To change those entities to 2-D entities, use the 3-D Line Convert macro (available from the DataCAD Web site and on the accompanying CD-ROM).
6. Go to the **Toolbox** and select the **3DLINCVT.DCX** macro.
7. Choose the **LyrsOn** option. All the entities are now 2-D.

Importing AutoCAD Drawings into DataCAD

This process is just as easy as the export process. You don't have to import a DWG file into a blank DataCAD drawing file, but things seem to go smoother if you do. To import AutoCAD drawings into DataCAD, follow these steps:

1. Select **File/Import**. In the dialog box, navigate to the folder with the DWG file.
2. Select and **Open** the file.
3. In most cases, you will see a dialog box like Figure 25-4.
4. Select the appropriate options (explained below) and then press OK.
5. The DWG file is imported into DataCAD.

Line Type Assignments

When the DWG file is imported, DataCAD reads all the AutoCAD linetype names in the file and displays them at the left side of the box. At the right

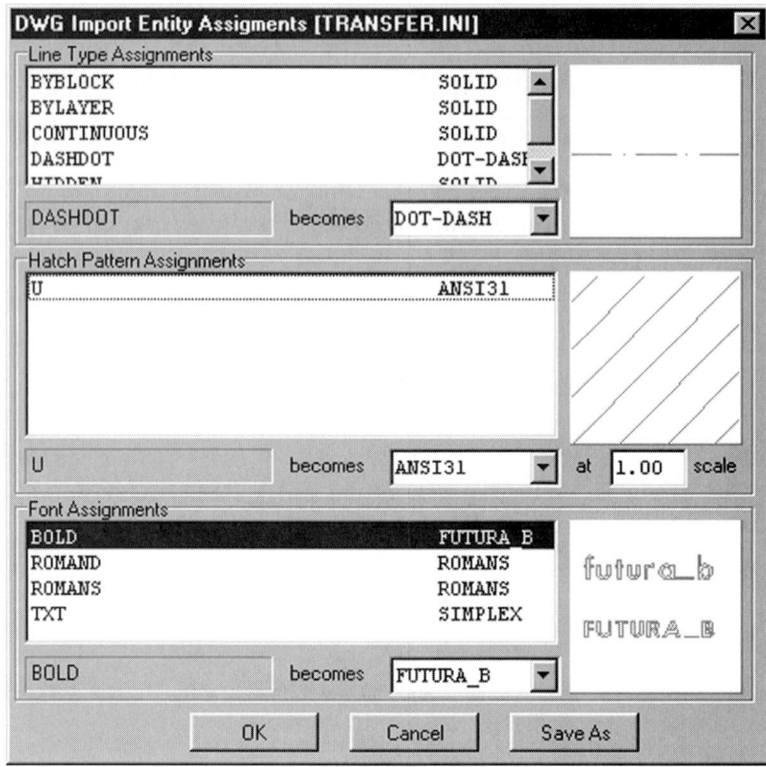

Figure 25-4 The DWG/DXF import dialog box

side of the box are the DataCAD linetypes assigned to each of the AutoCAD linetypes. If no match is found between the name of an AutoCAD linetype and a DataCAD linetype, the Solid linetype is automatically assigned.

If you want to reassign one of DataCAD's linetypes to one of the AutoCAD linetypes, follow these steps:

1. Highlight the AutoCAD linetype.
2. *Click* on the down arrow to the right of the **becomes** box. All of the linetypes in the DCADWIN.LIN file are displayed.
3. Select the DataCAD linetype to use. The selected linetype will be displayed in the preview window to the far right.
4. Oddly enough, the new selection will not display on the right side of the box until you highlight one of the other AutoCAD linetypes on the left.

The “Continuous” AutoCAD linetype is equivalent to DataCAD's “Solid” linetype.

Hatch Pattern Assignments

When the DWG file is imported, DataCAD reads all the AutoCAD associative hatch pattern names in the file and displays them in this box. If no match is found between the name of an AutoCAD hatch and a DataCAD hatch, the default hatch pattern is assigned. The default pattern is set in the **Tools/Program Preferences/DXF/DWG** tab in the **Hatch Pattern** box.

If you want to reassign one of DataCAD's hatch patterns to one of the AutoCAD hatches, perform these steps:

1. Highlight the AutoCAD hatch pattern.
2. *Click* on the down arrow to the right of the **becomes** box. All of the hatches in the DCADWIN.PAT file are displayed.
3. Select the DataCAD hatch pattern to use. The selected hatch will be displayed in the preview window to the far right.
4. Set the scale of the selected pattern by typing a value in the **At** . . . **Scale** box.

Font Assignments

When the DWG file is imported, DataCAD reads all the AutoCAD font names in the file and displays them at the left side of this box. At the right side of the box are the DataCAD fonts assigned to each of the AutoCAD fonts. If no match is found between the name of an AutoCAD font and a DataCAD font, the default font is assigned. The default font is set in the **Tools/Program Preferences/DXF/DWG** tab in the **Font Assignments** box.

If you want to reassign one of DataCAD's fonts to one of the AutoCAD fonts, perform these steps:

1. Highlight the AutoCAD font.
2. *Click* on the down arrow to the right of the **becomes** box. All of the fonts in the \CHR directory are displayed.
3. Select the DataCAD font to use. The selected font will be displayed in the preview window to the far right.
4. Oddly enough, the new selection will not display on the right side of the box until you highlight one of the other AutoCAD fonts on the left.

Note that some DataCAD fonts may have been created only for upper case letters. If you select one of these fonts none of the lower case letters can be displayed in DataCAD. If the selected font is created to handle both upper and lower case letters, then both will be displayed in the preview window. If you only see upper case letters in the preview, then it cannot handle lower case letters.

Save As

If you have just completed a complex series of reassignments of linetypes, hatches, and fonts, use the **Save As** button to save a new .INI file to the \SUP directory. Give the new .INI file a name and then select **Save**.

Let's say you saved an .INI file called CLIENT1.INI. This now becomes the active .INI file for all future DWG imports. All the settings you made will be remembered so that if DataCAD runs across linetypes, hatches, and fonts with the same names as the ones in the .INI file, the assignments you made will automatically be applied within the DWG Import dialog, saving you a lot of time.

To change the current .INI file, go to the **Tools/Program Preferences/ DXF/DWG** tab and pick a new .INI file in the **Translator Settings File** box.

AutoCAD Blocks

Blocks in AutoCAD are essentially the same as symbols in DataCAD with one important difference. AutoCAD's blocks support multiple layers, while DataCAD's symbols do not. Thus, when a block with multiple layers is imported into DataCAD, all the layers of the block are placed on the layer where the block is located.

Troubleshooting

The following section will cover common problems regarding importing AutoCAD drawings into DataCAD.

Bad Object If the import process stops and you get an error message saying that a "bad object" is found, there is a problem with a corrupted entity in the original AutoCAD file. In AutoCAD, the Purge and Audit commands should be run on the file to clear up the problem. Save the file and try to import it into DataCAD again.

Odd Characters After the file is imported, if you see a lot of %%U characters at the beginning of words, that is where AutoCAD has underlined the text. Since DataCAD does not support underlining text, you must manually erase the %%U characters (**Edit/Change/Text/Contents**). To underline the text, you would have to draw a line under it. You may see some other similarly out-of-place formatting characters in the drawing file, like these:

ACAD	DCAD
Degree	%%d
+/–	%%p
Diameter	%%c
Underline	%%u
$^{1}/_{2}$	≅8
$^{1}/_{4}$	≅4
$^{3}/_{4}$	≅w
$^{1}/_{8}$	≅2
$^{3}/_{8}$	≅6
$^{5}/_{8}$	≅f
$^{7}/_{8}$	≅r
$^{1}/_{16}$	≅1
$^{3}/_{16}$	≅3
$^{5}/_{16}$	≅5
$^{7}/_{16}$	≅7
$^{9}/_{16}$	≅9
$^{11}/_{16}$	≅q
$^{13}/_{16}$	≅e
$^{15}/_{16}$	≅t
Centerline	≅C or ≅c
PL	≅P
Plate	≅p
< (angle)	≅<
Perpendicular	≅d

Import and Export .DXF Drawing Files

.DXF stands for *Data eXchange File*. It is a relatively universal method of transferring 2-D and 3-D vector information between various graphic and CAD programs. The limitations of .DXF files have more to do with the .DXF specifications and less to do with DataCAD itself. .DXF files are meant to be generic and therefore do not interpret or transfer many higher order functions in DataCAD.

Two of the strongest reasons for exporting a .DXF file are as follows:

- Most any CAD program will read a .DXF file. It is such a simple, basic file format (far more so than a .DWG file) that it is the best way to ensure that nearly any CAD program will be able to import the file.
- Most any rendering program will read a 3-D .DXF file. It is by far the most common format among all rendering programs.

The primary reason *not* to export a .DXF file is that it is such a simple file format that much of the "smarts" of the CAD file are lost. See the following limitations for more information.

Much of the information here came from Bruce Kaplan, the author of the DXF PREP Cheap Tricks Ware macro.

The Bidirectional Limitations of .DXF Files

The .DXF file limitations are as follows:

- Only four fonts are directly compatible: SIMPLEX, COMPLEX, ROMANS, ROMANC.
- Only four line types will transfer: SOLID, DASHED, DOTTED, and DOT-DASH.
- The .DXF protocol only supports 3-D entities (polygons) with a maximum of four vertices and no thickness.

Limitations from DataCAD to DXF

From DataCAD to DXF, the following limitations exist due to the higher order functions of DataCAD that are not supported in AutoCAD or DXF.

DataCAD's DXF export utility will take care of these items for you without any intervention.

- Layer names must be all CAPITALS and contain only legal DOS characters.
- DataCAD's associative hatches do not transfer. The bounding polygon will, but no hatch will be displayed within it. The solution is to explode the associative hatch into non-associative hatching prior to creating the .DXF file.
- DataCAD's .DXF process will tessellate (bust up) the higher order 3-D entities (slabs, etc.) into individual polygon surfaces (no more than four vertices). The coincident edges of the coplanar polygons will be invisible to give the appearance of it being one whole 3-D entity.
- Nested symbols will not transfer.
- Single point polylines will cause an abort (usually caused by a user error).

Exporting a DataCAD Drawing to .DXF

The process can be accomplished by following these steps:

1. DataCAD's 2 $^1/_2$ D entities must be squashed to 2-D only. Do this by changing the Z-base and Z-height of all entities to zero.
2. Unused symbol definitions should be purged. Do this by going to **Utility/Directry** and selecting **Purge Symbols**.
3. Convert all text to a .DXF-compatible font, like Romans.
4. Run the *Layer Utility* (LyrUtil) macro and save all the layers you want to keep (use **Save All** or **Save On**).
5. Create a new blank drawing and load the saved layers back in with the **LyrUtil** macro (**LyrLoad**). Steps 4 and 5 will help ensure a clean file with fewer potential problems.
6. Select **File/Export/Dxf/All Layers** to export all the layers in the drawing file, or select **File/Export/Dxf/On Layers** to export only the layers that are currently turned on in the drawing file.
7. A dialog box will appear (see Figure 25-5).

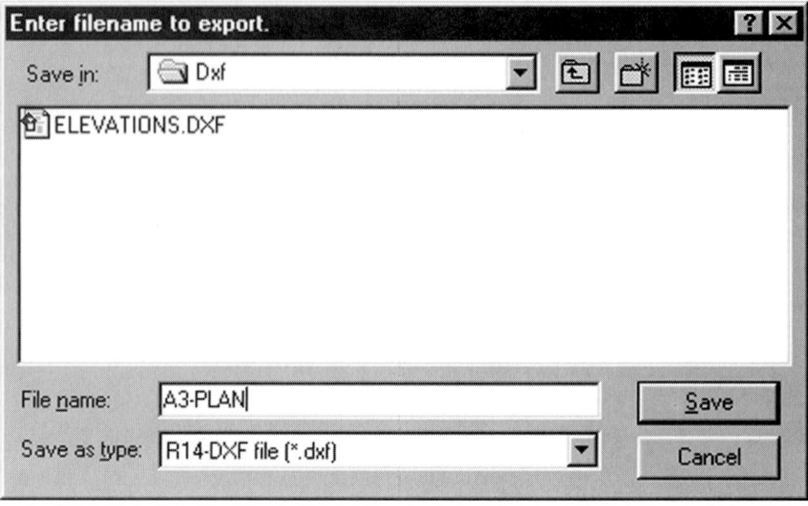

Figure 25-5
The DWG/DXF export dialog box

8. Navigate to the folder where you want to save the new drawing. By default, DataCAD selects the \XFER directory. In the above example it's the \DXF directory.
9. In the **Save as type:** box, *click* on the down arrow to select which .DXF version you want to save to. You can select from R12 through R2000.
10. In the **File name:** box, type the name of the file to be saved. You do not need to type the .DXF file extension. By default, DataCAD uses the name of the current .DC5 drawing file.

NOTE: *As stated earlier, when naming a .DXF file, keep in mind that not all programs can read long filenames or special characters like spaces, periods and slashes. For instance, AutoCAD 13 and earlier cannot read long file names. Keep the name simple and within eight characters long.*

11. *Click* on **Save**. DataCAD will create the .DXF drawing file. In most cases, this will happen fairly quickly, but if you have a large drawing with a lot of individual entities or one with a large amount of hatching, this process could take a few minutes.

Importing a .DXF File to DataCAD

The DXF format is such a simple, basic file format that generally anything, short of corrupt entities, could be imported by DataCAD. The import process therefore is incredibly simple:

1. You don't have to import a .DXF file into a blank DataCAD drawing file, but things seem to go smoother if you do.
2. Select **File/Import**. In the dialog box, navigate to the folder with the .DXF file.
3. Select and **Open** the file.
4. In most cases, you will see a dialog box like Figure 25-6.

   Select the appropriate options (explained earlier in the DWG Import section) and then press **OK**.
5. The .DXF file is imported into DataCAD.

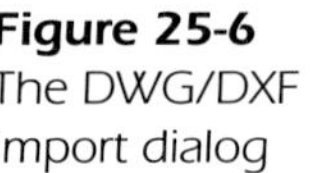

Figure 25-6 The DWG/DXF import dialog

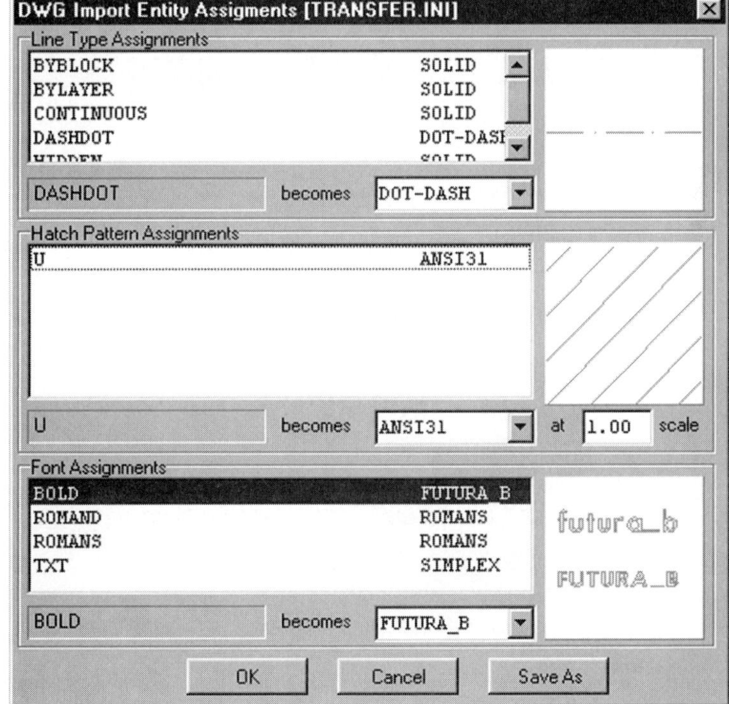

DXF/DWG Options

This information comes from the DataCAD **Help** file and relates to the **DXF/DWG** tab of the **Tools/Program Preferences** dialog box, shown in Figure 25-7.

These options let you choose an import method, translator default settings, and how linetypes, colors, and 3D entities will be exported.

Import Method

In the DXF/DWG tab, you can choose between two import methods:

- **Interactively assign . . . :** This allows you to interactively (manually) assign fonts, linetypes, and hatch patterns and is the default method. The DWG Import Entity Assignments dialog box, shown in Figure 29-6, is displayed during the import process only when the interactive option is checked.

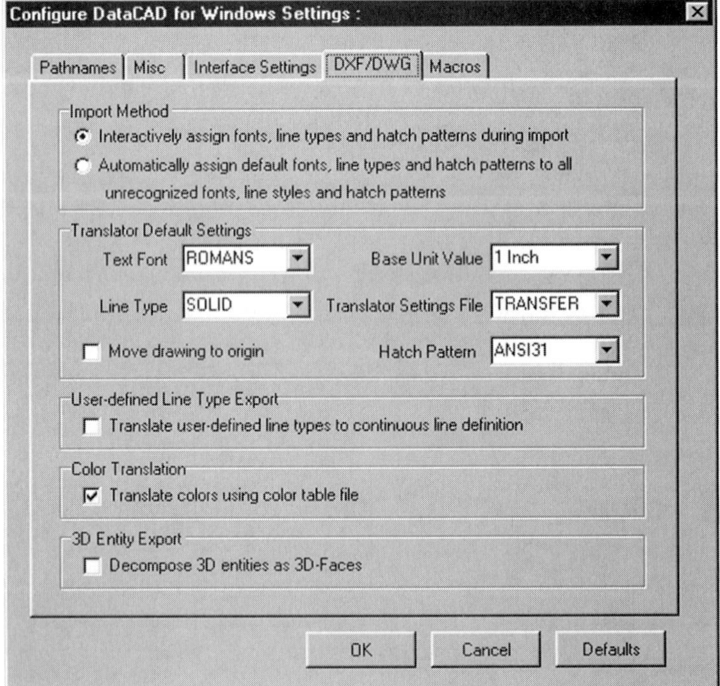

Figure 25-7
The DXF/DWG tab

- **Automatically assign . . . :** This option will automatically assign default fonts, linetypes, and hatch patterns to all unrecognized fonts, line styles, and hatch patterns. These defaults are selected in the Translator Default Settings section of the DXF/DWG tab.

Translator Default Settings

Several Translator Default Settings can be changed. These defaults will be used when DataCAD has no equivalent for an existing entity in the DXF or DWG file. To change any of these settings, *click* on the arrow to the right of the drop-down box and select another setting.

Text Font The current list of DataCAD fonts in the \CHR directory is available by *clicking* on the arrow to the right of the box. The selected font will be the default font used by DataCAD when imported AutoCAD text includes font names that don't match any of the DataCAD font names in your \CHR folder.

Line Type The current list of DataCAD linetypes in the DCADWIN. LIN file in the \SUP directory is available by *clicking* on the arrow to the right of the box. The selected linetype will be the default linetype used by DataCAD when imported AutoCAD linetypes include those with names that don't match any of the DataCAD linetypes.

Hatch Pattern The current list of DataCAD hatches in the DCADWIN. PAT file in the \SUP directory is available by *clicking* on the arrow to the right of the box. The selected hatch pattern will be the default hatch pattern used by DataCAD when imported associative AutoCAD hatches include those with names that don't match any of the DataCAD hatches.

Move Drawing to Origin This will force the imported .DWG drawing to be centered on the absolute zero point of the DataCAD drawing file.

Base Unit Value AutoCAD uses units of scale, rather than true architectural or engineering scales. If the DWG file you receive comes from an office that uses feet/inches, and you use feet/inches in DataCAD, then the default **base unit value** of one Inch will ensure that all the entities in the imported DWG file are at the correct size.

However, if the DWG file you receive comes from an office that uses millimeters, the imported entities will be very, very small and will need to be scaled up to the proper size (25.4 mm are in an inch, so you would need to use the **Enlarge** command to enlarge all the entities by a factor of 25.4).

To get around this problem, you have five base unit value options:

- One Inch
- One Foot
- One Millimeter
- One Centimeter
- One Meter

Set the value according to the base units used by the office that sent you the DWG file. If the office uses imperial units (feet/inches) and the base unit is one Inch, you would set the base unit value to one Inch. If the DWG file uses millimeters as its base units, then select the one Millimeter value.

Translator Settings File Each time the DWG Import Entity Assignments dialog box is accessed during an import, you have the option to save all the current settings in that dialog box to a file with an .INI file extension in the \SUP directory (refer to the previous section). All saved .INI files are available for selection here in the Translator Settings File dialog box. The .INI file selected here will automatically be used when the **Import Method/Automatically assign** . . . option is selected. The settings in the .INI file are also applied to an interactive import. The default translator file is called Transfer.ini.

User-Defined Line Type Exports

During export, the line definitions and shape codes for DataCAD linetypes are stored in a .SHX file. If you export a DataCAD drawing named MYDRAWING.DC5, the translator will create two files: MYDRAWING.DWG and MYDRAWING.SHX. In order for the linetypes in the DWG file to display properly, the .SHX file must be placed in the \SUPPORT subdirectory in the AutoCAD directory.

If you're not concerned with linetypes displaying as you drew them, you can check the Export Linetypes toggle in the **Tools/Program Preferences** dialog box to convert all user-defined linetypes (except Solid, Dotted,

Dashed, and Dot-Dash) to Continuous when exporting a DataCAD drawing to a DWG or DXF file. In this case, a shape file will not be created because none is needed. Instead, all user-defined linetypes in MYDRAWING.DC5 will be permanently translated to the Continuous linetype. The Export Linetypes toggle is off by default.

If you choose to export a DataCAD (.DC5) file to a Release 12 DWG/DXF file and the Export Linetypes toggle is off, all user-defined linetypes in DataCAD will be displayed as Continuous in the DWG/DXF file, since AutoCAD 12 doesn't support complex line shapes. These DataCAD linetypes, however, are not permanently changed to the Continuous line shape. Instead, they retain the DataCAD linetype name and merely display as Continuous in AutoCAD. When you import the file back into DataCAD, the proper linetypes will once again be displayed.

Color Translation

This is on by default and creates a table (.TBL) file so that DataCAD colors translate properly when exported to or imported from a DWG/DXF file. This file is created automatically during translation and requires no input from you. It is given the same filename as the currently selected RGB file and is saved to the \SUP subdirectory.

3D Entity Export

The 3D Entity Export option located at the bottom of the DXF/DWG tab of the **Tools/Program Preferences** dialog box is unchecked (off) by default. Any 3-D entities in a DataCAD (.DC5) file will be translated as a polyface mesh when exported to a .DXF or .DWG file. This polyface mesh is a single entity, but by checking the "Decompose 3D Entities as 3D Face" option, each facet of a 3-D entity will be turned into an individual polyface. Thus, a single sphere in DataCAD can be translated into many separate entities arranged in a sphere shape in a DWG file. Checking the Decompose option can make translation faster, and the translated entities will look the same after a shade or hidden line removal. However, editing may be more difficult since these translated 3-D entities are made up of many smaller and separate entities.

Recreating MSP Layouts in AutoCAD

A number of our recent clients have been other architects who have needed substantial help with construction drawings. They all use AutoCAD and expect at least some level of usability from our CAD files when turned over to them at the end of the project. Most obstacles can be overcome, but the biggest problem is in page layout. We make extensive use of *Clip Cubes* (CCs) and MSP, none of which translate to AutoCAD. Thus, at the end of the project we export our .DC5 drawings to .DWG format, open the drawing in AutoCAD, and use PS (the equivalent of MSP sheets in DataCAD), Viewport (the equivalent of CCs in DataCAD), and XREF features to recreate our MSP layouts. What follows here are the basic, condensed steps for creating PS drawing sheets in AutoCAD.

DataCAD Phase

To prepare a DataCAD drawing for export to AutoCAD, follow these steps:

1. Save a copy of the DataCAD drawing file. You may want to apply certain actions to the file (refer to the section about exporting DWG files) before exporting (like exploding the title block). You don't want these changes applied to your original file.
2. Turn on the layers for each detail in the drawing file. Write down the names of the layers that belong to each detail so that you will know which layers to turn on when recreating the details in AutoCAD. We use the DWGDOC (Drawing Documentation) macro, available from Cheap Tricks Ware to list all the layers for each detail. Just turn on the layers for each detail and then run the macro, which will give you a print-out of only the layers that are turned on.
3. Since AutoCAD also has XREFs, plan to export a .DWG file of each DataCAD.DC5 master file and XREF file:
   a. In each master file (see Chapter 13, "External Reference Files (XREFs)," for an explanation), turn off the layers that have XREFs on them. If you don't, the XREF entities will be exported with the master file.
   b. Export only the **activated layers**. Be sure to save the file in the correct AutoCAD version.

c. Export each of the XREF .DC5 drawing files using **All Layers**. Be sure to save the file in the correct AutoCAD version.

AutoCAD Phase (R13/R14)

After you create the exported .DWG files, open AutoCAD and follow these steps:

1. Open the drawing in AutoCAD.
   a. Type **Z** (for Zoom) and then press **Enter**.
   b. Type **E** (for extents) and then press **Enter**.
2. Save the drawing.
3. If this is one of the master file drawings, then you need to attach any XREFs that belong with this drawing:
   a. Type **xref** and then press **Enter**.
   b. Type **A** (or Attach) and then press **Enter** (Attaching in AutoCAD is like Inserting in DataCAD).
   c. Locate and select the drawing to be attached.
4. You are prompted for the insertion point. Type in **0,0,0**.
   a. Accept the default X scale factor of **1**.
   b. Accept the default Y scale factor of **1**.
   c. Accept the default rotation angle of **0**.
5. Just as in DataCAD, an XREF in AutoCAD is attached as a single entity, so you can move it as a group if you need to relocate it after attaching it.

Paper Space (Similar to MSP in DataCAD)

To recreate your DataCAD MSP layouts in AutoCAD's Paper Space, follow these steps:

1. You have several choices for accessing PS:
   - In the pull-down menu, go to **View/Floating Model Space**.
   - At the bottom of the drawing screen, *double-click* on **Model**. It will change to **Paper**.
   - Type **Tilemode** and then press **Enter**. Type **0** and then press **Enter.**
2. Type **visretain** and then press **Enter**. Type **1** and then press **Enter**. This step is very important. If you don't do this, then none of the

settings you make in the viewports will be retained. You only have to do this once in a drawing file.

3. Type **psltscale** and then press **Enter**. Type **0** and then press **Enter**. Like Step 2, you only have to do this once in a drawing file.

Create Viewports (Similar to CCs in DataCAD)

1. Zoom out prior to making a viewport. It is best to be at a screen scale that is at least 50 feet or more across. If you don't, after making the viewport and setting a scale you may have a difficult time finding the drawing entities in a small viewport.
2. Type **mview** and then **Enter**.
3. Draw a large box called a *viewport*. All the entities of the drawing file will be displayed in the viewport.
4. You must create a new viewport for each detail on the sheet. Each viewport can have its own scale, just as each MSP detail can.

Setting Layers in a Viewport

1. To work inside the viewport, change back to Model Space:
   a. To toggle between Paper Space and Model Space, *double-click* on Paper at the bottom of the screen. Model will then appear in place of Paper and you will be in the viewport.
   b. If there is more than one viewport, just *click* inside the viewport you want to be in and you will enter it.
2. Go to the Layer menu. *Click* on **Select All**. In the middle of the right side of the dialog box, *click* on **Frz** (freeze) next to the **Cur VP** (current viewport). Frozen layers are not displayed in the viewport. **Do not use the large buttons at top right of the dialog. These affect the whole drawing, not just the viewport.**
3. Select **Clear All**. All the layers will be deselected.
4. Now select only the layers that are to be on in this viewport.
5. Select **Thw** (thaw). Thawed layers will be displayed in the viewport.
6. Select **OK** to accept all the settings.

Setting the Scale of a Viewport

1. To scale the viewport to the correct scale you must be in Paper Space, not in the viewport (Model Space).
2. Change to Paper Space now.
3. Type **mvsetup** and then press **Enter**.
4. Type **S** (scale) and then press **Enter**. You are prompted to select the objects.
5. Select the viewport you want to scale and then press **Enter**.
6. You will be prompted for the number of Paper Space units. Press **Enter** to accept the default value of **1.0**.
7. You will then be prompted for the number of Model Space Units. The factor you enter will depend on the scale of the detail. Here are a few common factors:

Detail Scale	Viewport Factor
12” = 1’-0” (full scale)	1
6” = 1’-0”	2
3” = 1’-0”	4
1/2 “ = 1’-0”	8
3/4 “ = 1’-0”	16
1/2 “ = 1’-0”	24
3/8 “ = 1’-0”	32
1/4 “ = 1’-0”	48
1/8 “ = 1’-0”	96
1/16 “ = 1’-0”	192

Working with Viewports

1. To enlarge or reduce a viewport, type **stretch**. Draw a bounding box around the corner of the viewport that you want to stretch.
2. To move a viewport, type **move**. Select the viewport and move it.

3. The four lines that make up a viewport will be visible when you print the drawing. To keep the viewports from being visible, perform these steps:
   a. Create a new layer called *viewports*.
   b. Type **change** and then press **Enter**.
   c. Select all the viewports in the drawing and then press **Enter**.
   d. Type **P** (properties) and then press **Enter**.
   e. Type **LA** (layers) and then press **Enter**.
   f. Type **viewports** (as the layer name to change to). Press Enter twice.
   g. Finally, go to the Layer menu and turn off the viewports layer.

   If you place a viewport (called viewport 1) completely within another viewport (called vieport 2), you will not be able to access viewport 1 anymore. If you must place viewport 1 inside viewport 2, make sure that one side of viewport 1 is located just outside viewport 2. That way you can still access viewport 1 by selecting the area outside of viewport 2.

CHAPTER 26

Networking

Networking Concepts

In order to get the most out of DataCAD or any CAD program, any office with two or more CAD stations should be networked together. My expertise lies with DataCAD, not computer networking, so this chapter should be considered simply a primer for creating a network. The information here was put together while I was creating the network for my own office.

A *server* is simply a computer that stores common files that are accessed by all the other computers on the network, which are known as *clients*. A network can be set up so that the server is one person's CAD station, or the server can be a dedicated machine. The dedicated machine works best. Using a CAD station as a server will cause that computer to slow down, especially when there is a lot of activity over the network.

Here is a partial list of the items that you will want to place on the network server and share with other computers on the network:

- CAD files
- Template and symbol files
- Default drawing files
- Plot files
- Layer files
- .DWG and .DXF files
- Word processing documents
- Spreadsheets (for door, window, and finish schedules)
- Product specifications
- Graphic .BMP files

While I was installing the network in my office, I got some very useful additional information from a Web site at **www.helmig.com/j_helmig/faq.htm**. There you will find some very good graphical examples of the steps outlined in this chapter.

Hardware

The first choice you will have to make is which kind of network hardware to purchase. The store that you buy from should be able to help you get all of the correct supplies. A typical network looks like the wheel of a bicycle. All the computers on the network form the rim of the wheel. Each of the computers are then connected with cables, like spokes, to a hub in the center. The hub is a little box that ties all of the computers together, routes the data between them, and amplifies the signals as required.

Here is a partial and simplified list of the components you will need:

- *A network card for each computer on the network* Save yourself a lot of potential headaches by buying identical network cards. You will want to get cards that use twisted-pair wires (this looks almost identical to standard telephone wire, but it is not) with RJ-45 connectors, such as 10baseT/UTP or 100baseTX/100baseT4 (which is faster and more expensive).
- *Network hub* Make sure you get enough capacity to handle future expansion of the network.
- *Network cable* 10baseT requires standard UTP cable and 100baseT requires UTP cat 5 cable.

Networking in Windows

The basic steps to setting up a network are as follows:

1. Install the network cards
2. Decide on a common Workgroup Name for the network.
3. Set up each computer with a Computer Name.

Step 1: Installing the Network Cards

Now we'll break down each of the previous steps into more specific tasks:

1. Make sure your card is properly installed in the computer.
2. Reboot Windows. If Windows detects your card, follow the setup procedures. If it does not, go to **Control Panel/Add New Hardware/ Network Adapters/NEXT**. Choose your card from the available options. Have your Windows CD ready in case it needs to find the driver there.
3. If the card installs with no conflicts, you're all set. If conflicts occur with the IRQs and so on, you will have to go to **Control Panel/ System**, find which IRQs are available, and then reset your card. Then go back to Step 3 and start again.

Once the card is installed in each computer, the rest is not too difficult, just a bit confusing.

Step 2: Decide on a Common Workgroup Name

This is the easiest step of all, since you don't have to set up any hardware or software. All you have to do is decide on a name. A workgroup is like your house and its address. The parents and kids (each with a name) all live in their house at 100 Main Street. In this example, their workgroup name is 100 Main Street, and the computer names are the names of the kids. Anyone who does not live in that house cannot gain access without a key from the head of the household. Your network is like the house and the family. As long as each user is part of the same workgroup, they can all access the same network. We chose "WMA" as our workgroup name. Just choose something simple for your own.

Step 3: Set Up Each Computer on the Network

As mentioned earlier, a network can be set up so that the server is one person's CAD station, or the server can be a dedicated machine. Having a dedicated server makes everything that much easier to set up, and if the server is running on Windows NT, that makes it even easier.

Here are the generic steps for setting up each computer on a Windows 95/98 network:

1. Most of your networking options will be set from **My Computer/ Control Panel/Network**, so go there now.
2. *Click* on **Identification**:
   a. Give a name to the computer you are working on. This is the *Computer Name*. If the computer is dedicated to one user, you can give it a name like MIKE or ALPHA. Each computer must have a unique name.
   b. Assign a name to the workgroup (as described in the previous section). This is a *Workgroup Name*. This must be the same name on all the computers on the network (though you could have multiple workgroups in a larger office).

   c. Now give a description for the computer. This can be anything (like "Mike's computer"), but the best solution is to use the same name as the Computer Name from Step a.
   d. As an example, my computer's Computer Name is ALPHA, its Workgroup Name is WMA, and its Description is ALPHA.
3. *Click* on the **Access Control** tab. Check the **Share-level access control** box. This enables all computers to share all resources, such as hard drives or printers. You can still password-protect certain files, directories, drives, and so on later.
4. *Click* on the **Configuration** tab. This is the difficult part:
   a. *Click* on **ADD**.
   b. Go to **Client** and add "Client for Microsoft Networks."
   c. Go to **Adapter** and add your adapter card if it's not already there. It should install a couple of protocols. If not, go to **Protocol**, choose **Add/Microsoft**, and then choose the **IPX/SPX-Compatible Protocol** for your adapter card. Next, choose the **NetBEUI-Compatible Protocol** for your adapter card.
5. Now *click* on **Add/Service/Microsoft**.
6. Go back to the **Configuration** tab and *click* on **File and Print Sharing**. Check both boxes if you want to share both. Go to each item in the "The following network components are installed" window and check that **File and Print Sharing** is set for each one.
7. *Click* on **OK** to get out of the Network setup.
8. Repeat Steps 1 through 7 for all the computers on the network. Then restart all the machines.

To enable hard drive, file, and printer sharing, you need to take two more steps. First, you need to enable *sharing* for everything you want to share and, second, you need to *map* physical hard drives to a network drive letter for each computer.

Sharing

Before data can be accessed between computers on the network, each computer to be accessed must give the other computers permission to access that data, which by now you're already aware is called *sharing*. You can share a complete disk drive or just certain directories. For this example, let's assume that one computer (called SERVER) is acting as a server for all

Figure 26-1
A shared drive

the other computers on the network and that its D:\ drive contains the information to be shared with the other computers. To allow this sharing, you must follow these steps:

1. On the SERVER computer, go to **My Computer**, *right-click* on the D:\ drive icon, and *click* on **Sharing**.
2. Check the **Shared As:** circle.
3. In the **Share Name:** box, type in a descriptive name, such as SERVER FILES.
4. Under **Access Type:**, check the **Full** circle.
5. Select **OK** to accept all the settings.

To indicate that the drive is now shared, the drive icon has been changed to have a hand appearing under the drive (the hand is "giving" access to the data), as shown in Figure 26-1. Once a system has shared something, it becomes visible in the **Network Neighborhood** on the Windows desktop.

Mapping

After setting up data sharing, information on the shared drive is now accessible to everyone on the network. You could leave it at this, but one more step should be taken in order to make accessing specific data that much easier. The basic steps are as follows:

1. On the SERVER drive that is shared, create a new folder for each data type to be shared and move all the appropriate data to those folders.
2. Use drive *mapping* on all the other computers to make each of the new folders on the server look like separate hard drives (called *virtual drives*).

If all of that just went over your head, don't worry. Here is what it means. In our office, we have the following virtual drives set up:

F: de**F**ault drawings (.DC5)

G: drawin**G** files (.DC5)

L: **L**ayer files

P: **P**lot files

S: **S**ymbol files

T: **T**emplate files

X: .d**X**f & .dwg files

Y: stik**Y**back files (files created with the third-party Sticky Back macro)

For instance, all of our DataCAD drawing files are stored in a folder called Drawings on the server. That folder has been mapped as a virtual drive, however, so that when I access drawing files from my computer, it "thinks" that it is accessing the G:\ hard drive on the SERVER computer. Using this as an example, let's see how we accomplished this.

Although you don't have to use the exact same letters that we have, we highly recommend that you make virtual drives for all those file types (of course, you don't need the Sticky Back drive if you don't use that macro). You may not immediately understand why you should divide your files into separate virtual drives, but you most certainly will once you begin to use them. It may take you an hour or so to set them all up and copy the appropriate files to the new directories, but it will be well worth the effort.

To create a virtual drive, follow these steps:

1. On the server's D:\ drive (the one we set up to be shared earlier), create a new folder called Drawings. Place all the DataCAD .DC5 drawing files in that folder.
2. On the client, open **Network Neighborhood**.
3. *Double-click* on the SERVER FILES icon (the name we gave the shared D:\ drive earlier).
4. Locate the Drawings folder and *right-click* on it.
5. Select **Map Network Drive**. A dialog box appears.
6. In the **Drive:** box, select the G: drive letter.
7. The **Path:** should already be filled in with the path to the folder on the server. If not, type in the path yourself (see Figure 26-2).
8. Check the **Reconnect at logon** box so that the new virtual drive will be accessible every time the client computer is started.
9. *Click* on **OK** to accept all the new settings. The virtual drive is now displayed in **My Computer** with a network icon, as shown in Figure 26-3.

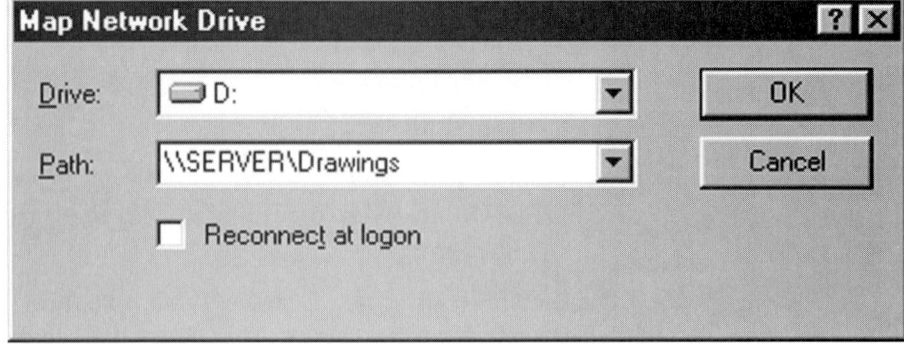

Figure 26-2 The drive and path to the folder

Figure 26-3 The virtual drive now has a network icon.

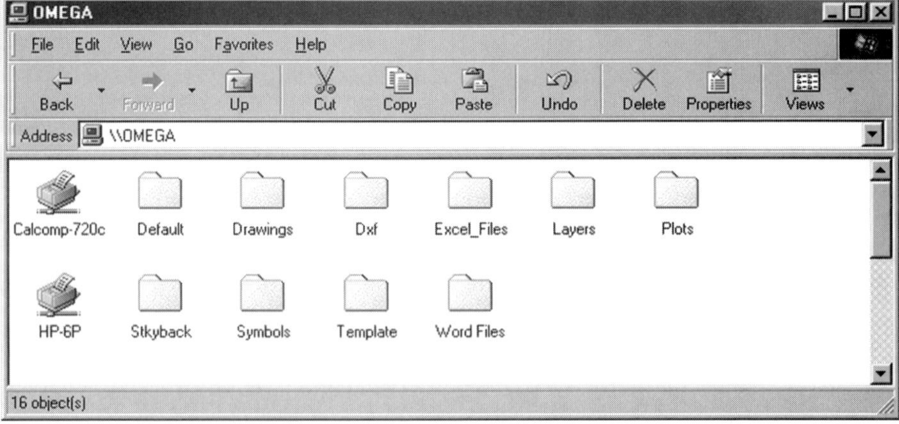

Figure 26-4 10 shared folders, and 2 shared printers on the server

An Example from My Own Office

Figure 26-4 shows what I see on my own computer at work when I open up the icon for our network server in Network Neighborhood.

In this example, the name of our server computer is OMEGA and you can see all of the folders that are shared between OMEGA and the network. Most of these folders are in turn mapped to a virtual drive in **My Computer** so that it looks like Figure 26-5.

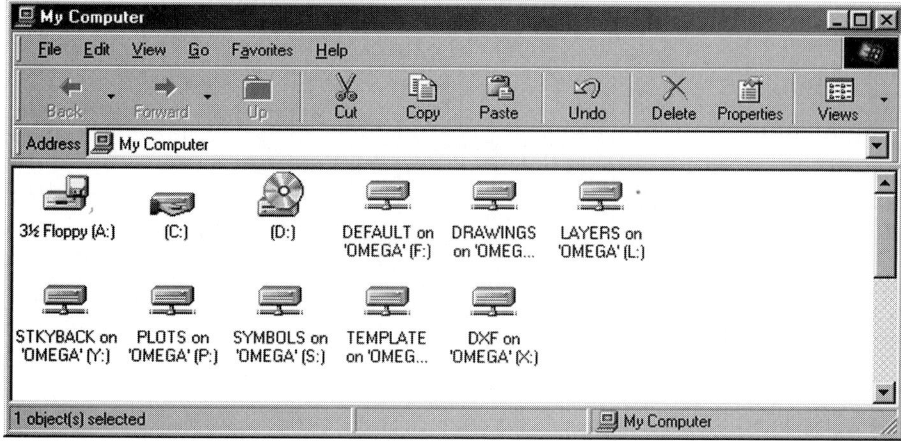

Figure 26-5 Eight mapped virtual drives on the client

Here you can see that besides my computer's own floppy drive, its C:\ hard drive (which is shared), and its CD-ROM drive, I have eight virtual drives, all mapped to the OMEGA server.

Mapping Drives on a Workstation Server

If you are going to use one of your Windows 95/98 CAD workstations as the network server, then you will find that if you try to follow the previous directions, you will get an error message when you try to map a folder with a virtual drive letter. That's because Windows 95/98 will not let your computer map virtual drives to itself, but an old DOS trick can help you get around this limitation. If the workstation is running Windows NT, however, you can map virtual drives to itself, so you won't have to revert to using this trick. You only have to do this on the workstation that will be acting as the server:

1. With a text editor like Notepad, open the Autoexec.bat file in your C:\ directory.
2. Add this line to the file:

   ```
   SUBST G: D:\drawings
   ```

   This tells the computer to map the D:\drawings folder as a virtual hard drive called drive G:. In Explorer, if you click on the G: drive, you will actually be seeing the contents of the D:\drawings folder.

3. Save the Autoexec.bat file and then reboot your computer.
4. After the computer restarts, go to Explorer and scroll down the list of folders and drives. Near the bottom you will now see a new one called XXXXX(G:), where XXXXX is the name of the D:\ drive of the server.

NOTE: *SUBST is a DOS command, and DOS can only read file names of up to eight characters in length. Thus, I could not use the SUBST command to map a folder called D:\Templates, since Templates is nine characters in length. To use the SUBST command, you would need to rename the folder to something shorter, such as Template or TPL. You also cannot use blank spaces in the folder name, such as TPL File.*

Caveats

In this example, on the workstation server the list of drives you see in Explorer will include a D:\ drive and a G:\ drive (plus whatever other drives you might have, such as a CD-ROM or ZIP drive). Notice that the directory tree for D:\ will be IDENTICAL to the one for G:\.

When you back up files on the workstation server, remember that you do **not** back up the virtual G:\ drive, since it is really just the D:\Drawings folder in disguise. Just back up the D:\ drive like you normally would.

Printers/Plotters on the Network

Like data files, to access printers and plotters over a network, we need to make each printer and plotter a shared resource and then create a connection to them from each computer that needs access to them. These printers do not have to be connected to the server; they could be connected to one or more client computers on the network, but once again, the system will be faster and more efficient if the output devices are connected to a dedicated machine.

One thing that cannot be overemphasized is the need for you to have the most up-to-date printer/plotter software driver for each printer/plotter. Even if you just purchased the device, there may already be a more up-to-date software driver available from the company's Internet Web site. You will probably have more trouble getting your network printers to work than you will any other aspect of your computer network. Printers and their drivers are notoriously finicky, and manufacturers are constantly fixing and improving their drivers to get them to work better. Often they will also add new features in the process, so check out the manufacturer's Web site right away and make sure you have the latest drivers.

We will be using the word *printer* to refer to either a printer or a plotter. Even though you are accessing the printer on another computer, you must still have the software drivers for each network printer installed on each client computer. This is because the client computer is actually processing all of the print data. The print server is simply receiving the information and printing it out. This print server is the computer that has the printer or plotter physically connected to it. The client in this case is any computer that will access the printer or plotter on the print server.

Setting Up the Print Server

To set up the print server, follow these steps:

1. Install the printer driver on the print server first. When asked during the setup, "How is this printer attached to your computer?", *click* on the **Local Printer** option.
2. Go to **My Computer/Control Panel** and open the **Network** applet.
3. Make sure "File and printer sharing for Microsoft Networks" is installed as one of the network components. If it is not, use the **Add** button to add it.
4. Now *click* on **File and Print Sharing**. Both boxes should be checked. *Click* on **OK** (see Figure 26-6).
5. *Click* **OK** to close the **Network** applet.
6. In **My Computer/Printers**, *right-click* on the printer to be shared. Select the **Sharing** option.
7. Check the **Shared As** circle.
8. In the **Share Name** box, give the printer a unique name that can be identified by everyone on the network.
9. If you want, you can also type in a **Comment**. *Right-click* on the word Comment if you want a description of what it is used for.
10. *Click* on **OK** to accept all the settings (see Figure 26-7).
11. The printer icon will change to display a hand underneath it, indicating it is now a shared resource (see Figure 26-8).

Notice that in this example, the Share Name, Comment, and the name under the printer icon are all different. Unless you renamed it earlier, the

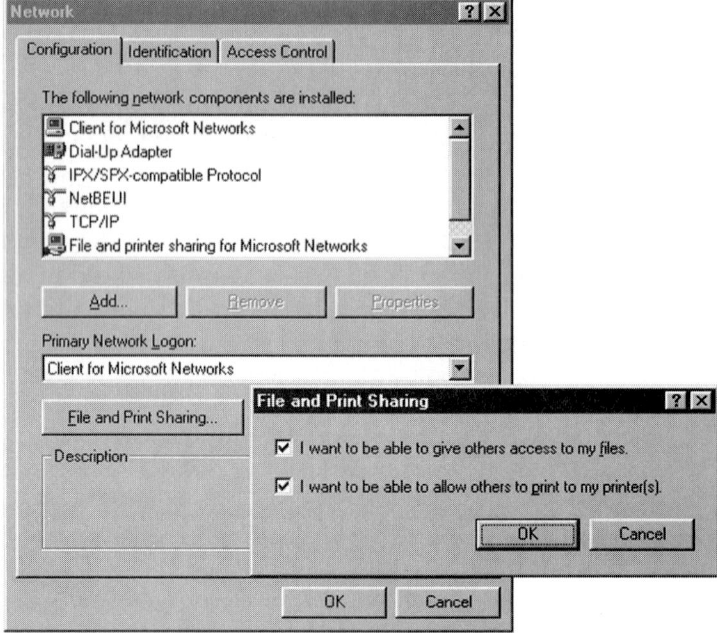

Figure 26-6 Activating file and print sharing

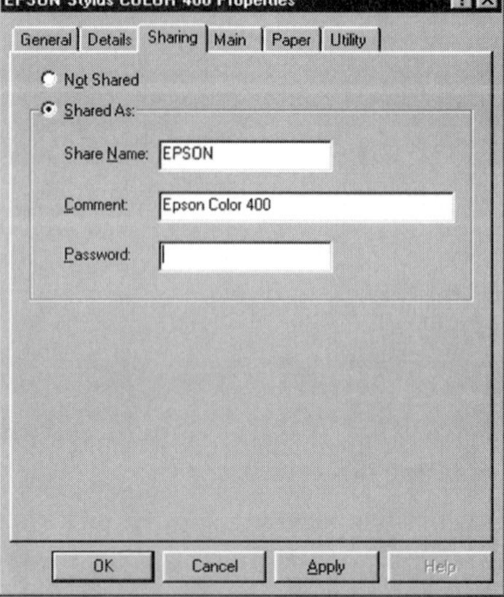

Figure 26-7 Printer sharing options

Figure 26-8 A shared printer icon

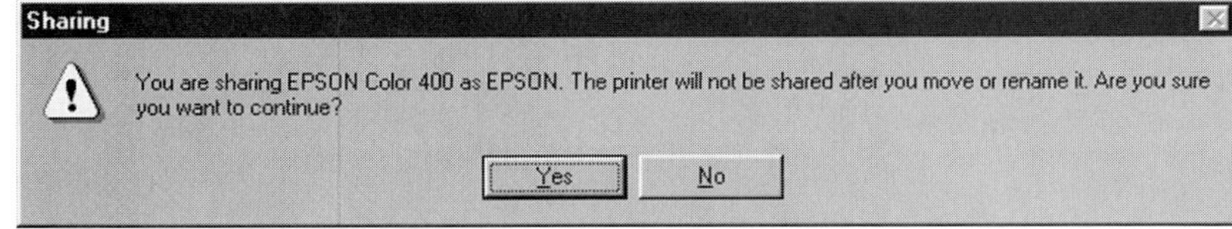

Figure 26-9
A name change error message

name under the icon is the name given when the printer driver was installed. If you change this name after setting up your printer sharing, then you will get an error message as shown in Figure 26-9.

If you rename the printer, you will have to follow all the Steps 6 through 10 all over again for every computer that was sharing that printer. So if you think you may want to rename the printer, do it before setting it up across the network.

Setting up the Client

To set up the client, follow these steps:

1. On the client computer, go to **My Computer/Printers** and select **Add Printer**. The Windows Add Printer Wizard dialog will appear. *Click* on **Next** to continue the installation.
2. When asked, "How is this printer attached to your computer?", *click* on the **Network Printer** option and then *click* on **Next**.
3. Either enter the **Network path** or **queue name** if you know it, or use the **Browse** button to locate the network printer under Network Neighborhood.
4. The dialog box also asks, "Do you print from MS-DOS based programs?" Unless you print from DOS programs, select the **No** option.
5. *Click* on **Next**.
6. Now you are prompted for a **Printer name**. This is the name that you will see in the **Printers** folder on your own computer. This name does not need to match any of the other names, so you can name it whatever you like. Just make sure it is adequately descriptive.
7. In the same dialog box, you are also asked if you want this printer to become the default printer. If you want this printer to be the one that is automatically selected by all your Windows programs, then check **Yes**. Otherwise, check **No**.
8. *Click* on **Next**.

9. The final dialog box will ask, "Would you like to print a test page?" Select **Yes** and then *click* on **Finish**. As long as everything is installed properly, when the installation process is completed, Windows will print a test page from the printer.

Print Spooling

To make printing faster, you need to tell the printer driver to *spool* all the print jobs. With spooling, your computer stores the print information and dishes it out to the printer as necessary. Without print spooling, the print information is sent directly to the printer, causing your computer to pause while the information is slowly transferred to the printer. Your new printer is probably already set up to use spooling, but you should check to make sure:

1. In the **Printers** folder, *right-click* on the new printer icon and select the Details tab.
2. Select the first option: "Spool print jobs so program finishes printing faster."
3. Select the third option: "Start printing after first page is spooled."
4. *Click* on **OK** to close the Spool Settings (see Figure 26-10).

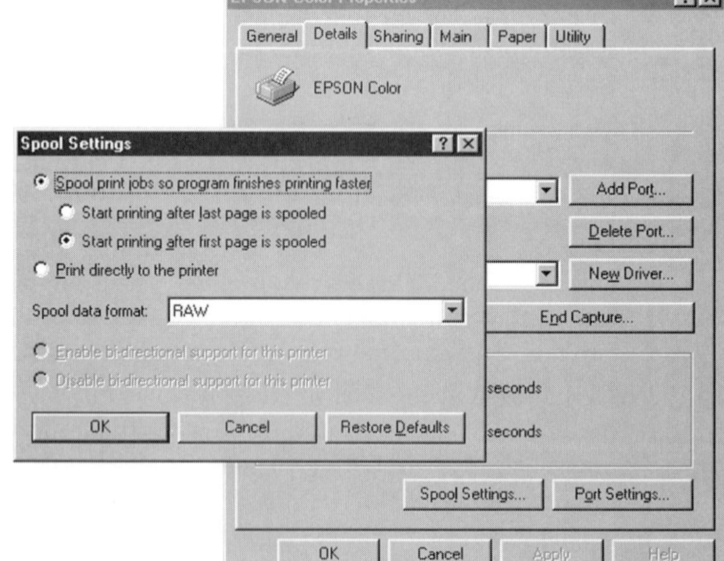

Figure 26-10
Printer spool settings

5. Now go to the adjacent **Port Settings** and make sure the following are checked:
    a. "Spool MS DOS print jobs"
    b. "Check port state before printing"
6. Press **OK** to close the printer dialog and accept all of the settings.

SUMMARY

Choosing the correct hardware and setting up an efficient network consists of more processes than we can cover here, but this should give you a good working knowledge. Any decent book on setting up a network should fill in the rest of the blanks for you. Of course, you could always hire a network professional to set things up as well. Whether you do it yourself or hire a professional, be sure to create the virtual drives that were described earlier. Doing so is what will really make your network useful.

CHAPTER 27

DataCAD Resources

This chapter lists numerous locations for support and information relating to the DataCAD software program including Web pages, dealers, and DataCAD user groups. Keep in mind that due to the continually changing nature of the Internet some of the Web pages listed here may have changed their addresses or are no longer in service.

Web Pages and Third-Party Support

DataCAD Home Page

`www.datacad.com`

This is the home page of DataCAD LLC, the company that brings you DataCAD. Here you will find general information and announcements, free downloads, powerful add-on macros, DataCAD dealers, a photo gallery, and technical support information. If you go to the Site Map section of the page, you will find all the following topics:

Gallery This section includes images of projects by various design professionals using DataCAD. It also proves links to Web sites of architects, builders, and designers using DataCAD.

Resources This area contains various A/E/C Web sites related to architecture, design, construction, and CADD.

Sales Here you will find DataCAD's own online store where you can purchase the DataCAD program, various add-ons, training and support materials, and even DataCAD T-shirts. This section also offers information on U.S. and international DataCAD dealers and support personnel, as well as U.S. educational dealers of DataCAD.

News The News section contains links to articles about DataCAD and the people who use it. Available here are links to news about DataCAD, present and past. Archives of past DataCAD press releases are also maintained, along with testimonials from various DataCAD users.

Products The Products section offers information about the latest release of DataCAD, including a list of the latest features. In this area, you can also learn about the following topics:

- The DC Viewer program that comes on the DataCAD CD-ROM.
- The latest *Architectural Graphics Standards* reference book and companion CD-ROM.
- Royalty-free clip-art ImageCELs for use in your 3-D renderings. These are photographs of people, automobiles, plants, doors, and other miscellaneous architecturally-related images specially edited for use in photorealistic renderings.
- The bundled Cheap Tricks Ware CD-ROMs available through DataCAD LLC.
- The DataCAD training tapes available through DataCAD LLC.
- The new DataCAD Estimator program (for cost estimating) that comes on the DataCAD CD-ROM.
- The DataCAD technical support certificate, entitling you to extended technical phone support from DataCAD LLC.
- Various international fonts available for free from the DataCAD LLC Web site (also available on the CD-ROM that comes with this book).

Events Here you can find out about upcoming events featuring DataCAD LLC and other DataCAD-related resources.

Employment If you are looking for a job, you can check here for current U.S. and worldwide listings. You can also post your resumé here for prospective employers to read. If you are looking to hire someone, you can post your advertisement here and review any of the current on-line resumés.

Support In this section, you can read about the various technical support options available to registered DataCAD users. Several great full-color technical bulletins about how to use various DataCAD products and features are included here. This is simply a must-visit page. Some of the topics include the following:

- Hardware key issues.
- Using the DataCAD estimator. This is some of the best basic information you can find.
- Converting 2-D drawings to a 3-D model.
- Frequently Asked Questions about DataCAD.
- Basic information about DataCAD templates and symbols.
- Importing and exporting .DWG drawing files.

- A great primer about polygon normals and their impact on 3-D rendering. If you do any 3-D rendering, you need to read this bulletin.
- A wonderful how-to for creating fully rendered models with the Renderize rendering program that comes with DataCAD. Get the whole manual boiled down to a few pages of concise information.
- Another great how-to for creating an animated walk-through of your 3-D model in Renderize.
- The *Multi-Scale Plotting* (MSP) feature of DataCAD, which enables you to create drawing sheets containing details of differing scales. This is one of DataCAD's most powerful features. You owe it to yourself to learn it.
- Printing your drawings from DataCAD using the new interface in DataCAD.
- Various issues relating specifically to the old DOS version of DataCAD.

Online Support Here you can access the *DataCAD Developer Network* (DDN), where you can get related information about programming with DataCAD's own *DataCAD Applications Language* (DCAL) programming language. You can also download sample programming code for a variety of fully-functioning free macros. Information also includes fonts, hatching, linetypes, and icon toolbars. Also available on the DDN site is a series of free macros that are not available from the main DataCAD download page.

The Online Support section also provides a list of current DCAL programming developers. If you have a macro that you need to have created or if you just have an idea for one, these are the guys to talk to.

In this section, you can also learn about the free DataCAD *Boston Users' Group* (DBUG) online forum. If you use DataCAD you will want to be a member of this group. You can get help and offer information among hundreds of worldwide DataCAD users. This is one of the best resources for DataCAD users, period.

About Us Here you can get information, addresses, and phone numbers for the DataCAD LLC company and staff.

Contact Us This section provides information on how to contact various DataCAD LLC departments via phone, mail, or e-mail. It also offers contact information for educational, institutional and international sales.

DBUG, Cheap Tricks, and DataCAD Links

Evan Shu of Shu Associates, Inc. is a practicing architect in Massachusetts and a long-time user of DataCAD, but he wears many other hats as well. He has also been running the DataCAD Boston Users' Group (DBUG) for more than 10 years; is the editor of the monthly *Cheap Tricks* DataCAD newsletter; and is the owner and operator of the Cheap Tricks Ware software company, a third-party seller of inexpensive DataCAD-related software.

The *Cheap Tricks* newsletter was a great source of information for us as we wrote this book. Evan Shu was kind enough to let us use some of his past articles for the basis of a few sections of this book.

You can find links to all the previously mentioned resources, plus many others, on his DataCAD Links Web page at **`www.world.std.com/~eshu/links.htm`**. There you will find dozens of links to various DataCAD resources, including the Web pages of other DataCAD users who offer free DataCAD software, DataCAD tutorials, and lots of general DataCAD information. This section lists some of these Web sites.

Cheap Tricks Ware

`www.world.std.com/~eshu/ctw/ctw.htm`
Shu Associates, Inc.
120 Trenton Street
Melrose, MA 02176-3714

Cheap Tricks Ware is unarguably the single most important software resource for the DataCAD user. Here you will find affordable add-on macros (mini-programs that work inside DataCAD), ready-made symbols, hatches, linetypes, shareware, and text-based information relating to DataCAD. Much of the software here costs only $5 (U.S.), with nearly everything else ranging from $10 to $30. For your convenience, you can even place credit card orders via the Internet. Software is submitted by a multitude of DataCAD users and programmers worldwide.

Cheap Tricks DataCAD Newsletter

`www.world.std.com/~eshu/cheap.htm`

At this site you can order a subscription to the *Cheap Tricks* newsletter. It is a monthly grass-roots newsletter with every DataCAD-related topic you can think of, including 2-D drafting, 3-D modeling, rendering, office procedures and standards, hardware and software information, networking, and much more. The publisher, Evan Shu, is himself a practicing architect using DataCAD on a daily basis. His wit and wisdom have become a staple for DataCAD users.

Cheap Tricks newsletter Reprints

`www.world.std.com/~eshu/cheap/reprints.htm`

At this site you can order reprints of many past years of the *Cheap Tricks* newsletter. A huge wealth of information is to be had here.

DBUG Meeting Schedule

`www.world.std.com/~eshu/dbug.htm`

DBUG is arguably the most active DataCAD user group in the world. If you live in the New England area or will be visiting the area, then drop on by. Meetings are usually held on or about the third Wednesday of the month.

BAM Compugraphyx, Inc.

1025 Estates Blvd.
Trenton, NJ 08690

`www.datacadsales.com`

BAMNET is a service of BAM Compugraphyx, Inc. They help those who use DataCAD, whether products were purchased from them or not. This Web site provides informational technical support and is a place to find answers to questions. You can also purchase DataCAD from this site and not pay shipping costs. Some of the information here includes DataCAD software sales and installation and configuration issues. A DBUG support archive can also be found here, as well.

AECNET

`cyburbia.ap.buffalo.edu/pairc`

This site proclaims to offer “Internet Resources for the Built Environment.”

Architectural CADD Consultants

`www.architecturalcadd.com`

This is another must-see Web site for DataCAD users. Here you will find information about DataCAD tutorials, seminars, reviews, and consulting in CAD; 2-D productivity; 3-D CADD modeling; imaging; rendering; animation; the Architectural CADD Shootout; and DataCAD training videos (including DataCAD for those who Know AutoCAD). Check out the new DataCAD CD-ROM-based tutorials, which are very well produced and worth a look.

Networth Computer Solutions

`www.wolfenet.com/~networth`

This page is dedicated to the user-development of DataCAD. Users are welcome to post questions and solutions for public use to NetWorth Computer Solutions, who offer this as a public resource. They encourage authors of macros, linetypes, hatches, and so on to post ZIPed files. All data posted is assumed upon receipt to be freeware, although authors may post shareware if they want.

Unique Software

`www.nxi.com/billd`

12 Chapel Street
Freeport, ME 04032

Bill D'Amico of Unique Software is one of the premier programmers of DataCAD macros. Some of these include the Blocker Space Planning Macro, the Touchup 3-D Hatching Macro, the TPLLIB Template Library Macro, and the 3-D Power Tools Macro. He is also the programmer responsible for developing many of the new DataCAD program features such as Display List, PDF output, Batch Plotting, and Multiple Undo and Redo.

Alchemy Mindworks Inc.

`www.mindworkshop.com/alchemy/alchemy.html`
`www.mindworkshop.com/alchemy/gww.html`

Alchemy Mindworks probably doesn't even know what DataCAD is, but their downloadable shareware, especially *Graphics Workshop for Windows*

(GWW), is highly recommended for graphics chores. GWW is invaluable for converting between various graphics formats and provides instantly viewable thumbnails of all your graphics images. As shareware, it is included on the CD-ROM at the back of this book.

JASC, Inc.

`www.jasc.com`

Paint Shop Pro is the latest graphics paint program from JASC, Inc. It is a great program to use for creating and editing DataCAD toolbar icons. As shareware, it is included on the CD-ROM at the back of this book.

Lightscape Rendering Software

`www.lightscape.com`

Lightscape is a powerful rendering program that many DataCAD users are currently using. You can export your DataCAD 3-D models to Lightscape via the .DXF format. From there, you can add various materials and lighting effects to achieve photographic quality images.

AutoCAD Shareware Clearinghouse

`cadalog.com/index.html`

You may think it odd to include an AutoCAD resource in this book, but if you check out this and other AutoCAD sites you will see that many utilities often deal with graphics in general and CAD topics specifically, some of which will help you make better use of DataCAD.

Three More Web Pages

Besides the sites that have already been mentioned, three other Web sites that can be found on Evan Shu's Links page are worth specifically mentioning here.

WMA's DataCAD How-To's and Information

www.tiac.net/users/wmitrop/wma/wmahome.htm

This is the home page of my own architectural office. We have an entire section dedicated to DataCAD how-to's and general DataCAD information. The positive comments I received in response to this site are what prompted me to write this book.

The DataCAD Roundhouse

www.pe.net/~jhorecka/DCADLinks.html

James Horecka is a registered architect practicing in California. He has been using DataCAD for many years and has created a Web page for other DataCAD users. Here is the description of his site as it is described on his opening web page:

> The DataCAD Roundhouse features DataCAD, the Architectural CADD software that I use daily as an architect. (CADD is an acronym for Computer Aided Design and Drafting.) I produce nearly all of my architectural work with DataCAD, from the initial project research phase, site and building design and development, all the way through detailed construction documents and display graphics.

For well over a year now, James has been posting a DataCAD Tip of the Day to the DBUG e-mail forum. On his Web page you will find an archive of all of the past tips. You will also find lots of other tidbits of DataCAD information along with more links to DataCAD-related Web sites. Here is a partial list:

- Southern California Users of DataCAD
- TenLinks.com
- Tip of the Day Archive
- DataCAD speed tips
- The MoolBox maneuver
- Things to come
- vRamDir
- Keyboard shortcuts

- Desktop icons
- CalComp TechJET 720 Model 5436 plotter
- Text manipulation

Tony's DataCAD Pages

www.homestead.com/blasio/datacad.html

This is another helpful Web page with lots of free stuff and general DataCAD information. Here is Tony's own description of his page:

> These are my pages dedicated to DataCAD. On these pages you will find numerous free downloads and tips & tricks for using what I feel is the best Architectural CAD Software on the market. I hope you like what you find here and I encourage everyone to donate anything you want to these pages. I don't know everything about DataCAD, but if all the users kick in a little knowledge then this site may become a source of quite a bit of knowledge and information! Here you can find DataCAD tips & tricks, free downloads, links and more!
>
> Topics you will find here include the following:

- DataCAD macro reviews
- Tony's dcad.mcr File
- Toolbar icons
- Custom toolbars
- Templates
- Hatches
- Linetypes
- Tips and tricks
- DataCAD links
- Keyboard macro archive
- DataCAD chat
- Sample images
- Miscellaneous DataCAD files
- Default drawing files

DataCAD Applications Language (DCAL)

DataCAD comes with its own programming language called DCAL. It is a very powerful language that will enable you to create custom **Toolbox** macros for your own use or for sale to others. You can read more about it in Chapter 21, "Techno-Files." If you are interested in pursuing it, here are some resources that can help you out.

DCAL Programming Language

Every DataCAD CD-ROM comes with the compiler for the DCAL programming language along with numerous DCAL programming examples. If you haven't already done so, you can install all the DCAL programming material from the CD-ROM by running the installation program and choosing to install only the DCAL material.

DCAL Reference Manual

Every DataCAD CD-ROM also comes with the complete reference documentation for the DCAL programming language. The document has basic formatting, but if you get into DCAL programming, you will want to get a better formatted version. You can get this from the DDN Web site at **`www.datacad.com/dcal/frame.htm`** or you can get it from the CD-ROM that comes with this book. It is very nicely formatted and is available in Microsoft Word format, but it may not contain a few of the latest DCAL commands since some were to be added after this book was already in publication, so check the DataCAD Web site or the DataCAD CD for the most up-to-date changes.

DataCAD Developer Network (DDN) Web Page

`www.datacad.com/dcal/frame.htm`

This web site is the best online resource for DCAL programming. You can get the full source code for dozens of DCAL macros along with the fully compiled macros. You will also find contact information for other DCAL programmers, all of whom are ready, willing, and able to help you.

APPENDIX

Linetypes

These are the standard DataCAD linetypes that come with the basic program. A chart of these linetypes can be printed from the CD-ROM.

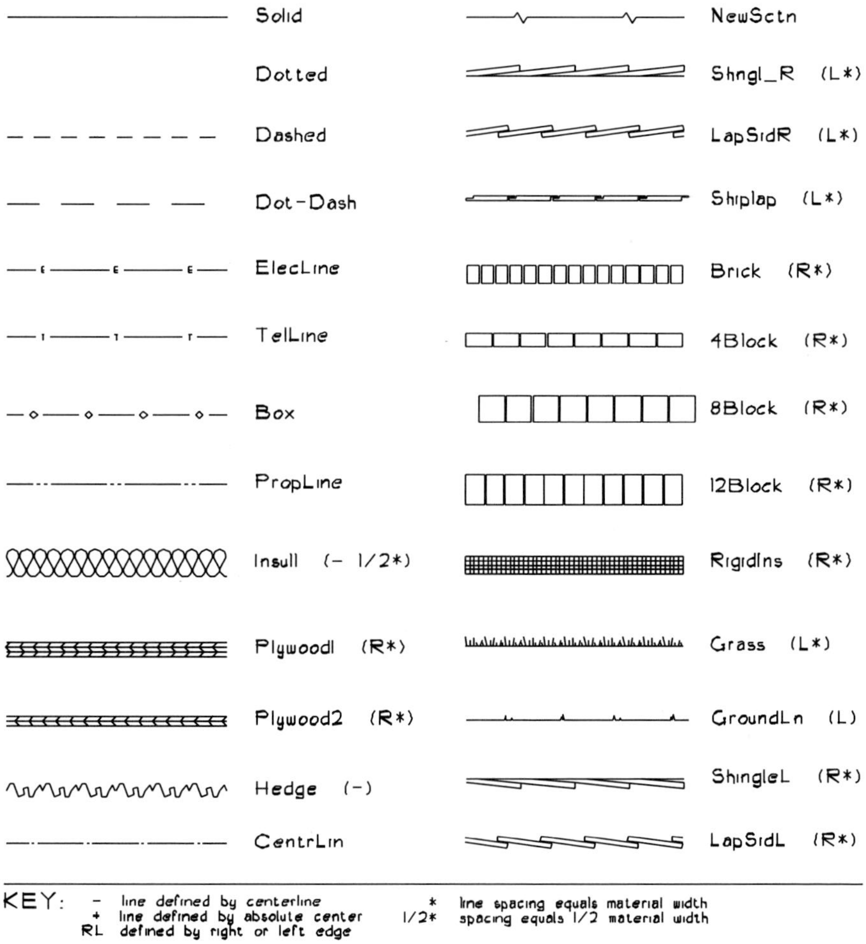

Linetype	Linetype
Solid	NewSctn
Dotted	Shngl_R (L*)
Dashed	LapSidR (L*)
Dot-Dash	Shiplap (L*)
ElecLine	Brick (R*)
TelLine	4Block (R*)
Box	8Block (R*)
PropLine	12Block (R*)
Insull (– 1/2*)	RigidIns (R*)
Plywood1 (R*)	Grass (L*)
Plywood2 (R*)	GroundLn (L)
Hedge (–)	ShingleL (R*)
CentrLin	LapSidL (R*)

KEY:
- – line defined by centerline
- + line defined by absolute center
- RL defined by right or left edge
- * line spacing equals material width
- 1/2* spacing equals 1/2 material width

Hatches

These are the standard DataCAD hatch patterns that come with the basic program, along with numerous additional patterns available through Cheap Tricks Ware. A chart of these hatch patterns can be printed from the CD-ROM. The indicated scales are typical scales that we often use for each of these hatches.

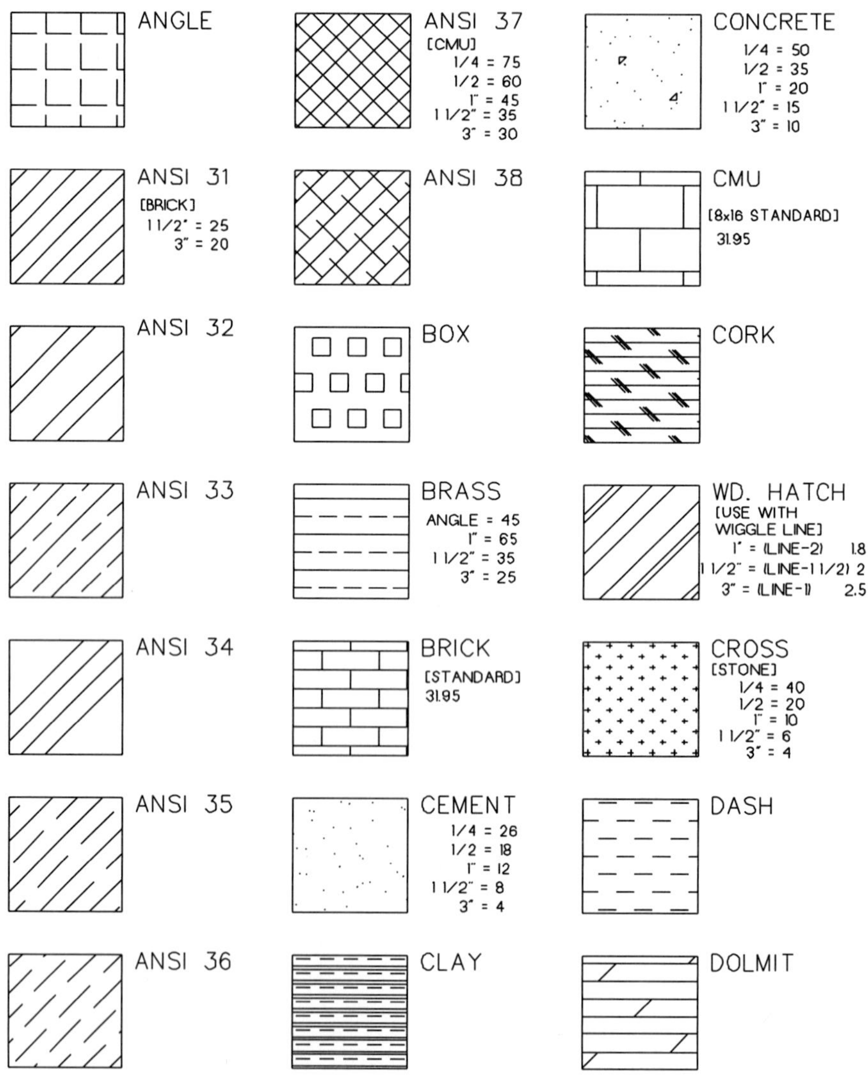

DOTS

HONEY
[GRAVEL]
1/4 = 60
1/2 = 30
1" = 15
1 1/2" = 10
3" = 5

PLAST.

EARTH
1/8 = 475
1/4 = 325
1/2 = 185
1" = 100
1 1/2" = 55
3" = 35

HOUND

PLASTI

ESCHER

INSUL.

SACNER

FLEX

LINE
[4" CLAPBDS]
SCALE = 128
[6" CLAPBDS]
SCALE = 192

SQUARE

GROSS

MUDST

STARS

GRATE

NET

STEEL
1/2 = 95
1" = 65
1 1/2" = 35
3" = 25

HEX

NET 3

TRANS

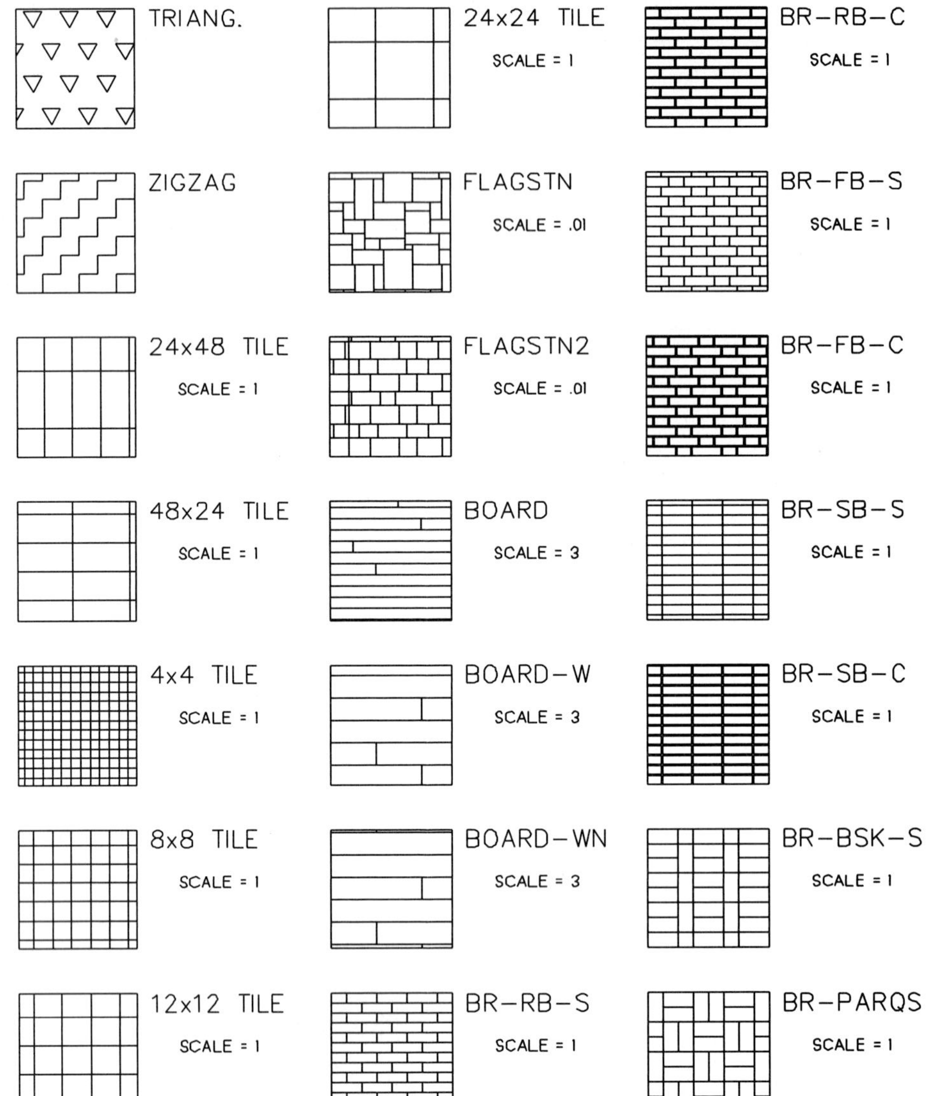
TRIANG.
24x24 TILE
SCALE = 1
BR–RB–C
SCALE = 1
ZIGZAG
FLAGSTN
SCALE = .01
BR–FB–S
SCALE = 1
24x48 TILE
SCALE = 1
FLAGSTN2
SCALE = .01
BR–FB–C
SCALE = 1
48x24 TILE
SCALE = 1
BOARD
SCALE = 3
BR–SB–S
SCALE = 1
4x4 TILE
SCALE = 1
BOARD–W
SCALE = 3
BR–SB–C
SCALE = 1
8x8 TILE
SCALE = 1
BOARD–WN
SCALE = 3
BR–BSK–S
SCALE = 1
12x12 TILE
SCALE = 1
BR–RB–S
SCALE = 1
BR–PARQS
SCALE = 1

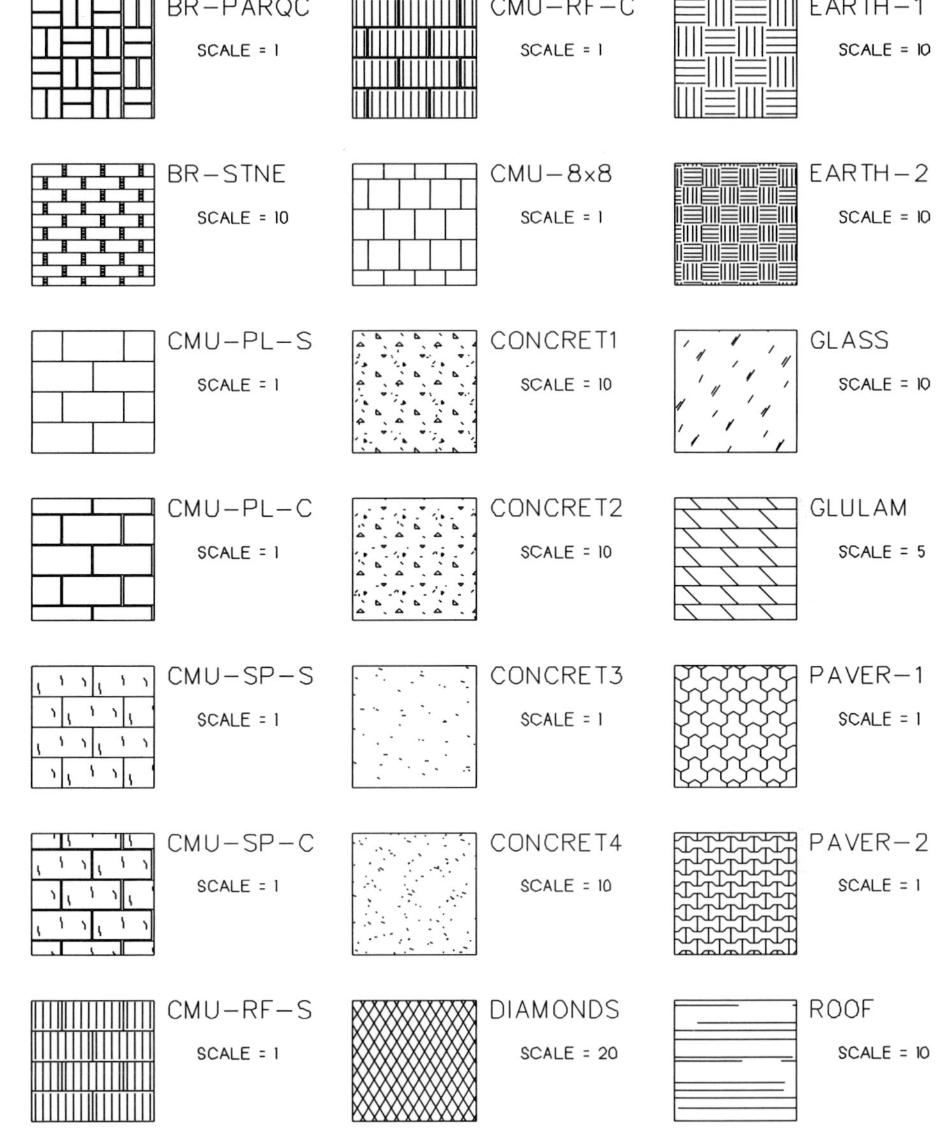
BR-PARQC
SCALE = 1
CMU-RF-C
SCALE = 1
EARTH-1
SCALE = 10
BR-STNE
SCALE = 10
CMU-8x8
SCALE = 1
EARTH-2
SCALE = 10
CMU-PL-S
SCALE = 1
CONCRET1
SCALE = 10
GLASS
SCALE = 10
CMU-PL-C
SCALE = 1
CONCRET2
SCALE = 10
GLULAM
SCALE = 5
CMU-SP-S
SCALE = 1
CONCRET3
SCALE = 1
PAVER-1
SCALE = 1
CMU-SP-C
SCALE = 1
CONCRET4
SCALE = 10
PAVER-2
SCALE = 1
CMU-RF-S
SCALE = 1
DIAMONDS
SCALE = 20
ROOF
SCALE = 10

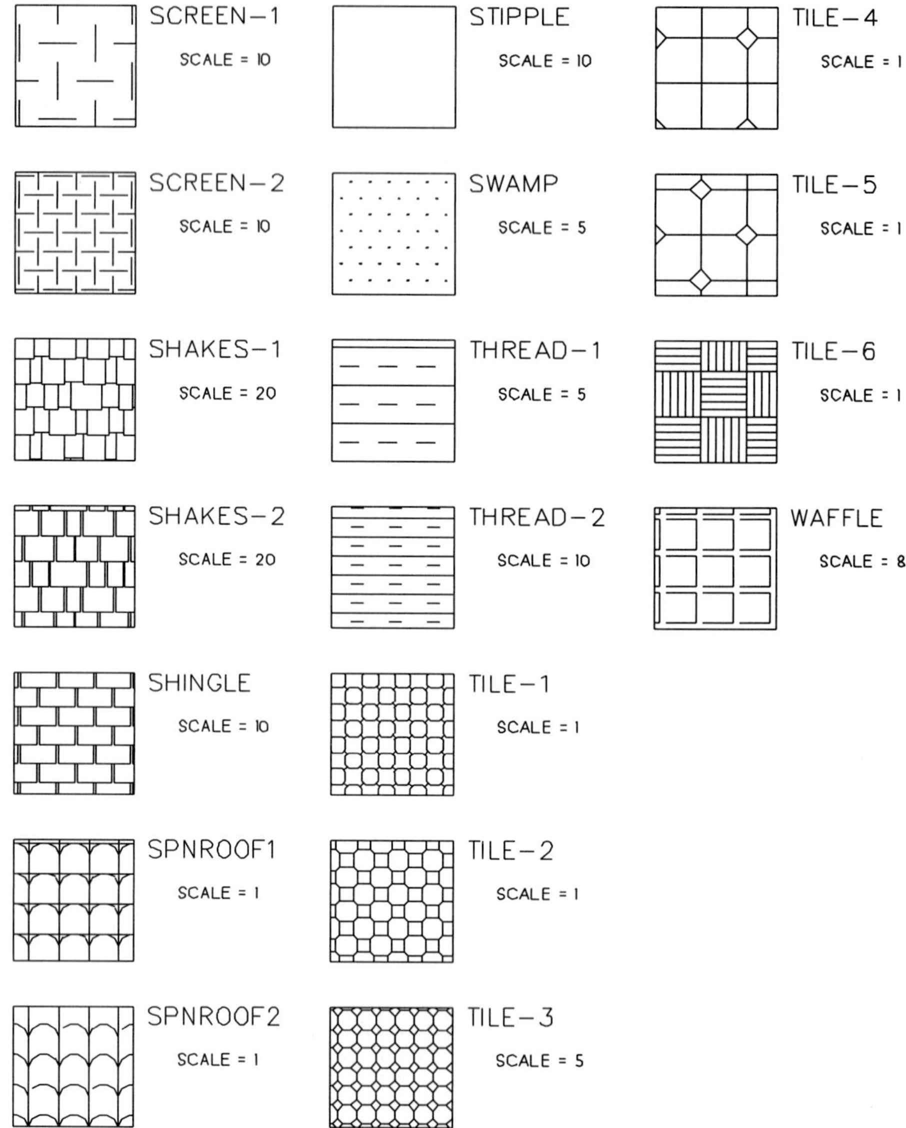
SCREEN-1
SCALE = 10
STIPPLE
SCALE = 10
TILE-4
SCALE = 1
SCREEN-2
SCALE = 10
SWAMP
SCALE = 5
TILE-5
SCALE = 1
SHAKES-1
SCALE = 20
THREAD-1
SCALE = 5
TILE-6
SCALE = 1
SHAKES-2
SCALE = 20
THREAD-2
SCALE = 10
WAFFLE
SCALE = 8
SHINGLE
SCALE = 10
TILE-1
SCALE = 1
SPNROOF1
SCALE = 1
TILE-2
SCALE = 1
SPNROOF2
SCALE = 1
TILE-3
SCALE = 5

Fonts

These are the standard DataCAD fonts that come with the basic program. A chart of these fonts can be printed from the CD-ROM.

ARCDR2CP	ABCDEFG abcdefg 1234567890
ARCWY2FW	ABCDEFG abcdefg 1234567890
ARCWY2GP	ABCDEFG abcdefg 1234567890
ARCWY2HC	ABCDEFG abcdefg 1234567890
ARCWY2LC	ABCDEFG abcdefg 1234567890
ARCWY2TX	ABCDEFG abcdefg 1234567890
AVG_TP	ABCDEFG abcdefg 1234567890
BLOCK	ABCDEFG ABCDEFG 1234567890
COMPLEX	ABCDEFG abcdefg 1234567890
HLV_BP	**ABCDEFG abcdefg 1234567890**
HLV_LP	ABCDEFG abcdefg 1234567890
HLV_MP	**ABCDEFG abcdefg 1234567890**
HLV_TM	ABCDEFG abcdefg 1234567890
HLV_TP	ABCDEFG abcdefg 1234567890
HLV_TPX	ABCDEFG abcdefg 1234567890
MCG_TP	ABCDEFG abcdefg 1234567890
ORIG	ABCDEFG abcdefg 1234567890
ORIG2	ABCDEFG abcdefg 1234567890
ROMAN	ABCDEFG abcdefg 1234567890
ROMAN2	ABCDEFG abcdefg 1234567890
ROMANC	**ABCDEFG abcdefg 1234567890**
ROMANS	ABCDEFG abcdefg 1234567890
SIMPLEX	ABCDEFG abcdefg 1234567890

Extended Character Codes

Here are the standard ANSII characters, numbers 1 through 127, for the English character set, for use with keyboard shortcuts and icon toolbars.

value	name	value	char	value	char	value	char
0	NUL	32	space	64	@	96	`
1	SOH	33	!	65	A	97	a
2	STX	34	"	66	B	98	b
3	ETX	35	#	67	C	99	c
4	EOT	36	$	68	D	100	d
5	ENQ	37	%	69	E	101	e
6	ACK	38	&	70	F	102	f
7	BEL	39	'	71	G	103	g
8	BS	40	(	72	H	104	h
9	HT	41	)	73	I	105	i
10	LF	42	*	74	J	106	j
11	VT	43	+	75	K	107	k
12	FF	44	,	76	L	108	l
13	CR	45	-	77	M	109	m
14	SO	46	.	78	N	110	n
15	SI	47	/	79	O	111	o
16	DLE	48	0	80	P	112	p
17	DC1	49	1	81	Q	113	q
18	DC2	50	2	82	R	114	r
19	DC3	51	3	83	S	115	s
20	DC4	52	4	84	T	116	t
21	NAK	53	5	85	U	117	u
22	SYN	54	6	86	V	118	v
23	ETB	55	7	87	W	119	w
24	CAN	56	8	88	X	120	x
25	EM	57	9	89	Y	121	y
26	SUB	58	:	90	Z	122	z
27	ESC	59	;	91	[	123	{
28	FS	60	<	92	\	124	|
29	GS	61	=	93	]	125	}
30	RS	62	>	94	^	126	≅
31	US	63	?	95	_	127	del

These are the extended ANSII characters, numbers 128 through 255, for the English character set.

128	Ç	171	½	214	╓
129	ü	172	¼	215	╫
130	é	173	¡	216	╪
131	â	174	«	217	┘
132	ä	175	»	218	┌
133	à	176	░	219	█
134	å	177	▒	220	▄
135	ç	178	▓	221	▌
136	ê	179	│	222	▐
137	ë	180	┤	223	▀
138	è	181	╡	224	α
139	ï	182	╢	225	β
140	î	183	╖	226	Γ
141	ì	184	╕	227	π
142	Ä	185	╣	228	Σ
143	Å	186	║	229	σ
144	É	187	╗	230	µ
145	æ	188	╝	231	τ
146	Æ	189	╜	232	Φ
147	ô	190	╛	233	Θ
148	ö	191	┐	234	Ω
149	ò	192	└	235	δ
150	û	193	┴	236	∞
151	ù	194	┬	237	ø
152	ÿ	195	├	238	∈
153	Ö	196	─	239	∩
154	Ü	197	┼	240	≡
155	¢	198	╞	241	±
156	£	199	╟	242	≥
157	¥	200	╚	243	≤
158	₧	201	╔	244	⌠
159	ƒ	202	╩	245	⌡
160	á	203	╦	246	÷
161	í	204	╠	247	≈
162	ó	205	═	248	°
163	ú	206	╬	249	∙
164	ñ	207	╧	250	·
165	Ñ	208	╨	251	√
166	ª	209	╤	252	^{n}
167	º	210	╥	253	^{2}
168	¿	211	╙	254	■
169	⌐	212	╘	255	
170	¬	213	╒		

Favorite Keyboard Macros

This chart represents the current **Alt+letter** keyboard shortcuts that we use in our office. The chart is available in Word for Windows format on the CD-ROM.

Shortcut	Function	Notes
Alt+A	Arrows Menu	Starts drawing arrows (red leader) on the current layer
Alt+B	Scale	Goes to Plotter/Scale menu; selects scale for use with Text Scale, etc.
Alt+C	Clip Cube Menu	Goes to Clip Cube Menu
Alt+D	Mltlyout Details	Goes to the MltLyout Menu and displays the current MSP sheet
Alt+E	Turn Off Layer by Entity	Places cursor on entity which is on the layer you want to turn off. That layer will turn off.
Alt+F	Perpendicular Snap	Toggles perpendicular snapping ON/OFF
Alt+G	3D GotoView Menu	Goes to 3D GotoView Menu
Alt+H	Selection Sets—by Color	Clears previous selection set 8, makes it active and prompts you to mask by color
Alt+I	Identify, Set—All	Places cursor on line to identify, then invokes the macro. Matches line type, spacing color & layer
Alt+J	Delete Associative Hatch	Makes hatching non-associative by exploding via 3D Explode. Keeps all origins correct. Default=area.
Alt+K	Group Function	Standard DCAD shortcut, selects entities to group together
Alt+L	Line Type Menu	Puts you in the DCAD linetype menu
Alt+M	Layout in Multi-Scale-Plot	Puts you in Multi-Scale-Plotting to lay out the current detail
Alt+N	Not Used	Not Used
Alt+O	Explode Dimensions	Explodes associative dimensions to non-associative. Default is by "area."
Alt+P	Plotter Menu	Standard DCAD shortcut, selects plotter options
Alt+Q	Update Gotoview 1	Saves current view as temporary ("temp") GotoView #1. (Assumes "temp" is always first GTV.)

Shortcut	Function	Notes
Alt+R	1-line Trim by Entity	Places cursor on line to trim to, then invokes macro. You will be prompted for the outside of the trim line.
Alt+S	Stretch Menu	Accesses the 2D Stretch menu.
Alt+T	Text Menu	Takes you to the Text menu
Alt+U	Unlock All Layers	Runs the UNLOCK macro and unlocks all layers, on or off
Alt+V	Match Current Line Settings	Selects entities. They will be changed to match all current line settings. (often used after ALT+I)
Alt+W	Weld Function	Standard DCAD shortcut, selects lines to weld together
Alt+X	Edit Text by Entity	Places cursor on text entity to edit text contents
Alt+Y	Create and Name New Layer	Creates new layer and prompts you to name it
Alt+Z	Save 3D GotoView	Goes to 3D GotoView and prompts you to enter a name

Italics denote a keyboard shortcut which requires cursor placement *before* invoking the macro.

And here is the DCADWIN.MCR file (in the \SUP directory) that defines each of the previous shortcuts:

```
A^;^S7^F1^S0^:^F3^F7^F8^S0^S0^S6^S2^
B^:^S5^F4^
C^;^S8^S9^S7^
D^:^S5^S1^S8^S0^
E^:^F3^F1^S8^$^S0^S0^;^
F^:^S3^F8^;^
G^:^F2^
H^;^S2^F4^F8^F5^F0^F1^F8^S0^F2^F8^F3^F0^F2^S9^
I^;^S1^$^S8^S0^
J^;^S8^F9^F3^F8
K^;^F8^
L^;^S7
M^:^S5^S1^S3^
```

```
N^
O^:^F8^F1^S7^F3^
P^:^S5^
Q^:^F2^S6^F1^S0^
R^;^F9^F4^F1^
S^;^F5^
T^;^S6
U^M^UNLOCK$^F0^S0^S0^
V^;^F0^S6^S7^S6^
W^;^F9^F7^
X^;^F0^S8^F1^S2^$^
Y^:^F3^F9^S0^F6^S7^S7^S7^S7^S7^S7^S7^S7^S7^S5^S4^S3^S2^
   S1^F0^F9^F8^F7^F6^F5^F4^F3^F2^F1
Z^V^S5^S5^
```

DWG/DXF Translation Files

This list is based on the DataCAD HELP file, and describes how DWG, DXF, and DC5 entities translate between these file fomats.

Importing from AutoCAD to DataCAD (and Back to AutoCAD)

AUTOCAD	DATACAD	AUTOCAD
2D Polyline (parallel to X, Y plane)	Polyline	2D Polyline (Rel. 13 or 14) Polyline (Rel. 12)
2D Polyline (not parallel to X, Y plane)	Polyline Symbol ∞	2D Polyline (Rel. 13 or 14) Polyline (Rel. 12)
2D Polyline (with width assigned to any vertices)	Polyline β	2D Polyline (Rel. 13 or 14) Polyline (Rel. 12)
3Dface	Polygon	3Dface
3D Polyline (with varying Z values)	3D Line	Line

AUTOCAD	DATACAD	AUTOCAD
3D Polyline (planar—no varying Z values)	Polyline (parallel to X, Y plane) Polyline Symbol ∞	2D Polyline
Arc (parallel to X, Y plane)	Arc	Arc
Arc (not parallel to X, Y plane; thickness = 0)	3D Arc	Arc
Arc (not parallel to X, Y plane; thickness ≠ 0)	Cylinder β	Arc
Attribute Definition	Text β (constant data is saved as attributes; variable data is lost)	Attribute Definition
Block Φ	Symbol	Block
Circle (parallel to X, Y plane)	Circle	Circle
Circle (not parallel to X, Y plane; thickness = 0)	3D Arc β	Circle
Circle (not parallel to X, Y plane; thickness ≠ 0)	Cylinder β	Circle
Dimensions	Associative Dimensions	Dimensions
Dimensions (angular, radius, or diameter dimensions and leaders)	Lines and text	Lines and text
Donut (parallel to X, Y plane)	Polyline β	Donut
Donut (not parallel to X, Y plane)	Polyline Symbol β∞	Donut
Ellipse (parallel to X, Y plane)	Ellipse	Ellipse (Rel. 13 or 14) Lines (Rel. 12)
Ellipse (not parallel to X, Y plane)	Ellipse Symbol ∞	Ellipse (Rel. 13 or 14) Lines (Rel. 12)
Hatch	Associative Hatch (if hatch boundary is LW polyline parallel to X, Y plane) Associative Hatch Symbol ∞ (if hatch boundary is LW polyline but not parallel to X, Y plane) Lines (if hatch boundary includes any other boundary types besides LW polyline) Lines and Text	Lines

AUTOCAD	DATACAD	AUTOCAD
Leaders	Lines and Text	Lines and Text
Light-Weight Polyline	(See 2D Polyline)	
Line (parallel to X, Y plane)	Line	Line
Line (not parallel to X, Y plane; thickness = 0)	3D Line	Line
Line (not parallel to X, Y plane; thickness ≠ 0)	Polygon β	Line (if coordinates unchanged) Line, with changed thickness (if polygon stretched along the axis of thickness) Polygon (if coordinates changed in directions other than axis of thickness)
MText	Text	Text
Mesh	Polygons	Polygons
Point (with no thickness)	Point	Point
Point (with thickness)	3D Line βΩ	Point
Polyface Mesh	Polygons	Polygons
Spline	Points β	Spline (Rel. 13 or 14)
Solid	Polygons, triangulated	3D Faces
Text, single line or multi-line	Text	Text

β AutoCAD entity is imported into DataCAD as an entity with attributes. If the entity is edited in DataCAD, upon export it will return to its original (AutoCAD) entity type, incorporating the changes you made in DataCAD. Exceptions to this are exploding an entity, which permanently changes the entity type.

∞ AutoCAD entity is imported into DataCAD as a symbol. These symbols are like any other in DataCAD; they cannot be edited unless they are first exploded. If you explode an entity in 3D space, however, it will flatten to the X, Y plane. For instance, a Donut in a DWG file that is not parallel to the X, Y plane will be imported into DataCAD as a Polyline Symbol. If this symbol is not exploded, it will maintain its position when exported back to a DWG file. If the symbol is exploded, however, it will flatten to the X, Y plane, even when exported back to a DWG file.

Ω Point shape is lost in translation.

Φ Entities on different layers that comprise of a single block will be imported into DataCAD as a symbol on the layer where the insertion is located; that is, all entities for that Block will be moved to the insertion layer. These entities will remain on this layer when exported back to a DXF of DWG file. Also, unnamed blocks in AutoCAD will be given a unique, numeric name when imported into DataCAD; this name will be saved with the block when exported.

There are a few AutoCAD entities that cannot be read, displayed, or otherwise saved during translation. These include: XREFs, viewports, ACIS data (Body), Rays (XLines), Ole Frames, Ole2 Frames, Tolerances, Images, Regions, Proxys, and named views.

Exporting from DataCAD to AutoCAD (and Back to DataCAD)

DATACAD	AUTOCAD	DATACAD
3D Arc	Arc	Arc
3D Line	Line	3D Line (if not parallel to X, Y plane) Line (if parallel to X, Y plane)
Arc	Arc	Arc
B-Spline	Polyline	Polyline
Bezier	Polyline	Polyline
Circle	Circle	Circle
Cone	Polyface Mesh 3D Faces (if Decompose in Program Preferences dialog is on during export)	Polygons
Contour	Polyline	Polyline
Cylinder	Polyface Mesh 3D Faces (if Decompose in Program Preferences dialog is on during export)	Polygons
Dimensions (Associative)	Dimensions	Dimensions (Associative)
Ellipse	Ellipse (Rel. 13 or 14) Lines (Rel. 12)	Ellipse
Hatch (Associative)	Lines	Lines
Line	Line	Line
Mesh Surface	Polyface Mesh 3D Faces (if Decompose in Program Preferences dialog is on during export)	Polygons

DATACAD	AUTOCAD	DATACAD
Point	Point	Point
Polygon (with 3–4 vertices and no voids)	3D Face	Polygon
Polygon (with 5 or more vertices and/or voids)	Triangulated into multiple 3D Faces	Multiple 3-sided polygons
Polyline	Lightweight Polyline (Rel. 14) Polyline (< Rel. 14)	Polyline
Sphere	Polyface Mesh 3D Faces (if Decompose in Program Preferences dialog is on during export)	Polygons
Surface of Revolution	Polyface Mesh 3D Faces (if Decompose in Program Preferences dialog is on during export)	Polygons
Symbol	Block	Symbol
Text	Text	Text
Torus	Polyface Mesh 3D Faces (if Decompose in Program Preferences dialog is on during export)	Polygons
Truncated Cone	Polyface Mesh 3D Faces (if Decompose in Program Preferences dialog is on during export)	Polygons

INDEX

A

B

D

F

G

H

K

L

M

P

Q

R

S

T

U

V

W

X

Z

What's On the CD-ROM?

The interface for this CD-ROM will run in your favorite Internet Web browser, such as Netscape Navigator or Microsoft Internet Explorer. Start by opening the **HOME.HTM** page, then simply click on the appropriate buttons to find and access the bonus material described in the text of the book. Here is a quick description of what you will find in each section:

PROGRAMS
Includes the demo version of the DataCAD 9 program, various shareware programs, and a couple of DataCAD utilities written by Michael Smith.

MACROS
Includes all the free DataCAD Toolbox macros currently available on the DataCAD LLC Web site, as well as some macros available ONLY on this CD-ROM!

CUSTOM
Here you will find various items from keyboard shortcuts to icon toolbars to customize DataCAD 9 to your liking.

CHAPTERS
This is where you will find the sample DataCAD drawing files described in various chapters of this book.

REFERENCE
This section contains various reference materials including tables and charts describing the many keyboard shortcuts available within DataCAD.

DCAL
If you would like to program your own DataCAD Toolbox macros, this is where you can find various references to *DataCAD Applications Language* (DCAL), including the complete documentation to the DCAL programming code.

GALLERY
A collection of 2D and 3D images and animations.

SOFTWARE AND INFORMATION LICENSE

The software and information on this diskette (collectively referred to as the "Product") are the property of The McGraw-Hill Companies, Inc. ("McGraw-Hill") and are protected by both United States copyright law and international copyright treaty provision. You must treat this Product just like a book, except that you may copy it into a computer to be used and you may make archival copies of the Products for the sole purpose of backing up our software and protecting your investment from loss.

By saying "just like a book," McGraw-Hill means, for example, that the Product may be used by any number of people and may be freely moved from one computer location to another, so long as there is no possibility of the Product (or any part of the Product) being used at one location or on one computer while it is being used at another. Just as a book cannot be read by two different people in two different places at the same time, neither can the Product be used by two different people in two different places at the same time (unless, of course, McGraw-Hill's rights are being violated).

McGraw-Hill reserves the right to alter or modify the contents of the Product at any time.

This agreement is effective until terminated. The Agreement will terminate automatically without notice if you fail to comply with any provisions of this Agreement. In the event of termination by reason of your breach, you will destroy or erase all copies of the Product installed on any computer system or made for backup purposes and shall expunge the Product from your data storage facilities.

LIMITED WARRANTY

McGraw-Hill warrants the physical diskette(s) enclosed herein to be free of defects in materials and workmanship for a period of sixty days from the purchase date. If McGraw-Hill receives written notification within the warranty period of defects in materials or workmanship, and such notification is determined by McGraw-Hill to be correct, McGraw-Hill will replace the defective diskette(s). Send request to:

Customer Service
McGraw-Hill
Gahanna Industrial Park
860 Taylor Station Road
Blacklick, OH 43004-9615

The entire and exclusive liability and remedy for breach of this Limited Warranty shall be limited to replacement of defective diskette(s) and shall not include or extend any claim for or right to cover any other damages, including but not limited to, loss of profit, data, or use of the software, or special, incidental, or consequential damages or other similar claims, even if McGraw-Hill has been specifically advised as to the possibility of such damages. In no event will McGraw-Hill's liability for any damages to you or any other person ever exceed the lower of suggested list price or actual price paid for the license to use the Product, regardless of any form of the claim.

THE McGRAW-HILL COMPANIES, INC. SPECIFICALLY DISCLAIMS ALL OTHER WARRANTIES, EXPRESS OR IMPLIED, INCLUDING BUT NOT LIMITED TO, ANY IMPLIED WARRANTY OF MERCHANTABILITY OR FITNESS FOR A PARTICULAR PURPOSE. Specifically, McGraw-Hill makes no representation or warranty that the Product is fit for any particular purpose and any implied warranty of merchantability is limited to the sixty day duration of the Limited Warranty covering the physical diskette(s) only (and not the software or information) and is otherwise expressly and specifically disclaimed.

This Limited Warranty gives you specific legal rights; you may have others which may vary from state to state. Some states do not allow the exclusion of incidental or consequential damages, or the limitation on how long an implied warranty lasts, so some of the above may not apply to you.

This Agreement constitutes the entire agreement between the parties relating to use of the Product. The terms of any purchase order shall have no effect on the terms of this Agreement. Failure of McGraw-Hill to insist at any time on strict compliance with this Agreement shall not constitute a waiver of any rights under this Agreement. This Agreement shall be construed and governed in accordance with the laws of New York. If any provision of this Agreement is held to be contrary to law, that provision will be enforced to the maximum extent permissible and the remaining provisions will remain in force and effect.